Finite Element Methods (Part 1)

Handbook of Numerical Analysis

General Editors:

P.G. Ciarlet

Analyse Numérique, Tour 55–65
Université Pierre et Marie Curie
4 Place Jussieu
75005 PARIS, France

J.L. Lions

Collège de France
Place Marcelin Berthelot
75005 PARIS, France

ELSEVIER
AMSTERDAM · BOSTON · HEIDELBERG · LONDON · NEW YORK · OXFORD
PARIS · SAN DIEGO · SAN FRANCISCO · SINGAPORE · SYDNEY · TOKYO

Volume II

Finite Element Methods
(Part 1)

ELSEVIER
AMSTERDAM · BOSTON · HEIDELBERG · LONDON · NEW YORK · OXFORD
PARIS · SAN DIEGO · SAN FRANCISCO · SINGAPORE · SYDNEY · TOKYO

ELSEVIER SCIENCE B.V.
Sara Burgerhartstraat 25
P.O. Box 211, 1000 AE Amsterdam. The Netherlands

First edition 1991
Second impression 1993
Third impression 2003

Library of Congress Cataloging in Publication Data
(Revised for Vol. 2)

Handbook of numerical analysis.
Includes bibliographical references and indexes.
Contents: v. 1. Finite difference methods (pt. 1):
Solutions of equations in $\mathbb{R}^n$ (pt.1)–v.2. Finite element methods (pt. 1)
ISBN 0-444-70365-9 (v. 2)
1. Numerical analysis. I. Ciarlet, Philippe G. II. Lions, Jacques Louis.
QA297.H287 1989 519.4 89-23314
ISBN 0-444-70366-7 (v.1)

ISBN: 0 444 70365 9

Transferred to digital printing 2006
Printed and bound by Antony Rowe Ltd, Eastbourne

General Preface

During the past decades, giant needs for ever more sophisticated mathematical models and increasingly complex and extensive computer simulations have arisen. In this fashion, two indissociable activities, *mathematical modeling* and *computer simulation*, have gained a major status in all aspects of science, technology, and industry.

In order that these two sciences be established on the safest possible grounds, mathematical rigor is indispensable. For this reason, two companion sciences, *Numerical Analysis* and *Scientific Software*, have emerged as essential steps for validating the mathematical models and the computer simulations that are based on them.

Numerical Analysis is here understood as the part of *Mathematics* that describes and analyzes all the numerical schemes that are used on computers; its objective consists in obtaining a clear, precise, and faithful, representation of all the "information" contained in a mathematical model; as such, it is the natural extension of more classical tools, such as analytic solutions, special transforms, functional analysis, as well as stability and asymptotic analysis.

The various volumes comprising the *Handbook of Numerical Analysis* will thoroughly cover all the major aspects of Numerical Analysis, by presenting accessible and in-depth surveys, which include the most recent trends.

More precisely, the Handbook will cover the *basic methods of Numerical Analysis*, gathered under the following general headings:

- Solution of Equations in $\mathbb{R}^n$,
- Finite Difference Methods,
- Finite Element Methods,
- Techniques of Scientific Computing,
- Optimization Theory and Systems Science.

It will also cover the *numerical solution of actual problems of contemporary interest in Applied Mathematics*, gathered under the following general headings:

- Numerical Methods for Fluids,
- Numerical Methods for Solids,
- Specific Applications.

"Specific Applications" include: Meteorology, Seismology, Petroleum Mechanics, Celestial Mechanics, etc.

Each heading is covered by several *articles*, each of which being devoted to a specialized, but to some extent "independent", topic. Each article contains a thorough description and a mathematical analysis of the various methods in actual use, whose practical performances may be illustrated by significant numerical examples.

Since the Handbook is basically expository in nature, only the most basic results are usually proved in detail, while less important, or technical, results may be only stated or commented upon (in which case specific references for their proofs are systematically provided). In the same spirit, only a "selective" bibliography is appended whenever the roughest counts indicate that the reference list of an article should comprise several thousands items if it were to be exhaustive.

Volumes are numbered by capital Roman numerals (as Vol. I, Vol. II, etc.), according to their *chronological appearance.*

Since all the articles pertaining to a given *heading* may not be simultaneously available at a given time, a given heading usually appears in more than one volume; for instance, if articles devoted to the heading "Solution of Equations in $\mathbb{R}^n$" appear in Volumes I and III, these volumes will include "Solution of Equations in $\mathbb{R}^n$ (Part 1)" and "Solution of Equations in $\mathbb{R}^n$ (Part 2)" in their respective titles. Naturally, all the headings dealt with within a given volume appear in its title; for instance, the complete title of Volume I is "Finite Difference Methods (Part 1)—Solution of Equations in $\mathbb{R}^n$ (Part 1)".

Each article is subdivided into *sections*, which are numbered consecutively throughout the article by *Arabic numerals*, as Section 1, Section 2, . . . , Section 14, etc. Within a given section, *formulas, theorems, remarks, and figures*, have their own independent numberings; for instance, within Section 14, formulas are numbered consecutively as (14.1), (14.2), etc., theorems are numbered consecutively as Theorem 14.1, Theorem 14.2, etc. For the sake of clarity, the article is also subdivided into *chapters*, numbered consecutively throughout the article by *capital Roman numerals;* for instance, Chapter I comprises Sections 1 to 9, Chapter II comprises Sections 10 to 16, etc.

P.G. CIARLET
J.L. LIONS
May 1989

Contents of Volume II

Contents of the Handbook

Finite Element Methods (Part 1)

Finite Elements: An Introduction

J. Tinsley Oden

Finite elements; perhaps no other family of approximation methods has had a greater impact on the theory and practice of numerical methods during the twentieth century. Finite element methods have now been used in virtually every conceivable area of engineering that can make use of models of nature characterized by partial differential equations. There are dozens of textbooks, monographs, handbooks, memoirs, and journals devoted to its further study; numerous conferences, symposia, and workshops on various aspects of finite element methodology are held regularly throughout the world. There exist easily over one hundred thousand references on finite elements today, and this number is growing exponentially with further revelations of the power and versatility of the method. Today, finite element methodology is making significant inroads into fields in which many thought were outside its realm; for example, computational fluid dynamics. In time, finite element methods may assume a position in this area of comparable or greater importance than classical difference schemes which have long dominated the subject.

Why finite elements?

A natural question that one may ask is: why have finite element methods been so popular in both the engineering and mathematical community? There is also the question, do finite element methods possess properties that will continue to make them attractive choices of methods to solve difficult problems in physics and engineering?

In answering these questions, one must first point to the fact that finite element methods are based on the weak, variational, formulation of boundary and initial value problems. This is a critical property, not only because it provides a proper setting for the existence of very irregular solutions to differential equations (e.g. distributions), but also because the solution appears in the integral of a quantity over

HANDBOOK OF NUMERICAL ANALYSIS, VOL. II
Finite Element Methods (Part 1)
Edited by P.G. Ciarlet and J.L. Lions

a domain. The simple fact that the integral of a measurable function over an arbitrary domain can be broken up into the sum of integrals over an arbitrary collection of almost disjoint subdomains whose union is the original domain, is a vital observation in finite element theory. Because of it, the analysis of a problem can literally be made locally, over a typical subdomain, and by making the subdomain sufficiently small one can argue that polynomial functions of various degrees are adequate for representing the local behavior of the solution. This summability of integrals is exploited in every finite element program. It allows the analysts to focus their attention on a typical finite element domain and to develop an approximation independent of the ultimate location of that element in the final mesh.

The simple integral property also has important implications in physics and in most problems in continuum mechanics. Indeed, the classical balance laws of mechanics are global, in the sense that they are integral laws applying to a given mass of material, a fluid or solid. From the onset, only regularity of the primitive variables sufficient for these global conservation laws to make sense is needed. Moreover, since these laws are supposed to be fundamental axioms of physics, they must hold over every finite portion of the material: every finite element of the continuum. Thus once again, one is encouraged to think of approximate methods defined by integral formulations over typical pieces of a continuum to be studied.

These rather primitive properties of finite elements lead to some of its most important features:

(1) *Arbitrary geometries.* The method is essentially geometry-free. In principle, finite element methods can be applied to domains of arbitrary shape and with quite arbitrary boundary conditions.

(2) *Unstructured meshes.* While there is still much prejudice in the numerical analysis literature toward the use of coordinate-dependent algorithms and mesh generators, there is nothing intrinsic in finite element methodology that requires such devices. Indeed, finite element methods by their nature lead to unstructured meshes. This means, in principle, analysts can place finite elements anywhere they please. They may thus model the most complex types of geometries in nature and physics, ranging from the complex cross-sections of biological tissues to the exterior of aircraft to internal flows in turbo machinery, without strong use of a global fixed coordinate frame.

(3) *Robustness.* It is well known that in finite element methods the contributions of local approximations over individual elements are assembled together in a systematic way to arrive at a global approximation of a solution to a partial differential equation. Generally, this leads to schemes which are stable in appropriate norms, and, moreover, insensitive to singularities or distortions of the mesh, in sharp contrast to classical difference methods. There are notable exceptions to this, of course, and these exceptions have been the subject of some of the most important works in finite element theory. But, by and large, the direct use of Galerkin or Petrov–Galerkin methods to derive finite element methods leads to conservative and stable algorithms, for most classes of problems in mechanics and mathematical physics.

(4) *Mathematical foundation.* Because of the extensive work on the mathematical foundations done during the seventies and eighties, finite elements now enjoy a rich and solid mathematical basis. The availability of methods to determine a priori and a posteriori estimates provides a vital part of the theory of finite elements, and makes it possible to lift the analysis of important engineering and physical problems above the traditional empiricism prevalent in many numerical and experimental studies.

These properties are intrinsic to finite element methods and continue to make these methods among the most attractive for solving complex problems.

They represent the most desirable properties of any numerical scheme designed to handle real-world problems. Moreover, the basic features of finite element methodology provide an ideal setting for innovative use of modern supercomputing architectures, particularly parallel processing. For these reasons, it is certain that finite element concepts will continue to occupy an important role in applications and in research on the numerical solution of partial differential equations.

The early history

When did finite elements begin? It is difficult to trace the origins of finite element methods because of a basic problem in defining precisely what constitutes a "finite element method". To most mathematicians, it is a method of piecewise polynomial approximation and, therefore, its origins are frequently traced to the appendix of a paper by COURANT [1943] in which piecewise linear approximations of the Dirichlet problem over a network of triangles is discussed. Also, the "interpretation of finite differences" by PÓLYA [1952] is regarded as embodying piecewise polynomial approximation aspects of finite elements.

On the other hand, the approximation of variational problems on a mesh of triangles goes back much further: 92 years. In 1851, SCHELLBACH [1851] proposed a finite-element-like solution to Plateau's problem of determining the surface S of minimum area enclosed by a given closed curve. Schellbach used an approximation S_h of S by a mesh of triangles over which the surface was represented by piecewise linear functions, and he then obtained an approximation to the solution to Plateau's problem by minimizing S_h with respect to the coordinates of hexagons formed by six elements (see WILLIAMSON [1980]). Not quite the conventional finite element approach, but certainly as much a finite element technique as that of Courant.

Some say that there is even an earlier work that uses some of the ideas underlying finite element methods: Gottfried Leibniz himself employed a piecewise linear approximation of the Brachistochrone problem proposed by Johann Bernoulli in 1696 (see the historical volume, LEIBNIZ [1962]). With the help of his newly developed calculus tools, Leibniz derived the governing differential equation for the problem, the solution of which is a cycloid. However, most would agree that to credit this work as a finite element approximation is somewhat stretching the point. Leibniz had no intention of approximating a differential equation; rather, his purpose was to derive one. Two and a half centuries later it was realized that useful approximations of differential equations could be determined by not necessarily taking infinitesimal elements as in the calculus, but by keeping the elements finite in size. This idea is, in fact, the basis of the term "finite element".

There is also some difference in the process of laying a mesh of triangles over a domain on the one hand and generating the domain of approximation by piecing together triangles on the other. While these processes may look the same in some cases, they may differ dramatically in how the boundary conditions are imposed. Thus, neither Schellbach nor Courant, nor for that matter Synge who used triangular meshes many years later, were particularly careful as to how boundary conditions were to be imposed or as to how the boundary of the domain was to be modeled by elements, issues that are now recognized as an important feature of finite element methodologies. If a finite element method is one in which a global approximation of a partial differential equation is built up from a sequence of local approximations over subdomains, then credit must go back to the early papers of HRENNIKOFF [1941], and perhaps beyond, who chose to solve plane elasticity problems by breaking up the domain of the displacements into little finite pieces, over which the stiffnesses were approximated using bars, beams, and spring elements. A similar "lattice analogy" was used by MCHENRY [1943]. While these works are draped in the most primitive physical terms, it is nevertheless clear that the methods involve some sort of crude piecewise linear or piecewise cubic approximation over rectangular cells. Miraculously, the methods also seem to be convergent.

To the average practitioner who uses them, finite elements are much more than a method of piecewise polynomial approximation. The whole process of partitioning of domains, assembling elements, applying loads and boundary conditions, and, of course, along with it, local polynomial approximation, are all components of the finite element method.

If this is so, then one must acknowledge the early papers of Gabriel Kron who developed his "tensor analysis of networks" in 1939 and applied his "method of tearing" and "network analysis" to the generation of global systems from large numbers of individual components in the 1940s and 1950s (KRON [1939]; see also KRON [1953]). Of course, Kron never necessarily regarded his method as one of approximating partial differential equations; rather, the properties of each component were regarded as exactly specified, and the issue was an algebraic one of connecting them all appropriately together.

In the early 1950s, ARGYRIS [1954] began to put these ideas together into what some call a primitive finite element method: he extended and generalized the combinatoric methods of Kron and other ideas that were being developed in the literature on system theory at the time, and added to it variational methods of approximation, a fundamental step toward true finite element methodology.

Around the same time, SYNGE [1957] described his "method of the hypercircle" in which he also spoke of piecewise linear approximations on triangular meshes, but not in a rich variational setting and not in a way in which approximations were built by either partitioning a domain into triangles or assembling triangles to approximate a domain (indeed Synge's treatment of boundary conditions was clearly not in the spirit of finite elements, even though he was keenly aware of the importance of convergence criteria and of the "angle condition" for triangles, later studied in some depth by others).

It must be noted that during the mid-1950s there were a number of independent studies underway which made use of "matrix methods" for the analysis of aircraft structures. A principal contributor to this methodology was LEVY [1953] who introduced the "direct stiffness method" wherein he approximated the structural behavior of aircraft wings using assemblies of box beams, torsion boxes, rods and shear panels. These assuredly represent some sort of crude local polynomial approximation in the same spirit as the Hrennikoff and McHenry approaches. The direct stiffness method of Levy had a great impact on the structural analysis of aircraft, and aircraft companies throughout the United States began to adopt and apply some variant of this method or of the methods of Argyris to complex aircraft structural analyses. During this same period, similar structural analysis methods were being developed and used in Europe, particularly in England, and one must mention in this regard the work of TAIG [1961] in which shear lag in aircraft wing panels was approximated using basically a bilinear finite element method of approximation. Similar element-like approximations were used in many aircraft industries as components in various matrix methods of structural analyses. Thus the precedent was established for piecewise approximations of some kind by the mid-1950s.

To a large segment of the engineering community, the work representing the beginning of finite elements was that contained in the pioneering paper of TURNER, CLOUGH, MARTIN and TOPP [1956] in which a genuine attempt was made at both a local approximation (of the partial differential equations of linear elasticity) and the use of assembly strategies essential to finite element methodology. It is interesting that in this paper local element properties were derived without the use of variational principles. It was not until 1960 that CLOUGH [1960] actually dubbed these techniques as "finite element methods" in a landmark paper on the analysis of linear plane elasticity problems.

The 1960s were the formative years of finite element methods. Once it was perceived by the engineering community that useful finite element methods could be derived from variational principles, variationally based methods significantly dominated all the literature for almost a decade. If an operator was unsymmetric, it was thought that the solution of the associated problem was beyond the scope of finite elements, since it did not lend itself to a traditional extremum variational approximation in the spirit of Rayleigh and Ritz.

From 1960 to 1965, a variety of finite element methods were proposed. Many were primitive and unorthodox; some were innovative and successful. During this time, a variety of attempts at solving the biharmonic equation for plate bending problems were proposed which employed piecewise polynomial approximations, but did not provide the essentials for convergence. This led to the concern of some as to whether the method was indeed applicable to such problems. On the other hand, it was clear that classical Fourier series solutions of plate problems were, under appropriate conditions, convergent and could be fit together in an assemblage of rectangular components (ODEN [1962]) and, thus, a form of "spectral finite element methods" was introduced early in the study of such problems. However, such high-order schemes never received serious attention in this period, as it was felt that piecewise polynomial approximations could be developed which did give satisfactory results. It was not until the mid- to late 1960s that papers on bicubic spline approximations by BOGNER, FOX, and SCHMIT [1966] and BIRKHOFF, SCHULTZ, and VARGA [1968] provided successful polynomial finite element approximations for these classes of problems.

Many workers in the field feel that the famous Dayton conferences on finite elements (at the Air Force Flight Dynamics Laboratory in Dayton, Ohio, USA) represented landmarks in the development of the field (see PRZEMIENIECKI et al. [1966]). Held in 1965, 1968, 1970, these meetings brought specialists from all over the world to discuss their latest triumphs and failures, and the pages of the proceedings, particularly the earlier volumes, were filled with remarkable and innovative accomplishments from a technical community just beginning to learn the richness and power of this new collection of ideas. In these volumes one can find many of the premier papers of now well-known methods. In the first volume alone one can find mixed finite element methods (HERRMANN [1966]), Hermite approximations (PESTEL [1966]), C^1-bicubic approximations (BOGNER, FOX and SCHMIT [1966]), hybrid methods (PIAN [1966]) and other contributions. In later volumes, further assaults on nonlinear problems and special element formulations can be found.

Near the end of the sixties and early seventies there finally emerged the realization that the method could be applied to unsymmetric operators without difficulty and thus problems in fluid mechanics were brought within the realm of application of finite element methods; in particular, finite element models of the full Navier–Stokes equations were first presented during this period (ODEN [1969], ODEN and SOMOGYI [1968], ODEN [1970]).

The early textbook by ZIENKIEWICZ and CHEUNG [1967] did much to popularize the method with the practicing engineering community. However, the most important factor leading to the rise in popularity during the late 1960s and early 1970s was not purely the publication of special formulations and algorithms, but the fact that the method was being very successfully used to solve difficult engineering problems. Much of the technology used during this period was due to Bruce Irons, who with his colleagues and students developed a multitude of techniques for the successful implementation of finite elements. These included the frontal solution technique (IRONS [1970]), the patch test (IRONS and RAZZAQUE [1972]), isoparametric elements (ERGATOUDIS, IRONS and ZIENKIEWICZ [1966]), and numerical integration schemes (IRONS [1966]) and many more. The scope of finite element applications in the 1970s would have been significantly diminished without these contributions.

The mathematical theory

The mathematical theory of finite elements was slow to emerge from this caldron of activity. The beginning works on the mathematical theory of finite elements were understandably concerned with one-dimensional elliptic problems and used many of the tools and jargon of Ritz methods, interpolation, and variational differences. An early work in this line was the paper of VARGA [1966] which dealt with "Hermite interpolation-type Ritz methods" for two-point boundary value problems. We also mention in this regard the paper of BIRKHOFF, DE BOOR, SCHWARTZ and WENDROFF [1966] on "Rayleigh–Ritz approximation by piecewise cubic polynomials". This is certainly one of the first papers to deal with the issue of convergence of finite element methods, although some papers on variational differences yielded similar results but did not focus on the piecewise polynomial features of finite elements. The work of KANG FENG [1965], published in Chinese (a copy of which I have not been able to acquire for review) may fall into this category and is sometimes noted as relevant to the convergence of finite element methods.

The mathematical theory of finite elements for two-dimensional and higher-dimensional problems began in 1968 and several papers were published that year on the subject. One of the first papers in this period to address the problem of convergence of a finite method in a rigorous way and in which a priori error estimates for bilinear approximations of a problem in plane elasticity are obtained, is the often overlooked paper of JOHNSON and MCLAY [1968], which appeared in the *Journal of Applied Mechanics*. This paper correctly developed error estimates in energy norms, and even attempted to characterize the deterioration of convergence rates due to corner singularities. In the same year there appeared the first of two important papers by OGENESJAN and RUCHOVEC [1968, 1969] in the Russian literature, in which "variational difference schemes" were proposed for linear second-order elliptic problems in two-dimensional domains. These works dealt with the estimates of the rate of convergence of variational difference schemes.

Also in 1968 there appeared the important mathematical paper of ZLÁMAL [1968] in which a detailed analysis of interpolation properties of a class of triangular elements and their application to second-order and fourth-order linear elliptic boundary value problems is discussed. This paper attracted the interest of a large segment of the numerical analysis community and several very good mathematicians began to work on finite element methodologies. The paper by Zlámal also stands apart from other multidimensional finite element papers of this era since it represented a departure of studies of tensor products of polynomials on rectangular domains and provided an approach toward approximations in general polygonal domains. In the same year, CIARLET [1968] published a rigorous proof of convergence of piecewise linear finite element approximation of a class of linear two-point boundary value problems and proved L^{∞} estimates using a discrete maximum principle. We also mention the work of OLIVEIRA [1968] on convergence of finite element methods which established correct rates of convergence for certain problems in appropriate energy norms.

A year later, SCHULTZ [1969] presented error estimates for "Rayleigh–Ritz–Galerkin methods" for multidimensional problems. Two years later, SCHULTZ [1971] published L^2 error bounds for these types of methods.

By 1972, finite element methods had emerged as an important new area of numerical analysis in applied mathematics. Mathematical conferences were held on the subject on a regular basis, and there began to appear a rich volume of literature on mathematical aspects of the method applied to elliptic problems, eigenvalue problems, and parabolic problems. A conference of special significance in this period was held at the University of Maryland in 1972 and featured a penetrating series of lectures by Ivo Babuška (see Babuška and Aziz [1972]) and several important mathematical papers by leading specialists in the mathematics of finite elements, all collected in the volume edited by Aziz [1972].

One unfamiliar with aspects of the history of finite elements may be led to the erroneous conclusion that the method of finite elements emerged from the growing wealth of information on partial differential equations, weak solutions of boundary value problems, Sobolev spaces, and the associated approximation theory for elliptic variational boundary value problems. This is a natural mistake, because the seeds for the modern theory of partial differential equations were sown about the same time as those for the development of modern finite element methods, but in an entirely different garden.

In the late 1940s, Laurent Schwartz was putting together his theory of distributions around a decade after the notion of generalized functions and their use in partial differential equations appeared in the pioneering work of S.L. Sobolev. A long list of other names could be added to the list of contributors to the modern theory of partial differential equations, but that is not our purpose here. Rather, we must only note that the rich mathematical theory of partial differential equations which began in the 1940s and 1950s, blossomed in the 1960s, and is now an integral part of the foundations of not only partial differential equations but also approximation theory, grew independently and parallel to the development of finite element methods for almost two decades. There was important work during this period on the foundations of variational methods of approximation, typified by the early work of Lions [1955] and by the French school in the early 1960s; but, while this work did concern itself with the systematic development of mathematical results that would ultimately prove to be vital to the development of finite element methods, it did not focus on the specific aspects of existing and already successful finite element concepts. It was, perhaps, an unavoidable occurrence, that in the late 1960s these two independent subjects, finite element methodology and the theory of approximation of partial differential equations via functional analysis methods, united in an inseparable way, so much so that it is difficult to appreciate the fact that they were ever separate.

The 1970s must mark the decade of the mathematics of finite elements. During this period, great strides were made in determining a priori error estimates for a variety of finite element methods, for linear elliptic boundary value problems, for eigenvalue problems, and certain classes of linear and nonlinear parabolic problems; also, some preliminary work on finite element applications to hyperbolic equations was done. It is both inappropriate and perhaps impossible to provide an adequate survey of this large volume of literature, but it is possible to present an albeit biased reference to some of the major works along the way.

An important component in the theory of finite elements is an interpolation theory: how well can a given finite element method approximate functions of a given class locally over a typical finite element? A great deal was known about this subject from the literature on approximation theory and spline analysis, but its particularization to finite elements involves technical difficulties. One can find results on finite element interpolation in a number of early papers, including those of ZLÁMAL [1968], BIRKHOFF [1969], SCHULTZ [1969], BRAMBLE and ZLÁMAL [1970], BABUŠKA [1970, 1971], and BABUŠKA and AZIZ [1972]. But the elegant work on Lagrange and Hermite interpolations of finite elements by CIARLET and RAVIART [1972a] must stand as a very important contribution to this vital aspect of finite element theory.

A landmark work on the mathematics of finite elements appeared in 1972 in the remarkably comprehensive and penetrating memoir of BABUŠKA and AZIZ [1972] on the mathematical foundations of finite element methods. Here one can find interwoven with the theory of Sobolev spaces and elliptic problems, general results on approximation theory that have direct bearing on finite element methods. It was known that Cea's lemma (CEA [1964]) established that the approximation error in a Galerkin approximation of a variational boundary value problem is bounded by the so-called interpolation error; that is, the distance in an appropriate energy norm from the solution of the problem to the subspace of approximations. Indeed, it was this fact that made the results on interpolation theory using piecewise polynomials of particular interest in finite element methods. In the work of BABUŠKA [1971] and BABUŠKA and AZIZ [1972], this framework was dramatically enlarged by Babuška's introduction of the so-called "INF-SUP" condition. This condition is encountered in the characterization of coerciveness of bilinear forms occuring in elliptic boundary value problems. The characterization of this "INF-SUP" condition for the discrete finite element approximation embodies in it the essential elements for studying the stability in convergence of finite element methods. BREZZI [1974] developed an equivalent condition for studying constrained elliptic problems and these conditions provide for a unified approach to the study of qualitative properties, including rates of convergence, of broad classes of finite element methods.

The fundamental work of NITSCHE [1970] on L^∞ estimates for general classes of linear elliptic problems must stand out as one of the most important contributions of the seventies. STRANG [1972], in an important communication, pointed out "variational crimes", inherent in many finite element methods, such as improper numerical quadrature, the use of nonconforming elements, improper satisfaction of boundary conditions, etc., all common practices in applications, but all frequently leading to exceptable numerical schemes.

In the same year, CIARLET and RAVIART [1972b, c] also contributed penetrating studies of these issues. Many of the advances of the 1970s drew upon earlier results on variational methods of approximation based on the Ritz method and finite differences; for example the fundamental Aubin–Nitsche method for lifting the order of convergence to lower Sobolev norms (see AUBIN [1967] and NITSCHE [1963]; see also OGENESJAN and RUCHOVEC [1969]) used such results. In 1974, the important paper of BREZZI [1974] mentioned earlier, used such earlier results on saddle point problems and laid the groundwork for a multitude of papers on problems with constraints and on the stability of various finite element procedures. While convergence of special types of finite element strategies such as mixed methods and hybrid methods had been attempted in the early 1970s (e.g. ODEN [1972]), the Brezzi results, and the methods of Babuška for constrained problems, provided a general framework for studying virtually all mixed and hybrid finite elements (e.g. RAVIART [1975], RAVIART and THOMAS [1977], BABUŠKA, ODEN and LEE [1977]).

The first textbook on mathematical properties of finite element methods was the popular book of STRANG and FIX [1973]. A book on an introduction to the mathematical theory of finite elements was published soon after by ODEN and REDDY [1976] and the well-known treatise on the finite element method for elliptic problems by CIARLET [1978] appeared two years later.

The penetrating work of NITSCHE and SCHATZ [1974] on interior estimates and SCHATZ and WAHLBIN [1978] on L^∞ estimates and singular problems represented notable contributions to the growing mathematical theory of finite elements. The important work of DOUGLAS and DUPONT (e.g. [1970, 1973]; DUPONT [1973]) on finite element methods for parabolic problems and hyperbolic problems must be mentioned along with the idea of elliptic projections of WHEELER [1973] which provided a useful technique for deriving error bounds for time-dependent problems.

The 1970s also represented a decade in which the generality of finite element methods began to be appreciated over a large portion of the mathematics and scientific community, and it was during this period that significant applications to highly nonlinear problems were made. The fact that very general nonlinear phenomena in continuum mechanics, including problems of finite deformation of solids and of flow of viscous fluids could be modeled by finite elements and solved on existing computers was demonstrated in the early seventies (e.g. ODEN [1972]), and, by the end of that decade, several "general purpose" finite element programs were in use by engineers to treat broad classes of nonlinear problems in solid mechanics and heat transfer. The mathematical theory for nonlinear problems also was advanced in this period, and the important work of FALK [1974] on finite element approximations of variational inequalities should be mentioned.

It is not too inaccurate to say that by 1980, a solid foundation for the mathematical theory of finite elements for linear problems had been established and that significant advances in both theory and application into nonlinear problems existed. The open questions that remain are difficult ones and their solution will require a good understanding of the mathematical properties of the method. The works collected in this volume should not only provide a summary of important results and approaches to mathematical issues related to finite elements, but also they should provide a useful starting point for further research.

References

ARGYRIS, J.H. (1954), Energy theorems and structural analysis, *Aircraft Engrg.* **26**, 347–356; 383–387; 394.

ARGYRIS, J.H. (1955), Energy theorems and structural analysis, *Aircraft Engrg.* **27**, 42–58; 80–94; 125–134; 145–158.

ARGYRIS, J.H. (1966), Continua and discontinua, in: *Proceedings Conference on Matrix Methods in Structural Mechanics*, Wright-Patterson AFB, Dayton, OH, 11–190.

AUBIN, J.P. (1967), Behavior of the error of the approximate solutions of boundary-value problems for linear elliptic operators by Galerkin's method and finite differences, *Ann. Scuola Norm. Pisa (3)* **21**, 599–637.

AZIZ, A.K., ed. (1972), *The Mathematical Foundations of the Finite Element Method with Applications to Partial Differential Equations* (Academic Press, New York).

BABUŠKA, I. (1970), Finite element methods for domains with corners, *Computing* **6**, 264–273.

BABUŠKA, I. (1971), Error bounds for the finite element method, *Numer. Math.* **16**, 322–333.

BABUŠKA, I. and A.K. AZIZ (1972), Survey lectures on the mathematical foundation of the finite element method, in: A.K. Aziz, ed. *The Mathematical Foundations of the Finite Element Method with Applications to Partial Differential Equations* (Academic Press, New York) 5–359.

BABUŠKA, I., J.T. ODEN and J.K. LEE (1977), Mixed-hybrid finite element approximations of second-order elliptic boundary-value problems, *Comput. Methods Appl. Mech. Engrg.* **11**, 175–206.

BIRKHOFF, G. (1969), Piecewise bicubic interpolation and approximation in polygons, in: I.J. Schoenberg, ed., *Approximations with Special Emphasis on Spline Functions* (Academic Press, New York) 85–121.

BIRKHOFF, G., C. DE BOOR, M.H. SCHULTZ and B. WENDROFF (1966), Rayleigh–Ritz approximation by piecewise cubic polynomials, *SIAM J. Numer. Anal.* **3**, 188–203.

BIRKHOFF, G., M.H. SCHULTZ and R.S. VARGA (1968), Piecewise Hermite interpolation in one and two variables with applications to partial differential equations, *Numer. Math.* **11**, 232–256.

BOGNER, F.K., R.L. FOX and L.A. SCHMIT Jr (1966), The generation of interelement, compatible stiffness and mass matrices by the use of interpolation formulas, in: *Proceedings Conference on Matrix Methods in Structural Mechanics*, Wright-Patterson AFB, Dayton, OH, 397–444.

BRAMBLE, J.H. and M. ZLÁMAL (1970), Triangular element in the finite element method, *Math. Comp.* **24** (112), 809–820.

BREZZI, F. (1974), On the existence, uniqueness, and approximation of saddle-point problems arising from Lagrange multipliers, *Rev. Française d'Automat. Inform. Rech. Opér.* **8-R2**, 129–151.

CEA, J. (1964), Approximation variationnelle des problems aux limites, *Ann. Inst. Fourier* (Grenoble) **14**, 345–444.

CIARLET, P.G. (1968), An $O(h^2)$ method for a non-smooth boundary-value problem, *Aequationes Math.* **2**, 39–49.

CIARLET, P.G. (1978), *The Finite Element Method for Elliptic Problems* (North-Holland, Amsterdam).

CIARLET, P.G. and P.A. RAVIART (1972a), General Lagrange and Hermite interpolation in $\mathbb{R}^n$ with applications to the finite elment method, *Arch. Rational Mech. Anal.* **46**, 177–199.

CIARLET, P.G. and P.A. RAVIART (1972b), Interpolation theory over curved elements with applications to finite element methods, *Comput. Methods Appl. Mech. Engrg.* **1**, 217–249.

CIARLET, P.G. and P.A. RAVIART (1972c), The combined effect of curved boundaries and numerical integration in isoparametric finite element methods, in: A.K. Aziz, ed., *The Mathematical Foundations of the Finite Element Method with Applications to Partial Differential Equations* (Academic Press, New York) 409–474.

CLOUGH, R.W. (1960), The finite element method in plane stress analysis, in: *Proceedings 2nd ASCE Conference on Electronic Computation*, Pittsburgh, PA.

COURANT, R. (1943), Variational methods for the solution of problems of equilibrium and vibration, *Bull. Amer. Math. Soc.* **49**, 1–23.

DOUGLAS, J. and T. DUPONT (1970), Galerkin methods for parabolic problems, *SIAM J. Numer. Anal.* **7**, 575–626.

DOUGLAS, J. and T. DUPONT (1973), Superconvergence for Galerkin methods for the two-point boundary problem via local projections, *Numer. Math.* **21**, 220–228.

DUPONT, T. (1973), L^2-estimates for Galerkin methods for second-order hyperbolic equations, *SIAM J. Numer. Anal.* **10**, 880–889.

ERGATOUDIS, I., B.M. IRONS and O.C. ZIENKIEWICZ (1966), Curved isoparametric quadrilateral finite elements, *Internat. J. Solids Structures* **4**, 31–42.

FALK, S.R. (1974), Error estimates for the approximation of a class of variational inequalities, *Math. Comp.* **28**, 963–971.

HERRMANN, L.R. (1966), A bending analysis for plates, in: *Proceedings Conference on Matrix Methods in Structural Mechanics*, Wright-Patterson AFB, Dayton, OH, 577.

HRENNIKOFF, H. (1941), Solutions of problems in elasticity by the framework method, *J. Appl. Mech.*, A169–175.

IRONS, B. (1966), Engineering applications of numerical integration in stiffness methods, *AIAA J.* **4**, 2035–3037

IRONS, B. (1970), A frontal solution program for finite element analysis, *Internat. J. Numer. Methods Engrg.* **2**, 5–32.

IRONS, B. and A. RAZZAQUE (1972), Experience with the patch test for convergence of finite elements, in: A.K. Aziz, Ed., *The Mathematical Foundations of the Finite Element Method with Applications to Partial Differential Equations* (Academic Press, New York) 557–587.

JOHNSON Jr, M.W. and R.W. MCLAY (1968), Convergence of the finite element method in the theory of elasticity, *J. Appl. Mech. E*, **35**, 274–278.

KANG, FENG (1965), A difference formulation based on the variational principle, *Appl. Math. Comput. Math.* **2**, 238–162 (in Chinese).

KRON, G. (1939), *Tensor Analysis of Networks* (Wiley, New York).

KRON, G. (1953), A set of principles to interconnect the solutions of physical systems, *J. Appl. Phys.* **24**, 965–980.

LEIBNIZ, G. (1962), *G.W. Leibniz Mathematische Schriften*, C. Gerhardt, ed. (G. Olms Verlagsbuchhandlung) 290–293.

LEVY, S. (1953), Structural analysis and influence coefficients for delta wings, *J. Aeronaut. Sci.* **20**.

LIONS, J. (1955), Problèmes aux limites en théorie des distributions, *Acta Math.* **94**, 13–153.

MCHENRY, D. (1943), A lattice analogy for the solution of plane stress problems, *J. Inst. Civ. Engrg.* **21**, 59–82.

NITSCHE, J.A. (1963), Ein Kriterium für die Quasi-Optimalität des Ritzschen Verfahrens, *Numer. Math.* **2**, 346–348.

NITSCHE, J.A. (1970), Lineare Spline-Funktionen und die Methoden von Ritz für elliptische Randwertprobleme, *Arch. Rational Mech. Anal.* **36**, 348–355.

NITSCHE, J.A. and A.H. SCHATZ (1974), Interior estimates for Ritz–Galerkin methods, *Math. Comp.* **28**, 937–958.

ODEN, J.T. (1962), Plate beam structures, Dissertation, Oklahoma State University, Stillwater, OK.

ODEN, J.T. (1969), A general theory of finite elements, II: Applications, *Internat. J. Numer. Methods Engrg.* **1**, 247–259.

ODEN, J.T. (1970), A finite element analogue of the Navier–Stokes equations, *J. Engrg. Mech. Div. ASCE* **96** (EM 4).

ODEN, J.T. (1972), *Finite Elements of Nonlinear Continua* (McGraw-Hill, New York).

ODEN, J.T. and J.N. REDDY (1976), *An Introduction to the Mathematical Theory of Finite Elements* (Wiley-Interscience, New York).

ODEN, J.T. and D. SOMOGYI (1968), Finite element applications in fluid dynamics, *J. Engrg. Mech. Div. ASCE* **95** (EM 4), 821–826.

OGENESJAN, L.A. and L.A. RUCHOVEC (1968), Variational-difference schemes for linear second-order elliptic equations in a two-dimensional region with piecewise smooth boundary, *U.S.S.R. Comput. Math. and Math. Phys.* **8** (1), 129–152.

OGENESJAN, L.A. and L.A. RUCHOVEC (1969), Study of the rate of convergence of variational difference schemes for second-order elliptic equations in a two-dimensional field with a smooth boundary, *U.S.S.R. Comput. Math. and Math. Phys.* **9** (5), 158–183.

OLIVEIRA, E.R. de Arantes e. (1968), Theoretical foundation of the finite element method, *Internat. J. Solids Structures* **4**, 926–952.

PESTEL, E. (1966), Dynamic stiffness matrix formulation by means of Hermitian polynomials, in: *Proceedings Conference on Matrix Methods in Structural Mechanics*, Wright-Patterson AFB, Dayton, OH, 479–502.

PIAN, T.H.H. (1966), Element stiffness matrices for boundary compatibility and for prescribed stresses, in: *Proceedings Conference on Matrix Methods in Structural Mechanics*, Wright-Patterson AFB, Dayton, OH, 455–478.

PÓLYA, G. (1952), Sur une interprétation de la méthode des différences finies qui peut fournir des bornes supérieures ou inférieures, *C.R. Acad. Sci. Paris* **235**, 995–997.

PRZEMIENIECKI, J.S., R.M. BADER, W.F. BOZICH, J.R. JOHNSON and W.J. MYKYTOW, eds. (1966), *Proceedings Conference on Matrix Methods in Structural Mechanics*, Wright-Patterson AFB, Dayton, OH.

RAVIART, P.A. (1975), Hybrid methods for solving 2nd-order elliptic problems, in: J.H.H. Miller, ed., *Topics in Numerical Analysis* (Academic Press, New York) 141–155.

RAVIART, P.A. and J.M. THOMAS (1977), A mixed finite element method for 2nd-order elliptic problems, in: *Proceedings Symposium on the Mathematical Aspects of the Finite Element Methods*, Rome.

SCHATZ, A.H. and L.B. WAHLBIN (1977), Interior maximum norm estimates for finite element methods, *Math. Comp.* **31**, 414–442.

SCHATZ, A.H. and L.B. WAHLBIN (1978), Maximum norm estimates in the finite element method on polygonal domains, Part I, *Math. Comp.* **32**, 73–109.

SCHELLBACH, K. (1851), Probleme der Variationsrechnung, *J. Reine Angew. Math.* **41**, 293–363.

SCHULTZ, M.H. (1969a), L^∞-multivariate approximation theory, *SIAM J. Numer. Anal.* **6**, 161–183.

SCHULTZ, M.H. (1969b), Rayleigh–Ritz–Galerkin methods for multi-dimensional problems, *SIAM J. Numer. Anal.* **6**, 523–538.

SCHULTZ, M.H. (1971), L^2 error bounds for the Rayleigh–Ritz–Galerkin method, *SIAM J. Numer. Anal.* **8**, 737–748.

STRANG, G. (1972), Variational crimes in the finite element method, in: A.K. Aziz, ed., *The Mathematical Foundations of the Finite Element Method with Applications to Partial Differential Equations* (Academic Press, New York).

STRANG, G. and G. FIX (1973), *An Analysis of the Finite Element Method* (Prentice-Hall, Englewood Cliffs, NJ).

SYNGE, J.L. (1957), *The Hypercircle Method in Mathematical Physics* (Cambridge University Press, Cambridge).

TAIG, I.C. (1961), Structural analysis by the matrix displacement method, English Electrial Aviation Ltd. Report, S-O-17.

TURNER, M.J., R.W. CLOUGH, H.C. MARTIN and L.J. TOPP (1956), Stiffness and deflection analysis of complex structures, *J. Aero. Sci.* **23**, 805–823.

VARGA, R.S. (1966), Hermite interpolation-type Ritz methods for two-point boundary value problems, J.H. Bramble, ed., *Numerical Solution of Partial Differential Equations* (Academic Press, New York).

WHEELER, M.F. (1973), A-priori L^2-error estimates for Galerkin approximations to parabolic partial differential equations, *SIAM J. Numer. Anal.* **11**, 723–759.

WILLIAMSON, F. (1980), A historical note on the finite element method, *Internat. J. Numer. Methods Engrg.* **15**, 930–934.

ZIENKIEWICZ, O.C. and Y.K. CHEUNG (1967), *The Finite Element Method in Structural and Continuum Mechanics* (McGraw-Hill, New York).

ZLÁMAL, M. (1968), On the finite element method, *Numer. Math.* **12**, 394–409.

Basic Error Estimates for Elliptic Problems

P.G. Ciarlet

Analyse Numérique, Tour 55–65
Université Pierre et Marie Curie
4, Place Jussieu
75005 Paris, France

HANDBOOK OF NUMERICAL ANALYSIS, VOL. II
Finite Element Methods (Part 1)
Edited by P.G. Ciarlet and J.L. Lions

Contents

Preface

The *objectives* of this article are to give a thorough *mathematical description* of the finite element method applied to *linear elliptic problems* of the *second* and *fourth order* that typically arise in *linearized elasticity* (including the "almost linear" problems modeled by variational inequalities) and to establish the corresponding "*global*" *error estimates*; the "interior" error estimates are established in the next article by Wahlbin.

The *prerequisites* consist essentially in a good knowledge of analysis, functional analysis, and a certain familiarity with Sobolev spaces and linear elliptic partial differential equations. Apart from these, the article is essentially self-contained.

The *main topics covered* are the following (more detailed informations are provided in the introductions to each chapter):

- Description and mathematical analysis of various problems of linearized elasticity, such as the membrane and plate problems, the boundary value problem of three-dimensional elasticity, the obstacle problem (Chapter I).
- Description of the conforming finite elements currently used for approximating second- and fourth-order problems, including composite and singular elements in the latter case (Chapters II and VII).
- Derivation of the fundamental error estimates in the H^1-norm, L^2-norm, and L^∞-norm, for conforming finite element methods applied to second-order problems, including detailed analyses of the discrete maximum principle, and of the method of weighted norms of Nitsche (Chapter III).
- Derivation of the error estimates in the H^1-norm for the obstacle problem (Chapter III).
- Description of finite element methods with numerical integration for second-order problems, and derivation of the corresponding error estimates in the H^1-norm (Chapter IV).
- Description of nonconforming finite element methods for second- and fourth-order problems, and derivation of the corresponding error estimates in the discrete H^1- and H^2-norm (Chapters V and VII).
- Description of the combined use of isoparametric finite elements and of isoparametric numerical integration for second-order problems posed over domains with curved boundaries, and derivation of the corresponding error estimates in the H^1-norm (Chapter VI).
- Derivation of the error estimates in the H^2-norm for the polynomial, composite, and singular, finite elements used for solving fourth-order problems (Chapter VII).

In addition, various relevant extensions, refinements, etc., of these estimates have been mentioned, with appropriate references to the existing literature.

This article is a revised, updated, and enlarged edition of those parts of my book, *The Finite Element Method for Elliptic Problems* (North-Holland, Amsterdam, 1978) that are relevant here. Although I have added more than 230 items to the 316 references that I have kept from this book, I have made no attempt to compile an exhaustive bibliography. I however hope that the present bibliography is "reasonably complete".

The various finite elements described in this article have been named according to the most common usages in the engineering literature. As a result, the terminology adopted in this article often departs strikingly from that used in my book of 1978. For instance, what I then called a "triangle of type (1)" or a "rectangle of type (2)" are now called a "linear triangle" or a "biquadratic rectangle". Since the readers of this article are perfectly aware that a triangle may be isoceles, but certainly not genuinely "linear", or that a rectangle may be square, but certainly not genuinely "biquadratic", I have blithely committed these serious *abus de langage*, which clearly convey more information than the names that I had originally chosen.

CHAPTER I

Elliptic Boundary Value Problems

Introduction

Many problems in linearized elasticity are modeled by a minimization problem of the following form: The unknown u, which is the *displacement* of a *mechanical system*, satisfies

$$u \in U \quad \text{and} \quad J(u) = \inf_{v \in U} J(v),$$

where the set U of *admissible displacements* is a closed convex subset of a Hilbert space V, and the *energy* J of the system takes the form

$$J(v) = \tfrac{1}{2}a(v, v) - l(v),$$

where $a(\cdot, \cdot)$ is a symmetric bilinear form and l a linear form, both defined and continuous over the space V. In Section 1, we first prove a general existence result (Theorem 1.1) for such minimization problems, the main assumptions being the completeness of the space V and the *V-ellipticity* of the bilinear form. We also describe other formulations of the same problem (Theorem 1.2), which are its *variational formulations.* When the bilinear form is not symmetric, these formulations make up *variational problems* on their own. For such problems, we give an existence theorem when $U = V$ (Theorem 1.3), which is the celebrated *Lax–Milgram lemma.*

All these problems are called *abstract problems* inasmuch as they represent an "abstract" formulation which is common to many examples, such as those examined in this chapter.

The analysis made in Section 1 shows that a candidate for the space V must have the following properties: It must be complete on the one hand, and it must be such that the expression $J(v)$ is well defined for all functions $v \in V$ on the other hand (V is a "space of finite energy"). The *Sobolev spaces* fulfill these requirements. After briefly mentioning in Section 2 some of their basic properties (other properties will be introduced in later sections as needed), we examine in Sections 3 and 4 specific examples that fit in the abstract setting of Section 1, such as the *membrane problem*, the *obstacle problem*, the *clamped plate problem*, and the *boundary value problem of linearized elasticity*, which is by far the most important example. Indeed, even though throughout this article we will often find it convenient to work with the simpler looking problems described at the beginning of Section 3, it must not be forgotten that these are essentially convenient *model problems* for the boundary value problem of linearized elasticity.

For each of the examples, we establish in particular the *V-ellipticity* of the associated bilinear form, and, using various *Green's formulae* in Sobolev spaces, we show that when solving these problems, one solves, at least *formally*, *elliptic boundary value problems of the second and fourth order*.

1. Abstract minimization problems, variational inequalities and the Lax–Milgram lemma

All functions and vector spaces considered in this article are real.

Let there be given a normed vector space V with norm $\|\cdot\|$, a *continuous* bilinear form $a(\cdot,\cdot)\colon V\times V\to\mathbb{R}$, a *continuous* linear form $l\colon V\to\mathbb{R}$ and a nonempty subset U of the space V. With these data we associate an *abstract minimization problem*: Find an element u such that

$$u\in U,\qquad J(u)=\inf_{v\in U} J(v), \tag{1.1}$$

where the functional $J\colon V\to\mathbb{R}$ is defined by

$$J\colon v\in V\to J(v)=\tfrac{1}{2}a(v,v)-l(v). \tag{1.2}$$

As regards existence and uniqueness properties of the solution of this problem, the following result is essential.

THEOREM 1.1. *Assume in addition that* (i) *the space V is complete,* (ii) *U is a closed convex subset of V,* (iii) *the bilinear form $a(\cdot,\cdot)$ is symmetric and* (iv) *the bilinear form is V-elliptic, in the sense that there exists a constant α such that*

$$\begin{aligned}&\alpha>0,\\ &\alpha\|v\|^2\leqslant a(v,v)\quad \textit{for all } v\in V.\end{aligned} \tag{1.3}$$

Then the abstract minimization problem (1.1) *has one and only one solution.*

PROOF. The bilinear form $a(\cdot,\cdot)$ is an inner product over the space V, and the associated norm is equivalent to the given norm $\|\cdot\|$. Thus the space V is a Hilbert space when it is equipped with this inner product. By the Riesz representation theorem, there exists an element $\sigma l\in V$ such that

$$l(v)=a(\sigma l,v)\quad\text{for all } v\in V,$$

so that, taking again into account the symmetry of the bilinear form, we may rewrite the functional as

$$J(v)=\tfrac{1}{2}a(v,v)-a(\sigma l,v)=\tfrac{1}{2}a(v-\sigma l,v-\sigma l)-\tfrac{1}{2}a(\sigma l,\sigma l).$$

Hence solving the abstract minimization problem amounts to minimizing the distance between the element σl and the set U, with respect to the norm $\sqrt{a(\cdot,\cdot)}$. Consequently, the solution is simply the projection of the element σl onto the set U, with respect to the inner product $a(\cdot,\cdot)$. By the projection theorem, such a projection exists and is unique, since U is a closed convex subset of the space V. □

Next, we give equivalent formulations of this problem.

THEOREM 1.2. *An element u is the solution of the abstract minimization problem* (1.1) *if and only if it satisfies the relations*

$$\begin{aligned} &u \in U, \\ &a(u, v-u) \geqslant l(v-u) \quad \textit{for all } v \in U, \end{aligned} \tag{1.4}$$

in the general case, or

$$\begin{aligned} &u \in U, \\ &a(u, v) \geqslant l(v) \quad \textit{for all } v \in U, \\ &a(u, u) = l(u), \end{aligned} \tag{1.5}$$

if U is a closed convex cone with vertex 0, *or*

$$\begin{aligned} &u \in U, \\ &a(u, v) = l(v) \quad \textit{for all } v \in U, \end{aligned} \tag{1.6}$$

if U is a closed subspace.

PROOF. The projection u of σl onto U is completely characterized by the relations

$$\begin{aligned} &u \in U, \\ &a(\sigma l - u, v-u) \leqslant 0 \quad \text{for all } v \in U, \end{aligned} \tag{1.7}$$

the geometrical interpretation of the last inequalities being that the angle between the vectors $(\sigma l - u)$ and $(v-u)$ is obtuse (Fig. 1.1) for all $v \in U$. These inequalities may

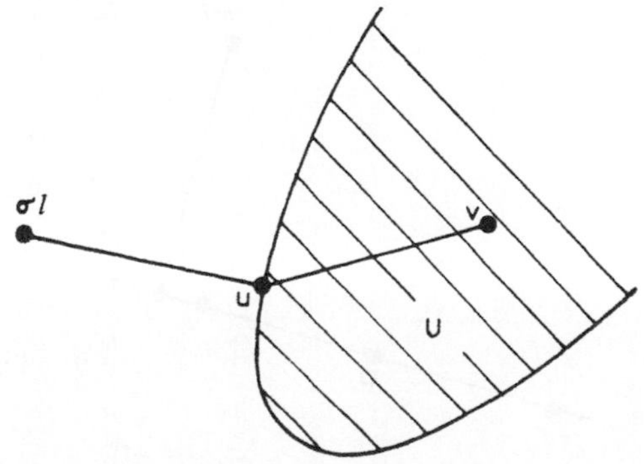

FIG. 1.1. In a Hilbert space with inner product $a(\cdot,\cdot)$, the projection u of σl onto the closed convex set U is characterized by the variational inequalities $a(\sigma l - u, v-u) \leqslant 0$ for all $v \in U$.

be written as

$$a(u, v-u) \geqslant a(\sigma l, v-u) = l(v-u) \quad \text{for all } v \in U,$$

which proves relations (1.4).

Assume next that U is a closed convex cone with vertex 0. Then the point $(u+v)$ belongs to the set U whenever the point v belongs to the set U (Fig. 1.2). Therefore,

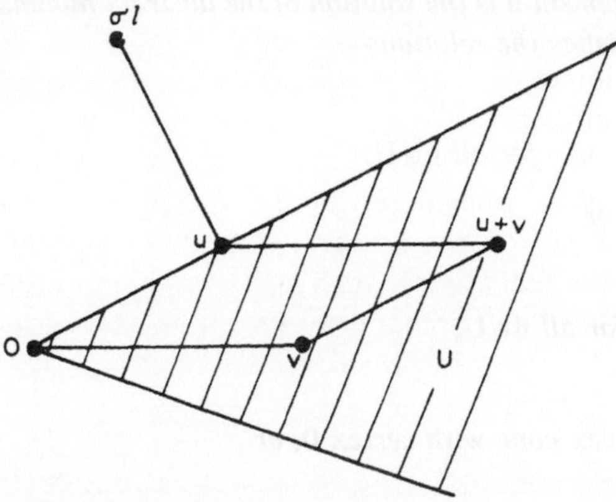

FIG. 1.2. In a Hilbert space with inner product $a(\cdot,\cdot)$, the projection u of σl onto a closed convex cone U with vertex 0 is characterized by the equation $a(u-\sigma l, u)=0$ and by the variational inequalities $a(u-\sigma l, v) \geqslant 0$ for all $v \in U$.

upon replacing v by $(u+v)$ in inequalities (1.4), we obtain the inequalities

$$a(u, v) \geqslant l(v) \quad \text{for all } v \in U,$$

so that, in particular, $a(u, u) \geqslant l(u)$. Letting $v=0$ in (1.4), we obtain $a(u, u) \leqslant l(u)$, and thus relations (1.5) are proved. The converse is clear.

If U is a subspace (Fig. 1.3), then inequalities (1.5) written with v and $-v$ yield

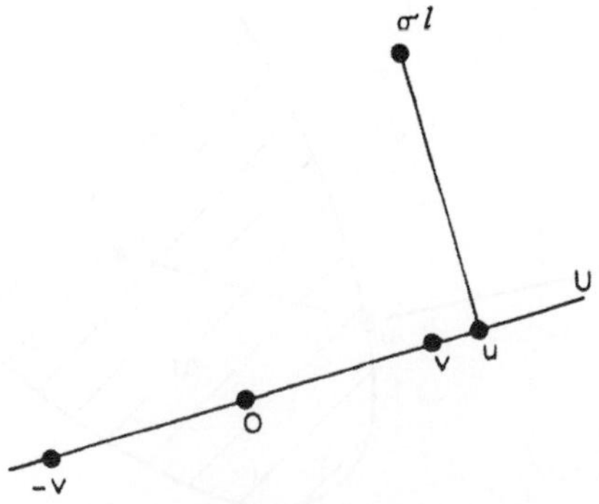

FIG. 1.3. In a Hilbert space with inner product $a(\cdot,\cdot)$, the projection u of σl onto a closed subspace U is characterized by the variational equations $a(\sigma l-u, v)=0$ for all $v \in U$.

$a(u, v) \geqslant l(v)$ and $a(u, v) \leqslant l(v)$ for all $v \in U$, from which relations (1.6) follow. Again the converse is clear. □

REMARK 1.1. Since the projection mapping is linear if and only if the subset U is a subspace, it follows that *problems associated with variational inequalities are generally nonlinear*, the linearity or nonlinearity being that of the mapping $l \in V' \to u \in V$, where V' is the dual space of V, all other data being fixed. One should not forget, however, that if the resulting problem is linear when one minimizes over a subspace, this is also because the functional is *quadratic*, i.e., it is of the form (1.2). The minimization of more general functionals over a subspace would correspond to nonlinear problems.

The various equivalent formulations of the minimization problem (1.1) *given in Theorems* 1.1 *and* 1.2 *may be also interpreted from the point of view of differential calculus*, as follows. We first observe that the functional J is differentiable at every point $u \in V$, the action of its Fréchet derivative $J'(u) \in V'$ on an arbitrary element $v \in V$ being given by

$$J'(u)v = a(u, v) - l(v). \tag{1.8}$$

Let then u be the solution of the minimization problem (1.1) and let $v = u + w$ be any point of the convex set U. Since the points $(u + \theta w)$ belong to the set U for all $\theta \in [0, 1]$ (Fig. 1.4), we have, by definition of the derivative $J(u)$,

$$0 \leqslant J(u + \theta w) - J(u) = \theta J'(u)w + \theta \| w \| \varepsilon(\theta)$$

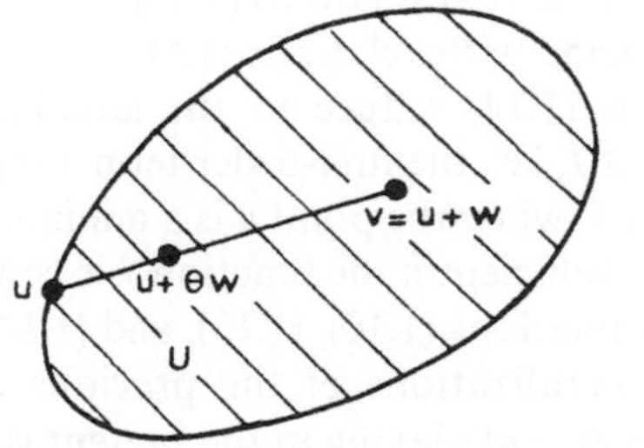

FIG. 1.4. If u belongs to a convex set U and if $J(u) = \inf_{u \in U} J(v)$, then for each $v = (u + w) \in U$, $J(u + \theta w) - J(u) \geqslant 0$ for all $0 \leqslant \theta \leqslant 1$.

for all $\theta \in [0, 1]$, with $\lim_{\theta \to 0} \varepsilon(\theta) = 0$. As a consequence, we necessarily have

$$J'(u)w \geqslant 0, \tag{1.9}$$

since otherwise the difference $J(u + \theta w) - J(u)$ would be < 0 for θ small enough. Using (1.8), we may rewrite inequality (1.9) as

$$J'(u)w = J'(u)(v - u) = a(u, v - u) - l(v - u) \geqslant 0,$$

which is precisely (1.4). Conversely, assume that we have found an element $u \in U$ such that

$$J'(u)(v-u) \geqslant 0 \quad \text{for all } v \in U. \tag{1.10}$$

The second derivative $J''(u) \in \mathscr{L}_2(V; \mathbb{R})$ of the functional J is independent of $u \in V$ and its action on arbitrary elements $v \in V$ and $w \in V$ is given by

$$J''(u)(v, w) = a(v, w). \tag{1.11}$$

Thus, for any point $v = u + w$ belonging to the set U, an application of Taylor's formula yields

$$J(u+w) - J(u) = J'(u)(w) + \tfrac{1}{2}a(w, w) \geqslant \tfrac{1}{2}\alpha \| w \|^2, \tag{1.12}$$

which shows that u is a solution of problem (1.1). We have $J(v) - J(u) > 0$ unless $v = u$ so that we see once again that the solution is unique.

Arguing as in the proof of Theorem 1.2, we then easily verify that inequalities (1.10) are equivalent to the relations

$$J'(u)v \geqslant 0, \quad J'(u)u = 0 \qquad \text{for all } v \in U, \tag{1.13}$$

when U is a convex cone with vertex 0, and that they reduce to

$$J'(u)v = 0 \quad \text{for all } v \in U, \tag{1.14}$$

when U is a subspace. Notice that relations (1.13) coincide with relations (1.5), and that relations (1.14) coincide with relations (1.6).

When $U = V$, relations (1.14) reduce to the familiar condition that the *first variation* of the functional J, i.e., the first-order term $J'(u)w$ in the Taylor expansion (1.12), vanishes for all $w \in V$ when the point u is a minimum of the function $J: V \to \mathbb{R}$, this condition being also sufficient if the function J is convex, as is the case here. By means of the equivalent relations (1.10), (1.13), and (1.14), relations (1.4), (1.5), and (1.6) thus appear as generalizations of the previous condition, the expression $a(u, v-u) - f(v-u) = J'(u)(v-u)$ playing in the present situation the role of the first variation of the functional J *relative to the convex set* U. It is in this sense that the formulations of Theorem 1.2 are called *variational*.

More precisely, the characterizations (1.4), (1.5), and (1.6) are called *variational formulations* of the original minimization problem, the equations (1.6) are called *variational equations*, and the inequalities of (1.4) and (1.5) are called *variational inequalities*.

Without making explicit reference to the functional J, we can also define various *abstract variational problems*: Find an element u such that

$$\begin{aligned} &u \in U, \\ &a(u, v-u) \geqslant l(v-u) \quad \text{for all } v \in U, \end{aligned} \tag{1.15}$$

in the general case, or, find an element u such that

$$\begin{aligned} &u \in U, \\ &a(u, v) \geqslant l(v) \quad \text{for all } v \in U, \\ &a(u, u) = l(u), \end{aligned} \tag{1.16}$$

if U is a cone with vertex 0, or, finally, find an element u such that

$$\begin{aligned} &u \in U, \\ &a(u, v) = l(v) \quad \text{for all } v \in V, \end{aligned} \tag{1.17}$$

if U is a subspace. By Theorem 1.1, each of these problems has one and only one solution if the space V is complete, if the subset U of V is closed and convex, and if the bilinear form is V-elliptic, continuous, and symmetric. *If the assumption of symmetry of the bilinear form is dropped, the above variational problems still have one and only one solution if the space V is a Hilbert space, but there is no longer an associated minimization problem.* Here we shall confine ourselves to the case where $U = V$.

THEOREM 1.3. (Lax–Milgram lemma). *Let V be a Hilbert space, let $a(\cdot,\cdot)\colon V \times V \to \mathbb{R}$ be a continuous V-elliptic bilinear form, and let $l\colon V \to \mathbb{R}$ be a continuous linear form.*

Then the abstract variational problem: Find an element u such that

$$\begin{aligned} &u \in V, \\ &a(u, v) = l(v) \quad \textit{for all } v \in V, \end{aligned} \tag{1.18}$$

has one and only one solution.

PROOF. Let M be a constant such that

$$|a(u, v)| \leqslant M \, \|u\| \, \|v\| \quad \text{for all } u, v \in V. \tag{1.19}$$

For each $u \in V$, the linear form $v \in V \to a(u, v)$ is continuous and thus there exists a unique element $Au \in V'$ (V' is the dual space of V) such that

$$a(u, v) = Au(v) \quad \text{for all } v \in V. \tag{1.20}$$

Denoting by $\|\cdot\|'$ the norm in the space V', we have

$$\|Au\|' = \sup_{v \in V} \frac{|Au(v)|}{\|v\|} \leqslant M \, \|u\|.$$

Consequently, the linear mapping $A\colon V \to V'$ is continuous, with

$$\|A\|_{\mathscr{L}(V;V')} \leqslant M. \tag{1.21}$$

Let $\tau\colon V' \to V$ denote the Riesz mapping which is such that, by definition,

$$l(v) = ((\tau l, v)) \quad \text{for all } l \in V' \text{ and all } v \in V, \tag{1.22}$$

where $((\cdot,\cdot))$ denotes the inner- product in the space V. Then solving the variational problem (1.18) is equivalent to solving the equation $\tau A u = \tau l$. We will show that this equation has one and only one solution by showing that, for appropriate values of a parameter $\rho>0$, the affine mapping

$$v\in V\to v-\rho(\tau Av-\tau l)\in V \tag{1.23}$$

is a contraction. To see this, we observe that

$$\begin{aligned}\|v-\rho\tau Av\|^2&=\|v\|^2-2\rho((\tau Av,v))+\rho^2\|\tau Av\|^2\\&\leqslant(1-2\rho\alpha+\rho^2M^2)\|v\|^2,\end{aligned}$$

since, by inequalities (1.3) and (1.21),

$$((\tau Av,v))=Av(v)=a(v,v)\geqslant\alpha\|v\|^2,$$
$$\|\tau Av\|=\|Av\|'\leqslant\|A\|\,\|v\|\leqslant M\|v\|.$$

Therefore the mapping defined in (1.23) is a contraction whenever the number ρ belongs to the interval $]0,2\alpha/M^2[$ and the proof is complete. □

REMARK 1.2. It follows from the previous proof that the mapping $A\colon V\to V'$ is onto. Since

$$\alpha\|u\|^2\leqslant a(u,u)=l(u)\leqslant\|l\|'\,\|u\|,$$

the mapping A has a continuous inverse A^{-1}, with

$$\|A^{-1}\|_{\mathscr{L}(V';V)}\leqslant 1/\alpha.$$

Therefore the variational problem (1.18) is *well-posed* in the sense that its solution *exists, is unique*, and *depends continuously on the data f* (all other data being equal).

More generally, one can show that, if u_1 and u_2 are solutions of problem (1.15) corresponding to linear forms l_1 and l_2, then

$$\|u_1-u_2\|\leqslant\frac{1}{\alpha}\|l_1-l_2\|'.$$

The original reference of the Lax–Milgram lemma is LAX and MILGRAM [1954]. Our proof follows the method of LIONS and STAMPACCHIA [1967], where it is applied to the general variational problem (1.15), and where the case of semipositive-definite bilinear forms is also considered; STAMPACCHIA [1964] had the original proof in this case. For constructive existence proofs and additional references, see also GLOWINSKI, LIONS and TRÉMOLIÈRES [1976a]. We also mention that Babuška (BABUŠKA and AZIZ [1972, Theorem 5.2.1]) has extended the Lax–Milgram lemma to the case of bilinear forms defined on a product of two distinct Hilbert spaces.

2. The Sobolev spaces $H^m(\Omega)$ and Green's formulae

For treatments of differential calculus with Fréchet derivatives, the reader may consult AVEZ [1983], CARTAN [1967], DIEUDONNÉ [1967], SCHWARTZ [1967]. For the theory of distributions and its applications to partial differential equations, see

SCHWARTZ [1966]. Other references are TRÈVES [1967], SHILOV [1968], VO-KHAC KHOAN [1972a, 1972b], CHOQUET-BRUHAT [1973], HÖRMANDER [1983]. The Hilbertian Sobolev spaces $H^m(\Omega)$ are studied in LIONS and MAGENES [1968, Chapter 1], DAUTRAY and LIONS [1984, Chapter 4] (references on the more general Sobolev spaces $W^{m,p}(\Omega)$ are given in Section 14).

Let us first briefly recall some results from differential calculus. Let there be given two normed vector spaces X and Y and a function v: $A\to Y$, where A is a subset of X. If the function is k times differentiable at a point $a\in A$, we shall denote by $D^k v(a)$, or simply by $Dv(a)$ if $k=1$, its kth *Fréchet derivative*. Note that we also use the alternate notations $Dv(a)=v'(a)$ and $D^2v(a)=v''(a)$. The kth derivative $D^k v(a)$ is a symmetric element of the space $\mathscr{L}_k(X;Y)$, and its norm is given by

$$\|D^k v(a)\| = \sup_{\substack{\|h_i\|\leqslant 1\\ 1\leqslant i\leqslant k}} \|D^k v(a)(h_1, h_2, \ldots, h_k)\|.$$

In the special case where $X=\mathbb{R}^n$ and $Y=\mathbb{R}$, let e_i, $1\leqslant i\leqslant n$, denote the canonical basis vectors of $\mathbb{R}^n$. Then the usual *partial derivatives* are given by

$$\partial_i v(a) = Dv(a)e_i,$$
$$\partial_{ij} v(a) = D^2 v(a)(e_i, e_j),$$
$$\partial_{ijk} v(a) = D^3 v(a)(e_i, e_j, e_k), \quad \text{etc.},$$

and occasionally, we use the notation $\nabla v(a)$, or $\boldsymbol{\nabla} v(a)$, to denote the *gradient* of the function v at the point a, i.e., the vector in $\mathbb{R}^n$ whose components are the partial derivatives $\partial_i v(a)$, $1\leqslant i\leqslant n$.

We also use the *multi-index notation* for denoting the partial derivatives: Given a multi-index $\alpha=(\alpha_1, \alpha_2, \ldots, \alpha_n)\in\mathbb{N}^n$, we let $|\alpha|=\sum_{i=1}^n \alpha_i$. Then the partial derivative $\partial^\alpha v(a)$ is the result of the application of the $|\alpha|$th derivative $D^{|\alpha|}v(a)$ to any $|\alpha|$-vector of $(\mathbb{R}^n)^{|\alpha|}$ where each vector e_i occurs α_i times, $1\leqslant i\leqslant n$. For instance, if $n=3$, we have $\partial_1 v(a)=\partial^{(1,0,0)}v(a)$, $\partial_{123}v(a)=\partial^{(1,1,1)}v(a)$, $\partial_{111}v(a)=\partial^{(3,0,0)}v(a), \ldots$ Clearly, there exist constants $C(m,n)$ such that for any partial derivative $\partial^\alpha v(a)$ with $|\alpha|=m$ and any function v,

$$|\partial^\alpha v(a)| \leqslant \|D^m v(a)\| \leqslant C(m,n) \max_{|\alpha|=m} |\partial^\alpha v(a)|$$

(unless otherwise specified, it is understood that the space $\mathbb{R}^n$ is equipped with the Euclidean norm).

As a rule, we represent by symbols such as $D^k v$, v'', $\partial_i v$, $\partial^\alpha v, \ldots$, the *functions* associated with any derivative or partial derivative.

When $h_1=h_2=\cdots=h_k=h$, we simply write

$$D^k v(a)(h_1, h_2, \ldots, h_k) = D^k v(a)h^k.$$

Thus, given a real-valued function v, *Taylor's formula of order* k is written as

$$v(a+h) = v(a) + \sum_{l=1}^{k} \frac{1}{l!} D^l v(a)h^l + \frac{1}{(k+1)!} D^{k+1} v(a+\theta h)h^{k+1},$$

for some $\theta\in\,]0,1[$ (whenever such a formula applies).

Given a bounded open subset Ω in $\mathbb{R}^n$, the space $\mathscr{D}(\Omega)$ consists of all indefinitely differentiable functions $v: \Omega \to \mathbb{R}$ with compact support.

For each integer $m \geqslant 0$, the *Sobolev space* $H^m(\Omega)$ consists of those functions $v \in L^2(\Omega)$ for which the partial derivatives $\partial^\alpha v$ in the distributional sense with $|\alpha| \leqslant m$, belong to the space $L^2(\Omega)$ for all $|\alpha| \leqslant m$, i.e., for each multi-index α with $|\alpha| \leqslant m$, there exists a function $\partial^\alpha v \in L^2(\Omega)$ that satisfies

$$\int_\Omega \partial^\alpha v \phi \, dx = (-1)^{|\alpha|} \int_\Omega v \partial^\alpha \phi \, dx \quad \text{for all } \phi \in \mathscr{D}(\Omega). \tag{2.1}$$

Equipped with the norm

$$v \to \|v\|_{m,\Omega} = \Bigg(\sum_{|\alpha| \leqslant m} \int_\Omega |\partial^\alpha v|^2 \, dx \Bigg)^{1/2},$$

the space $H^m(\Omega)$ is a Hilbert space. We shall also make frequent use of the *semi-norm*

$$|v|_{m,\Omega} = \Bigg(\sum_{|\alpha| = m} \int_\Omega |\partial^\alpha v|^2 \, dx \Bigg)^{1/2}.$$

We define the *Sobolev space*

$$H_0^m(\Omega) = \overline{\mathscr{D}(\Omega)},$$

the closure being understood in the sense of the norm $\|\cdot\|_{m,\Omega}$. When the set Ω is bounded, *the Poincaré–Friedrichs inequality holds*: there exists a constant $C(\Omega)$ such that

$$|v|_{0,\Omega} \leqslant C(\Omega) |v|_{1,\Omega} \quad \text{for all } v \in H_0^1(\Omega). \tag{2.2}$$

Therefore, *when the set Ω is bounded*, the seminorm $|\cdot|_{m,\Omega}$ is a norm over the space $H_0^m(\Omega)$, equivalent to the norm $\|\cdot\|_{m,\Omega}$ (see also Theorem 3.1).

Following NEČAS [1967], we say that an open set Ω has a *Lipschitz-continuous boundary* Γ if the following conditions are fulfilled: There exist constants $\alpha > 0$ and $\beta > 0$, and a finite number of local coordinate systems and local maps a_r, $1 \leqslant r \leqslant R$, that are Lipschitz-continuous on their respective domains of definitions $\{\hat{x}^r \in \mathbb{R}^{n-1}; |\hat{x}^r| \leqslant \alpha\}$, such that (Fig. 2.1):

$$\Gamma = \bigcup_{r=1}^{R} \{(x_1^r, \hat{x}^r); x_1^r = a_r(\hat{x}^r), |\hat{x}^r| < \alpha\},$$

$$\{(x_1^r, \hat{x}^r); a_r(\hat{x}^r) < x_1^r < a_r(\hat{x}^r) + \beta, |\hat{x}^r| < \alpha\} \subset \Omega, \quad 1 \leqslant r \leqslant R,$$

$$\{(x_1^r, \hat{x}^r); a_r(\hat{x}^r) - \beta < x_1^r < a_r(\hat{x}^r), |\hat{x}^r| < \alpha\} \subset \mathbb{R}^n - \bar{\Omega}, \quad 1 \leqslant r \leqslant R,$$

where $\hat{x}^r = (x_2^r, \ldots, x_n^r)$, and $|\hat{x}^r| < \alpha$ stands for $|x_i^r| < \alpha$, $2 \leqslant i \leqslant n$.

Occasionally, we shall also need the following definitions: A boundary is *of class* $\mathscr{X}$ if the functions $a_r: |\hat{x}^r| \leqslant \alpha \to \mathbb{R}$ are of class $\mathscr{X}$ (such as $\mathscr{C}^m$ or $\mathscr{C}^{m,\alpha}$), and a boundary

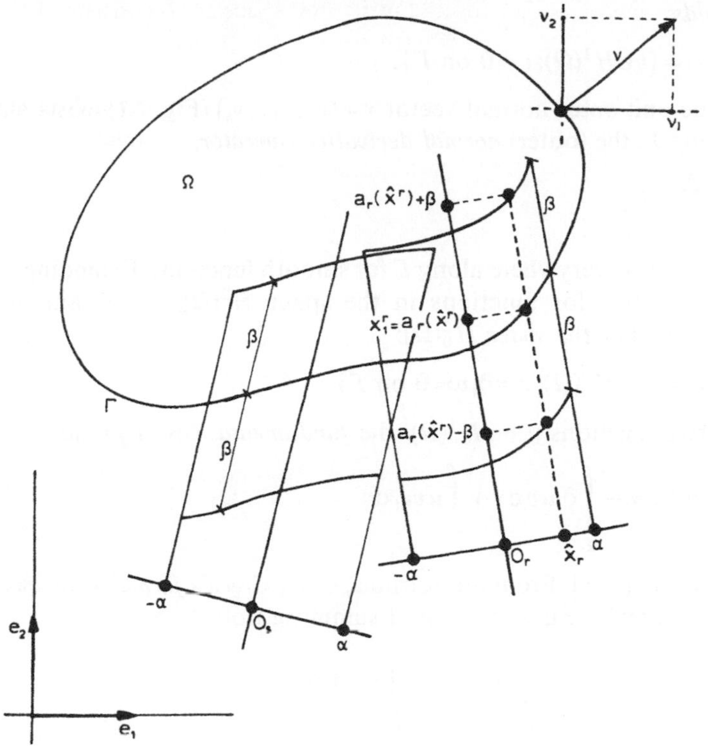

FIG. 2.1. A domain in $\mathbb{R}^2$ is an open, bounded, connected subset with a Lipschitz-continuous boundary.

is said to be *sufficiently smooth* if it is of class $\mathscr{C}^m$, or $\mathscr{C}^{m,\alpha}$, for sufficiently high values of m, or m and α.

A *domain* in $\mathbb{R}^n$ is an open, bounded, connected subset with a Lipschitz-continuous boundary. This definition is particularly well suited for our subsequent purposes, in that it allows the consideration of domains with "corners" or "edges", such as polyhedra.

In the remaining part of this section, it will be always understood that Ω is a domain in $\mathbb{R}^n$. In particular then, a superficial measure, which we shall denote $d\gamma$, can be defined along the boundary, so that it makes sense to consider the spaces $L^2(\Gamma)$, whose norm shall be denoted $\|\cdot\|_{L^2(\Gamma)}$.

Then it can be proved that there exists a constant $C(\Omega)$ such that

$$\|v\|_{L^2(\Gamma)} \leqslant C(\Omega)\|v\|_{1,\Omega} \quad \text{for all } v \in \mathscr{C}^\infty(\bar{\Omega}). \tag{2.3}$$

Since $\{\mathscr{C}^\infty(\bar{\Omega})\}^- = H^1(\Omega)$ if Ω is a domain, the closure being taken with respect to the norm $\|\cdot\|_{1,\Omega}$, there exists a continuous linear mapping $\operatorname{tr}: v \in H^1(\Omega) \to \operatorname{tr} v \in L^2(\Gamma)$, which is called the *trace operator*. Note however that, when no confusion should arise, we shall simply write $\operatorname{tr} v = v$. The following *characterization of the space*

$H_0^1(\Omega)$ holds:

$$H_0^1(\Omega)=\{v\in H^1(\Omega);\, v=0 \text{ on } \Gamma\}.$$

Since the unit outer normal vector $\nu=(\nu_1,\dots,\nu_n)$ (Fig. 2.1) exists almost everywhere along Γ, the (outer) *normal derivative operator*,

$$\partial_\nu=\sum_{i=1}^{n}\nu_i\partial_i,$$

is defined almost everywhere along Γ for smooth functions. Extending its definition to $\partial_\nu=\Sigma_{i=1}^n \nu_i \operatorname{tr} \partial_i$ for functions in the space $H^2(\Omega)$, we obtain the following *characterization of the space $H_0^2(\Omega)$*:

$$H_0^2(\Omega)=\{v\in H^2(\Omega);\, v=\partial_\nu v=0 \text{ on } \Gamma\}.$$

Given two functions $u, v\in H^1(\Omega)$, the *fundamental Green formula*

$$\int_\Omega u\partial_i v\,\mathrm{d}x=-\int_\Omega \partial_i u\, v\,\mathrm{d}x+\int_\Gamma u v\nu_i\,\mathrm{d}\gamma \tag{2.4}$$

holds for any $i\in[1,n]$. From this formula, other *Green's formulae* are easily deduced. For example, replacing u by $\partial_i u$ and summing from 1 to n, we get

$$\int_\Omega \sum_{i=1}^{n}\partial_i u\partial_i v\,\mathrm{d}x=-\int_\Omega \Delta u\, v\,\mathrm{d}x+\int_\Gamma \partial_\nu u\, v\,\mathrm{d}\gamma \tag{2.5}$$

for all $u\in H^2(\Omega)$, $v\in H^1(\Omega)$. As a consequence, we obtain by subtraction:

$$\int_\Omega (u\Delta v-\Delta u\, v)\,\mathrm{d}x=\int_\Gamma (u\partial_\nu v-\partial_\nu u\, v)\,\mathrm{d}\gamma \tag{2.6}$$

for all $u, v\in H^2(\Omega)$. Replacing u by Δu in formula (2.6), we obtain

$$\int_\Omega \Delta u\Delta v\,\mathrm{d}x=\int_\Omega \Delta^2 u\, v\,\mathrm{d}x-\int_\Gamma \partial_\nu\Delta u\, v\,\mathrm{d}\gamma+\int_\Gamma \Delta u\partial_\nu v\,\mathrm{d}\gamma \tag{2.7}$$

for all $u\in H^4(\Omega)$, $v\in H^2(\Omega)$. As another application of formula (2.4), let us prove the relation

$$|\Delta v|_{0,\Omega}=|v|_{2,\Omega} \quad \text{for all } v\in H_0^2(\Omega), \tag{2.8}$$

which implies that, *over the space $H_0^2(\Omega)$, the seminorm $v\to|\Delta v|_{0,\Omega}$ is a norm, equivalent to the norm $\|\cdot\|_{2,\Omega}$*: We have, by definition,

$$|v|_{2,\Omega}^2=\int_\Omega\Big\{\sum_i(\partial_{ii}v)^2+\sum_{i\neq j}(\partial_{ij}v)^2\Big\}\,\mathrm{d}x,$$

$$|\Delta v|_{0,\Omega}^2=\int_\Omega\Big\{\sum_i(\partial_{ii}v)^2+\sum_{i\neq j}\partial_{ii}v\partial_{jj}v\Big\}\,\mathrm{d}x.$$

Clearly, it suffices to prove relations (2.8) for all functions $v \in \mathcal{D}(\Omega)$. But for such functions, we have

$$\int_\Omega (\partial_{ij}v)^2 \, dx = -\int_\Omega \partial_i v \partial_{ijj} v \, dx = \int_\Omega \partial_{ii} v \partial_{jj} v \, dx,$$

as two applications of Green's formula (2.4) show, and thus (2.8) is proved.

For $n = 2$, let $\tau = (\tau_1, \tau_2)$ denote the unit tangential vector along the boundary Γ, oriented in the usual way. In addition to the normal derivative operator ∂_ν, we introduce the differential operators $\partial_\tau, \partial_{\nu\tau}, \partial_{\tau\tau}$ defined by

$$\partial_\tau v(a) = Dv(a)\tau = \sum_{i=1}^{2} \tau_i \partial_i v(a),$$

$$\partial_{\nu\tau} v(a) = D^2 v(a)(\nu, \tau) = \sum_{i,j=1}^{2} \nu_i \tau_j \partial_{ij} v(a),$$

$$\partial_{\tau\tau} v(a) = D^2 v(a)(\tau, \tau) = \sum_{i,j=1}^{2} \tau_i \tau_j \partial_{ij} v(a).$$

Note that $\partial_{\tau\tau} v$ does *not* coincide in general with the second derivative of the function v considered as a function of the curvilinear abscissa along the boundary. Then one can show that another Green's formula, viz.,

$$\int_\Omega \{2\partial_{12}u\partial_{12}v - \partial_{11}u\partial_{22}v - \partial_{22}u\partial_{11}v\} \, dx = \int_\Gamma \{-\partial_{\tau\tau}u\partial_\nu v + \partial_{\nu\tau}u\partial_\tau v\} \, d\gamma, \tag{2.9}$$

holds for all functions $u \in H^3(\Omega)$, $v \in H^2(\Omega)$.

3. Examples of second-order boundary value problems: The membrane problem, the boundary value problem of linearized elasticity and the obstacle problem

We next proceed to examine several *examples of minimization problems* that fall in the abstract setting of Section 1. According to the analysis made in this section, we thus need to specify for each example the space V, a subset U of the space V, a bilinear form $a(\cdot,\cdot)$: $V \times V \to \mathbb{R}$, and a linear form l: $V \to \mathbb{R}$. In all the examples, *the set Ω is a domain in $\mathbb{R}^n$*.

EXAMPLE 3.1. The first example corresponds to the following data:

$$V = U = H_0^1(\Omega),$$

$$a(u, v) = \int_\Omega \left(\sum_{i=1}^{n} \partial_i u \partial_i v + buv \right) dx, \tag{3.1}$$

$$l(v) = \int_\Omega fv \, dx,$$

and the following assumptions are made on the functions b and f:

$$b\in L^{\infty}(\Omega), \qquad b\geqslant 0 \quad \text{a.e. on } \Omega, \qquad f\in L^2(\Omega). \tag{3.2}$$

To begin with, it is clear that the symmetric bilinear form $a(\cdot,\cdot)$ is continuous since for all $u, v\in H^1(\Omega)$,

$$\begin{aligned}|a(u,v)| &\leqslant \sum_{i=1}^{n} |\partial_i u|_{0,\Omega}|\partial_i v|_{0,\Omega}+|b|_{0,\infty,\Omega}|u|_{0,\Omega}|v|_{0,\Omega}\\ &\leqslant \max\{1,|b|_{0,\infty,\Omega}\}\,\|u\|_{1,\Omega}\|v\|_{1,\Omega},\end{aligned}$$

where $|\cdot|_{0,\Omega}$ and $|\cdot|_{0,\infty,\Omega}$ denote the norms of the space $L^2(\Omega)$ and $L^{\infty}(\Omega)$ respectively. Further, it is $H_0^1(\Omega)$-elliptic since, for all $v\in H^1(\Omega)$,

$$a(v,v)\geqslant \int_\Omega \sum_{i=1}^{n} (\partial_i v)^2\,dx=|v|_{1,\Omega}^2$$

(by the Poincaré–Friedrichs inequality (2.2), the seminorm $|\cdot|_{1,\Omega}$ is a norm over the space $H_0^1(\Omega)$, equivalent to the norm $\|\cdot\|_{1,\Omega}$). Next, the linear form l is continuous since for all $v\in H^1(\Omega)$,

$$|l(v)|\leqslant |f|_{0,\Omega}|v|_{0,\Omega}\leqslant |f|_{0,\Omega}\|v\|_{1,\Omega}.$$

Therefore, by Theorem 1.1, there exists a unique function $u\in H_0^1(\Omega)$ that minimizes the functional

$$J: v\to J(v)=\tfrac{1}{2}\int_\Omega\left\{\sum_{i=1}^{n}(\partial_i v)^2+bv^2\right\}dx-\int_\Omega fv\,dx \tag{3.3}$$

over the space $H_0^1(\Omega)$, or equivalently, by Theorem 1.2, that satisfies the variational equations

$$\int_\Omega\left\{\sum_{i=1}^{n}\partial_i u\partial_i v+buv\right\}dx=\int_\Omega fv\,dx \quad \text{for all } v\in H_0^1(\Omega). \tag{3.4}$$

Using these equations, we now show that *we are also solving a partial differential equation in the distributional sense*. More specifically, let $\mathscr{D}'(\Omega)$ denote the *space of distributions over the set* Ω, i.e., the dual space of the space $\mathscr{D}(\Omega)$, equipped with the Schwartz topology, and let $\langle\cdot,\cdot\rangle$ denote the duality pairing between the spaces $\mathscr{D}'(\Omega)$ and $\mathscr{D}(\Omega)$. As usual *we identify a function g that is locally integrable over Ω with the distribution* $g: \phi\in\mathscr{D}(\Omega)\to\int_\Omega g\phi\,dx$.

Since the inclusion

$$\mathscr{D}(\Omega)\subset V=H_0^1(\Omega)$$

holds, the variational equations (3.4) are satisfied for all functions $v\in\mathscr{D}(\Omega)$. Therefore, by definition of the differentiation of distributions, we may write

$$a(u,\phi)=\sum_{i=1}^{n}\langle\partial_i u,\partial_i\phi\rangle+\langle bu,\phi\rangle=\langle -\Delta u+bu,\phi\rangle \quad \text{for all } \phi\in\mathscr{D}(\Omega).$$

Since $l(\phi)=\langle f,\phi\rangle$ for all $\phi\in\mathscr{D}(\Omega)$, it follows from the above relations that u is a solution of the partial differential equation $-\Delta u+bu=f$ in $\mathscr{D}'(\Omega)$.

To sum up, the solution u of the minimization (or variational) problem associated with the data (3.1) is also a solution of the problem: *Find a distribution* $u\in\mathscr{D}'(\Omega)$ *such that*

$$\begin{aligned}&u\in H_0^1(\Omega),\\&-\Delta u+bu=f\quad\text{in }\mathscr{D}'(\Omega),\end{aligned}\tag{3.5}$$

and conversely, if a distribution u satisfies (3.5), it is a solution of the original problem. To see this, we observe that the equalities

$$a(u,\phi)=\langle-\Delta u+bu,\phi\rangle=\langle f,\phi\rangle=l\phi)\quad\text{for all }\phi\in\mathscr{D}(\Omega)$$

hold in fact for all functions $\phi\in H_0^1(\Omega)$ since $\mathscr{D}(\Omega)$ is a dense subspace of the space $H_0^1(\Omega)$.

Remembering that the functions in the space $H_0^1(\Omega)$ have a vanishing trace along Γ, we shall say that we have *formally solved* the *associated boundary value problem*

$$-\Delta u+bu=f\quad\text{in }\Omega,\qquad u=0\quad\text{on }\Gamma.\tag{3.6}$$

The terminology "boundary value problem" refers to the fact that the values of the unknown function u (as in this example) or of some differential operator acting on the unknown function u (as in some of the subsequent examples) are specified along the boundary of the set Ω.

Problem (3.6) is called a *homogeneous Dirichlet problem* for the operator $u\to-\Delta u+bu$, since it is *formally* posed exactly as in the classical sense where, typically, one would seek a solution in the space $\mathscr{C}^0(\bar\Omega)\cap\mathscr{C}^2(\Omega)$. Actually, when the data are sufficiently smooth, it can be proved (but this is not easy) that the solution of (3.4) is also a solution of (3.6) in the classical sense. Nevertheless, one should keep in mind that, in general, nothing guarantees that the partial differential equation $-\Delta u+bu=f$ in Ω can be given a sense otherwise than in the space $\mathscr{D}'(\Omega)$. Likewise the boundary condition $u=0$ on Γ cannot be understood in general otherwise than in the sense of a vanishing trace, or even in no sense at all if the set Ω were "only" supposed to be bounded.

A modification of the linear form of (3.1) will lead us to the second example.

EXAMPLE 3.2. Let the space V and the bilinear form be as in (3.1), and let the linear form be defined by

$$l(v)=\int_\Omega fv\,\mathrm{d}x-a(u_0,v),\tag{3.7}$$

where the functions f and b satisfy assumptions (3.2) and u_0 is a given function in the space $H^1(\Omega)$. Proceeding as before, we could likewise show that these data correspond to the formal solution of the *nonhomogeneous Dirichlet problem for the operator* $u \to -\Delta u + bu$, viz.,

$$-\Delta u + bu = f \quad \text{in } \Omega, \qquad u = u_0 \quad \text{on } \Gamma. \tag{3.8}$$

When $b=0$ and $n=2$, this problem is called the *membrane problem*: It arises in linearized elasticity when one considers the problem of finding the *equilibrium position of an elastic membrane*, with tension τ, under the action of a vertical force of density $F=\tau f$, and whose vertical displacement u is equal to a known function u_0 along the boundary Γ (cf. Fig. 3.1).

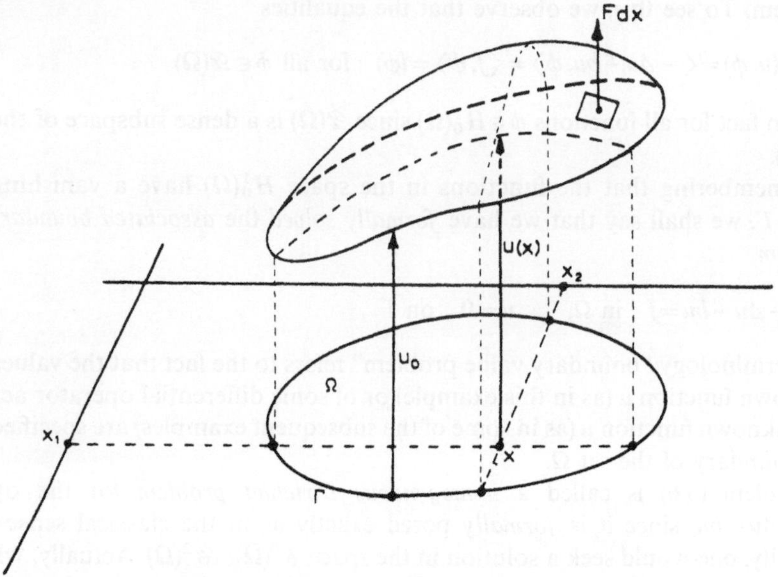

FIG. 3.1. The membrane problem: the unknown function u: $\Omega \subset \mathbb{R}^2 \to \mathbb{R}$ represents the vertical displacement of a membrane subjected to a vertical force of density F per unit area.

EXAMPLE 3.3. The third example corresponds to the following data:

$$V = U = H^1(\Omega),$$

$$a(u,v) = \int_\Omega \left(\sum_{i=1}^n \partial_i u \partial_i v + buv \right) \mathrm{d}x, \tag{3.9}$$

$$l(v) = \int_\Omega fv \, \mathrm{d}x + \int_\Gamma gv \, \mathrm{d}\gamma,$$

and the following assumptions are made on the functions a, f and g:

$$b \in L^\infty(\Omega), \qquad b \geqslant b_0 > 0 \quad \text{a.e. on } \Omega, \qquad f \in L^2(\Omega), \qquad g \in L^2(\Gamma) \tag{3.10}$$

for some constant b_0. Hence both the space V and the linear form l differ from those of Example 3.1.

The bilinear form is $H^1(\Omega)$-elliptic since $a(v,v) \geqslant \min\{1, b_0\}\, \|v\|_{1,\Omega}^2$ for all $v \in H^1(\Omega)$. The linear form $v \in H^1(\Omega) \to \int_\Gamma gv\, \mathrm{d}\gamma$ is continuous since by inequality (2.3),

$$\left|\int_\Gamma gv\, \mathrm{d}\gamma\right| \leqslant \|g\|_{L^2(\Gamma)} \|v\|_{L^2(\Gamma)} \leqslant C(\Omega) \|g\|_{L^2(\Gamma)} \|v\|_{1,\Omega}.$$

Therefore there exists a unique function $u \in H^1(\Omega)$ that minimizes the functional

$$J: v \to J(v) = \tfrac{1}{2} \int_\Omega \left\{ \sum_{i=1}^n (\partial_i v)^2 + bv^2 \right\} \mathrm{d}x - \int_\Omega fv\, \mathrm{d}x - \int_\Gamma gv\, \mathrm{d}\gamma,$$

over the space $H^1(\Omega)$ or, equivalently, such that

$$\int_\Omega \left\{ \sum_{i=1}^n \partial_i u \partial_i v + buv \right\} \mathrm{d}x = \int_\Omega fv\, \mathrm{d}x + \int_\Gamma gv\, \mathrm{d}\gamma \quad \text{for all } v \in H^1(\Omega). \tag{3.11}$$

Because $\mathscr{D}(\Omega)$ is a subspace of the space $H^1(\Omega)$, an argument similar to the one used in Example 3.1 shows that u is also a solution of the partial differential equation $-\Delta u + bu = f$ in $\mathscr{D}'(\Omega)$. Hence we have

$$\int_\Omega (-\Delta u + bu) v\, \mathrm{d}x = a(u,v) - \int_\Gamma gv\, \mathrm{d}\gamma \quad \text{for all } v \in H^1(\Omega).$$

To sum up, the solution u of the minimization (or variational) problem associated with the data (3.9) is also a solution of the problem: *Find a distribution $u \in \mathscr{D}'(\Omega)$ such that*

$$\begin{aligned} &u \in H^1(\Omega), \qquad -\Delta u + bu = f \ \text{ in } \mathscr{D}'(\Omega), \\ &\int_\Omega (-\Delta u + bu) v\, \mathrm{d}x = a(u,v) - \int_\Gamma gv\, \mathrm{d}\gamma \quad \text{for all } v \in H^1(\Omega), \end{aligned} \tag{3.12}$$

and, conversely, if a distribution u is a solution of problem (3.12), it is clearly a solution of the variational equations (3.11).

If we assume additional smoothness on the solution, the second relations in (3.12) shows that the solution satisfies a *boundary condition*: If the solution u is in the space $H^2(\Omega)$, for example, an application of Green's formula (2.5) shows that, for all $v \in H^1(\Omega)$,

$$\begin{aligned} a(u,v) &= \int_\Omega (-\Delta u + bu) v\, \mathrm{d}x + \int_\Gamma \partial_\nu u\, v\, \mathrm{d}\gamma \\ &= \int_\Omega fv\, \mathrm{d}x + \int_\Gamma gv\, \mathrm{d}\gamma. \end{aligned} \tag{3.13}$$

Therefore the conjunction of relations (3.12) and (3.13) implies that

$$\int_\Gamma \partial_\nu u v \, d\gamma = \int_\Gamma g v \, d\gamma \quad \text{for all } v \in H^1(\Omega), \tag{3.14}$$

and from these, one deduces that $\partial_\nu u = g$ on Γ.

Consequently, we shall say that we have *formally solved* the *associated boundary value problem*:

$$-\Delta u + bu = f \quad \text{in } \Omega, \qquad \partial_\nu u = g \quad \text{on } \Gamma, \tag{3.15}$$

which is called a *nonhomogeneous Neumann problem* if $g \neq 0$, or a *homogeneous Neumann problem* if $g = 0$, for the operator $u \to -\Delta u + bu$.

REMARK 3.1. Without using differentiation of distributions, we may use Green's formula (2.5) to obtain the partial differential equation since

$$a(u, \phi) = \int_\Omega (-\Delta u + au)\phi \, dx = \langle -\Delta u + au, \phi \rangle \quad \text{for all } \phi \in \mathcal{D}(\Omega).$$

This is not a coincidence: The definition of differentiation for distributions is precisely based upon the fundamental Green formula (2.4).

If $b = 0$, *the bilinear form* $a(\cdot,\cdot)$ *of* (3.9) *is no longer* $H^1(\Omega)$*-elliptic*. We can however circumvent this difficulty by using the following result (which is a special case of a result proved later; cf. Theorem 14.1): Let $P_0(\Omega)$ denote the space of all constant functions over Ω. Then *the seminorm* $|\cdot|_{1,\Omega}$ *is a norm over the quotient space* $H^1(\Omega)/P_0(\Omega)$, *equivalent to the quotient norm*. This observation leads us to our fourth example.

EXAMPLE 3.4. Let $\dot{w} \in H^1(\Omega)/P_0(\Omega)$ denote the equivalence class of an arbitrary element $w \in H^1(\Omega)$, and let:

$$V = U = H^1(\Omega)/P_0(\Omega),$$

$$a(\dot{u}, \dot{v}) = \int_\Omega \sum_{i=1}^n \partial_i u \partial_i v \, dx, \tag{3.16}$$

$$l(\dot{v}) = \int_\Omega f v \, dx + \int_\Gamma g v \, d\gamma.$$

The bilinear form $a(\cdot,\cdot)$ *is well defined and continuous over the quotient space* $V = H^1(\Omega)/P_0(\Omega)$, *and further, it is now* V*-elliptic* by the aforesaid result. If (and only if)

$$\int_\Omega f \, dx + \int_\Gamma g \, d\gamma = 0, \tag{3.17}$$

the linear form l is also well defined and continuous over the quotient space.

Arguing as before, we find that the variational problem associated with the data (3.16) corresponds to the *formal solution* of the associated boundary value problem

$$-\Delta u = f \quad \text{in } \Omega, \qquad \partial_\nu u = g \quad \text{on } \Gamma, \tag{3.18}$$

which is another *nonhomogeneous Neumann problem* for the operator $-\Delta$.

REMARK 3.2. The necessity of the relation (3.17) also follows a posteriori from (3.18) and from the formula $\int_\Omega \Delta u \, dx = \int_\Gamma \partial_\nu u \, d\gamma$, obtained by letting $v=1$ in Green's formula (2.5).

EXAMPLE 3.5. In the fifth example, we extend in two directions the previous examples: First the associated partial differential equation will have nonconstant coefficients and secondly, the bilinear form will not be necessarily symmetric so that the Lax–Milgram lemma (Theorem 1.3) will be needed for proving the existence of a solution. The data are the following:

$$V = U = \{v \in H^1(\Omega); v = 0 \text{ on } \Gamma_0\},$$

$$a(u, v) = \int_\Omega \left\{ \sum_{i,j=1}^n a_{ij} \partial_i u \partial_j v + buv \right\} dx, \tag{3.19}$$

$$l(v) = \int_\Omega fv \, dx + \int_{\Gamma_1} gv \, d\gamma,$$

where $\Gamma_0 = \Gamma - \Gamma_1$ is a $d\gamma$-measurable subset of the boundary Γ whose $d\gamma$-measure is >0, and the functions a_{ij}, b and f satisfy the following assumptions:

$$\begin{aligned} & a_{ij} \in L^\infty(\Omega), \quad 1 \leqslant i, j \leqslant n, \\ & b \in L^\infty(\Omega), \quad b \geqslant 0 \quad \text{a.e. on } \Omega, \\ & f \in L^2(\Omega), \quad g \in L^2(\Gamma_1), \end{aligned} \tag{3.20}$$

and there exists a constant β such that

$$\beta > 0, \qquad \sum_{i,j=1}^n a_{ij}\xi_i\xi_j \geqslant \beta \sum_{i=1}^n \xi_i^2 \quad \text{for all } \xi_i,\ 1 \leqslant i \leqslant n, \text{ a.e. on } \Omega. \tag{3.21}$$

The V-ellipticity of the bilinear form of (3.19) will be a consequence of the following result.

THEOREM 3.1. *Let Ω be a domain in $\mathbb{R}^n$. Then the space V defined in* (3.19) *is a closed subspace of $H^1(\Omega)$. If the* $d\gamma$*-measure of Γ_0 is >0, the seminorm $|\cdot|_{1,\Omega}$ is a norm over the space V, equivalent to the norm $\|\cdot\|_{1,\Omega}$.*

PROOF. Let (v_k) be a sequence of functions in the space V that converges to an element $v \in H^1(\Omega)$. Since the sequence $(\operatorname{tr} v_k)$ converges to $\operatorname{tr} v$ in the space $L^2(\Gamma)$ (cf. inequalities (2.3)), it contains a subsequence that converges almost everywhere to tr v and thus tr $v=0$ a.e. on Γ_0. This implies that the function v belongs to the space V.

Next, let us show that $|\cdot|_{1,\Omega}$ is a norm over the space V. Let v be a function in the space V that satisfies $|v|_{1,\Omega}=0$. Then it is a constant function by virtue of the connectedness of the set Ω; thus its trace is a constant function that takes the same value, and this value is zero since the trace vanishes on the set Γ_0, whose dγ-measure is >0.

Finally, assume that the two norms $|\cdot|_{1,\Omega}$ and $\|\cdot\|_{1,\Omega}$ are not equivalent over the space V. Then there exists a sequence (v_k) of functions $v_k \in V$ such that

$$\|v_k\|_{1,\Omega}=1 \quad \text{for all } k, \qquad \lim_{k\to\infty} |v_k|_{1,\Omega}=0.$$

By Rellich's theorem, any bounded sequence in the space $H^1(\Omega)$ contains a subsequence that converges in $L^2(\Omega)$, so that there exists a sequence (v_l) of functions $v_l \in V$ that converges in the space $L^2(\Omega)$. Since $\lim_{l\to\infty} |v_l|_{1,\Omega} = 0$, on the other hand, the sequence (v_l) is a Cauchy sequence in the complete space V, and therefore it converges with respect to the norm $\|\cdot\|_{1,\Omega}$ to an element $v \in V$.

Since $|v|_{1,\Omega}=\lim_{l\to\infty}|v_l|_{1,\Omega}=0$, we deduce that $v=0$, which is in contradiction with the equalities $\|v_k\|_{1,\Omega}=1$ for all k. □

From this theorem, we infer that the bilinear form of (3.19) is V-elliptic since we have $a(v,v)\geqslant \beta |v|^2_{1,\Omega}$ for all $v\in H^1(\Omega)$ by inequalities (3.20) and (3.21).

By the Lax–Milgram lemma (Theorem 1.3), there exists a unique function $u\in V$ that satisfies the variational equations

$$\int_\Omega \left\{ \sum_{i,j=1}^n a_{ij}\partial_i u \partial_j v + buv \right\} \mathrm{d}x = \int_\Omega fv\,\mathrm{d}x + \int_{\Gamma_1} gv\,\mathrm{d}\gamma \quad \text{for all } v\in V. \tag{3.22}$$

Referring once again to formula (2.4), we obtain another *Green's formula*:

$$\int_\Omega \sum_{i,j=1}^n a_{ij}\partial_i u \partial_j v\,\mathrm{d}x = -\int_\Omega \sum_{i,j=1}^n \partial_j(a_{ij}\partial_i u)v\,\mathrm{d}x + \int_\Gamma \sum_{i,j=1}^n a_{ij}\partial_i u\, \nu_j\,\mathrm{d}\gamma, \tag{3.23}$$

valid for all functions $u\in H^2(\Omega)$, $v\in H^1(\Omega)$, provided the functions a_{ij} are smooth enough, e.g., in the space $\mathscr{C}^1(\bar\Omega)$, so that the functions $a_{ij}\partial_i u$ belong to the space $H^1(\Omega)$. Using (3.23) we conclude that we have formally solved the *associated boundary value problem*

$$\begin{aligned} &-\sum_{i,j=1}^n \partial_j(a_{ij}\partial_i u)+bu=f \quad \text{in } \Omega,\\ &u=0 \quad \text{on } \Gamma_0,\\ &\sum_{i,j=1}^n a_{ij}\nu_j\partial_i u = g \quad \text{on } \Gamma_1, \end{aligned} \tag{3.24}$$

which is called a *homogeneous mixed problem if* $g=0$, or a *nonhomogeneous mixed problem* if $g\neq 0$, for the operator

$$u \to -\sum_{i,j=1}^{n} \partial_j(a_{ij}\partial_i u)+bu. \tag{3.25}$$

Notice that condition (3.21) is the classical *ellipticity condition* for an operator such as that of (3.25), which is accordingly called an *elliptic operator* when this condition is fulfilled. The boundary operator

$$u \to \sum_{i,j=1}^{n} a_{ij}\nu_j\partial_i u$$

is called the *conormal derivative operator* associated with the operator of (3.25).

If $\Gamma=\Gamma_0$, or if $\Gamma=\Gamma_1$, we have *formally solved* a *homogeneous Dirichlet problem*, or a *homogeneous* or a *nonhomogeneous Neumann problem* respectively, for the operator of (3.25) (in the second case, we would require an inequality such as $b \geqslant b_0 > 0$ a.e. on Ω to get existence).

We now come to the sixth example, which is by far the most important.

EXAMPLE 3.6. Let Ω be a domain in $\mathbb{R}^3$. We define the space

$$V=U=\{\boldsymbol{v}=(v_1,v_2,v_3)\in (H^1(\Omega))^3;\ v_i=0 \text{ on } \Gamma_0,\ 1\leqslant i\leqslant 3\}, \tag{3.26}$$

where Γ_0 is a dγ-measurable subset of Γ, whose dγ-measure is >0. The space V, which is a closed subspace of $(H^1(\Omega))^3$, is equipped with the product norm

$$\boldsymbol{v}=(v_1,v_2,v_3)\to \|\boldsymbol{v}\|_{1,\Omega}=\left(\sum_{i=1}^{3}\|v_i\|_{1,\Omega}^2\right)^{1/2}. \tag{3.27}$$

For any $\boldsymbol{v}=(v_1,v_2,v_3)\in (H^1(\Omega))^3$, let

$$\varepsilon_{ij}(\boldsymbol{v})=\varepsilon_{ji}(\boldsymbol{v})=\tfrac{1}{2}(\partial_j v_i+\partial_i v_j),\quad 1\leqslant i,j\leqslant 3, \tag{3.28}$$

and

$$\sigma_{ij}(\boldsymbol{v})=\sigma_{ji}(\boldsymbol{v})=\lambda\left(\sum_{k=1}^{3}\varepsilon_{kk}(\boldsymbol{v})\right)\delta_{ij}+2\mu\varepsilon_{ij}(\boldsymbol{v}),\quad 1\leqslant i,j\leqslant 3, \tag{3.29}$$

where δ_{ij} is the Kronecker's symbol, and λ and μ are two constants that satisfy

$$\lambda>0,\qquad \mu>0. \tag{3.30}$$

We then define the symmetric bilinear form

$$a(\boldsymbol{u},\boldsymbol{v})=\int_\Omega \sum_{i,j=1}^{3}\sigma_{ij}(\boldsymbol{u})\varepsilon_{ij}(\boldsymbol{v})\,\mathrm{d}x$$

$$=\int_\Omega\left\{\lambda \operatorname{div}\boldsymbol{u}\operatorname{div}\boldsymbol{v}+2\mu\sum_{i,j=1}^{3}\varepsilon_{ij}(\boldsymbol{u})\varepsilon_{ij}(\boldsymbol{v})\right\}\mathrm{d}x, \tag{3.31}$$

and the linear form

$$l(\boldsymbol{v}) = \int_\Omega \boldsymbol{f}\cdot\boldsymbol{v}\,\mathrm{d}x + \int_{\Gamma_1} \boldsymbol{g}\cdot\boldsymbol{v}\,\mathrm{d}\gamma$$

$$= \int_\Omega \sum_{i=1}^{3} f_i v_i\,\mathrm{d}x + \int_{\Gamma_1} \sum_{i=1}^{3} g_i v_i\,\mathrm{d}\gamma, \tag{3.32}$$

where $\boldsymbol{f}=(f_1,f_2,f_3)\in(L^2(\Omega))^3$ and $\boldsymbol{g}=(g_1,g_2,g_3)\in(L^2(\Gamma_1))^3$, with $\Gamma_1=\Gamma-\Gamma_0$.

It is clear that these bilinear and linear forms are continuous over the space V. To prove the V-ellipticity of the bilinear form (Theorem 3.2), one needs *Korn's inequality*: There exists a constant $C(\Omega)$ such that, for all $\boldsymbol{v}=(v_1,v_2,v_3)\in(H^1(\Omega))^3$,

$$\|\boldsymbol{v}\|_{1,\Omega} \leqslant C(\Omega)\left(\sum_{i,j=1}^{3} |\varepsilon_{ij}(\boldsymbol{v})|_{0,\Omega}^2 + \sum_{i=1}^{3} |v_i|_{0,\Omega}^2\right)^{1/2}. \tag{3.33}$$

This inequality is difficult to establish; its proof may be found in DUVAUT and LIONS [1972, Chapter 3, Section 3.3], or in FICHERA [1972a, Section 12].

THEOREM 3.2. *Let Ω be a domain in $\mathbb{R}^3$, and let Γ_0 denote a* dγ*-measurable subset of Γ, whose* dγ*-measure is >0. Then the mapping*

$$\boldsymbol{v}=(v_1,v_2,v_3)\to\left\{\sum_{i,j,1}^{3} |\varepsilon_{ij}(\boldsymbol{v})|_{0,\Omega}^2\right\}^{1/2} \tag{3.34}$$

is a norm over the space V of (3.26), *equivalent to the product norm* (3.27).

PROOF. The proof is similar to that of Theorem 3.1, and for this reason will be only sketched (for details, see e.g. DUVAUT and LIONS [1972] or CIARLET [1988, Section 6.3]). One first shows that a function $\boldsymbol{v}\in(H^1(\Omega))^3$ that satisfies $\varepsilon_{ij}(\boldsymbol{v})=0$ in Ω, $1\leqslant i,j\leqslant 3$, is of the form $\boldsymbol{v}: x\to\boldsymbol{v}(x)=\boldsymbol{a}+\boldsymbol{b}\times\mathbf{0}x$, for some constant vectors $\boldsymbol{a}\in\mathbb{R}^3$ and $\boldsymbol{b}\in\mathbb{R}^3$. Using Korn's inequality (3.33), one then shows that the norm defined in (3.34) is equivalent to the product norm. □

The V-ellipticity of the bilinear form (3.31) is therefore a consequence of inequalities (3.30), since by (3.31):

$$a(\boldsymbol{v},\boldsymbol{v}) \geqslant 2\mu \sum_{i,j=1}^{3} |\varepsilon_{ij}(\boldsymbol{v})|_{0,\Omega}^2.$$

We thus conclude that there exists a unique function $\boldsymbol{u}\in V$ that minimizes the functional

$$J(\boldsymbol{v}) = \tfrac{1}{2}\int_\Omega \left\{\lambda(\operatorname{div}(\boldsymbol{v}))^2 + 2\mu\sum_{i,j=1}^{3}(\varepsilon_{ij}(\boldsymbol{v}))^2\right\}\mathrm{d}x - \left(\int_\Omega \boldsymbol{f}\cdot\boldsymbol{v}\,\mathrm{d}x + \int_{\Gamma_1}\boldsymbol{g}\cdot\boldsymbol{v}\,\mathrm{d}\gamma\right) \tag{3.35}$$

over the space V or, equivalently, that satisfies

$$\int_\Omega \sum_{i,j=1}^{3} \sigma_{ij}(\boldsymbol{u})\varepsilon_{ij}(\boldsymbol{v})\,\mathrm{d}x = \int_\Omega \boldsymbol{f}\cdot\boldsymbol{v}\,\mathrm{d}x + \int_{\Gamma_1} \boldsymbol{g}\cdot\boldsymbol{v}\,\mathrm{d}\gamma \quad \text{for all } \boldsymbol{v}\in \boldsymbol{V}. \tag{3.36}$$

Since relations (3.36) are satisfied by all functions $\boldsymbol{v}\in(\mathscr{D}(\Omega))^3$, they could yield the associated partial differential equations. However, as was pointed out in Remark 3.1, it is equivalent to use Green's formulae, which in addition have the advantage of simultaneously yielding the associated boundary conditions (cf. (3.40)).

Using Green's formula (2.4), we obtain, for all $\boldsymbol{u}\in(H^2(\Omega))^3$ and all $\boldsymbol{v}\in(H^1(\Omega))^3$:

$$\int_\Omega \sigma_{ij}(\boldsymbol{u})\partial_j v_i\,\mathrm{d}x = -\int_\Omega (\partial_j\sigma_{ij}(\boldsymbol{u}))v_i\,\mathrm{d}x + \int_\Gamma \sigma_{ij}(\boldsymbol{u})v_i\nu_j\,\mathrm{d}\gamma,$$

so that, using definitions (3.28) and (3.29) we have proved that the following Green's formula holds:

$$\int_\Omega \sum_{i,j=1}^{3} \sigma_{ij}(\boldsymbol{u})\varepsilon_{ij}(\boldsymbol{v})\,\mathrm{d}x$$
$$= \int_\Omega \sum_{i=1}^{3}\left(-\sum_{j=1}^{3}\partial_j\sigma_{ij}(\boldsymbol{u})\right)v_i\,\mathrm{d}x + \int_\Gamma \sum_{i=1}^{3}\left(\sum_{j=1}^{3}\sigma_{ij}(\boldsymbol{u})\nu_j\right)v_i\,\mathrm{d}\gamma, \tag{3.37}$$

for all functions $\boldsymbol{u}\in(H^2(\Omega))^3$ and $\boldsymbol{v}\in(H^1(\Omega))^3$.

Arguing as in the previous examples, we find that we are formally solving the equations

$$-\sum_{j=1}^{3}\partial_j\sigma_{ij}(\boldsymbol{u}) = f_i, \quad 1\leqslant i\leqslant 3. \tag{3.38}$$

These equations are sometimes written in vector form as

$$-\mu\Delta\boldsymbol{u} - (\lambda+\mu)\,\mathbf{grad}\,\mathrm{div}\,\boldsymbol{u} = \boldsymbol{f} \quad \text{in } \Omega, \tag{3.39}$$

which is obtained from (3.38) simply by using relations (3.29).

Taking equations (3.38) into account, we find that the variational equations (3.36) reduce to

$$\int_{\Gamma_1} \sum_{i=1}^{3}\left(\sum_{j=1}^{3}\sigma_{ij}(\boldsymbol{u})\nu_j\right)v_i\,\mathrm{d}\gamma = \int_{\Gamma_1} \sum_{i=1}^{3} g_i v_i\,\mathrm{d}\gamma \quad \text{for all } \boldsymbol{v}\in \boldsymbol{V},$$

since $\boldsymbol{v}=\mathbf{0}$ on $\Gamma_0=\Gamma-\Gamma_1$.

To sum up, we have formally solved the following *associated boundary value problem*:

$$\begin{aligned} &-\mu\Delta \boldsymbol{u}-(\lambda+\mu)\,\mathbf{grad}\,\mathrm{div}\,\boldsymbol{u}=\boldsymbol{f} \quad \text{in } \Omega,\\ &\boldsymbol{u}=\mathbf{0} \quad \text{on } \Gamma_0,\\ &\sum_{j=1}^{3}\sigma_{ij}(\boldsymbol{u})\nu_j=g_i \quad \text{on } \Gamma_1, \quad 1\leqslant i\leqslant 3, \end{aligned} \tag{3.40}$$

which is called the *boundary value problem of linearized elasticity*: it is a linearization of the boundary value problem of nonlinear three-dimensional, elasticity that describes the equilibrium of an elastic homogeneous, isotropic body (Fig. 3.2) that occupies the set $\bar{\Omega}$ in the absence of forces; the vector $\boldsymbol{u}$ denotes the *displacement* of the points of $\bar{\Omega}$ under the influence of given forces (as usual, the scale for the displacements is distorted in the figure). The body $\bar{\Omega}$ cannot move along Γ_0, and along Γ_1, surface forces of density $\boldsymbol{g}$ are given. In addition, a volumic force, of density $\boldsymbol{f}$, is prescribed inside the body $\bar{\Omega}$. For details about three-dimensional elasticity and the validity of this linearization, see e.g. CIARLET [1988].

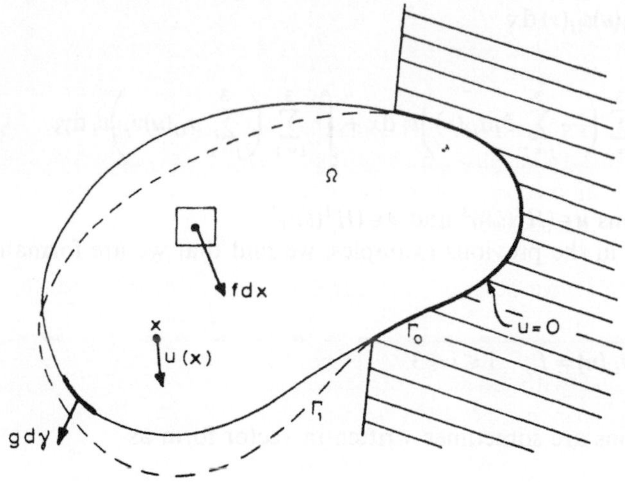

FIG. 3.2. The boundary value problem of linearized elasticity: the unknown vector field $\boldsymbol{u}:\bar{\Omega}\to\mathbb{R}^3$ represents the displacement vector of a homogeneous, isotropic, elastic body occuping the set $\bar{\Omega}$ in the absence of forces, and subjected to volumic forces of density $\boldsymbol{f}:\Omega\to\mathbb{R}^3$ and to surface forces of density $\boldsymbol{g}:\Gamma_1\to\mathbb{R}^3$.

Then we recognize in $(\varepsilon_{ij}(\boldsymbol{u}))$ the *linearized strain tensor* while $(\sigma_{ij}(\boldsymbol{u}))$ is the *linearized stress tensor*, the linear relationship (3.29) between the linearized tensors being known in linearized elasticity as *Hooke's law*. The constants λ and μ are the *Lamé coefficients* of the material that constitutes the body; experimental evidence indicates that they satisfy inequalities (3.30) for actual materials.

REMARK 3.3. It can be shown that the bilinear form of (3.31) remains elliptic under the weaker assumptions $3\lambda+2\mu>0$, $\mu>0$.

The variational equations (3.36) represent in linearized elasticity the *principle of virtual work*, valid for all *kinematically admissible displacements* $\boldsymbol{v}$, i.e., that satisfy the boundary condition $\boldsymbol{v}=\boldsymbol{0}$ on Γ_0.

REMARK 3.4. It is interesting to notice that the strict positiveness of the dγ-measure of Γ_0 has a physical interpretation: It is intuitively clear that if the dγ-measure of Γ_0 were zero, the body would be "free" and therefore there could not exist an equilibrium position in general.

The functional J of (3.35) represents in linearized elasticity the *total energy* of the body. It is the sum of the *strain energy*:

$$\tfrac{1}{2}\int\left\{\lambda(\operatorname{div}\boldsymbol{v})^2+2\mu\sum_{i,j=1}^{3}(\varepsilon_{ij}(\boldsymbol{v}))^2\right\}\mathrm{d}x,$$

and of the *potential energy of the exterior forces*:

$$-\left(\int_\Omega \boldsymbol{f}\cdot\boldsymbol{v}\,\mathrm{d}x+\int_{\Gamma_1}\boldsymbol{g}\cdot\boldsymbol{v}\,\mathrm{d}\gamma\right).$$

Example 3.6 is the most crucial one, not only because it has obviously many applications, but also because its variational formulation, described here, is basically responsible for the invention of the finite element method by engineers.

We conclude this section by studying various problems posed in terms of *variational inequalities*, i.e., where the set U is *not* a vector space. To begin this second series of examples, we describe an interesting variant of the membrane problem.

EXAMPLE 3.7. The *obstacle problem* consists in finding the equilibrium position of an elastic membrane, with tension τ, which, as before (cf. Fig. 3.1), passes through the boundary Γ of an open set Ω of the horizontal plane and is subjected to the action of a vertical force of density $F=\tau f$, but which, *in addition*, must lie over an "*obstacle*" which is represented by a function $\chi\colon\bar{\Omega}\to\mathbb{R}$, as illustrated in Fig. 3.3. This seventh example is thus naturally associated with the following data:

$$\begin{aligned}&V=H_0^1(\Omega),\quad \Omega\subset\mathbb{R}^2,\\&U=\{v\in H_0^1(\Omega);\, v\geqslant\chi \text{ a.e. in } \Omega\},\\&a(u,v)=\int_\Omega\nabla u\cdot\nabla v\,\mathrm{d}x,\qquad l(v)=\int_\Omega fv\,\mathrm{d}x.\end{aligned}\tag{3.41}$$

For definiteness, we shall make the following assumptions on the functions χ and f:

$$\chi\in H^2(\Omega),\qquad \chi\leqslant 0\quad\text{on }\Gamma,\qquad f\in L^2(\Omega).\tag{3.42}$$

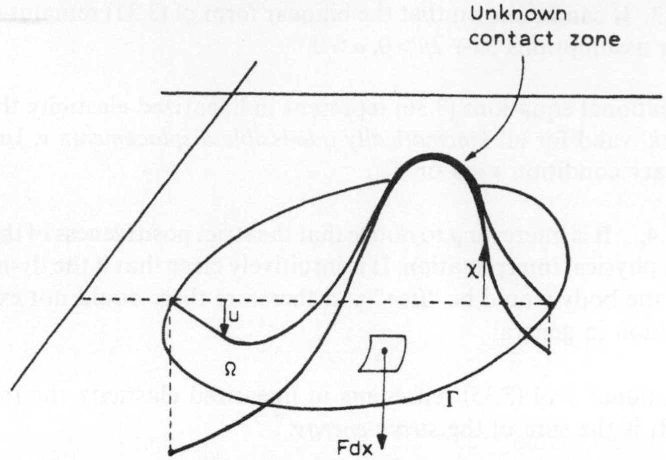

FIG. 3.3. The obstacle problem: the membrane must lie over an "obstacle", which is represented by a function $\chi:\bar{\Omega}\rightarrow\mathbb{R}$.

The set U, which is not empty by virtue of the second assumption of (3.42), is easily seen to be convex. To show that it is closed, it suffices to observe that every convergent sequence in the space $L^2(\Omega)$ contains an a.e. pointwise convergent subsequence.

Thus we may apply Theorem 1.1: There exists a unique function $u\in U$ that minimizes the *membrane energy*

$$J: v\rightarrow J(v)=\tfrac{1}{2}\int_\Omega \|\nabla v\|^2\,\mathrm{d}x-\int_\Omega fv\,\mathrm{d}x \tag{3.43}$$

over the set U, and it is also the unique solution of the *variational inequalities*

$$\int_\Omega \nabla u\cdot\nabla(v-u)\,\mathrm{d}x\geqslant\int_\Omega f(v-u)\,\mathrm{d}x\quad\text{for all } v\in U, \tag{3.44}$$

by Theorem 1.2.

By contrast with the solution of the membrane problem, *the solution of the obstacle problem is not smooth in general, even if the data are very smooth.* To be convinced of this phenomenon, consider the one-dimensional analog with $f=0$, as shown in Fig. 3.4. In this case, the solution is affine in the region where it does not touch the obstacle and consequently, whatever the smoothness of the function χ, the second derivatives will have discontinuities at points such as ξ and η. Therefore the solution u is "only" in the space $H^2(I)$.

These results carry over to the two-dimensional case, but they are of course much less easy to prove. For example, it is known that if the function χ satisfies the assumptions of (3.42), $f=0$, and $\bar{\Omega}$ is a convex polygon, the solution u belongs to the space $H_0^1(\Omega)\cap H^2(\Omega)$. If the set $\bar{\Omega}$ is convex with a boundary of class $\mathscr{C}^2$ and

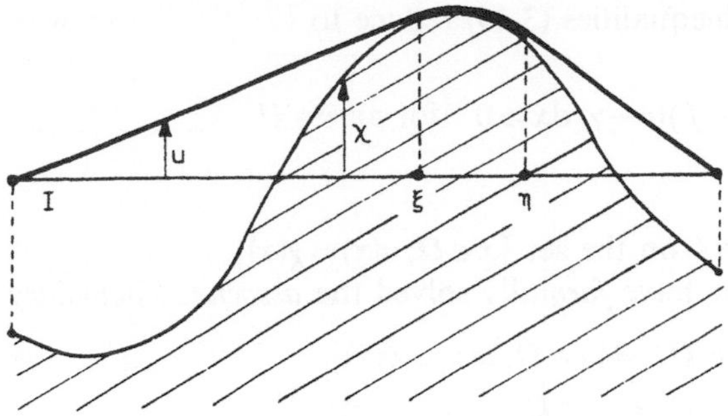

FIG. 3.4. The one-dimensional analog of the obstacle problem, with $f=0$.

assumptions (3.42) hold then we have again $u \in H_0^1(\Omega) \cap H^2(\Omega)$. In both cases, the norm $\|u\|_{2,\Omega}$ can be estimated in terms of the norms $\|\chi\|_{2,\Omega}$ and $|f|_{0,\Omega}$ of the data. These results are proved in BREZIS and STAMPACCHIA [1968] and LEWY and STAMPACCHIA [1969].

In order to interpret the variational inequalities (3.44) as a boundary value problem, assume that the solution u is in the space $H^2(\Omega)$. Hence the function $(u-\chi)$ is in the space $\mathscr{C}^0(\bar{\Omega})$, since $\chi \in H^2(\Omega)$ by the first assumption of (3.42), and $H^2(\Omega) \subset \mathscr{C}^0(\bar{\Omega})$ when $n = 2$. Let $x \in \Omega$ be such that $u(x) > \chi(x)$. Then there exist $\delta > 0$ and $r > 0$ such that

$$B_r(x) = \{y \in \mathbb{R}^2; \|y-x\| \leqslant r\} \subset \Omega,$$
$$u(y) \geqslant \chi(y) + \delta \quad \text{for all } y \in B_r(x).$$

Consequently, if $\varphi \in \mathscr{D}(\Omega)$ is such that $\operatorname{supp} \varphi \subset B_r(x)$, there exists $\varepsilon_0(\varphi) > 0$ such that

$$v_\varepsilon = (u + \varepsilon\varphi) \in U \quad \text{for all } |\varepsilon| \leqslant \varepsilon_0(\varphi).$$

Using the functions v_ε, $|\varepsilon| \leqslant \varepsilon_0(\varphi)$, in the variational inequalities (3.44), we find that

$$\varepsilon\left\{\int_\Omega \nabla u \cdot \nabla \varphi \, dx - \int_\Omega f\varphi \, dx\right\} \geqslant 0 \quad \text{for all } |\varepsilon| \leqslant \varepsilon_0(\varphi).$$

Therefore,

$$\int_\Omega \nabla u \cdot \nabla \varphi \, dx - \int_\Omega f\varphi \, dx = 0$$

for all $\varphi \in \mathscr{D}(\Omega)$ such that $\operatorname{supp} \varphi \subset B_r(x)$.

From Green's formula (2.5), we thus infer that $-\Delta u = f$ in the open ball $\{y \in \mathbb{R}^2; \|y-x\| < r\}$. Hence $-\Delta u = f$ in the open set $\Omega^+ = \{x \in \Omega; u(x) > \chi(x)\}$.

Using the same Green formula and the relation $-\Delta u = f$ in Ω^+, we next find that the variational inequalities (3.44) reduce to

$$\int_{\Omega-\Omega^+} (-\Delta u - f)(v-\chi)\,dx \geqslant 0 \quad \text{for all } v \in U,$$

and thus $-\Delta u \geqslant f$ on the set $\{x \in \Omega;\, u(x) = \chi(x)\}$.

To sum up, we have *formally* solved the *associated boundary value problem*:

$$\begin{aligned} &-\Delta u = f && \text{in } \Omega^+ = \{x \in \Omega;\, u(x) > \chi(x)\},\\ &-\Delta u \geqslant f && \text{in } \Omega^0 = \{x \in \Omega;\, u(x) = \chi(x)\},\\ &u \geqslant \chi && \text{in } \bar{\Omega}, \quad u = 0 \quad \text{on } \Gamma. \end{aligned} \tag{3.45}$$

As was already observed (cf. Remark 1.1), this is a *nonlinear* problem, since the set U of (3.41) is *not* a vector space. In the same vein, notice that *the region where the membrane touches the obstacle, i.e., the set* Ω^0, *is not known in advance*.

We may also view the boundary value problem (3.45) as a *free boundary problem*, where the *free boundary* $\Gamma^* = \partial\Omega^+ \cap \partial\Omega^0$ is one of the unknowns. In such a formulation, it is customary to adjoin to (3.45) two *transmission conditions*:

$$\left.\begin{aligned} &\operatorname{tr}(u|_{\Omega^+}) = \operatorname{tr}(u|_{\Omega^0})\\ &\operatorname{tr}\partial_\nu(u|_{\Omega^+}) = \operatorname{tr}\partial_\nu(u|_{\Omega^0}) \end{aligned}\right\} \text{ on } \Gamma^*, \tag{3.46}$$

which simply reflect here the assumption that $u \in H^2(\Omega)$. Observe, however, that these transmission conditions make sense only if Γ^* is smooth enough.

EXAMPLE 3.8. This example, which is closely related to Example 3.7, corresponds to the following data:

$$\begin{aligned} &V = H^1(\Omega),\\ &U = \{v \in H^1(\Omega);\, v \geqslant 0 \ d\gamma\text{-a.e. on } \Gamma\},\\ &a(u, v) = \int_\Omega \left\{\sum_{i=1}^n \partial_i u \partial_i v + buv\right\} dx,\\ &l(v) = \int_\Omega fv\,dx, \end{aligned} \tag{3.47}$$

where the functions b and f satisfy the same assumptions as in (3.10). Arguing as before, we easily see that the corresponding variational problem amounts to *formally* solving the *associated boundary value problem*:

$$\begin{aligned} &-\Delta u + bu = f \quad \text{in } \Omega,\\ &u \geqslant 0 \quad \text{on } \Gamma,\\ &\partial_\nu u \geqslant 0 \quad \text{on } \Gamma, \qquad u\partial_\nu u = 0 \quad \text{on } \Gamma. \end{aligned} \tag{3.48}$$

This problem falls in the class of *Signorini's problems*, named after Signorini [1933], i.e., problems where some boundary conditions take the form of inequalities, such as those appearing in (3.48). For extensive discussions of such problems, see DUVAUT and LIONS [1972b], FICHERA [1972], NEČAS [1975], GLOWINSKI, LIONS and TRÉMOLIÈRES [1976a, 1976b], NEČAS and HLAVÁČEK [1981], GLOWINSKI [1984].

EXAMPLE 3.9. Another problem modeled by *variational inequalities* is the *elastic-plastic torsion problem*, which arises in the following situation: Consider a cylindrical thin rod with a simply connected cross section $\bar{\Omega} \subset \mathbb{R}^2$, subjected to a torsion around the axis supporting the vector e_3. The torsion angle τ per unit length is assumed to be constant throughout the length of the rod (cf. Fig. 3.5, where the vertical scale should be considerably increased).

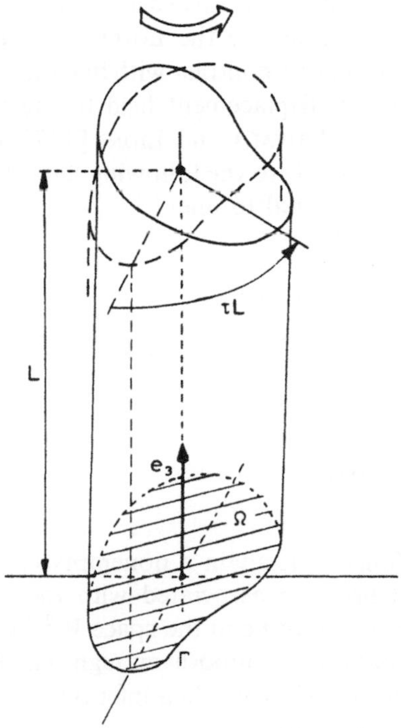

FIG. 3.5. The elastic-plastic torsion problem: A cylindrical rod is subjected to a constant torsion angle τ per unit length along the vertical direction.

Let us first assume that this problem is modeled by *linearized elasticity*. Then certain simplifying assumptions (the weight of the rod is neglected among other things) imply that the components $\sigma_{11}, \sigma_{12}, \sigma_{22}$ and σ_{33} of the stress tensor vanish everywhere in the rod, while the components σ_{13} and σ_{23}, which are functions of x_1 and x_2 only, are such that

$$\sigma_{13} = 2\mu\tau\partial_2 u, \qquad \sigma_{23} = -2\mu\tau\partial_1 u,$$

where μ is the second Lamé coefficient of the constitutive material of the rod, and the *stress function* u satisfies

$$-\Delta u = 1 \quad \text{in } \Omega, \qquad u = 0 \quad \text{on } \Gamma.$$

Therefore the function u minimizes the functional

$$J: v \to J(v) = \int_\Omega \|\nabla v\|^2 \,\mathrm{d}x - \int_\Omega f v \,\mathrm{d}x, \quad \text{with } f = 1,$$

over the space $H_0^1(\Omega)$.

If the *plasticity* of the material is also taken into account, the stresses cannot take arbitrary large values. A particular mathematical model of this restriction is the *von Mises criterion*, which reduces in this case to the condition that the quantity $(|\sigma_{13}|^2 + |\sigma_{23}|^2)^{1/2}$, and consequently the norm $\|\nabla u\|$, cannot exceed a certain constant. Notice, however, that by contrast with linearized elasticity, it is no longer straightforward to relate the displacement field to the stress field, as shown for instance by the discussion in DUVAUT and LIONS [1972, Chapter 5, Section 6].

This problem thus corresponds to the following data (for definiteness, the upper bound on $\|\nabla u\|$ has been set equal to one):

$$\begin{aligned}
&V = H_0^1(\Omega), \quad n = 2,\\
&U = \{v \in H_0^1(\Omega);\ \|\nabla v\| \leqslant 1 \text{ a.e. in } \Omega\},\\
&a(u, v) = \int_\Omega \nabla u \cdot \nabla v \,\mathrm{d}x. \qquad\qquad (3.49)\\
&l(v) = \int_\Omega f v \,\mathrm{d}x, \quad f \in L^2(\Omega),
\end{aligned}$$

with $f = 1$ in this case. Since U is again a nonempty closed convex subset of the space V, the variational problem associated with the data (3.49) has a unique solution u, which can be shown to be in the space $W^{2,p}(\Omega) \cap H_0^1(\Omega)$ for $1 < p < \infty$ if $f \in L^p(\Omega)$ and the boundary Γ is smooth enough; cf. BREZIS and STAMPACCHIA [1968]. Using Green's formula (2.5), we then infer that we have *formally* solved the *associated boundary value problem*:

$$\begin{aligned}
-\Delta u &= f \quad \text{in } \Omega^* = \{x \in \Omega;\ \|\nabla u(x)\| < 1\},\\
\|\nabla u\| &= 1 \quad \text{in } \Omega - \Omega^*, \qquad\qquad (3.50)\\
u &= 0 \quad \text{on } \Gamma.
\end{aligned}$$

As in the case of the obstacle problem, we may view (3.50) as a *free boundary value problem* and accordingly, adjoin to (3.50) appropriate *transmission conditions* at the interface between the sets Ω^* and $(\Omega - \Omega^*)$. More details about the elastic-plastic torsion problem are found in LANCHON [1972], GLOWINSKI and LANCHON [1973]. In NEČAS and HLAVÁČEK [1981], more general elastic-plastic problems are also considered.

Thorough treatments of the variational formulations of linear elliptic boundary value problems are given in LIONS [1962], AGMON [1965], NEČAS [1967], LIONS and MAGENES [1968], Vo-Khac KHOAN [1972b]. Shorter accounts are given in AUBIN [1972], BABUŠKA and AZIZ [1972], ODEN and REDDY [1976a]. More specialized treatments, particularly for general nonlinear problems, are LADYŽENSKAJA and URAL'CEVA [1968], LIONS [1969], EKELAND and TEMAM [1974]. For regularity results, see GRISVARD [1976, 1985, 1987], KONDRAT'EV [1967], BLUM and RANNACHER [1980]. For more classically oriented treatments, see for example BERS, JOHN and SCHECHTER [1964], COURANT and HILBERT [1953, 1962], GILBARG and TRUDINGER [1983], MIRANDA [1970], STAKGOLD [1968].

As introductions to classical linearized elasticity, see LANDAU and LIFCHITZ [1967], GERMAIN [1972], FRAEIJS DE VEUBEKE [1979], VALID [1977], BAMBERGER [1981], GERMAIN [1986a, 1986b]. Mathematical treatments of problems of linearized elasticity, including their variational formulations, are given in KNOPS and PAYNE [1971], GURTIN [1972], FICHERA [1972a, 1972b], DUVAUT and LIONS [1972], VILLAGIO [1972], ODEN and REDDY [1976b], NEČAS and HLAVÁČEK [1981]. The derivation of linearized elasticity from "genuine", nonlinear, elasticity, is described in CIARLET [1987, Section 6.2].

For detailed expositions of variational inequalities, and their applications to linearized elasticity, see in particular DUVAUT and LIONS [1972], FICHERA [1972b], GLOWINSKI, LIONS and TRÉMOLIÈRES [1976], BAIOCCHI and CAPELO [1978], KINDERLEHRER and STAMPACCHIA [1980], NEČAS and HLAVÁČEK [1981], GLOWINSKI [1984], PANAGIOTOPOULOS [1985], RODRIGUES [1987].

4. Examples of fourth-order boundary value problems: The biharmonic problem and the plate problems

Whereas in the preceding examples the spaces V were subspaces of the Sobolev space $H^1(\Omega)$, we consider in the last examples Sobolev spaces that involve second-order derivatives. To begin with, we consider the minimization problem that corresponds to the following data (Ω is again a domain in $\mathbb{R}^n$):

$$\begin{aligned} &V=U=H_0^2(\Omega),\\ &a(u,v)=\int_\Omega \Delta u \Delta v\,\mathrm{d}x, \qquad (4.1)\\ &l(v)=\int_\Omega fv\,\mathrm{d}x,\quad f\in L^2(\Omega). \end{aligned}$$

Since the mapping $v\to|\Delta v|_{0,\Omega}$ is a norm over the space $H_0^2(\Omega)$, as we showed in (2.8), the bilinear form is $H_0^2(\Omega)$-elliptic. Thus there exists a unique function $u\in H_0^2(\Omega)$ that minimizes the functional

$$J: v\to J(v)=\tfrac{1}{2}\int_\Omega |\Delta v|^2\,\mathrm{d}x-\int_\Omega fv\,\mathrm{d}x \qquad (4.2)$$

over the space $H_0^2(\Omega)$ or, equivalently, that satisfies the variational equations

$$\int_\Omega \Delta u \Delta v \, dx = \int_\Omega fv \, dx \quad \text{for all } v \in H_0^2(\Omega). \tag{4.3}$$

Using Green's formula (2.7):

$$\int_\Omega \Delta u \Delta v \, dx = \int_\Omega \Delta^2 u v \, dx - \int_\Gamma \partial_\nu \Delta u v \, d\gamma + \int_\Gamma \Delta u \partial_\nu v \, d\gamma,$$

we find that we have formally solved the following *homogeneous Dirichlet problem for the biharmonic operator* Δ^2:

$$\Delta^2 u = f \quad \text{in } \Omega, \qquad u = \partial_\nu u = 0 \quad \text{on } \Gamma. \tag{4.4}$$

This problem is, in particular, a mathematical model for a specific class of problems in fluid mechanics: It can be shown (see e.g. CIARLET [1978, p. 280ff.]) that the solution of the *Stokes problem* for an incompressible viscous fluid in a simply connected domain $\Omega \subset \mathbb{R}^2$ may be reduced to the solution of (4.4), where the unknown u is an appropriate *stream function.*

Finally, we let, for $n=2$,

$$V = U = H_0^2(\Omega),$$

$$\begin{aligned} a(u,v) &= \int_\Omega \{\Delta u \Delta v + (1-\nu)(2\partial_{12}u\partial_{12}v - \partial_{11}u\partial_{22}v - \partial_{22}u\partial_{11}v)\}\, dx \\ &= \int_\Omega \{\nu \Delta u \Delta v + (1-\nu)(\partial_{11}u\partial_{11}v + \partial_{22}u\partial_{22}v + 2\partial_{12}u\partial_{12}v)\}\, dx, \end{aligned} \tag{4.5}$$

$$l(v) = \int_\Omega fv \, dx, \quad f \in L^2(\Omega).$$

These data correspond to the variational formulation of the *clamped plate problem in linearized elasticity*: The unknown u represents the vertical displacement of a plate of constant thickness e under the action of a transverse force, of density $F = \frac{1}{12}Ee^3 f/(1-\sigma^2)$ per unit area. The constants $E = \mu(3\lambda + 2\mu)/(\lambda + \mu)$ and $\nu = \frac{1}{2}\lambda/(\lambda + \mu)$ are respectively the *Young modulus* and the *Poisson coefficient* of the elastic material constituting the plate, λ and μ being the Lamé coefficients of the same material. When $f=0$, the plate lies in the plane of coordinates (x_1, x_2) (cf. Fig. 4.1). The condition $u = \partial_\nu u = 0$ contained in the definition of the space $H_0^2(\Omega)$ take into account the fact that the plate is *clamped.*

It is worth noticing that the expressions found in the variational formulation (4.5) of the clamped plate problem can be obtained by applying a "limit" analysis (when the thickness of the plate approaches zero) to the variational formulation of the boundary value problem of three-dimensional elasticity (cf. (3.31) and (3.32)).

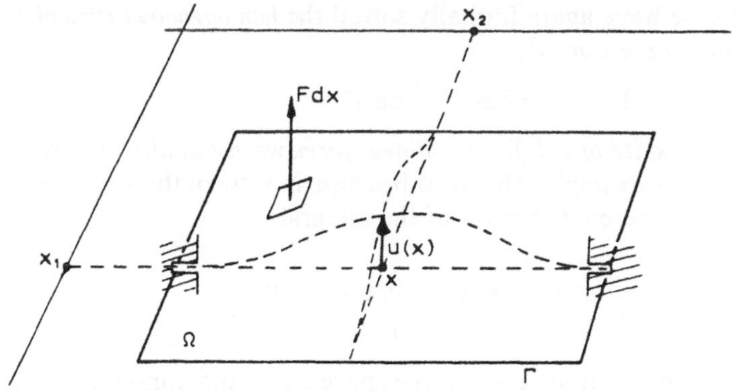

FIG. 4.1. The clamped plate problem: the unknown $u:\bar{\Omega}\subset\mathbb{R}^2\to\mathbb{R}$ represents the vertical displacement of a clamped plate occupying the set Ω in the absence of applied forces, and subjected to a vertical force of density F per unit area.

This process results in a simpler problem, in that there are now only two independent variables. However, this advantage is compensated by the fact that second partial derivatives are now present in the bilinear form. This will yield a fourth-order partial differential equation; see (4.8).

The Poisson coefficient ν satisfies the inequalities $0<\nu<\frac{1}{2}$; hence the bilinear form is $H_0^2(\Omega)$-elliptic, since we have

$$a(v,v)=\nu|\Delta v|_{0,\Omega}^2+(1-\nu)|v|_{2,\Omega}^2 \quad \text{for all } v\in H^2(\Omega).$$

Thus, there exists a unique function $u\in H_0^2(\Omega)$ that minimizes the *total energy of the plate*:

$$J(v)=\tfrac{1}{2}\int_\Omega \{|\Delta v|^2+2(1-\nu)((\partial_{12}v)^2-\partial_{11}v\partial_{22}v)\}\,\mathrm{d}x-\int_\Omega fv\,\mathrm{d}x, \tag{4.6}$$

over the space $H_0^2(\Omega)$ or, equivalently, that solves the variational equations

$$\int_\Omega \{\Delta u\Delta v+(1-\nu)(2\partial_{12}u\partial_{12}v-\partial_{11}u\partial_{22}v-\partial_{22}u\partial_{11}v)\}\,\mathrm{d}x=\int_\Omega fv\,\mathrm{d}x$$

$$\text{for all } v\in H_0^2(\Omega). \tag{4.7}$$

Using Green's formulae (2.7) and (2.9):

$$\int_\Omega \Delta u\Delta v\,\mathrm{d}x=\int_\Omega \Delta^2 uv\,\mathrm{d}x-\int_\Gamma \partial_\nu\Delta uv\,\mathrm{d}\gamma+\int_\Gamma \Delta u\partial_\nu v\,\mathrm{d}\gamma,$$

$$\int_\Omega \{2\partial_{12}u\partial_{12}v-\partial_{11}u\partial_{22}v-\partial_{22}u\partial_{11}v\}\,\mathrm{d}x=\int_\Gamma \{-\partial_{\tau\tau}u\partial_\nu v+\partial_{\nu\tau}u\partial_\tau v\}\,\mathrm{d}\gamma,$$

we find that we have again formally solved the *homogeneous Dirichlet problem for the biharmonic operator* Δ^2:

$$\Delta^2 u = f \quad \text{in } \Omega, \qquad u = \partial_\nu u = 0 \quad \text{on } \Gamma. \tag{4.8}$$

Therefore, *in spite of a different bilinear form, we eventually find the same problem as in the previous example.* This is so because, in view of the second Green formula which we used, the contribution of the integral

$$\int_\Omega (1-\nu)\{2\partial_{12}u\partial_{12}v - \partial_{11}u\partial_{22}v - \partial_{22}u\partial_{11}v\}\,\mathrm{d}x$$

is zero when the functions v are in the space $\mathscr{D}(\Omega)$, and consequently in its closure $H_0^2(\Omega)$. Thus, the partial differential equation is still $\Delta^2 u = f$ in Ω. However *different boundary conditions would result from another choice for the space* V.

To distinguish the two problems, we shall refer to a fourth-order problem corresponding to the functional of (4.2) as a *biharmonic problem,* while we shall refer to a fourth-order problem corresponding to the functional of (4.6) as a *plate problem.*

Other *plate problems* are also of interest. For instance, if we let

$$V = \{v \in H^2(\Omega);\, v = \partial_\nu v = 0 \text{ on } \Gamma_0,\ v = 0 \text{ on } \Gamma_1\} \tag{4.9}$$

where Γ_0 and Γ_1 are two disjoint dγ-measurable subsets of Γ, with dγ-meas $\Gamma_0 > 0$, and the bilinear and linear form are as in (4.5), we find that the plate is subjected to a boundary condition of *simple support* along Γ_1, and to a boundary condition of *free edge* along $\Gamma - \{\Gamma_0 \cup \Gamma_1\}$. For details about such boundary conditions, which are quite commonly encountered in practical situations, see, e.g., GERMAIN [1986b, p. 86], CIARLET [1990].

REMARK 4.1. The *membrane problem,* which we have already described (cf. Fig. 3.1), and the *plate problems,* which we just described, can be derived from the boundary value problem of *three-dimensional* linearized elasticity described in Section 3 through a systematic "limit" process, which has recently received considerable attention; see in particular CIARLET and DESTUYNDER [1979], CIARLET and KESAVAN [1980], CIARLET and RABIER [1980], DESTUYNDER [1981, 1986], RAOULT [1985], and the book by CIARLET [1990]. In particular, the problem is reduced by this process to a problem in two variables with only one unknown function (the "vertical" displacement).

To sum up, we have examined in Sections 3 and 4 various minimization or variational problems whose associated bilinear form is V-elliptic on an appropriate Hilbert space V. For this reason these minimization or variational problems, as well as their associated boundary value problems, are called *elliptic boundary value problems.* In the same spirit such problems are said to be *second-order problems,* or *fourth-order problems,* when the associated partial differential equation is of order two or four, respectively.

Finally, one should recall that even though the association between the two formulations may be formal, it is possible to prove, *under appropriate smoothness assumptions on the data*, that a solution of any of the variational problems considered here is also a solution in the classical sense of the associated boundary value problem.

REMARK 4.2. In this article, one could conceivably omit all reference to the associated boundary value problems, inasmuch as the finite element method is based only on the variational formulations. Note that, by contrast, finite difference methods are most often derived from the boundary value problems themselves.

formulations may be formal, it is possible to prove, under appropriate smoothness assumptions on the data, that a solution of any of the variational problems considered here is also a solution in the classical sense of the associated boundary value problem.

Remark 4.2. In this article, one could conceivably omit all reference to the associated boundary value problems inasmuch as the finite element method is based only on the variational formulations. Note that, by contrast, finite difference methods are most often derived from the boundary value problems themselves.

CHAPTER II

Introduction to the Finite Element Method

Introduction

This chapter is an introduction to the *finite element method*, which is a very powerful means of approximating the solutions of second-order or fourth-order problems posed in variational form over a space V. A well-known approach for approximating such problems is *Galerkin's method*, which consists in defining similar problems, called *discrete problems*, over finite-dimensional subspaces V_h of the space V. Then a *conforming finite element method*, i.e., a *finite element method in its simplest form*, is a Galerkin method characterized by *three basic aspects* in the construction of the space V_h: First, a *triangulation* $\mathcal{T}_h$ is established over the set $\bar{\Omega}$, i.e., the set $\bar{\Omega}$ is written as a finite union of *polyhedra* $T \in \mathcal{T}_h$. Secondly, the functions $v_h \in V_h$ are *piecewise polynomials*, in the sense that for each $T \in \mathcal{T}_h$, the spaces $P_T = \{v_h|_T;\, v_h \in V_h\}$ consist of polynomials. Thirdly, there should exist a basis in the space V_h whose functions have *small supports*. These three basic aspects are discussed in Section 5, where we also give simple criteria that insure the validity of inclusions such as $V_h \subset H^1(\Omega)$, or $V_h \subset H^2(\Omega)$ (Theorems 5.1 and 5.2). We also briefly indicate how these three basic aspects still pervade the more general finite element methods that will be subsequently described.

In Sections 6–9 we describe various *examples of finite elements*, which are either n-simplices (*simplicial* finite elements) or n-rectangles (*rectangular* finite elements), in which either all *degrees of freedom* are point values (*Lagrange* finite elements), or some degrees of freedom are directional derivatives (*Hermite* finite elements). These finite elements yield either the inclusion $V_h \subset H^1(\Omega)$ (finite elements *of class* $\mathscr{C}^0$) or the inclusion $V_h \subset H^2(\Omega)$ (finite elements *of class* $\mathscr{C}^1$) when they are assembled in a finite element space X_h.

In Sections 10–12, we give general definitions pertaining to *finite elements* and *finite element spaces*, and we discuss their various properties. Of particular importance are the notion of an *affine family* (T, P_T, Σ_T) of finite elements (Section 11), where all the finite elements in the family can be obtained as images through affine mappings of a single *reference finite element*, and the notion of P_T-*interpolation operators*; in particular, a basic relationship between these two notions is proved in Theorem 11.1. The P_T-interpolation operator and its global counterpart, the X_h-*interpolation operator*, both play a fundamental role in the interpolation theory in Sobolev spaces that will be developed in the next chapter. We also show how to impose *boundary conditions* on functions in finite element spaces. In Section 12, we also briefly indicate why a particular finite element should be preferred to another one in practical computations.

Finally, in Section 13, we define the *convergence* and the *order of convergence* for a family of discrete problems. In this respect, *Céa's lemma* (Theorem 13.1) is crucial: The *error* $\|u-u_h\|$, i.e., the distance (measured in the norm of the space V) between the solution u of the original problem and the solution u_h of the discrete problem, is, up to a constant independent of the space V_h, bounded above by the distance $\inf_{v_h \in V_h} \|u-v_h\|$ between the function u and the subspace V_h. This is a particularly important result, since all subsequent convergence results will be essentially variations on this theme!

The finite element method was first conceived in a paper by COURANT [1943], but the importance of this contribution was ignored at that time. Then the engineers independently re-invented the method in the early fifties: The earliest references generally quoted in the engineering literature are those of ARGYRIS [1954–1955], TURNER, CLOUGH, MARTIN and TOPP [1956]. The name of the method was proposed by CLOUGH [1960]. Historical accounts on the development of the method from the engineering point of view are given in ODEN [1972], and ZIENKIEWICZ [1973], and in the introductory article of J.T. Oden.

Many books have been written on the manifold aspects of the numerical implementation of the finite element method. We mention in particular the successive editions of the celebrated book by ZIENKIEWICZ [1971], the pioneering book of ODEN [1972] (whose value was probably not fully recognized when it appeared), and the other successful books of NORRIE and DE VRIES [1973], GALLAGHER [1975], IRONS and AHMAD [1979], ZIENKIEWICZ and MORGAN [1983], BATHE [1982], DHATT and TOUZOT [1984], KARDESTUNCER and NORRIE [1987], ARGYRIS and MLEJNEK [1986–1988].

Nice blends of computational *and* mathematical aspects are found in the pioneering books of STRANG and FIX [1973] and in the later books of ODEN and CAREY [1981–1984], AXELSSON and BARKER [1984], WAIT and MITCHELL [1985], and HUGHES [1987].

Mathematical aspects are the main themes in ODEN and REDDY [1976a], RAVIART and THOMAS [1983], JOHNSON [1987], and, at a more advanced level, in BABUŠKA and AZIZ [1972] and CIARLET [1978].

Among the many more specialized books on this method, we mention in particular GEORGE [1986] for the fascinating topic of automatic generation of triangulations, BERNADOU and BOISSERIE [1982] for shell problems, THOMASSET [1981], TEMAM [1984] and GIRAULT and RAVIART [1986] for the Navier–Stokes equations, and GLOWINSKI [1984] for nonlinear variational problems.

5. The three basic aspects of the finite element method

Consider the linear abstract variational problem: Find $u \in V$ such that

$$a(u, v) = l(v) \quad \text{for all } v \in V, \tag{5.1}$$

where the space V, the bilinear form $a(\cdot,\cdot)$, and the linear form l are assumed to satisfy the assumptions of the Lax–Milgram lemma (Theorem 1.3). Then the *Galerkin method* for approximating the solution of such a problem consists in defining similar problems in *finite-dimensional subspaces* of the space V. More specifically, with any finite-dimensional subspace V_h of V, we associate the *discrete problem*: Find $u_h \in V_h$ such that

$$a(u_h, v_h) = l(v_h) \quad \text{for all } v_h \in V_h. \tag{5.2}$$

Applying the Lax–Milgram lemma, we infer that such a problem has one and only one solution u_h, which we shall call a *discrete solution*. If the bilinear form is symmetric, the discrete solution is also characterized by the property (Theorem 1.2):

$$J(u_h) = \inf_{v_h \in V_h} J(v_h), \tag{5.3}$$

where the functional J is given by $J(v) = \frac{1}{2}a(v, v) - f(v)$. This alternate definition of the discrete solution is known as the *Ritz method*.

Let us henceforth assume that the abstract variational problem (5.1) *corresponds to a second-order or to a fourth-order elliptic boundary value problem posed over an open subset Ω of $\mathbb{R}^n$, with a Lipschitz-continuous boundary Γ*. Typical examples of such problems have been studied in Sections 3 and 4.

In order to apply the Galerkin, or the Ritz, method, we face, by definition, the problem of constructing finite-dimensional subspaces V_h of spaces V such as $H_0^1(\Omega), H^1(\Omega), H_0^2(\Omega), \ldots$. The *finite element method in its simplest form* is a specific process of constructing subspaces V_h, which are then called *finite element spaces*: This construction is characterized by *three basic aspects*, which for convenience shall be recorded as (FEM1), (FEM2) and (FEM3), respectively, and which shall be described in this section.

(FEM1) The *first basic aspect*, and certainly the *most characteristic*, is that a *triangulation* $\mathcal{T}_h$ is established over the set $\bar{\Omega}$, i.e., the set $\bar{\Omega}$ is subdivided into a finite number of subsets T, in such a way that the following properties are satisfied:

($\mathcal{T}_h$1) For each $T \in \mathcal{T}_h$, the set T is closed and its interior $\mathring{T}$ is nonempty and connected.

($\mathcal{T}_h$2) For each $T \in \mathcal{T}_h$, the boundary ∂T is Lipschitz-continuous.

($\mathcal{T}_h$3) $\bar{\Omega} = \bigcup_{T \in \mathcal{T}_h} T$.

($\mathcal{T}_h$4) For each distinct $T_1, T_2 \in \mathcal{T}_h$, one has $\mathring{T}_1 \cap \mathring{T}_2 = \emptyset$.

REMARK 5.1. A fifth condition ($\mathcal{T}_h$5), relating "adjacent" sets T, will be later introduced.

Once a triangulation $\mathcal{T}_h$ is established over the set $\bar{\Omega}$, one defines a *finite element space* X_h through a specific process, which will be illustrated by many examples in this chapter. All we need to know for the moment is that X_h is a finite-dimensional space of functions defined over the set $\bar{\Omega}$. Hence we ignore at this stage instances of finite element spaces whose "functions" may have two definitions across "adjacent" sets T (see Section 12).

Given a finite element space X_h, we define the *finite-dimensional spaces*

$$P_T = \{v_h|_T;\ v_h \in X_h\},$$

spanned by the restrictions $v_h|_T$ of the functions $v_h \in X_h$ to the sets $T \in \mathcal{T}_h$. Since our aim is to approximate the solutions of problems posed in spaces such as $H^1(\Omega)$ or $H^2(\Omega)$, our first task consists in obtaining sufficient conditions that guarantee that the inclusion $X_h \subset H^1(\Omega)$, or the inclusion $X_h \subset H^2(\Omega)$, holds (converses of these results hold, as we shall show in Theorems 30.1 and 49.1).

REMARK 5.2. Here and subsequently, we shall systematically use the improper notation $H^m(T)$ in lieu of $H^m(\mathring{T})$, for the sake of notational brevity.

THEOREM 5.1. *Assume that the inclusions $P_T \subset H^1(T)$ for all $T \in \mathcal{T}_h$ and $X_h \subset \mathscr{C}^0(\bar{\Omega})$ hold. Then the inclusions*

$$X_h \subset H^1(\Omega),$$

$$X_{0h} = \{v_h \in X_h;\ v_h = 0 \text{ on } \Gamma\} \subset H_0^1(\Omega),$$

hold.

PROOF. Let a function $v \in X_h$ be given. Since $X_h \subset \mathscr{C}^0(\bar{\Omega})$, v is in the space $L^2(\Omega)$. Therefore, by definition of the space $H^1(\Omega)$, it remains to find functions $v_i \in L^2(\Omega)$ such that

$$\int_\Omega v_i \phi \,\mathrm{d}x = -\int_\Omega v \partial_i \phi \,\mathrm{d}x \quad \text{for all } \phi \in \mathscr{D}(\Omega).$$

For each i, a natural candidate is the function whose restriction to each finite element T is the function $\partial_i(v|_T)$. Since by assumption each set T has a Lipschitz-continuous boundary ∂T, we may apply Green's formula (2.4): For each $T \in \mathcal{T}_h$,

$$\int_T \partial_i(v|_T)\phi \,\mathrm{d}x = -\int_T v|_T \partial_i \phi \,\mathrm{d}x + \int_{\partial T} v|_T \phi \nu_{i,T} \,\mathrm{d}\gamma,$$

where $\nu_{i,T}$ is the ith component of the unit outer normal vector along ∂T. By summing over all finite elements, we obtain

$$\int_\Omega v_i \phi \,\mathrm{d}x = -\int_\Omega v \partial_i \phi \,\mathrm{d}x + \sum_{T \in \mathcal{T}_h} \int_{\partial T} v|_T \phi \nu_{i,T} \,\mathrm{d}\gamma,$$

and the proof follows since the sum

$$\sum_{T \in \mathcal{T}_h} \int_{\partial T} v|_T \phi \nu_{i,T} \,\mathrm{d}\gamma$$

vanishes: Either a portion of ∂T is a portion of the boundary Γ of Ω in which case $\phi = 0$ along this portion, or the contribution of adjacent elements is zero.

The boundary Γ being Lipschitz-continuous by assumption, the second inclusion follows from the characterization (cf. Section 2)

$$H_0^1(\Omega)=\{v\in H^1(\Omega);\, v=0 \text{ on } \Gamma\}. \quad \square$$

If a space X_h fulfills the assumptions of Theorem 5.1, we shall use the finite element space $V_h = X_{0h}$ if we are solving a second-order homogeneous Dirichlet problem, or the space $V_h = X_h$ if we are solving a second-order homogeneous or non-homogeneous Neumann problem.

The proof of the next theorem is similar to that of Theorem 5.1 and, for this reason, is omitted.

THEOREM 5.2. *Assume that the inclusions $P_T \subset H^2(T)$ for all $T\in\mathcal{T}_h$, and $X_h\subset\mathscr{C}^1(\bar{\Omega})$, hold. Then the inclusions*

$$X_h\subset H^2(\Omega),$$
$$X_{0h}=\{v_h\in X_h;\, v_h=0 \text{ on } \Gamma\}\subset H^2(\Omega)\cap H_0^1(\Omega),$$
$$X_{00h}=\{v_h\in X_h;\, v_h=\partial_\nu v_h=0 \text{ on } \Gamma\}\subset H_0^2(\Omega),$$

hold.

Thus if we are to solve a simply supported plate problem, or a clamped plate problem, we shall use the finite element space $V_h = X_{0h}$, or the finite element space $V_h = X_{00h}$, respectively, as given in the previous theorem.

Let us return to the description of the finite element method.

(FEM2) The *second basic aspect* of the finite element method is that the spaces P_T, $T\in\mathcal{T}_h$, contain *polynomials*, or, at least, contain functions that are "close to" polynomials.

At this stage, we cannot be too specific about the underlying reasons for this aspect of the method but at least, we can say that (i) it is the key to all convergence results as we shall see and (ii) it yields simple computations of the coefficients of the resulting linear system (5.4).

Let us now briefly examine how the discrete problem (5.2) is solved in practice. Let $(w_k)_{k=1}^M$ be a *basis* in the space V_h. Then the solution $u_h=\Sigma_{k=1}^M \zeta_k w_k$ of problem (5.2) is such that the vector $(\zeta_1,\zeta_2,\ldots,\zeta_M)$ is solution of the *linear system*

$$\sum_{k=1}^{M} a(w_k,w_m)\zeta_k = l(w_m), \quad 1\leqslant m\leqslant M, \tag{5.4}$$

whose matrix is always *invertible*, since the bilinear form, being assumed to be V-elliptic, is a fortiori V_h-elliptic. By reference to the data associated with the boundary value problem of linearized elasticity (Section 3), the matrix $(a(w_k,w_m))$ and the vector $(l(w_m))$ are often called the *stiffness matrix* and the *load vector*, respectively.

In the choice of the basis $(w_k)_{k=1}^M$, it is of paramount importance, from a numerical standpoint, that *the resulting matrix possess as many zeros as possible.*

For *all* the examples that were considered in Section 3 the coefficients $a(w_k, w_m)$ are *integrals* of a specific form: For instance, in the case of Example 3.1, one has

$$a(w_k, w_m) = \int_\Omega \left(\sum_{i=1}^{n} \partial_i w_k \partial_i w_m + b w_k w_m \right) dx,$$

so that a coefficient $a(w_k, w_m)$ vanishes whenever the dx-measure of the intersection of the supports of the basis functions w_k and w_m is zero. As a consequence, we next state:

(FEM3) The *third basic aspect* of the finite element method is that there exists at least one "canonical" basis in the space V_h whose corresponding basis functions have *supports that are as "small" as possible*, it being implicitly understood that these basis functions can be easily described.

Note that when the bilinear form is symmetric, the matrix $(a(w_k, w_m))$ is *symmetric and positive-definite*, a fact of considerable practical importance for the numerical solution of the linear system (5.4). By contrast, this is not usually the case for standard finite difference methods, except for domains with special geometries, such as rectangular domains.

Since another noticeable practical feature of the matrix $(a(w_k, w_m))$ is its *sparsity*, one could conceivably, assuming again the symmetry of the bilinear form, start out with any given basis, and, using the Gram–Schmidt orthonormalization procedure, construct a new basis $(w_k^*)_{k=1}^M$ that is orthonormal with respect to the inner product $a(\cdot,\cdot)$. This is indeed an efficient way of getting a sparse matrix since the corresponding matrix $a(w_k^*, w_l^*)$ is the identity matrix! However, this process is not recommended from a *practical* standpoint: For comparable computing times, it yields worse results than the solution by standard methods of the linear system corresponding to the "canonical" basis.

It was mentioned at the beginning of this section that the three basic aspects were characteristic of the finite element method *in its simplest form*. Indeed, there are *more general finite element methods:*

(i) One may start out with *more general variational problems*, such as variational inequalities (see Section 3) or different variational formulations such as *mixed*, or *hybrid*, formulations (see the article by Roberts and Thomas in this volume), or *boundary formulations* (see the article by Nédélec in a later volume), etc.

(ii) The space V_h, in which one seeks the discrete solution, may no longer be a subspace of the space V. This may happen when the boundary of the set Ω is curved, for instance. Then it cannot be exactly triangulated in general by standard finite elements and thus it is replaced by an approximate set Ω_h (see Section 38). This also happens when the functions in the space V_h lack the proper continuity across adjacent finite elements (see the "nonconforming" methods described in Section 30 and Section 49).

(iii) Finally, the bilinear form and the linear form may be approximated. This is the case for instance when numerical integration is used for computing the coefficients of the linear system (5.4) (see Sections 25 and 39).

Nevertheless, it is characteristic of all these more general finite element methods that the three basic aspects, possibly in some more elaborate form, constitute their underlying principles. In order to establish a distinction with these more general methods, we shall reserve the terminology *conforming finite element methods* for the finite element methods described at the beginning of this section, i.e., for which the space V_h is a subspace of the space V, and the bilinear form and the linear forms of the discrete problem are identical to the original ones.

Let us therefore assume, as in the rest of this chapter, that we are using a *conforming finite element method* for solving a second-order or a fourth-order boundary value problem. We first summarize the various requirements that a finite element space X_h must satisfy, according to the considerations that we have made so far:

(i) Such a space is associated with a *triangulation* $\mathcal{T}_h$ of the set $\bar{\Omega}$, i.e., $\bar{\Omega} = \bigcup_{T\in\mathcal{T}_h} T$ (FEM1), and for each $T \in \mathcal{T}_h$, the space

$$P_T = \{v_h|_T;\ v_h \in X_h\} \tag{5.5}$$

should consist of functions that are polynomials or "nearly polynomials" (FEM2).

(ii) By Theorems 5.1 and 5.2, *inclusions such as* $X_h \subset \mathscr{C}^0(\bar{\Omega})$ *or* $X_h \subset \mathscr{C}^1(\bar{\Omega})$ should hold, depending upon whether we are solving a second-order or a fourth-order problem. For the time being, we shall ignore boundary conditions, which we shall take into account in Section 12.

(iii) Finally, we must check that *there exists one canonical basis in the space* X_h, *whose functions have "small" supports, and which are easy to describe* (FEM3).

In this chapter we shall describe various finite elements that are all *polyedra* in $\mathbb{R}^n$, and which are for this reason sometimes called *straight finite elements*. By virtue of the relation $\bar{\Omega} = \bigcup_{T\in\mathcal{T}_h} T$, we shall therefore restrict ourselves in this chapter to problems that are posed over a set $\bar{\Omega}$ that is itself a *polyhedron*, in which case we shall say that the set Ω is *polygonal.*

6. Examples of simplicial finite elements and their associated finite element spaces

We begin, in Sections 6–8, by examining examples of finite element spaces X_h that satisfy the inclusion $X_h \subset \mathscr{C}^0(\bar{\Omega})$, and which are *the most commonly used by engineers for solving second-order problems arising in linearized elasticity with conforming finite element methods.* As shown in Sections 3 and 4 such problems are posed in open subsets of either $\mathbb{R}^2$ or $\mathbb{R}^3$. Hence the value to be assigned in practice to the dimension n in the forthcoming examples is either 2 or 3.

We equip the space $\mathbb{R}^n$ with its canonical basis $(e_i)_{i=1}^n$. For each integer $k \geqslant 0$, we denote by P_k the space of all polynomials of degree $\leqslant k$ in the variables $x_1, x_2, \ldots, x_n$, i.e., a polynomial $p \in P_k$ is of the form

$$p\colon x = (x_1, x_2, \ldots, x_n) \in \mathbb{R}^n \to p(x) = \sum_{\sum_{i=1}^n \alpha_i \leqslant k} \gamma_{\alpha_1\alpha_2\cdots\alpha_n} x_1^{\alpha_1} x_1^{\alpha_2} \cdots x_n^{\alpha_n},$$

for appropriate coefficients $\gamma_{\alpha_1\alpha_2\cdots\alpha_n}$, or equivalently, if the multi-index notation is used,

$$p: x \in \mathbb{R}^n \to p(x) = \sum_{|\alpha| \leqslant k} \gamma_\alpha x^\alpha.$$

The dimension of the space P_k is given by

$$\dim P_k = \binom{n+k}{k}. \tag{6.1}$$

If Φ is a space of functions defined over $\mathbb{R}^n$, and if A is any subset of $\mathbb{R}^n$, we denote by $\Phi(A)$ the space formed by the restrictions to the set A of the functions in the space Φ. Thus, for instance, we let

$$P_k(A) = \{p|_A; p \in P_k\}. \tag{6.2}$$

Notice that the *dimension of the space $P_k(A)$ is the same as that of the space $P_k = P_k(\mathbb{R}^n)$ if the interior of the set A is nonempty.*

An *n-simplex* in $\mathbb{R}^n$ is the convex hull T of $(n+1)$ points $a_j = (a_{ij})_{i=1}^n \in \mathbb{R}^n$, which are called the *vertices* of the n-simplex, and which, by definition, are such that the matrix

$$A = \begin{pmatrix} a_{11} & a_{12} & \cdots & a_{1,n+1} \\ a_{21} & a_{22} & \cdots & a_{2,n+1} \\ \vdots & \vdots & & \vdots \\ a_{n1} & a_{n2} & \cdots & a_{n,n+1} \\ 1 & 1 & \cdots & 1 \end{pmatrix} \tag{6.3}$$

is *invertible* (equivalently, the $(n+1)$ points a_j are not contained in a hyperplane). Thus, one has

$$T = \left\{ x = \sum_{j=1}^{n+1} \lambda_j a_j;\ 0 \leqslant \lambda_j \leqslant 1, 1 \leqslant j \leqslant n+1, \sum_{j=1}^{n+1} \lambda_j = 1 \right\}. \tag{6.4}$$

Notice that *a 2-simplex is a triangle* and that *a 3-simplex is a tetrahedron.*

REMARK 6.1. If the matrix A of (6.3) is singular, the set T defined by (6.4) is still the convex hull of the points a_j, $1 \leqslant j \leqslant n+1$, but it is not necessarily an m-simplex for some $m < n$.

For any integer m with $0 \leqslant m \leqslant n$, an *m-face* of the n-simplex T is any m-simplex whose $(m+1)$ vertices are also vertices of T. In particular, any $(n-1)$-face is simply called a *face*, any 1-face is called an *edge*, or a *side*.

The *barycentric coordinates* $\lambda_j = \lambda_j(x)$, $1 \leqslant j \leqslant n+1$, of any point $x = (x_i)_{i=1}^n \in \mathbb{R}^n$ with respect to the $(n+1)$ vertices a_j are the (unique) solutions of the linear system

$$\sum_{j=1}^{n+1} a_{ij}\lambda_j = x_i, \quad 1 \leqslant i \leqslant n, \qquad \sum_{j=1}^{n+1} \lambda_j = 1, \tag{6.5}$$

whose matrix is precisely the matrix A of (6.3). By inspecting the linear system (6.5), one sees that the *barycentric coordinates of a point* $x \in \mathbb{R}^n$ *are affine functions of the coordinates* $x_1, x_2, \ldots, x_n$ of x (i.e., they belong to the space P_1):

$$\lambda_i = \sum_{j=1}^{n} b_{ij} x_j + b_{in+1}, \quad 1 \leqslant i \leqslant n+1, \tag{6.6}$$

where the matrix $B = (b_{ij})$ is the inverse of the matrix A.

REMARK 6.2. In the engineering literature, the barycentric coordinates are often called *area coordinates* if $n=2$, or *volume coordinates* if $n=3$.

The *barycenter, or center of gravity*, of an n-simplex T is the point of T whose all barycentric coordinates are equal (to $1/(n+1)$).

To describe our first finite element, we need to prove that *a polynomial* $p: x \to \sum_{|\alpha| \leqslant 1} \gamma_\alpha x^\alpha$ *of degree* 1 *is uniquely determined by its values at the* $(n+1)$ *vertices* a_j *of any n-simplex in* $\mathbb{R}^n$. To see this, it suffices to show that the linear system

$$\sum_{|\alpha| \leqslant 1} \gamma_\alpha (a_j)^\alpha = \mu_j, \quad 1 \leqslant j \leqslant n+1,$$

has one and only one solution $(\gamma_\alpha, |\alpha| \leqslant 1)$ for all right-hand sides μ_j, $1 \leqslant j \leqslant n+1$. Since

$$\dim P_1 = \operatorname{card}\left(\bigcup_{j=1}^{n+1} \{a_j\} \right) = n+1,$$

the matrix of this linear system is *square*, and therefore it suffices to prove either *uniqueness* or *existence*. In this case, the existence is clear: The barycentric coordinates verify $\lambda_i(a_j) = \delta_{ij}$, $1 \leqslant i, j \leqslant n+1$, and thus the polynomial

$$x \in \mathbb{R}^n \to \sum_{i=1}^{n+1} \mu_i \lambda_i(x)$$

has the desired interpolation property. As a consequence, we have proved the *identity*

$$p = \sum_{i=1}^{n+1} p(a_i) \lambda_i \quad \text{for all } p \in P_1. \tag{6.7}$$

Although we shall not repeat this kind of argument in the sequel, it will be often implicitly used.

Having thus completely determined a polynomial $p \in P_1$ by its values $p(a_i)$, $1 \leqslant i \leqslant n+1$, we can now define the simplest "paradigm of all finite elements", which we shall call *linear n-simplex*: The set T is an n-simplex with vertices a_i, $1 \leqslant i \leqslant n+1$, the space P_T is the space $P_1(T)$, and the *degrees of freedom of the finite element*, i.e., those parameters that uniquely define a function in the space P_T, consist of the values at the vertices. Denoting by Σ_T the corresponding *set of degrees of freedom*, we shall write symbolically

$$\Sigma_T = \{ p(a_i) \colon 1 \leqslant i \leqslant n+1 \}.$$

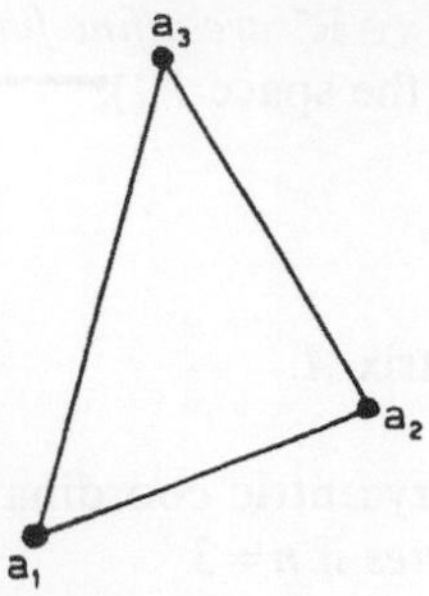

linear triangle, or
Courant's triangle,
$\dim P_T = 3$

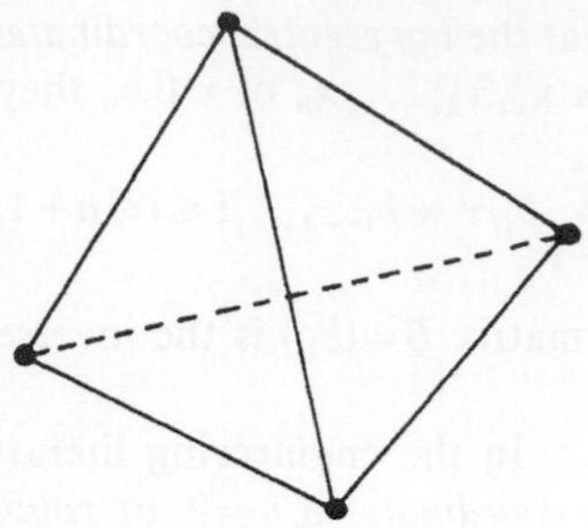

linear tetrahedron
$\dim P_T = 4$

linear n-simplex
$P_T = P_1(T)$, $\quad \dim P_T = (n+1)$
$\Sigma_T = \{p(a_i) : 1 \leqslant i \leqslant n+1\}$

FIG. 6.1

In Fig. 6.1, we have recorded the main characteristics of this finite element for arbitrary n, along with the figures in the special cases $n=2$ and 3. Note that the linear triangle ($n=2$) is also known as *Courant's triangle*, after COURANT [1943].

Let us call $a_{ij} = \frac{1}{2}(a_j + a_j)$, $1 \leqslant i < j \leqslant n+1$, the midpoints of the edges of the n-simplex T. Observing that $\lambda_k(a_{ij}) = \frac{1}{2}(\delta_{ki} + \delta_{kj})$, $1 \leqslant i < j \leqslant n+1, 1 \leqslant k \leqslant n+1$, and that

$$\dim P_2(T) = \operatorname{card}\left(\bigcup_{i=1}^{n+1} \{a_i\} \right) \cup \left(\bigcup_{1 \leqslant i < j \leqslant n+1} \{a_{ij}\} \right),$$

we obtain the identity (here and subsequently, indices $i, j, k, \ldots$, are always assumed to take all possible values in the set $\{1, 2, \ldots, n\}$ whenever this fact is not specified):

$$p = \sum_i \lambda_i(2\lambda_i - 1)p(a_i) + \sum_{i<j} 4\lambda_i\lambda_j p(a_{ij}) \quad \text{for all } p \in P_2. \tag{6.8}$$

This identity allows us to define a finite element, called the *quadratic n-simplex*: the space P_T is $P_2(T)$, and the set Σ_T consists of the values at the vertices and at the midpoints of the edges (Fig. 6.2).

Let $a_{iij} = \frac{1}{3}(2a_i + a_j)$ for $i \neq j$, and $a_{ijk} = \frac{1}{3}(a_i + a_j + a_k)$ for $i<j<k$. From the identity

$$p = \sum_i \tfrac{1}{2}\lambda_i(3\lambda_i - 1)(3\lambda_i - 2)p(a_i) + \sum_{i \neq j} \tfrac{9}{2}\lambda_i\lambda_j(3\lambda_i - 1)p(a_{iij}) + \sum_{i<j<k} 27\lambda_i\lambda_j\lambda_k p(a_{ijk})$$
$$\text{for all } p \in P_3, \tag{6.9}$$

we deduce the definition of the *cubic n-simplex* (Fig. 6.3).

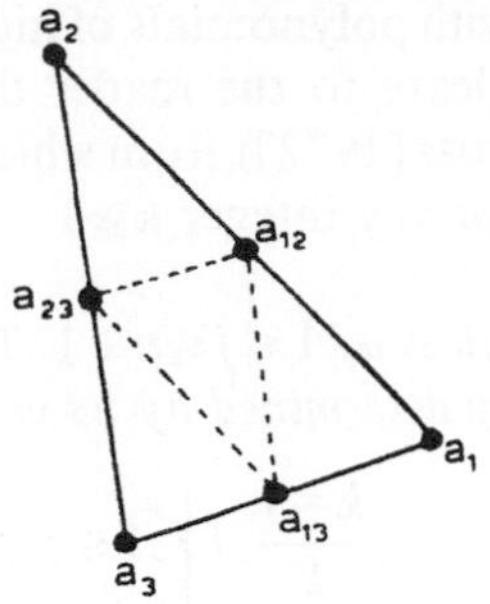

quadratic triangle
$\dim P_T = 6$

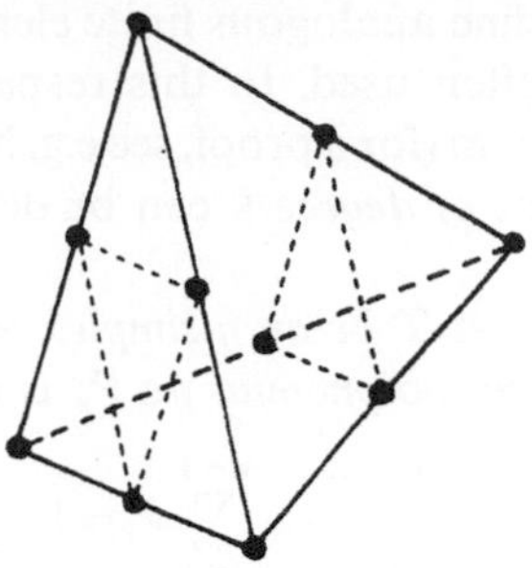

quadratic tetrahedron
$\dim P_T = 10$

quadratic n-simplex
$P_T = P_2(T)$, $\dim P_T = \frac{1}{2}(n+1)(n+2)$
$\Sigma_T = \{p(a_i)\colon 1 \leqslant i \leqslant n+1;\ p(a_{ij})\colon 1 \leqslant i < j \leqslant n+1\}$

FIG. 6.2.

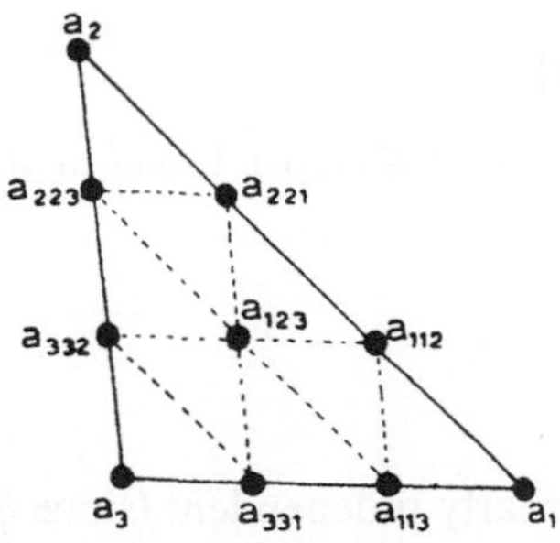

cubic triangle
$\dim P_T = 10$

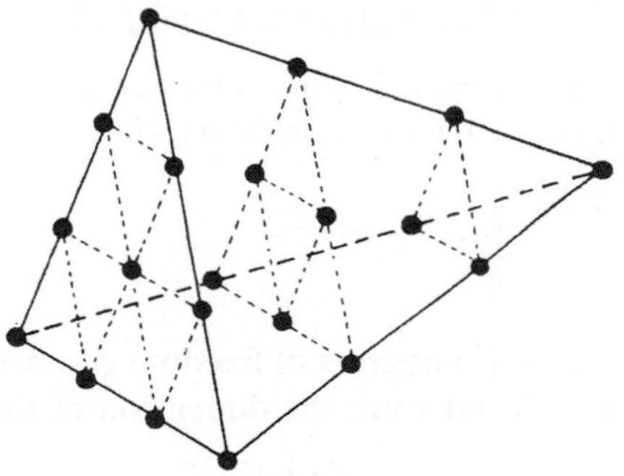

cubic tetrahedron
$\dim P_T = 20$

cubic n-simplex
$P_T = P_3(T)$, $\dim P_T = \frac{1}{6}(n+1)(n+2)(n+3)$
$\Sigma_T = \{p(a_i)\colon 1 \leqslant i \leqslant n+1;\ p(a_{iij})\colon 1 \leqslant i, j \leqslant n+1, i \neq j;$ $p(a_{ijk})\colon 1 \leqslant i < j < k \leqslant n+1\}$

FIG. 6.3.

One may define analogous finite elements with polynomials of higher degree, but they are not often used. In this respect, we leave to the reader the proof of the following theorem (for a proof, see e.g. NICOLAIDES [1972]), from which the definition of the *n-simplex of degree k* can be derived for any integer $k \geq 1$.

THEOREM 6.1. *Let T be an n-simplex with vertices* a_j, $1 \leq j \leq n+1$. *Then for a given integer* $k \geq 1$, *any polynomial* $p \in P_k$ *is uniquely determined by its values on the set*

$$L_k(T) = \left\{ x = \sum_{j=1}^{n+1} \lambda_j a_j;\ \sum_{j=1}^{n+1} \lambda_j = 1, \lambda_j \in \left\{0, \frac{1}{k}, \ldots, \frac{k-1}{k}, 1\right\}, 1 \leq j \leq n+1 \right\}. \quad (6.10)$$

REMARK 6.3. The set $L_k(T)$ is called the *principal lattice of order k* of the *n*-simplex *T*.

Let us now examine how the degrees of freedom $p(a_{ijk})$ can be eliminated in the definition of the cubic *n*-simplex. This elimination yields a new finite element, which is often preferred by engineers in actual computations, and whose definition is based on the following result:

THEOREM 6.2. *For each triple* (i, j, k) *with* $i < j < k$, *let*

$$\phi_{ijk}(p) = 12p(a_{ijk}) + 2 \sum_{l=i,j,k} p(a_l) - 3 \sum_{\substack{l,m=i,j,k \\ l \neq m}} p(a_{llm}). \quad (6.11)$$

Then any polynomial in the space

$$P'_3 = \{p \in P_3;\ \phi_{ijk}(p) = 0, 1 \leq i < j < k \leq n+1\} \quad (6.12)$$

is uniquely determined by its values at the vertices a_i, $1 \leq i \leq n+1$, *and at the points* a_{iij}, $1 \leq i, j \leq n+1, i \neq j$. *In addition, the inclusion*

$$P_2 \subset P'_3 \quad (6.13)$$

holds.

PROOF. The $\binom{n+1}{3}$ degrees of freedom ϕ_{ijk} are linearly independent (since $\phi_{ijk}(p) = 12p(a_{ijk}) + \cdots$) and thus, the dimension of the space P'_3 is

$$\dim P'_3 = \dim P_3 - \binom{n+1}{3} = (n+1)^2,$$

i.e., precisely the number of degrees of freedom. Using identity (6.9), we thus obtain the identity

$$p = \sum_i \left(\tfrac{1}{2}\lambda_i(3\lambda_i - 1)(3\lambda_i - 2) - \tfrac{27}{6} \sum_{\substack{k,j \neq i \\ j < i}} \lambda_i \lambda_j \lambda_k \right) p(a_i)$$

$$+ \sum_{i \neq j} \left(\tfrac{9}{2}\lambda_i \lambda_j (3\lambda_i - 1) + \tfrac{27}{4} \sum_{k \neq i,j} \lambda_i \lambda_j \lambda_k \right) p(a_{iij}) \quad \text{for all } p \in P'_3, \quad (6.14)$$

and the first part of the theorem is proved.

To prove that the inclusion (6.13) holds, let p be a polynomial of degree $\leqslant 2$ and let $A \in \mathscr{L}_2(\mathbb{R}^n; \mathbb{R})$ be its second derivative (which is constant). From the expansions

$$p(a_l) = p(a_{ijk}) + Dp(a_{ijk})(a_l - a_{ijk}) + \tfrac{1}{2}A(a_l - a_{ijk})^2, \quad l \in I,$$

valid for any triple $I = \{i, j, k\}$ with $i < j < k$, we deduce

$$\sum_{l \in I} p(a_l) = 3p(a_{ijk}) + \tfrac{1}{2} \sum_{l \in I} A(a_l - a_{ijk})^2,$$

since $\Sigma_{l \in I}(a_l - a_{ijk}) = 0$. Likewise, from the expansions

$$p(a_{llm}) = p(a_{ijk}) + Dp(a_{ijk})(a_{llm} - a_{ijk}) + \tfrac{1}{2}A(a_{llm} - a_{ijk})^2,$$
$$l, m \in I, \quad l \neq m,$$

we deduce

$$\sum_{\substack{l,m \in I \\ l \neq m}} p(a_{llm}) = 6p(a_{ijk}) + \tfrac{1}{2} \sum_{\substack{l,m \in I \\ l \neq m}} A(a_{llm} - a_{ijk})^2,$$

taking into account that

$$a_{ijk} = \tfrac{1}{2}(a_{iij} + a_{kkj}) = \tfrac{1}{2}(a_{jjk} + a_{iik}) = \tfrac{1}{2}(a_{kki} + a_{jji}).$$

Because A is a bilinear mapping, and because

$$a_{llm} - a_{ijk} = \tfrac{1}{3}(2(a_l - a_{ijk}) + (a_m - a_{ijk})),$$

we can write

$$\sum_{l \in I} A(a_l - a_{ijk})^2 - \tfrac{3}{2} \sum_{\substack{l,m \in I \\ l \neq m}} A(a_{llm} - a_{ijk})^2 = -\tfrac{2}{3} A\Big(\sum_{l \in I}(a_l - a_{ijk})\Big)^2 = 0,$$

and the proof is complete. □

From Theorem 6.1, we deduce the definition of the *reduced cubic n-simplex* (Fig. 6.4).

We next describe the construction of the *finite element spaces* associated with anyone of the finite elements that have been described so far. Since an n-simplex has a nonempty interior and a Lipschitz-continuous boundary, conditions $(\mathscr{T}_h 1)$ and $(\mathscr{T}_h 2)$ of Section 5 are automatically satisfied. In order that a triangulation $\mathscr{T}_h$ made up of n-simplices $T \in \mathscr{T}_h$ also satisfy conditions $(\mathscr{T}_h 3)$ and $(\mathscr{T}_h 4)$, we shall assemble them in such a way that $\bar{\Omega} = \bigcup_{T \in \mathscr{T}_h} T$ and that distinct n-simplices have piecewise disjoint interiors. Then *the first basic aspect* (FEM1) *of Section* 5 *is satisfied*. In view of satisfying inclusions such as $X_h \subset \mathscr{C}^0(\bar{\Omega})$ and $X_h \subset \mathscr{C}^1(\bar{\Omega})$, we shall however impose a *fifth condition on a triangulation made up of n-simplices*, viz. (cf. Theorems 5.1 and 5.2, and Theorem 6.3 below)

$(\mathscr{T}_h 5)$ *Any face of any n-simplex T_1 in the triangulation is either a subset of the boundary Γ, or a face of another n-simplex T_2 in the triangulation.*

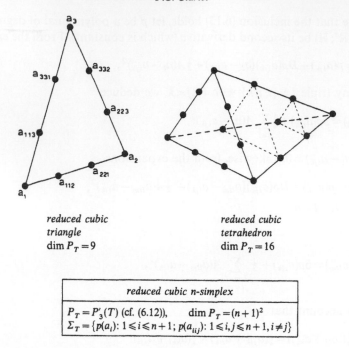

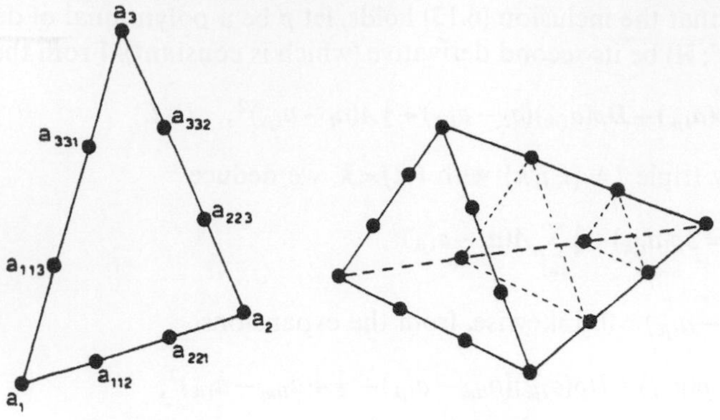

reduced cubic triangle
dim $P_T=9$

reduced cubic tetrahedron
dim $P_T=16$

reduced cubic n-simplex
$P_T=P_3'(T)$ (cf. (6.12)), dim $P_T=(n+1)^2$
$\Sigma_T=\{p(a_i)\colon 1\leqslant i\leqslant n+1;\ p(a_{iij})\colon 1\leqslant i,j\leqslant n+1, i\neq j\}$

FIG. 6.4.

In the second case, the n-simplices T_1 and T_2 are said to be *adjacent*. An example of a triangulation for $n=2$ is given in Fig. 6.5 while Fig. 6.6 shows an example of a triangulation that violates condition ($\mathcal{T}_h 5$), since the intersection of T_1 and T_2 is not an edge of T_2.

We then associate with any triangulation $\mathcal{T}_h$ that satisfies conditions ($\mathcal{T}_h 1$) to ($\mathcal{T}_h 5$) and with each type of finite element, a *finite element space* X_h, whose functions $v_h\colon \bar{\Omega}\to\mathbb{R}$ are constructed as follows:

With linear n-simplices, a function $v_h\in X_h$ is (i) such that each restriction $v_h|_T$ is in the space $P_T=P_1(T)$ for each $T\in\mathcal{T}_h$ and (ii) completely determined by its values at all the vertices of the triangulation. Likewise, with quadratic n-simplices, a function of X_h is (i) in the space $P_T=P_2(T)$ for each $T\in\mathcal{T}_h$ and (ii) completely determined by its values at all the vertices and all the mid-points of the edges of the triangulation. Similar constructions hold for cubic, or reduced cubic, n-simplices.

In all cases, a function v_h in the space X_h is thus determined by *degrees of freedom* $b\to v_h(b)$, which make up a set of the form

$$\Sigma_h=\{v_h(b)\colon b\in N_h\}, \tag{6.15}$$

where N_h is a finite subset of $\bar{\Omega}$. The set Σ_h is the *set of degrees of freedom of the finite element space* X_h.

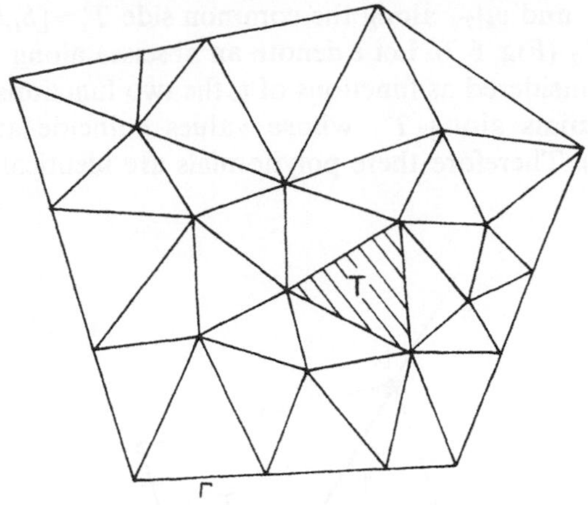

FIG. 6.5. An example of a triangulation of a polygonal set in $\mathbb{R}^2$.

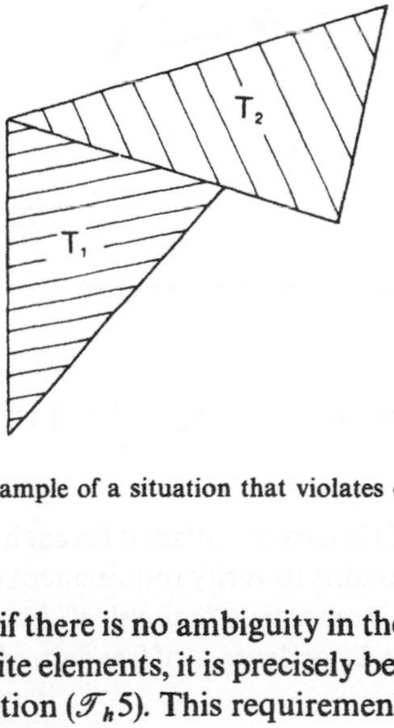

FIG. 6.6. An example of a situation that violates condition $(\mathcal{T}_h 5)$.

One should observe that if there is no ambiguity in the definition of the degrees of freedom across adjacent finite elements, it is precisely because we have required that triangulations satisfy condition $(\mathcal{T}_h 5)$. This requirement also plays a crucial role in the proof of the following result.

THEOREM 6.3. *Let X_h be a finite element space associated with n-simplices of degree k for some integer $k \geqslant 1$ or with reduced cubic n-simplices. Then the inclusion*

$$X_h \subset \mathscr{C}^0(\bar{\Omega}) \cap H^1(\Omega)$$

holds.

PROOF. We shall give the proof in the case $n=2$ and for quadratic triangles; the other cases are similarly treated. Given a function v_h in the space X_h, consider the two functions $v_h|_{T_1}$ and $v_h|_{T_2}$ along the common side $T' = [b_i, b_j]$ of two adjacent triangles T_1 and T_2 (Fig. 6.7). Let t denote an abscissa along the axis containing the segment T'. Considered as functions of t, the two functions $v_h|_{T_1}$ and $v_h|_{T_2}$ are quadratic polynomials along T', whose values coincide at the three points $b_i, b_j, b_{ij} = \frac{1}{2}(b_i + b_j)$. Therefore these polynomials are identical, and the inclusion

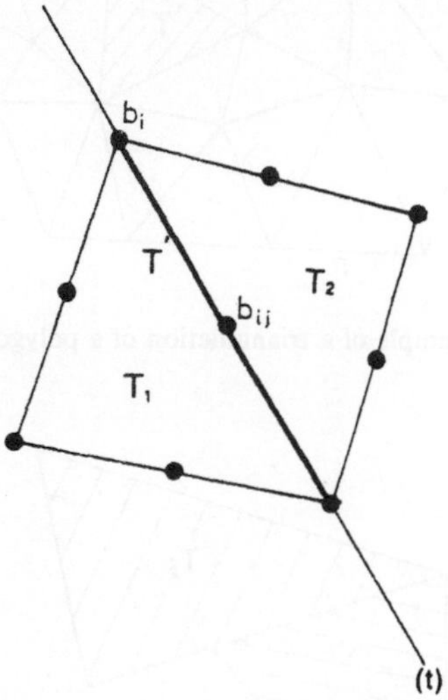

FIG. 6.7. Quadratic triangles assembled in a triangulation.

$X_h \subset \mathscr{C}^0(\bar{\Omega})$ holds. Hence the inclusion $X_h \subset H^1(\Omega)$ is a consequence of Theorem 5.1. □

Since requirement (FEM2) is clearly satisfied (in each case, the spaces P_T, $T \in \mathscr{T}_h$, consist of polynomials), it remains to verify requirement (FEM3), i.e., that *each one of these finite element spaces X_h possesses a "canonical" basis whose functions have small supports*. In each case, the set Σ_h of degrees of freedom of the spaces is of the form (cf. (6.15)):

$$\Sigma_h = \{v(b_k) \colon 1 \leqslant k \leqslant M\}. \tag{6.16}$$

If we define functions w_k, $1 \leqslant k \leqslant M$, by the conditions

$$w_k \in X_h, \qquad w_k(b_l) = \delta_{kl}, \quad 1 \leqslant k, l \leqslant M, \tag{6.17}$$

it is seen that (i) such functions form a basis of the space X_h and that (ii) they have "small" supports. In Fig. 6.8, we have represented the three types of supports that are encountered when cubic triangles are employed, for instance.

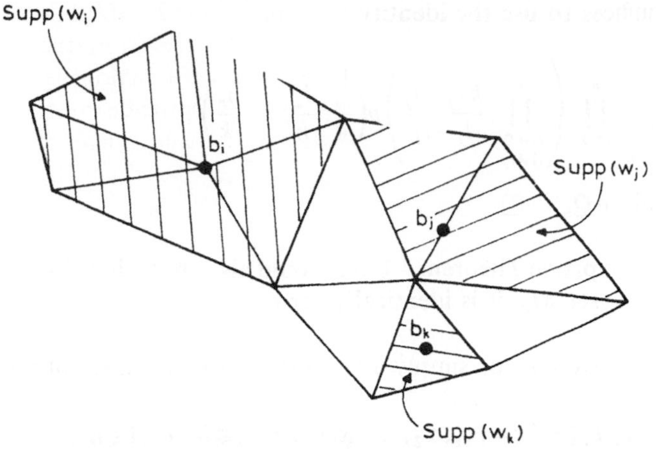

FIG. 6.8. Supports of the basis functions associated with cubic triangles.

7. Examples of rectangular finite elements and their associated finite element spaces

Before we turn to a second category of finite elements, we need a few definitions. For each integer $k \geqslant 0$, we denote by Q_k the space of all polynomials that are of degree $\leqslant k$ with respect to each one of the n variables $x_1, x_2, \ldots, x_n$, i.e., a polynomial $p \in Q_k$ is of the form

$$p: x=(x_1, x_2, \ldots, x_n) \in \mathbb{R}^n \rightarrow p(x)=\sum_{\substack{\alpha_i \leqslant k, \\ 1 \leqslant i \leqslant n}} \gamma_{\alpha_1 \alpha_2 \cdots \alpha_n} x_1^{\alpha_1} x_2^{\alpha_2} \cdots x_n^{\alpha_n},$$

for appropriate coefficients $\gamma_{\alpha_1 \alpha_2 \cdots \alpha_n}$. The dimension of the space Q_k is given by

$$\dim Q_k=(k+1)^n, \tag{7.1}$$

and the inclusions

$$P_k \subset Q_k \subset P_{nk} \tag{7.2}$$

hold. Notice that the dimension of the space $Q_k(A)$ is the same as that of the space $Q_k=Q_k(\mathbb{R}^n)$ as long as the interior of the set $A \subset \mathbb{R}^n$ is not empty.

THEOREM 7.1. *A polynomial $p \in Q_k$ is uniquely determined by its values on the set*

$$M_k=\left\{x=\left(\frac{i_1}{k}, \frac{i_2}{k}, \ldots, \frac{i_n}{k}\right) \in \mathbb{R}^n;\ i_j \in \{0,1, \ldots, k\},\ 1 \leqslant j \leqslant n\right\}. \tag{7.3}$$

PROOF. It suffices to use the identity

$$p=\sum_{\substack{0\leqslant i_j\leqslant k\\ 1\leqslant j\leqslant n}}\prod_{j=1}^{n}\left(\prod_{\substack{i'_j=0\\ i'_j\neq i_j}}^{k}\frac{kx_j-i'_j}{i_j-i'_j}\right)p\left(\frac{i_1}{k},\frac{i_2}{k},\ldots,\frac{i_n}{k}\right) \tag{7.4}$$

for all $p\in Q_k$. □

REMARK 7.1. To prove Theorem 7.1, one could also show that if a polynomial of Q_k vanishes on the set M_k, it is identically zero.

In $\mathbb{R}^n$, an *n-rectangle*, or simply a *rectangle* if $n=2$, is a set of the form

$$T=\prod_{i=1}^{n}[a_i,b_i]=\{x=(x_1,x_2,\ldots,x_n);\ a_i\leqslant x_i\leqslant b_i,\ 1\leqslant i\leqslant n\}, \tag{7.5}$$

with $-\infty<a_i<b_i<+\infty$ for each i, i.e., it is a product of compact intervals with nonempty interiors; in particular, the *unit hypercube* $[0,1]^n$ is an n-rectangle. A *face* of an n-rectangle T is any one of the sets

$$\{a_j\}\times\prod_{\substack{i=1\\ i\neq j}}^{n}[a_i,b_i]\quad\text{or}\quad\{b_j\}\times\prod_{\substack{i=1\\ i\neq j}}^{n}[a_i,b_i],\quad 1\leqslant j\leqslant n,$$

while an *edge* of T, also called a *side*, is any one of the sets

$$[a_j,b_j]\times\prod_{\substack{i=1\\ i\neq j}}^{n}\{c_i\},$$

with $c_i=a_i$ or b_i, $1\leqslant i\leqslant n, i\neq j, 1\leqslant j\leqslant n$. A *vertex* of T is any point $x=(x_1,x_2,\ldots,x_n)$ of T with $x_i=a_i$ or b_i, $1\leqslant i\leqslant n$.

Given an n-rectangle T, there exists a (nonuniquely determined) *diagonal affine mapping*, i.e., of the form $F_T(x)=B_Tx+b_T$, where B_T is an $n\times n$ diagonal matrix and b_T is a vector in $\mathbb{R}^n$, such that

$$T=F_T([0,1]^n).$$

Observing that the set M_k of (7.3) is a subset of the unit hypercube $[0,1]^n$, we then infer from Theorem 7.1 that a polonomial $p\in Q_k$ is also uniquely determined by its values on the subset

$$M_k(T)=F_T(M_k) \tag{7.6}$$

of the n-rectangle T; from this observation, we deduce the definition of finite elements called *n-rectangles of degree k*.

Just as in the case of n-simplices, the values $k=1,2$ or 3 are the most commonly encountered. In Figs. 7.1, 7.2 and 7.3, the corresponding elements are represented for $n=2$ and 3, their most common names are indicated, and the numbering of the nodes, i.e., of the points occurring in the sets of degrees of freedom, is indicated for

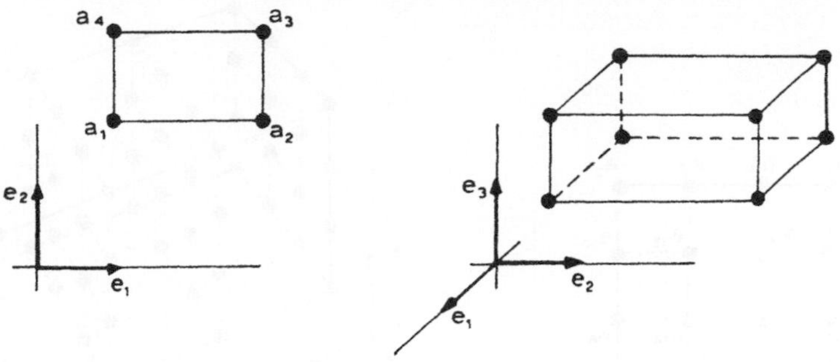

bilinear, or *4-node*, *rectangle*
dim $P_T = 4$

3-rectangle of degree 2
dim $P_T = 8$

n-rectangle of degree 2
$P_T = Q_1(T)$, dim $P_T = 2^n$ $\Sigma_T = \{p(a)\colon a \in M_1(T)\}$ (cf. (7.6))

FIG. 7.1.

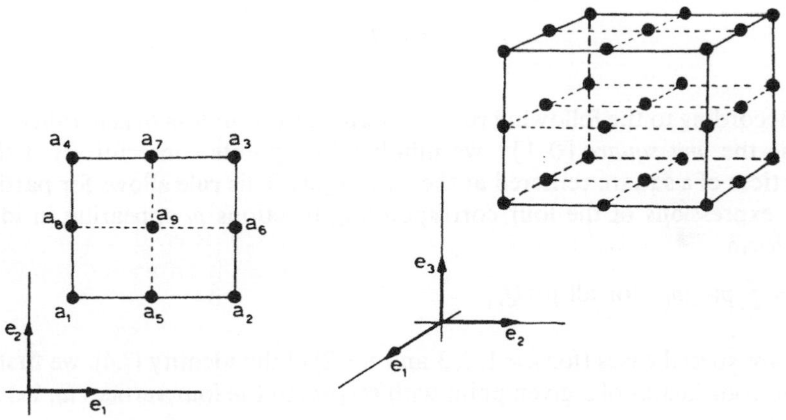

biquadratic, or *9-node*, *rectangle*
dim $P_T = 9$

3-rectangle of degree 3
dim $P_T = 27$

n-rectangle of degree 3
$P_T = Q_2(T)$, dim $P_T = 3^n$ $\Sigma_T = \{p(a)\colon a \in M_2(T)\}$ (cf. (7.6))

FIG. 7.2.

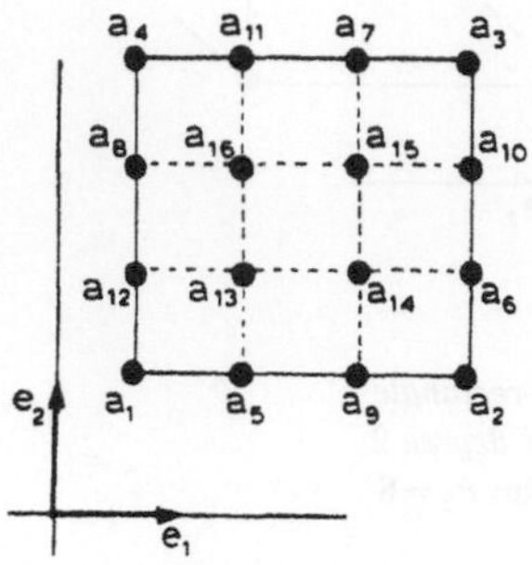

bicubic rectangle
dim $P_T=16$

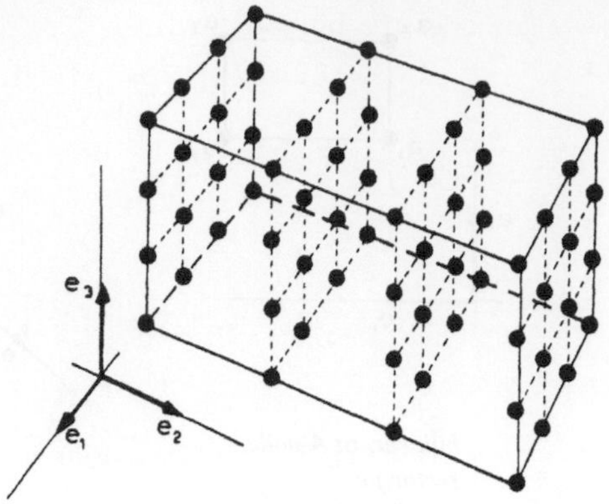

3-rectangle of degree 4
dim $P_T=64$

n-rectangle of degree 4
$P_T=Q_3(T)$, dim $P=4^n$ $\Sigma_T=\{p(a): a\in M_3(T)\}$ (cf. (7.6))

FIG. 7.3.

$n=2$, according to the following rule: Assuming, without loss of generality, that the set T is the *unit square* $[0,1]^2$, we number four points consecutively if they are the vertices of a square centered at the point $(\frac{1}{2},\frac{1}{2})$. This rule allows for particularly simple expressions of the four corresponding functions p_i appearing in identities of the form

$$p=\sum_i p(a_i)p_i \quad \text{for all } p\in Q_k,$$

which are special cases (for $k=1,2,3$ and $n=2$) of the identity (7.4): we first notice that the coordinates of a given point with respect to the four vertices a_i, $1\leqslant i\leqslant 4$, of the unit square are

$$(x_1,x_2),\quad (x_2,1-x_1),\quad (1-x_1,1-x_2),\quad (1-x_2,x_1),$$

respectively. Then, if we introduce the variables

$$x_3=1-x_1,\qquad x_4=1-x_2, \tag{7.7}$$

it can be easily checked that the four functions p_i corresponding to such four consecutively numbered points a_i are obtained through circular permutations of the variables x_1,x_2,x_3,x_4 (such permutations correspond to rotations of $+\frac{1}{2}\pi$ around the point $(\frac{1}{2},\frac{1}{2})$). Let us illustrate this fact by means of examples:

If we consider the biline... [0, 1]^2), we have the identity

$$p=\sum_{i=1}^{4} p(a_i)p_i \quad \text{for all } p\in Q_1,$$

with

$$p_1=(1-x_1)(1-x_2), \qquad p_2=x_1(1-x_2),$$
$$p_3=x_1x_2, \qquad p_4=(1-x_1)x_2.$$

We may thus condense these expressions as

$$p_1=x_3x_4, \quad \dots \tag{7.8}$$

Likewise, if we consider the biquadratic unit square, we have the identity

$$p=\sum_{i=1}^{9} p(a_i)p_i \quad \text{for all } p\in Q_2,$$

with

$$\begin{aligned} &p_1=x_3(2x_3-1)x_4(2x_4-1), \quad \dots\\ &p_5=-4x_3(x_3-1)x_4(2x_4-1), \quad \dots\\ &p_9=16x_1x_2x_3x_4, \end{aligned} \tag{7.9}$$

using the above rule. Finally, if we consider the bicubic unit square, we have

$$\begin{aligned} &p_1=\tfrac{1}{4}x_3(3x_3-1)(3x_3-2)x_4(3x_4-1)(3x_4-2), \quad \dots\\ &p_5=-\tfrac{9}{4}x_3(3x_3-1)(x_3-1)x_4(3x_4-1)(3x_4-2), \quad \dots\\ &p_9=\tfrac{9}{4}x_3(3x_3-2)(x_3-1)x_4(3x_4-1)(3x_4-2), \quad \dots\\ &p_{13}=\tfrac{81}{4}x_3(3x_3-1)(x_3-1)x_4(3x_4-1)(x_4-1), \quad \dots . \end{aligned} \tag{7.10}$$

REMARK 7.2. The inconsistency for the notations a_i, $5\leqslant i\leqslant 9$, between the biquadratic and bicubic rectangles avoids the introduction of a new letter.

We now define finite elements similar to the biquadratic or bicubic rectangles, but in which the values at the interior nodes are no longer degrees of freedom (for simplicity, we restrict ourselves to the case $n=2$). The existence of these finite elements is a consequence of the following two theorems.

THEOREM 7.2. *Let the points a_i, $1\leqslant i\leqslant 9$, be as in Fig. 7.2. Then any polynomial in the space*

$$Q_2'=\left\{p\in Q_2;\ 4p(a_9)+\sum_{i=1}^{4} p(a_i)-2\sum_{i=5}^{8} p(a_i)=0\right\} \tag{7.11}$$

is uniquely determined by its value at the points a_i, $1 \leqslant i \leqslant 8$. In addition, the inclusion

$$P_2 \subset Q_2' \tag{7.12}$$

holds.

PROOF. The first part of the proof is similar to the first part of the proof of Theorem 6.2. In particular we have the identity

$$p = \sum_{i=1}^{8} p(a_i)p_i \quad \text{for all } p \in Q_2',$$

with

$$\begin{aligned} p_1 &= x_3x_4(2x_3+2x_4-3), \quad \ldots \\ p_5 &= -4x_3x_4(x_3-1), \quad \ldots. \end{aligned} \tag{7.13}$$

To prove the inclusion (7.12), let p be a polynomial of degree 2, and let A denote its (constant) second derivative. From the expansions

$$p(a_i) = p(a_9) + Dp(a_9)(a_i - a_9) + \tfrac{1}{2}A(a_1 - a_9)^2, \quad 1 \leqslant i \leqslant 8,$$

we deduce

$$\sum_{i=1}^{4} p(a_i) = 4p(a_9) + \tfrac{1}{2}\sum_{i=1}^{4} A(a_i - a_9)^2,$$

$$\sum_{i=5}^{8} p(a_i) = 4p(a_9) + \tfrac{1}{2}\sum_{i=5}^{8} A(a_i - a_9)^2,$$

since

$$\sum_{i=1}^{4} (a_i - a_9) = \sum_{i=5}^{8} (a_i - a_9) = 0.$$

Because the mapping A is bilinear, and because $a_5 = \frac{1}{2}(a_1 + a_2), \ldots,$ we obtain

$$\sum_{i=5}^{8} A(a_i - a_9)^2 = \tfrac{1}{2}\sum_{i=1}^{4} A(a_i - a_9)^2.$$

Combining the previous relations, we deduce that

$$4p(a_9) + \sum_{i=1}^{4} p(a_i) - 2\sum_{i=5}^{8} p(a_i) = 0,$$

and the proof is complete. □

THEOREM 7.3. *Let the points a_i, $1 \leqslant i \leqslant 16$, be as in Fig. 7.3. Define the space*

$$Q_3' = \{p \in Q_3;\quad \psi_i(p) = 0,\ 1 \leqslant i \leqslant 4\}, \tag{7.14}$$

where

$$\psi_1(p)=9p(a_{13})+4p(a_1)+2p(a_2)+p(a_3)+2p(a_4)$$
$$-6p(a_5)-3p(a_6)-3p(a_{11})-6p(a_{12}), \tag{7.15}$$

and $\psi_2(p)$, $\psi_3(p)$, *and* $\psi_4(p)$ *are derived by circular permutations in the sets*

$$\bigcup_{i=1}^{4}\{a_i\},\quad \bigcup_{i=5}^{8}\{a_i\},\quad \bigcup_{i=9}^{12}\{a_i\},\quad \bigcup_{i=13}^{16}\{a_i\}.$$

Then any polynomial in the space Q_3' *is uniquely determined by its values at the points* a_i, $1\leqslant i\leqslant 12$. *In addition, the inclusion*

$$P_3\subset Q_3' \tag{7.16}$$

holds.

PROOF. The proof is similar to that of Theorem 7.2 and for this reason, is omitted. We shall only record here the identity

$$p=\sum_{i=1}^{12} p(a_i)p_i \quad \text{for all } p\in Q_3',$$

with

$$p_1=x_3x_4(1+\tfrac{9}{2}x_3(x_3-1)+\tfrac{9}{2}x_4(x_4-1)), \quad \cdots$$
$$p_5=-\tfrac{9}{2}x_3(x_3-1)(3x_3-1)x_4, \quad \cdots \tag{7.17}$$
$$p_9=\tfrac{9}{2}x_3(x_3-1)(3x_3-2)x_4, \quad \cdots.$$

From these two theorems we derive the definition of the *reduced*, or 8-*node*, *biquadratic rectangle* (Fig. 7.4) and of the *reduced*, or 12-*node*, *bicubic rectangle* (Fig. 7.5); they are also called the biquadratic, and bicubic, rectangles of the *Serendipity family*, as a reminder of the *ingenuity* that their discovery indeed required! Other

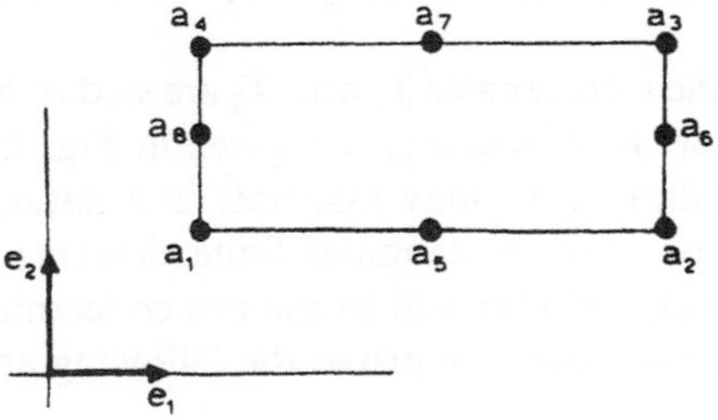

reduced, or 8-*node*, *biquadratic rectangle*	
$P_T=Q_2'(T)$ (cf. (7.11)), $\Sigma_T=\{p(a_i)\colon 1\leqslant i\leqslant 8\}$	$\dim P_T=8$

FIG. 7.4.

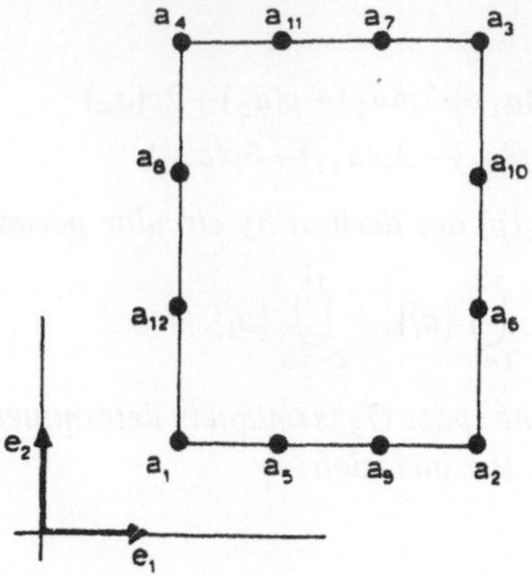

reduced, or *12-node*, *bicubic rectangle*
$P_T = Q_3'(T)$ (cf. (7.14)), dim $P_T = 12$ $\Sigma_T = \{p(a_i)\colon 1 \leqslant i \leqslant 12\}$

FIG. 7.5.

examples of *Serendipity finite elements* are found in ZIENKIEWICZ [1971, pp. 108, 121, 126], in particular for $n=3$. Their basis functions (and those of other rectangular finite elements) are found in EL-ZAFRANY and COOKSON [1986]. See also ZLÁMAL [1973c] for an interesting approach to such finite elements.

If it happens that the set $\bar{\Omega} \subset \mathbb{R}^n$ is *rectangular*, i.e., it is either an n-rectangle or a finite union of n-rectangles, it can be conveniently "triangulated" by finite elements which are themselves n-rectangles: The fifth condition $(\mathcal{T}_h 5)$ imposed on such a triangulation now reads:

$(\mathcal{T}_h 5)$ *Any face of any n-rectangle T_1 in the triangulation is either a subset of the boundary Γ, or a face of another n-rectangle T_2 in the triangulation.*

In the second case, the n-rectangles T_1 and T_2 are said to be *adjacent*. An example of a triangulation made up of rectangles is given in Fig. 7.6.

With such a triangulation, we may associate in a natural way a *finite element space* X_h with each type of the rectangular finite elements that we just described. Since the discussion is almost identical to the one concerning n-simplices, we shall be very brief. In particular, one can prove the following analog of Theorem 6.3.

THEOREM 7.4. *Let X_h be a finite element space associated with n-rectangles of degree k for some integer $k \geqslant 1$, or with reduced biquadratic, or reduced bicubic, rectangles. Then the inclusion*

$$X_h \subset \mathscr{C}^0(\bar{\Omega}) \cap H^1(\Omega) \tag{7.18}$$

holds.

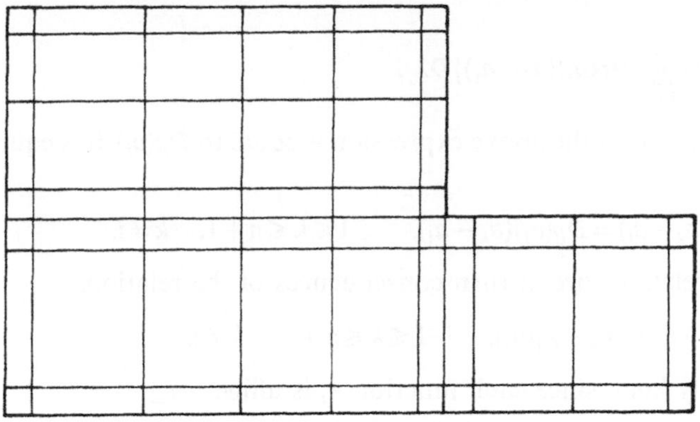

FIG. 7.6. A triangulation made up with rectangles.

Finally, as in the case of simplicial finite elements (Section 6), it is easily seen that such finite element spaces possess bases whose functions have "small" supports.

8. Examples of finite elements with derivatives as degrees of freedom and their associated finite element spaces

So far, the degrees of freedom of each finite element have been "point values", i.e., of the form $p(a)$, for some points $a \in T$. We now introduce finite elements in which some degrees of freedom are partial derivatives, or, more generally *directional derivatives*, i.e., expressions such as $Dp(a)b$, $D^2p(a)(b, c), \ldots$, where b, c are vectors in $\mathbb{R}^n$.

The first example of this type of finite element is based on the following theorem.

THEOREM 8.1. *Let T be an n-simplex with vertices a_i, $1 \leqslant i \leqslant n+1$, and let $a_{ijk} = \frac{1}{3}(a_i + a_j + a_k)$, $1 \leqslant i < j < k \leqslant n+1$. Then any polynomial in the space P_3 is uniquely determined by its values at the vertices a_i, $1 \leqslant i \leqslant n+1$, by the values of its n first partial derivatives at the vertices a_i, $1 \leqslant i \leqslant n+1$, and by its values at the points a_{ijk}, $1 \leqslant i < j < k \leqslant n+1$.*

PROOF. The argument relies as usual on an identity:

$$p = \sum_i \Bigg(-2\lambda_i^3 + 3\lambda_i^2 - 7\lambda_i \sum_{\substack{j<k \\ j,k \neq i}} \lambda_j \lambda_k \Bigg) p(a_i) + 27 \sum_{i<j<k} \lambda_i \lambda_j \lambda_k p(a_{ijk})$$

$$+ \sum_{i \neq j} \lambda_i \lambda_j (2\lambda_i + \lambda_j - 1) Dp(a_i)(a_j - a_i) \quad \text{for all } p \in P_3. \tag{8.1}$$

To prove this identity, one needs in particular an expression for the derivatives of the barycentric coordinates, in order to show that $Dp(a_i) = D\tilde{p}(a_i)$, $1 \leqslant i \leqslant n+1$, where $\tilde{p}$ denotes the right-hand side of (8.1). By differentiating the polynomial $\tilde{p}$, we obtain

$$D\tilde{p}(a_i)=\sum_{j\neq i}\{Dp(a_i)(a_j-a_i)\}D\lambda_j.$$

Hence, to show that the above expression is equal to $Dp(a_i)$, it is equivalent to show that

$$D\tilde{p}(a_i)(a_k-a_i)=Dp(a_i)(a_k-a_i), \qquad 1\leqslant k\leqslant n+1, \quad k\neq i.$$

These last relations are in turn consequences of the relations

$$D\lambda_j(a_k-a_i)=\delta_{jk}-\lambda_j(a_i), \qquad 1\leqslant k\leqslant n+1, \quad k\neq i,$$

which clearly hold, since each function λ_j is affine. □

From this theorem, we deduce the definition of a finite element, which is called the cubic *Hermite n-simplex* (Fig. 8.1), where the directional derivatives $Dp(a_i)(a_j-a_i)$ are degrees of freedom. Of course, the knowledge of these n directional derivatives at a vertex a_i is equivalent to the knowledge of the first derivative $Dp(a_i)$. Such a knowledge is indicated graphically by one small circle, or sphere, centered at the point a_i. Since the first derivative $Dp(a_i)$ is equally well determind by the partial derivatives $\partial_j p(a_i)$, $1\leqslant j\leqslant n$, another possible set of degrees of freedom for this element is the set Σ'_T indicated in Fig. 8.1.

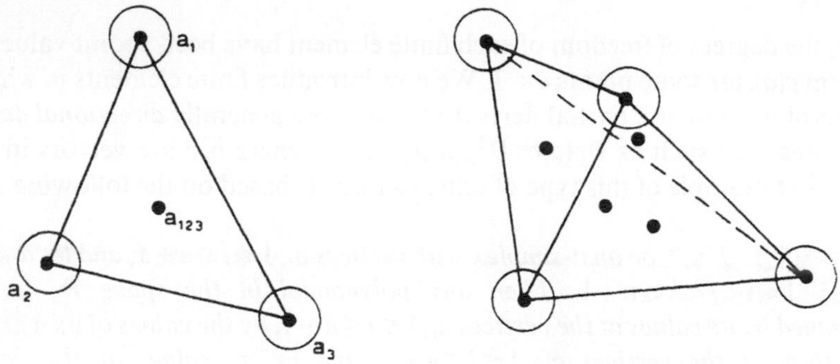

cubic
Hermite triangle
dim $P_T=10$

cubic
Hermite tetrahedron
dim $P_T=20$

cubic Hermite n-simplex
$P_T=P_3(T)$, dim $P_T=\frac{1}{6}(n+1)(n+2)(n+3)$
$\Sigma_T=\{p(a_i)\colon 1\leqslant i\leqslant n+1;\ p(a_{ijk})\colon 1\leqslant i<j<k\leqslant n+1;\ Dp(a_i)(a_j-a_i)\colon 1\leqslant i,j\leqslant n+1, j\neq i\}$
$\Sigma'_T=\{p(a_i)\colon 1\leqslant i\leqslant n+1;\ p(a_{ijk})\colon 1\leqslant i<j<k\leqslant n+1;\ \partial_j p(a_i)\colon 1\leqslant i\leqslant n+1,\ 1\leqslant j\leqslant n\}$

FIG. 8.1

By the same method that led us to the reduced cubic n-simplex, we can eliminate the degrees of freedom $p(a_{ijk})$, $i<j<k$. This "reduction" relies on the following theorem, whose proof is similar to that of Theorem 6.2:

THEOREM 8.2. *For each triple (i,j,k) with $i<j<k$, let*

$$\psi_{ijk}(p)=6p(a_{ijk})-2\sum_{l=i,j,k} p(a_l)+\sum_{l=i,j,k} Dp(a_l)(a_l-a_{ijk}). \tag{8.2}$$

Then any polynomial in the space

$$P_3''=\{p\in P_3;\ \psi_{ijk}(p)=0,\ 1\leqslant i<j<k\leqslant n+1\} \tag{8.3}$$

is uniquely determined by its values and the values of its n first partial derivatives at the vertices a_i, $1\leqslant i\leqslant n+1$. In addition, the inclusion

$$P_2\subset P_3''$$

holds.

From this theorem, we deduce the definition of the *reduced cubic Hermite n-simplex* which, for $n=2$, is also called the *Zienkiewicz triangle* (Fig. 8.2), after BAZELEY, CHEUNG, IRONS and ZIENKIEWICZ [1965].

Given a triangulation $\mathcal{T}_h$ made up of n-simplices, we associate in a natural way a finite element space X_h with either type of finite elements. For instance assume we

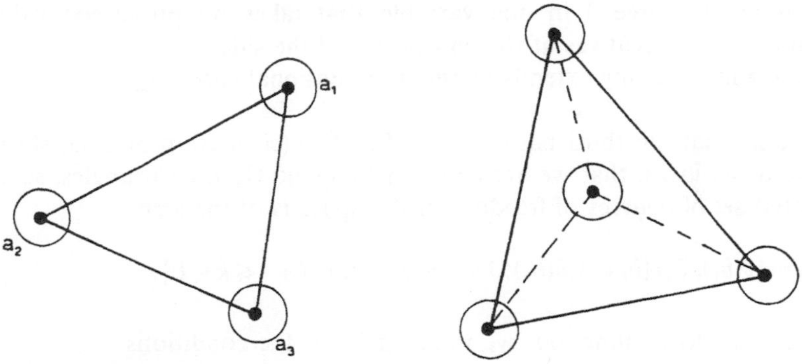

Zienkiewicz triangle, or *reduced cubic Hermite triangle* dim $P_T=9$

reduced cubic Hermite tetrahedron dim $P_T=16$

reduced cubic Hermite n-simplex
$P_T=P_3''\,(T)$(*cf.* (8.3)), dim $P_T=(n+1)^2$
$\Sigma_T=\{p(a_i)\colon 1\leqslant i\leqslant n+1;\ Dp(a_i)(a_j-a_i)\colon 1\leqslant i,j\leqslant n+1,\ i\neq j\}$,
$\Sigma_T'=\{p(a_i)\colon 1\leqslant i\leqslant n+1;\ \partial_j p(a_i)\colon 1\leqslant i\leqslant n+1,\ 1\leqslant j\leqslant n\}$.

FIG. 8.2

are using cubic Hermite n-simplices, the case of reduced cubic Hermite n-simplices being similar. Then a function v_h is in the space X_h if (i) each restriction $v_h|_T$ is in the space $P_T = P_3(T)$ for each $T \in \mathcal{T}_h$, and (ii) it is defined by its values at all the vertices of the triangulation, by its values at the centers of gravity of all triangles found as 2-faces of the n-simplices $T \in \mathcal{T}_n$, and by the values of its n first partial derivatives at all the vertices of the triangulation. The corresponding set of degrees of freedom of the space X_h is thus of the form

$$\Sigma_h = \{v_h(b): b \in N_v \cup N_c,\ \partial_j v_h(b),\ b \in N_v,\ 1 \leqslant j \leqslant n\}, \tag{8.4}$$

where N_v denotes the set of all the vertices of the n-simplices of the triangulation and N_c denotes the set of all centers of gravity of all 2-faces of the n-simplices found in the triangulation. Note that requirement ($\mathcal{T}_h 5$) (cf. Section 6) again insures that *the degrees of freedom are unambiguously defined across adjacent finite elements*; this requirement is also the basis for the following theorem.

THEOREM 8.3. *Let X_h be the finite element space associated with cubic, or reduced cubic, Hermite n-simplices. Then the inclusion*

$$X_h \subset \mathscr{C}^0(\bar{\Omega}) \cap H^1(\Omega) \tag{8.5}$$

holds.

PROOF. As in Theorem 6.3, it suffices to prove the inclusion $X_h \subset \mathscr{C}^0(\bar{\Omega})$. Assume $n=2$, then along any side common to two adjacent triangles, there is a unique polynomial of degree 3 in one variable that takes on prescribed values and prescribed first derivatives at the end points of the side.

This argument easily extends to the n-dimensional case. □

To verify that the third basic aspect (FEM 3) (cf. Section 5) is satisfied, let us assume, to fix ideas, that we are considering cubic Hermite triangles, so that the associated set of degrees of freedom of the space is of the form

$$\Sigma_h = \{v(b_k),\ \partial_1 v(b_k),\ \partial_2 v(b_k): 1 \leqslant k \leqslant J;\ v(b_k): J+1 \leqslant k \leqslant L\}. \tag{8.6}$$

Then if we define functions $w_k, w_k^1, w_k^2 \in X_h$ by the conditions

$$\begin{aligned}
&w_k(b_l) = \delta_{kl}, \quad 1 \leqslant k, l \leqslant L,\\
&\partial_1 w_k(b_l) = \partial_2 w_k(b_l) = 0, \quad 1 \leqslant k \leqslant L, \quad 1 \leqslant l \leqslant J,\\
&w_k^1(b_l) = 0, \quad 1 \leqslant k \leqslant J, \quad 1 \leqslant l \leqslant L,\\
&\partial_1 w_k^1(b_l) = \delta_{kl}, \quad \partial_2 w_k^1(b_l) = 0, \quad 1 \leqslant k, l \leqslant J,\\
&w_k^2(b_l) = 0, \quad 1 \leqslant k \leqslant J, \quad 1 \leqslant l \leqslant L,\\
&\partial_1 w_k^2(b_l) = 0, \quad \partial_2 w_k^2(b_l) = \delta_{kl}, \quad 1 \leqslant k, l \leqslant J,
\end{aligned} \tag{8.7}$$

it is easily seen that these functions form a basis of the space X_h and that they have "small" supports.

9. Examples of finite elements for fourth-order problems and their associated finite element spaces

Finally, we examine some examples of finite elements whose associated finite element spaces satisfy the inclusion $X_h \subset \mathscr{C}^1(\bar{\Omega})$, and which may therefore be used for solving fourth-order problems if the inclusions $P_T \subset H^2(T)$, $T \in \mathscr{T}_h$, also hold; cf. Theorem 5.2. It is legitimate to restrict ourselves to the case where $n=2$, in view of the examples given in Section 4. Our first example is based on the following result.

THEOREM 9.1. *Let T be a triangle with vertices a_i, $1 \leqslant i \leqslant 3$, and let $a_{ij} = \frac{1}{2}(a_i + a_j)$, $1 \leqslant i < j \leqslant 3$, denote the midpoints of the sides. Then any polynomial p of degree 5 is uniquely determined by the following set of 21 degrees of freedom:*

$$\Sigma_T = \{\partial^\alpha p(a_i) : |\alpha| \leqslant 2, 1 \leqslant i \leqslant 3;\ \partial_\nu p(a_{ij}) : 1 \leqslant i < j \leqslant 3\}, \tag{9.1}$$

where ∂_ν denote the normal derivative operator along the boundary of T.

PROOF. Given a set of degrees of freedom, finding the corresponding polynomial of degree 5 amounts to solving a linear system with a *square* matrix, for which existence and uniqueness for all right-hand sides are equivalent properties, as we already observed. We shall prove the latter property, i.e., that any polynomial $p \in P_5$ such that

$$\partial^\alpha p(a_i) = 0, \quad |\alpha| \leqslant 2, \quad 1 \leqslant i \leqslant 3,$$
$$\partial_\nu p(a_{ij}) = 0, \quad 1 \leqslant i < j \leqslant 3,$$

is identically zero.

Let t denote an abscissa along the axis that contains the side $T' = [a_1, a_2]$. Then the restriction $p|_{T'}$, considered as a function q of t, is a polynomial of degree 5, and q satisfies

$$q(a_1) = q'(a_1) = q''(a_1) = q(a_2) = q'(a_2) = q''(a_2) = 0,$$

since, if τ is a unit vector on the axis containing the side T', we have

$$q'(a_1) = \partial_\tau p(a_1), \qquad q''(a_1) = \partial_{\tau\tau} p(a_1), \quad \ldots;$$

hence $q = 0$.

Likewise, considered as a function r of t, the normal derivative $\partial_\nu p$ along T' is a polynomial of degree 4, and r satisfies

$$r(a_1) = r'(a_1) = r(a_{12}) = r(a_2) = r'(a_2) = 0,$$

since

$$r(a_1) = \partial_\nu p(a_1), \qquad r'(a_1) = \partial_{\nu\tau} p(a_1), \qquad r(a_{12}) = \partial_\nu p(a_{12}), \quad \ldots;$$

hence $r = 0$.

Since we have $\partial_\tau p = 0$ along T' ($p = 0$ along T'), we have proved that p and its first derivative Dp vanish identically along T'. We now show that this implies that the polynomial λ_3^2 is a factor of p: After using an appropriate affine mapping if necessary, we may assume without loss of generality that $\lambda_3(x_1,x_2)=x_1$. We can write

$$p(x_1,x_2)=\sum_{i=0}^{5} x_1^i p_i(x_2)$$

where $p_i, 0\leqslant i\leqslant 5$, are polynomials of degree $(5-i)$ in the variable x_2. Therefore

$$p(0,x_2)=p_0(x_2)=0 \quad \text{for all } x_2\in\mathbb{R},$$
$$\partial_1 p(0,x_2)=p_1(x_2)=0 \quad \text{for all } x_2\in\mathbb{R},$$

which proves our assertion.

Similar arguments hold for the other sides, and we find that the polynomial $(\lambda_1^2\lambda_2^2\lambda_3^2)$ is a factor of p. Since the λ_i are polynomials of degree 1 that do not reduce to constants, it necessarily follows that $p=0$. □

With Theorem 9.1 we can define a finite element, the $\mathscr{C}^1$-*quintic triangle*, also called the 21-*degree of freedom triangle*, or the *Argyris triangle* (Fig. 9.1), after ARGYRIS, FRIED and SCHARPF [1968]. Its basis functions may be found in BERNADOU

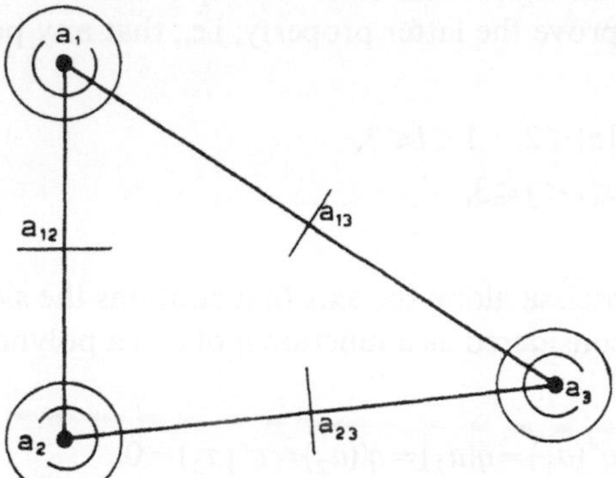

Argyris triangle, or $\mathscr{C}^1$-quintic triangle, or 21-degree of freedom triangle
$P_T=P_5(T)$, $\dim P_T=21$
$\Sigma_T=\{p(a_i),\partial_1 p(a_i),\partial_2 p(a_i),\partial_{11}p(a_i),\partial_{12}p(a_i),\partial_{22}p(a_i)\colon 1\leqslant i\leqslant 3;$ $\partial_\nu p(a_{ij})\colon 1\leqslant i<j\leqslant 3\}$
$\Sigma_T'=\{p(a_i)\colon 1\leqslant i\leqslant 3; Dp(a_i)(a_j-a_i)\colon 1\leqslant i,j\leqslant 3, j\neq i;$ $D^2p(a_i)(a_j-a_i,a_k-a_i)\colon 1\leqslant i,j,k\leqslant 3, j\neq i, k\neq i;$ $\partial_\nu p(a_{ij})\colon 1\leqslant i<j\leqslant 3\}$
$\Sigma_T''=\{p(a_i), Dp(a_i)(a_{i-1}-a_i), Dp(a_i)(a_{i+1}-a_i)\colon 1\leqslant i\leqslant 3;$ $D^2p(a_i)(a_{j+1}-a_j)^2\colon 1\leqslant i,j\leqslant 3; Dp(a_{ij})\nu_k\colon \{i,j,k\}=\{1,2,3\}, i<j\}$.

FIG. 9.1

and BOISSERIE [1982, p. 71]. We also note that Theorem 3 of ZLÁMAL [1968] yields an alternate proof to Theorem 9.1.

Figure 9.1 is self-explanatory as regards the graphical symbols used for representing the various degrees of freedom. We observe that at each vertex a_i, the first and second derivatives $Dp(a_i)$ and $D^2p(a_i)$ are known. With this observation in mind, we see that other possible definitions for the set of degrees of freedom are the sets Σ'_T and Σ''_T indicated in Fig. 9.1. In the expression of the set Σ''_T, the indices are numbered modulo 3, and each vector v_i, $1 \leqslant i \leqslant 3$, is the height originating at the point a_i.

It may be desirable to dispose of the degrees of freedom $\partial_\nu p(a_{ij})$, $1 \leqslant i < j \leqslant 3$. This reduction will be a consequence of the following result.

THEOREM 9.2. *Any polynomial in the space*

$$P'_5(T) = \{p \in P_5(T);\ \partial_\nu p \in P_3(T') \text{ for each side } T' \text{ of } T\} \tag{9.2}$$

is uniquely determined by the following set of 18 *degrees of freedom*:

$$\Sigma_T = \{\partial^\alpha p(a_i) : |\alpha| \leqslant 2,\ 1 \leqslant i \leqslant 3\}. \tag{9.3}$$

The space $P'_5(T)$ *satisfies the inclusion*

$$P_4(T) \subset P'_5(T). \tag{9.4}$$

PROOF. By writing $\partial_\nu p \in P_3(T')$ in definition (9.2), it is of course meant that, considered as a function of an abscissa along an axis containing the side T', the normal derivative $\partial_\nu p$ is a polynomial of degree 3. The inclusion (9.4) being obvious, it remains to prove the first part of the theorem.

To begin with, we prove a preliminary result: *Let* $T' = [a_i, a_j]$ *be a segment in* $\mathbb{R}^n$, *with midpoint* a_{ij}, *and let* v *be a function such that* $v|_{T'} \in P_4(T')$. *Then we have* $v|_{T'} \in P_3(T')$ *if and only if* $\chi_{ij}(v) = 0$, *where*

$$\chi_{ij}(v) = 4(v(a_i) + v(a_j)) - 8v(a_{ij}) + Dv(a_i)(a_j - a_i) + Dv(a_j)(a_i - a_j). \tag{9.5}$$

To see this, let, for any $x \in T'$, $\alpha_4 = D^4v(x)\tau^4$, where τ is a unit vector along T', so that α_4 is a constant. Then we have

$$\begin{aligned} v(a_i) &= v(a_{ij}) + Dv(a_{ij})(a_i - a_{ij}) + \tfrac{1}{2}D^2v(a_{ij})(a_i - a_{ij})^2 \\ &\quad + \tfrac{1}{6}D^3v(a_{ij})(a_i - a_{ij})^3 + \tfrac{1}{24}\alpha_4 \|a_i - a_{ij}\|^4, \\ v(a_j) &= v(a_{ij}) + Dv(a_{ij})(a_j - a_{ij}) + \tfrac{1}{2}D^2v(a_{ij})(a_j - a_{ij})^2 \\ &\quad + \tfrac{1}{6}D^3v(a_{ij})(a_j - a_{ij})^3 + \tfrac{1}{24}\alpha_4 \|a_j - a_{ij}\|^4, \end{aligned}$$

from which we deduce $(a_i - a_{ij} = -(a_j - a_{ij}))$:

$$\begin{aligned} v(a_i) + v(a_j) &= 2v(a_{ij}) + \tfrac{1}{2}\{D^2v(a_{ij})(a_i - a_{ij})^2 + D^2v(a_{ij})(a_j - a_{ij})^2\} \\ &\quad + \tfrac{1}{24}\alpha_4\{\|a_i - a_{ij}\|^4 + \|a_j - a_{ij}\|^4\}. \end{aligned}$$

Likewise,

$$Dv(a_i)(a_i-a_{ij})=D^2v(a_{ij})(a_i-a_{ij})^2+\tfrac{1}{2}D^3v(a_{ij})(a_i-a_{ij})^3+\tfrac{1}{6}\alpha_4\|a_i-a_{ij}\|^4,$$

$$Dv(a_j)(a_j-a_{ij})=D^2v(a_{ij})(a_j-a_{ij})^2+\tfrac{1}{2}D^3v(a_{ij})(a_j-a_{ij})^3+\tfrac{1}{6}\alpha_4\|a_j-a_{ij}\|^4,$$

and therefore,

$$D^2v(a_{ij})(a_i-a_{ij})^2+D^2v(a_{ij})(a_j-a_{ij})^2$$
$$=Dv(a_i)(a_i-a_{ij})+Dv(a_j)(a_j-a_{ij})-\tfrac{1}{6}\alpha_4\{\|a_i-a_{ij}\|^4+\|a_j-a_{ij}\|^4\}.$$

Combining our previous relations, we get

$$2v(a_{ij})=v(a_i)+v(a_j)+\tfrac{1}{4}\{Dv(a_i)(a_j-a_i)+Dv(a_j)(a_i-a_j)\}+\tfrac{1}{96}\alpha_4\|a_i-a_j\|^4,$$

and the assertion is proved.

As a consequence of this preliminary result, the space $P'_5(T)$ may be also defined as

$$P'_5(T)=\{p\in P_5(T);\ \chi_{ij}(\partial_\nu p)=0,\ 1\leqslant i<j\leqslant 3\},\tag{9.6}$$

i.e., we have characterized the space $P'_5(T)$ by the property that each normal derivative $\partial_\nu p(a_{ij})$ is expressed as a linear combination of the parameters $\partial^\alpha p(a_i)$, $\partial^\alpha p(a_j)$, $|\alpha|=1,2$. Then the proof is completed by combining the usual argument with the result of Theorem 9.1. □

From Theorem 9.2, we deduce the definition of a finite element, called the *reduced $\mathscr{C}^1$-quintic triangle*, or the 18-*degree of freedom triangle*, or the *Bell triangle*, after BELL [1969]; see Fig. 9.2, where we have indicated three possible sets of degrees of freedom that parallel those of the Argyris triangle.

REMARK 9.1. The Argyris and Bell triangles should be also attributed to FELIPPA [1966], who described them for the first time.

Given a triangulation made up of triangles, we associate a finite element space X_h with either type of finite elements. We leave it to the reader to derive the associated set of degrees of freedom of the space X_h and to check that the canonical basis is again composed of functions with "small" support. We shall only prove the following result.

THEOREM 9.3. *Let X_h be the finite element space associated with Argyris triangles or Bell triangles. Then the inclusion*

$$X_h\subset\mathscr{C}^1(\bar{\Omega})\cap H^2(\Omega)\tag{9.7}$$

holds.

PROOF. By Theorem 5.2, it suffices to show that the inclusion $X_h\subset\mathscr{C}^1(\bar{\Omega})$ holds.

Let T_1 and T_2 be two adjacent triangles with a common side $T'=[b_i,b_j]$ (Fig. 9.3) and let v_h be a function in the space X_h constructed with Argyris triangles. Considered as functions of an abscissa t along an axis containing the side T', the

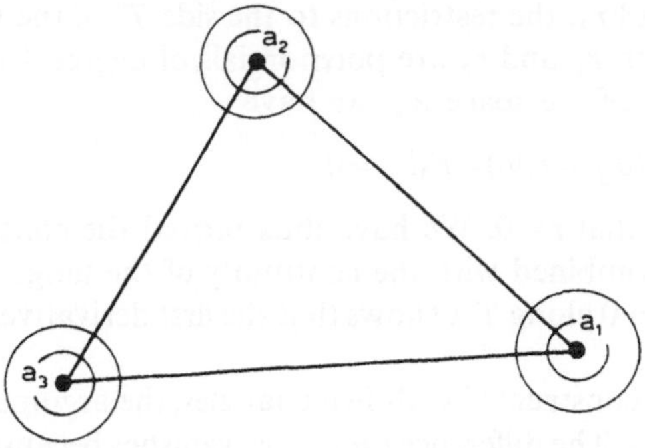

Bell triangle, or *reduced* $\mathscr{C}^1$*-quintic triangle,* or 18*-degree of freedom triangle*
$P_T = P'_5(T)$ (cf. (9.2)), $\dim P_T = 18$
$\Sigma_T = \{p(a_i), \partial_1 p(a_i), \partial_2 p(a_i), \partial_{11} p(a_i), \partial_{12} p(a_i), \partial_{22} p(a_i): 1 \leqslant i \leqslant 3\}$
$\Sigma'_T = \{p(a_i): 1 \leqslant i \leqslant 3; Dp(a_i)(a_j - a_i): 1 \leqslant i, j \leqslant 3, j \neq i;$ $D^2 p(a_i)(a_j - a_i, a_k - a_i): 1 \leqslant i, j, k \leqslant 3, j \neq i, k \neq i\}$
$\Sigma''_T = \{p(a_i), Dp(a_i)(a_{i-1} - a_i), Dp(a_i)(a_{i+1} - a_i): 1 \leqslant i \leqslant 3;$ $D^2 p(a_i)(a_{j+1} - a_j)^2 : 1 \leqslant i, j \leqslant 3\}$

FIG. 9.2.

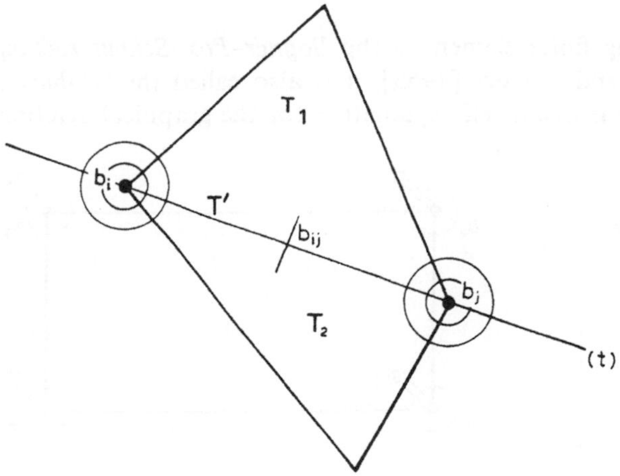

FIG. 9.3. Two adjacent Argyris triangles.

functions $v_h|_{T_1}$ and $v_h|_{T_2}$ are along T' polynomials of degree 5 in the variable t. Call these polynomials q_1 and q_2. Since, by definition of the space X_h, we have

$$q(b_i) = q'(b_i) = q''(b_i) = q(b_j) = q'(b_j) = q''(b_j) = 0,$$

with $q = q_1 - q_2$, it follows that $q = 0$ and hence the inclusion $V_h \subset \mathscr{C}^0(\bar{\Omega})$ holds.

Likewise, call r_1 and r_2, the restrictions to the side T' of the functions $\partial_\nu v_h|_{T_1}$ and $-\partial_\nu v_h|_{T_2}$. Then both r_1 and r_2 are polynomials of degree 4 in the variable t and, again by definition of the space X_h, we have

$$r(b_i)=r'(b_i)=r(b_{ij})=r(b_j)=r'(b_j)=0,$$

with $r=r_1-r_2$, so that $r=0$. We have thus proved the continuity of the normal derivative which, combined with the continuity of the tangential derivative ($q=0$ along T' implies $q'=0$ along T'), shows that the first derivatives are also continuous on $\bar{\Omega}$.

If the space X_h is constructed with Bell triangles, the argument is identical for the difference $q=q_1-q_2$. The difference $r=r_1-r_2$ vanishes because it is a polynomial of degree 3 in the variable t which is such that

$$r(b_i)=r'(b_i)=r(b_j)=r'(b_j)=0. \quad \square$$

To conclude, we give one instance of a rectangular finite element which may be used for solving fourth-order problems posed over rectangular domains. Its existence depends upon the following theorem, whose proof offers no difficulties.

THEOREM 9.4. *Let T denote a rectangle with vertices a_i, $1\leqslant i\leqslant 4$. Then a polynomial $p\in Q_3$ is uniquely determined by the following set of degrees of freedom:*

$$\Sigma_T=\{p(a_i), \partial_1 p(a_i), \partial_2 p(a_i), \partial_{12} p(a_i): 1\leqslant i\leqslant 4\}. \tag{9.8}$$

The resulting finite element is the *Bogner–Fox–Schmit rectangle* named after BOGNER, FOX and SCHMIT [1965]; it is also called the *$\mathscr{C}^1$-bicubic rectangle*; see Fig. 9.4, which is again self-explanatory for the graphical symbols.

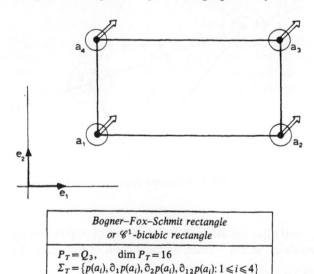

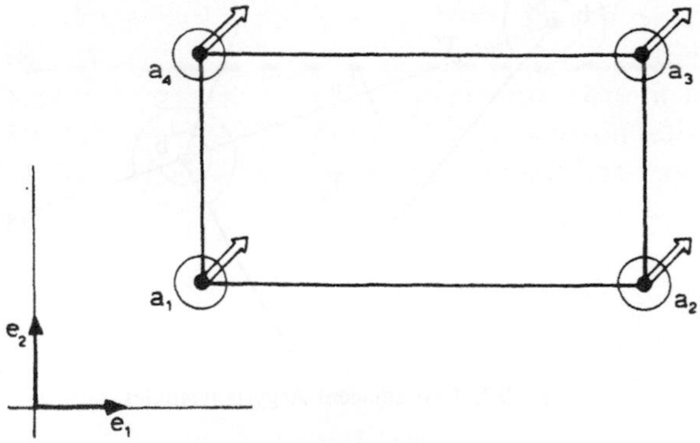

Bogner–Fox–Schmit rectangle   *or $\mathscr{C}^1$-bicubic rectangle*
$P_T=Q_3$, dim $P_T=16$ $\Sigma_T=\{p(a_i), \partial_1 p(a_i), \partial_2 p(a_i), \partial_{12} p(a_i): 1\leqslant i\leqslant 4\}$

FIG. 9.4.

The proof of the next result is similar to that of Theorem 9.3.

THEOREM 9.5. *Let X_h be the finite element space associated with Bogner–Fox–Schmit rectangles. Then the inclusion*

$$X_h \subset \mathscr{C}^1(\bar{\Omega}) \cap H^2(\Omega) \tag{9.9}$$

holds.

Finally, one easily checks, using the standard construction, that a finite element space constructed with any one of the last three finite elements indeed possesses canonical bases whose functions have "small" supports.

REMARK 9.2. Other examples of finite elements yielding the inclusion $X_h \subset \mathscr{C}^1(\bar{\Omega}) \cap H^2(\Omega)$ will be given in Sections 45–47.

Whereas it is fairly easy to conceive finite element spaces contained in $\mathscr{C}^0(\bar{\Omega})$, the construction of finite element spaces contained in $\mathscr{C}^1(\bar{\Omega})$ is less obvious, as shown by the three examples described in this section (and by the even more intricate examples that will be given in Chapter VII; see also the discussion in ZIENKIEWICZ [1971, Section 10.3]). This observation has been justified by the beautiful *result of* ŽENIŠEK [1973, 1974], who has proved the following: Let $n=2$, let X_h be a finite element space where all the sets T are *triangles*, and where the spaces P_T are spaces of *polynomials*, i.e., there exists some integer l such that the inclusions $P_T \subset P_l(T)$ hold for all $T \in \mathscr{T}_h$ (therefore finite elements of class $\mathscr{C}^1$ using "singular functions", or of "composite" type, as described in Chapter VII are excluded from the present analysis). Then, *for any integer $m \geqslant 0$, the inclusion $X_h \subset \mathscr{C}^m(\bar{\Omega})$ implies that, at each vertex b of the triangulation, the linear forms $v_h \to \partial^\alpha v_h(b)$ are degrees of freedom of the space X_h for all $|\alpha| \leqslant 2m$.* As a corollary, the inequality $l \geqslant 4m+1$ holds (the proof of the corollary is simple, but the proof of the first result is by no means trivial).

Thus for instance the particular choice $m=1$ shows that the Bell triangle is optimal for fourth-order problems, since the dimension (18) of the space $P'_5(T)$ of (9.2) is the smallest possible for conforming finite element methods using piecewise polynomial spaces and triangles. Note that ŽENÍŠEK [1972] has also extended his results to the case of higher dimensions. See also LE MÉHAUTÉ [1984] for further extensions.

Finite element spaces whose functions are piecewise polynomials, and which are contained in $\mathscr{C}^m(\bar{\Omega})$, have also been studied by BARNHILL and GREGORY [1975b], DÉLÈZE and GOËL [1976], MORGAN and SCOTT [1975], SCOTT [1974], STRANG [1973, 1974].

10. Finite elements as triples (T, P, Σ) and their associated P_T-interpolation operators Π_T

Motivated by the previous examples, we are now in a position to give a general definition, first proposed by CIARLET [1975, p. 61]: A *finite element* in $\mathbb{R}^n$ is a triple

(T, P, Σ) where:

(i) T is a closed subset of $\mathbb{R}^n$ with a nonempty interior and a Lipschitz-continuous boundary;

(ii) P is a finite-dimensional space of real-valued functions defined over the set T; we let $N = \dim P$;

(iii) Σ is a set of N linear forms $\phi_i, 1 \leqslant i \leqslant N$, defined over the space P and, *by definition*, it is assumed that the set Σ is *P-unisolvent*, in the following sense: given any real scalars $\alpha_i, 1 \leqslant i \leqslant N$, there exists a unique function $p \in P$ that satisfies

$$\phi_i(p) = \alpha_i, \quad 1 \leqslant i \leqslant N. \tag{10.1}$$

Of course, this implies that *the N linear forms ϕ_i are linearly independent.*

In particular, there exist N functions $p_i \in P, 1 \leqslant i \leqslant N$, that satisfy

$$\phi_j(p_i) = \delta_{ij}, \quad 1 \leqslant j \leqslant N, \tag{10.2}$$

and the following identity holds:

$$p = \sum_{i=1}^{N} \phi_i(p) p_i \quad \text{for all } p \in P. \tag{10.3}$$

The linear forms $\phi_i, 1 \leqslant i \leqslant N$, are called the *degrees of freedom of the finite element*, and the functions $p_i, 1 \leqslant i \leqslant N$, are called the *basis functions of the finite element.*

Whenever we find it convenient, we shall use the notations $P_T, \Sigma_T, \phi_{i,T}$ and p_{i}, in lieu of P, Σ, ϕ_i and p_i.

REMARK 10.1. The basis functions are also called the *shape functions* in the engineering literature.

REMARK 10.2. We shall see later that, in practice, the linear forms ϕ_i are in fact defined over spaces that are larger than P, but we ignore this fact for the time being.

REMARK 10.3. The set T itself is often called a *finite element*; this is clearly an *abus de langage*!

REMARK 10.4. The P-unisolvence of the set Σ is equivalent to the fact that the N linear forms ϕ_i form a basis in the dual space of P. One may then view the bases $(\phi_i)_{i=1}^N$ and $(p_i)_{i=1}^N$ as being *dual bases*, in the algebraic sense (cf. in particular identity (10.3)).

In the light of the definition of a finite element, let us briefly review the examples given in the previous sections.

We have seen examples for which the set T is either an n-simplex, in which case the finite element is said to be *simplicial*, or *triangular* if $n = 2$, or *tetrahedral* if $n = 3$, or an n-rectangle in $\mathbb{R}^n$, in which case the finite element is said to be *rectangular*. As we already mentioned, these are all special cases of *straight finite elements*, i.e., for which the set T is a polyhedron in $\mathbb{R}^n$. Other polygonal shapes are found in practice, such as *quadrilaterals* (see Sections 36 and 46) or "*prismatic*" *finite elements* (cf. Fig. 12.1). We shall also describe (Section 36) "*curved*" *finite elements*, i.e., whose boundaries are composed of "curved" faces.

The main characteristic of the various spaces P encountered in the examples is that *they all contain a "full" polynomial space* $P_k(T)$ for some integer $k \geq 1$, a property that will be shown in the subsequent chapters to be crucial as far as *convergence* properties are concerned.

In all the previously described examples, the degrees of freedom were of some of the following forms:

$$p \to p(a_i^0),$$
$$p \to Dp(a_i^1)\xi_{ik}^1, \tag{10.4}$$
$$p \to D^2 p(a_i^2)(\xi_{ik}^2, \xi_{il}^2),$$

where the points $a_i^r, r = 0, 1, 2$, belong to the finite element, and the nonzero vectors ξ_{ik}^1, ξ_{ik}^2, ξ_{il}^2 are either "attached to the geometry" of the finite element (as in $Dp(a_i)(a_j - a_i)$, $\partial_\nu p(a_{ij}), \ldots$) or are fixed vectors of $\mathbb{R}^n$ (as in $\partial_i p(a_j)$, $\partial_{ij} p(a_k)$). The points $a_i^r, r = 0,1,2$, are called the *nodes of the finite element*.

Whereas only directional derivatives of order 1 or 2 occurred in the examples, one could conceivably consider degrees of freedom that would be partial derivatives of arbitrarily high order, but these are seldom used in practice. As we shall see later (cf. Sections 32 and 50) there are however practical instances of degrees of freedom that are not attached to nodes: They are instead *averages* (over the finite element or over one of its faces) of some partial derivative.

When all the degrees of freedom of a finite element are of the form $p \to p(a_i)$, we say that the associated finite element is a *Lagrange finite element*, while if at least one directional derivative occurs as a degree of freedom, the associated finite element is said to be a *Hermite finite element*.

As the examples in the previous sections have shown, there are essentially *two methods for proving that a given set* Σ *of degrees of freedom is P-unisolvent*: After it has been checked that $\dim P = \text{card}(\Sigma)$, one either exhibits the basis functions, or one shows that if all the degrees of freedom are set equal to zero, then the only corresponding function in the space P is identically zero. Note that we have used the first method for all the examples, except for the Argyris triangle.

Given a finite element (T, P, Σ) and a function $v: T \to \mathbb{R}$ sufficiently smooth, so that the degrees of freedom $\phi_i(v)$, $1 \leq i \leq N$, are well defined, we let

$$\Pi v = \sum_{i=1}^{N} \phi_i(v) p_i \tag{10.5}$$

denote the *P-interpolant* of the function v; the P-interpolant is also denoted $\Pi_T v$ if necessary. Since the set Σ is P-unisolvent, the P-interpolant is also the unique function that satisfies

$$\Pi v \in P, \qquad \phi_i(\Pi v) = \phi_i(v), \quad 1 \leq i \leq N. \tag{10.6}$$

Whenever the degrees of freedom are of the form (10.4), let s denote the maximal order of derivatives occurring in the definition of the set Σ. Since, for all the finite elements described so far, the inclusion $P\subset\mathscr{C}^s(T)$ holds, we shall consider that *the domain* $\operatorname{dom}\Pi$ of the *P-interpolation operator* Π is the space

$$\operatorname{dom}\Pi=\mathscr{C}^s(T). \tag{10.7}$$

This being the case, it follows that the *P-interpolation operator reduces to the identity over the space* $P\subset\operatorname{dom}\Pi$, i.e.,

$$\Pi p=p \quad \text{for all } p\in P. \tag{10.8}$$

In order that the P-interpolation operator be unambiguously defined, it is therefore necessary that the forms ϕ_i, which are a priori only defined on the space P, be also defined on the space $\mathscr{C}^s(T)$. To see this, assume again that the space P is contained in the space $\mathscr{C}^s(T)$. Then if a degree of freedom were only defined over the space P, it would have infinitely many extensions to the space $\mathscr{C}^s(T)$. Let us give one simple example of such a phenomenon: Let T be an n-simplex with barycenter a. Then the linear form

$$p\in\mathscr{C}^0(T)\to\frac{1}{\operatorname{meas}(T)}\int_T p\,\mathrm{d}x$$

is one possible extension of the form $p\in P_1(T)\to p(a)$.

Of course, these considerations are usually omitted: when a degree of freedom such as $\partial_i p(a_j)$ is considered for instance, it is implicitly understood that this linear form is the usual one, i.e., it is defined over the space $\mathscr{C}^1(T)$, but it is not any one of its other possible extensions from the space P to the space $\mathscr{C}^1(T)$. For another illustration of this circumstance, see the description of *Wilson's brick* in Section 32.

Whereas for a Lagrange finite element, the set of degrees of freedom is unambiguously defined (indeed, it can be conveniently identified with the set of nodes), there are always several possible definitions for the degrees of freedom of a Hermite finite element which correspond to the "same" finite element. More precisely, we shall say that *two finite elements* (T,P,Σ) *and* (S,Q,E) *are equal* if we have

$$T=S,\qquad P=Q,\qquad \Pi_T=\Pi_S. \tag{10.9}$$

To illustrate this point, consider the reduced cubic Hermite n-simplex. Two possible sets of degrees of freedom for this element are (cf. Fig. 8.2):

$$\Sigma=\{p(a_i)\colon 1\leqslant i\leqslant n+1;\, Dp(a_i)(a_j-a_i)\colon 1\leqslant i,j\leqslant n+1, i\neq j\},$$
$$\Sigma'=\{p(a_i)\colon 1\leqslant i\leqslant n+1;\, \partial_k p(a_i)\colon 1\leqslant i\leqslant n+1, 1\leqslant k\leqslant n\}.$$

Let us denote by Π and Π' the corresponding $P_3''(T)$-interpolation operators. Then, for any function $v\in\mathscr{C}^1(T)=\operatorname{dom}\Pi=\operatorname{dom}\Pi'$, we have, with self-explanatory notations,

$$\Pi v = \sum_i v(a_i)p_i + \sum_{i,j} Dv(a_i)(a_j - a_i)p_{ij},$$

$$\Pi' v = \sum_i v(a_i)p'_i + \sum_{i,k} \partial_k v(a_i)p'_{ik},$$

and

$$Dv(a_i)(a_j - a_i) = \sum_{k=1}^{n} \mu_{ijk}\partial_k v(a_i)$$

for appropriate coefficients μ_{ijk} for each pair (i,j). To conclude that $\Pi = \Pi'$, it suffices to observe that for each polynomial $p \in P''_3(T)$, one also has $Dp(a_i)(a_j - a_i) = \sum_{k=1}^{n} \mu_{ijk}\partial_k p(a_i)$ with the *same* coefficients μ_{ijk}.

11. Affine families of finite elements

We now come to an essential idea, which we shall first illustrate by means of an example. *Suppose that we are given a family of quadratic triangles* (T, P_T, Σ_T), *and that we want to describe such a family as simply as possible.*

Let $\hat{T}$ be a triangle with vertices $\hat{a}_i$, and midpoints of the sides $\hat{a}_{ij} = \frac{1}{2}(\hat{a}_i + \hat{a}_j)$, $1 \leq i < j \leq 3$, and let $\hat{\Sigma} = \{p(\hat{a}_i)\colon 1 \leq i \leq 3;\ p(\hat{a}_{ij})\colon 1 \leq i < j \leq 3\}$, so that the triple $(\hat{T}, \hat{P}, \hat{\Sigma})$ with $\hat{P} = P_2(\hat{T})$ is also a quadratic triangle. Then for each finite element (T, P_T, Σ_T) in the family (Fig. 11.1), there exists a unique *invertible affine mapping*

$$F_T\colon \hat{x} \in \mathbb{R}^2 \to F_T(\hat{x}) = B_T\hat{x} + b_T,$$

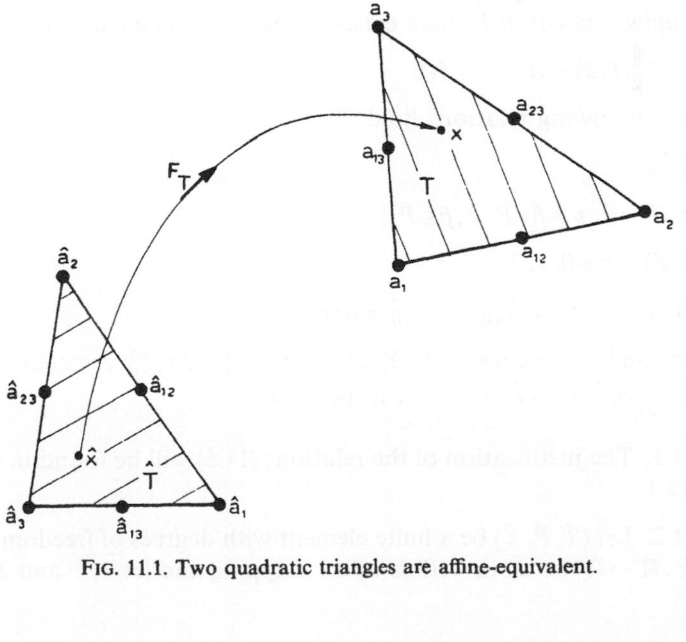

FIG. 11.1. Two quadratic triangles are affine-equivalent.

where B_T is an invertible 2×2 matrix and b_T is a vector of $\mathbb{R}^2$, such that

$$F_T(\hat{a}_i)=a_i, \quad 1\leqslant i\leqslant 3.$$

Then it automatically follows that

$$F_T(\hat{a}_{ij})=a_{ij}, \quad 1\leqslant i<j\leqslant 3,$$

since the property for a point to be the midpoint of a segment is preserved by an affine mapping (likewise, the points that we called a_{iij} or a_{ijk} keep their geometrical property through an affine mapping).

Once we have established a bijection $\hat{x}\in\hat{T}\to x=F_T(\hat{x})\in T$ between the points of the sets $\hat{T}$ and T, it is natural to associate the space

$$Q_T=\{p:T\to\mathbb{R};\, p=\hat{p}\circ F_T^{-1},\hat{p}\in\hat{P}\}$$

with the space P. Then it automatically follows that

$$Q_T=P_2(T)=P_T,$$

because the mapping F_T is affine.

In other words, rather than prescribing such a family by the data T, P_T and Σ_T, it suffices to give one *reference finite element* $(\hat{T},\hat{\Sigma},\hat{P})$ and to specify the affine mappings F_T. Then the generic finite element (T,P_T,Σ_T) in the family is such that

$$T=F_T(\hat{T}),$$
$$P_T=\{p:T\to\mathbb{R};\, p=\hat{p}\circ F^{-1},\hat{p}\in\hat{P}\},$$
$$\Sigma_T=\{p(F_T(\hat{a}_i)): 1\leqslant i\leqslant 3;\, p(F_T(\hat{a}_{ij})): 1\leqslant i<j\leqslant 3\}.$$

With this example in mind, we are in a position to give the general definition: Two finite elements $(\hat{T},\hat{P},\hat{\Sigma})$ and (T,P,Σ), with degrees of freedom of the form (10.4) are said to be *affine-equivalent* if there exists an *invertible affine mapping*:

$$F:\hat{x}\in\mathbb{R}^n\to F(\hat{x})=B\hat{x}+b\in\mathbb{R}^n, \tag{11.1}$$

such that the following relations hold:

$$T=F(\hat{T}), \tag{11.2}$$

$$P=\{p:T\to\mathbb{R};\, p=\hat{p}\circ F^{-1},\hat{p}\in\hat{P}\}, \tag{11.3}$$

$$a_i^r=F(\hat{a}_i^r), \quad r=0,1,2, \tag{11.4}$$

$$\xi_{ik}^1=B\hat{\xi}_{ik}^1, \qquad \xi_{ik}^2=B\hat{\xi}_{ik}^2, \qquad \xi_{il}^2=B\hat{\xi}_{il}^2, \tag{11.5}$$

whenever the nodes a_i^r, respectively $\hat{a}_i^r$, and vectors $\xi_{ik}^1, \xi_{ik}^2, \xi_{il}^2$, respectively $\hat{\xi}_{ik}^1, \hat{\xi}_{ik}^2, \hat{\xi}_{il}^2$, occur in the definition of the set Σ, respectively $\hat{\Sigma}$.

REMARK 11.1. The justification of the relations (11.5) will be found in the proof of Theorem 11.1.

REMARK 11.2. Let $(\hat{T},\hat{P},\hat{\Sigma})$ be a finite element with degrees of freedom of the form (10.4), let $F:\mathbb{R}^n\to\mathbb{R}^n$ be an invertible affine mapping and let T, P and Σ be defined through relations (11.2)–(11.5). Then it is easily seen that the triple (T, P, Σ) is a finite element. This observation thus provides a systematic means of constructing finite elements. For instance, let $(\hat{T}, \hat{P}, \hat{\Sigma})$ be an n-rectangle of degree k. Then finite elements whose associated sets T are parallelograms for $n=2$, parallelepipeds for $n=3,\ldots$, can be constructed in this fashion (such finite elements are seldom used in practice however). The real merit of this observation lies in that mappings F more general than affine mappings may be allowed: This is the basis for defining an important class of "curved" finite elements (cf. Section 35).

With this definition of affine-equivalent finite elements in mind, let us review the examples given so far (the reader should check the details of the various assertions that follow). To begin with, it is clear that two n-simplices of the same degree $k \geqslant 1$ are affine equivalent, and that this is also the case for reduced cubic n-simplices, in view of the definition (6.12) of the associated space P_T. Likewise, two n-rectangles of the same degree $k \geqslant 1$, or two reduced biquadratic, or bicubic rectangles are affine equivalent through diagonal affine mappings. In other words, *two Lagrange finite elements of any one of the types considered so far are affine-equivalent.*

As regards Hermite finite elements, the situation is less simple. Consider for example two cubic Hermite n-simplices with sets of degrees of freedom of the form Σ_T (Fig. 8.1). Then it is clear that they are affine-equivalent because the relations

$$a_j - a_i = F(\hat{a}_j) - F(\hat{a}_i) = B(\hat{a}_j - \hat{a}_i), \qquad 1 \leqslant i, j \leqslant n+1, \quad j \neq i,$$

hold among other things. However, had we chosen sets of degrees of freedom in the form Σ'_T, it would not have been clear to decide whether the two finite elements were affine-equivalent, and yet these two sets of degrees of freedom correspond to the *same* finite element, as we already pointed out.

The same analysis and conclusion apply to the reduced cubic Hermite n-simplex, or to the Bogner–Fox–Schmit rectangle. In the latter case, it suffices to observe that this rectangular finite element can also be defined by the following set of degrees of freedom (the index i being counted modulo 4)

$$\begin{aligned}\Sigma'_T = \{ & p(a_i), Dp(a_i)(a_{i-1} - a_i), Dp(a_i)(a_{i+1} - a_i), \\ & D^2 p(a_i)(a_{i-1} - a_i, a_{i+1} - a_i) \colon 1 \leqslant i \leqslant 4\},\end{aligned} \tag{11.6}$$

whose degrees of freedom satisfy relations (11.4) and (11.5).

There are counter-examples. For instance, consider a finite element in $\mathbb{R}^2$ where some degrees of freedom are normal derivatives at some boundary nodes. Then two such finite elements are not in general affine equivalent: The property for a vector to be normal to a side is not in general preserved through an affine mapping. Thus *two Argyris triangles are not affine-equivalent in general,* except (for instance) if they happen to be both equilateral triangles. In the same vein, it can be shown that *two Bell triangles are not affine-equivalent in general,* because condition (11.3) is violated.

Whenever two finite elements are affine-equivalent, we shall systematically use the correspondences

$$\hat{x} \in \hat{T} \to x = F(\hat{x}) \in T, \tag{11.7}$$

$$\hat{p} \in \hat{P} \to p = \hat{p} \circ F^{-1} \in P, \tag{11.8}$$

between the points $\hat{x} \in \hat{T}$ and $x \in T$, and the functions $\hat{p} \in \hat{P}$ and $p \in P$. Notice that these correspondences imply that

$$\hat{p}(\hat{x}) = p(x) \quad \text{for all } \hat{x} \in \hat{T}, \quad \hat{p} \in \hat{P}. \tag{11.9}$$

We next prove a crucial relationship between the $\hat{P}$-interpolation operator $\hat{\Pi}$ and the P-interpolation operator Π associated with two affine-equivalent finite elements $(\hat{T}, \hat{P}, \hat{\Sigma})$ and (T, P, Σ). This relationship is a consequence of the fact that the basis functions are also in the correspondence (11.8), as we now show:

THEOREM 11.1. *Let $(\hat{T}, \hat{P}, \hat{\Sigma})$ and (T, P, Σ) be two affine-equivalent finite elements with degrees of freedom of the form (10.4), and let $\hat{p}_i$, $1 \leqslant i \leqslant N$, denote the basis functions of the finite element $(\hat{T}, \hat{P}, \hat{\Sigma})$. Then the functions $p_i = \hat{p}_i \circ F^{-1}$, $1 \leqslant i \leqslant N$, are the basis functions of the finite element (T, P, Σ).*

In addition, the associated $\hat{P}$-interpolation operator $\hat{\Pi}$ and the P-interpolation operator Π satisfy

$$\{\Pi v\}\hat{} = \hat{\Pi}\hat{v}, \tag{11.10}$$

for all functions $\hat{v} \in \operatorname{dom} \hat{\Pi}$ and $v \in \operatorname{dom} \Pi$ associated in the correspondence

$$\hat{v} \in \operatorname{dom} \hat{\Pi} \to v = \hat{v} \circ F^{-1} \in \operatorname{dom} \Pi. \tag{11.11}$$

PROOF. The P-interpolation operator Π is of the following form (the notations are self-explanatory):

$$\Pi v = \sum_i v(a_i^0) p_i^0 + \sum_{i,k} \{Dv(a_i^1)\xi_{ik}^1\} p_{ik}^1 + \sum_{i,k,l} \{D^2 v(a_i^2)(\xi_{ik}^2, \xi_{il}^2)\} p_{ikl}^2.$$

Using the chain rule, we obtain

$$\begin{aligned} Dv(a_i^1)\xi_{ik}^1 &= Dv(F(\hat{a}_i^1))B\hat{\xi}_{ik}^1 = Dv(F(\hat{a}_i^1))DF(\hat{a}_i^1)\hat{\xi}_{ik}^1 \\ &= D(v \cdot F)(\hat{a}_i^1)\hat{\xi}_{ik}^1 = D\hat{v}(\hat{a}_i^1)\hat{\xi}_{ik}^1, \end{aligned}$$

and

$$\begin{aligned} D^2 v(a_i^2)(\xi_{ik}^2, \xi_{il}^2) &= D^2 v(F(\hat{a}_i^2))(B\hat{\xi}_{ik}^2, B\hat{\xi}_{il}^2) = D^2 v(F(\hat{a}_i^2))(DF(\hat{a}_i^2)\hat{\xi}_{ik}^2, DF(\hat{a}_i^2)\hat{\xi}_{il}^2) \\ &= D^2(v \cdot F)(\hat{a}_i^2)(\hat{\xi}_{ik}^2, \hat{\xi}_{il}^2) = D^2\hat{v}(\hat{a}_i^2)(\hat{\xi}_{ik}^2, \hat{\xi}_{il}^2), \end{aligned}$$

since $D^2 F = 0$. Thus we also have

$$\Pi v = \sum_i \hat{v}(\hat{a}_i^0) p_i^0 + \sum_{i,k} \{D\hat{v}(\hat{a}_i^1)\hat{\xi}_{ik}^1\} p_{ik}^1 + \sum_{i,k,l} \{D^2\hat{v}(\hat{a}_i^2)(\hat{\xi}_{ik}^2, \hat{\xi}_{il}^2)\} p_{ikl}^2$$

from which we deduce, using the correspondence (11.8),

$$\{\Pi v\}\hat{} = \sum_i \hat{v}(\hat{a}_i^0)\hat{p}_i^0 + \sum_{i,k} \{D\hat{v}(\hat{a}_i^1)\hat{\xi}_{ik}^1\}\hat{p}_{ik}^1 + \sum_{i,k,l} \{D^2\hat{v}(\hat{a}_i^2)(\hat{\xi}_{ik}^2, \hat{\xi}_{il}^2)\}\hat{p}_{ikl}^2.$$

Using functions $v \in P$ in this relation, we infer that the functions $\hat{p}_i^0, \hat{p}_{ik}^1, \hat{p}_{ik}^2$ are the basis functions of the finite element $(\hat{T}, \hat{P}, \hat{\Sigma})$, by virtue of identity (10.8). Using this result, we conclude that the function $\{\Pi v\}\hat{}$ is equal to the function $\hat{\Pi}\hat{v}$, by definition of the $\hat{P}$-interpolation operator $\hat{\Pi}$. □

REMARK 11.3. The basis functions of each simplicial finite element (T, P, Σ) described in Sections 6 and 8 are polynomial functions of the barycentric coordinates attached to the n-simplex T. The invariance of the barycentric coordinates by the affine mapping F shows directly in this case that the basis functions are in the correspondence (11.8).

It is easy to verify that the conclusions of Theorem 11.1 hold in the more general situation where the sets of degrees of freedom are given as $\hat{\Sigma} = \{\hat{\phi}_i : 1 \leqslant i \leqslant N\}$ and $\Sigma = \{\phi_i : 1 \leqslant i \leqslant N\}$, provided the degrees of freedom satisfy

$$\hat{\phi}_i(\hat{v}) = \phi_i(v), \quad 1 \leqslant i \leqslant N, \quad \text{for all } \hat{v} \in \operatorname{dom} \hat{\Pi}, \tag{11.12}$$

and, in essence, the proof of Theorem 11.1 consisted in showing that these relations are satisfied when the degrees of freedom are of the form (10.4).

REMARK 11.4. In Section 32, we shall encounter an instance of a degree of freedom, which is *not* of the form (10.4) but which satisfies relations (11.12).

A family of finite elements is called an *affine family* if all its finite elements are affine-equivalent to a single finite element $(\hat{T}, \hat{P}, \hat{\Sigma})$, which is called the *reference finite element* of the family (the finite element $(\hat{T}, \hat{P}, \hat{\Sigma})$ need not belong itself to the family).

If an affine family consists of *simplicial finite elements*, a customary choice for the set $\hat{T}$ is the *unit n-simplex*, whose vertices are

$$\begin{aligned}
\hat{a}_1 &= (1, 0, \ldots, 0), \\
\hat{a}_2 &= (0, 1, 0, \ldots, 0), \\
&\vdots \\
\hat{a}_n &= (0, \ldots, 0, 1), \\
\hat{a}_{n+1} &= (0, 0, \ldots, 0).
\end{aligned}$$

In this case, the barycentric coordinates take the simple form

$$\lambda_i = x_i, \quad 1 \leqslant i \leqslant n, \qquad \lambda_{n+1} = 1 - \sum_{i=1}^{n} x_i.$$

If an affine family consists of *rectangular finite elements*, customary choices for the set $\hat{T}$ are the unit hypercube $[0, 1]^n$, or the hypercube $[-1, 1]^n$.

The concept of an affine family of finite elements is of crucial importance, for the following reasons:

(i) In practical computations, *most of the work involved in the computation of the coefficients of the linear system* (5.4) *is performed on a reference finite element, not on a generic finite element.* This point will be further illustrated in Sections 25 and 39.

(ii) For such affine families, a fairly elegant *interpolation theory* can be developed (Section 16), which is in turn the basis of *convergence theorems.*

(iii) Even when a family of finite elements of a given type is not an affine family, it is generally associated in a natural way with an affine family whose "intermediate" role is essential. For example, when we shall study in Section 45 the interpolation properties of the Argyris triangle, an important step will consist in introducing a slightly different finite element, the $\mathscr{C}^0$-quintic Hermite triangle, which *can* be imbedded in an affine family. In the same fashion, we shall consider in Section 37 the "isoparametric" families of curved finite elements essentially as perturbations of affine families.

12. General properties of finite element spaces

Our next task is to give a precise description of the *construction of a finite element space* with finite elements (T, P_T, Σ_T), $T \in \mathscr{T}_h$. For the sake of simplicity, we shall restrict ourselves to the case where the finite elements are all *polygonal*, so that the set $\bar{\Omega} = \bigcup_{T \in \mathscr{T}_h} T$ is necessarily *polygonal*, and to the case where the finite elements are all of *Lagrange type.* These restrictions will allow us to avoid technical difficulties, such as appropriately defining a "face" of a nonpolygonal finite element, or explicitly stating the compatibility conditions that degrees of freedom of adjacent Hermite finite elements should satisfy.

There are indeed polygonal finite elements that are used in actual computations, and which are neither n-simplices nor n-rectangles. Of course, such finite elements are not just arbitrary polygonal domains. Rather they are adapted to special circumstances: Thus, if the domain $\bar{\Omega}$ is a cylindrical domain in $\mathbb{R}^3$, it might be appropriate to use *prismatic finite elements*, an example of which is given in Fig. 12.1:

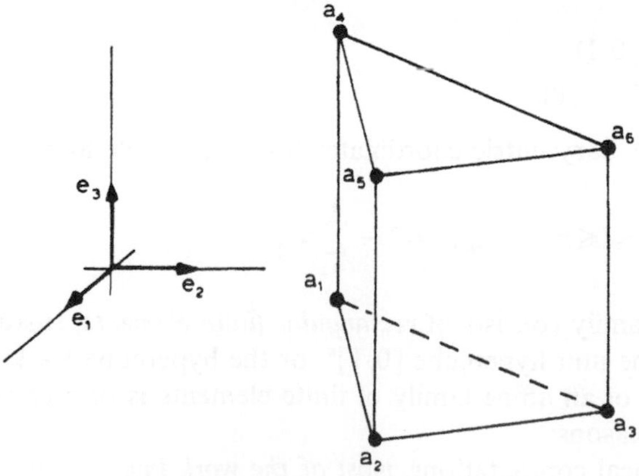

FIG. 12.1. An example of a prismatic finite element.

The space P is the tensor product of affine polynomials in x_1, x_2 by the space of affine polynomials in x_3, i.e., a function p in the space P is of the form

$$p(x_1, x_2, x_3) = \gamma_1 + \gamma_2 x_1 + \gamma_3 x_2 + \gamma_4 x_3 + \gamma_5 x_1 x_3 + \gamma_6 x_2 x_3.$$

Returning to the general case, we shall assume that each polygonal set T has a nonempty interior, and that the interiors of distincts sets $T \in \mathcal{T}_h$ are pairwise disjoint, so that requirements $(\mathcal{T}_h 1)$–$(\mathcal{T}_h 4)$ of Section 5 are satisfied (a polygonal domain has a Lipschitz-continuous boundary). A portion T' of the boundary of T is a *face* if it is a connected subset of an affine hyperplane of $\mathbb{R}^n$ with a nonempty interior relatively to this hyperplane.

In order to unambiguously define the functions of the finite element space, we ask that the finite elements satisfy the following requirement, which generalizes condition $(\mathcal{T}_h 5)$ already required for n-simplices and n-rectangles (Section 6 and 7):

$(\mathcal{T}_h 5)$ *Any face of a finite element T_1 is either a face of another finite element T_2, in which case the finite elements T_1 and T_2 are said to be adjacent, or it is a portion of the boundary of the set Ω.*

Finally, we ask that the sets of degrees of freedom of adjacent finite elements shall be related as follows: Whenever (T^l, P^l, Σ^l), with $\Sigma^l = \{p(a_i^l): 1 \leqslant i \leqslant N^l\}$, $l = 1, 2$, are two adjacent finite elements, then

$$\left(\bigcup_{i=1}^{N^1} \{a_i^1\}\right) \cap T^2 = \left(\bigcup_{i=1}^{N^2} \{a_i^2\}\right) \cap T^2. \tag{12.1}$$

We define the set

$$N_h = \bigcup_{T \in \mathcal{T}_h} N_T \tag{12.2}$$

where N_T denotes the set of nodes of the finite element (T, P_T, Σ_T). For each $b \in N_h$, we let $T(\lambda)$, $\lambda \in \Lambda(b)$, denote all those finite elements for which b is a node. Then the associated *finite element space* X_h is defined by

$$X_h = \Big\{ v = (v_T)_{T \in \mathcal{T}_h} \in \prod_{T \in \mathcal{T}_h} P_T;$$

$$v_{T(\lambda)}(b) = v_{T(\mu)}(b) \text{ for all } b \in N_h \text{ and all } \lambda, \mu \in \Lambda(b) \Big\}.$$

and a function in the space X_h, is uniquely determined by the set

$$\Sigma_h = \{v(b): b \in N_h\}, \tag{12.3}$$

which is called the *set of degrees of freedom of the finite element space*.

Since X_h is a priori defined as a subspace of the product space $\Pi_{T \in \mathcal{T}_h} P_T$, it is thus realized that an element $v \in X_h$ is not in general a function defined over the set $\bar{\Omega}$, since it need not have a unique definition along faces common to adjacent finite elements. It is nevertheless customary to say that the elements of the space X_h are "functions" over $\bar{\Omega}$, which, by virtue of assumption (12.1) are at least continuous at all nodes common to adjacent finite elements (the inclusions $P_T \subset \mathscr{C}^0(T)$, $T \in \mathscr{T}_h$, usually hold in practice). It is also a usual practice to consider the functions v_T, $T \in \mathscr{T}_h$, as being the *restrictions* to the finite elements T of the function $v \in X_h$, just as if v were an "ordinary" function defined over the set $\bar{\Omega}$. This is why we shall use the alternate notation $v|_T = v_T$.

If it happens, however, that for each function $v \in X_h$, the restrictions $v|_{T_1}$ and $v|_{T_2}$ coincide along the face common to any pair of adjacent finite elements T_1 and T_2, then the function v can indeed be identified with a function defined over the set $\bar{\Omega}$.

Although this last property is satisfied by all the finite element spaces that we have constructed so far, it is by no means necessary. Following CROUZEIX and RAVIART [1973], let us consider for example the finite element space constructed with the following finite element (T, P, Σ): The set T is an n-simplex with vertices a_j, $1 \leqslant j \leqslant n+1$, the space P is the space $P_1(T)$ and the set of degrees of freedom is the set $\Sigma_T = \{p(c_i)\colon 1 \leqslant i \leqslant n+1\}$, where for each i the point c_i is the barycenter of the face that does not contain the point a_i, i.e.,

$$c_i = \frac{1}{n} \sum_{\substack{j=1 \\ j \neq i}}^{n+1} a_j, \quad 1 \leqslant i \leqslant n+1.$$

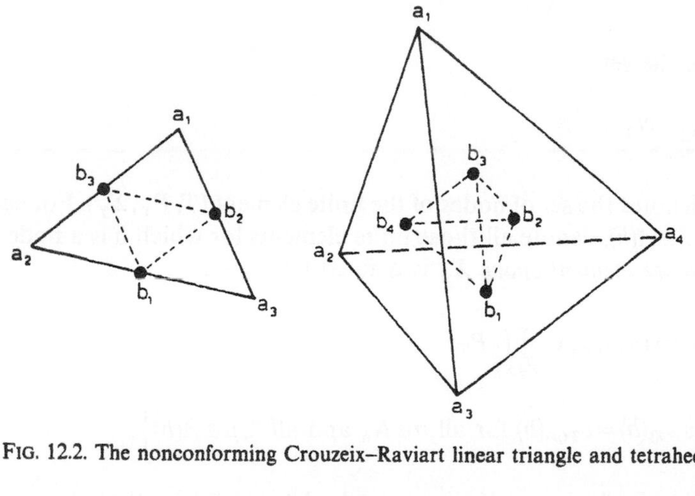

FIG. 12.2. The nonconforming Crouzeix–Raviart linear triangle and tetrahedron.

The resulting finite element is called the *nonconforming Crouzeix–Raviart linear n-simplex*; cf. Fig. 12.2 for the special cases $n=2$ and 3.

To show that the set Σ_T is $P_1(T)$-unisolvent, it suffices to observe that the points $c_i = (c_{ji})_{j=1}^n$, $1 \leqslant i \leqslant n+1$, are also the vertices of an n-simplex: If we let C denote the

$(n+1)\times(n+1)$ matrix defined by

$$C=\begin{pmatrix} c_{11} & c_{12} & \cdots & c_{1,n+1} \\ c_{21} & c_{22} & \cdots & c_{2,n+1} \\ \vdots & \vdots & & \vdots \\ c_{n1} & c_{n2} & \cdots & c_{n,n+1} \\ 1 & 1 & \cdots & 1 \end{pmatrix},$$

it is easily verified that $\det C=(-1/n)^n \det A$, where A is the matrix of (6.3) and thus $\det C\neq 0$. One may also notice that the functions

$$p_i=1-n\lambda_i,\quad 1\leqslant i\leqslant n+1,$$

are the associated basis functions. Then it is clear that the functions of the corresponding finite element space generally have two definitions along faces common to adjacent finite elements, except at the centroids of these faces.

REMARK 12.1. This triangular finite element has no "rectangular counterpart"! For let T be the unit square and let the points a_i, $5\leqslant i\leqslant 8$, be as in Fig. 7.2. Then the set $\{p(a_i): 5\leqslant i\leqslant 8\}$ is *not* $Q_1(T)$-unisolvent (consider the polynomial $p(x_1,x_2)=(x_1-\frac{1}{2})(x_2-\frac{1}{2})$).

In the same vein, let T be a triangle with vertices a_i, $1\leqslant i\leqslant 3$, and let $a_{iij}=\frac{1}{3}(2a_i+a_j)$. Then the set $\{p(a_{iij}): 1\leqslant i,j\leqslant 3, i\neq j\}$ is *not* $P_2(T)$-unisolvent.

All the previous considerations can be extended so as to include the case of finite element spaces constructed with Hermite finite elements: The only difficulties are technical; for instance, extending the compatibility conditions (12.1) to general Hermite finite elements is easy in its principle, yet it involves cumbersome notational devices. We shall simply point out that it is usually necessary to choose between various possible sets of degrees of freedom corresponding to the same finite element, so as to unambiguously define a *set* Σ_h *of degrees of freedom* of the corresponding finite element space. These considerations have been illustrated at various places in Sections 8 and 9.

When the degrees of freedom of all the finite elements encountered in a finite element space are of the forms (10.4), the degrees of freedom of the finite element space are of the following forms:

$$\begin{aligned} &v\to v(b_j^0),\\ &v\to Dv(b_j^1)\eta_{jk}^1, \\ &v\to D^2v(b_j^2)(\eta_{jk}^2,\eta_{jl}^2), \end{aligned}\tag{12.4}$$

where the points b_j^r, $r=0,1,2$, are called the *nodes of the finite element space.* If we write the set Σ_h as

$$\Sigma_h=\{\phi_{j,h}: 1\leqslant j\leqslant M\},\tag{12.5}$$

then the *basis functions* w_j, $1\leqslant j\leqslant M$, *of the finite element space* are defined by the relations

$$\begin{aligned} &w_j \in X_h, \\ &\phi_{i,h}(w_j) = \delta_{ij}, \quad 1 \leqslant i \leqslant M. \end{aligned} \tag{12.6}$$

It is easy to verify on each example that the *basis functions of the finite element space are constructed by appropriately assembling the basis functions of the finite elements*. More specifically, let $\phi_h \in \Sigma_h$ be a degree of freedom of one of the forms (12.4), let b be the associated node, and let T_λ, $\lambda \in \Lambda(b)$, denote all the finite elements of $\mathcal{T}_h$ that contain b as a node (see Fig. 12.3, where bilinear rectangles are considered). For each $\lambda \in \Lambda(b)$, let p_λ denote the basis function of the finite element T_λ

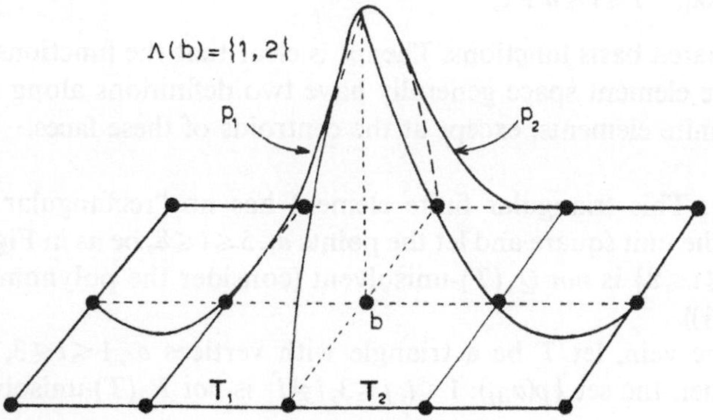

FIG. 12.3. Construction of a basis function of the finite element space associated with bilinear rectangles.

associated with the restriction of ϕ_h to T_λ. Then the function $w \in X_h$ defined by

$$w = \begin{cases} p_\lambda & \text{over } T_\lambda, \lambda \in \Lambda(b), \\ 0 & \text{elsewhere,} \end{cases} \tag{12.7}$$

is the basis function of the space X_n associated with the degree of freedom ϕ_h.

As a practical consequence, *requirement* (FEM3) (existence of a basis of X_h whose functions have "small supports") *is always satisfied by the examples*. The reader should refer to Fig. 6.8 where it was shown on an example that the basis functions constructed in this fashion have indeed "small" supports. The "worst" case concerns a basis function attached to a vertex, say b, of the triangulation. In this case, the corresponding support is the union of those finite elements that have b as a vertex. In most commonly encountered triangulations in the plane, the number of such finite elements is very low (six or seven, for example).

Let there be given a finite element space X_h with a set of degrees of freedom of the form (12.5). With any function $v : \bar{\Omega} \to \mathbb{R}$, sufficiently smooth in order that the degrees of freedom $\phi_{j,h}(v)$, $1 \leqslant j \leqslant M$, be well defined, we associate the function

$$\Pi_h v = \sum_{j=1}^{M} \phi_{j,h}(v) w_j, \tag{12.8}$$

where the functions w_j are the basis functions defined in (12.6). The function $\Pi_h v$, which is called the X_h-*interpolant* of the function v, is equivalently characterized by the conditions

$$\begin{aligned}&\Pi_h v \in X_h,\\&\phi_{j,h}(\Pi_h v)=\phi_{j,h}(v),\quad 1\leqslant j\leqslant M.\end{aligned}\tag{12.9}$$

If s denotes the maximal order of directional derivatives found in the finite elements (T, P_T, Σ_T), $\in \mathcal{T}_h$, we shall consider, in accordance with (10.7), that the domain dom Π_h of the X_h-*interpolation operator* Π_h is the space

$$\operatorname{dom}\Pi_h=\mathscr{C}^s(\bar{\Omega}).\tag{12.10}$$

It might be helpful to keep in mind the following tableau (Fig. 12.4), where we have listed the main "global" (i.e., on $\bar{\Omega}$) versus "local" (i.e., on a generic finite element (T, P, Σ)) notations, definitions and correspondences.

"Global" definitions	"Local" definitions
$\bar{\Omega}$	(T, P, Σ): finite element
Boundary of the set Ω: Γ	∂T: boundary of T; T': side, or face, of T
Triangulation of the set $\bar{\Omega}$: $\mathcal{T}_h$	
Finite element space: X_h	P or $P_T=\{v\|_T;\, v\in X_h\}$
Generic function of X_h: v or v_h	p or p_T: generic function of P_T
Set of degrees of freedom of X_h: Σ_h	Σ or Σ_T: set of degrees of freedom of (T, P, Σ)
Degrees of freedom of X_h: ϕ_h or $\phi_{j,h}$, $1\leqslant j\leqslant M$	ϕ_i or $\phi_{i,T}$, $1\leqslant i\leqslant N$: degrees of freedom of (T, P, Σ)
Basis functions of X_h: w_j, $1\leqslant j\leqslant M$	p_i or $p_{i,T}$, $1\leqslant i\leqslant N$: basis functions of (T, P, Σ)
Nodes of X_h: b_j	$a_i, a_{ij}, \ldots$: nodes of (T, P, Σ)
X_h-interpolation operator: Π_h	Π or Π_T: P_T-interpolation operator

FIG. 12.4. "Global" definitions, pertaining to a finite element space X_h, and "local" definitions, pertaining to a finite element (T, P, Σ).

We next state a relationship of paramount importance between the "global" interpolation operator Π_h and the "local" interpolation operators Π_T.

THEOREM 12.1. *Let v be any function in the space* dom Π_h. *Then the restrictions $v|_T$ belong to the spaces* dom Π_T, *and we have*

$$(\Pi_h v)|_T=\Pi_T(v|_T)\quad \textit{for all } T\in\mathcal{T}_h.\tag{12.11}$$

PROOF. These relations are direct consequences of the way in which the set Σ_h is derived from the sets Σ_T, $T\in\mathcal{T}_h$. □

It has always been assumed thus far that all the finite elements (T, P_T, Σ_T), $T\in\mathcal{T}_h$, used in the definition of a finite element space are all *of the same type*. By this, we mean that, for instance, the finite elements are all quadratic n-simplices, or that the finite elements are all Argyris triangles, etc. If this is the case, we shall say that any finite element (T, P_T, Σ_T), $T \in \mathcal{T}_h$, is the *generic* finite element of the finite element space. The next two definitions are of particular importance, in view of Theorems 5.1 and 5.2.

We say that a finite element (T, P_T, Σ_T) is *of class* $\mathscr{C}^0$ if (i) the inclusion $P_T \subset \mathscr{C}^0(T)$ holds and (ii) whenever it is the generic finite element of a triangulation and T_1 and T_2 are two adjacent finite elements, the restrictions $v_h|_{T_1}$ and $v_h|_{T_2}$ coincide along the face common to T_1 and T_2 for any function v_h in the corresponding finite element space. Hence the inclusion $X_h \subset \mathscr{C}^0(\bar{\Omega})$ holds in this case.

Likewise, we say that a finite element (T, P_T, Σ_T) of a given type is *of class* $\mathscr{C}^1$ if (i) the inclusion $P_T \subset \mathscr{C}^1(T)$ holds and (ii) whenever it is the generic finite element of a triangulation and T_1 and T_2 are two adjacent finite elements, for any function v_h in the corresponding finite element space the restrictions $v_h|_{T_1}$ and $v_h|_{T_2}$ coincide along the face T' common to T_1 and T_2 and the outer normal derivatives satisfy $\partial_\nu v_h|_{T_1} + \partial_\nu v_h|_{T_2} = 0$ along T'. Hence the inclusion $X_h \subset \mathscr{C}^1(\bar{\Omega})$ holds in this case.

Thus all the finite elements seen in Sections 6–8 are of class $\mathscr{C}^0$, the Argyris and Bell triangles and the Bogner–Fox–Schmit rectangle (Section 9) are of class $\mathscr{C}^1$. There are also finite elements that are not of class $\mathscr{C}^0$, such as the nonconforming Crouzeix–Raviart linear n-simplex (cf. Fig. 12.2).

One may also combine finite elements of *different types*, provided some compatibility conditions are satisfied along faces that are common to adjacent finite elements, in such a way that a function in the resulting finite element space X_h is still unambigously defined on the one hand, and an inclusion such as $X_h \subset \mathscr{C}^0(\bar{\Omega})$ (for example) holds on the other hand: For instance, one may combine n-simplices of degree k with n-rectangles of degree k; another example is considered in Fig. 12.5.

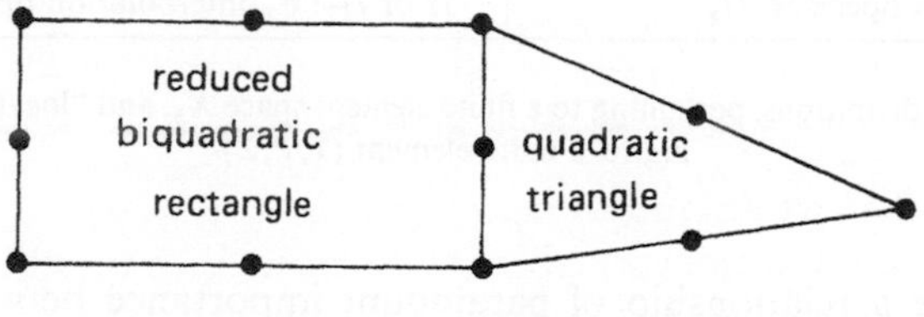

FIG. 12.5. Combining finite elements of different types in a finite element space.

The last topic we wish to examine in this section is: *How are boundary conditions taken into account in a finite element space?* Again, we shall essentially concentrate on examples.

Let X_h be a finite element space whose generic finite element is one of the following: n-simplex of degree k, $k \geq 1$, reduced cubic n-simplex, n-rectangle of degree k, $k \geq 1$, reduced biquadratic or bicubic rectangle. Then the inclusion $X_h \subset \mathscr{C}^0(\bar{\Omega}) \cap H^1(\Omega)$ holds (Theorems 6.3 and 7.4) and thus the inclusion

$$X_{0h} = \{v_h \in X_h;\ v_h|_\Gamma = 0\} \subset H_0^1(\Omega) \tag{12.12}$$

holds. In each case, it is easily verified that a sufficient (and clearly necessary) condition for a function $v_h \in X_h$ to vanish along Γ is that *it vanishes at all the boundary nodes*, i.e., those nodes of the space X_h that are on the boundary Γ. In other words, if we let N_h denote the set of nodes of the space X_h, the finite element space X_{0h} of (12.12) is simply given by

$$X_{0h} = \{v_h \in X_h;\ v_h(b) = 0 \text{ for all } b \in N_h \cap \Gamma\}. \tag{12.13}$$

When Hermite finite elements are used, the situation is less simple. Let us consider for example cubic, or reduced cubic, Hermite n-simplices. For each boundary node $b \in N_h \cap \Gamma$, we let $\tau_\gamma(b)$, $\gamma \in \Gamma(b)$, denote a maximal set of linearly independent vectors in $\mathbb{R}^n$ with the property that the points $(b + \tau_\gamma(b))$, $\gamma \in \Gamma(b)$, belong to the boundary Γ. Then the space X_{0h} of (12.12) is given in this case by

$$\begin{aligned} X_{0h} = \{v_h \in X_h;\ & v_h(b) = 0 \text{ for all } b \in N_h \cap \Gamma, \\ & Dv_h(b)\tau_\gamma(b) = 0 \text{ for all } b \in N_h \cap \Gamma \text{ and all } \gamma \in \Gamma(b)\}. \end{aligned} \tag{12.14}$$

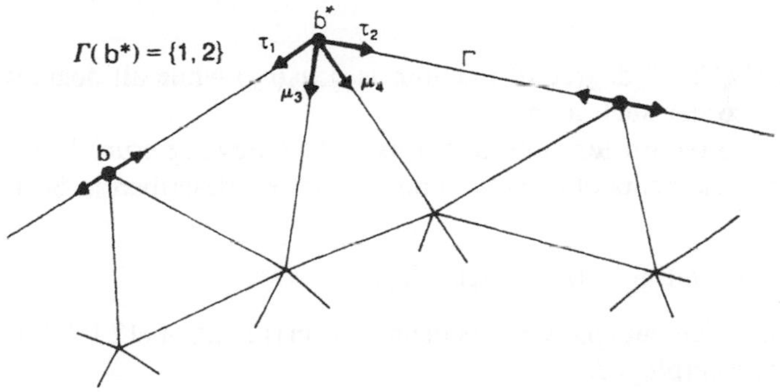

FIG. 12.6. When cubic, or reduced cubic, Hermite triangles are used, the values $v_h(b)$, $v_h(b^*)$, and the directional derivatives $Dv_h(b)\tau_1$, $Dv_h(b^*)\tau_1$, $Dv_h(b^*)\tau_2$ must vanish in order that $v_h = 0$ on Γ.

As an illustration, we have indicated in Fig. 12.6 the directional derivatives that must be set equal to zero along a specific portion of the boundary of a polygonal set in $\mathbb{R}^2$, when cubic or reduced cubic, Hermite triangles are used. In particular, one should observe that at a corner, such as b^*, the directional derivatives $Dv_h(b^*)\mu_3$ and $Dv_h(b^*)\mu_4$ must necessarily vanish.

If we next assume that the inclusion $X_h \subset \mathscr{C}^1(\bar{\Omega}) \cap H^2(\Omega)$ holds, it follows that we have the inclusions

$$X_{0h} = \{v_h \in X_h;\ v_h|_\Gamma = 0\} \subset H^2(\Omega) \cap H_0^1(\Omega), \tag{12.15}$$

$$X_{00h} = \{v_h \in X_h;\ v_h|_\Gamma = \partial_\nu v_h|_\Gamma = 0\} \subset H_0^2(\Omega), \tag{12.16}$$

so that we are facing the problem of constructing such spaces X_{0h} and X_{00h}. Again they are obtained by canceling appropriate values and directional derivatives at boundary nodes. As an example, we have indicated in Fig. 12.7 all the directional derivatives that must be set equal to zero when Argyris triangles are used and the second inclusion (12.16) is needed. It should be realized that at a boundary node such

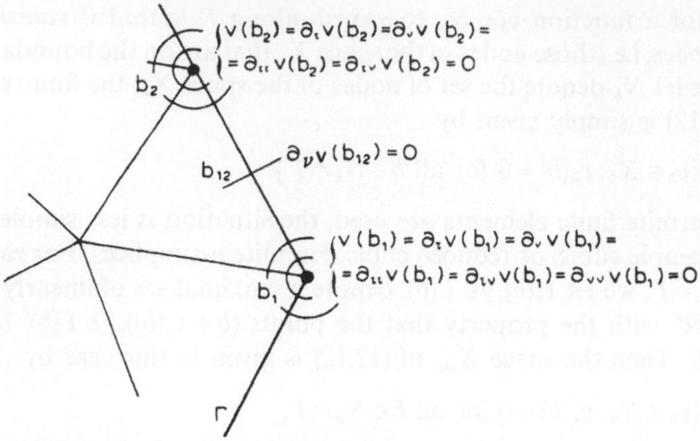

FIG. 12.7. Directional derivatives that must vanish in order that $v=\partial_\nu v=0$ on Γ when Argyris triangles are used.

as b_2, the only "free" degree of freedom is $\partial_{\nu\nu}v_h(b_2)$ while all degrees of freedom are zero at a corner such as b_1.

We shall also record for subsequent uses the following crucial properties:

(i) All finite elements of class $\mathscr{C}^0$ and of class $\mathscr{C}^1$ described in Sections 6–9 are such that

$$v\in \operatorname{dom} \Pi_h \text{ and } v|_\Gamma=0 \;\Rightarrow\; \Pi_h v\in X_{0h}, \tag{12.17}$$

where the finite element space X_{0h} is defined as in (12.12), or (12.15) if finite elements of class $\mathscr{C}^1$ are employed.

(ii) All finite elements of class $\mathscr{C}^1$ described in Sections 6–9 have the property that

$$v\in \operatorname{dom} \Pi_h \text{ and } v|_\Gamma=\partial_\nu v|_\Gamma=0 \;\Rightarrow\; \Pi_h v\in X_{00h}, \tag{12.18}$$

where the finite element space X_{00h} is defined as in (12.16).

Notice that it is clearly possible to extend the previous considerations so as to include the case where boundary conditions are imposed only over a *portion* Γ_0 of the boundary Γ, provided such a portion Γ_0 is exactly the union of some faces of the finite elements found in the triangulation.

REMARK 12.2. Let us briefly show how some "*global*" *properties*, i.e., of a finite element space, may be in fact derived from a "*local*" *property*, i.e., of a generic finite element. For simplicity, we restrict ourselves to Lagrange finite elements, leaving the case of Hermite finite elements to the reader.

Let (T, P, Σ) be a Lagrange finite element, with N as its set of nodes; hence the set of degrees of freedom is of the form $\Sigma=\{p(a)\colon a\in N\}$. If T' is any face of the set T, we let

$$\Sigma|_{T'}=\{p(a)\colon a\in N\cap T'\},$$
$$P(T')=\{p|_{T'};\ T'\to\mathbb{R}\colon p\in P\}.$$

Then all the Lagrange finite elements heretofore described, except the nonconforming Crouzeix–Raviart linear triangle, have the property that *for each one of their faces T', the set $\Sigma|_{T'}$ is $P(T')$-unisolvent*. This crucial "local" property has the following easily established consequences:

(i) The P_T-interpolant of a function $v \in \operatorname{dom} \Pi_T$ which vanishes along a face T' also vanishes along T'. As a consequence, the "global" property (12.17) holds.

(ii) Let $\phi: p \in P \to p(a)$ be one of the degrees of freedom of the finite element and let p be the associated basis function. Then the basis function p vanishes along any face that does not contain the node a. The way in which the basis functions of the space X_h are constructed from the basis functions of each finite element (cf. (12.7)) then implies the "global" property that the basis functions of the space X_h have indeed small supports (FEM3).

(iii) Assume in addition that, for each pair (T_1, T_2) of adjacent finite elements found in a triangulation, one has $P_{T_1}|_{T'} = P_{T_2}|_{T'}$ along the common face T', and that the inclusions $P_T \subset \mathscr{C}^0(T)$, $T \in \mathscr{T}_h$, hold. Then the "global" inclusion $X_h \subset \mathscr{C}^0(\bar{\Omega})$ holds.

The *choice* of a finite element for solving a given problem usually relies on the following considerations:

(i) The finite element must be well adapted to the *geometry* of the problem, and yet be easy to assemble. For example, assembling tetrahedra is not an easy task and thus prismatic finite elements (see Fig. 12.1 for an example) are usually preferred whenever possible. In this respect, see the discussion in ZIENKIEWICZ [1971, Chapter 6]. Geometrical considerations may also justify the choice of curved finite elements (cf. Chapter VI) instead of straight finite elements for some "particularly curved" domains.

(ii) The finite element must satisfy appropriate *continuity requirements*: They must be of class $\mathscr{C}^0$, or of class $\mathscr{C}^1$, if we are to solve a second-order, or a fourth-order, problem by a conforming method.

(iii) We shall see that a mathematical proof of *convergence* requires (among other things) the inclusions $P_1(T) \subset P_T$, $T \in \mathscr{T}_h$, for second-order problems, and the inclusions $P_2(T) \subset P_T$, $T \in \mathscr{T}_h$, for fourth-order problems. Incidentally, the engineers were well aware of these conditions, which they discovered empirically, long before the mathematicians undertook the numerical analysis of the finite element method!

(iv) The coefficients of the resulting *linear system* (5.4) should be easy to compute on the one hand, and the linear system should be as easy as possible to solve on the other hand. We shall very briefly discuss here these fundamental practical aspects of the finite element method, by simply recording two simple guidelines that tend to reduce certain computational difficulties:

A *first guideline* is that, if possible, the sets of degrees of freedom associated with a given node in the triangulation be all alike, so as to avoid different instructions depending on which node is considered. This explains for instance why reduced cubic Hermite n-simplices may be preferred to cubic Hermite n-simplices, or why Bell triangles may be preferred to Argyris triangles, even though there is in both cases a decrease of one in the order of convergence, as we shall see; in addition, such choices slightly reduce the dimension of the resulting linear system.

A *second guideline* is that each node of the finite element space should be common to the greatest possible number of finite elements. For example, it is easily realized that, for a given triangulation, cubic Hermite triangles lead to a smaller linear system than cubic triangles; yet the asymptotic rates of convergence are the same, as we shall see.

(v) In addition, miscellaneous aspects may be considered. For instance, one may argue that cubic Hermite triangles introduce artificial "constraints" (the continuity of the first derivatives at the vertices) on the one hand, but on the other hand, this is an advantage when solving the boundary value problem of linearized elasticity, if one needs to compute the stresses (cf. (3.29)) at the vertices. Likewise, one may argue that the use of Argyris triangles for solving a plate problem introduces artificial "constraints" (the continuity of the second derivatives at the vertices and "extra" boundary conditions as shown in Fig. 12.7) on the one hand, but on the other hand, this is an advantage if one needs to compute the "bending moments" at the vertices (the bending moments are linear expressions involving the second partial derivatives of the solution), etc.

For more details, the reader is referred to the books listed in the Introduction of this chapter.

13. General considerations on the convergence of finite element methods and Céa's lemma

Whereas up to now, our discussion has been concerned with *one* discrete problem, we shall now consider *families* of discrete problems. More specifically, assume that we are approximating the solution u of the variational equations

$$a(u, v) = l(v) \quad \text{for all } v \in V, \tag{13.1}$$

where the space V, the bilinear form $a(\cdot,\cdot)$, and the linear form l, satisfy the assumptions of the Lax–Milgram lemma (Theorem 1.3). Confining ourselves to the case of *conforming finite element methods*, we consider a family (V_h) of subspaces of the space V, where it is understood that h is the *defining parameter of the family* and that *h approaches zero* (the parameter h will be given a specific meaning in Section 17).

With each finite element space V_h is associated the discrete solution u_h, which satisfies

$$a(u_h, v_h) = l(v_h) \quad \text{for all } v_h \in V_h. \tag{13.2}$$

Then we shall say that the associated family of discrete problems is *convergent* or, equivalently, that *convergence* holds, if, for *any* problem of the form (13.1) posed in the space V, we have

$$\lim_{h \to 0} \| u - u_h \| = 0, \tag{13.3}$$

where $\|\cdot\|$ denotes the norm in the space V.

We are therefore interested in giving sufficient conditions for convergence and, as a first result in this direction, we have the following basic *abstract error estimate.*

THEOREM 13.1 (Céa's lemma). *There exists a constant C independent of the subspace V_h such that*

$$\|u-u_h\| \leqslant C \inf_{v_h \in V_h} \|u-v_h\|. \tag{13.4}$$

Consequently, a sufficient condition for convergence is that there exists a family (V_h) of subspaces of the space V such that, for each $u \in V$,

$$\lim_{h \to 0} \inf_{v_h \in V_h} \|u-v_h\| = 0. \tag{13.5}$$

PROOF. Let w_h be an arbitrary element in V_h. It follows from (13.1) and (13.2) that $a(u-u_h, w_h)=0$. Using the same constants α, M as in (1.3) and (1.19), we have, for any $v_h \in V_h$,

$$\begin{aligned}\alpha \|u-u_h\|^2 &\leqslant a(u-u_h, u-u_h)\\ &= a(u-u_h, u-v_h) \leqslant M \|u-u_h\| \, \|v-v_h\|,\end{aligned}$$

and the conclusion follows with $C = M/\alpha$. □

Céa's lemma is named after CÉA [1964, Proposition 3.1], who proved it in the symmetric case. It was independently rediscovered by VARGA [1966a], and extended to the nonsymmetric case by BIRKHOFF, SCHULTZ and VARGA [1968, Theorem 13].

Note that, when the bilinear form is *symmetric*, there is a remarkable interpretation of the discrete solution. Since $a(u-u_h, w_h)=0$ for all $w_h \in V_h$, it follows that *u_h is the projection over V_h of the exact solution u, with respect to the inner product* $a(\cdot,\cdot)$. Therefore, we have in this case:

$$a(u-u_h, u-u_h) = \inf_{v_h \in V_h} a(u-v_h, u-v_h).$$

Using the V-ellipticity and the continuity of the bilinear form, we deduce

$$\|u-u_h\| \leqslant \sqrt{(M/\alpha)} \inf_{v_h \in V_h} \|u-v_h\|.$$

Thus we have obtained a "better" constant than in the proof of Theorem 13.1, since we necessarily have $M \geqslant \alpha$.

The simple, yet crucial, inequality (13.4) shows that the problem of estimating the *error* $\|u-u_h\|$ is reduced to a problem in approximation theory, i.e., to evaluate the *distance* $d(u, V_h) = \inf_{v_h \in V_h} \|u-v_h\|$ between a function $u \in V$ and a subspace $V_h \subset V$. This explains why this problem will be a central theme of the next chapter, where we shall essentially prove results of the following type: Assuming appropriate smoothness on the function u, we shall show that the distance $d(u, V_h)$ is itself bounded by a constant (which usually involves norms of higher order derivatives of the function u) times h^β, for some exponent $\beta > 0$. We shall therefore obtain the additional information that, for a smooth enough solution u, there exists a constant $C(u)$ independent of h such that

$$\|u-u_h\| \leqslant C(u)h^{\beta}. \tag{13.6}$$

If this is the case, we shall say that the *order of convergence* is β, or equivalently, that we have an $O(h^{\beta})$ *convergence*, and we shall simply write

$$\|u-u_h\| = O(h^{\beta}). \tag{13.7}$$

Using more elaborated techniques, we shall also evaluate the difference $(u-u_h)$ in other norms than the norm of the space V (which is either the $\|\cdot\|_{1,\Omega}$ or the $\|\cdot\|_{2,\Omega}$ norm), such as the $|\cdot|_{0,\Omega}$ or $|\cdot|_{0,\infty,\Omega}$ norms (cf. Sections 19 and 21 respectively), and we shall also call *errors* the corresponding norms $|u-u_h|_{0,\Omega}$ or $|u-u_h|_{0,\infty,\Omega}$.

Whereas a mathematician is generally satisfied with a sufficient condition for convergence such as that of Theorem 13.1, this condition rightly appears as a philosophical matter to many an engineer, who is much more concerned in getting even a rough estimate of the error for a *given* space V_h: For practical problems, one chooses often one, sometimes two, seldom more, subspaces V_h, but certainly not an infinite family. *In other words, the parameter h never approaches zero in practice*!

CHAPTER III

Finite Element Methods for Second-Order Problems: The Basic Error Estimates

Introduction

In this chapter, we estimate various norms of the difference $(u-u_h)$, where $u \in V$ is the solution of a second-order boundary value problem and $u_h \in V_h$ is the discrete solution obtained in a subspace V_h of V.

The first error estimate is based on Céa's lemma (Theorem 13.1): it consists in estimating the distance $\inf_{v_h \in V_h} \|u-v_h\|_{1,\Omega}$ by means of the X_h-interpolant $\Pi_h u$ of the solution u. This gives the error estimate

$$\|u-u_h\|_{1,\Omega} \leqslant C\|u-\Pi_h u\|_{1,\Omega}.$$

Since we shall assume in this chapter that the set $\bar{\Omega}$ is polygonal, it can be written as a union $\bar{\Omega}=\bigcup_{T\in\mathscr{T}_h} T$ of polygonal sets T and straight finite elements (T, P_T, Σ_T), $T \in \mathscr{T}_h$, can thus be used for constructing subspaces $V_h \subset V$, as indicated in Chapter II. Hence the corresponding finite element method is *conforming*.

Taking into account that we are using the norm $\|\cdot\|_{1,\Omega}$ and that $(\Pi_h u)|_T = \Pi_T u$ for all $T \in \mathscr{T}_h$ (Theorem 12.1), we can write

$$\|u-\Pi_h u\|_{1,\Omega} = \left\{ \sum_{T\in\mathscr{T}_h} \|u-\Pi_T u\|_{1,T}^2 \right\}^{1/2}.$$

Therefore, the problem of finding an estimate of the error $\|u-u_h\|_{1,\Omega}$ is reduced to the problem of estimating the "*local*" *interpolation errors* $\|u-\Pi_T u\|_{1,T}$. The solution of such "local" interpolation problems is the object of Sections 15 and 16, where in view of other future needs, we shall also estimate more general norms or seminorms of the difference $(u-\Pi_T u)$.

A typical, and crucial, result in this direction is that, for a finite element (T, P_T, Σ_T) that can be imbedded in an *affine family* and whose P_T-interpolation operator *leaves invariant all polynomials of degree* $\leqslant k$ (equivalently, the inclusions $P_k(T) \subset P_T$ hold), there exists a constant C independent of T such that

$$|v-\Pi_T v|_{m,T} \leqslant C\frac{h_T^{k+1}}{\rho_T^m}|v|_{k+1,T}, \quad 0\leqslant m\leqslant k+1 \quad \text{for all } v \in H^{k+1}(T),$$

where

h_T = diameter of T,

ρ_T = supremum of the diameters of the spheres inscribed in T.

Such a result is proved (in a more general form) in Theorem 16.1.

One key idea for obtaining such an estimate consists in first obtaining it over a *reference finite element* and then to convert it into an estimate valid for any affine-equivalent finite element.

Another key idea is to use a *basic result about Sobolev spaces, due to Deny and Lions, which pervades the mathematical analysis of the finite element method*: Over the quotient space $H^{k+1}(\Omega)/P_k(\Omega)$, the seminorm $|\cdot|_{k+1,\Omega}$ is a norm equivalent to the quotient norm. This result is proved in Theorem 14.1, for the more general Sobolev spaces $W^{m,p}(\Omega)$.

In practice, one often considers a *regular family* of finite elements, in the sense that the diameters h_T approach zero, and that there exists a constant σ independent of T such that $h_T \leqslant \sigma \rho_T$. For such a regular family, the interpolation error estimate becomes (Theorem 16.2)

$$|v-\Pi_T v|_{m,T} = \mathrm{O}(h_T^{k+1-m}), \quad 0 \leqslant m \leqslant k+1, \quad \text{if } v \in H^{k+1}(T).$$

Hence, using Céa's lemma, we obtain in this fashion the error estimate (Theorem 18.1)

$$\|u-u_h\|_{1,\Omega} \leqslant C\|u-\Pi_h u\|_{1,\Omega} = \mathrm{O}(h^k),$$

with $h = \max_{T\in\mathcal{T}_h} h$, if $u \in H^{k+1}(\Omega)$.

We also mention in passing the *inverse inequalities* (Theorem 17.2), a "technical" property of finite element spaces that is often useful in the process of getting error estimates.

The range of applicability of the above error estimate is however limited since it requires that the solution u be smooth enough in order that its X_h-interpolant be well defined and that it be in the space $H^{k+1}(\Omega)$. Fortunately, we show in Theorem 18.2 that, under the minimal assumptions that the solution u is in the space $H^1(\Omega)$ and that the spaces P_T contain the space $P_1(T)$, $T \in \mathcal{T}_h$, convergence still holds, i.e., one has $\lim_{h\to 0}\|u-u_h\|_{1,\Omega} = 0$.

Next, using a method due to Aubin and Nitsche (cf. the Aubin–Nitsche lemma (Theorem 19.1)), we show that there is in most cases an improvement in the estimate of the error $|u-u_h|_{0,\Omega}$, in the sense that (Theorem 19.2)

$$|u-u_h|_{0,\Omega} = \mathrm{O}(h^{k+1}).$$

We then turn to various ways of estimating the error $|u-u_h|_{0,\infty,\Omega}$. After a first estimate, based on the estimate of the error $|u-u_h|_{0,\Omega}$ combined with an inverse inequality (cf. Theorem 19.3), we consider in Section 20 the situation where the discrete problem satisfies a *discrete maximum principle*, in the sense that

$$f \leqslant 0 \Rightarrow \max_{x\in\bar{\Omega}} u_h(x) \leqslant \max\{0, \max_{x\in\Gamma} u_h(x)\},$$

where f denotes the function appearing in the right-hand side of the partial differential equation of the associated boundary value problem.

In the case of the operator $(-\Delta u + au)$ with $a \geqslant 0$ and $n=2$, it is shown (Theorem 20.1) that the discrete maximum principle holds for h small enough if there exists $\varepsilon > 0$ such that all the angles of all the triangles found in all the triangulations are $\leqslant(\frac{1}{2}\pi - \varepsilon)$; if $a=0$, it suffices that the angles of the triangles be $\leqslant \frac{1}{2}\pi$ (Theorem 20.2). Returning to the general case, we show that when the discrete problems satisfy a maximum principle, one has (Theorem 21.5):

$$\lim_{h\to 0} |u-u_h|_{0,\infty,\Omega} = 0, \quad \text{if } u \in W^{1,p}(\Omega) \text{ with } p>n,$$

$$|u-u_h|_{0,\infty,\Omega} = \mathrm{O}(h), \quad \text{if } u \in W^{2,p}(\Omega) \text{ with } 2p>n.$$

We then follow in Section 22 the penetrating *method of weighted norms of Nitsche*, who has shown that, if $u \in W^{k+1,\infty}(\Omega)$,

$$|u-u_h|_{0,\infty,\Omega} = \begin{cases} \mathrm{O}(h^{2-\varepsilon}) \text{ for any } \varepsilon>0, & \text{if } k=1, \\ \mathrm{O}(h^{k+1}), & \text{if } k\geqslant 2, \end{cases}$$

$$\|u-u_h\|_{1,\infty,\Omega} = \begin{cases} \mathrm{O}(h^{1-\varepsilon}) \text{ for any } \varepsilon>0, & \text{if } k=1, \\ \mathrm{O}(h^{k}), & \text{if } k\geqslant 2, \end{cases}$$

where $|\cdot|_{0,\infty,\Omega}$ and $\|\cdot\|_{1,\infty,\Omega}$ represent the norms of the spaces $L^\infty(\Omega)$ and $W^{1,\infty}(\Omega)$, respectively. These error estimates are established in Theorem 22.7 for $k=1$.

It is worth pointing out that *all the error estimates found in this chapter are optimal* in the sense that, with the same regularity assumptions on the function u, one gets the same asymptotic estimates (or "almost" the same for the norms $|\cdot|_{0,\infty,\Omega}$ and $\|\cdot\|_{1,\infty,\Omega}$ when $k=1$) when the discrete solution $u_h \in V_h$ is replaced by the X_h-interpolant $\Pi_h u \in V_h$.

In Section 23, we consider the finite element approximation of problems posed as *variational inequalities*, and in particular, we consider the approximation of the obstacle problem: Following an ingenious method due to Falk, we show in this case that the discrete solutions obtained with linear triangles satisfy (Theorem 23.2)

$$\|u-u_h\|_{1,\Omega} = \mathrm{O}(h).$$

This result is itself a consequence of an abstract error estimate (Theorem 23.1), valid for general variational inequalities.

Finally, we give in Section 24 a review of various extensions of the basic error estimates established in this chapter.

It is only in the sixties that mathematicians, notably MIKHLIN [1964, 1971], showed real interest in the analysis of the Galerkin and Ritz methods. Although they were not aware of the engineers' contributions, it is interesting to notice that the approximate methods that they studied resembled more and more the finite element method, as exemplified by the basic contributions of CÉA [1964], VARGA [1966b] (for the one-dimensional case), BIRKHOFF, SCHULTZ and VARGA [1968] (for the multi-dimensional, but still tensor product, case), FRIEDRICHS and KELLER [1966], where a finite difference method was analyzed as a finite element method on special triangulations. Then the outbreak came with the paper of ZLÁMAL [1968], which is generally regarded as the first mathematical error analysis of the "general" finite element method as we know it today. Other "historical" references on the numerical analysis of the method are found in the introductory article of Oden.

14. The Sobolev spaces $W^{m,p}(\Omega)$ and the quotient space $W^{k+1,p}(\Omega)/P_k(\Omega)$

For general references on Sobolev spaces, see ADAMS [1975], LIONS [1962], NEČAS [1967]. An excellent introduction is given in BREZIS [1983, Chapter 9].

Throughout this section, Ω denotes a domain in $\mathbb{R}^n$, i.e., a bounded open connected subset of $\mathbb{R}^n$ with a Lipschitz-continuous boundary. For any integer $m \geqslant 0$, and any number p satisfying $1 \leqslant p \leqslant \infty$, the *Sobolev space* $W^{m,p}(\Omega)$ consists of those functions $v \in L^p(\Omega)$ for which all partial derivatives $\partial^\alpha v$ (in the distribution sense) with $|\alpha| \leqslant m$ belong to the space $L^p(\Omega)$. Equipped with the *norm*

$$\|v\|_{m,p,\Omega} = \begin{cases} \left\{ \sum_{|\alpha| \leqslant m} \int_\Omega |\partial^\alpha v|^p \, dx \right\}^{1/p}, & \text{if } 1 \leqslant p < \infty, \\ \max_{|\alpha| \leqslant m} \left\{ \operatorname{ess\,sup}_{x \in \Omega} |\partial^\alpha v(x)| \right\}, & \text{if } p = \infty, \end{cases} \tag{14.1}$$

the space $W^{m,p}(\Omega)$ is a Banach space. We shall also use the *seminorms*

$$|v|_{m,p,\Omega} = \begin{cases} \left\{ \sum_{|\alpha| = m} \int_\Omega |\partial^\alpha v|^p \, dx \right\}^{1/p}, & \text{if } 1 \leqslant p < \infty, \\ \max_{|\alpha| = m} \left\{ \operatorname{ess\,sup}_{x \in \Omega} |\partial^\alpha v(x)| \right\}, & \text{if } p = \infty. \end{cases} \tag{14.2}$$

The *Sobolev space* $W_0^{m,p}(\Omega)$ is the closure of the space $\mathcal{D}(\Omega)$ in the space $W^{m,p}(\Omega)$.

Given a subset A of $\mathbb{R}^n$ and given a function $v \in \mathscr{C}^m(A)$, the notations $\|v\|_{m,\infty,A}$ and $|v|_{m,\infty,A}$ also denote the norm $\max_{|\alpha| \leqslant m} \sup_{x \in A} |\partial^\alpha v(x)|$ and the seminorm $\max_{|\alpha| = m} \sup_{x \in A} |\partial^\alpha v(x)|$, respectively. Notice that

$$W^{m,2}(\Omega) = H^m(\Omega), \qquad W_0^{m,2}(\Omega) = H_0^m(\Omega),$$
$$\|\cdot\|_{m,2,\Omega} = \|\cdot\|_{m,\Omega}, \qquad |\cdot|_{m,2,\Omega} = |\cdot|_{m,\Omega}.$$

We now record some basic properties of the Sobolev spaces that will often be used. In what follows, the notation $X \hookrightarrow Y$ indicates that the normed linear space X is contained in the normed linear space Y with a continuous injection, and the notation $X \Subset Y$ indicates that this injection is compact. Finally, for any integer $m \geqslant 0$ and any number $\alpha \in]0, 1]$, $\mathscr{C}^{m,\alpha}(\bar{\Omega})$ denotes the space of all functions in $\mathscr{C}^m(\bar{\Omega})$ whose mth derivatives satisfy a Hölder's condition with exponent α. Equipped with the norm

$$\|v\|_{\mathscr{C}^{m,\alpha}} = \|v\|_{m,\infty,\Omega} + \max_{|\beta|=m} \sup_{\substack{x,y\in\Omega \\ x\neq y}} \frac{|\partial^\beta v(x) - \partial^\beta v(y)|}{\|x-y\|^\alpha},$$

where $\|\cdot\|$ denotes the Euclidean norm in $\mathbb{R}^n$, the space $C^{m,\alpha}(\bar{\Omega})$ is a Banach space.

By the *Sobolev imbedding theorems*, the following inclusions hold, for all integers $m \geqslant 0$ and for all numbers p with $1 \leqslant p \leqslant \infty$,

$$W^{m,p}(\Omega) \hookrightarrow \begin{cases} L^{p^*}(\Omega) \quad \text{with } \dfrac{1}{p^*} = \dfrac{1}{p} - \dfrac{m}{n}, & \text{if } m < \dfrac{n}{p}, \\ L^q(\Omega) \quad \text{for all } q \in [1, \infty[, & \text{if } m = \dfrac{n}{p}, \\ \mathscr{C}^{0,m-(n/p)}(\bar{\Omega}), & \text{if } \dfrac{n}{p} < m < \dfrac{n}{p} + 1, \\ \mathscr{C}^{0,\alpha}(\bar{\Omega}) \quad \text{for all } 0 < \alpha < 1, & \text{if } m = \dfrac{n}{p} + 1, \\ \mathscr{C}^{0,1}(\bar{\Omega}), & \text{if } \dfrac{n}{p} + 1 < m. \end{cases} \tag{14.3}$$

By the *Kondrasov theorems*, the compact injections

$$W^{m,p}(\Omega) \Subset \begin{cases} L^q(\Omega) \text{ for all } 1 \leqslant q < p^* \text{ with } \dfrac{1}{p^*} = \dfrac{1}{p} - \dfrac{m}{n}, & \text{if } m < \dfrac{n}{p}, \\ L^q(\Omega) \text{ for all } q \in [1, \infty[, & \text{if } m = \dfrac{n}{p}, \\ \mathscr{C}^0(\bar{\Omega}), & \text{if } \dfrac{n}{p} < m, \end{cases} \tag{14.4}$$

hold for all $1 \leqslant p \leqslant \infty$. The compact injection

$$H^1(\Omega) \Subset L^2(\Omega)$$

which holds for any n by the Kondrasov theorems, is known as *Rellich's theorem*. Notice that analogous inclusions can be derived by "translating" the orders of derivations. Thus for instance, one has $W^{m+r,p}(\Omega) \hookrightarrow W^{r,p^*}(\Omega)$ if $m < (n/p)$, etc.

We also note that, for $1 \leqslant p < \infty$, one has the important *density property*:

$$\{\mathscr{C}^\infty(\bar{\Omega})\}^- = W^{m,p}(\Omega).$$

REMARK 14.1. The assumption that Ω is a domain in $\mathbb{R}^n$ is not necessarily needed for proving some of the above properties. For example, one can derive the compact inclusion $W^{1,p}(\Omega) \Subset L^q(\Omega)$ for all $1 \leqslant q < p^*$, or the above density property, under weaker assumptions.

Since an open set Ω with a Lipschitz-continuous boundary is bounded, the space $P_k(\Omega)$ is contained in the space $W^{k+1,p}(\Omega)$, and thus it makes sense to consider the *quotient space* $W^{k+1,p}(\Omega)/P_k(\Omega)$. This space is a Banach space when it is equipped with the *quotient norm*

$$\dot{v} \in W^{k+1,p}(\Omega)/P_k(\Omega) \to \|\dot{v}\|_{k+1,p,\Omega} = \inf_{p \in P_k(\Omega)} \|v+p\|_{k+1,p,\Omega}, \tag{14.5}$$

where

$$\dot{v} = \{w \in W^{k+1,p}(\Omega);\ (w-v) \in P_k(\Omega)\} \tag{14.6}$$

denotes the equivalence class of the element $v \in W^{k+1,p}(\Omega)$.

Then the mapping

$$\dot{v} \in W^{k+1,p}(\Omega)/P_k(\Omega) \to |\dot{v}|_{k+1,p,\Omega} = |v|_{k+1,p,\Omega} \tag{14.7}$$

is a priori only a seminorm on the quotient space $W^{k+1,p}(\Omega)/P_k(\Omega)$, which satisfies the inequality

$$|\dot{v}|_{k+1,p,\Omega} \leqslant \|\dot{v}\|_{k+1,p,\Omega} \quad \text{for all } \dot{v} \in W^{k+1,p}(\Omega)/P_k(\Omega). \tag{14.8}$$

To see this, observe that, for any polynomial $p \in P_k(\Omega)$,

$$\|v+p\|_{k+1,p,\Omega} = \{|v|^p_{k+1,p,\Omega} + \|v+p\|^p_{k,p,\Omega}\}^{1/p} \geqslant |v|_{k+1,p,\Omega},$$

with the standard modification for $p=\infty$. It is a fundamental result that *the mapping defined in* (14.7) *is in fact a norm over the quotient space, equivalent to the quotient norm* (14.5). This result was first proved by DENY and LIONS [1953–1954].

THEOREM 14.1. *Let $k \geqslant 0$ be an integer, and let $p \in [1,\infty]$. There exists a constant $C(\Omega)$ such that*

$$\inf_{p \in P_k(\Omega)} \|v+p\|_{k+1,p,\Omega} \leqslant C(\Omega)|v|_{k+1,p,\Omega} \quad \textit{for all } v \in W^{k+1,p}(\Omega) \tag{14.9}$$

and consequently, such that

$$\|\dot{v}\|_{k+1,p,\Omega} \leqslant C(\Omega)|\dot{v}|_{k+1,p,\Omega} \quad \textit{for all } \dot{v} \in W^{k+1,p}(\Omega)/P_k(\Omega). \tag{14.10}$$

PROOF. Let $N = \dim P_k(\Omega)$ and let f_i, $1 \leqslant i \leqslant N$, be a basis of the dual space of $P_k(\Omega)$. Thus, by the Hahn–Banach extension theorem, there exist continuous linear forms over the space $W^{k+1,p}(\Omega)$, again denoted f_i, $1 \leqslant i \leqslant N$, such that $f_i(p)=0$, $1 \leqslant i \leqslant N$, implies $p=0$ if $p \in P_k(\Omega)$. We will show that there exists a constant $C(\Omega)$ such that

$$\|v\|_{k+1,p,\Omega} \leqslant C(\Omega)\left(|v|_{k+1,p,\Omega} + \sum_{i=1}^N |f_i(v)|\right) \quad \text{for all } v \in W^{k+1,p}(\Omega). \tag{14.11}$$

Inequality (14.9) will then be a consequence of inequality (14.11): Given any function $v \in W^{k+1,p}(\Omega)$, let $q \in P_k(\Omega)$ be such that $f_i(v+q)=0$, $1 \leqslant i \leqslant N$. Then by (14.11),

$$\inf_{p \in P_k(\Omega)} \|v+p\|_{k+1,p,\Omega} \leqslant \|v+q\|_{k+1,p,\Omega} \leqslant C(\Omega)|v|_{k+1,p,\Omega},$$

which proves (14.9). If inequality (14.11) is false, there exists a sequence (v_l) of functions $v_l \in W^{k+1,p}(\Omega)$, such that

$$\|v_l\|_{k+1,p,\Omega} = 1 \quad \text{for all } l \geqslant 1,$$
$$\lim_{l\to\infty} \left\{ |v_l|_{k+1,p,\Omega} + \sum_{i=1}^{N} |f_i(v_l)| \right\} = 0. \tag{14.12}$$

Since the sequence (v_l) is bounded in $W^{k+1,p}(\Omega)$, there exists a subsequence, again denoted (v_l) for notational convenience, that converges in the space $W^{k,p}(\Omega)$ (this follows from the Kondrasov or Rellich theorems for $1 \leqslant p < \infty$ and from Ascoli's theorem for $p=\infty$). Since

$$\lim_{l\to\infty} |v_l|_{k+1,p,\Omega} = 0,$$

by (14.12), and since the space $W^{k+1,p}(\Omega)$ is complete, the sequence (v_l) converges in the space $W^{k+1,p}(\Omega)$.

The limit v of this sequence is such that

$$|\partial^\alpha v|_{0,p,\Omega} = \lim_{l\to\infty} |\partial^\alpha v_l|_{0,p,\Omega} = 0 \quad \text{for all } \alpha \text{ with } |\alpha| = k+1,$$

and thus $\partial^\alpha v = 0$ for all multi-indices α with $|\alpha| = k+1$. Since a domain is connected by assumption, it follows from distribution theory (see SCHWARTZ [1966, p. 60] that the function v is a polynomial of degree $\leqslant k$. Using (14.12), we have

$$f_i(v) = \lim_{l\to\infty} f_i(v_l) = 0;$$

hence we conclude that $v=0$ since $v \in P_k(\Omega)$. But this contradicts the equality $\|v_l\|_{k+1,p,\Omega} = 1$ for all l. □

15. Estimate of the seminorms $|v - \Pi_T v|_{m,q,T}$ for polynomial-preserving operators Π_T

One of our main objectives in this chapter is to estimate the *interpolation errors* $|v - \Pi_T v|_{m,q,T}$ and $\|v - \Pi_T v\|_{m,q,T}$, where Π_T is the P_T-interpolation operator associated with a finite element (T, P_T, Σ_T). At other places, however, we shall need similar estimates, but in situations where Π_T is a more general polynomial-preserving operator, i.e., Π_T is not necessarily an interpolation operator. This is why we shall develop an error theory valid also for such general operators; we follow here CIARLET and RAVIART [1972a].

To begin with, we need a definition: We say that two open subsets Ω and $\hat{\Omega}$ of $\mathbb{R}^n$ are *affine-equivalent* if there exists an invertible affine mapping

$$F : \hat{x} \in \mathbb{R}^n \to F(\hat{x}) = B\hat{x} + b \in \mathbb{R}^n \tag{15.1}$$

such that

$$\Omega = F(\hat{\Omega}). \tag{15.2}$$

As in the case of affine-equivalent finite elements (compare with (11.1) and (11.3)), the correspondences

$$\hat{x} \in \hat{\Omega} \to x = F(\hat{x}) \in \Omega, \tag{15.3}$$

$$(\hat{v} : \hat{\Omega} \to \mathbb{R}) \to (v = \hat{v} \circ F^{-1} : \Omega \to \mathbb{R}), \tag{15.4}$$

hold between the points $\hat{x} \in \hat{\Omega}$ and $x \in \Omega$, and between functions defined over the set $\hat{\Omega}$ and the set Ω. Notice that we have

$$\hat{v}(\hat{x}) = v(x) \tag{15.5}$$

for all points $\hat{x}$ and x in the correspondence (15.3) and all functions $\hat{v}$, v in the correspondence (15.4).

REMARK 15.1. If the functions v and $\hat{v}$ are defined only almost everywhere (as in the next theorem for instance), it is understood that relation (15.5) needs to hold for almost all points $\hat{x} \in \hat{\Omega}$, and thus for almost all points $x \in \Omega$.

We first show that, if Ω and $\hat{\Omega}$ are affine-equivalent, any Sobolev seminorm $|\cdot|_{m,p,\Omega}$ (cf. (14.2)) is bounded above and below by the corresponding seminorm $|\cdot|_{m,p,\hat{\Omega}}$, times factors that depend on the matrix B of (15.1), and on the numbers m and p. Here and subsequently, $\|\cdot\|$ stands for both the Euclidean norm in $\mathbb{R}^n$ and for the associated matrix norm.

THEOREM 15.1. *Let Ω and $\hat{\Omega}$ be two affine-equivalent open domains in $\mathbb{R}^n$. If a function v belongs to the space $W^{m,p}(\Omega)$ for some integer $m \geq 0$ and some number $p \in [1, \infty]$, the function $\hat{v} = v \cdot F$ belongs to the space $W^{m,p}(\hat{\Omega})$; in addition, there exists a constant $C = C(m, n)$ such that*

$$|\hat{v}|_{m,p,\hat{\Omega}} \leq C \|B\|^m |\det B|^{-1/p} |v|_{m,p,\Omega} \quad \textit{for all } v \in W^{m,p}(\Omega), \tag{15.6}$$

where B is the matrix of (15.1). *Analogously, one has*

$$|v|_{m,p,\Omega} \leq C \|B^{-1}\|^m |\det B|^{1/p} |\hat{v}|_{m,p,\hat{\Omega}} \quad \textit{for all } \hat{v} \in W^{m,p}(\hat{\Omega}). \tag{15.7}$$

PROOF. (i) Let us first assume that the function v belongs to the space $\mathscr{C}^m(\bar{\Omega})$, so that the function $\hat{v}$ belongs to the space $\mathscr{C}^m(\bar{\hat{\Omega}})$.

Since, for any multi-index α with $|\alpha| = m$, one has

$$\partial^\alpha \hat{v}(\hat{x}) = D^m \hat{v}(\hat{x})(e_{\alpha_1}, e_{\alpha_2}, \ldots, e_{\alpha_m})$$

where the vectors e_{α_i}, $1 \leq i \leq m$, are some of the basis vectors of $\mathbb{R}^n$, we deduce that

$$|\partial^\alpha \hat{v}(\hat{x})| \leq \|D^m \hat{v}(\hat{x})\| = \sup_{\substack{\|\xi_i\| \leq 1 \\ 1 \leq i \leq m}} |D^m \hat{v}(\hat{x})(\xi_1, \xi_2, \ldots, \xi_m)|.$$

Consequently, we obtain

$$|\hat{v}|_{m,p,\hat{\Omega}} = \left\{ \int_{\hat{\Omega}} \sum_{|\alpha| = m} |\partial^\alpha \hat{v}(\hat{x})|^p \, d\hat{x} \right\}^{1/p} \leq C_1(m, n) \left\{ \int_{\hat{\Omega}} \|D^m \hat{v}(\hat{x})\|^p \, d\hat{x} \right\}^{1/p}, \tag{15.8}$$

where the constant $C_1(m, n)$ may be chosen as

$$C_1(m,n)=\sup_{1\leqslant p}(\operatorname{card}\{\alpha\in\mathbb{N}^m\colon |\alpha|=m\})^{1/p}.$$

Using the chain rule, we note that, for any vectors $\xi_i\in\mathbb{R}^n$, $1\leqslant i\leqslant m$,

$$D^m\hat{v}(\hat{x})(\xi_1,\xi_2,\ldots,\xi_n)=D^m v(x)(B\xi_1,B\xi_2,\ldots,B\xi_m),$$

so that

$$\|D^m\hat{v}(\hat{x})\|\leqslant\|D^m v(x)\|\;\|B\|^m,$$

and, therefore,

$$\int_{\hat{\Omega}}\|D^m\hat{v}(\hat{x})\|^p\,\mathrm{d}\hat{x}\leqslant\|B\|^{mp}\int_{\hat{\Omega}}\|D^m v(F(\hat{x}))\|^p\,\mathrm{d}\hat{x}. \tag{15.9}$$

Using the formula of change of variables in multiple integrals, we get

$$\int_{\hat{\Omega}}\|D^m v(F(\hat{x}))\|^p\,\mathrm{d}\hat{x}=|\det B^{-1}|\int_{\hat{\Omega}}\|D^m v(x)\|^p\,\mathrm{d}x, \tag{15.10}$$

and since there exists a constant $C_2(m, n)$ such that

$$\|D^m v(x)\|\leqslant C_2(m,n)\max_{|\alpha|=m}|\partial^\alpha v(x)|,$$

we obtain

$$\left\{\int_{\Omega}\|D^m v(x)\|^p\,\mathrm{d}x\right\}^{1/p}\leqslant C_2(m,n)|v|_{m,p,\Omega}. \tag{15.11}$$

Inequality (15.6) is then a consequence of inequalities (15.8)–(15.11).

(ii) To complete the proof when $p\neq\infty$, it remains to use the continuity of the linear operator $\imath\colon v\in\mathscr{C}^m(\bar{\Omega})\to\hat{v}\in W^{m,p}(\hat{\Omega})$ with respect to the norms $\|\cdot\|_{m,p,\Omega}$ and $\|\cdot\|_{m,p,\hat{\Omega}}$, the denseness of the space $\mathscr{C}^m(\bar{\Omega})$ in the space $W^{m,p}(\Omega)$, and the definition of the (unique) extension of the mapping $\imath$ to the space $W^{m,p}(\Omega)$.

(iii) Let us finally consider the case $p=\infty$. A function $v\in W^{m,\infty}(\Omega)$ belongs to the spaces $W^{m,p}(\Omega)$ for all $p<\infty$ (recall that a domain is bounded). Therefore, by (ii), the function $\hat{v}$ belongs to the spaces $W^{m,p}(\hat{\Omega})$ for all $p<\infty$, and there exists a constant $C(m, n)$ such that

$$\begin{aligned}|\partial^\alpha\hat{v}|_{0,p,\hat{\Omega}}&\leqslant|\hat{v}|_{|\alpha|,p,\hat{\Omega}}\\&\leqslant C(m,n)\|B\|^{|\alpha|}\sup_{1\leqslant p}|\det B|^{-1/p}\|v\|_{m,p,\Omega}\end{aligned}$$

for all $p\geqslant 1$ and for all multi-indices $\alpha\in\mathbb{N}^m$ such that $|\alpha|\leqslant m$. Since this upper bound on the seminorm $|\partial^\alpha v|_{0,p,\hat{\Omega}}$ is independent of the number p, it follows that, for each $|\alpha|\leqslant m$, the function $\partial^\alpha\hat{v}$ is in the space $L^\infty(\hat{\Omega})$ for each $|\alpha|\leqslant m$. Consequently, the function $\hat{v}$ belongs to the space $W^{m,\infty}(\hat{\Omega})$. To conclude, it suffices to use inequality

(15.7), which holds for all $p\geqslant 1$, in conjunction with the property that, for any function $w\in L^{\infty}(\Omega)$,

$$|w|_{0,\infty,\Omega} = \lim_{p\to\infty} |w|_{0,p,\Omega}.$$

(iv) Inequality (15.7) is proved in a similar fashion. □

In order to apply Theorem 15.1 it is desirable to evaluate the norms $\|B\|$ and $\|B^{-1}\|$ in terms of simple geometric quantities attached to the sets Ω and $\hat{\Omega}$. This is the object of the next theorem, where we use the following notations:

$$h=\operatorname{diam}(\Omega), \qquad \hat{h}=\operatorname{diam}(\hat{\Omega}), \tag{15.12}$$

$$\begin{aligned} \rho&=\sup\{\operatorname{diam}(S);\ S \text{ is a ball contained in } \Omega\},\\ \hat{\rho}&=\sup\{\operatorname{diam}(\hat{S});\ \hat{S} \text{ is a ball contained in } \hat{\Omega}\}. \end{aligned} \tag{15.13}$$

THEOREM 15.2. *Let $F: \hat{x}\in\mathbb{R}^n\to F(\hat{x})=B\hat{x}+b\in\mathbb{R}^n$ be an affine mapping, and let $\hat{\Omega}$ and $\Omega=F(\hat{\Omega})$ be two affine-equivalent bounded open subsets of $\mathbb{R}^n$. Then the upper bounds*

$$\|B\|\leqslant h/\hat{\rho}, \qquad \|B^{-1}\|\leqslant \hat{h}/\rho \tag{15.14}$$

hold.

PROOF. We may write

$$\|B\| = \frac{1}{\hat{\rho}} \sup_{\|\xi\|=\hat{\rho}} \|B\xi\|.$$

Given a vector ξ satisfying $\|\xi\|=\hat{\rho}$, there exist two points $\hat{y}, \hat{z}\in(\hat{\Omega})^-$ such that $\hat{y}-\hat{z}=\xi$, by definition of $\hat{\rho}$ (Fig. 15.1). Since $B\xi=F(\hat{y})-F(\hat{z})$ with $F(\hat{y})\in\bar{\Omega}$, $F(\hat{z})\in\bar{\Omega}$, we deduce that $\|B\xi\|\leqslant h$, and thus the first inequality of (15.14) is proved. The other inequality is proved in a similar fashion. □

We are now in a position to prove an important property of *polynomial-preserving operators*, i.e., that leave invariant some space $P_k(\hat{\Omega})$ (cf. (15.16)).

THEOREM 15.3. *Assume that, for some integers $k\geqslant 0$ and $m\geqslant 0$ and some numbers $p, q\in[1,\infty]$, $W^{k+1,p}(\hat{\Omega})$ and $W^{m,q}(\hat{\Omega})$ are two Sobolev spaces that satisfy*

$$W^{k+1,p}(\hat{\Omega}) \hookrightarrow W^{m,q}(\hat{\Omega}), \tag{15.15}$$

and let $\hat{\Pi}\in\mathscr{L}(W^{k+1,p}(\hat{\Omega}); W^{m,q}(\hat{\Omega}))$ be a mapping that satisfies

$$\hat{\Pi}\hat{p}=\hat{p} \quad \text{for all } \hat{p}\in P_k(\hat{\Omega}). \tag{15.16}$$

For any open set Ω that is affine-equivalent to the set $\hat{\Omega}$, let the mapping Π_Ω be defined by

$$\{\Pi_\Omega v\}\hat{} = \hat{\Pi}\hat{v}, \tag{15.17}$$

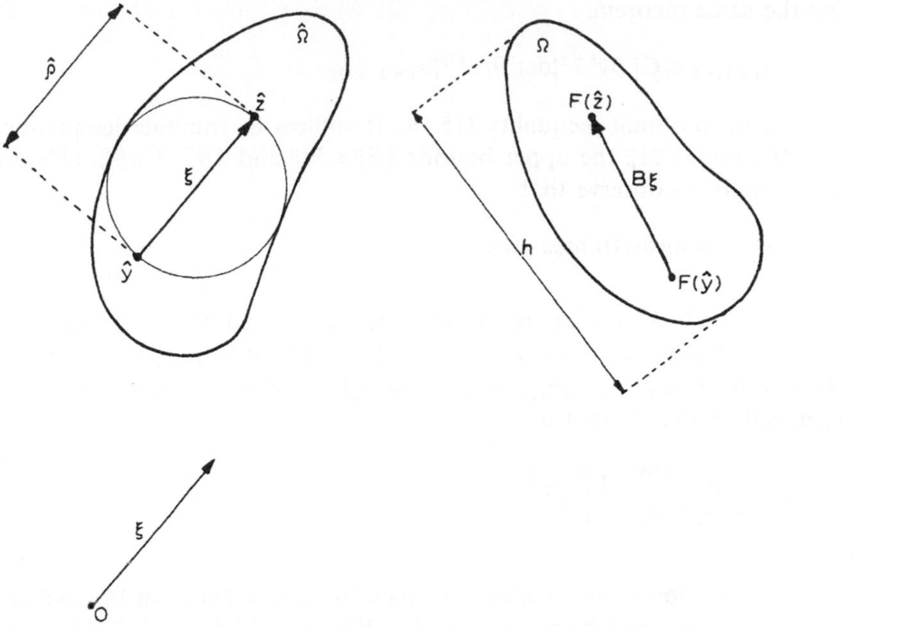

FIG. 15.1. Examples of affine-equivalent open subsets in $\mathbb{R}^2$. Two open subsets $\hat{\Omega}$ and Ω are affine-equivalent if there exists an affine mapping $F: \hat{x} \in \mathbb{R}^n \to (B\hat{x}+b) \in \mathbb{R}^n$ such that $\Omega = F(\hat{\Omega})$.

for all functions $\hat{v} \in W^{k+1,p}(\hat{\Omega})$ *and* $v \in W^{k+1,p}(\Omega)$ *in the correspondence* (15.4). *Then there exists a constant* $C(\hat{\Pi}, \hat{\Omega})$ *such that, for all affine-equivalent sets* Ω,

$$|v - \Pi_\Omega v|_{m,q,\Omega} \leqslant C(\hat{\Pi}, \hat{\Omega})\,\{\text{meas}(\Omega)\}^{1/q - 1/p} \frac{h^{k+1}}{\rho^m} |v|_{k+1,p,\Omega}, \tag{15.18}$$

for all $v \in W^{k+1,p}(\Omega)$,

where h *and* ρ *are defined as in* (15.12) *and* (15.13).

PROOF. Using the polynomial invariance (15.16), we obtain the identity

$$\hat{v} - \hat{\Pi}\hat{v} = (I - \hat{\Pi})(\hat{v} + \hat{p}) \quad \text{for all } \hat{v} \in W^{k+1,p}(\hat{\Omega}), \quad \hat{p} \in P_k(\hat{\Omega}),$$

where I denotes the identity mapping from $W^{k+1,p}(\hat{\Omega})$ into $W^{m,q}(\hat{\Omega})$, which is continuous by (15.15). From this identity we deduce that

$$\begin{aligned} |\hat{v} - \hat{\Pi}\hat{v}|_{m,q,\hat{\Omega}} &\leqslant \|I - \hat{\Pi}\|_{\mathscr{L}(W^{k+1,p}(\hat{\Omega}); W^{m,q}(\hat{\Omega}))} \inf_{\hat{p} \in P_k(\hat{\Omega})} \|\hat{v} + \hat{p}\|_{k+1,p,\hat{\Omega}} \\ &\leqslant C(\hat{\Pi}, \hat{\Omega}) |\hat{v}|_{k+1,p,\hat{\Omega}}, \end{aligned} \tag{15.19}$$

by Theorem 14.1. It follows from relation (15.17) that $\hat{v} - \hat{\Pi}\hat{v} = \{v - \Pi_\Omega v\}\hat{}$, and therefore an application of Theorem 15.1 yields

$$|v - \Pi_\Omega v|_{m,q,\Omega} \leqslant C \|B^{-1}\|^m |\det B|^{1/q} |\hat{v} - \hat{\Pi}\hat{v}|_{m,q,\hat{\Omega}}. \tag{15.20}$$

By the same theorem,

$$|\hat{v}|_{k+1,p,\hat{\Omega}} \leqslant C\|B\|^{k+1}|\det B|^{-1/p}|v|_{k+1,p,\Omega}, \tag{15.21}$$

and thus, to obtain inequality (15.18), it suffices to combine inequalities (15.19), (15.20) and (15.21), the upper bounds $\|B\| \leqslant h/\hat{\rho}$ and $\|B^{-1}\| \leqslant \hat{h}/\rho$ (Theorem 15.2), and, finally, to observe that

$$|\det B| = \text{meas}(\Omega)/\text{meas}(\hat{\Omega}). \qquad \square$$

REMARK 15.2. A similar result can be established for polynomial-preserving operators that leave invariant the space $Q_k(\Omega)$ instead of the space $P_k(\Omega)$. In this case, the seminorm $|v|_{k+1,p,\Omega}$ appearing in the right-hand side of inequality (15.18) can be replaced by the seminorm

$$v \to \left\{ \sum_{i=1}^{n} \int_\Omega \left| \frac{\partial^{k+1} v}{\partial x_i^{k+1}} \right|^p \mathrm{d}x \right\}^{1/p}.$$

This result follows from an abstract generalization of Theorem 14.1, which is due to Luc Tartar, and which may be found in BREZZI and MARINI [1975], or in CIARLET [1978, Exercise 3.1.1].

16. Estimate of the interpolation errors $|v - \Pi_T v|_{m,q,T}$ for an affine family of finite elements

In Theorem 15.3, let us now choose as a particular polynomial-preserving operator Π_T the P_T-interpolation operator associated with a finite element (T, P_T, Σ_T) that is affine-equivalent to a finite element $(\hat{T}, \hat{P}, \hat{\Sigma})$. We obtain in this fashion the following result, which is essentially Theorem 6 of CIARLET and RAVIART [1972a].

THEOREM 16.1. *Let $(\hat{T}, \hat{P}, \hat{\Sigma})$ be a finite element, and let s denotes the greatest order of the partial derivatives occurring in the definition of $\hat{\Sigma}$. Assume that for some integers $m \geqslant 0$ and $k \geqslant 0$ and for some numbers $p, q \in [1, \infty]$, the following inclusions hold:*

$$W^{k+1,p}(\hat{T}) \hookrightarrow \mathscr{C}^s(\hat{T}), \tag{16.1}$$

$$W^{k+1,p}(\hat{T}) \hookrightarrow W^{m,q}(\hat{T}), \tag{16.2}$$

$$P_k(\hat{T}) \subset \hat{P} \subset W^{m,q}(\hat{T}). \tag{16.3}$$

Then there exists a constant $C(\hat{T}, \hat{P}, \hat{\Sigma})$ such that, for all affine-equivalent finite elements (T, P, Σ), and all functions $v \in W^{k+1,p}(T)$,

$$|v - \Pi_T v|_{m,q,T} \leqslant C(\hat{T}, \hat{P}, \hat{\Sigma})\{\text{meas}(T)\}^{1/q - 1/p} \frac{h_T^{k+1}}{\rho_T^m} |v|_{k+1,p,T}, \tag{16.4}$$

where $\Pi_T v$ denotes the P_T-interpolant of the function v, and

$$\begin{aligned} &\operatorname{meas}(T) = \mathrm{d}x\text{-}measure\ of\ T,\\ &h_T = \operatorname{diam}(T), \qquad\qquad (16.5)\\ &\rho_T = \sup\{\operatorname{diam}(S);\ S \text{ is a ball contained in } T\}. \end{aligned}$$

PROOF. The inclusion $P_k(\hat{T}) \subset \hat{P}$ in conjunction with the fact that the $\hat{P}$-interpolation operator $\hat{\Pi}$ reduces to the identity over the space $\hat{P}$ (cf. (10.8)) implies that

$$\hat{\Pi}\hat{p} = \hat{p} \quad \text{for all } \hat{p} \in P_k(\hat{T}). \tag{16.6}$$

Let $\hat{v}$ be a function in the space $W^{k+1,p}(\hat{T})$. Then $\hat{v}$ belongs to the space $\operatorname{dom} \hat{\Pi} = \mathscr{C}^s(\hat{T})$ (cf. (10.7)) since the inclusion $W^{k+1,p}(\hat{T}) \subset \mathscr{C}^s(\hat{T})$ holds. For definiteness, let us assume that $s=2$ (recall that in practice, $s=0$, 1 or 2) so that the $\hat{P}$-interpolant of the function $\hat{v}$ takes the form

$$\hat{\Pi}\hat{v} = \sum_i \hat{v}(\hat{a}_i^0)\hat{p}_i^0 + \sum_{i,k} \{D\hat{v}(\hat{a}_i^1)\hat{\xi}_{i,k}^1\}\hat{p}_{ik}^1 + \sum_{i,k,l} \{D^2\hat{v}(a_i^2)(\hat{\xi}_{ik}^2, \hat{\xi}_{il}^2)\}\,\hat{p}_{ikl}^2. \tag{16.7}$$

We now show that the linear mapping $\hat{\Pi}$: $W^{k+1,p}(\hat{T}) \to W^{m,q}(\hat{T})$ defined in this fashion (recall that $\hat{\Pi}\hat{v} \in \hat{P}$ and that $\hat{P} \subset W^{m,q}(\hat{T})$ by (16.3)) is continuous. To see this, we observe that, by (16.7),

$$\begin{aligned} \|\hat{\Pi}\hat{v}\|_{m,q,\hat{T}} &\leqslant \sum_i |\hat{v}(\hat{a}_i^0)|\,\|\hat{p}_i^0\|_{m,q,\hat{T}} + \sum_{i,k} |\{D\hat{v}(\hat{a}_i^1)\hat{\xi}_{ik}^1\}|\,\|\hat{p}_{ik}^1\|_{m,q,\hat{T}}\\ &\quad + \sum_{i,k,l} |\{D^l\hat{v}(\hat{a}_i^2)(\hat{\xi}_{ik}^2, \hat{\xi}_{il}^2)\}|\,\|\hat{p}_{ikl}^2\|_{m,q,\hat{T}}\\ &\leqslant C(\|\hat{p}_i^0\|_{m,q,\hat{T}}, \|\hat{\xi}_{ik}^1\|\,\|\hat{p}_{ik}^1\|_{m,q,\hat{T}}, \|\hat{\xi}_{ik}^2\|\,\|\hat{\xi}_{il}^2\|\,\|\hat{p}_{ikl}^2\|_{m,q,\hat{T}})\|\hat{v}\|_{2,\infty,\hat{T}}, \end{aligned}$$

and thus, by (16.1)

$$\|\hat{\Pi}\hat{v}\|_{m,q,\hat{T}} \leqslant C(\hat{T}, \hat{P}, \hat{\Sigma})\,\|\hat{v}\|_{k+1,p,\hat{T}}.$$

Since the P_T- and $\hat{P}$-interpolation operators are related through the correspondence

$$\{\Pi_T v\}\hat{} = \hat{\Pi}\hat{v} \quad \text{for all } v \in \operatorname{dom} \Pi_T$$

(cf. (11.10)), we may apply Theorem 15.3. Hence inequality (16.4) is just inequality (15.18) adapted to the present situation. □

REMARK 16.1. The factor $\{\operatorname{meas}(T)\}^{1/q-1/p}$ found in (16.4) may also be expressed in terms of the parameters h_T and ρ_T by means of the inequalities

$$\{\Pi_T v\} = \hat{\Pi}\hat{v} \quad \text{for all } v \in \operatorname{dom} \Pi_T$$

where β_n denotes the dx-measure of the unit ball in $\mathbb{R}^n$.

Theorem 16.1 thus gives us an estimate of the interpolation error $|v - \Pi_T v|_{m,q,T}$, which is valid for all finite elements that are affine-equivalent to the same finite element $(\hat{T}, \hat{P}, \hat{\Sigma})$. In other words, *the estimate of Theorem* 16.1 *is valid for all finite elements of an affine family*, according to the definition given in Section 11.

We now show that it is possible to dispose of the parameter ρ_T in the estimate (16.4) provided we restrict ourselves to finite elements that do not become "flat". More specifically, we shall say that a family of finite elements (T, P_T, Σ_T) is *regular* if the following two conditions are satisfied (for notational brevity, T is viewed here as the parameter that defines the family):

(1) there exists a constant σ such that

$$\frac{h_T}{\rho_T} \leqslant \sigma \quad \text{for all } T; \tag{16.8}$$

(2) the family (h_T) is bounded and 0 is its unique accumulation point; by an *abus criant de notation*, we shall record the latter properties as

$$h_T \to 0. \tag{16.9}$$

When an *affine family* is *regular*, the interpolation error estimate of Theorem 16.1 can be immediately converted into simple estimates of the *norms* $\|v - \Pi_T v\|_{m,q,T}$.

THEOREM 16.2. *Let there be given a regular affine family of finite elements* (T, P_T, Σ_T) *whose reference finite element* $(\hat{T}, \hat{P}, \hat{\Sigma})$ *satisfies assumptions* (16.1), (16.2) *and* (16.3). *Then there exists a constant* $C(\hat{T}, \hat{P}, \hat{\Sigma})$ *such that, for all finite elements* (T, P_T, Σ_T) *in the family and all functions* $v \in W^{k+1,p}(T)$,

$$\|v - \Pi_T v\|_{m,q,T} \leqslant C(\hat{T}, \hat{P}, \hat{\Sigma}) \{\text{meas}(T)\}^{1/q - 1/p} h_T^{k+1-m} |v|_{k+1,p,T}. \tag{16.10}$$

REMARK 16.2. Only the boundedness of the diameters h_T is used here, in conjunction with inequality (16.8).

For a family of *triangular* finite elements, condition (16.8) is equivalent to *Zlámal's condition* (ZLÁMAL [1968]) that there exists a constant θ_0 such that

$$\theta_T \geqslant \theta_0 > 0 \quad \text{for all } T, \tag{16.11}$$

where for each triangle T, θ_T denotes the smallest angle of T.

In order to get a more concrete understanding of the estimates of Theorem 16.2 we have recorded in Fig. 16.1 estimates of the interpolation error $\|v - \Pi_T v\|_{m,T}$ $(p = q = 2)$ for various finite elements that can be imbedded in regular affine families.

Note that, if the function v lacks the "optimal" regularity assumed in Fig. 16.1 ("optimal", in the sense that it yields the highest possible exponent of h_T in the interpolation error estimate), weaker estimates may still hold, *provided the* P_T*-interpolant is still defined*. Assume for instance that we are considering cubic triangles, or cubic tetrahedra, and that the function v is only in the space $H^2(T)$. Since $H^2(T) \hookrightarrow \mathscr{C}^0(T)$ for $n \leqslant 3$, Theorems 16.1 and 16.2 can still be applied with $k = 1$; this yields $\|v - \Pi_T v\|_{m,T} = O(h_T^{2-m})$ for $0 \leqslant m \leqslant 2$. If we were considering Hermite cubic triangles, or Hermite cubic tetrahedra, however, these theorems could *not* be applied, since the space $H^2(T)$ is not contained in the space $\mathscr{C}^1(T)$ (except for $n = 1$).

$\Vert v-\Pi_T v\Vert_{m,T}$	$O(h_T^{2-m})$ $0\leqslant m\leqslant 2$ $(k=1)$	$O(h_T^{3-m})$ $0\leqslant m\leqslant 3$ $(k=2)$		$O(h_T^{4-m})$ $0\leqslant m\leqslant 4$ $(k=3)$		
Regularity of the function v	$H^2(T)$	$H^3(T)$		$H^4(T)$		
Upper bound on the dimension n, to insure that $H^{k+1}(T)\hookrightarrow\mathscr{C}^s(T)$	$n\leqslant 3$ $(s=0)$	$n\leqslant 5$ $(s=0)$	$n\leqslant 3$ $(s=1)$	$n\leqslant 7$ $(s=0)$	$n\leqslant 5$ $(s=1)$	$n\leqslant 3$ $(s=2)$

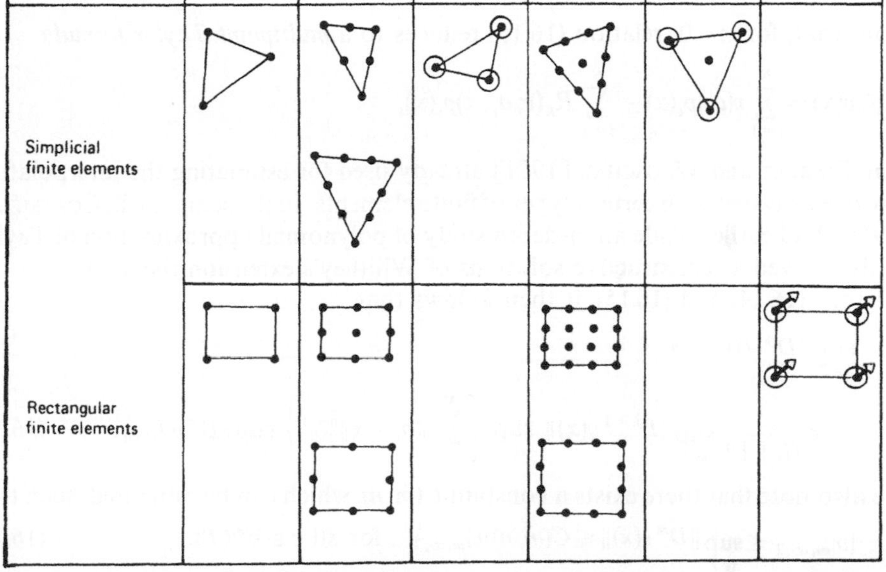

FIG. 16.1. Examples of interpolation error estimates $\Vert v-\Pi_T v\Vert_{m,T}$ for finite elements that are imbedded in regular affine families.

It is worth mentioning at this point that a different, and more constructive, approach to interpolation theory for affine-equivalent finite elements can be also developed, along the following lines: Let (T, P, Σ) be a Lagrange finite element such that the inclusion

$$P_k(T)\subset P\subset\mathscr{C}^k(T) \tag{16.12}$$

holds for some integer $k\geqslant 0$, and let there be given a function $v\in\mathscr{C}^{k+1}(T)$. If a and x are two points in the set T (assumed here to be convex), *Taylor's formula with integral remainder* gives us:

$$v(a)=v(x)+Dv(x)(a-x)+\cdots+\frac{1}{k!}D^k v(x)(a-x)^k+R_k(v;a,x), \tag{16.13}$$

where

$$R_k(v; a, x) = \frac{1}{k!} \int_0^1 (1-t)^k D^{k+1} v(ta + (1-t)x)(a-x)^{k+1}\, dt. \tag{16.14}$$

Let $\Sigma = \{p(a_i): 1 \leqslant i \leqslant N\}$, and let p_i, $1 \leqslant i \leqslant N$, denote the associated basis functions. It can then be shown (cf. CIARLET and RAVIART [1972a]) that the associated P-interpolation operator satisfies

$$D^m(\Pi v - v)(x) = \sum_{i=1}^{N} R_k(v; a_i, x) D^m p_i(x), \quad 0 \leqslant m \leqslant k, \quad \text{for all } x \in T. \tag{16.15}$$

Note that, for $m=0$, relation (16.15) reduces to a *multipoint Taylor formula*

$$v(x) = \sum_{i=1}^{N} v(a_i) p_i(x) - \sum_{i=1}^{N} R_k(v; a_i, x) p_i(x),$$

that CIARLET and WAGSCHAL [1971] already used for estimating the interpolation errors associated with certain types of finite elements. In the same spirit, COATMÉLEC [1966] had earlier made an in-depth study of polynomial approximation of Taylor fields, viewed as constructive solutions of Whitney's extension theorem.

From (16.14) and (16.15), it then follows that

$$\sup_{x \in T} \|D^m(\Pi v - v)(x)\| \leqslant \frac{1}{(k+1)!} \sup_{x \in T} \|D^{k+1} v(x)\| \sup_{x \in T} \left\{ \sum_{i=1}^{N} \|a_i - x\|^{k+1} \right\} \sup_{x \in T} \|D^m p_i(x)\|. \tag{16.16}$$

We also note that there exists a constant $C(m, n)$, which can be estimated, such that

$$|v|_{m,\infty,T} \leqslant \sup_{x \in T} \|D^m v(x)\| \leqslant C(m, n) |v|_{m,\infty,T} \quad \text{for all } v \in \mathscr{C}^m(T), \tag{16.17}$$

and that, if $(\hat{T}, \hat{P}, \hat{\Sigma})$ denotes an affine-equivalent finite element, we have

$$\|D^m p_i(x)\| \leqslant \|D^m \hat{p}_i(\hat{x})\| \, \|B^{-1}\|^m \tag{16.18}$$

(cf. the proof of Theorem 15.1). Hence (16.16)–(16.18) yield a *sharp estimate of the constant* $C(\hat{T}, \hat{P}, \hat{\Sigma})$ *that appears in the inequality*

$$|v - \Pi v|_{m,\infty,T} \leqslant C(\hat{T}, \hat{P}, \hat{\Sigma})(h_T^{k+1}/\rho_T^m) |v|_{k+1,\infty,T}, \tag{16.19}$$

which is nothing but the special case $p = q = \infty$ of inequality (16.4). As shown in CIARLET and RAVIART [1972a], this analysis can be also extended to affine-equivalent *Hermite* finite elements of the type considered in Theorem 16.1.

Using estimates of the norms $|R_k(v; a, \cdot)|_{0,p,T}$, ARCANGÉLI and GOUT [1976] and GOUT [1977] have likewise obtained sharp estimates of the constant $C(\hat{T}, \hat{P}, \hat{\Sigma})$ that appears in the inequality (16.4) when $p = q$. For instance, they have obtained the following estimates: For linear triangles,

$$|v - \Pi_T v|_{m,T} \leqslant 3(h_T^2/\rho_T^m) |v|_{2,T} \quad \text{for } m = 0, 1;$$

for quadratic triangles,

$$|v-\Pi_T v|_{0,T} \leqslant 2h_T^3 |v|_{3,T},$$

$$|v-\Pi_T v|_{1,T} \leqslant 6(h_T^3/\rho_T)|v|_{3,T},$$

$$|v-\Pi_T v|_{2,T} \leqslant 9(h_T^3/\rho_T^2)|v|_{3,T}.$$

17. Interpolation and approximation properties of finite element spaces

We are given a polygonal domain $\Omega \subset \mathbb{R}^n$, a family $(\mathcal{T}_h)$ of triangulations $\bar{\Omega} = \bigcup_{T\in\mathcal{T}_h} T$ and an associated family (X_h) of finite element spaces made up of finite elements $(T, P_T, \Sigma_T)_{T\in\mathcal{T}_h}$. We shall consistently make *three basic assumptions* about the triangulations and the finite elements, according to the following definitions:

(H1) A *family of triangulations* $\mathcal{T}_h$ *is regular* if there exists a constant σ such that

$$\frac{h_T}{\rho_T} \leqslant \sigma \quad \text{for all } T \in \bigcup_h \mathcal{T}_h, \tag{17.1}$$

and if the quantity

$$h = \max_{T\in\mathcal{T}_h} h_T \tag{17.2}$$

approaches zero.

In other words, *the family formed by the finite elements* (T, P_T, Σ_T), $T \in \bigcup_h \mathcal{T}_h$, *is a regular family of finite elements*, in the sense understood in Section 16.

REMARK 17.1. There is of course an *abus de notation* about h, which was first considered as a defining parameter of both families $(\mathcal{T}_h)$ and (X_h), and which is now specifically defined in (17.2).

(H2) *All the finite elements* (T, P_T, Σ_T), $T \in \bigcup_h \mathcal{T}_h$, *are affine-equivalent to a single reference finite element* $(\hat{T}, \hat{P}, \hat{\Sigma})$. In other words, the family (T, P_T, Σ_T), $T \in \bigcup_h \mathcal{T}_h$, is an affine family of finite elements, in the sense understood in Section 11.

(H3) *All the finite elements* (T, P_T, Σ_T), $T \in \bigcup_h \mathcal{T}_h$, *are of class* $\mathscr{C}^0$, in the sense understood in Section 12.

We first prove an approximation property of the associated family of finite element spaces V_h (Theorem 17.1), from which we will later derive an estimate for the error $\|u-u_h\|_{1,\Omega}$ (Theorem 18.1). Whenever no confusion should arise, we use in the sequel the same letter C to denote various constants which are not necessarily the same at their various occurrences. It is an easy exercise, which is left to the reader, to derive more general interpolation error estimates, obtained by replacing (17.3)–(17.4) by the more general assumptions (16.1)–(16.3).

THEOREM 17.1. *In addition to* (H1), (H2) *and* (H3), *assume that there exist integers k and l with $0 \leqslant l \leqslant k$, such that the following inclusions are satisfied*:

$$P_k(\hat{T}) \subset \hat{P} \subset H^l(\hat{T}), \tag{17.3}$$

$$H^{k+1}(\hat{T}) \hookrightarrow \mathscr{C}^s(\hat{T}), \tag{17.4}$$

where s is the maximal order of partial derivatives occurring in the definition of the set $\hat{\Sigma}$.

Then there exists a constant C independent of h such that, for all functions $v \in H^{k+1}(\Omega) \cap V$,

$$\|v - \Pi_h v\|_{m,\Omega} \leqslant C h^{k+1-m} |v|_{k+1,\Omega}, \quad 0 \leqslant m \leqslant \min\{1, l\}, \tag{17.5}$$

$$\left\{ \sum_{T \in \mathscr{T}_h} \|v - \Pi_h v\|^2_{m,T} \right\}^{1/2} \leqslant C h^{k+1-m} |v|_{k+1,\Omega}, \quad 2 \leqslant m \leqslant \min\{k+1, l\}, \tag{17.6}$$

where $\Pi_h v \in V_h$ is the X_h-interpolant of the function v.

PROOF. Applying Theorem 16.2 with $p = q = 2$, we obtain

$$\|v - \Pi_T v\|_{m,T} \leqslant C h_T^{k+1-m} |v|_{k+1,T}, \quad 0 \leqslant m \leqslant \min\{k+1, l\}.$$

Using the relations $(\Pi_h v)|_T = \Pi_T(v|_T)$, $T \in \mathscr{T}_h$ (cf. (12.11)) and the inequalities $h_T \leqslant h$, $T \in \mathscr{T}_h$ (cf. (17.2)), we get

$$\begin{aligned} \left\{ \sum_{T \in \mathscr{T}_h} \|v - \Pi_h v\|^2_{m,T} \right\}^{1/2} &\leqslant C h^{k+1-m} \left\{ \sum_{T \in \mathscr{T}_h} |v|^2_{k+1,T} \right\}^{1/2} \\ &= C h^{k+1-m} |v|_{k+1,\Omega}, \quad 0 \leqslant m \leqslant \min\{k+1, l\}. \end{aligned}$$

Thus inequalities (17.6) are proved, and inequalities (17.5) likewise follow, since

$$\left\{ \sum_{T \in \mathscr{T}_h} \|v - \Pi_h v\|^2_{m,T} \right\}^{1/2} = \|v - \Pi_h v\|_{m,\Omega},$$

for $m = 0$ and for $m = 1$ (when $l \geqslant 1$), and since the inclusions $\hat{P} \subset H^1(\hat{T})$ and $X_h \subset \mathscr{C}^0(\bar{\Omega})$ imply $X_h \subset H^1(\Omega)$ (Theorem 5.1). □

REMARK 17.2. Analogous interpolation error estimates hold if the function v is only in the spaces $\mathscr{C}^s(\bar{\Omega}) \cap \Pi_{T \in \mathscr{T}_h} H^{k+1}(T) \cap V$. It suffices to replace the seminorm $|v|_{k+1,\Omega}$ by the seminorm $\{\Sigma_{T \in \mathscr{T}_h} |v|^2_{k+1,T}\}^{1/2}$ in the right-hand sides of inequalities (17.5) and (17.6).

While the approximation properties of finite element spaces obtained in Theorem 17.1 rely on the interpolation theory developed in Section 16, a different, "*interpolation-free*", *approximation theory* is also possible. Let us describe such an approach, due to CLÉMENT [1975], and further generalized by BERNARDI [1986]: The objective is to construct an operator $r_h : v \in H^l(\Omega) \to r_h v \in X_h$, whose approximation properties are similar to those of the X_h-interpolation operator Π_h, even when the X_h-interpolant $\Pi_h v$ is not defined (for example, when $l = 0$).

To fix ideas, consider a finite element space X_h made up of linear triangles (the analysis can be extended to triangles of degree k). With each vertex b_i, $1 \leqslant i \leqslant M$, of the triangulation, we associate as usual the basis function $w_i \in X_h$ defined by the relations $w_i(b_j) = \delta_{ij}$, $1 \leqslant j \leqslant M$, and we let

$$S_i = \operatorname{supp} w_i. \tag{17.7}$$

Given a function $v \in L^2(\Omega)$, let $P_i v$ denote the $L^2(S_i)$-projection of v on the subspace $P_1(S_i)$, $1 \leqslant i \leqslant M$, i.e., $P_i v$ satisfies

$$\begin{aligned} &P_i v \in P_1(S_i), \\ &\int_{S_i} (v - P_i v) p \,\mathrm{d}x = 0 \quad \text{for all } p \in P_1(S_i), \end{aligned} \tag{17.8}$$

and let the operator r_h be defined by

$$r_h : v \in L^2(\Omega) \to r_h v = \sum_{i=1}^{M} P_i v(b_i) w_i \in X_h. \tag{17.9}$$

Then the operators $r_h : L^2(\Omega) \to X_h$ associated with a regular family of triangulations satisfy the following approximation properties (as usual, C denotes various constants independent of h): For all functions $v \in L^2(\Omega)$,

$$\lim_{h \to 0} |v - r_h v|_{0,\Omega} = 0, \qquad |v - r_h v|_{0,\Omega} \leqslant C |v|_{0,\Omega}; \tag{17.10}$$

for all functions $v \in H^1(\Omega)$,

$$\lim_{h \to 0} |v - r_h v|_{1,\Omega} = 0, \qquad |v - r_h v|_{m,\Omega} \leqslant C h^{1-m} |v|_{1,\Omega}, \quad m = 0,1; \tag{17.11}$$

for all functions $v \in H^2(\Omega)$,

$$|v - r_h v|_{m,\Omega} \leqslant C h^{2-m} |v|_{2,\Omega}, \quad m = 0, 1, \qquad \left\{ \sum_{T \in \mathcal{T}_h} |v - r_h v|_{2,\Omega}^2 \right\}^{1/2} \leqslant C |v|_{2,\Omega}. \tag{17.12}$$

There are other ways of defining an "interpolation-free" approximation theory for finite element spaces; see HILBERT [1973], PINI [1974], and, in particular, STRANG [1972a], who cleverly adapts to finite element spaces the *regularization by convolution* procedure. See also the interesting "*nonlocal*" *finite element* proposed by SCOTT [1976c].

There has been a considerable interest in *interpolation theory* and *approximation theory* in several variables during the past decades, one reason behind this interest being the need of such theories for studying the convergence of finite element methods. Special mention must be made of the pioneering works of PÓLYA [1952] and SYNGE [1957], who considered what we call here bilinear rectangles and linear triangles, respectively.

The "classical" approach consists in obtaining error estimates in $\mathscr{C}^m$-norms. In this direction, see the contributions of BARNHILL and GREGORY [1976b], BARNHILL and WHITEMAN [1973], BIRKHOFF [1971, 1972], BIRKHOFF, SCHULTZ and VARGA [1968], CARLSON and HALL [1973], CIARLET and RAVIART [1972a], CIARLET and

WAGSCHAL [1971], COATMÉLEC [1966], LEAF and KAPER [1974], NICOLAIDES [1972, 1973], NIELSON [1973], SCHULTZ [1969, 1973], STRANG [1971, 1972a], ŽENÍŠEK [1970, 1972, 1973], ZLÁMAL [1968, 1970].

Although in most cases a special role is played by the canonical Cartesian coordinates, a more powerful coordinate-free approach, using *Fréchet derivatives*, can be developed, as in COATMÉLEC [1966], CIARLET and WAGSCHAL [1971], CIARLET and RAVIART [1972a], where, as noted earlier in Section 16, the interpolation error estimates are obtained as corollaries of *multipoint Taylor formulae*. See also LE MÉHAUTÉ [1981, 1984], APPRATO, ARCANGÉLI and MANZANILLA [1987], for further extensions. Another frequently used tool is the *kernel theorem* of SARD [1963].

Some authors have obtained realistic *estimates of the constants* that appear in the interpolation error estimates. See ARCANGÉLI and GOUT [1976], ATTÉIA [1977], BARNHILL and WHITEMAN [1973], GOUT [1977], MEINGUET [1975, 1977, 1978, 1979, 1981, 1984], MEINGUET and DESCLOUX [1977], SANCHEZ and ARCANGÉLI [1984].

The approach in Sobolev spaces that we have followed here has been given much attention. In this respect, we quote the pioneering contributions of BRAMBLE and HILBERT [1970, 1971] and BRAMBLE and ZLÁMAL [1970]. Other relevant references are AUBIN [1967a, 1967b, 1968a, 1968b], BABUŠKA [1970b, 1972b], BIRKHOFF, SCHULTZ and VARGA [1968], BRAMBLE [1970], CIARLET and RAVIART [1972a], FIX and STRANG [1969], DI GUGLIELMO [1970], HEDSTROM and VARGA [1971], KOUKAL [1973], NITSCHE [1969, 1970], SCHULTZ [1969], VARGA [1971].

Interesting connections between interpolation theory and *spline theory* can be found in ATTÉIA [1975], MANSFIELD [1972b], NIELSON [1973], SABLONNIÈRE [1987], BOATTIN [1988] and, especially, DUCHON [1976, 1977].

The dependence of the interpolation error estimates upon the *geometry* of the element (through the parameters h_T and ρ_T) generalized Zlámal's condition, as given in ZLÁMAL [1968, 1970], and the "uniformity condition" of STRANG [1972a]. JAMET [1976a] has also shown that, for some finite elements at least, the regularity condition given in (16.8) can be replaced by a less stringent one; the same condition has been simultaneously and independently found in a special case by BABUŠKA and AZIZ [1976]. If triangles are considered, this condition states that no angle of the triangle should approach π in the limit while by the present analysis no angle should approach 0 in the limit. Incidentally, this was already observed by SYNGE [1957].

The classical Jackson–Bernstein–Zygmund theory for *trigonometric approximation* has been extended to finite element approximations by WIDLUND [1977], who has shown that approximation error estimates of a certain order with respect to h imply conversely that the approximated function has some specific smoothness.

Another constructive approach to approximation theory, based on appropriately averaged Taylor expansions, has been used by DUPONT and SCOTT [1978, 1980], who notably obtained *error estimates in fractional order Sobolev spaces*; see also SANCHEZ and ARCANGÉLI [1984].

The *Banach space interpolation method* of LIONS and PEETRE [1964] has been used by BRAMBLE and SCOTT [1978] and SCOTT [1979], who have shown that approximation estimates in the different norms of a same "Banach scale" may be all derived from the estimates expressed in the "highest-order" norm.

To conclude this section, we record another basic assumption on triangulations, which was not needed thus far, but which is crucial for establishing some refined error estimates (see in particular Theorem 19.3 and Section 22):

(H4) A family of triangulations $\mathcal{T}_h$ satisfies an *inverse assumption* if there exists a constant ν such that

$$\frac{h}{h_T} \leqslant \nu \quad \text{for all } T \in \bigcup_h \mathcal{T}_h. \tag{17.13}$$

A regular family of triangulations that satisfies an inverse assumption (i.e., (H1) and (H4) are satisfied) is called *quasi-uniform*. For such families, we are able to estimate the equivalence constants between Sobolev seminorms (we recall that σ is the constant that appears in the regularity assumption; cf. (17.1)).

THEOREM 17.2. *Let there be given a family of triangulations that satisfies hypotheses* (H1), (H2) *and* (H4), *and let there be given two pairs* (l, r) *and* (m, q) *with* $l, m \geqslant 0$ *and* $(r, q) \in [1, \infty]$ *such that*

$$l \leqslant m \quad \text{and} \quad \hat{P} \subset W^{l,r}(\hat{T}) \cap W^{m,q}(\hat{T}). \tag{17.14}$$

Then there exists a constant $C = C(\sigma, \nu, l, r, m, q)$ *such that*

$$\Big\{\sum_{T\in\mathcal{T}_h} |v_h|^q_{m,q,T}\Big\}^{1/q} \leqslant \frac{C}{(h^n)^{\max\{0,1/r-1/q\}} h^{m-l}} \Big\{\sum_{T\in\mathcal{T}_h} |v_h|^r_{l,r,T}\Big\}^{1/r} \quad \text{for all } v_h \in X_h, \tag{17.15}$$

if $p, q < \infty$, *with*

$$\max_{T\in\mathcal{T}_h} |v_h|_{m,\infty,T} \quad \text{in lieu of} \quad \Big\{\sum_{T\in\mathcal{T}_h} |v_h|^q_{m,q,T}\Big\}^{1/q}, \quad \text{if } q = \infty,$$

$$\max_{T\in\mathcal{T}_h} |v_h|_{l,\infty,T} \quad \text{in lieu of} \quad \Big\{\sum_{T\in\mathcal{T}_h} |v_h|^r_{l,r,T}\Big\}^{1/r}, \quad \text{if } r = \infty.$$

PROOF. Given a function $v_h \in X_h$ and a finite element $T \in \mathcal{T}_h$, we have by assumption (H2) and by Theorem 15.1,

$$\begin{aligned} |\hat{v}_T|_{l,r,\hat{T}} &\leqslant C \|B_T\|^l |\det B_T|^{-1/r} |v_h|_{l,r,T}, \\ |v_h|_{m,q,T} &\leqslant C \|B_T^{-1}\|^m |\det B_T|^{1/q} |\hat{v}_T|_{m,q,\hat{T}}, \end{aligned} \tag{17.16}$$

where the function $\hat{v}_T$ is in the standard correspondence with the function $v_h|_T$. Define the space

$$\hat{N} = \{\hat{p} \in \hat{P} : |\hat{p}|_{l,r,\hat{T}} = 0\} = \begin{cases} \{0\}, & \text{if } l = 0, \\ \hat{P} \cap P_{l-1}(\hat{T}), & \text{if } l \geqslant 1. \end{cases}$$

Since $l \leqslant m$ by assumption, the implication

$$\hat{p} \in \hat{N} \Rightarrow |\hat{p}|_{m,q,\hat{T}} = 0$$

holds, and therefore the mapping

$$\hat{p} \in \hat{P}/\hat{N} \to \|\dot{\hat{p}}\|_{m,q,\hat{T}} = \inf_{\hat{s} \in \hat{N}} |\hat{p} - \hat{s}|_{m,q,\hat{T}}$$

is a norm over the quotient space $\hat{P}/\hat{N}$. Since this quotient space is finite-dimensional, this norm is equivalent to the quotient norm $\|\cdot\|_{l,r,\hat{T}}$ and therefore there exists a constant $\hat{C} = \hat{C}(l, r, m, q)$ such that

$$|\hat{p}|_{m,q,\hat{T}} = \|\dot{\hat{p}}\|_{m,q,\hat{T}} \leqslant \hat{C} \|\dot{\hat{p}}\|_{l,r,\hat{T}} = \hat{C} |\hat{p}|_{l,r,\hat{T}} \quad \text{for all } \hat{p} \in \hat{P}. \tag{17.17}$$

Taking into account assumptions (H1) and (H4), we obtain from inequalities (17.16), (17.17), and from Theorem 15.2,

$$|v_h|_{m,q,T} \leqslant C(\sigma, \nu) \frac{(h^n)^{1/q - 1/r}}{h^{m-l}} |v_h|_{l,r,T}. \tag{17.18}$$

Assume first that $q = \infty$, so that there exists $T_0 \in \mathscr{T}_h$ such that

$$\max_{T \in \mathscr{T}_h} |v_h|_{m,\infty,T} = |v_h|_{m,\infty,T_0} \leqslant C \frac{(h^n)^{-1/r}}{h^{m-l}} |v_h|_{l,r,T_0} \leqslant C \frac{(h^n)^{-1/r}}{h^{m-l}} |v_h|_{l,r,\Omega}$$

by (17.18). Assume next that $q < \infty$. We deduce from inequality (17.18) that

$$\left\{ \sum_{T \in \mathscr{T}_h} |v_h|^q_{m,q,T} \right\}^{1/q} \leqslant C \frac{(h^n)^{1/q - 1/r}}{h^{m-l}} \left\{ \sum_{T \in \mathscr{T}_h} |v_h|^q_{l,r,T} \right\}^{1/q}.$$

Then we distinguish three cases:

(i) $r \leqslant q$, so that

$$\left\{ \sum_{T \in \mathscr{T}_h} |v_h|^q_{l,r,T} \right\}^{1/q} \leqslant \left\{ \sum_{T \in \mathscr{T}_h} |v_h|^r_{l,r,T} \right\}^{1/r}$$

by Jensen's inequality;

(ii) $q < r < \infty$, so that

$$\left\{ \sum_{T \in \mathscr{T}_h} |v_h|^q_{l,r,T} \right\}^{1/q} \leqslant \mathscr{K}_h^{1/q - 1/r} \left\{ \sum_{T \in \mathscr{T}_h} |v_h|^r_{l,r,T} \right\}^{1/r}$$

with

$$\mathscr{K}_h = \operatorname{card} \mathscr{T}_h \leqslant C(\sigma, \nu)/h^n,$$

by Hölder's inequality;

(iii) $r = \infty$, in which case

$$\left\{ \sum_{T \in \mathscr{T}_h} |v_h|^q_{l,\infty,T} \right\}^{1/q} \leqslant \mathscr{K}_h^{1/q} \max_{T \in \mathscr{T}_h} |v_h|_{l,\infty,T},$$

and inequality (17.15) is proved in all cases. □

Inequalities of the form (17.15) may be immediately converted into inequalities involving the seminorms $|\cdot|_{m,q,\Omega}$ or $|\cdot|_{l,r,\Omega}$ if it so happens that the inclusions $X_h \subset W^{m,q}(\Omega)$ or $X_h \subset W^{l,r}(\Omega)$ hold. For example, let us assume that assumption (H3) is also satisfied and that the inclusion $\hat{P} \subset H^1(\hat{T})$ holds, so that the inclusion $X_h \subset \mathscr{C}^0(\bar{\Omega}) \cap H^1(\Omega)$ holds. Then we have

$$|v_h|_{0,\infty,\Omega} \leqslant \frac{C}{h^{n/2}} |v_h|_{0,\Omega} \quad \text{for all } v_h \in X_h, \tag{17.19}$$

$$|v_h|_{1,\Omega} \leqslant \frac{C}{h} |v_h|_{0,\Omega} \quad \text{for all } v_h \in X_h, \text{ etc.} \tag{17.20}$$

If hypothesis (H3) is satisfied and if the inclusion $\hat{P} \subset W^{1,\infty}(\hat{T})$ holds, then we get similarly

$$|v_h|_{1,\infty,\Omega} \leqslant \frac{C}{h} |v_h|_{0,\infty,\Omega} \quad \text{for all } v_h \in X_h, \text{ etc.} \tag{17.21}$$

Clearly, similar inequalities between *norms* can be also derived from these inequalities. For instance, we deduce from (17.20) that

$$\|v_h\|_{1,\Omega} \leqslant \frac{C}{h} |v_h|_{0,\Omega} \quad \text{for all } v_h \in X_h. \tag{17.22}$$

Inequalities such as (17.19)–(17.22), and the more general inequality (17.15), bear the generic name of *inverse inequalities* (whence the terminology "inverse assumption" in (H4)). Inverse inequalities can be likewise established which involve other seminorms or norms: For instance, let there be given a family of triangulations that satisfies (H2) and (H4), and assume that $\hat{P} \subset \mathscr{C}^0(\hat{T})$. Then for each $p \in [1, \infty]$, there exists a constant $C(p)$ such that

$$\|v_h\|_{L^p(\Gamma)} \leqslant \frac{C(p)}{h^{1/p}} |v_h|_{0,p,\Omega}, \quad \text{etc.} \tag{17.23}$$

Inverse inequalities have appeared at many places; see notably DESCLOUX [1973].

REMARK 17.3. Usually, inverse inequalities have no "continuous" counterpart (consider e.g. inequalities (17.19)–(17.23)), as reflected by a factor in the right-hand sides that tends to infinity as h tends to zero.

18. Estimate of the error $\|u-u_h\|_{1,\Omega}$ when the solution u is smooth and sufficient conditions for $\lim_{h\to 0}\|u-u_h\|_{1,\Omega}=0$ when $u \in H^1(\Omega)$

Let there be given a second-order boundary value problem, posed over a space V that satisfies the usual inclusions $H_0^1(\Omega) \subset V = \bar{V} \subset H^1(\Omega)$. A basic assumption throughout the remainder of this chapter will be that the set $\bar{\Omega}$ is *polygonal*, essentially because such an assumption allows us to *exactly* cover the set $\bar{\Omega}$ with polygonal finite elements. Then with any such finite element and with any triangulation $\mathscr{T}_h$ *of* $\bar{\Omega}$, we associate a finite element space X_h, and we define an appropriate subspace V_h of X_h which takes into account the boundary conditions contained in the definition of the space V (cf. Section 12). Since V_h is then included in the space V, we are using a *conforming finite element method.*

Another basic assumption will be that *the space V_h contains the X_h-interpolant of the solution u* of the boundary value problem; in this respect, see Section 12, where the special cases $V_h = X_h \subset V = H^1(\Omega)$ and $V_h = X_{0h} \subset V = H_0^1(\Omega)$ have been thoroughly discussed. If the space V is of the form

$$V = \{v \in H^1(\Omega);\ v = 0 \text{ on } \Gamma_0\},$$

and if the subset Γ_0 of Γ can be exactly covered by a union of faces of finite elements, it is easily seen that the space

$$V_h = \{v_h \in X_h;\ v_h = 0 \text{ on } \Gamma_0\}$$

still contains the X_h-interpolant of u. By contrast, this is not true if we are approximating a nonhomogeneous Dirichlet problem; such a problem requires specific approximations, which will be briefly described in Section 24.

We now prove our first convergence result; we recall that assumptions (H1), (H2), (H3) have been defined in Section 17.

THEOREM 18.1. *In addition to* (H1), (H2) *and* (H3), *assume that there exists an integer $k \geqslant 1$ such that the following inclusions are satisfied:*

$$P_k(\hat{T}) \subset \hat{P} \subset H^1(\hat{T}), \tag{18.1}$$

$$H^{k+1}(\hat{T}) \hookrightarrow \mathscr{C}^s(\hat{T}), \tag{18.2}$$

where s is the maximal order of partial derivatives occurring in the definition of the set $\hat{\Sigma}$.

Then if the solution $u \in V$ of the variational problem is also in the space $H^{k+1}(\Omega)$, there exists a constant C independent of h such that

$$\|u - u_h\|_{1,\Omega} \leqslant Ch^k |u|_{k+1,\Omega}, \tag{18.3}$$

where $u_h \in V_h$ is the discrete solution.

PROOF. It suffices to use inequality (17.5) with $v = u$ and $m = 1$, in conjunction with Céa's lemma (Theorem 13.1); this yields

$$\|u - u_h\|_{1,\Omega} \leqslant C \inf_{v_h \in V_h} \|u - v_h\|_{1,\Omega} \leqslant C \|u - \Pi_h u\|_{1,\Omega}. \qquad \square$$

The error estimate $\|u - u_h\|_{1,\Omega} \leqslant Ch^k|u|_{k+1,\Omega}$ of Theorem 18.1 has been established under the assumptions that the solution u is *sufficiently smooth* ($u \in H^{k+1}(\Omega)$ for some $k \geqslant 1$) and that *the X_h-interpolant $\Pi_h u$ exists* (cf. the inclusion $H^{k+1}(\hat{T}) \hookrightarrow \mathscr{C}^s(\hat{T})$, which is satisfied if $k > \frac{1}{2}n - 1 + s$). If these hypotheses are not valid, it is still possible to prove the convergence of the method if the solution u "only" belongs to the space $H^1(\Omega)$ and if the "minimal" inclusions (18.4) below hold. One should notice that neither the inclusions (18.4) nor the assumption $s \leqslant 1$ in the next theorem are restrictive *in practice* for second-order problems.

THEOREM 18.2. *In addition to* (H1), (H2) *and* (H3), *assume that the inclusions*

$$P_1(\hat{T}) \subset \hat{P} \subset H^1(\hat{T}) \tag{18.4}$$

are satisfied, and that there are no directional derivatives of order $\geqslant 2$ *in the set* $\hat{\Sigma}$. *Then*

$$\lim_{h\to 0} \|u-u_h\|_{1,\Omega}=0. \tag{18.5}$$

PROOF. Define the space

$$\mathscr{V} = W^{2,\infty}(\Omega) \cap V. \tag{18.6}$$

Since the inclusions (18.1) and

$$W^{2,\infty}(\hat{K}) \hookrightarrow \mathscr{C}^s(\hat{K}), \quad s=0 \text{ or } 1, \qquad W^{2,\infty}(\hat{K}) \hookrightarrow H^1(\hat{K}),$$

hold, we may apply Theorem 16.2 with $k=1$, $p=\infty$, $m=1$, $q=2$: There exists a constant C such that

$$\|v-\Pi_T v\|_{1,T} \leqslant C\{\text{meas}(T)\}^{1/2} h_T |v|_{2,\infty,T} \quad \text{for all } v \in \mathscr{V}.$$

Therefore,

$$\|v-\Pi_h v\|_{1,\Omega} = \left\{\sum_{T\in\mathscr{T}_h} \|v-\Pi_T v\|_{1,T}^2\right\}^{1/2} \leqslant Ch\{\text{meas}(\Omega)\}^{1/2} |v|_{2,\infty,\Omega},$$

and thus we have proved that

$$\lim_{h\to 0} \|v-\Pi_h v\|_{1,\Omega}=0. \tag{18.7}$$

For all h and all $v \in \mathscr{V}$, we can write

$$\|u-\Pi_h v\|_{1,\Omega} \leqslant \|u-v\|_{1,\Omega} + \|v-\Pi_h v\|_{1,\Omega}. \tag{18.8}$$

Given the solution $u \in V$ and any number $\varepsilon > 0$, there exists a function $v_\varepsilon \in \mathscr{V}$ that satisfies the inequality $\|u-v_\varepsilon\|_{1,\Omega} \leqslant \frac{1}{2}\varepsilon$, since the space $\mathscr{V}$ is dense in the space V. Then by (18.7) there exists an $h_0(\varepsilon)$ such that $\|v_\varepsilon - \Pi_h v_\varepsilon\|_{1,\Omega} \leqslant \frac{1}{2}\varepsilon$ for all $h \leqslant h_0(\varepsilon)$. Hence $\|u-\Pi_h v_\varepsilon\|_{1,\Omega} \leqslant \varepsilon$ by (18.8), and thus

$$\lim_{h\to 0} \inf_{v_h \in V_h} \|u-v_h\| = 0.$$

Then the conclusion again follows from Céa's lemma (Theorem 13.1). □

REMARK 18.1. An inspection of the above proof shows that the choice (18.6) for the space $\mathscr{V}$ is the result of the following requirements: On the one hand $\mathscr{V}$ must to be dense in the space V; on the other hand the value $k=1$ is needed in order to apply Theorem 16.2 so as to obtain property (18.7) with the assumption $P_1(\hat{T}) \subset \hat{P}$. Therefore the space $\mathscr{V}$ must contain derivatives of order 2 (this condition limits in turn the admissible values of s to 0 and 1) and consequently one is naturally led to the choice (18.6). In fact, any space of the form $\mathscr{V} = W^{2,p}(\Omega) \cap V$ with p sufficiently large, would have also been acceptable, as one may verify.

REMARK 18.2. While we have used here a "density argument", based on the interpolation theory of Section 16, an "interpolation-free" approach is also possible, since the estimates of (17.11), used in conjunction with Céa's lemma, provide another proof of Theorem 18.2.

REMARK 18.3. (i) Using a priori estimates in various norms on the solution (cf. e.g. NEČAS [1967] and KONDRAT'EV [1967]), it is possible to get error estimates that depend solely on the *data* of the problem. See BRAMBLE and ZLÁMAL [1970], NITSCHE [1970], OGANESJAN and RUKHOVETS [1969].

(ii) In the case of the equation $-\Delta u=f$ over a rectangle, BARNHILL and GREGORY [1976a] obtain theoretical values for the *constants* that appear in the error estimate, and these values are realistic, as shown in BARNHILL, BROWN, MCQUEEN and MITCHELL [1976].

(iii) Error estimates expressed in *mesh-dependent norms* have been obtained by BABUŠKA and OSBORN [1980].

(iv) Using the theory of *n-widths*, BABUŠKA and AZIZ [1972, Section 6.4] have discussed whether the estimate of the error $\|u-u_h\|_{1,\Omega}$ is indeed the best possible.

19. Estimate of the error $|u-u_h|_{0,\Omega}$ when u is smooth and the Aubin–Nitsche lemma, first estimate of the error $|u-u_h|_{0,\infty,\Omega}$

In Theorem 18.1 we have given assumptions that insure that $\|u-u_h\|_{1,\Omega}=\mathrm{O}(h^k)$; hence the error $|u-u_h|_{0,\Omega}$ is at least of the same order. We now show that, under mild additional assumptions, one has in fact $|u-u_h|_{0,\Omega}=\mathrm{O}(h^{k+1})$. To this end, we begin by defining an abstract setting adapted to this type of improved error estimates:

Let there be given a normed vector space V, with norm $\|\cdot\|$, and a Hilbert space with norm $|\cdot|$ and inner product $(\cdot,\cdot)$, such that

$$\bar{V}=H \quad \text{and} \quad V\hookrightarrow H$$

(in the applications we have in mind, we shall have typically $H_0^1(\Omega)\subset V\subset H^1(\Omega)$ and $H=L^2(\Omega)$). Our first observation is that, *if the space H is identified with its dual space, the space H may then be identified with a subspace of the dual space V' of V.* To see this, let $f\in H$ be given. Since $V\subset H$ with a continuous injection $\imath$, we have

$$|(f,v)|\leqslant|f|\,|v|\leqslant\|\imath\|\,|f|\,\|v\| \quad \text{for all } v\in V,$$

and thus the mapping $v\in V\to(f,v)$ defines an element $\tilde{f}\in V'$. The mapping $f\in H\to\tilde{f}\in V'$ is an injection for if $(f,v)=0$ for all $v\in V$, then $(f,v)=0$ for all $v\in H$ since V is dense in H, and thus $f=0$. We may therefore identify f and $\tilde{f}$, i.e., we shall write

$$(f,v)=f(v) \quad \text{for all } f\in H,\quad v\in V. \tag{19.1}$$

REMARK 19.1. Some care should be exercised when making such identifications, in this respect see the discussion given by BREZIS [1983, p. 81].

We next prove an *abstract error estimate.* Making the same assumptions as in the Lax–Milgram lemma (Theorem 1.3), we let as usual $u \in V$ and $u_h \in V_h$ denote the solutions of the variational problems

$$a(u, v) = l(v) \qquad \text{for all } v \in V, \tag{19.2}$$

$$a(u_h, v_h) = l(v_h) \quad \text{for all } v_h \in V_h. \tag{19.3}$$

respectively. We recall that M denotes an upper bound for the norm of the bilinear form $a(\cdot,\cdot)$ (cf. (1.19)).

THEOREM 19.1 (Aubin–Nitsche lemma). *Let the assumptions and notations be as in the Lax–Milgram lemma. In addition let H be a Hilbert space, with norm $|\cdot|$ and inner product $(\cdot,\cdot)$, such that*

$$\bar{V} = H \quad \text{and} \quad V \hookrightarrow H. \tag{19.4}$$

Then one has

$$|u-u_h| \leqslant M\|u-u_h\| \left(\sup_{g\in H} \left\{ \frac{1}{|g|} \inf_{\varphi_h\in V_h} \|\varphi_g - \varphi_h\| \right\} \right), \tag{19.5}$$

where, for any $g \in H$, $\varphi_g \in V$ is the unique solution of the variational problem:

$$a(v, \varphi_g) = (g, v) \quad \text{for all } v \in V. \tag{19.6}$$

PROOF. To estimate $|u-u_h|$, we shall use the characterization

$$|u-u_h| = \sup_{g\in H} |(g, u-u_h)|/|g|. \tag{19.7}$$

Using the identification (19.1) we can solve problem (19.6) for all $g \in H$ (the proof is exactly the same as that of the Lax–Milgram lemma). Since $(u-u_h)$ is an element of the space V, we have in particular

$$a(u-u_h, \varphi_g) = (g, u-u_h)$$

on the one hand, and we have

$$a(u-u_h, \varphi_h) = 0 \quad \text{for all } \varphi_h \in V_h$$

on the other. Combining these relations, we obtain

$$(g, u-u_h) = a(u-u_h, \varphi_g - \varphi_h) \quad \text{for all } \varphi_h \in V_h,$$

and therefore,

$$|(g, u-u_h)| \leqslant M\|u-u_h\| \inf_{\varphi_h\in V_h} \|\varphi_g - \varphi_h\|. \tag{19.8}$$

Inequality (19.5) is thus a consequence of the characterization (19.7) and of inequality (19.8) □

An inspection of the above proof shows that φ_g has to be the solution of (19.6), i.e., of a problem where the arguments are *interchanged* in the bilinear form. Problem

(19.6) is a special case of the following variational problem: Given any element $g \in V'$, find an element $\varphi \in V$ such that

$$a(v, \varphi) = g(v) \quad \text{for all } v \in V. \tag{19.9}$$

Such a problem is called the *adjoint problem* of problem (19.2). When solving the variational problem (19.2) amounts to solving, at least formally, a second-order boundary value problem (i.e., when $V \subset H^1(\Omega)$), it is easily verified that solving its adjoint problem (19.9) also amounts to solving a second-order boundary value problem, which is in general different, however (unless of course the bilinear form is symmetric).

As we shall see, the abstract error estimate of Theorem 19.1 will yield an improvement of the order of the estimate of the error $|u-u_h|_{0,\Omega}$, but only for a restricted class of second-order problems, which we now define: A second-order boundary value problem whose variational formulation is (19.2), or (19.9), is said to be *regular* if, for *any* $f \in L^2(\Omega)$, or for *any* $g \in L^2(\Omega)$, the corresponding solution u_f, or u_g, is in the space $H^2(\Omega) \cap V$.

It then follows from the closed graph theorem (for a proof, see e.g. BREZIS [1983, p. 20]) that there exists a constant C such that

$$\|u_f\|_{2,\Omega} \leqslant C|f|_{0,\Omega} \quad \text{for all } f \in L^2(\Omega), \tag{19.10}$$

$$\|\varphi_g\|_{2,\Omega} \leqslant C|g|_{0,\Omega} \quad \text{for all } g \in L^2(\Omega). \tag{19.11}$$

REMARK 19.2. Consider problem (19.2) for instance. Without the assumption of regularity, we simply know that (use Remark 1.2 and the identification (19.1)):

$$\alpha\|u_f\|_{1,\Omega} \leqslant \|f\|^* = \sup_{v\in V} \frac{|f(v)|}{\|v\|_{1,\Omega}} = \sup_{v\in V} \frac{|\int fv \,\mathrm{d}x|}{\|v\|_{1,\Omega}} \leqslant |f|_{0,\Omega}$$

for all $f \in L^2(\Omega)$.

For problems with smooth data, *the assumption of regularity is essentially a restriction on the possibility of "mixing" different boundary conditions*. For instance, while the Dirichlet ($\Gamma_0 = \Gamma$) and Neumann ($\Gamma_0 = \emptyset$) problems associated with the data of (3.19) are regular if Ω is convex and if the functions a_{ij} and a are sufficiently smooth, the mixed problem associated with these data is not regular in general.

We are now in a position to estimate the error $|u-u_h|_{0,\Omega}$.

THEOREM 19.2. *In addition to* (H1), (H2) *and* (H3) (*cf. Section* 17), *assume that* $s=0$, *that the dimension* n *is* $\leqslant 3$, *and that there exists an integer* $k \geqslant 1$ *such that the solution* u *is in the space* $H^{k+1}(\Omega)$ *and such that the inclusions*

$$P_k(\hat{T}) \subset \hat{P} \subset H^1(\hat{T}) \tag{19.12}$$

hold.

Then if the adjoint problem is regular, there exists a constant C *independent of h such that*

$$|u-u_h|_{0,\Omega}\leqslant Ch^{k+1}|u|_{k+1,\Omega}. \tag{19.13}$$

PROOF. Since $n\leqslant 3$, *the inclusion* $H^2(\hat{T})\hookrightarrow \mathscr{C}^0(\hat{T})$ holds (if $s=1$, the inclusion $H^2(\hat{T})\hookrightarrow \mathscr{C}^1(\hat{T})$ holds only if $n=1$; this is why we have restricted ourselves to the case $s=0$). Applying Theorem 17.1 and inequality (19.11) we obtain, for each $g\in H=L^2(\Omega)$,

$$\inf_{\varphi_h\in V_h}\|\varphi_g-\varphi_h\|_{1,\Omega}\leqslant\|\varphi_g-\Pi_h\varphi_g\|_{1,\Omega}\leqslant Ch|\varphi_g|_{2,\Omega}\leqslant Ch|g|_{0,\Omega}.$$

Combining this inequality with inequality (19.5) then gives

$$|u-u_h|_{0,\Omega}\leqslant Ch\|u-u_h\|_{1,\Omega},$$

and it remains to use inequality (18.3) of Theorem 18.1. □

The results of Theorems 19.1 and 19.2 were established independently by AUBIN [1967b] and NITSCHE [1968], and also by OGANESJAN and RUKHOVETS [1969]. See also KIKUCHI [1975c] for a generalization.

The asymptotic error estimates obtained in Theorems 18.1 and 19.2 are optimal, in the sense that the orders of convergence are the same as if the discrete solution u_h were replaced by the X_h-interpolant of the function u: Compare (18.3) and (19.13) with (17.5) for $m=1$ and $m=0$.

Consequently, Fig. 16.1 is also useful for getting a practical appraisal of the upper bounds of Theorems 18.1 and 19.2. For instance, one gets $\|u-u_h\|_{m,\Omega}=O(h^{2-m})$, $m=0,1$, with linear n-simplices or with bilinear rectangles, $\|u-u_h\|_{m,\Omega}=O(h^{3-m})$, $m=0,1$, with quadratic, or reduced cubic, n-simplices or with biquadratic, or reduced biquadratic, rectangles, etc. Nevertheless, the higher the order of convergence, the higher the assumed regularity of the solution, and this observation limits considerably the practical value of such estimates. For example, let us assume that we are using cubic n-simplices but that the solution is "only" in the space $H^2(\Omega)$: Then the application of Theorems 18.1 and 19.2 with $k=1$ only implies that $\|u-u_h\|_{m,\Omega}=O(h^{2-m})$, $m=0,1$.

If the solution is not very smooth, special techniques may be applied that yield "local" error estimates which may nevertheless be of a high order. This fascinating aspect of the error analysis for finite element methods is treated in depth in the next article by Wahlbin.

We finally show that, if the family of triangulations satisfies an inverse assumption, the estimate of the error $|u-u_h|_{0,\Omega}$ established in Theorem 19.2 may be in turn used to derive a *first estimate of the error* $|u-u_h|_{0,\infty,\Omega}$ (other estimates of the same error will be given in Sections 21 and 22).

THEOREM 19.3. *In addition to* (H1), (H2), (H3), *and* (H4) (*cf. Section* 17), *assume that* $s=0$, *that the dimension n is* $\leqslant 3$, *and that there exists an integer* $k\geqslant 1$ *such that the* *solution u is in the space $H^{k+1}(\Omega)$ and such that the inclusions*

$$P_k(\hat{T}) \subset \hat{P} \subset H^1(\hat{T}) \tag{19.14}$$

hold.

Then if the adjoint problem is regular, there exists a constant C independent of h such that

$$|u-u_h|_{0,\infty,\Omega} \leqslant Ch^{k+1-n/2}|u|_{k+1,\Omega}. \tag{19.15}$$

PROOF. Write

$$|u-u_h|_{0,\infty,\Omega} \leqslant |u_h-\Pi_h u|_{0,\infty,\Omega} + |u-\Pi_h u|_{0,\infty,\Omega}. \tag{19.16}$$

We first infer from Theorem 16.2 that (by assumption (H3), we also have $\hat{P} \subset L^\infty(\hat{T})$):

$$|u-\Pi_h u|_{0,\infty,T} \leqslant Ch_T^{k+1-n/2}|u|_{k+1,T} \quad \text{for all } T \in \bigcup_h \mathcal{T}_h;$$

hence

$$|u-\Pi_h u|_{0,\infty,\Omega} \leqslant Ch^{k+1-n/2}|u|_{k+1,\Omega}. \tag{19.17}$$

We then infer from the inverse inequality (17.15) with $l=m=0$, $q=\infty$, $r=2$, from the error estimate (19.13), and from Theorem 16.2, that

$$\begin{aligned} |u_h-\Pi_h u|_{0,\infty,\Omega} &\leqslant Ch^{-n/2}|u_h-\Pi_h u|_{0,\Omega} \\ &\leqslant Ch^{-n/2}\{|u-u_h|_{0,\Omega}+|u-\Pi_h u|_{0,\Omega}\} \\ &\leqslant Ch^{k+1-n/2}|u|_{k+1,\Omega}, \end{aligned} \tag{19.18}$$

and the error estimate (19.15) follows by combining inequalities (19.16), (19.17), and (19.18). □

20. Discrete maximum principle in finite element spaces

Let Ω be a polyhedral domain in $\mathbb{R}^n$, with boundary Γ, and consider the *model problem*

$$-\Delta u + bu = f \quad \text{in } \Omega, \qquad u = u_0 \quad \text{on } \Gamma, \tag{20.1}$$

where b, f and u_0 are given functions, which are assumed to be sufficiently smooth for the time being (specific smoothness assumptions will be given later on). We also assume that

$$b \geqslant 0 \quad \text{in } \Omega. \tag{20.2}$$

If we approximate the solution u of (20.1) by using finite element spaces made up of *linear n-simplices*, it follows from Theorem 19.3 that

$$|u-u_h|_{0,\infty,\Omega} = \begin{cases} O(h) & \text{if } u \in H^2(\Omega), \quad n=2, \\ O(\sqrt{h}) & \text{if } u \in H^2(\Omega), \quad n=3, \end{cases} \tag{20.3}$$

if the adjoint problem is regular and if the associated family of triangulations is quasi-uniform. Our next objective is to show that, for a general class of second-order Dirichlet problems (cf. (20.16)), which include (20.1) as a special case, one can improve the estimates (20.3) if the finite element spaces satisfy a *discrete maximum principle* as introduced by CIARLET [1970a, 1971]; it is the discrete analog of the well-known "continuous" maximum principle (cf. PROTTER and WEINBERGER [1967], SPERB [1981]). More specifically, we shall show in Section 21 that, if this is the case,

$$\lim_{h\to 0} |u-u_h|_{0,\infty,\Omega}=0 \quad \text{if } u\in W^{1,p}(\Omega), \quad n<p, \tag{20.4}$$

$$|u-u_h|_{0,\infty,\Omega}=O(h) \quad \text{if } u\in W^{2,p}(\Omega), \quad n<2p. \tag{20.5}$$

The discrete maximum principle has been extensively used for proving convergence and error estimates for classical finite difference schemes (cf. for example VARGA [1966a], BRAMBLE, HUBBARD and THOMÉE [1969], CIARLET [1970b], and the references therein). In this case the solution of a problem such as that of (20.1) is supposed to be *classical*, in the sense that it has continuous derivatives of sufficiently high order.

If the continuous problem is now approached from a *variational* viewpoint, then a maximum principle for the continuous problem can be derived, as in ARONSZAJN and SMITH [1957], and a parallel approach can also be taken up for the finite element method (cf. CIARLET [1970a]).

We use here the discrete analog of another approach, due to STAMPACCHIA [1965], where the essential step consists in obtaining a priori bounds in $L^\infty(\Omega)$ for the solution of the discrete problem (cf. Theorem 21.4). A similar approach has been used in LEBAUD [1969], and LEBAUD and RAVIART [1969] for finite difference schemes of variational type. We also mention that NITSCHE [1970] has also obtained, by a different approach, an estimate of the form (20.5) when Ω is convex and $n=p=2$.

Therefore, it is somehow a satisfactory situation to realize that *for three well-known approximation schemes for solving second-order Dirichlet problems (classical finite differences, variational finite differences, finite element methods), the existence of a maximum principle for the discrete problem implies the possibility of obtaining uniform convergence of the approximate solutions to the exact solution.*

We now describe the variational problem that we will be approximating: Let Ω denote as before a polyhedral domain in $\mathbb{R}^n$ with $\Gamma=\partial\Omega$. We are given functions

$$\begin{aligned} &a_{kl}\in L^\infty(\Omega), \quad 1\leqslant k,l\leqslant n,\\ &a_k\in L^\infty(\Omega), \quad 1\leqslant k\leqslant n,\\ &b\in L^\infty(\Omega), \end{aligned} \tag{20.6}$$

we let

$$V=H_0^1(\Omega) \tag{20.7}$$

and we consider the bilinear form

$$a(u, v)=\int_\Omega \left\{ \sum_{k,l=1}^{n} a_{kl}\partial_l u \partial_k v + \sum_{k=1}^{n} a_k \partial_k u\, v + buv \right\} dx. \tag{20.8}$$

By (20.6), this bilinear form is continuous over the space $V \times V$. Furthermore, we shall assume that the bilinear form is V-elliptic, i.e., that there exists a constant $\alpha > 0$ such that

$$a(v, v) \geqslant \alpha \|v\|_{1,\Omega}^2 \quad \text{for all } v \in V. \tag{20.9}$$

For instance, this is the case if the functions a_{kl} satisfy the condition (3.21), if $a_k = 0$ and if $b \geqslant 0$.

Let $V' = H^{-1}(\Omega)$ denote the dual space of the space V. Given any function $u_0 \in H^1(\Omega)$ and any element $l \in V'$, there exists, by the Lax–Milgram lemma (Theorem 1.3), one and only one function $u \in H^1(\Omega)$ such that

$$a(u, v) = l(v) \quad \text{for all } v \in V, \tag{20.10}$$

$$u - u_0 \in V. \tag{20.11}$$

Suppose that we are given functions

$$f_k \in L^p(\Omega), \quad 0 \leqslant k \leqslant n, \tag{20.12}$$

$$u_0 \in W^{1,p}(\Omega), \tag{20.13}$$

with

$$2 \leqslant n < p, \tag{20.14}$$

so that the inclusion $W^{1,p}(\Omega) \hookrightarrow \mathscr{C}^0(\bar{\Omega})$ holds, in particular. Then the linear form l defined by

$$l: v \in V \to l(v) = \int_\Omega \left\{ f_0 v + \sum_{k=1}^{n} f_k \frac{\partial v}{\partial x_k} \right\} dx \tag{20.15}$$

belongs to the space V' since condition (20.14) implies that $f_k \in L^2(\Omega)$, $0 \leqslant k \leqslant n$. Moreover, the function u_0 belongs to the space $\mathscr{C}^0(\bar{\Omega})$ so that its restriction to Γ (still denoted u_0) is well defined as a function in the space $\mathscr{C}^0(\Gamma)$.

Using Green's formula, we find that solving the variational problem (20.10)–(20.11) amounts to formally solving the boundary value problem

$$\begin{aligned} &-\sum_{k,l=1}^{n} \partial_k(a_{kl}\partial_l u) + \sum_{k=1}^{n} a_k \partial_k u + bu = f_0 - \sum_{k=1}^{n} \partial_k f_k \quad \text{in } \Omega, \\ &u = u_0 \quad \text{on } \Gamma, \end{aligned} \tag{20.16}$$

which is a *nonhomogeneous Dirichlet problem for the operator*

$$L: u \to Lu = -\sum_{k,l=1}^{n} \partial_k(a_{kl}\partial_l u) + \sum_{k=1}^{n} a_k \partial_k u + bu. \tag{20.17}$$

With all the above assumptions, it can then be proved (cf. STAMPACCHIA [1965, Theorems 4.1 and 4.2]) that the solution u of (20.10)–(20.11) lies in the space $H^1(\Omega)\cap L^\infty(\Omega)$, and that

$$|u|_{0,\infty,\Omega}\leqslant \|u_0\|_{L^\infty(\Gamma)}+C\left\{\sum_{k=0}^{n}|f_k|_{0,p,\Omega}\right\}, \tag{20.18}$$

for some constant C independent of the particular functions $u_0\in W^{1,p}(\Omega)$, $f_k\in L^p(\Omega)$, $0\leqslant k\leqslant n$.

REMARK 20.1. It would be simpler to let $f_k=0$, $1\leqslant k\leqslant n$, but we shall need later on (cf. the proof of Theorem 21.5) the possibility of having nonzero functions f_k.

With a triangulation $\mathcal{T}_h$ of the set $\bar{\Omega}$, made up of n-simplices $T\in\mathcal{T}_h$, we associate as usual the finite element spaces

$$\begin{aligned} X_h&=\{v_h\in\mathscr{C}^0(\bar{\Omega});\ v_{h|T}\in P_1(T) \text{ for all } T\in\mathcal{T}_h\},\\ V_h&=\{v_h\in X_h;\ v_h=0 \text{ on } \Gamma\}. \end{aligned} \tag{20.19}$$

We let b_i, $1\leqslant i\leqslant N$, and b_i, $N+1\leqslant i\leqslant N+M$, denote the vertices of the triangulation $\mathcal{T}_h$ that belong to Ω, and to Γ, respectively, and we let w_i, $1\leqslant i\leqslant N+M$, denote the functions of X_h that satisfy

$$w_i(b_j)=\delta_{ij},\quad 1\leqslant i,j\leqslant N+M, \tag{20.20}$$

i.e., the functions w_i, $1\leqslant i\leqslant N$, or w_i, $1\leqslant i\leqslant N+M$, form a basis of V_h, or of X_h. Given the function u_0 of (20.13), we let

$$u_{0h}=\sum_{i=N+1}^{N+M}u_0(b_i)w_i, \tag{20.21}$$

so that $u_{0h}\in X_h$; then the *discrete problem* consists of finding a function $u_h\in X_h$ such that

$$a(u_h,v_h)=l(v_h)\quad \text{for all } v_h\in V_h, \tag{20.22}$$

$$u_h-u_{0h}\in V_h. \tag{20.23}$$

Clearly, this problem has a unique solution. If we look for a solution u_h of (20.22)–(20.23) in the form

$$u_h=u_{0h}+\sum_{i=1}^{N}\xi_i w_i\quad \text{with } \xi_i=u_h(b_i),\quad 1\leqslant i\leqslant N, \tag{20.24}$$

condition (20.22) is automatically satisfied, and the vector $\xi=(\xi_1,\xi_2,\ldots,\xi_N)$ is the solution of the linear system

$$A\xi=\beta, \tag{20.25}$$

with

$$A=(a_{ij}),\quad a_{ij}=a(w_j,w_i),\qquad 1\leqslant i\leqslant N,\quad 1\leqslant j\leqslant N+M, \tag{20.26}$$

$$\beta=(\beta_1,\beta_2,\ldots,\beta_N),\qquad \beta_i=l(w_i)-\sum_{j=N+1}^{n+M} u_0(b_j)a_{ij}\quad 1\leqslant i\leqslant N. \tag{20.27}$$

We then say that the discrete problem (20.22)–(20.23) is of *nonnegative type* if the matrix A of (20.26) is irreducibly diagonally dominant (VARGA [1962, p. 23]), and the coefficients a_{ij} defined in (20.26) satisfy

$$a_{ij}\leqslant 0\qquad \text{for } i\neq j,\quad 1\leqslant i\leqslant N,\quad 1\leqslant j\leqslant N+M, \tag{20.28}$$

$$\sum_{j=1}^{N+M} a_{ij}\geqslant 0,\quad 1\leqslant i\leqslant N. \tag{20.29}$$

Note that these conditions are reminiscent of similar conditions for discrete operators associated with finite difference schemes for solving (20.16). To be more specific assume that $f_0\leqslant 0$ and $f_k=0$, $1\leqslant k\leqslant n$, in (20.15) so that $l(w_i)=\int_\Omega f_0 w_i\,dx\leqslant 0$, $1\leqslant i\leqslant N$, since each basis function w_i is $\geqslant 0$. Then we have (cf. CIARLET [1970b, Theorem 3]):

$$\max\{\xi_i;\,1\leqslant i\leqslant N\}\leqslant\max\{0,\,\max\{u_0(b_j);\,N+1\leqslant j\leqslant N+M\}\}, \tag{20.30}$$

an inequality that can be equivalently written as

$$\max_{\Omega} u_h\leqslant \max_{\Gamma}\{0,\,\max\{u_{0h}\}\}. \tag{20.31}$$

Following CIARLET [1970a], we shall then say that the discrete problem (20.22)–(20.23) satisfies a *discrete maximum principle*, in the sense that inequality (20.31) holds when $l(w_i)\leqslant 0$, $1\leqslant i\leqslant N$; note that this inequality is nothing but the discrete analog of the maximum principle for the continuous problem (cf. STAMPACCHIA [1965, Theorem 3.6]).

Following CIARLET [1971], we next give an example of a discrete problem that satisfies the discrete maximum principle. Let

$$a(u,v)=\int_\Omega\left\{\sum_{k=1}^{n}\partial_k u\partial_k v+buv\right\}dx, \tag{20.32}$$

with

$$b\in L^\infty(\Omega),\qquad b\geqslant 0\quad \text{in } \Omega, \tag{20.33}$$

so that the operator L of (20.17) reduces to that of (20.1), viz.,

$$Lu=-\Delta u+bu. \tag{20.34}$$

Given an n-simplex T of the triangulation $\mathcal{T}_h$, let a_r, $1\leqslant r\leqslant n+1$, denote its vertices and let λ_r, $1\leqslant r\leqslant n+1$, denote the barycentric coordinates of a point $x\in T$ with respect to the points a_r. With each n-simplex T, we associate the parameter

$$\sigma_T=\max_{r\neq s}\{\cos(D\lambda_r,D\lambda_s)\}, \tag{20.35}$$

with

$$D\lambda_r = (\partial_1 \lambda_r, \partial_2 \lambda_r, \ldots, \partial_n \lambda_r), \quad 1 \leqslant r \leqslant n+1,$$

and

$$\cos(D\lambda_r, D\lambda_s) = \frac{D\lambda_r \cdot D\lambda_s}{\| D\lambda_r \| \, \| D\lambda_s \|}, \quad 1 \leqslant r, s \leqslant n+1,$$

where $\cdot$ and $\| \cdot \|$ respectively denote the Euclidean scalar product and Euclidean norm in $\mathbb{R}^n$. Finally, with each triangulation $\mathcal{T}_h$, we associate the parameter

$$\sigma_h = \max_{T \in \mathcal{T}_h} \sigma_T. \tag{20.36}$$

THEOREM 20.1. *Let the bilinear form be as in* (20.32)–(20.33) *with* $b \neq 0$, *and consider the finite element approximation* (20.22)–(20.33) *of the corresponding variational problem* (20.10)–(20.11). *Given a sequence of triangulations* $\mathcal{T}_h$ *for which*

$$h = \max_{T \in \mathcal{T}_h} h_T$$

approaches 0, *the discrete problem satisfies the discrete maximum principle for h small enough if there exists a constant* $\sigma_0 = \sigma_0(b)$ *independent of h such that*

$$\sigma_h \leqslant \sigma_0 < 0 \quad \textit{for all } h. \tag{20.37}$$

If $b = 0$, *it suffices that*

$$\sigma_h \leqslant 0 \quad \textit{for all } h. \tag{20.38}$$

PROOF. Consider an n-simplex $T \in \mathcal{T}_h$ and a basis function w_i. Then either $T \subset \operatorname{supp} w_i$, which implies that the point b_i is one of the vertices, say a_r, of T, so that $w_i|_T = \lambda_r$; or $T \not\subset \operatorname{supp} w_i$, in which case $w_i|_T = 0$. Therefore, for $j \neq i$, the coefficient $a_{ij} = a(w_j, w_i)$ reduces to a finite sum of integrals of the form

$$\alpha_{rs} = \int_T \left\{ \sum_{k=1}^{n} \partial_k \lambda_r \partial_k \lambda_s + b \lambda_r \lambda_s \right\} \mathrm{d}x,$$

and the indices r and s are always different when they occur in the sum. Since the functions λ_r are affine and satisfy $0 \leqslant \lambda_r \leqslant 1, 1 \leqslant r \leqslant n$, we have

$$\alpha_{rs} \leqslant (D\lambda_r \cdot D\lambda_s + |b|_{0,\infty,\Omega}) \operatorname{meas}(T) \leqslant \left(\frac{\sigma_0}{h^2} + |b|_{0,\infty,\Omega} \right) \operatorname{meas}(T),$$

since $1/h \leqslant \| D\lambda_r \|, 1 \leqslant r \leqslant n+1$. Thus, we have

$$\alpha_{rs} \leqslant 0 \quad \text{for all } r \neq s,$$

for h small enough if condition (20.37) is satisfied, or for all h if $b = 0$; this shows that conditions (20.28) are satisfied.

From the definition of the basis functions, it follows that $\Sigma_{j=1}^{N+M} w_j = 1$ over $\bar{\Omega}$, and therefore

$$\sum_{j=1}^{N+M} a_{ij} = \int_\Omega b w_i \, dx \geqslant 0, \quad 1 \leqslant i \leqslant N,$$

since $b \geqslant 0$ by (20.33). Hence conditions (20.29) are also satisfied. Since the matrix (a_{ij}) is irreducibly diagonally dominant, the proof is complete. □

As observed by CIARLET [1971], there is a simple geometrical intepretation of condition (20.37) when $n=2$:

THEOREM 20.2. *Let the assumptions be as in Theorem 20.1. When $n=2$, the discrete problem satisfies the discrete maximum principle for h small enough if there exists $\varepsilon > 0$ such that for all h, all the angles of the triangles of $\mathcal{T}_h$ are $\leqslant \frac{1}{2}\pi - \varepsilon$;* if $b=0$, *it suffices that all the angles of the triangles of $\mathcal{T}_h$ be $\leqslant \frac{1}{2}\pi$.*

Further information on the discrete maximum principle in finite element spaces are found in LORENZ [1977], HOHN and MITTELMANN [1981], RUAS SANTOS [1982].

21. Estimates of the error $|u-u_h|_{0,\infty,\Omega}$ when $u \in W^{1,p}(\Omega)$, $n<p$, or when $u \in W^{2,p}(\Omega)$, $n<2p$, when the discrete maximum principle holds

We follow here CIARLET and RAVIART [1973]. To begin with, we prove various technical results (Theorems 21.1, 21.2, 21.3).

THEOREM 21.1 *Let T be an n-simplex in $\mathbb{R}^n$, with vertices a_r, $1 \leqslant r \leqslant n+1$, and let a function $v \in P_1(T)$ be $\geqslant 0$ on T. Then for any p with $p \geqslant 1$, there exists a constant $C_1 > 0$, independent of T and v, such that*

$$C_1 \operatorname{meas}(T) \sum_{r=1}^{n+1} \{v(a_r)\}^p \leqslant |v|_{0,p,T}^p. \tag{21.1}$$

PROOF. Let $\lambda_r(x)$ denote the barycentric coordinates of a point $x \in T$ with respect to the points a_r, $1 \leqslant r \leqslant n+1$; observe that $\lambda_r \geqslant 0$ over T. Since the function $v = \sum_{r=1}^{n+1} v(a_r)\lambda_r$ is $\geqslant 0$ on T, we have, by Jensen's inequality,

$$|v|_{0,p,T}^p = \int_T \left\{ \sum_{r=1}^{n+1} v(a_r)\lambda_r(x) \right\}^p dx \geqslant \sum_{r=1}^{n+1} \{v(a_r)\}^p \int_T \{\lambda_r(x)\}^p \, dx.$$

Now let $\hat{T}$ be a fixed n-simplex of $\mathbb{R}^n$ with vertices $\hat{a}_r$, $1 \leqslant r \leqslant n+1$. Then there exists an invertible matrix B of order n and a vector $b \in \mathbb{R}^n$ such that the set T is the image of the set $\hat{T}$ through the mapping $F: \hat{x} \to F(\hat{x}) = B\hat{x} + b$, and this mapping can be so chosen that $F(\hat{a}_r) = a_r$, $1 \leqslant r \leqslant n+1$. Therefore, denoting by $\hat{\lambda}_r$, $1 \leqslant r \leqslant n+1$, the barycentric coordinates of a point $\hat{x} \in \hat{T}$ with respect to the points $\hat{a}_r$, we obtain

$$\int_T \{\lambda_r(x)\}^p \, dx = \int_{\hat{T}} \{\lambda_r(B\hat{x}+b)\}^p |\det DF(\hat{x})| \, d\hat{x} = \frac{\text{meas}(T)}{\text{meas}(\hat{T})} \int_{\hat{T}} \{\hat{\lambda}_r(\hat{x})\}^p \, d\hat{x},$$

since the barycentric coordinates are invariant through the affine mapping F. Thus, inequality (21.1) holds, with

$$C_1 = \{\text{meas}(\hat{T})\}^{-1} \min \left\{ \int_{\hat{T}} \hat{\lambda}_r(\hat{x})^p \, d\hat{x};\ 1 \leqslant r \leqslant n+1 \right\}. \qquad \square$$

THEOREM 21.2. *Let there be given a discrete problem of nonnegative type (cf. (20.28)–(20.29)). Let ξ_i, $1 \leqslant i \leqslant N+M$, be given real numbers and let $\alpha \in \mathbb{R}$ be given such that*

$$\max\{0, \max\{\xi_i;\ N+1 \leqslant i \leqslant N+M\}\} \leqslant \alpha. \tag{21.2}$$

Then if we let

$$\eta_i = \min\{\alpha, \xi_i\}, \quad 1 \leqslant i \leqslant N+M, \tag{21.3}$$

we have

$$\sum_{i=1}^{N} \sum_{j=1}^{N+M} a_{ij}(\xi_i - \eta_i)\eta_j \geqslant 0. \tag{21.4}$$

PROOF. Let $I = \{1 \leqslant i \leqslant N;\ \xi_i > \alpha\}$ and $J = \{1 \leqslant i \leqslant N+M;\ \xi_i \leqslant \alpha\}$. From (21.2), it follows that $I \cup J = \{1, 2, \ldots, N+M\}$. Next, it is easily verified that

$$\sum_{i=1}^{N} \sum_{j=1}^{N+M} a_{ij}(\xi_i - \eta_i)\eta_j = \alpha \sum_{i \in I} (\xi_i - \alpha) \sum_{j=1}^{N+M} a_{ij} + \sum_{i \in I} \sum_{j \in J} a_{ij}(\xi_i - \alpha)(\xi_j - \alpha),$$

and thus this quantity is $\geqslant 0$: the first term is $\geqslant 0$ since $\alpha \geqslant 0$ by (21.2), $(\xi_i - \alpha) > 0$ for $i \in I$ and $\Sigma_{j=1}^{N+M} a_{ij} \geqslant 0$ by (20.29), and the second term is also $\geqslant 0$ since for $i \in I$ and $j \in J$, we have $a_{ij} \leqslant 0$ by (20.28) ($I \cap J = \emptyset$), $(\xi_i - \alpha) > 0$ for $i \in I$ and $(\xi_j - \alpha) \leqslant 0$ for $j \in J$. $\square$

THEOREM 21.3. *Let there be given a discrete problem of nonnegative type. Given any function $u_h \in X_h$ and any real number α, let $u_{h,\alpha}$ denote the function of X_h that satisfies*

$$u_{h,\alpha}(b_i) = \min\{\alpha, u_h(b_i)\}, \quad 1 \leqslant i \leqslant N+M. \tag{21.5}$$

Then, if

$$\max\{0, \max\{u_h(b_i);\ N+1 \leqslant i \leqslant N+M\}\} \leqslant \alpha, \tag{21.6}$$

the function

$$v_{h,\alpha} = u_h - u_{h,\alpha} \tag{21.7}$$

belongs to V_h, and

$$a(v_{h,\alpha}, v_{h,\alpha}) \leqslant a(u_{h,\alpha}, v_{h,\alpha}). \tag{21.8}$$

PROOF. We have $u_h(b_i)=u_{h,\alpha}(b_i)$ for $N+1\leqslant i\leqslant N+M$ by (21.6); hence $v_{h,\alpha}$ belongs to V_h. Next, we may write

$$a(u_h, v_{h,\alpha})=a(v_{h,\alpha}, v_{h,\alpha})+a(u_{h,\alpha}, u_h-u_{h,\alpha}),$$

and thus, we have to prove that

$$a(u_{h,\alpha}, u_h-u_{h,\alpha})\geqslant 0.$$

If we let $\xi_i=u_h(b_i)$ and $\eta_i=u_{h,\alpha}(b_i)$, $1\leqslant i\leqslant N+M$, the last expression can be rewritten as

$$a(u_{h,\alpha}, u_h-u_{h,\alpha})=\sum_{i=1}^{N}\sum_{j=1}^{N+M} a_{ij}(\xi_i-\eta_i)\eta_j,$$

and this quantity is $\geqslant 0$ by Theorem 21.2. □

Following CIARLET and RAVIART [1973, Theorem 1], we now prove that the norms $|u_h|_{0,\infty,\Omega}$ *are bounded independently of h*. This crucial step towards estimating the error $|u-u_h|_{0,\infty,\Omega}$ is here nothing but the discrete analog of the a priori estimate (20.18) for the continuous problem. A similar estimate has been established by SCHATZ [1980].

THEOREM 21.4. *Let the assumptions on the continuous problem be as in Section 20. Then there exists a constant C, which is the same for all the discrete problems of nonnegative type, such that*

$$|u_h|_{0,\infty,\Omega}\leqslant \|u_0\|_{L^\infty(\Gamma)}+C\sum_{k=0}^{n}|f_k|_{0,p,\Omega}, \tag{21.9}$$

where $u_h\in X_h$ *is the solution of the discrete problem* (20.22)–(20.23).

PROOF. Let

$$a_{0h}=\max\{0,\max_{\Gamma} u_{0h}\}=\max\{0,\max\{u_h(b_i);\ N+1\leqslant i\leqslant N+M\}\},$$

and let α be any real number with

$$\alpha\geqslant a_{0h}.$$

By Theorem 21.3, the function $v_{h,\alpha}=u_h-u_{h,\alpha}$ belongs to V_h, and moreover

$$a(v_{h,\alpha}, v_{h,\alpha})\leqslant a(u_h, v_{h,\alpha})=(f, v_{h,\alpha}). \tag{21.10}$$

Since $v_{h,\alpha}\in V_h\subset H^1(\Omega)$, the functions $v_{h,\alpha}$ and $\partial_k v_{h,\alpha}$, $1\leqslant k\leqslant n$, belong to $L^{p'}(\Omega)$ with p' defined by $1/p+1/p'=1$ ($p'<2$ by (20.14)); thus, by Hölder's inequality, we obtain

$$l(v_{h,\alpha})\leqslant |f_0|_{0,p,\Omega}|v_{h,\alpha}|_{0,p',\Omega}+\sum_{k=1}^{n}|f_k|_{0,p,\Omega}|\partial_k v_{h,\alpha}|_{0,p',\Omega}. \tag{21.11}$$

Let

$$E(\alpha)=\{x\in\Omega;\ v_{h,\alpha}(x)>0\}. \tag{21.12}$$

Then $v_{h,\alpha}=\partial_k v_{h,\alpha}=0$, $1\leqslant k\leqslant n$, in $\Omega-\overline{E(\alpha)}$, since $\overline{E(\alpha)}$ is a union of n-simplices of $\mathcal{T}_h$ ($v_{h,\alpha}\geqslant 0$ over Ω); hence using again Hölder's inequality, we obtain

$$|v_{h,\alpha}|_{0,p',\Omega}=|v_{h,\alpha}|_{0,p',E(\alpha)}\leqslant|v_{h,\alpha}|_{0,E(\alpha)}\{\text{meas}(E(\alpha))\}^{(1/2)-(1/p)}.$$

From this inequality, similar inequalities for $\partial_k v_{h,\alpha}$, $1\leqslant k\leqslant n$, and inequality (21.11), we get

$$l(v_{h,\alpha})\leqslant(n+1)\left\{\sum_{k=0}^{n}|f_k|_{0,p,\Omega}\right\}\|v_{h,\alpha}\|_{1,\Omega}\{\text{meas}(E(\alpha))\}^{(1/2)-(1/p)};\tag{21.13}$$

from inequalities (20.9), (21.8), (21.10), and (21.13), we obtain

$$\alpha\|v_{h,\alpha}\|_{1,\Omega}\leqslant(n+1)\left\{\sum_{k=0}^{n}|f_k|_{0,p,\Omega}\right\}\{\text{meas}(E(\alpha))\}^{(1/2)-(1/p)}.\tag{21.14}$$

From the Sobolev imbedding theorem (14.3), we infer that

$$H^1(\Omega)\hookrightarrow L^{2^*}(\Omega)\quad\text{with}\begin{cases}1/2^*=\frac{1}{2}-1/n, & \text{if } n>2,\\ 2^*=\text{any real number }\geqslant 1, & \text{if } n=2;\end{cases}$$

hence inequality (21.14) implies that there exists a constant C_2 independent of h such that

$$|v_{h,\alpha}|_{0,2^*,\Omega}\leqslant C_2\left\{\sum_{k=0}^{n}|f_k|_{0,p,\Omega}\right\}\{\text{meas}(E(\alpha))\}^{(1/2)-(1/p)}.\tag{21.15}$$

Next, let β be any real number that satisfies

$$\beta>\alpha,$$

and let $E(\beta)$ be defined as in (21.12). Using Theorem 21.1, we obtain

$$\begin{aligned}|v_{h,\alpha}|_{0,2^*,\Omega}^{2^*}&=\sum_{T\in E(\alpha)}\int_T\{v_{h,\alpha}(x)\}^{2^*}\,dx\\&\geqslant C_1\sum_{b_i\in E(\alpha)}\{v_{h,\alpha}(b_i)\}^{2^*}\text{meas}(\text{supp}\,w_i)\\&\geqslant C_1(\beta-\alpha)^{2^*}\sum_{b_i\in E(\beta)}\text{meas}(\text{supp}\,w_i)=C_1(\beta-\alpha)^{2^*}\text{meas}(E(\beta)).\end{aligned}\tag{21.16}$$

Inequalities (21.15) and (21.16) imply that the function

$$\phi:\alpha\geqslant\alpha_{0h}\to\phi(\alpha)=\text{meas}(E(\alpha))$$

is $\geqslant 0$ and nondecreasing on the interval $[\alpha_{0h},+\infty[$, and that, for any $\beta>\alpha\geqslant\alpha_{0h}$,

$$\phi(\beta)\leqslant\frac{C_3}{(\beta-\alpha)^{2^*}}\phi(\alpha)^{\nu},$$

with

$$C_3=C_1^{-1}\left\{C_2\sum_{k=0}^{n}|f_k|_{0,p,\Omega}\right\}^{2^*},\qquad \nu=2^*((1/2)-(1/p)).$$

Therefore, applying a result of STAMPACCHIA [1965, Lemma 4.1], we obtain

$$\phi(\alpha_{0h}+\{2^{\nu/(\nu-1)}C_3(\phi(\alpha_{0h}))^{\nu-1}\}^{1/2^*})=0,$$

which means that, for all $x\in\bar{\Omega}$,

$$\begin{aligned} u_h(x) &\leqslant \alpha_{0h}+\{2^{\nu/(\nu-1)}C_3(\phi(\alpha_{0h}))^{\nu-1}\}^{1/2^*} \\ &\leqslant \max\{0, \max_{\Gamma} u_0\}+C\sum_{k=0}^{n}|f_k|_{0,p,\Omega}, \end{aligned} \tag{21.17}$$

with

$$C=C_1^{-1/2^*}C_2 2^{\nu/2^*(\nu-1)}(\text{meas}(\Omega))^{2^*(\nu-1)}$$

(since the piecewise affine function u_{0h} interpolates the function u_0 on Γ, we have $\alpha_{0h}\leqslant\max\{0,\max_\Gamma u_0\}$). The conclusion of the theorem follows by observing that we can similarly prove an inequality opposite to that of (21.17). □

We are now in a position to prove the main result of this section, due to CIARLET and RAVIART [1973, Theorem 2].

THEOREM 21.5. *Let the assumptions on the continuous problem be as in Section 20, and consider a family of discrete problems of nonnegative type, associated with a regular family of triangulations. Then*

$$\lim_{h\to 0}|u-u_h|_{0,\infty,\Omega}=0, \qquad \text{if } u\in W^{1,p}(\Omega),\quad n<p, \tag{21.18}$$

$$|u-u_h|_{0,\infty,\Omega}\leqslant Ch|u|_{2,p,\Omega}, \quad \text{if } u\in W^{2,p}(\Omega),\quad n<2p. \tag{21.19}$$

PROOF. For each h, let $\Pi_h u$ denote the X_h-interpolant of the solution u, i.e., the unique function in X_h that satisfies $\Pi_h u(b_i)=u(b_i)$, $1\leqslant i\leqslant N+M$. Since the function $(u_h-\Pi_h u)$ belongs to the space V_h, we infer from (20.10) and (20.22) that

$$a(u_h-\Pi_h u, v_h)=a(u-\Pi_h u, v_h) \quad \text{for all } v_h\in V_h.$$

Hence we may write

$$a(u_n-\Pi_h u, v_h)=\int_\Omega\left(f_0 v_h+\sum_{k=1}^{n}f_k\partial_k v_h\right)\mathrm{d}x \quad \text{for all } v_h\in V_h,$$

with

$$f_0=b(u-\Pi_h u)+\sum_{k=1}^{n}a_k\partial_k(u-\Pi_h u),$$

$$f_k=\sum_{l=1}^{n}a_{kl}\partial_l(u-\Pi_h u),\quad 1\leqslant k\leqslant n.$$

The assumptions (20.6) and the fact that $u\in W^{1,p}(\Omega)$ imply that $f_k\in L^p(\Omega)$, $0\leqslant k\leqslant n$. Hence we may apply Theorem 21.4; this shows that there exists a constant

C independent of h and u such that

$$|u_h - \Pi_h u|_{0,\infty,\Omega} \leqslant C\|u - \Pi_h u\|_{1,p,\Omega},$$

and thus

$$|u - u_h|_{0,\infty,\Omega} \leqslant (1+C)\|u - \Pi_h u\|_{1,p,\Omega}. \tag{21.20}$$

If $u \in W^{2,p}(\Omega)$ with $n < 2p$, then

$$\|u - \Pi_h u\|_{1,p,\Omega} \leqslant Ch|u|_{2,p,\Omega} \tag{21.21}$$

by Theorem 16.2, and inequality (21.19) is proved when $n < 2p$. Assume next that $n < p$. By Theorem 16.2 again, there exists a constant C independent of h such that

$$\|v - \Pi_h v\|_{1,p,\Omega} \leqslant C|v|_{1,p,\Omega} \quad \text{for all } v \in W^{1,p}(\Omega), \quad n < p,$$

and thus

$$\|\Pi_h v\|_{1,p,\Omega} \leqslant (1+C)\|v\|_{1,p,\Omega} \quad \text{for all } v \in W^{1,p}(\Omega), \quad n < p. \tag{21.22}$$

Let $\varepsilon > 0$ be given. Since the space $W^{2,p}(\Omega)$ is dense in the space $W^{1,p}(\Omega)$ and since inequality (21.22) holds, there exist $u_\varepsilon \in W^{2,p}(\Omega)$ such that

$$\|u - u_\varepsilon\|_{1,p,\Omega} \leqslant \tfrac{1}{3}\varepsilon, \qquad \|\Pi_h(u - u_\varepsilon)\|_{1,p,\Omega} \leqslant \tfrac{1}{3}\varepsilon. \tag{21.23}$$

Next, by (21.21), there exists $h_0(\varepsilon)$ such that

$$\|u_\varepsilon - \Pi_h u_\varepsilon\|_{1,p,\Omega} \leqslant \tfrac{1}{3}\varepsilon \quad \text{for all } h \leqslant h_0(\varepsilon). \tag{21.24}$$

Hence (21.23) and (21.24) combined imply that

$$\|u - \Pi_h u\|_{1,p,\Omega} \leqslant \varepsilon \quad \text{for all } h \leqslant h_0(\varepsilon), \tag{21.25}$$

and (21.18) follows from (21.20) and (21.25). □

22. Estimates of the errors $|u - u_h|_{0,\infty,\Omega}$ and $|u - u_h|_{1,\infty,\Omega}$ when $u \in W^{2,\infty}(\Omega)$ and Nitsche's method of weighted norms

Our objective in this section is again to study the *uniform convergence of the finite element method when linear n-simplices are used.* We shall however use a technique quite different from that of the previous section, which relied on a discrete maximum principle. We follow instead the particularly penetrating *method of weighted norms of Nitsche,* which yields *optimal orders of convergence* if the solution is sufficiently smooth, i.e., if $u \in W^{2,\infty}(\Omega)$ in the present case.

For ease of exposition, we shall simply consider the homogeneous Dirichlet problem for the operator $-\Delta$ in Ω, whose variational formulation corresponds to the following data:

$$\begin{aligned} &V = H_0^1(\Omega), \\ &a(u, v) = \int_\Omega \nabla v \cdot \nabla v \,\mathrm{d}x, \qquad l(v) = \int_\Omega fv \,\mathrm{d}x, \quad f \in L^2(\Omega). \end{aligned} \tag{22.1}$$

Assuming that $\bar{\Omega}$ is a *convex polygonal* subset of $\mathbb{R}^2$, we shall restrict ourselves to finite element spaces X_h which are made up of linear triangles; hence the corresponding discrete problems are posed in the spaces $V_h = \{v_h \in X_h;\, v_h = 0 \text{ on } \Gamma\}$. Extensions to triangles of higher degree and to higher dimensions are indicated at the end of this section.

We shall assume once and for all that we are given a *regular* family of triangulations of the set $\bar{\Omega}$ that also satisfies an *inverse assumption*, i.e., that there exist two constants σ and ν such that

$$\begin{aligned} &h_T/\rho_T \leqslant \sigma, \\ &h/h_T \leqslant \nu \quad \text{for all } T \in \bigcup_h \mathscr{T}_h. \end{aligned} \tag{22.2}$$

The main tool that Nitsche used in his study of the errors $|u-u_h|_{0,\infty,\Omega}$ and $|u-u_h|_{1,\infty,\Omega}$ is the consideration of appropriate *weighted norms and seminorms*. Accordingly, the first part of this section will be devoted to the study of those properties of such seminorms that are of interest for our subsequent analysis; these properties are the object of Theorems 22.1–22.4.

Given a *weight function* ϕ, i.e., a function that satisfies

$$\phi \in L^\infty(\Omega), \qquad \phi \geqslant 0 \quad \text{a.e. on } \Omega, \tag{22.3}$$

we define, for each integer $m \geqslant 0$, the *weighted seminorms*

$$v \in H^m(\Omega) \to |v|_{\phi;m,\Omega} = \left\{ \int_\Omega \phi \sum_{|\beta|=m} |\partial^\beta v|^2 \,\mathrm{d}x \right\}^{1/2}. \tag{22.4}$$

To begin with, we observe that, if the function ϕ^{-1} exists and is also in the space $L^\infty(\Omega)$, an application of Cauchy–Schwarz inequality gives

$$a(u, v) \leqslant |u|_{\phi^\alpha;1,\Omega} |v|_{\phi^{-\alpha};1,\Omega} \quad \text{for all } u, v \in H^1(\Omega), \quad \alpha \in \mathbb{R}. \tag{22.5}$$

Departing from the general case, we shall in fact concentrate our subsequent study on weighted seminorms of the particular type $|\cdot|_{\phi^\alpha;m,\Omega}$, $\alpha \in \mathbb{R}$, where the function ϕ is of the form (22.7) below. Our first task is to extend to such weighted seminorms the property that there exists a constant c_1, solely dependent upon the set Ω, such that

$$|v|_{2,\Omega} \leqslant c_1 |\Delta v|_{0,\Omega} \quad \text{for all } v \in H_0^1(\Omega) \cap H^2(\Omega). \tag{22.6}$$

We recall that such an inequality follows from the $H^2(\Omega)$-regularity of the solution u of problem (22.1) when $f \in L^2(\Omega)$ and $\bar{\Omega}$ is convex.

THEOREM 22.1. *There exists a constant $C_1 = C_1(\Omega)$ such that, for all functions ϕ of the form*

$$\begin{aligned} &\phi : x \in \bar{\Omega} \to \phi(x) = \frac{1}{\|x-\bar{x}\|^2 + \theta^2}, \\ &\theta > 0, \quad \bar{x} = (\bar{x}_1, \bar{x}_2) \in \mathbb{R}^2, \end{aligned} \tag{22.7}$$

we have

$$|v|^2_{\phi^{-1};2,\Omega} \leqslant C_1(|\Delta v|^2_{\phi^{-1};0,\Omega} + |v|^2_{1,\Omega}) \quad \textit{for all } v \in H^1_0(\Omega) \cap H^2(\Omega). \tag{22.8}$$

PROOF. Let v be an arbitrary function in the space $H^1_0(\Omega) \cap H^2(\Omega)$. Then the function

$$w = (x_1 - \bar{x}_1)v$$

also belongs to the space $H^1_0(\Omega) \cap H^2(\Omega)$, and

$$(x_1 - \bar{x}_1)\partial_{11}v = \partial_{11}w - 2\partial_1 v,$$
$$(x_1 - \bar{x}_1)\partial_{12}v = \partial_{12}w - \partial_2 v,$$
$$(x_1 - \bar{x}_1)\partial_{22}v = \partial_{22}w,$$
$$\Delta w = (x_1 - \bar{x}_1)\Delta v + 2\partial_1 v.$$

Using these relations and inequality (22.6), we find that there exists a constant c_2 such that

$$\begin{aligned}\int_\Omega (x_1 - \bar{x}_1)^2 \sum_{|\beta|=2} |\partial^\beta v|^2 \, dx &\leqslant 2c_1^2 |\Delta w|^2_{0,\Omega} + 8|v|^2_{1,\Omega} \\ &\leqslant c_2 \left\{ \int_\Omega (x_1 - \bar{x}_1)^2 (\Delta v)^2 \, dx + |v|^2_{1,\Omega} \right\}.\end{aligned}$$

Since we have likewise

$$\int_\Omega (x_2 - \bar{x}_2)^2 \sum_{|\beta|=2} |\partial^\beta v|^2 \, dx \leqslant c_2 \left\{ \int_\Omega (x_2 - \bar{x}_2)^2 (\Delta v)^2 \, dx + |v|^2_{1,\Omega} \right\},$$

we eventually obtain

$$\begin{aligned}|v|^2_{\phi^{-1};2,\Omega} &= \int_\Omega ((x_1 - \bar{x}_1)^2 + (x_2 - \bar{x}_2)^2 + \theta^2) \sum_{|\beta|=2} |\partial^\beta v|^2 \, dx \\ &\leqslant \max\{2c_2, c_1^2\}(|\Delta v|^2_{\phi^{-1};0,\Omega} + |v|^2_{1,\Omega}),\end{aligned}$$

and the proof is complete. □

As exemplified by the above computations, we shall depart in this section from our usual practice of letting the same letter C denote various constants, not necessarily the same in their various occurrences. This is due not only to the unusually large number of such constants which we shall come across, but also, and essentially, to their sometimes intricate interdependence. Therefore, constants will be numbered and, in addition, their dependence on other quantities will be made explicit when necessary. However the possible dependence on the set Ω and on the constants σ and ν of (22.2) will be systematically omitted. While we shall use capital letters $C_i, i \geqslant 1$, for constants occurring in important inequalities, lower-case letters $c_i, i \geqslant 1$, will rather be reserved for intermediate computations.

In the next two theorems, we examine the relationships between the weighted seminorms $|\cdot|_{\phi^\alpha;m,\Omega}$ (when the function ϕ is as in (22.7)) and the standard seminorms $|\cdot|_{m,\infty,\Omega}$. Such relationships will play a crucial role in the derivation of the eventual error estimates.

THEOREM 22.2. *For each number $\alpha > 1$ and each integer $m \geqslant 0$, there exists a constant $C_2(\alpha, m)$ such that, for all functions ϕ of the form*

$$\phi: x \in \bar{\Omega} \to \phi(x) = \frac{1}{\|x - \bar{x}\|^2 + \theta^2}, \qquad \theta > 0, \quad \bar{x} \in \bar{\Omega}, \tag{22.9}$$

we have

$$|v|_{\phi^\alpha;m,\Omega} \leqslant C_2(\alpha, m) \frac{1}{\theta^{\alpha-1}} |v|_{m,\infty,\Omega} \quad \text{for all } v \in W^{m,\infty}(\Omega). \tag{22.10}$$

For each number $\beta \in]0, 1[$, and each integer $m \geqslant 0$, there exists a constant $C_3(\beta, m)$ such that, for all functions ϕ of the form (22.9), *we have*

$$|v|_{\phi;m,\Omega} \leqslant C_3(\beta, m) |\ln \theta|^{1/2} |v|_{m,\infty,\Omega} \quad \text{for all } v \in W^{m,\infty}(\Omega), \quad \theta \leqslant \beta. \tag{22.11}$$

PROOF. Clearly, one has

$$|v|_{\phi^\alpha;m,\Omega} \leqslant c_3(m) \left\{ \int_\Omega \phi^\alpha \, dx \right\}^{1/2} |v|_{m,\infty,\Omega}.$$

Let next $\delta = \operatorname{diam} \Omega$, so that

$$\int_\Omega \phi^\alpha \, dx \leqslant \int_{B(\bar{x};\delta)} \phi^\alpha \, dx = 2\pi \int_0^\delta \frac{\tau \, d\tau}{(\tau^2 + \theta^2)^\alpha}.$$

If $\alpha > 1$, we write

$$\int_0^\delta \frac{\tau \, d\tau}{(\tau^2 + \theta^2)^\alpha} \leqslant \int_0^\infty \frac{\tau \, d\tau}{(\tau^2 + \theta^2)^\alpha} = \frac{1}{2(\alpha - 1)\theta^{2(\alpha-1)}},$$

and inequality (22.10) is proved with $C_2(\alpha, m) = c_3(m)\, \pi/(\alpha - 1)^{1/2}$. If $\alpha = 1$, we have for $\theta \leqslant \beta < 1$,

$$\begin{aligned} \int_0^\delta \frac{\tau \, d\tau}{\tau^2 + \theta^2} &= |\ln \theta| + \tfrac{1}{2} \ln(\theta^2 + \delta^2) \\ &\leqslant |\ln \theta| + \tfrac{1}{2} \ln(1 + \delta^2) \leqslant c_4(\beta) |\ln \theta|, \end{aligned}$$

with

$$c_4(\beta)=1+\frac{\ln(1+\delta^2)}{2|\ln\beta|},$$

and inequality (22.11) is proved with $C_3(\beta,m)=c_3(m)(2\pi c_4(\beta))^{1/2}$. □

We next obtain inequalities in the opposite direction. In order that they be useful for our subsequent purposes, however, we shall establish these inequalities only for functions in the finite element space X_h, and further, we shall restrict ourselves to weight functions of the form ϕ or ϕ^2, with ϕ as in (22.9) for which (i) the parameter θ cannot approach zero too rapidly when h approaches zero (cf. (22.13)), and for which (ii) the points $\bar{x}$ depend upon the particular function $v_h\in X_h$ under consideration (cf. (22.14) and (22.16)).

THEOREM 22.3. *Assume that for each h, we are given a function ϕ_h of the form*

$$\phi_h: x\in\bar{\Omega}\to\phi_h(x)=\frac{1}{\|x-x_h\|^2+\theta_h^2},\quad x_h\in\bar{\Omega}, \tag{22.12}$$

and that there exists a number γ such that

$$\gamma>0,\qquad \theta_h\geqslant\gamma h\quad\text{for all } h. \tag{22.13}$$

Then there exist constants $C_4(\gamma)$ and $C_5(\gamma)$ with the following properties: Let $v_h\in X_h$ be given. If the point $x_h\in\bar{\Omega}$ in (22.12) *is chosen in such a way that*

$$|v_h(x_h)|=|v_h|_{0,\infty,\Omega}, \tag{22.14}$$

then

$$|v_h|_{0,\infty,\Omega}\leqslant C_4(\gamma)(\theta_h^2/h)|v_h|_{\phi_h^2;0,\Omega}. \tag{22.15}$$

If the point $x_h\in\bar{\Omega}$ in (22.12) *is chosen in such a way that*

$$\max\{|\partial_1 v_h(x_h)|,|\partial_2 v_h(x_h)|\}=|v_h|_{1,\infty,\Omega}, \tag{22.16}$$

then

$$|v_h|_{1,\infty,\Omega}\leqslant C_5(\gamma)(\theta_h/h)|v_h|_{\phi_h;1,\Omega}. \tag{22.17}$$

PROOF. (i) Let v_h be an arbitrary function in the space X_h, and let the point x_h be chosen as in (22.14). Then there exists a constant c_5 such that

$$\begin{aligned}|v_h|_{0,\infty,\Omega}-|v_h(x)|&=|v_h(x_h)|-|v_h(x)|\leqslant|v_h(x_h)-v_h(x)|\\&\leqslant\sqrt{2}|v_h|_{1,\infty,\Omega}\|x-x_h\|\leqslant(c_5/h)|v_h|_{0,\infty,\Omega}\|x-x_h\|,\end{aligned}$$

for all $x\in\bar{\Omega}$ (in the last inequality we have used the fact that the family of triangulations satisfies an inverse assumption; cf. Theorem 17.2). In other words,

$$|v_h(x)|\geqslant(1-(c_5/h)\|x-x_h\|)|v_h|_{0,\infty,\Omega}\quad\text{for all } x\in\bar{\Omega},$$

and consequently ($B(a;r)=\{x\in\mathbb{R}^2;\ \|x-a\|\leqslant r\}$),

$$|v_h|^2_{\phi_h^2;0,\Omega}\geqslant|v_h|^2_{0,\infty,\Omega}\int\limits_{\bar\Omega\cap B(x_h;h/2c_5)}\left\{\frac{1-(c^5/h)\|x-x_h\|}{\|x-x_h\|^2+\theta_h^2}\right\}^2\mathrm{d}x.$$

The set $\bar\Omega$ being polygonal, there exists a constant c_6 such that

$$\operatorname{meas}\left\{\bar\Omega\cap B\left(x_h;\frac{h}{2c_5}\right)\right\}\geqslant c_6\left(\frac{h}{2c_5}\right)^2,$$

and we also have

$$1-\frac{c_5}{h}\|x-x_h\|\geqslant\tfrac12\quad\text{for all } x\in B\left(x_h;\frac{h}{2c_5}\right),$$

and

$$\frac{1}{\|x-x_h\|^2+\theta_h^2}\geqslant\frac{1}{\dfrac{h^2}{4c_5^2}+\theta_h^2}\geqslant\frac{1}{\theta_h^2\left(1+\dfrac{1}{4c_5^2\gamma^2}\right)}\quad\text{for all } x\in B\left(x_h;\frac{h}{2c_5}\right)$$

by assumption (22.13). Combining the previous inequalities, we obtain an inequality of the form (22.15), with

$$C_4(\gamma)=\frac{4c_5}{\sqrt{c_6}}\left(1+\frac{1}{4c_5^2\gamma^2}\right).$$

(ii) Let v_h be an arbitrary function in the space X_h, let the point x_h be chosen as in (22.16), and let $T_h\in\mathscr{T}_h$ denote a triangle that contains the point x_h. Since the gradient ∇v_h is constant over the set T_h, we deduce

$$|v_h|^2_{\phi_h;1,\Omega}\geqslant|v_h|^2_{1,\infty,\Omega}\int\limits_{T_h}\frac{\mathrm{d}x}{\|x-x_h\|^2+\phi_h^2}.$$

Combining this inequality and the inequalities

$$\operatorname{meas}(T_h)\geqslant c_7(\sigma,\nu)h^2,$$

$$\frac{1}{\|x-x_h\|^2+\theta_h^2}\geqslant\frac{1}{\theta_h^2(1+1/\gamma^2)}\quad\text{for all } x\in T_h,$$

we obtain an inequality of the form (22.17) with

$$C_5(\gamma)=\sqrt{(1+1/\gamma^2)/c_7}.\qquad\square$$

To conclude this analysis of weighted seminorms, we examine in the next theorem the *interpolation error estimates expressed with the seminorms* $|\cdot|_{\phi_h^\alpha;m,\Omega}$, where, for each h, the function ϕ_h is of the form (22.12). The conclusion (cf. (22.20)) is that *the error estimates are exactly the same as in the case of the usual seminorms* $|\cdot|_{m,\Omega}$, *provided the parameter θ_h does not approach zero too rapidly with h* (cf. (22.19)). Notice, however, that if the behavior of the function θ_h can be "at best" linear as in the previous theorem, the constant that appears in inequality (22.19) is not arbitrary, by contrast with the constant γ which appeared in inequality (22.13). Finally, observe that no restriction is imposed upon the points x_h.

THEOREM 22.4. *Assume that for each h, we are given a function ϕ_h of the form*

$$\phi_h: x \in \bar{\Omega} \to \phi_h(x) = \frac{1}{\|x-x_h\|^2+\theta_h^2}, \quad x_h \in \bar{\Omega}. \tag{22.18}$$

Then there exists a constant C_6, and for each $\alpha \in \mathbb{R}-\{0\}$, there exist constants $C_7(\alpha)>0$ and $C_8(\alpha)$ such that, if

$$\theta_h \geqslant C_7(\alpha)h \quad \textit{for all } h, \tag{22.19}$$

the following estimates hold:

$$|v-\Pi_h v|_{\phi_h^\alpha;m,\Omega} \leqslant C_6 h^{2-m}|v|_{\phi_h^\alpha;2,\Omega}, \quad m=0,1, \quad \textit{for all } v \in H^2(\Omega), \tag{22.20}$$

and

$$\begin{aligned} &|\phi_h^\alpha v_h - \Pi_h(\phi_h^\alpha v_h)|_{\phi_h^{-\alpha};1,\Omega} \\ &\quad \leqslant C_8(\alpha)(h/\theta_h)(|v_h|_{\phi_h^{\alpha+1};0,\Omega}+|v_h|_{\phi_h^\alpha;1,\Omega}) \quad \textit{for all } v_h \in X_h. \end{aligned} \tag{22.21}$$

PROOF. (i) There exists a constant c_8 such that

$$|v-\Pi_T v|_{m,T} \leqslant c_8 h^{2-m}|v|_{2,T}, \quad m=0,1,$$
$$\text{for all } v \in H^2(\Omega), \quad T \in \mathcal{T}_h.$$

Next, we have

$$|v-\Pi_T v|_{\phi_h^\alpha;m,T} \leqslant (\phi_h^\alpha(\bar{x}_T))^{1/2}|v-\Pi_T v|_{m,T},$$
$$|v|_{2,T} \leqslant (\phi_h^\alpha(x_T))^{-1/2}|v|_{\phi_h^\alpha;2,T},$$

if, for each $T \in \mathcal{T}_h$, the points $x_T \in T$ and $\bar{x}_T \in T$ are chosen in such a way that

$$0 < \phi_h^\alpha(x_T) = \inf_{x \in T} \phi_h^\alpha(x_T), \qquad \phi_h^\alpha(\bar{x}_T) = \sup_{x \in T} \phi_h^\alpha(x).$$

Since

$$\frac{\partial_i(\phi_h^\alpha)(x)}{\phi_h^\alpha(x)} = -2\alpha \frac{(x_i-x_{hi})}{\|x-x_h\|^2+\theta_h^2}, \quad i=1,2,$$

we obtain

$$\sup_{x \in \bar{\Omega}} \frac{|\partial_i(\phi_h^\alpha)(x)|}{\phi_h^\alpha(x)} \leqslant \sup_{x \in \bar{\Omega}} \frac{\|D(\phi_h^\alpha)(x)\|}{\phi_h^\alpha(x)} \leqslant 2|\alpha| \sup_{x \in \bar{\Omega}} \frac{\|x-x_h\|}{\|x-x_h\|^2+\theta_h^2} \leqslant \frac{|\alpha|}{\theta_h},$$

and therefore

$$\phi_h^\alpha(\bar{x}_T) \leqslant \phi_h^\alpha(x_T) + |\alpha|(h/\theta_h)\phi_h^\alpha(\bar{x}_T).$$

Consequently, if we let

$$C_7(\alpha)=2|\alpha|,$$

so that

$$\theta_h \geqslant C_7(\alpha)h$$

implies

$$\phi_h^\alpha(\bar{x}_T)/\phi_h^\alpha(x_T)\leqslant 2 \quad \text{for all } T\in\mathcal{T}_h,$$

the conjunction of the above inequalities yields inequality (22.20) with $C_6=\sqrt{2}c_8$.

(ii) Since the function $\phi_h^\alpha v_h$ is in the space $\mathscr{C}^0(\bar{\Omega})=\operatorname{dom}\Pi_h$ and since the restrictions $\phi_h^\alpha v_h|_T$ belong to the space $H^2(T)$ for all $T\in\mathcal{T}_h$, the same argument as in (i) shows that

$$|\phi_h^\alpha v_h-\Pi_T(\phi_h^\alpha v_h)|_{\phi_h^{-\alpha};1,\Omega}\leqslant C_7 h\left\{\sum_{T\in\mathcal{T}_h}|\phi_h^\alpha v_h|^2_{\phi_h^{-\alpha};2,T}\right\}^{1/2}.$$

We have (recall that $v_h|_T\in P_1(T)$ for all $T\in\mathcal{T}_h$)

$$\partial_{ij}(\phi_h^\alpha v_h)=(\partial_{ij}\phi_h^\alpha)v_h+(\partial_i\phi_h^\alpha)\partial_j v_h+(\partial_j\phi_h^\alpha)\partial_i v_h \quad \text{in any } T\in\mathcal{T}_h,$$

and

$$\frac{\partial_{ij}(\phi_h^\alpha)(x)}{\phi_h^\alpha(x)}=4\alpha(\alpha+1)\frac{(x_i-x_{hi})(x_j-x_{hj})}{(\|x-x_h\|^2+\theta_h^2)^2}-2\alpha\frac{\delta_{ij}}{\|x-x_h\|^2+\theta_h^2}$$

for all $x\in\bar{\Omega}$.

Hence, using the inequalities

$$\frac{\|x-x_h\|^2}{(\|x-x_h\|^2+\theta_h^2)^2}\leqslant\frac{\phi_h^{1/2}(x)}{2\theta_h}, \qquad \frac{1}{\|x-x_h\|^2+\theta_h^2}\leqslant\frac{\phi_h^{1/2}(x)}{\theta_h},$$

we deduce that

$$|\partial_{ij}(\phi_h^\alpha)(x)|\leqslant(2/\theta_h)(|\alpha^2+\alpha|+|\alpha|)\phi_h^{\alpha+1/2}(x) \quad \text{for all } x\in\bar{\Omega}.$$

Using the above inequalities and the inequality (cf. (i))

$$|\partial_i(\phi_h^\alpha)(x)|\leqslant(|\alpha|/\theta_h)\phi_h^\alpha(x) \quad \text{for all } x\in\bar{\Omega},$$

we conclude that there exists a constant $c_9(\alpha)$ such that, in any $T\in\mathcal{T}_h$,

$$\phi_h^{-\alpha}(x)\sum_{i,j=1}^{2}|\partial_{ij}(\phi_h^\alpha v_h)(x)|^2$$

$$\leqslant\frac{c_9(\alpha)}{\theta_h^2}\left(\phi_h^{\alpha+1}(x)|v_h(x)|^2+\phi_h^\alpha(x)\sum_{i=1}^{2}|\partial_i v_h(x)|^2\right) \quad \text{for all } x\in T,$$

and thus,

$$\sum_{T\in\mathcal{T}_h} |\phi_h^{\alpha} v_h|^2_{\phi_h^{-\alpha};2,T}$$

$$= \sum_{T\in\mathcal{T}_h} \int_T \phi_h^{-\alpha} \sum_{i,j=1}^{2} |\partial_{ij}(\phi_h^{\alpha} v_h)(x)|^2 \, dx \leqslant \frac{c_9(\alpha)}{\theta_h^2} (|v_h|^2_{\phi_h^{\alpha+1};0,\Omega} + |v_h|^2_{\phi_h^{\alpha};1,\Omega}).$$

Therefore we have proved inequality (22.21), with

$$C_8(\alpha) = C_7\sqrt{c_9(\alpha)}. \quad \square$$

After the above preliminaries, we now come to the central object of this section, i.e., the estimate of the errors $|u-u_h|_{0,\infty,\Omega}$ and $|u-u_h|_{1,\infty,\Omega}$ via *the method of weighted norms of Nitsche*. The analysis will comprise *three stages*. In the first stage (cf. the next theorem), we consider for each h *the projection operator*

$$P_h: v \in H_0^1(\Omega) \to P_h v \in V_h \tag{22.22}$$

associated with the inner product $a(\cdot,\cdot)$ of (22.1), which is therefore defined for each $v \in H_0^1(\Omega)$ by the relations

$$\begin{aligned} &P_h v \in V_h, \\ &a(v - P_h v, w_h) = 0 \quad \text{for all } w_h \in V_h. \end{aligned} \tag{22.23}$$

Thus we have in particular $u_h = P_h u$, where u_h is the discrete solution found in the space V_h and u is the solution of the problem defined in (22.1). We shall then show that for an appropriate choice of the parameters θ_h in the functions ϕ_h (cf. (22.25) and (22.26) below), *the mappings P_h are bounded independently of h when both spaces $H_0^1(\Omega)$ and V_h are equipped with the weighted norm*

$$v \to (|v|^2_{\phi_h^2;0,\Omega} + |v|^2_{\phi_h;1,\Omega})^{1/2}. \tag{22.24}$$

THEOREM 22.5. *Assume that for each h, we are given a function ϕ_h of the form*

$$\phi_h: x \in \bar{\Omega} \to \phi_h(x) = \frac{1}{\|x - x_h\|^2 + \theta_h^2}, \quad x_h \in \bar{\Omega}. \tag{22.25}$$

Then there exist three constants $h_0 \in]0, 1[$, $C_9 > 0$, and C_{10}, such that, if

$$\theta_h = C_9 h |\ln h|^{1/2} \quad \textit{for all } h, \tag{22.26}$$

the following inequalities hold for all $h \leqslant h_0$:

$$|P_h v|_{\phi_h^2;0,\Omega} + |P_h v|_{\phi_h;1,\Omega} \leqslant C_{10}(|v|_{\phi_h^2;0,\Omega} + |v|_{\phi_h;1,\Omega}) \quad \textit{for all } v \in H^1(\Omega). \tag{22.27}$$

PROOF. For convenience, the proof will be divided in four steps.

(i) *There exist two constants* C_{11} *and* C_{12} *such that, if*

$$\theta_h \geqslant C_{11} h \quad \textit{for all } h, \tag{22.28}$$

then

$$|P_h v|^2_{\phi_h;1,\Omega} \leqslant C_{12}(|P_h v|^2_{\phi_h^2;0,\Omega} + |v|^2_{\phi_h;1,\Omega}) \quad \textit{for all } v \in H_0^1(\Omega). \tag{22.29}$$

For brevity, let

$$v_h = P_h v.$$

Since

$$|v_h|^2_{\phi_h;1,\Omega} = a(v_h, \phi_h v_h) + \tfrac{1}{2} \int_\Omega \Delta\phi_h v_h^2 \, dx$$

and

$$\Delta\phi_h(x) = \frac{4(\|x - x_h\|^2 - \theta_h^2)}{\|x - x_h\|^2 + \theta_h^2} \phi_h^2(x) \leqslant 4\phi_h^2(x) \quad \text{for all } x \in \bar{\Omega},$$

we deduce that

$$|v_h|^2_{\phi_h;1,\Omega} \leqslant a(v_h, \phi_h v_h) + 2|v_h|^2_{\phi_h^2;0,\Omega}. \tag{22.30}$$

Using relations (22.23) we can write

$$a(v_h, \phi_h v_h) = a(v_h - v, \phi_h v_h - \Pi_h(\phi_h v_h)) + a(v, \phi_h v_h), \tag{22.31}$$

and an application of inequality (22.5) with $\alpha = 1$ shows that

$$|a(v_h - v, \phi_h v_h - \Pi_h(\phi_h v_h))| \leqslant (|v|_{\phi_h;1,\Omega} + |v_h|_{\phi_h;1,\Omega})|\phi_h v_h - \Pi_h(\phi_h v_h)|_{\phi_h^{-1};1,\Omega}.$$

We next infer from Theorem 22.4 that, if

$$\theta_h \geqslant c_{10} h \quad \text{with } c_{10} = C_7(1) \quad \text{for all } h, \tag{22.32}$$

then (cf. inequality (22.21) with $\alpha = 1$)

$$|\phi_h v_h - \Pi_h(\phi_h v_h)|_{\phi_h^{-1};1,\Omega} \leqslant c_{11}(h/\theta_h)(|v_h|_{\phi_h^2;0,\Omega} + |v_h|_{\phi_h;1,\Omega})$$

with $c_{11} = C_8(1)$.

Combining the previous inequalities, we find that, for $\theta_h \geqslant c_{10} h$,

$$|a(v_h - v, \phi_h v_h - \Pi_h(\phi_h v_h))|$$
$$\leqslant c_{11}(h/\theta_h)(|v|_{\phi_h;1,\Omega} + |v_h|_{\phi_h;1,\Omega})(|v_h|_{\phi_h^2;0,\Omega} + |v_h|_{\phi_h;1,\Omega}). \tag{22.33}$$

By another application of inequality (22.5) with $\alpha = 1$, we obtain

$$a(v, \phi_h v_h) \leqslant |v|_{\phi_h;1,\Omega} |\phi_h v_h|_{\phi_h^{-1};1,\Omega}. \tag{22.34}$$

Since

$$\sum_{i=1}^{2} |\partial_i \phi_h(x)|^2 \leqslant 4\phi_h^3(x) \quad \text{for all } x \in \bar{\Omega},$$

we find that there exists a constant c_{12} such that

$$|\phi_h v_h|_{\phi_h^{-1};1,\Omega} \leqslant c_{12}(|v_h|_{\phi_h^2;0,\Omega} + |v_h|_{\phi_h;1,\Omega}). \tag{22.35}$$

Combining relations (22.30)–(22.35), we find that, for $\theta_h \geqslant c_{10} h$,

$$|v_h|^2_{\phi_h;1,\Omega} \leqslant 2|v_h|^2_{\phi_h^2;0,\Omega} + c_{12}|v|_{\phi_h;1,\Omega}(|v_h|_{\phi_h^2;0,\Omega} + |v_h|_{\phi_h;1,\Omega})$$
$$+ c_{11}(h/\theta_h)(|v|_{\phi_h;1,\Omega} + |v_h|_{\phi_h;1,\Omega})(|v_h|_{\phi_h^2;0,\Omega} + |v_h|_{\phi_h;1,\Omega}),$$

i.e., we have found an inequality of the form

$$A^2 \leqslant 2C^2 + c_{12}B(A+C) + c_{11}(h/\theta_h)(A+B)(A+C).$$

Assuming that

$$\theta_h \geqslant 2c_{11} h \quad \text{for all } h, \tag{22.36}$$

we get

$$A^2 \leqslant 4C^2 + (1+2c_{12})BC + A((1+2c_{12})B + C)$$
$$\leqslant 4C^2 + (\tfrac{1}{2} + c_{12})(B^2 + C^2) + \tfrac{1}{2}A^2 + (1+2c_{12})^2 B^2 + C^2,$$

and therefore step (i) is proved with (cf. (22.32) and (22.36))

$$C_{11} = \max(c_{10}, 2c_{11}) \tag{22.37}$$

in relation (22.28), and

$$C_{12} = \max\{11 + 2c_{12}, (1+2c_{12})(3+4c_{12})\} \tag{22.38}$$

in relation (22.29).

(ii) *Assume that $\theta_h \geqslant C_{11} h$, where the constant C_{11} has been determined in step* (i). *Then there exists a constant C_{13} such that*

$$|P_h v|^2_{\phi_h^2;0,\Omega} + |P_h v|^2_{\phi_h;1,\Omega}$$
$$\leqslant C_{13}(|v|^2_{\phi_h^2;0,\Omega} + |v|^2_{\phi_h;1,\Omega} + h^2|\psi_h|^2_{\phi_h^{-1};2,\Omega}) \quad \textit{for all } v \in H_0^1(\Omega), \tag{22.39}$$

where, for each h, $\psi_h = \psi_h(v)$ is the solution of the variational problem:

$$\psi_h \in H_0^1(\Omega),$$

$$\int_\Omega \nabla\psi_h \cdot \nabla w \, dx = \int_\Omega \phi_h^2 (P_h v) w \, dx \quad \textit{for all } w \in H_0^1(\Omega). \tag{22.40}$$

Notice that, *because the set Ω is assumed to be convex, the function ψ_h is in the space* $H^2(\Omega)$, and therefore it is legitimate to consider the seminorm $|\cdot|_{\phi_h^{-1};2,\Omega}$ in inequality (22.39).

Using the definition of the function ψ_h, and letting again $v_h = P_h v$, we can write

$$|v_h|^2_{\phi_h^2;0,\Omega} = a(v_h - v, \psi_h - \Pi_h\psi_h) + \int_\Omega \phi_h^2 v_h v \,\mathrm{d}x. \tag{22.41}$$

By applying inequality (22.5) with $\alpha=1$ and inequality (22.20) with $\alpha=-1$ (this is possible because we assume $\theta_h \geqslant C_{11}h$ and $C_{11} \geqslant c_{10} = C_7(-1)$; cf. (22.32) and (22.37)), we obtain

$$|a(v_h - v, \psi_h - \Pi_h\psi_h)| \leqslant C_7 h(|v_h|_{\phi_h;1,\Omega} + |v|_{\phi_h;1,\Omega})|\psi_h|_{\phi_h^{-1};2,\Omega}. \tag{22.42}$$

Next we have

$$\int_\Omega \phi_h^2 v_h v\,\mathrm{d}x \leqslant |v_h|_{\phi_h^2;0,\Omega}|v|_{\phi_h^2;0,\Omega} \leqslant \tfrac{1}{2}(|v_h|^2_{\phi_h^2;0,\Omega} + |v|^2_{\phi_h^2;0,\Omega}), \tag{22.43}$$

so that, by combining relations (22.41), (22.42), and (22.43), we obtain the inequality

$$|v_h|^2_{\phi_h^2;0,\Omega} \leqslant C_7 h(|v_h|_{\phi_h;1,\Omega} + |v|_{\phi_h;1,\Omega})|\psi_h|_{\phi_h^{-1};2,\Omega} + \tfrac{1}{2}(|v_h|^2_{\phi_h^2;0,\Omega} + |v|^2_{\phi_h^2;0,\Omega}),$$

which in turn implies the inequality

$$|v_h|^2_{\phi_h^2;0,\Omega} \leqslant \delta|v_h|^2_{\phi_h;1,\Omega} + |v|^2_{\phi_h^2;0,\Omega} + |v|^2_{\phi_h;1,\Omega} + (1+1/\delta)C_7^2h^2|\psi_h|^2_{\phi_h^{-1};2,\Omega}$$

$$\text{for any } \delta > 0. \tag{22.44}$$

Let then $\delta = 1/(3C_{12})$ in inequality (22.44), where C_{12} is the constant appearing in inequality (22.29). Then inequality (22.44) added to inequality (22.29) times the factor $2/(3C_{12})$ yields

$$\begin{aligned}&\tfrac{1}{3}|v_h|^2_{\phi_h^2;0,\Omega} + (1/(3C_{12}))|v_h|^2_{\phi_h;1,\Omega}\\ &\quad \leqslant |v|^2_{\phi_h^2;0,\Omega} + \tfrac{5}{3}|v|^2_{\phi_h;1,\Omega} + (1+3C_{12})C_7^2h^2|\psi_h|^2_{\phi_h^{-1};2,\Omega},\end{aligned}$$

i.e., an inequality of the form (22.39).

(iii) *Given any number $\theta_0 \in\,]0,1[$, there exists a constant $C_{14}(\theta_0)$ such that*

$$|\psi_h|^2_{\phi_h^{-1};2,\Omega} \leqslant C_{14}(\theta_0)\frac{|\ln\theta_h|}{\theta_h^2}|P_h v|^2_{\phi_h^2;0,\Omega} \quad \textit{for all } \theta_h \in\,]0,\theta_0]. \tag{22.45}$$

Since $-\Delta\psi_h = \phi_h^2 v_h$ (recall that $v_h = P_h v$), we have

$$|\Delta\psi_h|_{\phi_h^{-1};0,\Omega} = |v_h|_{\phi_h^3;0,\Omega},$$

and consequently, by Theorem 22.1,

$$|\psi_h|^2_{\phi_h^{-1};2,\Omega} \leqslant C_1(|v_h|^2_{\phi_h^3;0,\Omega} + |\psi_h|^2_{1,\Omega}). \tag{22.46}$$

Since $\phi_h(x) \leqslant 1/\theta_h^2$ for all $x \in \bar\Omega$, we first find that

$$|v_h|^2_{\phi_h^3;0,\Omega} \leqslant \frac{1}{\theta_h^2}|v_h|^2_{\phi_h^2;0,\Omega}. \tag{22.47}$$

To take care of the other term that appears in the right-hand side of inequality (22.46), we shall prove that, *for each number $\theta_0 \in]0, 1[$, there exists a constant $c_{13}(\theta_0)$ such that, for all functions ϕ of the form*

$$\phi : x \in \bar{\Omega} \to \phi(x) = \frac{1}{\|x-\bar{x}\|^2+\theta^2}, \qquad \bar{x} \in \bar{\Omega}, \quad 0<\theta\leqslant\theta_0, \tag{22.48}$$

we have

$$|\psi|_{1,\Omega}^2 \leqslant c_{13}(\theta_0)\frac{|\ln\theta|}{\theta^2}|\Delta\psi|^2_{\phi^{-2};0,\Omega} \quad \text{for all } \psi\in H_0^1(\Omega)\cap H^2(\Omega); \tag{22.49}$$

the proof given here of inequality (22.49) is due to RANNACHER [1977].

Taking into account that

$$|\Delta\psi_h|_{\phi_h^{-2};0,\Omega} = |v_h|_{\phi_h^2;0,\Omega},$$

and applying inequalities (22.49) with $\psi=\psi_h$ and $\phi=\phi_h$, (22.46), and (22.47), we then find an inequality of the form (22.45), with

$$C_{14}(\theta_0) = C_1\left(\frac{1}{|\ln\theta_0|}+c_{13}(\theta_0)\right). \tag{22.50}$$

It therefore remains to prove relation (22.49) (another method for proving the same relation is given in NITSCHE [1977]). Given an arbitrary function $\psi\in H_0^1(\Omega)\cap H^2(\Omega)$, we have

$$\begin{aligned} |\psi|_{1,\Omega}^2 &= -\int_\Omega \psi\Delta\psi\,\mathrm{d}x \leqslant |\Delta\psi|_{\phi_h^{-2};0,\Omega}|\psi|_{\phi_h^2;0,\Omega} \\ &\leqslant \frac{|\ln\theta|}{2\theta^2}|\Delta\psi|^2_{\phi^{-2};0,\Omega} + \frac{\theta^2}{2|\ln\theta|}|\psi|^2_{\phi^2;0,\Omega}. \end{aligned} \tag{22.51}$$

Let then G denote the Green function associated with the operator $-\Delta$ in Ω and the boundary condition $v=0$ on Γ, so that

$$\begin{aligned} |\psi|^2_{\phi^2;0,\Omega} &= \int_\Omega \phi^2(x)\left|\int_\Omega G(x,\xi)\Delta\psi(\xi)\,\mathrm{d}\xi\right|^2 \mathrm{d}x \\ &\leqslant \int_\Omega \phi^{-2}(\xi)|\Delta\psi(\xi)|^2\left(\int_\Omega \phi^2(x)G(x,\xi)\left\{\int_\Omega \phi^2(\eta)G(x,\eta)\,\mathrm{d}\eta\right\}\mathrm{d}x\right)\mathrm{d}\xi. \end{aligned} \tag{22.52}$$

There exists a constant c_{14} such that (cf. for example STAKGOLD [1968, p. 143])

$$0\leqslant G(x,y)\leqslant c_{14}(1+|\ln\|x-y\||) \quad \text{for all } x,y\in\bar{\Omega}, \quad x\neq y. \tag{22.53}$$

Using this inequality, we proceed to show that for arbitrary points $x, \bar{x}\in\Omega$ and for any number θ with $0<\theta\leqslant\theta_0<1$, there exists a constant $c_{15}(\theta_0)$ such that

$$\int_\Omega \phi^2(\eta)G(x,\eta)\,d\eta=\int_\Omega \frac{G(x,\eta)}{(\|\eta-\bar{x}\|^2+\theta^2)^2}\,d\eta\leqslant c_{15}(\theta_0)\frac{|\ln\theta|}{\theta^2}. \tag{22.54}$$

To see this, write

$$\int_\Omega \frac{|\ln\|x-\eta\||}{(\|\eta-\bar{x}\|^2+\theta^2)^2}\,d\eta=\sum_{\lambda=1}^{3}\int_{\Omega_\lambda}\frac{|\ln\|x-\eta\||}{(\|\eta-\bar{x}\|^2+\theta^2)^2}\,d\eta,$$

where

$$\Omega_1=\Omega_1(x,\theta)=\{\eta\in\Omega;\ \|\eta-x\|\leqslant\theta\},$$
$$\Omega_2=\Omega_2(x,\theta)=\{\eta\in\Omega;\ \theta\leqslant\|\eta-x\|\leqslant 1\},$$
$$\Omega_3=\Omega_3(x)=\{\eta\in\Omega;\ 1\leqslant\|\eta-x\|\}.$$

We then obtain the following inequalities (observe that the last two inequalities make sense only if the sets Ω_2 and Ω_3 are not empty, and that we have $\operatorname{diam}\Omega\geqslant 1$ if the set Ω_3 is not empty):

$$\int_{\Omega_1}\frac{|\ln\|x-\eta\||}{(\|\eta-\bar{x}\|^2+\theta^2)^2}\,d\eta$$
$$\leqslant-\frac{1}{\theta^4}\int_{\Omega_1}\ln\|\eta-x\|\,d\eta=-\frac{1}{\theta^4}\int_{B(0;\theta)}\ln\|\xi\|\,d\xi$$
$$=\frac{\pi}{\theta^2}(\tfrac12-\ln\theta)\leqslant\pi\left(1+\frac{1}{2|\ln\theta_0|}\right)\frac{|\ln\theta|}{\theta^2},$$

$$\int_{\Omega_2}\frac{|\ln\|x-\eta\||}{(\|\eta-\bar{x}\|^2+\theta^2)^2}\,d\eta$$
$$\leqslant|\ln\theta|\int_{\Omega_2}\frac{d\eta}{(\|\eta-\bar{x}\|^2+\theta^2)^2}$$
$$\leqslant|\ln\theta|\int_{\mathbb{R}^2}\frac{d\eta}{(\|\eta-\bar{x}\|^2+\theta^2)^2}=\frac{|\ln\theta|}{\theta^2}\int_{\mathbb{R}^2}\frac{d\xi}{(1+\|\xi\|^2)^2}=\pi\frac{|\ln\theta|}{\theta^2},$$

$$\int_{\Omega_3}\frac{|\ln\|x-\eta\||}{(\|\eta-\bar{x}\|^2+\theta^2)^2}\,d\eta$$
$$\leqslant\ln(\operatorname{diam}\Omega)\int_{\Omega_3}\frac{d\eta}{(\|\eta-\bar{x}\|^2+\theta^2)^2}\leqslant\ln(\operatorname{diam}\Omega)\int_{\mathbb{R}^2}\frac{d\eta}{(\|\eta-\bar{x}\|^2+\theta^2)^2}$$
$$=\frac{\pi}{\theta^2}\ln(\operatorname{diam}\Omega)\leqslant\left(\frac{\pi\ln(\operatorname{diam}\Omega)}{|\ln\theta_0|}\right)\frac{|\ln\theta|}{\theta^2}.$$

Consequently, inequality (22.54) is proved, with

$$c_{15}(\theta_0)=\pi c_{14}\left\{\frac{3}{|\ln\theta_0|}+\frac{1}{2|\ln\theta_0|^2}+2+\frac{1+2\ln(\operatorname{diam}\Omega)}{2|\ln\theta_0|}\right\}.$$

Using this inequality, it is easy to verify that there exists a constant $c_{15}(\theta_0)$ such that

$$\int_\Omega \phi^2(\eta)G(x,\eta)\,\mathrm{d}\eta \leqslant c_{15}(\theta_0)\,\frac{|\ln\theta|}{\theta^2}.$$

Then inequalities (22.51)–(22.54) together imply inequality (22.49) with

$$c_{13}(\theta_0)=\tfrac{1}{2}(1+c_{15}^2(\theta_0)).$$

(iv) *It remains to combine the results of steps* (ii) *and* (iii): We have determined constants C_{11}, C_{13} and $C_{14}(\theta_0)$ for each $\theta_0\in\,]0,1[$ such that (cf. inequalities (22.39) and (22.45))

$$C_{11}h\leqslant\theta_h\leqslant\theta_0<1$$
$$\Rightarrow\ |P_hv|^2_{\phi_h^2;0,\Omega}+|P_hv|^2_{\phi_h;1,\Omega}$$
$$\leqslant C_{13}(|v|^2_{\phi_h^2;0,\Omega}+|v|^2_{\phi_h;1,\Omega})+C_{13}C_{14}(\theta_0)\,\frac{|\ln\theta_h|h^2}{\theta_h^2}\,|P_hv|^2_{\phi_h^2;0,\Omega}. \tag{22.55}$$

Let for example $\theta_0=\frac{1}{2}$ and let

$$\theta_h=C_9h|\ln h|^{1/2}\quad\text{with } C_9=2(C_{13}C_{14}(\tfrac{1}{2}))^{1/2}. \tag{22.56}$$

Then there exists a number $h_0\in\,]0,1[$ such that

$$h\leqslant h_0\ \Rightarrow\ \begin{cases}C_{11}h\leqslant\theta_h\leqslant\frac{1}{2}=\theta_0,\\ |\ln\theta_h|\leqslant 2|\ln h|.\end{cases} \tag{22.57}$$

This being the case, we have found an inequality of type (22.27) with

$$C_{10}=2\sqrt{C_{13}}.\qquad\square \tag{22.58}$$

We next develop the *second stage* of this analysis. Using the inequalities (cf. Theorems 22.2 and 22.3) relating the seminorms $|\cdot|_{m,\infty,\Omega}$, $m=0,1$, and the weighted seminorms that appear in inequality (22.27) we show in the next theorem that *the projection operators* P_h of (22.22), *considered as acting from the subspace* $H_0^1(\Omega)\cap W^{1,\infty}(\Omega)$ *of the space* $H_0^1(\Omega)$ *onto the space* V_h, *are bounded independently of* h *when the space* $H_0^1(\Omega)\cap W^{1,\infty}(\Omega)$ *is equipped with the norm*

$$v\to|v|_{0,\infty,\Omega}+h|\ln h|\,|v|_{1,\infty,\Omega} \tag{22.59}$$

and the space V_h *is equipped with the norm*

$$v\to|\ln h|^{-1/2}|v|_{0,\infty,\Omega}+h|v|_{1,\infty,\Omega}. \tag{22.60}$$

REMARK 22.1. Such norms may be viewed as "weighted $W^{1,\infty}(\Omega)$-like" norms.

THEOREM 22.6. *Let the constant $h_0>0$ be as in Theorem 22.5. Then there exists a constant C_{15} such that, for all $h\leqslant h_0$,*

$$|\ln h|^{-1/2}|P_h v|_{0,\infty,\Omega}+h|P_h v|_{1,\infty,\Omega}\leqslant C_{15}(|v|_{0,\infty,\Omega}+h|\ln h|\,|v|_{1,\infty,\Omega})$$
$$\text{for all } v\in H_0^1(\Omega)\cap W^{1,\infty}(\Omega). \tag{22.61}$$

PROOF. Let there be given a function v in the space $H_0^1(\Omega)\cap W^{1,\infty}(\Omega)$. For each $h\leqslant h_0$, we define the function

$$\phi_{0h}: x\in\bar{\Omega}\to\phi_{0h}(x)=\frac{1}{\|x-x_h^0\|^2+\theta_h^2}, \tag{22.62}$$

with

$$|P_h v(x_h^0)|=|P_h v|_{0,\infty,\Omega},\qquad \theta_h=C_9 h|\ln h|^{1/2}, \tag{22.63}$$

where $h_0>0$ and C_9 are the constants found in Theorem 22.5. Since $\theta_h\geqslant C_{11}h$ for $h\leqslant h_0$ (cf. (22.57)), we may apply inequality (22.15): We find in this fashion that there exists a constant

$$c_{16}=C_4(C_{11}) \tag{22.64}$$

such that

$$|P_h v|_{0,\infty,\Omega}\leqslant c_{16}(\theta_h^2/h)|P_v v|_{\phi_{0h}^2;0,\Omega}. \tag{22.65}$$

By inequality (22.27),

$$|P_h v|_{\phi_{0h}^2;0,\Omega}\leqslant C_{10}(|v|_{\phi_{0h}^2;0,\Omega}+|v|_{\phi_{0h};1,\Omega}), \tag{22.66}$$

and by inequalities (22.10) and (22.11), there exists a constant ($\theta_h\leqslant\theta_0=\frac{1}{2}$ for $h\leqslant h_0$; cf. (22.57))

$$c_{17}=\max\{C_2(2,0),\ C_3(\tfrac{1}{2},1)\} \tag{22.67}$$

such that

$$|v|_{\phi_{0h}^2;0,\Omega}+|v|_{\phi_{0h};1,\Omega}\leqslant c_{17}\left(\frac{1}{\theta_h}|v|_{0,\infty,\Omega}+|\ln\theta_h|^{1/2}|v|_{1,\infty,\Omega}\right). \tag{22.68}$$

Combining inequalities (22.65) to (22.68), we find that

$$|P_h v|_{0,\infty,\Omega}\leqslant C_{10}c_{16}c_{17}\left(\frac{\theta_h}{h}|v|_{0,\infty,\Omega}+\frac{\theta_h^2|\ln\theta_h|^{1/2}}{h}|v|_{1,\infty,\Omega}\right).$$

Using the relations $\theta_h=C_9 h|\ln h|^{1/2}$ (cf. (22.26)) and the inequality $|\ln\theta_h|\leqslant 2|\ln h|$ (cf. (22.57)), we eventually find that, for all $h\leqslant h_0$,

$$|\ln h|^{-1/2}|P_h v|_{0,\infty,\Omega}\leqslant c_{18}(|v|_{0,\infty,\Omega}+h|\ln h|\,|v|_{1,\infty,\Omega}), \tag{22.69}$$

with

$$c_{18}=C_{10}c_{16}c_{17}\max\{C_9,\sqrt{2}C_9^2\}. \tag{22.70}$$

Likewise, for each $h \leqslant h_0$, define the function

$$\phi_{1h} : x \in \bar{\Omega} \to \phi_{=h}(x) = \frac{1}{\|x - x_h^1\|^2 + \theta_h^2}, \tag{22.71}$$

with

$$\max\{|\partial_1 P_h v(x_h^1)|, |\partial_2 P_h v(x_h^1)|\} = |v_h|_{1,\infty,\Omega}, \qquad \theta_h = C_9 h |\ln h|^{1/2}. \tag{22.72}$$

Then inequality (22.17) shows that there exists a constant

$$c_{19} = C_5(C_{11}) \tag{22.73}$$

such that

$$|P_h v|_{1,\infty,\Omega} \leqslant c_{19}(\theta_h/h)|P_h v|_{\phi_{1h};1,\Omega}; \tag{22.74}$$

moreover, by inequality (22.27),

$$|P_h v|_{\phi_{1h};1,\Omega} \leqslant C_{10}(|v|_{\phi_{1h}^2;0,\Omega} + |v|_{\phi_{1h};1,\Omega}). \tag{22.75}$$

Then, arguing as before, we find that for all $h \leqslant h_0$,

$$h|P_h v|_{1,\infty,\Omega} \leqslant c_{20}(|v|_{0,\infty,\Omega} + h|\ln h|\, |v|_{1,\infty,\Omega}) \tag{22.76}$$

with

$$c_{20} = C_{10} c_{17} c_{19} \max\{1, \sqrt{2} C_9\}. \tag{22.77}$$

The conjunction of inequalities (22.69) and (22.76) implies that inequality (22.61) holds, with

$$C_{15} = c_{18} + c_{20}. \qquad \square$$

REMARK 22.2. In Theorem 22.5, the behavior of θ_h as a function of h is bounded below by a constant times $(h|\ln h|^{1/2})$. The key to the success of the present argument is that such a function of h tends nevertheless sufficiently rapidly vers zero with h so as to produce the right factors (as functions of h) in inequalities (22.69) and (22.76).

In the *third, and final, stage* of this study, we show that the uniform boundedness of the projection mappings P_h (which we just established) yields in turn the desired error estimates (recall that the discrete solution u_h is nothing but the projection $P_h u$ of the solution u).

THEOREM 22.7. *Assume that the solution $u \in H_0^1(\Omega)$ of the boundary value problem associated with the data (22.1) is also in the space $W^{2,\infty}(\Omega)$. Then there exists a constant C independent of h such that*

$$|u - u_h|_{0,\infty,\Omega} \leqslant C h^2 |\ln h|^{3/2} |u|_{2,\infty,\Omega}, \tag{22.78}$$

$$|u - u_h|_{1,\infty,\Omega} \leqslant C h |\ln h|\, |u|_{2,\infty,\Omega}. \tag{22.79}$$

PROOF. The norm of the identity mapping acting from the space $H_0^1(\Omega) \cap W^{1,\infty}(\Omega)$

equipped with the norm of (22.59) into the same space, but equipped with the norm of (22.60), is bounded above by $|\ln h_0|^{-1/2}$ for all $h \leqslant h_0' = \min\{h_0, 1/e\}$. We next have the identity

$$u - u_h = u - P_h u = (I - P_h)(u - v_h) \quad \text{for all } v_h \in V_h,$$

so that we infer from Theorem 22.6 that, for all $h \leqslant h_0'$,

$$|\ln h|^{-1/2} |u - u_h|_{0,\infty,\Omega} + h|u - u_h|_{1,\infty,\Omega}$$
$$\leqslant (|\ln h_0|^{-1/2} + C_{15}) \inf_{v_h \in V_h} (|u - v_h|_{0,\infty,\Omega} + h|\ln h|\,|u - v_h|_{1,\infty,\Omega}).$$

Since there exists a constant c_{21} such that

$$\inf_{v_h \in V_h} (|u - v_h|_{0,\infty,\Omega} + h|\ln h|\,|u - v_h|_{1,\infty,\Omega}) \leqslant c_{21} h^2 |\ln h|\,|u|_{2,\infty,\Omega},$$

inequalities (22.78) and (22.79) follow with

$$C = c_{21}(|\ln h_0|^{-1/2} + c_{15}). \qquad \square$$

In fact, the error estimate of (22.78) is not optimal: NITSCHE [1977] has obtained the improved error bound

$$|u - u_h|_{0,\infty,\Omega} \leqslant Ch^2 |\ln h|\,|u|_{2,\infty,\Omega}, \tag{22.80}$$

at the expense of a technical refinement in the argument, special to linear triangles. Observe, however, that the discrepancy between the error estimates (22.78) and (22.80) is in a sense insignificant: Both estimates yield an $O(h^{2-\varepsilon})$ convergence for any $\varepsilon > 0$.

To conclude, it is worth pointing out that all the essential features of *Nitsche's method of weighted norms* have been presented: Indeed, the extension to more general cases proceeds along the same lines. In particular, the use of *higher-order polynomial spaces* (i.e., $P_T = P_k(T)$ for some $k \geqslant 2$, n arbitrary) yields a simplification in that the "$|\ln h|$" term present for $k = 1$ disappears in the norms then considered. Thus inequality (22.61) is replaced by an inequality of the simpler form (cf. NITSCHE [1975])

$$|P_h v|_{0,\infty,\Omega} + h|P_h v|_{1,\infty,\Omega} \leqslant C(|v|_{0,\infty,\Omega} + h|v|_{1,\infty,\Omega}). \tag{22.81}$$

Such inequalities are obtained after inequalities reminiscent of that of (22.27) have been established for appropriate weighted norms of the form $|\cdot|_{\phi_h^{\alpha+1};0,\Omega} + |\cdot|_{\phi_h^{\alpha};1,\Omega}$, $\frac{1}{2}n < \alpha < \frac{1}{2}n + 1$, with functions ϕ_h again defined as in (22.25).

The "$|\ln h|$ factor" in the estimate of the error $|u - u_h|_{0,\infty,\Omega}$ for linear triangles has been the subject of various conjectures as to whether it should appear or not, although Claes Johnson (cf. SCOTT [1976a, p. 684]), JESPERSSEN [1978], FRIED [1980] had pointed out that it is unavoidable, the reason being that piecewise linear polynomials cannot approximate sufficiently well a Green's function. The controversy was settled by HAVERKAMP [1984], who showed by means of an example that the estimate no longer holds if $|\ln h|$ is replaced by $o(|\ln h|)$.

The subject of uniform convergence of finite element methods has a long story. In one dimension, an $O(h^2)$ uniform convergence for piecewise linear approximations was first obtained by CIARLET [1968]. Further results were then established by NITSCHE [1969], CIARLET and VARGA [1970], DOUGLAS and DUPONT [1973, 1976b], DOUGLAS, DUPONT and WAHLBIN [1975b], NATTERER [1977]. For special types of triangulations in higher dimensions, see BRAMBLE, NITSCHE and SCHATZ [1975], BRAMBLE and SCHATZ [1976], BRAMBLE and THOMÉE [1974], DOUGLAS, DUPONT and WHEELER [1974b], NATTERER [1975b].

The first contribution to the general case is that of NITSCHE [1970]. Then CIARLET and RAVIART [1973] improved the analysis of Nitsche by using a *discrete maximum principle*, as shown in Sections 20 and 21, where various references on this approach are found.

Then NATTERER [1975a], NITSCHE [1975, 1976b, 1977] and SCOTT [1976a] obtained simultaneously optimal (or nearly optimal) orders of convergence. The greatest generality is achieved by the method of weighted norms of NITSCHE [1975] (which we have followed in this section; see also NITSCHE [1977, 1981b]), and in the work of SCOTT [1976a]. While weighted Sobolev norms are also introduced by Natterer, Scott's main tool is a careful analysis of the *approximation* of the *Green's function*. Nitsche's method has been combined with a duality argument using a regularized Green's function, by FREHSE and RANNACHER [1976, 1978], and RANNACHER and SCOTT [1982]. Nitsche's method has also been extended to more general second-order boundary value problems, by RANNACHER [1976b]; to the obstacle problem (cf. the next section); to higher-order triangles, by RAUGEL [1978b]; to Orlicz normes, by DURÁN [1987]; to plate problems, by RANNACHER [1976a].

The uniform boundedness in appropriate norms of particular Hilbertian projections, on which Nitsche's argument is essentially based, was also noticed by DOUGLAS, DUPONT and WAHLBIN [1975a], who have established through a different approach the boundedness in the norms $|\cdot|_{0,q,\Omega}$, $1 \leqslant q \leqslant \infty$, of the $L^2(\Omega)$-projections onto certain finite element spaces; see also CROUZEIX and THOMÉE [1987]. The boundedness in $L^\infty(\Omega)$ of the $H^1(\Omega)$-projections have been established by SCHATZ [1980], SCHATZ and WAHLBIN [1982], SUZUKI and FUJITA [1986]; and in $W_0^{1,p}(\Omega)$, by RANNACHER and SCOTT [1982], who then obtained, by an interpolation argument, estimates of the error $|u-u_h|_{0,p,\Omega}$, $2 \leqslant p < \infty$, and $|u-u_h|_{1,p,\Omega}$, $1 < p \leqslant \infty$.

23. Estimate of the error $\|u-u_h\|_{1,\Omega}$ for the obstacle problem and Falk's method

To conclude this chapter, we consider the finite element approximation of problems posed as *variational inequalities*. Following an analysis due to FALK [1974], we shall first prove an abstract error estimate (Theorem 23.1), which is valid for a general class of approximation schemes for variational inequalities of the form (23.1) below, and then we shall apply this result to the finite element approximation of the *obstacle problem* (Theorem 23.2).

The *abstract setting* is the following: Let V be a Hilbert space, with norm $\|\cdot\|$, let $a(\cdot,\cdot): V \times V \to \mathbb{R}$ be a continuous, symmetric and V-elliptic bilinear form (with the usual V-ellipticity and continuity constants α and M), let $l: V \to \mathbb{R}$ be a continuous linear form, and let U be a nonempty closed convex subset of V. Then there is a unique element u that satisfies (cf. Theorem 1.2)

$$\begin{aligned} &u \in U, \\ &a(u, v-u) \geqslant l(v-u) \quad \text{for all } v \in U. \end{aligned} \tag{23.1}$$

Let V_h be a finite-dimensional subspace of the space V and let U_h be a nonempty closed convex subset of V_h. Observe that, in general, *the set U_h is not a subset of U*. Then, quite naturally, the *discrete problem* consists in finding an element u_h such that

$$\begin{aligned} &u_h \in U_h, \\ &a(u_h, v_h-u_h) \geqslant l(v_h-u_h) \quad \text{for all } v_h \in U_h, \end{aligned} \tag{23.2}$$

and, again by Theorem 1.2, this problem has a unique solution u_h.

In the proof of the next theorem, we shall use the mapping $A \in \mathscr{L}(V; V')$ defined by the relations

$$Av(w) = a(v, w) \quad \text{for all } v, w \in V, \tag{23.3}$$

and which was already used in the proof of Theorem 1.3. Notice that in the present situation we do *not* have $Au = l$ in general, as in the case of the linear problem $(U = V)$. Also, we shall consider a Hilbert space H, with the norm $|\cdot|$ and inner product $(\cdot, \cdot)$, such that

$$\bar{V} = H \quad \text{and} \quad V \hookrightarrow H. \tag{23.4}$$

The space H will be identified with its dual, so that it may be in turn identified with a subspace of the dual space of V, as we showed in Section 19.

We now estimate the error $\|u-u_h\|$.

THEOREM 23.1. *Assume that $l(v) = (f, v)$ for some $f \in H$ and that*

$$Au \in H. \tag{23.5}$$

Then there exists a constant C independent of the subspace V_h and of the set U_h such that

$$\|u-u_h\| \leqslant C\left\{ \inf_{v_h \in U_h} \{\|u-v_h\|^2 + |Au-f|\,|u-v_h|\} + |Au-f| \inf_{v \in U} |u_h - v| \right\}^{1/2}. \tag{23.6}$$

PROOF. We have

$$\alpha\|u-u_h\|^2 \leqslant a(u-u_h, u-u_h) = a(u,u) + a(u_h,u_h) - a(u,u_h) - a(u_h,u),$$

and thus, by (23.1) and (23.2),

$$a(u,u)\leqslant a(u,v)+f(u-v) \qquad \text{for all } v\in U,$$
$$a(u_h,u_h)\leqslant a(u_h,v_h)+f(u_h-v_h) \quad \text{for all } v_h\in U_h.$$

Therefore we deduce that, for all $v\in U$ and all $v_h\in U_h$,

$$\begin{aligned}\alpha\|u-u_h\|^2 &\leqslant a(u,v-u_h)+a(u_h,v_h-u)+f(u-v)+f(u_h-v_h)\\ &=a(u,v-u_h)-f(v-u_h)+a(u,v_h-u)-f(v_h-u)+a(u_h-u,v_h-u)\\ &=(f-Au,u-v_h)+(f-Au,u_h-v)+a(u-u_h,u-v_h).\end{aligned}$$

We thus have, for all $v\in U$ and all $v_h\in U_h$,

$$\alpha\|u-u_h\|^2\leqslant|f-Au|\{|u-v_h|+|u_h-v|\}+M\|u-u_h\|\,\|u-v_h\|.$$

Since

$$\|u-u_h\|\,\|u-v_h\|\leqslant\frac{1}{2}\left\{\frac{\alpha}{M}\|u-u_h\|^2+\frac{M}{\alpha}\|u-v_h\|^2\right\},$$

we obtain, upon combining the two previous inequalities,

$$\tfrac{1}{2}\alpha\|u-u_h\|^2\leqslant|f-Au|\{|u-v_h|+|u_h-v|\}+\tfrac{1}{2}(M^2/\alpha)\|u-v_h\|^2, \tag{23.7}$$

from which inequality (23.6) follows. □

Several comments are in order about this theorem:

(i) The proof has been given in such a way that it includes the case where the bilinear form is not symmetric.

(ii) If $U=V$ then $Au-f=0$, so that, with the natural choice $U_h=V_h$, the error estimate of (23.6) reduces to the familiar error estimate of Céa's lemma (Theorem 13.1).

(iii) If the inclusion $U_h\subset U$ holds, then the term $\inf_{v\in U}|u_h-v|$ (which can be expected to be the harder to evaluate) vanishes in the error estimate. This is not the case, however, of the finite element approximation of the obstacle problem which we shall describe.

(iv) Had we not introduced the space H in our argument, we would have found, instead of inequality (23.7), the inequality

$$\tfrac{1}{2}\alpha\|u-u_h\|^2\leqslant\|Au-l\|'\{\|u-v_h\|+\|u_h-v\|\}+\tfrac{1}{2}(M^2/\alpha)\|u-v_h\|^2, \tag{23.8}$$

where $\|\cdot\|'$ denotes as usual the norm of the dual space of V. However inequality (23.8) is likely to yield a poorer order of convergence, since the term $\inf_{v_h\in U_h}|u-v_h|$ can be reasonably anticipated to be of a higher order than the term $\inf_{v_h\in U_h}\|u-v_h\|$. This observation will be confirmed in the proof of Theorem 23.2.

We next apply the abstract error estimate of Theorem 23.1 to the finite element approximation of the *obstacle problem*; this problem corresponds to the following data (cf. (3.41))

$$\begin{aligned} &V=H_0^1(\Omega), \quad \Omega\subset\mathbb{R}^2,\\ &U=\{v\in H_0^1(\Omega); v\geqslant\chi \text{ a.e. in } \Omega\}, \qquad \chi\in H^2(\Omega), \quad \chi\leqslant 0 \quad \text{on } \Gamma,\\ &a(u,v)=\int_\Omega \nabla u\cdot\nabla v\,dx,\\ &l(v)=\int_\Omega fv\,dx, \quad f\in L^2(\Omega). \end{aligned} \tag{23.9}$$

We shall assume that $\bar{\Omega}$ is a *polygonal domain* (so that $\bar{\Omega}$ can be exactly covered by triangulations made up of triangles; cf. Remark 23.1 for the case of a domain with a curved boundary), and that the solution u is in the space $H^2(\Omega)$. Notice that while this is a reasonable regularity assumption, it would be unrealistic to assume that $u\in H^3(\Omega)$ for instance (cf. the discussion given in Section 3). Finally, observe that since $H^2(\Omega)\subset\mathscr{C}^0(\bar{\Omega})$, the point values of the function χ are well defined.

With a triangulation $\mathscr{T}_h$ of the set $\bar{\Omega}=\bigcup_{T\in\mathscr{T}_h}T$, we associate the finite element space X_h whose generic finite element is the linear triangle, and we let

$$V_h=\{v_h\in X_h;\ v_h=0 \text{ on } \Gamma\}=X_{0h}, \tag{23.10}$$

and

$$U_h=\{v_h\in V_h;\ v_h(b)\geqslant\chi(b) \text{ for all } b\in\mathscr{N}_h\}, \tag{23.11}$$

where $\mathscr{N}_h$ denotes the set of nodes of the space X_h.

Notice that *the set U_h is not in general contained in the set U*, as shown in Fig. 23.1 in the one-dimensional case.

Let us now apply the abstract error estimate of Theorem 23.1.

THEOREM 23.2. *Assume that the solution u is the space $H^2(\Omega)$. Then, for any regular family of triangulations, there exists a constant $C(u,f,\chi)$ independent of h such that*

$$\|u-u_h\|_{1,\Omega}\leqslant C(u,f,\chi)h. \tag{23.12}$$

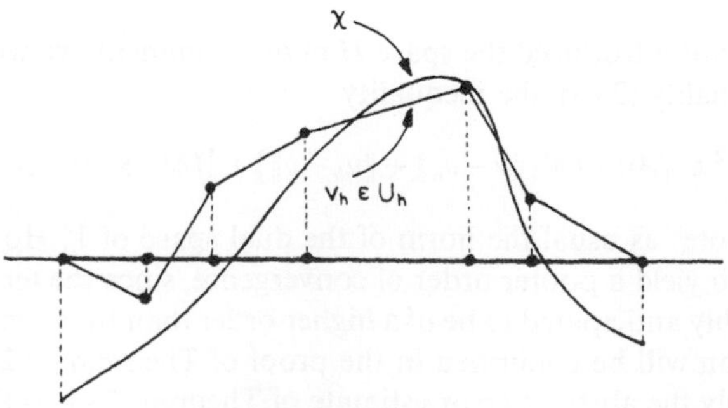

FIG. 23.1. The set $U_h=\{v_h\in V_h;\ v_h(b)\geqslant\chi(b) \text{ for all } b\in\mathscr{N}_h\}$ is not in general a subset of the set U.

PROOF. We apply Theorem 23.1 with

$$H = L^2(\Omega),$$

so that we need to verify that $Au \in L^2(\Omega)$ ($f \in L^2(\Omega)$ by assumption). Since the solution u is assumed to be in the space $H^2(\Omega)$, we have

$$Au(v) = \int_\Omega \nabla u \cdot \nabla v \, dx = -\int_\Omega \Delta u \, v \, dx \quad \text{for all } v \in V,$$

and thus

$$|Au(v)| \leqslant |\Delta u|_{0,\Omega} |v|_{0,\Omega} \quad \text{for all } v \in V.$$

Hence Au is indeed an element of the space H.

Let $\Pi_h u$ denote as usual the X_h-interpolant of the function u, which is in the space V_h. Since

$$\Pi_h u(b) = u(b) \geqslant \chi(b) \quad \text{for all } b \in \mathcal{N}_h,$$

it is also an element of the set U_h. Thus,

$$\begin{aligned} &\inf_{v_h \in U_h} \{\|u - v_h\|_{1,\Omega}^2 + |Au - f|_{0,\Omega} |u - v_h|_{0,\Omega}\} \\ &\leqslant \|u - \Pi_h u\|_{1,\Omega}^2 + \{|\Delta u|_{0,\Omega} + |f|_{0,\Omega}\} |u - u_h|_{0,\Omega} \\ &\leqslant C(|u|_{2,\Omega}^2 + \{|\Delta u|_{0,\Omega} + |f|_{0,\Omega}\}) u|_{2,\Omega}) h^2. \end{aligned} \tag{23.13}$$

In order to estimate the term $\inf_{v \in U} |u_h - v|_{0,\Omega}$, it is convenient to introduce the function (Fig. 23.2)

$$u_h^* = \max\{u_h, \chi\}. \tag{23.14}$$

Both functions u_h and χ being in the space $H^1(\Omega)$, their maximum u_h^* is also in $H^1(\Omega)$ (this is a nontrivial fact, whose proof may be found in LEWY and STAMPACCHIA [1969, p. 169]), and the assumption $\chi \leqslant 0$ on Γ further implies that $u_h^* \in H_0^1(\Omega)$. Finally, $u_h^* \geqslant \chi$ in Ω (by construction; cf. (23.14)) and thus $u_h^* \in U$. Let

$$\Lambda_h = \{x \in \Omega;\, u_h < \chi\},$$

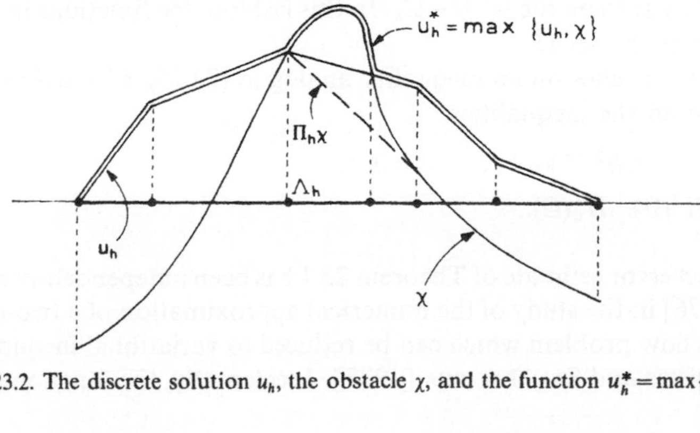

FIG. 23.2. The discrete solution u_h, the obstacle χ, and the function $u_h^* = \max\{u_h, \chi\}$.

so that

$$|u_h - u_h^*|_{0,\Omega}^2 = \int_{\Lambda_h} |u_h - \chi|^2 \, dx,$$

since $u_h - u_h^* = 0$ on $\Omega - \Lambda_h$. Since the X_h-interpolant $\Pi_h \chi$ of the function χ satisfies

$$u_h(b) \geqslant \chi(b) = \Pi_h \chi(b) \quad \text{for all } b \in \mathcal{N}_h,$$

it follows that

$$u_h - \Pi_h \chi \geqslant 0 \quad \text{in } \Omega,$$

since we are using piecewise affine functions. Consequently

$$0 < |(\chi - u_h)(x)| = (\chi - u_h)(x) \leqslant (\chi - \Pi_h \chi)(x) = |(\chi - \Pi_h \chi)(x)| \quad \text{for all } x \in \Lambda_h,$$

and thus

$$|u_h - u_h^*|_{0,\Omega}^2 = \int_{\Lambda_h} |u_h - \chi|^2 \, dx \leqslant \int_{\Lambda_h} |\chi - \Pi_h \chi|^2 \, dx \leqslant |\chi - \Pi_h \chi|_{0,\Omega}^2.$$

Therefore,

$$\inf_{v \in U} |u_h - v|_{0,\Omega} \leqslant |u_h - u_h^*|_{0,\Omega} \leqslant |\chi - \Pi_h \chi|_{0,\Omega} \leqslant C |\chi|_{2,\Omega} h^2, \tag{23.15}$$

and the conclusion follows from inequalities (23.13) and (23.15). □

REMARK 23.1. As shown by FALK [1975], the error estimate (23.12) of Theorem 23.2 holds unchanged in the following situation: The set Ω is convex with a sufficiently smooth boundary (so that $u \in H^2(\Omega)$). Let $\bar{\Omega}_h = \bigcup_{T \in \mathcal{T}_h} T$ denote a triangulation made up of triangles, in such a way that all the vertices of $\mathcal{T}_h$ that are on the boundary of the set Ω_h are also on Γ.

Let then X_h denote the associated finite element space, whose generic element is the linear triangle, and let X_{0h} denote the subspace of X_h whose functions vanish on the boundary of the set Ω_h. The space V_h then consists of the functions in the space X_{0h} extended by zero on the set $\bar{\Omega} - \bar{\Omega}_h$. In this fashion, the functions in the space V_h are defined over the set $\bar{\Omega}$.

The proof then relies on an inequality analog to (23.15), adapted to the present situation, and on the inequalities

$$\|u\|_{m,\Omega - \Omega_h} \leqslant C h^{2-m} \|u\|_{2,\Omega}, \quad m = 0, 1,$$
$$\text{if } u \in H^2(\Omega) \cap H_0^1(\Omega).$$

The abstract error estimate of Theorem 23.1 has been independently rediscovered by ROUX [1976] in the study of the numerical approximation of a two-dimensional compressible flow problem which can be reduced to variational inequalities by the method of BREZIS and STAMPACCHIA [1973]. Incidentally, the functional setting for this problem is interesting in itself in that the corresponding space V is a weighted Sobolev space, and the domain of definition of its functions is unbounded. The same problem is similarly studied via variational inequalities by CIAVALDINI and TOURNEMINE [1977], who have extended the abstract error estimate of Theorem 23.1 so as to include the case where the bilinear and linear forms are approximated, through the process of numerical integration. The related problem of a linearly elastic body resting on a support has been thoroughly studied by FREMOND [1971, 1972].

The error estimate of Theorem 23.2 has also been established by MOSCO and STRANG [1974]; see also MOSCO and SCARPINI [1975]. BREZZI, HAGER and RAVIART [1976] have given another proof of Theorem 23.2. They have also shown that $\|u-u_h\|_{1,\Omega}=O(h^{3/2-\varepsilon})$, $\varepsilon>0$ arbitrarily small, when quadratic triangles are used. NATTERER [1975a] has studied the error in the norm $|\cdot|_{0,\Omega}$, using an argument based on the Aubin–Nitsche lemma. For another approach, see BERGER [1976].

NITSCHE [1977] has been able to apply his method of weighted norms to this problem. In this fashion, he obtains an estimate of the form

$$|u-u_h^*|_{0,\infty,\Omega}\leqslant Ch^2|\ln h|(\|u\|_{2,\infty,\Omega}+\|\psi\|_{2,\infty,\Omega}).$$

However, the corresponding discrete solution u_h^* is found in the subset $U_h^*=U\cap X_{0h}$, instead of the present subset U_h. Then various estimates of the error $|u-u_h|_{0,\infty,\Omega}$ have been obtained, by BAIOCCHI [1977], FINZI–VITA [1982], CORTEY–DUMONT [1983].

The *elastic-plastic torsion problem*, which we described in Section 3, is another instance of a problem modeled by variational inequalities: The space V, the bilinear form $a(\cdot,\cdot)$, and the linear form l, are as in (23.9), while the set U is of the form

$$U=\{v\in H_0^1(\Omega);\ \|\nabla v\|\leqslant 1 \text{ a.e. in } \Omega\}. \tag{23.16}$$

Assuming again that $\bar{\Omega}$ is a polygonal domain, we let

$$U_h=\{v_h\in V_h;\ \|\nabla v_h\|\leqslant 1 \text{ a.e. in } \Omega\}, \tag{23.17}$$

the space V_h being defined as in (23.10). Hence the *inclusion*

$$U_h\subset U \tag{23.18}$$

is satisfied in this case. This simplification is compensated, however, by the fact that *the X_h-interpolant of a function of the set $U\cap\mathscr{C}^0(\bar{\Omega})$ is not necessarily contained in the set U_h*. It can be shown, however, that if $u\in W^{2,p}(\Omega)$ for some $p>2$, there exist numbers $\varepsilon(h)>0$ such that

$$\lim_{h\to 0}\varepsilon(h)\to 0,\qquad (1+\varepsilon(h))^{-1}\Pi_h u\in U_h.$$

This property implies in turn that (cf. GLOWINSKI [1984, p. 52]):

$$\|u-u_h\|_{1,\Omega}=O(h^{1/2-1/p}).$$

The elastic-plastic torsion problem has been extensively studied in LANCHON [1972]. Using techniques from duality theory, FALK and MERCIER [1977] have constructed a finite element method that directly yields an approximation of the stresses σ_{13} and σ_{23} with an O(h) convergence in the norm $|\cdot|_{0,\Omega}$, a particularly appropriate estimate for this problem, where a direct knowledge of the stresses is more important than a knowledge of the stress function. For related results, see MERCIER [1975a, 1975b], GABAY and MERCIER [1976], and BREZZI, JOHNSON and MERCIER [1977], where elasto-plastic plates are considered.

A third type of problem that reduces to variational inequalities occurs with sets U of the form

$$U=\{v\in H^1(\Omega);\ v\geqslant\psi \text{ a.e. on } \Gamma\}.$$

Such problems with unilateral constraints occur in particular in elasticity; they are then called *Signorini problems* (cf. (3.47)–(3.48)). A finite element approximation of such problems is studied in SCARPINI and VIVALDI [1977].

An extension of the present setting consists in looking for the solution u of variational inequalities of the form (see DUVAUT and LIONS [1972]):

$$a(u, v-u)+j(v)-j(u)\geqslant f(v-u) \quad \text{for all } v\in U,$$

where $j\colon V\to\mathbb{R}$ is a *nondifferentiable* functional. Such problems are found in particular in the study of *Bingham flows*, with $j(v)=\int_\Omega\|\nabla v\|\,dx$. Their finite element approximations have been analyzed in BRISTEAU [1975, chapter 2], FORTIN [1972], GLOWINSKI [1975].

Extensive treatments of variational inequalities and of their approximations are found in GLOWINSKI, LIONS and TRÉMOLIÈRES [1976a, 1976b] and in GLOWINSKI [1984].

24. Additional references

There is an enormous literature on the numerical analysis and implementation of the finite element method; our purpose in this section is simply to list some relevant references in various domains that shall not be discussed in this article.

We first mention that error analyses have often been performed on *variants of the finite element method* that are interesting by themselves. In this direction, see notably AUBIN [1967b, 1972], BABUŠKA [1970b, 1971a, 1971b, 1972b, 1974a], FIX and STRANG [1969], DI GUGLIELMO [1971], STRANG [1971], STRANG and FIX [1971], ROSE [1975], MOCK [1976]. More specifically, *"nonuniform" error estimates* have been obtained by BABUŠKA and KELLOGG [1975] and HELFRICH [1976]; *indefinite bilinear forms* are considered by CLÉMENT [1974], SCHATZ [1974]; HOPPE [1973] has suggested using *piecewise harmonic polynomials*, and his idea has been justified by RABIER [1977]. We also mention the very challenging and promising field of study opened by WERSCHULZ [1982], who has analyzed the finite element method from the viewpoint of *computational optimality*.

There are various ways of treating *nonhomogeneous Dirichlet boundary conditions*. The most straightforward method consists in interpolating the boundary condition at the boundary nodes: See AUBIN [1972], STRANG and FIX [1973, Section 4.4],

THOMÉE [1973a]. Lagrange multipliers may also be used, as in BABUŠKA [1973a], PITKÄRANTA [1979], BRAMBLE [1981], as well as penalty techniques, as in BABUŠKA [1973b], UTKU and CAREY [1982], SHI [1984f]. *Neumann problems* are considered by MOLCHANOV and GALBA [1985], while truly *mixed problems* are treated by CHOU and WANG [1979], ŽENÍŠEK [1987].

The *computation of gradients* of solutions is often of great importance, for example in the stress analysis of elastic structures. For this aspect, and the related question of *superconvergence*, see DOUGLAS and DUPONT [1973, 1974], DOUGLAS, DUPONT and WHEELER [1974a, 1974b], BRAMBLE and SCHATZ [1976, 1977], THOMÉE [1977], ZLÁMAL [1977, 1978], LESAINT and ZLÁMAL [1979], LOUIS [1979], KŘÍŽEK and NEITTAANMÄKI [1984], BABUŠKA and MILLER [1984a, 1984b], BAKKER [1984], CIAVALDINI and CROUZEIX [1985], LEVINE [1985], WHEELER and WHITEMAN [1987], and the article of Wahlbin.

Superconvergence may be also obtained by applying the *Richardson extrapolation procedure* to an appropriate *asymptotic expansion of the finite element solution*. After the pioneering work of LIN, LU and SHEN [1983], such asymptotic expansions have been studied by LIN and LU [1984a, 1984b], LIN and ZHU [1984, 1985, 1986], BLUM, LIN and RANNACHER [1986], NAKAO [1987].

Since the solution of realistic boundary value problems usually presents *singularities* at some portions of the domain $\bar{\Omega}$, "*interior*" or "*local*" *error estimates* have proved to be an invaluable tool for handling such situations, where the "global" error estimates obtained in this chapter are of no avail. In this respect, see the pioneering contributions of NITSCHE [1972a], NITSCHE and SCHATZ [1974], BRAMBLE and THOMÉE [1974], BRAMBLE, NITSCHE and SCHATZ [1975], DESCLOUX [1975, 1977], DESCLOUX and NASSIF [1977], DOUGLAS and DUPONT [1976], SCHATZ and WAHLBIN [1977, 1978, 1979, 1981], and the article of Wahlbin.

More generally, for problems where the solutions present *singularities* due to *corners*, to *changes in boundary conditions*, to *singular right-hand sides* (such as the Dirac distribution), or to *coefficients that present singularities*, see FIX [1969], CIARLET, NATTERER and VARGA [1970], WAIT and MITCHELL [1971], BABUŠKA [1970a, 1972a, 1974b, 1976], DAILEY and PIERCE [1972], BABUŠKA and ROSENZWEIG [1972], FRIED and YANG [1972], VEIDINGER [1972], BARNHILL and WHITEMAN [1973, 1975], CROUZEIX and THOMAS [1973], FIX, GULATI and WAKOFF [1973], STRANG and FIX [1973, Chapter 8], SCOTT [1973b], NITSCHE [1976a], SCHATZ and WAHLBIN [1976], THATCHER [1976], JESPERSSEN [1978], RAUGEL [1978a, 1978b], SCHREIBER [1980], WHITEMAN and AKIN [1980], BENDALLI [1981], MERCIER and RAUGEL [1982], WHITEMAN [1982], BLUM and DOBROWOLSKI [1983], ERIKSSON and THOMÉE [1984], WAHLBIN [1984], ERIKSSON [1985], CASAS [1985a, 1985b, 1985c], BABUŠKA and OSBORN [1986], LI [1986], FRENCH [1987], LI, MATHON and SERMER [1987], and the article of Wahlbin.

CHAPTER IV

The Effect of Numerical Integration for Second-Order Problems

Introduction

Up to now, we have considered finite elements methods that are *conforming*, in the sense that the space V_h is a subspace of the space V, and the bilinear form and the linear form used in the definition of the discrete problem are identical to those of the original problem. In this and the next two chapters, we study various violations of this "conformity", which are constantly used in everyday computations. To begin with, we examine in this chapter the effect of *numerical integration*.

Assuming as before that the domain $\bar{\Omega}$ is polygonal and that the inclusion $V_h \subset V$ still holds, we use in addition a *quadrature scheme* for computing the coefficients of the resulting linear system. Each such coefficient being of the form

$$\sum_{T \in \mathcal{T}_h} \int_T \varphi(x)\, dx,$$

the integrals $\int_T \varphi(x)\, dx$, $T \in \mathcal{T}_h$, are approximated by finite sums of the form $\sum_{l=1}^{L} \omega_{l,T}\, \varphi(b_{l,T})$, whose *weights* $\omega_{l,T}$ and *nodes* $b_{l,T} \in T$ are derived from a single *quadrature formula* defined over a reference finite element. Examples of useful quadrature formulae are given in Section 25. This approximation thus yields an *approximate bilinear form* $a_h(\cdot,\cdot)$ and an *approximate linear form* $l_h(\cdot)$, which are defined over the space V_h, but *not* over the space V.

Our study of this kind of approximation follows a general pattern that is also common to the two other methods described in Chapters V and VI. First, we prove an *abstract error estimate*, the *first Strang lemma* (cf. Theorem 26.1), which relies on the crucial assumption that the approximate bilinear forms are *uniformly V_h-elliptic*, in the sense that there exists a constant $\tilde{\alpha} > 0$ independent of h such that $a_h(v_h, v_h) \geqslant \tilde{\alpha} \|v_h\|^2$ for all $v_h \in V_h$. This is why we next examine (Theorem 27.1) under which assumptions on the quadrature scheme over the reference finite element this property is true.

The abstract error estimate of Theorem 26.1 generalizes Céa's lemma: In the right-hand side of the inequality, there appear two additional *consistency errors*, which measure the quality of the approximation of the bilinear form and of the linear form, respectively.

We are then in a position to study the convergence of such methods. More precisely, we shall essentially concentrate on the following problem: *Find sufficient conditions that insure that the order of convergence in the absence of numerical integration is unaltered by the effect of numerical integration.* Restricting ourselves for simplicity to the case where $P_T = P_k(T)$ for all $T \in \mathcal{T}_h$, our main result in this direction (Theorem 29.1) is that one still has

$$\|u - u_h\|_{1,\Omega} = O(h^k),$$

provided the quadrature formula is exact for all polynomials of degree $(2k-2)$. The proof of this result depends in particular on the *Bramble–Hilbert lemma* (Theorem 28.1), which is a useful tool for handling linear functionals that vanish on polynomial subspaces. In the present case, it is repeatedly used in the derivation of the consistency error estimates (Theorems 28.2 and 28.3).

25. The effect of numerical integration and examples of numerical quadrature schemes

Throughout this chapter, we shall assume that we are solving the second-order boundary value problem that corresponds to the following data:

$$\begin{aligned} &V = H_0^1(\Omega), \\ &a(u, v) = \int_\Omega \sum_{i,j=1}^n a_{ij} \partial_i u \partial_j v \, dx, \\ &l(v) = \int_\Omega f v \, dx, \end{aligned} \tag{25.1}$$

where $\bar{\Omega}$ is a polygonal domain in $\mathbb{R}^n$, and the functions $a_{ij} \in L^\infty(\Omega)$ and $f \in L^2(\Omega)$ are assumed to be *everywhere* defined over $\bar{\Omega}$. We shall also assume that the *ellipticity condition* is satisfied i.e., that there exists β such that

$$\begin{aligned} &\beta > 0, \\ &\sum_{i,j=1}^n a_{ij}(x)\xi_i\xi_j \geqslant \beta \sum_{i=1}^n \xi_i^2 \quad \text{for all } x \in \bar{\Omega} \text{ and all } \xi_i,\ 1 \leqslant i \leqslant n. \end{aligned} \tag{25.2}$$

Hence the bilinear form of (25.1) is $H_0^1(\Omega)$-elliptic.

Solving this problem amounts to formally solving (cf. (3.24)) the homogeneous Dirichlet problem for the operator $u \to -\sum_{i,j=1}^n \partial_j(a_{ij}\partial_i u)$, viz.,

$$\begin{aligned} &-\sum_{i,j=1}^n \partial_j(a_{ij}\partial_i u) = f \quad \text{in } \Omega, \\ &u = 0 \quad \text{on } \Gamma. \end{aligned} \tag{25.3}$$

Indications for handling more general operators, such as $u \to -\Sigma_{i,j=1}^{n} \partial_j(a_{ij}\partial_i u)+bu$, are given in Section 29.

We are given a family of finite element spaces X_h made up of finite elements (T, P_T, Σ_T) $T \in \mathcal{T}_h$, where $\mathcal{T}_h$ are triangulations of the set $\bar{\Omega}$ (because the set $\bar{\Omega}$ is assumed to be polygonal, it can be exactly covered by triangulations), and we define as usual the spaces $V_h = \{v_h \in X_h;\, v_h = 0 \text{ on } \Gamma\}$.

The assumptions made throughout this chapter about the triangulations and the finite elements are the same as in Section 17. Let us briefly record these assumptions for convenience:

(H1) The associated family of triangulations is regular.

(H2) All the finite elements (T, P_T, Σ_T), $T \in \bigcup_h \mathcal{T}_h$, are affine-equivalent to a single reference finite element $(\hat{T}, \hat{P}, \hat{\Sigma})$.

(H3) All the finite elements (T, P_T, Σ_T), $T \in \bigcup_h \mathcal{T}_h$, are of class $\mathscr{C}^0$.

As a consequence, the inclusions $X_h \subset H^1(\Omega)$ and $V_h \subset H_0^1(\Omega)$ hold, as long as the inclusion $\hat{P} \subset H^1(\hat{T})$ holds.

Given a finite element space V_h, solving the corresponding discrete problem amounts to finding the coefficients ζ_k, $1 \leqslant k \leqslant M$, of the expansion $u_h = \Sigma_{k=1}^{M} \zeta_k w_k$ of the discrete solution u_h over the basis functions w_k, $1 \leqslant k \leqslant M$, of the space V_h. These coefficients are solutions of the linear system (cf. (5.4))

$$\sum_{k=1}^{M} a(w_k, w_m)\zeta_k = l(w_m), \quad 1 \leqslant m \leqslant M, \tag{25.4}$$

where, according to (25.1),

$$a(w_k, w_m) = \sum_{T \in \mathcal{T}_h} \int_T \sum_{i,j=1}^{n} a_{ij} \partial_i w_k \partial_j w_m \, dx, \tag{25.5}$$

$$l(w_m) = \sum_{T \in \mathcal{T}_h} \int_T f w_m \, dx. \tag{25.6}$$

In practice, even if the functions a_{ij}, f have simple analytical expressions, the integrals $\int_T \cdots dx$ that appear in (25.5) and (25.6) are seldom computed exactly. Instead, they are approximated through the process of *numerical integration*, which we now describe:

Let $\int_T \varphi(x)\, dx$ denote any one of the integrals appearing in (25.5) or (25.6), and let $F_T: \hat{x} \in \hat{T} \to F_T(\hat{x}) = B_T\hat{x} + b_T$ be the invertible affine mapping that maps $\hat{T}$ onto T. Assuming, without loss of generality, that the (constant) Jacobian of the mapping F_T is >0, we have

$$\int_T \varphi(x)\, dx = (\det B_T) \int_{\hat{T}} \hat{\varphi}(\hat{x})\, d\hat{x}, \tag{25.7}$$

where the functions φ and $\hat{\varphi}$ are in the usual correspondence, i.e., $\varphi(x) = \hat{\varphi}(\hat{x})$ for all

$x = F_T(\hat{x})$, $\hat{x} \in \hat{T}$. Therefore, computing the integral $\int_T \varphi(x)\,dx$ amounts to computing the integral $\int_{\hat{T}} \hat{\varphi}(\hat{x})\,d\hat{x}$.

Then a *numerical quadrature scheme over the set* $\hat{T}$ consists in approximating the integral $\int_{\hat{T}} \hat{\varphi}(\hat{x})\,d\hat{x}$ by a finite sum of the form $\sum_{l=1}^{L} \hat{\omega}_l \hat{\varphi}(\hat{b}_l)$, an approximation that we shall symbolically record as

$$\int_{\hat{T}} \hat{\varphi}(\hat{x})\,d\hat{x} \sim \sum_{l=1}^{L} \hat{\omega}_l \hat{\varphi}(\hat{b}_l). \tag{25.8}$$

The numbers $\hat{\omega}_l$ are called the *weights*, and the points $\hat{b}_l$ are called the *nodes*, of the *quadrature formula* $\sum_{l=1}^{L} \hat{\omega}_l \hat{\varphi}(\hat{b}_l)$. For simplicity, we shall only consider in the sequel examples for which *the nodes belong to the set* $\hat{T}$ *and the weights are* >0. Nodes outside the set $\hat{T}$ and negative weights are not excluded in principle, but, as expected, they generally result in quadrature schemes that behave poorly in actual computations. Hence these situations will not be considered here.

From (25.7) and (25.8), we infer that *the quadrature scheme over the set* $\hat{T}$ *automatically induces a quadrature scheme over the set* T, namely,

$$\int_T \varphi(x)\,dx \sim \sum_{l=1}^{L} \omega_{l,T} \varphi(b_{l,T}), \tag{25.9}$$

where the *weights* $\omega_{l,T}$ and *nodes* $b_{l,T}$ are defined by

$$\omega_{l,T} = (\det B_T)\hat{\omega}_l, \quad b_{l,T} = F_T(\hat{b}_l), \qquad 1 \leqslant l \leqslant L. \tag{25.10}$$

Accordingly, we introduce the *quadrature error functionals*

$$E_T(\varphi) = \int_T \varphi(x)\,dx - \sum_{l=1}^{L} \omega_{l,T} \varphi(b_{l,T}), \tag{25.11}$$

$$\hat{E}(\hat{\varphi}) = \int_{\hat{T}} \hat{\varphi}(\hat{x})\,d\hat{x} - \sum_{l=1}^{L} \hat{\omega}_l \hat{\varphi}(\hat{b}_l), \tag{25.12}$$

which are related by

$$E_T(\varphi) = (\det B_T)\hat{E}(\hat{\varphi}). \tag{25.13}$$

Let us now give a few examples of *often used quadrature formulae*, which, by (25.10), need only be defined over the *reference set* $\hat{T}$. Notice that *each scheme preserves some space of polynomials*; this polynomial invariance will subsequently play a crucial role in the error estimates.

More precisely, given a space $\hat{\Phi}$ of functions $\hat{\varphi}$ defined over the set $\hat{T}$, we shall say that the quadrature scheme is *exact for the space* $\hat{\Phi}$, *or exact for the functions* $\hat{\varphi} \in \hat{\Phi}$, if $\hat{E}(\hat{\varphi}) = 0$ for all $\hat{\varphi} \in \hat{\Phi}$.

Before we turn to the examples, let us mention a useful relation: Let T be an n-simplex, and let $\lambda_i(x)$, $1 \leqslant i \leqslant n+1$, denote the barycentric coordinates of a point x with respect to the vertices of the n-simplex. Then for any integers $\alpha_i \geqslant 0$,

$1 \leqslant i \leqslant n+1$, one has

$$\int_T \lambda_1^{\alpha_1}(x)\lambda_2^{\alpha_2}(x)\cdots\lambda_{n+1}^{\alpha_{n+1}}(x)\,\mathrm{d}x = \frac{\alpha_1!\alpha_2!\cdots\alpha_{n+1}!n!}{(\alpha_1+\alpha_2+\cdots+\alpha_{n+1}+n)!}\,\mathrm{meas}(T). \tag{25.14}$$

Let $\hat{T}$ be an n-simplex with vertices $\hat{a}_i$, $1 \leqslant i \leqslant n+1$, and barycenter

$$\hat{a} = \frac{1}{(n+1)}\sum_{i=1}^{n+1} \hat{a}_i$$

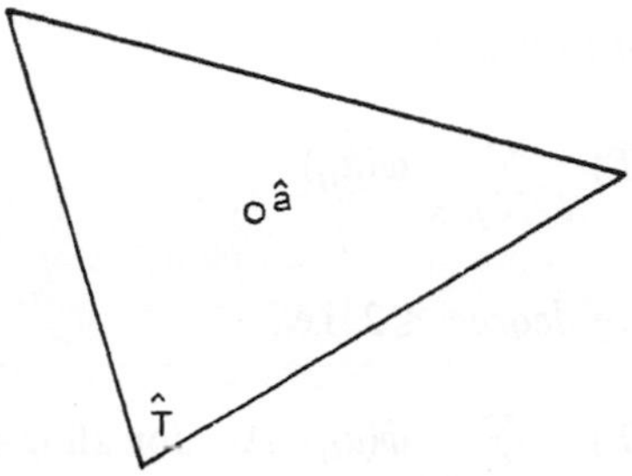

FIG. 25.1. The numerical quadrature scheme $\int_{\hat{T}}\hat{\varphi}(\hat{x})\,\mathrm{d}\hat{x} \sim \mathrm{meas}(\hat{T})\hat{\varphi}(\hat{a})$ is exact if $\hat{\varphi} \in P_1(\hat{T})$.

(Fig. 25.1). Then the *quadrature scheme*

$$\int_{\hat{T}} \hat{\varphi}(\hat{x})\,\mathrm{d}\hat{x} \sim \mathrm{meas}(\hat{T})\hat{\varphi}(\hat{a}) \tag{25.15}$$

is exact for polynomials of degree $\leqslant 1$, i.e.,

$$\int_{\hat{T}} \hat{\varphi}(\hat{x})\,\mathrm{d}\hat{x} - \mathrm{meas}(\hat{T})\hat{\varphi}(\hat{a}) = 0 \quad \text{for all } \hat{\varphi} \in P_1(\hat{T}). \tag{25.16}$$

To see this, let $\hat{\varphi} = \sum_{i=1}^{n+1}\hat{\varphi}(\hat{a}_i)\hat{\lambda}_i$ be any polynomial of degree $\leqslant 1$. Using the relations

$$(n+1)\int_{\hat{T}} \hat{\lambda}_i(\hat{x})\,\mathrm{d}\hat{x} = \mathrm{meas}(\hat{T}), \quad 1 \leqslant i \leqslant n+1$$

(cf. (25.14)), we obtain

$$\int_{\hat{T}} \hat{\varphi}(\hat{x})\,\mathrm{d}\hat{x} = \frac{\mathrm{meas}(\hat{T})}{n+1}\sum_{i=1}^{n+1}\hat{\varphi}(\hat{a}_i) = \mathrm{meas}(\hat{T})\hat{\varphi}(\hat{a}).$$

Let $n=2$ and let $\hat{T}$ be a triangle with midpoints of the sides $\hat{a}_{ij}$, $1 \leqslant i \leqslant j \leqslant 3$ (Fig.

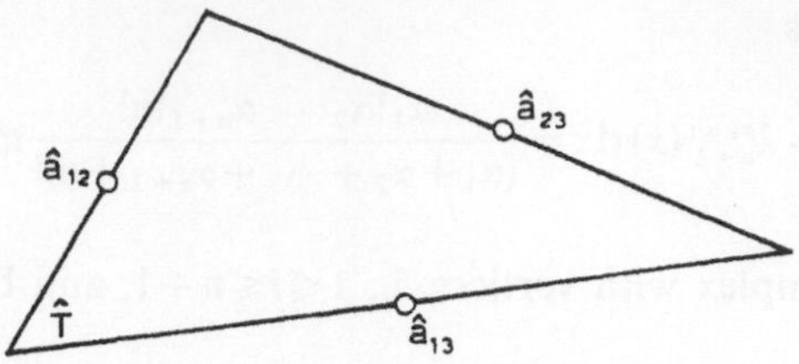

FIG. 25.2. The numerical quadrature scheme $\int_{\hat{T}}\hat{\varphi}(\hat{x})\,d\hat{x}\sim\frac{1}{3}\operatorname{meas}(\hat{T})\sum_{1\leqslant i<j\leqslant 3}\hat{\varphi}(\hat{a}_{ij})$ is exact if $\hat{\varphi}\in P_2(\hat{T})$.

25.2). Then *the quadrature scheme*

$$\int_{\hat{T}}\hat{\varphi}(\hat{x})\,d\hat{x}\sim\tfrac{1}{3}\operatorname{meas}(\hat{T})\sum_{1\leqslant i<j\leqslant 3}\hat{\varphi}(\hat{a}_{ij}) \tag{25.17}$$

is exact for polynomials of degree $\leqslant 2$, i.e.,

$$\int_{\hat{T}}\hat{\varphi}(\hat{x})\,d\hat{x}-\tfrac{1}{3}\operatorname{meas}(\hat{T})\sum_{1\leqslant i<j\leqslant 3}\hat{\varphi}(\hat{a}_{ij})=0\quad\text{for all }\hat{\varphi}\in P_2(\hat{T}). \tag{25.18}$$

To see this, let $\hat{\Pi}$ denote the $P_2(\hat{T})$-interpolation operator associated with the set $\hat{\Sigma}=\{p(\hat{a}_i)\colon 1\leqslant i\leqslant 3;\ p(\hat{a}_{ij})\colon 1\leqslant i<j\leqslant 3\}$. Then using (25.14), one can show that

$$\tfrac{1}{3}\operatorname{meas}(\hat{T})\sum_{1\leqslant i<j\leqslant 3}\hat{\varphi}(\hat{a}_{ij})=\int_{\hat{T}}\hat{\Pi}\hat{\varphi}(\hat{x})\,d\hat{x},$$

and the assertion is proved.

Finally, let $n=2$ and let $\hat{T}$ be a triangle with vertices $\hat{a}_i$, $1\leqslant i\leqslant 3$, with midpoints of the sides $\hat{a}_{ij}$, $1\leqslant i<j\leqslant 3$, and with barycenter $\hat{a}_{123}$ (Fig. 25.3). Then *the quadrature scheme*

$$\int_{\hat{T}}\hat{\varphi}(\hat{x})\,d\hat{x}\sim\tfrac{1}{60}\operatorname{meas}(\hat{T})\left\{3\sum_{i=1}^{3}\hat{\varphi}(\hat{a}_i)+8\sum_{1\leqslant i<j\leqslant 3}\hat{\varphi}(\hat{a}_{ij})+27\hat{\varphi}(\hat{a}_{123})\right\} \tag{25.19}$$

is exact for polynomials of degree $\leqslant 3$, i.e.,

$$\int_{\hat{T}}\hat{\varphi}(\hat{x})\,d\hat{x}-\tfrac{1}{60}\operatorname{meas}(\hat{T})\left\{3\sum_{i=1}^{3}\hat{\varphi}(\hat{a}_i)+8\sum_{1\leqslant i<j\leqslant 3}\hat{\varphi}(\hat{a}_{ij})+27\hat{\varphi}(\hat{a}_{123})\right\}=0$$
$$\text{for all }\hat{\varphi}\in P_3(\hat{T}). \tag{25.20}$$

To see this, one first observes that the set

$$\hat{\Sigma}=\{\hat{p}(\hat{a}_i)\colon 1\leqslant i\leqslant 3;\hat{p}(\hat{a}_{ij})\colon 1\leqslant i<j\leqslant 3;\ \hat{p}(\hat{a}_{123})\}$$

is $\hat{P}$-unisolvent, where

$$\hat{P}=P_2(\hat{T})\oplus V\{\hat{\lambda}_1\hat{\lambda}_2\hat{\lambda}_3\}.$$

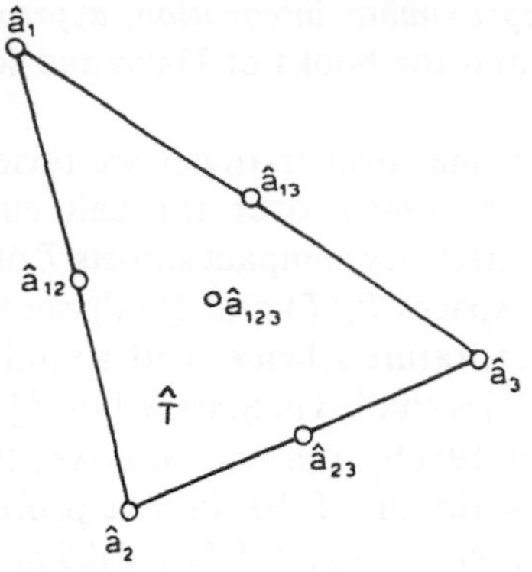

FIG. 25.3. The numerical quadrature scheme

$$\int_{\hat{T}} \hat{\varphi}(\hat{x})\,\mathrm{d}\hat{x} \sim \tfrac{1}{60}\operatorname{meas}(\hat{T})\left\{3\sum_{i=1}^{3}\hat{\varphi}(\hat{a}_i)+8\sum_{1\leqslant i<j\leqslant 3}\hat{\varphi}(\hat{a}_{ij})+27\hat{\varphi}(\hat{a}_{123})\right\}$$

is exact if $\hat{\varphi}\in P_3(\hat{T})$.

This fact, combined with (25.14), then implies that the quadrature scheme of (25.19) can also be written

$$\int_{\hat{T}} \hat{\varphi}(\hat{x})\,\mathrm{d}\hat{x} \sim \int_{\hat{T}} \hat{\Pi}\,\hat{\varphi}(\hat{x})\,\mathrm{d}\hat{x},$$

where $\hat{\Pi}$ is the $\hat{P}$-interpolation operator. From this, one deduces that the quadrature scheme of (25.19) is exact for the space $P_3(\hat{T})$, but not for the space $P_4(\hat{T})$.

Let us finally consider the case where $\hat{T}$ is an n-rectangle. It is well known that for each integer $k\geqslant 0$, there exist $(k+1)$ points $b_i\in[0,1]$ and $(k+1)$ weights $\omega_i>0$, $1\leqslant i\leqslant k+1$, such that the quadrature scheme

$$\int_{[0,1]} \varphi(x)\,\mathrm{d}x \sim \sum_{i=1}^{k+1}\omega_i\varphi(b_i) \tag{25.21}$$

is exact for the space $P_{2k+1}([0,1])$ (see e.g. CROUZEIX and MIGNOT [1984, Theorem 2.9]). This particular quadrature formula is known as the *Gauss–Legendre formula.* Then it is easy to infer from this result *that the numerical quadrature scheme*

$$\int_{[0,1]^n} \varphi(x)\,\mathrm{d}x \sim \sum_{\substack{i_j=1\\ 1\leqslant j\leqslant n}}^{k+1} (\omega_{i_1}\omega_{i_2}\cdots\omega_{i_n})\varphi(b_{i_1},b_{i_2},\ldots,b_{i_n}) \tag{25.22}$$

is exact for the space $Q_{2k+1}([0,1]^n)$. This result thus provides examples of *quadrature schemes over n-rectangles.*

Examples of numerical quadrature schemes used in actual finite element computations are found in the book of ZIENKIEWICZ [1971, Section 8.10]. For general introductions to the subject of numerical integration, also known as:

numerical quadrature, approximate integration, approximate quadrature, see the survey of HABER [1970], and the books of DAVIS and RABINOWITZ [1975], STROUD [1971].

Existing quadrature schemes over triangles are reviewed in DUNAVANT [1985]; over tetrahedra, in KEAST [1986]; over the unit cube, in DUNAVANT [1986]. Quadrature schemes over arbitrary compact subsets T of $\mathbb{R}^n$ that are exact for spaces intermediate between the spaces $P_k(T)$ and $Q_k(T)$ are studied in GUESSAB [1986]. The problem of finding quadrature schemes with a minimum number of nodes that are exact for the space $P_k(\hat{T})$ is studied in SCHMID [1978], and, for the space $Q_k(\hat{T})$, in GOUT and GUESSAB [1986a, 1986b]. See also GUESSAB [1987] for further extensions.

Let us return to the description of the discrete problem. Instead of solving the linear system (25.4) with the coefficients (25.5) and (25.6), all integrals $\int_T \cdots \mathrm{d}x$ will be computed using a quadrature scheme given on the set $\hat{T}$. In other words, we are solving the *modified linear system*

$$\sum_{k=1}^{M} a_h(w_k, w_m)\zeta_k = l_h(w_m), \quad 1 \leqslant m \leqslant M, \tag{25.23}$$

where (compare with (25.5) and (25.6) respectively)

$$a_h(w_k, w_m) = \sum_{T \in \mathcal{T}_h} \sum_{l=1}^{L} \omega_{l,T} \sum_{i,j=1}^{n} (a_{ij}\partial_i w_k \partial_j w_m)(b_{l,T}), \tag{25.24}$$

$$l_h(w_m) = \sum_{T \in \mathcal{T}_h} \sum_{l=1}^{L} \omega_{l,T}(f w_m)(b_{l,T}). \tag{25.25}$$

REMARK 25.1. Conceivably, *different* quadrature formulae could be used for approximating the coefficients $a(w_k, w_m)$ on the one hand, and the coefficients $l(w_m)$ on the other hand. However, our final result (Theorem 29.1) will show that this is not necessary.

For our subsequent analysis, it will be more convenient to consider the following equivalent formulation of the *discrete problem*, represented by the linear system (25.23). We seek a *discrete solution* $u_h \in V_h$ that satisfies

$$a_h(u_h, v_h) = l_h(v_h) \quad \text{for all } v_h \in V_h, \tag{25.26}$$

where, for all functions $u_h, v_h \in V_h$, the bilinear form a_h and the linear form f_h are respectively given by

$$a_h(u_h, v_h) = \sum_{T \in \mathcal{T}_h} \sum_{l=1}^{L} \omega_{l,T} \sum_{i,j=1}^{n} (a_{ij}\partial_i u_h \partial_j v_h)(b_{l,T}), \tag{25.27}$$

$$l_h(v_h) = \sum_{T \in \mathcal{T}_h} \sum_{l=1}^{L} \omega_{l,T}(f v_h)(b_{l,T}). \tag{25.28}$$

Note that the expressions (25.27) and (25.28) show why *the functions a_{ij} and f need* *to be defined everywhere over the set* $\bar{\Omega}$, since the nodes $b_{l,T}$ may be arbitrarily located in $\bar{\Omega}$.

Also, in order that definition (25.27) make sense, it is necessary that, over each set T, the first partial derivatives of the functions in the space X_h should have unambiguously defined extensions to the boundary of T, should some node $b_{l,T}$ be situated on the boundary of T. If this node coincides with a node b_{l,T^*} belonging to an adjacent set T^*, it should be clear that the values to be assigned to the derivatives $\partial_i v_h(b_{l,T})$ and $\partial_i v_h(b_{l,T^*})$ are thus generally different.

Since the definition of the discrete problem requires the knowledge of the values of the functions a_{ij} and f only at a finite number of points of $\bar{\Omega}$, the discrete problem is in this sense quite reminiscent of a discrete problem obtained by a finite difference method. In fact, this is even true to the extent that *most classical finite difference schemes can be exactly interpreted as finite element methods with specific finite element spaces and specific quadrature schemes.*

To be more specific, let $\bar{\Omega}=[0, I\rho]\times[0, J\rho]$ where I and J are integers and ρ is a strictly positive number, let $\mathcal{T}_h$ be a triangulation of the set $\bar{\Omega}$ made up of bilinear rectangles, with sets T of the form

$$[i\rho,(i+1)\rho]\times[j\rho,(j+1)\rho], \qquad 0\leqslant i\leqslant I-1, \quad 0\leqslant j\leqslant J-1,$$

and let U_{ij} denote the unknown, so far denoted ζ_k, corresponding to the kth node (ih, jh), $1\leqslant i\leqslant I-1$, $1\leqslant j\leqslant J-1$. We further assume that the bilinear form is of the form

$$a(u,v)=\int_\Omega \sum_{i=1}^n \partial_i u\partial_i v\,\mathrm{d}x,$$

i.e., that the corresponding partial differential equation is the Poisson equation $-\Delta u=f$ in Ω, and we only consider nodes $(i\rho, j\rho)$ that are at least two squares away from the boundary of the set Ω, i.e., for which $2\leqslant i\leqslant I-2$, $2\leqslant j\leqslant J-2$.

Then, in the absence of numerical integration, the expression $\Sigma_{k=1}^M a(w_k, w_m)\zeta_k$ corresponding to the mth node $(i\rho, j\rho)$ is, up to a constant factor, given by the expression

$$\begin{aligned}8U_{ij}-(&U_{i+1,j}+U_{i+1,j+1}+U_{i,j+1}+U_{i-1,j+1}+U_{i-1,j}\\&+U_{i-1,j-1}+U_{i,j-1}+U_{i+1,j-1}).\end{aligned}$$

Assume next that we are using the following numerical quadrature scheme over the reference square $\hat{T}=[0,1]^2$:

$$\int_{[0,1]^2}\hat{\varphi}(\hat{x})\,\mathrm{d}\hat{x}\sim\tfrac{1}{4}(\hat{\varphi}(0,0)+\hat{\varphi}(0,1)+\hat{\varphi}(1,1)+\hat{\varphi}(1,0)),$$

which is exact for the space $Q_1(\hat{T})$. Then the equality $\Sigma_{k=1}^M a_h(w_k, w_m)\zeta_k=l_h(w_m)$ becomes

$$4U_{ij}-(U_{i+1,j}+U_{i,j+1}+U_{i-1,j}+U_{i,j-1})=\rho^2 f(i\rho, j\rho),$$

i.e., it is exactly the standard five-point difference approximation to Poisson's equation. It is interesting to notice that it is generally impossible to derive this scheme from a finite element method without numerical integration (BIRKHOFF and GULATI [1974]).

More general comparisons between finite element methods, with or without numerical integration, and finite difference methods, are found in BIRKHOFF and GULATI [1974], TOMLIN [1972], WALSH [1971].

26. Abstract error estimate and the first Strang lemma

To sum up, we started out with a standard variational problem: Find $u \in V$ such that, for all $v \in V$, $a(u, v) = l(v)$, where the space V, the forms $a(\cdot,\cdot)$ and $l(\cdot)$ satisfy the assumptions of the Lax–Milgram lemma. Then given a finite-dimensional subspace V_h of the space V, the discrete problem consists in finding $u_h \in V_h$ such that, for all $v_h, \in V_h$, $a_h(u_h, v_h) = l_h(v_h)$, where $a_h(\cdot,\cdot)$ is a bilinear form defined over the space V_h and $l_h(\cdot)$ is a linear form defined over the space V_h.

Notice that, in the present case, *the forms $a_h(\cdot,\cdot)$ and $l_h(\cdot)$ are not defined on the space V*, since the point values are not defined in general for functions in the space $H^1(\Omega)$.

Our first task is to prove an *abstract error estimate* adapted to the above abstract setting, but first we need some definitions: We shall refer to $a_h(\cdot,\cdot)$ as an *approximate bilinear form* and to $l_h(\cdot)$ as an *approximate linear form*. Denoting by $\|\cdot\|$ the norm of the space V, we shall say that approximate bilinear forms $a_h(\cdot,\cdot): V_h \times V_h \to \mathbb{R}$, associated with a family of subspaces V_h of the space V, are *uniformly V_h-elliptic* if there exists $\tilde{\alpha}$ such that

$$\begin{aligned} &\tilde{\alpha} > 0, \\ &\tilde{\alpha}\|v_h\|^2 \leqslant a_h(v_h, v_h) \quad \text{for all } v_h \in V_h \text{ and all } h. \end{aligned} \tag{26.1}$$

In particular then, *the constant $\tilde{\alpha}$ is independent of the subspace V_h*. Notice that such an assumption implies in particular the existence of the discrete solutions.

The following error estimate is due to STRANG [1972b].

THEOREM 26.1 (First Strang lemma). *Consider a family of discrete problems whose associated approximate bilinear forms are uniformly V_h-elliptic.*

Then there exists a constant C independent of the space V_h such that

$$\begin{aligned} \|u-u_h\| \leqslant C\bigg(&\inf_{v_h \in V_h} \left\{ \|u-v_h\| + \sup_{w_h \in V_h} \frac{|a(v_h, w_h) - a_h(v_h, w_h)|}{\|w_h\|} \right\} \\ &+ \sup_{w_h \in V_h} \frac{|l(w_h) - l_h(w_h)|}{\|w_h\|} \bigg). \end{aligned} \tag{26.2}$$

PROOF. Let v_h be an arbitrary element in the space V_h. Using the assumption of uniform V_h-ellipticity, we may write:

$$\begin{aligned}\tilde{\alpha}\|u_h-v_h\|^2 &\leqslant a_h(u_h-v_h, u_h-v_h)\\ &= a(u-v_h, u_h-v_h)+\{a(v_h, u_h-v_h)-a_h(v_h, u_h-v_h)\}\\ &\quad +\{l_h(u_h-v_h)-l(u_h-v_h)\},\end{aligned}$$

and thus, by the continuity of the bilinear form $a(\cdot,\cdot)$,

$$\begin{aligned}\tilde{\alpha}\|u_h-v_h\| &\leqslant M\|u-v_h\|+\frac{|a(v_h, u_h-v_h)-a_h(v_h, u_h-v_h)|}{\|u_h-v_h\|}+\frac{|l_h(u_h-v_h)-l(u_h-v_h)|}{\|u_h-v_h\|}\\ &\leqslant M\|u-v_h\|+\sup_{w_h\in V_h}\frac{|a(v_h, w_h)-a_h(v_h, w_h)|}{\|w_h\|}+\sup_{w_h\in V_h}\frac{|l_h(w_h)-l(w_h)|}{\|w_h\|}.\end{aligned}$$

Combining the above inequality with the triangular inequality

$$\|u-u_h\|\leqslant\|u-v_h\|+\|u_h-v_h\|$$

and taking the infimum with respect to $v_h\in V_h$, we find inequality (26.2). □

Note that the abstract error estimate (26.2) generalizes the abstract error estimate established in Céa's lemma (Theorem 13.1) for conforming finite element methods, since, in the absence of numerical integration, we would have $a_h(\cdot,\cdot)=a(\cdot,\cdot)$ and $l_h(\cdot)=l(\cdot)$.

REMARK 26.1. One can similarly obtain an abstract error estimate that generalizes the Aubin–Nitsche lemma (Theorem 19.1) in the present setting. Let H be a Hilbert space such that $\bar{V}=H$ with $V\hookrightarrow H$. Then one can show that

$$\begin{aligned}|u-u_h|\leqslant\sup_{g\in H}\frac{1}{|g|}\inf_{\varphi_h\in V_h}&\{M\|u-u_h\|\,\|\varphi_g-\varphi_h\|\\ &+|a(u_h, \varphi_h)-a_h(u_h, \varphi_h)|+|l(\varphi_h)-l_h(\varphi_h)|\},\end{aligned}$$

where $|\cdot|$ denotes the norm in the space H, and for each $g\in H$, the function $\varphi_g\in V$ is the unique solution of the variational problem

$$a(v, \varphi_g)=(g, v)\quad\text{for all } v\in V,$$

where $(\cdot,\cdot)$ denotes the inner product in H.

27. Uniform V_h-ellipticity of the approximate bilinear forms

We now give sufficient conditions on a quadrature scheme that insure that the approximate bilinear forms are uniformly V_h-elliptic: Notice in particular that in the next theorem assumptions (27.2) and (27.3) exhibit the *relationship that should exist between the reference finite element* $(\hat{T}, \hat{P}, \hat{\Sigma})$ *and the quadrature scheme defined on* $\hat{T}$. The next proof is based on, and generalizes, an idea of Strang (cf. STRANG and FIX [1973, Section 4-3]).

THEOREM 27.1. *Let there be given a quadrature scheme*

$$\int_{\hat{T}} \hat{\varphi}(\hat{x})\,\mathrm{d}\hat{x} \sim \sum_{l=1}^{L} \hat{\omega}_l \hat{\varphi}(\hat{b}_l), \qquad \hat{\omega}_l > 0, \quad 1 \leqslant l \leqslant L,$$

over the reference finite element $(\hat{T}, \hat{P}, \hat{\Sigma})$, and assume that there exists an integer $k' \geqslant 1$ such that:

$$\hat{P} \subset P_{k'}(\hat{T}) \tag{27.1}$$

on the one hand, and either

$$\bigcup_{l=1}^{L} \{\hat{b}_l\} \text{ contains a } P_{k'-1}(\hat{T})\text{-unisolvent set,} \tag{27.2}$$

or

$$\int_{\hat{T}} \hat{\varphi}(\hat{x})\,\mathrm{d}\hat{x} = \sum_{l=1}^{L} \hat{\omega}_l \hat{\varphi}(\hat{b}_l) \quad \text{for all } \hat{\varphi} \in P_{2k'-2}(\hat{T}), \tag{27.3}$$

or both (27.2) *and* (27.3) *hold, on the other hand.*

Then there exists a constant $\tilde{\alpha} > 0$ independent of h such that, for all approximate bilinear forms of the form (25.27) *and all spaces V_h,*

$$\tilde{\alpha} |v_h|_{1,\Omega}^2 \leqslant a_h(v_h, v_h) \quad \text{for all } v_h \in V_h. \tag{27.4}$$

PROOF. (i) Let us first assume that the union $\bigcup_{l=1}^{L} \{\hat{b}_l\}$ contains a $P_{k'-1}(\hat{T})$-unisolvent subset. Using the positivity of the weights, we find that

$$\hat{p} \in \hat{P} \text{ and } \sum_{l=1}^{L} \hat{\omega}_l \sum_{i=1}^{n} (\partial_i \hat{p}(\hat{b}_l))^2 = 0$$
$$\Rightarrow \partial_i \hat{p}(\hat{b}_l) = 0, \qquad 1 \leqslant i \leqslant n, \quad 1 \leqslant l \leqslant L.$$

For each $i \in [1, n]$, the function $\partial_i \hat{p}$ is in the space $P_{k'-1}(\hat{T})$ by assumption (27.1) and thus it is identically zero since it vanishes on a $P_{k'-1}(\hat{T})$-unisolvent subset, by assumption (27.2). As a consequence, the mapping

$$\hat{p} \rightarrow \left\{ \sum_{l=1}^{L} \hat{\omega}_l \sum_{i=1}^{n} |\partial_i \hat{p}(\hat{b}_l)|^2 \right\}^{1/2}$$

defines a norm over the quotient space $\hat{P}/P_0(\hat{T})$. Since the mapping $\hat{p} \rightarrow |\hat{p}|_{1,\hat{T}}$ is also a norm over this space and since this space is finite-dimensional, *there exists a constant $\hat{C} > 0$ such that*

$$\hat{C} |\hat{p}|_{1,\hat{T}}^2 \leqslant \sum_{l=1}^{L} \hat{\omega}_l \sum_{i=1}^{n} |\partial_i \hat{p}(\hat{b}_l)|^2 \quad \text{for all } \hat{p} \in \hat{P}. \tag{27.5}$$

If we instead assume that the quadrature scheme is exact for the space $P_{2k'-2}(\hat{T})$, inequality (27.5) becomes an equality with $\hat{C} = 1$, since the function $\sum_{i=1}^{n} |\partial_i \hat{p}|^2$ belongs to the space $P_{2k'-2}(\hat{T})$ for all $\hat{p} \in \hat{P}$ and since $\sum_{l=1}^{L} \hat{\omega}_l \sum_{i=1}^{n} (\partial_i \hat{p}(\hat{b}_l))^2$ is precisely the quadrature formula that corresponds to the integral

$$\int_{\hat{T}} \sum_{i=1}^{n} |\partial_i \hat{p}|^2 \,\mathrm{d}\hat{x} = |\hat{p}|_{1,\hat{T}}^2.$$

(ii) Let us next consider the approximation of one of the integrals

$$\int_T \sum_{i,j=1}^{n} a_{ij}\partial_i v_h \partial_j v_h \,\mathrm{d}x.$$

Let $v_h|_T = p_T$, and let $\hat{p}_T \in \hat{P}$ be the function associated with p_T through the usual correspondence $\hat{x} \in \hat{T} \to F_T(\hat{x}) = B_T\hat{x} + b_T = x \in T$. Using the ellipticity condition (25.2) and the positivity of the weights, we can write

$$\begin{aligned}
&\sum_{l=1}^{L} \omega_{l,T} \sum_{i,j=1}^{n} (a_{ij}\partial_i v_h \partial_j v_h)(b_{l,T}) \\
&\quad = \sum_{l=1}^{L} \omega_{l,T} \sum_{i,j=1}^{n} (a_{ij}\partial_i p_T \partial_j p_T)(b_{l,T}) \\
&\quad \geqslant \beta \sum_{l=1}^{L} \omega_{l,T} \sum_{i=1}^{n} |\partial_i p(b_{l,T})|^2.
\end{aligned} \tag{27.6}$$

Observe that $\Sigma_{i=1}^{n} |\partial_i p_T(b_{l,T})|^2$ is the square of the Euclidean norm $\|\cdot\|$ of the vector $Dp_T(b_{l,T})$. Since $|D\hat{p}_T(\hat{b}_l)\| \leqslant \|B_T\| \, \|Dp_T(b_{l,T})\|$ (for all $\xi \in \mathbb{R}^n$, we have $D\hat{p}(\hat{b}_l)\xi = Dp(b_{l,T})(B_T\xi)$), we can write, using relations (25.10) and (27.5) and Theorem 15.1,

$$\begin{aligned}
&\sum_{l=1}^{L} \omega_{l,T} \sum_{i=1}^{n} |\partial_i p_T(b_{l,T})|^2 \\
&\quad \geqslant \|B_T\|^{-2} \sum_{l=1}^{L} \omega_{l,T} \sum_{i=1}^{n} |\partial_i \hat{p}_T(\hat{b}_l)|^2 \\
&\quad = (\det B_T)\|B_T\|^{-2} \sum_{l=1}^{L} \hat{\omega}_l \sum_{i=1}^{n} |\partial_i \hat{p}_T(\hat{b}_l)|^2 \\
&\quad \geqslant \hat{C}(\det B_T)\|B_T\|^{-2}|\hat{p}_T|_{1,\hat{T}}^2 \geqslant \hat{C}\{\|B_T\| \, \|B_T^{-1}\|\}^{-2}|p_T|_{1,T}^2.
\end{aligned} \tag{27.7}$$

Since the family of triangulations is regular by assumption, there exists a constant C independent of $T \in \mathcal{T}_h$ and h such that

$$\|B_T\| \, \|B_T^{-1}\| \leqslant \frac{\hat{h}}{\hat{\rho}} \frac{h_T}{\rho_T} \leqslant C. \tag{27.8}$$

Combining inequalities (27.6), (27.7) and (27.8), we find that *there exists a constant* $\tilde{\alpha} > 0$ *independent of* $T \in \mathcal{T}_h$ *and* h *such that*

$$\sum_{l=1}^{L} \omega_{l,T} \sum_{i,j=1}^{n} (a_{ij}\partial_i v_h \partial_j v_h)(b_{l,T}) \geqslant \tilde{\alpha}|v_h|_{1,T}^2 \quad \text{for all } v_h \in V_h. \tag{27.9}$$

(iii) It is then easy to conclude: Using inequalities (27.9) for all $T \in \mathcal{T}_h$, we obtain

$$a_h(v_h, v_h) = \sum_{T \in \mathcal{T}_h} \sum_{l=1}^{L} \omega_{l,T} \sum_{i,j=1}^{n} (a_{ij} \partial_i v_h \partial_j v_h)(b_{l,T})$$

$$\geqslant \tilde{\alpha} \sum_{T \in \mathcal{T}_h} |v_h|_{1,T}^2 = \tilde{\alpha} |v_h|_{1,\Omega}^2 \quad \text{for all } v_h \in V_h. \qquad \square$$

REMARK 27.1. The expressions $\Sigma_{l=1}^{L} \hat{\omega}_l \Sigma_{i=1}^{n} |\partial_i \hat{p}_T(\hat{b}_l)|^2$ are precisely the approximations we get when we apply the quadrature scheme to the integrals $|\hat{p}_T|_{1,\hat{T}}^2$, which in turn correspond to the model problem $-\Delta u = f$ in Ω, $u = 0$ on Γ. This observation is the basis for assumptions (27.2) or (27.3), which in essence insure that the mapping

$$\hat{p} \rightarrow \left\{ \sum_{l=1}^{L} \hat{\omega}_l \sum_{i=1}^{n} |\partial_i \hat{p}(\hat{b}_l)|^2 \right\}^{1/2}$$

is a norm over the quotient space $\hat{P}/P_0(\hat{T})$.

Let us apply this theorem to the examples of quadrature schemes given in Section 25:

– If $(\hat{T}, \hat{P}, \hat{\Sigma})$ is a linear n-simplex ($\hat{P} = P_1(\hat{T})$ and thus $k' = 1$), we may use the quadrature scheme of (25.15) since $\{\hat{a}\}$ is a $P_0(\hat{T})$-unisolvent set.

– If $(\hat{T}, \hat{P}, \hat{\Sigma})$ is a quadratic triangle ($\hat{P} = P_2(\hat{T})$ and thus $k' = 2$), we may use the quadrature scheme of (25.17) since $\bigcup_{i<j}\{\hat{a}_{ij}\}$ is a $P_1(\hat{T})$-unisolvent set.

Notice that in both cases, both assumptions (27.2) and (27.3) are satisfied.

– If $(\hat{T}, \hat{P}, \hat{\Sigma})$ is a cubic or reduced cubic triangle ($\hat{P} \subset P_3(\hat{T})$ and thus $k' = 3$), we may use the quadrature scheme of (25.19) since the set of numerical integration nodes (strictly) contains the $P_2(\hat{T})$-unisolvent subset $(\bigcup_i\{\hat{a}_i\}) \cup (\bigcup_{i<j}\{\hat{a}_{ij}\})$. Note however that the quadrature scheme is *not* exact for the space $P_4(\hat{T})$; hence assumption (27.3) is not satisfied in this case.

28. Consistency error estimates and the Bramble–Hilbert lemma

Now that the question of uniform V_h-ellipticity has been taken care of, we can turn to the problem of estimating the various terms appearing in the right-hand side of inequality (26.2). For the sake of clarity, we shall essentially concentrate on one special case (which nevertheless displays all the characteristic properties of the general case), namely the case where

$$\hat{P} = P_k(\hat{T}) \tag{28.1}$$

for some integer $k \geqslant 1$. Indications for handling the cases where $P_k(\hat{T}) \subset \hat{P} \subset P_{k'}(\hat{T})$ or where $P_k(\hat{T}) \subset \hat{P} \subset Q_k(\hat{T})$ are given at the end of Section 29.

If the solution u belongs to the space $H^{k+1}(\Omega)$ and if its X_h-interpolant is well defined, we infer from (28.1) that (cf. Theorem 17.1)

$$\inf_{v_h \in V_h} \|u - v_h\|_{1,\Omega} \leqslant \|u - \Pi_h u\|_{1,\Omega} \leqslant Ch^k |u|_{k+1,\Omega}. \tag{28.2}$$

Thus we have an $O(h^k)$ convergence in the absence of numerical integration.

Our basic objective is to give *sufficient conditions on the quadrature scheme that insure that the effect of numerical integration does not decrease this order of convergence.*

REMARK 28.1. This criterion for appraising the required quality of the quadrature scheme is to some extent arbitrary, but at least it is unambiguously defined. Surprisingly, the results that shall be obtained in this fashion are nevertheless quite similar to the conclusions that the engineers had empirically drawn from their numerical experience.

Let us assume that the approximate bilinear forms are uniformly V_h-elliptic, so that we may apply the abstract error estimate of Theorem 26.1. In view of (26.2) and (28.2), our aim is to obtain *consistency error estimates* of the form

$$\sup_{w_h \in V_h} \frac{|a(\Pi_h u, w_h) - a_h(\Pi_h u, w_h)|}{\|w_h\|_{1,\Omega}} \leqslant C(a_{ij}, u) h^k, \tag{28.3}$$

$$\sup_{w_h \in V_h} \frac{|l(w_h) - l_h(w_h)|}{\|w_h\|_{1,\Omega}} \leqslant C(f) h^k. \tag{28.4}$$

Notice that, in the usual terminology of numerical analysis, the uniform ellipticity condition appears as a *stability condition*, while the conditions (implied by the above error estimates)

$$\lim_{h\to 0} \sup_{w_h \in V_h} \frac{|a(\Pi_h u, w_h) - a_h(\Pi_h u, w_h)|}{\|w_h\|_{1,\Omega}} = 0,$$

$$\lim_{h\to 0} \sup_{w_h \in V_h} \frac{|l(w_h) - l_h(w_h)|}{\|w_h\|_{1,\Omega}} = 0,$$

appear as *consistency conditions*. This is why we call *consistency errors* the two terms of the form $\sup_{w_h \in V_h}\{\dots\}$ appearing in the left-hand sides of inequalities (28.3) and (28.4). By definition of the quadrature error functionals $E_T(\cdot)$ of (25.11), we have, for all $w_h \in V_h$,

$$a(\Pi_h u, w_h) - a_h(\Pi_h u, w_h) = \sum_{T \in \mathscr{T}_h} E_T\left(\sum_{i,j=1}^{n} a_{ij}\partial_i(\Pi_h u)\partial_j w_h\right), \tag{28.5}$$

$$l(w_h) - l_h(w_h) = \sum_{T \in \mathscr{T}_h} E_T(f w_h). \tag{28.6}$$

It turns out that we shall obtain (Theorems 28.2 and 28.3) "local" *quadrature error estimates* of the form

$$|E_T(a_{ij}\partial_i p'\partial_j p)| \leqslant C(a_{ij}|_T; \partial_i p')h_T^k |\partial_j p|_{0,T} \quad \text{for all } p', p \in P_T, \tag{28.7}$$

$$|E_T(fp)| \leqslant C(f|_T)h_T^k \|p\|_{1,T} \quad \text{for all } p \in P_T, \tag{28.8}$$

from which the "global" consistency error estimates (28.3) and (28.4) are deduced by an application of the Cauchy–Schwarz inequality; this is possible only because the constants $C(a_{ij}|_T; \partial_i p')$ and $C(f|_T)$ appearing in the above inequalities are of an appropriate form.

To begin with, we prove a preliminary result, due to BRAMBLE and HILBERT [1970]. It is a very useful tool for getting error estimates, not only in numerical integration, but also in interpolation theory, where it provides an alternate approach to that based (as here) on Theorem 15.3.

THEOREM 28.1 (Bramble–Hilbert lemma). *Let Ω be a domain in $\mathbb{R}^n$; let $k \geqslant 0$ be an integer, let p be a number satisfying $1 \leqslant p \leqslant \infty$, and let l be a continuous linear form on the space $W^{k+1,p}(\Omega)$ with the property that*

$$l(p) = 0 \quad \textit{for all } p \in P_k(\Omega). \tag{28.9}$$

Then there exists a constant $C(\Omega)$ such that

$$|l(v)| \leqslant C(\Omega)\|l\|'_{k+1,p,\Omega}|v|_{k+1,p,\Omega} \quad \textit{for all } v \in W^{k+1,p}(\Omega), \tag{28.10}$$

where $\|\cdot\|'_{k+1,p,\Omega}$ is the norm in the dual space of $W^{k+1,p}(\Omega)$.

PROOF. Let v be any function in the space $W^{k+1,p}(\Omega)$. Since by assumption, $l(v) = l(v+p)$ for all $p \in P_k(\Omega)$, we may write

$$|l(v)| = |l(v+p)| \leqslant \|l\|'_{k+1,p,\Omega}\|v+p\|_{k+1,p,\Omega} \quad \text{for all } p \in P_k(\Omega),$$

and thus

$$|l(v)| \leqslant \|l\|'_{k+1,p,\Omega} \inf_{p \in P_k(\Omega)} \|v+p\|_{k+1,p,\Omega}.$$

The conclusion follows by Theorem 14.1. □

In the sequel, we shall often use the following result: *Let the functions $\varphi \in W^{m,q}(\Omega)$, and $w \in W^{m,\infty}(\Omega)$ be given. Then the function φw belongs to the space $W^{m,q}(\Omega)$, and*

$$|\varphi w|_{m,q,\Omega} \leqslant C \sum_{j=0}^{m} |\varphi|_{m-j,q,\Omega}|w|_{j,\infty,\Omega}, \tag{28.11}$$

for some constant C solely dependent upon the integers m and n; thus C is in particular independent of the set Ω. To prove this, we use the formula

$$\partial^\alpha(\varphi w) = \sum_{j=0}^{m} \sum_{\substack{|\beta|=j \\ \beta+\beta'=\alpha}} \partial^\beta w \partial^{\beta'}\varphi \quad \text{for all } \alpha \text{ with } |\alpha| = m,$$

in conjunction with inequalities of the form

$$\left|\sum_{\lambda=1}^{\Lambda} a_\lambda f_\lambda\right|_{0,q,\Omega} \leqslant \sum_{\lambda=1}^{\Lambda} |a_\lambda|_{0,\infty,\Omega}|f_\lambda|_{0,q,\Omega}.$$

From now on in this chapter, we essentially follow RAVIART [1972] and CIARLET and RAVIART [1972c].

THEOREM 28.2. *Assume that, for some integer* $k \geqslant 1$,

$$\hat{P} = P_k(\hat{T}), \tag{28.12}$$

$$\hat{E}(\hat{\varphi}) = 0 \quad \textit{for all } \hat{\varphi} \in P_{2k-2}(\hat{T}). \tag{28.13}$$

Then there exists a constant C independent of $T \in \mathcal{T}_h$ *and h such that*

$$\begin{aligned} |E_T(a\partial_i p'\partial_j p)| &\leqslant Ch_T^k \|a\|_{k,\infty,T}\|\partial_i p'\|_{k-1,T}|\partial_j p|_{0,T} \\ &\leqslant Ch_T^k \|a\|_{k,\infty,T}\|p'\|_{k,T}|p|_{1,T} \end{aligned} \tag{28.14}$$

for all $a \in W^{k,\infty}(T)$, $p \in P_k(T)$, $p' \in P_k(T)$.

PROOF. We shall get an error estimate for the expression $E_T(avw)$ with $a \in W^{k,\infty}(T)$, $v \in P_{k-1}(T)$, $w \in P_{k-1}(T)$. From (25.13), we infer that

$$E_T(avw) = (\det B_T)\hat{E}(\hat{a}\hat{v}\hat{w}), \tag{28.15}$$

with $\hat{a} \in W^{k,\infty}(\hat{T})$, $\hat{v} \in P_{k-1}(\hat{T})$, $\hat{w} \in P_{k-1}(\hat{T})$. For any $\hat{w} \in P_{k-1}(\hat{T})$ and any $\hat{\varphi} \in W^{k,\infty}(\hat{T})$, we have ($W^{k,\infty}(\hat{T}) \hookrightarrow \mathscr{C}^0(\hat{T})$ since $k \geqslant 1$)

$$\begin{aligned} |\hat{E}(\hat{\varphi}\hat{w})| &= \left|\int_{\hat{T}} \hat{\varphi}\hat{w}\,\mathrm{d}\hat{x} - \sum_{l=1}^{L} \hat{\omega}_l(\hat{\varphi}\hat{w})(\hat{b}_l)\right| \\ &\leqslant \hat{C}|\hat{\varphi}\hat{w}|_{0,\infty,\hat{T}} \leqslant \hat{C}|\hat{\varphi}|_{0,\infty,\hat{T}}|\hat{w}|_{0,\infty,\hat{T}}, \end{aligned}$$

where, here and subsequently, the letter $\hat{C}$ represents various constants solely dependent upon the reference finite element. Since $|\hat{\varphi}|_{0,\infty,\hat{T}} \leqslant \|\hat{\varphi}\|_{k,\infty,\hat{T}}$, and since all norms are equivalent on the finite-dimensional space $P_{k-1}(\hat{T})$, we deduce that

$$|\hat{E}(\hat{\varphi}\hat{w})| \leqslant \hat{C}\|\hat{\varphi}\|_{k,\infty,\hat{T}}|\hat{w}|_{0,\hat{T}}.$$

Thus, for a given $\hat{w} \in P_{k-1}(\hat{T})$, the linear form

$$\hat{\varphi} \in W^{k,\infty}(\hat{T}) \rightarrow \hat{E}(\hat{\varphi}\hat{w})$$

is continuous with norm $\leqslant \hat{C}|\hat{w}|_{0,\hat{T}}$ on the one hand, and it vanishes over the space $P_{k-1}(\hat{T})$ on the other hand, by assumption (28.13). Therefore, using the Bramble–Hilbert lemma (Theorem 28.1), there exists a constant $\hat{C}$ such that

$$|\hat{E}(\hat{\varphi}\hat{w})| \leqslant \hat{C}|\hat{\varphi}|_{k,\infty,\hat{T}}|\hat{w}|_{0,\hat{T}} \quad \text{for all } \hat{\varphi} \in W^{k,\infty}(\hat{T}), \quad \hat{w} \in P_{k-1}(\hat{T}).$$

Next, let $\hat{\varphi} = \hat{a}\hat{v}$ with $\hat{a} \in W^{k,\infty}(\hat{T})$, $\hat{v} \in P_{k-1}(\hat{T})$. Using (28.11) and taking into account that $|\hat{v}|_{k,\infty,\hat{T}}=0$, we get

$$|\hat{\varphi}|_{k,\infty,\hat{T}}=|\hat{a}\hat{v}|_{k,\infty,\hat{T}}$$
$$\leqslant \hat{C}\sum_{j=0}^{k-1}|\hat{a}|_{k-j,\infty,\hat{T}}|\hat{v}|_{j,\infty,\hat{T}}\leqslant \hat{C}\sum_{j=0}^{k-1}|\hat{a}|_{k-j,\infty,\hat{T}}|\hat{v}|_{j,\hat{T}};$$

in the last inequality, we have again used the equivalence of norms over the finite-dimensional space $P_{k-1}(\hat{T})$. Therefore, we obtain

$$|\hat{E}(\hat{a}\hat{v}\hat{w})|\leqslant \hat{C}\left\{\sum_{j=0}^{k-1}|\hat{a}|_{k-j,\infty,\hat{T}}|\hat{v}|_{j,\hat{T}}\right\}|\hat{w}|_{0,\hat{T}} \tag{28.16}$$
$$\text{for all } \hat{a}\in W^{k,\infty}(\hat{T}),\quad \hat{v}\in P_{k-1}(\hat{T}),\quad \hat{w}\in P_{k-1}(\hat{T}).$$

Then it suffices to use the inequalities (cf. Theorems 15.1 and 15.2)

$$|\hat{a}|_{k-j,\infty,\hat{T}}\leqslant \hat{C}h_T^{k-j}|a|_{k-j,\infty,T},\quad 0\leqslant j\leqslant k-1,$$
$$|\hat{v}|_{j,\hat{T}}\leqslant \hat{C}h_T^{j}(\det B_T)^{-1/2}|v|_{j,T},\quad 0\leqslant j\leqslant k-1,$$
$$|\hat{w}|_{0,\hat{T}}\leqslant \hat{C}(\det B_T)^{-1/2}|w|_{0,T},$$

in conjunction with relations (28.15) and (28.16). We obtain in this fashion:

$$|E_T(avw)|\leqslant \hat{C}h_T^{k}\left\{\sum_{j=0}^{k-1}|a|_{k-j,\infty,T}|v|_{j,T}\right\}|w|_{0,T}$$
$$\leqslant Ch_T^{k}\|a\|_{k,\infty,T}\|v\|_{k-1,T}|w|_{0,T},$$
$$\text{for all } a\in W^{k,\infty}(T),\quad v\in P_{k-1}(T),\quad w\in P_{k-1}(T),$$

and we conclude the proof by replacing v by $\partial_i p'$ and w by $\partial_j p$ in the last inequality. □

REMARK 28.2. Let us indicate why a *direct* application of the Bramble–Hilbert lemma to the quadrature error functionals $E_T(\cdot)$ does *not* yield the proper estimate. Let us assume that, for some integer $l\geqslant 0$,

$$\hat{E}(\hat{\varphi})=0\quad\text{for all } \hat{\varphi}\in P_l(\hat{T}),$$

and let $r\in[1,\infty]$ be such that the inclusion $W^{l+1,r}(\hat{T})\hookrightarrow \mathscr{C}^0(\hat{T})$ holds; hence we have

$$|\hat{E}(\hat{\varphi})|\leqslant \hat{C}|\hat{\varphi}|_{0,\infty,\hat{T}}\leqslant \hat{C}\|\hat{\varphi}\|_{l+1,r,\hat{T}}\quad\text{for all } \hat{\varphi}\in W^{l+1,r}(\hat{T}),$$

and thus assumption (28.13), together with the Bramble–Hilbert lemma, implies that

$$|\hat{E}(\hat{\varphi})|\leqslant \hat{C}|\hat{\varphi}|_{l+1,r,\hat{T}}\quad\text{for all } \hat{\varphi}\in W^{l+1,r}(\hat{T}).$$

Let us then replace $\hat{\varphi}$ by the product $\hat{a}\hat{v}\hat{w}$, with a sufficiently smooth function $\hat{a}$, $\hat{v}\in P_{k-1}(\hat{T})$, $\hat{w}\in P_{k-1}(\hat{T})$. Using inequalities of the form (28.11) and the equivalence of norms over the space $P_{k-1}(\hat{T})$, we automatically get *all* the seminorms $|w|_{j,T}$, $0\leqslant j\leqslant \min\{l+1,k-1\}$, in the right-hand side of the final inequality, whereas only the seminorm $|w|_{0,T}$ should appear.

The reader should notice that the ideas involved in the proof of the previous theorem are quite reminiscent of those involved in the proof of Theorem 15.3. In both cases, the central idea is to apply the fundamental result of Theorem 14.1 (in the disguised form of the Bramble–Hilbert lemma in the present case) over the reference finite element and then to use the standard inequalities to go from T to $\hat{T}$, and back. The same analogies also hold for our next result.

THEOREM 28.3. *Assume that, for some integer* $k\geqslant 1$,

$$\hat{P}=P_k(\hat{T}), \tag{28.17}$$

$$\hat{E}(\hat{\varphi})=0 \quad \text{for all } \hat{\varphi}\in P_{2k-2}(\hat{T}), \tag{28.18}$$

and let $q\in[1,\infty]$ *be a number that satisfies the inequality*

$$k-(n/q)>0. \tag{28.19}$$

Then there exists a constant C *independent of* $T\in\mathscr{T}_h$ *and* h *such that*

$$\begin{aligned}&|E_T(fp)|\leqslant Ch_T^k\{\operatorname{meas}(T)\}^{1/2-1/q}\|f\|_{k,q,T}\|p\|_{1,T}\\&\quad\text{for all } f\in W^{k,q}(T),\quad p\in P_k(T).\end{aligned} \tag{28.20}$$

PROOF. For any $f\in W^{k,q}(T)$ and any $p\in P_k(T)$, we have

$$E_T(fp)=(\det B_T)\hat{E}(\hat{f}\hat{p}), \tag{28.21}$$

with $\hat{f}\in W^{k,q}(\hat{T})$, $\hat{p}\in P_k(\hat{T})$. Let us write

$$\hat{E}(\hat{f}\hat{p})=\hat{E}(\hat{f}\hat{\Pi}\hat{p})+\hat{E}(\hat{f}(\hat{p}-\hat{\Pi}\hat{p})), \tag{28.22}$$

where $\hat{\Pi}$ *is the orthogonal projection in the space* $L^2(\hat{T})$ *onto the subspace* $P_1(\hat{T})$.

(i) *Let us first estimate* $\hat{E}(\hat{f}\hat{\Pi}\hat{p})$. For all $\hat{\psi}\in W^{k,q}(\hat{T})$, we have

$$|\hat{E}(\hat{\psi})|\leqslant\hat{C}|\hat{\psi}|_{0,\infty,\hat{T}}\leqslant\hat{C}\|\hat{\psi}\|_{k,q,\hat{T}},$$

since inequality (28.19) implies that the inclusion $W^{k,q}(\hat{T})\hookrightarrow\mathscr{C}^0(\hat{T})$ holds, and, in addition, $\hat{E}(\hat{\psi})=0$ for all $\hat{\psi}\in P_{k-1}(\hat{T})$, by virtue of assumption (28.18) (therefore, this assumption is not fully used at this stage, unless $k=1$). Using the Bramble–Hilbert lemma, we obtain

$$|\hat{E}(\hat{\psi})|\leqslant\hat{C}|\hat{\psi}|_{k,q,\hat{T}}\quad\text{for all } \hat{\psi}\in W^{k,q}(\hat{T}).$$

In particular, let $\hat{\psi}=\hat{f}\hat{\Pi}\hat{p}$ with $\hat{f}\in W^{k,q}(\hat{T})$, $\hat{p}\in P_k(\hat{T})$. Using inequality (28.11), we find:

$$|\hat{f}\hat{\Pi}\hat{p}|_{k,q,\hat{T}}\leqslant\hat{C}\{|\hat{f}|_{k,q,\hat{T}}|\hat{\Pi}\hat{p}|_{0,\infty,\hat{T}}+|\hat{f}|_{k-1,q,\hat{T}}|\hat{\Pi}\hat{p}|_{1,\infty,\hat{T}}\},$$

since all seminorms $|\hat{\Pi}\hat{p}|_{l,\infty,\hat{T}}$ vanish for $l\geqslant 2$ ($\hat{\Pi}\hat{p}\in P_1(\hat{T})$). Using the equivalence of norms over the finite-dimensional space $P_1(\hat{T})$, we get

$$|\hat{f}\hat{\Pi}\hat{p}|_{k,q,\hat{T}}\leqslant\hat{C}\{|\hat{f}|_{k,q,\hat{T}}|\hat{\Pi}\hat{p}|_{0,\hat{T}}+|\hat{f}|_{k-1,q,\hat{T}}|\hat{\Pi}\hat{p}|_{1,\hat{T}}\}.$$

Further we have

$$|\hat{\Pi}\hat{p}|_{0,\hat{T}} \leqslant |\hat{p}|_{0,\hat{T}},$$

since $\hat{\Pi}$ is a projection operator, and

$$|\hat{\Pi}\hat{p}|_{1,\hat{T}} \leqslant |\hat{p}-\hat{\Pi}\hat{p}|_{1,\hat{T}} + |\hat{p}|_{1,\hat{T}}.$$

Since the mapping $\hat{\Pi}$ leaves the space $P_0(\hat{T})$ invariant, there exists (cf. Theorem 15.3) a constant $\hat{C}$ such that

$$|\hat{p}-\hat{\Pi}\hat{p}|_{1,\hat{T}} \leqslant \hat{C}|\hat{p}|_{1,\hat{T}}.$$

Thus, upon combining all our previous inequalities, we have found a constant $\hat{C}$ such that

$$|\hat{E}(\hat{f}\hat{\Pi}\hat{p})| \leqslant \hat{C}\{|\hat{f}|_{k,q,\hat{T}}|\hat{p}|_{0,\hat{T}} + |\hat{f}|_{k-1,q,\hat{T}}|\hat{p}|_{1,\hat{T}}\}$$
$$\text{for all } \hat{f} \in W^{k,q}(\hat{T}), \quad \hat{p} \in P_k(\hat{T}). \tag{28.23}$$

(ii) *Let us next estimate* $\hat{E}(\hat{f}(\hat{p}-\hat{\Pi}\hat{p}))$. We observe that if $k=1$, the difference $(\hat{p}-\hat{\Pi}\hat{p})$ vanishes and therefore, we may henceforth assume that $k \geqslant 2$. This being the case, there exists a number $\rho \in [1, +\infty]$ such that the inclusions

$$W^{k,q}(\hat{T}) \hookrightarrow W^{k-1,\rho}(\hat{T}) \hookrightarrow \mathscr{C}^0(\hat{T})$$

hold.

To see this, consider first the case where $1 \leqslant q < n$, and define a number ρ by letting $1/\rho = 1/q - 1/n$, so that the inclusion $W^{1,q}(\hat{T}) \hookrightarrow L^\rho(\hat{T})$ (and consequently the inclusion $W^{k,q}(\hat{T}) \hookrightarrow W^{k-1,\rho}(\hat{T})$) holds. Then the inclusion $W^{k-1,\rho}(\hat{T}) \hookrightarrow \mathscr{C}^0(\hat{T})$ also holds because $k-1-(n/\rho) = k-(n/q) > 0$ by (28.19).

Consider next the case where $n \leqslant q$. Then either $n < q$ and the inclusion $W^{1,q}(\hat{T}) \hookrightarrow L^\rho(\hat{T})$ holds for all $\rho \in [1, \infty]$, or $n = q$ and the same inclusion holds for all (finite) $\rho \geqslant 1$, so that in both cases the inclusion $W^{k,q}(\hat{T}) \hookrightarrow W^{k-1,\rho}(\hat{T})$ holds for all $\rho \geqslant 1$. Since in this part (ii) we assume $k \geqslant 2$, it suffices to choose ρ large enough so that $k-1-(n/\rho) > 0$ and then the inclusion $W^{k-1,\rho}(\hat{T}) \hookrightarrow \mathscr{C}^0(\hat{T})$ holds.

Using now familiar arguments, we eventually find that

$$|\hat{E}(\hat{f}(\hat{p}-\hat{\Pi}\hat{p}))| \leqslant \hat{C}|\hat{f}(\hat{p}-\hat{\Pi}\hat{p})|_{0,\infty,\hat{T}} \leqslant \hat{C}|\hat{f}|_{0,\infty,\hat{T}}|\hat{p}-\hat{\Pi}\hat{p}|_{0,\infty,\hat{T}}$$
$$\leqslant \hat{C}\|\hat{f}\|_{k-1,\rho,\hat{T}}|\hat{p}-\hat{\Pi}\hat{p}|_{0,\infty,\hat{T}} \quad \text{for all } \hat{f} \in W^{k-1,\rho}(\hat{T}), \quad \hat{p} \in P_k(\hat{T}).$$

Thus for a given $\hat{p} \in P_k(\hat{T})$, the linear form

$$\hat{f} \in W^{k-1,\rho}(\hat{T}) \to \hat{E}(\hat{f}(\hat{p}-\hat{\Pi}\hat{p}))$$

is continuous, with norm $\leqslant \hat{C}|\hat{p}-\hat{\Pi}\hat{p}|_{0,\infty,\hat{T}}$, and it vanishes over the space $P_{k-2}(\hat{T})$ (notice that, by contrast with step (i), the "full" assumption (28.18) is used here). Another application of the Bramble–Hilbert lemma shows that

$$|\hat{E}(\hat{f}(\hat{p}-\hat{\Pi}\hat{p}))| \leqslant C|\hat{f}|_{k-1,\rho,\hat{T}}|\hat{p}-\hat{\Pi}\hat{p}|_{0,\hat{T}} \quad \text{for all } \hat{f} \in W^{k-1,\rho}(\hat{T}), \quad \hat{p} \in P_k(\hat{T}).$$

Since the operator $\hat{\Pi}$ leaves the space $P_0(\hat{T})$ invariant, we have, again by Theorem 15.3,

$$|\hat{p} - \hat{\Pi}\hat{p}|_{0,\hat{T}} \leqslant \hat{C}|\hat{p}|_{1,\hat{T}}.$$

We also have

$$|\hat{g}|_{0,\rho,\hat{T}} \leqslant \hat{C}\{|\hat{g}|_{0,q,\hat{T}} + |\hat{g}|_{1,q,\hat{T}}\} \quad \text{for all } \hat{g} \in W^{1,q}(\hat{T}),$$

since the inclusion $W^{1,q}(\hat{T}) \hookrightarrow L^\rho(\hat{T})$ holds, and thus,

$$|\hat{f}|_{k-1,\rho,\hat{T}} \leqslant \hat{C}\{|\hat{f}|_{k-1,q,\hat{T}} + |\hat{f}|_{k,q,\hat{T}}\} \quad \text{for all } \hat{f} \in W^{k,q}(\hat{T}).$$

Combining all our previous inequalities, we obtain:

$$|\hat{E}(\hat{f}(\hat{p} - \hat{\Pi}\hat{p}))| \leqslant \hat{C}\{|\hat{f}|_{k-1,q,\hat{T}} + |\hat{f}|_{k,q,\hat{T}}\}|\hat{p}|_{1,\hat{T}}$$
$$\text{for all } \hat{f} \in W^{k,q}(\hat{T}), \quad \hat{p} \in P_k(\hat{T}). \tag{28.24}$$

(iii) The proof is completed by combining inequalities (28.21), (28.22), (28.23), (28.24), and

$$|\hat{f}|_{k-j,q,\hat{T}} \leqslant \hat{C}h_T^{k-j}(\det B_T)^{-1/q}|f|_{k-j,q,T}, \quad j = 0, 1,$$
$$|\hat{p}|_{j,\hat{T}} \leqslant \hat{C}h_T^j(\det B_T)^{-1/2}|p|_{j,T}, \quad\quad j = 0, 1. \quad \square$$

REMARK 28.3. Several comments are in order about this proof.

(i) There always exists a number q that satisfies inequality (28.19). In particular, the choice $q = \infty$ is possible in all cases.

(ii) As in Theorem 28.2, a direct application of the Bramble–Hilbert lemma yields unwanted norms in the right-hand side of the final inequality, which should be of the form $|E_T(fp)| \leqslant \cdots \|p\|_{1,\hat{T}}$ (cf. Remark 28.2).

(iii) Why did we have to introduce the projection $\hat{\Pi}$? Arguing otherwise as in part (ii) of the proof, we would find either $|\hat{E}(\hat{f}\hat{p})| \leqslant \hat{C}|\hat{f}|_{k-1,\rho,\hat{T}}|\hat{p}|_{0,\hat{T}}$, or $|\hat{E}(\hat{f}\hat{p})| \leqslant \hat{C}|\hat{f}|_{k-1,\rho,\hat{T}}\|\hat{p}\|_{1,\hat{T}}$: In both cases, there would be a loss of one in the exponent of h_T.

(iv) Since in both steps (i) and (ii) of the proof, only the invariance of the space $P_0(\hat{T})$ through the projection operator is used, why is it not enough to consider the orthogonal projection $\hat{\Pi}_0$ from the space $L^2(\hat{T})$ onto the subspace $P_0(\hat{T})$? It is true that, if $k \geqslant 2$, the whole argument holds with $\hat{\Pi}_0$ instead of $\hat{\Pi}$. If $k = 1$ however, part (i) of the proof yields the inequality $|\hat{E}(\hat{f}\hat{\Pi}_0\hat{p})| \leqslant \hat{C}|\hat{f}|_{k,q,\hat{T}}|\hat{p}|_{0,\hat{T}}$, which is perfectly admissible; but then part (ii) of the proof is needed, since it is necessary to estimate the quantity $\hat{E}(\hat{f}(\hat{p} - \hat{\Pi}_0\hat{p}))$ for $\hat{p} \in P_1(\hat{T})$. It is however impossible to find a space $W^{0,\rho}(\hat{T}) = L^\rho(\hat{T})$ that would be contained in the space $\mathscr{C}^0(\hat{T})$. Thus it is simply to avoid two distinct proofs (one with $\hat{\Pi}$ if $k = 1$, another one with $\hat{\Pi}_0$ if $k \geqslant 2$) that we have used the single mapping $\hat{\Pi}$.

(v) Why is it necessary to introduce the intermediate space $W^{k-1,\rho}(\hat{T})$? For all $\hat{p} \in P_k(\hat{T})$, the function $(\hat{p} - \hat{\Pi}\hat{p})$ is also a polynomial of degree $\leqslant k$. Since, on the other hand, the quadrature scheme is exact for polynomials of degree $\leqslant (2k-2)$, the application of the Bramble–Hilbert lemma to the linear form $\hat{f} \rightarrow \hat{E}(\hat{f}(\hat{p} - \hat{\Pi}\hat{p}))$ necessitates that the function $\hat{f}$ be taken in a Sobolev space that involves derivatives up to and including the order $(k-1)$, and no more.

For studies of numerical integration along the lines developed here, see also MANSFIELD [1971, 1972a]). In ARCANGÉLI and GOUT [1976] and MEINGUET [1975], the constants appearing in the quadrature error estimates are evaluated.

29. Estimate of the error $\|u-u_h\|_{1,\Omega}$

Combining the previous theorems, we obtain the main result of this chapter, which is due to RAVIART [1972] and CIARLET and RAVIART [1972c].

THEOREM 29.1. *In addition to* (H1), (H2) *and* (H3), *assume that there exists an integer* $k\geqslant 1$ *such that the following relations are satisfied*:

$$\hat{P}=P_k(\hat{T}), \tag{29.1}$$

$$H^{k+1}(\hat{T})\hookrightarrow \mathscr{C}^s(\hat{T}), \tag{29.2}$$

where s *is the maximal order of partial derivatives occurring in the definition of the set* $\hat{\Sigma}$, *that the weights of the quadrature schemes are* >0, *and that*

$$\hat{E}(\hat{\varphi})=0 \quad \text{for all } \hat{\varphi}\in P_{2k-2}(\hat{T}). \tag{29.3}$$

If the solution $u\in H_0^1(\Omega)$ *of the variational problem corresponding to the data* (25.1) *belongs to the space* $H^{k+1}(\Omega)$, *if* $a_{ij}\in W^{k,\infty}(\Omega)$, $1\leqslant i,j\leqslant n$, *and if* $f\in W^{k,q}(\Omega)$ *for some number* $q\geqslant 2$ *with* $k>n/q$, *there exists a constant* C *independent of* h *such that*

$$\|u-u_h\|_{1,\Omega}\leqslant Ch^k\left\{|u|_{k,+1,\Omega}+\sum_{i,j=1}^{n}\|a_{ij}\|_{k,\infty,\Omega}\|u\|_{k+1,\Omega}+\|f\|_{k,q,\Omega}\right\}. \tag{29.4}$$

PROOF. The assumptions (29.1) and (29.2) imply that (Theorem 17.1)

$$\|u-\Pi_h u\|_{1,\Omega}\leqslant Ch^k|u|_{k+1,\Omega},$$

where, here and subsequently, C stands for a constant independent of h.

Using (28.5), Theorem 28.2, and the Cauchy–Schwarz inequality, we obtain, for any $w_h\in V_h$,

$$\begin{aligned}
&|a(\Pi_h u, w_h)-a_h(\Pi_h u, w_h)|\\
&\leqslant \sum_{T\in\mathscr{T}_h}\sum_{i,j=1}^{n}|E_T(a_{ij}\partial_i(\Pi_h u|_T)\partial_j(w_{h|T}))|\\
&\leqslant C\sum_{T\in\mathscr{T}_h}h_T^k\sum_{i,j=1}^{n}\|a_{ij}\|_{k,\infty,T}\|\Pi_h u\|_{k,T}|w_h|_{1,T}\\
&\leqslant Ch^k\left(\sum_{i,j=1}^{n}\|a_{ij}\|_{k,\infty,\Omega}\right)\left\{\sum_{T\in\mathscr{T}_h}\|\Pi_h u\|_{k,T}^2\right\}^{1/2}|w_h|_{1,\Omega}.
\end{aligned}$$

By Theorem 17.1 we have

$$\left\{\sum_{T\in\mathcal{T}_h}\|\Pi_h u\|_{k,T}^2\right\}^{1/2}\leqslant\|u\|_{k,\Omega}+\left\{\sum_{T\in\mathcal{T}_h}\|u-\Pi_h u\|_{k,T}^2\right\}^{1/2}$$
$$\leqslant\|u\|_{k,\Omega}+Ch|u|_{k+1,\Omega}\leqslant C\|u\|_{k+1,\Omega},$$

and thus,

$$\inf_{v_h\in V_h}\left\{\|u-v_h\|_{1,\Omega}+\sup_{w_h\in V_h}\frac{|a(v_h,w_h)-a_h(v_h,w_h)|}{\|w_h\|_{1,\Omega}}\right\}$$
$$\leqslant\|u-\Pi_h u\|_{1,\Omega}+\sup_{w_h\in V_h}\frac{|a(\Pi_h u,w_h)-a_h(\Pi_h u,w_h)|}{\|w_h\|_{1,\Omega}}$$
$$\leqslant Ch^k\left\{|u|_{k+1,\Omega}+\sum_{i,j=1}^{n}\|a_{ij}\|_{k,\infty,\Omega}\|u\|_{k+1,\Omega}\right\}.$$

Combining likewise (28.6) and Theorem 28.3, we obtain

$$|l(w_h)-l_h(w_h)|\leqslant\sum_{T\in\mathcal{T}_h}|E_T(fw_h)|$$
$$\leqslant C\sum_{T\in\mathcal{T}_h}h_T^k\{\text{meas}(T)\}^{1/2-1/q}\|f\|_{k,q,T}\|w_h\|_{1,T}$$
$$\leqslant Ch^k\{\text{meas}(\Omega)\}^{1/2-1/q}\|f\|_{k,q,\Omega}\|w_h\|_{1,\Omega}.$$

In the last inequality, we have used the inequality

$$\sum_T|a_Tb_Tc_T|\leqslant\left\{\sum_T|a_T|^\alpha\right\}^{1/\alpha}\left\{\sum_T|b_T|^\beta\right\}^{1/\beta}\left\{\sum_T|c_T|^\gamma\right\}^{1/\gamma},$$

valid for any numbers $\alpha\geqslant1$, $\beta\geqslant1$, $\gamma\geqslant1$ that satisfy

$$\frac{1}{\alpha}+\frac{1}{\beta}+\frac{1}{\gamma}=1.$$

Here, $1/\alpha=\frac{1}{2}-1/q$, $\beta=q$, $\gamma=2$ (this is why the assumption $q\geqslant2$ was needed). Consequently, we obtain

$$\sup_{w_h\in V_h}\frac{|l(w_h)-l_h(w_h)|}{\|w_h\|_{1,\Omega}}\leqslant Ch^k\{\text{meas}(\Omega)\}^{1/2-1/q}\|f\|_{k,q,\Omega}.$$

To complete the proof, it remains to use the abstract error estimate of Theorem 26.1; we may indeed apply this result since, by virtue of assumptions (29.1) and (29.3), the approximate bilinear forms are uniformly V_h-elliptic, by Theorem 27.1. □

REMARK 29.1. When $\hat{P}=P_k(\hat{T})$, the condition that the quadrature scheme be exact for the space $P_{2k-2}(\hat{T})$ has a simple interpretation: It means that *all integrals* $\int_T a_{ij}\partial_i u_h\partial_j v_h\,\text{d}x$ *are exactly computed when all the functions* a_{ij} *are constant.* To see this, notice that

$$\int_T \partial_i p' \partial_j p \, dx = \int_{\hat{T}} (\det B_T) \{\partial_i p'\}^\wedge \{\partial_j p\}^\wedge \, d\hat{x} \quad \text{for all } p', p \in P_T,$$

with

$$\det B_T = \text{constant}, \qquad \{\partial_i p'\}^\wedge \in P_{k-1}(\hat{T}), \qquad \{\partial_j p\}^\wedge \in P_{k-1}(\hat{T}).$$

Let us now illustrate the error estimate of Theorem 29.1 by some applications:

– If we use linear n-simplices, we still get $\|u-u_h\|_{1,\Omega} = O(h)$ provided we use a quadrature scheme exact for constant functions, such as that of (25.15).

– If we use quadratic triangles, we still get $\|u-u_h\|_{1,\Omega} = O(h^2)$ provided we use a quadrature scheme exact for polynomials of degree $\leqslant 2$, such as that of (25.18).

– If we use cubic triangles, it is necessary to use a quadrature scheme exact for polynomials of degree $\leqslant 4$, in order to preserve the error estimate $\|u-u_h\|_{1,\Omega} = O(h^3)$, etc.

Let us next briefly mention various possible extensions of Theorem 29.1. For example, one might wish to analyze the effect of numerical integration for the *homogeneous Neumann problem* corresponding to the following data:

$$V = H^1(\Omega),$$

$$a(u,v) = \int_\Omega \left\{ \sum_{i,j=1}^n a_{ij} \partial_i u \partial_j v + buv \right\} dx, \tag{29.5}$$

$$l(v) = \int_\Omega fv \, dx,$$

where the functions $a_{ij} \in L^\infty(\Omega)$ and $f \in L^2(\Omega)$ satisfy the same assumptions as in (25.1), the function b is defined everywhere over the set $\bar{\Omega}$ and there exists b_0 such that

$$\begin{aligned} &b_0 > 0, \\ &b(x) \geqslant b_0 > 0 \quad \text{for all } x \in \bar{\Omega}. \end{aligned} \tag{29.6}$$

Thus the discrete problem corresponds to the approximate bilinear form

$$\begin{aligned} a_h(u_h, v_h) = &\sum_{T \in \mathcal{T}_h} \sum_{l=1}^L \omega_{l,T} \sum_{i,j=1}^n (a_{ij} \partial_i u_h \partial_j v_h)(b_{l,T}) \\ &+ \sum_{T \in \mathcal{T}_h} \sum_{l=1}^L \omega_{l,T} (b u_h v_h)(b_{l,T}). \end{aligned} \tag{29.7}$$

With the same assumptions as in Theorem 27.1, one can then show that there exists a constant $\tilde{\alpha} > 0$ such that

$$\tilde{\alpha} \|v_h\|_{1,\Omega}^2 \leqslant a_h(v_h, v_h) \quad \text{for all } v_h \in V_h \text{ and all } h,$$

i.e., that the approximate bilinear forms are uniformly V_h-elliptic. Assume next that, for some integer $k \geqslant 1$,

$$\hat{P} = P_k(\hat{T}),$$
$$\hat{E}(\hat{\varphi}) = 0 \quad \text{for all } \hat{\varphi} \in P_{2k-2}(\hat{T}).$$

Then, the same arguments as in the proof of Theorem 28.2 imply that there exists a constant C independent of $T \in \mathscr{T}_h$ and of h such that

$$|E_T(ap'p)| \leqslant Ch_T^k \|a\|_{k,\infty,T} \|p'\|_{k,T} \|p\|_{1,T}$$
$$\text{for all } a \in W^{k,\infty}(T), \quad p \in P_k(T), \quad p' \in P_k(T).$$

From these results the analogue of Theorem 29.1 then easily follows.

A second extension due to Ciarlet and Raviart (unpublished) consists in studying the case where *the space* $\hat{P}$ *satisfies the inclusions*

$$P_k(\hat{T}) \subset \hat{P} \subset P_{k'}(\hat{T}). \tag{29.8}$$

instead of the equality $\hat{P} = P_k(\hat{T})$ assumed thus far. We first observe that the question of uniform V_h-ellipticity is already settled by Theorem 27.1. Theorems 28.2 and 28.3 can then be extended to this case if the quadrature scheme is exact for the space $P_{k+k'-2}(\hat{T})$; these results then imply that the analogue of Theorem 29.1 holds if all the weights are >0, the union $\bigcup_{l=1}^{L} \{\hat{b}_l\}$ contains a $P_{k'-1}(\hat{T})$-unisolvent subset and the quadrature scheme is exact for the space $P_{k+k'-2}(\hat{T})$. This shows in particular that reduced cubic triangles may be used in conjunction with the quadrature scheme of (25.19).

Finally, consider the case where *the space* $\hat{P}$ *satisfies the inclusions*

$$P_k(\hat{T}) \subset \hat{P} \subset Q_k(\hat{T}), \tag{29.9}$$

as is the case for rectangular finite elements.

Assuming again the positivity of the weights, one can show that the approximate bilinear forms are uniformly V_h-elliptic if the union $\bigcup_{l=1}^{L} \{\hat{b}_l\}$ contains a $Q_k(\hat{T}) \cap P_{nk-1}(\hat{T})$-unisolvent subset; the analogues of Theorem 28.2 and 28.3 likewise hold if the quadrature scheme is exact for the space $Q_{2k-1}(\hat{T})$. These results then imply that the analogue of Theorem 29.1 holds if all the weights are >0, if the union $\bigcup_{l=1}^{L} \{\hat{b}_l\}$ contains a $Q_k(\hat{T}) \cap P_{nk-1}(\hat{T})$-unisolvent subset, and if the quadrature scheme is exact for the space $Q_{2k-1}(\hat{T})$ (an example of such a scheme has been given in (25.22)).

As a consequence, and contrary to the case where $\hat{P} = P_k(\hat{T})$ (cf. Remark 29.1), it is no longer necessary to exactly compute the integrals $\int_T a_{ij} \partial_i u_h \partial_j v_h \, dx$ when the coefficients a_{ij} are constant functions.

This last extension is again due to Ciarlet and Raviart. Using the abstract error estimate mentioned in Remark 26.1, Ciarlet and Raviart have also established that, if the assumptions of Theorems 19.2 and 29.1 simultaneously hold, then

$$|u - u_h|_{0,\Omega} = O(h^{k+1})$$

if the quadrature scheme is exact for the space $P_{2k-2}(\hat{T})$ if $k \geq 2$, or if the quadrature scheme is exact for the space $P_1(\hat{T})$ if $k=1$.

An interesting extension of Theorem 29.1, based on appropriate *discrete Poincaré–Friedrichs inequalities* (cf. ŽENÍŠEK [1981a]), has been given by ŽENÍŠEK [1987], who has established convergence under the "minimal" assumption that the solution u is in the space $H^1(\Omega)$.

Quadrature schemes of order lower than those indicated by the theory are often employed in practice, in order to reduce computational costs: This procedure of *underintegration* sometimes leads to various numerical instabilities; in particular, *hourglass instabilities* often develop, in that "spurious oscillations" are superposed on an otherwise smooth solution. This interesting phenomenon is described and studied in BELYTSCHKO and ONG [1984], JACQUOTTE and ODEN [1984], JACQUOTTE [1985], JACQUOTTE, ODEN and BECKER [1986], KOH and KIKUCHI [1987], WISSMANN, BECKER and MÖLLER [1987].

For other references concerning the effect of numerical integration, see BABUŠKA and AZIZ [1972, Chapter 9], FIX [1972a, 1972b], HERBOLD [1968] where this problem was studied for the first time, HERBOLD, SCHULTZ and VARGA [1969], HERBOLD and VARGA [1972], ODEN and REDDY [1976a, Section 8.8], SCHULTZ [1972], STRANG and FIX [1973, Section 4.3].

CHAPTER V

Nonconforming Finite Element Methods for Second-Order Problems

Introduction

We study in this chapter a first instance of finite element methods for which the spaces V_h are *not* contained in the space V. This violation of the inclusion $V_h \subset V$ results of the use of finite elements that are not of class $\mathscr{C}^0$, i.e., that are not continuous across adjacent finite elements, so that the inclusion $V_h \subset H^1(\Omega)$ is not satisfied (Theorem 30.1). The terminology "*nonconforming finite element method*" is specifically reserved for this type of "nonconformity" (likewise, nonconforming methods for fourth-order problems use finite elements that are not of class $\mathscr{C}^1$; cf. Section 49).

For definiteness, we assume throughout this chapter that we are solving a homogeneous Dirichlet problem posed over a polygonal domain $\bar{\Omega}$. Then the discrete problem consists in finding a function $u_h \in V_h$ such that, for all $v_h \in V_h$, $a_h(u_h, v_h) = l(v_h)$, where the *approximate bilinear form* $a_h(\cdot,\cdot)$ is defined by

$$a_h(\cdot,\cdot) = \sum_{T \in \mathscr{T}_h} \int_T \{\cdots\}\, dx,$$

the integrand $\{\cdots\}$ being the same as in the bilinear form that is used in the definition of the original problem. The linear form $l(\cdot)$ need not be approximated since the inclusion $V_h \subset L^2(\Omega)$ holds.

Assuming that the mapping

$$v_h \in V_h \to \| v_h \|_h = \left\{ \sum_{T \in \mathscr{T}_h} |v_h|_{1,T}^2 \right\}^{1/2}$$

is a norm over the space V_h, we prove an abstract error estimate, the *second Strang lemma* (Theorem 31.1), where a *consistency error term* is added to the expected term $\inf_{v_h \in V_h} \| u - v_h \|_h$. Just as in the case of numerical integration (cf. Section 26), this result holds under the assumption that the approximate bilinear forms are *uniformly V_h-elliptic*, in the sense that there exists a constant $\tilde{\alpha} > 0$ independent of h such that $a_h(v_h, v_h) \geqslant \tilde{\alpha} \| v_h \|_h^2$ for all $v_h \in V_h$.

We then proceed to describe in Section 32 a three-dimensional "nonconforming" finite element, known as *Wilson's brick*, which has gained some popularity among engineers for solving the boundary value problem of linearized elasticity. Apart from being nonconforming, this finite element presents the added theoretical interest that some of its degrees of freedom are of a form not yet encountered. This is why we need to adapt to this finite element the standard interpolation error analysis (Theorem 32.1).

Next, using a "*bilinear lemma*" which extends the Bramble–Hilbert lemma to bilinear forms (Theorem 33.1), we analyze the consistency error (Theorem 34.1); we show in this fashion that

$$\|u-u_h\|_h=\left\{\sum_{T\in\mathcal{T}_h}|u-u_h|_{1,T}^2\right\}^{1/2}=\mathrm{O}(h),$$

if the solution u is in the space $H^2(\Omega)$. We also establish the connection between the convergence of such nonconforming finite element methods and the *patch test* of Irons.

30. Nonconforming methods

Assume, as in Chapter IV, that we are solving the second-order boundary value problem that corresponds to the following data:

$$\begin{aligned} &V=H_0^1(\Omega),\\ &a(u,v)=\int_\Omega \sum_{i,j=1}^n a_{ij}\partial_i u\partial_j v\,\mathrm{d}x, \qquad (30.1)\\ &l(v)=\int_\Omega fv\,\mathrm{d}x. \end{aligned}$$

At this descriptive stage, the only assumptions that we need to record are that

$$a_{ij}\in L^\infty(\Omega),\quad 1\leqslant i,j\leqslant n,\qquad f\in L^2(\Omega), \tag{30.2}$$

and that the set $\bar\Omega$ is a *polygonal* subset of $\mathbb{R}^n$. As in the preceding chapter, this last assumption insures that the set $\bar\Omega$ can be exactly covered with triangulations.

Assume next that, with any triangulation $\bar\Omega=\bigcup_{T\in\mathcal{T}_h}T$ of the set $\bar\Omega$, we associate a finite element space X_h whose generic finite element is not of class $\mathscr{C}^0$. Then the space X_h will not be contained in the space $H^1(\Omega)$, as we now show (note that the next theorem is the converse of Theorem 5.1).

THEOREM 30.1. *Assume that the inclusions* $P_T\subset\mathscr{C}^0(T)$ *for all* $T\in\mathcal{T}_h$ *and* $X_h\subset H^1(\Omega)$ *hold. Then the inclusion*

$$X_h\subset\mathscr{C}^0(\bar\Omega)$$

holds.

PROOF. Assume that the conclusion is false. Then there exists a function $v \in X_h$, there exist two adjacent finite elements $T(1)$ and $T(2)$, and there exists a nonempty open set $O \subset T(1) \cup T(2)$ such that (for instance)

$$(v|_{T(1)} - v|_{T(2)}) > 0 \quad \text{along } T' \cap O, \tag{30.3}$$

where T' is the face common to $T(1)$ and $T(2)$. Let $\varphi \in \mathscr{D}(\Omega)$, $\varphi > 0$ with supp $\varphi \subset O$. Using Green's formula (2.4), we obtain:

$$\begin{aligned}\int_\Omega \partial_i v\, \varphi\, \mathrm{d}x &= \sum_{\lambda=1,2} \int_{T(\lambda)} \partial_i v\, \varphi\, \mathrm{d}x\\ &= -\sum_{\lambda=1,2} \int_{T(\lambda)} v \partial_i \varphi\, \mathrm{d}x + \sum_{\lambda=1,2} \int_{\partial T(\lambda)} v|_{T(\lambda)} \varphi \nu_{i,T(\lambda)}\, \mathrm{d}\gamma\\ &= -\int_\Omega v \partial_i \varphi\, \mathrm{d}x + \int_{T'} (v|_{T(1)} - v|_{T(2)}) \varphi \nu_{i,T(1)}\, \mathrm{d}\gamma,\end{aligned}$$

and thus we reach a contradiction since the integral along T' should be >0 by (30.3). □

More specifically, let us assume that the inclusions

$$P_T \subset H^1(T) \quad \text{for all } T \in \mathscr{T}_h, \tag{30.4}$$

hold; hence at least the inclusion

$$X_h \subset L^2(\Omega) \tag{30.5}$$

holds. We then define a subspace X_{0h} of X_h that takes as well as possible into account the boundary condition $v = 0$ along the boundary Γ of Ω. For example, if the generic finite element is a Lagrange element, all degrees of freedom are set equal to zero at the boundary nodes. But, again because the finite element is not of class $\mathscr{C}^0$ (cf. Remark 12.2), *the functions in the space X_{0h} will in general vanish only at the boundary nodes.*

In order to define a discrete problem over the space $V_h = X_{0h}$, we observe that, if the linear form l is still defined over the space V_h by virtue of the inclusion (30.5), this is not the case of the bilinear form $a(\cdot,\cdot)$. To obviate this difficulty, we define, in view of (30.1) and (30.4), the *approximate bilinear form*

$$a_h(u_h, v_h) = \sum_{T \in \mathscr{T}_h} \int_T \sum_{i,j=1}^n a_{ij} \partial_i u_h \partial_j v_h\, \mathrm{d}x, \tag{30.6}$$

and the *discrete problem* consists in finding a function $u_h \in V_h$ such that

$$a_h(u_h, v_h) = l(v_h) \quad \text{for all } v_h \in V_h. \tag{30.7}$$

We shall call any such finite element approximation of a second-order boundary value problem a *nonconforming finite element method*. By extension, we shall call any generic finite element used in such a method a *nonconforming finite element*.

31. Abstract error estimate and the second Strang lemma

To begin with, we must equip the space V_h with a norm. Since $|\cdot|_{1,\Omega}$ is a norm over the space $V=H_0^1(\Omega)$, a natural candidate is the mapping

$$v_h \to \|v_h\|_h = \left\{ \sum_{T\in\mathcal{T}_h} |v_h|_{1,T}^2 \right\}^{1/2} \tag{31.1}$$

which is a priori only a *seminorm* over the space V_h. Thus, given a specific nonconforming finite element, our first task is to check that the mapping of (31.1) is indeed a *norm* on the space V_h.

After this, we shall be interested in finding sufficient conditions that guarantee that the approximate bilinear forms of (30.6) are *uniformly V_h-elliptic*, in the sense that there exists a constant $\tilde{\alpha}$ such that

$$\begin{aligned} &\tilde{\alpha} > 0, \\ &\tilde{\alpha}\|v_h\|_h^2 \leqslant a_h(v_h, v_h) \quad \text{for all } v_h \in V_h \text{ and all } h. \end{aligned} \tag{31.2}$$

This condition, which clearly implies the existence and uniqueness of the solution of the discrete problem, is also a crucial assumption for obtaining the abstract error estimate (Theorem 31.1).

It follows from the definition (31.1) of the norm $\|\cdot\|_h$ that the *approximate bilinear forms* (30.6) *are uniformly V_h-elliptic if the functions a_{ij} satisfy the usual ellipticity condition*, i.e., if there exists a constant β such that

$$\begin{aligned} &\beta > 0, \\ &\sum_{i,j=1}^{n} a_{ij}(x)\xi_i\xi_j \geqslant \beta \sum_{i=1}^{n} \xi_i^2 \\ &\quad \text{for all } (\xi_1, \xi_2, \ldots, \xi_n) \in \mathbb{R}^n \text{ and for almost all } x \in \Omega. \end{aligned} \tag{31.3}$$

From now on, we shall consider that the domain of definition of both the approximate bilinear form of (30.7) and the seminorm of (31.1) is the space $(V_h + V)$. This being the case, notice that

$$a_h(v, v) = a(v, v), \quad \|v\|_h = |v|_{1,\Omega} \quad \text{for all } v \in V. \tag{31.4}$$

Also, the first assumptions (30.2) imply that there exists a constant $\tilde{M}$ independent of the space V_h such that

$$|a_h(u, v)| \leqslant \tilde{M}\|u\|_h\|v\|_h \quad \text{for all } u, v \in (V_h + V). \tag{31.5}$$

The next result is due to STRANG [1972b].

THEOREM 31.1 (Second Strang lemma). *Consider a family of discrete problems whose associated approximate bilinear forms are uniformly V_h-elliptic.*

Then there exists a constant C independent of the subspace V_h such that

$$\|u-u_h\|_h \leqslant C\left(\inf_{v_h\in V_h} \|u-v_h\|_h + \sup_{w_h\in V_h} \frac{|a_h(u,w_h)-l(w_h)|}{\|w_h\|_h}\right). \tag{31.6}$$

PROOF. Let v_h be an arbitrary element in the space V_h. Using the uniform V_h-ellipticity and the continuity of the bilinear forms a_h (cf. (31.2) and (31.5)) and the definition (30.7) of the discrete problem, we may write

$$\begin{aligned}\tilde{\alpha}\|u_h-v_h\|_h^2 &\leqslant a_h(u_h-v_h,u_h-v_h)\\ &= a_h(u-v_h,u_h-v_h)+\{l(u_h-v_h)-a_h(u,u_h-v_h)\},\end{aligned}$$

from which we deduce

$$\begin{aligned}\tilde{\alpha}\|u_h-v_h\|_h &\leqslant \tilde{M}\|u-v_h\|_h + \frac{|l(u_h-v_h)-a_h(u,u_h-v_h)|}{\|u_h-v_h\|_h}\\ &\leqslant \tilde{M}\|u-v_h\|_h + \sup_{w_h\in V_h}\frac{|l(w_h)-a_h(u,w_h)|}{\|w_h\|_h}.\end{aligned}$$

Inequality (31.6) then follows from this inequality and the triangular inequality

$$\|u-u_h\|_h \leqslant \|u-v_h\|_h + \|u_h-v_h\|_h. \qquad \square$$

Note that the error estimate (31.6) indeed generalizes the error estimate established in Céa's lemma (Theorem 13.1) for conforming methods, since the difference $\{l(w_h)-a_h(u,w_h)\}$ vanishes for all $w_h\in V_h$ when the space V_h is contained in the space V.

REMARK 31.1. An abstract error estimate can also be established that eventually yields an estimate of the error $|u-u_h|_{0,\Omega}$: Let H be a Hilbert space such that $\bar{V}=H$, $V\hookrightarrow H$ and $V_h\subset H$ for all h, and let, for all $u, v\in(V_h+V)$,

$$D_h(u,v)=a_h(u,v)-f(v).$$

Finally, assume that the bilinear form is symmetric. Then the abstract error estimate of the Aubin–Nitsche lemma (Theorem 19.1) is replaced in the present situation by the following inequality:

$$|u-u_h|\leqslant \sup_{g\in H}\frac{1}{|g|}\inf_{\varphi_h\in V_h}\{\tilde{M}\|u-u_h\|\,\|\varphi_g-\varphi_h\|+|D_h(u,\varphi_g-\varphi_h)|+|D_h(\varphi_g,u-u_h)|\},$$

where $|\cdot|$ denotes the norm in H, and for each $g\in H$, $\varphi_g\in V$ denotes the unique solution of the variational problem

$$a(v,\varphi_g)=(g,v)\quad \text{for all } v\in V,$$

where $(\cdot,\cdot)$ denotes the inner product in H.

This abstract error estimate is found in NITSCHE [1974] and LASCAUX and LESAINT [1975].

32. An example of a nonconforming finite element: Wilson's brick

We now describe an example of a nonconforming finite element, named *Wilson's brick* after WILSON, TAYLOR, DOHERTY and GHABUSSI [1973]. It is used for approximating the solution of linearized elasticity problems posed over three-dimensional rectangular domains.

REMARK 32.1. A two-dimensional analog of this finite element, known as *Wilson's rectangle*, is used for approximating problems in "plane elasticity". This finite element is studied in LESAINT [1976].

Wilson's brick is an example of a *rectangular* finite element in $\mathbb{R}^3$: The set T is a 3-rectangle, whose vertices will be denoted a_i, $1 \leqslant i \leqslant 8$ (Fig. 32.1).

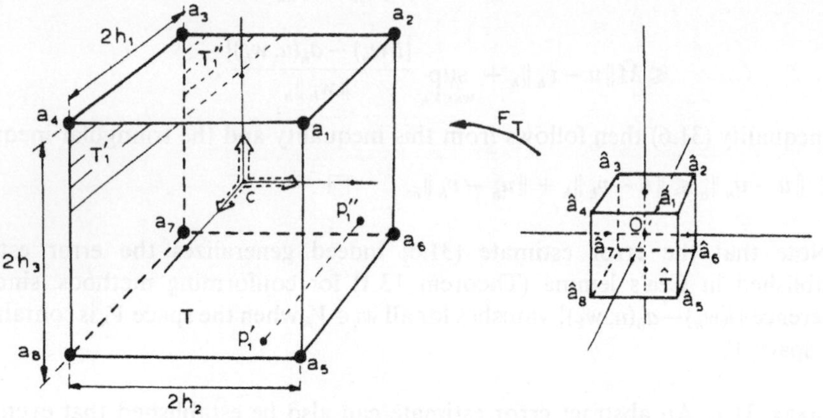

Wilson's brick, $n=3$	
$P_T = Q_1(T) \oplus \vee\{x_j^2;\ 1 \leqslant j \leqslant 3\}$,	$\dim(P_T)=11$
$\Sigma_T = \left\{ p(a_i)\text{: } 1 \leqslant i \leqslant 8;\ \dfrac{h_j^2}{h_1h_2h_3} \displaystyle\int_T \partial_{jj} p \,\mathrm{d}x\text{: } 1 \leqslant j \leqslant 3 \right\}$	

FIG. 32.1.

The space P_T is the space $P_2(T)$ to which are added linear combinations of the function $(x_1x_2x_3)$; equivalently, the space P_T is the space $Q_1(T)$ to which have been added linear combinations of the three functions x_j^2, $1 \leqslant j \leqslant 3$. We shall therefore record this definition by writing

$$P_T = P_2(T) \oplus \vee\{x_1x_2x_3\} = Q_1(T) \oplus \vee\{x_j^2;\ 1 \leqslant j \leqslant 3\}. \tag{32.1}$$

Notice that the inclusions

$$P_2(T) \subset P_T, \qquad Q_1(T) \subset P_T \tag{32.2}$$

hold and that

$$\dim(P_T)=11. \tag{32.3}$$

We next show that *the values* $p(a_i)$, $1\leqslant i\leqslant 8$, *at the vertices, together with the values of the* (constant) *second derivatives* $\partial_{jj}p$, $1\leqslant j\leqslant 3$, *form a* P_T*-unisolvent set*. To see this, it suffices to check the validity of the following identity: For all functions $\hat{p}\in P_{\hat{T}}$, with $\hat{T}=[-1,+1]^3$, one has

$$\begin{aligned}\hat{p}=&\tfrac{1}{8}(1+x_1)(1+x_2)\{(1+x_3)\hat{p}(\hat{a}_1)+(1-x_3)\hat{p}(\hat{a}_5)\}\\&+\tfrac{1}{8}(1-x_1)(1+x_2)\{(1+x_3)\hat{p}(\hat{a}_2)+(1-x_3)\hat{p}(\hat{a}_6)\}\\&+\tfrac{1}{8}(1-x_1)(1-x_2)\{(1+x_3)\hat{p}(\hat{a}_3)+(1-x_3)\hat{p}(\hat{a}_7)\}\\&+\tfrac{1}{8}(1+x_1)(1-x_2)\{(1+x_3)\hat{p}(\hat{a}_4)+(1-x_3)\hat{p}(\hat{a}_8)\}\\&+\tfrac{1}{2}(x_1^2-1)\partial_{11}\hat{p}+\tfrac{1}{2}(x_2^2-1)\partial_{22}\hat{p}+\tfrac{1}{2}(x_3^2-1)\partial_{33}\hat{p}.\end{aligned} \tag{32.4}$$

Therefore if we denote by $c=\frac{1}{8}\Sigma_{i=1}^{8}a_i$ the center of the finite element T, we are naturally tempted to define the following set of degrees of freedom:

$$\Xi_T=\{p(a_i)\colon 1\leqslant i\leqslant 8;\ \partial_{jj}p(c)\colon 1\leqslant j\leqslant 3\}, \tag{32.5}$$

whose degrees of freedom are all in a familiar form. Of course, nothing obliges us to attach the last three degrees of freedom to the particular point c (except perhaps an aesthetical reason of symmetry), since the second derivatives $\partial_{jj}p$, $1\leqslant j\leqslant 3$, are constant for any function $p\in P_T$.

Keeping this last property in mind, we may also choose for degrees of freedom the *averages* $\int_T\partial_{jj}p\,\mathrm{d}x$, $1\leqslant j\leqslant 3$, and we shall indeed show that this choice is more appropriate for our subsequent purposes. For the time being, we observe that *such degrees of freedom are of a new type*, although they are still linear forms over the space $\mathscr{C}^2(T)$ as indeed they should be, to comply with the general definition of a degree of freedom given in Section 10.

Notice that since any function $p\in P_T$ satisfies

$$\partial_{jj}p(c)=\frac{1}{8h_1h_2h_3}\int_T\partial_{jj}p\,\mathrm{d}x,\quad 1\leqslant j\leqslant 3, \tag{32.6}$$

where $2h_j$, $1\leqslant j\leqslant 3$, denote the lengths of the sides as indicated in Fig. 32.1, the two types of degrees of freedom are interchangeable over the space P_T. However, relations (32.6) do not hold in general for arbitrary functions in the space $\mathscr{C}^2(T)$; in other words we obtain in this fashion two *different* finite elements (cf. Remark 32.2 below; also, this is an instance of a phenomenon that was mentioned in Section 10).

Let us then equip Wilson's brick with degrees of freedom of the form (32.6). Our next objective is to extend the definition of affine-equivalence so that Wilson's bricks can be imbedded in an affine family, the reference finite element being in this case the hypercube $\hat{T}=[-1,+1]^3$. To do this, it suffices, according to equations (11.12), to write the degrees of freedom in such a way that if we have the identity

$$\hat{p}=\sum_{i=1}^{8}\hat{p}(\hat{a}_i)\hat{p}_i+\sum_{j=1}^{3}\hat{\phi}_j(\hat{p})\hat{q}_j\quad\text{for all }\hat{p}\in P_{\hat{T}}, \tag{32.7}$$

then we also have the identity

$$p=\sum_{i=1}^{8} p(a_i)p_i+\sum_{j=1}^{3} \phi_j(p)q_j \quad \text{for all } p\in P_T, \tag{32.8}$$

where the basis functions $\hat{p}_i$ and p_i, respectively $\hat{q}_j$ and q_j, are in the usual correspondence (11.9), and $\hat{\phi}_j$ and ϕ_j, $1\leqslant j\leqslant 3$, denote the degrees of freedom of the form $\int_T \partial_{jj}p\,\mathrm{d}x$, attached to the sets $\hat{T}$ and T, respectively. Using (32.4), we easily deduce that any function p in the space P_T satisfies the following identity, where c_i, $1\leqslant i\leqslant 3$, denote the coordinates of the point c:

$$p=\frac{1}{8}\left(1+\frac{(x_1-c_1)}{h_1}\right)\left(1+\frac{(x_2-c_2)}{h_2}\right)\left\{\left(1+\frac{(x_3-c_3)}{h_3}\right)p(a_1)+\left(1-\frac{(x_3-c_3)}{h_3}\right)p(a_5)\right\}$$
$$+\cdots+\sum_{j=1}^{3}\frac{1}{16}\left\{\left(\frac{x_j-c_j}{h_j}\right)^2-1\right\}\frac{h_j^2}{h_1h_2h_3}\int_T \partial_{jj}p\,\mathrm{d}x. \tag{32.9}$$

After inspecting (32.8) and (32.9), we find that the proper choices for ϕ_j and q_j are:

$$\phi_j(p)=\frac{h_j^2}{h_1h_2h_3}\int_T \partial_{jj}p\,\mathrm{d}x,\quad q_j=\frac{1}{16}\left(\left\{\frac{x_j-c_j}{h_j}\right\}^2-1\right),\qquad 1\leqslant j\leqslant 3. \tag{32.10}$$

These choices insure that the following relations hold:

$$\begin{aligned} &p_i(a_k)=\delta_{ik}, && 1\leqslant i,k\leqslant 8,\\ &q_j(a_i)=0, && 1\leqslant i\leqslant 8,\quad 1\leqslant j\leqslant 3,\\ &\phi_j(p_i)=0, && 1\leqslant i\leqslant 8,\quad 1\leqslant j\leqslant 3,\\ &\phi_l(q_j)=\delta_{lj}, && 1\leqslant j,l\leqslant 3. \end{aligned} \tag{32.11}$$

Consequently, we shall henceforth consider that *the set of degrees of freedom of Wilson's brick* is

$$\Sigma_T=\left\{p(a_i)\colon 1\leqslant i\leqslant 8;\ \frac{h_j^2}{h_1h_2h_3}\int_T \partial_{jj}p\,\mathrm{d}x\colon 1\leqslant j\leqslant 3\right\}. \tag{32.12}$$

Notice that we could drop the multiplicative factors $h_j^2/(h_1h_2h_3)$ in the last degrees of freedom without changing the definition of the finite element.

According to definition (10.6), the associated *P_T-interpolation operator Π_T* is such that, for any sufficiently smooth function $v: T\to\mathbb{R}$, the function $\Pi_T v$ belongs to the space P_T and is uniquely determined by the conditions

$$\begin{aligned} &\Pi_T v(a_i)=v(a_i), && 1\leqslant i\leqslant 8,\\ &\phi_j(\Pi_T v)=\phi_j(v), && 1\leqslant j\leqslant 3. \end{aligned} \tag{32.13}$$

Notice that the last three conditions can also be written as

$$\int_T \partial_{jj}(\Pi_T v)\,\mathrm{d}x = \int_T \partial_{jj} v\,\mathrm{d}x, \quad 1 \leqslant j \leqslant 3. \tag{32.14}$$

By construction, the P_T-interpolation operator satisfies

$$\{\Pi_T v\}\hat{} = \Pi_{\hat{T}} \hat{v} \tag{32.15}$$

for functions v and $\hat{v}$ in the usual correspondence $\hat{v} \to v = \hat{v} \cdot F^{-1}$. Also, by virtue of the first inclusion (32.2), we have

$$\Pi_{\hat{T}} \hat{p} = \hat{p} \quad \text{for all } \hat{p} \in P_2(\hat{T}). \tag{32.16}$$

REMARK 32.2. According to definition (10.9), the finite elements (T, P_T, Ξ_T) and (T, P_T, Σ_T) (cf. (32.5) and (32.12)) are *not* identical since the associated interpolation operators do not coincide over the space $\mathscr{C}^2(T)$ (we momentarily ignore that the domain of the interpolation operator corresponding to the set Σ_T is wider; see below).

We are now in a position to explain the definite advantage of choosing the forms ϕ_j as degrees of freedom, rather than the point values $\partial_{jj} p(c)$. On the one hand, the basic properties (32.15) and (32.16) of the interpolation operator are unaltered, but on the other hand, *the interpolation operator Π_T has a wider domain*: Whereas in the first case, one is led to assume that the function $v : T \to \mathbb{R}$ is twice differentiable over T in order to define its P_T-interpolant, in the second case the P_T-interpolant is well defined for functions "only" in the space $H^2(T)$ (which is contained in the space $\mathscr{C}^0(T)$ for $n=3$). This property will later avoid unnecessary restrictions on the smoothness of the solution u of our original problem (cf. Theorem 34.1).

Although the larger Sobolev space over which the P_T-interpolant is defined is the space $W^{2,p}(T)$ for $p > \frac{3}{2}$, we shall consider for simplicity that

$$\operatorname{dom} \Pi_T = H^2(T). \tag{32.17}$$

In the next theorem, we estimate the interpolation errors $|v - \Pi_T v|_{m,T}$. The notations h_T and ρ_T represent the usual geometrical parameters (cf. (16.5)).

THEOREM 32.1. *There exists a constant C such that, for all Wilson's bricks (T, P_T, Σ_T),*

$$|v - \Pi_T v|_{m,T} \leqslant C \frac{h_T^l}{\rho_T^m} |v|_{l,T}, \qquad 0 \leqslant m \leqslant l, \quad l = 2, 3, \quad \text{for all } v \in H^l(T). \tag{32.18}$$

PROOF. By an argument similar to that used in the proof of Theorem 16.1, it can be checked that the mapping

$$\Pi_{\hat{T}} : H^l(\hat{T}) \subset H^2(\hat{T}) = \operatorname{dom} \Pi_{\hat{T}} \to H^m(\hat{T})$$

is continuous for $0 \leqslant m \leqslant l$, $l = 2$ or 3. It then suffices to combine this fact with relations (32.15) and (32.16), and to apply Theorem 15.3. □

Let $\bar{\Omega}$ be a rectangular domain in $\mathbb{R}^3$, so that it may be covered by triangulations $\mathcal{T}_h$ composed of 3-rectangles. We then let X_h denote the finite element space whose functions v_h have the following properties:

(i) For each $T \in \mathcal{T}_h$, the restrictions $v_h|_T$ belong to the space P_T defined in (32.1).

(ii) Each function $v_h \in X_h$ is defined by its values at all the vertices and by the averages

$$\int_T \partial_{jj} v_h|_T \, \mathrm{d}x, \quad 1 \leqslant j \leqslant 3, \qquad T \in \mathcal{T}_h.$$

Since the basis functions q_j given in (32.10) do not vanish on the boundary of Wilson's brick, *this element is not of class* $\mathscr{C}^0$; *hence the space* X_h *is not contained in the space* $H^1(\Omega)$, by Theorem 30.1. Continuity is however guaranteed at the vertices of the triangulations, since the functions q_j vanish at all nodes of Wilson's brick (cf. (32.11)).

Finally, we let $V_h = X_{0h}$, where X_{0h} denotes the space of all functions $v_h \in X_h$ that vanish at the boundary nodes. For the same reasons as before, the functions in the space X_{0h} do not vanish along the boundary Γ, but they vanish at the boundary nodes.

According to the discussion made in Section 31, we need first to verify that the mapping $\|\cdot\|_h$ defined in (31.1) is indeed a norm over the space V_h.

THEOREM 32.2. *The mapping*

$$v_h \to \|v_h\|_h = \left\{ \sum_{T \in \mathcal{T}_h} |v_h|_{1,T}^2 \right\}^{1/2} \tag{32.19}$$

is a norm over the finite element space V_h *constructed with Wilson's bricks.*

PROOF. Let v_h be a function in the space V_h that satisfies

$$\|v_h\|_h = \left\{ \sum_{T \in \mathcal{T}_h} |v_{h|T}|_{1,T}^2 \right\}^{1/2} = 0.$$

Then each polynomial $v_h|_T$ is a constant function. Hence $\partial_{jj}(v_h|_T) = 0$, $1 \leqslant j \leqslant 3$, in each $T \in \mathcal{T}_h$, on the one hand. On the other, the function $v_h : \bar{\Omega} \to \mathbb{R}$ must be equal to the same constant over all $T \in \mathcal{T}_h$, since it is continuous at all the vertices and thus, it is identically zero since it vanishes at the boundary nodes. □

In order to simpify the exposition, *we shall henceforth assume that the bilinear form of* (30.1) *is*

$$a(u, v) = \int_\Omega \sum_{i=1}^{3} \partial_i u \partial_i v \, \mathrm{d}x, \tag{32.20}$$

i.e., the corresponding boundary value problem is a homogeneous Dirichlet problem for the operator $-\Delta$. The extension of the subsequent analysis to more general bilinear forms, such as

$$a(u, v) = \int_\Omega \left\{ \sum_{i,j=1}^n a_{ij}\partial_i u \partial_j v + buv \right\} dx,$$

offers no particular difficulties (other than technical). With the choice (32.20), the *uniform V_h-ellipticity of the approximate bilinear forms* is a consequence of the identity

$$\|v_h\|_h^2 = a_h(v_h, v_h) \quad \text{for all } v_h \in V_h. \tag{32.21}$$

This being the case, we may apply the abstract error estimate of Theorem 31.1. The first term, $\inf_{v_h \in V_h} \|u - v_h\|_h$, is easily taken care of: Assuming that we consider a family of discrete problems associated with a regular family of triangulations, and assuming that the solution u is in the space $H^2(\Omega)$, we deduce from Theorem 32.1 that

$$\inf_{v_h \in V_h} \|u - v_h\|_h \leqslant \left\{ \sum_{T \in \mathcal{T}_h} |u - \Pi_T u|_{1,T}^2 \right\}^{1/2} \leqslant Ch|u|_{2,\Omega}. \tag{32.22}$$

Notice that the derivation of this interpolation error estimate makes a crucial use of the familiar implication (cf. (12.17))

$$v \in \operatorname{dom} \Pi_h = H^2(\Omega) \text{ and } v_{|\Gamma} = 0 \;\Rightarrow\; \Pi_h v \in X_{0h},$$

where Π_h is the X_h-interpolation operator.

REMARK 32.3. Of course, we could assume that $u \in H^3(\Omega)$, thus getting an $O(h^2)$ estimate instead of (32.22). However the eventual gain is nil because the other term in the right-hand side of inequality (31.6) is of order h, whatever the additional smoothness of the solution may be. Besides, we recall that the weaker assumption $u \in H^2(\Omega)$ is more realistic: One does not have a smoother solution in general on convex polygonal domains.

33. Consistency error estimate and the bilinear lemma

The other term,

$$\sup_{w_h \in V_h} \frac{|a_h(u, w_h) - l(w_h)|}{\|w_h\|_h},$$

which appears in inequality (31.6), is a *consistency error term* due to the "nonconformity" of the method. Consequently, a sufficient condition for convergence is the *consistency condition*:

$$\lim_{h\to 0} \sup_{w_h \in V_h} \frac{|a_h(u, w_h) - l(w_h)|}{\|w_h\|_h} = 0.$$

Our next objective is thus *to estimate this consistency error term*, through a careful analysis of the difference

$$D_h(u, w_h) = a_h(u, w_h) - l(w_h), \quad w_h \in V_h \tag{33.1}$$

Since $-\Delta u = f$, we can write for any function $w_h \in V_h$,

$$D_h(u, w_h) = \sum_{T \in \mathcal{T}_h} \int_T \sum_{i=1}^{3} \partial_i u \partial_i w_h \, dx - \int_\Omega f w_h \, dx.$$

$$= \sum_{T \in \mathcal{T}_h} \int_T \left\{ \sum_{i=1}^{3} \partial_i u \partial_i w_h \, dx + \Delta u \, w_h \right\} dx, \tag{33.2}$$

i.e., we have obtained *one* decomposition of the form

$$D_h(u, w_h) = \sum_{T \in \mathcal{T}_h} D_T(u|_T, w_h|_T), \tag{33.3}$$

where, for each $T \in \mathcal{T}_h$, the mapping $D_T(\cdot, \cdot)$ is a bilinear form defined over the space $H^2(T) \times P_T$. Ignoring for the time being that such a decomposition is not unique (we shall return to this crucial point later), let us assume that, for one decomposition of the form (33.3), we can show that there exists a constant C independent of $T \in \mathcal{T}_h$ and h such that

$$|D_T(v, p)| \leqslant C h_T |v|_{2,T} |p|_{1,T} \quad \text{for all } v \in H^2(T), \quad p \in P_T. \tag{33.4}$$

Then an application of Cauchy–Schwarz inequality yields

$$|D_h(u, w_h)| \leqslant C h |u|_{2,\Omega} \|w_h\|_h, \tag{33.5}$$

and therefore we obtain

$$\sup_{w_h \in V_h} \frac{|a_h(u, w_h) - l(w_h)|}{\|w_h\|_h} \leqslant C h |u|_{2,\Omega}, \tag{33.6}$$

i.e., *an estimate of the same order as that of* (32.22).

For proving estimates such as those of (33.4), the following result turns out to be useful. It plays with respect to bilinear forms the role played by the Bramble–Hilbert lemma (Theorem 28.1) with respect to linear forms. For this reason, we shall refer to this result, due to CIARLET [1974a], as the "*bilinear lemma*".

THEOREM 33.1. *Let Ω be a domain in $\mathbb{R}^n$, let $k \geqslant 0$, $l \geqslant 0$ be integers, let $p, q \in [1, +\infty]$, let W be a space that satisfies the inclusions*

$$P_l(\Omega) \subset W \subset W^{l+1}(\Omega), \tag{33.7}$$

and finally, let b be a continuous bilinear form over the space $W^{k+1,p}(\Omega) \times W$ (the space W is equipped with the norm $\|\cdot\|_{l+1,q,\Omega}$) that satisfies

$$b(p, w) = 0 \quad \textit{for all } p \in P_k(\Omega), \quad w \in W, \tag{33.8}$$

$$b(v, q) = 0 \quad \textit{for all } v \in W^{k+1,p}(\Omega), \quad q \in P_l(\Omega). \tag{33.9}$$

Then there exists a constant $C(\Omega)$ *such that*

$$|b(v,w)| \leq C(\Omega)\|b\|\,|v|_{k+1,p,\Omega}|w|_{l+1,q,\Omega} \quad \text{for all } v \in W^{k+1,p}(\Omega), \quad w \in W, \tag{33.10}$$

where $\|b\|$ *denotes the norm of the bilinear form b in the space* $\mathscr{L}_2(W^{k+1,p}(\Omega)\times W;\mathbb{R})$.

PROOF. For each function $w \in W$, the linear form $b(\cdot,w): v \in W^{k+1,p}(\Omega) \to b(v,w)$ is continuous and it vanishes over the space $P_k(\Omega)$, by (33.8). Thus, by the Bramble–Hilbert lemma, there exists a constant $C_1(\Omega)$ such that

$$|b(v,w)| \leq C_1(\Omega)\|b(\cdot,w)\|'_{k+1,p,\Omega}|v|_{k+1,p,\Omega} \quad \text{for all } v \in W^{k+1,p}(\Omega). \tag{33.11}$$

Using (33.9),we may write $b(v,w)=b(v,w+q)$ for all $q \in P_l(\Omega)$, and thus

$$|b(v,w)| = |b(v,w+q)| \leq \|b\|\,\|v\|_{k+1,p,\Omega}\|w+q\|_{l+1,q,\Omega}.$$

Therefore, by Theorem 14.1,

$$\begin{aligned}|b(v,w)| &\leq \|b\|\,\|v\|_{k+1,p,\Omega} \inf_{q \in P_l(\Omega)} \|w+q\|_{l+1,q,\Omega}\\ &\leq C_2(\Omega)\|b\|\,\|v\|_{k+1,p,\Omega}|w|_{l+1,q,\Omega} \quad \text{for all } v \in W^{k+1,p}(\Omega),\ w \in W,\end{aligned}$$

Consequently,

$$\|b(\cdot,w)\|'_{k+1,p,\Omega} = \sup_{v \in W^{k+1,p}(\Omega)} \frac{|b(v,w)|}{\|v\|_{k+1,p,\Omega}} \leq C_2(\Omega)\|b\|\,|w|_{l+1,q,\Omega}, \tag{33.12}$$

and inequality (33.10) follows from inequalities (33.11) and (33.12). □

34. Estimate of the error $\{\sum_{T\in\mathscr{T}_h}|u-u_h|^2_{1,T}\}^{1/2}$ for Wilson's brick and the patch test

We now prove the main result of this chapter.

THEOREM 34.1. *Assume that the solution u is in the space* $H^2(\Omega)$. *Then, for any regular family of triangulations, there exists a constant C independent of h such that*

$$\|u-u_h\|_h = \left\{\sum_{T\in\mathscr{T}_h} |u-u_h|^2_{1,T}\right\}^{1/2} \leq Ch|u|_{2,\Omega}. \tag{34.1}$$

PROOF. The central idea of the proof is to apply the *bilinear lemma* to each term $D_T(u,w_h)$ occurring in a decomposition of the expression $D_h(u,w_h)$ of the form (33.3). Some care has to be exercised, however: From (33.2), an obvious choice for the bilinear forms D_T is

$$(v,p) \in H^2(T)\times P_T \to \int_T \left\{\sum_{i=1}^{3} \partial_i v \partial_i p + \Delta v\, p\right\} \mathrm{d}x = \int_{\partial T} \partial_{\nu,T} v\, p \,\mathrm{d}\gamma,$$

where $\partial_{\nu,T}$ denotes the normal derivative operator along the boundary ∂T of the set T. It is quickly realized, however, that there are not "enough" polynomial invariances at our disposal in such bilinear forms D_T in order to eventually obtain estimates of the form (33.4). Fortunately, there is another possible decomposition of the form (33.3) which will yield the right estimates. The key idea is *to obtain the desired additional "local" polynomial invariances from a "global" polynomial invariance*, as we now show.

Let Y_h denote the finite element space whose generic finite element is the bilinear rectangle. In other words, for each $T\in\mathcal{T}_h$, the restrictions $v_h|_T$ span the space $Q_1(T)$, and each function $v_h\in Y_h$ is defined by its values at all the vertices of the triangulation. Next, let $W_h=Y_{0h}$ denote the space of all functions $v_h\in Y_h$ that vanish at all the boundary nodes, so that *the inclusion*

$$W_h\subset\mathscr{C}^0(\bar{\Omega})\cap H_0^1(\Omega)$$

holds. Notice also that the second inclusion of (32.2) implies that the inclusions

$$Y_h\subset X_h,\qquad Y_{0h}=W_h\subset X_{0h}=V_h \tag{34.2}$$

hold. Finally, let $D_h\colon H^2(\Omega)\times X_h\to\mathbb{R}$ denote from now on the function defined according to the second expression found in (33.2), viz.,

$$D_h(v,w_h)=\sum_{T\in\mathcal{T}_h}\int_T\left\{\sum_{i=1}^{3}\partial_i v\partial_i w_h+\Delta v\,w_h\right\}\mathrm{d}x. \tag{34.3}$$

Then

$$D_h(v,w_h)=0\quad\text{for all } v\in H^2(\Omega),\quad w_h\in W_h. \tag{34.4}$$

For any function $w_h\in X_h$, let $\Lambda_h w_h$ denote the unique function in the space Y_h that takes the same values as w_h at all the vertices of the triangulation. Then, for each $T\in\mathcal{T}_h$, $\Lambda_h w_{h|T}=\Lambda_T(w_{h|T})$, where Λ_T denotes the corresponding $Q_1(T)$-interpolation operator, and the function $\Lambda_h w_h$ belongs to the space $W_h=Y_{0h}$ if the function w_h belongs to the space $V_h=X_{0h}$. Using the definition (34.3), we deduce that

$$D_h(v,w_h)=D_h(v,w_h-\Lambda_h w_h)\quad\text{for all } v\in H^2(\Omega),\quad w_h\in V_h, \tag{34.5}$$

so that another possible decomposition of the difference $D_h(\cdot,\cdot)$ of (34.3) consists in writing

$$D_h(v,w_h)=\sum_{T\in\mathcal{T}_h}D_T(v,w_h)\quad\text{for all } v\in H^2(\Omega),\quad w_h\in V_h, \tag{34.6}$$

where the bilinear forms $D_T(\cdot,\cdot)$ are now given by (compare with (33.2)):

$$D_T(v,p)=\int_{\partial T}\partial_{\nu,T}v(p-\Lambda_T p)\,\mathrm{d}\gamma\quad\text{for all } v\in H^2(T),\quad p\in P_T. \tag{34.7}$$

We observe that, by definition of the operator Λ_T, we have

$$D_T(v,p)=0\quad\text{for all } v\in H^2(T),\quad p\in Q_1(T), \tag{34.8}$$

and thus we get a *first polynomial invariance*.

To obtain a polynomial invariance with respect to the first argument v, assume that the function v belongs to the space $P_1(T)$. Then the expression $D_T(v, p)$ is a sum of three terms, each of which is, up to a constant multiplicative factor, the difference between integrals of the expression $(p-\Lambda_T p)$ over opposite faces. Consider one such term, say (with the notation of Fig. 32.1):

$$\delta_1 = \int_{T_1'} (p-\Lambda_T p)\,\mathrm{d}x_2\,\mathrm{d}x_3 - \int_{T_1''} (p-\Lambda_T p)\,\mathrm{d}x_2\,\mathrm{d}x_3. \tag{34.9}$$

Using the properties of the interpolation operator Λ_T, the identity (32.9), and the equations $\partial_{jj}(\Lambda_h p)=0$, $1 \leqslant j \leqslant 3$, we deduce that

$$p-\Lambda_T p = \sum_{j=1}^{3} \frac{1}{16}\left(\left\{\frac{x_j-c_j}{h_j}\right\}^2 - 1\right) \frac{h_j^2}{h_1 h_2 h_3} \int_T \partial_{jj} p\,\mathrm{d}x.$$

Since the function $(\{(x_1-c_1)/h_1\}^2-1)$ vanishes along the faces T_1' and T_1'', and since the functions $(\{(x_j-c_j)/h_j\}^2-1)$, $j=2, 3$, take on the same values at the points P_1' and P_1'' (cf. Fig. 32.1), we conclude that $\delta_1=0$. The other analogous terms vanish for the same reasons. Consequently, we obtain a *second polynomial invariance*:

$$D_T(v, p)=0 \quad \text{for all } v\in P_1(T), \quad p\in P_T. \tag{34.10}$$

Each expression $D_T(v, p)$ found in (34.6) is of the form

$$D_T(v, p) = \sum_{j=1}^{3} \Delta_{j,T}(v, p), \tag{34.11}$$

where

$$\Delta_{1,T}(v,p) = \int_{T_1'} \partial_1 v(p-\Lambda_T p)\,\mathrm{d}x_2\mathrm{d}x_3 - \int_{T_1''} \partial_1 v(p-\Lambda_T p)\,\mathrm{d}x_2\,\mathrm{d}x_3, \tag{34.12}$$

and the expressions $\Delta_{2,T}(v, p)$ and $\Delta_{3,T}(v, p)$ are analogously defined. Using the standard correspondences $\hat{v} \to v$ between the functions $\hat{v} : \hat{T} \to \mathbb{R}$ and $v : T \to \mathbb{R}$, we note that

$$\Delta_{1,T}(v, p) = (h_2 h_3/h_1)\Delta_{1,\hat{T}}(\hat{v}, \hat{p}), \quad \text{etc.} \tag{34.13}$$

Since each function $\Delta_{j,\hat{T}}$, $j\in\{1, 2, 3\}$, satisfies the following polynomial invariances:

$$\begin{aligned} &\Delta_{j,\hat{T}}(\hat{v}, \hat{p})=0 \quad \text{for all } \hat{v}\in H^2(\hat{T}), \quad \hat{p}\in P_0(\hat{T}), \\ &\Delta_{j,\hat{T}}(\hat{v}, \hat{p})=0 \quad \text{for all } \hat{v}\in P_1(\hat{T}), \quad \hat{p}\in P_{\hat{T}}, \end{aligned} \tag{34.14}$$

we deduce from the bilinear lemma (Theorem 33.1) that there exists a constant $\hat{C}$ such that

$$|\Delta_{j,\hat{T}}(\hat{v}, \hat{p})| \leqslant \hat{C}|\hat{v}|_{2,\hat{T}}|\hat{p}|_{1,\hat{T}} \quad \text{for all } \hat{v}\in H^2(\hat{T}), \quad \hat{p}\in \hat{P}_{\hat{T}}. \tag{34.15}$$

Using Theorem 15.1 and the regularity of the family of triangulations, we find that there exist constants C such that

$$|\hat{v}|_{2,\hat{T}} \leqslant C\|B_T\|^2|\det B_T|^{-1/2}|v|_{2,T} \leqslant Ch_T^{1/2}|v|_{2,T}, \tag{34.16}$$

$$|\hat{p}|_{1,\hat{T}} \leqslant C\|B_T\|\,|\det B_T|^{-1/2}|p|_{1,T} \leqslant Ch_T^{-1/2}|p|_{1,T}, \tag{34.17}$$

so that, upon combining (34.11), (34.13), (34.15), (34.16), and (34.17), we eventually find that there exists a constant C such that

$$|D_T(v,p)| \leqslant Ch_T|v|_{2,T}|p|_{1,T} \quad \text{for all } T \in \bigcup_h \mathscr{T}_h, \quad v \in H^2(T).$$

This last inequality is of the desired form (33.4) and therefore the proof is complete. □

REMARK 34.1. Loosely speaking, one may think of the space W_h introduced in the proof as representing the "conforming" part of the otherwise "nonconforming" space V_h (cf. (34.4)).

In analyzing the consistency error, we have followed the method described in CIARLET [1974a] for studying nonconforming methods: The main idea consists in obtaining *two polynomial invariances* in the functions $D_T(\cdot,\cdot)$, in such a way that an application of the *bilinear lemma* yields the desired consistency error estimates. For the specific application of this method to Wilson's brick, we have extended to the three-dimensional case the analysis that LESAINT [1976] has made for *Wilson's rectangle.* As already mentioned (cf. Remark 32.1), Lesaint has considered the use of this element for approximating problems of "plane elasticity", for which he was able to show the uniform ellipticity of the corresponding approximate bilinear forms. In this fashion, Lesaint obtains an O(h) convergence in the norm $\|\cdot\|_h$. Also, the idea of introducing the degrees of freedom $\int_T \partial_{jj}p\,dx$ is due to Lesaint. More recently, SHI [1988] has established that the O(h) convergence is optimal.

Using the abstract error estimate mentioned in Remark 31.1, LESAINT [1976] has also shown that, if the solution u is in the space $H^2(\Omega)$,

$$|u-u_h|_{0,\Omega} \leqslant Ch^2|u|_{2,\Omega}. \tag{34.18}$$

It is worth pointing out that, by contrast with (34.14), the "full" available polynomial invariances are used in the derivation of the error estimate (34.18).

The second polynomial invariance (34.10) is related to the *patch test*, a famous milestone in the history of the finite element method. It follows from (34.6) and (34.10) that

$$D_h(p,w_h)=0 \quad \text{for all } p \in P_1(\bar{\Omega}), \quad w_h \in V_h.$$

Hence if the function w_h is a basis function $w_i \in V$, whose support is thus a *patch* $\mathscr{P}_i$, i.e., a union of finite elements $T \in \mathscr{T}_h$, it follows that

$$D_h(p,w_i)=0 \quad \text{for all } p \in P_1(\mathscr{P}_i) \tag{34.19}$$

This is an instance of the celebrated *Irons patch test*, which Irons (cf. BAZEKEY, CHEUNG, IRONS and ZIENKIEWICZ [1965], IRONS and RAZZAQUE [1972a]) was the first to empirically recognize as a condition for getting convergence of a nonconforming finite element method. The theoretical importace of the patch test was immediately recognized in the pioneering work of Gilbert Strang on the mathematical analysis of nonconforming methods (cf. STRANG [1972b], and STRANG and FIX [1973, Section 4.2]).

The patch test and, more generally, the assessment of nonconforming methods have received considerable attention in the engineering literature, as illustrated by the contributions of WILSON, TAYLOR, DOHERTY and GHABUSSI [1973], FRAEIJS DE VEUBEKE [1974], CAREY [1976], TAYLOR, BERSFORD and WILSON [1976], OLIVEIRA [1977], SANDER and BECKERS [1977], SAMUELSSON [1979].

STUMMEL [1979, 1980a] has shown that the following *generalized patch test* is a necessary and sufficient condition for the convergence to 0 of the consistency error term arising in the approximation of a $2m$th-order elliptic problem over $\Omega \subset \mathbb{R}^n$ by a nonconforming method:

$$v_h \in V_h \text{ and } \|v_h\| \leqslant M \text{ for all } h \;\Rightarrow\; \lim_{h\to 0} \sum_{T\in\mathcal{T}_h} \int_{\partial T} \varphi\partial^\alpha v_h \nu_{i,T}\,\mathrm{d}\gamma = 0 \tag{34.20}$$

for any $M > 0$, for all $|\alpha| \leqslant m-1$, for all $\varphi \in \mathcal{D}(\mathbb{R}^n)$, and for all $i = 1, 2, \ldots, n$, where $\nu_{i,T}$ denotes as usual the ith component of the unit outer normal vector along ∂T.

This generalized patch test is thus independent of the particular operator considered (in particular, nonconstant coefficients are allowed), it holds for operators of arbitrary orders, and, finally, it yields convergence (without orders, however).

At the same time, STUMMEL [1980b] has questioned the validity of the "traditional" patch test, and since then, a controversy has developed about which patch test should be preferred! In this direction, see ROBINSON [1982], IRONS and LOIKKANEN [1983], SHI [1984c, 1984d, 1987a, 1989], RAZZAQUE [1986], TAYLOR, SIMO, ZIENKIEWICZ and CHAN [1986].

Among the other studies of nonconforming methods, we mention RACHFORD and WHEELER [1974], NITSCHE [1974], CÉA [1976], KANG [1979], SHI [1985]. The *nonconforming Crouzeix–Raviart linear triangle* (cf. Fig. 12.2), proposed and analyzed by CROUZEIX and RAVIART [1973], has proved to be very useful for studying two-dimensional incompressible flows; this has prompted FORTIN and SOULIE [1983] and FORTIN [1985] to propose related nonconforming quadratic triangles and tetrahedra. Estimates of the error $|u-u_h|_{0,\infty,\Omega}$ have been obtained by GASTALDI and NOCHETTO [1987].

A "quadrilateral version" of Wilson's brick, called the *quadrilateral Wilson's element*, has been proposed by TAYLOR, BERSFORD and WILSON [1976], and its convergence has been studied by LESAINT and ZLÁMAL [1980] and SHI [1984a]. Another nonconforming quadrilateral element, introduced by SANDER and BECKERS [1977], has been studied by SHI [1984b].

Finally, we mention that nonconforming finite elements may be also viewed as special cases of "primal-hybrid" finite elements. See Section 19 of the article by Roberts and Thomas.

CHAPTER VI

Finite Element Methods for Second-Order Problems Posed over Curved Domains

Introduction

We studied in Chapter V a first violation of the inclusion $V_h \subset V$. Another violation of the inclusion $V_h \subset V$ occurs in the approximation of a boundary value problem posed over a domain $\bar{\Omega}$ with a *curved* boundary Γ (i.e., the set $\bar{\Omega}$ is no longer assumed to be polygonal). In this case, the set $\bar{\Omega}$ is usually approximated by two types of finite elements: The finite elements of the first type are *straight*, i.e., they have plane faces, and they are typically used "inside" $\bar{\Omega}$. The finite elements of the second type have at least one "curved" face, and they are typically used so as to approximate "as well as possible" the boundary Γ.

In Sections 35 and 36, we describe one way of generating finite elements of the second type, the *isoparametric finite elements*, which are quite commonly used in actual computations. The key idea underlying their conception is the generalization of the notion of affine-equivalence: Let there be given a Lagrange finite element $(\hat{T}, \hat{P}, \{\hat{p}(\hat{a}_i): 1 \leqslant i \leqslant N\})$ in $\mathbb{R}^n$ and let $F: \hat{x} \in \hat{T} \to F(\hat{x}) = (F_i(\hat{x}))_{i=1}^n \in \mathbb{R}^n$ be a mapping such that $F_i \in \hat{P}$, $1 \leqslant i \leqslant n$. Then the triple

$$(T = F(\hat{T}), P = \{p = \hat{p} \circ F^{-1}; \hat{p} \in \hat{P}\}, \{p(a_i = F(\hat{a}_i)): 1 \leqslant i \leqslant N\})$$

is also a Lagrange finite element (Theorem 35.1), and two cases can be distinguished:

(i) The mapping F is *affine*, i.e., $F_i \in P_1(\hat{T})$, $1 \leqslant i \leqslant n$; therefore the finite elements (T, P, Σ) and $(\hat{T}, \hat{P}, \hat{\Sigma})$ are affine-equivalent.

(ii) Otherwise, the finite element (T, P, Σ) is said to be *isoparametric*, and *isoparametrically equivalent* to the finite element $(\hat{T}, \hat{P}, \hat{\Sigma})$. If $(\hat{T}, \hat{P}, \hat{\Sigma})$ is a standard straight finite element, it is easily seen in the second case that the boundary of the set T is curved in general. This fact is illustrated by several examples.

We then develop an interpolation theory adapted to this type of finite element. In our analysis, however, we shall restrict ourselves to the *isoparametric quadratic n-simplex*: it is one of the simplest isoparametric finite element, yet it retains all the characteristic features of more general such elements. For an *isoparametric family* (T, P_T, Σ_T) of quadratic n-simplices, we show (Theorem 37.3) that the Π_T-interpolant of a function v satisfies inequalities of the form

$$|v-\Pi_T v|_{m,T} \leqslant C h_T^{3-m} \|v\|_{3,T}, \quad 0 \leqslant m \leqslant 3,$$

where $h_T = \text{diam}(T)$. This estimate, which is of the same form as in the case of affine families (cf. Section 16) is established under the crucial assumption that *the "isoparametric" mappings F_T do not deviate too much from affine mappings*; in addition, the family is also assumed to be regular, in a sense that generalizes the regularity of affine families.

Even if we use isoparametric finite elements $T \in \mathcal{T}_h$ to "triangulate" a set $\bar{\Omega}$, the boundary of the set $\Omega_h = \bigcup_{T \in \mathcal{T}_h} T$ is usually not identical to the boundary Γ, even though it may be much closer to Γ than if we used straight finite elements. Since the domain of definition of the functions in the resulting finite element space V_h is the set $\bar{\Omega}_h$, the space V_h is thus *not* contained in the space V; therefore both the bilinear form and the linear form need to be approximated.

In order to be in as realistic a situation as possible, we then study in Section 39 the *simultaneous* effects of such an *approximation of the domain* $\bar{\Omega}$ and of *isoparametric numerical integration.* As in Chapter IV, this last approximation amounts to use a quadrature formula over a reference finite element $\hat{T}$ for computing all the integrals $\int_T \varphi(x)\,dx$ appearing in the linear system, via the isoparametric mappings $F_T: \hat{T} \to T$, $T \in \mathcal{T}_h$. Restricting ourselves again to isoparametric quadratic n-simplices for simplicity, we show (Theorem 43.1) that, if the quadrature formula over the set $\hat{T}$ is exact for polynomials of degree 2, we have

$$\|\tilde{u} - u_h\|_{1,\Omega_h} = \mathrm{O}(h^2),$$

where $\tilde{u}$ is an extension of the solution of the given boundary value problem to the set Ω_h (in general $\bar{\Omega}_h \not\subset \bar{\Omega}$), and $h = \max_{T \in \mathcal{T}_h} h_T$. This error estimate is obtained through the familiar process: We first prove *an abstract error estimate* (Theorem 40.1), valid under a *uniform V_h-ellipticity assumption* of the approximate bilinear forms. Then we use the interpolation theory developed in Section 37 for evaluating the term $\|\tilde{u} - \Pi_h \tilde{u}\|_{1,\Omega_h}$ (Theorem 42.1) and finally, we estimate the two *consistency errors* (Theorems 42.2 and 42.3; these results largely depend on related results of Chapter IV). It is precisely in these last estimates that a remarkable conclusion arises: *In order to retain the* $\mathrm{O}(h^2)$ *convergence, it is not necessary to use more sophisticated quadrature schemes for approximating the integrals when isoparametric, instead of straight, finite elements are used.*

35. Isoparametric families of finite elements

Our first task consists in extending the notions of affine-equivalence and affine families that we discussed in Section 11. There, we saw how to generate finite elements through affine maps, a construction that will now be generalized in Theorem 35.1 for more general mappings F. We begin by considering *Lagrange finite elements.*

THEOREM 35.1 *Let $(\hat{T}, \hat{P}, \hat{\Sigma})$ be a Lagrange finite element in $\mathbb{R}^n$ with $\hat{\Sigma} = \{\hat{p}(\hat{a}_i): 1 \leqslant i \leqslant N\}$, let there be given a one-to-one mapping $F: \hat{x} \in \hat{T} \to (F_j(\hat{x}))_{j=1}^n \in \mathbb{R}^n$ that satisfies*

$$F_j \in \hat{P}, \quad 1 \leqslant j \leqslant n, \tag{35.1}$$

and let

$$\begin{aligned} &T = F(\hat{T}), \\ &P = \{p: T \to \mathbb{R};\ p = \hat{p} \circ F^{-1}, \hat{p} \in \hat{P}\}, \\ &\Sigma = \{p(F(\hat{a}_i)): 1 \leqslant i \leqslant N\}. \end{aligned} \tag{35.2}$$

Then the set Σ is P-unisolvent. Consequently, if T is a closed subset of $\mathbb{R}^n$ with a nonempty interior, the triple (T, P, Σ) is a Lagrange finite element.

PROOF. Let us establish the natural correspondences

$$\begin{aligned} &\hat{x} \in \hat{T} \to x = F(\hat{x}) \in T, \\ &\hat{p} \in \hat{P} \to p = \hat{p} \circ F^{-1} \in P, \end{aligned}$$

between the points in the sets $\hat{T}$ and T, and between the functions in the spaces $\hat{P}$ and P. If $\hat{p}_i$, $1 \leqslant i \leqslant N$, denote the basis functions of the finite element $(\hat{T}, \hat{P}, \hat{\Sigma})$, we have for all $p \in P$ and all $x \in T$,

$$p(x) = \hat{p}(\hat{x}) = \sum_{i=1}^{N} \hat{p}(\hat{a}_i)\hat{p}_i(\hat{x}) = \sum_{i=1}^{N} p(a_i)p_i(x),$$

i.e.,

$$p = \sum_{i=1}^{N} p(a_i)p_i \quad \text{for all } p \in P.$$

The functions p_i, $1 \leqslant i \leqslant N$, are linearly independent since $\sum_{i=1}^N \lambda_i p_i = 0$ implies $\sum_{i=1}^N \lambda_i \hat{p}_i = 0$ and thus $\lambda_i = 0$, $1 \leqslant i \leqslant N$. In other words, we have shown that the set Σ is P-unisolvent, which completes the proof. □

REMARK 35.1. If the mapping F is in addition assumed to be continuous, the set $T = F(\hat{T})$ has a nonempty interior: Since $\hat{T}$ has a nonempty interior (by definition of a finite element; cf. Section 10), this follows from the *invariance of domain theorem* (cf. DIEUDONNÉ [1982, Theorem 24.8.7]).

We shall henceforth use the following notation: To indicate that a mapping $F: \hat{x} \in \hat{T} \to F(\hat{x}) = (F_j(\hat{x}))_{j=1}^n \in \mathbb{R}^n$ satisfies relations (35.1), we write:

$$F \in (\hat{P})^n \Leftrightarrow F_j \in \hat{P}, \quad 1 \leqslant j \leqslant n.$$

Notice that the construction of Theorem 35.1 is indeed a *generalization of the construction that led to affine-equivalent finite elements,* because the inclusion $P_1(\hat{T}) \subset \hat{P}$ is satisfied by all the finite elements hitherto considered.

With Theorem 35.1 in mind, we now give several definitions: First, any finite element (T, P, Σ) constructed from another finite element $(\hat{T}, \hat{P}, \hat{\Sigma})$ through the process described in this theorem is called an *isoparametric finite element*, and the finite element (T, P, Σ) is said to be *isoparametrically equivalent* to the finite element $(\hat{T}, \hat{P}, \hat{\Sigma})$. Observe, however, that *the finite elements* $(\hat{T}, \hat{P}, \hat{\Sigma})$ *and* (T, P, Σ) *cannot in general be interchanged in this definition*, by contrast with the definition of affine-equivalence.

Next, we say that a family of finite elements (T, P_T, Σ_T) is an *isoparametric family* if all its elements are isoparametrically equivalent to a single finite element $(\hat{T}, \hat{P}, \hat{\Sigma})$, called the *reference finite element* of the family. In other words, for each T, there exists an *isoparametric mapping* $F_T: \hat{T} \to \mathbb{R}^n$, i.e., that satisfies

$$F_T \in (\hat{P})^n \quad \text{and} \quad F_T \text{ is one-to-one}, \tag{35.3}$$

such that

$$\begin{aligned} &T = F_T(\hat{T}), \\ &P_T = \{p: T \to \mathbb{R};\, p = \hat{p} \circ F_T^{-1},\, \hat{p} \in \hat{P}\}, \\ &\Sigma_T = \{p(F_T(\hat{a}_i)):\ 1 \leqslant i \leqslant N\}. \end{aligned} \tag{35.4}$$

As exemplified by the special case of affine-equivalent finite elements, one may consider a family of isoparametric finite elements whose associated mappings F_T all belong to the same space $(\hat{Q})^n$, where $\hat{Q}$ is a strict subspace of the space $\hat{P}$. Such finite elements are called *subparametric finite elements*. Examples are given in Fig. 36.4.

REMARK 35.2. The prefix “iso” in the adjective “isoparametric” refers to the fact that the same space $\hat{P}$ is used for defining both the finite element $(\hat{T}, \hat{P}, \hat{\Sigma})$ and the isoparametric mapping $F_T \in (\hat{P})^n$.

It is worth pointing out that, by contrast with the space P_T corresponding to an affine-equivalent finite element, the space P_T defined in (35.4) generally contains functions that are not polynomials, even when the space $\hat{P}$ consists of polynomials. However this complication is ignored in practical computation, inasmuch as *all the computations are performed on the set* $\hat{T}$, *not on the set* T. All that is needed is the knowledge of the mapping F_T, as we shall notably see in Section 39.

In practice, the isoparametric finite element found in Theorem 35.1 is not determined by a mapping F but, rather, by N distinct points a_i, $1 \leqslant i \leqslant N$, which in turn uniquely determine a mapping F satisfying

$$F \in (\hat{P})^n, \qquad F(\hat{a}_i) = a_i, \quad 1 \leqslant i \leqslant N. \tag{35.5}$$

Such a mapping is given by

$$F: \hat{x} \in \hat{T} \to F(\hat{x}) = \sum_{i=1}^{N} \hat{p}_i(\hat{x}) a_i, \tag{35.6}$$

as it is readily verified, and it is uniquely defined since the set $\hat{\Sigma}$ is $\hat{P}$-unisolvent and since, for each $j \in \{1, 2, \ldots, n\}$, we must have

$$F_j \in \hat{P}, \qquad F_j(\hat{a}_i) = a_{ji}, \quad 1 \leqslant i \leqslant N,$$

where we have let $a_i = (a_{ji})_{j=1}^n$. However, in the absence of additional assumptions, nothing garantees that the mapping $F: \hat{T} \to F(\hat{T})$ is *invertible*, and indeed this property will require a verification in each instance. Finally, notice that the points a_i are precisely the *nodes* of the finite element (T, P, Σ).

The main interest of isoparametric finite elements lies in the freedom in the choice of the points a_i, which in turn yields more general geometric shapes of sets T than the polygonal shapes considered up to now. As we shall show in Section 38, this property is crucial for getting a good approximation of a curved boundary.

Isoparametric Hermite finite elements can be likewise defined. Their construction relies on the following theorem whose proof, quite similar to that of Theorem 35.1, is omitted.

THEOREM 35.2. *Let $(\hat{T}, \hat{P}, \hat{\Sigma})$ be a Hermite finite element in $\mathbb{R}^n$, and assume that the order of directional derivatives occurring in the set $\hat{\Sigma}$ is one, i.e.,*

$$\hat{\Sigma} = \{\hat{\varphi}_i^0 : 1 \leqslant i \leqslant N_0;\ \hat{\varphi}_{ik}^1 : 1 \leqslant k \leqslant d_i, 1 \leqslant i \leqslant N_1\}, \tag{35.7}$$

and the degrees of freedom are of the following form:

$$\hat{\varphi}_i^0 : \hat{p} \to \hat{p}(\hat{a}_i^0), \qquad \hat{\varphi}_{ik}^1 : \hat{p} \to D\hat{p}(\hat{a}_i^1)\hat{\xi}_{ik}^1. \tag{35.8}$$

Let there be given a differentiable one-to-one mapping $F: \hat{T} \to \mathbb{R}^n$ that satisfies

$$F \in \hat{P}^n, \tag{35.9}$$

and let

$$\begin{aligned} &T = F(\hat{T}), \qquad P = \{p : T \to \mathbb{R};\ p = \hat{p} \circ F^{-1}, \hat{p} \in \hat{P}\}, \\ &\Sigma = \{\varphi_i^0 : 1 \leqslant i \leqslant N_0;\ \varphi_{ik}^1 : 1 \leqslant k \leqslant d_i, 1 \leqslant i \leqslant N_1\}, \\ &\varphi_i^0 : p \to p(a_i^0), \qquad \varphi_{ik}^1 : p \to Dp(a_i^1)\xi_{ik}^1, \end{aligned} \tag{35.10}$$

where

$$\begin{aligned} &a_i^0 = F(\hat{a}_i^0), \quad 1 \leqslant i \leqslant N_0, \\ &a_i^1 = F(\hat{a}_i^1), \quad 1 \leqslant i \leqslant N_1, \\ &\xi_{ik}^1 = DF(\hat{a}_i^1)\hat{\xi}_{ik}^1, \quad 1 \leqslant k \leqslant d_i, \quad 1 \leqslant i \leqslant N_1. \end{aligned} \tag{35.11}$$

Then the set Σ is P-unisolvent. Consequently, if T is a closed subset of $\mathbb{R}^n$ with a nonempty interior, the triple (T, P, Σ) is a Hermite finite element.

The finite element (T, P, Σ), which is called an *isoparametric Hermite finite element*, shares the same properties as the isoparametric Lagrange finite element constructed in Theorem 35.1: In particular, it generalizes the notion of affine-equivalent Hermite finite element; as in (35.6), the mapping F can be easily expressed in terms of the basis functions of the finite element $(\hat{T}, \hat{P}, \hat{\Sigma})$; the finite element (T, P, Σ) is completely determined once the finite element $(\hat{T}, \hat{P}, \hat{\Sigma})$, the points $\hat{a}_i^0$, and the vectors ξ_{ik}^1 are known (note that the points $\hat{a}_i^1$ cannot be arbitrarily chosen, however). For more details, cf. CIARLET and RAVIART [1972b].

36. Examples of isoparametric finite elements

Let us now describe several instances of commonly used isoparametric finite elements. For brevity, we shall give a detailed discussion only for our first example, the *isoparametric quadratic n-simplex*, for which the finite element $(\hat{T}, \hat{P}, \hat{\Sigma})$ is the quadratic n-simplex. Such an isoparametric finite element is uniquely determined once we are given $(n+1)$ points denoted a_i, $1 \leqslant i \leqslant n+1$, and $\frac{1}{2}n(n+1)$ points denoted a_{ij}, $1 \leqslant i < j \leqslant n+1$, since there exists a unique mapping F such that (cf. Theorem 35.1)

$$F \in (P_2(\hat{T}))^n,$$
$$F(\hat{a}_i) = a_i, \quad 1 \leqslant i \leqslant n+1,$$
$$F(\hat{a}_{ij}) = a_{ij}, \quad 1 \leqslant i < j \leqslant n+1.$$

This mapping is given by (cf. (6.8) and (35.6)):

$$F: \hat{x} \in \hat{T} \to F(\hat{x}) = \sum_i \lambda_i(\hat{x})(2\lambda_i(\hat{x}) - 1)a_i + \sum_{i<j} 4\lambda_i(\hat{x})\lambda_j(\hat{x})a_{ij}. \tag{36.1}$$

Observe that, if it so happens that the points a_{ij} are exactly the midpoints $\frac{1}{2}(a_i + a_j)$, then (by virtue of the uniqueness of the mapping F) the mapping F is "degenerate" and becomes affine. This consideration is illustrated in Fig. 36.1 for $n=2$, i.e., in the case of the *isoparametric quadratic triangle*.

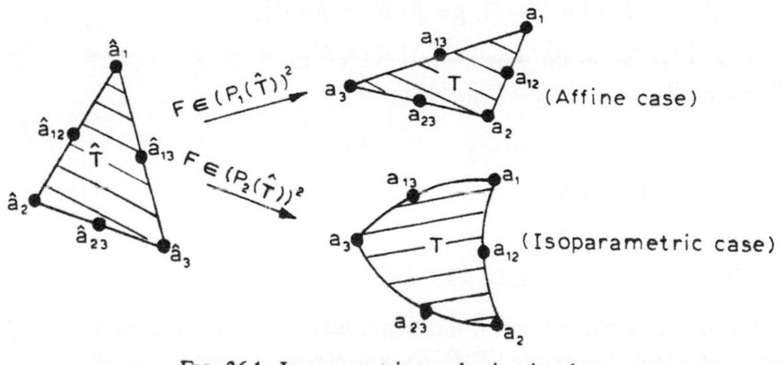

FIG. 36.1. Isoparametric quadratic triangle.

It is only later (Theorem 37.2) that we shall give sufficient conditions that guarantee the invertibility of the mapping F of (36.1), but at least we can already indicate that these conditions proceed from a natural idea: When $n=2$ (cf. Fig. 36.1), let us assume that the three vertices a_i, $1 \leqslant i \leqslant 3$, are the vertices of a nondegenerate triangle $\tilde{T}$. Then the mapping F: $\hat{T} \to T$ is invertible if the points a_{ij} are not "too far" from the actual midpoints $\frac{1}{2}(a_i + a_j)$ of the triangle $\hat{T}$ (for a counterexample, see Remark 36.1).

The *boundary* of the set $T = F(\hat{T})$ is composed of *faces*, which are the images $F(\hat{T}')$ of the faces $\hat{T}'$ of the n-simplex $\hat{T}$. Since each basis function $\hat{\varphi}$ of the quadratic n-simplex $(\hat{T}, P_2(\hat{T}), \hat{\Sigma})$ vanishes along any face of $\hat{T}$ that does not contain the node associated with $\hat{\varphi}$ (cf. Remark 12.2), we conclude that *each face of the isoparametric quadratic n-simplex is solely determined by the nodes through which it passes.* This property, which is common to all the isoparametric finite elements considered in this chapter (as the reader may check) allows the construction of triangulations made up of isoparametric finite elements (cf. Section 38).

REMARK 36.1. When $n = 2$, these considerations can be made more specific: Let a_i and $a_j, i \neq j$, be two "vertices" of an isoparametric quadratic triangle. Then the curved "side" joining these two points is an *arc of parabola* uniquely determined by the following conditions: It passes through the points a_i, a_j, a_{ij} and its asymptotic direction is parallel to the vector $a_{ij} - \frac{1}{2}(a_i + a_j)$.

With this result, it is easy to construct examples where the mapping F is not one-to-one. For instance, the mapping F corresponding to the following nodes is not invertible:

$$a_1 = (0, 0), \qquad a_2 = (2, 0), \qquad a_3 = (0, 2),$$
$$a_{12} = (1, 0), \qquad a_{13} = (1, 1), \qquad a_{23} = (0, 1).$$

REMARK 36.2. Let us verify that our description indeed coincides with the one used by the engineers; consider for instance the quadratic isoparametric triangle as described by FELIPPA and CLOUGH [1970, p. 224]: Given six points $a_i = (a_{1i}, a_{2i})$, $1 \leqslant i \leqslant 6$, in the plane (the points a_4, a_5 and a_6 play momentarily the role of the points that we usually call a_{12}, a_{23} and a_{13}, respectively), a "natural" coordinate system is defined, whereby the following relation (written in matrix form) should hold between the Cartesian coordinates x_1 and x_2 describing the finite element and the "new" coordinates λ_1, λ_2 and λ_3:

$$\begin{pmatrix} x_1 \\ x_2 \\ 1 \end{pmatrix} = \begin{pmatrix} a_{11} & a_{12} & a_{13} & a_{14} & a_{15} & a_{16} \\ a_{21} & a_{22} & a_{23} & a_{24} & a_{25} & a_{26} \\ 1 & 1 & 1 & 1 & 1 & 1 \end{pmatrix} \begin{pmatrix} \lambda_1(2\lambda_1 - 1) \\ \lambda_2(2\lambda_2 - 1) \\ \lambda_3(2\lambda_3 - 1) \\ 4\lambda_1\lambda_2 \\ 4\lambda_2\lambda_3 \\ 4\lambda_3\lambda_1 \end{pmatrix}.$$

Then we observe that the first two lines of this matrix equation precisely represent relation (36.1), with $F(\hat{x}) = (F_1(\hat{x}), F_2(\hat{x}))$ now denoted (x_1, x_2). The last line implies either $\lambda_1 + \lambda_2 + \lambda_3 = 1$ or $\lambda_1 + \lambda_2 + \lambda_3 = -\frac{1}{2}$, hence the solution $\lambda_1 + \lambda_2 + \lambda_3 = 1$ is the only one that is acceptable if we impose the restriction that $\lambda_i \geqslant 0$, $1 \leqslant i \leqslant 3$.

Therefore, the "natural" coordinates λ_1, λ_2 and λ_3 are nothing but the barycentric coordinates with respect to a fixed triangle $\hat{T}$, and the isoparametric finite element associated with the points a_i, $1 \leq i \leq 6$, is in this formulation the set of those points (x_1, x_2) given by the first two lines of the above matrix equation when the "natural" coordinates λ_i, also known as "*curvilinear*" *coordinates*, satisfy the inequalities $0 \leq \lambda_i \leq 1$, $1 \leq i \leq 3$, and the equality $\sum_{i=1}^{3} \lambda_i = 1$.

A general description of isoparametric finite elements along these lines is found in ZIENKIEWICZ [1971, Chapter 8]. The first references where such finite elements were found in the engineering literature are ARGYRIS and FRIED [1968] and ERGATOUDIS, IRONS and ZIENKIEWICZ [1968].

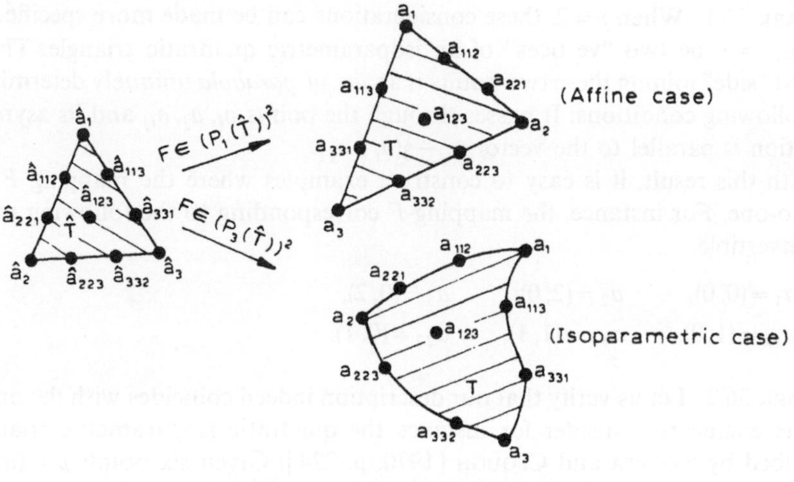

FIG. 36.2. Isoparametric cubic triangle.

We next consider the *isoparametric cubic n-simplex* (cf. Fig. 36.2 for $n=2$), whose associated mapping F is given by (cf. (6.9)):

$$F: \hat{x} \in \hat{T} \to F(\hat{x}) = \sum_i \tfrac{1}{2}\lambda_i(\hat{x})(3\lambda_i(\hat{x})-1)(3\lambda_i(\hat{x})-2)(a_i) + \sum_{i \neq j} \tfrac{9}{2}\lambda_i(\hat{x})\lambda_j(\hat{x})(3\lambda_i(\hat{x})-1)(a_{iij}) + \sum_{i<j<k} 27\lambda_i(\hat{x})\lambda_j(\hat{x})\lambda_k(\hat{x})a_{ijk}. \tag{36.2}$$

Observe that, if the points a_{ijk} play no role in the definition of the boundary of the set T, the space P_T depend on their positions.

We could similarly define the *isoparametric reduced cubic n-simplex* and, more generally, the *isoparametric n-simplex of degree k* for any integer $k \geq 1$. All these isoparametric finite elements are instances of *isoparametric simplicial*, or *triangular* if $n=2$, or *tetrahedral* if $n=3$, *finite elements* in the sense that they are isoparametrically equivalent to a finite element for which the set $\hat{T}$ is an n-simplex.

As an instance of *isoparametric Hermite triangular element*, we finally mention the *isoparametric cubic Hermite triangle* which, according to Theorem 35.2, is defined by three "vertices" a_i, $1 \leqslant i \leqslant 3$, two directions ξ_i, η_i at each point a_i, $1 \leqslant i \leqslant 3$, and a point a_{123} (cf. Fig. 36.3).

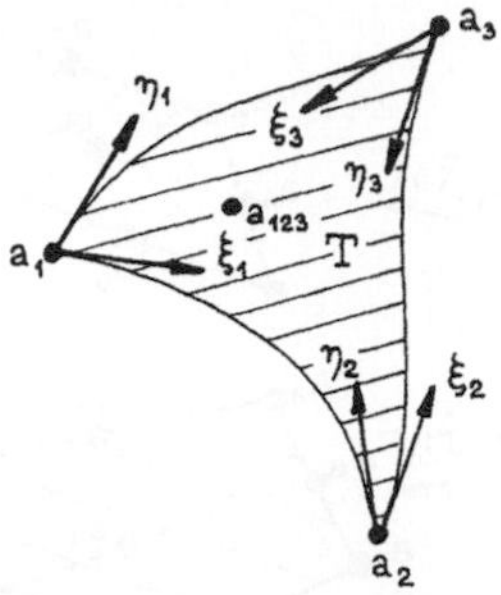

FIG. 36.3. Isoparametric cubic Hermite triangle.

We next describe some examples of *isoparametric rectangular finite elements*, in the sense that they are isoparametrically equivalent to a finite element for which the set $\hat{T}$ is an n-rectangle, for example the unit hypercube $\hat{T} = [0, 1]^n$. In this fashion we obtain the *isoparametric bilinear n-rectangle*, also known for $n = 2$ as the *isoparametric 4-node rectangle*, or as the *4-node quadrilateral* (cf. Fig. 36.4).

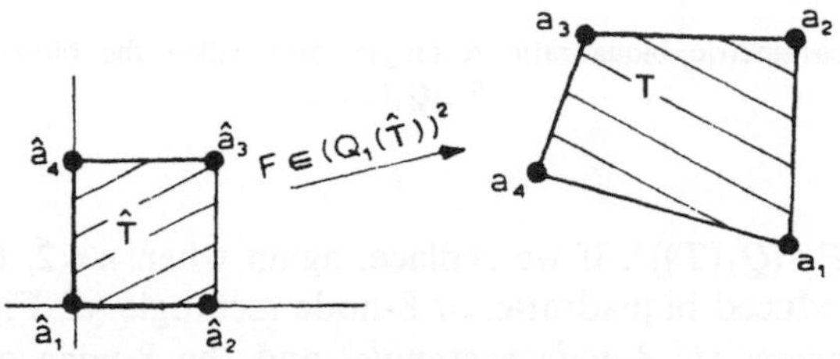

FIG. 36.4. The isoparametric bilinear rectangle, also called the isoparametric 4-node rectangle, or the 4-node quadrilateral.

For $n = 2$, this is an example of a true isoparametric finite element whose sides are not curved! This is so because the functions in the space $Q_1([0, 1]^2)$ are affine in the direction of each coordinate axis. However, this is special to dimension 2. If $n = 3$ for instance, the faces of the set T are portions of hyperbolic paraboloids and are therefore generally curved.

Another instance of isoparametric rectangular finite element is the *isoparametric n-rectangle of degree* 2, also known as the *isoparametric biquadratic rectangle* when $n = 2$. This last element is represented in Fig. 36.5, where various subparametric cases of interest are also indicated, notably the *biquadratic quadrilateral*, which

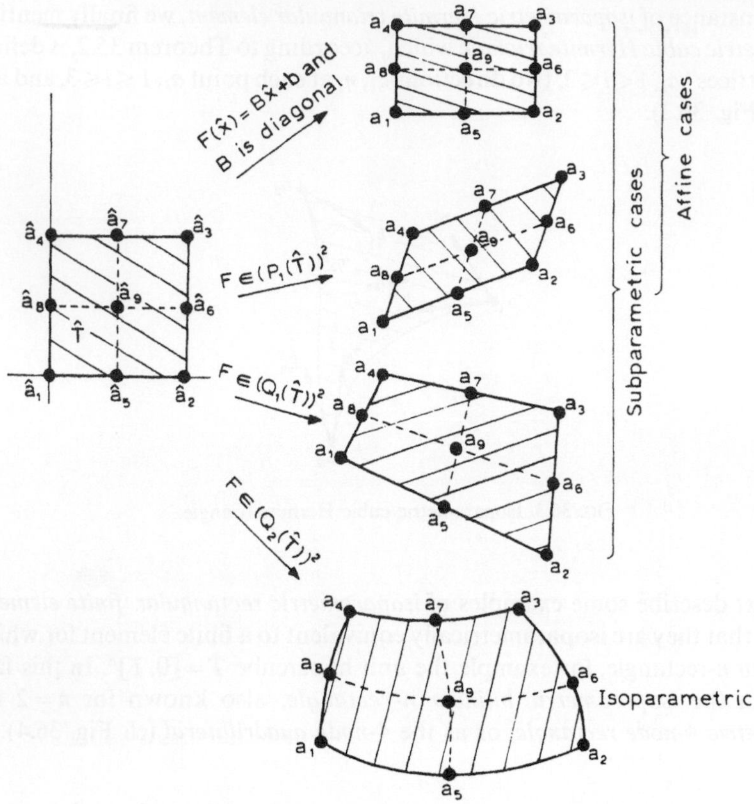

FIG. 36.5. The isoparametric biquadratic rectangle, also called the biquadratic quadrilateral if $F\in(Q_1(\hat{T}))^2$.

corresponds to $F\in(Q_1(T))^2$. If we replace, again when $n=2$, the reference finite element by the reduced biquadratic, or 8-node rectangle (cf. Fig. 7.4), we likewise obtain the *isoparametric* 8-*node rectangle*, and the 8-*node quadrilateral* when $F\in(Q_1(T))^2$.

The biquadratic quadrilateral and the 8-node quadrilateral provide instances of *quadrilateral finite elements*, that is, finite elements for which the set T is a quadrilateral.

When a finite element (T, P, Σ) is isoparametrically equivalent to a finite element $(\hat{T}, \hat{P}, \hat{\Sigma})$ through a mapping F, we shall systematically use the correspondences

$$\hat{x}\in\hat{T}\rightarrow x=F(\hat{x})\in T, \tag{36.3}$$

$$\hat{p}\in\hat{P}\rightarrow p=\hat{p}\circ F^{-1}\in P, \tag{36.4}$$

between the points in the sets $\hat{T}$ and T, and between the functions in the spaces $\hat{P}$ and P, respectively. We shall extend the correspondence (36.4) to more general functions defined over the sets $\hat{T}$ and T by letting

$$(\hat{v}: \hat{T}\to\mathbb{R})\to(v=\hat{v}\circ F^{-1}: T\to\mathbb{R}). \tag{36.5}$$

Then the associated $\hat{P}$-interpolation and P-interpolation operators $\hat{\Pi}$ and Π are such that

$$\{\Pi v\}\hat{} = \hat{\Pi}\hat{v} \quad \text{for all } \hat{v}\in \operatorname{dom}\hat{\Pi} = \mathscr{C}^0(\hat{T}), \tag{36.6}$$

provided $\hat{v}\in\operatorname{dom}\hat{\Pi}\Rightarrow v=\hat{v}\circ F^{-1}\in\operatorname{dom}\Pi=\mathscr{C}^0(T)$; hence this condition excludes situations where the mapping F^{-1} would not be continuous.

37. Estimates of the interpolation errors $\|v-\Pi_T v\|_{m,q,T}$ for an isoparametric family of finite elements

This section, which is based on CIARLET and RAVIART [1972b], is devoted to the derivation of an interpolation theory for isoparametric finite elements, i.e., we shall estimate the *interpolation errors* $\|v-\Pi_T v\|_{m,q,T}$ for finite elements (T, P, Σ) that are isoparametrically equivalent to a reference finite element $(\hat{T}, \hat{P}, \hat{\Sigma})$. This analysis is carried out in three stages, which parallel those used for affine-equivalent finite elements:

(i) Assuming the $\hat{P}$-interpolation operator $\hat{\Pi}$ leaves the space $P_k(\hat{T})$ invariant, we obtain by an argument similar to that used in Theorem 15.3 inequalities of the form

$$|\hat{v}-\hat{\Pi}\hat{v}|_{m,q,\hat{T}}\leqslant C(\hat{T},\hat{P},\hat{\Sigma})|\hat{v}|_{k+1,p,\hat{T}}. \tag{37.1}$$

Thus this step is the same as before.

(ii) We then examine how the seminorms occurring in (37.1) can be bounded above by analogous seminorms defined on T, and vice versa. Recall that for affine families, we found inequalities of the form (cf. Theorem 15.1)

$$|\hat{v}|_{m,p,\hat{T}}\leqslant C\|B\|^m|\det B^{-1}|^{1/p}|v|_{m,p,T}, \tag{37.2}$$

$$|v|_{m,p,T}\leqslant C\|B^{-1}\|^m|\det B|^{1/p}|\hat{v}|_{m,p,\hat{T}}, \tag{37.3}$$

with $F:\hat{x}\in\hat{T}\to F(\hat{x})=B\hat{x}+b$. In the present case, we shall find (Theorem 37.1) that the seminorms $|v|_{m,p,T}$ are bounded above not only in terms of the seminorm $|\hat{v}|_{m,p,\hat{T}}$, but instead in terms of *all* the seminorms $|\hat{v}|_{l,p,\hat{T}}$, $1\leqslant l\leqslant m$.

(iii) As in the affine case, where the quantities $\|B\|$, $|\det B^{-1}|$, $\|B^{-1}\|$, $|\det B|$, which appear in (37.2) and (37.3), were eventually estimated in terms of the geometrical parameters meas(T), h_T and ρ_T (cf. Theorem 15.2), we shall also estimate (Theorem 37.2) analogous quantities appearing in the more general inequalities found in step (ii) in terms of simple geometrical parameters attached to the finite element (T, P, Σ_T).

Thus there are essentially two new steps ((ii) and (iii)) to develop. For ease of exposition, we shall detail these new steps about one example only, the *isoparametric quadratic n-simplex*. Indications for handling more general isoparametric elements will be also given.

We shall use the following notations:

$$\begin{aligned} &J(F)(\hat{x})=\det(\partial_j F_i(\hat{x}))=\text{Jacobian of } F \text{ at } \hat{x},\\ &J(F^{-1})(x)=\{J(F^{-1})(x)\}^{-1}=\text{Jacobian of } F^{-1} \text{ at } x,\end{aligned} \tag{37.4}$$

$$\begin{aligned} &|F|_{l,\infty,\hat{T}}=\sup_{\hat{x}\in\hat{T}}\|D^l F(\hat{x})\|_{\mathscr{L}_l(\mathbb{R}^n;\,\mathbb{R}^n)},\\ &|F^{-1}|_{l,\infty,T}=\sup_{x\in T}\|D^l F^{-1}(x)\|_{\mathscr{L}_l(\mathbb{R}^n;\,\mathbb{R}^n)},\end{aligned} \tag{37.5}$$

whenever $F:\hat{T}\subset\mathbb{R}^n\to T=F(\hat{T})\subset\mathbb{R}^n$ is a sufficiently smooth mapping with a sufficiently smooth inverse $F^{-1}: T\to\hat{T}$. Notice that when the mapping F is of the form $F:\hat{x}\to F(\hat{x})=B\hat{x}+b$, then

$$\begin{aligned} &J(F)=\det B, && J(F^{-1})=\det B^{-1},\\ &|F|_{1,\infty,\hat{T}}=\|B\|, && |F^{-1}|_{1,\infty,T}=\|B^{-1}\|.\end{aligned}$$

Since we are considering isoparametric quadratic n-simplices, we shall apply inequality (37.1) with the values $m=0,1,2,3$ and $k+1=3$ only; thus we shall restrict ourselves to the seminorms $|\cdot|_{l,p,\Omega}$ with $0\leqslant l\leqslant 3$ in the next theorem. Notice that the following result is valid for general mappings F, i.e., it is irrelevant here that the mapping F be in the space $(\hat{P})^n$, with $\hat{P}$ attached to some finite element $(\hat{T},\hat{P},\hat{\Sigma})$.

THEOREM 37.1. *Let Ω and $\hat{\Omega}$ be two bounded open subsets of $\mathbb{R}^n$ such that $\Omega=F(\hat{\Omega})$, where F is a sufficiently smooth one-to-one mapping with a sufficiently smooth inverse $F^{-1}:\Omega\to\hat{\Omega}$.*

Then if a function $\hat{v}:\hat{\Omega}\to\mathbb{R}$ belongs to the space $W^{l,p}(\hat{\Omega})$ for some integer $l\geqslant 0$ and some number $p\in[1,\infty]$, the function $v=\hat{v}\circ F^{-1}:\Omega\to\mathbb{R}$ belongs to the space $W^{l,p}(\Omega)$ and, in addition, there exist constants C such that

$$|v|_{0,p,\Omega}\leqslant|J(F)|_{0,\infty,\hat{\Omega}}^{1/p}|\hat{v}|_{0,p,\hat{\Omega}} \quad \text{for all } \hat{v}\in L^p(\hat{\Omega}), \tag{37.6}$$

$$|v|_{1,p,\Omega}\leqslant C|J(F)|_{0,\infty,\hat{\Omega}}^{1/p}|F^{-1}|_{1,\infty,\Omega}|\hat{v}|_{1,p,\hat{\Omega}} \quad \text{for all } \hat{v}\in W^{1,p}(\hat{\Omega}), \tag{37.7}$$

$$\begin{aligned}|v|_{2,p,\Omega}\leqslant{}& C|J(F)|_{0,\infty,\hat{\Omega}}^{1/p}\{|F^{-1}|_{1,\infty,\Omega}^2|\hat{v}|_{2,p,\hat{\Omega}}+|F^{-1}|_{2,\infty,\Omega}|\hat{v}|_{1,p,\hat{\Omega}}\}\\ &\text{for all } \hat{v}\in W^{2,p}(\hat{\Omega}),\end{aligned} \tag{37.8}$$

$$\begin{aligned}|v|_{3,p,\Omega}\leqslant{}& C|J(F)|_{0,\infty,\hat{\Omega}}^{1/p}\{|F^{-1}|_{1,\infty,\Omega}^3|\hat{v}|_{3,p,\hat{\Omega}}\\ &+|F^{-1}|_{1,\infty,\Omega}|F^{-1}|_{2,\infty,\Omega}|\hat{v}|_{2,p,\hat{\Omega}}+|F^{-1}|_{3,\infty,\Omega}|\hat{v}|_{1,p,\hat{\Omega}}\}\\ &\text{for all } \hat{v}\in W^{3,p}(\hat{\Omega}).\end{aligned} \tag{37.9}$$

PROOF. Let $1\leqslant p<\infty$. As in Theorem 15.1, it suffices to prove inequalities (37.6)–(37.9) for smooth functions. Using the formula for change of variables in multiple integrals, we obtain

$$|v|_{0,p,\Omega}^p=\int_\Omega|v(x)|^p\,\mathrm{d}x=\int_\Omega|\hat{v}(F^{-1}(x))|^p\,\mathrm{d}x=\int_{\hat{\Omega}}|J(F)(\hat{x})||\hat{v}(\hat{x})|^p\,\mathrm{d}\hat{x}.$$

from which we deduce inequality (37.6).

Since $v = \hat{v} \circ F^{-1}$, we next infer that

$$Dv(x) = D\hat{v}(\hat{x}) \circ DF^{-1}(x) \quad \text{for all } x = F(\hat{x}),$$

and thus,

$$\| Dv(x) \| \leqslant | F^{-1} |_{1,\infty,\Omega} \| D\hat{v}(F^{-1}(x)) \| \quad \text{for all } x \in \Omega.$$

Consequently,

$$\begin{aligned}\int_{\Omega} \| Dv(x) \|^p \,\mathrm{d}x &\leqslant | F^{-1} |^p_{1,\infty,\Omega} \int_{\Omega} \| D\hat{v}(F^{-1}(x)) \|^p \,\mathrm{d}x \\ &= | F^{-1} |^p_{1,\infty,\Omega} \int_{\hat{\Omega}} | J(F)(\hat{x}) | \, \| D\hat{v}(\hat{x}) \|^p \,\mathrm{d}\hat{x} \\ &\leqslant | F^{-1} |^p_{1,\infty,\Omega} | J(F) |_{0,\infty,\hat{\Omega}} \int_{\hat{\Omega}} \| D\hat{v}(\hat{x}) \|^p \,\mathrm{d}\hat{x},\end{aligned}$$

and inequality (37.7) follows from this inequality and the equivalence between the seminorms (cf. (15.8) and (15.11))

$$v \to |v|_{m,p,\Omega} \quad \text{and} \quad v \to \left\{ \int_{\Omega} \| D^m v(x) \|^p \,\mathrm{d}x \right\}^{1/p}.$$

We likewise have for all $x \in \Omega$, $\xi_1 \in \mathbb{R}^n$, $\xi_2 \in \mathbb{R}^n$,

$$D^2 v(x)(\xi_1, \xi_2) = D\hat{v}(\hat{x})(D^2 F^{-1}(x)(\xi_1, \xi_2)) + D^2 \hat{v}(\hat{x})(DF^{-1}(x)\xi_1, DF^{-1}(x)\xi_2);$$

hence we obtain, for all $x = F(\hat{x}) \in \Omega$,

$$\| D^2 v(x) \| = \sup_{\substack{\|\xi_l\| \leqslant 1 \\ l=1,2}} | D^2 v(x)(\xi_1, \xi_2) | \leqslant | F^{-1} |_{2,\infty,\Omega} \| D\hat{v}(\hat{x}) \| + | F^{-1} |^2_{1,\infty,\Omega} \| D^2 \hat{v}(\hat{x}) \|.$$

Therefore,

$$\begin{aligned}\left\{ \int_{\Omega} \| D^2 v(x) \|^p \,\mathrm{d}x \right\}^{1/p} &\leqslant | F^{-1} |_{2,\infty,\Omega} \left\{ \int_{\Omega} \| D\hat{v}(F^{-1}(x)) \|^p \,\mathrm{d}x \right\}^{1/p} \\ &\quad + | F^{-1} |^2_{1,\infty,\Omega} \left\{ \int_{\Omega} \| D^2 \hat{v}(F^{-1}(x)) \|^p \,\mathrm{d}x \right\}^{1/p}.\end{aligned}$$

Since

$$\left\{ \int_{\Omega} \| D^l \hat{v}(F^{-1}(x)) \|^p \, dx \right\}^{1/p}$$
$$= \left\{ \int_{\hat{\Omega}} |J(F)(\hat{x})| \, \| D^l \hat{v}(\hat{x}) \|^p \, d\hat{x} \right\}^{1/p} \leqslant |J(F)|_{0,\infty,\hat{\Omega}}^{1/p} \left\{ \int_{\hat{\Omega}} \| D^l \hat{v}(\hat{x}) \|^p \, d\hat{x} \right\}^{1/p},$$

for any integer l, inequality (37.8) is proved.

Inequality (37.9) is proved analogously by using the following inequality:

$$\| D^3 v(x) \| \leqslant | F^{-1} |_{3,\infty,\Omega} \| D\hat{v}(\hat{x}) \| + 3 | F^{-1} |_{1,\infty,\Omega} | F^{-1} |_{2,\infty,\Omega} \| D^2 \hat{v}(\hat{x}) \|$$
$$+ | F^{-1} |_{1,\infty,\Omega}^3 \| D^3 \hat{v}(\hat{x}) \|$$

which holds for all $x = F(\hat{x}) \in \Omega$. □

To apply the previous theorem, we must next obtain estimates of the following quantities:

$$|J(F)|_{0,\infty,\hat{T}}, \qquad |J(F^{-1})|_{0,\infty,T},$$
$$|F|_{l,\infty,\hat{T}}, \qquad l = 1, 2, 3,$$
$$|F^{-1}|_{l,\infty,T}, \qquad l = 1, 2,$$

for an isoparametric quadratic n-simplex. To this end, the key idea is the following: Since the affine case is a special case of the isoparametric case, we may expect the same type of error bounds as in the affine case, provided the mapping F is not "too far" from the unique affine mapping $\tilde{F}$ that satisfies

$$\tilde{F}(\hat{a}_i) = a_i, \quad 1 \leqslant i \leqslant n+1. \tag{37.10}$$

Therefore we are naturally led to introduce the n-simplex

$$\tilde{T} = \tilde{F}(\hat{T}) \tag{37.11}$$

and the points (cf. Fig. 37.1 for $n = 2$)

$$\tilde{a}_{ij} = \tilde{F}(\hat{a}_{ij}), \quad 1 \leqslant i < j \leqslant n+1. \tag{37.12}$$

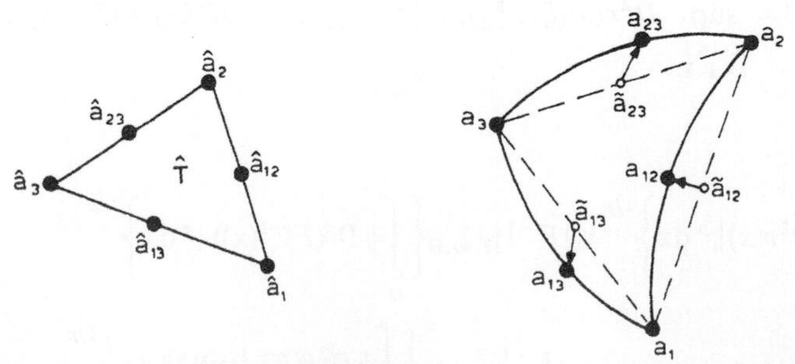

FIG. 37.1. The quadratic triangle $(\tilde{T}, \tilde{P}, \tilde{\Sigma})$ associated with an isoparametric quadratic triangle: $\tilde{T}$ has the same vertices as T; $\tilde{P} = P_2(\tilde{T})$, and $\tilde{\Sigma} = \{p(a_i)\colon 1 \leqslant i \leqslant 3;\ p(\tilde{a}_{ij})\colon 1 \leqslant i < j \leqslant 3\}$.

An inspection of Fig. 37.1 suggests that the vectors $(a_{ij}-\tilde{a}_{ij})$ serve as a good measure of the discrepancy between the mappings F and $\tilde{F}$: To see that this is indeed the case, let $\hat{p}_{ij}$ denote the basis functions attached to the node $\hat{a}_{ij}$ in the quadratic n-simplex $(\hat{T}, P_2(\hat{T}), \hat{\Sigma})$, where

$$\hat{\Sigma}=\{p(\hat{a}_i)\colon 1\leqslant i\leqslant n+1;\ p(\hat{a}_{ij})\colon 1\leqslant i<j\leqslant n+1\}.$$

Then we can write

$$F=\tilde{F}+\sum_{i<j}\hat{p}_{ij}(a_{ij}-\tilde{a}_{ij}), \tag{37.13}$$

since the mapping

$$G=\tilde{F}+\sum_{i<j}\hat{p}_{ij}(a_{ij}-\tilde{a}_{ij})$$

satisfies the relations

$$G\in(P_2(\hat{T}))^n,$$

$$G(\hat{a}_i)=a_i,\quad 1\leqslant i\leqslant n+1,$$

$$G(\hat{a}_{ij})=a_{ij},\quad 1\leqslant i<j\leqslant n+1,$$

which precisely characterize the unique isoparametric mapping F.

Let there be given an isoparametric family of quadratic n-simplices (T, P_T, Σ_T), each of which is determined by the data of vertices $a_{i,T}$, $1\leqslant i\leqslant n+1$, and points $a_{ij,T}$, $1\leqslant i<j\leqslant n+1$. As in (37.10) and (37.11), we let $\tilde{F}_T$ denote for each T the unique affine mapping that satisfies $\tilde{F}_T(\hat{a}_i)=a_{i,T}$, $1\leqslant i\leqslant n+1$, and we define the n-simplex $\tilde{T}=\tilde{F}_T(\hat{T})$. Finally, we let, for each $(\tilde{T})$,

$$h_T=\operatorname{diam}(\tilde{T}), \tag{37.14}$$

$$\rho_T=\text{diameter of the sphere inscribed in } \tilde{T}. \tag{37.15}$$

We shall then say that such an isoparametric family of quadratic n-simplices is *regular* if the following three conditions are simultaneously satisfied:

(i) There exists a constant σ such that

$$h_T/\rho_T\leqslant\sigma \quad \text{for all } T. \tag{37.16}$$

(ii) The quantities h_T approach zero.

(iii) We have

$$\|a_{ij,T}-\tilde{a}_{ij,T}\|=O(h_T^2),\quad 1\leqslant i<j\leqslant n+1, \tag{37.17}$$

where, for each T, $\tilde{a}_{ij,T}=\tilde{F}_T(\hat{a}_{ij})$.

REMARK 37.1. If the family is affine, condition (37.17) is automatically satisfied since $a_{ij,T}=\tilde{a}_{ij,T}$. In this case, the above definition thus reduces to that of a regular affine family as given in Section 16.

Although it is clear that condition (37.17) does insure that the mappings F_T and $\tilde{F}_T$ do not differ too much (cf. (37.13)), the condition that the vectors $(a_{ij,T}-\tilde{a}_{ij,T})$ have to be precisely of order $O(h_T^2)$ may seem arbitrary at this stage. As we shall show later (cf. Theorem 37.3), the basic justification of this assumption is that it yields the same interpolation error estimates as in the affine case.

To begin with, we show that this assumption allows us to obtain upper bounds of the various quantities found in the inequalities of Theorem 37.1 and that it also implies that the mappings F_T are invertible for h_T small enough (recall that the invertibility of the mapping F_T is part of the definition of an isoparametric family).

THEOREM 37.2. *Let there be given a regular isoparametric family of quadratic n-simplices. Then, provided h_T is small enough, the mappings $F_T\colon \hat{T}\to T=F_T(\hat{T})$ are one-to-one, their Jacobians $J(F_T)$ do not vanish, and there exist constants C such that*

$$|F_T|_{1,\infty,\hat{T}}\leqslant Ch_T, \qquad |F_T|_{2,\infty,\hat{T}}\leqslant Ch_T^2, \qquad |F_T|_{3,\infty,\hat{T}}=0, \tag{37.18}$$

$$|F_T^{-1}|_{1,\infty,T}\leqslant C/h_T, \qquad |F_T^{-1}|_{2,\infty,T}\leqslant C/h_T, \tag{37.19}$$

$$|J(F_T)|_{0,\infty,\hat{T}}\leqslant C\ \mathrm{meas}(\hat{T}), \qquad |J(F_T^{-1})|_{0,\infty,T}\leqslant C/\mathrm{meas}(\hat{T}). \tag{37.20}$$

PROOF. For notational convenience, we shall drop the index T throughout the proof. Using the decomposition (37.13) of the mapping F, we deduce that, for all $\hat{x}\in\hat{T}$,

$$DF(\hat{x})=D\tilde{F}(\hat{x})+E(\hat{x})=B+E(\hat{x}) \quad \text{with } E(\hat{x})=\sum_{i<j}(a_{ij}-\tilde{a}_{ij})D\hat{p}_{ij}(\hat{x}). \tag{37.21}$$

Since assumption (37.17) holds and since the basis functions $\hat{p}_{ij}$ are independent of T, we thus find that

$$\sup_{\hat{x}\in\hat{T}}\|E(\hat{x})\|\leqslant Ch^2 \tag{37.22}$$

(as usual the same letter C stands for various constants). Therefore

$$|F|_{1,\infty,\hat{T}}=\sup_{\hat{x}\in\hat{T}}\|DF(\hat{x})\|\leqslant\|B\|+\sup_{\hat{x}\in\hat{T}}\|E(\hat{x})\|\leqslant Ch,$$

since $\|B\|\leqslant Ch$ (cf. Theorem 15.2); likewise,

$$D^2F(\hat{x})=DE(\hat{x}),$$

since $D^2\tilde{F}=0$; hence arguing as before, we find that

$$\sup_{\hat{x}\in\hat{T}}\|DE(\hat{x})\|\leqslant Ch^2,$$

and thus

$$|F|_{2,\infty,\hat{T}}=\sup_{\hat{x}\in\hat{T}}\|D^2F(\hat{x})\|\leqslant Ch^2.$$

Hence all relations (37.18) are proved (the last one is clear, since $F\in(P_2(\hat{T}))^n$).

Considered as a function of its column vectors $\partial_j F(\hat{x})$, $1\leqslant j\leqslant n$, the determinant $J(F)(\hat{x})=\det(DF(\hat{x}))$ is a continuous multilinear mapping; therefore there exists a constant $C=C(n)$ such that

$$J(F)(\hat{x})\leqslant C\sum_{j=1}^{n}\|\partial_j F(\hat{x})\|\quad\text{for all } \hat{x}\in\hat{T}.$$

Since the inequality $|F|_{1,\infty,\hat{T}}\leqslant Ch$ proved earlier implies that $\sup_{\hat{x}\in\hat{T}}\|\partial_j F(\hat{x})\|\leqslant Ch$, $1\leqslant j\leqslant n$, it follows that

$$|J(F)|_{0,\infty,\hat{T}}=\sup_{\hat{x}\in\hat{T}}|J(F)(\hat{x})|\leqslant Ch^n\leqslant C\ \text{meas}(\tilde{T}),$$

and the first inequality of (37.20) is proved.

Because of assumption (37.16), the matrices B are all invertible; hence (37.21) may be rewritten as

$$DF(\hat{x})=B(I+B^{-1}E(\hat{x})).$$

Using the inequality $\|B^{-1}\|\leqslant C/h$ (cf. Theorem 15.2 and assumptions (37.16)) and inequality (37.22), we conclude that $\sup_{\hat{x}\in\hat{T}}\|B^{-1}E(\hat{x})\|\leqslant Ch$. Let then γ be a fixed number in the interval $]0,1[$. There exists $h_0>0$ such that

$$\sup_{\hat{x}\in\hat{T}}\|B^{-1}E(\hat{x})\|\leqslant\gamma\quad\text{for all } h\leqslant h_0;$$

hence for $h\leqslant h_0$, the operator $(I+B^{-1}E(\hat{x}))$ is invertible for each $\hat{x}\in\hat{T}$, and

$$\sup_{\hat{x}\in\hat{T}}\|(I+B^{-1}E(\hat{x}))^{-1}\|\leqslant\frac{1}{1-\gamma}.\tag{37.23}$$

This shows that the derivative $DF(\hat{x})$ is invertible for all $\hat{x}\in\hat{T}$, with

$$(DF(\hat{x}))^{-1}=(I+B^{-1}E(\hat{x}))^{-1}B^{-1}.\tag{37.24}$$

We next prove that *the mapping* $F\colon\hat{T}\to T$ *is invertible.* By the implicit function theorem, we can only deduce that the mapping F is *locally invertible*, i.e., in a sufficiently small neighborhood of each point of $\hat{T}$; this is why the global invertibility requires an additional analysis. Let then $\hat{x},\hat{y}\in\hat{T}$ be such that $F(\hat{x})=F(\hat{y})$. Since the set $\hat{T}$ is convex, Taylor formula yields

$$F(\hat{y})=F(\hat{x})+DF(\hat{x})(\hat{y}-\hat{x})+\tfrac{1}{2}A(\hat{y}-\hat{x})^2,$$

where $A\in\mathscr{L}_2(\mathbb{R}^n;\mathbb{R}^n)$ is the constant second derivative of the mapping F; hence

$$DF(\hat{x})(\hat{y}-\hat{x})=-\tfrac{1}{2}A(\hat{y}-\hat{x})^2=-\tfrac{1}{2}A(\hat{x}-\hat{y})^2=DF(\hat{y})(\hat{x}-\hat{y}),$$

and consequently,

$$(DF(\hat{x})+DF(\hat{y}))(\hat{y}-\hat{x})=0.$$

Since each component F_i of the mapping F is in the space $P_2(\hat{T})$, and since $(\tfrac{1}{2}(\hat{x}+\hat{y}))\in\hat{T}$,

$$\partial_j F_i\in P_1(\hat{T})\ \Rightarrow\ \partial_j F_i(\hat{x})+\partial_j F_i(\hat{y})=2\partial_j F_i(\tfrac{1}{2}(\hat{x}+\hat{y})),\quad 1\leqslant i,j\leqslant n,$$

i.e.,

$$0=(DF(\hat{x})+DF(\hat{y}))(\hat{y}-\hat{x})=2\ DF(\tfrac{1}{2}(\hat{x}+\hat{y}))(\hat{y}-\hat{x}).$$

Since the derivative $DF(\frac{1}{2}(\hat{x}+\hat{y}))$ is an invertible mapping in $\mathscr{L}(\mathbb{R}^n)$, we conclude that $\hat{x}=\hat{y}$.

We can now turn to the proof of inequalities (37.19) and (37.20). To begin with, we have

$$\{DF(\hat{x})\}^{-1}=DF^{-1}(x) \quad \text{for each } \hat{x}\in\hat{T},$$

and thus, by inequalities (37.23) and (37.24),

$$|F^{-1}|_{1,\infty,T}=\sup_{x\in T}\|DF^{-1}(x)\|\leqslant C/h,$$

which proves the first inequality of (37.19).

Next, let there be given smooth enough functions $F\colon\mathbb{R}^n\to\mathbb{R}^n$ and $G\colon\mathbb{R}^n\to\mathbb{R}^n$; then the function $H=G\circ F\colon\mathbb{R}^n\to\mathbb{R}^n$ is such that

$$D^2H(\hat{x})(\xi_1,\xi_2)=DG(x)(D^2F(\hat{x})(\xi_1,\xi_2))+D^2G(x)(DF(\hat{x})\xi_1,DF(\hat{x})\xi_2)$$
$$\text{for all } \xi_1,\xi_2\in\mathbb{R}^n.$$

If we apply this formula with $G=F^{-1}$, in which case $H=I$, we obtain

$$D^2F^{-1}(x)(DF(\hat{x})\xi_1,DF(\hat{x})\xi_2)=-DF^{-1}(x)(D^2F(\hat{x})(\xi_1,\xi_2))$$
$$\text{for all } x=F(\hat{x})\in\hat{T}.$$

Since the mapping $DF(\hat{x})\colon\mathbb{R}^n\to\mathbb{R}^n$ is invertible for each $x=F(\hat{x})\in T$, we infer that

$$D^2F^{-1}(x)(\eta_1,\eta_2)=-DF^{-1}(x)(D^2F(\hat{x})(DF^{-1}(x)\eta_1,DF^{-1}(x)\eta_2))$$
$$\text{for all } \eta_1,\eta_2\in\mathbb{R}^n,$$

and thus,

$$\|D^2F^{-1}(x)\|=\sup_{\substack{\|\eta_l\|\leqslant 1\\ l=1,2}}\|D^2F^{-1}(x)(\eta_1,\eta_2)\|\leqslant\|D^2F(\hat{x})\|\,\|DF^{-1}(x)\|^3.$$

Hence, using the second inequality of (37.18) and the first inequality (37.19), we conclude that

$$|F^{-1}|_{2,\infty,T}=\sup_{x\in T}\|D^2F^{-1}(x)\|\leqslant|F|_{2,\infty,\hat{T}}|F^{-1}|^3_{1,\infty,T}\leqslant C/h,$$

and the second inequality of (37.19) is proved.

Using (37.24), we can write

$$B=DF(\hat{x})(I+B^{-1}E(\hat{x}))^{-1} \quad \text{for all } \hat{x}\in\hat{T},$$

and thus, by (37.23),

$$|\det B|=|J(F)(\hat{x})|\,|\det(I+B^{-1}E(\hat{x}))^{-1}|\leqslant\frac{|J(F)(\hat{x})|}{(1-\gamma)^n} \quad \text{for all } \hat{x}\in\hat{T}.$$

Therefore, we deduce that

$$\frac{1}{|J(F^{-1})|_{0,\infty,T}} = \frac{1}{\sup_{x\in T}|J(F^{-1})(x)|} = \inf_{\hat{x}\in\hat{T}} J(F)(\hat{x})$$
$$\geqslant (1-\gamma)^n |\det B| \geqslant C\, \text{meas}(\tilde{T}),$$

and the second inequality of (37.20) is proved. □

The elegant proof of the invertibility of the mapping $F_T \in P_2(\hat{T})$ given here is due to Annie Raoult. In more general cases, it can often be derived from the following result, due to MEISTERS and OLECH [1963]: *Let $\hat{T}$ be a compact subset of $\mathbb{R}^n$ with a connected boundary $\partial\hat{T}$, and let there be given a mapping $F: \hat{T} \subset \mathbb{R}^n \to \mathbb{R}^n$ that satisfies the following assumptions:*

(i) *The mapping F can be extended to an open subset $\hat{\Omega}$ containing $\hat{T}$ in such a way that its extension (still denoted) F is in the space $\mathscr{C}^1(\hat{\Omega}; \mathbb{R}^n)$.*

(ii) *$J(F)(\hat{x}) > 0$ for all $\hat{x} \in \hat{T}$.*

(iii) *The restriction of F to $\partial\hat{T}$ is one-to-one.*

Then the mapping $F: \hat{T} \to \mathbb{R}^n$ is one-to-one.

Since it is usually straightforward to prove that the Jacobians $J(F)(\hat{x}) > 0$ for all $\hat{x} \in \hat{T}$, checking that the mapping F is one-to-one on $\hat{T}$ thus amounts to checking that it is one-to-one on $\partial\hat{T}$; when $n=2$ for instance, this is particularly easy since the boundary $\partial\hat{T}$ is a union of arcs of simple algebraic curves (cf. Remark 36.1). Other sufficient conditions of injectivity for such mappings F are given in CIARLET [1987, Section 5.5].

Let us now return to our discussion of the isoparametric quadratic triangle. Combining Theorems 37.1 and 37.2, we are in a position to prove an important result (compare with Theorem 16.2).

THEOREM 37.3. *Let there be given a regular isoparametric family of quadratic n-simplices (T, P_T, Σ_T), and let there be given an integer $m \geqslant 0$ and two numbers $p, q \in [1, \infty]$ such that the following inclusions hold:*

$$W^{3,p}(\hat{T}) \hookrightarrow \mathscr{C}^0(\hat{T}), \tag{37.25}$$
$$W^{3,p}(\hat{T}) \hookrightarrow W^{m,q}(\hat{T}), \tag{37.26}$$

where $(\hat{T}, \hat{P}, \hat{\Sigma})$ is the reference quadratic n-simplex of the family.

Then, if the diameters h_T are small enough, there exists a constant C such that, for all finite elements in the family, and all functions $v \in W^{3,p}(T)$,

$$\|v - \Pi_T v\|_{m,q,T} \leqslant C\{\text{meas}(\tilde{T})\}^{1/q-1/p} h_T^{3-m}\{|v|_{2,p,T} + |v|_{3,p,T}\}, \tag{37.27}$$

where, for each T, $\tilde{T}$ denotes the n-simplex that has the same vertices as T.

PROOF. The inclusion (37.25) guarantees the existence of the $\hat{P}$- and P_T-interpolation operators $\hat{\Pi}$ and Π_T, which satisfy relation (36.6). Combining the inequalities established in Theorems 37.1 and 37.2, we obtain

$$|v-\Pi_T v|_{0,q,T} \leqslant |J(F_T)|_{0,\infty,\hat{T}}^{1/q} |\hat{v}-\hat{\Pi}\hat{v}|_{0,q,\hat{T}}$$
$$\leqslant C\{\text{meas}(\tilde{T})\}^{1/q} |\hat{v}-\hat{\Pi}\hat{v}|_{0,q,\hat{T}},$$
$$|v-\Pi_T v|_{1,q,T} \leqslant C|J(F_T)|_{0,\infty,\hat{T}}^{1/q} |F_T^{-1}|_{1,\infty,T} |\hat{v}-\hat{\Pi}\hat{v}|_{1,q,\hat{T}}$$
$$\leqslant C\{\text{meas}(\tilde{T})\}^{1/q} \left\{\frac{1}{h_T} |\hat{v}-\hat{\Pi}\hat{v}|_{1,q,\hat{T}}\right\},$$
$$|v-\Pi_T v|_{2,q,T} \leqslant C|J(F_T)|_{0,\infty,\hat{T}}^{1/q} \{|F_T^{-1}|_{1,\infty,T}^2 |\hat{v}-\hat{\Pi}\hat{v}|_{2,q,\hat{T}} + |F_T^{-1}|_{2,\infty,T} |\hat{v}-\hat{\Pi}\hat{v}|_{1,q,\hat{T}}\}$$
$$\leqslant C\{\text{meas}(\tilde{T})\}^{1/q} \left\{\frac{1}{h_T^2} |\hat{v}-\hat{\Pi}\hat{v}|_{2,q,\hat{T}} + \frac{1}{h_T} |\hat{v}-\hat{\Pi}\hat{v}|_{1,q,\hat{T}}\right\}.$$

The relation $\hat{P}=P_2(\hat{T})$ and the inclusions (37.25) and (37.26) next imply, by Theorem 15.3, that there exists a constant C depending only on the set $\hat{T}$ such that, for all $\hat{v}\in W^{3,p}(\hat{T})$,

$$|\hat{v}-\hat{\Pi}\hat{v}|_{l,q,\hat{T}} \leqslant C|\hat{v}|_{3,p,\hat{T}}, \quad l \leqslant m,$$

and thus, after combining the above inequalities, we obtain

$$|v-\Pi_T v|_{m,q,T} \leqslant C\{\text{meas}(\tilde{T})\}^{1/q} \frac{1}{h_T^m} |\hat{v}|_{3,p,\hat{T}}.$$

Another application of Theorems 37.1 and 37.2 then yields:

$$|\hat{v}|_{3,p,\hat{T}} \leqslant C|J(F_T^{-1})|_{0,\infty,T}^{1/p} \{|F_T|_{1,\infty,\hat{T}}^3 |v|_{3,p,T} + |F_T|_{1,\infty,\hat{T}} |F_T|_{2,\infty,\hat{T}} |v|_{2,p,T} + |F_T|_{3,\infty,\hat{T}} |v|_{1,p,T}\}$$
$$\leqslant C\{\text{meas}(\tilde{T})\}^{-1/p} h_T^3 \{|v|_{2,p,T} + |v|_{3,p,T}\},$$

and thus inequality (37.27) is proved for the values $m=0,1,2$. The proof for $m=3$ proceeds along the same lines and, for this reason, is omitted. □

It is interesting to compare the estimate of the above theorem with the analogous estimates obtained for a regular *affine* family of quadratic n-simplices (cf. Theorem 16.2; in this case, $T=\tilde{T}$):

$$\|v-\Pi_T v\|_{m,q,T} \leqslant C\{\text{meas}(\tilde{T})\}^{1/q-1/p} h_T^{3-m} |v|_{3,p,T}.$$

We conclude that the two estimates coincide except for the additional seminorm $|v|_{2,p,T}$ (which appears when one differentiates a function composed with other than an affine function; cf. the end of the proof of Theorem 37.1). Also, the present estimates have been established under the additional assumption that the diameter h_T are sufficiently small, basically to insure the invertibility of the derivatives $DF_T(\hat{x})$, $\hat{x}\in\hat{T}$ (cf. the proof of Theorem 37.2).

REMARK 37.2. As in the case of affine families (cf. Remark 16.1), the parameter

$\text{meas}(\tilde{T})$ can be replaced by h_T^n in inequality (37.27), since it satisfies (cf. (37.16)) the inequalities

$$\beta_n \sigma^{-n} h_T^n \leqslant \text{meas}(\tilde{T}) \leqslant \beta_n h_T^n,$$

where β_n denotes the dx-measure of the unit ball in $\mathbb{R}^n$.

Similar analyses can be carried out for other isoparametric finite elements. Consider for instance the *isoparametric cubic n-simplex* (cf. Fig. 36.2 for $n=2$). If we introduce as in (37.10) the unique affine mapping $\tilde{F}_T$ that satisfies $\tilde{F}_T(\hat{a}_i) = a_i$, $1 \leqslant i \leqslant n+1$, we can obtain by an argument similar to that of Theorem 37.3 interpolation error estimates similar to those of the affine theory, viz.,

$$\| v - \Pi_T v \|_{m,q,T} \leqslant C\{\text{meas}(\tilde{T})\}^{1/q-1/p} h_T^{4-m} \| v \|_{4,p,T}, \quad 0 \leqslant m \leqslant 4, \tag{37.28}$$

provided we consider a *regular* isoparametric family of isoparametric cubic n-simplices, in the following sense (compare with (37.17)):

$$\| a_{iij,T} - \tilde{a}_{iij,T} \| = \text{O}(h_T^3), \quad 1 \leqslant i,j \leqslant n+1, \quad i \neq j, \tag{37.29}$$

$$\| a_{ijk,T} - \tilde{a}_{ijk,T} \| = \text{O}(h_T^3), \quad 1 \leqslant i < j < k \leqslant n+1, \tag{37.30}$$

where $\tilde{a}_{iij,T} = \tilde{F}_T(\hat{a}_{iij})$ and $\tilde{a}_{ijk,T} = \tilde{F}_T(\hat{a}_{ijk})$.

It is clear however that if the points $a_{iij,T}$ are taken from an actual boundary (as they would be in practice), condition (37.29) cannot be satisfied since in this situation one has at best $\| a_{iij,T} - \tilde{a}_{iij,T} \| = \text{O}(h_T^2)$. There is nevertheless one case where this difficulty can be circumvented: Assume that $n=2$ and that (cf. Fig. 37.2, where the

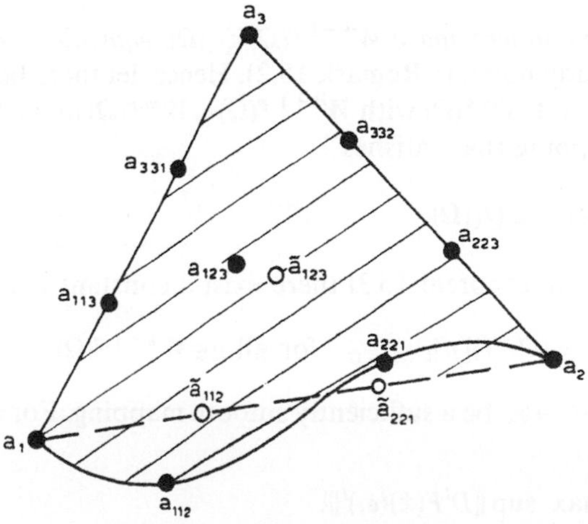

FIG. 37.2. The nodes a_{112} and a_{221} of the isoparametric cubic triangle may be chosen in such a way that $\| a_{112} - \tilde{a}_{112} \| = \text{O}(h^2)$, $\| a_{221} - \tilde{a}_{221} \| = \text{O}(h^2)$; yet if $v \in H^4(T)$, the interpolation error estimate is still $\| v - \Pi v \|_{m,T} = \text{O}(h^{4-m})$, $0 \leqslant m \leqslant 4$.

indices T have been dropped for convenience)

$$\begin{aligned}
&a_{331,T}=\tilde a_{331,T}, \qquad a_{113,T}=\tilde a_{113,T},\\
&a_{332,T}=\tilde a_{332,T}, \qquad a_{223,T}=\tilde a_{223,T},
\end{aligned}\tag{37.31}$$

$$\begin{aligned}
&\|a_{112,T}-\tilde a_{112,T}\| = \mathrm{O}(h_T^2), \quad \|a_{221,T}-\tilde a_{221,T}\| = \mathrm{O}(h_T^2),\\
&\|(a_{112,T}-\tilde a_{112,T})-(a_{221,T}-\tilde a_{221,T})\| = \mathrm{O}(h_T^3),\\
&\|4(a_{123,T}-\tilde a_{123,T})-(a_{112,T}-\tilde a_{112,T})-(a_{221,T}-\tilde a_{221,T})\| = \mathrm{O}(h_T^3).
\end{aligned}\tag{37.32}$$

Note that *assumptions* (37.32) *are now realistic*, in the sense that the points $a_{112,T}$ and $a_{221,T}$ can be actually chosen along a smooth boundary so as to fulfill the above conditions. Then it can be shown that *the interpolation error estimates of* (37.28) *still hold if the assumptions* (37.29)–(37.30) *are replaced by* (37.31)–(37.32). For details, see CIARLET and RAVIART [1972b], where the *isoparametric cubic Hermite triangle* (Fig. 36.3) is also considered.

If we turn to *isoparametric rectangular finite elements*, the situation is less simple. Of course, we could again consider this case as a perturbation of the affine case. But, as exemplified by Fig. 36.5, this would reduce the possible shapes to "nearly parallelograms". Hopefully, a new approach can be developed whereby the admissible shapes correspond to *mappings F_T that are perturbations of mappings $\tilde F_T$ in the space* $(Q_1(\hat T))^n$, instead of the space $(P_1(\hat T))^n$. Accordingly, a new theory, valid in particular for the 4-node quadrilateral, has to be developed, along the following lines:

(i) Let Ω be a domain in $\mathbb{R}^n$. Then for any integer $k\geqslant 1$ and any $p\in[1,\infty]$, *the seminorm*

$$v\to[v]_{k,p,\Omega}=\left\{\int_\Omega \sum_{i=1}^n |D^k v(x)(e_i)^k|^p\,\mathrm{d}x\right\}^{1/p}\tag{37.33}$$

is a norm over the quotient space $W^{k+1,p}(\Omega)/Q_k(\Omega)$, equivalent to the quotient norm (this fact was already noted in Remark 15.2). Hence, let there be given two Sobolev spaces $W^{k+1,p}(\Omega)$ and $W^{m,q}(\Omega)$ with $W^{k+1,p}(\Omega)\hookrightarrow W^{m,p}(\Omega)$ and let $\Pi\in\mathscr{L}(W^{k+1,p}(\Omega); W^{m,q}(\Omega))$ be a mapping that satisfies

$$\Pi q=q \quad \text{for all } q\in Q_k(\Omega).\tag{37.34}$$

Then (compare with Theorem 15.3) there exist a constant $C(\Omega,\Pi)$ such that

$$|v-\Pi v|_{m,q,\Omega}\leqslant C(\Omega,\Pi)[v]_{k+1,p,\Omega} \quad \text{for all } v\in W^{k+1,p}(\Omega).\tag{37.35}$$

(ii) Let $F:\hat\Omega\subset\mathbb{R}^n\to\mathbb{R}^n$ be a sufficiently smooth mapping. For any integer $l\geqslant 1$, we let

$$[F]_{l,\infty,\hat\Omega}=\max_{1\leqslant i\leqslant n}\sup_{\hat x\in\hat\Omega}\|D^l F(\hat x)(e_i)^l\|.\tag{37.36}$$

Then, if the same assumptions as in Theorem 37.1 hold, there exist constants C such that

$$[\hat{v}]_{2,p,\hat{\Omega}} \leqslant C|J(F^{-1})|_{0,\infty,\Omega}^{1/p}\{[F]_{1,\infty,\hat{\Omega}}^2|v|_{2,p,\Omega}+[F]_{2,\infty,\hat{\Omega}}|v|_{1,p,\Omega}\}$$
$$\text{for all } \hat{v}\in W^{2,p}(\hat{\Omega}), \tag{37.37}$$

$$[\hat{v}]_{3,p,\hat{\Omega}} \leqslant C|J(F^{-1})|_{0,\infty,\Omega}^{1/p}\{[F]_{1,\infty,\hat{\Omega}}^3|v|_{3,p,\Omega}+[F]_{1,\infty,\hat{\Omega}}[F]_{2,\infty,\hat{\Omega}}|v|_{2,p,\Omega}$$
$$+[F]_{3,\infty,\hat{\Omega}}|v|_{1,p,\Omega}\} \quad \text{for all } \hat{v}\in W^{3,p}(\hat{\Omega}). \tag{37.38}$$

(iii) Consider an isoparametric family of *4-node quadrilaterals* (cf. Fig. 36.4); and for each T, let

$$\begin{aligned} &h_T=\text{diam}(T),\\ &h'_T=\text{smallest length of the sides of } T,\\ &\gamma_T=\max\{|\cos\{(a_{i+1}-a_i)\cdot(a_{i-1}-a_i)\}|, \quad 1\leqslant i\leqslant 4(\text{mod } 4)\}. \end{aligned} \tag{37.39}$$

Such a family is said to be *regular* if all the sets T are convex, if there exist constants σ' and γ such that

$$h_T/h'_T\leqslant\sigma', \quad \gamma_T\leqslant\gamma<1 \qquad \text{for all } T. \tag{37.40}$$

and if the quantity h_T approaches zero. Note that condition (37.40) implies that the ratios h_T/ρ_T are bounded but that the converse is clearly false.

Then given a regular family, the mappings F_T: $\hat{T}=[0,1]^2\rightarrow T$ are one-to-one, and the following estimates hold (compare with Theorem 37.2)

$$[F_T]_{1,\infty,\hat{T}}\leqslant Ch_T, \qquad [F_T]_{2,\infty,\hat{T}}\leqslant Ch_T^2, \tag{37.41}$$

$$[F_T^{-1}]_{1,\infty,\hat{T}}\leqslant C/h_T, \tag{37.42}$$

$$|J(F_T)|_{0,\infty,\hat{T}}\leqslant Ch_T^2, \qquad |J(F_T^{-1})|_{0,\infty,T}\leqslant C/h_T^2. \tag{37.43}$$

(iv) All the above results combined then imply that the following *interpolation error estimate* holds: Assume that $W^{2,p}(\hat{T})\hookrightarrow\mathscr{C}^0(\hat{T})$ and that $W^{2,p}(\hat{T})\hookrightarrow W^{m,q}(\hat{T})$. Then there exists a constant C such that, for all $v\in W^{2,p}(T)$,

$$\|v-\Pi_T v\|_{m,q,T}\leqslant C\{h_T^2\}^{1/q-1/p}h_T^{2-m}\|v\|_{2,p,T}, \quad m=0,1. \tag{37.44}$$

(v) Consider next an isoparametric family of *biquadratic quadrilaterals* (cf. Fig. 36.5). For each $T=F(\hat{T})$, the mapping $F_T\in(Q_2(\hat{T}))^2$ is uniquely determined by nine points $a_{i,T}$, $1\leqslant i\leqslant 9$.

Let then $\tilde{F}_T$ denote the mapping uniquely determined by the conditions

$$\tilde{F}_T\in(Q_1(\hat{T}))^2, \qquad \tilde{F}_T(\hat{a}_i)=a_{i,T}, \quad 1\leqslant i\leqslant 4. \tag{37.45}$$

Then we say that the family is *regular* if the family of quadrilaterals $\tilde{T}=\tilde{F}_T(\hat{T})$ is regular, if conditions (37.40) hold, and if (compare with (37.17)):

$$\|a_{i,T}-\tilde{a}_{i,T}\|=\mathrm{O}(h_T^2), \quad 5\leqslant i\leqslant 9, \tag{37.46}$$

where $\tilde{a}_{i,T}=\tilde{F}_T(\hat{a}_i)$, $5\leqslant i\leqslant 9$.

If the family is regular, the mappings $F_T:\hat{T}=[0,1]^2\rightarrow T$ are one-to-one for h_T small enough, and the following estimates hold (compare again with Theorem 37.2):

$$[F_T]_{1,\infty,\hat{T}}\leqslant Ch_T, \qquad [F_T]_{2,\infty,\hat{T}}\leqslant Ch_T^2, \qquad [F_T]_{3,\infty,\hat{T}}=0, \tag{37.47}$$

$$|F_T^{-1}|_{1,\infty,T} \leqslant C/h_T, \quad |F_T^{-1}|_{2,\infty,T} \leqslant C/h_T, \tag{37.48}$$

$$|J(F_T)|_{0,\infty,T} \leqslant Ch_T^2, \quad |J(F_T^{-1})|_{0,\infty,T} \leqslant C/h_T^2, \tag{37.49}$$

and, finally, the following *interpolation error estimates* hold: Assume that $W^{3,p}(\hat{T}) \hookrightarrow \mathscr{C}^0(T)$ and that $W^{3,p}(\hat{T}) \hookrightarrow W^{m,q}(\hat{T})$. Then, there exists a constant C such that, for all $v \in W^{3,p}(T)$,

$$\|v - \Pi_T v\|_{m,q,T} \leqslant C\{h^2\}^{1/q-1/p} h_T^{3-m} \|v\|_{3,p,T}, \quad m=0,1,2. \tag{37.50}$$

A general interpolation theory for isoparametric rectangular finite elements, which comprises the above results as special cases, is given in CIARLET and RAVIART [1972b]. Significant improvements have then been obtained by JAMET [1976b], who obtained interpolation errors for quadrilateral elements degenerating into triangles.

Curved finite elements of other than isoparametric type have also been considered, notably by ZLÁMAL [1970, 1973a, 1973b, 1974] and SCOTT [1973a]. Both authors begin by constructing a curved face T' by approximating a smooth surface through an $(n-1)$-dimensional interpolation process. This interpolation serves to define a mapping F_T which in turn allows to define a finite element with T' as a curved face. Then the corresponding interpolation theory follows basically the same pattern as here. In particular, Scott constructs in this fashion a curved finite element that resembles the isoparametric cubic triangle and for which an interpolation theory can be developed which requires weaker assumptions than those indicated here. See also BERNARDI [1988].

WACHSPRESS [1971, 1973, 1975] has devised a clever class of *rational finite elements*, where the sets T are *quadrilaterals*, or more generally, *polygons*, and where the spaces P_T consist of *rational functions*. These finite elements of a new type have then been studied along the lines developed here by APPRATO and ARCANGÉLI [1979], APPRATO, ARCANGÉLI and GOUT [1979a, 1979b], GOUT [1979a, 1979b], LAYDI and LESAINT [1985].

Curved finite elements based on *blending function interpolation* have been advocated by GORDON and HALL [1973]; in this direction, see also CAVENDISH, GORDON and HALL [1976], BARNHILL [1975a, 1975b], ZACHARIAS and SUBBA RAO [1982]. Other relevant references are MCLEOD and MITCHELL [1972, 1975], LUKÁŠ [1974], MITCHELL [1976], MITCHELL and MARSHALL [1975], ARCANGÉLI and GOUT [1976], LEAF, KAPER and LINDEMAN [1976].

38. Approximation of a domain with a curved boundary with isoparametric finite elements

As in Chapters IV and V, we consider the homogeneous second-order Dirichlet problem that corresponds to the following data:

$$V = H_0^1(\Omega)$$

$$a(u,v) = \int_\Omega \sum_{i,j=1}^n a_{ij}\partial_i u \partial_j v \,\mathrm{d}x, \qquad l(v) = \int_\Omega fv \,\mathrm{d}x, \tag{38.1}$$

where Ω is a bounded open subset of $\mathbb{R}^n$ with a *curved* boundary Γ (the main novelty), and the functions $a_{ij} \in L^\infty(\Omega)$ and $f \in L^2(\Omega)$ are *everywhere* defined over the set $\bar{\Omega}$. We shall assume that the *ellipticity condition* holds, i.e., that there exists a constant β such that

$$\beta > 0,$$
$$\sum_{i,j=1}^{n} a_{ij}(x)\xi_i\xi_j \geqslant \beta \sum_{i=1}^{n} \xi_i^2 \quad \text{for all } x \in \bar{\Omega},\ (\xi_i) \in \mathbb{R}^n. \tag{38.2}$$

In this section, we describe in detail how to construct a *triangulation* of the set $\bar{\Omega}$ with isoparametric finite elements, a process that will lead us to the definition of a discrete problem *without* numerical integration; the definition of a discrete problem *with* numerical integration will be given in Section 39.

Our objective is to *assemble isoparametric finite elements* (T, P_T, Σ_T), $T \in \mathcal{T}_h$, *in such a way that the union* $\bar{\Omega}_h = \bigcup_{T\in\mathcal{T}_h} T$ *is a "good" approximation of the set* $\bar{\Omega}$. We assume that these elements form an isoparametric family, i.e., that they are all constructed from a single reference finite element $(\hat{T}, \hat{P}, \hat{\Sigma})$ through isoparametric mappings $F_T \in (\hat{P})^n$ (cf. Section 35). Since we shall assume for ease of exposition that $(\hat{T}, \hat{P}, \hat{\Sigma})$ is a *Lagrange element*, each mapping F_T is thus uniquely determined once the nodes of the finite element (T, P_T, Σ_T) are specified. These nodes, which will always be assumed to belong to the set $\bar{\Omega}$, must thus be chosen in such a way that $\mathcal{T}_h$ is a triangulation of the set $\bar{\Omega}_h$, in the sense that conditions $(\mathcal{T}_h 1)$–$(\mathcal{T}_h 4)$ set up in Section 5 hold, and that the boundary of the set $\bar{\Omega}_h$ "approximates at best" the boundary of Ω.

Finally, we shall restrict ourselves to finite elements that possess the following property:

> Each basis function $\hat{\varphi}$ of the reference finite element $(\hat{T}, \hat{P}, \hat{\Sigma})$ vanishes along any face of the set $\hat{T}$ that does not contain the node associated with $\hat{\varphi}$. (38.3)

As shown by the examples given in Section 36, this is not a restrictive assumption.

Of course, we shall take advantage of the isoparametric mappings F_T for getting a good approximation of the boundary Γ: By an appropriate choice of nodes along Γ, we construct finite elements with (at least) one curved face which should be "close" to Γ, at any rate closer than a straight face would be. Let us assume for definiteness that we are using *simplicial* finite elements. We then distinguish two cases: either the mapping F_T is affine, i.e., $F_T \in (P_1(\hat{T}))^n$, or the mapping F_T is "truly" isoparametric, i.e., $F_T \in (\hat{P})^n$ but $F_T \notin (\hat{P}_1(\hat{T}))^n$. The latter case applies in particular to "boundary" finite elements, while the former rather applies to "interior" finite elements. These considerations are illustrated in Fig. 38.1, where we consider quadratic triangles.

For computational simplicity, it is desirable to keep to a minimum the number of curved faces, and this is why, in general, only the "boundary" finite elements will have one curved face. However, all the subsequent analysis applies equally well to all

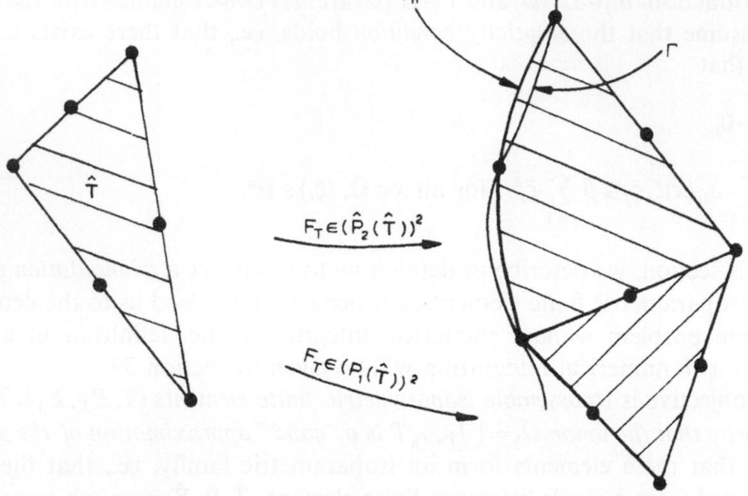

FIG. 38.1 "Affine" and "truly isoparametric" quadratic triangles.

possible cases, including those in which *all* the finite elements (T, P_T, Σ_T), $T \in \mathcal{T}_h$, are "truly" isoparametric.

Hence we must check in particular that the intersection of "adjacent" finite elements is indeed a face for both of them. In other words, *there should be no holes and no overlaps.* This is true because, by (35.6), the finite elements that satisfy (38.3) are such that *any one of their faces is solely determined by the nodes that are on it* (of course, the nodes that define a common face are assumed to be the same for two adjacent finite elements). As an illustration, we have represented in Fig. 38.2 three

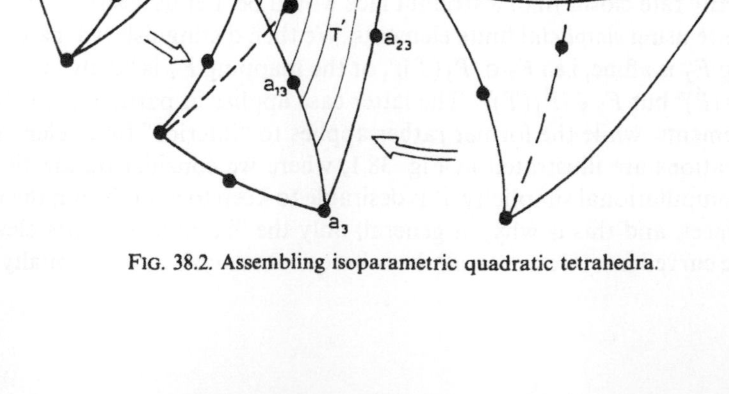

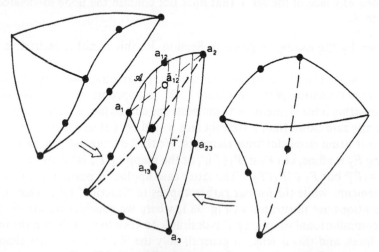

FIG. 38.2. Assembling isoparametric quadratic tetrahedra.

isoparametric quadratic tetrahedra "just before assembly": The face T' is completely determined by the nodes $a_1, a_2, a_3, a_{12}, a_{23}, a_{13}$, and the arc $\mathscr{A}$ is completely defined by the nodes a_1, a_2 and a_{12}.

Returning to the general case, we shall assume that *all the nodes that are used in the definition of the faces that approximate the boundary Γ are on Γ and that there are no other nodes on Γ.*

Because each face T' of an isoparametric finite element is necessarily of the form $T' = F_T(\hat{T}')$ with $F_T \in (\hat{P})^n$ and $\hat{T}'$ a face of $\hat{T}$, it is clear that *the boundary Γ_h of the set $\bar{\Omega}_h = \bigcup_{T\in\mathscr{T}_h} T$ does not coincide in general with the boundary Γ of the set Ω.* Nevertheless, we shall call $\mathscr{T}_h$ a *triangulation of the set $\bar{\Omega}$*, even though it should more appropriately be called a triangulation of the set $\bar{\Omega}_h$.

We then let X_h denote the *finite element space* whose functions $v_h: \bar{\Omega}_h \to \mathbb{R}$ are defined as follows:

(i) For each $T \in \mathscr{T}_h$, the restrictions $v_h|_T$ span the space

$$P_T = \{p : T \to \mathbb{R}; p = \hat{p} \circ F_T^{-1}, \hat{p} \in \hat{P}\}.$$

(ii) Over each $T \in \mathscr{T}_h$, the restrictions $v_h|_T$ are defined by their values at the nodes of the finite element (T, P_T, Σ_T).

If the functions of the space $\hat{P}$ are smooth enough, such a space X_h is contained in the space $\mathscr{C}^0(\bar{\Omega}_h)$ (by (38.3), a function in the space P is solely determined along a face T' of T by its values at the nodes that lie on T'), and consequently the inclusion $X_h \subset H^1(\Omega_h)$ holds by Theorem 5.1, where Ω_h denotes the interior of the set $\bar{\Omega}_h$.

We let X_{0h} denote the *subspace of X_h whose functions vanish at the boundary nodes*, i.e., those nodes that are on the boundary Γ. We recall that, by construction, these nodes are precisely those that are on the boundary Γ_h. Therefore another application of property (38.3) shows that the functions in the space X_{0h} vanish along the boundary Γ_h, and thus the inclusion

$$V_h = X_{0h} \subset H_0^1(\Omega_h) \tag{38.4}$$

holds.

Since we expect that the boundaries Γ_h and Γ are closer and closer as $h \to 0$, we shall henceforth assume that *there exists a bounded open set $\tilde{\Omega}$ such that*

$$\Omega \subset \tilde{\Omega} \quad \text{and} \quad \Omega_h \subset \tilde{\Omega} \tag{38.5}$$

for all the triangulations $\mathscr{T}_h$ that we shall consider.

Then the most natural definition of a *discrete problem* associated with the space V_h consists in finding a function $\tilde{u}_h \in V_h$ such that

$$\int_{\Omega_h} \sum_{i,j=1}^{n} \tilde{a}_{ij} \partial_i \tilde{u}_h \partial_j v_h \,\mathrm{d}x = \int_{\Omega_h} \tilde{f} v_h \,\mathrm{d}x \quad \text{for all } v_h \in V_h, \tag{38.6}$$

where the functions $\tilde{a}_{ij}$ and $\tilde{f}$ are some *extensions* of the functions a_{ij} and f to the set $\tilde{\Omega}$.

39. Isoparametric numerical integration

In spite of the simplicity and of the natural character of definition (38.6), several questions immediately arise: How should one choose between all possible extensions? How should one construct such extensions in practice? What is the dependence of the discrete solution $\tilde{u}_h$ upon these extensions? Surprisingly, it turns out that these ambiguities will be circumvented by taking into account the effect of *isoparametric numerical integration*:

As in Section 25, we assume that we have at our disposal a quadrature scheme over the set $\hat{T}$:

$$\int_{\hat{T}} \hat{\varphi}(\hat{x})\,\mathrm{d}\hat{x} \sim \sum_{l=1}^{L} \hat{\omega}_l \hat{\varphi}(\hat{b}_l), \qquad \hat{\omega}_l \in \mathbb{R}, \quad \hat{b}_l \in \hat{T}, \quad 1 \leqslant l \leqslant L. \tag{39.1}$$

Given two functions $\hat{\varphi}\colon \hat{T} \to \mathbb{R}$ and $\varphi : T = F_T(\hat{T}) \to \mathbb{R}$ in the usual correspondence $\hat{\varphi} \to \varphi = \hat{\varphi} \circ F_T^{-1}$, we have

$$\int_T \varphi(x)\,\mathrm{d}x = \int_{\hat{T}} \hat{\varphi}(\hat{x}) J(F_T)(\hat{x})\,\mathrm{d}\hat{x},$$

where the Jacobian $J(F_T)$ of the mapping F_T may be assumed without loss of generality to be >0 over the set $\hat{T}$. Therefore, *the quadrature scheme* (39.1) *over the reference element* $\hat{T}$ *automatically induces a quadrature scheme over the finite element* T (compare with (25.9) and (25.10)), viz.,

$$\int_T \varphi(x)\,\mathrm{d}x \sim \sum_{l=1}^{L} \omega_{l,T}\varphi(b_{l,T}), \tag{39.2}$$

with *weights* $\omega_{l,T}$ and *nodes* $b_{l,T}$ defined by

$$\omega_{l,T} = \hat{\omega}_l J(F_T)(\hat{b}_l), \quad b_{l,T} = F_T(\hat{b}_l), \qquad 1 \leqslant l \leqslant L. \tag{39.3}$$

Accordingly, we define the *quadrature error functionals*

$$E_T(\varphi) = \int_T \varphi(x)\,\mathrm{d}x - \sum_{l=1}^{L} \omega_{l,T}\varphi(b_{l,T}), \tag{39.4}$$

$$\hat{E}(\hat{\varphi}) = \int_{\hat{T}} \hat{\varphi}(\hat{x})\,\mathrm{d}\hat{x} - \sum_{l=1}^{L} \hat{\omega}_l \hat{\varphi}(\hat{b}_l), \tag{39.5}$$

which are related through the equation

$$E_T(\varphi) = \hat{E}(\hat{\varphi} J(F_T)). \tag{39.6}$$

Let us examine how isoparametric numerical integration affects the definition of the discrete problem (38.6). Assuming that the extensions $\tilde{a}_{ij}$ and $\tilde{f}$ are defined everywhere over the set $\{\tilde{\Omega}\}^-$, we now seek a *discrete solution* $u_h \in V_h$ that satisfies (compare with (25.27) and (25.28)):

$$\sum_{T\in\mathcal{T}_h}\sum_{l=1}^{L}\omega_{l,T}\sum_{i,j=1}^{n}(\tilde{a}_{ij}\partial_i u_h\partial_j v_h)(b_{l,T})=\sum_{T\in\mathcal{T}_h}\sum_{l=1}^{L}\omega_{l,T}(\tilde{f}v_h)(b_{l,T})$$

$$\text{for all } v_h\in V_h. \tag{39.7}$$

We then show that the extensions $\tilde{a}_{ij}$ and $\tilde{f}$ are not needed in the definition of the discrete problem (39.7) if all the quadrature nodes $b_{l,T}$, $1\leq l\leq L$, $T\in\mathcal{T}_h$, belong to the set $\bar{\Omega}$. To show that this is indeed a common circumstance, let us consider one typical example. Let $n=2$ and assume that we are using *isoparametric quadratic triangles* and that each node of the quadrature scheme over the set $\hat{T}$ either coincides with a node of the quadratic triangle T, or is in the interior int $\hat{T}$ of the set $\hat{T}$; as shown by the examples given in Section 25 (cf. Figs. 25.1, 25.2 and 25.3), this is a realistic situation.

To prove our assertion, it suffices to consider only a "boundary" finite element and, at this point, it becomes necessary to indicate *how the boundary nodes are actually chosen.* With the notations of Fig. 39.1, the point $a_{12,T}$ is chosen at the

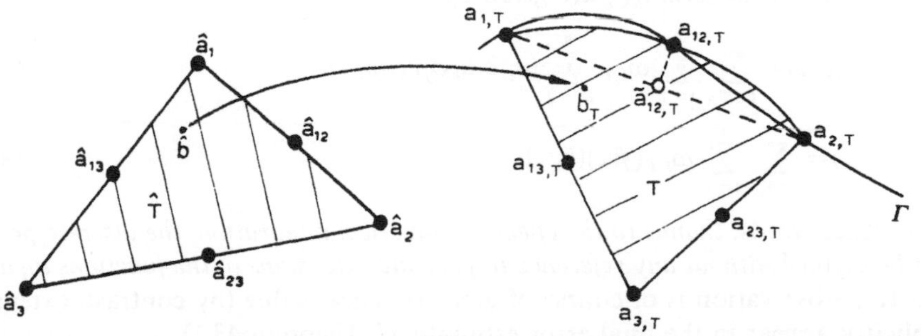

FIG. 39.1. The boundary node $a_{12,T}$ of an isoparametric quadratic triangle is chosen as the intersection of the boundary Γ with the perpendicular bissector of the segment $[a_{1,T}, a_{2,T}]$.

intersection between the boundary Γ and the line perpendicular to the segment $[a_{1,T}, a_{2,T}]$ passing through its midpoint $\tilde{a}_{12,T}=\frac{1}{2}(a_{1,T}+a_{2,T})$. *This choice has three important consequences:*

First, if the boundary Γ is smooth enough, we automatically have

$$\|a_{12,T}-\tilde{a}_{12,T}\|=\mathrm{O}(h_T^2), \tag{39.8}$$

where h_T is the diameter of the triangle with vertices $a_{i,T}$, $1\leq i\leq 3$. This estimate will insure that a family made up of such isoparametric quadratic triangles is regular in the sense understood in Section 37; we shall use this property in Theorem 42.1.

Secondly, *the image $b_T=F_T(\hat{b})$ of any point $\hat{b}\in$ int $\hat{T}$ belongs to the set $\bar{\Omega}\cap T$ provided h_T* is small enough.

Thirdly, it is clear that there exists a bounded open set $\tilde{\Omega}$ such that the inclusions (38.5) hold.

REMARK 39.1. This construction can be easily extended to an open set with a *piecewise smooth boundary*, i.e., a Lipschitz-continuous boundary composed of a finite number of smooth arcs, provided each intersection of adjacent arcs is a "vertex" of at least one isoparametric quadratic triangle.

REMARK 39.2. When $n=3$, a node such as a_{12} (cf. Fig. 38.2) may be chosen in such a way that the distance between the points $\tilde{a}_{12}$ and a_{12} is equal to the distance between the point $\tilde{a}_{12}$ and the boundary Γ.

Returning to the general case, we are therefore justified in *assuming from now on that the relations*

$$b_{l,T}=F_T(\hat{b}_l)\in\bar{\Omega},\quad 1\leqslant l\leqslant L,\quad \text{for all } T\in\mathcal{T}_h, \tag{39.9}$$

hold for all the triangulations $\mathcal{T}_h$ *to be considered*. This being the case, the *discrete problem* (39.7) consists in finding a discrete solution $u_h\in V_h$ such that

$$a_h(u_h,v_h)=l_h(v_h)\quad \text{for all } v_h\in V_h, \tag{39.10}$$

where, for all functions $u_h, v_h\in V_h$, the *approximate bilinear form* $a_h(\cdot,\cdot)$ and the *approximate linear form* $l_h(\cdot)$ are given by

$$a_h(u_h,v_h)=\sum_{T\in\mathcal{T}_h}\sum_{l=1}^{L}\omega_{l,T}\sum_{i,j=1}^{n}(a_{ij}\partial_i u_h\partial_j v_h)(b_{l,T}), \tag{39.11}$$

$$l_h(v_h)=\sum_{T\in\mathcal{T}_h}\sum_{l=1}^{L}\omega_{l,T}(fv_h)(b_{l,T}). \tag{39.12}$$

In other words, *thanks to the effect of numerical integration, the discrete problem can be defined without any reference to possible extensions of the functions* a_{ij} *and* f, and this observation is of course of great practical value (by contrast, extensions explicitly appear in the final error estimate; cf. Theorem 43.1).

Conceivably, *several* quadrature schemes over the reference finite element could be used, which vary according to which finite element is considered in the triangulation. In particular, one would naturally expect that more sophisticated schemes are needed for dealing with the "truly" isoparametric finite elements. Since our final result (Theorem 43.1) shows however that this is not necessary, we shall deliberately ignore this possibility (which, at this stage, would simply require notational modifications in (39.11) and (39.12)).

40. Abstract error estimate

The remainder of this chapter is based on CIARLET and RAVIART [1972c].

Given a family of discrete problems of the form (39.10), we shall say that the approximate bilinear forms $a_h(\cdot,\cdot)$ of (39.11) are *uniformly* V_h*-elliptic* if there exists a constant $\tilde{\alpha}$ such that

$$\begin{aligned}&\tilde{\alpha}>0,\\&\tilde{\alpha}\|v_h\|^2_{1,\Omega_h}\leqslant a_h(v_h,v_h)\quad \text{for all } v_h\in V_h \text{ and all } h.\end{aligned} \tag{40.1}$$

As usual, we first prove an *abstract error estimate*. The arbitrariness of the functions $\tilde{a}_{ij}$ and $\tilde{u}$ appearing in the next theorem is temporary: When this error estimate is actually applied, these functions will be *extensions* of the functions a_{ij} and u (cf. Theorem 43.1).

THEOREM 40.1. *Given an open set $\tilde{\Omega}$ that contains all the sets Ω_h, and given functions $\tilde{a}_{ij} \in L^\infty(\tilde{\Omega})$, let*

$$\tilde{a}_h(v, w) = \int_{\Omega_h} \sum_{i,j=1}^{n} \tilde{a}_{ij} \partial_i v \partial_j w \,\mathrm{d}x \quad \textit{for all } v, w \in H^1(\Omega_h), \tag{40.2}$$

and consider a family of discrete problems of the form (39.10), *whose associated approximate bilinear forms are uniformly V_h-elliptic.*

Then there exists a constant C independent of the space V_h such that

$$\|\tilde{u} - u_h\|_{1,\Omega_h} \leqslant C\left(\inf_{v_h \in V_h} \left\{ \|\tilde{u} - v_h\|_{1,\Omega_h} + \sup_{w_h \in V_h} \frac{|\tilde{a}_h(v_h, w_h) - a_h(v_h, w_h)|}{\|w_h\|_{1,\Omega_h}} \right\} + \sup_{w_h \in V_h} \frac{|\tilde{a}_h(\tilde{u}, w_h) - l_h(w_h)|}{\|w_h\|_{1,\Omega_h}} \right), \tag{40.3}$$

where $\tilde{u}$ is any function in the space $H^1(\tilde{\Omega})$, and u_h denotes for each h the solution of the discrete problem (39.10).

PROOF. The assumption of uniform V_h-ellipticity insures in particular that each discrete problem has a unique solution u_h. Also, there exists a constant $\tilde{M}$ independent of h such that

$$|\tilde{a}_h(v, w)| \leqslant \tilde{M} \|v\|_{1,\Omega_h} \|w\|_{1,\Omega_h} \quad \text{for all } v, w \in H^1(\Omega_h). \tag{40.4}$$

Let then v_h denote an arbitrary element in the space V_h. We have

$$\begin{aligned} \tilde{\alpha} \|u_h - v_h\|_{1,\Omega_h}^2 &\leqslant a_h(u_h - v_h, u_h - v_h) \\ &= \tilde{a}_h(\tilde{u} - v_h, u_h - v_h) + \{\tilde{a}_h(v_h, u_h - v_h) - a_h(v_h, u_h - v_h)\} \\ &\quad + \{l_h(u_h - v_h) - \tilde{a}_h(\tilde{u}, u_h - v_h)\}, \end{aligned}$$

so that, using (40.4),

$$\begin{aligned} &\tilde{\alpha} \|u_h - v_h\|_{1,\Omega_h} \\ &\quad \leqslant \tilde{M} \|\tilde{u} - v_h\|_{1,\Omega_h} + \frac{|\tilde{a}_h(v_h, u_h - v_h) - a_h(v_h, u_h - v_h)|}{\|u_h - v_h\|_{1,\Omega_h}} \\ &\qquad + \frac{|\tilde{a}_h(\tilde{u}, u_h - v_h) - l_h(u_h - v_h)|}{\|u_h - v_h\|_{1,\Omega_h}} \\ &\quad \leqslant \tilde{M} \|\tilde{u} - v_h\|_{1,\Omega_h} + \sup_{w_h \in V_h} \frac{|\tilde{a}_h(v_h, w_h) - a_h(v_h, w_h)|}{\|w_h\|_{1,\Omega_h}} + \sup_{w_h \in V_h} \frac{|\tilde{a}_h(\tilde{u}, w_h) - l_h(w_h)|}{\|w_h\|_{1,\Omega_h}}. \end{aligned}$$

Combining the above inequality with the triangular inequality

$$\|\tilde{u}-u_h\|_{1,\Omega_h} \leqslant \|\tilde{u}-v_h\|_{1,\Omega_h} + \|u_h-v_h\|_{1,\Omega_h},$$

and taking the infimum with respect to $v_h \in V_h$, we obtain inequality (40.3). □

The remainder of this chapter will thus be devoted to giving sufficient conditions that imply the uniform V_h-ellipticity of the approximate bilinear forms (Theorem 41.1) and to estimating the various terms that appear in the right-hand side of inequality (40.3). To keep the development within reasonable limits, we shall however restrict ourselves to finite element spaces made up of *isoparametric quadratic n-simplices.*

In Section 37, we have defined a *regular* family of isoparametric quadratic n-simplices (cf. (37.16) and (37.17)). Since this regularity will be a pervading assumption in the remainder of this chapter, it is crucial to notice that condition (37.17) is perfectly compatible with the construction of boundary elements (cf. (39.8) and Remark 39.2).

41. Uniform V_h-ellipticity of the approximate bilinear forms

It is remarkable that the following sufficient conditions for the uniform V_h-ellipticity are the same as for straight quadratic n-simplices (cf. Theorem 27.1 with $k'=2$).

THEOREM 41.1. *Let (V_h) be a family of finite element spaces made up of isoparametric quadratic n-simplices forming a regular family, and let there be given a quadrature scheme*

$$\int_{\hat{T}} \hat{\varphi}(\hat{x})\,\mathrm{d}\hat{x} \sim \sum_{l=1}^{L} \hat{\omega}_l \hat{\varphi}(\hat{b}_l), \qquad \hat{\omega}_l > 0, \quad 1 \leqslant l \leqslant L,$$

such that either the union $\bigcup_{l=1}^{L} \{\hat{b}_l\}$ contains a $P_1(\hat{T})$-unisolvent subset, or the quadrature scheme is exact for the space $P_2(\hat{T})$, or both these assumptions hold.

Then the associated approximate bilinear forms are uniformly V_h-elliptic, i.e., there exists a constant $\tilde{\alpha}$ such that

$$\begin{aligned} &\tilde{\alpha} > 0, \\ &\tilde{\alpha}\|v_h\|^2_{1,\Omega_h} \leqslant a_h(v_h, v_h) \quad \textit{for all } v_h \in V_h \textit{ and all } h. \end{aligned} \tag{41.1}$$

PROOF. (i) Arguing as in part (i) of the proof of Theorem 27.1 we find that there exists a constant $\hat{C} > 0$ such that

$$\hat{C}|\hat{p}|^2_{1,\hat{T}} \leqslant \sum_{l=1}^{L} \hat{\omega}_l \sum_{i=1}^{n} |\partial_i \hat{p}(\hat{b}_l)|^2 \quad \text{for all } \hat{p} \in \hat{P} = P_2(\hat{T}). \tag{41.2}$$

(ii) Given a finite element (T, P_T, Σ_T), $T \in \mathcal{T}_h$, and a function $v_h \in V_h$, let $p = v_h|_T$.

Using the ellipticity condition (38.2), we obtain

$$\sum_{l=1}^{L} \omega_{l,T} \sum_{i,j=1}^{n} (a_{ij}\partial_i v_h \partial_j v_h)(b_{l,T}) \geqslant \beta \sum_{l=1}^{L} \omega_{l,T} \sum_{i=1}^{n} |\partial_i p(b_{l,T})|^2, \tag{41.3}$$

and we recognize in the expression $\Sigma_{i=1}^{n} |\partial_i p(b_{l,T})|^2$ the square of the Euclidean norm $\|\cdot\|$ of the vector $Dp(b_{l,T})$.

Since F_T is invertible for h small enough by Theorem 37.2, we have $Dp(x) = D\hat{p}(\hat{x})DF_T^{-1}(x)$ for all $x = F_T(\hat{x})$, $\hat{x} \in \hat{T}$; here $Dp(x)$ and $D\hat{p}(\hat{x})$ are identified with the row vectors $(\partial_1 p(x), \ldots, \partial_n p(x))$ and $(\partial_1 \hat{p}(\hat{x}), \ldots, \partial_n \hat{p}(\hat{x}))$ respectively, and $DF_T^{-1}(x)$ is identified with the Jacobian matrix of the mapping F_T^{-1} at x. Using the inequality $\xi A A^{\mathrm{T}} \xi^{\mathrm{T}} \geqslant \|A^{-1}\|^{-2} \xi \xi^{\mathrm{T}}$ valid for any invertible matrix A and any row vector ξ (the superscript T denotes transposition), we obtain

$$\sum_{i=1}^{n} |\partial_i p(x)|^2 = Dp(x)Dp(x)^{\mathrm{T}} \geqslant \frac{1}{\|DF_T(\hat{x})\|^2} \sum_{i=1}^{n} |\partial_i \hat{p}(\hat{x})|^2 \quad \text{for all } x = F_T(\hat{x}) \in T. \tag{41.4}$$

Since $\omega_{l,T} = \hat{\omega}_l J(F_T)(\hat{b}_l)$ (cf. (39.3)) and since the weights $\hat{\omega}_l$ are positive, we deduce from (41.2) and (41.4) that:

$$\begin{aligned} &\sum_{l=1}^{L} \omega_{l,T} \sum_{i=1}^{n} |\partial_i p(b_{l,T})|^2 \\ &\geqslant \inf_{\hat{x} \in \hat{T}} \{J(F_T)(\hat{x})\} \inf_{\hat{x} \in \hat{T}} \left\{\frac{1}{\|DF_T(\hat{x})\|^2}\right\} \sum_{l=1}^{L} \hat{\omega}_l \sum_{i=1}^{n} |\partial_i \hat{p}(\hat{b}_l)|^2 \\ &\geqslant \hat{C} \frac{1}{|J(F_T^{-1})|_{0,\infty,T} |F_T|_{1,\infty,\hat{T}}^2} |\hat{p}|_{1,\hat{T}}^2, \end{aligned} \tag{41.5}$$

where, here and subsequently, we use the notations introduced in (37.5). By Theorem 37.1, there exists a constant C such that

$$|\hat{p}|_{1,\hat{T}} \geqslant C \frac{1}{|J(F_T)|_{0,\infty,\hat{T}}^{1/2} |F_T^{-1}|_{1,\infty,T}} |p|_{1,T} \quad \text{for all } p = \hat{p} \circ F_T^{-1}, \quad \hat{p} \in \hat{P}. \tag{41.6}$$

Hence, combining inequalities (41.3), (41.5) and (41.6), we obtain

$$\begin{aligned} &\sum_{l=1}^{L} \omega_{l,T} \sum_{i,j=1}^{n} (a_{ij}\partial_i v_h \partial_j v_h)(b_{l,T}) \\ &\geqslant \beta \hat{C} C^2 \frac{|v_h|_{1,T}^2}{|J(F_T^{-1})|_{0,\infty,T} |J(F_T)|_{0,\infty,T} \{|F_T|_{1,\infty,\hat{T}} |F_T^{-1}|_{1,\infty,T}\}^2}. \end{aligned} \tag{41.7}$$

If we next make use of the estimates established in Theorem 37.2, which we may apply since the isoparametric family is assumed to be regular, we find that the denominators appearing in the right-hand side of inequality (41.7) are uniformly bounded for all $T \in \mathcal{T}_h$, all $v_h \in V_h$ and all h. Therefore we have shown that there exists $\tilde{\alpha}'$ such that

$$\tilde{\alpha}'>0,$$

$$\sum_{l=1}^{L}\omega_{l,T}\sum_{i,j=1}^{n}(a_{ij}\partial_i v_h\partial_j v_h)(b_{l,T})\geqslant\tilde{\alpha}'|v_h|_{1,T}^2 \tag{41.8}$$

for all $T\in\mathcal{T}_h$, all $v_h\in V_h$, and all h.

(iii) Using inequality (41.8), we obtain

$$\begin{aligned}a_h(v_h,v_h)&=\sum_{T\in\mathcal{T}_h}\sum_{l=1}^{L}\omega_{l,T}\sum_{i,j=1}^{n}(a_{ij}\partial_i v_h\partial_j v_h)(b_{l,T})\\&\geqslant\tilde{\alpha}'\sum_{T\in\mathcal{T}_h}|v_h|_{1,T}^2=\tilde{\alpha}'|v_h|_{1,\Omega_h}^2\quad\text{for all } v_h\in V_h \text{ and all } h.\end{aligned} \tag{41.9}$$

Since all the sets Ω_h are contained in a single bounded open set $\tilde{\Omega}$ (cf. (38.5)), there exists a constant C independent of h such that

$$\|v\|_{1,\Omega_h}\leqslant C|v|_{1,\Omega_h}\quad\text{for all } v\in H_0^1(\Omega_h). \tag{41.10}$$

To see this, it suffices to apply the Poincaré–Friedrichs inequality over the set $\tilde{\Omega}$ to the function $\tilde{v}\in H_0^1(\tilde{\Omega})$ that equals an arbitrary function $v\in H_0^1(\Omega)$ on Ω_h and that otherwise vanishes on $\tilde{\Omega}-\Omega_h$.

Inequality (41.1) is then a consequence of inequalities (41.9) and (41.10). □

42. Interpolation and consistency error estimates

With the finite element space X_h constructed in Section 38, we associate the *X_h-interpolation operator* Π_h, whose definition is the natural extension of the definition given in Section 12 in the case of straight finite elements: The *X_h-interpolant* $\Pi_h v$ of a function $v\in\operatorname{dom}\Pi_h=\mathscr{C}^0(\bar{\Omega}_h)$ is the unique function that satisfies

$$\begin{aligned}&\Pi_h v\in X_h,\\&\Pi_h v(a_{i,T})=v(a_{i,T}),\quad 1\leqslant i\leqslant n+1,\quad\text{for all } T\in\mathcal{T}_h,\\&\Pi_h v(a_{ij,T})=v(a_{ij,T}),\quad 1\leqslant i<j\leqslant n+1,\quad\text{for all } T\in\mathcal{T}_h.\end{aligned} \tag{42.1}$$

In this fashion, it is clear that the relations

$$\Pi_h v|_T=\Pi_T v\quad\text{for all } T\in\mathcal{T}_h \tag{42.2}$$

again hold.

We now estimate various norms of the difference $(v-\Pi_h v)$. In particular, these estimates will subsequently allow us to obtain, for a specific choice of function $\tilde{u}$, an estimate of the first term in the right-hand side of the error estimate (40.3) when we choose $v_h=\Pi_h\tilde{u}$. As usual, the same letter C stands for various constants independent of h and of the various functions involved.

THEOREM 42.1. *Let (X_h) be a family of finite element spaces made up of isoparametric quadratic n-simplices forming a regular family and assume that $n \leqslant 5$.*

Then there exists a constant C independent of h such that

$$\| v-\Pi_h v\|_{m,\Omega_h} \leqslant Ch^{3-m} \| v\|_{3,\Omega_h}, \quad m=0,1, \quad \text{for all } v \in H^3(\tilde{\Omega}), \tag{42.3}$$

$$\left\{ \sum_{T\in\mathcal{T}_h} \| v-\Pi_T v\|_{m,T}^2 \right\}^{1/2} \leqslant Ch^{3-m} \| v\|_{3,\Omega_h}, \quad m=2,3, \quad \text{for all } v \in H^3(\tilde{\Omega}), \tag{42.4}$$

where

$$h = \max_{T\in\mathcal{T}_h} h_T, \tag{42.5}$$

and $\tilde{\Omega}$ is any open set such that the inclusions (38.5) *hold. We also have the implication*

$$v\in H^3(\tilde{\Omega}) \text{ and } v=0 \text{ on } \Gamma \;\Rightarrow\; \Pi_h v\in X_{0h}. \tag{42.6}$$

PROOF. Since $n \leqslant 5$, the inclusion $H^3(\hat{T}) \hookrightarrow \mathscr{C}^0(\hat{T})$ holds and thus we may apply Theorem 37.3: For all functions $v\in H^3(T)$,

$$\| v-\Pi_T v\|_{m,T} \leqslant Ch_T^{3-m}\{|v|_{2,T}+|v|_{3,T}\} \leqslant Ch_T^{3-m}\| v\|_{3,T}, \quad 0\leqslant m\leqslant 3,$$

and inequalities (42.3) and (42.4) follow from these inequalities and property (42.2).

If a function vanishes on Γ, its X_h-interpolant vanishes at all the nodes situated on Γ_h by construction; therefore it vanishes on the boundary Γ_h of the set $\bar{\Omega}_h = \bigcup_{T\in\mathcal{T}_h} T$, and implication (42.6) is thus proved. □

As in Chapter IV, the *consistency errors*

$$\sup_{w_h\in V_h} \frac{|\tilde{a}_h(v_h,w_h)-a_h(v_h,w_h)|}{\| w_h\|_{1,\Omega_h}} \quad \text{with } v_h\in V_h, \qquad \sup_{w_h\in V_h} \frac{|\tilde{a}_h(\tilde{u},w_h)-l_h(w_h)|}{\| w_h\|_{1,\Omega_h}},$$

which appeared in the abstract error estimate (40.3), will be estimated after similar, but "local", terms have been themselves estimated. Such "local" estimates are the object of the next two theorems (compare with Theorems 28.2 and 28.3). The quadrature error functionals $E_T(\cdot)$ and $\hat{E}(\cdot)$ have been defined in (39.4) and (39.5).

THEOREM 42.2. *Let there be given a regular isoparametric family of quadratic n-simplices (T, P_T, Σ_T), and let the quadrature scheme over the reference finite element be exact for the space $P_2(\hat{T})$, i.e.,*

$$\hat{E}(\hat{\varphi})=0 \quad \text{for all } \hat{\varphi}\in P_2(\hat{T}). \tag{42.7}$$

Then there exists a constant C independent of T such that

$$|E_T(a\partial_i p'\partial_j p)| \leqslant Ch_T^2\| a\|_{2,\infty,T}\| p'\|_{2,T}|p|_{1,T} \quad \text{for all } a\in W^{2,\infty}(T),\ p\in P_T,\ p'\in P_T. \tag{42.8}$$

PROOF. For notational convenience, the indices T are dropped throughout the proof.

(i) To begin with, we record some consequences of Theorem 37.2: Inequalities (37.18) imply that

$$|\partial_i F_k|_{0,\infty,\hat{T}} \leqslant Ch, \qquad 1 \leqslant i, k \leqslant n, \tag{42.9}$$

$$|\partial_{ij} F_k|_{0,\infty,\hat{T}} \leqslant Ch^2, \quad 1 \leqslant i, j, k \leqslant n, \tag{42.10}$$

and inequalities (37.20) imply that

$$|J(F)|_{0,\infty,\hat{T}} \leqslant Ch^n, \qquad |J(F^{-1})|_{0,\infty,T} = |J(F)^{-1}|_{0,\infty,\hat{T}} \leqslant C/h^n. \tag{42.11}$$

Next, we show that

$$\begin{aligned} &|\partial_i J(F)|_{0,\infty,\hat{T}} \leqslant Ch^{n+1}, \quad 1 \leqslant i \leqslant n, \\ &|\partial_{ij} J(F)|_{0,\infty,\hat{T}} \leqslant Ch^{n+2}, \quad 1 \leqslant i, j \leqslant n. \end{aligned} \tag{42.12}$$

Let us denote by $\partial_i F(\hat{x})$ and $\partial_{ij} F(\hat{x})$ the column vectors with components $\partial_i F_k(\hat{x})$, $1 \leqslant k \leqslant n$, and $\partial_{ij} F_k(\hat{x})$, $1 \leqslant k \leqslant n$, respectively. To prove the first inequalities of (42.12), we observe that, for any $\hat{x} \in \hat{T}$, we have

$$\partial_i J(F)(\hat{x}) = \sum_{j=1}^{n} \det(\partial_1 F(\hat{x}), \ldots, \partial_{j-1} F(\hat{x}), \partial_{ij} F(\hat{x}), \partial_{j+1} F(\hat{x}), \ldots, \partial_n F(\hat{x})),$$

and we use inequalities (42.9) and (42.10). The second inequalities of (42.12) are proved in a similar fashion (since $F \in (P_2(\hat{T}))^n$, the partial derivatives $\partial_{ijk} F$ vanish identically).

(ii) The expression to be estimated can be written as

$$E_T(a\partial_i p' \partial_j p) = \hat{E}(\hat{a}\{\partial_i p'\}\hat{}\,\{\partial_j p\}\hat{}\,J(F)). \tag{42.13}$$

Then it is clear that, by contrast with the affine case, the functions $\{\partial_i p'\}^\wedge$ and $\{\partial_j p\}^\wedge$ no longer belong to the space $P_1(\hat{T})$ in general. This is why our first task is to determine the nature of these functions: Denoting by e_j the jth basis vector of $\mathbb{R}^n$, we have

$$\{\partial_j p\}\hat{}\,(\hat{x}) = \partial_j p(x) = Dp(x)e_j = D\hat{p}(\hat{x})DF^{-1}(x)e_j = D\hat{p}(\hat{x})(DF(\hat{x}))^{-1}e_j.$$

Noting that the vector $f_j = (DF(\hat{x}))^{-1}e_j$ is the solution of the linear system $DF(\hat{x})f_j = e_j$, we find that

$$\begin{aligned} \{\partial_j p\}\hat{}\,(\hat{x}) = {} & \{J(F)(\hat{x})\}^{-1} \\ & \times \sum_{k=1}^{n} \partial_k \hat{p}(\hat{x}) \det(\partial_1 F(\hat{x}), \ldots, \partial_{k-1} F(\hat{x}), e_j, \partial_{k+1} F(\hat{x}), \ldots, \partial_n F(\hat{x})). \end{aligned} \tag{42.14}$$

Consequently, the expression $\{\partial_j p\}\hat{}\,(\hat{x})J(F)(\hat{x})$ is a finite sum of terms of the form

$$\pm \partial_k \hat{p}(\hat{x}) \prod_{l \neq k} \partial_l F_{j(l)}(\hat{x}),$$

and likewise the expression $\{\partial_i p'\}\hat{}\,(\hat{x})$ is a finite sum of terms of the form

$$\pm \{J(F)(\hat{x})\}^{-1} \partial_r \hat{p}'(\hat{x}) \prod_{s \neq r} \partial_s F_{j(s)}(\hat{x}).$$

Using (42.13), we thus find that

$$E_T(a\partial_i p'\partial_j p) = \sum_{\substack{k,j(l),l\neq k\\ r,j(s),s\neq r}}' \pm \hat{E}\left(\{J(F)\}^{-1}\hat{a}\prod_{s\neq r}\partial_s F_{j(s)}\prod_{l\neq k}\partial_l F_{j(l)}\partial_r\hat{p}'\partial_k\hat{p}\right), \tag{42.15}$$

where the symbol Σ' simply reminds that the indices $j(l)$ and $j(s)$ do not take all possible values $1, 2, \ldots, n$.

(iii) We now take crucial advantage of the fact that the functions $\{\partial_i p'\}\hat{}$ and $\{\partial_j p\}\hat{}$ can be expressed in terms of the functions $\partial_k\hat{p}$, $1 \leqslant k \leqslant n$, which *do* belong to the space $P_1(\hat{T})$: Any one of the terms occurring in the sum (42.15) can be written as

$$\hat{E}\left(\{J(F)\}^{-1}\hat{a}\prod_{s\neq r}\partial_s F_{j(s)}\prod_{l\neq k}\partial_l F_{j(l)}\partial_r\hat{p}'\partial_k\hat{p}\right) = \hat{E}(\hat{b}\hat{v}\hat{w}), \tag{42.16}$$

with

$$\begin{aligned} &\hat{b} = \{J(F)\}^{-1}\hat{a}\prod_{s\neq r}\partial_s F_{j(s)}\prod_{l\neq k}\partial_l F_{j(l)} \in W^{2,\infty}(\hat{T}),\\ &\hat{v} = \partial_r\hat{p}' \in P_1(\hat{T}), \qquad \hat{w} = \partial_k\hat{p} \in P_1(\hat{T}), \end{aligned} \tag{42.17}$$

and consequently, inequality (28.16) with $k = 2$ yields

$$\begin{aligned} |\hat{E}(\hat{b}\hat{v}\hat{w})| &\leqslant C\{|\hat{b}|_{2,\infty,\hat{T}}|\hat{v}|_{0,\hat{T}} + |\hat{b}|_{1,\infty,\hat{T}}|\hat{v}|_{1,\hat{T}}\}|\hat{w}|_{0,\hat{T}}\\ &\leqslant C\{|\hat{b}|_{2,\infty,\hat{T}}|\hat{p}'|_{1,\hat{T}} + |\hat{b}|_{1,\infty,\hat{T}}|\hat{p}'|_{2,\hat{T}}\}|\hat{p}|_{1,\hat{T}}. \end{aligned} \tag{42.18}$$

It thus remains to express the various seminorms found in this inequality in terms of appropriate norms defined over the set T. Using Theorems 37.1 and 37.2, we obtain:

$$\begin{aligned} &|\hat{p}|_{l,\hat{T}} \leqslant Ch^{-n/2}h^l|p|_{l,T}, \qquad l = 0, 1,\\ &|\hat{p}'|_{l,\hat{T}} \leqslant Ch^{-n/2}h^l\|p'\|_{l,T}, \qquad l = 1, 2. \end{aligned} \tag{42.19}$$

Next, we have (cf. (28.11))

$$\begin{aligned} |\hat{b}|_{1,\infty,\hat{T}} &= \left|\{J(F)\}^{-1}\hat{a}\prod_{s\neq r}\partial_s F_{j(s)}\prod_{l\neq k}\partial_l F_{j(l)}\right|_{1,\infty,\hat{T}}\\ &\leqslant C\left\{|J(F)^{-1}|_{0,\infty,\hat{T}}|\hat{a}|_{1,\infty,\hat{T}}\left|\prod_{s\neq r}\partial_s F_{j(s)}\prod_{l\neq k}\partial_l F_{j(l)}\right|_{0,\infty,\hat{T}}\right.\\ &\quad + |J(F)^{-1}|_{0,\infty,\hat{T}}|\hat{a}|_{0,\infty,\hat{T}}\left|\prod_{s\neq r}\partial_s F_{j(s)}\prod_{l\neq k}\partial_l F_{j(l)}\right|_{1,\infty,\hat{T}}\\ &\quad \left. + |J(F)^{-1}|_{1,\infty,\hat{T}}|\hat{a}|_{0,\infty,\hat{T}}\left|\prod_{s\neq r}\partial_s F_{j(s)}\prod_{l\neq k}\partial_l F_{j(l)}\right|_{0,\infty,\hat{T}}\right\}, \end{aligned} \tag{42.20}$$

and we likewise have an analogous inequality for the seminorm $|\hat{b}|_{2,\infty,\hat{T}}$. Using inequalities (42.9) and (42.10), we obtain

$$\left|\prod_{s\neq r}\partial_s F_{j(s)}\prod_{l\neq k}\partial_l F_{j(l)}\right|_{\lambda,\infty,\hat{T}}\leqslant Ch^{2n-2+\lambda},\quad \lambda=0,1,2, \tag{42.21}$$

and, using inequalities (42.11) and (42.12), we obtain

$$|J(F)^{-1}|_{\mu,\infty,\hat{T}}\leqslant Ch^{\mu-n},\quad \mu=0,1,2. \tag{42.22}$$

Therefore, combining inequalities (42.20), (42.21), (42.22) with the inequalities (cf. Theorems 37.1 and 37.2)

$$|\hat{a}|_{\nu,\infty,\hat{T}}\leqslant Ch^{\nu}\|a\|_{\nu,\infty,T},\quad \nu=0,1,2, \tag{42.23}$$

we eventually find that

$$|\hat{b}|_{1,\infty,\hat{T}}\leqslant Ch^{n-1}\|a\|_{1,\infty,T}. \tag{42.24}$$

By a similar analysis, we would find that

$$|\hat{b}|_{2,\infty,\hat{T}}\leqslant Ch^{n}\|a\|_{2,\infty,T}. \tag{42.25}$$

Then the inequalities (42.18), (42.19), (42.24), (42.25) together with (42.16) imply that

$$\left|\hat{E}\left(\{J(F)\}^{-1}\hat{a}\prod_{s\neq r}\partial_s F_{j(s)}\prod_{l\neq k}\partial_l F_{j(l)}\partial_r\hat{p}'\partial_k\hat{p}\right)\right|\leqslant Ch^2\|a\|_{2,\infty,T}\|p'\|_{2,T}|p|_{1,T}. \tag{42.26}$$

Adding up inequalities (42.26), we find that the expression $E_T(a\partial_i p'\partial_j p)$ (cf. (42.15)) satisfies an inequality similar to that of (42.26), and the proof is complete. □

REMARK 42.1. With the same assumptions as in Theorem 42.2, one can show that the estimates

$$|E_T(a\partial_i p'\partial_j p)|\leqslant Ch_T\|a\|_{2,\infty,T}\|p'\|_{1,T}|p|_{1,T}$$

hold. From these, another proof of the uniform V_h-ellipticity of the approximate bilinear forms can be deduced (this type of argument was used notably by ZLÁMAL [1974]).

THEOREM 42.3. *Let there be given a regular isoparametric family of quadratic n-simplices (T, P_T, Σ_T), let the quadrature scheme over the reference finite element be such that*

$$\hat{E}(\hat{\varphi})=0\quad \text{for all } \hat{\varphi}\in P_2(\hat{T}), \tag{42.27}$$

and finally, let $q\in[1,\infty]$ be any number that satisfies the inequality

$$2-n/q>0. \tag{42.28}$$

Then there exists a constant C independent of T such that

$$|E_T(fp)|\leqslant Ch_T^2\{\text{meas}(\tilde{T})\}^{1/2-1/q}\|f\|_{2,q,T}\|p\|_{1,T}\quad \text{for all } f\in W^{2,q}(T),\quad p\in P_T, \tag{42.29}$$

where $\tilde{T}$ denotes for each T the n-simplex with the same vertices as those of T.

PROOF. For all $f \in W^{2,q}(T)$ and all $p \in P_T$, we have

$$E_T(fp) = \hat{E}(\hat{f}\hat{p}J(F)), \tag{42.30}$$

and it follows from the proof of Theorem 28.3 (cf. (28.23) and (28.24)) that there exists a constant C such that

$$|\hat{E}(\hat{g}\hat{p})| \leqslant C(\{|\hat{g}|_{1,q,\hat{T}} + |\hat{g}|_{2,q,\hat{T}}\}|\hat{p}|_{1,\hat{T}} + |\hat{g}|_{2,q,\hat{T}}|\hat{p}|_{0,\hat{T}})$$
$$\text{for all } \hat{g} \in W^{2,q}(\hat{T}), \quad \hat{p} \in P_2(\hat{T}). \tag{42.31}$$

Letting

$$\hat{g} = \hat{f}J(F) \tag{42.32}$$

in inequality (42.31) and making use of inequalities (42.11), (42.12), and

$$|\hat{f}|_{\mu,q,\hat{T}} \leqslant C\{\text{meas}(\tilde{T})\}^{-1/q}h^{\mu}\|f\|_{\mu,q,T}, \quad \mu = 0,1,2$$

(cf. Theorems 37.1 and 37.2), we obtain

$$|\hat{f}J(F)|_{l,q,\hat{T}} \leqslant C\left\{\sum_{j=0}^{l} |J(F)|_{j,\infty,\hat{T}}|\hat{f}|_{l-j,q,\hat{T}}\right\}$$
$$\leqslant C\{\text{meas}(\tilde{T})\}^{-1/q}h^{n+l}\|f\|_{l,q,T}, \quad l = 1,2.$$

These last inequalities, combined with relations (42.30), (42.31), (42.32) and the first inequalities of (42.19) with $l = 0, 1$, yield inequality (42.29). □

43. Estimate of the error $\|\tilde{u} - u_h\|_{1,\Omega_h}$

We are now in a position to prove the main result of this chapter, which should be compared with Theorem 29.1. We recall that u is the solution of the variational problem corresponding to the data (38.1). For references concerning the existence of extensions such as $\tilde{u}$ and $\tilde{a}_{ij}$ below, see LIONS [1965, Chapter 2], NEČAS [1967, Chapter 2].

THEOREM 43.1. *Let $n \leqslant 5$, let (V_h) be a family of finite element spaces made up of isoparametric quadratic n-simplices forming a regular family, and let there be given a quadrature scheme on the reference finite element that satisfies*

$$\hat{E}(\hat{\varphi}) = 0 \quad \textit{for all } \hat{\varphi} \in P_2(\hat{T}). \tag{43.1}$$

Let $\tilde{\Omega}$ be an open set such that the inclusions

$$\Omega \subset \tilde{\Omega}, \qquad \Omega_h \subset \tilde{\Omega} \quad \textit{for all } h \tag{43.2}$$

hold, and such that the functions u and a_{ij}, $1 \leqslant i,j \leqslant n$, possess extensions $\tilde{u}$ and

$\tilde{a}_{ij}$, $1 \leqslant i, j \leqslant n$, *that satisfy*

$$\tilde{u} \in H^3(\tilde{\Omega}), \qquad \tilde{a}_{ij} \in W^{2,\infty}(\tilde{\Omega}), \quad 1 \leqslant i, j \leqslant n, \tag{43.3}$$

$$\tilde{f} = \sum_{i,j=1}^{n} \partial_j(\tilde{a}_{ij}\partial_i \tilde{u}) \in W^{2,q}(\tilde{\Omega}) \quad \textit{for some } q \geqslant 2 > n/q. \tag{43.4}$$

Then there exists a constant C independent of h such that

$$\|\tilde{u} - u_h\|_{1,\Omega_h} \leqslant Ch^2 \left\{ \|\tilde{u}\|_{3,\Omega} + \sum_{i,j=1}^{n} \|\tilde{a}_{ij}\|_{2,\infty,\Omega} \|\tilde{u}\|_{3,\Omega} + \|\tilde{f}\|_{2,q,\Omega} \right\}, \tag{43.5}$$

where $h = \max_{T \in \mathcal{T}_h} h_T$.

PROOF. By Theorem 41.1, the approximate bilinear forms are uniformly V_h-elliptic and therefore, we can use the abstract error estimate (40.3) of Theorem 40.1.

(i) Since $n \leqslant 5$, the inclusion $H^3(\tilde{\Omega}) \hookrightarrow \mathscr{C}^0(\{\tilde{\Omega}\}^-)$ holds, and by Theorem 42.1, the function $\Pi_h \tilde{u}$ belongs to the space X_{0h} since $u = \tilde{u} = 0$ on the boundary Γ. We may thus let $v_h = \Pi_h \tilde{u}$ in the term $\inf_{v_h \in V_h}\{\ldots\}$ that appears in the abstract error estimate. In this fashion, we obtain

$$\|\tilde{u} - u_h\|_{1,\Omega_h} \leqslant C \left\{ \|\tilde{u} - \Pi_h \tilde{u}\|_{1,\Omega_h} + \sup_{w_h \in V_h} \frac{|\tilde{a}_h(\Pi_h \tilde{u}, w_h) - a_h(\Pi_h \tilde{u}, w_h)|}{\|w_h\|_{1,\Omega_h}} \right.$$

$$\left. + \sup_{w_h \in V_h} \frac{|\tilde{a}_h(\tilde{u}, w_h) - l_h(w_h)|}{\|w_h\|_{1,\Omega_h}} \right\}, \tag{43.6}$$

and, by Theorem 42.1, we know that

$$\|\tilde{u} - \Pi_h \tilde{u}\|_{1,\Omega_h} \leqslant Ch^2 \|\tilde{u}\|_{3,\Omega_h} \leqslant Ch^2 \|\tilde{u}\|_{3,\Omega}. \tag{43.7}$$

(ii) To estimate the two consistency errors, a specific choice must be made of the functions $\tilde{a}_{ij}$ that appear in the bilinear form $\tilde{a}_h(\cdot,\cdot)$: *We shall choose precisely the functions given in* (43.3). Notice that, since the inclusion $W^{2,\infty}(\tilde{\Omega}) \hookrightarrow \mathscr{C}^1(\{\tilde{\Omega}\}^-)$ holds, the functions $\tilde{a}_{ij}$ are in particular defined everywhere on the set $\tilde{\Omega}$. Then we have, for all $w_h \in V_h$,

$$\tilde{a}_h(\Pi_h \tilde{u}, w_h) - a_h(\Pi_h \tilde{u}, w_h)$$

$$= \int_{\Omega_h} \sum_{i,j=1}^{n} \tilde{a}_{ij} \partial_i \Pi_h \tilde{u} \partial_j w_h \, dx - \sum_{T \in \mathcal{T}_h} \sum_{l=1}^{L} \omega_{l,T} \sum_{i,j=1}^{n} (a_{ij} \partial_i \Pi_h \tilde{u} \partial_j w_h)(b_{l,T}),$$

and, since all the quadrature nodes $b_{l,T}$ belong to the set $\tilde{\Omega}$, we have $a_{ij}(b_{l,T}) = \tilde{a}_{ij}(b_{l,T})$. Consequently, we can rewrite the above expression as

$$\tilde{a}_h(\Pi_h \tilde{u}, w_h) - a_h(\Pi_h \tilde{u}, w_h) = \sum_{T \in \mathcal{T}_h} \sum_{i,j=1}^{n} E_T(\tilde{a}_{ij} \partial_i \Pi_h \tilde{u} \partial_j w_h).$$

Using the estimates of Theorem 42.2 and Cauchy–Schwarz inequality, we obtain

$$|\tilde{a}_h(\Pi_h\tilde{u}, w_h) - a_h(\Pi_h\tilde{u}, w_h)|$$

$$\leqslant C \sum_{T\in\mathcal{T}_h} h_T^2 \sum_{i,j=1}^{n} \|\tilde{a}_{ij}\|_{2,\infty,T}\|\Pi_T\tilde{u}\|_{2,T}|w_h|_{1,T}$$

$$\leqslant Ch^2\left\{\sum_{i,j=1}^{n} \|\tilde{a}_{ij}\|_{2,\infty,\tilde{\Omega}}\right\}\left\{\sum_{T\in\mathcal{T}_h} \|\Pi_T\tilde{u}\|_{2,T}^2\right\}^{1/2} |w_h|_{1,\Omega_h}.$$

Another application of Theorem 42.1 yields

$$\left\{\sum_{T\in\mathcal{T}_h} \|\Pi_T\tilde{u}\|_{2,T}^2\right\}^{1/2} \leqslant \|\tilde{u}\|_{2,\Omega_h} + \left\{\sum_{T\in\mathcal{T}_h} \|\tilde{u}-\Pi_T\tilde{u}\|_{2,T}^2\right\}^{1/2}$$

$$\leqslant \|\tilde{u}\|_{2,\Omega_h} + Ch\|\tilde{u}\|_{3,\Omega_h} \leqslant C\|\tilde{u}\|_{3,\tilde{\Omega}},$$

and thus, we have shown that

$$\sup_{w_h\in V_h} \frac{|\tilde{a}_h(\Pi_h\tilde{u}, w_h) - a_h(\Pi_h\tilde{u}, w_h)|}{\|w_h\|_{1,\Omega_h}} \leqslant Ch^2 \sum_{i,j=1}^{n} \|\tilde{a}_{ij}\|_{2,\infty,\tilde{\Omega}}\|\tilde{u}\|_{3,\tilde{\Omega}}. \tag{43.8}$$

(iii) Let us next estimate the numerator of the second consistency error. First it is easily verified that assumptions (43.3) imply in particular that the functions $(\tilde{a}_{ij}\partial_i\tilde{u})$ belong to the space $H^1(\tilde{\Omega})$. Therefore Green's formula yields

$$\tilde{a}_h(\tilde{u}, w_h) = \int_{\Omega_h} \sum_{i,j=1}^{n} \tilde{a}_{ij}\partial_i\tilde{u}\partial_j w_h \,dx$$

$$= -\int_{\Omega_h} \sum_{i,j=1}^{n} \partial_j(\tilde{a}_{ij}\partial_i\tilde{u})w_h \,dx = \int_{\Omega_h} \tilde{f}w_h \,dx \quad \text{for all } w_h\in V_h \subset H_0^1(\Omega_h).$$

Since

$$-\sum_{i,j=1}^{n} \partial_j(\tilde{a}_{ij}\partial_i\tilde{u}) = -\sum_{i,j=1}^{n} \partial_j(a_{ij}\partial_i u) = f \quad \text{on } \Omega,$$

the function $\tilde{f}$ given in (43.4) *is an extension of the function f*. Besides, using once again the fact that all integration nodes $b_{l,T}$ belong to the set $\tilde{\Omega}$, we infer that $f(b_{l,T}) = \tilde{f}(b_{l,T})$; consequently, we can write

$$\tilde{a}_h(\tilde{u}, w_h) - l_h(w_h) = \int_{\Omega_h} \tilde{f}w_h \,dx - \sum_{T\in\mathcal{T}_h} \sum_{l=1}^{L} \omega_{l,T}(fw_h)(b_{l,T})$$

$$= \sum_{T\in\mathcal{T}_h} E_T(\tilde{f}w_h),$$

and the estimates of Theorem 42.3 imply

$$|\tilde{a}_h(\tilde{u}, w_h) - l_h(w_h)| \leqslant C \sum_{T\in\mathcal{T}_h} h_T^2 \{\operatorname{meas}(\tilde{T})\}^{1/2-1/q} \|\tilde{f}\|_{2,q,T} \|w_h\|_{1,T}$$

$$\leqslant Ch^2 \left\{ \sum_{T\in\mathcal{T}_h} \operatorname{meas}(\tilde{T}) \right\}^{1/2-1/q} \|\tilde{f}\|_{2,q,\Omega_h} \|w_h\|_{1,\Omega_h}.$$

By construction, the interiors of the n-simplices do not overlap. Hence the quantity

$$\sum_{T\in\mathcal{T}_h} \operatorname{meas}(\tilde{T}) = \operatorname{meas}\left(\bigcup_{T\in\mathcal{T}_h} \tilde{T} \right)$$

is bounded independently of h. Thus, we have shown that

$$\sup_{w_h\in V_h} \frac{|\tilde{a}_h(u, w_h) - l_h(w_h)|}{\|w_h\|_{1,\Omega_h}} \leqslant Ch^2 \|\tilde{f}\|_{2,q,\Omega}, \tag{43.9}$$

and inequality (43.5) follows from inequalities (43.6), (43.7), (43.8) and (43.9). □

We have therefore reached a remarkable conclusion: *In order to retain the same order of convergence as in the case of a polygonal domain (when only straight finite elements are used), the same quadrature scheme should be used, whether it be for straight or for isoparametric finite elements.* Thus, if $n = 2$ for instance, we can use the quadrature scheme of (25.18), which is exact for polynomials of degree $\leqslant 2$.

REMARK 43.1. As expected, the order of convergence in the absence of numerical integration is the same, i.e., one has $\|\tilde{u} - \tilde{u}_h\|_{1,\Omega_h} = \mathrm{O}(h^2)$, where $\tilde{u}_h$ is now the solution of the discrete problem (38.6). The corresponding error analysis, which relies again on the abstract error estimate of Theorem 40.1, was carried out in SCOTT [1973a].

To complete this study, it remains to show that for a given domain with a curved boundary (irrespectively of whether or not numerical integration is used), isoparametric quadratic n-simplices do yield better estimates than their straight counterparts!

More specifically, assume that the set Ω is a bounded convex domain in $\mathbb{R}^2$ (Fig. 43.1). Given a triangulation $\mathcal{T}_h$ made up only of "straight" triangles as indicated in Fig. 43.1(a), let X_h denote the finite element space whose generic finite element is the quadratic triangle, and let $V_h = \{v_h \in X_h;\ v_h = 0 \text{ on } \Gamma_h\}$, where Γ_h is the boundary of the set $\bigcup_{T\in\mathcal{T}_h} T$.

Then one can show (cf. STRANG and BERGER [1971], THOMÉE [1973b]; see also STRANG and FIX [1973, chapter 4]) that

$$\|u - u_h\|_{1,\Omega_h} = \mathrm{O}(h^{3/2}),$$

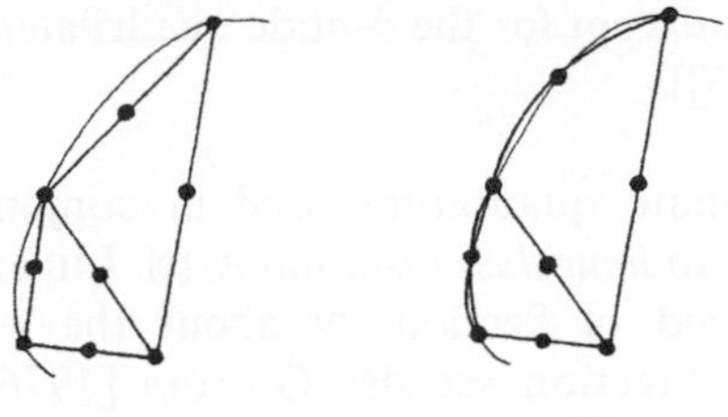

FIG. 43.1. Approximation of a domain in $\mathbb{R}^3$ with straight (a) or isoparametric (b) quadratic triangles.

where $u_h \in V_h$ now satisfies

$$\int_\Omega \sum_{i,j=1}^n a_{ij}\partial_i u_h \partial_j v_h \,dx = \int_\Omega f v_h \,dx \quad \text{for all } v_h \in V_h$$

(note that in this case, the X_h-interpolant of the solution u does *not* belong to the space V_h). This shows that *finite element spaces constructed with isoparametric quadratic triangles, as in* Fig. 43.1(b), *yields better error estimates.*

REMARK 43.2. By contrast with the case of straight finite elements (cf. Remark 29.1) *the integrals* $\int_T a_{ij}\partial_i u_h \partial_j v_h \,dx$ *are no longer computed exactly when the coefficients* a_{ij} *are constant functions.* To see this, let (T, P_T, Σ_T) be an isoparametric quadratic n-simplex; then we have

$$\int_T \partial_i p' \partial_j p \,dx = \int_T J(F)\{\partial_i p'\}\hat{}\{\partial_j p\}\hat{}\,d\hat{x} \quad \text{for all } p', p \in P_T,$$

and (cf. (42.14)),

$$\begin{aligned} J(F)(\hat{x})\{\partial_i p'\}\hat{}(\hat{x}) &= \sum_{k=1}^n \partial_k \hat{p}'(\hat{x}) \det(\partial_1 F(\hat{x}), \ldots, \partial_{k-1}F(\hat{x}), e_i, \partial_{k+1}F(\hat{x}), \ldots, \partial_n F(\hat{x})) \\ &= \{\text{polynomial of degree} \leqslant n \text{ in } \hat{x}\}, \\ \{\partial_j p\}\hat{}(x) &= \{J(F)(\hat{x})\}^{-1} \times \{\text{polynomial of degree} \leqslant n \text{ in } \hat{x}\}. \end{aligned}$$

Since

$$J(F)(\hat{x}) = \det(\partial_1 F(\hat{x}), \ldots, \partial_n F(\hat{x})) = \{\text{polynomial of degree} \leqslant n \text{ in } \hat{x}\},$$

it eventually follows that

$$\int_T \partial_i p' \partial_j p \,dx = \int_{\hat{T}} \frac{\{\text{polynomial of degree } \leqslant 2n \text{ in } \hat{x}\}}{\{\text{polynomial of degree } \leqslant n \text{ in } \hat{x}\}}\,d\hat{x}.$$

The exact computation of such integrals would thus require a quadrature scheme exact for rational functions of the form N/D with $N \in P_{2n}(\hat{T})$, $D \in P_n(\hat{T})$, and such schemes are not known, except for the 4-node quadrilateral (cf. BABU and PINDER [1984], MIZUKAMI [1986]).

REMARK 43.3. Approximate quadrature used in conjunction with the 4-node quadrilateral often leads to *hourglass instabilities* (cf. LIU and BELYTSCHKO [1984]), already noted at the end of Section 29 about the general phenomenon of *underintegration*. In this direction, see also GIRAULT [1976], who has nevertheless shown that 4-node quadrilaterals may be used in conjunction with a one-node quadrature scheme, even though the uniform V_h-ellipticity no longer holds.

As already mentioned, the error analysis developed in this chapter follows the general approach set up in CIARLET and RAVIART [1972c] (however it was thought at that time that more accurate schemes were needed for isoparametric elements), where an estimate of the error in the norm $|\cdot|_{0,\Omega}$ was also obtained. See also NEDOMA [1979] and LENOIR [1986], for further results in the same spirit, and, especially, WAHLBIN [1978], who establishes the *uniform convergence*.

An analogous study was made by ZLÁMAL [1974], who showed that, for two-dimensional curved elements for which $\hat{P} = P_k(\hat{T})$, k even, it is sufficient to use quadrature schemes exact for polynomials of degree $\leqslant 2k-2$, in order to retain the $O(h^k)$ convergence in the norm $\|\cdot\|_{1,\Omega_h}$. ZLÁMAL [1973b] has also evaluated the error in the absence of numerical integration; for complementary results, see VEIDINGER [1975]. Likewise, SCOTT [1973a] has shown that quadrature schemes of higher order of accuracy are not needed when curved finite elements are used. However, the finite elements considered by Zlámal and Scott are *not* of the isoparametric type as understood here. For such elements, a general theory is yet to be developed, in particular for quadrilateral finite elements.

Zlámal's method has been further improved by ŽENÍŠEK [1981a, 1981b, 1987], who was able in particular to handle various types of nonhomogeneous boundary conditions, then by LENOIR [1986], who extended it to an arbitrary space dimension, and finally by BERNARDI [1986, Chapter 6], who was in addition able to handle *nonconforming curved elements*. See also ČERMÁK [1983a, 1983b] for related ideas.

Alternate ways of handling Dirichlet problems posed over domains with curved boundaries have been proposed, which rely on various alterations of the bilinear form of the given problem. In this direction, we notably mention:

(i) penalty methods, as advocated by AUBIN [1969] and BABUŠKA [1973b], and later improved by KING [1976];

(ii) methods where the boundary condition is considered as a constraint and as such is treated via techniques from duality theory, as in BABUŠKA [1973a];

(iii) least square methods as proposed and studied in BRAMBLE and SCHATZ [1970, 1971], BRAMBLE and NITSCHE [1973], BAKER [1973];

(iv) methods where the domain is approximated by a polygonal domain, as in BRAMBLE, DUPONT and THOMÉE [1972];

(v) various methods proposed by NITSCHE [1971, 1976c].

Finite element approximations of *Neumann problems* posed over domains with curved boundaries have been studied by BARRETT and ELLIOTT [1984, 1987],

MOLCHANOV and GALBA [1985]. References about *nonconforming* isoparametric *quadrilateral* elements have been given at the end of Section 34. For additional references for the finite element approximation of boundary value problems posed over domains with curved boundaries, see BABUŠKA [1971b], BERGER [1973], BERGER, SCOTT and STRANG [1972], BLAIR [1976], BRAMBLE [1975], NITSCHE [1972b], SCOTT [1975], SHAH [1970], STRANG and BERGER [1971], STRANG and FIX [1973, Chapter 4], THOMÉE [1973a, 1973b].

We finally mention that, following the terminology of STRANG [1972b], we have perpetrated in the last three chapters three *variational crimes*, by using numerical integration, nonconforming finite elements, and approximate boundaries.

CHAPTER VII

Finite Element Methods for Fourth-Order Problems

Introduction

In this chapter, we study two commonly used finite element approximations of the plate problem on polygonal domains.

To begin with, we consider various *conforming methods*, which thus require *straight finite elements of class* $\mathscr{C}^1$. Although such finite elements cannot be imbedded in affine families in general, we show that, under very reasonable assumptions, they form *almost-affine families* (Section 44), in the sense that if the associated P_T-interpolation operator Π_T leaves invariant the space $P_k(T)$, there exists a constant C independent of T such that

$$|v-\Pi_T v|_{m,T} \leqslant C h_T^{k+1-m} |v|_{k+1,T} \quad \text{for all } v \in H^{k+1}(T),$$

for all integers $m \leqslant k+1$ for which $P_T \subset H^m(T)$. This property is shared not only by the finite elements of class $\mathscr{C}^1$ introduced in Section 9, such as the Argyris triangle (Section 45), but also by *composite finite elements*, such as the Hsieh–Clough–Tocher triangle (Section 46), and by *singular finite elements*, such as the singular Zienkiewicz triangle (Section 47).

For finite element spaces made up of such almost-affine families, we obtain (Theorem 48.1) error estimates of the form

$$\|u-u_h\|_{2,\Omega} \leqslant C\|u-\Pi_h u\|_{2,\Omega} = \mathrm{O}(h^{k-1}), \quad h = \max_{T\in\mathscr{T}_h} h_T,$$

by an application of Céa's lemma. We also show (Theorem 48.2) that the minimal assumptions "$u \in H^2(\Omega)$" and "$P_2(T) \subset P_T$, $T \in \mathscr{T}_h$" insure convergence, i.e., $\lim_{h\to 0} \|u-u_h\|_{2,\Omega} = 0$.

The actual implementation of conforming methods offers serious computational difficulties: Either the dimension of the "local" spaces P_T is fairly large (at least 18 for triangular polynomial elements) or the structure of the space P_T is complicated (cf. the Hsieh–Clough–Tocher triangle or the singular Zienkiewicz triangle for example). The basic source of these difficulties is of course the required continuity of the first-order partial derivatives across adjacent finite elements.

It is therefore tempting to relax this continuity requirement, and this results in *nonconforming methods*: One looks for a discrete solution in a finite element space V_h that is no longer contained in the space $H^2(\Omega)$, or not even in the space $H^1(\Omega)$ in some cases. The discrete solution then satisfies $a_h(u_h, v_h) = l(v_h)$ for all $v_h \in V_h$, where

$$a_h(\cdot,\cdot) = \sum_{T \in \mathcal{T}_h} \int_T \{\cdots\} \, dx,$$

the integrand $\{\cdots\}$ being the same as in the bilinear form of the original problem.

The analysis of such nonconforming methods follows exactly the same pattern as that of nonconforming methods for second order problems (cf. Chapter V). We concentrate our study here on one example, where the generic finite element is the *Adini–Clough–Melosh rectangle*, in which case we show that (Theorem 50.1)

$$\left\{ \sum_{T \in \mathcal{T}_h} |u - u_h|_{2,T}^2 \right\}^{1/2} = O(h),$$

if the solution u is in the space $H^3(\Omega)$.

The very important *mixed* and *hybrid finite element methods* for approximating plate problems, or other fourth-order problems, are treated in depth in the article by Roberts and Thomas.

There are numerous publications in the engineering literature about the various finite elements that can be employed for solving plate problems. Rather than giving a necessarily incomplete list, we refer the reader to the two "review" papers of BATOZ, BATHE and HO [1980] and HRABOK and HRUDEY [1984].

44. Conforming methods for fourth-order problems: Almost-affine families of finite elements

To begin with, we study several types of *conforming finite element methods* commonly used for approximating the solution of plate problems. For definiteness, we consider the *clamped plate problem*, which corresponds to the following data (cf. Section 4):

$$V = H_0^2(\Omega), \quad \Omega \subset \mathbb{R}^2,$$

$$a(u,v) = \int_\Omega \{\Delta u \Delta v + (1-\nu)(2\partial_{12}u\partial_{12}v - \partial_{11}u\partial_{22}v - \partial_{22}u\partial_{11}v)\} \, dx, \tag{44.1}$$

$$l(v) = \int_\Omega fv \, dx, \quad f \in L^2(\Omega),$$

where the constant ν (the Poisson coefficient of the material constituting the plate) lies in the interval $]0, \frac{1}{2}[$.

The conforming methods that we shall describe apply equally well to *any* fourth-order boundary value problem posed over a space V such as $H_0^2(\Omega)$, $H^2(\Omega) \cap H_0^1(\Omega)$, $H^2(\Omega)$, whose data $a(\cdot,\cdot)$ and $l(\cdot)$ satisfy the assumptions of the Lax–Milgram lemma.

For instance, we could likewise consider the simply supported plate or the biharmonic problem. By contrast, the *nonconforming methods* studied at the end of this chapter are specifically adapted to plate problems (cf. Remark 49.1).

We assume that the set Ω is polygonal, so that it may be covered by triangulations composed of polygonal sets. In order to develop a *conforming method*, we face the problem of constructing subspaces of the space $H^2(\Omega)$. Since the functions found in standard finite element spaces are "locally regular" ($P_T \subset H^2(T)$ for all $T \in \mathcal{T}_h$), this construction amounts in practice to finding finite element spaces X_h that satisfy the inclusion $X_h \subset \mathscr{C}^1(\bar{\Omega})$ (cf. Theorem 5.2), i.e., whose finite elements are of class $\mathscr{C}^1$.

REMARK 44.1. A thorough discussion about the use of finite elements of class $\mathscr{C}^1$ from the engineering viewpoint is given in ZIENKIEWICZ [1971, Chapter 10]. There, finite elements of class $\mathscr{C}^1$ are called *compatible*, while finite elements that are not of class $\mathscr{C}^1$ are called *incompatible*.

We have already described three finite elements that are of class $\mathscr{C}^1$, the *Argyris triangle* (cf. Fig. 9.1), the *Bell triangle* (cf. Fig. 9.2), and the *Bogner–Fox–Schmit rectangle* (cf. Fig. 9.4).

As we pointed out in Section 11, Argyris triangles or Bell triangles *cannot* be imbedded in affine families in general, because normal derivatives at some nodes are used either as degrees of freedom (for the Argyris triangle) or in the definition of the space P_T (for the Bell triangle). This is in general the rule for finite elements of class $\mathscr{C}^1$, but there are exceptions. For instance, the Bogner–Fox–Schmit rectangle is a rectangular finite element of class $\mathscr{C}^1$ that *can* be imbedded in an affine family.

Nevertheless, if most finite elements of class $\mathscr{C}^1$ do not form affine families, we shall show that their interpolation properties are quite similar to those of affine families, and it is this similarity that motivates the following definition (compare with Theorem 16.1).

DEFINITION 44.1. Consider a family of finite elements (T, P_T, Σ_T) of a given type, and let s denote the greatest order of partial derivatives occurring in the definition of the set Σ_T. Then such a family is said to be *almost-affine* if, for any integers $k, m \geqslant 0$ and any numbers $p, q \in [1, \infty]$ compatible with the following inclusions:

$$W^{k+1,p}(T) \hookrightarrow \mathscr{C}^s(T), \tag{44.2}$$

$$W^{k+1,p}(T) \hookrightarrow W^{m,q}(T), \tag{44.3}$$

$$P_k(T) \subset P_T \subset W^{m,q}(T), \tag{44.4}$$

there exists a constant C independent of T such that

$$\begin{aligned} &\| v - \Pi_T v \|_{m,q,T} \leqslant C\{\text{meas}(T)\}^{1/q - 1/p} h_T^{k+1-m} |v|_{k+1,p,T} \\ &\quad \text{for all } v \in W^{k+1,p}(T), \end{aligned} \tag{44.5}$$

where Π_T denotes as usual the associated P_T-interpolation operator, and $h_T = \text{diam}(T)$.

In order to simplify the exposition, we shall consider in the subsequent examples *only the highest possible value* of the integer k for which the inclusions $W^{k+1,p}(T) \hookrightarrow \mathscr{C}^s(T)$ and $P_k(T) \subset P_T$ are satisfied, but it is implicitly understood that any lower value of k compatible with these two inclusions is also admissible.

By definition, *a regular affine family is almost-affine*. In particular then, *a regular family of Bogner–Fox–Schmit rectangles* (T, P_T, Σ_T) (cf. Fig. 9.4) *is almost-affine*. We recall that the set T is a rectangle with vertices a_i, $1 \leqslant i \leqslant 4$, $P_T = Q_3(T)$, and

$$\Sigma_T = \{p(a_i), \partial_i p(a_i), \partial_2 p(a_i), \partial_{12} p(a_i) \colon 1 \leqslant i \leqslant 4\}.$$

Hence, *for all* $p \in]1, \infty]$ (the inequality $p > 1$ implies the inclusion $W^{4,p}(T) \hookrightarrow \mathscr{C}^2(T) = \operatorname{dom} \Pi_T$) *and all pairs* (m, q) *with* $m \geqslant 0$ *and* $q \in [1, \infty]$ *compatible with the inclusion*

$$W^{4,p}(T) \hookrightarrow W^{m,q}(T), \tag{44.6}$$

there exists a constant C independent of T such that

$$\| v - \Pi_T v \|_{m,q,T} \leqslant C \{\operatorname{meas}(T)\}^{1/q - 1/p} h_T^{4-m} |v|_{4,p,T} \quad \text{for all } v \in W^{4,p}(T). \tag{44.7}$$

REMARK 44.2. The Bogner–Fox–Schmit rectangle is not the only rectangular finite element of class $\mathscr{C}^1$ that may be used in practice. See for instance GOPALACHARYULU [1973].

45. Examples of polynomial finite elements of class $\mathscr{C}^1$

Let us next examine the *Argyris triangle* (cf. Fig. 9.1). We recall that this finite element is a triple (T, P_T, Σ_T) where the set T is a triangle with vertices a_i, $1 \leqslant i \leqslant 3$, and midpoints $a_{ij} = \frac{1}{2}(a_i + a_j)$, $1 \leqslant i < j \leqslant 3$, of the sides, the space P_T is the space $P_5(T)$, and the set Σ_T (whose $P_5(T)$-unisolvence has been proved in Theorem 9.1) can be chosen in the form

$$\Sigma_T = \{\partial^\alpha p(a_i) \colon 1 \leqslant i \leqslant 3, |\alpha| \leqslant 2;\ \partial_\nu p(a_{ij}) \colon 1 \leqslant i < j \leqslant 3\}.$$

The *basis functions* of the Argyris triangle are explicitly computed in BERNADOU and BOISSERIE [1982, pp. 69ff].

The first interpolation error estimates for the Argyris triangle are due to ZLÁMAL [1968], who obtained estimates of the norms $\| v - \Pi_T v \|_{\mathscr{C}^m(T)}$; the results and methods of Zlámal were later extended by ŽENÍŠEK [1970] to finite element spaces that satisfy the inclusion $X_h \subset \mathscr{C}^m(\bar{\Omega})$. BRAMBLE and ZLÁMAL [1970] then obtained estimates in Sobolev norms, which are contained in the next estimates. The method of proof that we follow here is due to CIARLET [1976].

THEOREM 45.1. *A regular family of Argyris triangles is almost-affine: For all* $p \in [1, \infty]$ *and all pairs* (m, q) *with* $m \geqslant 0$ *and* $q \in [1, \infty]$ *compatible with the inclusion*

$$W^{6,p}(T) \hookrightarrow W^{m,q}(T), \tag{45.1}$$

there exists a constant C independent of T such that

$$\| v-\Pi_T v \|_{m,q,T} \leqslant C\{\text{meas}(T)\}^{1/q-1/p} h_T^{6-m} |v|_{6,p,T} \quad \textit{for all } v\in W^{6,p}(T), \tag{45.2}$$

where Π_T denotes the associated $P_5(T)$-interpolation operator.

PROOF. The key idea is to introduce a finite element similar to the Argyris triangle, but which *can* be imbedded in an affine family, and which will play a crucial intermediary role in obtaining the interpolation error estimate. Since the degrees of freedom $\partial_\nu p(a_{ij})$, $1\leqslant i<j\leqslant 3$, are those that prevent the property of affine equivalence, we naturally introduce the *$\mathscr{C}^0$-quintic Hermite triangle*, as described in Fig. 45.1. For notational convenience, we shall henceforth denote by b_i the midpoint of the side that does not contain the vertex a_i, $1\leqslant i\leqslant 3$. Arguments similar to those used in the proof of Theorem 9.1 then show that the set Ξ_T is $P_5(T)$-unisolvent and that this is a finite element of class $\mathscr{C}^0$, but not of class $\mathscr{C}^1$.

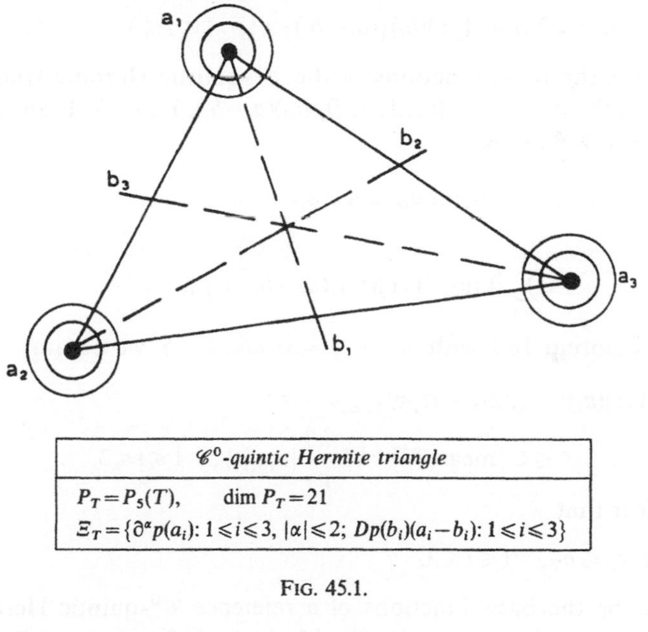

$\mathscr{C}^0$-quintic Hermite triangle
$P_T=P_5(T)$, dim $P_T=21$ $\Xi_T=\{\partial^\alpha p(a_i)\colon 1\leqslant i\leqslant 3,\ \lvert\alpha\rvert\leqslant 2;\ Dp(b_i)(a_i-b_i)\colon 1\leqslant i\leqslant 3\}$

FIG. 45.1.

In addition, it is clear that *two arbitrary $\mathscr{C}^0$-quintic Hermite triangles are affine-equivalent*. Therefore, if Λ_T denotes the associated $P_5(T)$-interpolation operator, for all $p\in[1,\infty]$ and all pairs (m,q) with $0\leqslant m\leqslant 6$ and $q\in[1,\infty]$ such that $W^{6,p}(T)\hookrightarrow W^{m,q}(T)$, there exists a constant C independent of T such that

$$|v-\Lambda_T v|_{m,q,K} \leqslant C\{\text{meas}(T)\}^{1/q-1/p} h_T^{6-m} |v|_{6,p,T} \quad \text{for all } v\in W^{6,p}(T). \tag{45.3}$$

It thus remains to estimate the seminorms $|\Pi_T v-\Lambda_T v|_{m,q,T}$, where v is any function in $W^{6,p}(T)$. The difference

$$\Delta=\Pi_T v-\Lambda_T v \tag{45.4}$$

is a polynomial of degree $\leqslant 5$ that satisfies

$$\partial^\alpha \Delta(a_i)=0, \quad |\alpha|\leqslant 2, \quad 1\leqslant i\leqslant 3, \tag{45.5}$$

since $\partial^\alpha \Pi_T v(a_i)=\partial^\alpha \Lambda_T v(a_i)=\partial^\alpha v(a_i), |\alpha|\leqslant 2, 1\leqslant i\leqslant 3$, and

$$\partial_\nu \Delta(b_i)=\partial_\nu(v-\Lambda_T v)(b_i), \quad 1\leqslant i\leqslant 3, \tag{45.6}$$

since $\partial_\nu \Pi_T v(b_i)=\partial_\nu v(b_i), 1\leqslant i\leqslant 3$.

For $1\leqslant i\leqslant 3$, let ν_i and τ_i be the unit outer normal and tangential vectors along the side opposite to the vertex a_i. Denoting by $\cdot$ the Euclidean inner product in $\mathbb{R}^2$, we can write, for $1\leqslant i\leqslant 3$,

$$D\Delta(b_i)(a_i-b_i)=\partial_\nu \Delta(b_i)\{(a_i-b_i)\cdot \nu_i\}, \tag{45.7}$$

since on the one hand $D\Delta(b_i)\nu_i=\partial_\nu \Delta(b_i)$, and since on the other $D\Delta(b_i)\tau_i=0$ as a consequence of relations (45.5), which imply that the difference Δ vanishes along each side of the triangle. Combining relations (45.6) and (45.7), we obtain

$$D\Delta(b_i)(a_i-b_i)=\partial_\nu(v-\Lambda_T v)(b_i)\{(a_i-b_i)\cdot \nu_i\}, \quad 1\leqslant i\leqslant 3. \tag{45.8}$$

Let q_i denote the basis functions of the $\mathscr{C}^0$-quintic Hermite triangle that are associated with the degrees of freedom $Dp(b_i)(a_i-b_i), 1\leqslant i\leqslant 3$. Then using relations (45.5) and (45.8), we can write

$$\begin{aligned}\Delta=\Pi_T v-\Lambda_T v&=\sum_{i=1}^{3}\{D\Delta(b_i)(a_i-b_i)\}q_i\\ &=\sum_{i=1}^{3}\partial_\nu(v-\Lambda_T v)(b_i)\{(a_i-b_i)\cdot\nu_i\}q_i.\end{aligned} \tag{45.9}$$

Applying Theorem 16.1 with $m=1$, $q=\infty$ and $k=5$, we obtain

$$\begin{aligned}|\partial_\nu(v-\Lambda_T v)(b_i)|&\leqslant\sqrt{2}|v-\Lambda_T v|_{1,\infty,T}\\ &\leqslant C\{\text{meas}(T)\}^{-1/p}\frac{h_T^6}{\rho_T}|v|_{6,p,T}, \quad 1\leqslant i\leqslant 3.\end{aligned} \tag{45.10}$$

Next, it is clear that

$$|(a_i-b_i)\cdot\nu_i|\leqslant h_T, \quad 1\leqslant i\leqslant 3. \tag{45.11}$$

Finally, let $\hat{q}_i$ be the basis functions of a reference $\mathscr{C}^0$-quintic Hermite triangle, associated in the usual correspondence with the basis functions q_i. From Theorems 15.1 and 15.2, we infer that

$$|q_i|_{m,q,T}\leqslant C\frac{\{\text{meas}(T)\}^{1/q}}{\rho_T^m}|\hat{q}_i|_{m,q,\hat{T}}. \tag{45.12}$$

Relations (45.9)–(45.12) then imply that

$$\begin{aligned}|\Pi_T v-\Lambda_T v|_{m,q,T}&\leqslant C\{\text{meas}(T)\}^{1/q-1/p}\frac{h_T^7}{\rho_T^{m+1}}|v|_{6,p,T}\\ &\leqslant C\{\text{meas}(T)\}^{1/q-1/p}h_T^{6-m}|v|_{6,p,T},\end{aligned} \tag{45.13}$$

since we are considering a regular family, and inequality (45.2) follows from inequalities (45.3) and (45.13). □

By a similar method, we could likewise show that *a regular family of Bell triangles* (cf. Fig. 9.2) *is almost-affine*, with the value $k=4$ in the corresponding inequality (44.5). The *basis functions of the Bell triangle* are explicitly computed in MITCHELL [1973], BARNHILL and FARIN [1981].

46. Examples of composite finite elements of class $\mathscr{C}^1$

For the first time in this article, we leave the realm of "purely polynomial" finite elements.

As we already pointed out in Section 9, Bell's triangle is optimal among triangular polynomial finite elements of class $\mathscr{C}^1$, since $\dim P_T \geqslant 18$ for such finite elements, as a consequence of *Ženíšek's result*. Therefore, a smaller dimension of the space P_T for triangular finite elements of class $\mathscr{C}^1$ requires that *functions other than polynomials* be used.

For example, one can use piecewise polynomials inside the set T, a process that is the basis for constructing *composite finite elements*, also named *macro-elements*. Or one can add some judiciously selected *rational functions* to a space of polynomials, a process that is the basis for constructing *singular finite elements* (singular in the sense that some functions in the space P_T or some of their derivatives become infinite, or are not defined at some points of T). We shall describe and study in detail one example of each type, and then briefly mention other examples of composite and singular finite elements.

The *Hsieh–Clough–Tocher triangle*, sometimes abbreviated as the *HCT triangle*, is defined as follows: The set T is a triangle subdivided into three triangles T_i with vertices a, a_{i+1}, a_{i+2}, $1 \leqslant i \leqslant 3$ (Fig. 46.1), where a is an arbitrary point in the interior of the set T (here and subsequently, the indices are counted modulo 3 when necessary). The space P_T and the set Σ_T are indicated in Fig. 46.1. For convenience, we again denote by b_i, $1 \leqslant i \leqslant 3$, the midpoint of the side that does not contain the vertex a_i. The Hsieh–Clough–Tocher triangle appeared in CLOUGH and TOCHER [1965]. It is also named after Hsieh who was the first to conceive in 1962 the idea of matching three polynomials in order to get a finite element of class $\mathscr{C}^1$.

Our first task is as usual to prove the P_T-unisolvence of the set Σ_T. Since $\dim P_3(T_i) = 10$, it is necessary to find 30 equations to define the three polynomials $p|_{T_i}$, $1 \leqslant i \leqslant 3$. First, it is easily seen that the degrees of freedom of the set Σ_T provide 21 equations. To see that the condition "$p \in \mathscr{C}^1(T)$" yields 9 additional equations, it suffices to express that the functions p, $\partial_1 p$ and $\partial_2 p$ are continuous at the point a (6 equations) and that the normal derivatives are continuous across the midpoints of the sides $[a, a_i]$ (3 equations). It therefore remains to show that the 30×30 matrix of the corresponding linear system is invertible, and this is the object of the next theorem.

The first proof of unisolvence that did not use the basis functions was given by

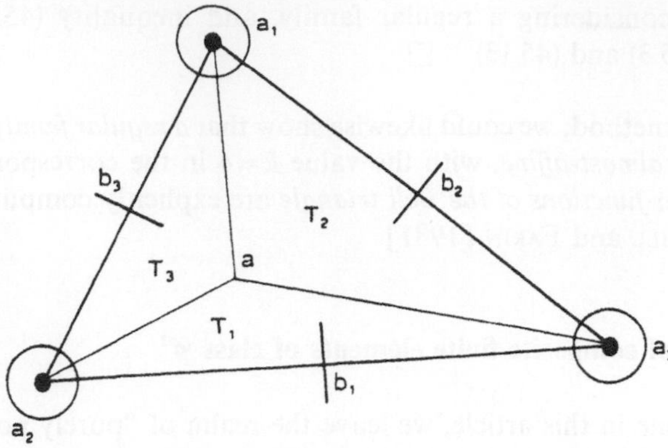

Hsieh–Clough–Tocher triangle	
$P_T=\{p\in\mathscr{C}^1(T); p\vert_{T_i}\in P_3(T_i),\ 1\leqslant i\leqslant 3\}$,	$\dim P_T=12$
$\Sigma_T=\{p(a_i), \partial_1 p(a_i), \partial_2 p(a_i), \partial_\nu p(b_i)\colon 1\leqslant i\leqslant 3\}$	

FIG. 46.1.

CIARLET [1974c]. The proof given here is due to PERCELL [1976]; see also DOUGLAS, DUPONT, PERCELL and SCOTT [1979]. Another proof consists in explicitly computing its *basis functions*, as in MITCHELL [1973], BERNADOU and HASSAN [1981], BERNADOU and BOISSERIE [1982, pp. 75ff].

THEOREM 46.1. *Let the triple (T, P_T, Σ_T) be as in Fig. 46.1. Then the set Σ_T is P_T-unisolvent. The resulting Hsieh–Clough–Tocher triangle is a finite element of class $\mathscr{C}^1$.*

PROOF. It suffices to show that a function p in the space P_T vanishes if

$$p(a_i)=\partial_1 p(a_i)=\partial_2 p(a_i)=\partial_\nu p(b_i)=0, \quad 1\leqslant i\leqslant 3. \tag{46.1}$$

For $1\leqslant i\leqslant 3$, let μ_i denote the unique function that satisfies

$$\mu_i\in P_1(T_i), \qquad \mu_i(a)=1, \qquad \mu_i(a_{i+1})=\mu_i(a_{i+2})=0.$$

Hence the function $\mu: T\to\mathbb{R}$ defined by

$$\mu|_{T_i}=\mu_i, \quad 1\leqslant i\leqslant 3,$$

is continuous. Since the function $p|_{T_i}$ is a polynomial of degree $\leqslant 3$ over each triangle T_i, assumptions (46.1) imply that there exist functions v_i such that

$$v_i\in P_1(T_i), \quad p|_{T_i}=v_i\mu_i^2, \qquad 1\leqslant i\leqslant 3.$$

Since the functions $p: T\to\mathbb{R}$ and $\mu: T\to\mathbb{R}$ are continuous, the function $v: T\to\mathbb{R}$

defined by

$$v|_{T_i} = v_i, \quad 1 \leqslant i \leqslant 3,$$

is also continuous (the function μ does not vanish in the interior $\mathring{T}$ of T).

On each segment $[a, a_{i+2}]$, the gradient ∇p, which is unambiguously defined since the function p is continuously differentiable, is given by either expression

$$\nabla p|_{[a,a_{i+2}]} = \begin{cases} (2v_i\mu_i \nabla\mu_i + \mu_i^2 \nabla v_i)|_{[a,a_{i+2}]}, \\ (2v_{i+1}\mu_{i+1}\nabla\mu_{i+1} + \mu_{i+1}^2 \nabla v_{i+1})|_{[a,a_{i+2}]}, \end{cases}$$

and thus

$$2v\nabla(\mu_{i+1} - \mu_i) + \mu\nabla(v_{i+1} - v_i) = 0 \quad \text{along } [a, a_{i+2}],$$

since $\mu \neq 0$ in $\mathring{T}$. Since $\mu(a_{i+2}) = 0$ and $\nabla(\mu_{i+1} - \mu_i) \neq 0$ (otherwise the lines $\mu_i = 0$ and $\mu_{i+1} = 0$ would be parallel), we conclude that $v(a_{i+2}) = v_i(a_{i+2}) = 0$. A similar argument shows that $v_i(a_{i+1}) = 0$. Consequently, each function $v_i \in P_1(T_i)$ is of the form

$$v_i = C_i \mu_i, \quad C_i = \text{constant},$$

and the continuity of the function v then implies that

$$v(a) = v_i(a) = C_i, \quad 1 \leqslant i \leqslant 3.$$

Denoting by C the common values of the constant C_i, we conclude that

$$v|_{T_i} = v_i = C\mu_i, \quad 1 \leqslant i \leqslant 3,$$

and therefore that

$$p|_{T_i} = C\mu_i^3,$$

whence

$$\nabla p|_{T_i} = 3C\mu_i^2 \nabla\mu_i, \quad 1 \leqslant i \leqslant 3.$$

Then the constant C is necessarily zero for otherwise the function p would not be continuously differentiable along the segment $[a, a_{i+2}]$ since $\nabla\mu_i \neq \nabla\mu_{i+1}$.

That the Hsieh–Clough–Tocher triangle is of class $\mathscr{C}^1$ follows by an argument analogous to the proof of Theorem 9.3. □

REMARK 46.1. The normal derivatives at the midpoint of the sides can be eliminated by requiring that the normal derivative vary linearly along the sides. This elimination results in a finite element of class $\mathscr{C}^1$ with $\dim P_T = 9$, which will be described at the end of this section.

REMARK 46.2. Triangular finite elements of class $\mathscr{C}^1$ analogous to the Hsieh–Clough–Tocher triangle, but which use polynomials of higher degree, may be also defined: For instance, let $T = \bigcup_{i=1}^3 T_i$ denote a similar subdivision of the triangle T, and let

$$P_T=\{p\in\mathscr{C}^1(K);\ p|_{T_i}\in P_4(T_i),\ 1\leqslant i\leqslant 3\},$$
$$\Sigma_T=\{p(a_i), \partial_1 p(a_i), \partial_2 p(a_i): 1\leqslant i\leqslant 3;\ p(a_{ij}): 1\leqslant i<j\leqslant 3;$$
$$\partial_\nu p(a_{iij}): 1\leqslant i,j\leqslant 3,\ i\neq j;\ p(a), \partial_1 p(a), \partial_2 p(a)\},$$

where $a_{ij}=\frac{1}{2}(a_i+a_j)$, $a_{iij}=\frac{1}{3}(2a_i+a_j)$. Then the set Σ_T is P_T-unisolvent (cf. PERCELL [1976]).

Two reasons prevent the Hsieh–Clough–Tocher triangle from being imbedded in an affine family: As for the Argyris triangle, one reason is the occurrence of the normal derivatives $\partial_\nu p(b_i)$ as degrees of freedom; the other reason is that the point a may be allowed to vary inside the set T. This is why we must adapt to this element the notion of a regular family:

We shall say that a family of Hsieh–Clough–Tocher triangles (T, P_T, Σ_T) is *regular* if the following three conditions are simultaneously satisfied:

(i) There exists a constant σ such that $h_T/\rho_T\leqslant\sigma$ for all T.

(ii) The quantities h_T approach zero.

(iii) Let $\hat{T}$ be a fixed triangle with vertices $\hat{a}_i$, $1\leqslant i\leqslant 3$. For each triangle T with vertices $a_{i,T}$, $1\leqslant i\leqslant 3$, let F_T denote the unique affine mapping that satisfies $F_T(\hat{a}_i)=a_{i,T}$, $1\leqslant i\leqslant 3$. Then (Fig. 46.2) all the points $\hat{a}_T=F_T^{-1}(a_T)$ must belong to a fixed compact subset $\hat{B}$ of the interior of the triangle $\hat{T}$ (the compact subset $\hat{B}$ may vary from one regular family to another).

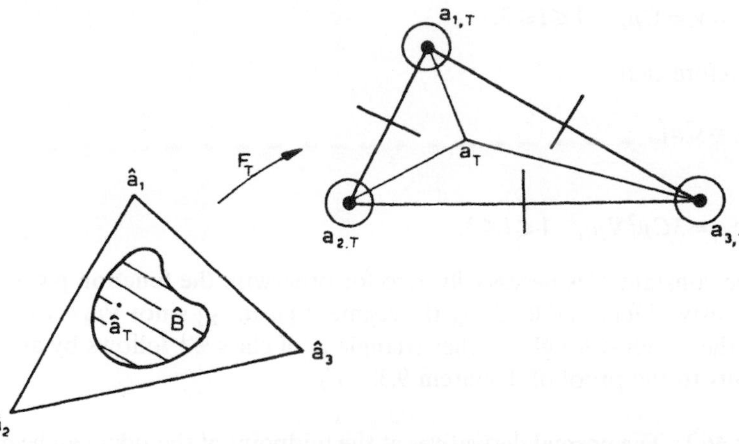

FIG. 46.2. In a regular family of Hsieh–Clough–Tocher triangles, all the points $\hat{a}_T=F_T^{-1}(a_T)$ belong to a fixed compact subset $\hat{B}$ of int T.

REMARK 46.3. Conditions (i) and (ii) are those that define a regular family of finite elements (cf. Section 16). Condition (iii) expresses precisely in which sense the points a_T may vary inside the triangle T, so as to guarantee that the family under consideration is almost-affine (cf. the next theorem).

The next interpolation error estimates are due to CIARLET [1974c].

THEOREM 46.2. *A regular family of Hsieh–Clough–Tocher triangles is almost-affine: For all* $p \in [1, \infty]$ *and all pairs* (m, q) *with* $m \geq 0$ *and* $q \in [1, \infty]$ *such that*

$$W^{4,p}(T) \hookrightarrow W^{m,q}(T), \qquad P_T \subset W^{m,q}(T), \tag{46.2}$$

there exists a constant C independent of T such that

$$\| v - \Pi_T v \|_{m,q,T} \leq C \{\operatorname{meas}(T)\}^{1/q - 1/p} h_T^{4-m} |v|_{4,p,T} \quad \text{for all } v \in W^{4,p}(T). \tag{46.3}$$

PROOF. We first observe that the inclusion $W^{4,p}(T) \hookrightarrow \mathscr{C}^1(T) = \operatorname{dom} \Pi_T$ holds for all $p \geq 1$. The proof of the theorem comprises three steps.

(1) As in the proof of Theorem 45.1, we shall introduce an "intermediary" finite element which is similar to the Hsieh–Clough–Tocher triangle, but which *can* be imbedded in an affine family. To this end, we replace in the set Σ_T the normal derivatives by appropriate directional derivatives, and we restrict the position of the points a_T, as follows: With each Hsieh–Clough–Tocher triangle (T, P_T, Σ_T), we associate the finite element (T, P_T, Ξ_T), where

$$\Xi_T = \{ p(a_i), Dp(a_i)(a_{i+1} - a_i), Dp(a_i)(a_{i-1} - a_i), Dp(b_i)(a - b_i) \colon 1 \leq i \leq 3 \} \tag{46.4}$$

(the proof of the P_T-unisolvence of the set Ξ_T is similar to that of the P_T-unisolvence of the set Σ_T as given in Theorem 46.1). We denote by Λ_T the P_T-interpolation operator associated with each finite element (T, P_T, Ξ_T).

For each point $\hat{a} \in \hat{B}$, let $\mathscr{T}(\hat{a})$ denote the (possibly empty) subfamily of Hsieh–Clough–Tocher triangles for which $a_T = F_T(\hat{a})$. Then, for each $\hat{a} \in \hat{B}$, the subfamily (T, P_T, Ξ_T), $T \in \mathscr{T}(\hat{a})$, is affine. Consequently, for all pairs (m, q) compatible with the inclusions (46.2), the inclusion

$$P_3(T) \subset P_T \tag{46.5}$$

implies that there exists a constant $C(\hat{a}, \hat{T})$ such that

$$|v - \Lambda_T v|_{m,q,T} \leq C(\hat{a}, \hat{T}) \{\operatorname{meas}(T)\}^{1/q - 1/p} h_T^{4-m} |v|_{4,p,T} \quad \text{for all } v \in W^{4,p}(T) \text{ and all } T \in \mathscr{T}(\hat{a}). \tag{46.6}$$

(2) We next show that, when the points $\hat{a}$ vary in the compact set $\hat{B}$, the constants $C(\hat{a}, \hat{T})$ appearing in the last inequality are bounded. To prove this, we recall that in the proof of Theorem 15.3, we found that these constants are of the form (cf. inequality (15.19)):

$$C(\hat{a}, \hat{T}) = C(\hat{T}) \| I - \hat{\Lambda}(\hat{a}) \|_{\mathscr{L}(W^{4,p}(\hat{T}); W^{m,q}(\hat{T}))}, \tag{46.7}$$

where, for each $\hat{a} \in \hat{B}$, $\hat{\Lambda}(\hat{a})$ denotes the $P_{\hat{T}}$-interpolation operator associated with the corresponding reference finite element $(\hat{T}, \hat{P}(\hat{a}), \Xi(\hat{a}))$.

The $P_{\hat{T}}$-interpolant $\hat{\Lambda}(\hat{a})\hat{v}$ of a function $\hat{v} \in W^{4,p}(\hat{T})$ is given by (the notation should be self-explanatory):

$$\hat{\Lambda}(\hat{a})\hat{v} = \sum_{i=1}^{3} \hat{v}(\hat{a}_i)\hat{p}_i(\hat{a},\cdot) + \sum_{\substack{i=1\\ |j-i|=1}}^{3} \{D\hat{v}(\hat{a}_i)(\hat{a}_j - \hat{a}_i)\}\hat{q}_{ij}(\hat{a},\cdot)$$

$$+ \sum_{i=1}^{3} \{D\hat{v}(\hat{b}_i)(\hat{a} - \hat{b}_i)\}\hat{r}_i(\hat{a},\cdot), \tag{46.8}$$

and there exist constants $C(\hat{T})$ independent of $\hat{a}$ such that

$$|\hat{v}(\hat{a}_i)| \leqslant |\hat{v}|_{0,\infty,\hat{T}} \leqslant C(\hat{T})\|\hat{v}\|_{4,p,\hat{T}},$$

$$\left.\begin{array}{l}|\{D\hat{v}(\hat{a}_i)(\hat{a}_j-\hat{a}_i)\}|\\ |\{D\hat{v}(\hat{b}_i)(\hat{a}-\hat{b}_i)\}|\end{array}\right\}\leqslant\sqrt{2}\,\mathrm{diam}(\hat{T})|\hat{v}|_{1,\infty,\hat{T}}\leqslant C(\hat{T})\|\hat{v}\|_{4,p,\hat{T}}. \tag{46.9}$$

Let us then consider the norm $\|\cdot\|_{m,q,T}$ of any one of the basis functions $\hat{p}_i(\hat{a},\cdot)$, $\hat{q}_{ij}(\hat{a},\cdot)$ and $\hat{r}_i(\hat{a},\cdot)$. On each one of the triangles $\hat{T}_i(\hat{a})$, $1 \leqslant i \leqslant 3$, that subdivide the triangle $\hat{T}$, the restriction of any one of these basis functions is a polynomial of degree $\leqslant 3$, whose coefficients are obtained through the solution of a linear system with an invertible matrix (the set $\hat{\Xi}(\hat{a})$ is $\hat{P}(\hat{a})$-unisolvent as long as the point $\hat{a}$ belongs to the interior of the set $\hat{T}$). This matrix depends continuously on the point $\hat{a}$ since its coefficients are polynomial functions of the coordinates of the point $\hat{a}$. Consequently, each basis function is in turn a continuous function of the point $\hat{a}$ and there exists a constant $\hat{C}$ such that

$$\sup_{\hat{a}\in\hat{B}} \{\|\hat{p}_i(\hat{a},\cdot)\|_{m,q,\hat{T}}, \|\hat{q}_{ij}(\hat{a},\cdot)\|_{m,q,\hat{T}}, \|\hat{r}_i(\hat{a},\cdot)\|_{m,q,\hat{T}}\} \leqslant \hat{C}, \tag{46.10}$$

since the set $\hat{B}$ is compact. Hence it follows from relations (46.7) to (46.10) that

$$\sup_{\hat{a}\in\hat{B}} C(\hat{a},\hat{T}) = C(\hat{B},\hat{T}) < \infty.$$

Combining this result with inequality (46.6), we obtain

$$|v - \Lambda_T v|_{m,q,T} \leqslant C(\hat{B},\hat{T})\{\mathrm{meas}(T)\}^{1/q-1/p} h_T^{4-m}|v|_{4,p,T}$$
$$\text{for all } v \in W^{4,p}(T) \text{ and all } T. \tag{46.11}$$

(3) By an argument similar to that used in the proof of Theorem 45.1 (cf. (45.9)), we find that

$$\Pi_T v - \Lambda_T v = \sum_{i=1} \partial_\nu(v - \Lambda_T v)(b_i)\{(a - b_i)\cdot\nu_i\}r_i, \tag{46.12}$$

where the functions r_i, $1 \leqslant i \leqslant 3$, are the basis functions associated with the degrees of freedom $\{Dp(b_i)(a - b_i)\}$ in the finite element (T, P_T, Ξ_T). Applying Theorem 16.1 with $m = 1$, $q = \infty$ and $k = 3$, we find that

$$|\partial_\nu(v - \Lambda_T v)(b_i)| \leqslant \sqrt{2}|v - \Lambda_T v|_{1,\infty,T} \leqslant C\{\mathrm{meas}(T)\}^{-1/p}\frac{h_T^4}{\rho_T}|v|_{4,p,T}. \tag{46.13}$$

We next have

$$|\{(a-b_i)\cdot v_i\}| \leqslant h_T, \tag{46.14}$$

$$|r_i|_{m,q,T} \leqslant C \frac{\{\text{meas}(T)\}^{1/q}}{\rho_T^m} |\hat{r}_i|_{m,q,\hat{T}}, \tag{46.15}$$

and we deduce from relations (46.12) to (46.15) that

$$\begin{aligned} |\Pi_T v - \Lambda_T v|_{m,q,T} &\leqslant C\{\text{meas}(T)\}^{1/q-1/p} \frac{h_T^5}{\rho_T^{m+1}} |v|_{4,p,T} \\ &\leqslant C\{\text{meas}(T)\}^{1/q-1/p} h_T^{4-m} |v|_{4,p,T}. \end{aligned} \tag{46.16}$$

Then the proof is completed by combining this inequality with inequality (46.11). □

When $q=2$, it is easily seen that the highest admissible value for the integer m compatible with the inclusion $P_T \subset H^m(T)$ is $m=2$ (that the integer m is at least 2 follows from an application of Theorem 5.2 to the partitioned triangle $T = \bigcup_{i=1}^3 T_i$; that the integer m cannot exceed 2 can be seen by an argument that will be used later; cf. Theorem 49.1). Hence this is the first instance of a *restriction on the possible inclusions* $P_T \subset W^{m,q}(T)$ (the other finite elements described in this and the next sections provide further instances of this restriction). Fortunately, the inclusion $P_T \subset H^2(T)$ is precisely that which is needed to insure convergence for fourth-order problems (cf. Section 48).

Let us now describe the "reduced" finite element that was announced in Remark 46.1. The *reduced Hsieh–Clough–Tocher triangle* is a triangular finite element whose corresponding space P_T and set Σ_T are indicated in Fig. 46.3. Combining the

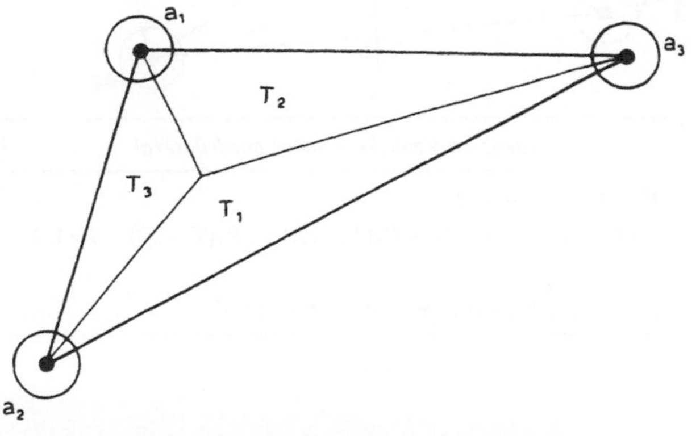

reduced Hsieh–Clough–Tocher triangle

$P_T = \{p \in \mathscr{C}^1(T);\ p|_{T_i} \in P_3(T_i),\ 1 \leqslant i \leqslant 3,\ \partial_\nu p \in P_1(T') \text{ for each side } T' \text{ of } T\}$
dim $P_T = 9$
$\Sigma_T = \{p(a_i), \partial_1 p(a_i), \partial_2 p(a_i) : 1 \leqslant i \leqslant 3\}$

FIG. 46.3.

methods found in the proofs of Theorems 9.2, 46.1, and 46.2, we could then show that the set Σ_T is P_T-unisolvent, and that *a regular family of reduced Hsieh–Clough–Tocher triangles is almost-affine,* with the value $k=2$ in the corresponding interpolation error estimates (44.5); the definition of a regular family is *verbatim* that of Hsieh–Clough–Tocher triangles.

The *basis functions* of the reduced Hsieh–Clough–Tocher triangle are explicitly computed in KIKUCHI [1975c], BERNADOU and HASSAN [1981], BERNADOU and BOISSERIE [1982, pp. 77ff].

The Hsieh–Clough–Tocher triangle, the reduced one, or more general *composite triangular finite elements,* also play a crucial role for constructing smooth *surface interpolants* with specified degrees of freedoms at arbitrary scattered points in the plane; in this direction, see notably POWELL and SABIN [1977], SABLONNIÈRE [1980, 1984, 1985], BARNHILL and FARIN [1981]. A three-dimensional *Hsieh–Clough–Tocher tetrahedron* has been proposed by ALFELD [1984].

Composite finite elements are not necessarily triangular. Consider for instance the *Fraeijs de Veubeke–Sander quadrilateral* (T, P_T, Σ_T): The set T is a convex non-degenerate quadrilateral, and the space P_T and Σ_T are as in Fig. 46.4, where T_1

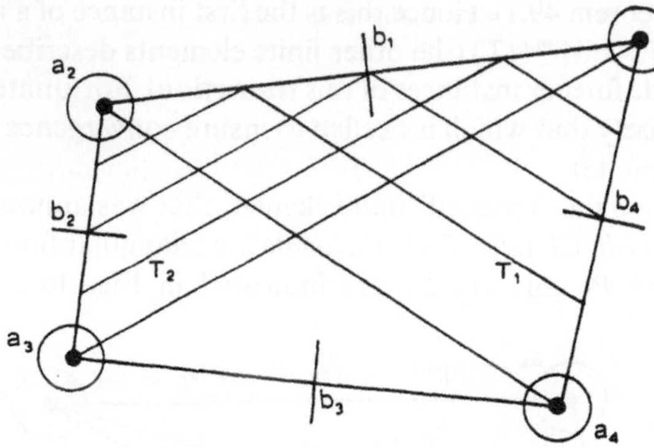

Fraeijs de Veubeke–Sander quadrilateral
$P_T=R_1(T)+R_2(T)$
$R_i(T)=\{p\in\mathscr{C}^1(T);\ p\|_{T_i}\in P_3(T_i),\ p\|_{T-T_i}\in P_3(T-T_i)\},\quad i=1,2$
$\dim P_T=16$
$\Sigma_T=\{p(a_i), \partial_1 p(a_i), \partial_2 p(a_i), \partial_\nu p(b_i): 1\leqslant i\leqslant 4\}$

FIG. 46.4.

denotes the subtriangle with vertices a_1, a_2, a_4, T_2 denotes the subtriangle with vertices a_1, a_2, a_3, and b_i denotes the midpoint of $[a_i, a_{i+1}]$. The P_T-unisolvence of the set Σ_T has been established by CIAVALDINI and NÉDÉLEC [1974].

We shall say that a family of Fraeijs de Veubeke–Sander quadrilaterals is *regular* if it is a regular family of finite elements in the usual sense and if, in addition, the following condition is satisfied: For each quadrilateral T in the family, let F_T denote the unique affine mapping that satisfies $F_T(0)=a_T$, $F_T(\hat{a}_1)=a_{1,T}$ and $F_T(\hat{a}_2)=a_{2,T}$, where a_T is the intersection of the two diagonals of the quadrilateral T, and where $\hat{a}_1=(1,0)$, $\hat{a}_2=(0,1)$ (cf. Fig. 46.5). Then there must exist compact intervals $\hat{I}_3$ and $\hat{I}_4$ contained in the half-axes

$$\{(x_1,x_2)\in \mathbb{R}^2;\, x_1<0,\ x_2=0\} \text{ and } \{(x_1,x_2)\in \mathbb{R}^2;\, x_1=0,\ x_2<0\}$$

respectively, such that all the points $\hat{a}_{j,T}=F_T^{-1}(a_{j,T})$ belong to the intervals $\hat{I}_j$, for $j=3$ and 4. In other words, the quadrilaterals $F_T^{-1}(T)$ must lie "in between" the two "extremal" quadrilaterals $\hat{T}_0$ and $\hat{T}_1$ indicated in Fig. 46.5.

CIAVALDINI and NÉDÉLEC [1974] have shown that *a regular family of Fraeijs de*

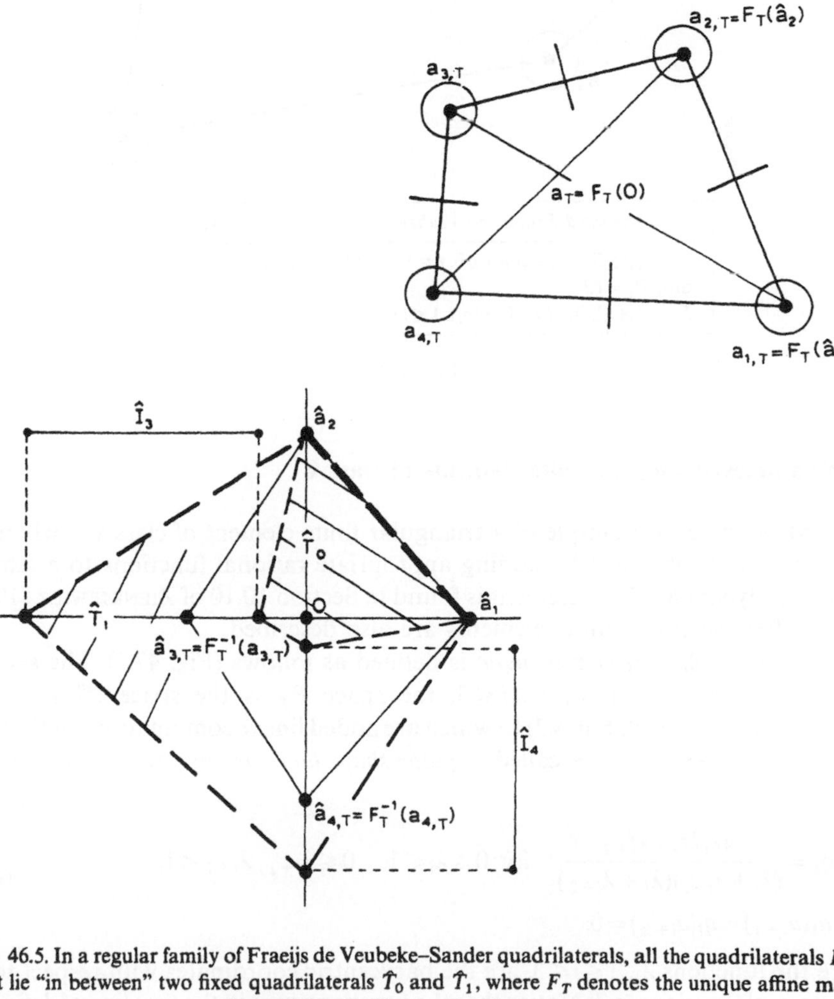

FIG. 46.5. In a regular family of Fraeijs de Veubeke–Sander quadrilaterals, all the quadrilaterals $F_T^{-1}(T)$ must lie "in between" two fixed quadrilaterals $\hat{T}_0$ and $\hat{T}_1$, where F_T denotes the unique affine mapping that satisfies $F_T(0)=a_T$, $F_T(\hat{a}_i)=a_{i,T}$, $i=1,2$.

Veubeke–Sander quadrilaterals is almost affine, with the value $k=3$ in the corresponding interpolation error estimates (44.5).

A similar analysis (unisolvence, interpolation error estimates) can be carried out for the *reduced Fraeijs de Veubeke–Sander quadrilateral*, whose characteristics are indicated in Fig. 46.6 (the spaces $R_1(T)$ and $R_2(T)$ are defined in Fig. 46.4).

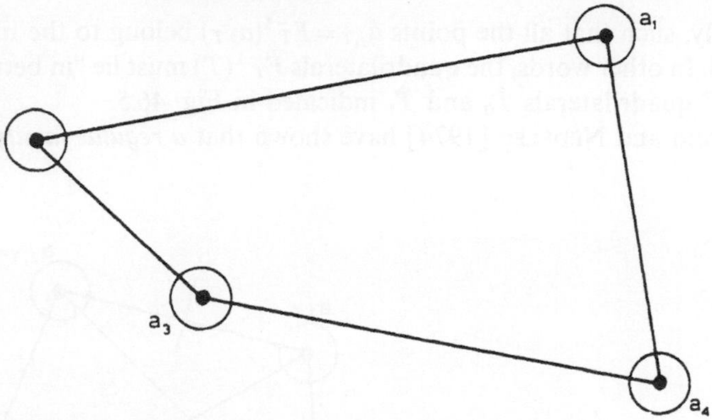

reduced Fraeijs de Veubeke–Sander quadrilateral
$P_T=\{p\in R_1(T)+R_2(T);\ \partial_\nu p\in P_1(T')$ along each side T' of $T\}$ $\dim P_T=12$ $\Sigma_T=\{p(a_i), \partial_1 p(a_i), \partial_2 p(a_i): 1\leqslant i\leqslant 4\}$

FIG. 46.6.

47. Examples of singular finite elements of class $\mathscr{C}^1$

We next describe an example of a triangular finite element of class $\mathscr{C}^1$, where the $\mathscr{C}^1$-continuity is obtained by adding appropriate rational functions to a familiar space of polynomials. This element is found in Section 10.10 of ZIENKIEWICZ [1971], where other "singular" finite elements are also described.

The *singular Zienkiewicz triangle* is defined as follows (Fig. 47.1): The set T is a triangle with vertices a_i, $1\leqslant i\leqslant 3$, the space P_T is the space $P_3''(T)$ of the Zienkiewicz triangle (cf. Fig. 8.2) to which are added linear combinations of the three functions q_i: $T\to\mathbb{R}$, $1\leqslant i\leqslant 3$, called *singular shape functions* by ZIENKIEWICZ [1971], and defined by

$$q_i=\frac{4\lambda_i\lambda_{i+1}^2\lambda_{i+2}^2}{(\lambda_i+\lambda_{i+1})(\lambda_i+\lambda_{i+2})}\quad\text{for } 0\leqslant\lambda_i\leqslant 1,\quad 0\leqslant\lambda_{i+1},\lambda_{i+2}<1, \qquad (47.1)$$
$$q_i(a_{i+1})=q_i(a_{i+2})=0,$$

where the functions λ_i, $1\leqslant i\leqslant 3$, are the barycentric coordinates with respect to the vertices a_i of the triangle T. Notice that the function given in the first line of definition

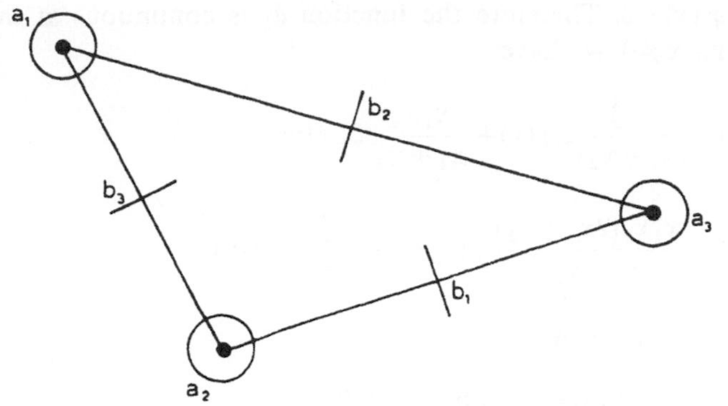

singular Zienkiewicz triangle
$P_T = P_3''(T) \oplus \bigvee_{i=1}^{3} \left\{ \dfrac{4\lambda_i\lambda_{i+1}^2\lambda_{i+2}^2}{(\lambda_i+\lambda_{i+1})(\lambda_i+\lambda_{i+2})} \right\}$ (cf. (8.3) and (47.1))
$\dim P_T = 12$ $\Sigma_T = \{p(a_i), \partial_1 p(a_i), \partial_2 p(a_i), \partial_\nu p(b_i) \colon 1 \leqslant i \leqslant 3\}$

FIG. 47.1.

(47.1) is *not* defined for $\lambda_i+\lambda_{i+1}=0$ or for $\lambda_i+\lambda_{i+2}=0$, i.e., for $\lambda_{i+2}=1$ or $\lambda_{i+1}=1$; this is why values must be assigned to the function q_i at the vertices a_{i+1} and a_{i+2}. Finally, the set Σ_T is the same as the set Σ_T of the Hsieh–Clough–Tocher triangle.

As usual, we begin by examining the question of unisolvence. Observe that this finite element is an instance where the validity of the inclusion $P_T \subset \mathscr{C}^1(T)$, which is part of the definition of a finite element of class $\mathscr{C}^1$, and of the inclusion $P_T \subset H^2(T)$ requires a proof.

THEOREM 47.1. *Let the triple* (T, P_T, Σ_T) *be as in Fig. 47.1. Then the set* Σ_T *is* P_T*-unisolvent. The resulting singular Zienkiewicz triangle is a finite element of class* $\mathscr{C}^1$*, and the inclusion* $P_T \subset H^2(T)$ *holds.*

PROOF. (1) To begin with, we verify that the inclusions $P_T \subset \mathscr{C}^1(T)$ and $P_T \subset H^2(T)$ hold. Since such inclusions are invariant through affine transformations, we may consider the case where the set T is the unit triangle with vertices $\hat{a}_1=(1,0)$, $\hat{a}_2=(0,1)$, and $\hat{a}_3=(0,0)$. Then it suffices to study the behavior of the function $\hat{q}_1\colon \hat{T}\to\mathbb{R}$ in a neighborhood of the origin in $\hat{T}$. We have

$$\begin{aligned} \hat{q}_1(x) &= \frac{x_1x_2^2}{(x_1+x_2)} f(x) \quad \text{for } x \neq 0, \\ \hat{q}_1(0) &= 0, \end{aligned} \tag{47.2}$$

where the function $f(x)=4(1-x_1-x_2)^2/(1-x_2)$ and its derivatives have no singularity at the origin. Since $\lim_{x_1,x_2\to 0^+} x_1x_2^2/(x_1+x_2)=0$, we deduce that

$\lim_{x_1,x_2\to 0^+}\hat{q}_1(x)=0$. Therefore the function $\hat{q}_1$ is continuous at the origin. For $x_1, x_2\geqslant 0$ and $x\neq 0$, we have:

$$\begin{aligned}\partial_1\hat{q}_1(x)&=\frac{x_2^3}{(x_1+x_2)^2}f(x)+\frac{x_1x_2^2}{x_1+x_2}\partial_1 f(x),\\ \partial_2\hat{q}_1(x)&=\frac{x_1x_2(2x_1+x_2)}{(x_1+x_2)^2}f(x)+\frac{x_1x_2^2}{x_1+x_2}\partial_2 f(x),\end{aligned}\tag{47.3}$$

and thus we conclude that

$$\lim_{x_1,x_2\to 0^+}\partial_j\hat{q}_1(x)=0=\partial_j\hat{q}_1(0),\quad j=1,2,\tag{47.4}$$

which proves that the function $\hat{q}_1$ is continuously differentiable at the origin.

Arguing analogously with the vertex $\hat{a}_2$, and next with the functions $\hat{q}_2$ and $\hat{q}_3$, we conclude that the inclusion

$$P_T\subset\mathscr{C}^1(T)$$

holds. This inclusion implies the inclusion $P_T\subset H^1(T)$ and thus, in order to obtain the inclusion $P_T\subset H^2(T)$, it remains to show that the second partial derivatives of the function $\hat{q}_1$ are square integrable around the origin. For $x\neq 0$, we can write

$$\begin{aligned}\partial_{11}\hat{q}_1(x)&=-\frac{2x_2^3}{(x_1+x_2)^3}f(x)+g_{11}(x),\\ \partial_{12}\hat{q}_1(x)&=\frac{x_2^2(3x_1+x_2)}{(x_1+x_2)^3}f(x)+g_{12}(x),\\ \partial_{22}\hat{q}_1(x)&=\frac{2x_1^3}{(x_1+x_2)^3}f(x)+g_{22}(x),\end{aligned}\tag{47.5}$$

where the functions g_{11}, g_{12} and g_{22} are continuous around the origin. Since the three functions factoring the function $f(x)$ are bounded on the set $\hat{T}$, the inclusion

$$P_T\subset H^2(T)$$

thus follows.

(2) The inclusion $P_T\subset\mathscr{C}^1(T)$ proved in (1) guarantees that the degrees of freedom of the set Σ_T are well-defined for the functions in the space P_T. The P_T-unisolvence of the set Σ_T will be a straightforward consequence of the P_T-unisolvence of the set

$$\Xi_T=\{p(a_i), Dp(a_i)(a_j-a_i), Dp(b_i)(a_i-b_i)\colon 1\leqslant i,j\leqslant 3,\quad |j-i|=1\},\tag{47.6}$$

which we first establish.

Let p_i, $1\leqslant i\leqslant 3$, and p_{ij}, $1\leqslant i,j\leqslant 3$, $|j-i|=1$, denote the basis functions of the space $P_3''(T)$ as given in (8.3). By definition, they satisfy

$$p_i(a_k)=\delta_{ik},\qquad Dp_i(a_k)(a_l-a_k)=0,\tag{47.7}$$

for $1 \leqslant i,k,l \leqslant 3$, $|k-l|=1$, and

$$p_{ij}(a_k)=0, \qquad Dp_{ij}(a_k)(a_l-a_k)=\delta_{ik}\delta_{jl}, \tag{47.8}$$

for $1 \leqslant i,j,k,l \leqslant 3$, $|j-i|=|k-l|=1$. We shall show that they also satisfy

$$Dp_i(b_k)(a_k-b_k)=-\tfrac{1}{4}+\tfrac{3}{4}\delta_{ik}, \quad 1 \leqslant i,k \leqslant 3, \tag{47.9}$$

$$Dp_{ij}(b_k)(a_k-b_k)=-\tfrac{1}{4}+\tfrac{3}{8}\delta_{ik}+\tfrac{5}{8}\delta_{jk}, \qquad 1 \leqslant i,j,k \leqslant 3, \quad |j-i|=1. \tag{47.10}$$

To this end, we first compute the directional derivatives $Dp(b_i)(a_i-b_i)$ of a function $p: T \to \mathbb{R}$ expressed in terms of barycentric coordinates. Let $p(x_1,x_2)=q(\lambda_1,\lambda_2,\lambda_3)$ be such a function. Denoting by $B=(b_{ij})$ the inverse matrix of the matrix A of (6.3) for $n=2$, we find that

$$\partial_j p=\sum_{k=1}^{3} \partial_k q \partial_j \lambda_k=\sum_{k=1}^{3} b_{kj}\partial_k q, \quad j=1,2.$$

Let us compute for cxample the quantity

$$Dp(b_1)(a_1-b_1)=\sum_{j=1}^{2}\sum_{k=1}^{3} b_{kj}\partial_k q(0,\tfrac{1}{2},\tfrac{1}{2})(a_{j1}-\tfrac{1}{2}(a_{j2}+a_{j3})),$$

where a_{ji}, $j=1,2$, denote the coordinates of the vertex a_i. By definition of the matrices B and A,

$$\sum_{j=1}^{2} b_{kj}a_{jl}=\delta_{kl}-b_{k3}, \quad 1 \leqslant k,l \leqslant 3,$$

and thus

$$Dp(b_1)(a_1-b_1)=\partial_1 q(0,\tfrac{1}{2},\tfrac{1}{2})-\tfrac{1}{2}\{\partial_2 q(0,\tfrac{1}{2},\tfrac{1}{2})+\partial_3 q(0,\tfrac{1}{2},\tfrac{1}{2})\}. \tag{47.11}$$

Then relations (47.9) and (47.10) follow from this relation and analogous relations for $Dp(b_i)(a_i-b_i)$, $i=2,3$, combined with the following expressions of the basis functions p_i and p_{ij}, which are easily derived from relations (8.1) and (8.2):

$$p_i=-2\lambda_i^3+3\lambda_i^2+2\lambda_1\lambda_2\lambda_3, \tag{47.12}$$

$$p_{ij}=\tfrac{1}{2}\lambda_i\lambda_j(\lambda_i-\lambda_j+1). \tag{47.13}$$

On the other hand, the functions q_i as defined in (47.1) satisfy

$$\begin{aligned} &q_i(a_k)=0, \quad 1 \leqslant i,k \leqslant 3, \\ &Dq_i(a_k)(a_l-a_k)=0, \qquad 1 \leqslant i,k,l \leqslant 3, \quad |k-l|=1, \\ &Dq_i(b_k)(a_k-b_k)=\delta_{ik}, \quad 1 \leqslant i,k \leqslant 3. \end{aligned} \tag{47.14}$$

The second relations follow from (47.4); the last ones follow from another application of relation (47.11) and similar ones.

Then relations (47.7)–(47.14) imply together that the functions (which all belong to the space P_T):

$$\begin{aligned}&\Big(p_i+\tfrac{1}{4}\Big\{-2q_i+\sum_{|j-i|=1} q_j\Big\}\Big),\quad 1\leqslant i\leqslant 3,\\ &(p_{ij}-\tfrac{1}{8}\{q_i+3q_j-2q_l\}),\qquad 1\leqslant i,j,l\leqslant 3,\quad \{i,j,l\}=\{1,2,3\},\qquad (47.15)\\ &q_i,\quad 1\leqslant i\leqslant 3,\end{aligned}$$

form a basis of the space P_T that corresponds to the degrees of freedom of the set Ξ_T of (47.6). Hence Ξ_T is P_T-unisolvent.

It remains to establish that the set Σ_T is also P_T-unisolvent. To prove this, we make the following observation: Along each side T' of the triangle T, the restrictions $p|_{T'}, p\in P_T$, are polynomials of degree $\leqslant 3$ in one variable, while the restrictions $Dp(\cdot)\xi|_{T'}, p\in P_T$, of any directional derivatives are polynomials of degree $\leqslant 2$ in one variable. This is clearly true for the functions in the space $P_3''(T)$, and it is a straightforward consequence of the definition for the functions q_i. Notice that this property implies in particular that this finite element is of class $\mathscr{C}^1$.

Let then $p\in P_T$ be a function that satisfies

$$p(a_i)=\partial_1 p(a_i)=\partial_2 p(a_i)=\partial_\nu p(b_i)=0,\quad 1\leqslant i\leqslant 3.$$

The conjunction of these relations and of the above property implies that the normal derivative and the tangential derivative vanish along any side of the triangle T. Consequently, the directional derivatives $Dp(b_i)(a_i-b_i)$, $1\leqslant i\leqslant 3$, vanish, and therefore the function p is identically zero since the set Ξ_T is P_T-unisolvent. □

REMARK 47.1. The second partial derivatives of the basis function $\hat{q}_1$ as given in (47.5) are not defined at the origin. More specifically, an easy computation shows that, for each slope $t>0$,

$$\lim_{\substack{x_2=tx_1\\ x_1\to 0^+}} \partial_{11}\hat{q}_1(x)=\frac{-8t^3}{(1+t)^3}.$$

This phenomenon is also noted by ZIENKIEWICZ [1971, p. 199], who observes that "second-order derivatives have non-unique values at nodes."

While this observation does not prevent the function q_1 from being in the space $\mathscr{C}^1(T)\cap H^2(T)$, it may nevertheless cause computational difficulties; in particular, very accurate quadrature schemes should be used in conjunction with this element. IRONS and RAZZAQUE [1972b] and RAZZAQUE [1973] obviate this difficulty by "smoothing" the second derivatives.

We next establish the interpolation error estimates.

THEOREM 47.2. *A regular family of singular Zienkiewicz triangles is almost-affine*: *For all* $p \in\ ]1, \infty]$ *and all pairs* (m, q) *with* $m \geq 0$ *and* $q \in [1, \infty]$ *such that*

$$W^{3,p}(T) \hookrightarrow W^{m,q}(T), \quad (47.16)$$
$$P_T \subset W^{m,q}(T),$$

there exists a constant C independent of T such that

$$\|v - \Pi_T v\|_{m,q,T} \leq C\{\text{meas}(T)\}^{1/q - 1/p} h_T^{3-m} |v|_{3,p,T} \quad \textit{for all } v \in W^{3,p}(T). \quad (47.17)$$

PROOF. We simply give some indications. The inequalities (47.17) are first established for the "intermediary" finite element (T, P_T, Ξ_T) with Ξ_T as in (47.6), which *can* be imbedded in an affine family, and which satisfies

$$P_2(T) \subset P_T$$

(the inequality $p > 1$ guarantees that the inclusion $W^{3,p}(T) \hookrightarrow \mathscr{C}^1(T) = \text{dom}\ \Pi_T$ holds). These estimates are then extended to the Zienkiewicz triangle by the same device as in the proofs of Theorems 45.1 or 46.2. □

The normal derivatives at the midpoints of the sides can be again eliminated from the set of degrees of freedom, by requiring that the normal derivative vary linearly along each side. We obtain in this fashion the *reduced singular Zienkiewicz triangle*, whose corresponding space P_T and set Σ_T are indicated in Fig. 47.2.

Arguing as in Theorems 47.1 and 47.2, we could then show that the set Σ_T is P_T-unisolvent and that *a regular family of reduced singular Zienkiewicz triangles is almost-affine*, with the value $k = 2$ in the corresponding interpolation error estimates (44.5).

We next describe yet another way of adding rational functions to a polynomial space so as to obtain a singular finite element of class $\mathscr{C}^1$.

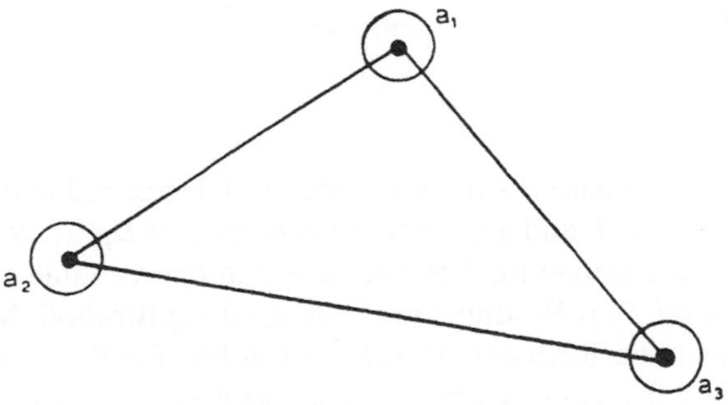

reduced singular Zienkiewicz triangle
$P_T = \{p \in P_3''(T) \oplus \bigvee_{i=1}^{3} \{q_i\};\ \partial_\nu p \in P_1(T') \text{ for each side } T' \text{ of } T\}$ (cf. (8.3) and (47.1)) $\dim P_T = 9$ $\Sigma_T = \{p(a_i), \partial_1 p(a_i), \partial_2 p(a_i) : 1 \leq i \leq 3\}$

FIG. 47.2.

Let T be a triangle, and let $R_3(T)$ denote the space of *tricubic polynomials*, i.e., polynomials whose restrictions along each parallel to any side of T are polynomials of degree $\leqslant 3$ in one variable. One can then show that the space $R_3(T)$ coincides with the space $P_3(T)$ to which are added linear combinations of the three functions $\lambda_1^2\lambda_2\lambda_3$, $\lambda_1\lambda_2^2\lambda_3$ and $\lambda_1\lambda_2\lambda_3^2$ (which are not linearly independent, however) and that $\dim R_3(T)=12$. This space was introduced by BIRKHOFF [1971].

Following BIRKHOFF and MANSFIELD [1974], we then define the *Birkhoff–Mansfield triangle* as indicated in Fig. 47.3 (as usual, $\partial_{\nu\tau}p(b_i)=D^2p(b_i)(\nu,\tau)$ where τ is the unit tangential vector at the point b_i). The following properties can then be established:

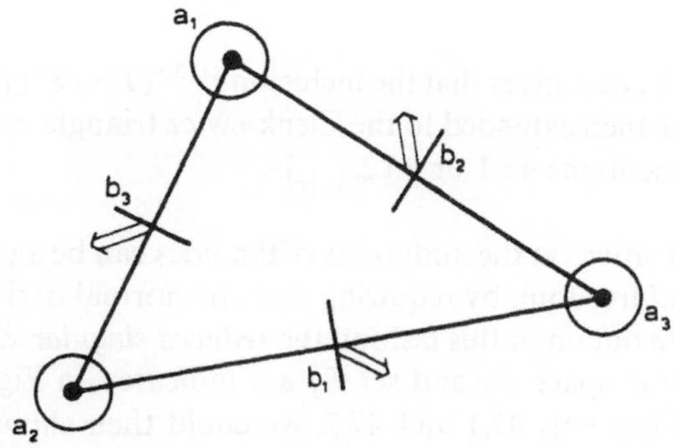

Birkhoff–Mansfield triangle
$P_T=R_3(T)\oplus \bigvee_{i=1}^{3}\left\{\dfrac{\lambda_i^2\lambda_{i+1}}{\lambda_i+\lambda_{i+1}}\right\}$, $\quad \dim P_T=15$
$\Sigma_T=\{p(a_i), \partial_1 p(a_i), \partial_2 p(a_i), \partial_\nu p(b_i), \partial_{\nu\tau}p(b_i): 1\leqslant i\leqslant 3\}$

FIG. 47.3.

The functions in the space P_T are polynomials of degree $\leqslant 3$ in one variable along each side of the triangle T, and any directional derivative $Dp(\cdot)\xi$, where ξ is any fixed vector in $\mathbb{R}^2$, is also a polynomial of degree $\leqslant 3$ in one variable along each side of the triangle T; the set Σ_T is P_T-unisolvent; the resulting Birkhoff–Mansfield triangle is of class $\mathscr{C}^1$, and the inclusion $P_T\subset H^2(T)$ holds. Finally, *a regular family of Birkhoff–Mansfield triangles is almost-affine*, with the value $k=3$ in the corresponding interpolation error estimates (44.5).

A similar analysis (unisolvence and interpolation error estimate) can be carried out for the *reduced Birkhoff–Mansfield triangle*, whose characteristics are indicated in Fig. 47.4. Other ways of adding rational functions are mentioned in MANSFIELD [1974, 1978], DUPUIS and GOËL [1970]. *Boolean sum interpolation theory* can also be used to derive *blending polynomial interpolants*, which interpolate a function $v\in\mathscr{C}^m(T)$ and all its derivatives of order $\leqslant m$ on the (possibly curved) boundary of

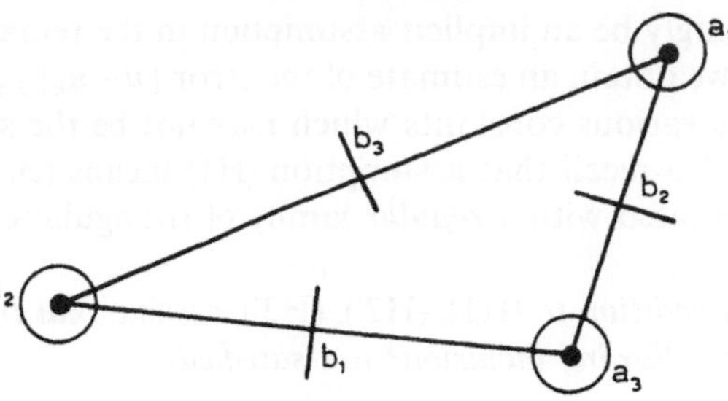

reduced Birkhoff–Mansfield triangle
$P_T=\{P\in R_3(T)\oplus \bigvee_{i=1}^{3}\left\{\frac{\lambda_i^2\lambda_{i+1}}{\lambda_i+\lambda_{i+1}}\right\} Dp(\cdot)\xi\in P_2(K')$ for each side T' of T and all $\xi\in\mathbb{R}^2\}$ $\dim P_T=12$ $\Sigma_T=\{p(a_i), \partial_1 p(a_i), \partial_2 p(a_i), \partial_\nu p(b_i): 1\leqslant i\leqslant 3\}$

FIG. 47.4.

a triangle T. In this direction, see BARNHILL [1975a, 1975b], BARNHILL, BIRKHOFF and GORDON [1973], BARNHILL and GREGORY [1975a, 1975b].

48. Estimates of the error $\|u-u_h\|_{2,\Omega}$ for finite elements of class $\mathscr{C}^1$

Let us now return to the finite element approximation of a fourth-order problem, such as the clamped plate problem (44.1), to fix ideas. We consider a family of finite element spaces X_h, made up with finite elements (T, P_T, Σ_T) of a given type, which satisfy the following assumptions:

(H2′) The family (T, P_T, Σ_T), $T\in\bigcup_h \mathscr{T}_h$, is almost-affine.

(H3′) All finite elements (T, P_T, Σ_T), $T\in\bigcup_h \mathscr{T}_h$, are of class $\mathscr{C}^1$.

Note that (H2′) and (H3′) parallel the assumptions (H2) and (H3) made in Section 17.

If we assume, as in the subsequent theorems, that the inclusions $P_T\subset H^2(T)$ hold, the inclusion $X_h\subset H^2(\Omega)$ then follows from hypothesis (H3′). This being the case, let

$$V_h=X_{00h}=\{v_h\in X_h;\ v_h=\partial_\nu v_h=0 \text{ on } \Gamma\}. \tag{48.1}$$

Note further that the X_h-interpolation operator associated with any one of the finite elements of class $\mathscr{C}^1$ that we have so far considered satisfy the implication

$$v\in \operatorname{dom}\Pi_h,\qquad v=\partial_\nu v=0 \text{ on } \Gamma \;\Rightarrow\; \Pi_h v\in X_{00h}, \tag{48.2}$$

which will accordingly be an implicit assumption in the remainder of this section.

To begin with, we obtain an estimate of the error $\|u-u_h\|_{2,\Omega}$. As usual, the same letter C represents various constants which may not be the same in their various occurrences. We also recall that assumption (H1) means (cf. Section 17) that the spaces V_h are associated with a *regular* family of triangulations.

THEOREM 48.1. *In addition to* (H1), (H2′), (H3′), *assume that there exists an integer* $k \geqslant 2$ *such that the following inclusions are satisfied*:

$$P_k(T) \subset P_T \subset H^2(T), \tag{48.3}$$

$$H^{k+1}(T) \hookrightarrow \mathscr{C}^s(T), \tag{48.4}$$

where s is the maximal order of partial derivatives occuring in the definition of the set Σ_T, *and let* $u_h \in V_h$ *denote the discrete solutions.*

Then if the solution $u \in H_0^2(\Omega)$ *of the clamped plate problem is also in the space* $H^{k+1}(\Omega)$, *there exists a constant C independent of h such that*

$$\|u-u_h\|_{2,\Omega} \leqslant Ch^{k-1}|u|_{k+1,\Omega}. \tag{48.5}$$

PROOF. Using Céa's lemma (Theorem 13.1), inequality (44.5), and relation (48.2), we obtain

$$\begin{aligned}\|u-u_h\|_{2,\Omega} &\leqslant C \inf_{v_h \in V_h} \|u-v_h\|_{2,\Omega} \\ &\leqslant C\|u-\Pi_h u\|_{2,\Omega} = C\Big\{\sum_{T\in\mathscr{T}_h} \|u-\Pi_T u\|_{2,T}^2\Big\}^{1/2} \\ &\leqslant Ch^{k-1}\Big\{\sum_{T\in\mathscr{T}_h} |u|_{k+1,T}^2\Big\}^{1/2} = Ch^{k-1}|u|_{k+1,\Omega}. \qquad \square\end{aligned}$$

From this theorem, the least assumptions that insure an $O(h)$ convergence of the error $\|u-u_h\|_{2,\Omega}$ are the inclusions $P_2(T) \subset P_T$ on the one hand, and the $H^3(\Omega)$-regularity of the solution u of the plate problem on the other hand. Notice that it is remarkable that this regularity result holds precisely if the right-hand side f is in the space $L^2(\Omega)$, and if $\bar{\Omega}$ is a convex polygon, an assumption often satisfied by plates. Therefore, since one cannot expect better regularity *in general*, the choice $P_T = P_2(T)$ seems optimal from the point of view of convergence. However, by Ženíšek's result (Section 9), this choice is not compatible with the inclusion $X_h \subset \mathscr{C}^1(\bar{\Omega})$ for triangular "polynomial" finite elements, and thus either higher-degree polynomials must be used, or one should use "nonpolynomial" finite elements such as singular or composite elements in order to decrease the dimension of the space P_T. These considerations are illustrated in the following tableau (Fig. 48.1), where we have summarized the application of Theorem 48.1 to various finite elements of class $\mathscr{C}^1$.

One should notice that, if the reduced Hsieh–Clough–Tocher triangle and the reduced singular Zienkiewicz triangle are optimal in that the dimension of the corresponding spaces P_T is the smallest, this reduction in the dimension of the spaces

Finite element	dim P_T	$P_k(T) \subset P_T$	$\|u-u_h\|_{2,\Omega}$	Assumed regularity of the solution
Argyris triangle	21	$P_5(T)=P_T$	$O(h^4)$	$u \in H^6(\Omega)$
Bell triangle	18	$P_4(T) \subset P_T$	$O(h^3)$	$u \in H^5(\Omega)$
Bogner–Fox–Schmit rectangle	16	$P_3(T) \subset P_T$	$O(h^2)$	$u \in H^4(\Omega)$
Hsieh–Clough–Tocher triangle	12	$P_3(T) \subset P_T$	$O(h^2)$	$u \in H^4(\Omega)$
Reduced Hsieh–Clough–Tocher triangle	9	$P_2(T) \subset P_T$	$O(h)$	$u \in H^3(\Omega)$
Singular Zienkiewicz triangle	12	$P_2(T) \subset P_T$	$O(h)$	$u \in H^3(\Omega)$
Reduced singular Zienkiewicz triangle	9	$P_2(T) \subset P_T$	$O(h)$	$u \in H^3(\Omega)$

FIG. 48.1. Orders of convergence obtained with various finite elements of class $\mathscr{C}^1$.

P_T is obtained at the expense of an increased complexity in the *structure* of the functions $p \in P_T$.

Notice that in order to get an $O(h^{k+1})$ estimate of the error $|u-u_h|_{0,\Omega}$ based on the Aubin–Nitsche lemma (Theorem 19.1), it would be necessary to assume that, for *any* $g \in L^2(\Omega)$, the corresponding solution φ_g of the plate problem belongs to the space $H^4(\Omega) \cap H_0^2(\Omega)$ and that there exists a constant C such that $\|\varphi_g\|_{4,\Omega} \leqslant C|g|_{0,\Omega}$ for *all* $g \in L^2(\Omega)$. However, *this regularity no longer holds on convex polygonal domains in general*; cf. KONDRAT'EV [1967], GRISVARD [1985]. It is true only if the boundary Γ is sufficiently smooth: For example, this is the case if the boundary Γ is of class $\mathscr{C}^4$. But then this regularity of the boundary becomes incompatible with our assumption that $\bar{\Omega}$ be a polygonal set!

We next give minimal assumptions (cf. (48.6)) that guarantee *convergence*.

THEOREM 48.2. *In addition to* (H1), (H2′), *and* (H3′), *assume that the inclusions*

$$P_2(T) \subset P_T \subset H^2(T) \tag{48.6}$$

are satisfied, and that the maximal order s *of partial derivatives found in the set* Σ_T *is* $\leqslant 2$. *Then*

$$\lim_{h \to 0} \|u-u_h\|_{2,\Omega}=0. \tag{48.7}$$

PROOF. The argument is the same as in the proof of Theorem 18.2 and, for this reason, will be only sketched. Using inequality (44.5) with $k=2$, $p=\infty$, $m=2$ and $q=2$, one first shows that the space

$$\mathscr{V}=W^{3,\infty}(\Omega) \cap H_0^2(\Omega)$$

is dense in the space $H_0^2(\Omega)$, and this result is used in conjunction with the inequality

$$\inf_{v_h \in V_h} \|u-v_h\|_{2,\Omega} \leqslant \|u-v\|_{2,\Omega} + \|v-\Pi_h v\|_{2,\Omega}$$

which holds for any function $v \in \mathscr{V}$. □

There are further aspects of the finite element approximation of the plate problem or of more general fourth-order problems that will not be covered here. Let us simply mention some relevant references: RANNACHER [1976a] has obtained *estimates of the error* $|u-u_h|_{0,\infty,\Omega}$. The effect of *numerical integration* is analyzed in BERNADOU and DUCATEL [1978]. The approximation of fourth-order problems on *domains with curved boundaries* is considered by MANSFIELD [1978], who handles in addition the effect of *numerical integration*. Her approach parallels that given in CIARLET and RAVIART [1972c] for second-order problems (cf. Chapter VI). Curved isoparametric finite elements of a new type are suggested by ROBINSON [1973]. In the case of the *simply supported plate problem*, we mention the *Babuška paradox* (cf. BABUŠKA [1963]; see also BIRKHOFF [1969]): Contrary to second-order problems, no convergent approximation may be found if the curved boundary is replaced by a polygonal domain. For, assume Ω is a circular domain; then the exact solutions on a sequence of polygonal domains approaching the circle do *not* converge to the solution on the circle!

Additional references concerning *curved boundaries*, or *different boundary conditions* for the plate problems, are NITSCHE [1971, 1972b], CHERNUKA, COWPER, LINDBERG and OLSON [1972], and the survey of SCOTT [1976b].

Finite element approximation of *variational inequalities of order four* are considered by GLOWINSKI [1975, 1984], GLOWINSKI, MARINI and VIDRASCU [1984]. See also GLOWINSKI, LIONS and TRÉMOLIÈRES [1976b, Chapter 4].

49. A nonconforming finite element for the plate problem: The Adini–Clough–Melosh rectangle

The general approach followed in this section is that of CIARLET [1974a, 1974b]. In LASCAUX and LESAINT [1975], a thorough study is made not only of the Adini–Clough–Melosh rectangle, but of other nonconforming finite elements for the plate problem, such as the *Zienkiewicz triangle*, *Morley's triangle*, and various instances of *Fraeijs de Veubeke triangles* (these last finite elements are briefly described at the end of this section); the convergence of the Adini–Clough–Melosh rectangle has also been studied by KIKUCHI [1975b, 1976] and MIYOSHI [1972].

Curved nonconforming elements have been considered by BARNHILL and BROWN [1975]. Nonconforming elements for the plate problem are extensively discussed from an engineering viewpoint in ZIENKIEWICZ [1971, Chapter 10]. Although the references given in Chapter V were more specifically concerned with second-order problems, some of them are also relevant in the present situation, notably CÉA [1976], NITSCHE [1974], OLIVEIRA [1977].

We first give the general definition of a *nonconforming method for solving the clamped plate problem*, which corresponds to the data (44.1). Assuming that the set $\bar{\Omega}$ is polygonal, so that it may be exactly covered with triangulations, we construct a finite element space X_h whose generic finite element (T, P_T, Σ_T) is not of class $\mathscr{C}^1$. Using the same arguments as in Theorem 30.1, one can then prove the following result, which shows that in this case, the space X_h cannot be a subspace of $H^2(\Omega)$ (note that this theorem is the converse of Theorem 5.2).

THEOREM 49.1 *Assume that the inclusions $P_T \subset \mathscr{C}^1(T)$ for all $T \in \mathscr{T}_h$ and $X_h \subset H^2(\Omega)$ hold. Then the inclusion*

$$X_h \subset \mathscr{C}^1(\bar{\Omega})$$

holds.

Let us henceforth assume that we are using finite elements (T, P_T, Σ_T) that satisfy

$$P_T \subset H^2(T) \quad \text{for all } T \in \mathscr{T}_h. \tag{49.1}$$

Then we have in particular

$$X_h \subset L^2(\Omega). \tag{49.2}$$

After defining an appropriate subspace X_{00h} of X_h, which takes the boundary conditions $v = \partial_\nu v = 0$ along Γ into account as well as possible (this will be illustrated on one example), we define the *approximate bilinear form*:

$$\begin{aligned} a_h(u_h, v_h) &= \sum_{T \in \mathscr{T}_h} \int_T \{\Delta u_h \Delta v_h + (1-\nu)(2\partial_{12} u_h \partial_{12} v_h - \partial_{11} u_h \partial_{22} v_h - \partial_{22} u_h \partial_{11} v_h)\} \,\mathrm{d}x \\ &= \sum_{T \in \mathscr{T}_h} \int_T \{\nu \Delta u_h \Delta v_h + (1-\nu)(\partial_{11} u_h \partial_{11} v_h + \partial_{22} u_h \partial_{22} v_h + 2\partial_{12} u_h \partial_{12} v_h)\} \,\mathrm{d}x. \end{aligned} \tag{49.3}$$

Observe that this definition is justified by the inclusions (49.1). Then the *discrete problem* consists in finding a function $u_h \in V_h = X_{00h}$ such that

$$a_h(u_h, v_h) = l(v_h) \quad \text{for all } v_h \in V_h \tag{49.4}$$

(the linear form need not be approximated in view of the inclusion (49.2)). Since $|\cdot|_{2,\Omega}$ is a norm over the space $V = H_0^2(\Omega)$, we are naturally led to introduce the *seminorm*

$$v_h \to \| v_h \|_h = \left\{ \sum_{T \in \mathscr{T}_h} |v_h|_{2,T}^2 \right\}^{1/2} \tag{49.5}$$

over the space V_h. We also extend the domains of definition of the mappings $a_h(\cdot, \cdot)$ and $\|\cdot\|_h$ to the space $(V_h + V)$. Thus there exists a constant $\tilde{M}$ independent of the space V_h such that

$$|a_h(u, v)| \leqslant \tilde{M} \| u \|_h \| v \|_h \quad \text{for all } u, v \in (V_h + V). \qquad (4$$

In the ensuing analysis, we shall essentially concentrate our study on one exam of a *nonconforming finite element method for solving the plate problem*, which ma use of the *Adini–Clough–Melosh rectangle*, sometimes abbreviated as the *A(rectangle.*

This element, due to ADINI and CLOUGH [1961] and MELOSH [1963], correspo to the following data: The set R is a rectangle whose vertices a_i, $1 \leqslant i \leqslant 4$, are coun

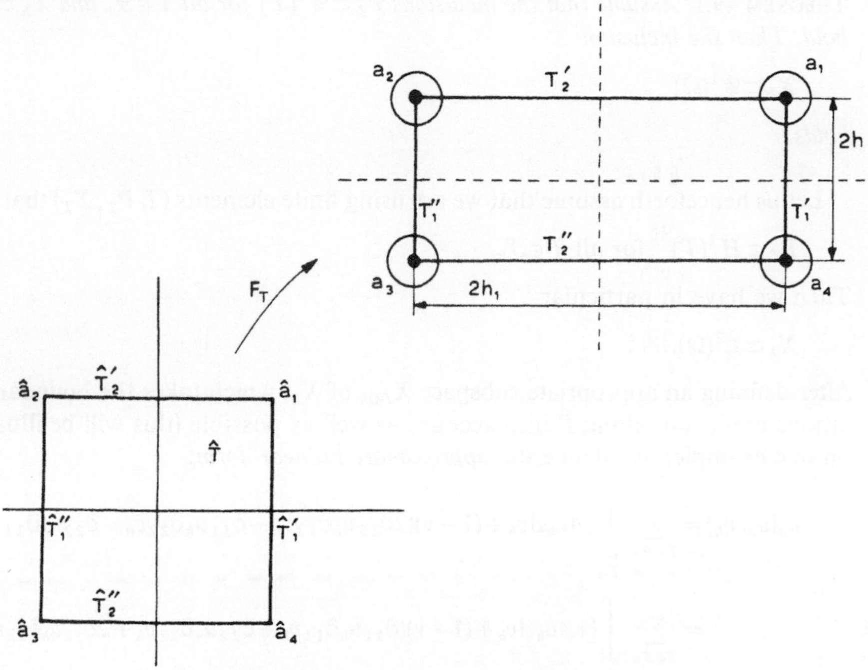

Adini–Clough–Melosh rectangle
$P_T = P_3(T) \oplus \mathrm{V}\{x_1 x_2^3, x_1^3 x_2\}$, dim $P_T = 12$ $\Sigma_T = \{p(a_i), \partial_1 p(a_i), \partial_2 p(a_i)\colon 1 \leqslant i \leqslant 4\}$

FIG. 49.1.

as in Fig. 49.1. The space P_T consists of all polynomials of the form

$$p\colon x = (x_1, x_2) \to p(x) = \sum_{\alpha_1 + \alpha_2 \leqslant 3} \gamma_{\alpha_1 \alpha_2} x_1^{\alpha_1} x_2^{\alpha_2} + \gamma_{13} x_1 x_2^3 + \gamma_{31} x_1^3 x_2,$$

i.e.,

$$P_T = P_3(T) \oplus \mathrm{V}\{x_1 x_2^3, x_1^3 x_2\}; \qquad (4$$

hence the inclusion

$$P_3(T) \subset P_T \tag{49.8}$$

holds, and

$$\dim P_T = 12. \tag{49.9}$$

Finally, the set Σ_T is given by

$$\Sigma_T = \{p(a_i), \partial_1 p(a_i), \partial_2 p(a_i) \colon 1 \leqslant i \leqslant 4\}. \tag{49.10}$$

Let us assume without loss of generality that the set T is the square $\hat{T} = [-1, +1]^2$, in which case it is easily verified that

$$\hat{p} = \sum_{i=1} \hat{p}(\hat{a}_i)\hat{p}_i + \sum_{|j-i|=1\,(\mathrm{mod}\,4)} D\hat{p}(\hat{a}_i)(\hat{a}_j - \hat{a}_i)\hat{p}_{ij} \quad \text{for all } \hat{p} \in P_{\hat{T}}, \tag{49.11}$$

with

$$\begin{aligned} &\hat{p}_1(x) = \tfrac{1}{4}(1+x_1)(1+x_2)(1+\tfrac{1}{2}(x_1+x_2) - \tfrac{1}{2}(x_1^2+x_2^2)),\\ &\hat{p}_{12}(x) = \tfrac{1}{8}(1+x_1)(1+x_2)^2(1-x_2),\\ &\hat{p}_{14}(x) = \tfrac{1}{8}(1+x_2)(1+x_1)^2(1-x_1), \quad \text{etc.} \end{aligned} \tag{49.12}$$

Therefore this identity implies that *the set Σ_T is P_T-unisolvent.*

Let us assume that the set $\bar{\Omega}$ is rectangular, so that it may be covered by triangulations made up of rectangles. With such a triangulation $\mathcal{T}_h$, we associate a finite element space X_h whose functions v_h are defined as follows: (i) for each rectangle $T \in \mathcal{T}_h$, the restrictions $v_h|_T$ span the space P_T of (49.7); (ii) each function $v_h \in X_h$ is defined by its values and the values of its first derivatives at all the vertices of the triangulation.

Along each side T' of a rectangle T, the restrictions $p|_{T'}$, $p \in P_T$ are polynomials of degree $\leqslant 3$ in one variable. Since such polynomials are uniquely determined by their values and the values of their first derivative at the end-points of T', we conclude that the *Adini–Clough–Melosh rectangle is a finite element of class $\mathscr{C}^0$. It is not of class $\mathscr{C}^1$*, however: Along the side $T_1' = [a_4, a_1]$ (for instance; cf. Fig. 49.1), the normal derivative is a polynomial of degree $\leqslant 3$ in the variable x_2 on the one hand, and on the other the only degrees of freedom that are available for specifying the normal derivative along the side T_1' are its values at the two end-points.

We then let $V_h = X_{00h}$, where X_{00h} denotes the space of all functions $v_h \in X_h$ that satisfy $v_h(b) = \partial_1 v_h(b) = \partial_2 v_h(b) = 0$ at all the boundary nodes b. Then the functions $v_h \in V_h$ vanish along the boundary Γ; their derivatives $\partial_\nu v_h$ vanish at all the boundary nodes, but not along the entire boundary Γ, in general. To sum up, we have constructed a finite element space V_h whose functions v_h satisfy:

$$\begin{aligned} &v_h \in H_0^1(\Omega) \cap C^0(\bar{\Omega}),\ v_h|_T \in H^2(T) \quad \text{for all } T \in \mathcal{T}_h,\\ &\partial_\nu v_h(b) = 0 \quad \text{at the boundary nodes.} \end{aligned} \tag{49.13}$$

Observe that the associated X_h-interpolation operator Π_h is such that

$$v \in H_0^2(\Omega) \cap \operatorname{dom} \Pi_h \;\Rightarrow\; \Pi_h v \in X_{00h} = V_h. \tag{49.14}$$

Hence this implication holds in particular for functions in the space $H_0^2(\Omega)\cap H^3(\Omega)$, since $H^3(\Omega)\subset\mathscr{C}^1(\bar{\Omega})=\operatorname{dom}\Pi_h$.

Prior to the error analysis, we must examine whether the mapping $\|\cdot\|_h$ of (49.5) is indeed a norm.

THEOREM 49.2. *The mapping*

$$v_h\to\|v_h\|_h=\left\{\sum_{T\in\mathscr{T}_h}|v_h|_{2,T}^2\right\}^{1/2}$$

is a norm over the space V_h.

PROOF. Let v_h be a function in the space V_h such that $\|v_h\|_h=0$. Then the functions $\partial_j(v_h|_T)$, $j=1,2$, are constant over each rectangle $T\in\mathscr{T}_h$. Since they are continuous at the vertices, the functions $\partial_j v_h$, $j=1,2$, are therefore constant over the set $\bar{\Omega}$, and since they vanish at the boundary nodes, they are identically zero. Thus the function $v_h\in V_h$ is identically zero, as a consequence of the inclusion $V_h\subset H_0^1(\Omega)\cap\mathscr{C}^0(\Omega)$. □

Notice that the *approximate bilinear forms* $a_h(\cdot,\cdot)$ *are uniformly* V_h*-elliptic*, since

$$(1-\nu)\|v_h\|_h^2\leqslant a_h(v_h,v_h)\quad\text{for all } v_h\in V_h, \tag{49.15}$$

by (49.3), and the Poisson coefficient ν lies in the interval $]0,\tfrac{1}{2}[$.

REMARK 49.1. Had we tried to use nonconforming finite element methods for the biharmonic problem, in which case the approximate bilinear form reduces to $\Sigma_{T\in\mathscr{T}_h}\int_T\Delta u_h\Delta v_h\,dx$, the uniform V_h-ellipticity is no longer automatic, and this is essentially why we restrict ourselves to plate problems. By contrast, the conforming methods described in the previous section apply equally well to any fourth-order elliptic boundary value problem.

50. Estimate of the error $\{\Sigma_{T\in\mathscr{T}_h}|u-u_h|_{2,T}^2\}^{1/2}$ for the Adini–Clough–Melosh rectangle

We shall assume in what follows that the spaces V_h are associated with a *regular family of triangulations*, and that the *solution u is in the space* $H^3(\Omega)\cap H_0^2(\Omega)$; this is

$$\|u-u_h\|_h\leqslant C\left(\inf_{v_h\in V_h}\|u-v_h\|_h+\sup_{w_h\in V_h}\frac{|a_h(u,w_h)-l(w_h)|}{\|w_h\|_h}\right). \tag{50.1}$$

We shall assume in what follows that the spaces V_h are associated with a *regular family of triangulations*, and that the *solution u is in the space* $H^3(\Omega)\cap H_0^2(\Omega)$; this is a reasonable regularity assumption, which holds in particular if $f\in L^2(\Omega)$ and $\bar{\Omega}$ is a convex polygon, i.e., a rectangle in the present case. Since any family of

Adini–Clough–Melosh rectangles is affine, we obtain

$$\inf_{v_h\in V_h}\|u-v_h\|_h\leqslant\Big\{\sum_{T\in\mathcal{T}_h}|u-\Pi_T u|_{2,T}^2\Big\}^{1/2}\leqslant Ch|u|_{3,\Omega}, \tag{50.2}$$

and this estimate takes care of the first term in the right-hand side of inequality (50.1). The estimate of the second term, i.e., the *consistency error estimate*, rests on a careful decomposition of the difference

$$D_h(u,w_h)=a_h(u,w_h)-f(w_h),\quad w_h\in V_h. \tag{50.3}$$

Let us first show that the term $l(w_h)=\int_\Omega fw_h\,\mathrm{d}x$ can be rewritten as

$$l(w_h)=-\int_\Omega \nabla(\Delta u)\cdot\nabla w_h\,\mathrm{d}x\quad\text{for all } w_h\in V_h \tag{50.4}$$

(this equality clearly holds if $u\in H^4(\Omega)\cap H_0^2(\Omega)$, in which case $l(w_h)=\int_\Omega \Delta^2u\,w_h\,\mathrm{d}x$, but we only assume here that $u\in H^3(\Omega)\cap H_0^2(\Omega)$). To see this, let $w_h\in V_h$ be given, and let (w_h^k) be a sequence of functions $w_h^k\in\mathcal{D}(\Omega)$ such that $\lim_{k\to\infty}\|w_h^k-w_h\|_{1,\Omega}=0$ (recall that $w_h\in V_h\subset H_0^1(\Omega)$). By making use of Green's formulae (2.5) and (2.9), we obtain for all integers k,

$$\int_\Omega \Delta u\Delta w_h^k\,\mathrm{d}x=-\int_\Omega\nabla(\Delta u)\cdot\nabla w_h^k\,\mathrm{d}x,$$

$$\int_\Omega\{2\partial_{12}u\partial_{12}w_h^k-\partial_{11}u\partial_{22}w_h^k-\partial_{22}u\partial_{11}w_h^k\}\,\mathrm{d}x=0,$$

since $\partial_\nu w_h^k=\partial_\tau w_h^k=0$ along Γ, and thus, by definition of the abstract problem (cf. (44.1)),

$$\int_\Omega fw_h^k\,\mathrm{d}x=-\int_\Omega\nabla(\Delta u)\cdot\nabla w_h^k\,\mathrm{d}x.$$

Therefore,

$$\int_\Omega fw_h\,\mathrm{d}x=\lim_{k\to\infty}\int_\Omega fw_h^k\,\mathrm{d}x=\lim_{k\to\infty}\int_\Omega\Delta u\Delta w_h^k\,\mathrm{d}x$$

$$=\lim_{k\to\infty}\Big\{-\int_\Omega\nabla(\Delta u)\cdot\nabla w_h^k\,\mathrm{d}x\Big\}=-\int_\Omega\nabla(\Delta u)\cdot\nabla w_h\,\mathrm{d}x,$$

and equality (50.4) is proved.

Using the same Green formulae as above, we obtain with the same notation as in (2.9),

$$\int_T \{\Delta u \Delta w_h + (1-\nu)(2\partial_{12}u\partial_{12}w_h - \partial_{11}u\partial_{22}w_h - \partial_{22}u\partial_{11}w_h)\}\,dx$$

$$= -\int_T \nabla(\Delta u)\cdot\nabla w_h\,dx + \int_{\partial T} \Delta u\,\partial_{\nu,T}w_h\,d\gamma$$

$$+(1-\nu)\int_{\partial T}\{-\partial_{\tau\tau,T}u\partial_{\nu,T}w_h + \partial_{\nu\tau,T}u\partial_{\tau,T}w_h\}\,d\gamma$$

for all $T\in\mathcal{T}_h$ and all $w_h\in V_h$.

When these expressions are added as in the approximate bilinear form of (49.3), we find that

$$\sum_{T\in\mathcal{T}_h}\left\{-\int_T \nabla(\Delta u)\cdot\nabla w_h\,dx\right\} = -\int_\Omega \nabla(\Delta u)\cdot\nabla w_h\,dx = l(w_h),$$

using the inclusion $V_h\subset H^1(\Omega)$ and relation (50.4). Let us next show that

$$\sum_{T\in\mathcal{T}_h}\int_{\partial T}\partial_{\nu\tau,T}u\partial_{\tau,T}(w_h|_T)\,d\gamma = 0.$$

To this end, consider separately the situation where $T'\subset\partial T$ is a side common to two adjacent rectangles T_1 and T_2, and the situation where $T'\subset\partial T$ is a portion of the boundary Γ. In the first case the two corresponding integrals cancel because $u\in H^3(\Omega)$ and $w_h\in C^0(\bar{\Omega})$, and in the second case the integral vanishes because $w_h=0$ along Γ.

To sum up, we have found that

$$D_h(u, w_h) = a_h(u, w_h) - l(w_h)$$

$$= \sum_{T\in\mathcal{T}_h}\int_{\partial T}(\Delta u - (1-\nu)\partial_{\tau\tau,T}u)\partial_{\nu,T}(w_{h|T})\,d\gamma \quad \text{for all } w_h\in V_h, \tag{50.5}$$

i.e., we have obtained *one* decomposition of the expression $D_h(u, w_h)$ as

$$D_h(u, w_h) = \sum_{T\in\mathcal{T}_h} D_T(u|_T, w_h|_T),$$

where each mapping $D_T(\cdot,\cdot)$ is a bilinear form over the space $H^3(T)\times P_T$. As in the proof of Theorem 34.1, the key idea consists in obtaining another decomposition of the form (50.5) (cf. (50.8)), which takes into account the "conforming" part of the first-order partial derivatives of the functions in the space V_h (for a related idea, cf. Remark 34.1). This will in turn allow us to obtain appropriate estimates of the difference $D_h(u, w_h)$, as we now show.

THEOREM 50.1. *Assume that the solution u of the plate problem is in the space* $H_0^2(\Omega)\cap H^3(\Omega)$. *Then, for any regular family of triangulations, there exists a constant C independent of h such that*

$$\|u-u_h\|_h=\left\{\sum_{T\in\mathcal{T}_h}|u-u_h|_{2,T}^2\right\}^{1/2}\leqslant Ch|u|_{3,\Omega},\tag{50.6}$$

where u_h *denotes the discrete solution found with Adini–Clough–Melosh rectangles.*

PROOF. After inspecting the decomposition (50.5), we are naturally led to study the bilinear form

$$\begin{aligned}D_h(\cdot,\cdot):(v,w_h)&\in H^3(\Omega)\times V_h\\ &\to D_h(v,w_h)=\sum_{T\in\mathcal{T}_h}\int_{\partial T}(\Delta v-(1-\nu)\partial_{\tau\tau,T}v)\partial_{\nu,T}(w_{h|T})\,\mathrm{d}\gamma\\ &=D_h^1(v,\partial_1 w_h)+D_h^2(v,\partial_2 w_h),\end{aligned}\tag{50.7}$$

with

$$\begin{aligned}D_h^j(v,\partial_j w_h)=\sum_{T\in\mathcal{T}_h}\Bigg\{&\int_{T'_j}(\Delta v-(1-\nu)\partial_{\tau\tau,T}v)\partial_j(w_{h|T})\,\mathrm{d}\gamma\\ &-\int_{T''_j}(\Delta v-(1-\nu)\partial_{\tau\tau,T}v)\partial_j(w_{h|T})\,\mathrm{d}\gamma\Bigg\},\quad j=1,2,\end{aligned}$$

where, for each $T\in\mathcal{T}_h$, the sides T'_j and $T''_j, j=1,2$, are defined as in Fig. 49.1.

For each triangulation $\mathcal{T}_h$, let Y_h denote the finite element space whose generic finite element is the bilinear rectangle, and let $Z_h=Y_{0h}$ denote the space of all functions $w_h\in Y_h$ that vanish at the boundary nodes. Clearly, the inclusion

$$Z_h\subset\mathscr{C}^0(\bar\Omega)\cap H_0^1(\Omega)$$

implies that

$$D_h^j(v,z_h)=0,\quad j=1,2,\quad\text{for all } v\in H^3(\Omega),\quad z_h\in Z_h,$$

where

$$\begin{aligned}D_h^j(v,z_h)=\sum_{T\in\mathcal{T}_h}\Bigg\{&\int_{T'_j}(\Delta v-(1-\nu)\partial_{\tau\tau,T}v)z_h\,\mathrm{d}\gamma\\ &-\int_{T''_j}(\Delta v-(1-\nu)\partial_{\tau\tau,T}v)z_h\,\mathrm{d}\gamma\Bigg\},\quad j=1,2.\end{aligned}$$

Consequently, if Λ_T denotes for each $T\in\mathcal{T}_h$ the $Q_1(T)$-interpolation operator associated with the bilinear rectangle, we can also write

$$D_h(v, w_h) = \sum_{T\in\mathcal{T}_h} D_T(v, w_h) \quad \text{for all } (v, w_h) \in H^3(\Omega)\times V_h, \tag{50.8}$$

where, for each $T\in\mathcal{T}_h$, the bilinear form $D_T(\cdot,\cdot)$ is given by

$$D_T(v, p) = \Delta_{1,T}(v, \partial_1 p) + \Delta_{2,T}(v, \partial_2 p) \quad \text{for all } (v, p)\in H^3(T)\times P_T, \tag{50.9}$$

with

$$\begin{aligned}\Delta_{j,T}(v, \partial_j p) = &\int_{T_j'} (\Delta v - (1-\nu)\partial_{\tau\tau,T} v)(\partial_j p - \Lambda_T \partial_j p)\, \mathrm{d}\gamma \\ &- \int_{T_j''} (\Delta v - (1-\nu)\partial_{\tau\tau,T} v)(\partial_j p - \Lambda_T \partial_j p)\, \mathrm{d}\gamma, \quad j=1,2.\end{aligned} \tag{50.10}$$

Using the definition of the operator Λ_T, we find a *first polynomial invariance*:

$$\Delta_{j,T}(v, q) = 0, \quad j=1,2, \quad \text{for all } v\in H^3(T), \quad q\in Q_1(T), \tag{50.11}$$

with

$$\begin{aligned}\Delta_{j,T}(v, q) = &\int_{T_j'} (\Delta v - (1-\nu)\partial_{\tau\tau,T} v)(q - \Lambda_T q)\, \mathrm{d}\gamma \\ &- \int_{T_j''} (\Delta v - (1-\nu)\partial_{\tau\tau,T} v)(q - \Lambda_T q)\, \mathrm{d}\gamma, \quad j=1,2.\end{aligned}$$

We next establish the *second polynomial invariance*:

$$\Delta_{j,T}(v, q) = 0, \quad j=1,2, \quad \text{for all } v\in P_2(T), \quad q\in \partial_j P_T, \tag{50.12}$$

where the spaces

$$\partial_j P_T = \{\partial_j p;\ p\in P_T\}, \quad j=1,2, \tag{50.13}$$

both contain the space $Q_1(T)$. To see this, it suffices to show that

$$\int_{T_j'} (q - \Lambda_T q)\, \mathrm{d}\gamma = \int_{T_j''} (q - \Lambda_T q)\, \mathrm{d}\gamma, \quad j=1,2 \quad \text{for all } q\in\partial_j P_T. \tag{50.14}$$

Let us prove this equality for $j=1$, for instance. Each function $q\in\partial_1 P_T$ is of the form

$$q = \gamma_0(x_1) + \gamma_1(x_1)x_2 + \gamma_2 x_2^2 + \gamma_3 x_2^3,$$

where γ_0 and γ_1 are polynomials of degree $\leqslant 2$ in the variable x_1. Given any function

r defined on a side T', let $\lambda_{T'}(r)$ denote the linear function along T' that assumes the same values as the function r at the end points of T'. Then we have

$$(q-\Lambda_T q)|_{T_1'}(x_2)=\gamma_2 x_2^2+\gamma_3 x_2^3-\lambda_{T_1'}(\gamma_2 x_2^2+\gamma_3 x_2^3),$$

and therefore,

$$(q-\Lambda_T q)|_{T_1'}(x_2)=(q-\Lambda_T q)|_{T_1''}(x_2),$$

which proves (50.14). Consequently, the polynomial invariance of (50.12) holds.

To estimate the quantities $\Delta_{j,T}(v, \partial_j p)$ of (50.10), it suffices to estimate the similar expressions

$$\delta_{j,T}(\varphi,q)=\int_{T_j'}\varphi(q-\Lambda_T q)\,\mathrm{d}\gamma-\int_{T_j''}\varphi(q-\Lambda_T q)\,\mathrm{d}\gamma \tag{50.15}$$

for $\varphi\in H^1(T), q\in\partial_j P_T, j=1,2$. Using the standard correspondences between the functions $\hat{v}:\hat{T}\to\mathbb{R}$ and $v:T\to\mathbb{R}$, we obtain

$$\delta_{1,T}(\varphi,q)=h_2\delta_{1,\hat{T}}(\hat{\varphi},\hat{q}),\qquad \delta_{2,T}(\varphi,q)=h_1\delta_{2,\hat{T}}(\hat{\varphi},\hat{q}); \tag{50.16}$$

we shall also take into account the fact that a function $\hat{q}$ belongs to the space $\partial_j P_{\hat{T}}$ when the function q belongs to the space $\partial_j P_T$.

Corresponding to the polynomial invariances (50.11) and (50.12), we now have:

$$\begin{aligned}&\delta_{j,\hat{T}}(\hat{\varphi},\hat{q})=0\quad\text{for all }\hat{\varphi}\in H^1(\hat{T}),\quad \hat{q}\in P_0(\hat{T}),\\ &\delta_{j,\hat{T}}(\hat{\varphi},\hat{q})=0\quad\text{for all }\hat{\varphi}\in P_0(\hat{T}),\quad \hat{q}\in\partial_j P_{\hat{T}}.\end{aligned} \tag{50.17}$$

Hence if we equip the spaces $\partial_j P_T$ with the norm $\|\cdot\|_{1,T}$, we obtain

$$|\delta_{j,\hat{T}}(\hat{\varphi},\hat{q})|\leqslant\hat{C}\|\hat{\varphi}\|_{L^2(\partial\hat{T})}\|\hat{q}\|_{L^2(\partial\hat{T})}\leqslant\hat{C}\|\hat{\varphi}\|_{1,\hat{T}}\|\hat{q}\|_{1,\hat{T}}$$
$$\text{for all }\hat{\varphi}\in H^1(\hat{T}),\quad \hat{q}\in\partial_j P_{\hat{T}},$$

and thus each bilinear form $\delta_{j,\hat{T}}(\cdot,\cdot)$ is continuous over the space $H^1(\hat{T})\times\partial_j P_{\hat{T}}$. The bilinear lemma (Theorem 33.1) then implies that there exists another constant $\hat{C}$ such that

$$|\delta_{j,\hat{T}}(\hat{\varphi},\hat{q})|\leqslant\hat{C}|\hat{\varphi}|_{1,\hat{T}}|\hat{q}|_{1,\hat{T}}\quad\text{for all }\hat{\varphi}\in H^1(\hat{T}),\quad \hat{q}\in\partial_j P_{\hat{T}}. \tag{50.18}$$

By Theorem 15.1 and the regularity assumption, there exists a constant C such that

$$|\hat{\varphi}|_{1,\hat{T}}\leqslant C|\varphi|_{1,T},\qquad |\hat{q}|_{1,\hat{T}}\leqslant C|q|_{1,T}. \tag{50.19}$$

Combining relations (50.16), (50.18) and (50.19), we conclude that

$$\begin{aligned}&|\delta_{j,T}(\varphi,q)|\leqslant Ch|\varphi|_{1,T}|q|_{1,T}\\ &\quad\text{for all }\varphi\in H^1(T),\quad q\in\partial_j P_T,\quad j=1,2.\end{aligned} \tag{50.20}$$

Let then $v\in H^3(T)$ and $p\in P_T$ be two given functions; hence the functions $\varphi=\Delta v-(1-\nu)\partial_{22}v$ and $q=\partial_1 p$ belong to the spaces $H^1(T)$ and $\partial_1 P_T$, respectively.

Then we have

$$|\Delta_{1,T}(v,p)| = |\delta_{1,T}(\Delta v - (1-\nu)\partial_{22}v, \partial_1 p)| \leqslant Ch_T|v|_{3,T}|p|_{2,T}.$$

Thus, after analogously estimating the term $|\Delta_{2,T}(v,p)|$, we obtain

$$|D_T(v,p)| \leqslant \sum_{j=1}^{2} |\Delta_{j,T}(v,\partial_j p)| \leqslant Ch_T|v|_{3,T}|p|_{2,T}$$

$$\text{for all } v\in H^3(T), \quad p\in P_T.$$

In this fashion we are able to estimate the second term in the abstract error estimate (50.1): We find that, for all $w_h \in V_h$,

$$|a_h(u,w_h) - l(w_h)| \leqslant \sum_{T\in\mathcal{T}_h} |D_T(u,w_h)| \leqslant Ch|u|_{3,\Omega}\|w_h\|_h,$$

and the proof is complete. □

Using the abstract error estimate mentioned in Remark 31.1, LASCAUX and LESAINT [1975] have shown that the error estimate expressed with the norm $\|\cdot\|_{1,\Omega}$ is

$$\|u-u_h\|_{1,\Omega} \leqslant Ch^2|u|_{3,\Omega},$$

i.e., it is of a higher order, as expected. The same authors have also shown that *the error estimate* (50.6) *can be improved when all the rectangles* $T\in\mathcal{T}_h$ *are equal*, in the sense that in this case,

$$\|u-u_h\|_h \leqslant Ch^2|u|_{4,\Omega},$$

if the solution u is in the space $H^4(\Omega)$.

Another popular nonconforming finite element for solving the plate problem is the *Zienkiewicz triangle* (cf. BAZELEY, CHEUNG, IRONS and ZIENKIEWICZ [1965]) which was described in Section 8 (cf. Fig. 8.2). Through a refinement of the argument used in the proof of Theorem 50.1, LASCAUX and LESAINT [1975] have shown that the needed polynomial invariances hold if and only if *all sides of all the triangles found in the triangulation are parallel to three directions only*, in which case one gets

$$\|u-u_h\|_h \leqslant Ch|u|_{3,\Omega}, \qquad \|u-u_h\|_{1,\Omega} \leqslant Ch^2|u|_{3,\Omega},$$

if the solution u is the space $H^3(\Omega)$. This is therefore an answer to the *Union Jack problem*: As pointed out in ZIENKIEWICZ [1971, pp. 188–189], the engineers had empirically discovered that the configuration of Fig. 50.1(a) systematically yields poorer results than the configuration of Fig. 50.1(b).

The reason why the degree of freedom $p(a_{123})$ (which is normally found in the cubic Hermite triangle) should be eliminated is that the presence of the associated basis function $\lambda_1\lambda_2\lambda_3$ (cf. (8.1)) destroys the required polynomial invariances. See also SHI [1984e, 1987b] for a study of the Zienkiewicz triangle from the viewpoint of Stummel's generalized patch test.

A modified Zienkiewicz triangle has also been proposed by HANSSEN, SYVERTSEN and BERGAN [1978], and further developed by ARGYRIS, HAASE and MLEJNEK [1980,

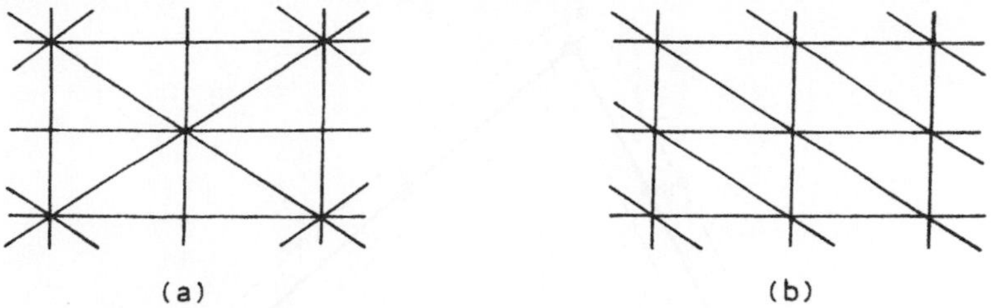

FIG. 50.1. The Union Jack problem: If the nonconforming Zienkiewicz triangles are arranged as in (a), the method does not converge; if the triangles are arranged as in (b), the method converges.

1982]. Then SHI [1987] has shown that the associated nonconforming finite element method is convergent, without any geometrical restriction imposed on the triangulations.

Whereas the Adini–Clough–Melosh rectangle and the Zienkiewicz triangle yield finite element spaces that satisfy the inclusion $V_h \subset \mathscr{C}^0(\bar{\Omega}) \cap H_0^1(\Omega)$, *there exist convergent nonconforming finite elements for the plate problem that are not even of class* $\mathscr{C}^0$. Here we shall simply describe two such elements and mention their convergence properties, referring the reader to LASCAUX and LESAINT [1975] for complete proofs. See also RUAS [1988] for extensions to biharmonic equations in $\mathbb{R}^n$, $n \geqslant 3$.

The first element is named *Morley triangle*, after MORLEY [1968]; its characteristics are indicated in Fig. 50.2. Details about its implementation are found in PEISKER and BRAESS [1987].

The second element, named *Fraeijs de Veubeke triangle* after FRAEIJS DE VEUBEKE [1974]), is an example of a finite element where some degrees of freedom are *averages*

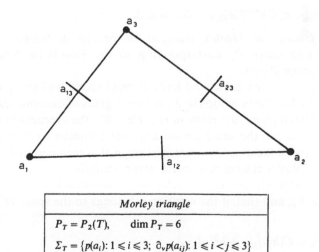

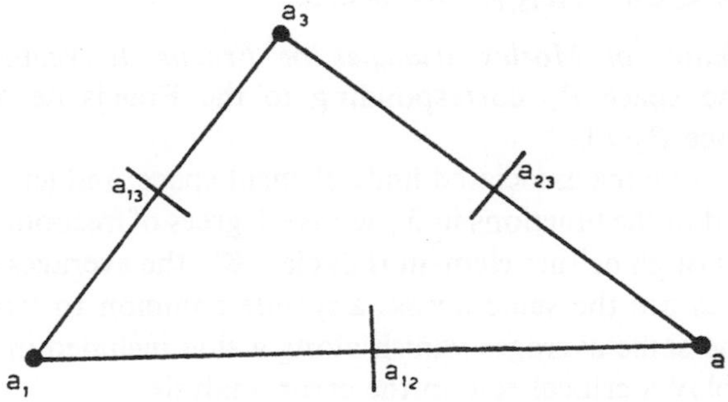

Morley triangle
$P_T = P_2(T)$, $\dim P_T = 6$
$\Sigma_T = \{p(a_i)\colon 1 \leqslant i \leqslant 3;\ \partial_\nu p(a_{ij})\colon 1 \leqslant i < j \leqslant 3\}$

FIG. 50.2.

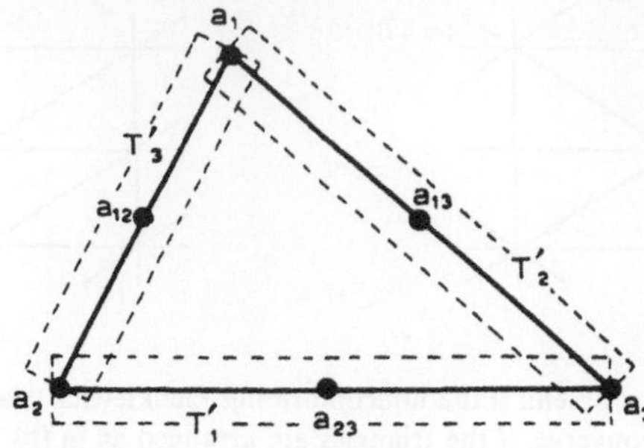

Fraeijs de Veubeke triangle
$P_T=\{p\in P_3(T);\ \phi(p)=0\},\quad \dim P_T=9$
$\phi(p)=27p(a_{123})-\sum_{i=1}^{3}p(a_i)-8\sum_{1\leqslant i<j\leqslant 3}p(a_{ij})+3\sum_{i=1}^{3}\frac{1}{\lvert T_i'\rvert}\int_{T_i'}Dp(\cdot)(a_{123}-a_i)\,\mathrm{d}\gamma$
$\Sigma_T=\left\{p(a_i)\colon 1\leqslant i\leqslant 3;\ p(a_{ij})\colon 1\leqslant i<j\leqslant 3;\ \frac{1}{\lvert T_i'\rvert}\int_{T_i'}\partial_\nu p\,\mathrm{d}\gamma\colon 1\leqslant i\leqslant 3\right\}.$

FIG. 50.3.

(another similar instance is Wilson's brick; cf. Section 32). Its characteristics are indicated in Fig. 50.3 where for each i, $|T_i'|$ denotes the length of the side T_i'.

One can then prove that each set Σ_T is P_T-unisolvent, and that, for regular families and for any $v\in H^3(T)\subset \operatorname{dom}\Pi_T$,

$$|v-\Pi_T v|_{m,T}\leqslant Ch^{3-m}|v|_{3,T},\quad 0\leqslant m\leqslant 3,$$

i.e., *regular families of Morley triangles or Fraeijs de Veubeke triangles are almost-affine* (the space P_T corresponding to the Fraeijs de Veubeke triangle contains the space $P_2(T)$).

Let next X_h denote the associated finite element space, and let $V_h=X_{00h}$, where X_{00h} is composed of the functions in X_h whose degrees of freedom vanish along the boundary Γ. Although neither element is of class $\mathscr{C}^0$, the averages of the first-order partial derivatives are the same across any side common to two adjacent finite elements, and the same averages vanish along a side included in Γ in both cases; these last facts play a critical role in the error analysis.

It can then be established that, in each case, the seminorm $\|\cdot\|_h$ of (49.5) is a norm over the space V_h, and that, if the solution u belongs to the space $H^4(\Omega)$, the error estimate

$$\|u-u_h\|_h\leqslant C(h|u|_{3,\Omega}+h^2|u|_{4,\Omega})$$

holds. Note that, contrary to the Zienkiewicz triangle, *no restriction need to be imposed here on the geometry of the triangulations in order to get convergence.*

The decomposition (50.5) is here replaced by

$$D_h(u, w_h) = a_h(u, w_h)$$
$$= \sum_{T\in\mathcal{T}_h} \int_{\partial T} (\Delta u - (1-\nu)\partial_{\tau\tau,T} u)\partial_{\nu,T}(w_h|_T)\,\mathrm{d}\gamma$$
$$- \sum_{T\in\mathcal{T}_h} \int_{\partial T} \{(\partial_{\nu,T}\Delta u)(w_h|_T) + (1-\nu)\partial_{\nu\tau,T} u \partial_{\tau,T}(w_h|_T)\}\,\mathrm{d}\gamma$$

for all $w_h \in V_h$,

and the key idea is again to subtract off appropriate "conforming" parts in the above expression. Then it is possible to apply the bilinear lemma, one side at a time rather than one element at a time, as in the case of Wilson's brick or the Adini–Clough–Melosh rectangle.

There are alternate ways of defining *nonconforming methods for plate problems.* For example, let us assume that we are given a finite element space V_h that satisfies only the inclusion $V_h \subset \mathscr{C}^0(\bar{\Omega}) \cap H_0^1(\Omega)$. If we assume as usual that the functions in the spaces P_T are smooth, the conformity would require the additional conditions that $\partial_\nu(v_h|_{T(1)}) + \partial_\nu(v_h|_{T(2)}) = 0$ along any side T' common to two adjacent finite elements $T(1)$ and $T(2)$, and that $\partial_\nu v_h = 0$ along Γ. If these conditions cannot be exactly fulfilled, they may be considered as *constraints*, and accordingly, they may be dealt with either by a *penalty method* or by *duality techniques* (for a general introduction to these techniques, see e.g. CIARLET [1983]).

In the first approach, one minimizes a functional of the form

$$J_h^*(v_h) = \tfrac{1}{2} a_h(v_h, v_h) - l(v_h) + \frac{1}{\varepsilon(h)} \Phi(v_h),$$

where

$$\Phi(v_h) = \sum_{\substack{T(1),T(2)\in\mathcal{T}_h \\ T(1)\neq T(2)}} \int_{T(1)\cap T(2)} \{\partial_\nu(v_h|_{T(1)}) + \partial_\nu(v_h|_{T(2)})\}^2\,\mathrm{d}\gamma + \int_\Gamma (\partial_\nu v_h)^2\,\mathrm{d}\gamma,$$

and $\varepsilon(\cdot)$ is a function of h that approaches zero as $h \to 0$, e.g., $\varepsilon(h) = Ch^\sigma$, $C > 0$, where the exponent $\sigma > 0$ is chosen so as to maximize the order of convergence. A method of this type has been studied by BABUŠKA and ZLÁMAL [1973], who showed that cubic Hermite triangles used in this fashion yield the error estimates

$$\| u - u_h \|_h \leqslant C\sqrt{h} \| u \|_{3,\Omega}, \quad \| u - u_h \|_{1,\Omega} \leqslant Ch \| u \|_{3,\Omega},$$

if $u \in H^3(\Omega)$ and $\varepsilon(h) = Ch^2$, and

$$\| u - u_h \|_h \leqslant Ch \| u \|_{4,\Omega}, \qquad \| u - u_h \|_{1,\Omega} \leqslant Ch \| u \|_{4,\Omega},$$

if $u \in H^4(\Omega)$ and $\varepsilon(h) = Ch^3$. These authors have applied this penalty method to the biharmonic problem instead of the plate problem. Such techniques are actually used in practice: see ZIENKIEWICZ [1974].

The second approach consists in introducing an appropriate *Lagrangian*. This is advocated for example by HARVEY and KELSEY [1971], who also use the cubic Hermite triangle, but for solving the plate problem.

The uniform convergence of nonconforming finite element methods for plate problems is studied in RANNACHER [1977].

References

ADAMS, R.A. (1975), *Sobolev Spaces* (Academic Press, New York).

ADINI, A. and R.W. CLOUGH (1961), Analysis of plate bending by the finite element method, NSF Rept. G.7337.

AGMON, S. (1965), *Lectures on Elliptic Boundary Value Problems* (Van Nostrand, Princeton, NJ).

ALFELD, P. (1984), A trivariate Clough–Tocher scheme for tetrahedral data, *Comput. Aided Geom. Design* **1**, 169–181.

APPRATO, D. and R. ARCANGÉLI (1979), Approximation d'un problème aux limites elliptique d'ordre deux par éléments finis rationnels de Wachspress avec intégration numérique, *RAIRO Anal. Numér.* **13**, 3–20.

APPRATO, D., R. ARCANGÉLI and J.L. GOUT (1979a), Sur les éléments finis rationnels de Wachspress, *Numer. Math.* **32**, 247–270.

APPRATO, D., R. ARCANGÉLI and J.L. GOUT (1979b), Rational interpolation of Wachspress error estimates, *Comput. Math. Appl.* **5**, 329–336.

APPRATO, D., R. ARCANGÉLI and R. MANZANILLA (1987), Sur la construction de surfaces de class C^k à partir d'un grand nombre de données de Lagrange, *Modélisation Math. Anal. Numér.* **21**, 529–555.

ARCANGÉLI, R. and J.L. GOUT (1976), Sur l'évaluation de l'erreur d'interpolation de Lagrange dans un ouvert de $\mathbb{R}^n$, *Rev. Française Automat. Informat. Recherche Opérationnelle Sér. Rouge Anal. Numér.* **10**, 5–27.

ARGYRIS, J.H. (1954–1955), Energy theorems and structural analysis, Part I: General theory, *Aircraft Engrg.* **26**, 347–356, 383–387, 394; **27**, 42–58, 80–94, 125–134 (also published as a book, Butterworths Scientific Publications, London, 1960).

ARGYRIS, J.H. and I. FRIED (1968), The LUMINA element for the matrix displacement method (Lagrangian interpolation), *Aero. J. Roy. Aero. Soc.* **72**, 514–517.

ARGYRIS, J.H., I. FRIED and D.W. SCHARPF (1968), The TUBA family of plate elements for the matrix displacement method, *Aero. J. Roy. Aero. Soc.* **72**, 701–709.

ARGYRIS, J.H., M. HAASE and H.P. MLEJNEK (1982), On an unconventional but natural formation of a stiffness matrix, *Comput. Methods Appl. Mech. Engrg.* **22**, 1–22.

ARGYRIS, J.H., M. HAASE and H.P. MLEJNEK (1980), Some considerations on the natural approach, *Comput. Methods Appl. Mech. Engrg.* **30**, 335–346.

ARGYRIS, J.H. and H.P. MLEJNEK (1986–1988), *Die Methode der Finiten Elemente, Band 1: Verschiebungsmethode in der Statik; Band 2: Kraft und gemischte Methoden, Nicht-linearitäten; Band 3: Einführung in die Dynamik* (Vieweg, Braunschweig) (English translation in preparation; North-Holland, Amsterdam).

ARONSZAJN, N. and K.T. SMITH (1957), Characterization of positive reproducing kernels; Application to Green's functions, *Amer. J. Math.* **79**, 611–622.

ATTÉIA, M. (1975), Fonctions "spline" et méthode d'éléments finis, *Rev. Française Automat. Informat. Recherche Opérationnelle Sér. Rouge Anal. Numér.* **R-2**, 13–40.

ATTÉIA, M. (1977), Evaluation de l'erreur dans la méthode des éléments finis, *Numer. Math.* **28**, 295–306.

AUBIN, J.P. (1967a), Approximation des espaces de distributions et des opérateurs différentiels, *Mémoire* **12**, *Bull. Soc. Math. France.*

AUBIN, J.P. (1967b), Behavior of the error of the approximate solutions of boundary value problems for linear elliptic operators by Galerkin's and finite difference methods, *Ann. Scuola Norm. Sup. Pisa* **21**, 599–637.

AUBIN, J.P. (1968a), Evaluation des erreurs de troncature des approximations des espaces de Sobolev, *J. Math. Anal. Appl.* **21**, 356–368.

AUBIN, J.P. (1968b), Interpolation et approximations optimales et "spline functions", *J. Math. Anal. Appl.* **24**, 1–24.

AUBIN, J.P. (1969), Approximation des problèmes aux limites non homogènes et régularité de la convergence, *Calcolo* **6**, 117–139.

AUBIN, J.P. (1972), *Approximation of Elliptic Boundary-Value Problems* (Wiley-Interscience, New York).

AVEZ, A. (1983), *Calcul Différentiel* (Masson, Paris).

AXELSSON, O. and V.A. BARKER (1984), *Finite Element Solution of Boundary Value Problems: Theory and Computation* (Academic Press, New York).

BABU, D.K. and G.F. PINDER (1984), Analytical integration formulae for linear isoparametric finite elements, *Internat. J. Numer. Methods Engrg.* **20**, 1153–1166.

BABUŠKA, I. (1963), The theory of small changes in the domain of existence in the theory of partial differential equations and its applications, in: *Differential Equations and Their Applications* (Academic Press, New York) 13–26.

BABUŠKA, I. (1970a), Finite element methods for domains with corners, *Computing* **6**, 264–273.

BABUŠKA, I. (1970b), Approximation by hill functions, *Comment. Math. Univ. Carolin.* **11**, 787–811.

BABUŠKA, I. (1971a), The rate of convergence for the finite element method, *SIAM J. Numer. Anal.* **8**, 304–315.

BABUŠKA, I. (1971b), Error bounds for the finite element method, *Numer. Math.* **16**, 322–333.

BABUŠKA, I. (1972a), A finite element scheme for domains with corners, *Numer. Math.* **20**, 1–21.

BABUŠKA, I. (1972b), Approximation by hill functions II, *Comment. Math. Univ. Carolin.* **13**, 1–22.

BABUŠKA, I. (1973a), The finite element method with Lagrangian multipliers, *Numer. Math.* **20**, 179–192.

BABUŠKA, I. (1973b), The finite element method with penalty, *Math. Comp.* **27**, 221–228.

BABUŠKA, I. (1974a), Method of weak elements, Technical Note BN-809, University of Maryland, College Park, MD.

BABUŠKA, I. (1974b), Solution of problems with interfaces and singularities, in: C. DE BOOR, ed., *Mathematical Aspects of Finite Elements in Partial Differential Equations*, (Academic Press, New York) 213–277.

BABUŠKA, I. (1976), Singularities problem in the finite element method, in: K.-J. BATHE, J.T. ODEN and W. WUNDERLICH, eds., *Formulation and Computational Algorithms in Finite Element Analysis* (MIT Press, Cambridge, MA) 748–792.

BABUŠKA, I. and A.K. AZIZ (1972), Survey lectures on the mathematical foundations of the finite element method, in: A.K. AZIZ, ed., *The Mathematical Foundations of the Finite Element Method with Applications to Partial Differential Equations* (Academic Press, New York) 3–359.

BABUŠKA, I. and A.K. AZIZ (1976), On the angle condition in the finite element method, *SIAM J. Numer. Anal.* **13**, 214–226.

BABUŠKA, I. and R.B. KELLOGG (1975), Nonuniform error estimates for the finite element method, *SIAM J. Numer. Anal.* **12**, 868–875.

BABUŠKA, I. and A. MILLER (1984a), The post-processing approach in the finite element method, Part 1: Calculation of displacements, stresses and other higher derivatives of the displacements, *Internat. J. Numer. Methods Engrg.* **20**, 1085–1109.

BABUŠKA, I. and A. MILLER (1984b), The post-processing approach in the finite element method, Part 2: The calculation of stress intensity factors, *Internat. J. Numer. Methods Engrg.* **20**, 1111–1129.

BABUŠKA, I. and J. OSBORN (1980), Analysis of finite element methods for second order boundary value problems using mesh dependent norms, *Numer. Math.* **34**, 41–62.

BABUŠKA, I. and J. OSBORN (1986), Finite element methods for the solution of problems with rough input data, in: P. GRISVARD, W. WENDLAND and J.R. WHITEMAN, eds., *Singularities and Constructive Methods for Their Treatment*, Lecture Notes in Mathematics **1121** (Springer, Berlin) 1–18.

BABUŠKA, I. and M.B. ROSENZWEIG (1972), A finite element scheme for domains with corners, *Numer. Math.* **20**, 1–21.

BABUŠKA, I. and M. ZLÁMAL (1973), Nonconforming elements in the finite element method with penalty, *SIAM J. Numer. Anal.* **10**, 863–875.

BAIOCCHI, C. (1977), Estimations d'erreur dans L^∞ pour les inéquations à obstacle, in: I. GALLIGANI and E. MAGENES, eds., *Mathematical Aspects of Finite Element Methods*, Lecture Notes in Mathematics **606** (Springer, Berlin) 27–34.

BAIOCCHI, C. and A. CAPELO (1978), *Disequazioni Variazionali e Quasi Variazionali: Applicazioni a Problemi di Frontiera Libera* (2 Vols.) (Pitagora, Bologna). English translation: *Variational and Quasivariational Inequalities: Applications to Free Boundary Problems* (Wiley-Interscience, New York, 1984).

BAKER, G.A. (1973), Simplified proofs of error estimates for the least squares method for Dirichlet's problem, *Math. Comp.* **27**, 229–235.

BAKKER, M. (1984), One-dimensional Galerkin methods and superconvergence at interior nodal points, *SIAM J. Numer. Anal.* **21**, 101–110.

BAMBERGER, Y. (1981), *Mécanique de l'Ingénieur II: Milieux Déformables* (Hermann, Paris).

BARNHILL, R.E. (1975a), Blending function finite elements for curved boundaries, in: J.R. WHITEMAN, ed., *Proceedings Conference on the Mathematics of Finite Elements and Applications* (Academic Press, London) 59–66.

BARNHILL, R.E. (1975b), Blending function interpolation: A survey and some new results, in: *Numerische Methoden der Approximationstheorie, Band 3* (Oberwolfach) 43–89.

BARNHILL, R.E., G. BIRKHOFF and W.J. GORDON (1973), Smooth interpolation in triangles, *J. Approximation Theory* **8**, 114–128.

BARNHILL, R.E. and J.H. BROWN (1975), Curved nonconforming elements for plate problems, Rep. No. 8, University of Dundee.

BARNHILL, R.E., J.H. BROWN, N. MCQUEEN and A.R. MITCHELL (1976), Computable finite element error bounds for Poisson's equation, *Internat. J. Numer. Methods Engrg.* **11**, 593–597.

BARNHILL, R.E. and G. FARIN (1981), C^1 quintic interpolation over triangles: Two explicit representations, *Internat. J. Numer. Methods Engrg.* **17**, 1763–1778.

BARNHILL, R.E. and J.A. GREGORY (1975a), Compatible smooth interpolation in triangles, *J. Approximation Theory* **15**, 214–225.

BARNHILL, R.E. and J.A. GREGORY (1975b), Polynomial interpolation to boundary data on triangles, *Math. Comp.* **29**, 726–735.

BARNHILL, R.E. and J.A. GREGORY (1976a), Sard kernel theorems on triangular domains with application to finite element error bounds, *Numer. Math.* **25**, 215–229.

BARNHILL, R.E. and J.A. GREGORY (1976b), Interpolation remainder theory from Taylor expansions on triangles, *Numer. Math.* **25**, 401–408.

BARNHILL, R.E. and J.R. WHITEMAN (1973), Error analysis of finite element methods with triangles for elliptic boundary value problems, in: J.R. WHITEMAN, ed., *The Mathematics of Finite Elements and Applications* (Academic Press, London) 83–112.

BARNHILL, R.E. and J.R. WHITEMAN (1975), Error analysis of Galerkin methods for Dirichlet problems containing boundary singularities, *J. Inst. Math. Appl.* **15**, 121–125.

BARRETT, J.W. and C.M. ELLIOTT (1984), A finite element method for solving elliptic equations with Neumann data on a curved boundary using unfitted meshes, *IMA J. Numer. Anal.* **4**, 309–325.

BARRETT, J.W. and C.M. ELLIOTT (1987), A practical finite element approximation of a semi-definite Neumann problem on a curved domain, *Numer. Math.* **51**, 23–36.

BATHE, K.-J. (1982), *Finite Element Procedures in Engineering Analysis* (Prentice-Hall, Englewood Cliffs, NJ).

BATOZ, J.-L., K.-J. BATHE and L-W. HO (1980), A study of three-node triangular plate bending elements, *Internat. J. Numer. Methods Engrg.* **15**, 1771–1812.

BAZELEY, G.P., Y.K. CHEUNG, B.M. IRONS and O.C. ZIENKIEWICZ (1965), Triangular elements in bending: Conforming and nonconforming solutions, in: *Proceedings Conference on Matrix Methods in Structural Mechanics*, Wright Patterson A.F.B., Dayton, OH, 547–576.

BELL, K. (1969), A refined triangular plate bending element, *Internat. J. Numer. Methods Engrg.* **1**, 101–122.

BELYTSCHKO, T. and J.S.-J. ONG (1984), A consistent control of spurious singular modes in the 9-node Lagrange element for the Laplace and Mindlin plate equations, *Comput. Methods Appl. Mech. Engrg.* **44**, 269–295.

BENDALLI, A. (1981), Approximation of a degenerate boundary value problem by a finite element method, *RAIRO Anal. Numér.* **15**, 87–99.

BERGER, A.E. (1973), L^2-error estimates for finite elements with interpolated boundary conditions, *Numer. Math.* **21**, 345–349.

BERGER, A.E. (1976), The truncation method for the solution of a class of variational inequalities, *Rev. Française Automat. Informat. Recherche Opérationnelle Sér. Rouge Anal. Numér.* **10**, 29–42.

BERGER, A.E., R. SCOTT and G. STRANG (1972), Approximate boundary conditions in the finite element method, in: *Symposia Mathematica* **10** (Academic Press, New York) 295–313.

BERNADOU, M. and J.M. BOISSERIE (1982), *The Finite Element Method in Thin Shell Theory: Application to Arch Dam Simulation* (Birkhaüser, Boston, MA).

BERNADOU, M. and Y. DUCATEL (1978), Méthodes d'éléments finis avec intégration numérique pour des problèmes elliptiques du quatrième ordre, *RAIRO Anal. Numér.* **12**, 3–26.

BERNADOU, M. and K. HASSAN (1981), Basis functions for general Hsieh–Clough–Tocher triangles, complete or reduced, *Internat. J. Numer. Methods Engrg.* **17**, 784–789.

BERNARDI, C. (1986), *Contributions à l'Analyse Numérique de Problèmes Non Linéaires*, Doctoral Dissertation, Université Pierre et Marie Curie, Paris, France.

BERNARDI, C. (1988), Optimal finite element interpolation on curved domains, Technical Rept. R88008, Laboratoire d'Analyse Numérique, Université Pierre et Marie Curie, Paris, France.

BERS, L., F. JOHN and M. SCHECHTER (1964), *Partial Differential Equations* (Wiley, New York).

BIRKHOFF, G. (1969), Piecewise bicubic interpolation and approximation in polygons, in: I.J. SCHOENBERG, ed., *Approximation with Special Emphasis on Spline Functions* (Academic Press, New York) 185–221.

BIRKHOFF, G. (1971), Tricubic polynomial interpolation, *Proc. Nat. Acad. Sci. U.S.A.* **68**, 1162–1164.

BIRKHOFF, G. (1972), Piecewise analytic interpolation and approximation in triangulated polygons, in: A.K. AZIZ, ed., *The Mathematical Foundations of the Finite Element Method with Applications to Partial Differential Equations* (Academic Press, New York) 363–385.

BIRKHOFF, G. and S. GULATI (1974), Optimal few-point discretizations of linear source problems, *SIAM J. Numer. Anal.* **11**, 700–728.

BIRKHOFF, G. and L. MANSFIELD (1974), Compatible triangular finite elements, *J. Math. Anal. Appl.* **47**, 531–553.

BIRKHOFF, G., M.H. SCHULTZ and R.S. VARGA (1968), Piecewise Hermite interpolation in one and two variables with applications to partial differential equations, *Numer. Math.* **11**, 232–256.

BLAIR, J.J. (1976), Higher order approximations to the boundary conditions for the finite element method, *Math. Comp.* **30**, 250–262.

BLUM, H. and M. DOBROWOLSKI (1983), On finite element methods for elliptic equations on domains with corners, *Computing* **28**, 53–63.

BLUM, H., Q. LIN and R. RANNACHER (1986), Asymptotic error expansion and Richardson extrapolation for linear finite elements, *Numer. Math.* **49**, 11–37.

BLUM, H. and R. RANNACHER (1980), On the boundary value problem of the biharmonic operator on domains with angular corners, *Math. Methods Appl. Sci.* **2**, 556–581.

BOATTIN, M. (1988), Approximation en coordonnées barycentriques généralisées, Thèse, Université Paul Sabatier, Toulouse, France.

BOGNER, F.K., R.L. FOX and L.A. SCHMIT (1965), The generation of interelement compatible stiffness and mass matrices by the use of interpolation formulas, in: *Proceedings Conference on Matrix Methods in Structural Mechanics*, Wright Patterson A.F.B., Dayton, OH, 397–444.

BRAMBLE, J.H. (1970), *Variational Methods for the Numerical Solution of Elliptic Problems* (Chalmers Institute of Technology, Göteborg).

BRAMBLE, J.H. (1975), A survey of some finite element methods proposed for treating the Dirichlet problem, *Adv. Math.* **16**, 187–196.

BRAMBLE, J.H. (1981), The Lagrange multiplier method for Dirichlet's problem, *Math. Comp.* **37**, 1–12.

BRAMBLE, J.H., T. DUPONT and V. THOMÉE (1972), Projection methods for Dirichlet's problem in approximating polygonal domains with boundary-value corrections, *Math. Comp.* **26**, 869–879.

BRAMBLE, J.H. and S.R. HILBERT (1970), Estimation of linear functionals on Sobolev spaces with application to Fourier transforms and spline interpolation, *SIAM J. Numer. Anal.* **7**, 113–124.

BRAMBLE, J.H. and S.R. HILBERT (1971), Bounds for a class of linear functionals with applications to Hermite interpolation, *Numer. Math.* **16**, 362–369.

BRAMBLE, J.H., B.E. HUBBARD and V. THOMÉE (1969), Convergence estimates for essentially positive type discrete problems, *Math. Comp.* **23**, 695–710.

BRAMBLE, J.H. and J.A. NITSCHE (1973), A generalized Ritz-least-squares method for Dirichlet problems, *SIAM J. Numer. Anal.* **10**, 81–93.

BRAMBLE, J.H., J.A. NITSCHE and A.H. SCHATZ (1975), Maximum-norm interior estimates for Ritz-Galerkin methods, *Math. Comp.* **29**, 677–688.

BRAMBLE, J.H. and A.H. SCHATZ (1970), Rayleigh–Ritz–Galerkin methods for Dirichlet's problem using subspaces without boundary conditions, *Comm. Pure Appl. Math.* **23**, 653–675.

BRAMBLE, J.H. and A.H. SCHATZ (1971), Least squares methods for 2*m*th order elliptic boundary-value problems, *Math. Comp.* **25**, 1–32.

BRAMBLE, J.H. and A.H. SCHATZ (1976), Estimates for spline projections, *Rev. Française Automat. Informat. Recherche Opérationnelle Sér. Rouge Anal. Numér.* **10**, 5–37.

BRAMBLE, J.H. and A.H. SCHATZ (1977), Higher order local accuracy by averaging in the finite element method, in: C. DE BOOR, ed., *Mathematical Aspects of Finite Elements in Partial Differential Equations* (Academic Press, New York), 1–14.

BRAMBLE, J.H. and R. SCOTT (1978), Simultaneous approximation in scales of Banach spaces, *Math. Comp.* **32**, 947–954.

BRAMBLE, J.H. and V. THOMÉE (1974), Interior maximum norm estimates for some simple finite element methods, *Rev. Française Automat. Informat. Recherche Opérationnelle Sér. Rouge Anal. Numér.* **R-2**, 5–18.

BRAMBLE, J.H. and M. ZLÁMAL (1970), Triangular elements in the finite element method, *Math. Comp.* **24**, 809–820.

BREZIS, H. (1983), *Analyse Fonctionnelle* (Masson, Paris).

BREZIS, H. and G. STAMPACCHIA (1968), Sur la régularité de la solution d'inéquations elliptiques, *Bull. Soc. Math. France* **96**, 153–180.

BREZIS, H. and G. STAMPACCHIA (1973), Une nouvelle méthode pour l'étude d'écoulements stationnaires, *C.R. Acad. Sci. Paris Sér A* **276**, 129–132.

BREZZI, F., W.W. HAGER and P.-A. RAVIART (1977), Error estimates for the finite element solution of variational inequalities, Part 1: Primal theory, *Numer. Math.* **28**, 431–443.

BREZZI, F., C. JOHNSON and B. MERCIER (1977), Analysis of a mixed finite element method for elasto-plastic plates, *Math. Comp.* **31**, 809–817.

BREZZI, F. and L.D. MARINI (1975), On the numerical solution of plate bending problems by hybrid methods, *Rev. Française Automat. Informat. Recherche Opérationnelle Sér. Rouge Anal. Numér.* **R-3**, 5–50.

BREZZI, F. and P.-A. RAVIART (1976), Mixed finite element methods for 4th order elliptic equations, Rapport interne No. 9, Centre de Mathématiques Appliquées, Ecole Polytechnique, Palaiseau.

BRISTEAU, M.-O. (1975), Application de la méthode des eléments finis à la résolution numérique d'inéquations variationnelles d'evolution de type Bingham, Doctoral Thesis (3ème Cycle), Université Pierre et Marie Curie, Paris, France.

CAREY, G.F. (1976), An analysis of finite element equations and mesh subdivisions, *Comput. Methods Appl. Mech. Engrg.* **9**, 165–179.

CARLSON, R.E. and C.A. HALL (1973), Error bounds for bicubic spline interpolation, *J. Approximation Theory* **4**, 41–47.

CARTAN, H. (1967), *Calcul Différentiel* (Hermann, Paris).

CASAS, E. (1985), L^2 estimates for the finite element method for the Dirichlet problem with singular data, *Numer. Math.* **47**, 627–632.

CAVENDISH, J.C., W.J. GORDON and C.A. HALL (1976), Ritz–Galerkin approximations in blending function spaces, *Numer. Math.* **26**, 155–178.

CÉA, J. (1964), Approximation variationnelle des problèmes aux limites, *Ann. Inst. Fourier (Grenoble)* **14**, 345–444.

CÉA, J. (1976), Approximation variationnelle; convergence des éléments finis; un test, in: *Journées Eléments Finis* (Université de Rennes, Rennes).

Cermák, L. (1983a), The finite element solution of second order elliptic problems with the Newton boundary condition, *Apl. Mat.* **28**, 430–456.

Čermák, L. (1983b), A note on a discrete form of Friedrichs' inequality, *Apl. Mat.* **28**, 457–466.

Chernuka, M.W., G.R. Cowper, G.M. Lindberg and M.D. Olson (1972), Finite element analysis of plates with curved edges, *Internat. J. Numer. Methods Engrg.* **4**, 49–65.

Choquet-Bruhat, Y. (1973), *Distributions, Théorie et Problèmes* (Masson, Paris).

Chou, S.-I. and C.-C. Wang (1979), Error estimates of finite element approximations for problems in linear elasticity, Part 1: Problems in elastostatics, *Arch. Rational Mech. Anal.* **72**, 41–60.

Ciarlet, P.G. (1968), An $O(h^2)$ method for a non-smooth boundary value problem, *Aequationes Math.* **2**, 39–49.

Ciarlet, P.G. (1970a), Discrete variational Green's function, I: *Aequationes Math.* **4**, 74–82.

Ciarlet, P.G. (1970b), Discrete maximum principle for finite-difference operators, *Aequationes Math.* **4**, 338–352.

Ciarlet, P.G. (1971), Fonctions de Green discrètes et principe du maximum discret, Doctoral Dissertation, Université de Paris VI, Paris, France.

Ciarlet, P.G. (1974a), Conforming and nonconforming finite element methods for solving the plate problem, in: G.A. Watson, ed., *Conference on the Numerical Solution of Differential Equations*, Lecture Notes in Mathematics **363** (Springer, Berlin) 21–31.

Ciarlet, P.G. (1974b), Quelques méthodes d'éléments finis pour le problème d'une plaque encastrée, in: R. Glowinski and J.L. Lions, eds., *Computing Methods in Applied Sciences and Engineering*, Lecture Notes in Computer Science **10** (Springer, Berlin) 156–176.

Ciarlet, P.G. (1974c), Sur l'élément de Clough et Tocher, *Rev. Française Automat. Informat. Recherche Opérationnelle Sér. Rouge Anal. Numér.* **R-2**, 19–27.

Ciarlet, P.G. (1975), Lectures on the finite element method, Tata Institute of Fundamental Research, Bombay.

Ciarlet, P.G. (1976), *Numerical Analysis of the Finite Element Method*, Séminaire de Mathématiques Supérieures (Presses de l'Université de Montréal, Montréal, Que.).

Ciarlet, P.G. (1978), *The Finite Element Method for Elliptic Problems* (North-Holland, Amsterdam).

Ciarlet, P.G. (1983), *Introduction à l'Analyse Numérique Matricielle et à l'Optimisation* (Masson, Paris); English translation: *Introduction to Numerical Linear Algebra and Optimization* (Cambridge University Press, Cambridge 1989).

Ciarlet, P.G. (1988), *Mathematical Elasticity* I: *Three-Dimensional Elasticity* (North-Holland, Amsterdam).

Ciarlet, P.G. (1990), *Mathematical Elasticity* II: *Lower-Dimensional Theories of Plates and Rods* (North-Holland, Amsterdam).

Ciarlet, P.G. and P. Destuynder (1979), A justification of the two-dimensional linea· · .te model, *J. Mécanique* **18**, 315–344.

Ciarlet, P.G. and S. Kesavan (1980), Two-dimensional approximation of three-dime..onal eigenvalue problems in plate theory, *Comput. Methods Appl. Mech. Engrg.* **26**, 149–172.

Ciarlet, P.G., F. Natterer and R.S. Varga (1970), Numerical methods of high-order accuracy for singular nonlinear boundary value problems, *Numer. Math.* **15**, 87–99.

Ciarlet, P.G. and P. Rabier (1980), *Les Equations de von Kármán*, Lecture Notes in Mathematics **826** (Springer, Berlin).

Ciarlet, P.G. and P.-A. Raviart (1972a), General Lagrange and Hermite interpolation in R^n with applications to finite element methods, *Arch. Rational Mech. Anal.* **46**, 177–199.

Ciarlet, P.G. and P.-A. Raviart (1972b), Interpolation theory over curved elements, with applications to finite element methods, *Comput. Methods Appl. Mech. Engrg.* **1**, 217–249.

Ciarlet, P.G. and P.-A. Raviart (1972c), The combined effect of curved boundaries and numerical integration in isoparametric finite element methods, in: A.K. Aziz, ed., *The Mathematical Foundations of the Finite Element Method with Applications to Partial Differential Equations* (Academic Press, New York) 409–474.

Ciarlet, P.G. and P.-A. Raviart (1973), Maximum principle and uniform convergence for the finite element method, *Comput. Methods Appl. Mech. Engrg.* **2**, 17–31.

CIARLET, P.G. and R.S. VARGA (1970), Discrete variational Green's function, II: One dimensional problem, *Numer. Math.* **16**, 115–128.

CIARLET, P.G. and C. WAGSCHAL (1971), Multipoint Taylor formulas and applications to the finite element method, *Numer. Math.* **17**, 84–100.

CIAVALDINI, J.F. and M. CROUZEIX (1985), A finite element method scheme for one dimensional elliptic equations with high super convergence at the nodes, *Numer. Math.* **46**, 417–427.

CIAVALDINI, J.F. and J.C. NÉDÉLEC (1974), Sur l'élément de Fraeijs de Veubeke et Sander, *Rev. Française Automat. Informat. Recherche Opérationnelle Sér. Rouge Anal. Numér.* **R-2**, 29–45.

CIAVALDINI, J.F. and G. TOURNEMINE (1977), A finite element method to compute stationary steady state flows in the hodograph plane, *J. Indian Math. Soc.* **41**, 69–89.

CLÉMENT, P. (1974), Méthode des éléments finis appliquée à des problèmes variationnels de type indéfini. Doctoral Thesis, Ecole Polytechnique Fédérale de Lausanne, Lausanne, Switzerland.

CLÉMENT, P. (1975), Approximation by finite element functions using local regularization, *Rev. Française Automat. Informat. Recherche Opérationnelle Sér. Rouge Anal. Numér.* **R-2**, 77–84.

CLOUGH, R.W. (1960), The finite element method in plane stress analysis, in: *Proceedings Second ASCE Conference on Electronic Computation*, Pittsburgh, PA.

CLOUGH, R.W. and J.L. TOCHER (1965), Finite element stiffness matrices for analysis of plates in bending, in: *Proceedings Conference on Matrix Methods in Structural Mechanics*, Wright Patterson A.F.B., Dayton, OH, 515–545.

COATMÉLEC, C. (1966), Approximation et interpolation des fonctions différentiables de plusieurs variables, *Ann. Sci. Ecole Norm. Sup.* **83**, 271–341.

CORTEY-DUMONT, P. (1983), Contribution à l'approximation des inéquations variationelles en norme L^∞, *C.R. Acad. Sci. Paris Sér. I. Math.* **296**, 753–756.

COURANT, R. (1943), Variational methods for the solution of problems of equilibrium and vibrations, *Bull. Amer. Math. Soc.* **49**, 1–23.

COURANT, R. and D. HILBERT (1953), *Methods of Mathematical Physics*, I (Interscience, New York).

COURANT, R. and D. HILBERT (1962), *Methods of Mathematical Physics*, II (Interscience, New York).

CROUZEIX, M. and A. MIGNOT (1984), *Analyse Numérique des Equations Différentielles* (Masson, Paris).

CROUZEIX, M. and P.-A. RAVIART (1973), Conforming and nonconforming finite element methods for solving the stationary Stokes equations I, *Rev. Française Automat. Informat. Recherche Opérationnelle Sér. Rouge Anal. Numér.* **R-3**, 33–76.

CROUZEIX, M. and J.M. THOMAS (1973), Eléments finis et problèmes elliptiques dégénérés, *Rev. Française Automat. Informat. Recherche Opérationnelle Sér. Rouge Anal. Numér.* **R-3**, 77–104.

CROUZEIX, M. and V. THOMÉE (1987), The stability in L_p and W_p^1 of the L_2 projection onto finite element function spaces, *Math. Comp.* **48**, 521–532.

DAILEY, J.W. and J.G. PIERCE (1972), Error bounds for the Galerkin method applied to singular and nonsingular boundary value problems, *Numer. Math.* **19**, 266–282.

DAUTRAY, R. and J.-L. LIONS (1984), *Analyse Mathématique et Calcul Numérique pour les Sciences et les Techniques*, Collection du Commissariat à l'Energie Atomique (Masson, Paris).

DAVIS, P.J. and P. RABINOWITZ (1975), *Methods of Numerical Integration* (Academic Press, New York).

DÉLÈZE, M. and J.-J. GOËL (1976), Tétraèdre comme élément fini de classe C^1, à seize paramètres, contenant les polynômes de degré deux, Report, Institut de Mathématiques, Université de Fribourg, Fribourg, France.

DENY, J. and J.L. LIONS (1953–1954), Les espaces du type de Beppo Levi, *Ann. Inst. Fourier (Grenoble)* **5**, 305–370.

DESCLOUX, J. (1973), Two basic properties of finite elements, Report, Département de Mathématiques, Ecole Polytechnique Fédérale de Lausanne, Switzerland.

DESCLOUX, J. (1975), Interior regularity and local convergence of Galerkin finite element approximations for elliptic equations, in: J.J.H. MILLER, ed., *Topics in Numerical Analysis* II (Academic Press, New York) 27–41.

DESCLOUX, J. (1977), Interior L^∞ estimates for Galerkin approximations of elliptic equations, in: *Proceedings Symposium on the Mathematical Aspects of the Finite Element Methods*, Rome, Italy.

DESCLOUX, J. and N. NASSIF (1977), Interior L^∞ estimates for finite element approximations of solutions of elliptic equations, in: I. GALLIGANI and E. MAGENES, eds., *Mathematical Aspects of Finite Element Methods*, Lecture Notes in Mathematics **606** (Springer, Berlin) 56–63.

DESTUYNDER, P. (1981), Comparaison entre les modèles tridimensionnels et bidimensionnels de plaques en élasticité, *RAIRO Anal. Numér.* **15**, 331–369.

DESTUYNDER, P. (1986), *Une Théorie Asymptotique des Plaques Minces en Elasticité Linéaire* (Masson, Paris).

DHATT, G. and G. TOUZOT (1984), *The Finite Element Method Displayed* (Wiley, New York).

DIEUDONNÉ, J. (1967), *Fondements de l'Analyse Moderne* (Gauthier-Villars, Paris).

DIEUDONNÉ, J. (1982), *Eléments d'Analyse* **9** (Gauthier-Villars, Paris).

DI GUGLIELMO, F. (1971), Résolution approchée de problèmes aux limites elliptiques par des schémas aux éléments finis à plusieurs fonctions arbitraires, *Calcolo* **8**, 185–213.

DI GUGLIELMO, F. (1971), Résolution approchée de problèmes aux limites elliptiques par des schémas aux éléments finis à plusieurs fonctions arbitraires, *Calcolo* **8**, 185–213.

DOUGLAS Jr, J. and T. DUPONT (1973), Superconvergence for Galerkin Methods for the two point boundary problem via local projections, *Numer. Math.* **21**, 270–278.

DOUGLAS, Jr, J. and T. DUPONT (1974), Galerkin approximations for the two point boundary problem using continuous, piecewise polynomial spaces, *Numer. Math.* **22**, 99–109.

DOUGLAS Jr, J. and T. DUPONT (1976), Interior penalty procedures for elliptic and parabolic Galerkin methods, in: R. GLOWINSKI and J.L. LIONS, ed., *Computing Methods in Applied Sciences*, Lecture Notes in Physics **58** (Springer, Berlin) 207–216.

DOUGLAS Jr, J., T. DUPONT, P. PERCELL and R. SCOTT (1979), A family of C^1 finite elements with optimal approximation properties for various Galerkin methods for 2nd and 4th order problems, *Rev. Française Automat. Informat. Recherche Opérationelle Sér. Rouge Anal. Numér.* **R-13**, 227–255.

DOUGLAS Jr, J., T. DUPONT and L. WAHLBIN (1975a), The stability in L^q of the L^2-projection into finite element function spaces, *Numer. Math.* **23**, 193–197.

DOUGLAS Jr, J., T. DUPONT and L. WAHLBIN (1975b), Optimal L_∞ error estimates for Galerkin approximations to solutions of two point boundary value problems, *Math. Comp.* **29**, 475–483.

DOUGLAS Jr, J., T. DUPONT and M.F. WHEELER (1974a), A Galerkin procedure for approximating the flux on the boundary for elliptic and parabolic boundary value problems, *Rev. Française Automat. Informat. Recherche Opérationnelle Sér. Rouge Anal. Numér.* **R-2**, 47–59.

DOUGLAS Jr, J., T. DUPONT and M.F. WHEELER (1974b), An L^∞ estimate and a super-convergence result for a Galerkin method for elliptic equations based on tensor products of piecewise polynomials, *Rev. Française Automat. Informat. Recherche Opérationnelle Sér. Rouge Anal. Numér.* **R-2**, 61–66.

DUCHON, J. (1976), Interpolation des fonctions de deux variables suivant le principe de la flexion des plaques minces, *Rev. Française Automat. Informat. Recherche Opérationelle Sér. Rough Anal. Numér.* **10** (12), 5–12.

DUCHON, J. (1977), Splines minimizing rotation: invariant semi-norms in Sobolev spaces in: *Constructive Theory of Functions of Several Variables*, Lecture Notes in Mathematics **571** (Springer, Berlin) 85–100.

DUNAVANT, D.A. (1985), High degree efficient symmetrical Gaussian quadrature rules for the triangle, *Internat. J. Numer. Meth. Engrg.* **21**, 1129–1148.

DUNAVANT, D.A. (1986), Efficient symmetrical cubature rules for complete polynomials of high degree over the unit cube, *Internat. J. Numer. Meth. Engrg.* **23**, 397–407.

DUPONT, T. and R. SCOTT (1978), Constructive polynomial approximation in Sobolov spaces, in: C. DE BOOR and G. GOLUB, eds., *Recent Advances in Numerical Analysis* (Academic Press, New York) 31–44.

DUPONT, T. and R. SCOTT (1980), Polynomial approximation of functions in Sobolov spaces, *Math. Comp.* **34**, 441–463.

DUPUIS, G. and J.-J. GOËL (1970), Finite elements with a high degree of regularity, *Internat. J. Numer. Methods Engrg.* **2**, 563–577.

DURÁN, R.G. (1987), Quasi-optimal estimates for finite element approximations using Orlicz norms, *Math. Comp.* **49**, 17–23.

DUVAUT, G. and J.L. LIONS (1972), *Les Inéquations en Mécanique et en Physique* (Dunod, Paris).

EKELAND, I. and R. TEMAM (1974), *Analyse Convexe et Problèmes Variationnels* (Dunod, Paris).

EL-ZAFRANY, A. and R.A. COOKSON (1986), Derivation of Lagrangian and Hermitian shape functions for quadrilateral finite elements, *Internat. J. Numer. Meth. Engrg.* **23**, 1939–1958.

ERGATOUDIS, I., B.M. IRONS and O.C. ZIENKIEWICZ (1968), Curved, isoparametric, "quadrilateral" elements for finite element analysis, *Internat. J. Solids and Structures* **4**, 31–42.

ERIKSSON, K. (1985a), Improved accuracy by adapted mesh-refinements in the finite element method, *Math. Comp.* **44**, 321–343.

ERIKSSON, K. (1985b), Finite element methods of optimal order for problems with singular data, *Math. Comp.* **44**, 345–360.

ERIKSSON, K. (1985c), Higher order local rate of convergence by mesh refinement in the finite element method, *Math. Comp.* **45**, 109–142.

ERIKSSON, K. and V. THOMÉE (1984), Galerkin methods for singular boundary value problems in one space dimension, *Math. Comp.* **42**, 345–367.

FALK, R.S. (1974), Error estimates for the approximation of a class of variational inequalities, *Math. Comp.* **28**, 963–971.

FALK, R.S. (1975), Approximation of an elliptic boundary value problem with unilateral constraints, *Rev. Française Automat. Informat. Recherche Opérationnelle Sér. Rouge Anal. Numér.* **R-2**, 5–12.

FALK, R.S. and B. MERCIER (1977), Error estimates for elasto-plastic problems, *Rev. Française Automat. Informat. Recherche Opérationnelle Sér. Rouge Anal. Numér.* **11**, 135–144.

FELIPPA, C.A. (1966), Refined finite element analysis of linear and nonlinear two-dimensional structures, Doctoral Thesis, University of California, Berkeley, CA.

FELIPPA, C.A. and R.W. CLOUGH (1970), The finite element method in solid mechanics, in: G. BIRKHOFF and R.S. VARGA, eds., *Numerical Solution of Field Problems in Continuum Mechanics* (American Mathematical Society, Providence, RI) 210–252.

FICHERA, G. (1972a), Existence theorems in elasticity, *Handbuch der Physik* VIa/2 (Springer, Berlin) 347–389.

FICHERA, G. (1972b), Boundary value problems of elasticity with unilateral constraints, *Handbuch der Physik* VIa/2 (Springer, Berlin) 391–424.

FINZI-VITA, S. (1982), L^∞-error estimates for variational inequalities with Hölder continuous obstacle, *RAIRO Anal. Numér.* **16**, 27–37.

FIX, G.J. (1969), Higher-order Rayleigh-Ritz approximations, *J. Math. Mech.* **18**, 645–658.

FIX, G.J. (1972a), On the effects of quadrature in the finite element method, in: J.T. ODEN, R.W. CLOUGH and Y. YAMAMOTO, eds., *Advances in Computational Methods in Structural Mechanics and Design* (The University of Alabama Press, Huntsville, AL) 55–68.

FIX, G.J. (1972b), Effects of quadrature errors in finite element approximation of steady state, eigenvalue and parabolic problems, in: A.K. AZIZ, ed., *The Mathematical Foundations of the Finite Element Method with Applications to Partial Differential Equations* (Academic Press, New York) 525–556.

FIX, G.J., S. GULATI and G.I. WAKOFF (1973), On the use of singular functions with finite element approximations, *J. Comput. Phys.* **13**, 209–228.

FIX, G.J. and G. STRANG (1969), Fourier analysis of the finite element method in Ritz–Galerkin theory, *Stud. Appl. Math.* **48**, 265–273.

FORTIN, M. (1972), Calcul numérique des ecoulements des fluides de Bingham et des fluides Newtoniens incompressibles par des méthodes d'eléments finis, Doctoral Thesis, Université de Paris VI, France.

FORTIN, M. (1985), A three-dimensional quadratic nonconforming element, *Numer. Math.* **46**, 269–279.

FORTIN, M. and M. SOULIE (1983), A non-conforming piecewise quadratic finite element on triangles, *Internat. J. Numer. Methods Engrg.* **19**, 505–520.

FRAEIJS DE VEUBEKE, B. (1979), *A Course in Elasticity* (Springer, New York).

FRAEIJS DE VEUBEKE, B. (1984), Variational principles and the patch test, *Internat. J. Numer. Methods Engrg.* **8**, 783–801.

FREHSE, J. and R. RANNACHER (1976), Eine L^1-Fehlerabschätzung für diskrete Grundlösungen in der Methode der finite elemente, *Bonner Math. Schriften* **89**, 92–114.

FREHSE, J. and R. RANNACHER, (1978), Asymptotic L^∞-error estimates for linear finite element approximations of quasilinear boundary value problems, *SIAM J. Numer. Anal.* **15**, 418–431.

FREMOND, M. (1971), Etude de structures visco-elastiques stratifées soumises à des charges harmoniques, et de solides elastiques reposant sur ces structures, Doctoral Thesis, Université Pierre et Marie Curie (Paris VI), Paris, France.

FREMOND, M. (1972), Utilisation de la dualité en élasticité. Compléments sur les énergies de Reissner.

Equilibre d'une dalle élastique reposant sur une structure stratifiée, Annales de l'Institut Technique du Bâtiment et des Travaux Publics, Supplément au No. **294**, 54–66.

FRENCH, D.A. (1987), The finite element method for a degenerate elliptic equation, *SIAM J. Numer. Anal.* **24**, 788–815.

FRIED, I. (1980), On the optimality of pointwise accuracy of the finite element solution, *Internat. J. Numer. Methods Engrg.* **15**, 451–456.

FRIED, I., YANG, S.K. (1972), Best finite elements distribution around a singularity, *AIAA J.* **10**, 1244–1246.

FRIEDRICHS, K.O. and H.B. KELLER (1966), A finite difference scheme for generalized Neumann problems, in: J.H. BRAMBLE, ed., *Numerical Solution of Partial Differential Equations* (Academic Press, New York) 1–19.

GABAY, D. and B. MERCIER (1976), A dual algorithm for the solution of nonlinear variational problems via finite element approximation, *Comput. Math. Appl.* **2**, 17–40.

GALLAGHER, H. (1975), *Finite Element Analysis: Fundamentals*, (Prentice-Hall, Englewood Cliffs, NJ).

GASTALDI, L. and R. NOCHETTO (1987), Optimal L^∞-error estimates for nonconforming and mixed finite element methods of lowest order, *Numer. Math.* **50**, 587–611.

GEORGE, P.L. (1986), MODULEF: Génération automatique de maillages, Collection Didactique, INRIA, Rocquencourt, France.

GERMAIN, P. (1972), *Mécanique des Milieux Continus* I (Masson, Paris).

GERMAIN, P. (1986a), Mécanique I, Ecole Polytechnique, Palaiseau, and Ellipses, Paris, France.

GERMAIN, P. (1986b), Mécanique II, Ecole Polytechnique, Palaiseau, and Ellipses, Paris, France.

GILBARG, D. and N.S. TRUDINGER (1983), *Elliptic Partial Differential Equations of Second Order* (Springer, New York, 2nd ed.).

GIRAULT, V. (1976), Nonelliptic approximation of a class of partial differential equations with Neumann boundary conditions, *Math. Comp.* **30**, 68–91.

GIRAULT, V. and P.-A. RAVIART (1986), *Finite Element Methods for Navier-Stokes Equations* (Springer, Berlin).

GLOWINSKI, R. (1984), *Numerical Methods for Nonlinear Variational Problems* (Springer, New York).

GLOWINSKI, R. and H. LANCHON (1973), Torsion élasto-plastique d'une barre cylindrique de section multi-connexe, *J. Mécanique* **12**, 151–171.

GLOWINSKI, R., J.-L. LIONS and R. TRÉMOLIÈRES (1976a), *Analyse Numérique des Inéquations Variationnelles 1: Théorie Générale, Premières Applications* (Dunod, Paris).

GLOWINSKI, R., J.-L. LIONS and R. TRÉMOLIÈRES (1976b), *Analyse Numérique des Inéquations Variationnelles 2: Applications aux Phénomènes Stationnaires et d'Evolution* (Dunod, Paris).

GLOWINSKI, R., L.D. MARINI and M. VIDRASCU (1984), Finite-element approximations and iterative solutions of a fourth-order elliptic variational inequality, *IMA J. Numer. Anal.* **4**, 127–167.

GOPALACHARYULU, S. (1973), A higher order conforming, rectangular plate element, *Internat. J. Numer. Methods Engrg.* **6**, 305–309.

GORDON, W.J. and C.A. HALL (1973), Transfinite element methods: Blending-function interpolation over arbitrary curved element domains, *Numer. Math.* **21**, 109–129.

GOUT, J.L. (1977), Estimation de l'erreur d'interpolation d'Hermite dans R^n. *Numer. Math.* **28**, 407–429.

GOUT, J.L. (1979a), Interpolation error estimates on Hermite rational "Wachspress type" 3rd-degree finite element, *Comput. Math. Appl.* **5**, 249–257.

GOUT, J.L. (1979b), Construction of an Hermite rational interpolation Wachspress type finite element, *Comput. Math. Appl.* **5**, 337–347.

GOUT, J.L. and A. GUESSAB (1986a), Sur les formules de quadrature numérique à nombre minimal de noeuds d'intégration, *Numer. Math.* **49**, 439–455.

GOUT, J.L. and A. GUESSAB (1986b), Exemples de formules de quadrature numérique à nombre minimal de noeuds sur des domaines à double symétrie axiale, *Modélisation Math. Anal. Numér.* **20**, 287–314.

GRISVARD, P. (1976), Behavior of the solutions of an elliptic boundary value problem in a polygonal or polyhedral domain, in: B. HUBBARD, ed., *Numerical Solution of Partial Differential Equations*, III (SYNSPADE 1975) (Academic Press, New York) 207–274.

GRISVARD, P. (1985), *Elliptic Problems in Nonsmooth Domains* (Pitman, Boston, MA).

GRISVARD, P. (1987), Singularities in elasticity theory, in: P.G. CIARLET and M. ROSEAU, eds., *Applications of Multiple Scaling in Mechanics* (Masson, Paris).

GUESSAB, A. (1986), Cubature formulae which are exact on spaces P intermediate between P_k and Q_k, *Numer. Math.* **49**, 561–576.

GUESSAB, A. (1987), Sur les formules de quadrature numérique dans $\mathbb{R}^n$ avec certains noeuds ayant une composante connue, *Appl. Anal.* **26**, 129–144.

GURTIN, M.E. (1972), The linear theory of elasticity, in: S. FLÜGGE and C. TRUESDELL, eds., *Handbuch der Physik* VIa/2 (Springer, Berlin) 1–295.

HABER, S. (1970), Numerical evaluation of multiple integrals, *SIAM Rev.* **12**, 481–526.

HANSSEN, L., T.G. SYVERTSEN and P.G. BERGAN (1978), Stiffness derivation based on element convergence requirements, in: J.R. WHITEMAN, ed., *The Mathematics of Finite Elements and Applications* (Academic Press, London) 83–96.

HARVEY, J.W. and S. KELSEY (1971), Triangular plate bending elements with enforced compatibility, *AIAA J.* **9**, 1023–1026.

HAVERKAMP, R. (1984), Eine Aussage zur L_∞-Stabilität und zur genauen Konvergenzordnung der H_0^1-Projektionen, *Numer. Math.* **44**, 393–405.

HEDSTROM, G.W. and R.S. VARGA (1971), Application of Besov spaces to spline approximation, *J. Approximation Theory* **4**, 295–327.

HELFRICH, H.-P. (1976), Charakterisierung des K-Funktionales zwischen Hilberträumen und nicht-uniforme Fehlerschranken, *Bonner Math. Schriften* **89**, 31–41.

HERBOLD, R.J. (1968), Consistent quadrature schemes for the numerical solution of boundary value problems by variational techniques, Doctoral Thesis, Case Western Reserve University, Cleveland, OH.

HERBOLD, R.J., M.H. SCHULTZ and R.S. VARGA (1969), The effect of quadrature errors in the numerical solution of boundary value problems by variational techniques, *Aequationes Math.* **3**, 247–270.

HERBOLD, R.J. and R.S. VARGA (1972), The effect of quadrature errors in the numerical solution of two-dimensional boundary value problems by variational techniques, *Aequationes Math.* **7**, 36–58.

HILBERT, S. (1973), A mollifier useful for approximations in Sobolev spaces and some applications to approximating solutions of differential equations, *Math. Comp.* **27**, 81–89.

HOHN, W. and H.D. MITTELMANN (1981), Some remarks on the discrete maximum principle for finite elements of higher order, *Computing* **27**, 145–154.

HOPPE, V. (1973), Finite elements with harmonic interpolation functions, in: J.R. WHITEMAN, ed., *The Mathematics of Finite Elements and Applications* (Academic Press, London) 131–142.

HÖRMANDER, L. (1983), *The Analysis of Partial Differential Equations* I, Grundlehren der Mathematischen Wissenschaften **256** (Springer, Berlin).

HRABOK, M.M. and T.M. HRUDEY (1984), A review and catalogue of plate bending finite elements, *Comput. & Structures* **19**, 479–495.

HUGHES, T.J.R. (1987), *The Finite Element Method: Linear Static and Dynamic Finite Element Analysis* (Prentice-Hall, Englewood Cliffs, NJ).

IRONS, B. and S. AHMAD (1979), *Techniques of Finite Elements* (Ellis Horwood, Chichester).

IRONS, B.M. and M.J. LOIKKANEN (1983), An engineers' defence of the patch test, *Internat. J. Numer. Methods Engrg.* **19**, 1391–1401.

IRONS, B.M. and A. RAZZAQUE (1972a), Experience with the patch test for convergence of finite elements, in: A.K. AZIZ, ed., *The Mathematical Foundations of the Finite Element Method with Applications to Partial Differential Equations* (Academic Press, New York) 557–587.

IRONS, B.M. and A. RAZZAQUE (1972b), Shape function formulations for elements other than displacement models, presented at the International Conference on Variational Methods in Engineering, Southampton, UK.

JACQUOTTE, O.-P. (1985), Stability, accuracy, and efficiency of some underintegrated methods in finite element computations, *Comput. Methods Appl. Mech. Engrg.* **50**, 275–293.

JACQUOTTE, O.-P. and J.T. ODEN (1984), Analysis of hourglass instabilities and control in underintegrated finite element methods, *Comput. Methods Appl. Mech. Engrg.* **44**, 339–363.

JACQUOTTE, O.-P., J.T. ODEN and E.B. BECKER (1986), Numerical control of the hourglass instability, *Internat. J. Numer. Methods Engrg.* **22**, 219–228.

JAMET, P. (1976a), Estimation d'erreur pour des éléments finis droits presque dégénéres, *Rev. Française Automat. Informat. Recherche Opérationnelle Sér. Rouge Anal. Numér.* **10**, 43–61.

JAMET, P. (1976b), Estimation de l'erreur d'interpolation dans un domaine variable et application aux éléments finis quadrilatéraux dégénérés, in: *Méthodes Numériques en Mathématiques Appliquées* (Presses de l'Université de Montréal, Montréal, Que.) 55–100.

JESPERSSEN, D. (1978), Ritz–Galerkin methods for singular boundary value problems, *SIAM J. Numer. Anal.* **15**, 813–834.

JOHNSON, C. (1987), *Numerical Solutions of Partial Differential Equations by the Finite Element Method* (Cambridge University Press, Cambridge).

KANG, F. (1979), On the theory of discontinuous finite elements, *Math. Numer. Sinica* **1**, 378–385.

KARDESTUNCER, H. and D.H. NORRIE, eds. (1987), *Finite Element Handbook* (McGraw-Hill, New York).

KEAST, P. (1986), Moderate-degree tetrahedal quadrature formulas, *Comput. Methods Appl. Mech. Engrg.* **55**, 339–348.

KHOAN, V.-K. (1972a), *Distributions, Analyse de Fourier, Opérateurs aux Dérivées Partielles* 1 (Vuibert, Paris).

KHOAN, V.-K. (1972b): *Distributions, Analyse de Fourier, Opérateurs aux Dérivées Partielles* 2 (Vuibert, Paris).

KIKUCHI, F. (1975a), Approximation in finite element models, Rept. No. 531, Institute of Space and Aeronautical Science, University of Tokyo.

KIKUCKI, F. (1975b), Convergence of the ACM finite element scheme for plate bending problems, *Publ. Res. Inst. Math. Sci., Kyoto Univ.* **11**, 247–265.

KIKUCHI, F. (1975c), On a finite element scheme based on the discrete Kirchoff assumption, *Numer. Math.* **24**, 211–231.

KIKUCKI, F. (1976), Theory and examples of partial approximation in the finite element method, *Internat. J. Numer. Methods Engrg.* **10**, 115–122.

KINDERLEHRER, D. and G. STAMPACCHIA (1980), *An Introduction to Variational Inequalities and Their Applications* (Academic Press, New York).

KING, J.T. (1976), New error bounds for the penalty method and extrapolation, *Numer. Math.* **23**, 153–165.

KNOPS, R.J. and L.E. PAYNE (1971), *Uniqueness Theorems in Linear Elasticity*, Springer Tracts in Natural Philosophy **19** (Springer, Berlin).

KOH, B.C. and N. KIKUCHI (1987), New improved hourglass control for bilinear and trilinear elements in anisotropic linear elasticity, *Comput. Methods Appl. Mech. Engrg.* **65**, 1–46.

KONDRAT'EV, V.A. (1967), Boundary value problems for elliptic equations in domains with conical or angular points, *Trudy Moskov. Mat. Obshch.* **16**, 209–292.

KOUKAL, S. (1973), Piecewise polynomial interpolations in the finite element method, *Apl. Math.* **18**, 146–160.

KŘÍŽEK, M. and P. NEITTAANMÄKI (1984), Superconvergence phenomenon in the finite element method arising from averaging gradients, *Numer. Math.* **45**, 105–116.

LADYŽENSKAJA, O.A. and N.N. URAL'CEVA (1968), *Linear and Quasilinear Elliptic Equations* (Academic Press, New York).

LANCHON, H. (1972), Torsion élastoplastique d'un arbre cylindrique de section simplement ou multiplement connexe, Doctoral Thesis, Université Pierre et Marie Curie (Paris VI), France.

LANDAU, L. and E. LIFCHITZ (1967), *Théorie de l'Elasticité* (Mir, Moscow).

LASCAUX, P. and P. LESAINT (1975), Some nonconforming finite elements for the plate bending problem, *Rev. Française Automat. Informat. Recherche Opérationnelle Sér. Rouge Anal. Numér* **R-1**, 9–53.

LAX, P.D. and A.N. MILGRAM (1954), Parabolic equations, in: Annals of Mathematics Studies **33** (Princeton University Press, Princeton, NJ) 167–190.

LAYDI, M. and P LESAINT (1985), Sur les éléments finis rationnels de Serendip de degré quelconque, *Numer. Math.* **46**, 175–187.

LEAF, G.K. and H.G. KAPER (1974), L^∞-error bounds for multivariate Lagrange approximation, *SIAM J. Numer. Anal.* **11**, 363–381.

LEAF, G.K., H.G. KAPER and A.J. LINDEMAN (1976), Interpolation and approximation properties of rational coordinates over quadrilaterals, *J. Approximation Theory* **16**, 1–15.

LEBAUD, G. (1969), Sur l'approximation de l'équation $Au = -\Sigma_{i=1}^{n} D_i(|D_i u|^{p-2} D_i u) = T$, *Rend. Mat.* **2**, 443–471.

LEBAUD, G. and P.-A. RAVIART (1969), Sur l'approximation du problème de Dirichlet pour les opérateurs elliptiques d'ordre 2, *Rend. Mat.* **2**, 507–562.

LE MÉHAUTÉ, A. (1981), Taylor interpolation of order n at the vertices of a triangle. Applications to Hermite interpolation and finite elements, in: *Approximation Theory and Applications* (Academic Press, New York) 171–185.

LE MÉHAUTÉ, A. (1984), Interpolation et approximation par des fonctions polynomiales par morceaux dans $\mathbb{R}^n$, Doctoral Dissertation, Université de Rennes, France.

LENOIR, M. (1986), Optimal isoparametric finite elements and error estimates for domains involving curved boundaries, *SIAM J. Numer. Anal.* **23**, 562–580.

LESAINT, P. (1976), On the convergence of Wilson's nonconforming element for solving the elastic problem, *Comput. Methods Appl. Mech. Engrg.* **7**, 1–16.

LESAINT, P. and M. ZLÁMAL (1979), Superconvergence of the gradient of finite element solutions, *RAIRO Anal. Numér.* **13**, 139–166.

LESAINT, P. and M. ZLÁMAL (1980), Convergence of the nonconforming Wilson element for arbitrary quadrilateral meshes, *Numer. Math.* **36**, 33–52.

LEVINE, N. (1985), Superconvergent recovery of the gradient from piecewise linear finite element approximations, *IMA J. Numer. Anal.* **5**, 407–427.

LEWY, H. and G. STAMPACCHIA (1969), On the regularity of the solution of a variational inequality, *Comm. Pure Appl. Math.* **22**, (1969) 153–188.

LI, Z.-C. (1986), A nonconforming combined method for solving Laplace's boundary value problems with singularities, *Numer. Math.* **49**, 475–497.

LI, Z.-C., R. MATHON and P. SERMER (1987), Boundary methods for solving elliptic problems with singularities and interfaces, *SIAM J. Numer. Anal.* **24**, 487–498.

LIN, Q. and T. LU (1984a), Asymptotic expansions for finite element approximation of elliptic problems on polygonal domains, in: R. GLOWINSKI and J.L. LIONS, eds., *Computing Methods in Applied Sciences and Engineering* VI (North-Holland, Amsterdam) 317–321.

LIN, Q. and T. LU (1984b), Asymptotic expansions for finite element eigenvalues and finite element solutions, *Bonner Math. Schriften* **158**, 1–10.

LIN, Q., T. LU and S. SHEN (1983), Asymptotic expansion for finite element approximations, Research Rept. IMS-11, Chengdu Branch of Academia Sinica.

LIN, Q. and Q. ZHU (1984), Asymptotic expansion for the derivative of finite elements, *J. Comput. Math* **2**, 361–363.

LIN, Q. and Q. ZHU (1985), Linear finite elements with high accuracy, *J. Comput. Math.* **3**, 115–133.

LIN, Q. and Q. ZHU (1986), Local asymptotic expansion and extrapolation for finite elements, *J. Comput. Math.* **4**, 263–265.

LIONS, J.L. (1962), *Problèmes aux Limites dans les Equations aux Dérivées Partielles* (Presses de l'Université de Montréal, Montréal, Que.).

LIONS, J.L. (1965), *Problèmes aux Limites dans les Equations aux Dérivées Partielles* (Presses de l'Université de Montréal, Montréal, Que.).

LIONS, J.L. (1969), *Quelques Méthodes de Résolution des Problèmes aux Limites Non Linéaires* (Dunod, Paris).

LIONS, J.L. and E. MAGENES (1968), *Problèmes aux Limites non Homogènes et Applications*, 1 (Dunod, Paris).

LIONS, J.L. and J. PEETRE (1964), Sur une classe d'espaces d'interpolation, *Publ. Math. Inst. Hautes Etudes Sci.* **19**, 5–68.

LIONS, J.L. and G. STAMPACCHIA (1967), Variational inequalities, *Comm. Pure Appl. Math.* **20**, 493–519.

LIU, W.K. and T. BELYTSCHKO (1984), Efficient linear and nonlinear heat conduction with a quadrilateral element, *Internat. J. Numer. Methods Engrg.* **20**, 931–948.

LORENZ, J. (1977), Zur Inversmonotonie diskreter Probleme, *Numer. Math.* **27**, 227–238.

LOUIS, A. (1979), Acceleration of convergence for finite element solutions of the Poisson equation, *Numer. Math.* **33**, 43–53.

LUKÁŠ, I.L. (1974), Curved boundary elements: General forms of polynomial mappings, in: J.T. ODEN et al., eds., *Computational Methods in Nonlinear Mechanics* (Texas Institute for Computational Mechanics, Austin, TX) 37–46.

MANSFIELD, L.E. (1971), On the optimal approximation of linear functionals in spaces of bivariate functions, *SIAM J. Numer. Anal.* **8**, 115–126.

MANSFIELD, L.E. (1972a), Optimal approximation and error bounds in spaces of bivariate functions, *J. Approximation Theory* **5**, 77–96.

MANSFIELD, L.E. (1972b), On the variational characterization and convergence of bivariate splines, *Numer. Math.* **20**, 99–114.

MANSFIELD, L.E. (1974), Higher order compatible triangular finite elements, *Numer. Math.* **22**, 89–97.

MANSFIELD, L.E. (1978), Approximation of the boundary in the finite element solution of fourth order problems, *SIAM J. Numer. Anal.* **15**, 568–579.

MCLEOD, R. and A.R. MITCHELL (1972), The construction of basis functions for curved elements in the finite element method, *J. Inst. Math. Appl.* **10**, 382–393.

MCLEOD, R. and A.R. MITCHELL (1975), The use of parabolic arcs in matching curved boundaries in the finite element method, *J. Inst. Math. Appl.* **16**, 239–246.

MEINGUET, J. (1975), Realistic estimates for generic constants in multivariate pointwise approximation, in: J.J.H. MILLER, ed., *Topics in Numerical Analysis* II (Academic Press, New York) 89–107.

MEINGUET, J. (1977), Structure et estimations de coefficients d'erreur, *RAIRO Anal. Numér.* **11**, 355–368.

MEINGUET, J. (1978), A practical method for estimating approximation errors in Sobolev spaces, in: D.C. HANDSCOMB, ed., *Multivariate Approximation* (Academic Press, London) 169–187.

MEINGUET, J. (1979), A convolution approach to multivariate representation formulas, in: W. SCHEMPP and K. ZELLER, eds., *Multivariate Approximation Theory* (Birkhäuser, Basel) 198–210.

MEINGUET, J. (1981), From Dirac distribution to multivariate representation formulas, in: Z. ZIEGLER, ed., *Approximation Theory and Applications* (Academic Press, New York) 225–248.

MEINGUET, J. (1982), Sharp "a priori" error bounds for polynomial approximation in Sobolev spaces, in: W. SCHEMPP and K. ZELLER, eds., *Multivariate Approximation Theory* II (Birkhäuser, Basel) 255–274.

MEINGUET, J. (1984), A practical method for obtaining a priori error bounds in point-wise and mean-square approximation problems, in: *Approximation Theory, Spline Functions, and Applications* (Reidel, Dordrecht, Netherlands) 97–125.

MEINGUET, J. and J. DESCLOUX (1977), An operator-theoretical approach to error estimation, *Numer. Math.* **27**, 307–326.

MEISTERS, G.H. and C. OLECH (1963), Locally one-to-one mappings and a classical theorem on Schlicht functions, *Duke Math. J.* **30**, 63–80.

MELOSH, R.J. (1963), Basis of derivation of matrices for the direct stiffness method, *AIAA J.* **1**, 1631–1637.

MERCIER, B. (1975a), Approximation par éléments finis et résolution, par un algorithme de pénalisation-dualité, d'un problème d'élasto-plasticité, *C.R. Acad. Sci. Paris, Sér. A* **280**, 287–290.

MERCIER, B. (1975b), Une méthode de résolution du problème des charges limites utilisant les fluides de Bingham, *C.R. Acad. Sci. Paris, Sér. A* **281**, 525–527.

MERCIER, B. and G. RAUGEL (1982), Solution of a boundary value problem in an axisymmetric domain by finite elements in r, z and Fourier series in θ, *RAIRO Anal. Numér.* **16**, 405–461.

MIKHLIN, S.G. (1964), *Variational Methods in Mathematical Physics* (Pergamon, Oxford) (original Russian edition: 1957).

MIKHLIN, S.G. (1971), *The Numerical Performance of Variational Methods* (Wolters-Noordhoff, Groningen, Netherlands).

MIRANDA, C. (1970), *Partial Differential Equations of Elliptic Type* (Springer, New York).

MITCHELL, A.R. (1973), An introduction to the mathematics of the finite element method, in: J.R. WHITEMAN, ed., *The Mathematics of Finite Elements and Applications* (Academic Press, London) 37–58.

MITCHELL, A.R. (1976), Basis functions for curved elements in the mathematical theory of finite elements in: J.R. WHITEMAN, ed., *The Mathematics of Finite Elements and Applications* II (Academic Press, London) 43–58.

MITCHELL, A.R. and J.A. MARSHALL (1975), Matching of essential boundary conditions in the finite element method, in: J.J.H. MILLER, ed., *Topics in Numerical Analysis* II (Academic Press, New York) 109–120.

MIYOSHI, T. (1972), Convergence of finite elements solutions represented by a non-conforming basis, *Kumamoto J. Sci.* (*Math.*) **9**, 11–20.

MIZUKAMI, A. (1986), Some integration formulas for a four-node isoparametric element, *Computer Methods Appl. Mech. Engrg.* **59**, 111–121.

MOCK, M.S. (1976), Projection methods with different trial and test spaces, *Math. Comp.* **30**, 400–416.

MOLCHANOV, I.N. and E.F. GALBA (1985), On finite element methods for the Neumann problem, *Numer. Math.* **46**, 587–598.

MOLCHANOV, I.N. and E.F. GALBA (1985), On finite element methods for the Neumann problem, *Numer. Math.* **46**, 587–598.

MORGAN, J. and R. SCOTT (1975), A nodal basis for C^1 piecewise polynomials of degree $n \geqslant 5$, *Math. Comp.* **29**, 736–740.

MORLEY, L.S.D. (1968), The triangular equilibrium element in the solution of plate bending problems, *Aero. Quart.* **19**, 149–169.

MOSCO, U. and F. SCARPINI (1975), Complementarity systems and approximations of variational inequalities, *Rev. Française Automat. Informat. Recherche Opérationnelle Sér. Rouge Anal. Numér.* **R-1**, 5–8.

MOSCO, U. and G. STRANG (1974), One-sided approximation and variational inequalities, *Bull. Amer. Math. Soc.* **80** (1974), 308–312.

NAKAO, M.T. (1987), Superconvergence of the gradient of Galerkin approximations for elliptic problems, *Modélisation Math. Anal. Numér.* **21**, 679–695.

NATTERER, F. (1975a), Über die punktweise Konvergenz finiter Elemente, *Numer. Math.* **25**, 67–77.

NATTERER, F. (1975b), Berechenbare Fehlerschranken für die Methode der Finiten Elemente, in: International Series of Numerical Mathematics **28** (Birkhäuser, Basel) 109–121.

NATTERER, F. (1977), Uniform convergence of Galerkin's method for splines on highly nonuniform meshes, *Math. Comp.* **31**, 457–468.

NEČAS, J. (1967), *Les Méthodes Directes en Théorie des Equations Elliptiques* (Masson, Paris).

NEČAS, J. (1975), On regularity of solutions to non-linear variational inequalities for second order elliptic systems, *Rend. Mat.* **8**, 481–498.

NEČAS, J. and I. HLAVÁČEK (1981), *Mathematical Theory of Elastic and Elasto-Plastic Bodies: An Introduction* (Elsevier, Amsterdam).

NEDOMA, J. (1979), The finite element solution of elliptic and parabolic equations using simplicial isoparametric elements, *RAIRO Anal. Numér.* **13**, 257–289.

NICOLAIDES, R.A. (1972), On a class of finite elements generated by Lagrange interpolation, *SIAM J. Numer. Anal.* **9**, 435–445.

NICOLAIDES, R.A. (1973), On a class of finite elements generated by Lagrange interpolation II, *SIAM J. Numer. Anal.* **10**, 182–189.

NIELSON, G.M. (1973), Bivariate spline functions and the approximation of linear functionals, *Numer. Math.* **21**, 138–160.

NITSCHE, J.A. (1968), Ein Kriterium für die Quasi-Optimalitat des Ritzchen Verfahrens, *Numer. Math.* **11**, 346–348.

NITSCHE, J.A. (1969), Orthogonalreihenentwicklung nach linearen Spline-Funktionen, *J. Approximation Theory* **2**, 66–78.

NITSCHE, J.A. (1970), Linear Spline-Funktionen und die Methoden von Ritz für elliptische Randwert-probleme, *Arch. Rational Mech. Anal.* **36**, 348–355.

NITSCHE, J.A. (1971), Über ein Variationsprinzip zur Lösung von Dirichlet-Problemen bei Verwendung von Teilräumen, die keinen Randbedingungen unterworfen sind, *Abh. Math. Sem. Univ. Hamburg* **36**, 9–15.

NITSCHE, J.A. (1972a), Interior error estimates of projection methods, in: *Proceedings EquaDiff 3*, Purkyně University, Brno, 235–239.

NITSCHE, J.A. (1972b), On Dirichlet problems using subspaces with nearly zero boundary conditions, in: A.K. AZIZ, ed., *The Mathematical Foundations of the Finite Element Method with Applications to Partial Differential Equations* (Academic Press, New York) 603–628.

NITSCHE, J.A. (1974), Convergence of nonconforming methods, in: C. DE BOOR, ed., *Mathematical Aspects of Finite Elements in Partial Differential Equations* (Academic Press, New York) 15–53.

NITSCHE, J.A. (1975), L_∞-convergence of finite element approximation, in: *Proceedings Second Conference on Finite Elements*, Rennes, France.

NITSCHE, J.A. (1976a), Der Einfluss von Randsingularitäten beim Ritzschen Verfahren, *Numer. Math.* **25**, 263–278.

NITSCHE, J.A. (1976b), Über L_∞-Abschätzungen von Projektionen auf finite Elemente, *Bonner Math. Schriften* **89**, 13–30.

NITSCHE, J.A. (1977), L_∞-convergence of finite element approximations, in: I. GALLIGANI and E. MAGENES, eds., *Mathematical Aspects of Finite Element Methods*, Lecture Notes in Mathematics **606** (Springer, Berlin) 261–274.

NITSCHE, J.A. (1981a), On Korn's second inequality, *RAIRO Anal. Numér.* **15**, 237–248.

NITSCHE, J.A. (1981b), Schauder estimates for finite element approximations of second order elliptic boundary value problems, in: *Proceedings, Special Year on Numerical Analysis*, College Park, MD, 290–343.

NITSCHE, J.A. and A.H. SCHATZ (1974), Interior estimates for Ritz–Galerkin methods, *Math. Comp.* **28**, 937–958.

NORRIE, D.H. and G. DE VRIES (1973), *The Finite Element Method: Fundamentals and Applications* (Academic Press, New York).

ODEN, J.T. (1972), *Finite Elements of Nonlinear Continua* (McGraw-Hill, New York).

ODEN, J.T. and G.F. CAREY (1981–1984), *Finite Elements* I: *An Introduction* (with E.B. BECKER); II: *A Second Course*; III: *Computational Aspects*; IV: *Mathematical Aspects*; V: *Special Problems in Solid Mechanics* (Prentice-Hall, Englewood Cliffs, NJ).

ODEN, J.T. and J.N. REDDY (1976a), *An Introduction to the Mathematical Theory of Finite Elements* (Wiley-Interscience, New York).

ODEN, J.T. and J.N. REDDY (1976b), *Variational Methods in Theoretical Mechanics* (Springer, Heidelberg).

OGANESJAN, L.A. and P.A. RUKHOVETS (1969), Investigation of the convergence rate of variational-difference schemes for elliptic second order equations in a two-dimensional domain with a smooth boundary, *Zh. Vychisl. Mat. i Mat. Fiz.* **9**, 1102–1120 (in Russian).

OLIVEIRA, E.R. DE ARANTES E (1977), The patch test and the general convergence criteria of the finite element method, *Internat. J. Solids Structures* **13**, 159–178.

PANAGIOTOPOULOS, P.D. (1985), *Inequality Problems in Mechanics* and *Applications* (Birkhäuser, Boston, MA).

PEISKER, P. and D. BRAESS (1987), A conjugate gradient method and a multigrid algorithm for Morley's finite element approximation of the biharmonic equation, *Numer. Math.* **50**, 567–586.

PERCELL, P. (1976), On cubic and quartic Clough–Tocher finite elements, *SIAM J. Numer. Anal.* **13**, 100–103.

PINI, F. (1974), Approximation by finite element functions using global regularization, Report, Département de Mathématiques, Ecole Polytechnique Fédérale de Lausanne, Switzerland.

PITKÄRANTA, J. (1979), Boundary subpaces for the finite element method with Lagrange multipliers, *Numer. Math.* **33**, 273–289.

PÓLYA, G. (1952), Sur une interprétation de la méthode des différences finies qui peut fournir des bornes supérieures ou inférieures, *C.R. Acad. Sci. Paris* **235**, 995–997.

POWELL, M.J.D. and M.A. SABIN (1977), Piecewise quadratic approximations on triangles, *ACM Trans. Math. Software* **3**, 316–325.

PROTTER, M. and H. WEINBERGER (1967), *Maximum Principles in Differential Equations*, (Prentice-Hall, Englewood Cliffs, NJ), also (Springer, Berlin, 1984).

RABIER, P. (1977), Interpolation harmonique, *Rev. Française Automat. Informat. Recherche Opérationnelle Sér. Rouge Anal. Numér.* **11**, 159–180.

RACHFORD Jr, H.H. and M.F. WHEELER (1974), An H^{-1}-Galerkin procedure for the two-point boundary value problem, in: C. DE BOOR, ed., *Mathematical Aspects of Finite Elements in Partial Differential Equations* (Academic Press, New York) 353–382.

RANNACHER, R. (1976a), Zur punktweisen Konvergenz der Methode der finiten Elemente beim Plattenproblem, *Manuscripta Math.* **19**, 401–416.

RANNACHER, R. (1976b), Zur L^∞-Konvergenz linearer finiter Elemente beim Dirichlet problem, *Math. Z.* **149**, 69–77.

RANNACHER, R. (1977), L^∞-Fehlerabschätzung für ein nicht konforme Finite-Elemente-Methode beim Plattenproblem, *Z. Angew. Math. Mech.* **57**, 247–249.

RANNACHER, R. and R. SCOTT (1982), Some optimal error estimates for piecewise linear finite element approximations, *Math. Comp.* **38**, 437–445

RAOULT, A. (1985), Construction d'un modèle d'évolution de plaques avec terme d'inertie de rotation, *Ann. Mat. Pura Appl.* **139**, 361–400.

RAUGEL, G. (1978a), Résolution numérique de problèmes elliptiques dans des domaines avec coins, *C.R. Acad. Sci. Paris Sér. A–B* **286**, 791–794.

RAUGEL, G. (1978b), Résolution numérique de problèmes elliptiques dans des domaines avec coins, Doctoral Dissertation, Université de Rennes, France.

RAVIART, P.-A. (1972), *Méthodes des Eléments Finis*, Lecture Notes, Laboratoire d'Analyse Numérique, Université Pierre et Marie Curie, Paris, France.

RAVIART, P.-A. and J.M. THOMAS (1983), *Introduction à l'Analyse Numérique des Equations aux Dérivées Partielles* (Masson, Paris).

RAZZAQUE, A. (1973), Program for triangular bending elements with derivatives smoothing, *Internat. J. Numer. Methods Engrg.* **6**, 333–343.

RAZZAQUE, A. (1986), The patch test for elements, *Internat. J. Numer. Methods Engrg.* **22**, 63–71.

ROBINSON, J. (1973), Basis for isoparametric stress elements, *Comput. Methods Appl. Mech. Engrg.* **2**, 43–63.

ROBINSON, J. (1982), The patch test: Is it or isn't it?, *Finite Element News* **1**, 30–34.

RODRIGUES, J.F. (1987), *Obstacle Problems in Mathematical Physics* (North-Holland, Amsterdam).

ROSE, M.E. (1975), Weak-element approximations to elliptic differential equations, *Numer. Math.* **24**, 185–204.

ROUX, J. (1976), Résolution numérique d'un problème d'écoulement subsonique de fluides compressibles, *Rev. Française Automat. Informat. Recherche Opérationnelle Sér. Rouge Anal. Numér.* **10**, (12), 31–50.

RUAS, V. (1988), A quadratic finite element method for solving biharmonic problems in $\mathbb{R}^n$, *Numer. Math.* **52**, 33–43.

RUAS SANTOS, V. (1982), On the strong maximum principle for some piecewise linear finite element approximate problems of non-positive type, *J. Fac. Sci. Univ. Tokyo, Sec. IA* **29**, 473–491.

SABLONNIÈRE, P. (1980), Interpolation d'Hermite par des surfaces de classe C^1 quadratiques par morceaux, in: *Actes du Deuxième Congrès International sur les Méthodes Numériques de l'Ingénieur* (Dunod, Paris) 175–185.

SABLONNIÈRE, P. (1984), Composite finite elements of class C^k, *J. Comput. Appl. Math.* **12–13**, 541–550.

SABLONNIÈRE, P. (1987), Error bounds for Hermite interpolation by quadratic splines on an α-triangulation, *IMA J. Numer. Anal.* **7**, 495–508.

SAMUELSSON, A. (1979), The global constant strain condition and the patch test, in: R. GLOWINSKI, E.Y. RODIN and O.C. ZIENKIEWICZ, eds., *Energy Methods in Finite Element Analysis* (Wiley, Chichester) 49–68.

SANCHEZ, A.M. and R. ARCANGÉLI (1984), Estimations des erreurs de meilleure approximation polynomiale et d'interpolation de Lagrange dans les espaces de Sobolev d'ordre non entier, *Numer. Math.* **45**, 301–321.

SANDER, G. and P. BECKERS (1977), The influence of the choice of connectors in the finite element method, *Internat. J. Numer. Methods Engrg.* **11**, 1491–1505.

SARD, A. (1963), *Linear Approximation*, Mathematical Surveys **9** (American Mathematical Society, Providence, RI).

SCARPINI, F. and VIVALDI, M.A. (1977), Error estimates for the approximation of some unilateral problems, *Rev. Française Automat. Informat. Recherche Opérationnelle Sér. Rouge Anal. Numér.* **11**, 197–208.

SCHATZ, A.H. (1974), An observation concerning Ritz-Galerkin methods with indefinite bilinear forms, *Math. Comp.* **28**, 959–962.

SCHATZ, A.H. (1980), A weak discrete maximum principle and stability of the finite element method in L^∞ on plane polygonal domains, *Math. Comp.* **34**, 77–91.

SCHATZ, A.H. and L.B. WAHLBIN (1976), Maximum norm error estimates in the finite element method for

Poisson equation on plane domains with corners, in: C.K. CHUI, G.G. LORENTZ and L.L. SCHUMAKER, eds., *Approximation Theory* II (Academic Press, New York) 541–547.

SCHATZ, A.H. and L.B. WAHLBIN (1977), Interior maximum norm estimates for finite element methods, *Math. Comp.* **31**, 414–442.

SCHATZ, A.H. and L.B. WAHLBIN (1978), Maximum norm estimates in the finite element method on plane polygonal domains, Part 1, *Math. Comp.* **32**, 73–109.

SCHATZ, A.H. and L.B. WAHLBIN (1979), Maximum norm estimates in the finite element method on plane polygonal domains, Part 2: Refinements, *Math. Comp*, **33**, 465–492.

SCHATZ, A.H. and L.B. WAHLBIN (1981), On a local asymptotic error estimate in finite elements and its use: numerical examples, in: R. VICHNEVETSKY and R.S. STEPLEMAN, eds., *Advances in Computer Methods for Partial Differential Equations* IV (IMACS, New Brunswick, NJ) 14–17.

SCHATZ, A.H. and L.B. WAHLBIN (1982), On the quasi-optimality in L_∞ of the H^1 projection into finite element spaces, *Math. Comp.* **38**, 1–22.

SCHMID, H.J. (1978), Cubature formulas with a minimal number of knots, *Numer. Math.* **31**, 281–297.

SCHREIBER, R. (1980), Finite element methods of high-order accuracy for singular two-point boundary value problems with nonsmooth solutions, *SIAM J. Numer. Anal.* **17**, 547–566.

SCHULTZ, M.H. (1969), L^∞-multivariate approximation theory, *SIAM J. Numer. Anal.* **6**, 161–183.

SCHULTZ, M.H. (1972), Quadrature–Galerkin approximations to solutions of elliptic differential equations, *Proc. Amer. Math. Soc.* **33**, 511–515.

SCHULTZ, M.H. (1973), Error bounds for a bivariate interpolation scheme, *J. Approximation Theory* **8**, 189–194.

SCHWARTZ, L. (1966), *Théorie des Distributions* (Hermann, Paris).

SCHWARTZ, L. (1967), *Cours d'Analyse* (Hermann, Paris).

SCOTT, R. (1973a), Finite element techniques for curved boundaries, Doctoral Thesis, MIT, Cambridge, MA.

SCOTT, R. (1973b), Finite element convergence for singular data, *Numer. Math.* **21**, 317–327.

SCOTT, R. (1974), C^1 continuity via constraints for 4th order problems, in: C. DE BOOR, ed., *Mathematical Aspects of Finite Elements in Partial Differential Equations* (Academic Press, New York) 171–193.

SCOTT, R. (1975), Interpolated boundary conditions in the finite element method, *SIAM J. Numer. Anal.* **12**, 404–427.

SCOTT, R. (1976a), Optimal L^∞ estimates for the finite element method on irregular meshes, *Math. Comp.* **30**, 681–697.

SCOTT, R. (1976b), A survey of displacement methods for the plate bending problem, in: K.-J. BATHE, J.T. ODEN and W. WUNDERLICH, eds., *Proceedings U.S.–Germany Symposium on Formulations and Computational Algorithms in Finite Element Analysis* (MIT Press, Cambridge, MA) 855–876.

SCOTT, R. (1976c), A hexagonal finite element with one degree-of-freedom per hexagon, *Internat. J. Numer. Methods Engrg.* **10**, 958–959.

SCOTT, R. (1979), Applications of Banach space interpolation to finite element theory, in: M.Z. NASHED, ed., *Functional Analysis Methods in Numerical Analysis*, Lecture Notes in Mathematics **701** (Springer, Berlin) 298–318.

SHAH, J.M. (1970), Two-dimensional polynomial splines, *Numer. Math.* **15**, 1–14.

SHI, Z. (1984a), A convergence condition for the quadrilateral Wilson element, *Numer. Math.* **44**, 349–361.

SHI, Z. (1984b), On the convergence properties of the quadrilateral elements of Sander and Beckers, *Math. Comp.* **42**, 493–504.

SHI, Z. (1984c), An explicit analysis of Stummel's patch test examples, *Internat. J. Numer. Methods Engrg.* **20**, 1233–1246.

SHI, Z. (1984d), Difficulties with Irons' patch test, *Z. Angew. Math. Mech.* **64**, T314–T315.

SHI, Z. (1984e), The generalized patch test for Zienkiewicz's triangles, *J. Comput. Math.* **2**, 279–286.

SHI, Z. (1984f), On the convergence rate of the boundary penalty method, *Internat. J. Numer. Methods Engrg.* **20**, 2027–2032.

SHI, Z. (1985), Convergence properties of two nonconforming finite elements, *Comput. Methods Appl. Mech. Engrg.* **48**, 123–137.

SHI, Z. (1986), The optimal order of convergence of Wilson's nonconforming element, *Math. Numer. Sinica* **8**, 159–163.

SHI, Z. (1987a), The F-E-M-test for convergence of nonconforming finite elements, *Math. Comp.* **49**, 391–405.

SHI, Z. (1987b), Convergence of the TRUNC plate element, *Comput. Methods Appl. Mech. Engrg.* **62**, 71–88.

SHI, Z. (1989), On Stummel's examples to the patch test, *Comput. Mech.* **5**, 81–87.

SHILOV, G.E. (1968), *Generalized Functions and Partial Differential Equations* (Gordon and Breach, New York).

SIGNORINI, A. (1933), Sopra alcune questioni di elastostatica, *Atti Soc. Ital. Prog. Sci.*

SPERB, R. (1981), *Maximum Principles and Their Applications* (Academic Press, New York).

STAKGOLD, I. (1968), *Boundary Value Problems of Mathematical Physics* II (MacMillan, New York).

STAMPACCHIA, G. (1964), Formes bilinéaires coercitives sur les ensembles convexes, *C.R. Acad. Sci. Paris Sér. A.* **258**, 4413–4416.

STAMPACCHIA, G. (1965), Le problème de Dirichlet pour les équations elliptiques du second ordre à coefficients discontinus, *Ann. Inst. Fourier (Grenoble)* **15**, 189–258.

STRANG, G. (1971), The finite element method and approximation theory, in: B.E. HUBBARD, ed., *Numerical Solutions of Partial Differential Equations* II (Academic Press, New York) 547–583.

STRANG, G. (1972a), Approximation in the finite element method, *Numer. Math.* **19**, 81–98.

STRANG, G. (1972b), Variational crimes in the finite element method, in: A.K. AZIZ, ed., *The Mathematical Foundations of the Finite Element Method with Applications to Partial Differential Equations* (Academic Press, New York) 689–710.

STRANG, G. (1973), Piecewise polynomials and the finite-element method, *Bull. Amer. Math. Soc.* **79**, 1128–1137.

STRANG, G. (1974), The dimension of piecewise polynomials, and one-sided approximation, in: G.A. WATSON, ed., *Conference on the Numerical Solution of Differential Equations*, Lecture Notes in Mathematics **363** (Springer, New York) 144–152.

STRANG, G. and A. BERGER (1971), The change in solution due to change in domain, in: *Proceedings AMS Symposium on Partial Differential Equations*, Berkeley, CA, 199–205.

STRANG, G. and G.J. FIX (1971), A Fourier analysis of the finite element method, in: *Proceedings CIME Summer School*, Cremonese, Rome.

STRANG, G. and G.J. FIX (1973), *An Analysis of the Finite Element Method* (Prentice-Hall, Englewood Cliffs, NJ).

STROUD, A.H. (1971), *Approximate Calculation of Multiple Integrals* (Prentice-Hall, Englewood Cliffs, NJ).

STUMMEL, F. (1979), The generalized patch test, *SIAM J. Numer. Anal.* **16**, 449–471.

STUMMEL, F. (1980a), Basic compactness properties of nonconforming and hybrid finite element spaces, *RAIRO Anal. Numér.* **4**, 81–115.

STUMMEL, F. (1980b), The limitations of the patch test, *Internat. J. Numer. Methods Engrg.* **15**, 177–188.

SUZUKI, T. and H. FUJITA (1986), A remark on the L^∞ bounds of the Ritz operator associated with a finite element approximation, *Numer. Math.* **49**, 529–544.

SYNGE, J.L. (1957), *The Hypercircle in Mathematical Physics* (Cambridge University Press, Cambridge).

TAYLOR, R.L., P.J. BERSFORD and E.L. WILSON (1976), A non-conforming element for stress analysis, *Internat. J. Numer. Methods Engrg.* **10**, 1211–1220.

TAYLOR, R.L., J.C. SIMO, O.C. ZIENKIEWICZ and A.C.H. CHAN (1986), The patch test: A condition for assessing FEM convergence, *Internat. J. Numer. Methods Engrg.* **22**, 39–62.

TEMAM, R. (1984), *Navier–Stokes Equations: Theory and Numerical Analysis* (North-Holland, Amsterdam, 3rd rev. ed.).

THATCHER, R.W. (1976), The use of infinite grid refinements at singularities in the solution of Laplace's equation, Numer. Math. **25**, 163–178.

THOMASSET, F. (1981), *Implementation of Finite Element Methods for Navier–Stokes Equations* (Springer, Heidelberg).

THOMÉE, V. (1973a), Approximate solution of Dirichlet's problem using approximating polygonal domain, in: J.J.H. MILLER, ed., *Topics in Numerical Analysis* (Academic Press, New York) 311–328.

THOMÉE, V. (1973b), Polygonal domain approximation in Dirichlet's problem, *J. Inst. Math. Appl.* **11**, 33–44.

THOMÉE, V. (1977), High-order local approximations to derivatives in the finite element method, *Math. Comp.* **31**, 652–660.

TOMLIN, G.R. (1972), An optimal successive overrelaxation technique for solving second order finite difference equations for triangular meshes, *Internat. J. Numer. Methods Engrg.* **5**, 25–39.

TRÈVES, F. (1967), *Topological Vector Spaces, Distributions and Kernels* (Academic Press, New York).

TURNER, M.J., R.W. CLOUGH, H.C. MARTIN and L.J. TOPP (1956), Stiffness and deflection analysis of complex structures, *J. Aero. Sci.* **23**, 805–823.

UTKU, M. and G.F. CAREY (1982), Boundary penalty techniques, *Comput. Methods Appl. Mech. Engrg.* **30**, 103–118.

VALID, R. (1977), *La Mécanique des Milieux Continus et le Calcul des Structures* (Eyralles, Paris). English translation: *Mechanics of Continuous Media and Analysis of Structures* (North-Holland, Amsterdam, 1981).

VARGA, R.S. (1962), *Matrix Iterative Analysis* (Prentice-Hall, Englewood Cliffs, NJ).

VARGA, R.S (1966a), On a discrete maximum principle, *SIAM J. Numer. Anal.* **3**, 355–359.

VARGA, R.S. (1966b), Hermite interpolation-type Ritz methods for two-point boundary value problems, in: J.H. BRAMBLE, ed., *Numerical Solution of Partial Differential Equations* (Academic Press, New York) 365–373.

VARGA, R.S. (1971), *Functional Analysis and Approximation Theory in Numerical Analysis*, Regional Conference Series in Applied Mathematics (SIAM, Philadelphia, PA).

VEIDINGER, L. (1972), On the order of convergence of the Rayleigh-Ritz method with piecewise linear trial functions, *Acta Math. Acad. Sci. Hungar.* **23**, 507–517.

VEIDINGER, L. (1975), On the order of convergence of a finite element method in regions with curved boundaries, *Acta Math. Acad. Sci. Hungar.* **26**, 419–431.

VILLAGIO, P. (1972), Energetic bounds in finite elasticity, *Arch. Rational Mech. Anal.* **45**, 282–293.

WACHSPRESS, E.L. (1971), A rational basis for function approximation, *J. Inst. Math. Appl.* **8**, 57–68.

WACHSPRESS, E.L. (1973), A rational basis for function approximation, II: Curved sides, *J. Inst. Math. Appl.* **11**, 83–104.

WACHSPRESS, E.L. (1975), *A Rational Finite Element Basis* (Academic Press, New York).

WAHLBIN, L.B. (1978), Maximum norm error estimates in the finite element method with isoparametric quadratic elements and numerical integration, *RAIRO Anal. Numér.* **12**, 173–202.

WAHLBIN, L.B. (1984), On the sharpness of certain local estimates for $\dot{H}^1$ projections into finite element spaces: influence of a reentrant corner, *Math. Comp.* **42**, 1–8.

WAIT, R. and A.R. MITCHELL (1971), Corner singularities in elliptic problems by finite element methods, *J. Comput. Phys.* **8**, 45–52.

WAIT, R. and A.R. MITCHELL (1985), *Finite Element Analysis and Applications* (Wiley, New York); rev. ed. of *The Finite Element Method in Partial Differential Equations* (Wiley, New York 1977).

WALSH, J. (1971), Finite-difference and finite-element methods of approximation, *Proc. Roy. Soc. London Ser. A* **323**, 155–165.

WERSCHULZ, A.G. (1982), Optimal error properties of finite element methods for second order elliptic Dirichlet problems, *Math. Comp.* **158**, 401–413.

WHEELER, M.F. and J.R. WHITEMAN (1987), Superconvergent recovery of gradients on subdomains from piecewise linear finite-element approximations, *Numer. Methods Partial Differential Equations* **3**, 65–82.

WHITEMAN, J.R. (1982), Finite elements for singularities in two- and three-dimensions, in: J.R. WHITEMAN, ed., *The Mathematics of Finite Elements and Applications* IV (Academic Press, London) 35–51.

WHITEMAN, J.R. and J.E. AKIN (1980), Finite elements, singularities, and fracture, in: J.R. WHITEMAN, ed., *The Mathematics of Finite Elements and Applications* III (Academic Press, London) 35–51.

WIDLUND, O. (1977), On best error bounds for approximation by piecewise polynomial functions, *Numer. Math.* **27**, 327–338.

WILSON, E.L., R.L. TAYLOR, W.P. DOHERTY and J. GHABUSSI (1973), Incompatible displacement models, in: S.T. FENVES et al., eds., *Numerical and Computational Methods in Structural Mechanics* (Academic Press, New York) 43–57.

WISSMANN, J.W., T. BECKER and H. MÖLLER (1987), Stabilization of the zero-energy modes of under-integrated isoparametric finite elements, *Comput. Mech.* **2**, 289–306.

ZACHARIAS, C. and V. SUBBA RAO (1982), Curved elements with polynomials of varying degree, *J. Math. Phys. Sci.* **16**, 223–230.

ŽENÍŠEK, A. (1970), Interpolation polynomials on the triangle, *Numer. Math.* **15**, 283–296.

ŽENÍŠEK, A. (1972), Hermite interpolation on simplexes in the finite element method, in *Proceedings EquaDiff 3*, Purkyně University, Brno, 271–277.

ŽENÍŠEK, A. (1973), Polynomial approximation on tetrahedrons in the finite element method, *J. Approximation Theory* **7**, 334–351.

ŽENÍŠEK, A. (1974), A general theorem on triangular finite $C^{(m)}$-elements, *Rev. Française Automat. Informat. Recherche Opérationnelle Sér. Rouge Anal. Numér.* **R-2**, 119–127.

ŽENÍŠEK, A. (1981a), Discrete form of Friedrichs' inequalities in the finite element method, *RAIRO Anal. Numér.* **15**, 265–286.

ŽENÍŠEK, A. (1981b), Nonhomogeneous boundary conditions and curved triangular finite elements, *Apl. Mat.* **26**, 121–141.

ŽENÍŠEK, A. (1987), How to avoid the use of Green's theorem in the Ciarlet–Raviart theory of variational crimes, *Modélisation Math. Anal. Numér.* **21**, 171–191.

ZIENKIEWICZ, O.C. (1971), *The Finite Element Method in Engineering Science* (McGraw-Hill, London).

ZIENKIEWICZ, O.C. (1973), Finite elements: The background story, in: J.R. WHITEMAN, ed., *The Mathematics of Finite Elements and Applications* (Academic Press, London) 1–35.

ZIENKIEWICZ, O.C. (1974), Constrained variational principles and penalty function methods in finite element analysis, in: G.A. WATSON, ed., *Conference on the Numerical Solution of Differential Equations*, Lecture Notes in Mathematics **363** (Springer, New York) 207–214.

ZIENKIEWICZ, O.C. and K. MORGAN (1983), *Finite Elements and Approximation* (Wiley, New York).

ZLÁMAL, M. (1968), On the finite element method, *Numer. Math.* **12**, 394–409.

ZLÁMAL, M. (1970), A finite element procedure of the second order of accuracy, *Numer. Math.* **16**, 394–402.

ZLÁMAL, M. (1973a), The finite element method in domains with curved boundaries, *Internat. J. Numer. Methods Engrg.* **5**, 367–373.

ZLÁMAL, M. (1973b), Curved elements in the finite element method, I. *SIAM J. Numer. Anal.* **10**, 229–240.

ZLÁMAL, M. (1973c), A remark on the "Serendipity family", *Internat. J. Numer. Methods Engrg.* **7**, 98–100.

ZLÁMAL, M. (1974), Curved elements in the finite element method, II, *SIAM J. Numer. Anal.* **11**, 347–362.

ZLÁMAL, M. (177), Some superconvergence results in the finite element method, in: I. GALLIGANI and E. MAGENES, eds., *Mathematical Aspects of Finite Element Methods*, Lecture Notes in Mathematics **606** (Springer, Berlin) 353–362.

ZLÁMAL, M. (1978), Superconvergence and reduced integration in the finite element method, *Math. Comp.* **32**, 663–685.

Glossary of Symbols

General notation

$\{x \in A; R(x)\}$: general notation for a set.

$v(\cdot), v(\cdot,\cdot), \ldots$: function v of one variable, two variables, ...

$v(\cdot, b)$: partial mapping $x \to v(x, b)$.

$\operatorname{supp} v = \operatorname{cl}\{x \in X;\ v(x) \neq 0\}$: support of a function v.

$\operatorname{osc}(v; A) = \sup_{x,y \in A} |v(x) - v(y)|$.

v_A or $v|_A$: restriction of a function v to the set A.

$P(A) = \{p|_A;\ p \in P\}$, where P is any space of functions defined over a domain that contains the set A.

$\operatorname{tr} v$ or simply v: trace of a function v.

$C(a)$, $C(a, b), \ldots$: any "constant" that depends solely on a, a and b, ...

$\operatorname{cl} A$ or $\bar{A}$: closure of a set A.

$\operatorname{int} A$ or $\mathring{A}$: interior of a set A.

∂A: boundary of a set A.

$\operatorname{card}(A)$: number of elements of a set A.

$\operatorname{diam}(A)$: diameter of a set A.

$X - A$: complement set of the subset A of the set X.

$\Rightarrow$: implies.

Derivatives and differential calculus

$Dv(a)$ or $v'(a)$: first (Fréchet) derivative of a function v at a point a.

$D^2 v(a)$ or $v''(a)$: second (Fréchet) derivative of a function v at a point a.

$D^k v(a)$ or $v^{(k)}(a)$: kth (Fréchet) derivative of a function v at a point a.

$D^k v(a) h^k = D^k v(a)(h_1, h_2, \ldots, h_k)$ if $h_1 = h_2 = \cdots = h_k = h$.

$$\left.\begin{aligned} \partial_i v(a) &= Dv(a) e_i \\ \partial_{ij} v(a) &= D^2 v(a)(e_i, e_j) \\ \partial_{ijk} v(a) &= D^3 v(a)(e_i, e_j, e_k) \end{aligned}\right\} \quad \text{also used for vector-valued functions.}$$

$J(F)(\hat{x}) = \det(\partial_j F_i(\hat{x}))$:

Jacobian of a mapping $F: \hat{x} \in \mathbb{R}^n \to F(\hat{x}) = (F_i(x))_{i=1}^n \in \mathbb{R}^n$ at $\hat{x}$.

$\operatorname{div} \boldsymbol{v} = \sum_{i=1}^n \partial_i v.$

$\nabla v(a) = (\partial_i v)_{i=1}^n$, also denoted $\nabla v(a)$ or grad $v(a)$.

$\Delta v = \sum_{i=1}^n \partial_{ii} v, \qquad \Delta \boldsymbol{v} = (\Delta v_i)_{i=1}^n.$

$|\alpha| = \sum_{i=1}^n \alpha_i$, for a multi-index $\alpha = (\alpha_1, \ldots, \alpha_n) \in \mathbb{N}^n$.

$\partial^\alpha v(a) = D^{|\alpha|} v(a)(\overbrace{e_1, \ldots, e_1}^{\alpha_1 \text{ times}}, \overbrace{e_2, \ldots, e_2}^{\alpha_2 \text{ times}}, \ldots, \overbrace{e_n, \ldots, e_n}^{\alpha_n \text{ times}}).$

$\nu = (\nu_1, \nu_2, \ldots, \nu_n)$: unit outer normal vector.

$\partial_\nu = \sum_{i=1}^n \nu_i \partial_i$: (outer) normal derivative operator.

$\tau = (\tau_1, \tau_2)$: unit tangential vector along the boundary of a plane domain.

$\partial_\tau v(a) = Dv(a)\tau = \sum_{i=1}^2 \tau_i \partial_i v(a).$

$\partial_{\nu\tau} v(a) = D^2 v(a)(\nu, \tau) = \sum_{i,j=1}^2 \nu_i \tau_j \partial_{ij} v(a).$

$\partial_{\tau\tau} v(a) = D^2 v(a)(\tau, \tau) = \sum_{i,j=1}^2 \tau_i \tau_j \partial_{ij} v(a).$

General notation for vector spaces

$\|\cdot\|$ or $\|\cdot\|_X$: norm (in the space X).

$|\cdot|$: seminorm.

$B_r(a) = \{x \in X;\ \|x - a\| \leqslant r\}$.

$B_r = B_r(0)$.

$\mathscr{L}(X; Y)$: space of continuous linear mappings from X into Y.

$\mathscr{L}(X) = \mathscr{L}(X; X)$.

$\mathscr{L}_k(X; Y)$: space of continuous k-linear mappings from X^k into Y.

X': dual of a space X.

$\|\cdot\|'$: norm in the space X'.

$\langle \cdot, \cdot \rangle$: duality pairing between a space and its dual.

$x + Y = \{x + y;\ y \in Y\}$.

$X + Y = \{x + y;\ x \in X, y \in Y\}$.

$X \oplus Y = \{x + y;\ x \in X,\ y \in Y\}$ when $X \cap Y = \{0\}$.

X/Y: quotient space of X by Y.

$\mathsf{V}\{e_\lambda, \lambda \in \Lambda\}$: vector space spanned by the vectors $e_\lambda, \lambda \in \Lambda$.

I: identity mapping.

$\hookrightarrow$: inclusion with continuous injection.

$\Subset$: inclusion with compact injection.

$\dim X$: dimension of the space X.

$\operatorname{Ker} A = \{x \in X;\ Ax = 0\}$.

$\operatorname{Im} A = \{Ax \in Y;\ x \in X\}$.

Notation for specific vector spaces

$(u, v) = \int_\Omega uv \,\mathrm{d}x$: inner product in $L^2(\Omega)$.

$(\boldsymbol{u}, \boldsymbol{v}) = \int_\Omega \boldsymbol{u} \cdot \boldsymbol{v} \,\mathrm{d}x$: inner product in $(L^2(\Omega))^n$.

$\mathscr{C}^m(A)$: space of real-valued, m times continuously differentiable functions on a subset A of $\mathbb{R}^n$.

$$\mathscr{C}^\infty(A) = \bigcap_{m=0}^{\infty} \mathscr{C}^m(A).$$

$$\mathscr{C}^{m,\alpha}(A) = \left\{ v \in \mathscr{C}^m(\bar{\Omega});\ \text{for each } \beta \text{ with } \beta = m,\ \sup_{\substack{x,y \in A \\ x \neq y}} \frac{|\partial^\beta v(x) - \partial^\beta v(y)|}{\|x - y\|^\alpha} < +\infty \right\}.$$

$$\|v\|_{\mathscr{C}^{m,\alpha}(A)} = \|v\|_{m,\infty,A} + \max_{|\beta| = m} \sup_{\substack{x,y \in A \\ x \neq y}} \frac{|\partial^\beta v(x) - \partial^\beta v(y)|}{\|x - y\|^\alpha}.$$

$\mathscr{D}(\Omega) = \{v \in \mathscr{C}^\infty(\Omega);\ \operatorname{supp} v \text{ is a compact subset of } \Omega\}$.

$\mathscr{D}'(\Omega)$; space of distributions over Ω.

$H^m(\Omega) = \{v \in L^2(\Omega);\ \partial^\alpha v \in L^2(\Omega) \text{ for all } \alpha \text{ with } |\alpha| \leqslant m\}$.

$H_0^m(\Omega)$: closure of $\mathscr{D}(\Omega)$ in $H^m(\Omega)$.

$$\|v\|_{m,\Omega} = \left\{ \sum_{|\alpha| \leqslant m} \int_\Omega |\partial^\alpha v|^2 \, dx \right\}^{1/2}.$$

$$|v|_{m,\Omega} = \left\{ \sum_{|\alpha| = m} \int_\Omega |\partial^\alpha v|^2 \, dx \right\}^{1/2}.$$

$$\|\boldsymbol{v}\|_{m,\Omega} = \left\{ \sum_{i=1}^n \|v_i\|_{m,\Omega}^2 \right\}^{1/2} \quad \text{for functions } \boldsymbol{v} = (v_i)_{i=1}^n \text{ in } (H^m(\Omega))^n.$$

$$|\boldsymbol{v}|_{m,\Omega} = \left\{ \sum_{i=1}^n |v_i|_{m,\Omega}^2 \right\}^{1/2} \quad \text{for functions } \boldsymbol{v} = (v_i)_{i=1}^n \text{ in } (H^m(\Omega))^n.$$

$W^{m,p}(\Omega) = \{v \in L^p(\Omega);\ \partial^\alpha v \in L^p(\Omega) \text{ for all } \alpha \text{ with } |\alpha| \leqslant m\}$.

$W_0^{m,p}(\Omega)$: closure of $\mathscr{D}(\Omega)$ in $W^{m,p}(\Omega)$.

$$\|v\|_{m,p,\Omega} = \left\{ \sum_{|\alpha| \leqslant m} \int_\Omega |\partial^\alpha v|^p \, dx \right\}^{1/p}, \quad 1 \leqslant p < \infty.$$

$$\|v\|_{m,\infty,\Omega} = \max_{|\alpha| \leqslant m} \left\{ \operatorname*{ess\,sup}_{x \in \Omega} |\partial^\alpha v(x)| \right\}.$$

$\|v\|'_{m,p,\Omega}$: norm in the dual space of $W^{m,p}(\Omega)$.

$$|v|_{m,p,\Omega} = \left\{ \sum_{|\alpha| = m} \int_\Omega |\partial^\alpha v|^p \, dx \right\}^{1/p}, \quad 1 \leqslant p < \infty.$$

$$|v|_{m,\infty,\Omega} = \max_{|\alpha| = m} \left\{ \operatorname*{ess\,sup}_{x \in \Omega} |\partial^\alpha v(x)| \right\}.$$

$$\left.\begin{aligned} &\dot{v} = \{w \in W^{k+1,p}(\Omega);\ (w - v) \in P_k(\Omega)\} \\ &\|\dot{v}\|_{k+1,p,\Omega} = \inf_{p \in P_k(\Omega)} \|v + p\|_{k+1,p,\Omega}, \quad v \in \dot{v} \\ &|\dot{v}|_{k+1,p,\Omega} = |v|_{k+1,p,\Omega}, \quad v \in \dot{v} \end{aligned}\right\} \text{notation in the quotient space } W^{k+1,p}(\Omega)/P_k(\Omega).$$

$$|v|_{\varphi;m,\Omega} = \left\{ \int_\Omega \varphi \sum_{|\beta| = m} |\partial^\beta v|^2 \, dx \right\}^{1/2} : \text{weighted seminorm.}$$

$$|v|_{m,\infty,T} = \sup_{x \in T} \|D^m v(x)\|_{\mathscr{L}_m(\mathbb{R}^n;\mathbb{R})} \quad \text{for } v: T \subset \mathbb{R}^n \to \mathbb{R}.$$

$|F|_{m,\infty,\hat{T}} = \sup_{\hat{x}\in\hat{T}} \|D^m F(\hat{x})\|_{\mathscr{L}_m(\mathbb{R}^n;\mathbb{R}^n)}$ for $F:\hat{T}\subset\mathbb{R}^n\to\mathbb{R}^n$.

$[F]_{m,\infty,\hat{T}} = \max_{1\leqslant i\leqslant n} \sup_{\hat{x}\in\hat{T}} \|D^m F(\hat{x})(e_i)^m\|$ for $F:\hat{T}\subset\mathbb{R}^n\to\mathbb{R}^n$.

$H^m(T), W^{m,p}(T), \|\cdot\|_{m,p,T},\ldots$:
alternate notation for $H^m(\mathring{T}), W^{m,p}(\mathring{T}), \|\cdot\|_{m,p,\mathring{T}},\ldots$, where T is a subset of $\mathbb{R}^n$ with interior $\mathring{T}$.

Elasticity

λ, μ: Lamé's coefficients of a material.

$E = \dfrac{\mu(3\lambda+2\mu)}{\lambda+\mu}$: Young's modulus.

$\nu = \dfrac{\lambda}{2(\lambda+\mu)}$: Poisson's coefficient.

$\varepsilon_{ij}(\boldsymbol{u}) = \frac{1}{2}(\partial_j u_i + \partial_i u_j)$: components of the linearized strain tensor.

$\sigma_{ij}(\boldsymbol{u})$: components of the stress tensor.

Spaces of polynomials

P_k: space of all polynomials in $x_1,\ldots,x_n$ of degree $\leqslant k$.

Q_k: space of all polynomials in $x_1,\ldots,x_n$, of degree $\leqslant k$ with respect to each variable $x_i, 1\leqslant i\leqslant n$.

Notation special to $\mathbb{R}^n$

$e_i, 1\leqslant i\leqslant n$: canonical basis of $\mathbb{R}^n$.

$\|v\| = \left\{\sum_{i=1}^n |v_i^2|\right\}^{1/2}$: Euclidean norm of the vector $v=(v_i)_{i=1}^n$.

$\|B\| = \sup_{v\in\mathbb{R}^n} \dfrac{\|Bv\|}{\|v\|}$: norm of the matrix B, induced by the Euclidean vector norm.

$\boldsymbol{a}\cdot\boldsymbol{b}$: Euclidean scalar product in $\mathbb{R}^n$ of the vectors $\boldsymbol{a}$ and $\boldsymbol{b}$.

$\boldsymbol{a}\times\boldsymbol{b}$: vector product of the vectors $\boldsymbol{a}$ and $\boldsymbol{b}$.

$\det B$: determinant of a square matrix B.

$\operatorname{meas}(A) = \int_A \mathrm{d}x$: $\mathrm{d}x$-measure of a set $A\subset\mathbb{R}^n$.

$d\gamma$: superficial measure along a Lipschitz-continuous boundary of an open subset of $\mathbb{R}^n$.

$\lambda_j = \lambda_j(x)$: barycentric coordinates of a point $x \in \mathbb{R}^n$, $1 \leqslant j \leqslant n+1$.

$a_{ij} = \frac{1}{2}(a_i + a_j)$, $i<j$.

$a_{iij} = \frac{1}{3}(2a_i + a_j)$, $i \neq j$.

$a_{ijk} = \frac{1}{3}(a_i + a_j + a_k)$, $i \neq j$, $j \neq k$, $k \neq i$.

Finite elements (most common notation)

(T, P, Σ) or (T, P_T, Σ_T): finite element.

$P = P_T$: space of functions p or $p_T: T \to \mathbb{R}$.

$\Sigma = \Sigma_T$: set of degrees of freedom of a finite element.

$\varphi_i = \varphi_{i,T}$, $1 \leqslant i \leqslant N$: degrees of freedom of a finite element.

$p_i = p_{i,T}$, $1 \leqslant i \leqslant N$: basis functions of a finite element.

η_T: set of nodes of a finite element.

$s = s_T$: maximal order of directional derivatives found in the set Σ.

$\Pi v = \Pi_T v$: P- or P_T-interpolant of a function v.

$\operatorname{dom} \Pi = \mathscr{C}^s(T)$.

$h_T = \operatorname{diam}(T)$.

$\rho_T = \sup\{\operatorname{diam}(S)$: S is a ball contained in $T\}$.

$\hat{x} \in \hat{T} \to x = F(\hat{x}) \in T$:
corresponding points of $\hat{T}$ and $T = F(\hat{T})$ (F: bijection).

$(\hat{v}: \hat{T} \to \mathbb{R}) \to (v = \hat{v} \cdot F^{-1}: T \to \mathbb{R})$:
corresponding functions defined over $\hat{T}$ and $T = F(\hat{T})$ (F: bijection).

$F \in (\hat{P})^n \Leftrightarrow F_i \in \hat{P}$, $1 \leqslant i \leqslant n$

$\tilde{T} = \tilde{F}(\hat{T})$, where $\tilde{F} \in (P_1(\hat{T}))^n$,
$\tilde{F}(\hat{a}_i) = a_i$, $1 \leqslant i \leqslant n+1$

$h_T = \operatorname{diam}(\tilde{T})$

ρ_T: diameter of the sphere inscribed in $\tilde{T}$

} for isoparametric simplicial elements.

$\int_T \varphi(x)\,dx \sim \sum_{l=1}^{L} \omega_l \varphi(b_l)$: quadrature formula with weights ω_l and nodes b_l.

$\hat{E}(\hat{\varphi}) = \int_{\hat{T}} \hat{\varphi}(\hat{x})\,d\hat{x} - \sum_{l=1}^{L} \hat{\omega}_l \hat{\varphi}(\hat{b}_l)$: quadrature error functional on $\hat{T}$.

$$E_T(\varphi)=\int_{\hat{T}}\varphi(x)\,\mathrm{d}x-\sum_{l=1}^{L}\omega_{l,T}\varphi(b_{l,T}):$$

quadrature error functional on $T=F_T(\hat{T})$, where
$\omega_{l,T}=\hat{\omega}_l J(F_T)(\hat{b}_l),\quad b_{l,T}=F_T(\hat{b}_l)$.

Finite element spaces (most common notation)

$\mathscr{T}_h$: triangulation of a set $\bar{\Omega}$.
X_h: finite element space without boundary conditions.
$X_{0h}=\{v_h\in X_h;\ v_h=0 \text{ on } \Gamma\}$.
$X_{00h}=\{v_h\in X_h;\ v_h=\partial_\nu v_h=0 \text{ on } \Gamma\}$.
V_h: finite element space with boundary conditions.
Σ_h: set of degrees of freedom of a finite element space X_h.
φ_h or φ_{kh}, $1\leqslant k\leqslant M$: degrees of freedom of a finite element space X_h.
$(w_k)_{k=1}^M$: basis in a finite element space X_h or V_h.
η_h: set of nodes of a finite element space X_h.
$\Pi_h v$: X_h-interpolant of a function v.
$\operatorname{dom}\Pi_h=\mathscr{C}^s(\bar{\Omega}),\quad s=\max_{T\in\mathscr{T}_h} s_T$.

Various sets of hypotheses concerning the finite element method

(FEM1) The set $\bar{\Omega}$ is triangulated.

(FEM2) The spaces P_T, $T\in\mathscr{T}_h$, contain polynomials or "nearly polynomials".

(FEM3) There exists a basis in the finite element space V_h whose functions have "small" supports.

($\mathscr{T}_h$1) Each set T is closed, $\mathring{T}\neq\emptyset$, and $\mathring{T}$ is connected.

($\mathscr{T}_h$2) Each boundary ∂T is Lipschitz-continuous.

($\mathscr{T}_h$3) $\bar{\Omega}=\bigcup_{T\in\mathscr{T}_h} T$.

($\mathscr{T}_h$4) $T_1\neq T_2\ \Rightarrow\ \mathring{T}_1\cap\mathring{T}_2=\emptyset$.

($\mathscr{T}_h$5) Condition on adjacent finite elements.

(H1) Regularity of a family of triangulations.

(H2) All finite elements (T, P_T, Σ_T), $T \in \bigcup_h \mathscr{T}_h$, are affine-equivalent to a single reference finite element.

(H3) All finite elements (T, P_T, Σ_T), $T \in \bigcup_h \mathscr{T}_h$, are of class $\mathscr{C}^0$.

(H4) The family of triangulations satisfies an inverse assumption.

(H2′) The family (T, P_T, Σ_T), $T \in \bigcup_h \mathscr{T}_h$, is almost-affine.

(H3′) All finite elements (T, P_T, Σ_T), $T \in \bigcup_h \mathscr{T}_h$, are of class $\mathscr{C}^1$.

Subject Index

An asterisk in the left margin indicates a specific finite element

Local Behavior in Finite Element Methods

Lars B. Wahlbin

Department of Mathematics
and Center for Applied Mathematics
White Hall
Cornell University
Ithaca, NY 14853, USA

Supported by the National Science Foundation, U.S.A. and by the Army Research Office through the Mathematical Sciences Institute at Cornell.

HANDBOOK OF NUMERICAL ANALYSIS, VOL. II
Finite Element Methods (Part 1)
Edited by P.G. Ciarlet and J.L. Lions

Contents

Note added in proof

This article reflects my knowledge as of December 1986. Only minor corrections have been made during proof-reading.

CHAPTER I

Introduction

1. Foreword

The purpose of this article is to survey what is mathematically known about local behavior in finite element projection methods. The purpose of this introductory chapter is to display some relevant questions, sidelights, and techniques, in very simple one-dimensional cases.

When facing various types of singularities one is often forced to consider the local behavior of an approximation method. This situation is eloquently summed up in the last section of OLIGER [1978, Section 5, pp. 105–106], in his article from an AMS short course in numerical analysis (italics mine):

> In closing, I want to point out an area of research activity which involves detailed mathematical analysis of the solutions of approximate methods. In our discussions up to this point we have only considered approximations of smooth solutions. However, many problems of interest do not have smooth solutions—they have contact discontinuities, shocks, etc. *It is necessary to know how our methods behave in the neighborhood of such discontinuities and whether or not the resulting effects are local or global.* That is, if something "funny" happens near a discontinuity, will this pollute our answers everywhere, or not? *Most of the studies regarding these questions are carried out on simple model equations....* This is an area of research activity where the tools of classical and modern analysis are providing significant results, where questions are abundant and answers are rare.

An archetypical question is the following: The problem at hand contains isolated singularities and we know, a priori, that our approximation method cannot resolve these singularities (to resolve them may be too costly, we do not know where the singularities are, or even what is) their nature. Assuming then that we have given up on resolving the singular behaviors, can we at last assert how good our approximation is in regions of smooth behavior? Can we precisely account for the spread of errors emanating at the singularities into smooth regions? And, can the analysis indicate an efficient way of resolving the singularities, e.g. by mesh refinement, inclusion of special functions mimicking a singularity, tracking of its unknown location, or by a posteriori processing? Regrettably, "... questions are abundant and answers are rare."

The term "pollution" may be given the following meaning: On a subdomain A of the basic domain, the function sought is smooth and, in principle, approximable to some optimal order $O(h^r)$. However, the projection method under consideration gives a *lower* rate of convergence on A.

Some theory is known. As an example the reader may ponder the following: Why is it that finite element solutions with unrefined meshes of a two-dimensional cracked torsion problem lead only to first order accuracy in function values away from the crack, regardless of the order of the finite elements used, whereas when approximating the Green's function in an uncracked problem (with a smooth boundary), while still not refining the meshes, we have essentially the best possible order of accuracy that the finite elements allow away from the point of singularity. After all, the Green's function singularity is worse than the crack singularity! This is a question that present theory can elucidate.

Furthermore, present theory can tell how to do an economical mesh refinement, e.g. in the cracked torsion problem, either a priori or adaptively a posteriori. The refinement suggested by theory often looks "dramatic" and therefore many an eyeball refinement ends up being not daring enough and, in the end, uneconomical. (Cf., e.g., Figs. 14.2 and 14.3 below.)

Another, rather distinct, question about local behavior is that of "superconvergence": Are there identifiable points where the rate of convergence can be proven to be better than in general? Or, is there some simple a posteriori processing that gives better rate of convergence at selected points? Again, answers are rare.

I now remind the reader that "most of the studies regarding these questions are carried out on simple model equations". Amplifying this point, I quote from KREISS and OLIGER [1973, Introduction]:

> We have developed many concepts and carried out many analyses by studying model equations. We find that most computational difficulties are linear effects and can be studied in simple situations where a detailed analysis is possible. We stress the importance of this technique. An adequate, rigorous analysis is usually practically impossible for the large nonlinear models; computational difficulties are apt to be wrongly ascribed. This can easily lead to a large and incorrect folklore which can steer future research in the wrong direction. There are certainly many examples of this in the past. This is not to say that there are no pitfalls inherent in this technique. Great care must be used in selecting model equations and in extrapolating conclusions to more complicated phenomena. However, this is an invaluable tool to isolate and analyze phenomena.

The impracticality of a complete, rigorous analysis in complicated situations is easily understood: You have to combine, say, four model situations (sharp front not resolved, numerical integration, approximate treatment of boundary conditions and approximate, fast, solution of matrix equations, e.g.). Each of the basic papers in model situations cover, say, 30 pages. A fully combined proof will end up being excessive in length.

When wishing to apply a typical result to a specific situation, then, the First Law of Applied Mathematics is in effect: *Nothing Fits.* The treatment of boundary conditions at curved boundaries is different, numerical integration is not taken into account, the result is proven for triangular elements but not for quadrilaterals, for strictly polynomial shape functions but not for isoparametric ones, the result is proven only for strictly interior subdomains but not for subdomains contiguous to the boundary of the basic domain, for Runge–Kutta time stepping methods but not for linear multistep methods, etc. (Sometimes results are given in a model situation which contains generically conflicting hypotheses, "for simplicity". A common case is that of a basic polygonal domain where the solution is as smooth as required, affectionately known as the "smooth polygonal" case. Another common pitfall is that of assuming

$$\min_{\chi \in V_h} \|u-\chi\|_{1,2,\Omega} \leqslant Ch^{r-1} \|u\|_{r,2,\Omega}$$

for general r, general curved boundaries and imposed essential boundary conditions.)

It is important to understand the basic techniques of proofs concerning local behavior. Once the underlying ideas are comprehended, one may extend the known results to a model situation which contains the salient features of interest. It is therefore my intention to elucidate the techniques of proof in this survey.

The rest of this Introduction will be devoted to a brief overview of results and techniques for Fourier series and spline projections with respect to L^2 and H^1 norms in one-space dimension. The aim is to contrast the three cases in order to appreciate the local nature of the spline projections, and also to give an introduction to important techniques for proving local results about finite element projections. We end with a section on identifiable superconvergence points in one dimension, and also an overview of the article.

I will, no doubt, commit errors of omission of relevant results, errors of misattributing credit, as well as factual errors in giving results and proofs. I would appreciate having such errors brought to my attention so that I can correct them in future editions of the Handbook.

2. Perspective: Local behavior in Fourier series

The finite element methods we shall consider are of Petrov–Galerkin, most often Galerkin, type. Thus, in simple cases, they are methods of projection with respect to some norm. To put the present survey in perspective, we shall briefly consider the local convergence in Fourier series; in Section 3 we shall contrast the results with those for spline-L^2 projections and, in Section 4, with $\mathring{H}_1$ projections.

The well-known controversy in the eighteenth century as to whether a plucked string may be represented by its Fourier series is, perhaps, the first example of a typical local problem in projection methods. For our present purposes it is appropriate to pose three questions:

Can the plucked string be represented by any trigonometric series? (2.1)

How does the Fourier series behave near the singularity? (2.2)

How does the Fourier series behave in smooth regions? (2.3)

In the finite element context the analogue of question (2.1) is frequently easy to settle. Various interpolants or quasi-interpolants are of a highly local nature which can or cannot be fitted to a particular singularity and, if they can, the question is answered in the positive. If the interpolant cannot fit the singularity, it is often easy enough to show that no function from the finite element space employed can. (Of course the word "fit" above implies questions of asymptotic convergence.)

The analogues of questions (2.2) and (2.3) are often much harder in the finite element context.

In the Fourier series context a high point in elucidating the local behavior came with Riemann's Habilitationsschrift, RIEMANN [1854], also known by his charming prelude that, although he knew that the Fourier coefficients were given by integrals, he did not quite know what an integral was, and hence desired to discuss that point. His localization principle asserts that, provided the function is globally in L^1 (in modern terms), convergence at a point is determined only by the behavior of the function approximated in any open (periodic) neighborhood of that point.

The questions (2.1) – (2.3) are not asked with sufficient precision for modern purposes. It is necessary to specify what error measure is used and, at least, what is then the asymptotic rate of convergence. Otherwise, the results would give little practical guidance to a computer user. (Some researchers impose more rigorous standards of precision, in particular, they take issue with the very concept of asymptotic rate of convergence. As an example I quote from FICHERA [1978, p. 10] in his critique of the investigations of BRAMBLE and OSBORN [1973]:

> Bramble and Osborn can only claim to have "ultimately" estimated the rate of convergence of $\mu_k^{(n)}$ towards μ. But the concept of an *ultimate* estimate (i.e. valid starting from some *unspecified* n_0) is, indeed, meaningless in numerical analysis!

In order not to end this article prematurely, asymptotic estimates with constants that are impossible to realistically secure, except by numerical experimentation, are admitted in this article. Such asymptotic estimates constitute an invaluable debugging tool, for example.)

We shall next consider the selection of an error measure in the context of local error behavior near a singularity. We shall do so in the situation of a simple example of a stepfunction. Let thus $u(x)$ denote the 2π-periodic function defined by

$$u(x)=\begin{cases}0 & \text{for } -\tfrac{1}{2}\pi\leqslant x\leqslant\tfrac{1}{2}\pi,\\ 1 & \text{for } -\pi\leqslant x<-\tfrac{1}{2}\pi \quad\text{or}\quad \tfrac{1}{2}\pi<x\leqslant\pi.\end{cases} \tag{2.4}$$

The Fourier series is

$$\frac{1}{2}-\frac{2}{\pi}(\cos x-\tfrac{1}{3}\cos 3x+\tfrac{1}{5}\cos 5x-\cdots) \tag{2.5}$$

and we let $u_n(x)$ stand for the nth truncated Fourier series. Let further V_n denote the trigonometric polynomials of degree n.

Let now A denote any open interval of $\Omega=(-\pi, \pi)$ containing the two rough points, $x=\pm\frac{1}{2}\pi$. Assuming that we desire to measure the error as the maximum over A, i.e., as $\|u-u_n\|_{\infty,A}$, it is clear that we do not have convergence, since even

$$\inf_{v_n\in V_n} \|u-v_n\|_{0,\infty,A}\geqslant\tfrac{1}{2}. \tag{2.6}$$

We try instead an error measure more forgiving to approximating discontinuous functions by continuous ones, let us say the Hausdorff graph measure. For the convenience of the reader I will recall its definition. First, for the discontinuous function such as $u(x)$ of (2.4), we define its augmented plane graph AGu by vertically filling in the jump discontinuities.

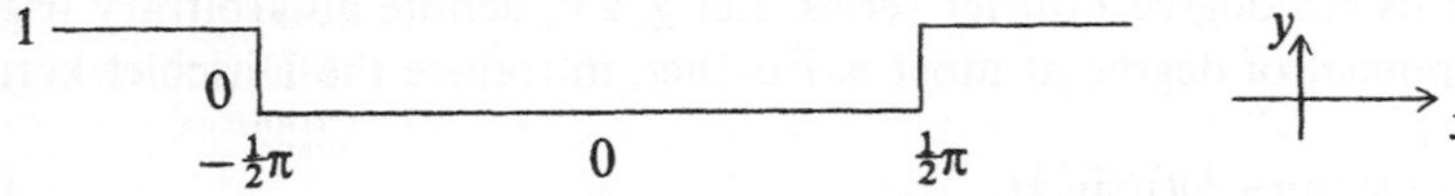

(AGu)

For a continuous function, the augmented graph is the ordinary one. For points $P=(x_P, y_P)$, $Q=(x_Q, y_Q)$ in the plane, set $\operatorname{dist}(P,Q)=\max(|x_P-x_Q|, |y_P-y_Q|)$. The Hausdorff graph measure of the distance between u and v over the interval A is then

$$H(u,v;A):=\max\left(\max_{P\in U}\min_{Q\in V}\operatorname{dist}(P,Q),\ \max_{Q\in V}\min_{P\in U}\operatorname{dist}(P,Q)\right) \tag{2.7}$$

where U denotes the portion of AGu over the x-interval A, and similarly for V. From SENDOV [1969] it is known that there exists $v_n\in V_n$ such that

$$H(u, v_n; \Omega)=\mathrm{O}\left(\frac{\ln n}{n}\right). \tag{2.8}$$

However, for the Fourier series u_n the classical Gibbs phenomenon tells us that for any open interval A containing $-\frac{1}{2}\pi$ or $\frac{1}{2}\pi$,

$$H(u, u_n; A)=\mathrm{O}(1) \tag{2.9}$$

and not better. Thus, approximation of this step function is in principle possible in the Hausdorff measure to order $(\ln n)/n$ but the Fourier series does not converge in Hausdorff measure near $\pm\frac{1}{2}\pi$.

Various error measures in which the Fourier series for the step function u does converge near the singular points include any L^p measure for $p<\infty$, any negative norm....

We are now advised that results concerning local behavior in regions including singularities may dramatically differ according to the error measure chosen.

In our finite element situation a more *typical* problem, at present, is to consider the error in smooth regions, away from singularities. In the Fourier series context, Riemann's localization principle tells us that we have pointwise convergence in such smooth regions (provided the function approximated is globally in L^1). However, consider the step function $u(x)$ of (2.4) and its Fourier series given in (2.5). The interval $A:=[-\frac{1}{4},\frac{1}{4}]$ is certainly an interval of smoothness for u; $u\equiv 0$ there! But, as is easily seen from (2.5),

$$\|u-u_n\|_{0,\infty,A}=O(n^{-1}) \tag{2.10}$$

and not better. In fact, it is easy to construct examples for which, in regions where the approximating function vanishes identically, its Fourier series converges arbitrarily slow, see WAHLBIN [1985, Example 6.1].

It is now a challenge to determine the underlying principles of how the global deportment of a function approximated determines the rate of convergence of its Fourier series in smooth regions. We shall first give an almost trivial result, cast in a form that can be readily compared with the corresponding result for spline-L^2 and $\mathring{H}^1$ projections (to be given in Sections 3 and 4).

To fix notation, let $u(x)$ denote a 2π-periodic function to be approximated and let $u_n(x)$ denote its nth degree Fourier series. Let $\chi_n\in V_n$ denote an arbitrary trigonometric polynomial of degree at most n. Further, introduce the Dirichlet kernel

$$D_n(t)=\frac{1}{2\pi}\sin((n+\tfrac{1}{2})t)/\sin\tfrac{1}{2}t. \tag{2.11}$$

Then, fixing a point x_0 of interest,

$$(u-u_n)(x_0)=(u-\chi_n)(x_0)+(\chi_n-u_n)(x_0). \tag{2.12}$$

Since the nth Fourier approximation of $\chi_n\in V_n$ is itself, we have as is well known,

$$(\chi_n-u_n)(x_0)=\int_{-\pi}^{\pi} D_n(x_0-t)(\chi_n-u)(t)\,dt. \tag{2.13}$$

At this point another question of precision enters: we wish to explain the rate of convergence of the Fourier series for u, at x_0, in terms of the local behavior of u around x_0, and the global (periodic) behavior of u. (Certainly, as argued above, both these factors need to be taken into account.) To make those two factors precise, introduce a parameter d, where (always interpreted in the periodic sense),

$$A_d=[x_0-d,x_0+d] \tag{2.14}$$

will be the interval around x_0 where local behavior of u will be taken into account. Assume for the moment that

$$d\geqslant n^{-1}. \tag{2.15}$$

Then from (2.13) (again in the periodic sense),

$$\begin{aligned}
&|(\chi_n-u_n)(x_0)|\\
&\quad\leqslant \int_{|x_0-t|\leqslant n^{-1}} |D_n(x_0-t)(\chi_n-u)(t)|\,dt+\int_{n^{-1}\leqslant|x_0-t|\leqslant d} |D_n(x_0-t)(\chi_n-u)(t)|\,dt\\
&\qquad+\int_{|x_0-t|\geqslant d} |D_n(x_0-t)(\chi_n-u)(t)|\,dt.
\end{aligned} \tag{2.16}$$

From (2.11) it is easy to estimate the Dirichlet kernel in the various regions involved in (2.16). In the first integral, $|D_n(x_0-t)|\leqslant Cn$, for the second,

$$\int\limits_{n^{-1}\leqslant|x_0-t|\leqslant d} |D_n(x_0-t)|\,dt\leqslant C\int_{n^{-1}}^{d}\frac{1}{x}\,dx\leqslant C\ln(nd).$$

Finally, in the third integral of (2.16), we use the estimate

$$|D_n(t-x_0)|\leqslant C/d$$

in that region. Thus, using the L^∞ norm for χ_n-u in the first two integrals in (2.16) and the L^1 norm in the third, and combining with (2.12), and letting $\Omega=[-\pi,\pi]$ denote the global interval,

$$|(u-u_n)(x_0)|$$
$$\leqslant(1+C|\ln(nd)|)\|u-\chi_n\|_{0,\infty,A_d}+(C/d)\|u-\chi_n\|_{0,1,\Omega},\quad \text{any } \chi_n\in V_n. \tag{2.17}$$

(It is trivial to extend this result to the case $d\leqslant n^{-1}$.)

For future reference we state (2.17) in precise form.

THEOREM 2.1. *Let u be a 2π-periodic function and let u_n be its nth-order Fourier approximation. There exists a constant C, independent of n, such that if $A_d=[x_0-d,x_0+d]$ (in the periodic sense), and $\Omega=[-\pi,\pi]$, for any χ_n a trigonometric polynomial of degree at most n,*

$$|(u-u_n)(x_0)|\leqslant(1+C|\ln(nd)|)\|u-\chi_n\|_{0,\infty,A_d}+(C/d)\|u-\chi_n\|_{0,1,\Omega}. \tag{2.17$'$}$$

We have here a trivial result which gives the rate of convergence for the Fourier series at a point x_0 in terms of how the basic function can be approximated by any trigonometric polynomial of degree n, viz., χ_n. We point out that the estimate (2.17) involves the local approximability of u on the neighboring domain A_d in a more severe error measure (viz., the maximum norm) than that which measures the global influence (viz., the global L^1 deviance).

Another, most important, feature of (2.17) is the following. If we try to be more lenient about taking the local behavior of u into account, that is, we decrease d, the more prominently the global L^1 term enters due to the factor Cd^{-1}. This balance between local and global influences to convergence is typical also of the finite element situation.

To use (2.17′) in order to predict the rate of convergence at x_0 in any specific situation, one must select a suitable trigonometric polynomial χ_n. In order to give a simple result, introduce the pth modulus of continuity in a Banach space B with step t,

$$\omega^p(u,t;B)=\sup_{|\tau|\leqslant t}\|\Delta_\tau^p u\|_B \tag{2.18}$$

where Δ_τ^p is the pth undivided difference,

$$\Delta_\tau^p u(x)=(T_\tau-I)^p u(x),\qquad T_\tau u(x)=u(x+\tau). \tag{2.19}$$

THEOREM 2.2. *Given natural numbers p and q, there exist a constant C such that the following holds: There exists $\chi_n \in V_n$ such that*

$$\|u-\chi_n\|_{0,\infty,A_d} \leqslant C\omega^p\left(u, \frac{C \ln n}{n}; L^\infty(A_{d+C(\ln n)/n})\right) + Cn^{-q}(\|u\|_{0,\infty,A_d} + \|u\|_{0,1,\Omega}) \tag{2.20}$$

and

$$\|u-\chi_n\|_{0,1,\Omega} \leqslant C\omega^p\left(u, \frac{C \ln n}{n}; L^1(\Omega)\right) + Cn^{-q}\|u\|_{0,1,\Omega}. \tag{2.21}$$

SKETCH OF PROOF. The proof follows traditional lines. We shall use a very localized trigonometric kernel $k_n(t)$ due to NOBLE [1954], cf. BARY [1964, Lemma, p. 270]. The kernel has the following properties: With a uniform constant $C = C(q)$,

$$|k_n(t)| \leqslant Cn/\ln n, \quad \text{all } t, \tag{2.22}$$

$$|k_n(t)| \leqslant Cn^{-q} \quad \text{for } C(\ln n)/n \leqslant |t| \leqslant \pi, \tag{2.23}$$

$$\int_{-\pi}^{\pi} k_n(t)\, dt = 1, \tag{2.24}$$

$$\int_{-\pi}^{\pi} |k_n(t)|\, dt \leqslant C. \tag{2.25}$$

Define then for any p,

$$\chi_n(x) = -\int_{-\pi}^{\pi} k_n(t) \sum_{l=1}^{p} \binom{p}{l} u(x+lt)\, dt \tag{2.26}$$

which is a trigonometric polynomial of degree n, cf. LORENTZ [1966, p. 58]. Then

$$u(x) - \chi_n(x) = \int_{-\pi}^{\pi} k_n(t) \Delta_t^p u(x)\, dt. \tag{2.27}$$

For $x \in A_d$,

$$|(u-\chi_n)(x)| \leqslant \int_{|t| \leqslant C(\ln n)/n} |k_n(t)||\Delta_t^p u(x)|\, dt + \int_{|t| \geqslant C(\ln n)/n} |k_n(t)||\Delta_t^p u(x)|\, dt$$

which by use of (2.22) and (2.23), (2.25) gives (2.20).

For (2.21),

$$\|u-\chi_n\|_{1,\Omega} \leqslant \int_{-\pi}^{\pi}\int_{-\pi}^{\pi} |k_n(t)\Delta_t^p u(x)|\,dx\,dt$$

$$= \int_{|t|\leqslant C(\ln n)/n} \cdots + \int_{|t|\geqslant C(\ln n)/n} \cdots$$

which gives (2.21) by use of (2.22), and (2.23). □

Combining Theorems 2.1 and 2.2, we see that in order to guarantee a rate of convergence of almost order p ($(\ln n)^{p+1}n^{-p}$ to be exact) for the Fourier series at x_0, it is necessary that u has essentially p derivatives in the L^∞ sense on an interval around x_0, but also that u has essentially p derivatives globally in the L^1 sense.

The reader may readily apply the two theorems to the stepfunction of (2.4), taking $x_0=0$ for instance. The prediction is, with $p=1$, $q=2$, that the convergence is as $O((\ln n)/n)$ which is sharp apart from the logarithmic factor.

3. Prologue: Local behavior in one-dimensional spline-L^2 projections. Exponential decay

Theorem 2.1 gives a result on how the local behavior in Fourier series is influenced by the mixture of local and global effects in a "best possible" approximation by trigonometric series. Our goal in this section is to give two "analogues" of Theorem 2.1. One, Theorem 3.1, is a rather straight analogue, but does display that spline-L^2 approximation is appreciably more local than Fourier series approximation is. Then, Theorem 3.3 gives a further localization, of considerable interest for our future purposes. (The "analogue" of Theorem 2.2, referring to straight approximation theory, is well known in the spline case and stated as Theorem 3.2 below merely for completeness; again, though, it is considerably more local than its Fourier counterpart.) To amplify this point, in every comparison spline-L^2 projection will turn out to be much more local than Fourier series, i.e., L^2 projection into trigonometric polynomials.

We shall use the fundamental techniques of NITSCHE and SCHATZ [1972] and since some essential ideas will recur throughout this article we shall give a fair amount of detail. Our aim is not, however, to give the best possible results; this would swell the exposition to unmanageable bounds.

We first need to establish some notation. Let $\bar{\Omega}=[-1,1]$, say, and let Υ_h denote a family of interval subdivisions of $\bar{\Omega}$,

$$\bar{\Omega} = \bigcup_{T_i\in\Upsilon_h} T_i, \quad T_i=[x_i, x_{i+1}],$$

$i=0,\ldots,N$, with $x_0=-1$, $x_{N+1}=1$. The parameter h can be thought of as (proportional to) N^{-1}. *Let* V_h denote spline spaces on Υ_h. With integers μ and r,

$$-1 \leqslant \mu \leqslant r-2, \tag{3.1}$$

we define

$$V_h = \{v \in \mathscr{C}^\mu(\bar{\Omega}): v \text{ is a polynomial of degree } r-1 \text{ on each interval } T_i\}. \tag{3.2}$$

Given a function u on Ω we let $u_h = P_h u \in V_h$ denote its L^2 projection into V_h so that

$$(u - P_h u, v) = 0 \quad \text{for } v \in V_h, \tag{3.3}$$

where $(f, g) = \int_\Omega fg$.

As a first indication that convergence may be more locally determined than in the case of Fourier series, consider the case $\mu = -1$ so that V_h consists of unconnected piecewise polynomials. Clearly then $P_h u$ is completely determined on each subinterval T_i by the behavior of u only on that T_i. At present we do not wish to consider local properties on a scale smaller than that of a mesh interval. The properties of L^2 projection into full polynomials on an interval, that is, Legendre series, have been extensively studied; a facile investigation of their local properties from the present points of view is contained in WAHLBIN [1985]; in the case of discontinuous splines this amounts to a "microlocal" analysis. (We shall return to such microlocal, or subscale, analysis only in connection with superconvergence points.)

For piecewise polynomials that are at least continuously connected across mesh points x_i, i.e., $\mu \geqslant 0$, $P_h u$ suffers some influence from u from outside intervals. To simplify our analysis of the situation (cf. references at the end of this section) we assume that our family of triangulations Υ_h is quasi-uniform, i.e., that there exist positive constants c and C independent of h such that

$$ch \leqslant \text{meas } T_i \leqslant Ch, \quad i = 0, \ldots, N. \tag{3.4}$$

The assumption (3.4) implies the following inverse property. Its proof is well known and can be found in any basic text on finite element methods.

LEMMA 3.1 (Inverse property). *There exists a constant C independent of h such that for any mesh interval* T_i,

$$\|v\|_{m,p,T_i} \leqslant Ch^{-m-(1/q-1/p)} \|v\|_{0,q,T_i} \quad \textit{for } v \in V_h, \quad 1 \leqslant q \leqslant p \leqslant \infty. \tag{3.5}$$

The next result represents the basic technical tool introduced by NITSCHE and SCHATZ [1972]. It is often referred to as "superapproximation" and is still the most important general tool. To describe it we need some notation. For intervals $A_0 \subseteq A_1 \subseteq \Omega$ set

$$\partial_<(A_0, A_1) = \text{dist}(\partial A_0 \backslash \partial\Omega, \partial A_1 \backslash \partial\Omega) \tag{3.6}$$

i.e., the minimum distance between endpoints of A_0 and A_1, not counting endpoints coinciding with $\partial\Omega$. Let further

$$\mathscr{C}^{\infty}_{<}(\bar{A})=\{v\in\mathscr{C}^{\infty}(\bar{A}): \partial_{<}(\operatorname{supp} v, A)>0\} \tag{3.7}$$

and

$$V_h^{<}(A)=\{v\in V_h: \partial_{<}(\operatorname{supp} v, A)>0\}. \tag{3.8}$$

LEMMA 3.2 ("Superapproximation"). *There exist constants c and C such that the following holds.*

Let $A_0\subseteq A_1\subseteq\Omega$ be intervals with $d=\partial_{<}(A_0, A_1)\geqslant ch$. Let $\omega\in\mathscr{C}^{\infty}_{<}(\bar{A}_0)$ with

$$\|\omega\|_{l,\infty,A_0}\leqslant \Lambda d^{-l},\quad l=0,\ldots, r. \tag{3.9}$$

Then for any $\chi\in V_h$ there exists $\psi\in V_h^{<}(A_1)$ such that

$$\|\omega\chi-\psi\|_{0,2,A_1}\leqslant C\Lambda(h/d)\|\chi\|_{0,2,A_0}. \tag{3.10}$$

PROOF. We first consider the case of $\mathscr{C}^0$ splines, i.e., $\mu=0$ in (3.2). We may then for any function $v\in\mathscr{C}(\bar{T}_i)$ construct its $(r-1)$st-degree polynomial interpolant $\Pi_h v$ on $\bar{T}_i$ using r equispaced interpolation points, including both endpoints of $\bar{T}_i$. By the Bramble–Hilbert lemma, BRAMBLE and HILBERT [1970, 1971],

$$\|v-\Pi_h v\|^2_{0,2,T_i}\leqslant Ch^{2r}\|v^{(r)}\|^2_{0,2,T_i}. \tag{3.11}$$

For $v\in\mathscr{C}(\bar{A}_0)$ the combination of those local interpolants result in a function in $V_h^{<}(A_1)$ provided c is large enough; we denote it still by $\Pi_h v$. Further with

$$\|v\|_{h;r,A_0}=\Bigg(\sum_{T_i\cap A_0\neq\emptyset}\|v\|^2_{r,2,T_i}\Bigg)^{1/2} \tag{3.12}$$

denoting the piecewise $H^r(A_0)$ norm, (3.11) gives

$$\|v-\Pi_h v\|^2_{0,2,A_1}\leqslant Ch^{2r}\|v\|^2_{h;r,A_0}. \tag{3.13}$$

To prove (3.10) we take $\psi=\Pi_h(\omega\chi)$; it remains to calculate $h^{2r}\|\omega\chi\|^2_{h;r,A_0}$. By Leibnitz' lemma and since $D^r\chi\equiv 0$ on each T_i, this can be estimated by a sum of terms of the form

$$h^{2r}\|(D^{\alpha}\omega)(D^{\beta}\chi)\|^2_{0,2,T_i},\qquad \alpha+\beta=r,\quad \beta\leqslant r-1.$$

By (3.9) and the inverse property of Lemma 3.1 each such term is estimated by

$$C\Lambda^2 d^{-2\alpha}h^{2(r-\beta)}\|\chi\|^2_{0,2,T_i}$$

and for $h/d\leqslant 1$, this quantity is bounded by

$$C\Lambda^2(h/d)^2\|\chi\|^2_{0,2,T_i}.$$

Summing over all T_i involved gives (3.10) in the case $\mu=0$.

The case of general μ is similar. An analogue of (3.13) is needed and to construct it we first recall the quasi-interpolation operator of DE BOOR and FIX [1973]: Set

$$\tilde{T}_i=[x_{\max(0,i-r+1)}, x_{\min(N+1,i+r)}]. \tag{3.14}$$

There exists a constant C and a family of operators $\Pi_h:H^r\to V_h$ such that

$$\|v-\Pi_h v\|^2_{0,2,T_i}\leqslant Ch^{2r}\|v^{(r)}\|^2_{0,2,T_i}. \tag{3.15}$$

Taking $v=\omega\chi$ we would need to convert the norm on the right of (3.15) into a piecewise norm. This is done by the following trick from DOUGLAS, DUPONT and WAHLBIN [1975a]: With each interior mesh point x_i associate the functions $v_{ik}(x)=(x-x_i)^k_+$, $k=\mu+1,\ldots,r-1$. Starting at left we determine a linear combination w of the v_{ik} so that $\omega\chi+w\in\mathscr{C}^{r-1}\cap H^r$. Set then $\psi=\Pi_h(\omega\chi+w)-w$ which is clearly in V_h. By (3.15),

$$\begin{aligned}\|\omega\chi-\psi\|^2_{0,2,T_i}&=\|(\omega\chi+w)-\Pi_h(\omega\chi+w)\|^2_{0,2,T_i}\\&\leqslant Ch^{2r}\|(\omega\chi+x)^{(r)}\|^2_{h;T_i}=Ch^{2r}\|(\omega\chi)^{(r)}\|^2_{h,T_i},\end{aligned}$$

and we have obtained a proper estimate involving only the piecewise r derivatives. The proof is concluded as before. □

We now wish to consider the error $(u-P_hu)(x_0)$. Stressing the analogy with the Fourier case we proceed as follows: For any $\chi_h\in V_h$,

$$(u-P_hu)(x_0)=(u-\chi_h)(x_0)+P_h(\chi_h-u)(x_0). \tag{3.16}$$

Assume that $x_0\in\bar{T}_i$. Let δ^0 be a polynomial of degree $r-1$ on T_i such that

$$\int_{T_i} v\delta^0\,\mathrm{d}x=v(x_0)\quad\text{for any } v \text{ a polynomial of degree } r-1. \tag{3.17}$$

Regard δ^0 as extended by zero to Ω. It is clear from simple considerations involving mapping T_i to a unit interval that

$$|\delta^0(x)|\leqslant Ch^{-1}\quad\text{for } x\in T_i. \tag{3.18}$$

Consider now the second term on the right of (3.16). We have

$$P_h(\chi_h-u)(x_0)=(P_h(\chi_h-u),\delta^0)=(\chi_h-u,P_h\delta^0). \tag{3.19}$$

Thus, $\delta^0_h:=P_h\delta^0\in V_h$ here takes the role of the Dirichlet kernel in the Fourier case.

We shall next give an estimate for δ^0_h. For future purposes we state a preliminary result in greater generality.

LEMMA 3.3. *There exist positive constants c_0, c_1 and C such that the following holds. Let $A\subseteq A_d\subseteq\Omega$ with $d=\eth_<(A,A_d)\geqslant c_0h$. Assume that $v_h\in V_h$ is such that*

$$(v_h,\chi_h)=0\quad\textit{for all } \chi_h\in V_h^<(A_d). \tag{3.20}$$

Then

$$\|v_h\|_{0,2,A}\leqslant Ce^{-c_1d/h}\|v_h\|_{0,2,A_d}. \tag{3.21}$$

PROOF. With the constant c from Lemma 3.2, let $ch=\delta$; let $A\subseteq A_\delta\subseteq A_{2\delta}$ with $\delta=\eth_<(A,A_\delta)=\eth_<(A_\delta,A_{2\delta})$. Let $\omega\in\mathscr{C}^\infty_<(A_\delta)$ be a nonnegative function with

$$\|\omega\|_{l,\infty}\leqslant\Lambda\delta^{-l},\quad l=0,\ldots,r, \tag{3.22}$$

and with

$$\omega \equiv 1 \quad \text{on } A. \tag{3.23}$$

Then

$$\| v_h \|^2_{0,2,A} \leqslant (v_h, \omega v_h). \tag{3.24}$$

Assuming now $A_{2\delta} \subseteq A_d$, we have $(v_h, \psi) = 0$ for any $\chi \in V_h^{<}(A_{2\delta})$ and thus

$$\| v_h \|^2_{0,2,A} \leqslant (v_h, \omega v_h - \chi). \tag{3.25}$$

Using Lemma 3.2,

$$\| v_h \|^2_{0,2,A} \leqslant C(h/\delta) \| v_h \|^2_{0,2,A_{2\delta}}. \tag{3.26}$$

The argument can clearly be repeated with domains $A_{2\delta} \subseteq A_{3\delta} \subseteq A_{4\delta} \subseteq \cdots \subseteq A_{2N\delta}$ as long as they are inside A_d, i.e., since $\delta = ch$,

$$N = \mathrm{O}(d/h). \tag{3.27}$$

Assuming, as we may, that $Ch/\delta < 1$, (3.26) leads to the result (3.21). □

REMARK 3.1. The quasi-uniform condition (3.4) is demanded in Lemma 3.3 only for those T_i that meet A_d.

Applying (3.21) to $v_h = \delta_h^0$, using that, by the inverse property of Lemma 3.1,

$$|\delta_h^0(x)| \leqslant Ch^{-1/2} \| \delta_h^0 \|_{0,2,T_j}, \quad x \in T_j \tag{3.28}$$

and that by the stability in L^2 and (3.18),

$$\| \delta_h^0 \|_{0,2,\Omega} \leqslant \| \delta^0 \|_{0,2,\Omega} = Ch^{-1/2}, \tag{3.29}$$

we have

LEMMA 3.4. *Assume that* (3.4) *holds. Then there exist constants c and C such that*

$$|\delta_h^0(x)| \leqslant C\mathrm{e}^{-c|x-x_0|/h} h^{-1}, \tag{3.30}$$

or, after a change of constants,

$$|\delta_h^0(x)| = \begin{cases} Ch^{-1}, & |x - x_0| \leqslant h, \\ \dfrac{C}{|x - x_0|} \mathrm{e}^{-c|x-x_0|/h}, & |x - x_0| \geqslant h. \end{cases} \tag{3.31}$$

Thus the kernel in spline-L^2 projection is considerably more localized than the Dirichlet kernel in Fourier series, for which the estimate corresponding to (3.31) is

$$|D_n(x - x_0)| \leqslant \begin{cases} Cn \quad, & |x - x_0| \leqslant n^{-1}, \\ \dfrac{C}{|x - x_0|}, & |x - x_0| \geqslant n^{-1}. \end{cases}$$

We note incidentally the following consequence of Lemma 3.4.

LEMMA 3.5. *Assume that* (3.4) *holds. Then the* L^2 *projection is stable in any* L^p *norm,* $1 \leqslant p \leqslant \infty$.

PROOF. From (3.30) it follows that $\| \delta_h^0 \|_{0,1,\Omega} \leqslant C$ with C independent of x_0. Thus,

$$|P_h u(x_0)| = |(u, \delta_h^0)| \leqslant \| u \|_{0,\infty,\Omega} C,$$

and P_h is stable in L^∞. By duality

$$\| P_h u \|_{1,\Omega} = \sup_{\substack{v \in L^\infty \\ \|v\|_{\infty,\Omega} = 1}} (P_h u, v)$$

and since $(P_h u, v) = (P_h u, P_h v) = (u, P_h v)$, the already established result gives

$$(P_h u, v) \leqslant \| u \|_{0,1,\Omega} C$$

so that P_h is also stable in L^1. By Riesz–Thorin's interpolation lemma it is then stable in any L^p space. □

We return now to (3.19) and use Lemma 3.4. Let $A_d = [x_0 - d, x_0 + d] \cap \Omega$; we obtain for $d \geqslant h$,

$$\begin{aligned} |P_h(\chi_h - u)(x_0)| &\leqslant \int_{A_d} |\chi_h - u| \, |\delta_h^0| + \int_{\Omega \setminus A_d} |\chi_h - u| \, |\delta_h^0| \\ &\leqslant C \| u - \chi_h \|_{0,\infty,A_d} + (C/d) \mathrm{e}^{-cd/h} \| u - \chi_h \|_{0,1,\Omega}. \end{aligned}$$

Combining this with (3.16) we have the following analogue of Theorem 2.1.

THEOREM 3.1 (Exponential decay in spline-L^2 projection). *Assume that the family of subdivisions of* Ω *is quasi-uniform,* (3.4), *and let* V_h *be given as in* (3.2). *There exist positive constants* c *and* C *such that the following holds. If* $A_d = [x_0 - d, x_0 + d] \cap \Omega$ *and* u_h *denotes the spline-*L^2 *projection of a function* u, *then for* χ_h *arbitrary in* V_h,

$$|(u - u_h)(x_0)| \leqslant C \| u - \chi_h \|_{0,\infty,A_d} + (C/d) \mathrm{e}^{-cd/h} \| u - \chi_h \|_{0,1,\Omega}. \tag{3.32}$$

Thus, in spline-L^2 projection the global influence is weighted down by an additional factor $\mathrm{e}^{-cd/h}$ compared to the global influence in Fourier series.

The analogue of Theorem 2.2 is well known for splines. The result of de Boor and Fix quoted in (3.15) holds in greater generality with respect to norms and rates of convergence less than r, and one may derive the following by use of well-known smoothing techniques, cf. STRANG [1972] and HILBERT [1973].

THEOREM 3.2. *Given a natural number* $p \leqslant r$ *there exists a constant* C *such that the following holds.*

There exists $\chi_h \in V_h$ *such that*

$$\| u - \chi_h \|_{0,\infty,A_d} \leqslant C h^p \| u^{(p)} \|_{0,\infty,A_{d+Ch}}, \tag{3.33}$$

$$\| u - \chi_h \|_{0,1,\Omega} \leqslant C \| u \|_{0,1,\Omega}. \tag{3.34}$$

We next give a result which "completely" localizes Theorem 3.1 to a subdomain. It assumes the quasi-uniformity condition (3.4) only for elements T_i in that subdomain; often this criterion is enough to meet practical situations involving systematic mesh refinements towards a known singularity. Although the technique is by no means the best possible in many situations for one-dimensional spline-L^2 projections (cf. the references at the end of this section) it is the way in which we shall proceed in multidimensional Galerkin methods for partial differential equations and so we take this opportunity to illustrate the technique in a situation where all the technical work is already done.

Assume thus that x_0 is given, $A_d=[x_0-d, x_0+d]\cap\Omega$ with $d\geqslant c_1 h$ (c_1 conveniently large) and assume that the family of subdivisions is quasi-uniform on A_d, i.e., corresponding to (3.4),

$$ch\leqslant \text{meas } T_i\leqslant Ch \quad \text{for those } T_i \text{ that intersect } A_d. \tag{3.35}$$

Assume also that

$$(u-u_h, \chi)=0 \quad \text{for } \chi\in V_h^<(A_d). \tag{3.36}$$

THEOREM 3.3 (Further localization of exponential decay in spline-L^2 projection). *Assuming* (3.35) *and* (3.36) *there exist constants* c, c_1 *and* C *such that the following holds for any* $\chi_h\in V_h$: *If* $d\geqslant c_1 h$, *then*

$$|(u-u_h)(x_0)|\leqslant C\|u-\chi_h\|_{0,\infty,A_d}+Cd^{-1/2}\mathrm{e}^{-cd/h}\|u-u_h\|_{0,2,A_d}. \tag{3.37}$$

Thus when applying this result the quantity $\|u-u_h\|_{0,2,A_d}$ has to be estimated. This is frequently easy to do to one's satisfaction, by standard techniques.

PROOF. Let $\tilde{P}_h$ denote the L^2 projection into $V_h^<(A_d)$. We have already seen, combining Remark 3.1 and Lemma 3.5, that $\tilde{P}_h$ is stable in L^∞. Thus,

$$(u-u_h)(x_0)=(u(x_0)-\tilde{P}_h u(x_0))+(\tilde{P}_h u(x_0)-P_h u(x_0)). \tag{3.38}$$

Here,

$$|u(x_0)-\tilde{P}_h u(x_0)|\leqslant C\|u\|_{0,\infty,A_d} \tag{3.39}$$

while $\tilde{P}_h u-P_h u$ satisfies (3.20) and hence by Lemma 3.3, cf. again Remark 3.1, after combining with the inverse estimate,

$$\begin{aligned}|\tilde{P}_h u(x_0)-P_h u(x_0)|&\leqslant Ch^{-1/2}\mathrm{e}^{-cd/h}\|\tilde{P}_h u-P_h u\|_{0,2,A_d}\\ &\leqslant Cd^{-1/2}\mathrm{e}^{-cd/h}\|\tilde{P}_h u-P_h u\|_{0,2,A_d}\end{aligned} \tag{3.40}$$

(after a change of constants).

Next,

$$\begin{aligned}&\|\tilde{P}_h u-P_h u\|_{0,2,A_d}\\ &\quad\leqslant\|\tilde{P}_h u-u\|_{0,2,A_d}+\|P_h u-u\|_{0,2,A_d}\\ &\quad\leqslant\|u\|_{0,2,A_d}+\|P_h u-u\|_{0,2,A_d}\leqslant d^{1/2}\|u\|_{0,\infty,A_d}+\|P_h u-u\|_{0,2,A_d}.\end{aligned}$$

Combining this with (3.38)–(3.40),

$$|(u-u_h)(x_0)| \leqslant C\|u\|_{0,\infty,A_d} + Cd^{-1/2}\mathrm{e}^{-cd/h}\|u-u_h\|_{0,2,A_d}.$$

Since $P_h\chi_h = \chi_h$ for any $\chi_h \in V_h$ we may add and subtract χ_h to arrive at (3.37). □

Our treatment above has followed the fundamental paper of NITSCHE and SCHATZ [1972]. In particular, *Lemma 3.2 is basic.* In the one-dimensional spline situation one may proceed via a careful analysis of the inverse of a banded matrix, cf. DEMKO [1977], to obtain similar results.

The decay estimates given above are frequently implicit in investigations concerning the L^p stability of the L^2 projection, also under relaxed mesh conditions as compared to quasi-uniformity. As examples we mention DESCLOUX [1972], DOUGLAS, DUPONT and WAHLBIN [1975a], and Demko's paper. We shall give Descloux's argument, as adapted by CROUZEIX and THOMÉE [1987a], in Section 8. For a survey of L^p stability of the L^2 projection under relaxed mesh conditions, see DE BOOR [1979] where references to original contributions in this regard are given.

4. Prologue continued: Local behavior in one-dimensional elliptic projections. Negative norms

Our goal in this section is not to prove the best possible results for one-dimensional elliptic spline projections but rather to introduce, in a simple setting, techniques and results that generalize to problems in more dimensions. Nevertheless, the results are often sharp also in one dimension, cf. Section 17.

(To prove "better", i.e., more localized, results in one dimension, the reader may contemplate integrating the results for the L^2 projection in Section 3, or, in the case of continuous splines, $\mu=0$, skip forward to Theorems 5.1 and 5.3 to "deduce" that the elliptic projection lies within $\mathrm{O}(h^{r+1})$ of a completely local interpolant on each mesh interval where the solution is smooth. Care is advised in carrying through such a program, to account for outside singularities, in particular for roughness emanating from unsmooth coefficients, cf. again Section 17.)

We shall first give an analogue of Theorem 3.3 (Theorem 4.1 below) and contrast the two. We then proceed to weaken the global influence, in Theorem 4.2.

The reader not familiar with negative norms is urged to contemplate Theorem 4.2, Lemma 4.4 and their interplay.

Let $x_0 \in \Omega = [-1, 1]$ be a point of interest and let $A_d = [x_0 - d, x_0 + d]$. For simplicity assume $A_d \subseteq \operatorname{Int}\Omega$. Let V_h be a spline space as in (3.1), (3.2) with $\mu \geqslant 0$ and let for $A \Subset \Omega$,

$$\mathring{V}_h(A) = \{v \in V_h \colon \operatorname{dist}(\operatorname{supp} v, A) > 0\}, \tag{4.1}$$

corresponding to (3.8) (for notational simplicity we do not consider the case when A, or A_d, may abut on the boundary of Ω).

Let now $u(x)$ be a function and $u_h \in V_h$ be such that

$$(u' - u_h', \chi') = 0 \quad \text{for all } \chi \in \mathring{V}_h(A_d). \tag{4.2}$$

We first observe that (4.2) makes sense also if u is merely continuous. For, integrating by parts over each element T_i intersecting A_d,

$$(u'-u_h',\chi')=\sum_{T_i\cap A_d\neq\emptyset}\left[-\int_{T_i}(u-u_h)\chi''\,\mathrm{d}x+(u-u_h)\chi'|_{\partial T_i}\right]. \tag{4.3}$$

Thus we may, perhaps, expect to obtain a result valid if u is merely continuous on A_d.

Let now $\mathring{P}_h$ denote a local elliptic projection into $\mathring{V}_h$, i.e., $\mathring{P}_h u\in\mathring{V}_h$ with

$$(u'-(\mathring{P}_h u)',\chi')=0 \quad \text{for all } \chi\in\mathring{V}_h. \tag{4.4}$$

Let further $\omega\in\mathscr{C}_0^\infty(A_{d/2})$ with

$$\omega(x)\equiv 1 \quad \text{on } A_{d/4}. \tag{4.5}$$

Then

$$(u-u_h)(x_0)=(\omega u-\mathring{P}_h\omega u)(x_0)+(\mathring{P}_h\omega u-u_h)(x_0). \tag{4.6}$$

We first argue that $\mathring{P}_h$ is stable in L^∞ for functions with support in $A_{d/2}$, $d\geqslant ch$. Assume for this that the family of triangulations is quasi-uniform on A_d, i.e., that (3.35) holds.

LEMMA 4.1. *There exists a constant C such that for v with* $\operatorname{supp} v\subseteq A_{d/2}$,

$$\|\mathring{P}_h v\|_{0,\infty,A_{d/2}}\leqslant C\|v\|_{0,\infty,A_{d/2}}. \tag{4.7}$$

PROOF. We have

$$\|\mathring{P}_h v\|_{0,\infty,A_{d/2}}=\sup_{\substack{w\in\mathscr{C}_0^\infty(A_{d/2})\\ \|w\|_{1,A_{d/2}}=1}}(\mathring{P}_h v,w). \tag{4.8}$$

For each such w, let W be the solution of

$$-W''=w \quad \text{in } A_d, \qquad W=0 \quad \text{on } \partial A_d. \tag{4.9}$$

Further, set $W_h=\mathring{P}_h W$. Then

$$\begin{aligned}(\mathring{P}_h v,w)&=((\mathring{P}_h v)',W')=((\mathring{P}_h v)',W_h')\\&=(v',W_h')=(v',(W_h-W)')+(v,w)\\&=\sum_i\left\{-\int_{T_i}v(W_h-W)''\,\mathrm{d}x+v(W_h-W)'|_{\partial T_i}\right\}+(v,w).\end{aligned} \tag{4.10}$$

A simple trace inequality establishes that (since the mesh is quasi-uniform on A_d), for $\bar{x}\in\bar{T}_i$,

$$|f(\bar{x})|\leqslant C\|f'\|_{0,1,T_i}+Ch^{-1}\|f\|_{0,1,T_i}. \tag{4.11}$$

(Use a cutoff function that isolates T_i and then the Fundamental Theorem of

Calculus.) Thus from (4.10), using (4.11) with $f=(W_h-W)'$,

$$|(\mathring{P}_h v, w)| \leqslant C\|v\|_{0,\infty,A_{d/2}}\{\|(W_h-W)''\|_{h;1,A_{d/2}}+h^{-1}\|(W_h-W)'\|_{0,1,A_{d/2}}+1\} \tag{4.12}$$

where the subscript h in the norm for the second derivatives of W_h-W denotes that we are operating piecewise on each element intersecting $A_{d/2}$.

Letting $\Pi_h W$ be, say, the piecewise linear interpolant of W, we have since then $(\Pi_h W)''=0$ piecewise,

$$\|(W_h-W)''\|_{h;1,A_{d/2}} \leqslant \|W''\|_{0,1,A_d}+\|(W_h-\Pi_h W)''\|_{h;1,A_{d/2}}. \tag{4.13}$$

Note that

$$\|W''\|_{0,1,A_d} \leqslant 1. \tag{4.14}$$

By inverse properties, assuming as we may here that $A_{d/2}$ is a mesh domain for notational simplicity,

$$\begin{aligned}\|(W_h-\Pi_h W)''\|_{h;1,A_{d/2}} &\leqslant Ch^{-1}\|(W_h-\Pi_h W)'\|_{0,1,A_{d/2}}\\ &\leqslant Ch^{-1}\|(W_h-W)'\|_{0,1,A_{d/2}}+Ch^{-1}\|(W-\Pi_h W)'\|_{0,1,A_{d/2}},\end{aligned} \tag{4.15}$$

and since, as is well known, cf. Theorem 3.2,

$$\|(W-\Pi_h W)'\|_{0,1,A_{d/2}} \leqslant Ch\|W''\|_{0,1,A_{d/2}} \leqslant Ch,$$

we have from (4.13)–(4.15),

$$\|(W_h-W)''\|_{h;1,A_{d/2}} \leqslant C+Ch^{-1}\|(W_h-W)'\|_{0,1,A_{d/2}}. \tag{4.16}$$

Thus, combining with (4.12),

$$|(\mathring{P}_h v, w)| \leqslant C\|v\|_{0,\infty,A_{d/2}}\{1+h^{-1}\|(W_h-W)'\|_{0,1,A_{d/2}}\}. \tag{4.17}$$

Under the circumstances, since $(W_h'-W', 1)=0$, W_h' is the L^2 projection of W' into a spline space with μ and r lowered by one. Thus by Lemma 3.5 (easily localized), for $\tilde{\chi}$ in that lower spline space,

$$\|W_h'-W'\|_{0,1,A_{d/2}} \leqslant C\min_{\tilde{\chi}}\|W'-\tilde{\chi}\|_{0,1,A_{d/2}} \tag{4.18}$$

and using again a variation of Theorem 3.2 for approximation,

$$\min_{\tilde{\chi}}\|W'-\tilde{\chi}\|_{0,1,A_d} \leqslant Ch\|W''\|_{0,1,A_d} \leqslant Ch. \tag{4.19}$$

By (4.17)–(4.19) then,

$$|(\mathring{P}_h v, w)| \leqslant C\|v\|_{0,\infty,A_{d/2}},$$

which in light of (4.8) proves (4.7). □

Returning now to (4.6) we have, using Lemma 4.1 with $v=\omega u$,

$$|(u-u_h)(x_0)|\leqslant C\|u\|_{0,\infty,A_{d/2}}+C|(\mathring{P}_h\omega u-u_h)(x_0)|. \tag{4.20}$$

We next come to a crucial part of our argument. Let

$$w_h:=\mathring{P}_h\omega u-u_h. \tag{4.21}$$

Then $w_h\in V_h$ and satisfies

$$(w_h',\chi')=0\quad\text{for all } \chi\in\mathring{V}_h(A_{d/4}), \tag{4.22}$$

since $\omega\equiv 1$ on $A_{d/4}$. Following NITSCHE and SCHATZ [1974] we may call w_h "discrete harmonic" in $A_{d/4}$.

Letting now $\tilde{\omega}$ with $\tilde{\omega}(x_0)=1$ be a suitable cutoff function isolating $A_{d/8}$, hence $\|\tilde{\omega}'\|_\infty\leqslant Cd^{-1}$, we have by Cauchy–Schwarz' inequality,

$$|w_h(x_0)|=\int_{-\infty}^{x_0}(\tilde{\omega}w_h)'\leqslant Cd^{-1/2}\|w_h\|_{0,2,A_{d/8}}+Cd^{1/2}\|w_h'\|_{0,2,A_{d/8}}. \tag{4.23}$$

The following is a fundamental result from NITSCHE and SCHATZ [1974]. Again "superapproximation", cf. Lemma 3.2, will figure in its proof but we shall merely note so at the relevant place.

LEMMA 4.2. *Assuming that the family of meshes is quasi-uniform on A_d, that $d\geqslant ch$, and that $w_h\in V_h$ satisfies* (4.22), *we have*

$$\|w_h'\|_{0,2,A_{d/8}}\leqslant cd^{-1}\|w_h\|_{0,2,A_{d/4}}. \tag{4.24}$$

PROOF. For simplicity in writing we set $\frac{1}{8}d=\delta$, $\frac{1}{4}d=2\delta$. Let here $\omega\in\mathscr{C}_0^\infty(A_{3\delta/2})$ be such that

$$\omega\equiv 1\quad\text{on } A_\delta, \tag{4.25}$$

$$\|\omega^{(l)}\|_{0,\infty}\leqslant\Lambda\delta^{-l},\quad l=0,\ldots,r. \tag{4.26}$$

We have

$$\begin{aligned}\|w_h'\|_{0,2,A_\delta}^2&\leqslant\|\omega w_h'\|_{0,2,A_{3\delta/2}}^2=(w_h',\omega^2w_h')\\&=(w_h',(\omega^2w_h)')-2(\omega w_h',\omega'w_h).\end{aligned} \tag{4.27}$$

Here,

$$\begin{aligned}(\omega w_h',\omega'w_h)&\leqslant C\delta^{-1}\|\omega w_h'\|_{0,2,A_{3\delta/2}}\|w_h\|_{0,2,A_{3\delta/2}}\\&\leqslant\tfrac{1}{2}\|\omega w_h'\|_{0,2,A_{3\delta/2}}^2+C\delta^{-2}\|w_h\|_{0,2,A_{3\delta/2}}^2.\end{aligned} \tag{4.28}$$

Further, by (4.22),

$$(w_h',(\omega^2w_h)')=(w_h',(\omega^2w_h-\chi)')\quad\text{for any } \chi\in\mathring{V}_h(A_{2\delta}). \tag{4.29}$$

An easy modification (left to the reader) of the superapproximation result of

Lemma 3.2 gives that there is a χ in $\mathring{V}_h(A_{3\delta/2})$ (slightly enlarged) such that

$$\|(\omega^2 w_h-\chi)'\|_{0,2,A_{3\delta/2}} \leqslant C(h/\delta)(\|w_h'\|_{0,2,A_{3\delta/2}}+\delta^{-1}\|w_h\|_{0,2,A_{3\delta/2}}). \tag{4.30}$$

Thus, from (4.29),

$$\begin{aligned}|(w_h', (\omega^2 w_h)')| &\leqslant C(h/\delta)(\|w_h'\|^2_{0,2,A_{3\delta/2}}+\delta^{-1}\|w_h'\|_{0,2,A_{3\delta/2}}\|w_h\|_{0,2,A_{3\delta/2}}\\ &\leqslant C(h/\delta)\|w_h'\|^2_{0,2,A_{3\delta/2}}+C\delta^{-2}\|w_h\|^2_{0,2,A_{3\delta/2}}.\end{aligned} \tag{4.31}$$

Combining (4.27) with (4.28) and (4.31),

$$\|w_h'\|^2_{0,2,A_\delta} \leqslant C(h/\delta)\|w_h'\|^2_{0,2,A_{3\delta/2}}+C\delta^{-2}\|w_h\|^2_{0,2,A_{3\delta/2}}$$

and repeating the argument once more with $\frac{3}{2}\delta$ and 2δ taking the role of δ and $\frac{3}{2}\delta$, respectively,

$$\|w_h'\|^2_{0,2,A_\delta} \leqslant C(h^2/\delta^2)\|w_h'\|^2_{0,2,A_{2\delta}}+C\delta^{-2}\|w_h\|^2_{0,2,A_{2\delta}}. \tag{4.32}$$

Assuming as we may (modulo shrinking 2δ somewhat) that $A_{2\delta}$ is a mesh domain we have by the inverse property Lemma 3.1 applied locally,

$$h^2\|w_h'\|^2_{0,2,A_{2\delta}} \leqslant C\|w_h\|^2_{0,2,A_{2\delta}}.$$

Inserting this into (4.32) we obtain the desired result (4.24). □

We now use Lemma 4.2 in equation (4.23). Thus,

$$|w_h(x_0)| \leqslant Cd^{-1/2}\|w_h\|_{0,2,A_{d/4}}. \tag{4.33}$$

Next use this with $w_h=\mathring{P}_h\omega u-u_h$ in (4.20) so that

$$|u-u_h(x_0)| \leqslant C\|u\|_{0,\infty,A_{d/2}}+Cd^{-1/2}\|\mathring{P}_h\omega u-u_h\|_{0,2,A_{d/4}}. \tag{4.34}$$

To further elaborate the last term of (4.34), since $\omega\equiv 1$ on $A_{d/4}$,

$$\|\mathring{P}_h\omega u-u_h\|_{0,2,A_{d/4}} \leqslant \|\mathring{P}_h\omega u-\omega u\|_{0,2,A_{d/4}}+\|u-u_h\|_{0,2,A_{d/4}}. \tag{4.35}$$

By Cauchy–Schwarz' inequality and Lemma 3.1,

$$\|\mathring{P}_h\omega u-\omega u\|_{0,2,A_{d/4}} \leqslant Cd^{1/2}\|\mathring{P}_h\omega u-\omega u\|_{0,\infty A_{d/4}} \leqslant Cd^{1/2}\|u\|_{0,\infty,A_{d/2}}. \tag{4.36}$$

We next combine (4.34)–(4.36), add and subtract a general $\chi\in V_h$, and have then proven the following result (changing the constant slightly). Recall that $A_d=[x_0-d, x_0+d]$.

THEOREM 4.1. *Assume that the family of triangulations is quasi-uniform on A_d, cf. (3.35). There exist constants c and C such that the following holds. Let $d\geqslant ch$ and $u_h\in V_h$ be such that $((u-u_h)', \psi')=0$ for all $\psi\in\mathring{V}_h(A_d)$. Then for any $\chi\in V_h$,*

$$|(u-u_h)(x_0)| \leqslant C\|u-\chi\|_{0,\infty,A_d}+Cd^{-1/2}\|u-u_h\|_{0,2,A_d}. \tag{4.37}$$

Comparing Theorem 4.1 with Theorem 3.3 one immediately notices the absence of the factor $e^{-cd/h}$ in (3.37). The form $\|u-u_h\|_{2,A_d}$ contains the global influences affecting the error at x_0; in light of the quasi-interpolant of DE BOOR and FIX [1973], cf. (3.15), the first term on the right of (4.37) is a very adequate representation of local influences to the error at x_0.

It is then clearly desirable to weaken the global term. This can be done by replacing the L^2 norm by a still weaker norm, viz., a negative norm in $H^{-s}(A_d)$, the dual of $\mathring{H}^s(A_d)$ over the pivot space L^2. Any s may be taken. We follow here the development of NITSCHE and SCHATZ [1974].

Thus define

$$\|v\|_{-s,2,A_d} = \sup_{\substack{w\in\mathscr{C}^0_\infty(A_d)\\ \|w\|_{s,2,A_d}=1}} (v,w). \tag{4.38}$$

If a function is appreciably smaller in a negative norm than in, say, the L^2 norm, this is frequently due to oscillations. As an example the reader may contemplate the functions $v_n(x)=\sin(nx)$, $n=1,2,\ldots$ on $[0,\pi]$. Here,

$$\|v_n\|_{-s,2}\leqslant Cn^{-s}.$$

We need to extend the inverse estimates of Lemma 3.1 somewhat to negative norms.

LEMMA 4.3. *There exists a constant C such that for A any mesh domain and $v\in V_h$,*

$$\|v\|_{0,2,A}\leqslant Ch^{-s}\|v\|_{-s,2,A}. \tag{4.39}$$

PROOF. Let $A=\bigcup T_i$, T_i mesh intervals. Then

$$\|v\|^2_{0,A}=\sum_i \|v\|^2_{0,T_i}.$$

A standard scaling argument establishes that

$$\|v\|^2_{0,T_i}\leqslant Ch^{-2s}\|v\|_{-s,2,T_i}.$$

The proof is then concluded by noting that for the negative norms we are considering,

$$\sum_i \|v\|^2_{-s,2,T_i}\leqslant \|v\|^2_{-s,2,A}. \qquad \square$$

REMARK 4.1. We do *not* claim the extended estimate

$$\|v\|_{-t,2,A}\leqslant Ch^{-s+t}\|v\|_{-s,2,A},\quad 0\leqslant t\leqslant s,$$

with C independent of A. Actually this is true in one dimension but we know of no proof in many dimensions except by the theory of interpolation spaces in which case C will depend on A.

To weaken the global term in (4.37) proceed as follows:

Break into the proof of Theorem 4.1 at the point (4.34) and consider the term

$$\|\mathring{P}_h\omega u-u_h\|_{0,2,A_{d/4}}.$$

Now $w_h=\mathring{P}_h\omega u-u_h$ is a function in $\mathring{V}_h$ which is discrete harmonic, i.e., satisfies

$$(w_h', \psi')=0 \text{ for } \psi\in \mathring{V}_h(A_{2d}) \tag{4.40}$$

(we change domain notation slightly at this point for simplicity).

LEMMA 4.4. *For* $w_h\in \mathring{V}_h$ *satisfying* (4.40),

$$\|w_h\|_{0,2,A_d}\leqslant Cd^{-s}\|w_h\|_{-s,2,A_{2d}}. \tag{4.41}$$

PROOF. Consider first the case of d of unit size. Set

$$A_0=A_d, \qquad A_1=A_{3d/2}, \qquad A_2=A_{2d}.$$

Let $\omega\in\mathscr{C}_0^\infty(A_{3d/2})$ with $\omega\equiv 1$ on A_d. Then, for any nonnegative integer l,

$$\|w_h\|_{-l,2,A_0}\leqslant \|\omega w_h\|_{-l,2,A_1}=\sup_{\substack{v\in\mathscr{C}_0^\infty(A_1)\\ \|v\|_{l,2}=1}} (\omega w_h, v). \tag{4.42}$$

For any such v, let V solve

$$\begin{aligned} -V''&=v \quad \text{on } A_1,\\ V&=0 \quad \text{on } \partial A_1. \end{aligned} \tag{4.43}$$

Then

$$\|V\|_{l+2,2,A_1}\leqslant \|v\|_{l,2,A_1}. \tag{4.44}$$

Now

$$\begin{aligned} (\omega w_h, v)&=((\omega w_h)', V')=(\omega' w_h, V')+(\omega w_h', V')\\ &=(\omega' w_h, V')+(w_h', (\omega V)'-\omega' V)\\ &=(\omega' w_h, V')-(\omega' w_h', V)+(w_h', (\omega V)')\\ &=(w_h, 2\omega' V'+\omega'' V)+(w_h', (\omega V)'-\psi_h') \end{aligned}$$

for any $\psi_h\in \mathring{V}(A_0)$. Thus, by approximation theory,

$$(\omega w_h, v)\leqslant C\|w_h\|_{-l-1,A_1}\|V\|_{l+2,A_1}+\|w_h\|_{1,A_2}Ch^\gamma\|V\|_{l+2,A_1}$$

where $\gamma=\min(l+1, r-1)$, or, in light of (4.44) and (4.42),

$$\|w_h\|_{-l,A_0}\leqslant C\|w_h\|_{-l-1,A_1}+Ch^\gamma\|w_h\|_{1,A_2}.$$

Now recalling Lemma 4.2 we obtain (changing notation for the domains),

$$\|w_h\|_{-l,A_0}\leqslant C\|w_h\|_{-l-1,A_2}+Ch^\gamma\|w_h\|_{0,A_2}. \tag{4.45}$$

For $l=0$, using Lemma 4.3, we have since $\gamma=1$ then (assuming A_2 a mesh domain for notational simplicity),

$$\|w_h\|_{0,A_0}\leqslant C\|w_h\|_{-1,A_2}.$$

The argument now proceeds by induction.

In the case of general d, transform the situation to unit size d. The mesh parameter for the spline space will then be replaced by h/d, and the factor d^{-s} in (4.41) results from this transformation. □

We now return to (4.34) and apply the above lemma, (4.41), to obtain (again changing notation)

$$\begin{aligned}|(u-u_h)(x_0)| &\leq C\|u\|_{0,\infty,A_{d/2}} + Cd^{-1/2-s}\|\mathring{P}_h\omega u - u_h\|_{-s,A_{d/4}} \\ &\leq C\|u\|_{0,\infty,A_{d/2}} + Cd^{-1/2-s}\|\mathring{P}_h\omega u - \omega u\|_{-s,A_{d/4}} \\ &\quad + Cd^{-1/2-s}\|u-u_h\|_{-s,A_{d/4}}. \end{aligned} \tag{4.46}$$

By Friedrich's inequality, $\|v\|_{0,A_{d/4}} \leq Cd^s\|v\|_{s,A_{d/4}}$ for $v \in \mathscr{C}_0^\infty(A_{d/4})$ and hence by duality,

$$\begin{aligned} &d^{-s-1/2}\|\mathring{P}_h\omega u - \omega u\|_{-s,A_{d/4}} \\ &\leq Cd^{-1/2}\|\mathring{P}_h\omega u - \omega u\|_{0,A_{d/4}} \\ &\leq C\|\mathring{P}_h\omega u - \omega u\|_{0,\infty,A_{d/4}} \leq C\|u\|_{0,\infty,A_{d/2}}. \end{aligned} \tag{4.47}$$

We thus obtain the following further elaboration of Theorem 4.1.

THEOREM 4.2. *Under the general hypotheses of Theorem* 4.1, *for any* $s \geq 0$, *for any* $\chi \in V_h$,

$$|(u-u_h)(x_0)| \leq C\|u-\chi\|_{0,\infty,A_d} + Cd^{-1/2-s}\|u-u_h\|_{-s,A_d} \tag{4.48}$$

(*where C now depends also on s*).

Theorem 4.2 holds if the elliptic projection is taken with respect to an equation with variable coefficients, i.e.,

$$(a(x)(u-u_h)', \psi') + (b(x)(u-u_h)', \psi) + (c(x)(u-u_h), \psi) = 0$$
$$\text{for } \psi \in \mathring{V}_h(A_d),$$

provided the coefficients are sufficiently smooth; we shall not give the arguments but refer to THOMÉE and WAHLBIN [1983, Lemma 3.4] for the counterpart of Lemma 4.1, i.e., L^∞ stability of the elliptic projection. (Although given there in a global setting it is not hard to localize the arguments.)

Theorem 4.2 is a one-dimensional version of a multidimensional result of SCHATZ and WAHLBIN [1977] which will be given later in this article. Some essential points of proof will recur, e.g. the L^∞ stability of a localized projection (very simple in the one-dimension situation) and the use of the energy-norm-based techniques of NITSCHE and SCHATZ [1974].

In the one-dimensional spline-L^2 projection there was room for improvement of the results with respect to constraints on the triangulations, mainly to reduce the quasi-uniformity constraint. Such is also the situation for one-dimensional elliptic projections. We refer to NATTERER [1977] and GÜSMAN [1981] for results which are very much related to Theorem 4.1.

Finally we remark on the use of Theorem 4.2. Pose the following question: What does it take in order to guarantee optimal rate of convergence, $O(h^r)$, at a point x_0? For the first term on the right of (4.48) it takes e.g. that $u\in\mathscr{C}^r(A_d)$. We then ask for

$$\|u-u_h\|_{-s,A_d}\leqslant Ch^r, \quad \text{some } s. \tag{4.49}$$

(We disregard the factor $d^{-s-1/2}$ here.)

A simple answer is as follows; we assume for simplicity that the meshes are globally quasi-uniform although this is by no means necessary.

LEMMA 4.5. *For* $1\leqslant s\leqslant r$,

$$\|u-u_h\|_{-(r-2),\Omega}\leqslant Ch^{r-1}\min_{\chi\in\mathscr{V}_h}\|u-\chi\|_{1,2,\Omega}\leqslant Ch^{r+s-2}\|u\|_{s,2,\Omega}. \tag{4.50}$$

PROOF. We have

$$\|u-u_h\|_{-(r-2),\Omega}=\sup_{\substack{v\in\mathscr{C}_0^\infty(\Omega)\\ \|v\|_{r-2,\Omega}=1}}(u-u_h,v).$$

For each such v, let

$$-w''=v \quad \text{in } \Omega, \qquad w(-1)=w(1)=0.$$

Then with w_h the projection of w,

$$\begin{aligned}(u-u_h,v)&=((u-u_h)',w')=((u-u_h)',(w-w_h)')\\ &=((u-\chi)',(w-w_h)')\leqslant\|u-\chi\|_{1,2,\Omega}Ch^{r-1}\|w\|_{r,2,\Omega}\end{aligned}$$

which leads to the desired result. □

Hence (4.49) and local optimal order follows if $u\in H^2(\Omega)$.

The result of this lemma can often be, and needs to be, improved. E.g., let $u=G^1$, the Green's function centered at x_1 for our two-point boundary value problem and consider $x_0\neq x_1$. Then G^1 is not in $H^2(\Omega)$ so that our theory does not predict optimal order at x_0. If x_1 is a mesh point and $\mu=0$ this is disturbing since, in our simple case, then $G^1-G_h^1\equiv 0$. A slightly different result is as follows.

LEMMA 4.6.

$$\|u-u_h\|_{-(r-1),\Omega}\leqslant Ch^{r-1}\min_{\chi\in\mathscr{V}_h}\|u-\chi\|_{1,1,\Omega}. \tag{4.51}$$

PROOF. As in the proof of the previous lemma we have with $\|v\|_{r-1,2,\Omega}=1$ now,

$$(u-u_h,v)=((u-\chi)',(w-w_h)')\leqslant\|u-\chi\|_{1,1,\Omega}\|w-w_h\|_{1,\infty,\Omega}.$$

We know (for quasi-uniform meshes) that the elliptic projection is stable in L^∞; Lemma 4.1 is simple to adapt to the case of the whole of Ω. Using then inverse estimates and approximation theory it is easy to see that the projection is stable in

$W^{1,\infty}$. Hence

$$\|w-w_h\|_{1,\infty,\Omega} \leqslant C \min_{\chi} \|w-\chi\|_{1,\infty,\Omega} \leqslant Ch^{r-1}\|w\|_{r,\infty,\Omega}$$

$$\leqslant Ch^{r-1}\|w\|_{r+1,1,\Omega} = Ch^{r-1}.$$

The desired result (4.51) is thus proven. □

Going back to the Green's function G^1 above, it is easily derived that

$$\min_{\chi \in \mathring{V}_h} \|G^1-\chi\|_{1,1,\Omega} \leqslant Ch$$

and thus G_h^1 approximates to optimal order h^r at $x_0 \neq x_1$. (Since we cannot expect better for variable smooth coefficients and we have a proof that clearly generalizes to that case, we stop.)

Negative norm estimates play important roles in finite element theory. As we have seen above they help explain how local roughness spreads into regions of smooth behavior. They are essential for results about convergence in finite element approximations for eigenvalues, see BRAMBLE and OSBORN [1973]. In the next section they will play a prominent role in connection with averaging to obtain higher rate of convergence. (In our multidimensional investigations later they will also be an important tool, but, alas, the counterparts of Lemmas 4.4 and 4.5 are seldom perfectly valid in practice, either because the domain or coefficients do not admit perfect duality estimates or, in smooth situations, because of slight "imperfections" in approximating curved boundaries.)

5. Prologue concluded: Superconvergence and local averaging in one-dimensional elliptic projections

In a smooth problem the best general pointwise approximation of function values is $O(h^r)$, with r as in (3.1), (3.2). In this section we consider two problems: (i) Can we find a priori identifiable points with higher order of accuracy, and (ii) can we obtain higher order of accuracy via (simple) postprocessing? For problem (i), that of superconvergent points, we first consider the case of continuous splines in two-point boundary value problems ($\mu=0$ in (3.1), (3.2)). Here identifiable superconvergence points for function values are found; the mesh points themselves are $O(h^{2r-2})$ and some $r-2$ specified interior points on each mesh interval are of order $O(h^{r+1})$. This result also holds for nonuniform meshes.

A second result about superconvergent points concerns Hermite cubics ($\mu=1$, $r=4$) for two-point boundary value problems. Here there are two identifiable superconvergent points on each mesh interval of $O(h^5)$ accuracy, but only on strictly interior subintervals and only for completely uniform meshes. As we look on mesh intervals T_i approaching the boundary, or we consider nonuniform meshes, we can still assert that $\int_{T_i}$ error$=0$ (in a very simple situation, see (5.18)) so that we know that the error vanishes at least at one point; however, no such point has ever been identified. This may illustrate that the problem of finding superconvergent points is challenging.

In each case mentioned above we first find superconvergent points for the derivatives and then the result is integrated. Since the derivative of the elliptic projection is the L^2 projection into a lower-order spline space we actually give results for L^2 projections too. (These latter results are classical for the case $\mu=0$, that is, for L^2 projections into full polynomial spaces.)

For problem (ii) above, that of a posteriori processing, we describe the salient points about the "K-operator" of BRAMBLE and SCHATZ [1977]. This easily implemented method gives $O(h^{2r-2})$ accuracy at points where the solution is smooth and the mesh is completely uniform in a neighborhood (provided the solution is somewhat smooth globally). We shall later return to the K-operator in several dimensions; it is the one practical method of some generality.

Let now $\Omega=[-1,1]$ be a basic domain, let $\mathring{V}_h$ denote a spline space with $v_h(-1)=0$ for all $v_h \in \mathring{V}_h$ and consider the problem of finding u s.t.

$$-(a(x)u')' = f(x) \quad \text{in } \Omega, \qquad u(-1)=u'(1)=0. \tag{5.1}$$

We assume $a(x) \geqslant a_0 > 0$ and that $a(x)$ and f are as smooth as required. (The form of (5.1) considered is for simplicity only.)

The projection (Galerkin approximation) $u_h \in \mathring{V}_h$ to (5.1) is given by

$$A(u-u_h, \chi) \equiv (a(x)(u-u_h)', \chi') = 0 \quad \text{for } \chi \in \mathring{V}_h. \tag{5.2}$$

Consider first the case of continuous splines, $\mu=0$ in (3.1), (3.2). Since (5.1) has a unique solution for every $f(x)$, it has a Green's function $G^0(x)=G(x,x_0)$ centered at x_0 such that

$$A(v, G^0) = v(x_0) \tag{5.3}$$

for each $v \in H^1$ with $v(-1)=0$. In particular,

$$(u-u_h)(x_0) = A(u-u_h, G^0). \tag{5.4}$$

Letting $G_h^0 \in \mathring{V}_h$ be the projection of G^0, by (5.2) and (5.4),

$$(u-u_h)(x_0) = A(u-u_h, G^0 - G_h^0). \tag{5.5}$$

Thus,

$$|(u-u_h)(x_0)| \leqslant C \| u-u_h \|_{1,2,\Omega} \| G^0 - G_h^0 \|_{1,2,\Omega}. \tag{5.6}$$

Since u is smooth we have, as is well known,

$$\| u-u_h \|_{1,2,\Omega} \leqslant Ch^{r-1}. \tag{5.7}$$

Further, as is a nice observation from DOUGLAS and DUPONT [1974], since G^0 has a break in its first derivative at x_0 but is smooth on both sides if x_0 is a mesh point, the continuous splines can mimic that break so that

$$\| G^0 - G_h^0 \|_{1,2,\Omega} \leqslant C \inf_{\chi \in \mathring{V}_h} \| G^0 - \chi \|_{1,2,\Omega} \leqslant Ch^{r-1}, \tag{5.8}$$

as is easily seen. Combining this with (5.6) and (5.7), we have the following result of Douglas and Dupont.

THEOREM 5.1 (Mesh point superconvergence in $\mathscr{C}^0$ splines). *Assume the problem* (5.1) *is smooth and that* $\mu=0$ *in* (3.1), (3.2). *Then for* x_i *a mesh point,*

$$|(u-u_h)(x_i)| \leqslant Ch^{2r-2}. \tag{5.9}$$

We continue with the case of continuous splines and search for superconvergent point(s) inside each element T_i. (The simple argument appears to be first published in CHEN [1979], cf. also BAKKER [1982, 1984].)

Let $e(x)=(u-u_h)(x)$ denote the error. Let $p(x)$ denote any function in $P_{r-2}(T_i)$, the polynomials of degree $r-2$ on $T_i=[x_i, x_{i+1}]$. Set with $p(x)\equiv 0$ outside T_i,

$$P(x)=\begin{cases} \displaystyle\int_{-1}^{x} p(t)\,\mathrm{d}t, & x \leqslant x_{i+1}, \\ P(x_{i+1}), & x > x_{i+1}. \end{cases} \tag{5.10}$$

Then $P(x)\in \mathring{V}_h$. By (5.2),

$$\int_{-1}^{1} a(x)\,e'(x)P'(x)=0,$$

or

$$\int_{T_i} ae'p=0 \quad \text{for any } p\in P_{r-2}(T_i). \tag{5.11}$$

We next Taylor expand ae' on T_i around the midpoint $x_{i+1/2}$. Writing the Taylor expansion in terms of Legendre polynomials $L_i(x)$ we have

$$ae' = c_0L_0(x)+c_1L_1(x)+\cdots+c_{r-1}L_{r-1}(x)+\mathrm{O}(h^r),$$

and, by (5.11), $c_0=c_1=\cdots=c_{r-2}=0$ so that

$$ae'(x)=c_{r-1}L_{r-1}(x)+\mathrm{O}(h^r). \tag{5.12}$$

Since $a(x)\geqslant a_0>0$ we have the following superconvergence result for derivatives, one order better than the global result.

THEOREM 5.2 (Interior superconvergence for derivatives in $\mathscr{C}^0$ splines). *Assume that the problem* (5.1) *is smooth and that* $\mu=0$ *in* (3.1), (3.2). *Then for* $\bar{x}$ *a zero of the* $(r-1)$*th Legendre polynomial on* T_i,

$$(u-u_h)'(\bar{x})=\mathrm{O}(h^r). \tag{5.13}$$

Finally, integrating from x_i,

$$\int_{x_i}^{x} ae' = \int_{x_i}^{x} (ae)' - \int_{x_i}^{x} a'e = ae(x) - ae(x_i) - \int_{x_i}^{x} a'e$$

or by (5.9) and (5.12), assuming

$$|e(x)| \leqslant Ch^r \tag{5.14}$$

(as is known in many cases), and that $r \geqslant 3$ so that $2r-2 \geqslant r+1$,

$$ae(x) = O(h^{r+1}) + c_{r-1} \int_{x_i}^{x} L_{r-1}(t)\,dt.$$

Using Legendre's differential equation one sees that the inside zeros of the integral on the right coincides with the inside zeros of L'_{r-1}.

Thus we have the following superconvergence result for the function values, again one order better than globally.

THEOREM 5.3 (Interior superconvergence in $\mathscr{C}^0$ splines). *Assume that the problem* (5.1) *is smooth, and that* $\mu=0$ *in* (3.1), (3.2). *Further assume* (5.14) *and that* $r \geqslant 3$. *Then for* $\tilde{x}$ *a zero of the derivative of the* $(r-1)$*th Legendre polynomial on* T_i,

$$(u-u_h)(\tilde{x}) = O(h^{r+1}). \tag{5.15}$$

Hence, in the case of merely continuous splines, we have identifiable superconvergence points on all mesh intervals, also for nonuniform meshes.

Without attempting to be systematic we next consider the example of Hermite cubics as an illustration. Further, the arguments will only be sketched. Assume for simplicity that $a \equiv 1$ in (5.1). With $e(x) = (u-u_h)(x)$, we shall first investigate $e'(x)$. Then u'_h is the L^2 projection P^0u' of u' into $\mathscr{C}^0$ quadratics, essentially. Set, for notational simplicity, $v = u'$, $v_h = P^0 v$. At a point x_0, let further $\tilde{v}_h$ be a function in the $\mathscr{C}^0$ quadratics that interpolates v at x_0. Assume also a uniform mesh. Then with δ^0 an approximate Dirac delta for quadratics centered at x_0, and with δ^0_h its L^2 projection into the continuous quadratics,

$$(v-v_h)(x_0) = (\tilde{v}_h - v_h)(x_0) = (\tilde{v}_h - v_h, \delta^0_h) = (\tilde{v}_h - v, \delta^0_h). \tag{5.16}$$

Lemma 3.4 gives us very fine control over the behavior of δ^0_h. In particular, if x_0 is interior to Ω the exponential decay suggests that, provided the mesh is completely uniform, δ^0_h is (almost) symmetric if x_0 is either a mesh point or halfway between two mesh points. If then $\tilde{v}_h - v \cong \text{const}\cdot(x-x_0)^3 + \cdots$, the third-order term will be cancelled out in (5.16) so that

$$(v-v_h)(x_0) = O(h^4),$$
$$x_0 \text{ interior mesh point or midpoint, uniform mesh.} \tag{5.17}$$

This is a superconvergence result for derivatives in Hermite cubic approximations to two-point boundary value problems. The approximation is one order better than globally at the above-mentioned points.

To translate (5.17) into a superconvergence result for function values, note first that

$$\int_{T_i} e = \int_{T_i} ex = 0. \tag{5.18}$$

For, if

$$-w'' = 1|_{T_i} \text{ or } x|_{T_i}, \quad 0 \text{ outside } T_i,$$

$$w(-1) = w'(1) = 0,$$

then $w \in \mathring{V}_h$ in the Hermite cubic case.

Thus,

$$\int_{T_i} e = -\int_{-1}^{1} ew'' = (e', w') = 0.$$

(We have assumed $a \equiv 1$ for simplicity.) This shows (5.18).

Writing now for $T_i = [\bar{x} - \frac{1}{2}h, \bar{x} + \frac{1}{2}h]$ with $\bar{x}$ the midpoint,

$$e(x) = A + B(x-\bar{x}) + C(x-\bar{x})^2 + D(x-\bar{x})^3 + E(x-\bar{x})^4 + O(h^5) \tag{5.19}$$

we know that $B = O(h^4)$ since $e'(\bar{x}) = O(h^4)$. Further, (5.18) leads to, when integrating (5.19) against $x - \bar{x}$, that $D = O(h^2)$. Thus,

$$e(x) = A + C(x-\bar{x})^2 + E(x-\bar{x})^4 + O(h^5). \tag{5.20}$$

Integrating this against 1, (5.18) gives

$$hA + \tfrac{2}{3}C(\tfrac{1}{2}h)^3 + \tfrac{2}{5}E(\tfrac{1}{2}h)^5 = O(h^6). \tag{5.21}$$

Further $e'(x) = O(h^4)$ at endpoints gives

$$2C \cdot \tfrac{1}{2}h + 4E(\tfrac{1}{2}h)^3 = O(h^4),$$

or,

$$C = -\tfrac{1}{2}Eh^2 + O(h^3). \tag{5.22}$$

Reporting (5.22) in (5.21),

$$A - \tfrac{1}{3}Eh^2 \tfrac{1}{8}h^2 + \tfrac{2}{5}E \tfrac{1}{32}h^4 = O(h^5)$$

so that

$$A = Eh^4 \frac{7}{15 \cdot 16} + O(h^5).$$

Thus, inserting this and (5.22) in (5.20),

$$e(x)=Eh^4\frac{7}{15\cdot 16}-\tfrac{1}{2}Eh^2(x-\bar{x})^2+E(x-\bar{x})^4+O(h^5)$$
$$=E\left(\tfrac{7}{240}h^4-\tfrac{1}{2}h^2(x-\bar{x})^2+(x-\bar{x})^4\right)+O(h^5).$$

We see from this that the superconvergent points for function values are given at the inside roots of the polynomial multiplying E, i.e., at

$$(x-\bar{x})/h=\pm(\tfrac{1}{4}-1/\sqrt{30})^{1/2}=\pm 0.2596648112\ldots. \tag{5.23}$$

We summarize the above investigation of superconvergence in Hermite cubics. The result extends to variable coefficients but, as numerical experimentation shows, the restriction to *interior intervals* and *completely uniform meshes* is essential. (This is in spite of the fact that, by (5.18), the error $e(x)$ changes sign on each T_i also for nonuniform meshes and T_i near ends. The point(s) where this happens in general have never been identified.) Thus we are dealing with a fairly fickle phenomenon here.

THEOREM 5.4 (Superconvergence in Hermite cubics). *Assume that the problem* (5.1) *is smooth and that* $\mu=1$, $r=4$ *in* (3.1), (3.2). *Further assume that the mesh is uniform, i.e.,* meas$(T_i)\equiv h$, *all i. Let* $A\Subset\Omega$ *be a fixed subinterval of* Ω. *Then for any* T_i *inside A*,

$$(u-u_h)'(\bar{x})=O(h^4),\quad \bar{x}=x_i,x_{i+1/2},x_{i+1}, \tag{5.24}$$

and

$$(u-u_h)(\tilde{x})=O(h^5),\quad \tilde{x}=x_{i+1/2}\pm h(\tfrac{1}{4}-1/\sqrt{30})^{1/2}. \tag{5.25}$$

The argument given extends to other $\mathscr{C}^1$ splines, since essentially only the symmetry (almost) of δ_h^0 and (5.18) were used.

In the remainder of this section we describe, in a simple case, the so-called K-operator of BRAMBLE and SCHATZ [1977]. This method extends to several dimensional problems. It applies to a very general class of locally uniform meshes and sometimes furnishes a pointwise accuracy of $O(h^{2r-2})$ by use of a local averaging operator which is a priori computed and depends only on the finite elements being used, not on h or on the variable coefficients of the second-order elliptic problem considered.

A point to note is that sometimes somewhere (in a smooth problem) an order of accuracy $O(h^{2r-2})$ is lurking, viz., if one measures the error in a negative norm $H^{-(r-2)}$, see Lemma 4.5.

Higher-order accuracy in a negative norm suggests oscillatory behavior of the error. BRAMBLE and SCHATZ found a local, simple and systematic way of averaging out those oscillations to obtain $O(h^{2r-2})$ convergence in smooth problems.

Consider, for simplicity in one part of the argument, the problem

$$-u''=f\quad\text{in } [-1,1]=\Omega,\qquad u(-1)=u'(1)=0. \tag{5.26}$$

Assume that the mesh is uniform (throughout the domain, for simplicity at present), i.e., that

$$\operatorname{meas}(T_i) \equiv h. \tag{5.27}$$

For each of the finite element spaces $\mathring{V}_h$ considered, we have for interior points, $x \in A \Subset \Omega$, for h sufficiently small, a local basis: There exist functions $\varphi_1, \ldots, \varphi_n$ independent of h which are piecewise polynomials with compact support such that any $\chi(x) \in \mathring{V}_h$ is of the form

$$\chi(x) = \sum_{j=1}^{m} \sum_{\alpha \in \mathbb{Z}} a_\alpha^j \varphi_j(x/h - \alpha), \quad x \in A. \tag{5.28}$$

A particular case is that of smoothest splines, $\mu = r-2$, in which we have the "classical" B splines. Since the construction of the K-operator is based on those, we describe them in more detail:

For t real, define

$$\psi(t) = \begin{cases} 1, & |t| \leqslant \frac{1}{2}, \\ 0, & |t| > \frac{1}{2}, \end{cases} \tag{5.29}$$

and for l an integer,

$$\psi_l(t) = \psi * \psi * \cdots * \psi, \quad \text{convolution } l \text{ times}. \tag{5.30}$$

Then ψ_l is the B spline basis function, a piecewise polynomial of degree l. In this case, $\mu = r-2 = l-1$, any $\chi \in \mathring{V}_h$ is of the form

$$\chi(x) = \sum_{\beta \in \mathbb{Z}} a_\beta \psi_l(x/h - \beta). \tag{5.31}$$

Consider now a kernel K_h defined in terms of the above B spline basis function as

$$K_h(x) = \sum_{\beta = -(r-2)}^{(r-2)} h^{-1} k_\beta \psi_{r-2}(x/h - \beta). \tag{5.32}$$

The coefficients k^β are solutions of a certain system of linear equations and tables are given in BRAMBLE and SCHATZ [1977]. The choice of k_β can be summed up in K_h being an approximate identity of order $2r-2$. Here we shall merely point out certain properties of K_h at the points of our development where they are used.

The idea is then to replace u_h, the Galerkin approximation, by

$$K_h * u_h(x) = h^{-1} \int K_h(x-y) u(y)\, \mathrm{d}y.$$

If, cf. (5.28), $u_h = \sum_j \sum_\alpha a_\alpha^j \varphi_j(x/h - \alpha)$, if $x = h\gamma$ is a mesh point in $A \Subset \Omega$ (so that resulting integrals never abut on the boundary on Ω), then

$$\begin{aligned}(K_h * u_h)(h\gamma) &= \sum_{j=1}^{m} \sum_{\alpha} a_\alpha^j \left[\sum_\beta k_\beta \int \psi_{r-2}(\gamma - \beta - \alpha - y) \varphi_j(y)\, \mathrm{d}y \right] \\ &= \sum_{j=1}^{m} \sum_{\gamma,\delta} a_{\gamma-\delta}^j d_\delta^j\end{aligned}$$

where

$$d_\delta^j = \sum_\beta k_\beta \int \psi_{r-2}(\delta-\beta-y)\varphi_j(y)\,dy.$$

Thus the values of $K_h * u_h$ at mesh points are given via a finite linear combination of the coefficients a_α^j for u_h in the basis φ_j. The finitely many nonzero d_δ^j may be computed a priori; they are independent of h, of the particular mesh point considered and depend only on the particular finite element function space used.

We proceed to give the salient points about why

$$u(x_0) - K_h * u_h(x_0) = O(h^{2r-2}), \quad x_0 \in A, \tag{5.33}$$

in smooth situations. Write

$$u - K_h * u_h = (u - K_h * u) + K_h * (u - u_h). \tag{5.34}$$

The choice of coefficients k_β in K_h has been made so that it is an approximate identity (with a highly localized kernel). Then, for $x_0 \in \operatorname{Int} A_0 \Subset A$,

$$|(u - K_h * u)(x_0)| \leqslant C h^{2r-2} \|u\|_{2r-2,\infty,A_0}. \tag{5.35}$$

It remains to consider $K_h * (u - u_h)$ in (5.34). For a general v, one has for $A_0 \Subset A_1$, see BRAMBLE and SCHATZ [1977, Lemma 4.2],

$$\|v\|_{0,A_0} \leqslant C \sum_{0 \leqslant j \leqslant r-2} \|D^j v\|_{-(r-2),2,A_1}. \tag{5.36}$$

A first crucial fact is that

$$D^j(K_h * v)(x) = \tilde{K}_h^{(j)} * \partial_h^j v(x) \tag{5.37}$$

where ∂_h^j is the jth centered divided difference. Here

$$\tilde{K}_h^{(j)}(x) = h^{-1} \sum_{\beta=-(r-2)}^{(r-2)} k^\beta \psi_{r-2-j}(x/h - \beta).$$

as is easily seen. It follows that

$$\|D^j K_h * v\|_{-(r-2),2,A_1} \leqslant C \|\partial_h^j v\|_{-(r-2),2,A_2} \quad \text{for } A_1 \Subset A_2.$$

Based on the above ideas and after some minor further technicalities involving Sobolev's inequality, Bramble and Schatz show in their Lemma 6.1 that

$$|(K_h * v)(x_0)| \leqslant C \sum_{j=0}^{r-1} \|\partial_h^j v\|_{-(r-2),2,A_0} + h^{r-2} \sum_{j=0}^{r-2} |\partial_h^j v|_{\infty,A_0}. \tag{5.38}$$

It remains to apply this with $v = e = u - u_h$. A second crucial aspect can now be appreciated. Namely, that on uniform meshes, difference quotients of e satisfies "the same" approximate equation as e does. In the present simple situation, i.e.,

$$(e', \chi') = 0 \quad \text{for } \chi \in \mathring{V}_h(A_0)$$

one has

$$(\partial_h^j e', \chi') = (e', \partial_h^j(\chi')) = (e', (\partial_h^j \chi)') = 0$$

since (if $A_0 \Subset \Omega$), $\partial_h^j \chi \in \mathring{V}_h$, cf. (5.28). (For variable coefficients there are some further technicalities.)

Returning to (5.38) and taking $v = e$, we have then for the two terms on the right, for $A_0 \Subset A_1$,

$$\|\partial_h^j e\|_{-(r-2),2,A_0} \leq Ch^{2r-2}(\|u\|_{r+j,2,A_1} + \|u\|_{r,2,\Omega}) \tag{5.39}$$

(after localization of Lemma 4.4) and, using the local result of Theorem 4.2,

$$\begin{aligned}\|\partial_h^j e\|_{0,\infty,A_0} &\leq C\|\partial_h^j u - \chi\|_{\infty,A_{1/2}} + C\|\partial_h^j e\|_{-(r-2),2,A_1} \\ &\leq Ch^r(\|u\|_{r+j,\infty,A_1} + \|u\|_{r,2,\Omega}).\end{aligned} \tag{5.40}$$

Inserting (5.39) and (5.40) into (5.38), and combining with (5.35) and then (5.34), we have the following theorem.

THEOREM 5.5. *Assume that* $x_0 \in \operatorname{Int} A \Subset \Omega$, *and that the mesh is uniform on A. There exists a constant C such that with r as in* (3.1), (3.2),

$$|(u - K_h * u_h)(x_0)| \leq Ch^{2r-2}(\|u\|_{2r-1,2,A} + \|u\|_{r,2,\Omega}). \tag{5.41}$$

We may summarize the most important points in the above derivation via the following mnemonic notation.

$$\begin{aligned}&e(x_0) \cong \text{Local approximability} + \|e\|_{-(r-2),2,\Omega} \quad \text{(Theorem 4.2)}, \\ &D^j(K_h * e) = \tilde{K}_h^{(j)} * \partial_h^j e, \\ &\partial_h^j(u_h) \cong (\partial_h^j u)_h \quad \text{for locally translation-invariant finite element spaces}, \\ &\|e\|_{-(r-2),2,\Omega} = O(h^{2r-2}) \quad \text{in smooth problems (Lemma 4.5)}.\end{aligned} \tag{5.42}$$

Note that, as a consequence of (5.40), difference quotients are better approximations to derivatives than the differentiated approximate solution itself. We shall return to this point in several dimensions in Section 30.

6. Overview: Capsule summaries of each section

In this section we give a capsule summary of each section of this article, except the Foreword and this section itself.

Chapter I. Introduction

2. Perspective: Local behavior in Fourier series

Local behavior in Fourier series is determined by local approximability and global influences. The global influences may be measured by best global L^1 approximation and are, in comparison to the finite element situation, quite severe.

3. Prologue: Local behavior in one-dimensional spline-L^2 projections. Exponential decay

Local behavior in spline-L^2 projections is determined by local approximability and exponentially decaying outside influences.

4. Prologue continued: Local behavior in one-dimensional elliptic projections. Negative norms

Local behavior in approximating second-order two-point boundary value problems is determined by local approximability and global influences. The global influences may be measured by the error in any negative norm, a "weak" measure.

5. Prologue concluded: Superconvergence and local averaging in one-dimensional elliptic projections

Superconvergence without postprocessing is a fickle phenomenon. Only the case of continuous splines is reasonably understood. With Hermite cubics, as an example, superconvergence occurs on uniform meshes but only *away from the boundary*. On uniform meshes the K-operator averages out oscillations and recovers the best error, hidden in negative norms.

Chapter II. Local behavior in finite element L^2 projections

7. Locally quasi-uniform meshes

As in the one-dimensional situation, local behavior is determined by local approximability and exponentially decaying outside influences.

8. Less regular meshes in two-dimensional Lagrangian elements

A more precise accounting of the outside influences is given.

Chapter III. Local estimates in the finite element method for second-order elliptic problems, with applications

9. Local estimates in energy and L^2 norms

Basic estimates for the local error in the above-mentioned norms are given. They depend on local approximability and global influences; the latter are measured in negative norms.

10. Local pointwise error estimates

The estimates again depend on local approximability and global influences, the latter measured in negative norms. The theory is not mature when estimating the pointwise error close to the boundary. The rest of this chapter consists of applications of the basic estimates of this section.

11. Problems with a rough right-hand side

The theory of Section 10 is applied away from rough spots. Careful duality estimates for the global influences show that, in smooth domains, the error in the finite element solution conforms to what one would expect from elementary approximation theory.

12. A model problem on a plane polygonal domain with a reentrant corner: Unrefined meshes

The basic theory of Section 10 is applied. On interior domains, suboptimal order estimates result, due to a failure of the duality argument in this case. This is in contrast to the results of Section 11.

13. Sharpness of the results of Section 12: Pollution

The results of Section 12 are shown to be sharp.

14. A priori and adaptive mesh refinements

The basic theory of Section 10 is applied to the case of a model problem on a polygonal domain with a reentrant corner. It is shown that mesh refinements guided by elementary approximation theory give optimal results. Thereafter the results of ERIKSSON and JOHNSON [1988] for an adaptive algorithm are described.

15. Stress intensity factors, and better a posteriori approximations via the Trigonometric Fitting Method

In the case of a model problem on a polygonal domain, the basic theory of Section 10 gives guidelines as how to recover the parameters in the singular expansion around a corner.

16. Local behavior when singular functions are added

Still in the model problem on a polygonal domain, it is shown how the local behavior when singular functions are added may be reduced to the "same" question without singular functions, already solved in Section 10.

17. Local behavior in problems with rough coefficients: Pollution

One-dimensional examples are given and, following the inverse technique of Section 13, severe pollution effects are exhibited.

18. An example of global pollution with numerical integration

Systematic integration errors lead to severe global pollution.

Chapter IV. A singularly perturbed elliptic to elliptic model problem

19. The model problem

Salient features of the singularities entering are described.

20. The finite element method and its local behavior in energy

A preliminary investigation, to be used in earnest in the next section.

21. Local pointwise error estimates
The pollution effects are not severe in this problem; influences of the singularities are locally confined.

22. Some concrete estimates using Theorem 21.2
The estimates of Section 21 are further elucidated in the neighborhood of typical singularities.

Chapter V. A singularly perturbed convection-dominated model problem

23. The model problem
The singularly features are elucidated, via proofs that will later be used in the finite element situation.

24. The streamline diffusion finite element method
The ordinary Galerkin method is worthless. Instead, the streamline diffusion method is introduced. It is modified with artificial crosswind diffusion.

25. Numerical crosswind and downwind spread
While downwind spread is always restricted to $Ch \ln(1/h)$, numerical crosswind spread is harder to analyze. We give an analysis and choose crosswind diffusion as to minimize crosswind spread according to these results. (According to numerical experiments these results are *not* sharp.)

26. The numerical crosswind spread: An alternative proof
We explain in more transparent detail how our minimizing crosswind result of Section 25 was arrived at. Also, we further localize the result.

27. Pointwise local error estimates
The tight control of numerical crosswind spread gained in Sections 25 and 26 allows decent pointwise error estimates. (They are, however, probably not sharp in practice.)

Chapter VI. Time-localized behavior in second-order parabolic problems

28. A linear model problem: Full advantage of the smoothing effect
Solving a second order parabolic problem with rough or incompatible initial data, but otherwise smooth data, the solution is smooth for positive time. Using the L^2 projection into V_h to approximate initial data, the semidiscrete time continuous Galerkin approximation is of optimal order for positive time in a linear problem, provided integrations in evaluating the L^2 projection are faithfully carried out. Systematic errors in evaluating L^2 projections or time-discretization in the presence of variable coefficients may drastically alter these optimal order results.

29. A nonlinear model problem: Restricted advantage of the smoothing effect
Still the solution is smooth for positive time but, due to nonlinear aliazing, the Galerkin solution does not take "full" advantage of this. The restricted result is shown to be sharp.

Chapter VII. Superconvergence

30. Difference quotients on translation-invariant meshes for approximating derivatives
On domains where the solution is smooth, *difference quotients* of the finite element solution provide an *optimal* $O(h^r)$ order approximation to derivatives of the continuous solution. (This is in contrast to what the *derivatives* of the finite element solution provide.) The result is subject to sufficient-order approximation globally in negative norms, which may be hampered by singularities in right-hand sides, coefficients, reentrant corners, or, more subtly, deficient approximation of smooth boundaries.

31. Higher-order local accuracy by averaging: The K-operator
On locally translation invariant meshes, a simple postprocessing of the finite element solution provides a superapproximating approximation (up to order h^{2r-2}), subject again to global influences measured in negative norms (which may again be subtly obstructed).

32. Miscellaneous results
Various identifiable superconvergence points for functions or gradients are described, in special situations. Various postprocessing methods, local and global, are also given. Again the results are subject to subtle obstructions, in particular from approximations of smooth boundaries.

CHAPTER II

Local Behavior in Finite Element L^2 Projections

7. Locally quasi-uniform meshes

When seeking finite element solutions of partial differential equations the L^2 projection frequently appears. E.g., in Lagrangian formulations the mesh moves and may, after some time, deteriorate so that it is necessary to rectify it. Then, how does one transfer data from the old mesh to the new? One possibility is to use the L^2 projection, cf. CHAVEZ [1983]. Another example occurs in the solution of parabolic problems where it has been found that taking the approximate initial data as the L^2 projection of the given data into a finite element space has some interesting properties, cf. Section 28, when initial data is not smooth, or incompatible. Also, a right-hand side of an equation is, in many finite element methods, replaced by its L^2 projection.

In this section we shall give the multidimensional counterpart of Theorem 3.3. Again our development follows NITSCHE and SCHATZ [1972] with some ideas from SCHATZ and WAHLBIN [1983] thrown in to show exponential decay.

In the one-dimensional case our spline spaces were characterized by two parameters, μ and r in (3.1) and (3.2). (Clearly, we could have let these parameters vary across, respectively on, each mesh interval.) In several dimensions there is a plethora of possible finite element spaces and the only practical way is to nail down a priori assumptions, letting the user of a particular finite element space worry about verifying these assumptions.

For the reader not familiar with applying (or deriving) local results, we point out that deciding on what these a priori assumptions should be is not easy: one may err on the side of giving too specific hypotheses which, say, only apply to Lagrange elements, or one may err in giving very general hypotheses which are almost impossible to verify for a nonexpert.

In the present section we shall formalize the superapproximation Lemma 3.2, as translated to the multidimensional case, as our basic assumption. (Various comments about this will be given later.) We now need to introduce notation.

Let Ω be a basic compact domain in $\mathbb{R}^N$ and for $A_0 \subseteq A_1 \subseteq \Omega$ set

$$\partial_<(A_0, A_1) = \operatorname{dist}(\partial A_0 \setminus \partial\Omega, \partial A_1 \setminus \partial\Omega). \tag{7.1}$$

Thus for instance $\partial_<(A_0, A_1) = d$ in the following situation

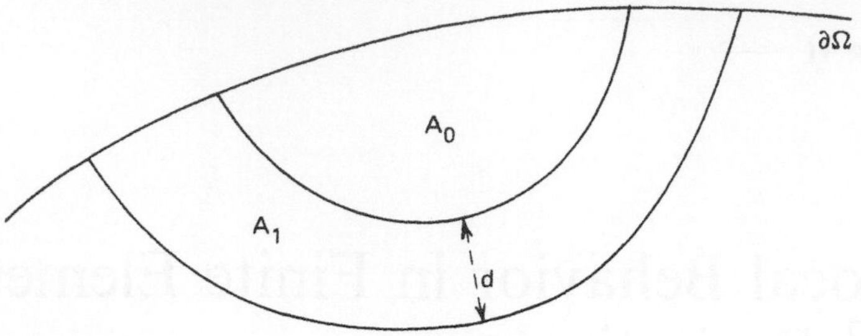

FIG. 7.1.

Further, let $\mathscr{C}^{\infty}_{<}(\bar{A}) = \{v \in \mathscr{C}^{\infty}(\bar{A}) : \partial_{<}(\operatorname{supp} v, A) > 0\}$ and

$$V_h^{<}(A) = \{v \in V_h : \partial_{<}(\operatorname{supp} v, A) > 0\} \tag{7.2}$$

where V_h is some basic family of finite element spaces on Ω.

ASSUMPTION 7.1 (Superapproximation). *There exist constants c and C and a number L such that the following holds: Let* $A_0 \subseteq A_1 \subseteq \Omega$ *with* $d = \partial_{<}(A_0, A_1) \geqslant ch$. *Let further* $\omega \in \mathscr{C}^{\infty}_{<}(\bar{A}_0)$ *with*

$$\|\omega\|_{l,\infty,A_0} \leqslant \Lambda d^{-l}, \quad l = 0, \ldots, L. \tag{7.3}$$

Then for any $\chi \in V_h$ *there exists* $\psi \in V_h^{<}(A_1)$ *such that*

$$\|\omega\chi - \psi\|_{0,2,A_1} \leqslant C\Lambda h/d \|\chi\|_{0,2,A_0}. \tag{7.4}$$

We have seen the basic ingredients necessary for verifying Assumption 7.1 in Lemma 3.2: It takes a local approximation operator with error governed by the local $\| \ \|_{r,2}$ norm, it takes that $D^r\chi \cong 0$ for $\chi \in V_h$ (e.g., for isoparametric elements, $D^r\chi$ is not exactly zero), and it takes a locally quasi-uniform mesh. In the case of tensor products, e.g., one needs to have a more precise error functional, cf. BRAMBLE and HILBERT [1970, 1971]. It is my experience that it is quite easy to verify Assumption 7.1 in any situation with a locally quasi-uniform mesh family.

The next result corresponds to Lemma 3.3. The proof is virtually the same but repeated here to bring out essential features.

LEMMA 7.1. *Assume that Assumption 7.1 holds. There exist positive constants* c_0, c_1 *and C such that the following holds. Let* $A_0 \subseteq A_1$ *with* $d = \partial_{<}(A_0, A_1) \geqslant c_0 h$. *Assume further that* $v_h \in V_h$ *is such that*

$$(v_h, \chi_h) = 0, \quad \text{all } \chi_h \in V_h^{<}(A_1). \tag{7.5}$$

Then

$$\|v_h\|_{0,2,A_0} \leqslant C e^{-c_1 d/h} \|v_h\|_{0,2,A_1}. \tag{7.6}$$

PROOF. Let $\delta = ch$, with c as in Assumption 7.1. Let further $A_0 \subseteq A_\delta \subseteq A_{2\delta}$ with

$\delta = \partial_<(A_0, A_\delta) = \partial_<(A_{\delta,2\delta})$. Let $\omega \in \mathscr{C}^\infty_<(A_\delta)$ be a nonnegative function with

$$\|\omega\|_{l,\infty} \leqslant \Lambda\delta^{-l}, \quad l=0,\ldots,L, \tag{7.7}$$

and with

$$\omega \equiv 1 \quad \text{on } A_0. \tag{7.8}$$

Then

$$\|v_h\|^2_{0,A_0} \leqslant (v_h, \omega v_h). \tag{7.9}$$

Assuming that $A_{2\delta} \subseteq A_1$, we have from (7.5),

$$\|v_h\|^2_{0,A_0} \leqslant (v_h, \omega v_h - \chi), \tag{7.10}$$

for $\chi \in V_h^<(A_{2\delta})$. Thus, via Assumption 7.1,

$$\|v_h\|^2_{0,A_0} \leqslant C(h/\delta)\|v_h\|^2_{0,A_{2\delta}}.$$

Repeat the argument with domains $A_{2\delta} \subseteq \cdots \subseteq A_{2N\delta}$ as long as they remain inside A_1, i.e., since $\delta = ch$, for $N = \mathrm{O}(d/h)$. Since $Ch/\delta < 1$ may be assumed, we obtain the result upon setting $Ch/\delta = \mathrm{e}^{-\mathrm{const}}$. □

Now consider $(u-u_h)(x_0)$ for $x_0 \in \Omega$. Assume that

$$(u-u_h, \chi) = 0 \quad \text{for } \chi \in V_h, \tag{7.11}$$

i.e., that u_h is the L^2 projection of u into V_h.

We next make an assumption that is trivial to verify in most situations.

ASSUMPTION 7.2. *There exists a constant C such that for any $x_0 \in \Omega$, there exists a function δ^0 with* $\operatorname{supp}(\delta^0) \in B_{Ch}(x_0)$ *and for* $1 \leqslant p \leqslant \infty$,

$$\|\delta^0\|_{0,p,\Omega} \leqslant Ch^{(1/p-1)N},$$

and such that

$$(v_h, \delta^0) = v_h(x_0) \quad \textit{for } v_h \in V_h. \tag{7.12}$$

Now, for any $\chi \in V_h$,

$$(u-u_h)(x_0) = (u-\chi)(x_0) + (\chi - u_h)(x_0). \tag{7.13}$$

Here,

$$(\chi - u_h)(x_0) = (\chi - u_h, \delta^0)_\Omega = (\chi - u_h, \delta^0_h)_\Omega \tag{7.14}$$

where δ^0_h denotes the $L^2(\Omega)$ projection of δ^0 into V_h.

We continue to parallel the development in Section 3. We are aiming for local results expressed in a pointwise fashion. Since the L^2 projection is inherently based on the L^2 inner product, cf. Lemma 7.1, it becomes necessary to have means of going between various L^p norms. We, therefore, formalize the inverse property of Lemma 3.1 in the multidimensional situation.

ASSUMPTION 7.3 (Inverse assumption). *There exists a constant C independent of h such that for any element* T_i,

$$\|v\|_{0,p,T_i} \leqslant Ch^{-N(1/q-1/p)}\|v\|_{0,q,T_i} \quad \text{for } v \in V_h, \quad 1 \leqslant q \leqslant p \leqslant \infty. \tag{7.15}$$

We can now give the counterpart of Lemma 3.4.

LEMMA 7.2. *Assume Assumptions* 7.1, 7.2, *and* 7.3. *Then there exist positive constants c and C such that*

$$|\delta_h^0(x)| \leqslant Ce^{-c|x-x_0|/h}/h^N. \tag{7.16}$$

PROOF. By Assumption 7.3,

$$|\delta_h^0(x)| \leqslant Ch^{-N/2}\|\delta_h^0\|_{0,T_j} \tag{7.17}$$

where $x \in T_j$. For $|x-x_0| \geqslant Ch$ we have from Lemma 7.1 that

$$\|\delta_h^0\|_{0,T_j} \leqslant Ce^{-c|x-x_0|/h}\|\delta_h^0\|_{0,\Omega} \tag{7.18}$$

and then by Assumption 7.2 and the obvious stability in L^2 of the L^2 projection,

$$\|\delta_h^0\|_{0,\Omega} \leqslant \|\delta^0\|_{0,\Omega} \leqslant Ch^{-N/2}. \tag{7.19}$$

The desired result (7.16) obtains. □

As in Lemma 3.5 we have the L^p stability of the L^2 projection. The proof is exactly the same.

LEMMA 7.3. *Let Assumptions* 7.1, 7.2, *and* 7.3 *hold. Then the* L^2 *projection is stable in any* L^p *norm*, $1 \leqslant p \leqslant \infty$.

The counterpart of Theorem 3.1 is now the following. Again the proof is the same.

THEOREM 7.1 (Exponential decay in finite element L^2 projections). *Let Assumptions* 7.1, 7.2, *and* 7.3 *hold. There exist positive constants c and C such that the following holds. If* $A_d = B(x_0, d) \cap \Omega$ *and* $u_h \in V_h$ *is the* L^2 *projection into* V_h, *then for* χ_h *arbitrary in* V_h,

$$|(u-u_h)(x_0)| < C\|u-\chi_h\|_{0,\infty,A_d} + Cd^{-N}e^{-cd/h}\|u-\chi_h\|_{0,1,\Omega}. \tag{7.20}$$

To utilize this result one has to make assumptions about local approximability, cf. Theorem 3.2. Since this is well known in standard finite element analysis, we leave it to the reader.

Finally in this section we will further localize Theorem 7.1 as in Theorem 3.3. Again let $A_d = B(x_0, d) \cap \Omega$ and let Assumptions 7.1, 7.2, and 7.3 hold whenever we are on A_{2d}. Assume also that

$$(u-u_h, \chi) = 0 \quad \text{for } \chi \in V_h^{<}(A_d). \tag{7.21}$$

THEOREM 7.2 (Further localization of exponential decay in finite element L^2 projection). *There exists a constant c_1 such that if $d \geqslant c_1 h$ and Assumptions 7.1, 7.2, and 7.3 hold on A_{2d}, and (7.21) holds, then with positive c and C independent of h, for any $\chi_h \in V_h$,*

$$|(u-u_h)(x_0)| \leqslant C\|u-\chi_h\|_{0,\infty,A_d} + Cd^{-N/2}e^{-cd/h}\|u-u_h\|_{0,A_d}. \tag{7.22}$$

SKETCH OF PROOF. The proof is so close to that of Theorem 3.3 that we shall merely sketch it. Let P_h denote the L^2 projection into V_h and $\tilde{P}_h$ the L^2 projection into $V_h^{<}(A_d)$. Then

$$(u-u_h)(x_0) = (u(x_0) - \tilde{P}_h u(x_0)) + (\tilde{P}_h u(x_0) - P_h u(x_0)). \tag{7.23}$$

By an obvious localization of Lemma 7.3,

$$|u(x_0) - \tilde{P}_h u(x_0)| \leqslant C\|u\|_{0,\infty,A_d}. \tag{7.24}$$

The quantity $(\tilde{P}_h - P_h)u$ satisfies (7.5) and hence from Lemma 7.1 (localized), combined with the inverse assumption,

$$\begin{aligned}|(\tilde{P}_h - P_h)u(x_0)| &\leqslant Ch^{-N/2}e^{-cd/h}\|(\tilde{P}_h - P_h)u\|_{0,A_d}\\ &\leqslant Cd^{-N/2}e^{-cd/h}\|(\tilde{P}_h - P_h)u\|_{0,A_d},\end{aligned} \tag{7.25}$$

after a change of constants. Then use the triangle inequality, and Cauchy–Schwarz' inequality, and again Lemma 7.3 (localized), obtaining

$$\begin{aligned}\|\tilde{P}_h u - P_h u\|_{0,A_d} &\leqslant \|u - \tilde{P}_h u\|_{0,A_d} + \|u - P_h u\|_{0,A_d}\\ &\leqslant Cd^{N/2}\|u - \tilde{P}_h u\|_{0,\infty,A_d} + \|u - P_h u\|_{0,A_d}\\ &\leqslant Cd^{N/2}\|u\|_{0,\infty,A_d} + \|u - P_h u\|_{0,A_d}.\end{aligned} \tag{7.26}$$

The combination of (7.23)–(7.26) leads to (7.22) for $\chi_h = 0$. Since $P_h \chi_h = \chi_h$ for $\chi_h \in V_h$ we may add and subtract χ_h to arrive at (7.22). □

8. Less regular meshes in two-dimensional Lagrangian elements

In Section 7 we gave Theorem 7.2, a result pertaining to the case when the family of meshes is quasi-uniform on the subregion of the full region Ω considered. Considering its proof, or the proofs of L^p stability of the L^2 projection as in DESCLOUX [1972], DOUGLAS, DUPONT and WAHLBIN [1975b], or the one-dimensional cases surveyed in DE BOOR [1979], one is led to consider relaxing the assumption of local quasi-uniformity of the meshes; exponential decay as d/h should be replaced by decay according to the number of elements which are removed from the basic point. Here we shall review the careful investigation of CROUZEIX and THOMÉE [1987a, Section 2] which is based on Descloux's technique.

We consider a bounded polygonal domain Ω in the plane. The family of triangulations $\mathcal{T}_h$ is into disjoint triangles T_i such that no vertex of any triangle lies on the interior of a side of another triangle. (This technically rules out many situations of a posteriori, or adaptive, mesh refinement.) The finite element spaces V_h will be of Lagrangian, i.e., $\mathscr{C}^0$, type:

$$V_h = \{v \in \mathscr{C}^0(\bar{\Omega}): v|_{T_i} \text{ is a polynomial of degree } r-1\}. \tag{8.1}$$

We introduce the relevant notation of Crouzeix and Thomée. Let T_0 be a fixed triangle. Let $R_j(T_0)$ be the set of triangles which are j triangles away from T_0, defined recursively as $R(T_0)=T_0$ and, for $j \geqslant 1$, $R_j(T_0)$ is the union of closed triangles in $\mathscr{T}_h$ which are not in $\bigcup_{l<j} R_l$ but which have at least one vertex in R_{j-1}. For $T \in R_j(T_0)$ we set the generalized distance between T and T_0 as

$$\mathscr{D}(T_0, T) := j. \tag{8.2}$$

Thus, for $T \neq T_0$, $\mathscr{D}(T_0, T)$ is the least integer j such that there exists a sequence of vertices $v_i, i=1,\ldots,j$, with v_1 a vertex of T_0, v_j a vertex of T, and v_i and v_{i+1} are vertices of the same triangle for $1 \leqslant i \leqslant j$. It follows that $\mathscr{D}(T, T_0) = \mathscr{D}(T_0, T)$.

Let further

$$a_T = \text{area of triangle } T. \tag{8.3}$$

The following is the basic result of Descloux, as further amended by Crouzeix and Thomée.

THEOREM 8.1. *Let $1 \leqslant p \leqslant \infty$. There exist positive constants c and C, depending only on r (the polynomial degree), and r and p, respectively, such that if* $\operatorname{supp} v_0 \subseteq T_0$, *if*

$$D_j = \bigcup_{l>j} R_l(T_0) = \bigcup_{\mathscr{D}(T_0,T)>j} T \tag{8.4}$$

and if $P_h v_0$ is the L^2 projection into V_h, then

$$\|P_h v_0\|_{0,2,D_j} \leqslant C e^{-cj} a_{T_0}^{1/2-1/p} \|v_0\|_{0,p}. \tag{8.5}$$

PROOF. We shall show that for some $\kappa > 0$,

$$\|P_h v_0\|_{0,D_j}^2 \leqslant \kappa \|P_h v_0\|_{0,R_j}^2 \quad \text{for } j \geqslant 1. \tag{8.6}$$

Assuming this for the moment, since $R_j = D_{j-1} \backslash D_j$, we have

$$Q_j \leqslant \kappa (Q_{j-1} - Q_j), \tag{8.7}$$

where

$$Q_j = \|P_h v_0\|_{0,D_j}^2. \tag{8.8}$$

Thus

$$Q_j \leqslant (\kappa/(1+\kappa)) Q_{j-1} \leqslant (\kappa/(\kappa+1))^j Q_0 \leqslant e^{-c2j} \|P_h v_0\|_0^2, \tag{8.9}$$

where $e^{-c}=(\kappa/(1+\kappa))^{1/2}$. Since $\operatorname{supp} v_0 \subseteq T_0$,

$$\|P_h v_0\|_0 = \max_{\chi \in V_h} \frac{(v_0, \chi)_{T_0}}{\|\chi\|_0}$$

$$\leqslant \max_{q \in \Pi_{r-1}(T_0)} \frac{(v_0, q)_{T_0}}{\|q\|_{0,T_0}} \leqslant \|v_0\|_{0,p,T_0} \max_{q \in \Pi_{r-1}} \frac{\|q\|_{0,p',T_0}}{\|q\|_{0,T_0}}$$

where p' is the conjugate index to p.

By a standard transformation to a reference triangle (as used in proving most inverse assumptions), with C depending only on r,

$$\|q\|_{0,p',T_0} \leqslant C a_{T_0}^{1/p'-1/2} \|q\|_{0,T_0}$$

whence

$$\|P_h v_0\|_0 \leqslant C a_{T_0}^{1/2-1/p} \|v_0\|_{0,p,T_0}. \tag{8.10}$$

Using this in (8.9), and taking square roots,

$$\|P_h v_0\|_{0,D_j} \leqslant e^{-cj} a_{T_0}^{1/2-1/p} \|v_0\|_{0,p,T_0},$$

which would show (8.5).

It remains to prove (8.6). We have

$$(P_h v_0, \chi) = 0 \quad \text{for } \chi \text{ in } V_h, \tag{8.11}$$

with support in $D_{j-1} = D_j \cup R_j$, $j \geqslant 1$. Let $w = P_h v_0$ and define, for any $w \in V_h$, a new function $\tilde{w}_j$ in V_h by setting

$$\begin{aligned} \tilde{w}_j &= w \quad \text{on } D_j, \\ \tilde{w}_j &= 0 \quad \text{on } \Omega \backslash D_{j-1}. \end{aligned} \tag{8.12}$$

To define $\tilde{w}_j$ on the triangles T in $R_j = D_j \backslash D_{j-1}$, introduce for such a triangle T the Lagrangian nodes, with barycentric coordinates $(i_1/(r-1), i_2/(r-1), i_3/(r-1))$ for i_1, i_2, i_3 nonnegative numbers.

Now set $\tilde{w}_j = w$ on all nodes that are interior to R_j (i.e., $\tilde{w}_j$ is cut down to zero on nodes on or inside the closure of $\bigcup_{l<j} R_l(T_0)$). Then setting $\chi = \tilde{w}_j$ in (8.11) we have with $w = P_h v_0$,

$$(w, \tilde{w}_j) = \|w\|_{0,D_j}^2 + (w, \tilde{w}_j)_{R_j} = 0, \tag{8.13}$$

so that

$$\|w\|_{0,D_j}^2 = -(w, \tilde{w}_j)_{R_j}. \tag{8.14}$$

In order to estimate the quantity on the right of (8.14), consider again a triangle T in R_j. Note that T has either only one vertex or one whole edge on R_{j-1}. For $q \in \Pi_{r-1}(T)$ we let $\tilde{q}_T \in \Pi_{r-1}(T)$ be the polynomial that vanishes at the Lagrangian nodal points of T in R_{j-1} and that agrees with q at other Lagrangian nodal points. Then

$$-(w, \tilde{w}_j)_T = -(w, \tilde{w}_T)_T. \tag{8.15}$$

We now have a localized problem: What happens on one triangle if we cut down a polynomial w to vanish on one vertex or one edge? Transformation to a reference triangle is then possible, Jacobian determinants cancel, etc. and we easily find that

$$|(w, \tilde{w}_T)_T| \leqslant \kappa \|w\|_{0,T}^2, \tag{8.16}$$

where κ only depends on r, the polynomial degree. Thus from (8.14), (8.15), (8.16), after summation,

$$\|w\|_{0,D_j}^2 \leqslant \kappa \|w\|_{0,R_j}^2.$$

This proves (8.6) and thus, as we have noted, completes the proof of the theorem. □

It should now be clear that we can select $v_0 = \delta^0$ as a polynomial of degree $r-1$ on $T_0 \ni x_0$ such that

$$(\chi, \delta^0)_{T_0} = \chi(x_0), \quad \text{all } \chi \in \Pi_{r-1}(T_0). \tag{8.17}$$

Further,

$$\|\delta^0\|_{0,1,T_0} \leqslant C \tag{8.18}$$

so that, cf. Assumption 7.2,

$$\|\delta^0\|_{0,p} \leqslant C a_{T_0}^{(1/p-1)}, \tag{8.19}$$

as follows by transforming the situation to a reference triangle.

Then as before, for u_h the L^2 projection into V_h of u, for $x_0 \in T_0$, $\chi_h \in V_h$,

$$(u-u_h)(x_0) = (u-\chi_h)(x_0) + (\chi_h - u_h)(x_0) \tag{8.20}$$

where

$$(\chi_h - u_h)(x_0) = (\chi_h - u_h, \delta^0) = (\chi_h - u, \delta_h^0)_\Omega \tag{8.21}$$

with δ_h^0 the L^2 projection of δ^0. By use of (8.19) and Theorem 8.1 results about the local behavior of $u-u_h$ can then be obtained. Clearly the precise expression of such results will depend (if phrased in norms other than L^2) on assumptions concerning the mesh family, e.g., triangle size on $R_j(T)$ as compared to T etc. We refrain from writing them up; the reader should have no problem in a situation where the mesh is, say, orderly refined. In CROUZEIX and THOMÉE [1987a] conditions are given that assure that the L^2 projection is stable in $L^p(\Omega)$ or $W^{1,p}(\Omega)$, thus opening the possibility for further localization, cf. Theorem 7.2.

The conditions given there will give the reader sufficient clues as how to proceed in a specific case.

CHAPTER III

Local Estimates in the Finite Element Method for Second-Order Elliptic Problems, with Applications

9. Local estimates in energy and L^2 norms

In this section we shall give local error estimates in energy and L^2 norms. We follow, by and large, the treatment in NITSCHE and SCHATZ [1974], cf. also NITSCHE [1972a] and DESCLOUX [1976]. Modifications are introduced as in SCHATZ and WAHLBIN [1977, 1982]. In particular, our statements follow these two later papers.

The results are not only models for later pointwise local error estimates but also provide basic technical tools in deriving these. Indeed, the local estimates in energy based norms give a fundamental tool in proving *any* L^p or $W^{1,p}$ error estimate as witnessed by the proofs in the following papers devoted to *global* estimates: NATTERER [1975], NITSCHE [1975, 1977], FREHSE and RANNACHER [1976], SCOTT [1976], RANNACHER and SCOTT [1982], and SCHATZ and WAHLBIN [1982].

The investigation in NITSCHE and SCHATZ [1974] is quite general in that it treats variable coefficient problems and a general set of finite element spaces, given via assuming suitable hypotheses. We wish to keep the present exposition in a less general setting as not to overburden it. The interested reader is referred to the original paper for the general case. On the other hand, the results of Nitsche and Schatz are not given for subdomains abutting on the boundary of the basic domain. We shall include this case, very important in practice, in our simplified setting.

To fix thoughts, let $\Omega \Subset \mathbb{R}^N$ be a basic domain and consider as a basic problem that of finding u such that

$$\begin{aligned} -\Delta u &= f \quad \text{on } \Omega \\ u &= 0 \quad \text{on } \partial\Omega. \end{aligned} \tag{9.1}$$

To approximate (9.1), let $\tau_h = \bigcup T_i$ be a family of triangulations of Ω_h into simplicial elements, possibly modified in the isoparametric fashion near the boundary. Although somewhat questionable in practice, we assume for simplicity that

$$\Omega_h \subseteq \Omega. \tag{9.2}$$

On these triangulations let V_h denote Lagrangian elements of order $r-1$ which vanish on $\partial\Omega_h$.

Let next

$$B \subseteq \Omega_h \subseteq \Omega \tag{9.3}$$

be a basic domain inside of which almost all our considerations will take place. Most often, B can be thought of as a disc intersected with Ω_h. Further, with notation as in (7.1) with Ω now replaced by Ω_h,

$$V_h^{<}(B) = \{v \in V_h : \partial_{<}(\operatorname{supp} v, B) > 0\}, \tag{9.4}$$

and similarly for $\mathscr{C}_{<}^{\infty}(\bar{B})$. Consider then any function $u_h \in V_h$ such that

$$D(u-u_h, \chi) \equiv \int_{\Omega_h} \nabla(u-u_h)\nabla\chi = 0 \quad \text{for } \chi \in V_h^{<}(B). \tag{9.5}$$

In particular, u_h could be the finite element solution to (9.1).

Furthermore, we asume that the family of triangulations is quasi-uniform on B, i.e., that there exist positive constants c and C such that

$$ch \leqslant \rho_T \leqslant \operatorname{diam}(T) \leqslant Ch \quad \text{for } T \cap B \neq \emptyset, \tag{9.6}$$

where ρ_T denotes the radius of the largest inscribed ball of T.

In this setting and seeing to it that the possible isoparametric modifications at the boundary are well behaved, cf. CIARLET [1978, 4.3], it is frequently easy to verify the following two properties which will be basic in our development. Since we do not wish to go into details about the isoparametric modifications the properties are stated as assumptions. (The properties are the analogues of Assumptions 7.1 and 7.3, but we restate them here since there are some minor changes.)

ASSUMPTION 9.1 (Superapproximation). *Let $A_0 \subseteq A_1 \subseteq B$ with $d = \partial_{<}(A_0, A_1) \geqslant ch$. Let $\omega \in \mathscr{C}_{<}^{\infty}(\bar{A}_0)$. Then for any $\chi \in V_h$ there exists $\psi \in V_h^{<}(A_1)$ with*

$$\|\omega\chi - \psi\|_{1,2,A_1} \leqslant Ch\|\chi\|_{1,2,A_0} \tag{9.7}$$

where $C = C(\omega)$.

ASSUMPTION 9.2 (Inverse assumption). *There exists a constant C independent of h such that for any element T_i meeting B,*

$$\|v\|_{l,p,T_i} \leqslant Ch^{-(l-m)-N(1/q-1/p)}\|v\|_{m,q,T_i} \tag{9.8}$$

for $v \in V_h$, $1 \leqslant q \leqslant p \leqslant \infty$, $l \geqslant m$. Further let

$$\|v\|_{-s,A} = \sup_{\substack{w \in \mathscr{C}_{<}^{\infty}(\bar{A}) \\ |w|_{s,A} = 1}} (v, w).$$

Then for $v \in V_h$ and A a mesh domain

$$\|v\|_{0,A} \leqslant Ch^{-s}\|v\|_{-s,A}$$

where C does not depend on A or v.

For this, cf. Lemma 4.3; the proof there easily extends to the case when A abuts on the boundary.

The following is our basic local energy estimate. Many ideas in its proof were sketched in a simple situation in Section 4.

THEOREM 9.1. *Let Assumptions* 9.1 *and* 9.2 *hold on* $B \subseteq \Omega_h$ *and let* $e = u - u_h$ *satisfy* (9.5). *Let* $A \subseteq B$ *with* $\partial_<(A, B) > 0$. *Then (for h sufficiently small),*

$$\|\nabla e\|_{0,A} \leqslant C \min_{\chi \in V_h} \|u - \chi\|_{1,B} + C\|e\|_{0,B} \tag{9.9}$$

where C depends on $\partial_<(A, B)$ *and the constants of the assumptions.*

The proof will rely on the following lemma.

LEMMA 9.1. *Let* $w_h \in V_h$ *be such that*

$$D(w_h, \chi) = 0 \quad \text{for } \chi \in V_h^<(A_1). \tag{9.10}$$

Then for $A \subseteq A_1$, $\partial_<(A, A_1) > 0$,

$$\|w_h\|_{1,A} \leqslant C\|w_h\|_{0,A_1}. \tag{9.11}$$

The proof of this lemma is postponed for the moment.

PROOF OF THEOREM 9.1. Let $A \subseteq A_1 \subseteq A_2 \subseteq B$ with $\partial_<(A, A_1)$, $\partial_<(A_1, A_2) > 0$. Let $\omega \in \mathscr{C}_<^\infty(A_2)$ with $\omega \equiv 1$ on A_1, and let P_1 denote the $H^1(\Omega)$ projection into V_h. Then on A,

$$u - u_h = [\omega u - P_1(\omega u)] + [P_1(\omega u) - u_h]. \tag{9.12}$$

Here, as is well known,

$$\|\nabla(\omega u - P_1(\omega u))\|_{\Omega_h} \leqslant C\|\omega u\|_{1,\Omega_h} \leqslant C\|u\|_{1,B}. \tag{9.13}$$

Set next $w_h = P_1(\omega u) - u_h$. Then w_h satisfies (9.10) and so by Lemma 3.1,

$$\|w_h\|_{1,A} \leqslant C\|w_h\|_{A_1} \leqslant \|\omega u - P_1(\omega u)\|_{A_1} + \|e\|_{A_1}. \tag{9.14}$$

By Poincaré's inequality (note that ωu can be assumed to vanish on a major portion of ∂A_1 even if $\omega u \neq 0$ on $\partial\Omega_h$), and using (9.13),

$$\|\omega u - P_1(\omega u)\|_{A_1} \leqslant \|\nabla(\omega u - P_1(\omega u))\|_{\Omega_h} \leqslant C\|u\|_{1,B}. \tag{9.15}$$

Combining (9.12)–(9.15) proves (9.9) with $\chi = 0$. Now add and subtract a $\chi \in V_h$. □

It remains to verify the lemma.

PROOF OF LEMMA 9.1. We first consider $\|\nabla w_h\|_{0,A}$. Let $\omega \in \mathscr{C}_<^\infty(A_1)$ with $\omega \equiv 1$ on A. Then, by (9.10),

$$\begin{aligned}\|\nabla w_h\|_{0,A}^2 &\leqslant (\nabla w_h, \omega \nabla w_h) = (\nabla w_h, \nabla(\omega w_h)) + (\nabla w_h, (\nabla\omega) w_h)\\ &= (\nabla w_h, \nabla(\omega w_h - \chi)) + \tfrac{1}{2}(w_h, (\Delta\omega) w_h).\end{aligned}$$

Thus, by the superapproximation hypothesis,

$$\|\nabla w_h\|_{0,A}^2 \leqslant Ch\|w_h\|_{1,A_1}^2 + \tfrac{1}{2}\|w_h\|_{0,A_1}^2$$

and so

$$\|w_h\|_{1,A} \leqslant Ch^{1/2}\|w_h\|_{1,A_1} + C\|w_h\|_{0,A_1}.$$

The argument may be repeated for $\|w_h\|_{1,A_1}$ on the right to yield with $A \subseteq A_2 \subseteq B$,

$$\|w_h\|_{1,A} \leqslant Ch\|w_h\|_{1,A_2} + C\|w_h\|_{0,A_2}. \tag{9.16}$$

In light of the inverse hypothesis this proves (9.11) after a change of notation. □

It is easy to trace through the proof how the constant in Theorem 9.1 varies if $d = \partial_<(A, B)$ becomes small. Another way of seeing the same thing is to consider domains of size comparable to $\partial_<(A, B)$, scale such up to unit size and then note (or, in general, make an assumption) that after scaling, the previous assumptions hold with h replaced by h/d. In the case of variable coefficients one would then have to be careful about how constants associated with them enter into the proof, cf. SCHATZ and WAHLBIN [1977, Lemma 3.1].

COROLLARY 9.1. *With* $d = \partial_<(A, B) \geqslant \text{const} \cdot h$, *(9.9) is replaced by*

$$\|\nabla e\|_{0,A} \leqslant C \min_{\chi \in V_h} (\|\nabla(u-\chi)\|_{0,B} + d^{-1}\|u-\chi\|_{0,B}) + Cd^{-1}\|e\|_{0,B}. \tag{9.17}$$

We next want to replace the L^2 norm on the right-hand side of (9.9) by a weaker, negative, norm,

$$\|e\|_{-s,B} = \sup_{\substack{v \in \mathscr{C}_<^\infty(B) \\ |v|_{s,B} = 1}} (e, v). \tag{9.18}$$

For this we shall need three additional assumptions. The first one amounts in practice to a very weak regularity condition on the boundary of Ω and is not needed if we consider only completely interior subdomains. It is needed for a duality argument.

ASSUMPTION 9.3. *The families of mesh boundaries* $\partial\Omega_h$ *are uniformly Lipschitz so that there exists a universal extension operator* E (cf., e.g. STEIN [1970]).

ASSUMPTION 9.4 (Locally smooth boundary). *Let* $A_1 \subseteq B$, $\partial_<(A_1, B) > 0$. *There exists* $\tilde{A}_1 \subseteq \Omega$ *such that* $\partial_<(\tilde{A}_1 \cap B, B) > 0$ *and such that if*

$$\begin{aligned} -\Delta V &= v \quad \text{in } \tilde{A}_1, \\ v &= 0 \quad \text{on } \partial\tilde{A}_1, \end{aligned} \tag{9.19}$$

then

$$\|V\|_{l+2,\tilde{A}_1} \leqslant C\|v\|_{l,\tilde{A}_1}, \quad l = 0, \ldots, s-1. \tag{9.20}$$

In order to verify Assumption 9.4 one would adjoin a surface smoothly to $\partial\Omega$, taking care that it goes between A_1 and B in Ω_h.

We shall also need a very weak approximation hypothesis.

ASSUMPTION 9.5 (Weak approximation). *There exists a constant C and a positive number γ such that the following holds. With $\tilde{A}_1$ as in Assumption 9.4, for V vanishing on $\partial\tilde{A}_1$,*

$$\min_{\chi\in V_h^{<}(\tilde{A}_1\cap\Omega_h)} \|V-\chi\|_{1,\tilde{A}_1\cap\Omega_h}\leqslant Ch^{\gamma}\|V\|_{2,\tilde{A}_1}.$$

In practice this is clear with $\gamma=\frac{1}{2}$.

We can then proceed essentially as in Section 4.

THEOREM 9.2. *With assumptions as in Theorem 9.1 and with Assumptions 9.3, 9.4, and 9.5,*

$$\|\nabla e\|_{1,A}\leqslant C\min_{\chi\in V_h}\|u-\chi\|_{1,B}+C\|e\|_{-s,B} \tag{9.21}$$

From the proof of Theorem 9.1 it is clear that Theorem 9.2 would follow from the following result which extends Lemma 9.1.

LEMMA 9.2. *Let $w_h\in V_h$ satisfy* (9.10). *Then*

$$\|w_h\|_A\leqslant C\|w_h\|_{-s,A_1}. \tag{9.22}$$

PROOF. Let $A\subseteq A_0\subseteq A_1$ and let $\omega\in\mathscr{C}^{\infty}_{<}(A_0)$ with $\omega\equiv 1$ on A. Then

$$\|w_h\|_{-l,A}=\sup_{\|v\|_{l,A}=1}(w_h,v). \tag{9.23}$$

For each fixed such v, let V solve (9.19) on a domain $\tilde{A}_0$, where v is replaced by Ev via the universal extension operator. Then

$$\begin{aligned}(w_h,v)&=(\omega w_h,v)=D(\omega w_h,V)\\&=\int w_h(2\nabla\omega\,\nabla V-(\Delta\omega)V)+\int\nabla w_h\,\nabla(\omega V-\chi)\end{aligned}$$

for arbitrary $\chi\in V_h^{<}(\tilde{A}_0\cap\Omega_h)$. Consequently by Assumption 9.5,

$$|(w_h,v)|\leqslant C(\|w_h\|_{-l-1,A_0}+\|w_h\|_{1,A_0}h^{\gamma})\|V\|_{l+2,A_0}$$

so that by Assumptions 9.3 and 9.4,

$$\|w_h\|_{-l,A}\leqslant C\|w_h\|_{-l-1,A_0}+Ch^{\gamma}\|w_h\|_{1,A_0}. \tag{9.24}$$

By Lemma 9.1 then, taking $l=0$, and changing notation for domains,

$$\|w_h\|_{0,A}\leqslant C\|w_h\|_{-1,A_0}+Ch^{\gamma}\|w_h\|_{0,A_0}.$$

The argument may now be repeated until the inverse assumption can be applied to the last term on the right. This would prove (9.22) for $s=1$. The argument then proceeds via induction in (9.24). □

Again the mapping as described for Corollary 9.1 gives:

COROLLARY 9.2. *With $d=\partial_<(A, B)\geq \text{const}\cdot h$, (9.21) is replaced by*

$$\|\nabla e\|_{0,A}\leq C \min_{\chi\in V_h} (\|\nabla(u-\chi)\|_{0,B}+d^{-1}\|u-\chi\|_{0,B}) \quad +Cd^{-1-s}\|e\|_{-s,B}. \tag{9.25}$$

We now turn to local error estimates in the L^2 norm. We cannot expect a "straight" analogue of Theorem 9.1 (or Theorem 9.2) since the H^1 projection is not in itself stable in L^2 for merely continuous elements, see BABUŠKA and OSBORN [1980, p. 58]. The traditional path to L^2 error estimates is via a duality argument and we shall follow this path. Note that in our previous duality arguments in this section, namely in the proof of Lemma 9.2, we worked opposite a function in V_h and thus the "skin layer" $\Omega\backslash\Omega_h$ never entered our considerations. In particular, all our results were stated in terms of domains which are subdomains of Ω_h. Since the H^1 projection into V_h does not see u in the skin layer this is correct.

In the argument to follow shortly the behavior of u in the local skin layer will enter through a boundary integral term.

We shall make the following modification of Assumption 9.5.

ASSUMPTION 9.6. *There exists a constant C such that the following holds: With $\tilde{A}_1$ as in Assumption 9.4, with V vanishing on $\partial\Omega\cap\tilde{A}_1$, and with $A_0\subseteq A_1$ where $\partial_<(A_0, A_1)>0$,*

$$\min_{\chi\in V_h^<(A_1)} \|V-\chi\|_{1,A_0}\leq Ch\|V\|_{2,\tilde{A}_1}. \tag{9.26}$$

In practice, this is clear provided $\text{dist}(\partial\Omega, \partial\Omega_h)\leq Ch^2$ locally.

THEOREM 9.3. *Let Assumptions 9.1–9.6 hold on $B\subseteq\Omega_h$ and let $e=u-u_h$ satisfy (9.5). Let $A\subseteq B$ with $\partial_<(A, B)>0$. Then (for h sufficiently small)*

$$\|e\|_{0,A}\leq Ch \min_{\chi\in v_h} \|u-\chi\|_{1,B}+C|u|_{-1/2,B\cap\partial\Omega_h}+C\|e\|_{-s,B}. \tag{9.27}$$

PROOF. Let $\tilde{A}$ be as in Assumption 9.4 and $A\subseteq A_0\subseteq\tilde{A}\cap\Omega_h$ with their $\partial_<$ distances positive (as can always be arranged). Let $\omega\in\mathscr{C}_<^\infty(A_0)$ with $\omega\equiv 1$ on A. We have

$$\|e\|_{-l,A}= \sup_{\|v\|_{l,A}=1} (\omega e, v). \tag{9.28}$$

For each fixed v, let

$$-\Delta V=Ev \quad \text{on } \tilde{A}, \qquad V=0 \quad \text{on } \partial\tilde{A}. \tag{9.29}$$

Then, as in the proof of Lemma 9.2, and since $u_h=0$ on $\partial\Omega_h$, for any $\chi\in V_h$,

$$(\omega e, v)_{(\tilde{A}\cap\Omega_h)} = -\int_{\tilde{A}\cap\Omega_h} \omega e \Delta V = \int_{\tilde{A}\cap\Omega_h} \nabla(\omega e)\nabla V + \int_{\partial\tilde{A}\cap\partial\Omega_h} \omega u(\partial V/\partial n)$$

$$= \int_{\tilde{A}\cap\Omega_h} e(2\nabla\omega\nabla V-(\Delta\omega)V) + \int_{\tilde{A}\cap\Omega_h} \nabla e\nabla(\omega V-\chi) + \int_{\partial\tilde{A}\cap\partial\Omega_h} \omega u(\partial V/\partial n).$$

Thus by our Assumptions 9.3 and 9.5 and by a trace inequality,

$$\|e\|_{-l,A} \leqslant C\|e\|_{-l-1,A_0} + Ch\|\nabla e\|_{0,A_0} + |u|_{-1/2,\partial A_0\cap\partial\Omega_h}.$$

The desired result now follows in the obvious fashion by use of Theorem 9.1 and induction.

It is left to the reader to formulate the obvious analogue of Corollary 9.1.

REMARK 9.1. As the reader has undoubtedly noticed, we have gone through some contortions in order to keep the domains entering in our statements inside Ω_h. Apart from the intellectual honesty of this (since only u as seen on Ω_h enters in the elliptic projection) it is of some practical importance.

A well-known example is that of quadratics in the plane where the triangulation is straight-edged with no isoparametric modifications at the boundary. Then for smooth functions vanishing on $\partial\Omega$ on a domain B abutting on a smooth boundary $\partial\Omega$, the error in $H^1(B)$ is typically $O(h)$ at best whereas the error in $H^1(B\cap\Omega_h)$ is $O(h^{3/2})$, see STRANG and FIX [1973, p. 195]. Likewise, for isoparametric modifications of polynomial order $r-1$, the $H^1(B)$ error is typically of order $O(h^{r/2})$ whereas the error in $H^1(B\cap\Omega_h)$ is of order $O(h^{r-1})$. (In certain cases it does happen that the error in $H^1(B)$ is of order $O(h^{r-1})$; the reader may, as an elementary but interesting calculus exercise, verify that this happens if quadratic plane isoparametrics are used on Ω the unit ball in $\mathbb{R}^2$; further, in this case $\Omega_h\subseteq\Omega$ when the usual halfpoint interpolation procedure is used. These two facts are connected: if $\Omega_h\subseteq\Omega$ should occur, then $\operatorname{dist}(\Omega,\Omega_h)=O(h^4)$ while if the boundaries of Ω_h and Ω interweave, then $\operatorname{dist}(\Omega,\Omega_h)=O(h^3)$ in general. In the latter case, however, the undershoot and overshoot areas cancel to order $O(h^4)$.)

The case of natural (Neumann) boundary conditions is considerably easier to treat. We leave it to the reader and shall later freely use the relevant anologues of the results above.

For local error estimates in mixed methods, see DOUGLAS and MILNER [1985].

10. Local pointwise error estimates

We shall first consider interior estimates, following SCHATZ and WAHLBIN [1977]. In order not to overburden the exposition we shall be somewhat cavalier in giving hypotheses and refer to the paper mentioned for exactness. (By now the reader should have appreciated the role of superapproximation.)

Let x_0 be a point in Ω, and let $B \Subset \Omega$ be a ball of radius d around x_0. With $A(u, v)$ the bilinear form associated with a uniformly elliptic second-order operator with smooth coefficients, assume that $u_h \in V_h(B)$ satisfies

$$A(u-u_h, \chi)=0, \quad \text{for } \chi \in \mathring{V}_h(B), \tag{10.1}$$

where $V_h(B)$ denotes the functions in V_h restricted to B, and

$$\mathring{V}_h(B)=\{\chi \in V_h(B): \operatorname{supp} \chi \Subset B\}. \tag{10.2}$$

Note that in this setting the boundary conditions on $\partial\Omega$ do not enter.

Assume further that the family of meshes is quasi-uniform on B and that the highest possible (the optimal) approximation order in L^p is $O(h^r)$, $r \geqslant 2$.

THEOREM 10.1. *Under the above hypotheses, there are constants c and C such that for $d \geqslant ch$, for $s \geqslant 0$,*

$$|(u-u_h)(x_0)| \leqslant C(\ln d/h)^{\bar{r}} \min_{\chi \in V_h} \|u-\chi\|_{0,\infty,B} + Cd^{-N/2-s}\|u-u_h\|_{-s,B}, \tag{10.3}$$

where

$$\bar{r}=\begin{cases} 0 & \text{for } r \geqslant 3, \\ 1 & \text{for } r=2. \end{cases} \tag{10.4}$$

For the necessity of the logarithmic factor, see HAVERKAMP [1984].

The result of Theorem 10.1 was known earlier than 1977 in various cases of perfectly regular interior meshes (uniform, translation invariant). We refer to BRAMBLE and THOMÉE [1974], BRAMBLE, NITSCHE and SCHATZ [1975] and BRAMBLE and SCHATZ [1976] for such results, often based on local Fourier analysis.

We proceed to describe the salient features in the proof. We consider only the case $d \cong 1$; the case of $d \ll 1$ follows by mapping B to a unit ball and replacing h by h/d. In a sense one relies on global L_∞ stability results for elliptic projections in a localized boundary value problem. For simplicity assume that the form A is coercive on B, cf. SCHATZ and WAHLBIN [1977, Appendix 1]. We then take the Neumann problem over B as our basic localized problem: Find $v \in H^1(B)$ such that

$$A(v, \varphi)=(f, \varphi) \quad \text{for } \varphi \in H^1(B). \tag{10.5}$$

Let then P_1 denote the elliptic projection into $V_h(B)$. We note that

$$\|v-P_1 v\|_{l,B} \leqslant Ch^{s-l}\|v\|_{s,B} \tag{10.6}$$

for $l=0, 1$, $2 \leqslant s \leqslant r$. This follows by extending v over ∂B and using ordinary approximation theory and duality arguments.

Let next $\omega \in \mathscr{C}_0^\infty(\frac{1}{2}B)$ with $\omega(x) \equiv 1$ on $\frac{1}{4}B$. Then

$$(u-u_h)(x_0)=(\omega u-P_1(\omega u))(x_0)+(P_1(\omega u)-u_h)(x_0). \tag{10.7}$$

The two essential steps are the following:

$$\|\omega u - P_1(\omega u)\|_{0,\infty,B/4} \leqslant C(\ln 1/h)^{\bar r}\, \|\omega u\|_{0,\infty,B} \tag{10.8}$$

and for the function $w_h = P_1(\omega u) - u_h \in V_h$ which is "discrete harmonic" on $\frac{1}{4}B$,

$$|w_h(x_0)| \leqslant C\|w_h\|_{1,B/8}. \tag{10.9}$$

One then uses Lemmas 9.1 and 9.2 to obtain

$$\begin{aligned} |w_h(x_0)| &\leqslant C\|w_h\|_{-s,B/4} \\ &\leqslant C\|P_1(wu) - wu\|_{-s,B/4} + C\|u - u_h\|_{-s,B/4}. \end{aligned} \tag{10.10}$$

Here, by (10.8),

$$\|\omega u - P_1(\omega u)\|_{-s,B/4} \leqslant C\|\omega u - P_1(\omega u)\|_{0,\infty,B/4} \leqslant C(\ln 1/h)^{\bar r}\, \|u\|_{0,\infty,B}. \tag{10.11}$$

Combining the above would give (10.3) for $\chi = 0$; then write $u - u_h = (u - \chi) + (\chi - u_h)$ to complete the proof.

The two basic building blocks are thus (10.8) and (10.9).

The stability result (10.8) is reminiscent of global stability results such as those referred to in the second paragraph of Section 9. We shall follow the ideas of FREHSE and RANNACHER [1976] and SCOTT [1976], in particular.

A nice thing now happens: The results on the error in L_1-based norms for smoothed Green's functions that are basic in the global investigations just mentioned also give a way of proving (10.9). Let us state this fundamental result:

LEMMA 10.1. *Let $T \Subset \frac{1}{4}B$ be an element. Let $\varphi \in \mathscr{C}_0^\infty(T)$ and let v and $v_h \in V_h(B)$ satisfy*

$$A(\psi, v) = (\psi, \varphi) \quad \text{for } \psi \in H^1(B), \tag{10.12}$$

$$A(\chi, v - v_h) = 0 \quad \text{for } \chi \in V_h(B). \tag{10.13}$$

Then

$$\|v - v_h\|_{1,1,B} \leqslant Ch^{N/2+1}(\ln 1/h)^{\bar r}\, \|\varphi\|_{0,T}, \tag{10.14}$$

and for $B_1 \Subset B$,

$$\|v - v_h\|_{h;2,1,B_1} \leqslant Ch^{N/2}(\ln 1/h)^{\bar r}\, \|\varphi\|_{0,T}. \tag{10.15}$$

Here $W^{h;2,1,h}$ is the piecewise $W^{2,1}$ norm.

Before indicating the proof of Lemma 10.1 in a simple case, let us show how (10.8) and (10.9) would follow.

For (10.8), let x_1 be a point in $\frac{1}{4}B$ and denote $\tilde u = \omega u$. If $x_1 \in T_1$ we have by the inverse property,

$$\begin{aligned} |P_1\tilde u(x_1)| &\leqslant Ch^{-N/2}\|P_1\tilde u\|_{0,T_1} \\ &\leqslant C\|\tilde u\|_{0,\infty,T_1} + Ch^{-N/2}\|\tilde u - P_1\tilde u\|_{0,T_1}. \end{aligned} \tag{10.16}$$

Here

$$\|\tilde{u}-P_1\tilde{u}\|_{0,T_1}=\sup_{\substack{\varphi\in\mathscr{C}_0^\infty(T_1)\\ \|\varphi\|_{0,T_1}=1}}(\tilde{u}-P_1\tilde{u},\varphi). \tag{10.17}$$

For each such φ, let u and v_h satisfy (10.12), (10.13). Then

$$(\tilde{u}-P_1\tilde{u},\varphi)=A(\tilde{u}-P_1\tilde{u},v)=A(\tilde{u}-P_1\tilde{u},v-v_h)=A(\tilde{u},v-v_h). \tag{10.18}$$

With L^* the adjoint elliptic operator associated with A, and with $\partial/\partial n$ denoting the conormal derivative,

$$A(\tilde{u},v-v_h)=\sum\int_{T_i}\tilde{u}L^*(v-v_h)+\sum\int_{\partial T_i}\tilde{u}\frac{\partial}{\partial n}(v-v_h). \tag{10.19}$$

The summation here is only over elements meeting supp $\tilde{u}$ and thus they are quasi-uniform. A simple trace inequality gives that

$$|f|_{L_1(\partial T_i)}\leqslant C(\|\nabla f\|_{L_1(T_i)}+h^{-1}\|f\|_{L_1(T_i)}). \tag{10.20}$$

Thus, using Lemma 10.1,

$$\begin{aligned}A(\tilde{u},v-v_h)&\leqslant C\|\tilde{u}\|_{0,\infty}(\|v-v_h\|_{h;2,1,B/4}+h^{-1}\|v-v_h\|_{1,1,B/4})\\&\leqslant C(\ln 1/h)^{\bar{r}}h^{N/2}\|\tilde{u}\|_{0,\infty}.\end{aligned} \tag{10.21}$$

Combining the above (10.8) obtains.

For (10.9), let $\eta_h\in V_h$ be such that $w_h\equiv\eta_h$ on $\frac{1}{16}B$ while supp $\eta_h\subseteq\frac{1}{8}B$ and $\|\eta_h\|_1\leqslant C\|w_h\|_{1,B/8}$. Such a cut-down is easily constructed, see SCHATZ and WAHLBIN [1977, Proposition 2.2]. Then if $x_0\in T_0$,

$$|w_h(x_0)|=|\eta_h(x_0)|\leqslant Ch^{-N/2}\|\eta_h\|_{0,T_0}=Ch^{-N/2}\sup_{\substack{\varphi\in\mathscr{C}_0^\infty(T_0)\\ \|\varphi\|_{0,T_0}=1}}(\eta_h,\varphi). \tag{10.22}$$

Again, for each such φ let v and v_h satisfy (10.12), (10.13). Then

$$(\eta_h,\varphi)=A(\eta_h,v)=A(\eta_h,v_h). \tag{10.23}$$

Let now $\chi_h\equiv v_h$ on $\frac{1}{16}B$ and supported in $\frac{1}{8}B$, while $\|v_h-\chi_h\|_{1,(B/8)\setminus(B/16)}\leqslant C\|v_h\|_{1,(B/8)\setminus(B/16)}$. Then since $\eta_h\equiv w_h$ on $\frac{1}{16}B$ and w_h is discrete harmonic,

$$(\eta_h,\varphi)=A(\eta_h,v_h-\chi_h)\leqslant C\|w_h\|_{1,B/8}\|v_h\|_{1,(B/8)\setminus(B/16)}. \tag{10.24}$$

The important point here is that we now only need to estimate v_h on an annulus $A=\frac{1}{8}B\setminus\frac{1}{16}B$ away from the "singular" point x_0. Then v_h is "discrete harmonic" on this annulus and thus by (an easy extension of) Lemmas 9.1 and 9.2,

$$\|v_h\|_{1,A}\leqslant C\|v_h\|_{1,1,B}. \tag{10.25}$$

Here, by Lemma 10.1,

$$\begin{aligned}\|v_h\|_{1,1,B} &\leqslant \|v-v_h\|_{1,1,B}+\|v\|_{1,1,B}\\ &\leqslant Ch^{N/2+1}(\ln 1/h)^{\bar r}+\|v\|_{1,1,B}.\end{aligned} \tag{10.26}$$

It is easily seen, cf. SCHATZ and WAHLBIN [1977, Lemma 4.2], that

$$\|v\|_{1,1,B}\leqslant C\|\varphi\|_{0,1,B}\leqslant Ch^{N/2}\|\varphi\|_{0,B}=Ch^{N/2}.$$

Collecting the above, the estimate (10.9) follows.

To complete this informal description of the proof of Theorem 2.1, one should indicate the proof of Lemma 10.1. First note that (10.15) follows from (10.14) by introducing a suitable interpolant to v, by using inverse properties, and by using well-known results about the continuous problem. (The use of inverse properties accounts for the restriction to $B_1 \Subset B$ since elements at ∂B may be very irregular.)

The result (10.14) is, for $N=2$, contained in e.g. SCOTT [1976]. While being essentially a global result and thus outside the scope of this article, the proof is a very nice application of the local H^1 estimates of Section 9 and thus we shall take this opportunity to give a very brief sketch in the case of $N=2$ and piecewise linear elements ($r=2$).

SKETCH OF PROOF OF LEMMA 10.1, (10.14), in a simple case. Let the element T contain the center of the unit circle B, taken as the origin for notational simplicity. Let A_j denote the annuli

$$A_j=\{x: 2^{-j-1}\leqslant|x|\leqslant 2^{-j}\}$$

and let J be the largest integer such that $2^{-j}\geqslant h$. Set $d_j=2^{-j}$, $A_h=B\setminus\bigcup_{j=0}^{J}A_j$ so that $T\subseteq A_h$, and let

$$A_j'=A_{j-1}\cup A_j\cup A_{j+1}.$$

Setting $e=v-v_h$ we have

$$\|e\|_{1,1,B}=\sum_{j=0}^{J}\|e\|_{1,1,A_j}+\|e\|_{1,1,A_h}. \tag{10.27}$$

By Cauchy–Schwarz' inequality and (10.6),

$$\|e\|_{1,1,A_h}\leqslant Ch\|e\|_{1,A_h}\leqslant Ch^2\|v\|_{2,B}\leqslant Ch^2\|\varphi\|_{0,T}. \tag{10.28}$$

Further, again using Cauchy–Schwarz, by the local H^1 estimates and approximation theory,

$$\|e\|_{1,1,A_j}\leqslant Cd_j\|e\|_{1,A_j}\leqslant Cd_j\{h\|v\|_{2,A_j'}+d_j^{-1}\|e\|_{0,A_j'}\}. \tag{10.29}$$

From well-known properties of the continuous problem, since v satisfies a homogeneous equation away from T,

$$\|v\|_{2,A_j}\leqslant Cd_j^{-1}\|v\|_{1,A_j'}\leqslant Cd_j^{-1}h\|\varphi\|_{0,T}.$$

Hence,

$$\|e\|_{1,1,A_j} \leqslant Ch^2 \|\varphi\|_{0,T} + C\|e\|_{0,A'_j}$$

so that combining with (10.28), and using again (10.6),

$$\begin{aligned}\|e\|_{1,1,B} &\leqslant C\sum_{j=0}^{J} h^2 \|\varphi\|_{0,T} + C\sum_{j=0}^{J} \|e\|_{0,A'_j}\\ &\leqslant C(\ln 1/h)h^2 \|\varphi\|_{0,T} + C(\ln 1/h)\|e\|_{0,B}\\ &\leqslant Ch^2(\ln 1/h)\|\varphi\|_{0,T}\end{aligned}$$

which is (10.14) in the present case. □

REMARK 10.1. One may inquire about the counterpart of Theorem 10.1 in $W^{1,\infty}$. The proof above can rather easily be adapted to cover that case. The crucial point is to estimate $\nabla w_h(x_0)$ for w_h discrete harmonic. Proceeding as in (10.22) one ends up with

$$\sup(\eta_h, \varphi_x) \quad (\text{or } \varphi_y)$$

on the right. For the duality argument, let $-\Delta v = \varphi_x$. Then proceed to (10.24) essentially as before. It may be assumed that φ behaves like a finite element function so that $\|\nabla\varphi\|_0 \leqslant Ch^{-1}\|\varphi\|_0$. Then on the right of (10.24)

$$\begin{aligned}\|v_h\|_{1,(B/8)\setminus(B/16)} &\leqslant C\|v_h\|_{0,1} \leqslant \|v - v_h\|_{0,1} + \|v\|_{0,1}\\ &\leqslant Ch^{N/2+1}(\ln 1/h)^{\bar r}\|\varphi_x\|_0 + Ch^{N/2}\|\varphi\|_0\\ &\leqslant Ch^{N/2}(\ln 1/h)^{\bar r}.\end{aligned}$$

This leads to the appropriate analogue of Theorem 10.1 in $W^{1,\infty}$. RANNACHER and SCOTT [1982] have shown that the logarithmic factor is not necessary for $r=2$.

It is now natural to ask about the analogue of Theorem 10.1 in case B abuts on the boundary. To my knowledge no systematic general investigation has been performed in the literature, not even in the Neumann case assuming $\Omega_h = \Omega$. The problem, apparently, lies with constructing a localized boundary value problem (analogous to the interior Neumann problem (10.5)) for which one can prove all that is needed. The problem should have (at least) H^2 regularity. (In the Neumann case, adjoin a surface smoothly to $\partial\Omega$ and consider a local Neumann problem.) However, for essential boundary conditions on $\partial\Omega$, the boundary conditions cannot, with present technology of proofs, be taken of Dirichlet type on the rest of the boundary of the localized problem since, as of today, proofs of L^∞ stability of elliptic projections require that the mesh boundary is within $O(h^2)$ of the exact one, cf. SCHATZ and WAHLBIN [1982]; such would not be the case on the rest of the boundary. In a very special case a mixed problem was used in SCHATZ and WAHLBIN [1978].

In extending Theorem 10.1 up to the boundary one is then reduced to applying various ad hoc methods. We shall give two such special methods in the following examples.

EXAMPLE 10.1 (*Reflection over a straight boundary*). Consider the Dirichlet form, in the plane for simplicity,

$$D(u, v) = \int u_x v_x + u_y v_y.$$

Let Γ be the boundary $x=0$ (note that rotations do not change the Dirichlet form). Let V_h be Lagrangian elements on triangles which vanish on Γ, and assume also that $u=0$ on Γ. Assume

$$D(u-u_h, \chi)=0 \quad \text{for } \chi \in V_h^{<}(B)$$

where B is a semidisc centered on Γ. We wish to estimate $u-u_h$ on $A \subseteq B$, $\partial_{<}(A, B)>0$, say A a semidisc concentric with B.

If one extends u oddly over Γ to $\tilde{u}$, u_h to $\tilde{u}_h$, and of course the mesh similarly, one has (for $\mathscr{C}^0$ or $\mathscr{C}^1$ finite elements)

$$D(\tilde{u}-\tilde{u}_h, \chi)=0 \quad \text{for } \chi \in \mathring{V}_h(\tilde{B})$$

since this is automatic for even functions in $\mathring{V}_h(\tilde{B})$. Thus, e.g.

$$\|\tilde{u}-\tilde{u}_h\|_{0,\infty,\tilde{A}} C(\ln 1/h)^{\bar{r}} \min_{\chi \in V_h(\tilde{B})} \|\tilde{u}-\chi\|_{0,\infty,\tilde{B}} + C\|\tilde{u}-\tilde{u}_h\|_{0,\tilde{B}}$$

so that, restoring now a general $d=\partial_{<}(A, B)$,

$$\|u-u_h\|_{\infty,A} \leqslant C(\ln d/h)^{\bar{r}} \min_{\chi \in V_h} \|u-\chi\|_{\infty,B} + Cd^{-N/2}\|u-u_h\|_{0,B}. \tag{10.30}$$

The example clearly generalizes to more dimensions, to Neumann problems, ...

EXAMPLE 10.2 (*A two-dimensional technique*). Let

$$A(u-u_h, \chi)=0 \quad \text{for } \chi \in V_h^{<}(B). \tag{10.31}$$

Assume now that one can find *some* boundary value problem containing the region of interest such that the elliptic projection P_1 thereon is (almost) stable in L^∞. This problem could e.g. be the basic problem with coefficients changed outside B as in SCHATZ and WAHLBIN [1983], or it could be the basic problem with the mesh changed so as to be globally quasi-uniform if this can be done, cf. SCHATZ and WAHLBIN [1982].

With ω a cutoff function isolating $A \subseteq A_1 \subseteq B$, i.e. $\omega \equiv 1$ on A_1, while $\partial_{<}(\operatorname{supp} \omega, B)>0$,

$$(u-u_h)|_A = (\omega u - P_1(\omega u)) + (P_1(\omega u) - u_h). \tag{10.32}$$

Then by assumption,

$$\begin{aligned} &\|\omega u - P_1(\omega u)\|_{0,\infty} \\ &\quad \leqslant C(\ln 1/h)^{\bar{r}} \|\omega u\|_{0,\infty} \leqslant C(\ln 1/h)^{\bar{r}} \|u\|_{0,\infty,B}. \end{aligned} \tag{10.33}$$

Now by (10.31), $w_h = P_1(\omega u) - u_h$ is "discrete harmonic" on A_1. In the two-dimensional case with quasi-uniform meshes one has the following variant of Sobolev's inequality,

$$\|\chi\|_{0,\infty,A} \leqslant C(\ln 1/h)^{1/2} \|\chi\|_{1,2,A'} \quad \text{for any } \chi \in V_h, \tag{10.34}$$

see WENDLAND [1979, Theorem 8.3.3] and SCHATZ, THOMÉE and WAHLBIN [1980, Lemma 1.1]. (To prove (10.34), express $\chi(x_0)$ in terms of a suitable Green's function, after extension over $\partial\Omega_h$ by zero.) Here $A \subseteq A' \subseteq A_1$.

Next use the discrete harmonic estimates of Lemmas 9.1 and 9.2, assuming the boundary is sufficiently smooth. Thus

$$\|w_h\|_{0,\infty,A} \leqslant C(\ln 1/h)^{1/2} \|w_h\|_{-s,A_1}. \tag{10.35}$$

Since

$$\|w_h\|_{-s,A_1} \leqslant \|\omega u - P_1(\omega u)\|_{-s,A_1} + \|u-u_h\|_{-s,A_1}$$

and $\|\omega u - P_1(\omega u)\|_{-s,A_1} \leqslant C(\ln 1/h)^{\bar{r}} \|u\|_{0,\infty,B}$, the combination of the above give, upon writing $u-u_h=(u-\chi)+(\chi-u_h)$, with general $d=\partial_<(A,B)$,

$$\|u-u_h\|_{0,\infty,A} \leqslant C(\ln 1/h)^{\bar{r}+1/2} \min_{\chi\in V_h} \|u-\chi\|_{0,\infty,B} + C(\ln d/h)^{1/2} d^{-1-s} \|u-u_h\|_{-s,B}. \tag{10.36}$$

Although a result lacking in elegance it is, at present, the best we can do in many situations close to the boundary.

In the rest of this chapter we shall give applications of the basic estimate (10.3).

11. Problems with a rough right-hand side

Here we assume that the basic second-order problem is uniformly elliptic on a smooth domain and has smooth coefficients. We first consider the (much simpler than essential boundary conditions) case of homogeneous natural Neumann conormal conditions. Thus,

$$\begin{aligned} &Lu = f \quad \text{in } \Omega, \\ &\frac{\partial u}{\partial n} = 0 \quad \text{on } \partial\Omega \quad (n \text{ conormal direction}) \end{aligned} \tag{11.1}$$

where f may be rough. We assume furthermore that the associated bilinear form A is coercive over $H^1(\Omega)$ so that (11.1) has a unique solution.

Let now V_h be finite element spaces and let x_0 be a point interior to Ω,

$$x_0 \in \operatorname{Int} \Omega. \tag{11.2}$$

Let further B be a ball of radius d centered at x_0 with $B \Subset \Omega$. (The case when x_0 is close to the boundary is left to the reader, cf. Examples 10.1 and 10.2.) Assuming that the meshes are quasi-uniform on B we have thus from Theorem 10.1, for u_h the elliptic projection into V_h,

$$|(u-u_h)(x_o)| \leqslant C(\ln 1/h)^{\bar{r}} \min_{\chi \in V_h} \|u-\chi\|_{0,\infty,B} + Cd^{-N/2-s}\|u-u_h\|_{-s,B}. \tag{11.3}$$

Here the influences on the error in the elliptic projection at x_0 are neatly separated into a local approximability term and a term in a weaker norm which accounts for all outside influences. Note that if one attempts to make the local term more local, i.e., decrease d, then the factor in front of the global influence term increases.

Assume now further that f is smooth in a neighborhood B' of B; then so is u and we assume, as is not too unreasonable in practice, that

$$\min_{\chi \in V_h} \|u-\chi\|_{0,\infty,B} \leqslant Ch^r \|u\|_{r,\infty,B'}. \tag{11.4}$$

To estimate the global influence in (11.3) one most often resorts to a duality argument. We give a very simple one here. In the present Neumann situation we assume (as is again not totally unreasonable in practice) that for P_1 the global projection into V_h,

$$\|v-P_1\|_{1,\Omega} \leqslant Ch^{s-1}\|v\|_{s,\Omega} \quad \text{for } 2 \leqslant s \leqslant r. \tag{11.5}$$

Now

$$\|u-u_h\|_{-(r-2),\Omega} = \sup_{\substack{\varphi \in \mathscr{C}_0^\infty(\Omega) \\ \|\varphi\|_{r,\Omega}=1}} (u-u_h, \varphi). \tag{11.6}$$

For each such φ, let v solve

$$\begin{aligned} L^*v &= \varphi \quad \text{in } \Omega, \\ \frac{\partial v}{\partial n} &= 0 \quad \text{on } \partial\Omega. \end{aligned} \tag{11.7}$$

Then with $v_h = P_1 v$, by use of (11.5),

$$\begin{aligned} (u-u_h, \varphi) &= A(u-u_h, v-v_h) = A(u-\chi, v-v_h) \\ &\leqslant C\|u-\chi\|_{1,\Omega}\|v-v_h\|_{1,\Omega} \\ &\leqslant C\|u-\chi\|_{1,\Omega} h^{r-1}\|v\|_{r,\Omega} \\ &\leqslant Ch^{r-1}\|u-\chi\|_{1,\Omega}. \end{aligned} \tag{11.8}$$

Provided $f \in L^2(\Omega)$ we have for a suitable χ in V_h,

$$\|u-\chi\|_{1,\Omega} \leqslant Ch\|u\|_{2,\Omega} \leqslant Ch\|f\|_{0,\Omega}. \tag{11.9}$$

We have thus proven the following local error estimate of optimal order, apart from the logarithmic factor.

THEOREM 11.1. *Consider* (11.1) *with L and $\partial\Omega$ smooth. Let x_0 be interior to Ω and the ball B of radius d also interior to Ω. Assume further that the meshes are quasi-uniform on B and that* (11.4) *and* (11.5) *hold.*

Let $u\in W^{r,\infty}(B)$ *and* $f\in L^2(\Omega)$. *Then*

$$|(u-u_h)(x_0)|\leqslant Ch^r(\ln d/h)^{\bar r}\|u\|_{r,\infty,B}+Ch^r\|f\|_{0,\Omega}. \tag{11.10}$$

It is sometimes necessary to perform the duality argument in (11.8) with considerably more care. As an example we consider interior approximation of the Green's function in the setting of (11.1). We continue to assume (11.4) and (11.5).

Let thus

$$y\in \operatorname{Int}\Omega \tag{11.11}$$

and the Green's function $G^{(y)}$ centered at y be defined by

$$A(G^{(y)},v)=v(y)\quad \text{for } v\in W^{1,\infty}(\Omega). \tag{11.12}$$

The approximate Green's function $G_h^{(y)}\in V_h$ is given as

$$A(G_h^{(y)},\chi)=\chi(y)\quad \text{for } \chi\in V_h. \tag{11.13}$$

We have the following result, SCHATZ and WAHLBIN [1977, Theorem 6.1].

THEOREM 11.2. *Let* $x,y\in\Omega_1\Subset\Omega$ *and assume in particular* (11.1), (11.4) *and* (11.5). *Then for* $|x-y|\geqslant Ch$,

$$|G^{(y)}(x)-G_h^{(y)}(x)|\leqslant Ch^r(\ln|x-y|/h)^{\bar r}/|x-y|^{N-2+r}, \tag{11.14}$$

and for $|x-y|\leqslant Ch$,

$$|G^{(y)}(x)-G_h^{(y)}(x)|\leqslant \begin{cases} C|\ln|x-y||, & N=2,\\ C|x-y|^{2-N}, & N>2.\end{cases} \tag{11.15}$$

Thus, for interior situations at least, the error in the Green's function is, apart from the logarithmic factor if $\bar r=2$, exactly as one would guess from pure approximation theory, (11.4). In the case of $|x-y|\leqslant Ch$ the approximate Green's function behaves no worse than the continuous one.

As noted in BRAMBLE and SCHATZ [1976], a use for such refined error estimates in the Green's function is as follows. For u_h the elliptic projection of u one has

$$(u-u_h)(y)=\int_{\operatorname{supp} f}(G^{(y)}(x)-G_h^{(y)}(x))f(x)\,dx. \tag{11.16}$$

Thus, provided $\operatorname{supp} f\Subset\Omega$ (otherwise one would have to do additional work, cf. Examples 10.1 and 10.2), the estimates of Theorem 11.2 will furnish error estimates for $u-u_h$ under very weak regularity assumptions on f. In this connection see also SCOTT [1973].

PROOF OF THEOREM 11.2. We shall content ourselves with showing (11.14) and refer to SCHATZ and WAHLBIN [1977, Theorem 6.1] for the estimate (11.5) (which has a less interesting proof). The proof of (11.14) involves a precise variant of the duality argument. Set $d=|x-y|$.

Since, as is well known, $\|G^{(y)}\|_{r,\infty,B} \leqslant Cd^{-N-r+2}$ we have by use of (an obvious modification of) Theorem 10.1, with $e^{(y)} = G^{(y)} - G_h^{(y)}$, with B of radius $\frac{1}{4}$d centered at x:

$$|e^{(y)}(x)| \leqslant Ch^r(\ln d/h)^{\bar{r}} d^{-N-r+2} + Cd^{-N-r+2} \|e^{(y)}\|_{2-r,1,B} \tag{11.17}$$

where

$$\|e^{(y)}\|_{2-r,1,\infty} = \sup_{\substack{\varphi \in \mathscr{C}_0^\infty(B) \\ \|\varphi\|_{r-2,\infty,B}=1}} (e^{(y)}, \varphi). \tag{11.18}$$

For each φ as above, let v be such that $A(w, v) = (w, \varphi)$ for $w \in H^1(\Omega)$. Then v is smooth and

$$(e^{(y)}, \varphi) = A(e^{(y)}, v) = v(y) - v_h(y) \tag{11.19}$$

with v_h the adjoint projection of v into V_h.

Since y is interior to Ω we may apply Theorem 10.1 and (11.4) so that

$$|(v - v_h)(y)| \leqslant C\{h^r(\ln d/h)^{\bar{r}} \|v\|_{r,\infty,\tilde{B}} + d^{-N/2} \|v - v_h\|_{0,\Omega}\} \tag{11.20}$$

where $\tilde{B}$ is centered at y with radius $\frac{1}{4}d$. Thus B and $\tilde{B}$ are separated by $\frac{1}{2}d$ and one may derive by use of the continuous Green's function that

$$\|v\|_{r,\infty,\tilde{B}} \leqslant C\|\varphi\|_{r-2,\infty,B} = C. \tag{11.21}$$

Further, by (11.5), and straight duality,

$$\begin{aligned} \|v - v_h\|_{0,\Omega} &\leqslant Ch^r \|\varphi\|_{r-2,B} \\ &\leqslant Ch^r d^{N/2} \|\varphi\|_{r-2,\infty,B} = Ch^r d^{N/2}. \end{aligned} \tag{11.22}$$

Hence, by (11.18)–(11.22),

$$\|e^{(y)}\|_{2-r,1,B} \leqslant Ch^r(\ln d/h)^{\bar{r}}.$$

Reporting this into (11.17) the desired result (11.14) obtains. □

REMARK 11.1. Note that the proof of Theorem 11.2 still holds if the mesh is refined in a "systematic" fashion towards the point y. We refer to ERIKSSON [1985a,b] for details.

We lastly consider the counterpart of Theorem 11.1 for essential boundary conditions, i.e.,

$$Lu = f \quad \text{in } \Omega, \qquad u = 0 \quad \text{on } \partial\Omega. \tag{11.23}$$

Now there are severe difficulties in performing even the simple duality argument given in the proof of Theorem 11.1. One may for simplicity assume that $\Omega_h \subseteq \Omega$ and the functions in V_h vanish on $\partial\Omega_h$. Typically, for isoparametric elements of order $r-1$ (after an imaginary shift of Ω as to be a superset of Ω_h), the distance between $\partial\Omega_h$ and $\partial\Omega$ is $O(h^r)$. We amplify one difficulty:

REMARK 11.2. In the above situation the counterpart of (11.5), i.e., for $v \in \mathring{H}^1(\Omega)$,

$$\inf_{\chi \in V_h} \| v-\chi \|_{1,\Omega} \leqslant Ch^{r-1} \| v \|_{r,\Omega}$$

does not in general hold. Indeed, Ω needs to be replaced by Ω_h on the left, since, in general,

$$\| v \|_{1,\Omega \setminus \Omega_h} = \mathrm{O}(h^{r/2})$$

and not better.

This leads to grief in the duality argument.

The correct assumptions to make are now to keep the interior high-order assumption (11.4), and to replace (11.5) by: For $v \in \mathring{H}^1(\Omega) \cap H^s(\Omega)$,

$$\min_{\chi \in V_h} \| v-\chi \|_{1,\Omega_h} \leqslant Ch^{s-1} \| v \|_{s,\Omega} \tag{11.24}$$

for $1 \leqslant s \leqslant r'$. E.g., for plane quadratic elements on straight-edged triangles, $r' = \frac{5}{2}$ ($r=3$ in (11.4) but globally up to the boundary, $r=2$ only). For isoparametric elements of basic polynomial order $r-1$, if $\Omega_h \subseteq \Omega$ (not always practical!), one has the distance between $\partial\Omega_h$ and $\partial\Omega$ of order $\mathrm{O}(h^r)$ and $r'=r$ in (11.24).

Furthermore one needs to formalize that

$$\operatorname{dist}(\partial\Omega_h, \partial\Omega) = \mathrm{O}(h^q) \tag{11.25}$$

and that $\partial\Omega_h$ is uniformly Lipschitz.

In the counterpart of Theorem 11.1 to be given we consider only d of unit size.

THEOREM 11.3. *Consider* (11.23) *with globally quasi-uniform meshes such that* (11.4), (11.24), *and* (11.25) *(with* $\Omega_h \subseteq \Omega$*) hold. Then for* x_0 *interior to* Ω, *if* $u \in W^{r,\infty}$ *in a neighborhood* B *of* x_0 *and* $f \in L^2(\Omega)$, *then*

$$|(u-u_h)(x_0)| \leqslant Ch^r (\ln 1/h)^{\bar{r}} \| u \|_{r,\infty,B} + C(h^{r'} + h^q) \| f \|_{0,\Omega}. \tag{11.26}$$

Note that the term h^q describing the size of the boundary perturbation *should* enter. Consider e.g. what happens even in the continuous problem if the boundary is perturbed by $\mathrm{O}(h^q)$ in one direction; cf. Remark 11.3 below for cases where the boundaries $\partial\Omega$ and $\partial\Omega_h$ interweave.

PROOF. We have by Theorem 10.1 ($d \cong 1$),

$$|(u-u_h)(x_0)| \leqslant C(\ln 1/h)^{\bar{r}} \min_{\chi \in V_h} \| u-\chi \|_{0,\infty,B/2} + C \| u-u_h \|_{-s,B}, \tag{11.27}$$

where $B \subseteq \Omega_h$. From (11.4),

$$\| u-\chi \|_{0,\infty,B/2} \leqslant Ch^r \| u \|_{0,\infty,B} \leqslant Ch^r. \tag{11.28}$$

Next we apply a duality argument,

$$\| u-u_h \|_{-s,B} = \sup_{\substack{\varphi \in \mathscr{C}_0^\infty(B) \\ \|\varphi\|_{s,B}=1}} (u-u_h, \varphi), \tag{11.29}$$

For each such φ, let $v\in \mathring{H}^1(\Omega)$ with

$$A(w, v)=(w, \varphi) \quad \text{for } w\in \mathring{H}^1(\Omega). \tag{11.30}$$

Then having u_h extended by zero to Ω,

$$(u-u_h, \varphi)=A(u-u_h, v)=A(u-u_h, v-v_h)=A(u-\chi, v-v_h) \quad \text{any } \chi\in V_h, \tag{11.31}$$

where v_h is the elliptic projection of v. In obvious notation, since $u=v=0$ on $\partial\Omega$,

$$\begin{aligned}(u-u_h, \varphi)&=A_{\Omega_h}(u-\chi, v-v_h)+A_{\Omega\setminus\Omega_h}(u, v)\\ &=A_{\Omega_h}(u-\chi, v-v_h)-\int\limits_{\Omega\setminus\Omega_h} fv-\int\limits_{\partial\Omega_h}\frac{\partial u}{\partial n}v.\end{aligned} \tag{11.32}$$

Thus, from (11.24) and (11.25), for s sufficiently large,

$$|(u-u_h, \varphi)|\leqslant Ch\|u\|_{2,\Omega}Ch^{r'-1}\|v\|_{r',\Omega}+Ch^q(\|f\|_{0,\Omega}+|u|_{1,\partial\Omega_h})$$

which, in case $\partial\Omega_h$ is uniformly Lipschitz, is bounded (by well-known trace inequalities and regularity in (11.23)), by

$$C(h^{r'}+h^q)\|f\|_{0,\Omega}.$$

From (11.27), (11.28) and the above the desired result (11.26) obtains. □

REMARK 11.3. There are some special cases in which the negative norm estimates in the proof of Theorem 11.3 can be appreciably improved on. This happens e.g. on square domains in the plane, and, for quadratic elements with a special technique for the essential boundary conditions, in SCOTT [1975]; cf. also e.g. NITSCHE [1972b] and BRAMBLE [1975]. In this connection, cf. Remark 9.1.

We may summarize the present investigations in saying that, in case singularities are introduced by a rough right-hand side only, the elliptic projection is *very locally determined* (at least in the interior).

12. A model problem on a plane polygonal domain with a reentrant corner: Unrefined meshes

Let Ω be a bounded polygonal domain in the plane (see Fig. 12.1), and consider the model problem of finding u on Ω such that

$$\begin{aligned}-\Delta u&=f \quad \text{in } \Omega,\\ u&=0 \quad \text{on } \Gamma=\partial\Omega.\end{aligned} \tag{12.1}$$

Assume (to isolate effects) that $f\in\mathscr{C}^\infty(\Omega)$.

Let $\tau_h=\bigcup T_i$ be a family of quasi-uniform (i.e., unrefined) triangulations of Ω ($\Omega_h\equiv\Omega$ is natural for a polygonal domain). That is, with ρ_i the radius of the largest inscribed disc of T_i,

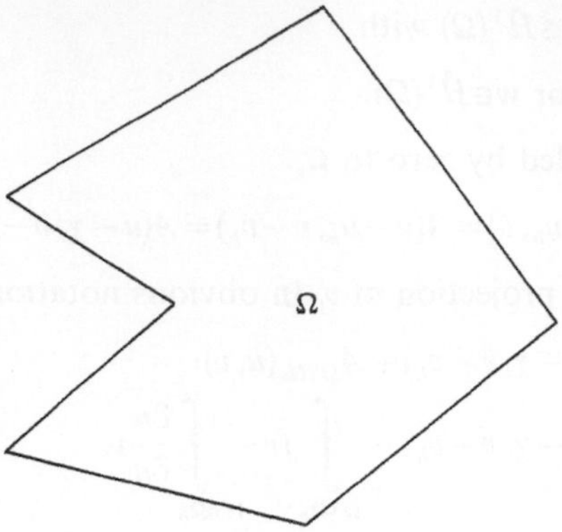

FIG. 12.1.

$$ch \leqslant \rho_i \leqslant \operatorname{diam}(T_i) \leqslant Ch \tag{12.2}$$

where the positive constants c and C are independent of h.

Let α_j denote the angle *interior* to Ω at each vertex v_j of Ω. We designate as the "worst" vertex that with α_j maximal, i.e.,

$$\beta_j \equiv \pi/\alpha_j, \quad \alpha_j \neq \pi, \tag{12.3}$$

minimal. Set that worst vertex as v_0 and

$$\beta = \pi/\alpha_0 < 1, \tag{12.4}$$

i.e., the worst corner is reentrant.

As is well known, cf. KONDRAT'EV [1967], KELLOGG [1971, 1972] and GRISVARD [1976, 1985], with polar coordinates r, θ centered at a fixed vertex v_j,

$$u(r, \theta) = a_{0,j} r^{\beta_j} \sin(\beta_j \theta) + \cdots \tag{12.5}$$

as r tends to zero. In general $a_{0,j}$ is not zero and thus the worst singularity we have to contend with occurs at v_0.

Consider now an interior fixed domain $A \Subset \Omega$. By the results of Section 10, for u_h the obvious projection into a suitable finite element space V_h on this unrefined mesh family, under rather standard assumptions, we have with $A \Subset A' \Subset \Omega$,

$$\| u - u_h \|_{0,\infty,A} \leqslant C(\ln 1/h)^{\bar{r}} \min_{\chi \in V_h} \| u - \chi \|_{0,\infty,A'} + C \| u - u_h \|_{0,\Omega}. \tag{12.6}$$

Since, if f is smooth, u is also smooth on A', with standard notation,

$$\min_{\chi \in V_h} \| u - \chi \|_{0,\infty,A'} \leqslant Ch^r. \tag{12.7}$$

To estimate the second term on the right of (12.6) we shall use a duality argument. Thus write

$$\| u - u_h \|_{0,\Omega} = \sup_{\substack{v \in \mathscr{C}_0^\infty(\Omega) \\ \|v\|_0 = 1}} (u - u_h, v). \tag{12.8}$$

For each such fixed v, let

$$-\Delta w = v \quad \text{in } \Omega, \qquad w = 0 \quad \text{on } \Gamma. \tag{12.9}$$

Then, cf. the papers referred to above, for $\beta < 1$,

$$\|w\|_{\beta+1-\varepsilon,\Omega} \leqslant C\|v\|_{0,\Omega}, \tag{12.10}$$

where now fractional order Sobolev spaces are employed. Hence,

$$\begin{aligned}(u-u_h, v) &= D(u-u_h, w) = D(u-u_h, w-w_h) \\ &\leqslant \|u-u_h\|_{1,\Omega}\|w-w_h\|_{1,\Omega}\end{aligned} \tag{12.11}$$

and by interpolation in standard approximation analysis, and by (12.10),

$$|(u-u_h, v)| \leqslant Ch^{2\beta-2\varepsilon}\|u\|_{\beta+1-\varepsilon,\Omega}\|w\|_{\beta+1-\varepsilon,\Omega} \leqslant Ch^{2\beta-2\varepsilon}. \tag{12.12}$$

Thus, noting that $2\beta < r$ in standard situations, we have:

THEOREM 12.1. *Let Ω be a polygonal domain with worst corner $\alpha_0 > \pi$ and set $\beta = \pi/\alpha_0 < 1$. Then on globally quasi-uniform meshes with standard finite element spaces ($r \geqslant 2$), for the solution u of* (12.1) *with smooth f and its elliptic projection u_h, on*

$$A \Subset \Omega, \tag{12.13}$$

we have for any $\varepsilon > 0$,

$$\|u-u_h\|_{0,\infty,A} \leqslant C_\varepsilon h^{2\beta-\varepsilon}. \tag{12.14}$$

We remark that, by use of Example 10.1, (12.14) holds for $A \subset \Omega$ as long as A is a fixed domain away from any vertex of Ω. Or, one may use Example 10.2 and the maximum norm stability of SCHATZ [1980b].

We shall next consider the corresponding estimate fine-tuned somewhat in that we consider the error at a point x_0 as it approaches the worst vertex v_0. We remind the reader that the mesh family is globally quasi-uniform, i.e., unrefined.

Set

$$\operatorname{dist}(x_0, v_0) = d \geqslant \text{const} \cdot h. \tag{12.15}$$

Assume still that $\beta < 1, r \geqslant 2$.

Letting

$$A_d = \{x : \tfrac{1}{4}d \leqslant |x-x_0| \leqslant 2d\} \tag{12.16}$$

we have from Section 10, cf. Example 10.1,

$$|(u-u_h)(x_0)| \leqslant C(\ln d/h)^{\bar{r}} \min_{\chi \in V_h} \|u-\chi\|_{0,\infty,A_d} + Cd^{-1}\|u-u_h\|_{0,A_d}. \tag{12.17}$$

For f smooth it is known, cf. (12.5) and our general references above, that

$$|D^r u(x)| \leqslant C|x-v_0|^{\beta-r}. \tag{12.18}$$

Thus, by standard approximation theory, for the first term on the right in (12.17),

$$(\ln d/h)^{\bar{r}} \min_{\chi \in V_h} \|u-\chi\|_{0,\infty,A_d} \leqslant C(\ln d/h)^{\bar{r}} h^r d^{\beta-r}. \tag{12.19}$$

For the second term on the right of (12.17), we employ a duality argument as in (12.8), (12.9) and (12.11). However, to obtain the precise dependence on d in our final result (Theorem 12.2 below) we shall (cf. the Green's function estimation in Section 11) need to perform a rather sophisticated duality argument.

The argument is based on the fact that for $p<(2/(2-\beta))$,

$$\|w\|_{2,p,\Omega} \leqslant C_p \|v\|_{0,p,\Omega}, \tag{12.20}$$

cf. MERIGOT [1971, 1972, 1974a, b].

An easy variation of the Bramble–Hilbert Lemma gives, for $1<p\leqslant 2$,

$$\min_{\chi \in V_h} \|\nabla(x-\chi)\|_{0,\Omega} \leqslant C_p h^{2-2/p} \|w\|_{2,p,\Omega}. \tag{12.21}$$

Thus, proceeding as in (12.8), (12.9), (12.11), for the second term in (12.17), for v in $\mathscr{C}_0^\infty(A_d)$ with $\|v\|_{0,A_d}=1$,

$$\begin{aligned} d^{-1}&\|u-u_h\|_{0,A_d} \\ &\leqslant d^{-1} D(u-u_h, w-w_h) \leqslant C d^{-1} h^{2(2-2/p)} \|u\|_{2,p,\Omega} \|w\|_{2,p,\Omega} \\ &\leqslant C(f) d^{-1} h^{2(2-2/p)} \|v\|_{0,p,A_d} \\ &\leqslant C(f) d^{-1} h^{2(2-2/p)} d^{(2-p)/p} \|v\|_{0,A_d} = C(f) h^{2(2-2/p)} d^{-(2-2/p)}. \end{aligned} \tag{12.22}$$

Since $p<2/(2-\beta)$, i.e., $2-2/p<\beta$, is allowed, we get for any $\varepsilon>0$,

$$d^{-1}\|u-u_h\|_{0,A_d} \leqslant C_\varepsilon h^{2\beta-\varepsilon} d^{-\beta-\varepsilon}. \tag{12.23}$$

(Recall that $f\in\mathscr{C}^\infty(\Omega)$.)

Comparing now the two terms involved on the right of (12.17), i.e., the two terms estimated in (12.19) and (12.23), respectively, for $d\geqslant h$, since $\beta<1$ and $r\geqslant 2$, we see that (for ε small enough) the term (12.23) is dominant. We have thus the following result:

THEOREM 12.2. *Let Ω be a polygonal domain with worst corner $\alpha_0>\pi$ and set $\beta=\pi/\alpha_0<1$. Then on globally quasi-uniform meshes with standard finite element spaces ($r\geqslant 2$), for the solution u of* (12.1) *with smooth f and its elliptic projection u_h, for $|x_0-v_0|=d\geqslant \text{const}\cdot h$, for any $\varepsilon>0$ we have*

$$|(u-u_h)(x_0)| \leqslant C_\varepsilon h^{2\beta-\varepsilon} d^{-\beta-\varepsilon}. \tag{12.24}$$

Note that for $d\simeq 1$, Theorems 12.1 and 12.2 coincide (upon taking into account Example 10.1).

We note that, in the basic estimates (12.6) and (12.17), the second terms measuring global influences (pollution effects) dominate over the first terms, the influence of local approximability. This is in decided contrast to the situation in Section 11 where roughness was introduced by nonsmooth right-hand sides in otherwise smooth problems.

13. Sharpness of the results of Section 12: Pollution

Considering unrefined (globally quasi-uniform) meshes, we found in Section 11 that, in model situations, the error in problems where roughness is introduced by an unsmooth right-hand side only, the error away from the singularity is almost of optimal order. In the situation of Section 12, where roughness is introduced by a reentrant corner while the right-hand side is smooth, the error estimate is decidedly not optimal, since $h^{2\beta}d^{-\beta} \gg h^r d^{\beta-r}$.

The estimates of Section 12 were dominated by the estimations of global influences.

In this section we show that the estimates of Section 12 are sharp in general for globally quasi-uniform meshes. This is often referred to as a "pollution effect". The arguments of this section are taken from WAHLBIN [1984], cf. DOBROWOLSKI [1981, Theorem 7.1].

We place ourselves in the situation of a plane bounded polygonal domain Ω with, for the worst vertex,

$$\beta = \pi/\alpha < 1. \tag{13.1}$$

Assume that with polar coordinates centered at the worst vertex,

$$u(r,\theta) = ar^{\beta}\sin(\beta\theta) + \cdots \tag{13.2}$$

where the "stress-intensity factor" is nonzero,

$$a \neq 0. \tag{13.3}$$

Assume further that there is an element T_0 of "size" h at the worst vertex. Then since, by a simple scaling argument,

$$\min_{\substack{\chi \text{ polynomial} \\ \text{of degree} \leqslant r-1}} \|r^{\beta}\sin(\beta\theta) - \chi\|_{1,2,T_0} \geqslant ch^{\beta} \tag{13.4}$$

we are justified in practical unrefined situations with (13.3) to assume that with $c>0$,

$$\min_{\chi \in V_h} \|u-\chi\|_{1,2,\Omega} \geqslant ch^{\beta}. \tag{13.5}$$

Our first result shows that the global influence terms considered in Section 12 can never be better than $O(h^{2\beta})$. Thus, they dominate over the local approximability influence. Precisely, we have as follows.

THEOREM 13.1. *Assume* (13.1), (13.3) *and* (13.5). *Let $A \subseteq \Omega$ be any subdomain of Ω and s any nonnegative number. Then for u_h the elliptic projection of u into V_h,*

$$\|u-u_h\|_{-s,A} \geqslant ch^{2\beta} \tag{13.6}$$

where $c = c(a, A, s) > 0$.

The result of Dobrowolski referred to above gives (13.6) for $s=0$ and $A=\Omega$.

PROOF. We shall give only the essential details and refer to WAHLBIN [1984, Theorem 1.1] for the full technical details. Consider any specific

$$u_0=a_0r^\beta\sin(\theta\beta)+\cdots$$

with $a_0\neq 0$. Setting

$$w=aa_0^{-1}u_0-u$$

then w is somewhat smoother than u_0 or u is, and by a standard duality argument, the error in the elliptic projection in w is $h^{2\beta}o(1)$. Thus it suffices to consider any specific u_0.

To construct our basic u_0, proceed as follows. Let $x_0\in\operatorname{Int}A$, and let $A_0\Subset A$ be an *annulus* centered at x_0. Denote $A_0=B_1\setminus B_0$ where $B_0\Subset B_1\Subset A$ are concentric discs. Let $\omega\in\mathscr{C}^\infty(\bar\Omega)$ with

$$\omega=\begin{cases}1 & \text{outside } B_1,\\ 0 & \text{inside } B_0.\end{cases}$$

Let further $G_0(x)$ be the Green's function for (12.1) with singularity at x_0 and set

$$u_0:=\omega G_0.$$

Then clearly $u_0\in\mathscr{C}^\infty(\operatorname{Int}\Omega)$, $u_0=0$ on $\partial\Omega$, and

$$\operatorname{supp}(\Delta u_0)\subseteq A_0\subseteq A. \tag{13.7}$$

An essential fact is that for this u_0, we have $a_0\neq 0$. To see this, use a conformal map z^β to locally straighten the boundary around the worst vertex v_0. Since u_0 is harmonic near v_0 in Ω, the transformed function $\tilde u_0$ is harmonic in the localized halfplane and vanishes on a piece of the real axis. Hence it is smooth and harmonic in a neighborhood of the origin by Schwarz' reflection principle and Weyl's lemma (or, by Schauder estimates). Thus, in new polar coordinates,

$$\tilde u_0=\sum_{i=1}^{\infty}A_i\rho^i\sin(i\phi)$$

for ρ small. Note now that by the maximum principle, $\tilde u_0$ is positive (for ρ small). Hence, A_1 is positive by the orthogonality of the $\sin(i\phi)$-functions on $[0,\pi]$. Transform back to original coordinates, where a_0 corresponds to A_1.

With $E_0=u_0-P_hu_0$, P_h the elliptic projection, we have by (13.5),

$$\begin{aligned}ch^{2\beta}&\leqslant\|E\|^2_{1,\Omega}=(\nabla E_0,\nabla(u_0-P_hu_0))=(\nabla E_0,\nabla u_0)\\&=-\int_A E_0(\Delta u_0)\leqslant\|E_0\|_{-s,A}\|\Delta u_0\|_{s,A}\end{aligned}$$

so that

$$\|E_0\|_{-s,A}\geqslant(c/\|\Delta u_0\|_{s,A})h^{2\beta}. \qquad\square$$

Our next result in this section is that Theorem 12.2 is sharp in the situation of unrefined (globally quasi-uniform) meshes. We refer to WAHLBIN [1984, Remark 1.1] for exact hypotheses.

THEOREM 13.2. *Assume* (13.1), (13.3) *and* (13.5). *Let, with* v_0 *the worst corner,*

$$A_d=\{x: d\leqslant |x-v_0|\leqslant 2d\}\cap\Omega.$$

Then for any $\delta>0$, *there exist positive constants c and* d_0 *such that for* $h^{1-\delta}\leqslant d\leqslant d_0$,

$$\|u-u_h\|_{0,\infty,A_d}\geqslant ch^{2\beta}d^{-\beta}. \tag{13.8}$$

SKETCH OF PROOF. See WAHLBIN [1984, Theorem 1.2] for all details. We shall only give the essential points.

Consider first a specific $u=u_d$ depending on d. Let $\omega_d\in\mathscr{C}^\infty(\bar{\Omega})$ with

$$\omega_d(x)=\begin{cases}1 & \text{for } |x-x_0|\leqslant d,\\ 0 & \text{for } |x-x_0|\geqslant 2d.\end{cases} \tag{13.9}$$

Then $\operatorname{supp}(\nabla\omega_d)\subseteq A_d$ and we may assume (by a scaling argument) that

$$\|\omega_d\|_{k,\infty}\leqslant C_kd^{-k}. \tag{13.10}$$

Define then a specific u_d,

$$u_d:=\omega_d r^\beta\sin(\beta\theta). \tag{13.11}$$

Then

$$\operatorname{supp}(\Delta u_d)\subseteq A_d. \tag{13.12}$$

Set next $E_d:=u_d-P_hu_d$, with P_h the elliptic projection. Since $d\gg h$ it is easy to see from (13.5) that

$$\|E_d\|_{1,\Omega}\geqslant ch^\beta. \tag{13.13}$$

Thus, as in the proof of Theorem 13.1, using (13.12),

$$ch^{2\beta}\leqslant\|E_d\|_{1,\Omega}^2=(\nabla E_d,\nabla(u_d-P_hu_d))=(\nabla E_d,\nabla u_d)$$

$$=-\int_{A_d}E_d(\Delta u_d)\leqslant\|E_d\|_{0,\infty,A_d}\|\Delta u_d\|_{0,1,A_d}.$$

By (13.10) and the general form of u_d, we easily have

$$\|\Delta u_d\|_{0,1,A_d}\leqslant Cd^\beta. \tag{13.14}$$

Hence,

$$\|E_d\|_{0,\infty,A_d}\geqslant ch^{2\beta}d^{-\beta}. \tag{13.15}$$

The rest of the proof consists in showing that the elliptic projection error on A_d in

$$w_d:=au_d-u$$

(for $u = ar^\beta \sin(\beta\theta) + \cdots$) is of higher order than $O(h^{2\beta} d^{-\beta})$. This follows by a variant of the arguments of Theorem 12.2, since w_d is "smoother"; the essential singularity at the worst vertex is knocked out. The reader can easily supply the arguments or consult WAHLBIN [1984] for details. □

The result of this section and Section 12 are easily extended to the situation where Dirichlet and Neumann conditions are mixed. The lasting message when compared to the results of Section 11 is, that for unrefined meshes, singularities introduced by rough domains (or abrupt change in boundary conditions) pollute throughout the domain whereas singularities introduced by rough right-hand sides are nicely confined to a small neighborhood of the singularity.

The results given here and proven in full detail in WAHLBIN [1984] have various precedents. The fact that the error is not better, anywhere, than $h^{2\beta}$ on an unrefined mesh has been computational folklore for about twenty years. Attempts to furnish a rigorous result were given in NITSCHE and SCHATZ [1974, Section 7, Example 4] for an L-shaped domain (the example is originally due to Babuška and Bramble), and in the already mentioned result by DOBROWOLSKY [1981, Theorem 7.1]. Another extremely interesting result on the structure of the error is given in NITSCHE [1976a]. The result cannot (apparently) explain the full pollution effect from reentrant corners; it is more successful in one-dimensional problems where singularities are introduced by degenerating coefficients, see NITSCHE [1976b], and SCHREIBER [1980, Section 6.2]. (Cf. also JESPERSEN [1978] and ERIKSSON and THOMÉE [1984].) (We shall not touch the question of degenerate elliptic problems in two dimensions. We refer to FRENCH [1987] for state of the art results.)

Let us also point out that in the case of finite difference methods for the problem (12.1), results analogous to Theorem 12.1 were given in LAASONEN [1967].

14. A priori and adaptive mesh refinements

We start by treating systematic a priori refinements.

We consider the model problem

$$\begin{aligned} -\Delta u &= f \quad \text{in } \Omega \Subset \mathbb{R}^2, \\ u &= 0 \qquad \text{on } \Gamma = \partial\Omega, \end{aligned} \tag{14.1}$$

where Ω is a bounded polygonal domain in the plane. We use polynomial finite elements of degree $r-1$ with $\Omega_h = \Omega$, as is natural on a polygonal domain. (For concreteness, the reader may consider Lagrangian triangular elements.) The pointwise results of Section 10 suggest that the resulting error in the finite element solutions are, by and large, determined by local approximability, provided we can handle the global influences.

We assume that $f \in \mathscr{C}^\infty(\bar{\Omega})$. Then as we approach a corner of interior angle α, u behaves as

$$u(r, \theta) = a_0 r^\beta \sin(\beta\theta) + \cdots \tag{14.2}$$

as the polar coordinate r centered at that vertex approaches zero, where $\beta=\pi/\alpha$.

Now think of h as the diameter of elements that are unit size away from vertices. As we approach a vertex v, on

$$\Omega_j=\{x: 2^{-j-1}\leqslant|x-v|\leqslant 2^{-j}\}, \tag{14.3}$$

$$d_j=2^{-j}, \tag{14.4}$$

we have

$$\min_{\chi\in V_h}\|u-\chi\|_{0,\infty,\Omega_j}\leqslant ch_j^r d_j^{\beta-r} \tag{14.5}$$

where h_j is the mesh parameter on Ω_j. We would like to make this local approximability equal to h^r, viz., what rules a unit distance away from the corners.

For technical reasons we shall need a slight "over-refinement". We proceed to detail our assumptions.

Consider a neighborhood of any corner v with interior angle α, and set $\beta=\pi/\alpha$. Let $\Omega_j, j=j_\alpha,\ldots,J_\alpha$ denote the domains in (14.3). With h the "interior h" (14.5) suggests for the local h_j on Ω_j, we should take $h_j=hd_j^{1-\beta/r}$. We assume that the family is quasi-uniform on Ω_j with, for some $\delta>0$,

$$h_j=hd_j^{1-\beta/r+\delta}. \tag{14.6}$$

Then the right-hand side in (14.5) can be replaced by $O(h^r)$, h being the interior mesh diameter.

This annular refinement is broken off at some point, viz. J_α, and setting for the "innermost" domain

$$\Omega_{I,\alpha}=\{x: |x-v|\leqslant 2^{-J_\alpha}\} \tag{14.7}$$

we take on this domain a quasi-uniform mesh with $h_I=h_{I,\alpha}$ where

$$h_I\simeq h^{r/\beta+\delta}. \tag{14.8}$$

Note that for h small enough, we have with any preassigned constant c_0, for $\varepsilon'>0$ small enough compared to δ,

$$h^{-\varepsilon'}h_j/d_j\leqslant c_0. \tag{14.9}$$

Here $j=I$ is allowed, with $d_I=2^{-J_\alpha}$.

The following is the main result of this section.

THEOREM 14.1. *With the above assumptions, for any $\varepsilon>0$ there exists a constant $C=C_\varepsilon$ such that*

$$\|u-u_h\|_{0,\infty,\Omega}\leqslant Ch^{r-\varepsilon}. \tag{14.10}$$

Before proving Theorem 14.1 we wish to investigate the total number of degrees of freedom asymptotically involved in such a mesh refinement. On each Ω_j, since we assume quasi-uniformity, we have by (14.6),

$$\#\text{ elements on }\Omega_j=O(\text{Area}(\Omega_j)/(h_j^2))=O(d_j^2/h_j^2)=h^{-2}O(d_j^{2\beta/r-2\delta}).$$

We may assume that $2\beta/r - 2\delta > 0$ and thus, summing a geometric series, the total number of elements involved is $O(h^{-2})$, i.e., not worse (in the asymptotic sense) than that involved in a globally quasi-uniform mesh family.

The argument in proving Theorem 14.1 below is based on SCHATZ and WAHLBIN [1978, 1979]. However, considerable simplifications are introduced which shorten the proof to about a quarter of its original length. (These simplifications have, apparently, never been published before. The main point is to use the estimates of Section 10 in a careful manner.)

The original papers by SCHATZ and WAHLBIN [1978, 1979] also contain various "bells and whistles": What happens if one underrefines; how to get $O(h^r)$ accuracy close to one corner while having a minimal refinement at other corners to keep the pollution effect under control, etc. We shall not treat those questions with our simplified technique but the reader wishing to do so should have no particular difficulties.

Naturally, before the question of mesh refinement was treated in the pointwise situation, it was considered in energy-based norms. We refer to BABUŠKA [1970], BABUŠKA and ROSENZWEIG [1973], THATCHER [1976], and RAUGEL [1978] for a representative survey. In the pointwise situation we also note ERIKSSON [1985b].

It is somewhat technical to construct mesh refinements as above while keeping quasi-uniformity on each Ω_j, cf. ERIKSSON [1985a, Section 6]. In SCHATZ and WAHLBIN [1978, 1979] it is also shown how the simpler case when elements are merely required to satisfy a maximum angle conditions works. Then the basic approximation theory of BABUŠKA and AZIZ [1976] and JAMET [1976] is applied. Tracing this case through the present development is easy.

The description of our mesh refinement is further elaborated in the one-dimensional situation in RICE [1969].

PROOF OF THEOREM 14.1. Consider the error on Ω_J close to the generic vertex v of angle α, and set $\beta = \pi/\alpha$. If Ω_J does not include the vertex, then we use Theorem 10.1 and Example 10.1 to obtain that

$$\|u-u_h\|_{0,\infty,\Omega_J} \leqslant C(\ln 1/h)^{\bar r} \min_{\chi\in V_h} \|u-\chi\|_{0,\infty,\Omega'_J} + C d_J^{-1} \|u-u_h\|_{0,\Omega'_J} \tag{14.11}$$

where

$$\Omega'_J = \Omega_{J-1} \cup \Omega_J \cup \Omega_{J+1}. \tag{14.12}$$

Assuming further that Ω'_J does not contain the vertex, we have by (14.6) and straightforward approximation theory,

$$\min \|u-\chi\|_{0,\infty,\Omega'_J} \leqslant Ch^r. \tag{14.13}$$

In case Ω'_J abuts on the vertex, which happens only if $J = I$ or $I-1$, we proceed as follows. We have by Sobolev's inequality on V_h, cf. Example 10.2,

$$\|u-u_h\|_{0,\infty,\Omega'_J} \leqslant \|u\|_{0,\infty,\Omega'_J} + \|u_h\|_{0,\infty,\Omega'_J} \leqslant Ch^r + C\|u_h\|_{1,\Omega''_J}$$

with $\Omega''_{\bar J}=(\Omega'_{\bar J})'$. Thus, in this case of being very close to a vertex, by the triangle inequality,

$$\|u-u_h\|_{0,\infty,\Omega'_{\bar J}} \leqslant Ch^r + \|u-u_h\|_{1,\Omega''_{\bar J}}. \tag{14.14}$$

We now proceed to consider the case when $\Omega'_{\bar J}$ does not contain a vertex. By (14.11) and (14.13), for any $\varepsilon>0$,

$$\|u-u_h\|_{0,\infty,\Omega'_{\bar J}} \leqslant Ch^{r-\varepsilon} + d_{\bar J}^{-1}\|u-u_h\|_{0,\Omega'_{\bar J}}. \tag{14.15}$$

Next in store is a duality argument. We have

$$\|u-u_h\|_{0,\Omega'_{\bar J}} = \sup_{\substack{v\in\mathscr{C}_0^\infty(\Omega''_{\bar J})\\ \|v\|_{0,\Omega'_{\bar J}}=1}} (u-u_h, v).$$

For each such v, let w solve

$$\begin{aligned} -\Delta w &= v \quad \text{in } \Omega, \\ w &= 0 \quad \text{on } \Gamma=\partial\Omega. \end{aligned}$$

Then

$$(u-u_h, v) = (\nabla(u-u_h), \nabla w) = (\nabla(u-u_h), \nabla(w-\chi))$$

for any $\chi\in V_h$. Thus, by Theorem 9.1, and Cauchy–Schwarz' inequality, for any $\psi\in V_h$,

$$\begin{aligned} d_{\bar J}^{-1}|(u-u_h, v)| &\leqslant d_{\bar J}^{-1}\sum_j \|\nabla(u-u_h)\|_{0,\Omega_j}\|\nabla(w-\chi)\|_{0,\Omega_j} \\ &\leqslant Cd_{\bar J}^{-1}\sum_j (\|\nabla(u-\psi)\|_{0,\Omega'_j} + d_j^{-1}\|u-u_h\|_{0,\Omega'_j})h_j|w|_{2,2,\Omega'_j} \\ &\quad + Cd_{\bar J}^{-1}\|\nabla(u-u_h)\|_{0,\Omega''_{\bar J}}\|\nabla(w-\chi)\|_{0,\Omega''_{\bar J}}. \end{aligned}$$

Note that, for Ω'_j not touching a vertex, and $j\neq\bar J\pm1$,

$$|w|_{2,2,\Omega'_j} \leqslant Cd_j^{-1}|w|_{1,2,\Omega''_j}$$

and since

$$\begin{aligned} \|\nabla w\|_{0,\Omega}^2 &= (\nabla w, \nabla w) = (w, v) = (w, v)_{\Omega_{\bar J}} \\ &\leqslant \|w\|_{0,p}\|v\|_{0,q,\Omega_{\bar J}}, \quad 1/p+1/q=1 \end{aligned}$$

we have by Sobolev's inequality, applied with $p\simeq\infty, q\simeq1$,

$$\|\nabla w\|_{0,\Omega} \leqslant C\|v\|_{q,\Omega_{\bar J}} \leqslant Cd_{\bar J}^{1-\varepsilon}\|v\|_{0,\Omega_{\bar J}} = Cd_{\bar J}^{1-\varepsilon}.$$

Hence,

$$|w|_{2,2,\Omega''_j} \leqslant Cd_j^{-1}d_{\bar J}^{1-\varepsilon}.$$

A separate argument easily establishes this also for $j=\bar J\pm1$. Thus, noting also that for a suitable ψ,

$$\|\nabla(u-\psi)\|_{0,\Omega_j'} \leqslant \begin{cases} Ch_j^{r-1} d_j^{\beta-r+1} \\ Ch^r, \quad \text{if } \Omega_j \text{ close to the vertex,} \end{cases}$$

we have from (14.15) et seq., for any $\varepsilon > 0$, using also Cauchy–Schwarz' inequality, and a simple estimate for $w-\chi$ on Ω_J'',

$$\begin{aligned} \|u-u_h\|_{0,\infty,\Omega_J} &\leqslant Ch^{r-\varepsilon} + d_J^{-\varepsilon} \sum_j h_j d_j^{-2} \|u-u_h\|_{0,\Omega_j'} + C\|u-u_h\|_{1,\Omega_J''} \\ &\leqslant Ch^{r-\varepsilon} + d_J^{-\varepsilon} \sum_j h_j d_j^{-1} \|u-u_h\|_{0,\infty,\Omega_j'} + C\|u-u_h\|_{1,\Omega_J''}. \end{aligned}$$

Combine this with (14.14). With

$$S = \sum_j h_j d_j^{-1} \|u-u_h\|_{0,\infty,\Omega_j} + \|u-u_h\|_{1,\Omega_I}, \tag{14.16}$$

which includes corresponding sums at each vertex, we have for any $\varepsilon' > 0$,

$$\|u-u_h\|_{0,\infty,\Omega_J} \leqslant Ch^{r-\varepsilon} + h^{-\varepsilon'} S. \tag{14.17}$$

Multiplying (14.17) (and (14.14)) by $h_J d_J^{-1}$ and summing, by use of (14.9), for any $c_0 < 1$, for h small,

$$S \leqslant Ch^{r-\varepsilon} + c_0 S$$

so that

$$S \leqslant Ch^{r-\varepsilon}.$$

Thus from (14.17)

$$\|u-u_h\|_{0,\infty,\Omega_J} \leqslant Ch^{r-\varepsilon}$$

which proves Theorem 14.1. □

We proceed to consider a posteriori adaptive mesh refinements following ERIKSSON and JOHNSON [1988]. For energy-based investigations, see e.g. BABUŠKA, MILLER and VOGELIUS [1983]. Considering piecewise linear elements one first guesses that on a triangulation $\tau_h = \bigcup T_i$, with $h_i = \operatorname{diam} T_i$,

$$\|e\|_{1,\infty,\Omega} \leqslant K \max_{T_i} h_i |u|_{2,\infty,T_i}. \tag{14.18}$$

We thus consider gradient errors (in which case for piecewise linears there is no pollution effect). The extent to which (14.18) is true is, of course, a main point. It is further assumed that the constant K can be estimated; it may depend on interpolation error estimates and estimates for discrete Green's functions which would then have to be calculated and estimated.

We next need to compute approximations to $D^\alpha u$, $|\alpha|=2$. Consider difference operators

$$D_H^\gamma v(x)=\frac{v(x\pm\gamma H)-v(x)}{H} \tag{14.19}$$

where $\gamma=(1,0)$ or $(0,1)$. Here $H=C_1h_T$ if $x\in T$ and C_1 is a sufficiently large constant. This type of difference quotient is then applied to $(u_h)_x$ or $(u_h)_y$. If x is close to the boundary $\chi\pm\gamma h$ is chosen so as to belong to Ω_h.

Specifying an error tolerance δ, the algorithm proceeds as follows:

Step 1. Start with an initial mesh $\tau_h=\tau_{\bar h}$ which is quasi-uniform with $c\delta\leqslant c\bar h\leqslant h_T\leqslant C\bar h$.

Step 2. Given a mesh τ_h compute the corresponding finite element solution $u_h\in V_h$.

Step 3. Compute the following quantity for each $x\in T\in\tau_h$:

$$D_H^2u_h(x)=\max\{|D_H^\gamma D^\alpha u_h(u)|, |\gamma|=|\alpha|=1, |x-y|\leqslant C_2\underline{h}\}$$

where $\underline{h}=\min_{T\in\tau_h}h_T$ and C_2 is a sufficiently large constant. Set

$$D_H^2(u_h;T)=\max_{x\in T} D_H^2u_h(x).$$

Step 4. If for all $K\in\tau_h$ we have

$$Kh_KD_H^2(u_h;T)\leqslant\delta,$$

then stop and accept the finite element solution u_h. If not, construct a new mesh $\hat\tau_h$ by "minimally" refining the old mesh τ_h so that for each $K\in\tau_h$,

$$Kh_{\hat T}D_H^2(u_h;T)\leqslant\delta \quad \text{for all } \hat T\in\hat\tau_h \text{ with } \hat T\subseteq T.$$

Then redefine $T_h=\hat T_h$ and return to Step 2.

Eriksson and Johnson proceed to show that their algorithm works in a model case which "mimics" a boundary singularity, except it is placed inside a smooth convex plane domain Ω. More exactly, let the origin be inside Ω and with $D^mu(x)=\max\{|D^\gamma u(x)|, |\gamma|=m\}$, for c and C positive,

$$c|x|^{\beta-2}\leqslant D^2u(x)\leqslant C|x|^{\beta-2}, \qquad D^3u(x)\leqslant C|x|^{\beta-3} \tag{14.20}$$

where $1<\beta<2$.

Using local error estimates for gradients in an essential way (we refer to their paper for details) they then prove:

THEOREM 14.2. *Suppose the exact solution u satisfies* (14.20). *Then the adaptive algorithm with initial quasi-uniform mesh $\bar h=\delta$ will produce a sequence of meshes T_h^n, $n=0,1,2,\ldots$ with corresponding finite element solutions u_h^n such that*

$$\|u-u_h^n\|_{1,\infty,\Omega_h}\leqslant C\delta^{1-(2-\beta)^{n+1}}.$$

Further, the mesh T_h^n will be "correctly refined" in the region $\{|x| \geqslant \bar{C}\underline{h}_{n-1}\}$ where $\underline{h}_{n-1}$ is the minimal stepsize of elements in T_h^{n-1} and given by

$$\underline{h}_{n-1} = c\delta^{\theta_n}, \quad \theta_n = (1-(2-\beta)^n)/(\beta-1)$$

and the region $\{|x| \leqslant \bar{C}\underline{h}_{n-1}\}$ will have a quasi-uniform mesh of size $\underline{h}_n$.

We conclude this section with a numerical illustration courtesy of Eriksson and Johnson. They applied the method to the mixed Dirichlet–Neumann problem

$$\begin{aligned} \Delta u &= 0 \quad \text{in } \Omega, \\ u &= 0 \quad \text{on } \Gamma_1^+ \cup \Gamma_2, \\ \frac{\partial u}{\partial n} &= u_0 \quad \text{on } \Gamma_1^-, \end{aligned}$$

where Ω is the semidisc $\{|x| \leqslant 1, x_2 \geqslant 0\}$,

$$\begin{aligned} &\Gamma_1^+ = \{x \in \Gamma : x_2 = 0, x_1 > 0\}, \qquad \Gamma_1^- = \{x \in \Gamma : x_2 = 0, x_1 < 0\}, \\ &\Gamma_2 = \{x \in \Gamma : x_2 > 0\}. \end{aligned}$$

The exact solution was taken as $u(x) = r^{2/3} \sin(\frac{2}{3}\theta)$ and since then ∇u is infinite at the origin, error control of the form

$$\|e\|_{\infty,\Omega} \leqslant \max_{T \in \tau} h_T^2 |u|_{2,\infty,T} \leqslant \delta$$

was used. Taking $\delta = 0.01$ the illustrated initial and final mesh and actual error distributions were obtained (see Figs. 14.1–14.4).

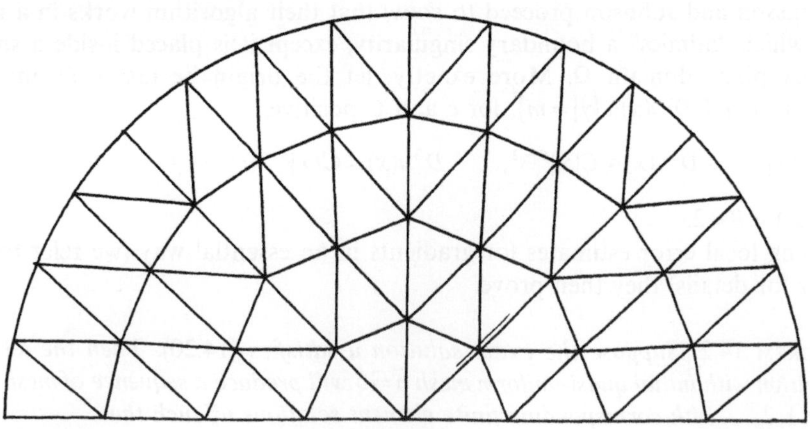

FIG. 14.1. Initial mesh.

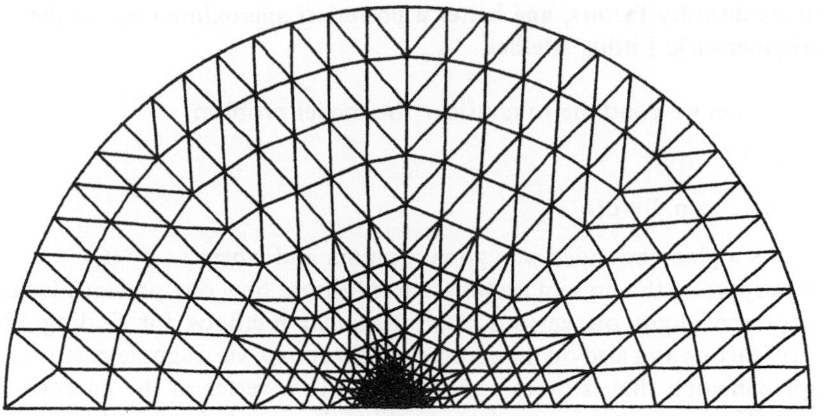

FIG. 14.2. Final mesh.

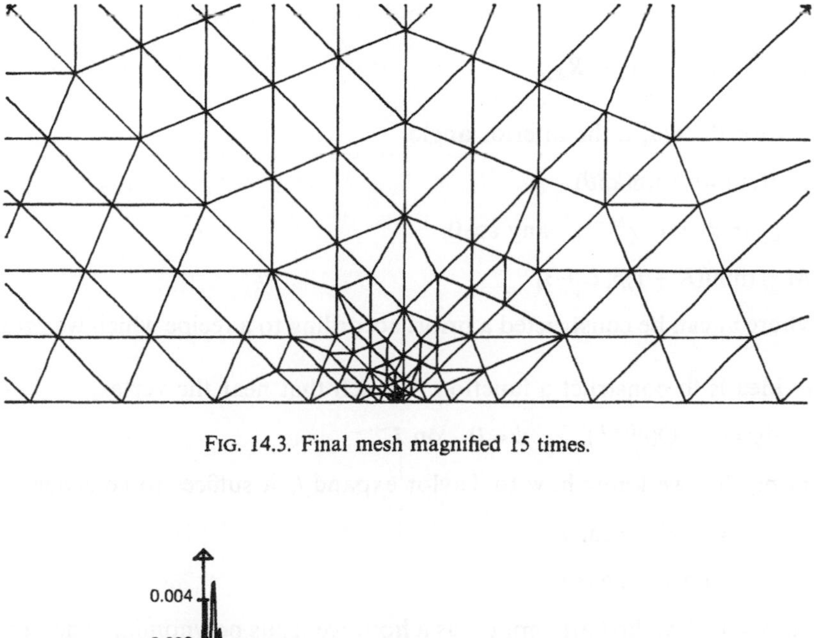

FIG. 14.3. Final mesh magnified 15 times.

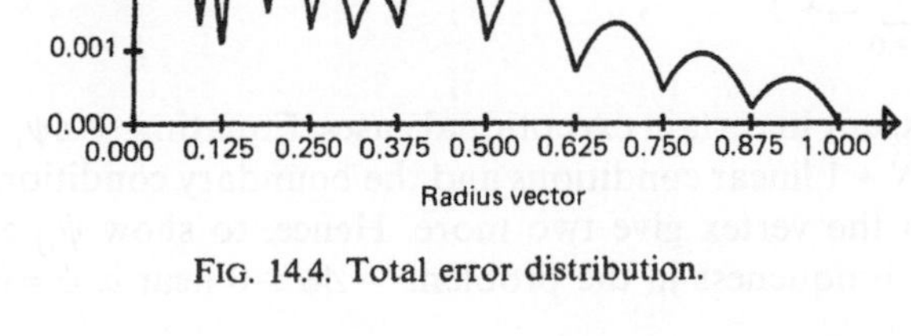

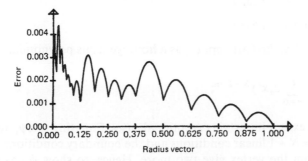

FIG. 14.4. Total error distribution.

15. Stress intensity factors, and better a posteriori approximations via the Trigonometric Fitting Method

In this section we continue to consider the model problem

$$-\Delta u = f \quad \text{in } \Omega, \tag{15.1}$$

$$u = 0 \qquad \text{on } \Gamma = \partial\Omega \tag{15.2}$$

where Ω is a plane bounded polygonal domain and where $f \in \mathscr{C}^\infty(\bar{\Omega})$.

The purpose of the present section is to illustrate how our previous local error estimates shed light on various computational methods for finding "stress-intensity" factors and also on the problem of obtaining, via a simple postprocessing, an approximation that is more accurate close to a vertex in the unrefined mesh situation. Thus our aim is *not* to describe *all* of the multitude of methods extant for finding stress intensity factors but rather to give applications of local error estimates.

We shall first need to describe in some detail the structure of u near a vertex, cf. GRISVARD [1985]. In terms of local polar coordinates (r, θ) centered at that vertex,

$$u = \sum_{k=1}^{K} a_k \mathscr{S}_k + v_L + R_{K,L} \tag{15.3}$$

where with $\beta = \pi/\alpha$, α the interior angle,

$$\mathscr{S}_k(r, \theta) = r^{k\beta} \sin(k\beta\theta), \tag{15.4}$$

$$|R_{K,L}(r, \theta)| \leqslant C_\varepsilon r^{M-\varepsilon}, \quad \text{any } \varepsilon > 0, \tag{15.5}$$

$$M = \min((K+1)\beta, L+3) \tag{15.6}$$

and where v_L can be constructed a priori according to a recipe which we proceed to give.

The idea is to construct a function ψ much that near the vertex,

$$-\Delta\psi = f + \mathrm{O}(r^{L+1}), \qquad \psi = 0 \quad \text{on } \Gamma. \tag{15.7}$$

Assuming that we know how to Taylor expand f, it suffices to consider

$$\begin{aligned} -\Delta\psi_{ij} &= x^i y^j \quad \text{near } v, \\ \psi_{ij} &= 0 \quad \text{on } \Gamma \quad \text{near } v. \end{aligned} \tag{15.8}$$

Setting $N = i + j$ we first attempt ψ_{ij} as a homogeneous polynomial of degree $N+2$,

$$\psi_{ij}(x, y) = \sum_{n=0}^{N+2} c_n x^n y^{N+2-n}; \tag{15.9}$$

this will work except in certain exceptional cases. Equating $-\Delta\psi_{ij}$ to $x^i y^j$ termwise gives $i + j + 1 = N + 1$ linear conditions and the boundary conditions on the two legs emanating from the vertex give two more. Hence, to show ψ_{ij} exists as above it suffices to show uniqueness in the problem $-\Delta\psi \equiv 0$ near v, $\psi = 0$ on Γ near v. In polar coordinates $\psi = r^{N+2} w(\theta)$ and thus

$$(N+2)^2 w(\theta) + w''(\theta) = 0$$

with $w(0) = w(\alpha) = 0$. Thus uniqueness follows unless $N+2$ is an eigenvalue for the two-point boundary value problem just described, i.e., unless

$$N+2 = S\beta, \quad S \text{ integer}. \tag{15.10}$$

If such is the case first seek (assume the $\theta = 0$ axis is the positive x-axis)

$$P_{ij} = y^2 \sum_{n=0}^{N} c_n x^n y^{N-n} \tag{15.11}$$

so that $-\Delta P_{ij} = x^i y^j$. This does have a unique solution since it corresponds, with $P_{ij} = r^{N+2} w(\theta)$, to the initial value problem

$$(N+2)^2 w(\theta) + w''(\theta) = 0, \qquad w(0) = w'(0) = 0.$$

But P_{ij} only satisfies the boundary condition on $\theta = 0$, not necessarily on $\theta = \alpha$. To adjust P_{ij}, let

$$\varphi_N := r^{N+2}(\ln r \sin((N+2)\theta) + \theta \cos((N+2)\theta)). \tag{15.12}$$

This φ_N is harmonic (being the real part of $z^{N+2} \ln z$), satisfies the Dirichlet condition for $\theta = 0$, while for $\theta = \alpha$ one has (S as in (15.10)),

$$\varphi_N = r^{N+2}(-1)^S \alpha \neq 0, \quad \theta = \alpha. \tag{15.13}$$

Thus adding a suitable multiple of φ_N to P_{ij} will furnish a desired ψ_{ij} for (15.8) also in the exceptional cases.

Setting $w = u - v_L$ we have near the vertex v,

$$\begin{aligned} -\Delta w &= t \quad \text{near } v, \\ w &= 0 \quad \text{on } \Gamma \quad \text{near } v, \end{aligned} \tag{15.14}$$

with

$$|t| \leqslant C r^{L+1}. \tag{15.15}$$

We shall next indicate how this leads to (15.3). Let us assume $\beta < 1$, this being the case of most interest. Straightening the boundary via a $Z = z^\beta$ transformation, $z = x + \mathrm{i}y = r\,\mathrm{e}^{\mathrm{i}\theta}$, $Z = \xi + \mathrm{i}\eta = \rho\,\mathrm{e}^{\mathrm{i}\varphi}$, since $\Delta_{xy} = 4\Delta_{\xi\eta}|(z^\beta)'|^2$, we have

$$\begin{aligned} -\Delta_{\xi\eta} \tilde{w} &= \tilde{t} \quad \text{for } \eta \geqslant 0, \quad |Z| \leqslant \rho_0, \\ \tilde{w}(\xi, 0) &= 0, \end{aligned} \tag{15.16}$$

where $\tilde{w}$ is the transformed function and

$$|D^\gamma_{\xi\eta} \tilde{t}| \leqslant C \rho^{(L+3)\alpha/\pi - 2 - |\gamma|} \tag{15.17}$$

so that $\tilde{t}$ belongs to the Hölder class $\mathscr{C}^{(L+3)\alpha/\pi - 2}$ on the closed half-disc. By classical Schauder estimates then,

$$\tilde{w} \in \mathscr{C}^{(L+3)\alpha/\pi - \varepsilon}, \quad \text{any } \varepsilon > 0, \tag{15.18}$$

and hence $\tilde{w}$ has a Taylor expansion

$$\tilde{w} = \sum_{i+j \leqslant \mathscr{K}} c_{ij} \xi^i \eta^j + \tilde{R} \tag{15.19}$$

where $\mathscr{K} = [(L+3)\alpha/\pi - \varepsilon]$, and $\tilde{R} \in \mathscr{C}^{(L+3)\alpha/\pi - \varepsilon}$ with

$$|D^\gamma \tilde{R}| \leqslant C\rho^{(L+3)\alpha/\pi - \varepsilon - |\gamma|}. \tag{15.20}$$

In polar coordinates ρ, φ,

$$\tilde{w} = \sum_{k \leqslant \mathscr{K}} \rho^k T_k(\varphi) + \tilde{R} \tag{15.21}$$

and so

$$-\Delta \tilde{w} = \sum_{k \leqslant \mathscr{K}} \rho^{k-2} [T_k(\varphi) k^2 + T_k''(\varphi)] + \mathrm{O}(\rho^{(L+3)\alpha/\pi - 2 - \varepsilon}). \tag{15.22}$$

By (15.16), (15.17) this equals $\tilde{f} = \mathrm{O}(\rho^{(L+3)\alpha/\pi - 2})$ and hence $T_k(\varphi)k^2 + T_k''(\varphi) = 0$, or,

$$T_k(\varphi) = a_k \sin(k\varphi) + b_k \cos(k\varphi).$$

By the boundary condition $\tilde{w}(\xi, 0) = 0$, $b_k = 0$ so that

$$T_k(\varphi) = a_k \sin(k\varphi). \tag{15.23}$$

Going back to original coordinates shows that (15.3) is satisfied.

From now on we assume that the function v_L in (15.3) is known in advance, for given L. Note that $L = -1$ is allowed, which means setting $v_L = 0$.

We now consider finite element solutions u_h in some family of finite element spaces V_h which satisfy the homogeneous Dirichlet conditions on Γ. We concentrate attention at the worst vertex, i.e., that of lowest β. We assume that near the vertex,

$$|(u - u_h)(r, \theta)| \leqslant CE(h, r) \qquad \text{for } r \leqslant r_0, \quad 0 \leqslant \theta \leqslant \pi \tag{15.24}$$

where $E(h, r)$ is a known function. As examples, in the situation where the mesh is globally quasi-uniform, we saw in Section 12 that $E(h, r) = h^{2\beta - \varepsilon} r^{-\beta - \varepsilon}$ may be taken, and we saw in Section 13 that this estimate is sharp. In a situation where suitable mesh refinements are performed we found in Section 14 that $E(h, r) = h^{s - \varepsilon}$, s being the order of the finite element space and h the interior element diameter.

The coefficient a_1 in (15.3), the "stress intensity factor" in linear elastic fracture mechanics, is often of some interest and we first consider how to recover it. This problem has an enormous literature associated with it and we can only give a very small sample of articles where further references are given: GALLAGHER [1975], HELLEN [1980], and many articles in the conference report GRISVARD, WENDLAND and WHITEMAN [1985].

The simplest method for finding a_1 goes under names such as the "Direct Method" or "Equation Substitution". By (15.3) we have for $\theta = \frac{1}{2}\alpha$, $\beta < 1$ and $L = -1$ (so that no preliminary evaluation of v_L needs be done),

$$a_1 = u(r, \tfrac{1}{2}\alpha) r^{-\beta} + \tilde{E}, \tag{15.25}$$

where $\tilde{E}(r) \leqslant C \max(r^{2\beta}, r^{2-\beta}(\ln 1/r))$ (since $\mathscr{S}_2(\alpha/2)=0$ and $v_{-1}=O(r^2(\ln 1/r))$). Thus with (15.24),

$$a_1 = u_h(r, \tfrac{1}{2}\alpha) r^{-\beta} + \{\tilde{E} + E(h,r) r^{-\beta}\}. \tag{15.26}$$

This relation gives some indication of what point r to choose to evaluate a_1. E.g., in an unrefined situation $E(h,r) \cong h^{2\beta} r^{-\beta}$ (modulo ε's) and if $3\beta < 2$, then $r \cong h^{1/2}$ should be taken. The resulting accuracy is $O(h^\beta)$.

The low accuracy of the "Direct Method" is well known in practice and the "Extrapolation Method" is one example of a method suggested to improve the accuracy. From (15.3) we have with v_3 known, for $\theta = \tfrac{1}{2}\alpha$,

$$u - v_3 = a_1 r^\beta - a_3 r^{3\beta} + O(r^{5\beta}) \tag{15.27}$$

so that if

$$F(r_0) = (u - v_3)(r_0, \tfrac{1}{2}\alpha)/r_0^\beta, \tag{15.28}$$

we have

$$F(r_0) = a_1 - a_3 r_0^{2\beta} + O(r_0^{4\beta}), \qquad F(2r_0) = a_1 - a_3 2^{2\beta} r_0^{2\beta} + O(r_0^{4\beta}), \tag{15.29}$$

so that extrapolating away a_3,

$$\frac{2^{2\beta} F(r_0) - F(2r_0)}{2^{2\beta} - 1} = a_1 + O(r_0^{4\beta}). \tag{15.30}$$

Replacing u by u_h in $F(r_0)$ it is obvious how to use the local error estimate (15.24) to obtain a rational choice for r_0.

Other methods proposed for finding a_1 include the "Energy Release Method", the "J-Integral Method of Rice" and various methods based on the Volkov–Masja–Plamenevski formula,

$$a_1 = (2\beta)^{-1} \int_\Omega f(r^{-\beta} \sin(\beta\theta) + H)$$

where H is harmonic with $H = -r^{-\beta} \sin(\beta\theta)$ on Γ. With apologies to many researchers for giving only a brief summary of the field we point out that in all cases local error estimates can be used to evaluate the resulting accuracy in approximating a_1.

We proceed now to consider methods in which, assuming v_L known, a fair number of coefficients a_k in (15.3) are identified so as to furnish a high-order a posteriori approximation to u close to the worst vertex, say, even if u_h was originally found only on an unrefined mesh with then only h^β accuracy close to the vertex. Clearly, we may go on with the extrapolation method based on distances $r_0, 2r_0, 3r_0$, etc., in the obvious analogues of (15.29), (15.30). However, in practice one then often finds oneself (after some preliminary analysis based on local error estimates to see what r_0 should be) using points outside Ω!

A more satisfactory method in practice is to stick with an optimal distance r_0 away from the vertex and try to get hold of approximations to $a_1, a_2, a_3 \ldots$ by using values

$u_h(r_0, \theta)$ on the part of the circle of radius r_0 around v meeting Ω. We call this method "Trigonometric Fitting". It was introduced in ROSSER [1975, 1977], cf. also SCHATZ and WAHLBIN [1981]. We proceed to describe this method.

We wish to arrange so that

$$\sum_{k=1}^{K} a_k^h \mathscr{S}_k(r_0, \theta) \cong (u_h - v_L)(r_0, \theta), \tag{15.31}$$

in some sense, in order to find approximations to a_k^h. To motivate our method, note that from (15.3) for any $I \geq K$,

$$\sum_{k=1}^{I} a_k \mathscr{S}_k(r_0, \theta) = ((u - v_L) - R_{I,L})(r_0, \theta) \tag{15.32}$$

where

$$|R_{I,L}| \leq C r_0^{\tilde{M} - \varepsilon} \tag{15.33}$$

with

$$\tilde{M} = \min((I+1)\beta, L+3). \tag{15.34}$$

Introduce now the notation

$$A_k = a_k r_0^{k\beta} \tag{15.35}$$

and similarly for A_k^h. Also set

$$\omega_k(\theta) = \sin(k\theta\beta). \tag{15.36}$$

Then from (15.32), by well-known relations for trigonometric functions,

$$A_k = \frac{2}{\alpha} \int_0^\alpha (u - v_L - R_{I,L})(r_0, \theta)\omega_k(\theta)\, \mathrm{d}\theta. \tag{15.37}$$

Employing the trapezoidal rule on $I+2$ points including the endpoints, we are led to defining

$$\begin{aligned} &A_k^h = \frac{2}{I+1} \sum_{i=1}^{I} (u_h - v_L)(r_0, \theta_i)\, \omega_k(\theta_i), \quad k = 1, \ldots, I, \\ &\theta_i = i\alpha/(I+1), \quad i = 1, \ldots, I. \end{aligned} \tag{15.38}$$

We first wish to consider the error in A_k. Setting

$$TF = \frac{1}{2(I+1)} \sum_{i=1}^{I} F(\theta_i) \tag{15.39}$$

we have from the above (since the trapezoidal rule is exact on $\omega_k \omega_l$),

$$A_k - A_k^h = T((u - u_h)(r_0, \cdot)\omega_k(\cdot)) - T(R_{I,L}\omega_k). \tag{15.40}$$

Consequently, with $\tilde{M}$ as in (15.34) and $E(h,r_0)$ as in (15.24), for any $\varepsilon>0$,

$$|A_k - A_k^h| \leqslant CE(h,r_0) + C_\varepsilon r_0^{\tilde{M}-\varepsilon}. \tag{15.41}$$

Note that $\tilde{M}$ depends on the number of interpolation points used, not on the number of coefficients desired to be found.

Assuming next, for simplicity, that we form the a posteriori approximation

$$u_h^* := \sum_{k=1}^{I} a_k^h \mathscr{S}_h + v_L \tag{15.42}$$

(with $K=I$ so that we use all our identified coefficients), then

$$(u-u_h^*)(r,\theta) = \sum_{k=1}^{I} (A_k - A_k^h)\mathscr{S}_k\left(\frac{r}{r_0}\right) + R_{I,L}. \tag{15.43}$$

Thus by (15.41),

$$|(u-u_h^*)(r,\theta)| \leqslant C(E(h,r_0) + C_\varepsilon r_0^{\tilde{M}-\varepsilon})\left(\frac{r}{r_0}\right)^\beta + C_\varepsilon r^{\tilde{M}-\varepsilon}. \tag{15.44}$$

It is now clear that (15.41) gives a clue as to how to choose a suitable r_0, based on local error estimates and $\tilde{M}$; note that $\tilde{M}$ includes some work expanded in identifying v_L if $\tilde{M}$ is high. E.g., if (modulo ε in our expressions below) we equalize the two terms on the right of (15.41), then in the unrefined mesh situation $E(h,r) \cong h^{2\beta} d^{-\beta}, r_0 = h^{2\beta/(\tilde{M}+\beta)}$, so that e.g.

$$|a_1 - a_1^h| \leqslant Ch^{2\beta(1-2\beta/(\tilde{M}+\beta))} \tag{15.45}$$

and, to be used only for $r \leqslant r_0$ of course,

$$|(u-u_h^*)(r,\theta)| \leqslant Ch^{2\beta(1-\beta/(\tilde{M}+\beta))}\left(\frac{r}{r_0}\right)^\beta. \tag{15.46}$$

Thus, for $\tilde{M}$ high, this trivial postprocessing in the unrefined situation gives an approximation to u that is almost $O(h^{2\beta})$ also as we approach the corner. Note that $O(h^{2\beta})$ is the inherent best possible order of accuracy, by the results of Section 13.

Various numerical experiments with the Trigonometric Fitting Method as compared to the Direct Method and the Extrapolation Method were reported in SCHATZ and WAHLBIN [1981] in the case of the L-shaped domain (and with an embarrassing misprint for v_L). For uniform $h=\frac{1}{20}$, the Direct Method (with optimized r_0) gave a relative error in a_1 of 17% (it gave 8% *un*optimized, i.e., with $r_0 \cong h$, indicating that asymptotic error estimates are suspect in inaccurate methods), the Extrapolation Method gave 1%, and the Trigonometric Fitting Method 0.2% for $I=2$, 0.1% for $I=3$ and higher. The predicted optimal r_0's from the theory were used, setting all unknown constants to 1 (and ε to 0). It thus appears that our local error estimates not only give a formidable insight in the general behavior of the approximations but can be quite useful in actual computations.

The brief description and analysis of the Trigonometric Fitting Method above is based on an unpublished investigation by Schatz, Whiteman and myself. The reader who wishes to do so should have no trouble in extending the method to nonhomogeneous boundary data, Neumann conditions, or mixed Dirichlet–Neumann conditions.

16. Local behavior when singular functions are added

We continue to consider the model problem (15.1), (15.2) on a plane polygonal domain. Concentrating attention at the corner v of maximum interior angle α, we have with $\beta=\pi/\alpha$, in polar coordinates,

$$u(r,\theta)=a_1\mathscr{S}_1+w, \tag{16.1}$$

where

$$\mathscr{S}_1(r,\theta)=r^{\beta}\sin(\beta\theta)\omega(r) \tag{16.2}$$

and where ω is a cutoff function isolating the corner. In particular, $\mathscr{S}_1=0$ on $\partial\Omega$.

A natural and well-known finite element procedure is to adjoin singular functions to the basic finite element space. For simplicity we shall only consider the case of adjoining *one*, the first, singular function. The generalization to inclusion of more singular functions (at various corners) is obvious.

Let thus V_h be a basic finite element space, satisfying the homogeneous Dirichlet conditions (15.2), and based on a *quasi-uniform* mesh family.

Set

$$\tilde{V}_h=\mathscr{S}_1\oplus V_h \tag{16.3}$$

and seek $u_h\in\tilde{V}_h$ so that

$$D(u-u_h,\chi)=\int_{\Omega}\nabla(u-u_h)\cdot\nabla\chi=0 \quad \text{for all } \chi\in\tilde{V}_h. \tag{16.4}$$

REMARK 16.1. To solve for u_h, one naturally utilizes the theory of bordered matrices. In fact, including k singular functions, $\tilde{V}_h=\mathscr{S}_1\oplus\mathscr{S}_2\oplus\cdots\oplus\mathscr{S}_k\oplus V_h$, so that

$$u_h=a_1^h\mathscr{S}_1+\cdots+a_k^h\mathscr{S}_k+w^h,\quad w^h\in V_h,$$

and letting $a_1^h,\ldots,a_k^h$ be the last variables adjoined to the (typically nodal) variables involved in $w^h\in V_h$, one has with S the stiffness matrix in V_h a matrix equation involving a matrix of the form

$$
\begin{array}{cc}
 & \begin{array}{cccccc} & & & \multicolumn{3}{c}{\overbrace{\hphantom{\times\quad\times\quad\times}}^{k}} \end{array} \\
\begin{array}{c} \\ \\ k\left\{\begin{array}{c} \\ \\ \\ \end{array}\right. \end{array} &
\left[\begin{array}{cccccc}
 & & & \times & \times & \times \\
 & S & & \times & \times & \times \\
\times & \times & \times & \times & \times & \times \\
\times & \times & \times & \times & \times & \times \\
\times & \times & \times & \times & \times & \times
\end{array}\right]
\end{array}
$$

By the theory of bordered matrices this is solved by $k+1$ application of S^{-1} (perhaps, once and for all LU decomposed or utilizing some other fast method) and then solving a (small) $k \times k$ system.

The basic fact which reduces the local behavior in $u-u_h$ as above to previously known cases is a representation of u_h due to SCHATZ [1980a], cf. DOBROWOLSKI [1981] for an exposé. Note that

$$u_h = a_1^h \mathscr{S}_1 + w^h \tag{16.5}$$

where $w^h \in V_h$.

THEOREM 16.1 (Representation). *Let P_h denote the $\mathring{H}^1(\Omega)$ projection into V_h, without singular functions. Then*

$$u-u_h = (a_1 - a_1^h)(\mathscr{S}_1 - P_h\mathscr{S}_1) + (w - P_h w). \tag{16.6}$$

PROOF. We first show that

$$u_h = P_h u + a_1^h(\mathscr{S}_1 - P_h\mathscr{S}_1). \tag{16.7}$$

For, with

$$w^h := u_h - a_1^h \mathscr{S}_1 \in V_h, \tag{16.8}$$

we have

$$D(w^h, \chi) = D(u_h, \chi) - a_1^h D(\mathscr{S}_1, \chi). \tag{16.9}$$

Considering only $\chi \in V_h$ here,

$$D(u_h, \chi) = (f, \chi) = D(u, \chi) = D(P_h u, \chi) \tag{16.10}$$

and

$$D(\mathscr{S}_1, \chi) = D(P_h\mathscr{S}_1, \chi). \tag{16.11}$$

Thus, by (16.9), for all $\chi \in V_h$,

$$D(w^h - P_h u + a_1^h P_h\mathscr{S}_1, \chi) = 0 \tag{16.12}$$

meaning that

$$w^h = P_h u - a_1^h P_h\mathscr{S}_1. \tag{16.13}$$

Hence

$$u_h = w^h + a_1^h\mathscr{S}_1 = P_h u + a_1^h(\mathscr{S}_1 - P_h\mathscr{S}_1) \tag{16.14}$$

which proves (16.7).

Continuing, by (16.7)

$$\begin{aligned} u-u_h &= a_1\mathcal{S}_1+w-(P_hu+a_1^h(\mathcal{S}_1-P_h\mathcal{S}_1)) \\ &= a_1\mathcal{S}_1+w-(P_h(a_1\mathcal{S}_1+w)+a_1^h(\mathcal{S}_1-P_h\mathcal{S}_1)) \\ &= a_1\mathcal{S}_1+w-a_1P_h\mathcal{S}_1-P_hw-a_1^h\mathcal{S}_1+a_1^hP_h\mathcal{S}_1 \\ &= (a_1-a_1^h)(\mathcal{S}_1-P_h\mathcal{S}_1)+(w-P_hw) \end{aligned} \tag{16.15}$$

proving (16.6). □

By our previous results we know how $\mathcal{S}-P_h\mathcal{S}_1$ behaves, and also, under suitable assumptions on w, how $w-P_hw$ behaves locally. Thus, the next order of business is clearly to get an estimate for $a_1-a_1^h$.

THEOREM 16.2.

$$a_1-a_1^h=\frac{D(w-P_hw,\mathcal{S}_1-P_h\mathcal{S}_1)}{\|\nabla(\mathcal{S}_1-P_h\mathcal{S}_1)\|_0^2}. \tag{16.16}$$

PROOF. By (16.6), since $D(u-u_h,\chi)=0$ for any $\chi\in\tilde{V}_h$, we have

$$D(w-P_hw-(a_1-a_1^h)(\mathcal{S}_1-P_h\mathcal{S}_1),\chi)=0 \quad \text{for any } \chi\in\tilde{V}_h. \tag{16.17}$$

Take here $\chi=\mathcal{S}_1-P_h\mathcal{S}_1\in\tilde{V}_h$ to show (16.16). □

Since we have assumed a quasi-uniform mesh family, we know, for $\beta<1$, that

$$\|\nabla(\mathcal{S}_1-P_h\mathcal{S}_1)\|_0^2\geqslant Ch^{2\beta}. \tag{16.18}$$

For the numerator NU in (16.16) we have with χ any approximation to w in V_h,

$$\mathrm{NU}=D(w-\chi,\mathcal{S}_1-P_h\mathcal{S}_1). \tag{16.19}$$

To get any further we have to make assumptions about w. A natural one (which can be bettered in many situations) is to assume, still with $\beta<1$,

$$w=Cr^{2\beta}\sin(\beta\theta)+\cdots \tag{16.20}$$

with corresponding behavior of derivatives.

Further, assume piecewise linear elements and $2\beta>1$.

To further estimate the numerator NU one should *not* merely estimate it as $\|w-\chi\|_1\|\mathcal{S}_1-P_h\mathcal{S}_1\|_1$; this is too rough. Instead, a more refined argument is called for; thus consider domains $\Omega_j=\{x: 2^{-j-1}\leqslant|x-v|\leqslant 2^{-j}\}$ towards the worst corner. Assuming suitable behavior at the other corners (we leave the details to the reader), with $d_j=2^{-j}$,

$$\begin{aligned} \mathrm{NU} &\lessapprox \sum_j \|\nabla(w-\chi)\|_{0,\Omega_j}\|\nabla(\mathcal{S}_1-P_h\mathcal{S})\|_{0,\Omega_j} \\ &\leqslant \sum_j hd_j^{2\beta-1}(\|\nabla(\mathcal{S}_1-\xi)\|_{0,\Omega_j'}+d_j^{-1}\|\mathcal{S}_1-P_h\mathcal{S}_1\|_{0,\Omega_j'}) \end{aligned} \tag{16.21}$$

by the local H^1 estimates of Section 9, and where $\Omega'_j = \Omega_{j-1} \cup \Omega_j \cup \Omega_{j+1}$. Thus,

$$\mathrm{NU} \lessapprox \sum_j h d_j^{2\beta-1}(h d_j^{\beta-1} + d_j^{-1} \|\mathscr{S}_1 - P_h \mathscr{S}_1\|_{0,\Omega'_j})$$
$$\leqslant Ch^2 \sum_j d_j^{3\beta-2} + Ch \sum_j d_j^{2\beta-1} d_j^{-1} \|\mathscr{S}_1 - P_h \mathscr{S}_1\|_{0,\Omega'_j}. \tag{16.22}$$

A local duality argument as in Section 12, (12.22) et seq. gives

$$d_j^{-1} \|\mathscr{S}_1 - P_h \mathscr{S}_1\|_{0,\Omega'_j} \leqslant C_\varepsilon h^{2\beta-\varepsilon} d_j^{-\beta-\varepsilon}, \tag{16.23}$$

so that from (16.22),

$$\mathrm{NU} \leqslant Ch^2 \sum_j d_j^{3\beta-2} + C_\varepsilon h^{1+2\beta-\varepsilon} \sum_j d_j^{\beta-1-\varepsilon}. \tag{16.24}$$

Summing this (to the highest J s.t. $2^{-J} \simeq h$, some technical details are left to the reader in this sketch), since $\beta < 1$,

$$\mathrm{NU} \leqslant \begin{cases} Ch^{3\beta-\varepsilon} & \text{for } \beta \leqslant \frac{2}{3}, \\ Ch^{2-\varepsilon} & \text{for } \beta > \frac{2}{3}. \end{cases} \tag{16.25}$$

Inserting this and (16.18) into (16.16), we have:

THEOREM 16.3. *For $\beta < 1$ with quasi-uniform piecewise linear elements, and assuming* (16.20) *for the behavior of w,*

$$|a_1 - a_1^h| \leqslant \begin{cases} C_\varepsilon h^{\beta-\varepsilon} & \text{for } \beta \leqslant \frac{2}{3}, \\ C_\varepsilon h^{2-2\beta-\varepsilon} & \text{for } 1 \geqslant \beta > \frac{2}{3}. \end{cases} \tag{16.26}$$

We comment on the above result. Of course, as β tends to 1 the solution becomes smooth and we would expect no identification of the nonexistent stress intensity factor. Otherwise, the identification of it is rather "flaky" in that it is growing in accuracy up to $\beta = \frac{2}{3}$ and then decreasing in accuracy. For $\beta < \frac{2}{3}$ the accuracy is the same as in the "Direct Method" of Section 15, i.e., worse than in the "Trigonometric Fitting Method". This flakiness led to the introduction of the so-called Dual Singular Function method, using $V_h \oplus \mathscr{S}_{-1}$, $\mathscr{S}_{-1} = r^{-\beta} \sin(\beta\theta)$ the dual singular function to $\mathscr{S}_1$, as test functions in order to pick out a_1^h in a sharper fashion, see DOBROWOLSKI [1981].

We have now (under certain assumptions) an estimate for $a_1 - a_1^h$. Using this in (16.6) of Theorem 16.1 the question of the local behavior of $u - u_h$ reduces to the already investigated case of the local behavior in the elliptic projection into V_h, the finite element space without singular function. The result will obviously depend on assumptions for w, and what occurs at the other corners. Again a rather careful local duality argument needs to be performed to estimate the last term in

$$|(w - P_h w)(r, \theta)| \leqslant Ch^2 r^{2\beta-2} + Cd^{-1} \|w - P_h w\|_{0,A_r}$$

but we leave this to the reader.

17. Local behavior in problems with rough coefficients: Pollution

Very little is known concerning the global effects of rough coefficients in a problem. To my knowledge no systematic investigations have been undertaken. Therefore, the present section will treat only simple two-point boundary value problems. The typical situation is that of a jump discontinuity in a coefficient at a point, and the globally quasi-uniform mesh *not* matched to that discontinuity.

As a first example, consider that of seeking $u(x)$, $0\leqslant x\leqslant 2$, such that

$$\begin{aligned}&-u''=f \quad \text{in } (0,1),\\ &-u''+u=f \quad \text{in } (1,2),\\ &u(0)=u(2)=0,\end{aligned} \tag{17.1}$$

with u demanded to be $\mathscr{C}^1$ across $x=1$.

We give an example. Let $I_0 \Subset (0,1)$ and $x_0\in I_0$. Let v be linear from 0 to x_0 (where it assumes the value $(\mathrm{e}-\mathrm{e}^{-1})+(1-x_0)(\mathrm{e}+\mathrm{e}^{-1})$, then of the form

$$v(x)=(\mathrm{e}-\mathrm{e}^{-1})+(1+x)(\mathrm{e}+\mathrm{e}^{-1}),\quad x_0\leqslant x\leqslant 1, \tag{17.2}$$

and finally

$$v(x)=\mathrm{e}^{2-x}-\mathrm{e}^{x-2},\quad 1\leqslant x\leqslant 2. \tag{17.3}$$

Thus v has a sharp break at x_0 but is $\mathscr{C}^1$ across $x=1$. Let further

$$\omega(x)=\begin{cases}1 & \text{outside } I_0,\\ 0 & \text{in a neighborhood of } x_0.\end{cases} \tag{17.4}$$

Set

$$u:=\omega v. \tag{17.5}$$

Then, since the singularity at x_0 in v has been killed off, u satisfies (17.1) above with f smooth and

$$\operatorname{supp} f\subseteq I_0. \tag{17.6}$$

Assume now that we are using uniform (or, quasi-uniform) meshes, but $x=1$ is not a mesh point but, say, always halfways between. For V_h any ordinary finite element space then,

$$\|u-\chi\|_{1,2}^2\geqslant ch^3,\quad \text{any } \chi\in V_h, \tag{17.7}$$

as is trivial to see since u has a sharp break across $x=1$.

Let

$$a(u,v)=\int_0^2 u'v'+\int_1^2 uv \tag{17.8}$$

and

$$E:=u-u_h \tag{17.9}$$

the error in the corresponding finite element solution. By (17.7) and (17.6),

$$ch^3 \leqslant a(E,E) = a(E,u) = -(E,f)_{I_0} \leqslant \|E\|_{\text{Any norm on } I_0} \|f\|_{\text{Dual norm}}. \tag{17.10}$$

Thus,

$$\|u-u_h\|_{\text{Any norm on } I_0} \geqslant c(I_0)h^3. \tag{17.11}$$

Hence, for $r \geqslant 4$, there is no chance that the Galerkin solution will approach optimal order on I_0, which is a domain away from the rough spot, $x=1$.

We next wish to match this negative result with a positive result. From the results of Section 4 we have for the general error $e=u-u_h$ in (17.1),

$$|e(x_0)| \leqslant C \min_{\chi \in V_h} \|u-\chi\|_{0,\infty,I_0} + Cd^{-1}\|e\|_{-1/2,I_0}. \tag{17.12}$$

Assume now for simplicity that

$$r \geqslant 3. \tag{17.13}$$

Employing the interpolation spaces $H^{k,\infty}$, based on the Sobolev spaces $H^k = W^{k,2}$ via the real interpolation method $[\ ]_{\theta,\infty}$, cf. BERGH and LÖFSTRÖM [1976], we prove the following:

CLAIM 17.1.

$$\text{If } f \in H^{1/2,\infty}, \text{ then } u \in H^{2.5,\infty}. \tag{17.14}$$

PROOF. Let χ_t, with $t>0$, be as follows:

$$\chi_t(x) = \begin{cases} 0 & \text{for } 0 \leqslant x \leqslant 1-t, \\ \text{linear} & \text{in } 1-t \leqslant x \leqslant 1, \\ 1 & \text{for } 1 \leqslant x \leqslant 2. \end{cases} \tag{17.15}$$

Let

$$-\tilde{u}'' = \chi_t u + f; \tag{17.16}$$

then with $\chi_{[1,2]}$ the characteristic function of $[1,2]$,

$$(u-\tilde{u})'' = (\chi_t - \chi_{[1,2]})u, \tag{17.17}$$

so that

$$\|u-\tilde{u}\|_{2,2} \leqslant \|u\|_{0,\infty} \|\chi_t - \chi_{[1,2]}\|_0 \leqslant Ct^{1/2}. \tag{17.18}$$

Next, split f into

$$f = f_0 + f_1 \tag{17.19}$$

where $f_0 \in L^2$, $f_1 \in H^1$. Correspondingly, let

$$\tilde{u} = \tilde{u}_0 + \tilde{u}_1 \tag{17.20}$$

where

$$-\tilde{u}_0''=f_0, \tag{17.21}$$

$$-\tilde{u}_1''=\chi_t u+f_1. \tag{17.22}$$

Then

$$\|\tilde{u}_0\|_{2,2}\leqslant C\|f_0\|_0, \tag{17.23}$$

and since

$$-\tilde{u}_1'''=(\chi_t)'u+\chi_t u'+f_1', \tag{17.24}$$

$$\|\tilde{u}_1\|_{3,2}\leqslant Ct^{-1/2}+\|f_1\|_{1,2}. \tag{17.25}$$

Thus,

$$\|u\|_{H^{2.5,\infty}}=\sup_{0<t<1}\frac{K(u,t;H^3,H^2)}{t^{1/2}} \tag{17.26}$$

where for the Lions–Petree K-functional,

$$\frac{K(u,t)}{t^{1/2}}\leqslant\frac{\|u-\tilde{u}-\tilde{u}_0\|_{H^2}+t\|\tilde{u}_1\|_{H^3}}{t^{1/2}}$$

$$\leqslant\frac{Ct^{1/2}+C\|f_0\|_{L^2}+Ct^{1/2}+Ct\|f_1\|_{H^1}}{t^{1/2}}$$

$$\leqslant C+C\frac{K(f,t;H^1,L^2)}{t^{1/2}} \tag{17.27}$$

and Claim 17.1 obtains. □

We may then employ an obvious duality argument in (17.12) to prove that (for $r\geqslant 3$, remember),

$$|e(x_0)|\leqslant Ch^3 \tag{17.28}$$

so that the lower bound (17.11) is complemented with an upper bound.

We next consider a pollution effect when a higher-order coefficient is rough. Let

$$a(x)=\begin{cases}1 & \text{for } 0\leqslant x\leqslant 1\\ 2 & \text{for } 1<x\leqslant 2\end{cases}, \tag{17.29}$$

and consider the problem

$$(au')'=f \quad \text{in } 0\leqslant x\leqslant 2, \qquad u(0)=u(2)=0. \tag{17.30}$$

Then since (for f smooth), u' has a jump discontinuity across $x=1$, we have

$$ch\leqslant\|u-u_h\|_{1,2}^2 \tag{17.31}$$

if $x=1$ is halfway between mesh points for a quasi-uniform mesh. Following the procedure in the first example above, it is easy to show that, for an example,

$$ch \leqslant \|E\|_{\text{Any norm on } I_0}. \tag{17.32}$$

Thus we have a rather severe pollution effect in this example.

We leave it to the reader to give a positive result in this case.

18. An example of global pollution with numerical integration

In general the influence of numerical integration of coefficients, right-hand sides, etc., is easy to trace through in smooth problems. (In my experience, the simplest way is to first use results concerning $u-u_h$ where u_h is a fictitious projection into V_h *without* numerical integration. Then consider the difference $u_h-u_h^*$ where u_h^* is obtained *with* numerical integration. The general principle should be clear from WAHLBIN [1978].)

Consider the problem of finding $u(x)$ such that with $0<\sigma<1$,

$$\begin{aligned} &u''=f=x_+^\sigma, \quad -1\leqslant x\leqslant 1,\\ &u(-1)=0=u(1), \end{aligned} \tag{18.1}$$

where

$$f=x_+^\sigma=\begin{cases} x^\sigma, & x\geqslant 0,\\ 0, & x\leqslant 0. \end{cases} \tag{18.2}$$

Use piecewise linear elements on a quasi-uniform mesh. To get a total numerical method we interpolate x_+^σ linearly (with a mesh point at 0) and use that in an exact fashion. (This is equivalent to some numerical integration.) Let f_I denote this linearly interpolated $f=x_+^\sigma$. Clearly, this is h^2 accurate away from the origin.

To elucidate the pollution ensuing from the above procedure, we need not even consider the finite element projection: it is enough to consider what would happen in the continuous problems when $f=x_+^\sigma$ is replaced by f_I.

Thus let $\tilde{u}$ be given by

$$\tilde{u}''=f_I, \quad -1\leqslant x\leqslant 1, \qquad \tilde{u}(-1)=0=\tilde{u}(1). \tag{18.3}$$

For the error $e=u-\tilde{u}$ then,

$$e''=f-f_I, \quad -1\leqslant x\leqslant 1, \qquad e(-1)=e(1)=0. \tag{18.4}$$

Since $x=0$ is a mesh point it is clear from the geometry of the situation that $f-f_I\geqslant 0$ for $0<\sigma<1$.

Now let x_0 be near $-\frac{1}{2}$, where, after all, both u and $\tilde{u}$ are linear. Then with G^0 the Green's function for the problem centered at x_0,

$$e(x_0)=\int_{-1}^{1} G^0(y)(f-f_I)(y)\,\mathrm{d}y. \tag{18.5}$$

Note that if $x_0 \simeq -\frac{1}{2}$, G^0 is positive for $0 \leqslant x \leqslant 1$. Thus, since, as already noted, $f-f_I \geqslant 0$, the error in $f-f_I$ rules $e(x_0)$. By the well-known error formula for the trapezoidal rule, for x_i, x_{i+1} mesh points, $i \geqslant 1$,

$$\int_{x_i}^{x_{i+1}} (f-f_I) = \left(\int_{x_i}^{x_{i+1}} f \right) - \tfrac{1}{2}h[f_{i+1}+f_i] = -\tfrac{1}{12}h^3 f''(\xi) \tag{18.6}$$

$$\geqslant \tfrac{1}{12}h^3 \sigma(1-\sigma) x_{i+1}^{\sigma-2}.$$

Thus, summing,

$$e(x_0) \geqslant Ch^{\sigma+1}, \quad x_0 \simeq -\tfrac{1}{2}, \tag{18.7}$$

giving a systematic suboptimal error away from the problematic point, i.e., a pollution effect.

Chapter IV

A Singularly Perturbed Elliptic to Elliptic Model Problem

19. The model problem

Consider the problem of finding $u=u(x;\varepsilon)$ such that

$$L_\varepsilon u \equiv -\varepsilon^2 \Delta u + a(x)u = f(x) \quad \text{in } \Omega, \tag{19.1}$$

where Ω is a *plane convex* polygon with straight edges. In general, Dirichlet boundary conditions lead to sharper boundary layers than Neumann conditions and hence we take

$$u=0 \quad \text{on } \partial\Omega \tag{19.2}$$

as our boundary conditions, selecting the harder case.

We wish to avoid the problem of turning points and assume therefore that with positive constants a_0, a_1,

$$0<a_0 \leqslant a(x) \leqslant a_1. \tag{19.3}$$

The limiting case of (19.1) is $u=f/a$. This is singularly perturbed in L^∞ but, e.g., regularly perturbed in L^2.

In the case that $\partial\Omega \in \mathscr{C}^{2,\alpha}$ for some $\alpha>0$, the precise behavior of boundary layers as $\varepsilon \to 0$ in (19.1), (19.2) is known, provided (19.3) holds and $a(x), f(x) \in \mathscr{C}^2$. Let $\partial(x)$ denote the distance from x to $\partial\Omega$ along the normal to $\partial\Omega$, let x' denote the point where the normal meets $\partial\Omega$, and let $\psi(x)$ be a smooth cutoff function isolating a layer around $\partial\Omega$. Set

$$u_\varepsilon(x) := \frac{f(x)}{a(x)} - \frac{f(x')}{a(x')} \exp\left(- \frac{\partial(x)}{\varepsilon} \sqrt{a(x')} \right) \psi(x). \tag{19.4}$$

The first term on the right is called the "regular" expansion, the second term the "boundary layer correction". The theory of such expansions can be found in Besjes [1975], Eckhaus [1973, Section 2.5.3; 1979, Section 7.1.1], and Lions [1973, Chapter II]. Using a "normal-tangential" coordinate system it is easy to see that

$L_\varepsilon(u-u_\varepsilon)=\mathrm{O}(\varepsilon)$. The maximum principle then gives

$$\|u-u_\varepsilon\|_{0,\infty,\Omega}\leqslant a_0^{-1}\mathrm{O}(\varepsilon)=C\varepsilon, \tag{19.5}$$

so that, within $\mathrm{O}(\varepsilon)$, in this case (19.4) describes the boundary layer behavior as $\varepsilon\to 0$. The thickness of the layer is thus $\mathrm{O}(\varepsilon \ln 1/\varepsilon)$, cf. BARANGER [1979].

In the case of less smooth boundary or coefficients, e.g. our basic case of a convex polygon, we cannot ascertain this precise behavior. For our future local error estimates in the finite element method we need some basic estimates for the continuous problem. For convenience we collect them here. These proofs utilize well-known regularity theory from, e.g., GRISVARD [1985]. The details are given in SCHATZ and WAHLBIN [1983, Section 2 and Appendix]; that paper is the basis for this whole chapter.

LEMMA 19.1. *There exists a constant C such that*

$$\|u\|_{0,\Omega}\leqslant C\|f\|_{0,\Omega}, \tag{19.6}$$

$$\|u\|_{1,\Omega}\leqslant C\varepsilon^{-1}\|f\|_{0,\Omega}, \tag{19.7}$$

and

$$\|u\|_{2,\Omega}\leqslant C\varepsilon^{-2}\|f\|_{0,\Omega}. \tag{19.8}$$

The next result concerns exponential decay in functions which satisfy the homogeneous equation in a subset of Ω.

LEMMA 19.2. *There exist positive constants c and C such that the following holds. Let $A\subseteq B$ with $d=\partial_<(A,B)>0$ and let $L_\varepsilon v=0$ on B. Then for $0<\varepsilon\leqslant 1$,*

$$\|v\|_{0,A}+d\|\nabla v\|_{0,A}\leqslant C\,\mathrm{e}^{-cd/\varepsilon}\|v\|_{0,B}, \tag{19.9}$$

$$\|\nabla v\|_{0,A}\leqslant \mathrm{e}^{-cd/\varepsilon}\|\nabla v\|_{0,B}, \tag{19.10}$$

and

$$\varepsilon^2\|v\|_{2,A}\leqslant C(\varepsilon^2 d^{-1}\|\nabla v\|_{0,B}+\|v\|_{0,B}). \tag{19.11}$$

20. The finite element method and its local behavior in energy

We consider the problem (19.1), (19.2) on a convex polygonal domain with a uniformly positive $a(x)$, (19.3). Let V_h be a finite element family satisfying the homogeneous Dirichlet conditions and based on a quasi-uniform family of partitions. For concreteness we may take $\Omega_h=\Omega$ and Lagrangian elements of order $r-1$ on triangular subdivisions. Then the properties listed in Chapters II and III, such as approximation, superapproximation, and inverse properties, hold.

Let

$$A_\varepsilon(v,w)=\varepsilon^2(\nabla v,\nabla w)+(av,w). \tag{20.1}$$

THEOREM 20.1. *There exist positive constants c_1, c_2 and C such that the following holds. Let $A \subseteq B \subseteq \Omega$ with $d = \partial_<(A, B) \geqslant c_1 h$. Let $v \in \mathring{H}^1(\Omega)$ and $v_h \in V_h$ be such that*

$$A_\varepsilon(v - v_h, \chi) = 0 \quad \text{for } \chi \in V_h^<(B). \tag{20.2}$$

Then

$$\begin{aligned}\|v - v_h\|_{1,A} \leqslant\ & C \min_{\chi \in V_h}(\|\nabla(v-\chi)\|_{0,B} + d^{-1}\|v - \chi\|_{0,B})\\ & + Cd^{-1} e^{-c_2 d/(\varepsilon + h)} \|v - v_h\|_{0,B}.\end{aligned} \tag{20.3}$$

The result (20.3) should be compared with exponential decay result in the L^2 projection, Theorem 7.2; note that $\varepsilon = 0$ "formally" corresponds to the L^2 projection. (In the next section we will give the pointwise result which is more perfectly analogous to Theorem 7.2.) For ε of unit size (20.3) is exactly Corollary 9.2 for $s = 0$. Thus the whole present chapter is, in a formal sense, intermediate between the limiting cases of the L^2 projection and an ordinary elliptic projection. In light of Chapters II and III, Sections 9 and 10, it is not surprising that we manage to give rather satisfactory results.

In order to prove Theorem 20.1 we shall need the following discrete version of Lemma 19.2, (19.9).

LEMMA 20.1. *With notation as in Theorem 20.1, if $A_\varepsilon(v_h, \chi) = 0$ for $\chi \in V_h^<(B)$, then*

$$\|v_h\|_{0,A} + d\|\nabla v_h\|_{0,A} \leqslant C e^{-c_2 d/(\varepsilon + h)} \|v_h\|_{0,B}. \tag{20.4}$$

PROOF. Let D_ρ and $D_{\rho+\delta}$ be two concentric discs of radii ρ and $\rho + \delta$, respectively, with center in A, and let $\omega \in \mathscr{C}_0^\infty(D_{\rho+\delta})$ be such that

$$\omega \equiv 1 \quad \text{on } D_\rho, \tag{20.5}$$

$$\|\omega\|_{\mathscr{C}^l(D_{\rho+\delta})} \leqslant C\delta^{-l}. \tag{20.6}$$

Let $B_r = \Omega \cap D_r$ and assume that

$$A_\varepsilon(v_h, \chi) = 0 \quad \text{for } \chi \in V_h^<(B_{\rho+\delta}) \tag{20.7}$$

and that

$$\delta \geqslant \tilde{c} h. \tag{20.8}$$

(The dependence of $\tilde{c}$ on various approximation and inverse constants is easy to trace but left to the reader.) We have now for any $\chi \in V_h^<(B_{\rho+\delta})$,

$$\begin{aligned}&\varepsilon^2 \|\omega \nabla v_h\|_0^2 + a_0 \|\omega v_h\|_0^2\\ &\quad \leqslant \varepsilon^2(\nabla v_h, \omega^2 \nabla v_h) + (a v_h, \omega^2 v_h)\\ &\quad = A_\varepsilon(v_h, \omega^2 v_h) - \varepsilon^2(\omega \nabla v_h, 2(\nabla\omega) v_h)\\ &\quad = A_\varepsilon(v_h, \omega^2 v_h - \chi) - \varepsilon^2(\omega \nabla v_h, 2(\nabla \omega) v_h).\end{aligned} \tag{20.9}$$

By "superapproximation" we find thus that

$$\begin{aligned}&\varepsilon^2\|\omega\nabla v_h\|_0^2+a_0\|\omega v_h\|_0^2\\&\quad\leqslant C\varepsilon^2\|\nabla v_h\|_{0,B_{\rho+\delta}}(h\delta^{-1}\|\nabla v_h\|_{0,B_{\rho+\delta}}+h\delta^{-2}\|v_h\|_{0,B_{\rho+\delta}})\\&\qquad+C\|v_h\|_{0,B_{\rho+\delta}}(h^2\delta^{-1}\|\nabla v_h\|_{0,B_{\rho+\delta}}h^2\delta^{-2}\|v_h\|_{0,B_{\rho+\delta}})\\&\qquad+C\varepsilon^2\delta^{-1}\|\omega\nabla v_h\|_0\|v_h\|_{0,B_{\rho+\delta}}.\end{aligned}\tag{20.10}$$

Since $\delta\geqslant\tilde{c}h$,

$$\begin{aligned}&\varepsilon^2h\delta^{-2}\|\nabla v_h\|_{0,B_{\rho+\delta}}\|v_h\|_{0,B_{\rho+\delta}}\\&\quad\leqslant C\varepsilon^2h\delta^{-1}\|\nabla v_h\|_{0,B_{\rho+\delta}}^2+C\varepsilon^2\delta^{-2}\|v_h\|_{0,B_{\rho+\delta}}^2.\end{aligned}\tag{20.11}$$

Further, by the inverse property, for $\delta\geqslant\tilde{c}h$ with $\tilde{c}$ large,

$$h^2\delta^{-1}\|v_h\|_{0,B_{\rho+\delta}}\|\nabla v_h\|_{0,B_{\rho+\delta}}\leqslant Ch\delta^{-1}\|v_h\|_{0,B_{\rho+2\delta}}^2.\tag{20.12}$$

Also,

$$C\varepsilon^2\delta^{-1}\|\omega\nabla v_h\|_0\|v_h\|_{0,B_{\rho+\delta}}\leqslant\tfrac12\varepsilon^2\|\omega\nabla v_h\|_0^2+C\varepsilon^2\delta^{-2}\|v_h\|_{0,B_{\rho+\delta}}^2.\tag{20.13}$$

Using (20.11)–(20.13) in (20.10),

$$\begin{aligned}&\tfrac12\varepsilon^2\|\nabla v_h\|_{0,B_\rho}^2+a_0\|v_h\|_{0,B_\rho}^2\\&\quad\leqslant C\varepsilon^2h\delta^{-1}\|\nabla v_h\|_{0,B_{\rho+\delta}}^2+C(\varepsilon^2\delta^{-2}+h\delta^{-1})\|v_h\|_{0,B_{\rho+2\delta}}^2.\end{aligned}\tag{20.14}$$

Iterating the gradient term on the right once more and using inverse properties,

$$\begin{aligned}&\varepsilon^2h\delta^{-1}\|\nabla v_h\|_{0,B_{\rho+\delta}}^2\\&\quad\leqslant h\delta^{-1}[C\varepsilon^2h\delta^{-1}\|\nabla v_h\|_{0,B_{\rho+2\delta}}^2+C(\varepsilon^2\delta^{-2}+h\delta^{-1})\|v_h\|_{0,B_{\rho+2\delta}}^2]\\&\quad\leqslant C\left(\frac{\varepsilon^2+h^2}{\delta^2}\right)\|v_h\|_{0,B_{\rho+3\delta}}^2.\end{aligned}\tag{20.15}$$

Similarly, iterating the last part of the L^2 term on the right of (20.14), and using (20.15),

$$\begin{aligned}h\delta^{-1}\|v_h\|_{0,B_{\rho+2\delta}}^2&\leqslant Ch\delta^{-1}(\varepsilon^2\delta^{-2}h\delta^{-1})\|v_h\|_{0,B_{\rho+4\delta}}^2\\&\leqslant C\left(\frac{\varepsilon^2+h^2}{\delta^2}\right)\|v_h\|_{0,B_{\rho+4\delta}}^2.\end{aligned}\tag{20.16}$$

Using (20.15) and (20.16) in (20.14), and changing notation slightly,

$$\tfrac12\varepsilon^2\|\nabla v_h\|_{0,B_\rho}^2+a_0\|v_h\|_{0,B_\rho}^2\leqslant C\left(\frac{\varepsilon^2+h^2}{\delta^2}\right)\|v_h\|_{0,B_{\rho+\delta}}^2.\tag{20.17}$$

Consequently, with K independent of ε, h, ρ or δ,

$$\|\nabla v_h\|_{0,B_\rho}\leqslant K\delta^{-1}\|v_h\|_{0,B_{\rho+\delta}}\quad\text{for }\varepsilon\geqslant h,\tag{20.18}$$

and

$$\|v_h\|_{0,B_\rho}\leqslant K\frac{\sqrt{\varepsilon^2+h^2}}{\delta}\|v_h\|_{0,B_{\rho+\delta}}.\tag{20.19}$$

(In the case $\varepsilon \leqslant h$ the inequality (20.18) has to be separately established. This is easy from (20.19) and inverse properties.)

We now choose δ so that

$$K\frac{\sqrt{\varepsilon^2+h^2}}{\delta}=\mathrm{e}^{-1}. \tag{20.20}$$

Then, iterating,

$$\| v_h \|_{0,B_\rho} \leqslant \mathrm{e}^{-N} \| v_h \|_{0,B_{\rho+N\delta}}. \tag{20.21}$$

We may do this for

$$N\delta \cong d, \tag{20.22}$$

or, from (20.20)

$$N \cong d/\delta = c_2 d/(\varepsilon + h). \tag{20.23}$$

By (20.21) (and squaring and summing) this proves half of (20.4) and combining with (20.18) ($\delta = d$ taken), the lemma is proven. □

We can now prove Theorem 20.1.

PROOF OF THEOREM 20.1. It suffices to verify the result with A and B replaced by B_ρ and $B_{\rho+\delta}$, $\delta \cong d \geqslant \tilde{c}h$, and then squaring and summing. Let $\omega \in \mathscr{C}_0^\infty(D_{\rho+2\delta})$ with

$$\omega \equiv 1 \quad \text{on } D_{\rho+\delta}, \qquad \| \omega \|_{\mathscr{C}^l(D_{\rho+2\delta})} \leqslant C\delta^{-l}. \tag{20.24}$$

With $\tilde{v} = \omega v$, $\tilde{v}_h = P_h^\varepsilon(\omega v)$, the projection w.r.t. the form A_ε, we have

$$\| v - v_h \|_{1,B_\rho} \leqslant \| \tilde{v} - \tilde{v}_h \|_1 + \| \tilde{v}_h - v_h \|_{1,B_\rho}. \tag{20.25}$$

Here, by Lemma 20.1, since

$$A_\varepsilon(\tilde{v}_h - v_h, \chi) = 0 \quad \text{for } \chi \in V_h^<(B_{\rho+\delta}),$$

we have

$$\begin{aligned} \| \tilde{v}_h - v_h \|_{1,B_\rho} &\leqslant C\delta^{-1}\, \mathrm{e}^{-c\delta/(\varepsilon+h)} \| \tilde{v}_h - v_h \|_{0,B_{\rho+\delta}} \\ &\leqslant C\delta^{-1}\, \mathrm{e}^{-c\delta/(\varepsilon+h)} (\| \tilde{v} - \tilde{v}_h \|_{0,B_{\rho+\delta}} + \| v - v_h \|_{0,B_{\rho+\delta}}). \end{aligned} \tag{20.26}$$

Hence,

$$\| v - v_h \|_{1,B_\rho} \leqslant \| \tilde{v} - \tilde{v}_h \|_1 + \delta^{-1} \| \tilde{v} - \tilde{v}_h \|_0 + C\delta^{-1}\, \mathrm{e}^{-c\delta/(\varepsilon+h)} \| v - v_h \|_{0,B_{\rho+\delta}}. \tag{20.27}$$

It remains to estimate the first two terms on the right of (20.27). For this we use the following lemma.

LEMMA 20.2 (Global energy estimates). *There exists a constant C such that*

$$\| \nabla(v - P_h^\varepsilon v) \|_0 \leqslant \begin{cases} C\| v \|_1, \\ Ch\| v \|_2, \end{cases} \tag{20.28}$$

and

$$\| v-P_h^{\varepsilon}v \|_0 \leqslant \begin{cases} Ch\| v\|_1, \\ Ch^2\| v\|_2. \end{cases} \tag{20.29}$$

Admitting this lemma for the moment,

$$\| \tilde{v}-\tilde{v}_h \|_1 \leqslant C\| \tilde{v}\|_1$$

and, for $\delta \cong d \geqslant ch$,

$$\delta^{-1}\| \tilde{v}-\tilde{v}_h \|_1 \leqslant Ch\delta^{-1}\| \tilde{v}\|_1 \leqslant C\| \tilde{v}\|_1.$$

Since

$$\| \tilde{v}\|_1 \leqslant C\| \nabla v\|_{0,B_{\rho+2\delta}} + c\delta^{-1}\| v\|_{0,B_{\rho+2\delta}},$$

we have from (20.27),

$$\begin{aligned}\| v-v_h \|_{1,B_\rho} \leqslant{}& C(\| \nabla v\|_{0,B_{\rho+\delta}} + c\delta^{-1}\| v\|_{0,B_{\rho+2\delta}}) \\ &+ C\delta^{-1}\,\mathrm{e}^{-c\delta/(\varepsilon+h)}\| v-v_h\|_{0,B_{\rho+2\delta}}\end{aligned} \tag{20.30}$$

which, upon squaring and summing, shows (20.3) with $\chi=0$. The whole of Theorem 20.1 now follows by writing $v-v_h=(v-\chi)-(v_h-\chi)$. □

It remains to show Lemma 20.2.

PROOF OF LEMMA 20.2. Set $e=v-P_h^{\varepsilon}v$. Then

$$\varepsilon^2\| \nabla e\|_0^2 + a_0\| e\|_0^2 \leqslant A_\varepsilon(e,e) = A_\varepsilon(e, v-\chi) \quad \text{for any } \chi\in V_h. \tag{20.31}$$

Hence,

$$\varepsilon^2\| \nabla e\|_0^2 + a_0\| e\|_0^2 \leqslant C\{\varepsilon^2\| \nabla(v-\chi)\|_0^2 + \| v-\chi\|_0^2\}. \tag{20.32}$$

We first treat the case $\varepsilon \geqslant ch$. Taking $\chi=P_h^1 v$ in (20.32) it is well known that $\| \nabla(v-\chi)\|_1 \leqslant Ch\| v\|_1$ and $\| v-\chi\|_0 \leqslant Ch\| v\|_1$ so that

$$\| \nabla e\|_0^2 \leqslant C\{\| v\|_1^2 + h^2\varepsilon^{-2}\| v\|_1^2\} \leqslant C\| v\|_1^2.$$

Hence, the first inequality of (20.28) obtains. Using instead that $\| \nabla(v-\chi)\|_1 \leqslant Ch\| v\|_2$ and $\| v-\chi\|_0 \leqslant Ch^2\| v\|_2$ the second part of (20.28) follows too.

We continue with a duality argument. Let w solve

$$L_\varepsilon w = e \quad \text{in } \Omega, \qquad w=0 \quad \text{on } \partial\Omega.$$

Then, for any $\chi\in V_h$,

$$\| e\|_0^2 = A_\varepsilon(w-\chi, e)$$

so that with a suitable χ,

$$\| e\|_0^2 \leqslant C\{\varepsilon^2\| \nabla e\|_0 h\| w\|_2 + C\| e\|_0 h^2\| w\|_2\}$$

and by use of Lemma 19.1,

$$\|e\|_0^2 \leqslant C\{h\|\nabla e\|_0\|e\|_0 + h^2\varepsilon^{-2}\|e\|_0^2\}.$$

Hence for $\varepsilon \geqslant \tilde{c}h$, $\tilde{c}$ large,

$$\|e\|_0 \leqslant Ch\|\nabla e\|_0$$

and (20.29) follows.

It remains to consider the case of $\varepsilon \leqslant \tilde{c}h$. By (20.32) with $\chi = P_h^1 v$ we have also

$$\|e\|_0^2 \leqslant C(\varepsilon^2 + h^2)\|v\|_1^2 \leqslant Ch^2\|v\|_1^2,$$

or

$$\|e\|_0^2 \leqslant C(\varepsilon^2 + h^2)h^2\|v\|_2^2 \leqslant Ch^4\|v\|_2^2,$$

establishing (20.29) in this case of $\varepsilon \leqslant \tilde{c}h$. For (20.28), use (20.29), a suitable intermediate $\chi \in V_h$, and the inverse property.

This proves Lemma 20.2. □

21. Local pointwise error estimates

As remarked in Section 10, to prove local pointwise estimates present technology of proofs requires a "related" projection on a domain, including the one of interest, for which one knows (almost) stability in L^∞. In the situation of Section 10 that problem was a local Neumann problem on a disc (and the approximation properties at the boundary of its domain never really "entered" since all functions could be extended to the outside and then approximated on full elements). Thereafter, we gave some examples of how to extend this to domains abutting on the boundary.

In the present setting we shall use the global P_h^ε projection *itself* as the fundamental stable process. Thereafter, localization will be done by cutoff functions and (the cheap way out!) use of the two-dimensional Sobolev's inequality on V_h. Clearly, this is extremely inelegant, leads to loss of unnecessary log $1/h$ factors, is restricted to two dimensions, etc., but it does, at present-day technology of proofs, lead to a *manageable* and *readable* proof. As usual at this stage of our survey, we shall be cavalier about exact assumptions. Consult Chapters II and III for more exact details, or, consult SCHATZ and WAHLBIN [1983]. Suffice it here to have in mind Lagrangian elements on a quasi-uniform family of triangulations so that all the relevant formalized assumptions listed in previous chapters hold.

We start by proving the global almost best approximation property of the P_h^ε projection into V_h based on the form A_ε, see (20.1).

THEOREM 21.1 (Global almost stability in L^∞). *Assume a quasi-uniform mesh family and that u is continuous. Then*

$$\|u - P_h^\varepsilon u\|_{0,\infty,\Omega} \leqslant \ln(C + \varepsilon/h) \min_{\chi \in V_h} \|u - \chi\|_{0,\infty,\Omega}. \tag{21.1}$$

PROOF. We write $u_h = P_h^\varepsilon u$. Let x_0 be a point in Ω, $x_0 \in \bar{T}_0$ an element. By the inverse property,

$$\begin{aligned}|(u-u_h)(x_0)| &\leqslant |u(x_0)| + |u_h(x_0)| \leqslant |u(x_0)| + Ch^{-1}\|u_h\|_{0,T_0} \\ &\leqslant C\|u\|_{0,\infty,\Omega} + Ch^{-1}\|u-u_h\|_{0,T_0}.\end{aligned} \tag{21.2}$$

Here,

$$h^{-1}\|u-u_h\|_{0,T_0} = h^{-1} \sup_{\substack{\phi\in\mathscr{C}_0^\infty(T_0)\\ \|\phi\|_0=1}} (u-u_h, \phi). \tag{21.3}$$

For each such fixed ϕ, let

$$L_\varepsilon v = \phi \quad \text{in } \Omega, \qquad v=0 \quad \text{on } \partial\Omega, \tag{21.4}$$

and set $v_h = P_h^\varepsilon v$. The following are simple results that we shall need:

$$\|v\|_0 + \|v_h\|_0 \leqslant \begin{cases} C, \\ Ch/\varepsilon \end{cases}. \tag{21.5}$$

$$\|\nabla v\|_0 + \|\nabla v_h\|_0 \leqslant h\varepsilon^{-2}\ln^{1/2}(C+\varepsilon/h). \tag{21.6}$$

To see these results, note first that

$$\varepsilon^2\|\nabla v\|_0^2 + a_0\|v\|_0^2 \leqslant A_\varepsilon(v,v) = (\phi, v), \tag{21.7}$$

and similarly for v_h, so that the first inequality in (21.5) obtains.

For the second piece of (21.5), consider first v itself. Letting $L_\varepsilon w = v/\|v\|_0$, $w=0$ on $\partial\Omega$, we have

$$\|v\|_0 = A_\varepsilon(v, w) = (\phi, w). \tag{21.8}$$

Sobolev's inequality in the form $\|w\|_{0,\infty} \leqslant C\|w\|_{2,2}^{1/2}\|w\|_0^{1/2}$ leads to

$$(\phi, w) \leqslant \|\phi\|_{0,1}\|w\|_{2,2}^{1/2}\|w\|^{1/2} \leqslant Ch\|\phi\|_0 C\varepsilon^{-1} = Ch\varepsilon^{-1}$$

by use of Cauchy–Schwarz' inequality and Lemma 19.1. This shows half of the second part of (21.5) and since, by Lemmas 20.2 and 19.1, $\|v-v_h\|_0 \leqslant Ch\|v\|_1 \leqslant Ch/\varepsilon$ all of (21.5) follows.

For (21.6), by the analogue (21.7),

$$\varepsilon^2\|v_h\|_1^2 \leqslant (\phi, v_h) \leqslant \|\phi\|_{0,1,T_0}\|v_h\|_{0,\infty,T_0}. \tag{21.9}$$

Let Ω^1 denote the intersection of a disc of radius $\max(Ch, \varepsilon)$ around x_0. By Cauchy–Schwarz' inequality and by Sobolev's lemma on V_h, see (10.34), and also by use of (21.5),

$$\begin{aligned}\varepsilon^2\|v_h\|_1^2 &\leqslant Ch\ln^{1/2}(C+\varepsilon/h)(\|v_h\|_{1,\Omega'} + \varepsilon^{-1}\|v_h\|_{0,\Omega'}) \\ &\leqslant Ch\ln^{1/2}(C+\varepsilon/h)\|v_h\|_1 + Ch^2\varepsilon^{-2}\ln^{1/2}(C+\varepsilon/h).\end{aligned} \tag{21.10}$$

By this,

$$\|v_h\|_1 \leqslant Ch\varepsilon^{-2}\ln^{1/2}(C+\varepsilon/h).$$

Since also, by Lemmas 20.2 and 19.1, $\|v-v_h\|_{H^1}\leqslant Ch\|v\|_{H^2}\leqslant Ch\varepsilon^{-2}$ we have completed the proof of (21.6).

Return now to (21.3):

$$\begin{aligned}
&(u-u_h,\phi)\\
&\quad=A_\varepsilon(u-u_h,v)=A_\varepsilon(u-u_h,v-v_h)=A_\varepsilon(u,v-v_h)\\
&\quad=-\varepsilon^2\sum_{T_i}\left(\int_{T_i}u\Delta(v-v_h)+\oint_{\partial T_i}u\frac{\partial}{\partial n}(v-v_h)\right)+(au,v-v_h)\\
&\quad\leqslant C\|u\|_{0,\infty}\{\varepsilon^2|v-v_h|_{2,1;h}+\varepsilon^2h^{-1}\|\nabla(v-v_h)\|_{0,1}+\|v-v_h\|_{0,1}\},
\end{aligned}\tag{21.11}$$

where we used a trace inequality on ∂T_i. The notation $|\cdot|_{2,1;h}$ means a piecewise norm. Here, adding and subtracting a suitable χ,

$$|v-v_h|_{2,1;h}\leqslant C\|v\|_{2,1}+Ch^{-1}\|v-v_h\|_{1,1}$$

so that

$$\begin{aligned}
&|h^{-1}(u-u_h,\phi)|\\
&\quad\leqslant C\|u\|_{0,\infty}\{\varepsilon^2h^{-1}\|v\|_{2,1}+\varepsilon^2h^{-2}\|v-v_h\|_{1,1}+h^{-1}\|v-v_h\|_{0,1}\}.
\end{aligned}\tag{21.12}$$

Let now

$$\Omega_j=\{x\in\Omega:2^{-j}\leqslant|x-x_0|\leqslant2^{-j+1}\}\tag{21.13}$$

and

$$\Omega'_j=\Omega_{j-1}\cup\Omega_j\cup\Omega_{j+1},\qquad \Omega''_j=\Omega'_{j-1}\cup\Omega'_j\cup\Omega'_{j+1}.$$

Assume for simplicity in writing that $\Omega=\overline{\bigcup_{j=1}^{\infty}\Omega_j}$. Let J_* be such that $2^{-J_*}\cong C_*h$ with C_* sufficiently large, to be determined, and let $\Omega_*=\{x\in\Omega: |x-x_0|\leqslant 2^{-J_*}\}$ so that $\Omega=(\bigcup_{j=1}^{J_*}\bar{\Omega}_j)\cup\Omega_*$. Set further $d_j=2^{-j}$ and $d_*=2^{-J_*}$, and

$$\sum{}^*\|v\|_{k,p,\Omega_j}=\sum_{j=1}^{J_*}\|v\|_{k,p,\Omega_j}+\|v\|_{k,p,\Omega_*}.\tag{21.14}$$

Returning now to (21.12) we estimate the three terms on the right. For the first we have by use of Cauchy–Schwarz' inequality, and Lemma 19.1, (19.8) and Lemma 19.2, (19.11),

$$\begin{aligned}
\varepsilon^2h^{-1}\|v\|_{2,1}&=\varepsilon^2h^{-1}\sum{}^*\|v\|_{2,1,\Omega_j}\\
&\leqslant C\varepsilon^2h^{-1}\sum{}^*d_j\|v\|_{2,2,\Omega_j}\\
&\leqslant C+C\varepsilon^2h^{-1}\sum_{j=1}^{J_*}\|\nabla v\|_{0,\Omega'_j}+Ch^{-1}\sum_{j=1}^{J_*}d_j\|v\|_{0,\Omega'_j}.
\end{aligned}\tag{21.15}$$

Note next that

$$\left(\sum_{j=1}^{J_*} e^{-2cd_j/\varepsilon}\right)^{1/2} \leqslant \ln^{1/2}(C+\varepsilon/h), \tag{21.16}$$

for, since $\exp(-2c2^{-x}/\varepsilon)$ is increasing with x we may estimate the sum inside the square root by $I=\int_1^{J_*+1}\exp(-2c2^{-x}/\varepsilon)\,dx$. Substituting $2^{-x}/\varepsilon=y$, since $2^{-J_*}\cong C_* h$,

$$I \leqslant \log_2 \int_{C_*h/4\varepsilon}^{1/2\varepsilon} e^{-2cy}\frac{dy}{y}$$

and the result readily follows.

By the exponential decay results of Lemma 19.2, (19.10), for C_* large, using also (21.6), (21.16),

$$\begin{aligned}
&\varepsilon^2 h^{-1}\sum_{j=1}^{J_*}\|\nabla v\|_{0,\Omega_j'}\\
&\leqslant C\varepsilon^2 h^{-1}\sum_{j=1}^{J_*} e^{-cd_j/\varepsilon}\|\nabla v\|_{0,\Omega_j'}\\
&\leqslant C\varepsilon^2 h^{-1}\left(\sum_{j=1}^{J_*} e^{-2cd_j/\varepsilon}\right)^{1/2}\|\nabla v\|_0 \leqslant \ln(C+\varepsilon/h).
\end{aligned} \tag{21.17}$$

Introducing the notation

$$\mathscr{S}\equiv h^{-1}\sum\nolimits^* d_j(\|v\|_{0,\Omega_j}+\|v_h\|_{0,\Omega_j}), \tag{21.18}$$

we have from (21.15) and (21.17),

$$\varepsilon^2 h^{-1}\|v\|_{1,2}\leqslant C\varepsilon^2 h^{-1}\sum\nolimits^* d_j\|v\|_{2,2,\Omega_j}\leqslant \ln(C+\varepsilon/h)+C\mathscr{S}. \tag{21.19}$$

For the term $h^{-1}\|v-v_h\|_1$ in (21.12) it is immediately bounded by $C\mathscr{S}$. Hence, by (21.12) and (21.19),

$$\begin{aligned}
&|h^{-1}(u-u_h\,\phi)|\\
&\leqslant C\|u\|_{0,\infty}(\ln(C+\varepsilon/h)+\mathscr{S}+\varepsilon^2h^{-2}\|v-v_h\|_{1,1}).
\end{aligned} \tag{21.20}$$

For the last term on the right we use the local energy norm error estimate of Section 19 and find, for any $\chi\in V_h$,

$$\begin{aligned}
\varepsilon^2h^{-2}\|v-v_h\|_{1,1} &= \varepsilon^2h^{-2}\sum\nolimits^*\|v-v_h\|_{1,1,\Omega_j}\\
&\leqslant C\varepsilon^2h^{-2}\sum\nolimits^* d_j\|v-v_h\|_{1,2,\Omega_j}\\
&\leqslant C\varepsilon^2h^{-2}\sum\nolimits^*(d_j\|v-\chi\|_{1,2,\Omega_j'}+\|v-\chi\|_{0,\Omega_j'})\\
&\quad + C\varepsilon^2h^{-2}\sum\nolimits^* e^{-cd_j/(\varepsilon+h)}\|v-v_h\|_{0,\Omega_j'}\\
&\equiv \textstyle\sum_1^*+\sum_2^*.
\end{aligned} \tag{21.21}$$

Here, by local approximation theory and by (21.19),

$$\textstyle\sum_1^* \leqslant C\varepsilon^2 h^{-1} \sum^* d_j \|v\|_{2,2,\Omega_j''} \leqslant \ln(C+\varepsilon/h) + C\mathscr{S}. \tag{21.22}$$

Further, by Lemma 20.2, (20.19) and Lemma 19.1, (19.8),

$$\begin{aligned}\textstyle\sum_2^* &\leqslant C\varepsilon^2 h^{-2}\Big(\sum^* \mathrm{e}^{-2cd_j/(\varepsilon+h)}\Big)^{1/2} \|v-v_h\|_0 \\ &\leqslant C\varepsilon^2 \ln^{1/2}(C+(\varepsilon+h)/h)\|v\|_{2,2} \\ &\leqslant C \ln^{1/2}(C+\varepsilon/h).\end{aligned} \tag{21.23}$$

Inserting (21.22) and (21.23) into (21.21), and then the result into (21.20),

$$|h^{-1}(u-u_h, \phi)| \leqslant C\|u\|_{0,\infty}(\ln(C+\varepsilon/h)+\mathscr{S}). \tag{21.24}$$

It remains to prove that

$$\mathscr{S} \equiv h^{-1}\textstyle\sum^* d_j(\|v\|_{0,\Omega_j} + \|v_h\|_{0,\Omega_j}) \leqslant C. \tag{21.25}$$

Admitting this for the moment,

$$|h^{-1}(u-u_h, \phi)| \leqslant C\|u\|_{0,\infty} \ln(C+\varepsilon/h)$$

and so

$$\|u-u_h\|_{0,\infty} \leqslant \ln(C+\varepsilon/h)\|u\|_{0,\infty}$$

and the theorem would follow upon adding and subtracting $\chi \in V_h$.

To verify (21.25) proceed as follows. Since, by (21.5), $\|v\| + \|v_h\| \leqslant C$ it suffices to estimate

$$h^{-1} \sum_{j=1}^{J_*} d_j(\|v\|_{0,\Omega_j} + \|v_h\|_{0,\Omega_j}).$$

For C_* large enough we invoke the exponential decay results of Lemmas 19.2 and 20.1. Thus,

$$\begin{aligned}\mathscr{S} &\leqslant C + Ch^{-1}\sum_{j=1}^{J_*} d_j \mathrm{e}^{-cd_j/\varepsilon}\|v\|_{0,\Omega_j'} + Ch^{-1}\sum_{j=1}^{J_*} d_j \mathrm{e}^{-cd_j/(\varepsilon+h)}\|v_h\|_{0,\Omega_j'} \\ &\leqslant C + Ch^{-1}\varepsilon\bigg(\sum_{j=1}^{J_*}\Big(\frac{d_j}{\varepsilon}\Big)^2 \mathrm{e}^{-2cd_j/\varepsilon}\bigg)^{1/2}\|v\|_0 \\ &\quad + Ch^{-1}(\varepsilon+h)\bigg(\sum_{j=1}^{J_*}\Big(\frac{d_j}{\varepsilon+h}\Big)^2 \mathrm{e}^{-2cd_j/(\varepsilon+h)}\bigg)^{1/2}\|v_h\|_0 \\ &\leqslant C + Ch^{-1}\varepsilon\|v\|_0 + Ch^{-1}(\varepsilon+h)\|v_h\|_0.\end{aligned} \tag{21.26}$$

Applying (21.5) and separating the cases $\varepsilon \geqslant h$ and $\varepsilon \leqslant h$ we deduce (21.25).

This completes the proof of Theorem 21.1. □

Note that the global estimate of Theorem 21.1 is of no use whatsoever if $\varepsilon \ll h$ and boundary layers are present; it then merely says that u_h is bounded! Hence, in this case, it is imperative to localize the result.

For the result to be stated in Theorem 21.2, we consider a disc D of radius d around a point x_0, and, if abutting on the boundary, its intersection with Ω. We assume that the mesh family is quasi-uniform on the whole of Ω, but, as is easily seen, it suffices that a mesh family quasi-uniform on D can be *extended* to a mesh family quasi-uniform on Ω. (However, this problem has not been given much thought in the literature.) Likewise, we assume that on the whole of Ω,

$$0<a_0\leqslant a(x)\leqslant a_1, \quad x\in\Omega; \tag{21.27}$$

if this holds for $x\in D$ it is quite easy to extend $a(x)$ "virtually" to the whole of Ω so that (21.27) is satisfied. Thus, under various circumstances, Theorem 21.2 can be further localized.

THEOREM 21.2 (A local pointwise error estimate). *Let the mesh family be quasi-uniform on the convex polygonal domain Ω, and let $0<a_0\leqslant a(x)\leqslant a_1$, $x\in\Omega$. There exist positive constants c_1, c_2 and C such that the following holds. Let u be a continuous function with $u=0$ on $\partial\Omega$. Let $u_h\in V_h$ be such that*

$$A_\varepsilon(u-u_h,\chi)=0 \quad \textit{for } \chi\in V_h^{<}(\Omega_d) \tag{21.28}$$

where $\Omega_d=D\cap\Omega$ with $d\geqslant c_1h$. Then

$$\begin{aligned}|(u-u_h)(x_0)|\leqslant C\ln^{1/2}(d/h)\{&\ln(C+\varepsilon/h)\min_{\chi\in V_h}\|u-\chi\|_{0,\infty,\Omega_d}\\ &+d^{-1}\mathrm{e}^{-c_2d/(\varepsilon+h)}\|u-u_h\|_{0,\Omega_d}\}.\end{aligned} \tag{21.29}$$

PROOF. Let D'' and D' be discs concentric with D of radii $\frac{1}{4}d$ and $\frac{1}{2}d$, respectively, and set $\Omega''=\Omega\cap D''$, $\Omega'=\Omega\cap D'$. Let $\omega\in\mathscr{C}_0^\infty(D)$ be such that

$$\omega\equiv1 \quad \text{on } D', \qquad \|\omega\|_{0,\infty,D}\leqslant1. \tag{21.30}$$

Set $\tilde{u}=\omega u$ and $\tilde{u}_h=P_h^\varepsilon(\omega u)$. Then

$$|(u-u_h)(x_0)|\leqslant|(\tilde{u}-\tilde{u}_h)(x_0)|+|(\tilde{u}_h-u_h)(x_0)|. \tag{21.31}$$

Here by Theorem 21.1,

$$\begin{aligned}|(\tilde{u}-\tilde{u}_h)(x_0)|&\leqslant\ln(C+\varepsilon/h)\|\tilde{u}\|_{0,\infty}\\ &\leqslant\ln(C+\varepsilon/h)\|u\|_{0,\infty,\Omega_d}.\end{aligned} \tag{21.32}$$

For the function $\tilde{u}_h-u_h$ in V_h we have by (21.30),

$$A_\varepsilon(\tilde{u}_h-u_h,\chi)=0 \quad \text{for } \chi\in V_h^{<}(\Omega'). \tag{21.33}$$

By Sobolev's inequality in V_h,

$$\begin{aligned}&|(\tilde{u}_h-u_h)(x_0)|\\ &\quad\leqslant C\ln^{1/2}(d/h)\{\|\tilde{u}_h-u_h\|_{1,\Omega''}+d^{-1}\|\tilde{u}_h-u_h\|_{0,\Omega''}\}.\end{aligned} \tag{21.34}$$

Hence by Lemma 20.1, and by (21.30),

$$\begin{aligned}
&|(\tilde{u}_h - u_h)(x_0)| \\
&\quad \leqslant C \ln^{1/2}(d/h) d^{-1} \{ \mathrm{e}^{-c_2 d/(\varepsilon + h)} \|\tilde{u}_h - u_h\|_{0,\Omega'} \} \\
&\quad \leqslant C \ln^{1/2}(d/h) d^{-1} \|\tilde{u}_h - \tilde{u}\|_{0,\Omega'} \\
&\qquad + C \ln^{1/2}(d/h) d^{-1} \mathrm{e}^{-c_2 d/(\varepsilon + h)} \|u - u_h\|_{0,\Omega'}. \qquad (21.35)
\end{aligned}$$

By Theorem 20.1 and Cauchy–Schwarz' inequality,

$$\begin{aligned}
&C \ln^{1/2}(d/h) d^{-1} \|\tilde{u}_h - \tilde{u}\|_{0,\Omega'} \\
&\quad \leqslant C \ln^{1/2}(d/h) \|\tilde{u}_h - \tilde{u}\|_{0,\infty,\Omega'} \\
&\quad \leqslant C \ln(C + \varepsilon/h) \ln^{1/2}(d/h) \|u\|_{0,\infty,\Omega_d}. \qquad (21.36)
\end{aligned}$$

Combining the above we have proven (21.29) with $\chi = 0$. Writing $u - u_h = (u - \chi) - (u - \chi)_h$ completes the proof of Theorem 21.2. □

22. Some concrete estimates using Theorem 21.2

The reader may easily extend the result of Theorem 21.2 to the case when Ω is a convex domain with a smooth boundary, and the approximating domain $\Omega_h \subseteq \Omega$ has a boundary within $O(h^2)$ distance of the boundary of Ω.

In this section we shall consider the case when the nonsmoothness in the homogeneous Dirichlet problem (19.1), (19.2) is occurring only in a boundary layer. Furthermore, we concentrate attention on the case of piecewise linears on a quasi-uniform straight-edged family of triangulations.

Thus, we make the following assumptions:

Local behavior of u. (22.1)

There exist positive constants C and c, independent of ε, such that with

$$d(x) = \operatorname{dist}(x, \partial\Omega), \qquad (22.2)$$

we have

$$\left| \left(\frac{\partial}{\partial x} \right)^{\alpha} u(x) \right| \leqslant C \left(1 + \frac{\mathrm{e}^{-cd(x)/\varepsilon}}{\varepsilon^2} \right) \quad \text{for } |\alpha| \leqslant 2. \qquad (22.3)$$

If data in the problem are smooth enough, this follows from BESJES [1975, Theorem 13].

A priori assumptions. (22.4)

There exists a constant C such that

$$\|f\|_{\infty,\Omega} \leqslant C. \qquad (22.5)$$

Finally, we make an a priori approximation assumption appropriate to the piecewise linear case.

Local approximation. (22.6)

There exist constants c and C such that the following holds. Given a function v with $v=0$ on $\partial\Omega$ there is $\chi\in V_h$ such that for T_i an element,

$$\|v-\chi\|_{\infty,T_i}\leqslant Ch^2\|v\|_{2,\infty,T_i'} \tag{22.7}$$

where $T_i'=(\bigcup T_j)\cap\Omega_h$ with the union taken over elements T_j within a ch distance of T_i.

Note that by the maximum principle, cf. SCHATZ and WAHLBIN [1983, Lemma A.1] for less smooth cases, with (22.5),

$$\|u\|_{0,\infty,\Omega}\leqslant a_0^{-1}\|f\|_{0,\infty,\Omega}\leqslant C. \tag{22.8}$$

Let now $e(x)=u(x)-u_h(x)$.

THEOREM 22.1. *Assume that the local result of Theorem 21.2 holds. Assume furthermore (22.1), (22.4) and (22.6) above. There exist positive constants c_1, c_2, c_3 and C, independent of h and ε, such that the following holds. Recall that*

$$\partial(x)=\operatorname{dist}(x,\partial\Omega). \tag{22.9}$$

(a) *If $\varepsilon\leqslant h$, then*

$$|e(x)|\leqslant C\ln^{1/2}(d/h)h^2 \quad \textit{for } d(x)\geqslant c_2h\ln(1/h), \tag{22.10}$$

and

$$|e(x)|\leqslant C \quad \textit{for } d(x)\leqslant c_2h\ln(1/h). \tag{22.11}$$

If furthermore $\varepsilon\leqslant c_1h\ln(1/h)$, then

$$|e(x)|\leqslant C\ln^{1/2}(d/h)(h^2+\mathrm{e}^{-c_3d/h}) \quad \textit{for } d(x)\geqslant c_2h. \tag{22.12}$$

(b) *If $\varepsilon\geqslant h$, then*

$$|e(x)|\leqslant\ln^{1/2}(d/h)\ln(C+\varepsilon/h)h^2 \quad \textit{for } d(x)\geqslant c_2\varepsilon\ln(1/\varepsilon), \tag{22.13}$$

and

$$|e(x)|\leqslant\ln(C+\varepsilon/h)h^2\varepsilon^{-2} \quad \textit{for any } d(x). \tag{22.14}$$

PROOF. The easier estimates are (22.11) and (22.14). From the global almost best approximation property of Theorem 21.1,

$$|e(x)|\leqslant\ln(C+\varepsilon/h)\min_{\chi\in V_h}\|u-\chi\|_{0,\infty}. \tag{22.15}$$

For (22.11), simply take $\chi\equiv0$ and use (22.8), and that $\varepsilon/h\leqslant1$.

For (22.14), using (22.7) and (22.3),

$$\min_{\chi\in V_h}\|u-\chi\|_{0,\infty}\leqslant Ch^2\|u\|_{2,\infty}\leqslant Ch^2\varepsilon^{-2} \tag{22.16}$$

and the desired results follow.

The remaining results are based on the local estimate (21.29) of Theorem 21.2. Taking there d equalling the present $d(x)=\operatorname{dist}(x,\partial\Omega)$, we have for $d\geqslant c'h$,

$$\begin{aligned}|e(x)|\leqslant C\ln^{1/2}(d/h)\{&\ln(C+\varepsilon/h)\min_{\chi\in V_h}\|u-\chi\|_{0,\infty,\Omega_d}\\&+Cd^{-1}\mathrm{e}^{-c''d/(\varepsilon+h)}\|u-u_h\|_{0,\Omega_d}\},\end{aligned} \tag{22.17}$$

where $\operatorname{dist}(\Omega_d,\partial\Omega)\geqslant\frac{1}{2}d'd$.

Here, by (22.6) and (22.1),

$$\min_{\chi\in V_h}\|u-\chi\|_{0,\infty,\Omega_d}\leqslant Ch^2(1+\mathrm{e}^{-cd/\varepsilon}\varepsilon^{-2})\leqslant C\ln^{1/2}(d/h)h^2 \tag{22.18}$$

under our assumptions on the relative sizes of ε, h and $d=d(x)$. Similarly,

$$d^{-1}\|u-u_h\|_{0,\Omega_d}\leqslant C\|u-u_h\|_{0,\infty}\leqslant\ln(C+\varepsilon/h).$$

This proves (22.10).

Similarly, easy arguments establish (22.12) and (22.13), see SCHATZ and WAHLBIN [1983, Theorem 10.1] for details. □

REMARK 22.1. Finally, we remark that the basic paper SCHATZ and WAHLBIN [1983] comes equipped with various bells and whistles, such as treating nonlinear problems, and giving very sharp estimates for $\|u-u_h\|$ under minimal assumptions on the smoothness of coefficients, and under minimal smoothness or compatibility assumptions on the right-hand side f. Also, Neumann conditions and nonhomogeneous conditions of Dirichlet of Neumann type are treated there. The present chapter is merely an excerpt introducing the reader to the basic ideas.

For one-dimensional results we refer the reader to NIIJIMA [1980a,b] and references therein.

Numerical illustrations can be found in the papers of Niijima and Schatz–Wahlbin. We give one here showing how boundary layer pollution abates. With $\varepsilon=0.001$, let

$$\begin{aligned}&-\varepsilon^2\Delta u+u=0\quad\text{in } 0\leqslant x,y\leqslant 1,\\&u(x,y)=\mathrm{e}^{-x/\varepsilon}+\varepsilon^{-y/\varepsilon}\quad\text{on the boundary,}\end{aligned}$$

which has the obvious exact solution. With triangular linear elements the mesh was essentially the product of a uniform subdivision in the y-direction with $h=0.05$ and a subdivision $\{x_{j+1}\}_0^{20}$ in the x-direction, $x_{j+1}=0.01j+0.002j^2$. Thus the x-mesh was slightly refined towards the origin. There were 361 interior nodes and 800 elements. We display the errors at mesh points in a subregion of the unit square (see Fig. 22.1, reproduced with the kind permission of the American Mathematical Society).

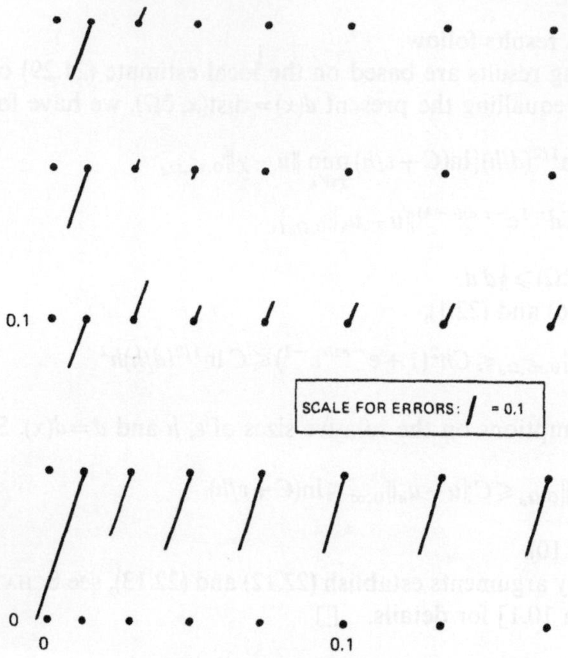

FIG. 22.1.

CHAPTER V

A Singularly Perturbed Convection-Dominated Model Problem

23. The model problem

Consider the problem of finding $u=u(x, y; \delta, \varepsilon)$ such that

$$L_{\delta,\varepsilon} \equiv -\delta u_{xx} - \varepsilon u_{yy} + u_x + u = f \quad \text{in } \Omega, \qquad u=0 \quad \text{on } \partial\Omega, \tag{23.1}$$

where Ω is a plane convex domain.

Clearly this problem has a unique solution for any (reasonable) f and it has a maximum principle so that (cf. as general references ECKHAUS [1972, 1973, 1979], ECKHAUS and DE JAEGER [1966], and LIONS [1973])

$$\|u\|_\infty \leqslant C\|f\|_\infty \tag{23.2}$$

where C does not depend on δ or ε.

We are interested in the case when

$$\varepsilon, \delta > 0 \text{ and small.} \tag{23.3}$$

The formal limiting case is the "hyperbolic" problem of finding u^0 such that

$$u_x^0 + u^0 = f \quad \text{in } \Omega, \qquad u^0 = 0 \quad \text{on } \Gamma_-, \tag{23.4}$$

where Γ_- is the inflow part of the convex boundary $\Gamma = \partial\Omega$ (see Fig. 23.1)

$$\Gamma_- = \text{closure}\{x \in \Gamma : \nu_x < 0\} \tag{23.5}$$

with $\nu = (\nu_x, \nu_y)$ the outward normal.

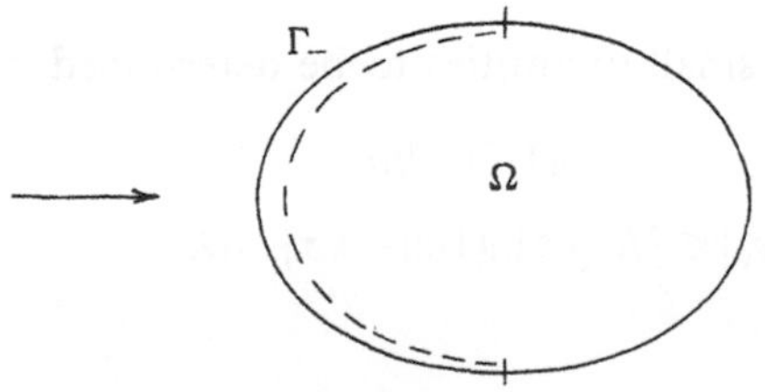

FIG. 23.1.

Clearly, at least as measured in L^∞, the problem is singularly perturbed as ε, δ tend to zero in the sense that

$$\|u^0-u\|_{0,\infty,\Omega}\not\to 0 \quad \text{as } \delta, \varepsilon\to 0^+. \tag{23.6}$$

Under various hypotheses the behavior in boundary layers of u is well described in the general references listed above. In order to highlight a comparison with the finite element situation we shall state and sketch an elementary proof of the following result. For simplicity of proof we assume that δ and ε are not completely independent, but that with some p,

$$\varepsilon\geqslant\delta^p \quad \text{or} \quad \delta\geqslant\varepsilon^p. \tag{23.7}$$

LEMMA 23.1. *Let $s>0$ be any number. There exist constants C and K such that the following holds. For any $(x_0, y_0)\in\Omega$, let*

$$\Omega_0=\{(x, y): x-x_0\leqslant K\delta\ln 1/\delta, |y-y_0|\leqslant K\sqrt{\varepsilon}\ln 1/\varepsilon\}\cap\Omega. \tag{23.8}$$

Then for u the solution of (23.1),

$$|u(x_0, y_0)|\leqslant C\|f\|_{0,\infty,\Omega_0}+C(\delta^s+\varepsilon^s)\|f\|_{0,\Omega}. \tag{23.9}$$

Thus, Lemma 23.1 rather sharply delineates how f influences the solution u at a point. Note that the region Ω_0 extends all the way upwind from (x_0, y_0). We say that the downwind spread is $\delta\ln 1/\delta$ and the crosswind spread $\sqrt{\varepsilon}\ln 1/\varepsilon$ in the continuous problem.

For the proof of Lemma 23.1 we shall use an auxiliary result in L^2. Since the proof is very analogous to the finite element situation, for further future elucidation and comparison, we state this result separately. We denote by L^* the adjoint to L.

LEMMA 23.2. *Let $L^*_{\delta,\varepsilon}v=0$ in $\Omega'_1=\Omega\backslash\Omega'_0$, Ω'_0 as in* (23.8) *with K replaced by $2K$. Then with $\Omega_1=\Omega\backslash\Omega_0\supseteq\Omega'_1$, setting*

$$[[v]]^2_D=\delta\|v_x\|^2_{0,D}+\varepsilon\|v_y\|^2_{0,D}+\|v\|^2_{0,D}, \tag{23.10}$$

we have

$$[[v]]_{\Omega'_1}\leqslant C(\delta^s+\varepsilon^s)[[v]]_{\Omega_1}. \tag{23.11}$$

PROOF. With ρ and σ small quantities to be determined, set

$$\begin{aligned}\Omega^k_0=\{(x, y): x-x_0\leqslant 2K\delta\ln(1/\delta)-k\rho,\\ |y-y_0|\leqslant 2K\sqrt{\varepsilon}\ln(1/\varepsilon)-k\sigma\}\cap\Omega.\end{aligned} \tag{23.12}$$

Let

$$\Omega^k_1=\Omega\backslash\Omega^k_0. \tag{23.13}$$

Note that $\Omega'_1=\Omega^0_1$.

Introduce now suitable cutoff functions to isolate Ω^k_1. In this "rectangular" situation we can find $\psi_k(x, y)$ such that $\psi_k \equiv 1$ on Ω_1^k, $\psi_k \equiv 0$ on Ω_0^{k+1},

$$\left|\frac{\partial}{\partial x}\psi_k\right| \leqslant C\rho^{-1}, \qquad \left|\frac{\partial}{\partial y}\psi_k\right| \leqslant C\sigma^{-1}, \tag{23.14}$$

and such that

$$\frac{\partial}{\partial x}\psi_k \geqslant 0. \tag{23.15}$$

Let further

$$B(v, \varphi) = \delta(v_x, \varphi_x) + \varepsilon(v_y, \varphi_y) + (v_x, \varphi) + (v, \varphi), \tag{23.16}$$

and with $\|v\|_k = \|v\|_{0,\Omega_1^k}$,

$$\mathscr{V}_k = \delta\|v_x\|_k^2 + \varepsilon\|v_y\|_k^2 + \|v\|_k^2. \tag{23.17}$$

Then, by (23.15),

$$\mathscr{V}_k \leqslant \delta(v_x, \psi_k v_x) + \varepsilon(v_y, \psi_k v_y) + (v, \psi_k v) + \tfrac{1}{2}\int_{\Omega_1^{k+1}\setminus\Omega_1^k} v^2(\psi_k)_x. \tag{23.18}$$

Here,

$$\delta(v_x, \psi_k v_x) = \delta(v_x, (\psi_k v)_x) - \delta(v_x, (\psi_k)_x v),$$
$$\varepsilon(v_y, \psi_k\, y) = \varepsilon(v_y, (\psi_k\, v)_y) - \varepsilon(v_y, (\psi_k)_y v),$$

and

$$\tfrac{1}{2}\int v^2(\psi_k)_x = -\int v_x(v\psi_k).$$

Thus,

$$\mathscr{V}_k \leqslant B(v, \psi_k v) - \delta(v_x, (\psi_k)_x v) - \varepsilon(v_y, (\psi_k)_y v)$$

and since $B(v, \psi_k v) = 0$ by assumption,

$$\mathscr{V}_k \leqslant -\delta(v_x, (\psi_k)_x v) - \varepsilon(v_y, (\psi_k)_y v). \tag{23.19}$$

By Cauchy–Schwarz' inequality and the arithmetic-geometric mean inequality, and by (23.14),

$$\begin{aligned}\delta(v_x, (\psi_k)_x v) &\leqslant \delta\|\sqrt{(\psi_k)_x}\, v_x\|_0 \|\sqrt{(\psi_k)_x}\, v\|_0 \\ &\leqslant C\delta\rho^{-1/2}\|v_x\|_{k+1}\|\sqrt{(\psi_k)_x}\, v\|_{0,\Omega_1^{k+1}\setminus\Omega_1^k} \\ &\leqslant \frac{\delta}{\mathrm{e}}\|v_x\|_{k+1}^2 + \frac{C\delta}{\rho}\int_{\Omega_1^{k+1}\setminus\Omega_1^k} (\psi_k)_x v^2.\end{aligned} \tag{23.20}$$

Further, by (23.14),

$$\varepsilon(v_y, (\psi_k)_y v) \leqslant \frac{\varepsilon}{\mathrm{e}}\|v_y\|_{k+1}^2 + C\frac{\varepsilon}{\sigma^2}\|v\|_{k+1}^2. \tag{23.21}$$

Select now

$$C\delta/\rho=\tfrac{1}{2} \tag{23.22}$$

so that the term $\int_{\Omega_1^{k+1}\setminus\Omega_1^k}(\psi_k)_x v^2$ cancels back in (23.18). If further

$$C\varepsilon/\sigma^2=\mathrm{e}^{-1}, \tag{23.23}$$

then it follows that

$$\mathscr{V}_k\leqslant \mathrm{e}^{-1}\mathscr{V}_{k+1}. \tag{23.24}$$

Iterating N times,

$$[[v]]_{\Omega_1'}\leqslant \mathrm{e}^{-N}[[v]]_{\Omega_1^N}. \tag{23.25}$$

Since $\rho\simeq\delta$ and $\sigma\simeq\sqrt{\varepsilon}$, by (23.21) and (23.22), if $K=K(s)$ is large enough we can do this iteration a number of times N which is logarithmic in the minimum of $\ln(1/\delta)$, $\ln(1/\varepsilon)$. Thus, (23.11) follows. □

PROOF OF LEMMA 23.1. Let $f=f_0+f_1$ where f_0 is supported on Ω_0, and correspondingly $u=u_0+u_1$. By the maximum principle,

$$\|u_0\|_{0,\infty}\leqslant C\|f_0\|_{0,\infty}\leqslant C\|f\|_{0,\infty,\Omega_0}. \tag{23.26}$$

For u_1, with G the Green's function centered at (x_0, y_0), so that

$$u_1(x_0,y_0)=B(u_1,G)=(f_1,G), \tag{23.27}$$

since G satisfies the conditions of Lemma 23.2 on the support of f_1, we have

$$|u_1(x_0,y_0)|\leqslant \|f_1\|_{0,\Omega}\|G\|_{0,\Omega_1'}\leqslant C\|f\|_{0,\Omega}(\delta^s+\varepsilon^s)[[G]]_{\Omega}. \tag{23.28}$$

By the simplifying assumption (23.7) and standard elliptic estimates, for some l,

$$[[G]]_{\Omega}\leqslant C\min[\delta^{-l},\varepsilon^{-l}]\|G\|_{0,\Omega}$$

and the desired result would, clearly, by (23.26) through (23.28), follow if we can show that

$$\|G\|_{0,\Omega}\leqslant C\varepsilon^{-1/4}. \tag{23.29}$$

To prove (23.29), let $\tilde{G}$ be the Green's function on the whole of $\mathbb{R}^2$ for $L_{\delta,\varepsilon}$. By the maximum principle , $\tilde{G}\geqslant G\geqslant 0$ so that it suffices to estimate $\|\tilde{G}\|_{\mathbb{R}^2}$. But

$$\mathscr{F}\tilde{G}(\xi,\eta)=(\delta\xi^2+\varepsilon\eta^2-\mathrm{i}\xi+1)^{-1}$$

and using Parseval's formula and elementary estimations the result obtains. (Considering e.g. the first quadrant in (ξ,η), subdivide it into six regions by $\eta\leqslant\varepsilon^{-1/2}$, $\eta\geqslant\varepsilon^{-1/2}$, $\xi\leqslant 1$, $\xi\geqslant 1$, $\xi\leqslant\delta^{-1}$, $\xi\geqslant\delta^{-1}$. In each region, estimate $|\mathscr{F}\tilde{G}(\xi,\eta)|^2$ accordingly as

$$1,\quad \xi^{-2},\quad \delta^{-2}\xi^{-4},\quad \varepsilon^{-2}\eta^{-4},\quad (\xi^2+\varepsilon^2\eta^4)^{-1},\quad (\delta^2\xi^4+\varepsilon^2\eta^4)^{-1}$$

and integrate.)

This proves Lemma 23.1. □

24. The streamline diffusion finite element method

Let V_h denote a typical family of finite element spaces, based on quasi-uniform mesh families and vanishing on the boundary of $\partial\Omega_h$. Our interest is in the singularly perturbed case and we assume that

$$\delta, \varepsilon \leqslant h. \tag{24.1}$$

Hence we do not seek to resolve boundary layers or other singularities but rather aim for a numerical method in which singularities do not pollute into regions where the solution is smooth.

The ordinary Galerkin method, i.e., seeking $u_h \in V_h$ such that

$$\delta((u_h)_x, \chi_x) + \varepsilon((u_h)_y, \chi_y) + ((u_h)_x, \chi) + (u_h, \chi) = (f, \chi) \quad \text{for } \chi \in V_h, \tag{24.2}$$

is well known for its severe pollution. This may be understood already from the one-dimensional analogue,

$$\begin{aligned} &-\delta u_{xx} + u_x + u = f \quad \text{on } I = [0, 1], \\ &u(0) = u(1) = 0, \end{aligned} \tag{24.3}$$

and the case of piecewise linear functions on a uniform mesh. Then the equations for nodal values correspond, essentially, to finite difference equations where u_x has been replaced by a centered difference quotient. Thus, in terms of nodal values u_j, the significant terms on the left are

$$\begin{aligned} &-\delta h^{-1}(u_{j+1} - 2u_j + u_{j-1}) - \tfrac{1}{2}(u_{j+1} - u_{j-1}) \\ &\quad + h(\tfrac{1}{6}u_{j+1} + \tfrac{1}{3}u_j + \tfrac{1}{6}u_{j-1}) = \cdots. \end{aligned} \tag{24.4}$$

The corresponding homogeneous three-term recurrence relation has fundamental solutions of the form r_1^j, r_2^j where

$$r_{1,2} = \frac{\frac{1}{6}h + \delta/h \pm \frac{1}{2}}{\frac{1}{2} - \delta/h - \frac{1}{6}h}. \tag{24.5}$$

If h and δ/h are small, one of the roots is close to -1 and hence contributes a highly oscillatory piece to the solution of (24.4).

In the simple one-dimensional situation, one well-known fixup is that of using upstream differencing for approximating u_x. In the finite element situation, a similar effect, in the one-dimensional problem (24.3), comes about by using a Petrov–Galerkin method, i.e., a method with *different* trial and test spaces. To whit, we seek $u_h \in V_h$ in the same trial space as before but use test functions $\chi + \frac{1}{2}h\chi_x$, $\chi \in V_h$. Thus,

$$\delta(u_h', \chi' + \tfrac{1}{2}h\chi'') + (u_h', \chi + \tfrac{1}{2}h\chi') + (u_h, \chi + \tfrac{1}{2}h\chi') = (f, \chi + \tfrac{1}{2}\chi') \quad \text{for } \chi \in V_h. \tag{24.6}$$

In the one-dimensional piecewise linear situation, the equations for nodal values are now

$$\begin{aligned} &-(\delta + \tfrac{1}{2}h)h^{-1}(u_{j+1} - 2u_j + u_{j-1}) - (1 - \tfrac{1}{2}h)(u_{j+1} - u_{j-1}) \\ &\quad + h(\tfrac{1}{6}u_{j+1} + \tfrac{1}{3}u_j + \tfrac{1}{6}u_{j-1}) = \cdots. \end{aligned} \tag{24.7}$$

(We have discarded the form $\delta h(u_h', \chi_h'')$ which does not make sense. Its formal order is $\delta h \leqslant h^2$.) Rewriting (24.7) to make it look like upwinding,

$$-\delta h^{-1}(u_{j+1}-2u_j+u_{j-1})+(u_{j-1}-u_j) \\ +h(\tfrac{1}{6}+\tfrac{1}{4})u_{j+1}+\tfrac{1}{3}u_j+(\tfrac{1}{6}-\tfrac{1}{4})u_{j-1})=\cdots \tag{24.8}$$

and we recognize the classical upwinding. (In case the test functions were taken as $\chi+\text{const}\cdot h\chi'$, with const >0, we would see that the method gets more information from upstream than from downstream.)

In the one-dimensional situation, note that if we can make sense of the terms $\delta h(u_h', \chi'')$, as is the case for $\mathscr{C}^1$-splines, e.g., we may keep them. Taking testfunctions $\chi+h\chi'$ for simplicity, by rearrangement, the equation analogous to (24.6) becomes

$$(\delta+h)(u_h', \chi')+(1-h)(u_h', \chi)+(u_h, \chi)+\delta h(u_h', \chi'') \\ =(f, \chi+h\chi') \tag{24.9}$$

which exhibits the feature of having diffusion of strength $(\delta+h)$ while being *exactly* satisfied for the real solution to the continuous problem (24.3). In short, we have a method of upwinding which is, at least formally, of *any* desired order of accuracy.

More elaboration on the one-dimensional situation can be found in WAHLBIN [1974a,b], where the rationale for using test functions $\chi+h\chi'$ was to obtain optimal order error estimates in hyperbolic problems, cf. DUPONT [1973]. As for other articles, we mention only CHRISTIE, GRIFFITHS, MITCHELL and ZIENKIEWICZ [1976], and LEONARD [1979]. The literature on numerical solution of one-dimensional convection-dominated flows is enormous.

Now back to the two-dimensional situation of the model equation (23.1). As is well known, the ordinary Galerkin method fails badly in practice as wild oscillations spreading into most of the domain occur when boundary layers are present, cf. the one-dimensional situation described above. A "classical" method of introducing artificial diffusion, or damping, is to find $u_h \in V_h$ such that

$$h((u_h)_x, \chi_x)+h((u_h)_y, \chi_y)+((u_h)_x, \chi)+(u_h, \chi)=(f, \chi) \\ \text{for } \chi \in V_h. \tag{24.10}$$

This method is, in practice, discarded due to excessive damping, in particular in the crosswind direction. Note also that the solution u of (23.1) would satisfy (24.10) only to within $O(h)$ accuracy in general, i.e., the higher diffusion is completely artificial, and the method is at best of first order.

The streamline diffusion method is a finite element method which combines formal high accuracy with decent stability properties. The method was introduced in the case of stationary problems by HUGHES and BROOKS [1979], cf. RAITHBY and TORRANCE [1974] for the finite difference situation.

The idea is to introduce extra diffusion only in the wind direction (= positive x-direction in our model). Following the ideas of the one-dimensional case already described, this can be done without changing the (formally) optimal order of the scheme, namely by using a Petrov–Galerkin method with test functions $\chi+h\chi_x$. Note that, in contrast, any crosswind (y-direction) diffusion has to be artificial.

Thus, in a formal sense, we consider the problem of finding $u_h \in V_h$ such that

$$\begin{aligned}&\delta((u_h)_x, (\chi+h\chi_x)_x)+\varepsilon((u_h)_y, (\chi+h\chi_x)_y)\\&\quad+((u_h)_x, \chi+h\chi_x)+(u_h, \chi+h\chi_x)=(f, \chi+h\chi_x).\end{aligned} \tag{24.11}$$

For simplicity below we shall only consider the case of

$$V_h \text{ piecewise linears on quasi-uniform triangulations.} \tag{24.12}$$

Then the terms $\delta h((u_h)_x, \chi_{xx})$ and $\varepsilon h((u_h)_y, \chi_{xy})$ do not make sense and we shall simply discard them. Furthermore, we shall allow artificial crosswind diffusion so that ε is, possibly, replaced by $\varepsilon_{\mathrm{mod}} \geqslant 0$. Thus, introducing the notation (motivated by some integration by parts in (24.11)),

$$B(v, \varphi) \equiv (\delta+h)(v_x, \varphi_x)+\varepsilon_{\mathrm{mod}}(v_y, \varphi_y)+(1-h)(v_x, \varphi)+(v, \varphi) \tag{24.13}$$

and

$$\tilde{\varphi}=\varphi+h\varphi_x, \tag{24.14}$$

we seek $u_h \in V_h$ such that

$$B(u_h, \chi)=(f, \tilde{\chi}) \quad \text{for } \chi \in V_h. \tag{24.15}$$

It is clear that (24.15) has a unique solution for any f. For, the homogeneous equation enjoys uniqueness, since if $B(u_h, \chi)=0$ we have upon taking $\chi=u_h$,

$$0=B(u_h, u_h)=(\delta+h)\|(u_h)_x\|_0^2+\varepsilon_{\mathrm{mod}}\|(u_h)_y\|_0^2+\|u_h\|_0^2 \tag{24.16}$$

so that $u_h \equiv 0$.

The modified crosswind diffusion $\varepsilon_{\mathrm{mod}}$ will be given as follows in the sequel. Let $0 \leqslant \varepsilon_{\mathrm{co}} \leqslant h$ be a crossover point, typically dependent on h, and set

$$\varepsilon_{\mathrm{mod}}=\begin{cases}\varepsilon_{\mathrm{co}}, & \text{if } \varepsilon<\varepsilon_{\mathrm{co}},\\ \varepsilon, & \text{if } \varepsilon \geqslant \varepsilon_{\mathrm{co}}.\end{cases} \tag{24.17}$$

Thus in the "classical" method (24.10), $\varepsilon_{\mathrm{co}}=h$, while for the "pure" streamline diffusion method, $\varepsilon_{\mathrm{co}}=0$. In our later theoretical development there will be a theoretically optimal $\varepsilon_{\mathrm{co}} \simeq h^{3/2}$, which may, however, not be optimal in computational practice.

We next summarize the deviations made from (23.1) in (24.15). Differentiating (23.1) with respect to x, multiplying by $-h$, adding to (23.1) itself, and integrating some by parts, it is easily found that the solution u of (23.1) satisfies

$$B(u, \varphi)=(f, \tilde{\varphi})+\operatorname{Per}(u, \varphi) \quad \text{for } \varphi \in \mathring{H}^1(\Omega), \tag{24.18}$$

with the perturbation form given as

$$\begin{aligned}\operatorname{Per}(u, \varphi)&=h(\delta u_{xx}+\varepsilon u_{yy}, \varphi_x)+\varepsilon_{\mathrm{per}}(u_{yy}, \varphi)\\&\equiv \operatorname{Per}_1(u, \varphi)+\operatorname{Per}_2(u, \varphi)\end{aligned} \tag{24.19}$$

where

$$\varepsilon_{\text{per}} = \begin{cases} 0 & \text{for } \varepsilon \geqslant \varepsilon_{\text{co}}, \\ -(\varepsilon_{\text{co}} - \varepsilon) & \text{for } \varepsilon < \varepsilon_{\text{co}}. \end{cases} \tag{24.20}$$

Formally, these perturbation terms are of order $h\delta + h\varepsilon + \varepsilon_{\text{per}}$.

We next give general references to the literature on streamline diffusion methods. As already mentioned, the germs are found in RAITHBY and TORRANCE [1974] and WAHLBIN [1974a,b], cf. also DENDY [1974]. The paper by HUGHES and BROOKS [1979] in the stationary situation is seminal. The problem of putting this computationally promising method on a sound mathematical basis was started in JOHNSON and NÄVERT [1981] and continued with extensions to, e.g. time-dependent problems and the zero diffusion limit of hyperbolic problems in NÄVERT [1982], JOHNSON, NÄVERT and PITKÄRANTA [1984], and JOHNSON and SARANEN [1986]. For computational aspects of the method in various situations, see BROOKS and HUGHES [1982] and also the series of papers summarily entitled “A new finite element formulation for computational fluid dynamics”, HUGHES, FRANCA, MALLET and MIZUKAMI (in various combinations) [1986]. In particular, we mention number IV of that series where an interesting nonlinear method for pouring on more streamline diffusion in regions of high gradients (but moderate streamline derivative for u_h so that in the case of downstream layers the numerical solution is clearly amiss) is given.

Our analysis in upcoming sections is based on JOHNSON, SCHATZ and WAHLBIN [1987], with due thanks to predecessors mentioned above. A local analysis for a hybrid-upwind finite element method is contained in RISCH [1986].

In the papers mentioned above, time dependent problems and the limiting hyperbolic case are sometimes treated, with local error estimates. The techniques of analysis are quite similar in the singularly perturbed convection-dominated case and in the pure hyperbolic case. However, in the pure hyperbolic case one also has recourse to “time-stepping” methods such as the discontinuous Galerkin method. We refer to LESAINT and RAVIART [1974] for the inception of the method, to e.g. RICHTER and FALK [1984] for further considerations and references, and to JOHNSON and PITKÄRANTA [1986] for an elucidation of the local behavior in the discontinuous Galerkin method.

Ending this section on a somewhat loose note, I wish to point out the problem of *crosswind* spread in the model problem (23.1) and its numerical analogue (24.15). At present, the question of numerical *downwind* spread of very sharp fronts is answered in a rather satisfactory fashion: It is limited to $O(h \ln 1/h)$. (Naturally, this is an asymptotic result and oscillations may not please the practical eye.) However, the numerical spread *crosswind* of fronts following characteristics (in the limiting zero diffusion hyperbolic case) is virtually unknown; we shall give some theoretical results, optimized by playing around with the artificial crosswind diffusion ε_{mod}. We warn the reader that the theoretical results do not match computational experience. (Of course, the theoretical results are too conservative.)

We end this section by displaying some graphs depicting crosswind spread (see Fig. 24.1). These graphs were computed by William Semper and are included here with his kind permission.

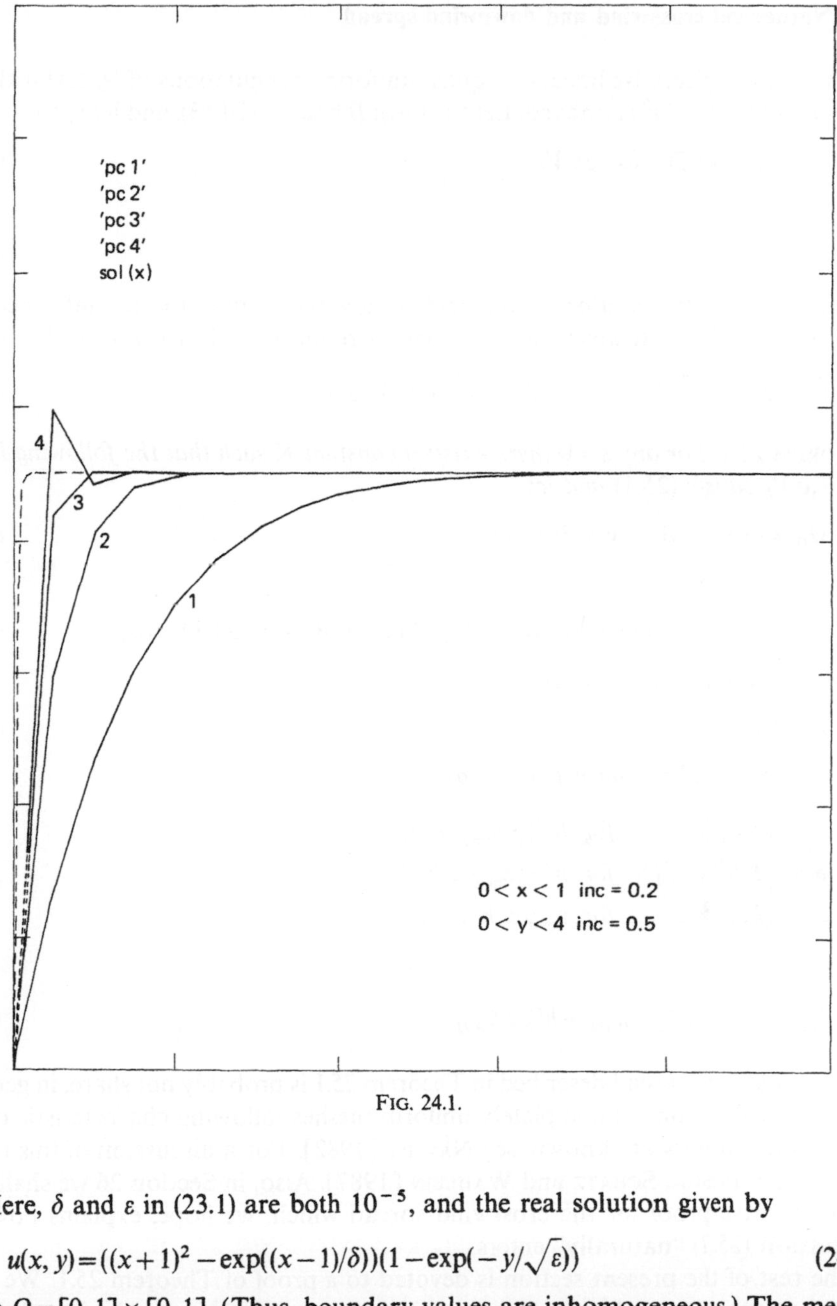

FIG. 24.1.

Here, δ and ε in (23.1) are both 10^{-5}, and the real solution given by

$$u(x, y) = ((x+1)^2 - \exp((x-1)/\delta))(1 - \exp(-y/\sqrt{\varepsilon})) \tag{24.21}$$

with $\Omega = [0, 1] \times [0, 1]$. (Thus, boundary values are inhomogeneous.) The method (24.15) was applied with $h = \frac{1}{20}$ with four-node quadrilaterals on a uniform mesh. A series of ε_{mod} were taken in (24.15), starting with $\varepsilon_{\text{mod}} = \frac{1}{2}h$ and then quartering it in each successive computation. The graphs depict a cut at $x = \frac{1}{2}$, with $u(\frac{1}{2}, y)$ and $u_h(\frac{1}{2}, y)$ sketched versus y for each modified ε_{mod}. (Of course, u is the same.)

25. Numerical crosswind and downwind spread

Let V_h denote piecewise linears on quasi-uniform triangulations of $\Omega_h \subseteq \Omega$ with $\partial\Omega_h$ and $\partial\Omega_h$ at most $O(h^2)$ removed. Let the form B be as in (24.13), and let $u_h \in V_h$ satisfy

$$B(u_h, \chi) = (f, \tilde{\chi}) \quad \text{for } \chi \in V_h, \tag{25.1}$$

where

$$\tilde{\chi} = \chi + h\chi_x. \tag{25.2}$$

The object of this section is to elucidate how the values of u_h are influenced by f, cf. Lemma 23.1. However, we start with a result in L^2-based norms. Let again

$$[[v]]_D = h^{1/2}\|v_x\|_{0,D}^2 + \varepsilon_{\mathrm{mod}}^{1/2}\|v_y\|_{0,D}^2 + \|v\|_{0,D}^2. \tag{25.3}$$

THEOREM 25.1. *For any $s>0$ there exists a constant K such that the following holds. Let $u_h \in V_h$ satisfy* (25.1) *and let*

$$\Omega_0 = \{x \leqslant A,\ B_1 \leqslant y \leqslant B_2\} \cap \Omega, \tag{25.4}$$

and

$$\Omega_0^+ = \{x \leqslant A + \rho\ln(1/h),\ B_1 - \sigma\ln(1/h) \leqslant y \leqslant B_2 + \sigma\ln(1/h)\} \cap \Omega, \tag{25.5}$$

where the downwind spread is

$$\rho = Kh \tag{25.6}$$

and the crosswind spread σ is given by

$$\sigma = \begin{cases} K\varepsilon_{\mathrm{mod}}^{1/2} & \text{for } h^{3/2} \leqslant \varepsilon_{\mathrm{mod}} \leqslant h, \\ Kh^{3/2}\varepsilon_{\mathrm{mod}}^{-1/2} & \text{for } h^2 \leqslant \varepsilon_{\mathrm{mod}} \leqslant h^{3/2}, \\ Kh^{1/2} & \text{for } \varepsilon_{\mathrm{mod}} \leqslant h^2. \end{cases} \tag{25.7}$$

Then

$$[[u_h]]_{\Omega_0} \leqslant K\|f\|_{0,\Omega_0^+} + h^s\|f\|_{0,\Omega}. \tag{25.8}$$

The crosswind spread described in Theorem 25.1 is probably not sharp, in general. In various situations of completely uniform meshes following characteristic directions, better results are known, see NÄVERT [1982]. For a discussion of this point, see again JOHNSON, SCHATZ and WAHLBIN [1987]. Also, in Section 26 we shall give an alternative proof for the crosswind spread which, we hope, explains how the expression (25.7) "naturally" enters.

The rest of the present section is devoted to a proof of Theorem 25.1. We note that the case of $\sigma = Kh^{1/2}$ is contained in NÄVERT [1982], as is the fact that the downwind spread is $\rho = Kh$ in all cases. Thus, the contribution of JOHNSON, SCHATZ and WAHLBIN [1987] is to fine-tune the situation when $h^2 \leqslant \varepsilon_{\mathrm{mod}} \leqslant h$. (This fine-tuning is, however, crucial in order to obtain pointwise estimates that are somewhat realistic.)

For notational simplicity in the proof of Theorem 25.1 we let $u=u_h$, $\varepsilon=\varepsilon_{\text{mod}}$ and replace the form B of (24.13) by

$$B(v,\varphi)=h(v_x,\varphi_x)+\varepsilon(v_y,\varphi_y)+(v_x,\varphi)+(v,\varphi). \tag{25.9}$$

Following JOHNSON and NÄVERT [1981] and NÄVERT [1982] we start by introducing a suitable cutoff function. Let $g(s)\in\mathscr{C}^2(-\infty,\infty)$ with $g(s)=|s|$ for $|s|\geqslant 1$ and set

$$\varphi(t)=\int_t^\infty \exp(-g(s))\,\mathrm{d}s. \tag{25.10}$$

Then, as is easily checked, there exist positive constants c and C such that

$$c\leqslant\varphi(t)\leqslant C \quad \text{for } t\leqslant 1, \tag{25.11}$$

$$\varphi(t)=\mathrm{e}^{-t} \quad \text{for } t>1, \tag{25.12}$$

$$\varphi'(t)\leqslant 0 \quad \text{all } t, \tag{25.13}$$

$$|\varphi'(t)|+|\varphi''(t)|\leqslant C|\varphi(t)| \quad \text{all } t, \tag{25.14}$$

$$|\varphi''(t)|\leqslant -C\varphi'(t) \quad \text{all } t. \tag{25.15}$$

Further, with the relative oscillation on a domain D defined as

$$\mathrm{RO}(D,v)=\max_{x\in D}|v(x)|/\min_{x\in D}|v(x)|, \tag{25.16}$$

on any interval I of length 1,

$$\mathrm{RO}(I,\varphi)+\mathrm{RO}(I,\varphi')\leqslant C. \tag{25.17}$$

Define then

$$\omega(x,y)=\varphi\left(\frac{x-A}{\rho}\right)\varphi\left(\frac{B_1-y}{\sigma}\right)\varphi\left(\frac{y-B_2}{\sigma}\right). \tag{25.18}$$

From the properties (25.11)–(25.17) above it follows that

$$\omega_x\leqslant 0, \tag{25.19}$$

and

$$|D_x^\alpha D_y^\beta\omega|\leqslant\begin{cases} C\rho^{-\alpha}\sigma^{-\beta}\omega & \text{for } \alpha+\beta\leqslant 2, \quad (25.20)\\ -C\rho^{-\alpha+1}\sigma^{-\beta}\omega_x & \text{for } \sigma\geqslant 1,\ \alpha+\beta\leqslant 2, \quad (25.21)\end{cases}$$

and that

$\mathrm{RO}(T,\omega)$ and $\mathrm{RO}(T,\omega_x)$ are bounded independently of h on any element T. (25.22)

From the stringently given properties above it is clear that Theorem 25.1 would follow from the following: Let

$$L\equiv\begin{cases}\max((\rho/h)^{1/2},\,h^{3/2}/(\sigma\varepsilon^{1/2}),\,\varepsilon/\sigma^2) & \text{for } h^2\leqslant\varepsilon\leqslant h, \quad (25.23)\\ \max((h/\rho)^{1/2},\,h^{1/2}\sigma^{-1}) & \text{for } \varepsilon\leqslant h^2. \quad (25.24)\end{cases}$$

Further set

$$Q(u) \equiv (h\|\omega u_x\|^2 + \varepsilon\|\omega u_y\|^2 + \|\omega u\|^2 + \|(\omega|\omega_x|)^{1/2}u\|^2)^{1/2}. \tag{25.25}$$

Then for L sufficiently small,

$$Q(u) \leqslant C\|\omega f\|_0. \tag{25.26}$$

In order to show (25.26) we rely on the following customized result on "superapproximation" in the present situation. Let

$$E = \omega^2 u - \mathrm{Int}(\omega^2 u) \tag{25.27}$$

where $u = u_h \in V_h$, ω is as above, and Int denotes the piecewise linear interpolant into V_h.

LEMMA 25.1. *There exists a constant C such that*

$$h\|\omega^{-1}\nabla E\|_0 + \|\omega^{-1}E\|_0 \leqslant Ch^{1/2}LQ(u). \tag{25.28}$$

We postpone the proof of this lemma until the end of this section and proceed with the main business, viz. to prove (25.26). Note first that

$$0 = ((\omega u)_x, \omega u) = (\omega_x u, \omega u) + (u_x, \omega^2 u). \tag{25.29}$$

Hence, since $B(u_h, \chi) = (f, \tilde{\chi})$ with $\chi = \mathrm{Int}(\omega^2 u) \in V_h$,

$$\begin{aligned}
Q^2(u) &\equiv h(\omega u_x, \omega u_x) + \varepsilon(\omega u_y, \omega u_y) + (\omega u, \omega u) - (\omega_x u, \omega y) \\
&= h(u_x, (\omega^2 u)_x - 2\omega\omega_x u) + \varepsilon(u_y, (\omega^2 u)_y - 2\omega\omega_y u) \\
&\quad + (u, \omega^2 u) + (u_x, \omega^2 u) \\
&= B(u, \omega^2 u) - 2h(\omega u_x, \omega_x u) - 2\varepsilon(\omega u_y, \omega_y u) \\
&= B(u, E) - (f, \tilde{E}) + (f, (\omega^2 u)^{\sim}) - 2h(\omega u_x, \omega_x u) - 2\varepsilon(\omega u_y, \omega_y u) \\
&\equiv I_1 + \cdots + I_5.
\end{aligned} \tag{25.30}$$

Here, by Lemma 25.1 and since $\varepsilon \leqslant h$,

$$\begin{aligned}
I_1 &= h(\omega u_x, \omega^{-1}E_x) + \varepsilon(\omega u_y, \omega^{-1}E_y) + (\omega u_x, \omega^{-1}E) + (\omega u, \omega^{-1}E) \\
&\leqslant [\|\omega u_x\|_0 + (\varepsilon/h)\|\omega u_y\|_0 + \|\omega u\|_0] \times [h\|\omega^{-1}\nabla E\|_0 + \|\omega^{-1}E\|_0] \\
&\leqslant [h^{1/2}\|\omega u_x\|_0 + \varepsilon^{1/2}\|\omega u_y\|_0(\varepsilon/h)^{1/2} + h^{1/2}\|\omega u\|_0]CLQ(u) \\
&\leqslant CLQ^2(u).
\end{aligned} \tag{25.31}$$

Again using Lemma 25.1,

$$I_2 \leqslant \|\omega f\|_0[\|\omega^{-1}E\|_0 + h\|\omega^{-1}\nabla E\|_0] \leqslant \|\omega f\|_0^2 + ChL^2Q^2(u). \tag{25.32}$$

Further, by (25.20) and since $\rho \geqslant h$,

$$\begin{aligned}
I_3 &\leqslant \|\omega f\|_0[\|\omega u\|_0 + 2h\|\omega_x u\|_0 + h\|\omega u_x\|_0] \\
&\leqslant \|\omega f\|_0[C\|\omega u\|_0 + h\|\omega u_x\|_0] \leqslant \tfrac{1}{2}Q^2(u) + C\|\omega f\|_0^2.
\end{aligned} \tag{25.33}$$

Next, by (25.21),

$$\begin{aligned} I_4 &\leqslant 2h\|\omega u_x\|_0\|\omega_x u\|_0 \leqslant \tfrac{1}{8}h\|\omega u_x\|_0^2 + Ch\|\omega_x u\|_0^2 \\ &\leqslant \tfrac{1}{8}h\|\omega u_x\|_0^2 + C(h/\rho)^{1/2}\|(\omega|\omega_x|)^{1/2}u\|_0^2 \\ &\leqslant (\tfrac{1}{8}+CL)Q^2(u). \end{aligned} \tag{25.34}$$

Finally, again using (25.20),

$$\begin{aligned} I_5 &\leqslant C\varepsilon\sigma^{-1}\|\omega u_y\|_0\|\omega u\|_0 \\ &\leqslant \tfrac{1}{8}\varepsilon\|\omega u_y\|_0^2 + C\varepsilon\sigma^{-2}\|\omega u\|_0^2 \leqslant (\tfrac{1}{8}+CL)Q^2(u). \end{aligned} \tag{25.35}$$

It is now clear that (25.26) will follow if L is small enough, and, as we have said, so would then Theorem 25.1 with K large enough.

It remains to prove Lemma 25.1.

PROOF OF LEMMA 25.1. On any triangle T we have by well known approximation theory,

$$h\|\nabla E\|_{0,T} + \|E\|_{0,T} \leqslant Ch^2 \sum_{|\gamma|=2} \|D^\gamma(\omega^2 u)\|_{0,T}. \tag{25.36}$$

Since $D^2 u = 0$ it follows by use of (25.20) that

$$\begin{aligned} |D_y^2(\omega^2 u)| &\leqslant |(\omega^2)_{yy}u| + 2|(\omega^2)_y u_y| \\ &\leqslant C\sigma^{-2}|\omega^2 u| + C\sigma^{-1}|\omega^2 u_y|. \end{aligned} \tag{25.37}$$

For the second mixed derivative,

$$D_x D_y(\omega^2 u) = 2\omega_x\omega_y u + 2\omega\omega_{xy}u + 2\omega\omega_x u_y + 2\omega\omega_y u_x, \tag{25.38}$$

we use (22.20) and (22.21) to arrive at

$$|D_x D_y(\omega^2 u)| \leqslant C\sigma^{-1}|\omega\omega_x u| + C|\omega\omega_x u_y| + C\sigma^{-1}|\omega^2 u_x|. \tag{25.39}$$

Similarly,

$$|D_x(\omega^2 u)| \leqslant C\rho^{-1}|\omega\omega_x u| + C|\omega\omega_x u_x|. \tag{25.40}$$

Inserting the above in (25.36) and employing (25.22),

$$\begin{aligned} &h\|\omega^{-1}\nabla E\|_{0,T} + \|\omega^{-1}E\|_{0,T} \\ &\quad\leqslant C[h^2\sigma^{-2}\|\omega u\|_{0,T} + h^2\sigma^{-1}\|\omega u_y\|_{0,T} + h^2\sigma^{-1}\|\omega_x u\|_{0,T} \\ &\qquad + h^2\|\omega_x u_y\|_{0,T} + h^2\sigma^{-1}\|\omega u_x\|_{0,T} + h^2\rho^{-1}\|\omega_x u\|_{0,T} \\ &\qquad + h^2\|\omega_x u_x\|_{0,T}]. \end{aligned} \tag{25.41}$$

We proceed to operate further on the last five terms. By (25.20) and since $\sigma \geqslant h$,

$$\begin{aligned} h^2\sigma^{-1}\|\omega_x u\|_{0,T} &\leqslant Ch^2\delta^{-1}\rho^{-1/2}\|(\omega|\omega_x|)^{1/2}u\|_{0,T} \\ &\leqslant Ch\rho^{-1/2}\|(\omega|\omega_x|)^{1/2}u\|_{0,T}. \end{aligned} \tag{25.42}$$

By inverse properties, and as above,

$$h^2\|\omega_y u_y\|_{0,T} \leqslant Ch\|\omega_x u\|_{0,T} \leqslant Ch\rho^{-1/2}\|(\omega|\omega_x|)^{1/2}u\|_{0,T}. \tag{25.43}$$

Further, since $\sigma \geqslant \rho$,

$$h^2 \sigma^{-1} \| \omega u_x \|_{0,T} \leqslant C h^2 \rho^{-1} \| \omega^2 u_x \|_{0,T}. \tag{25.44}$$

Finally, as for the first term treated above,

$$h^2 \rho^{-1} \| \omega_x u \|_{0,T} \leqslant C h \rho^{-1/2} \| (\omega |\omega_x|)^{1/2} u \|_{0,T} \tag{25.45}$$

and by (25.20),

$$h^2 \| \omega_x u_x \|_{0,T} \leqslant C h^2 \rho^{-1} \| \omega u_x \|_{0,T}. \tag{25.46}$$

Thus, and we consider now the case $\varepsilon \geqslant h^2$, by (25.41) and the above,

$$\begin{aligned} h \| \omega^{-1} \nabla E \|_{0,T} &+ \| \omega^{-1} E \|_{0,T} \\ &\leqslant C[h^2 \rho^{-1} \| \omega u_x \|_{0,T} + h^2 \sigma^{-1} \| \omega u_y \|_{0,T} + h^2 \sigma^{-2} \| \omega u \|_{0,T} \\ &\quad + h \rho^{-1/2} \| (\omega |\omega_x|)^{1/2} u \|_{0,T}] \\ &\leqslant C h^{1/2} [(h/\rho) h^{1/2} \| \omega u_x \|_{0,T} + (h^{3/2} / \sigma \varepsilon^{1/2}) \varepsilon^{1/2} \| \omega u_y \|_{0,T} \\ &\quad + (h^{3/2} / \sigma^2) \| \omega u \|_{0,T} + (h/\rho)^{1/2} \| (\omega |\omega_x|)^{1/2} u \|_{0,T}]. \end{aligned} \tag{25.47}$$

Since $L \geqslant h^{3/2} \sigma^{-2}$, in the case of $\varepsilon \geqslant h^2$, we obtain Lemma 25.1 upon squaring and summing over all elements.

In the case of $\varepsilon \leqslant h^2$, we replace the second term on the right of (25.41) via the inverse property,

$$h^2 \sigma^{-1} \| \omega u_y \|_{0,T} \leqslant C h \sigma^{-1} \| \omega u \|_{0,T} \tag{25.48}$$

and continue as before. □

26. The numerical crosswind spread: An alternative proof

In this section we will give another proof of the numerical crosswind spread result in Theorem 25.1. That theorem suffers from two deficiencies: The proof is rather hard to see through, in particular, the customized superapproximation property of Lemma 25.1 obscures how the expressions for the crosswind spread in (25.7) were arrived at. Secondly, in (25.8), the global term $h^s \| f \|_{\Omega}$, although multiplied by an arbitrary power h^s, may not be easy to estimate in very rough flows.

On the other hand, there is a reason why we presented the proof of Theorem 25.1: It is short, comparatively speaking.

Thus, in the present section, we shall only consider the question of crosswind spread and shall do so only if $f \equiv 0$ on the relevant domain. This brings the more elucidating and localized proof within manageable proportions.

Again we consider the notationally simplified situation of $u = u_h \in V_h$, $\varepsilon = \varepsilon_{\text{mod}}$, and

$$B(u, \chi) \equiv h(u_x, \chi_x) + \varepsilon(u_y, \chi_y) + (u_x, \chi) + (u, \chi). \tag{26.1}$$

As remarked above, for pedagogical reasons we present the following result elucidating the role of the crosswind spread parameters in (25.7).

THEOREM 26.1. *For any $s>0$ there exists a constant K such that the following holds. Let*

$$\Omega_0=\{B_1\leqslant y\leqslant B_2\}\cap\Omega \tag{26.2}$$

and

$$\Omega_0^+=\{B_1-\sigma\ln(1/h)\leqslant y\leqslant B_2+\sigma\ln(1/h)\}\cap\Omega, \tag{26.3}$$

where σ is as in (25.7).

Let $u\in V_h$ be such that

$$B(u,\chi)=0 \text{ for } \chi\in V_h \text{ with support in } \Omega_0^+. \tag{26.4}$$

Then

$$[[u]]_{\Omega_0}\leqslant Kh^s[[u]]_{\Omega_0^+}. \tag{26.5}$$

REMARK 26.1. The techniques employed give, when combined with a downwind spread analysis, that for u and Ω_0^+ as in Theorem 25.1,

$$[[u]]_{\Omega_0}\leqslant K\|f\|_{0,\Omega_0^+}+h^s\|u\|_{0,\Omega_0^+}.$$

The details for doing the complete analysis of this are lengthy, but not hard.

PROOF OF THEOREM 26.1. Let $\omega\equiv1$ on Ω_0, while $\omega(y)$ depends only on y, is supported within a σ-layer, and

$$|\omega^{(j)}(y)|\leqslant C\sigma^{-j},\quad j=0,1,2. \tag{26.6}$$

Then, since $((\omega u)_x,\omega u)=0$ in our case,

$$\begin{aligned}
[[u]]_{\Omega_0}&=h\|u_x\|^2_{0,\Omega_0}+\varepsilon\|u_y\|^2_{0,\Omega_0}+\|u\|^2_{0,\Omega_0}\\
&\leqslant h(\omega u_x,\omega u_x)+\varepsilon(\omega u_y,\omega u_y)+(\omega u,\omega u)+((\omega u)_x,\omega u)\\
&=h(u_x,(\omega^2u)_x)+\varepsilon(u_y,(\omega^2u)_y)+(u_x,\omega^2u)\\
&\quad+(u,\omega^2u)-2\varepsilon(u_y,\omega\omega_yu)\\
&=B(u,\omega^2u)-2\varepsilon(u_y,\omega\omega_yu)\\
&=B(u,\omega^2u-\chi)-2\varepsilon(u_y,\omega\omega_yu).
\end{aligned} \tag{26.7}$$

Thus, taking $\chi=\mathrm{Int}(\omega^2u)$, with Ω_0' the support of ω,

$$\begin{aligned}
[[u]]_{\Omega_0}&\leqslant C(h\|u_x\|_{0,\Omega_0'}+\varepsilon\|u_y\|_{0,\Omega_0'}+h\|u\|_{0,\Omega_0'})\\
&\quad\cdot h(\sigma^{-2}\|u\|_{0,\Omega_0'}+\sigma^{-1}\|u_x\|_{0,\Omega_0'}+\sigma^{-1}\|u_y\|_{0,\Omega_0'})\\
&\quad+|2\varepsilon(u_y,\omega\omega_yu)|.
\end{aligned} \tag{26.8}$$

We consider now the case of $h^2\leqslant\varepsilon\leqslant h$; in the case of $\varepsilon\leqslant h^2$ the y-derivatives of u occurring on the right of (26.8) are all replaced by u via inverse properties. (We leave the easy argument to the reader.)

Multiply out the parentheses in (26.8) to obtain ten terms to estimate. We retire them in order.

$$Ch\|u_x\|_{0,\Omega_0'}\cdot\frac{h}{\sigma^2}\|u\|_{0,\Omega_0'}\leqslant\frac{e^{-1}}{64}h\|u_x\|^2_{0,\Omega_0'}+C\frac{h^3}{\sigma^4}\|u\|^2_{0,\Omega_0'}. \tag{26.9}$$

$$Ch\|u_x\|_{0,\Omega_0'}\frac{h}{\sigma}\|u_x\|_{0,\Omega_0'}\leqslant C\frac{h}{\sigma}h\|u_x\|^2_{0,\Omega_0'}. \tag{26.10}$$

$$Ch\|u_x\|_{0,\Omega_0'}\frac{h}{\sigma}\|u_y\|_{0,\Omega_0'}\leqslant\frac{e^{-1}}{64}h\|u_x\|^2_{0,\Omega_0'}+\frac{Ch^3}{\sigma^2\varepsilon}(\varepsilon\|u_y\|^2_{0,\Omega_0'}). \tag{26.11}$$

$$C\varepsilon\|u_y\|_{0,\Omega_0'}\frac{h}{\sigma^2}\|u\|_{0,\Omega_0'}\leqslant\frac{e^{-1}}{64}\varepsilon\|u_y\|^2_{0,\Omega_0'}+C\frac{\varepsilon h^2}{\sigma^4}\|u\|^2_{0,\Omega_0'}. \tag{26.12}$$

$$C\varepsilon\|u_y\|_{0,\Omega_0'}\frac{h}{\sigma}\|u_x\|_{0,\Omega_0'}\leqslant\frac{e^{-1}}{64}\varepsilon\|u_y\|^2_{0,\Omega_0'}+\frac{C\varepsilon h}{\sigma^2}(h\|u_x\|^2_{0,\Omega_0'}). \tag{26.13}$$

$$C\varepsilon\|u_y\|_{0,\Omega_0'}\frac{h}{\sigma}\varepsilon\|u_y\|^2_{0,\Omega_0'}\leqslant C\frac{h}{\sigma}\varepsilon\|u_y\|^2_{0,\Omega_0'}. \tag{26.14}$$

The next three terms are easily combined as (assuming as we may that Ω_0' is a mesh domain so we can use inverse properties),

$$Ch^2\|u\|_{0,\Omega_0'}\left(\frac{1}{\sigma^2}\|u\|_{0,\Omega_0'}+\frac{1}{\sigma}\|u_x\|_{0,\Omega_0'}+\frac{1}{\sigma}\|u_y\|_{0,\Omega_0'}\right)$$
$$\leqslant C\frac{h}{\sigma}\|u\|^2_{0,\Omega_0'}. \tag{26.15}$$

Finally,

$$|2\varepsilon(u_y,\omega\omega_y u)|\leqslant\frac{e^{-1}}{64}\varepsilon\|u_y\|^2_{0,\Omega_0'}+\frac{\varepsilon}{\sigma^2}\|u\|^2_{0,\Omega_0'}. \tag{26.16}$$

Employing the above in (26.8), noting that $h^2/\sigma^4\leqslant h/\sigma^2$ and $\varepsilon h/\sigma^2\leqslant\varepsilon/\sigma^2$,

$$[[u]]^2_{\Omega_0}\leqslant[[u]]^2_{\Omega_0'}\cdot\left[\frac{e^{-1}}{8}+\frac{Ch}{\sigma}+\frac{Ch^3}{\sigma^2\varepsilon}+\frac{Ch^3}{\sigma^4}+C\frac{\varepsilon}{\sigma^2}\right]. \tag{26.17}$$

We now desire to choose σ so that (26.17) becomes

$$[[u]]^2_{\Omega_0}\leqslant e^{-1}[[u]]^2_{\Omega_0'}. \tag{26.18}$$

Then the argument could be iterated with Ω_0 replaced by Ω_0', Ω_0' by Ω_0^2, etc., a number N times, where $N=O(\ln 1/h)$. Clearly, (26.5) would follow.

To obtain (26.18), by (26.17) we need

h/σ small (26.19)

$h^3/\sigma^2\varepsilon$ small (26.20)

h^3/σ^4 small (26.21)

ε/σ^2 small, (26.22)

or

$$\sigma \gg /h, \qquad \sigma \gg /h^{3/2}\varepsilon^{-1/2}, \qquad \sigma \gg /h^{3/4}, \qquad \sigma \gg /\varepsilon^{1/2}. \tag{26.23}$$

An elementary investigation of these inequalities leads to the condition (25.7) in the case of $h^2 \leqslant \varepsilon \leqslant h$.

This proves Theorem 26.1 and has, we hope, clarified how the crosswind spread σ was chosen. □

27. Pointwise local error estimates

Compared with the situations in Chapters III and IV, in the convection-dominated case there are no results in general asserting L^∞ stability of the streamline diffusion method (or, any relevant localization of it). Thus, obtaining localized pointwise error estimates will *not* follow the pattern of Chapters III and IV. (Certain first-order accurate methods do enjoy L^∞ stability, cf. e.g. RISCH [1986].)

The investigations of JOHNSON and NÄVERT [1981] and NÄVERT [1982] give localized L^2 estimates of the following type.

Let

$$\Omega_0 = \{x \leqslant A, B_1 \leqslant y \leqslant B_2\} \cap \Omega \tag{27.1}$$

and

$$\Omega_0^+ = \{x \leqslant A + Kh\ln(1/h),\ B_1 - K\sqrt{h}\ln(1/h) \leqslant y \leqslant B_2 + K\sqrt{h}\ln(1/h)\} \cap \Omega. \tag{27.2}$$

Then if the streamline diffusion method is used with piecewise polynomials of degree $r-1$ and if u is smooth on Ω_0^+ (and the terms $h\delta(u_{xx}, \chi_x) + h\varepsilon(u_{yy}, \chi_x)$ are suitably retained in the form B, see the papers mentioned above for details), and $f \in L^2(\Omega)$,

$$\|u - u_h\|_{0,\Omega_0} \leqslant Ch^{r-1/2}. \tag{27.3}$$

In two-space dimensions it would then follow by Sobolev's lemma on V_h that, if u is smooth on Ω_0^+,

$$\|u - u_h\|_{0,\infty,\Omega_0} \leqslant C(\ln 1/h)^{1/2} h^{r-3/2}. \tag{27.4}$$

As already remarked, in the convection-dominated problem we have no known L^∞ stable projections to use. The only theoretical results improving those of (27.4) are those of JOHNSON, SCHATZ and WAHLBIN [1987] which we proceed to describe.

The results are based on a careful investigation of how the approximate Green's function behaves. For this it is essential to have tight control over crosswind spread. Looking back at the expression (25.7) for the crosswind spread, we shall only consider the choice of ε_{mod} that minimizes σ. Thus, choose

$$\varepsilon_{\text{mod}} = \max(\varepsilon, h^{3/2}), \tag{27.5}$$

so that

$$\sigma \leqslant \max(\varepsilon^{1/2}, h^{3/4}). \tag{27.6}$$

Of course, in making the choice (27.5) we have introduced artificial crosswind diffusion when $\varepsilon \leqslant h^{3/2}$ and thus destroyed any formal accuracy properties above $O(h^{3/2})$. We shall only consider the case of V_h being piecewise linears ($r=2$) on a family of quasi-uniform triangulations.

We point out that our results are probably not sharp, but they are the best that can be proven for general mesh families at the present level of technology of proof.

Let (x_0, y_0) be a point of Ω, let $\varepsilon_{co} = h^{3/2}$, cf. (27.5), and set

$$\Omega_0 = \{x \leqslant x_0 + 2Kh \ln(1/h),\ |y - y_0| \leqslant 2K\varepsilon_{\text{mod}}^{1/2} \ln(1/h)\} \cap \Omega, \tag{27.7}$$

where K is as in Theorem 26.1 with $s=6$. Let further

$$\|u\|_{\mathscr{C}^2(\tilde{\Omega}_0)} + \|\delta u_{xx} + \varepsilon u_{yy}\|_{L^1(\Omega)} + \|\nabla u\|_{L^1(\Omega)} + \|f\|_{L^2(\Omega)} \leqslant Q. \tag{27.8}$$

Then:

THEOREM 27.1.

$$|(u - u_h)(x_0, y_0)| \leqslant C h^{5/4} \ln^{3/2}(1/h), \tag{27.9}$$

where C depends on Q of (27.8) *and parameters in the quasi-uniformity of the triangulations.*

REMARK 27.1. Clearly, the result (27.9) misses being of optimal order by a factor of about $h^{-3/4}$. Compared to the L^2 result of $O(h^{3/2})$ it misses being "optimal" by a factor of about $h^{-1/4}$. This missing factor $h^{-1/4}$ will be further elucidated below.

We proceed to prove Theorem 27.1.

PROOF OF THEOREM 27.1. Let $G = G_h^{(x_0, y_0)} \in V_h$ be the discrete Green's function,

$$B(\chi, G) = \chi(x_0, y_0) \quad \text{for } \chi \in V_h. \tag{27.10}$$

With $P_0 u$ the L^2 projection into V_h we have, see (24.18) et seq. for notation,

$$\begin{aligned}(u_h - P_0 u)(x_0, y_0) &= B(u_h - P_0 u, G) = (f, \tilde{G}) - B(P_0 u, G)\\ &= (f, \tilde{G}) + \text{Per}(u, G) - B(P_0 u, G) - \text{Per}(u, G)\\ &= B(u - P_0 u, G) - \text{Per}(u, G).\end{aligned} \tag{27.11}$$

Let now $\Omega_0' \subseteq \Omega_0$ be as in (27.7) with $2K$ replaced by K. We claim that

$$\|G\|_{1,\infty,\Omega \setminus \Omega_0'} \leqslant C h^3. \tag{27.12}$$

To see this let δ_h be a linear function on the element T containing (x_0, y_0) such that $(\delta_h, \chi) = \chi(x_0, y_0)$ for χ linear on T, with δ_h extended to be zero outside T. Then (27.10) amounts to $B(\chi, G) = (\delta_h, \chi)$ for $\chi \in V_h$ so that by the counterpart of Theorem 25.1 (with the wind direction reversed),

$$\|G\|_{0,\Omega\setminus\Omega'_0} \leqslant Ch^6\|\delta_h\|_0. \tag{27.13}$$

Since the dimensions involved are much greater than h and since G vanishes outside Ω_h, we may assume that $\Omega\setminus\Omega'_0$ is a mesh domain. Since $\|\delta_h\| \leqslant Ch^{-1}$, (27.12) follows by inverse properties.

Let now $B_D(v, \varphi)$ denote that the integrations in the form B are extended only over the domain D. Then,

$$B_{\Omega\setminus\Omega'_0}(u-P_0u, G) \leqslant (\|u\|_{1,1}+\|P_0u\|_{1,1})\|G\|_{1,\infty,\Omega\setminus\Omega_0}. \tag{27.14}$$

Since by assumption $\|u\|_{1,1} \leqslant C$ and since (by inverse assumptions)

$$\|P_0u\|_{1,1} \leqslant C\|P_0u\|_{1,2} \leqslant Ch^{-1}\|P_0u\|_0 \leqslant Ch^{-1}\|u\|_0 \leqslant Ch^{-1}, \tag{27.15}$$

we get from (27.12),

$$B_{\Omega\setminus\Omega'_0}(u-P_0u, G) \leqslant Ch^2. \tag{27.16}$$

For the remaining part of B we have

$$\begin{aligned} &B_{\Omega'_0}(u-P_0, G)\\ &\quad= \int\limits_{\Omega'_0\cap\Omega_h} [(h+\delta)(u-P_0u)_xG_x+\varepsilon_{\text{mod}}(u-P_0u)_yG_y-(u-P_0u)G_x\\ &\qquad +(u-P_0u)G]\\ &\quad\leqslant C[h\|\nabla(u-P_0u)\|_{0,\infty,\Omega'_0\cap\Omega_h}+\|u-P_0u\|_{0,\infty,\Omega'_0\cap\Omega_h}]\cdot I \end{aligned} \tag{27.17}$$

where

$$I = \|G_x\|_{L^1(\Omega'_0)}+\varepsilon_{\text{mod}}h^{-1}\|G_y\|_{L^1(\Omega'_0)}+\|G\|_{L^1(\Omega'_0)}. \tag{27.18}$$

Since $\text{meas}(\Omega'_0) \leqslant C\varepsilon_{\text{mod}}^{1/2}\ln(1/h)$ we obtain by Cauchy–Schwarz' inequality (this is the point where it is important to have tight control over crosswind spread!),

$$I \leqslant C\varepsilon_{\text{mod}}^{1/4}\ln^{1/2}(1/h)\mathscr{I}, \tag{27.19}$$

with

$$\mathscr{I} = \|G_x\|_0+\varepsilon_{\text{mod}}h^{-1}\|G_y\|_0+\|G\|_0. \tag{27.20}$$

Hence, using the highly local properties of the L^2 projection from Chapter II,

$$B_{\Omega'_0}(u-P_0u, G) \leqslant Ch^2\varepsilon_{\text{mod}}^{1/2}\ln^{1/2}(1/h)\mathscr{I}. \tag{27.21}$$

We next estimate the perturbation form

$$\text{Per}(u, G) \equiv \text{Per}_1(u, G)+\text{Per}_2(u, G)$$

in (27.11). For the first part we have again using (27.12),

$$\begin{aligned} \text{Per}_1(u, G) &= h(\delta u_{xx}+\varepsilon u_{yy}, G_x)\\ &\leqslant Ch^2\|u\|_{\mathscr{C}^2(\Omega_0)}\|G_x\|_{L^1(\Omega_0)}+Ch\|\delta u_{xx}+\varepsilon u_{yy}\|_{L^1(\Omega)}h^3\\ &\leqslant Ch^2\varepsilon_{\text{mod}}^{1/4}\ln^{1/2}(1/h)\|G_x\|_0+Ch^4 \end{aligned} \tag{27.22}$$

and for the second, after integration by parts over $\Omega\backslash\Omega_0$,

$$\begin{aligned}\mathrm{Per}_2(u, G) &= \varepsilon_{\mathrm{per}}(u_{yy}, G)\\ &= \varepsilon_{\mathrm{per}}\left[(u_{yy}, G)_{\Omega_0} - \int_{\Omega\backslash\Omega_0} u_y G_y + \int_{\partial\Omega_0} u_y G\right]\\ &\leqslant C\varepsilon_{\mathrm{per}}\varepsilon_{\mathrm{mod}}^{1/4}\ln^{1/2}(1/h)\|G\|_0 + C\varepsilon_{\mathrm{per}}h^3. \end{aligned} \tag{27.23}$$

Collecting the above into (27.11),

$$\begin{aligned}&|(u-u_h)(x_0, y_0)|\\ &\leqslant Ch^2\varepsilon_{\mathrm{mod}}^{1/4}\ln^{1/2}(1/h)[\|G_x\|_0 + \varepsilon_{\mathrm{mod}}h^{-1}\|G_y\|_0 + \|G\|_0]\\ &\quad + C\varepsilon_{\mathrm{per}}\varepsilon_{\mathrm{mod}}^{1/4}\ln^{1/2}(1/h)\|G\|_0 + Ch^2. \end{aligned} \tag{27.24}$$

We shall next use the following lemma, the proof of which will be postponed.

LEMMA 27.1.

$$\|G_x\|_0 \leqslant Ch^{-3/4}\varepsilon_{\mathrm{mod}}^{-1/4}\ln(1/h), \tag{27.25}$$

$$\|G_y\|_0 \leqslant Ch^{-1/4}\varepsilon_{\mathrm{mod}}^{-3/4}\ln(1/h), \tag{27.26}$$

$$\|G\|_0 \leqslant Ch^{-1/4}\varepsilon_{\mathrm{mod}}^{-1/4}\ln(1/h). \tag{27.27}$$

Admitting this lemma for the moment, we have from (27.24) that

$$\begin{aligned}&|(u-u_h)(x_0, y_0)|\\ &\leqslant Ch^2\varepsilon_{\mathrm{mod}}^{1/4}\ln^{3/2}(1/h)[h^{-3/4}\varepsilon_{\mathrm{mod}}^{-1/4} + h^{-5/4}\varepsilon_{\mathrm{mod}}^{1/4} + h^{-1/4}\varepsilon_{\mathrm{mod}}^{-1/4}]\\ &\quad + C\varepsilon_{\mathrm{per}}\varepsilon_{\mathrm{mod}}^{1/4}\ln^{3/2}(1/h)h^{-1/4}\varepsilon_{\mathrm{mod}}^{1/4}\\ &\leqslant Ch^{5/4}\ln^{3/2}(1/h), \end{aligned} \tag{27.28}$$

since $h^{-5/4}\varepsilon_{\mathrm{mod}}^{1/4} \leqslant h^{-3/4}\varepsilon_{\mathrm{mod}}^{1/4}$ and since $\varepsilon_{\mathrm{per}} \leqslant h^{3/2}$ in our present case. This would conclude the proof of Theorem 27.1. □

It remains to prove Lemma 27.1, and for this we shall need the following variant of Sobolev's inequality.

LEMMA 27.2. *For* $v \in H_0^1(\Omega)$,

$$\|v\|_{0,p} \leqslant \|v_x\|_{0,2}^{1/2}\|v_y\|_{0,2}^{1/2}\cdot\tfrac{1}{4}p\,(\mathrm{meas}(\Omega))^{1/p}). \tag{27.29}$$

PROOF. Our proof is a minor modification of a standard proof of Sobolev's inequality. Let $w \in \mathscr{C}_0^\infty(\Omega)$. Then

$$|w(x, y)| \leqslant \tfrac{1}{2}\int |w_{x'}(x', y)|\,\mathrm{d}x' \quad \text{or} \quad |w(x, y)| \leqslant \tfrac{1}{2}\int |w_{y'}(x, y')|\,\mathrm{d}y'.$$

Thus,

$$|w(x,y)|^2 \leqslant \tfrac{1}{4}\int |w_{x'}(x',y)|\,\mathrm{d}x' \int |w_{y'}(x,y')|\,\mathrm{d}y'.$$

Integrating (and removing the primes),

$$\iint |w(x,y)|^2\,\mathrm{d}x\,\mathrm{d}y \leqslant \tfrac{1}{4}\left(\iint |w_x(x,y)|\,\mathrm{d}x\,\mathrm{d}y\right)\left(\iint |w_y(x,y)|\,\mathrm{d}x\,\mathrm{d}y\right).$$

Apply this to $w=|v|^p$ so that $Dw=p|v|^{p-1}\operatorname{sgn}(Dv)$. Using also Cauchy–Schwarz' inequality,

$$\iint |v|^{2p} \leqslant \tfrac{1}{4}p^2\left(\iint |v|^{p-1}|v_x|\right)\left(\iint |v|^{p-1}|v_y|\right)$$
$$\leqslant \tfrac{1}{4}p^2\,\|v_x\|_0\|v_y\|_0 \iint |v|^{2(p-1)}.$$

By Hölder's inequality,

$$\iint |v|^{2(p-1)} \leqslant \|v\|_{L^{2p}}^{2(p-1)}\,\mathrm{meas}(\Omega)^{1/p}$$

and hence

$$\|v\|_{L^{2p}}^{2p} \leqslant \|v_x\|_0\|v_y\|_0\|v\|_{L^{2p}}^{2(p-1)}\cdot\tfrac{1}{4}p^2\,\mathrm{meas}(\Omega)^{1/p}$$

or,

$$\|v\|_{L^{2p}}^{2} \leqslant \|v_x\|_0\|v_y\|_0\cdot\tfrac{1}{4}p^2\,\mathrm{meas}(\Omega)^{1/p}.$$

Changing $2p$ to p completes the proof of Lemma 27.2. □

We proceed now to prove Lemma 27.1.

PROOF OF LEMMA 27.1. We have

$$h\|G_x\|_0^2+\varepsilon_{\mathrm{mod}}\|G_y\|_0^2+\|G\|_0^2=B(G,G)=G(x_0,y_0). \tag{27.30}$$

By well-known inverse estimates, and then using Lemma 27.2,

$$G(x_0,y_0)\leqslant Ch^{-2/p}\|G\|_{0,p}\leqslant Ch^{-2/p}p\|G_x\|_0^{1/2}\|G_y\|_0^{1/2}. \tag{27.31}$$

Choosing $p=\ln(1/h)$ and $A=(h/\varepsilon_{\mathrm{mod}})^{1/4}$ below,

$$\begin{aligned} G(x_0,y_0) &\leqslant C\ln(1/h)\|G_x\|_0^{1/2}\|G_y\|_0^{1/2}\\ &\leqslant C\ln(1/h)(A\|G_x\|_0+A^{-1}\|G_y\|_0)\\ &\leqslant C\ln(1/h)(Ah^{-1/2}h^{1/2}\|G_x\|_0+A^{-1}\varepsilon_{\mathrm{mod}}^{-1/2}\varepsilon_{\mathrm{mod}}^{1/2}\|G_y\|_0)\\ &\leqslant \tfrac{1}{2}h\|G_x\|_0^2+\tfrac{1}{2}\varepsilon_{\mathrm{mod}}\|G_y\|_0^2+C\ln^2(1/h)(A^2h^{-1}+\varepsilon^{-1}A^{-2}). \end{aligned} \tag{27.32}$$

Hence, from (27.30),

$$h\|G_x\|_0^2+\varepsilon_{\text{mod}}\|G_y\|_0^2+\|G\|_0^2\leqslant C\ln^2(1/h)(h\varepsilon_{\text{mod}})^{-1/2} \tag{27.33}$$

which proves Lemma 27.1. □

We wish to end this chapter by pointing out the feeble nature of the results above. What we have stated above, for general meshes not uniform and not aligned to the characteristics of the limiting hyperbolic problem, exhausts the present knowledge of crosswind spread. Comparing the results of Section 23 for the continuous problem with those of the discrete problem, one is humbled. Compare e.g. (27.27) with (23.29). Numerical experiments suggest that the theoretical results are way too conservative.

As an example of the sorry state of affairs, let us return to the maximum principle (23.2) in the continuous problem. The best known result for the numerical scheme (for piecewise linears and when $\varepsilon_{\text{mod}}=\min(h^{3/2}, \varepsilon)$) is as follows, displaying a loss of $h^{-1/4}$, essentially.

THEOREM 27.2.

$$\|u_h\|_{0,\infty}\leqslant Ch^{-1/4}\ln^{3/2}(1/h)\|f\|_{0,\infty}. \tag{27.34}$$

PROOF. We have with G the discrete Green's function,

$$|u_h(x_0,y_0)|=|B(u_h,G)|=|(f,\tilde{G})|\leqslant|(f,\tilde{G})_{\Omega_0}|+|(f,\tilde{G})_{\Omega\backslash\Omega_0}|$$

with Ω_0 as in Theorem 27.1. Thus, by inverse estimates, assuming Ω_0 is a mesh domain,

$$|u_h(x_0,y_0)|\leqslant C\|f\|_{0,\infty,\Omega_0}\|G\|_{0,1,\Omega_0}+\|f\|_{0,\infty}\|G\|_{0,1,\Omega\backslash\Omega_0}.$$

We have seen that

$$\|G\|_{0,1,\Omega\backslash\Omega_0}\leqslant Ch^s,$$

while by Cauchy–Schwarz' inequality and Lemma 27.1,

$$\|G\|_{0,1,\Omega_0}\leqslant C\varepsilon_{\text{mod}}^{1/4}\ln^{1/2}(1/h)\|G\|_0\leqslant Ch^{-1/4}\ln^{3/2}(1/h).$$

This proves the theorem. □

Finally, we have already remarked that the present investigations do not follow the typical pattern from Chapters III and IV, in that there is no basic problem having pointwise stability to start from. Thus, perhaps, it is not surprising that in order to give even a weak global L^1 estimate in the presence of singularities one needs to use two *local* tools developed above.

As an example the reader may consider the global L^1 estimate of order almost $O(h^{1/2})$ given under realistic assumptions in JOHNSON, SCHATZ and WAHLBIN [1987, Corollary 3.6].

Note added in proof.

In the paper "Pointwise error estimates for a streamline diffusion finite element scheme" by K. Niijima (Preprint, Kyushu University, Fukuoka, 1988) the author replaces $h^{5/4}$ in (27.9) by $h^{11/8}$. The key is an improvement of the estimates of Lemma 27.1.

CHAPTER VI

Time-Localized Behavior in Second-Order Parabolic Problems

28. A linear model problem: Full advantage of the smoothing effect

We shall consider only the simplest model problem and shall be content with merely giving references to relevant extensions.

Thus, let Ω be a bounded domain in $\mathbb{R}^N$ with smooth boundary and consider the problem of finding $u(x, t)$, $x \in \Omega$, $t \geqslant 0$, such that

$$\begin{aligned} &u_t = \Delta u, && x \in \Omega, \quad t > 0, \\ &u(x, t) = 0, && x \in \partial\Omega, \\ &u(x, 0) = v(x), && x \in \Omega. \end{aligned} \tag{28.1}$$

In the case that v is smooth and compatible enough so that $u(t)$ is sufficiently smooth on the closed interval $\bar{I}$, then for $u_h(t)$ a semidiscrete finite element solution, with r the optimal order of approximation in V_h,

$$\|u(t) - u_h(t)\|_0 \leqslant Ch^r \|v\|_r, \quad t \in \bar{I},$$

for $v \in D((-\Delta)^{r/2})$ which thus requires smoothness and the compatibility conditions $\Delta^j v|_{\partial\Omega} = 0$ for $j < \frac{1}{2}r$.

The object of this section is to elucidate what happens if initial data v are not smooth or compatible enough for the above to hold.

As is well known, even if initial data $v(x)$ is rough, or incompatible, the solution $u(x, t)$ for $t > 0$ is smooth, viz., as smooth as $\partial\Omega$ allows. Assume in the rest of this section that $\partial\Omega$ is infinitely differentiable, for simplicity.

The question now is whether the finite element solution to (28.1) takes advantage of this smoothing property, i.e., whether in the case of rough (or incompatible) initial data v, the error in the Galerkin finite element solution is small for *positive* (and bounded) time.

Our analysis (indicated in a simple case) will be given only for the semidiscrete (continuous in time) finite element solution. The analysis is based on having a method for approximating the associated elliptic problem.

Let T denote the solution operator to the problem

$$-\Delta v = f \quad \text{in } \Omega, \qquad v = 0 \quad \text{on } \partial\Omega \tag{28.2}$$

so that $v=Tf$. Let $\{\lambda_j\}_1^\infty$ and $\{\varphi_j\}_1^\infty$ be the eigenvalues (in nondecreasing order) and L^2 orthonormal eigenfunctions for this problem. We can then introduce the associated $\dot{H}^s=\dot{H}^s(\Omega)$ spaces,

$$\|w\|_s=\left(\sum_{j=1}^{\infty}\lambda_j^s\beta_j^2\right)^{1/2}, \quad \beta_j=(w,\varphi_j). \tag{28.3}$$

Since for the solution of (28.1),

$$u(x,t)=\sum_{j=1}^{\infty}\beta_j e^{-\lambda_j t}\varphi_j(x), \quad \beta_j=(v,\varphi_j), \tag{28.4}$$

an exact statement of the smoothing property is that with $E(t)$ the solution operator to (28.1) so that $u(x,t)=(E(t)v)(x)$,

$$\|E(t)\|_{p,q}\leqslant Ct^{-(p-q)/2}, \quad p\geqslant q. \tag{28.5}$$

We assume now that we have a family of approximations $T_h: L_2\to V_h$ to the inverse of $(-\Delta)$ such that

T_h is self-adjoint, positive-semidefinite on L^2, and positive-definite on V_h, (28.6)

and

there is an integer $r\geqslant 2$ such that

$$\|T_h-T\|_{0,q}\leqslant Ch^{q+2} \quad \text{for } 0\leqslant q\leqslant r-2. \tag{28.7}$$

The integer r is, of course, the optimal L^2 error.

Note that equation (28.1) may be written as

$$D_t Tu+u=0, \qquad u(0)=v. \tag{28.8}$$

Correspondingly, we let the semidiscrete finite element solution be given by

$$D_t T_h u_h+u_h=0, \qquad u_h(0)=v_h \tag{28.9}$$

with $v_h\in V_h$ given. We shall assume that with P_0 the L^2 projection into V_h,

$$v_h=P_0 v. \tag{28.10}$$

We then have the following estimate showing that for positive time the semidiscrete error in approximating (28.1) is of optimal order in the linear problem. Here, $e_h(t)=(u_h-u)(t)$.

THEOREM 28.1. *Assume* (28.1), $\partial\Omega$ *smooth,* (28.6), (28.7) *and* (28.9), (28.10). *There exists a constant* $C(t^*)$ *such that for* $0<t\leqslant t^*$,

$$\|e_h(t)\|_0\leqslant Ch^r t^{-r/2}\|v\|_0. \tag{28.11}$$

Before proving Theorem 28.1 we give references to the literature. The fact that numerical solutions to parabolic problems take advantage of the smoothing property was noticed in the context of finite difference methods in JUNCOSA and YOUNG [1957]. The theory for difference schemes was further developed in KREISS, THOMÉE and WIDLUND [1970], LÖFSTRÖM [1970] and THOMÉE and WAHLBIN [1974], cf. BRENNER, THOMÉE and WAHLBIN [1975, Chapter 4] for a summary.

In the finite element situation Theorem 28.1 is due to HELFRICH [1974], cf. BLAIR [1970] for earlier thoughts in this direction. Our write up (with the formalism of the operators T and T_h) is taken from BRAMBLE, SCHATZ, THOMÉE and WAHLBIN [1977]. Extensions (with identical statement) of Theorem 28.1 to situations with time-dependent coefficients and non-self-adjoint elliptic forms are given in the basic paper of Helfrich and also in HUANG and THOMÉE [1981], LUSKIN and RANNACHER [1982a] and SAMMON [1982]. The extension to nonhomogeneous equations with a right-hand side smooth near the time of interest is in THOMÉE [1980].

The papers referenced above all give error estimates in spatial L^2 norms at a given time. For pointwise error estimates the translation technique of BRAMBLE, SCHATZ, THOMÉE and WAHLBIN [1977, Section 4] is easy to apply also for non-self-adjoint problems where the elliptic operator has time-dependent coefficients. Also, the estimates can be further localized in space, see BRAMBLE, SCHATZ, THOMÉE and WAHLBIN [1977, Section 5], cf. also NITSCHE [1979] and THOMÉE [1979]. Further, and sharper, pointwise error estimates with the smoothing property are given in SCHATZ, THOMÉE and WAHLBIN [1980] and in THOMÉE and WAHLBIN [1983].

For the analogues of Theorem 28.1 with time discretization, see HUANG and THOMÉE [1982], SAMMON [1983], which in particular contains a very thought-provoking observation concerning higher-order schemes, LUSKIN and RANNACHER [1982b], and for pointwise estimates, SCHATZ, THOMÉE and WAHLBIN [1980, Section 4].

An analogous use of a smoothing property in the case of the Euler–Poisson–Darboux equation is found in GENIS [1982].

We remark that in all references above (and in Theorem 28.1) *it is assumed that the L^2 projection of initial data, P_0v, is evaluated exactly. The effect of numerical integration brings the situation close to that of finite differences; this is thoroughly elucidated in* WAHLBIN [1980], *and shows that in order to retain the beneficial effects of the smoothing property one should, if possible, subtract off singularities in initial data v and treat the evaluation of the L^2 projection of those separately and accurately. I.e., in practice one needs a certain amount of data preparation to fully benefit from the parabolic smoothing effect.*

In the same spirit is the counterexample in SAMMON [1983] just referred to. He gives simple model numerical examples with time-dependent coefficients where smooth initial data leads to fourth-order convergence but nonsmooth initial data only gives second-order convergence for positive time. This again indicates that the semidiscrete results in the presence of perfect integration may not always carry over in practice.

Further, the reason why the choice $v_h = P_0v$ in (28.10) is appropriate in order to achieve the smoothing gain is explained in THOMÉE [1972] and WAHLBIN [1981].

We next give the proof of Theorem 28.1, following HELFRICH [1974] and BRAMBLE, SCHATZ, THOMÉE and WAHLBIN [1977].

PROOF OF THEOREM 28.1. Let $\mu_j=\lambda_j^{-1}$ and φ_j denote the eigenvalues and L^2 orthonormalized eigenfunctions of $T=(-\Delta)^{-1}$. Then with $\beta_j=(v,\varphi_j)$ the Fourier coefficients of initial data v,

$$u(t)=\sum_j \mathrm{e}^{-t/\mu_j}\beta_j\varphi_j(x). \tag{28.12}$$

Introducing the resolvent $R_z(T)=(z-T)^{-1}$ we shall use that the L^2 projection onto the eigenspace corresponding to the eigenvalue μ can be expressed as

$$\sum_{\mu_j=\mu}\beta_j\varphi_j(x)=\frac{1}{2\pi\mathrm{i}}\int_{\gamma_\mu}R_x(T)v\,\mathrm{d}z, \tag{28.13}$$

where γ_μ is a curve in the complex plane enclosing only this eigenvalue. It follows that

$$\sum_{j=1}^{J}\mathrm{e}^{-t/\mu_j}\beta_j\varphi_j=\frac{1}{2\pi\mathrm{i}}\int_{\Gamma_J}\mathrm{e}^{-t/z}R_z(T)v\,\mathrm{d}z, \tag{28.14}$$

where Γ_J is a curve enclosing only the eigenvalues $\mu_1,\dots\mu_J$ (for $\mu_{J+1}\neq\mu_J$). Taking limits, for $u(t)=E(t)v$ the solution of (28.1),

$$u(t)=E(t)v=\frac{1}{2\pi\mathrm{i}}\int_{\Gamma}\mathrm{e}^{-t/z}R_z(T)v\,\mathrm{d}z, \tag{28.15}$$

with Γ the positively oriented curve defined by $|\arg z|=\frac{1}{4}\pi$.

We note that since

$$R_z(T)v=\sum_j\frac{1}{(z-\mu_j)}\beta_j\varphi_j,$$

we have for $|\arg z|=\frac{1}{4}\pi$,

$$\|R_z(T)\|_{s,s}\leqslant|\mathrm{Im}\,z|^{-1}\leqslant\sqrt{2}|z|^{-1}, \tag{28.16}$$

and

$$\|R_z(T)\|_{s,s+2}\leqslant\sqrt{2}. \tag{28.17}$$

Corresponding to the eigenvalues and eigenfunctions μ_j and φ_j for T, let $\mu_{j,h}$ and $\varphi_{j,h}$ be those for the approximate elliptic solution operator T_h. Hence, for $u_h(t)$ the solution of the semidiscrete problem (28.9),

$$u_h(t)=\sum_{j=1}^{J_h}\mathrm{e}^{-t/\mu_{j,h}}\beta_{j,h}\varphi_{j,h} \tag{28.18}$$

where $\beta_{j,h}=(v_h,\varphi_{j,h})$. In particular, since $v_h=P_0v$, the L^2 projection of v, we have $\beta_{j,h}=(v,\varphi_{j,k})$ and we obtain as above, with $E_h(t)$ the discrete solution operator,

$$u_h(t)=E_h(t)v=\frac{1}{2\pi \mathrm{i}}\int_\Gamma \mathrm{e}^{-t/z}R_z(T_h)v\,\mathrm{d}z. \tag{28.19}$$

Since T_h is self-adjoint on L^2 we have as in (28.16),

$$\|R_z(T_h)\|_{0,0}\leqslant\sqrt{2}|z|^{-1}. \tag{28.20}$$

Combining (28.15) and (28.19) we have for $e_h(t)=u_h(t)-u(t)$,

$$e_h(t)=F_h(t)v\equiv E_h(t)v-E(t)v \tag{28.21}$$

where the error operator $F_h(t)$ is given as

$$F_h(t)=\frac{1}{2\pi \mathrm{i}}\int_\Gamma \mathrm{e}^{-t/z}(R_z(T_h)-R_z(T))\,\mathrm{d}z. \tag{28.22}$$

We next derive the following lemma.

LEMMA 28.1. *For* $0\leqslant q\leqslant r-2$,

$$\|R_z(T_h)-R_z(T)\|_{0,q}\leqslant Ch^{q+2}|z|^{-2}\quad \textit{for } z\in\Gamma. \tag{28.23}$$

PROOF. As a preliminary step we prove that if T_h satisfies (28.6), (28.7), then

$$\|R_z(T_h)-R(T)\|_{0,s}\leqslant Ch^s|z|^{-1}\quad \text{for } 0\leqslant s\leqslant r. \tag{28.24}$$

To see this we use the resolvent formula to write

$$R_z(T_h)-R_z(T)=R_z(T_h)(T_h-T)R_z(T) \tag{28.25}$$

so that by (28.20), (28.7) and (28.17),

$$\begin{aligned}&\|R_z(T_h)-R_z(T)\|_{0,r}\\&\quad\leqslant\|R_z(T_h)\|_{0,0}\|T_h-T\|_{0,r-2}\|R_z(T)\|_{r-2,r}\\&\quad\leqslant Ch^r|z|^{-1}\end{aligned} \tag{28.26}$$

which is the desired result (28.24) for $s=r$. Also, by (28.20) and (28.16),

$$\begin{aligned}&\|R_z(T_h)-R_z(T)\|_{0,0}\\&\quad\leqslant\|R_z(T_h)\|_{0,0}+\|R_z(T)\|_{0,0}\leqslant C|z|^{-1}.\end{aligned} \tag{28.27}$$

Our spaces defined by eigenfunction expansions are easily seen to be interpolation spaces (BERGH and LÖFSTRÖM [1976]) and thus (28.26) and (28.27) give (28.24) by interpolation.

In order to prove (28.23), we write, again using the resolvent formula,

$$\begin{aligned}&R_z(T_h)-R_z(T)\\&\quad=R_z(T)(T_h-T)R_z(T)+R_z(T)(T_h-T)(R_z(T_h)-R_z(T)).\end{aligned} \tag{28.28}$$

Hence, by (28.16) and the preliminary result (28.24),

$$\begin{aligned}&\| R_z(T_h)-R_z(T)\|_{0,q}\\&\quad\leqslant \| R_z(T)\|_{0,0}\| T_h-T\|_{0,q}\| R_z(T)\|_{q,q}\\&\qquad+\| R_z(T)\|_{0,0}\| T_h-T\|_{0,0}\| R_z(T_h)-R_z(T)\|_{0,q}\\&\quad\leqslant Ch^{q+2}|z|^{-2}\end{aligned}\tag{28.29}$$

which proves the desired result (28.23). □

Continuing with the proof of Theorem 28.1, we have

$$F_h(t)=\frac{1}{2\pi \mathrm{i}}\int_\Gamma \mathrm{e}^{-t/z}(R_z(T_h)-R_z(T))\,\mathrm{d}z$$

so that by the above Lemma 28.1,

$$\| F_h(t)\|_{0,q}\leqslant Ch^{q+2}\int_\Gamma |z|^{-2}\mathrm{e}^{-ct/|z|}\,\mathrm{d}|z|=Ch^{q+2}t^{-1}. \tag{28.30}$$

Writing next

$$F_h(t)=F_h(\tfrac12 t)E(\tfrac12 t)+E(\tfrac12 t)F_h(\tfrac12 t)-(F_h(\tfrac12 t))^2 \tag{28.31}$$

and noting that it suffices to consider $h^2t^{-2}\leqslant 1$, we have for the first term on the right, by (28.30) and (28.5),

$$\| F_h(\tfrac12 t)E(\tfrac12 t)\|_{0,0}\leqslant \| F_h(\tfrac12 t)\|_{0,r-2}\| E(\tfrac12 t)\|_{r-2,0}\leqslant Ch^r t^{-r/2}. \tag{28.32}$$

Since the second term on the right of (28.31) is the adjoint of the one just treated, its L^2 norm is the same. For the last term on the right of (28.31), by (28.30) for $q=0$,

$$\| (F_h(\tfrac12 t))^2\|_{0,0}\leqslant \| F_h(\tfrac12 t)\|_{0,0}^2\leqslant Ch^2t^{-1}\| F_h(\tfrac12 t)\|_{0,0}. \tag{28.33}$$

Thus, from (28.31)–(28.33),

$$\| F_h(t)\|_{0,0}\leqslant Ch^r t^{-r/2}+C(h^2t^{-1})\| F_h(\tfrac12 t)\|_{0,0}. \tag{28.34}$$

By repeated application s times,

$$\| F_h(t)\|_{0,0}\leqslant Ch^r t^{-r/2}+C(h^2t^{-1})^s\| F_h((\tfrac12)^s t)\|_{0,0} \tag{28.35}$$

from which the desired result (28.11) follows if we take $s\geqslant \frac12 r$. □

29. A nonlinear model problem: Restricted advantage of the smoothing effect

In this section we consider the semilinear problem of finding $u=u(x,t)$ such that

$$\begin{aligned}&u_t-\Delta u=f(u)\quad \text{in } \Omega\times I,\qquad I=(0,t^*],\\&u=0\qquad \text{on } \partial\Omega\times I,\\&u(0)=v\quad \text{in } \Omega.\end{aligned}\tag{29.1}$$

Here Ω is a bounded domain in $\mathbb{R}^N$ with a sufficiently smooth boundary. As always when solving a nonlinear problem we want some guarantee that the solution does not blow up in finite time, at least not inside I in our case. We assume thus, cf. SMOLLER [1983, Chapter 14, Section B], that there is an a priori bound B so that

$$|u(x,t)| \leqslant B, \qquad x \in \Omega, \quad t \in I. \tag{29.2}$$

In many situations, following the theory of invariant regions just referred to in Smoller's book, B is a priori computable. Thus we may artificially change $f(y)$ for $|y| > B$ and assume that f is a smooth function for $y \in \mathbb{R}$ with

$$|f(y)|, |f'(y)| \leqslant B \quad \text{for } y \in \mathbb{R}. \tag{29.3}$$

(Certain finite element situations where this a priori change is not necessary are given in THOMÉE and WAHLBIN [1975].)

For spatial discretization of (29.1) let $V_h \subset \mathring{H}^1(\Omega)$ be such that Theorem 28.1 holds in the linear homogeneous case. For simplicity in this outline, assume further that the basic method for the elliptic problem is the ordinary Galerkin method, i.e., that the semidiscrete solution $u_h : \bar{I} \to V_h$ is defined by

$$\begin{aligned} &(u_{h,t}, \chi) + (\nabla u_h, \nabla \chi) = (f(u_h), \chi) \quad \text{for } \chi \in V_h, \\ &u_h(0) = P_0 v = v_h, \end{aligned} \tag{29.4}$$

where $P_0 v$ is the L^2 projection of v into V_h.

In the case that v is smooth and compatible enough so that $u(t)$ is sufficiently smooth on the closed interval $\bar{I}$, then

$$\|u_h(t) - u(t)\|_0 \leqslant C(u, B) h^r \quad \text{for } t \in \bar{I}, \tag{29.5}$$

where r is the optimal order of approximation, cf. WHEELER [1973].

The object of this section is to elucidate the cases when initial data v is not smooth enough, or incompatible, so that (29.5) cannot be guaranteed to hold. We shall merely give an outline of the most surprising results of JOHNSON, LARSSON, THOMÉE and WAHLBIN [1987].

The main result is the following.

THEOREM 29.1. *Let* $|u(x,t)| \leqslant B$, $x \in \Omega$, $t \in \bar{I}$; *let* $v_h = P_0 v$ *and let* u_h *be given by* (29.4). *Then with* C *depending on* B, t^* *and approximation parameters for* V_h,

$$\|u_h(t) - u(t)\|_0 \leqslant C h^2 (t^{-1} + |\log(h^2/t)|), \quad t \in I. \tag{29.6}$$

PROOF. Simple energy arguments show that $u(t)$ and $u_h(t)$ are bounded in L^2 so that (29.6) trivially holds for $t \leqslant h^2$. With $E(t)$ the solution operator to the linear homogeneous problem (28.1) and $E_h(t)$ its semidiscrete counterpart, we have by Duhamel's principle,

$$u(t) = E(t)v + \int_0^t E(t-s) f(u(s)) \, \mathrm{d}s, \tag{29.7}$$

and

$$u_h(t)=E_h(t)v_h+\int_0^t E_h(t-s)P_0 f(u_h(s))\,\mathrm{d}s. \tag{29.8}$$

Hence with $F_h(t)=E_h(t)P_0-E(t)$ the error operator for the linear homogeneous equation, for $e_h(t)=u_h(t)-u(t)$ the error in the semilinear problems,

$$e_h(t)=F_h(t)v+\int_0^t E_h(t-s)[f(u_h(s))-f(u(s))]\,\mathrm{d}s+\int_0^t F_h(t-s)f(u(s))\,\mathrm{d}s. \tag{29.9}$$

From the proof of Theorem 28.1, see (28.30), it is seen that, regardless of the optimal order $r\geqslant 2$,

$$\|F_h(t)v\|_0\leqslant Ch^2t^{-1}\|v\|_0\leqslant CBh^2t^{-1}, \tag{29.10}$$

and

$$\|E_h(t)w\|_0\leqslant C\|w\|_0. \tag{29.11}$$

Thus, since $|f'|\leqslant B$, see (29.3), and since the theorem holds for $t<h^2$ as noted above,

$$\begin{aligned}\|e_h(t)\|_0&\leqslant Ch^2t^{-1}+C\left(\int_0^{h^2}+\int_{h^2}^t\right)\|e_h(s)\|_0\,\mathrm{d}s\\&\quad+\left(\int_0^{t-h^2}+\int_{t-h^2}^t\right)(\|F_h(t-s)f(u(s))\|_0)\,\mathrm{d}s\\&\leqslant Ch^2t^{-1}+Ch^2+C\int_{h^2}^t\|e_h(s)\|_0\,\mathrm{d}s+Ch^2\int_0^{t-h^2}\frac{\mathrm{d}s}{t-s}+Ch^2\\&\leqslant Ch^2t^{-1}+Ch^2\log(t/h^2)+C\int_{h^2}^t\|e_h(s)\|_0\,\mathrm{d}s.\end{aligned} \tag{29.12}$$

Letting $\varphi(t)=\int_{h^2}^t\|e_h(s)\|_0$ we thus have

$$\begin{aligned}&\varphi'(t)-C\varphi(t)\leqslant Ch^2t^{-1}+Ch^2\log(t/h^2)\quad\text{for } h^2\leqslant t\leqslant t^*,\\&\varphi(h^2)=0\end{aligned} \tag{29.13}$$

and hence,

$$\varphi(t)\leqslant C\int_{h^2}^t \mathrm{e}^{C(t-s)}(h^2s^{-1}+h^2\log(s/h^2))\,\mathrm{d}s\leqslant Ch^2\log(t/h^2). \tag{29.14}$$

Inserting this into (29.12) completes the proof. □

Note that if the optimal order of approximation is $r>2$, this is *not* reflected in Theorem 29.1.

The result of Theorem 29.1 is sharp in the following sense: An equation can be exhibited such that the following holds: If for *all* u with $|u(x,t)|\leqslant B$ for $x\in\Omega$, $t\in I$, we have for a fixed t_0, $0<t_0\leqslant t^*$,

$$\|u_h(t_0)-u(t_0)\|_0\leqslant C(t_0,B)h^{\alpha}, \tag{29.15}$$

then α *cannot* be greater than 2. This should be contrasted with the linear situation where, provided numerical integration is used carefully enough in evaluating the L^2 projection of initial data, WAHLBIN [1980], the error is of optimal order for positive time.

A complete counterexample in spline spaces is given in JOHNSON, LARSSON, THOMÉE and WAHLBIN [1987, Section 6]. However, it is easy to understand why the order α is restricted to 2 in (29.15), regardless of the optimal approximation order of the finite element spaces V_h. For, note that Theorem 29.1 would hold, with $h=(n-1)^{-1}$, if trigonometric polynomials of degree $(n-1)$ were used on $[-\pi,\pi]$ with periodic boundary conditions and v_h the Fourier series of v. (The proof of this is *exactly* as the proof of Theorem 29.1.) Further note that trigonometric approximation has unbounded optimal order r! Finally, note that Theorem 29.1 holds as well for systems.

Then take $u=(u_1,u_2)$ as

$$u_{1,t}=u_{1,xx}+f(u_2),\qquad u_{2,t}=u_{2,xx} \tag{29.16}$$

on $[-\pi,\pi]\times(0,\infty)$ where f is smooth with

$$f(y)=4y^2\quad\text{for } |y|\leqslant 1. \tag{29.17}$$

Further, take

$$u_1(0)=0,\qquad u_2(0)=\cos(nx) \tag{29.18}$$

so that by elementary checking,

$$\begin{aligned} u_2(x,t)&=\exp(-n^2t)\cos(nx),\\ u_1(x,t)&=\frac{1-\exp(-2n^2t)}{n^2}[1+\exp(-2n^2t)\cos(2nx)]. \end{aligned} \tag{29.19}$$

Using for approximation, trigonometric polynomials of degree $\leqslant(n-1)$ and, with Fourier series for initial data, the approximate solutions are identically zero. Thus, (29.19) describes the error which is, indeed, *exactly* of order $n^{-2}=O(h^2)$ for t positive.

Note that in (29.19), a high Fourier mode ($\cos(nx)$) has aliazed itself by nonlinear interaction into a low, indeed constant, mode (viz. $1/n^2$). Also note that the example is in complete contrast to the linear situation, regardless of whether the equation (29.16) is regarded as an analogue of the linear homogeneous or inhomogeneous case.

Further results are given in JOHNSON, LARSSON, THOMÉE and WAHLBIN [1987]. In particular, results when initial data v are somewhat smooth and compatible are derived, and also pointwise results in space, and results for gradients.

For further investigations concerning the smoothing effect in nonlinear problems we refer to HEYWOOD and RANNACHER [1988] for the Navier–Stokes equations; they also noted an upper bound in their proof for the rate of convergence with incompatible data similar to our counterexample. For time discretizations, see CROUZEIX and THOMÉE [1987b].

Interesting uses of Theorem 29.1 to the long time behavior of Galerkin approximations are given in LARSSON [1985] and HALE, LIN and RAUGEL [1988].

CHAPTER VII

Superconvergence

30. Difference quotients on translation-invariant meshes for approximating derivatives

This section is based on the fundamental investigation in NITSCHE and SCHATZ [1974, Section 6]. For simplicity we consider the model problem

$$-\Delta u = f \quad \text{in } \Omega, \qquad u=0 \quad \text{on } \partial\Omega. \tag{30.1}$$

In two or more space dimensions, results on superconvergence are restricted to uniform, or nearly uniform, meshes, cf. Section 5. In the present situation we consider an interior subdomain $A \Subset \Omega$ and let with V_h finite element spaces,

$$\mathring{V}_h(A) = \{\chi \in V_h : \operatorname{supp} \chi \subseteq \bar{A}\}. \tag{30.2}$$

The approximation $u_h \in V_h$ we are seeking is required to satisfy

$$(\nabla(u-u_h), \nabla\chi)=0 \quad \chi \in \mathring{V}_h(A_1), \tag{30.3}$$

where $A \Subset A_1 \Subset \Omega$. We let as usual r denote the optimal order of approximation in L^p spaces.

The mesh is uniform and translation-invariant on a neighborhood of A in the sense that with $T_h^\nu v(x) = v(x+\nu h)$, ν a multi-integer, with $A_1 \supseteq A$,

$$T_h^\nu \chi \in \mathring{V}_h(A_1) \quad \text{for } \chi \in \mathring{V}_h(A). \tag{30.4}$$

REMARK 30.1. Of course, the mesh parameter h may be reparametrized. Thus for this mesh configuration (see Fig. 30.1), we have translation invariance if h is replaced by $2h$ in the translation operators. Also, the translation invariance need not be with respect to the coordinate axes; in *any* direction with translation invariance we may consider corresponding difference quotients.

To approximate $D^\alpha u$, α a multi-integer, consider a difference operator

$$D_h^\alpha u = \sum_{|\nu| \leqslant M} C_{\nu\alpha}(h) T_h^\nu u, \tag{30.5}$$

for some finite M so that (30.4) holds for $|\nu| \leqslant M$.

We assume that

$$|(D^\alpha u - D_h^\alpha u)(x_0)| \leqslant Ch^r \|u\|_{|\alpha|+r,\infty,S(x_0,M)}, \tag{30.6}$$

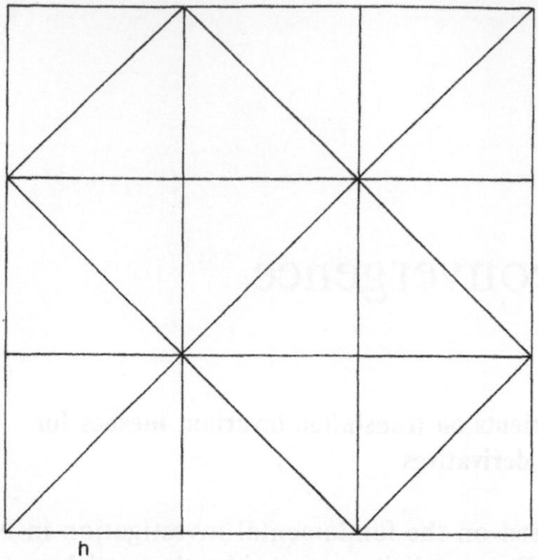

FIG. 30.1.

where

$$S(x_0, M)=\{x: |x_i - x_{0i}| \leqslant Mh\}, \tag{30.7}$$

so that the order of accuracy of the difference approximation matches that of the finite element space.

Set now

$$e = u - u_h. \tag{30.8}$$

The basic observation in NITSCHE and SCHATZ [1974] is that

$$(\nabla D_h^\alpha e, \nabla \chi)=0 \quad \text{for } \chi \in \mathring{V}_h(A). \tag{30.9}$$

For the form on the left equals $(\nabla e, \nabla(D_h^{\alpha *}\chi))$ and, by (30.4), $D_h^{\alpha *}\chi \in \mathring{V}_h(A_1)$ so that (30.3) applies.

The development in Nitsche and Schatz' paper now proceeds via local L^2 or energy estimates. In this presentation we shall consider pointwise estimates and make the assumption,

The pointwise local estimate of Theorem 10.1 holds. (30.10)

This assumption (30.10) is verified in SCHATZ and WAHLBIN [1977] for a large class of finite element spaces.

It follows from (30.9) with x_0 in A,

$$|D_h^\alpha e(x_0)| \leqslant C \ln(1/h)^{\bar{r}} \min_{\chi \in V_h} \|D_h^\alpha u - \chi\|_{0,\infty,A} + C\|D_h^\alpha u - \chi\|_{-s',A}. \tag{30.11}$$

Assuming that $u\in\mathscr{C}^{r+|\alpha|}(A_1)$, by approximation theory, and by (30.6),

$$\begin{aligned}&\min_{\chi\in V_h}\|D_h^\alpha u-\chi\|_{0,\infty,A}\\&\quad\leqslant\|(D_h^\alpha-D^\alpha)u\|_{0,\infty,A}+\min_{\chi\in V_h}\|D^\alpha u-\chi\|_{0,\infty,A}\\&\quad\leqslant Ch^r\|u\|_{0,\infty,|\alpha|+r,A_1}.\end{aligned}\tag{30.12}$$

Hence, again using (30.6),

$$\begin{aligned}&|(D^\alpha u-D_h^\alpha u_h)(x_0)|\\&\quad\leqslant|(D^\alpha-D_h^\alpha)u(x_0)|+|D_h^\alpha e(x_0)|\\&\quad\leqslant C\ln(1/h)^{\bar r}h^r+C\|D_h^\alpha e\|_{-s',A}.\end{aligned}\tag{30.13}$$

For the last term on the right of (30.13),

$$\|D_h^\alpha e\|_{-s',A}=\sup_{\substack{\|\varphi\|_{s',A}=1\\ \varphi\in\mathscr{C}_0^\infty(A)}}(D_h^\alpha e,\varphi).\tag{30.14}$$

For each such φ,

$$\begin{aligned}(D_h^\alpha e,\varphi)=(e,D_h^{\alpha*}\varphi)&\leqslant\|e\|_{-s,A_1}\|D_h^{\alpha*}\varphi\|_{s,A_1}\\&\leqslant C\|e\|_{-s,A_1}\|\varphi\|_{s',A_1}.\end{aligned}$$

e.g. for $s'=s+|\alpha|+\frac{1}{2}N+1$. Thus, from (30.13) and (30.14), we have the following theorem.

THEOREM 30.1. *Assume* (30.3), (30.4), (30.6), (30.10) *and that* $u\in\mathscr{C}^{r+|\alpha|}(A_1)$. *For any* s,

$$\begin{aligned}&|(D^\alpha u-D_h^\alpha u_h)(x_0)|\\&\quad\leqslant C\ln(1/h)^{\bar r}h^r\|u\|_{r+|\alpha|,\infty,A_1}+C\|e\|_{-s,A_1}.\end{aligned}\tag{30.15}$$

A similar estimate was given in BRAMBLE, NITSCHE and SCHATZ [1975].

The upshot of this is that, in the case of locally translation-invariant meshes, one should take for an approximation of a derivative $D^\alpha u$ a suitable *difference quotient*, not a derivative, of the finite element solution u_h. The error is then of the order governed by the minimum of the local approximability (h^r if $u\in\mathscr{C}^{r+|\alpha|}$ locally) and $u-u_h$ in any negative norm, the latter accounting for global influences such as e.g. pollution effects in problems with reentrant corners.

Note that in general, say for first derivatives, $\nabla u-\nabla u_h$ is only of order h^{r-1}. Thus, the result above is "superconvergent", the global effects permitting.

How to estimate $\|u-u_h\|_{-s,A_1}$ in various situations has been clarified in Chapter III. Clearly, extreme care is called for. As in the one-dimensional case of Chapter I, there may be singularities in data, such as right-hand sides or coefficients, preventing the negative norm from being of sufficiently high order to entail superconvergence. In the multidimensional case, the domain may also contain reentrant corners, i.e., be singular, so that superconvergence in the difference approximation to derivatives is obstructed, see Section 13. A more subtle point is that even if all data in the problem are smooth, the negative norm may not be of order h^r because a curved boundary is not fitted well enough. In the next section this problem is even more pronounced, since there higher-order approximations are involved, and we refer the reader forward to the first paragraph of Section 31 for further amplification of this point.

We further remark that the investigation in NITSCHE and SCHATZ [1974, Section 6] also covers variable smooth coefficients in the basic elliptic operator. It is easy to modify the above pointwise error estimate to cover that case.

A frequently rediscovered special case of Theorem 30.1 occurs in the plane case when the mesh is triangular and uniform on A, and translation-invariant (see Fig. 30.2). Consider piecewise linear finite elements on this mesh. To approximate u_x and u_y we take forward difference quotients,

$$\begin{aligned} u_x &\sim \partial_{x,h}u=(u(x+h,y)-u(x,y))/h,\\ u_y &\sim \partial_{y,h}u=(u(x,y+h)-u(x,y))/h. \end{aligned} \tag{30.16}$$

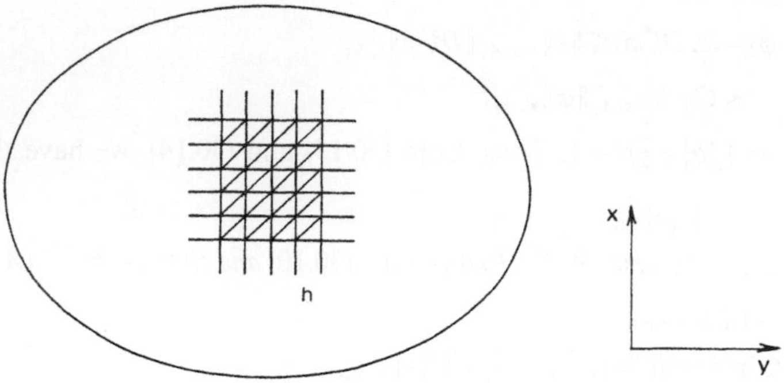

FIG. 30.2.

These should be considered as approximations for $u_x(x+\frac{1}{2}h, y)$ and $u_y(x, y+\frac{1}{2}h)$, respectively, so that they are of second-order accuracy.

Thus, if $u\in\mathscr{C}^3(A_1)$ we have by Theorem 30.1,

$$|(u_x(x_0+\tfrac{1}{2}h, y_0)-\partial_{x,h}u_h(x_0, y_0))|\leqslant C\ln(1/h)h^2+C\|u-u_h\|_{-s,A_1}, \tag{30.17}$$

and

$$|(u_y(x_0, y_0+\tfrac{1}{2}h)-\partial_{y,h}u_h(x_0, y_0))|\leqslant C\ln(1/h)h^2+C\|u-u_h\|_{-s,A_1}. \tag{30.18}$$

In particular, consider two adjoining triangles such as, e.g. in Fig. 30.3, and take $(x_0, y_0)_1$ in (30.17) and $(x_0, y_0)_2$ in (30.18) as indicated.

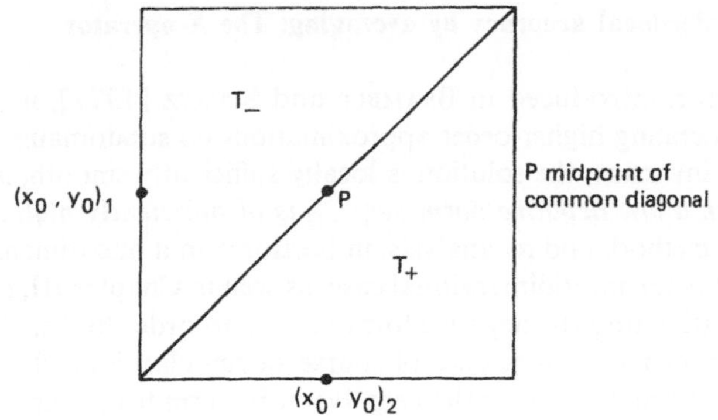

FIG. 30.3.

Since in the piecewise linear situation,

$$\begin{aligned}
&\partial_{x,h}u_h((x_0, y_0)_1)\\
&\quad=(u_h(x_0+h, y_0)-u_h(x_0, y_0))/h\\
&\quad=(u_h(x_0+h, y_0)-y_h(x_0+\tfrac{1}{2}h, y_0))/h+(u_h(x_0+\tfrac{1}{2}h, y_0)-u_h(x_0, y_0))/h\\
&\quad=\tfrac{1}{2}(u_h)_x T_- +\tfrac{1}{2}(u_h)_x T_+
\end{aligned} \tag{30.19}$$

and similarly for the y-difference with $(x_0, y_0)_2$, we have the following.

COROLLARY 30.1. *Let the triangular meshes be uniform and translation-invariant with piecewise linears on the plane interior domain A_1. Let $u \in \mathscr{C}^3(A_1)$. Then the average of ∇u_h taken over two neighboring triangles approximates ∇u at the midpoint of their common diagonal to order*

$$C\ln(1/h)h^2\|u\|_{3,\infty,A_1}+C\|u-u_h\|_{-s,A_1}. \tag{30.20}$$

for any s.

We leave it to the reader to account for how the distance between x_0 and ∂A_1 enters; the guidelines of Chapter III should suffice.

Extensions of the result of Corollary 30.1 to, for example, meshes which are "almost uniform", that is, perturbed $O(h^2)$ from the uniform situation, occur, e.g. in LIN, LU and SHEN [1983] and LIN and XU [1985]. In the latter paper it is shown how any domain can be triangulated into piecewise almost uniform meshes and how the averaging techniques apply in that case.

REMARK 30.2. Note that the method of taking difference quotients works up to straight boundaries in two dimensions, cf. Example 10.2, or for Poisson's equation in more dimensions, Example 10.1, for tangential derivatives provided the mesh is translation-invariant in the tangential direction. For approximation of the normal derivative at the boundary, one may try reflection but very few mesh configurations would work, cf. Remark 30.1 for one of the few.

31. Higher-order local accuracy by averaging: The *K*-operator

The *K*-operator, introduced in BRAMBLE and SCHATZ [1977], is a fairly general method of generating higher-order approximations on subdomains where the mesh is translation-invariant, the solution is locally sufficiently smooth, *and provided the global error in a low negative norm* $\|e\|_{-s,\Omega}$ *is of sufficiently high order.* We have described the method, and its analysis, in Section 5 in a one-dimensional case. We point out that in the multidimensional case, as seen in Chapter III, many things can go wrong in estimating the negative form $\|e\|_{-s,\Omega}$ to order higher than $O(h^r)$. As in the one-dimensional case there may, of course, be singularities in data away from the point of interest which prevents the negative norm term from being of higher order. However, in the multidimensional case the boundary of the domain may also be "singular", cf. Section 13. The most subtle point is, however, that even if all data in the problem are smooth, $\|e\|_{-s,\Omega}$ may still not be better than $O(h^r)$ because a curved boundary $\partial\Omega$ may not have been fitted well enough. A common pitfall is to assume that

$$\min_{\chi\in V_h} \|u-\chi\|_{1,\Omega}\leqslant Ch^{r-1}\|u\|_{r,\Omega}. \tag{31.1}$$

Then, in a smooth problem, a standard duality argument would give

$$\|e\|_{-(r-2),\Omega}\leqslant Ch^{2(r-1)}, \tag{31.2}$$

where for $r\geqslant 3$, $2(r-1)>r$. However, (31.1) does not hold in general.

Thus, when applying the results below concerning the *K*-operator one should be very careful.

We consider the problem

$$\begin{aligned} -\Delta u&=f \quad \text{in } \Omega\Subset\mathbb{R}^N,\\ u&=0 \quad \text{on } \partial\Omega \end{aligned} \tag{31.3}$$

for simplicity; the paper by Bramble and Schatz contains the case of variable smooth coefficients.

We assume that locally for $x\in A\Subset\Omega$,

$$\chi(x)=\sum_{j=1}^{m}\sum_{\alpha\in\mathbb{Z}^N} a_\beta^j\varphi_j(h^{-1}x-\alpha), \quad \chi\in V_h. \tag{31.4}$$

Thus, $\varphi_1,\dots,\varphi_m$ are the basic building blocks for the finite element space, and in adjoining elements the building blocks are merely translated. Let again r be the optimal order of approximation in L^p spaces.

As in Section 5, consider smoothest splines of degree $r-1$ and continuity $r-2$. With ψ_{r-2}^1 the one-dimensional basis function, let

$$K_h(x)=\prod_{m=1}^{M}\left(\sum_{\beta=-(r-2)}^{r-2} h^{-1}k_\beta\psi_{r-2}^1(h^{-1}x_m-\beta)\right), \tag{31.5}$$

where the k_β are as in Section 5, so that we have an approximate identity of order $2r-2$. (Tables are given in BRAMBLE and SCHATZ [1977].)

As in Section 5 it is easily seen that for u_h developed as in (31.4), $K_h * u_h$ at mesh points is given by a finite linear combination of a^j_β with coefficients d^j_δ that only depend on the finite element space used, not on h, or on the particular mesh point considered (or, on the coefficients of the elliptic operator).

The idea is now to consider $u - K_h * u_h$. We have with $e = u - u_h$,

$$u - K_h * u_h = (u - K_h * u) + K_h * e. \tag{31.6}$$

For $x_0 \in A$, since K_h is an approximate identity and localized,

$$|(u - K_h * u)(x_0)| \leqslant Ch^{2r-2} \|u\|_{2r-2,\infty,A}. \tag{31.7}$$

The rest of the argument now proceeds as in (5.36) et seq.: $K_h * e$ may be estimated in terms of negative norms of derivatives of $K_h * e$; derivatives of $K_h * e$ may be bounded by difference quotients of e. Thus BRAMBLE and SCHATZ [1977, Lemma 6.1],

$$|(K_h * e)(x_0)| \leqslant C\Bigg(\sum_{|\alpha| \leqslant N_0 + r - 2} \|\partial^\alpha_h e\|_{2-r,A} + h^{r-2} \sum_{|\alpha| \leqslant r-2} \|\partial^\alpha_h v\|_{0,\infty,A} \Bigg), \tag{31.8}$$

where $N_0 = [\frac{1}{2}N] + 1$. By localized duality estimates, cf. Section 9, and by Section 30 (assuming these results may be applied), we arrive at:

THEOREM 31.1. *Under the above assumptions, for $x_0 \in A \Subset \Omega$ and for any s,*

$$|(u - K_h * u_h)(x_0)| \leqslant Ch^{2r-2} \|u\|_{2r-2+N_0,2,A} + C\|e\|_{-s,A}. \tag{31.9}$$

Thus, for $r \geqslant 3$, we have an easily calculated candidate for a superconvergent approximation to u at interior nodes. Again we point out that one must be careful in estimating $\|e\|_{-s,A}$.

The technique above in deriving Theorem 31.1 includes a "Sobolev-loss" from (31.8). In the special case of tensor products of smoothest splines, it was shown in BRAMBLE and SCHATZ [1976, Theorem 8] how the first term on the right of (31.9) can be replaced by $Ch^{2r-2} \ln(1/h) \|u\|_{2r-2,\infty,A}$, a more satisfying result one would believe holds in general.

The approaches of Sections 30 and 31 can clearly be combined so that approximations to derivatives of higher order than $O(h^r)$ can be obtained by postprocessing, see e.g. THOMÉE [1977].

Similarly, both approaches can readily be adapted to parabolic problems, hyperbolic problems, Euler–Poisson–Darboux, Sobolev equations, etc., all time-dependent problems containing a spatial second-order elliptic operator. In this respect see e.g. BRAMBLE, SCHATZ, THOMÉE and WAHLBIN [1977, Sections 6 and 7]. Later advancements in the theory of approximations for time-dependent problems have, of course, yielded better results, see e.g. BALES [1984], GENIS [1984], and for the parabolic situation, THOMÉE [1984].

32. Miscellaneous results

In this section we shall describe some superconvergence results, frequently derived for special type of elements. A nice survey of superconvergence exists in KRIZEK and NEITTAANMÄKI [1987], which has a rather complete bibliography, including one-dimensional cases. However, our survey in this chapter has a different emphasis, in that I have stressed two *systematic* techniques in Sections 30 and 31 for translation-invariant meshes. Furthermore, the results were always given *pointwise*, always *localized*, although not systematically carried up to the boundary or done for piecewise almost uniform (or more general) meshes.

The results we are about to describe are frequently phrased so that they are dependent on global smoothness assumptions, and they are sometimes given in less satisfying error measures than the pointwise measure. Sometimes, they do apply up to boundaries.

We saw in Corollary 30.1 how the quite general theory for approximating derivatives by difference quotients lead to the fact that, in the uniform piecewise linear triangular situation, the average of the gradients of u_h in two neighboring triangles is an

$$O(h^2 \ln(1/h))\|u\|_{3,\infty,A_1} + \|e\|_{-s,A_1} \tag{32.1}$$

approximation to ∇u at the midpoint of the common side.

As noted by LEVINE [1985], this implies an averaging technique for approximating the gradient of u at centroids of triangles: Simply take the three edge-midpoint approximations already found and average them.

In the uniform piecewise linear situation, to approximate the gradient at a nodal point, by the theory of Section 30 simply use two suitable (following edges) centered difference quotients. Due to cancellations this method can be expressed as averaging the approximate gradient in all six surrounding elements, see KRIZEK and NEITTAANMÄKI [1984], where also a technique up to the boundary is given.

Clearly, the results of Sections 30 and 31 provide an almost inexhaustible source for deriving special formulae in special situations. Frequently, however, such special formulae were not first derived by the difference quotient or K-operator techniques. Indeed, in e.g. the piecewise linear situation a result by OGANESYAN and RUKHOVETS [1969, (3.18)], namely that for u_h^I the piecewise linear interpolant to u,

$$\|u_h - u_h^I\|_{H^1(\Omega_h)} \leqslant Ch^2 \|u\|_{H^3(\Omega)}, \tag{32.2}$$

appears to have served as inspiration to many researchers. Having (32.2) one may sink one's teeth into the explicitly and locally given u_h^I and come up with some averaging formula A such that $\nabla u - A\nabla u_h^I$ is of higher order at desired points. It is then automatic that $\nabla u - A\nabla u_h$ is of order h^2, at last in energy.

In the results mentioned so far the meshes are either completely uniform, a higher order perturbation of such a mesh, see LIN and XU [1985], or a smooth transformation of a uniform mesh, see LEVINE [1985]. (Levine's case thus falls into that of Lin and Xu.)

We shall next give two results which identify superconvergent points for *the function u itself*. Following first DOUGLAS, DUPONT and WHEELER [1974a], consider

$$\begin{aligned} -\Delta u &= f \quad \text{on } \Omega = \text{the plane rectangle } [0,1]\times[0,1] \\ u &= 0 \quad \text{on } \partial\Omega. \end{aligned} \tag{32.3}$$

Let V_h consist of tensor products of *continuous* piecewise polynomials of any order $r-1$; the meshes are quasi-uniform and rectangular but *not* necessarily uniform. They show that if u is globally smooth enough, then at the mesh points K,

$$|(u-u_h)(K)| \leqslant Ch^{r+1} \tag{32.4}$$

which is superconvergent to one extra order.

At the "opposite end of the scale", consider tensor products of *smoothest* splines, i.e., tensor products of one-dimensional splines with polynomial degree $r-1$ and continuity degree $r-2$. Then from BRAMBLE and SCHATZ [1976, Theorem 10] we have for the problem (32.3) (also generalized to $\mathbb{R}^N$) provided the meshes are *uniform*, and provided $r \geqslant 3$ is *odd*, for K an interior mesh point, with $e = u - u_h$,

$$|e(K)| \leqslant \begin{cases} h^4 \ln(1/h)\|u\|_{4,\infty,A} + \|e\|_{-s,A} & \text{for } r=3,\ N\geqslant 3, \\ h^{r+1}\|u\|_{r+1,\infty,A} + \|e\|_{-s,A} & \text{otherwise,} \end{cases} \tag{32.5}$$

which is again superconvergent to one extra order, the global influences permitting.

Turning now to identifiable superconvergence points for *the gradients*, we first describe the results of ZLAMAL [1977]. He considers also variable smooth coefficients in (32.3). The finite elements are (almost) rectangular quasi-uniform (but *not* necessarily uniform) eight-node quadrilaterals and he shows that under appropriate global smoothness assumptions, with G denoting the four Gauss points on each rectangle (i.e., the maps of $(\pm 1/\sqrt{3}, \pm 1/\sqrt{3})$ to each rectangle),

$$\left(\sum_G |\nabla(u-u_h)(G)|^2 h\right)^{1/2} \leqslant Ch^3. \tag{32.6}$$

Hence, in this global discrete energy norm we have superconvergence for gradients at Gauss points to one further order.

This result was generalized in LESAINT and ZLAMAL [1979] to any degree continuous (almost) rectangular tensor product of polynomials, still showing (in an average sense) one order of superconvergence at the appropriate Gauss points for gradients. Further, ZLAMAL [1978] gives similar results for quadratic and linear triangular elements. ANDREEV and LAZAROV [1988] show in the quadratic triangular case on almost uniform meshes that the Gauss points on triangle edges are points of superconvergence for the derivative in the direction of the side. Based on this they give an averaging scheme to produce superconvergence of gradients at element vertices.

NAKAO [1987] has shown how to obtain, via a certain local postprocessing, one order higher convergence *at nodes* for the gradient in the case of continuous tensor products on rectangular elements, in the case that the polynomial degree $r-1$ is *odd*.

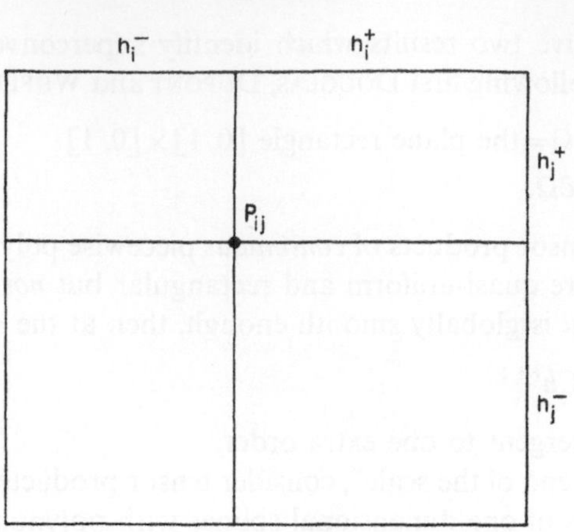

FIG. 32.1.

Consider the situation shown in Fig. 32.1. With $\alpha_i = h_i^-/h_i^+$, $\alpha_j = h_j^-/h_j^+$, set

$$\tilde{D}_{ix}u_h(P_{ij}) = \frac{1}{1+\alpha_i^r}\left\{\frac{\partial}{\partial x}u_h(x_i-, y_j) + \alpha_i^r\frac{\partial}{\partial x}u_h(x_i+, y_j)\right\},$$

$$\tilde{D}_{jy}u_h(P_{ij}) = \frac{1}{1+\alpha_j^r}\left\{\frac{\partial}{\partial y}u_h(x_i, y_j-) + \alpha_j^r\frac{\partial}{\partial y}u_h(x_i, y_j+)\right\}. \tag{32.7}$$

Letting K denote the knots P_{ij}, set

$$\tilde{\nabla}u_h(K) = (\tilde{D}_{ix}u_h, \tilde{D}_{jy}u_h). \tag{32.8}$$

If u is sufficiently smooth ($u \in W_p^{r+2}(\Omega)$, $p>2$, for example) then

$$|(\nabla u - \tilde{\nabla}u_h)(K)| \leqslant Ch^r, \tag{32.9}$$

again exhibiting one order of superconvergence.

We conclude this section by describing two more computationally complicated methods for obtaining higher-order approximations. The first one is a method due to WHEELER [1973] for obtaining a better approximation to the flux at the boundary of a domain (in the case of Dirichlet conditions).

To motivate the method, with $\gamma = (\partial u/\partial n)|_\Gamma$ we have by integration by parts in (32.3)

$$\int_{\partial\Omega} \gamma v = \int_{\Omega} (fv - \nabla u \cdot \nabla v), \quad \text{any } v \text{ smooth enough.} \tag{32.10}$$

Thus, it makes sense to approximate the normal flux γ by γ_h, where γ_h is given by

(and u_h is the already computed finite element approximation)

$$\int_{\partial\Omega_h} \gamma_h \chi = \int_{\Omega_h} (f\chi - \nabla u_h \cdot \nabla \chi) \quad \text{for } \chi \in V_h, \tag{32.11}$$

where V_h does *not* satisfy the essential boundary conditions.

Solving for γ_h amounts to solving for an L^2 projection on $\partial\Omega_h$ into $\tilde{V}_h|_{\partial\Omega_h}$, once u_h is computed.

In DOUGLAS, DUPONT and WHEELER [1974b] it is shown that (for Ω a rectangle, for u globally smooth enough, and for continuous tensor product elements), the approximation γ_h defined by (32.11) to $\gamma = (\partial u/\partial n)|_\Gamma$ is superconvergent in the sense that

$$\|\gamma_h - \gamma\|_{0,\infty,\partial\Omega} \leqslant Ch^r, \tag{32.12}$$

which is one order superconvergent.

We remark that methods such as the Lagrange multiplier method of BABUŠKA [1973] incorporate a separate approximation to γ in the basic finite element formulation. We refer to BRAMBLE [1981] for some results about this approximation.

Our second computationally more complicated method should probably be applied only if one desires a good approximation to u at only a few points inside Ω. It applies, however, to *any* mesh but requires exact knowledge of a fundamental solution. We follow the particular presentation of LOUIS [1979].

Let x_0 be a fixed point and

$$B(x_0, r) = \{x : |x - x_0| \leqslant r\} \subset \Omega. \tag{32.13}$$

Let

$$\omega(z) = -\frac{1}{2\pi} \ln|z| + \psi(z), \tag{32.14}$$

where $\psi(z) \in \mathscr{C}^\infty$ is chosen so that

$$\omega(r) = \partial\omega(r)/\partial n = 0. \tag{32.15}$$

For example, Louis gives the choice

$$\psi(x) = -\frac{1}{8\pi r^4}(|z|^4 - 4r^2|z|^2 + (3 - 4\ln r)r^4). \tag{32.16}$$

By Green's second formula,

$$u(x_0) = \int_{B(x_0,r)} (f(y)\omega(x_0 - y) + u(y)\Delta\psi(x_0 - y))\,\mathrm{d}y. \tag{32.17}$$

Having then $u_h \in V_h$ a basic finite element solution, define

$$\tilde{u}_h(x_0) = \int_{B(x_0,r)} (f(y)\omega(x_0 - y) + u_h(y)\Delta\psi(x_0 - y))\,\mathrm{d}y. \tag{32.18}$$

It is clear that (if the integral in (32.18) is evaluated exactly),

$$|(u-\tilde{u}_h)(x_0)| \leqslant C|||u-u_h|||_{-s,B(x_0,r)}, \tag{32.19}$$

where $|||\ |||$ denotes the dual norm to $H^s(\bar{B})$. Thus the approximation $\tilde{u}_h$ picks up whatever negative norm accuracy the finite element solution u_h enjoys. (As we have remarked at many places, one must be careful in evaluating negative norms. Boundary approximations may cause grief even on smooth domains.)

References

ANDREEV, A.B. and R.D. LAZAROV (1988), Superconvergence of the gradient for quadratic triangular finite elements. *Numer. Meths. Partial Differential Equations* **4**, 15–32.

BABUŠKA, I. (1970), Finite element methods for domains with corners, *Computing* **6**, 264–273.

BABUŠKA, I. (1973), The finite element method with Lagrangian multipliers, *Numer. Math.* **20**, 179–192.

BABUŠKA, I. and A.K. AZIZ (1976), On the angle condition in the finite element method, *SIAM J. Numer. Anal.* **13**, 214–226.

BABUŠKA, I., A. MILLER and M. VOGELIUS (1983), Adaptive methods and error estimation for elliptic problems of structural mechanics, in: I. BABUŠKA, J. CHANDRA and J.E. FLAHERTY, eds., *Adaptive Computational Methods for Partial Differential Equations* (SIAM, Philadelphia, PA) 53–73.

BABUŠKA, I. and J. OSBORN (1980), Analysis of finite element methods for second order boundary value problems using mesh dependent norms, *Numer. Math.* **34**, 41-62.

BABUŠKA, I. and M. ROSENZWEIG (1973), A finite element scheme for domains with corners, *Numer. Math.* **20**, 1–21.

BAKKER, M. (1982), A note on C^0 Galerkin methods for two-point boundary value problems, *Numer. Math.* **38**, 447–453.

BAKKER, M. (1984), One-dimensional Galerkin methods and superconvergence at interior nodal points, *SIAM J. Numer. Anal.* **21**, 101–110.

BALES, L. (1984), Semidiscrete and single step fully discrete approximations for second order hyperbolic equations with time-dependent coefficients, *Math. Comp.* **43**, 383–414.

BARANGER, J. (1979), On the thickness of the boundary layer in elliptic-elliptic singular perturbation problems, in: P.W. HEMKER and J.J.H. MILLER, eds., *Numerical Analysis of Singular Perturbation Problems* (Academic Press, New York) 395–400.

BARY, N.K. (1964), *A Treatise on Trigonometric Series* **2** (Macmillan, New York).

BERGH, J. and J. LÖFSTRÖM (1976), *Interpolation Spaces: An Introduction* (Springer, New York).

BESJES, J.G. (1975), Singular perturbation problems for linear elliptic differential operators of arbitrary order, I: Degeneration to elliptic operators, *J. Math. Anal. Appl.* **49**, 24–46.

BLAIR, J. (1970), Approximate solution of elliptic and parabolic boundary value problems, Thesis, University of California, Berkeley, CA.

BRAMBLE, J.H. (1975), A survey of some finite element methods proposed for treating the Dirichlet problem, *Adv. in Math.* **16**, 187–196.

BRAMBLE, J.H. (1981), The Lagrange multiplier method for Dirichlet's problem, *Math. Comp.* **37**, 1–12.

BRAMBLE, J.H. and S.R. HILBERT (1970), Estimation of linear functionals on Sobolev spaces with application to Fourier transforms and spline interpolation, *SIAM J. Numer. Anal.* **7**, 113–124.

BRAMBLE, J.H. and S.R. HILBERT (1971), Bounds for a class of linear functionals with application to Hermite interpolation, *Numer. Math.* **16**, 362–369.

BRAMBLE, J.H., J.A. NITSCHE and A.H. SCHATZ (1975), Maximum norm interior estimates for Ritz–Galerkin methods, *Math. Comp.* **29**, 677–688.

BRAMBLE, J.H. and J.E. OSBORN (1973), Rate of convergence estimates for non-selfadjoint eigenvalue approximation, *Math. Comp.* **27**, 525–549.

BRAMBLE, J.H. and A.H. SCHATZ (1976), Estimates for spline projections, *RAIRO Anal. Numér.* **10**, 5–37.

BRAMBLE, J.H. and A.H. SCHATZ (1977), Higher order local accuracy by averaging in the finite element method, *Math. Comp.* **31**, 94–111.

BRAMBLE, J.H., A.H. SCHATZ, V. THOMÉE and L.B. WAHLBIN (1977), Some convergence estimates for semidiscrete Galerkin type approximations for parabolic equations, *SIAM J. Numer. Anal.* **14**, 218–240.

BRAMBLE, J.H. and V. THOMÉE (1974), Interior maximum norm estimates for some simple finite element methods, *RAIRO Anal. Numér.* **8**, 5–18.

BRENNER, P., V. THOMÉE and L.B. WAHLBIN (1975), *Besov Spaces and Applications to Difference Methods for Initial Value Problems*, Lecture Notes in Mathematics **434** (Springer, Berlin).

BROOKS, A.N. and T.J.R. HUGHES (1982), Streamline upwind Petrov–Galerkin formulations for convection dominated flows with particular emphasis on the incompressible Navier–Stokes equations, *Comput. Meths. Appl. Mech. Engrg.* **32**, 199–259.

CHAVEZ, P. (1983), Automatic procedures in evolutionary finite element calculations: Restoration of deteriorated meshes, data transfer between meshes and mesh refinement, Thesis, Cornell University, Ithaca, NY.

CHEN, C.M. (1979), Superconvergent points of Galerkin's method for two-point boundary value problems, *J. Numer. Anal. Chinese Univ.* **1**, 73–79.

CHRISTIE, I., D.F. GRIFFITHS, A.R. MITCHELL and O.C. ZIENKIEWICZ (1976), Finite element methods for second order differential equations with significant first derivatives, *Internat. J. Numer. Meths. Engrg.* **10**, 1389–1396.

CIARLET, P.G. (1978), *The Finite Element Method for Elliptic Problems* (North-Holland, Amsterdam).

CROUZEIX, M. and V. THOMÉE (1987a), The stability in L_p and W_p^1 of the L_2 projection into finite element function spaces, *Math. Comp.* **48**, 521–532.

CROUZEIX, M. and V. THOMÉE (1987b), On the discretization in time of semilinear parabolic equations with nonsmooth initial data, *Math. Comp.* **49**, 359–377.

DE BOOR, C. (1979), On a max-bound for the least-squares spline approximant, in: Z. CIESIELSKI, ed., *Approximation and Function Spaces* (North-Holland, Amsterdam) 163–175.

DE BOOR, C. and G.J. FIX (1973), Spline approximation by quasiinterpolants, *J. Approx. Theory* **8**, 19–45.

DEMKO, S. (1977), Inverses of band matrices and local convergence of spline projections, *SIAM J. Numer. Anal.* **14**, 616–619.

DENDY, J.E. (1974), Two methods of Galerkin type achieving optimum L^2 accuracy for first order hyperbolics, *SIAM J. Numer. Anal.* **11**, 637–653.

DESCLOUX, J. (1972), On finite element matrices, *SIAM J. Numer. Anal.* **9**, 260–265.

DESCLOUX, J. (1976), Interior regularity and local convergence of Galerkin finite element approximations for elliptic equations, in: J. MILLER, ed., *Topics in Numerical Analysis* **2** (Academic Press, New York).

DOBROWOLSKI, M. (1981), Numerical approximation of elliptic interface and corner problems, Rheinische Friedrich-Wilhelms-Universität, Bonn.

DOUGLAS, J. and T. DUPONT (1974), Galerkin approximation for the two-point boundary problem using continuous piecewise polynomial spaces, *Numer. Math.* **22**, 99–109.

DOUGLAS, J., T. DUPONT and L.B. WAHLBIN (1975a), Optimal L_∞ error estimates for Galerkin approximations to solutions of two-point boundary value problems, *Math. Comp.* **39**, 475–483.

DOUGLAS, J., T. DUPONT and L. WAHLBIN (1975b), The stability in L^q of the L^2-projection into finite element function spaces, *Numer. Math.* **23**, 193–197.

DOUGLAS, J., T. DUPONT and M.F. WHEELER (1974a), An L^∞ estimate and a superconvergence result for a Galerkin method for elliptic equations based on tensor products of piecewise polynomials, *RAIRO Anal. Numér.* **8**, 61–66.

DOUGLAS, J., T. DUPONT and M.F. WHEELER (1974b), A Galerkin procedure for approximating the flux on the boundary for elliptic and parabolic boundary value problems, *RAIRO Anal. Numér.* **8**, 47–59.

DOUGLAS, J. and F. MILNER (1985), Interior and superconvergence estimates for mixed methods for second order elliptic problems, *RAIRO Modél. Math. Anal. Numér.* **19** (3), 397–428.

DUPONT, T. (1973), Galerkin methods for first order hyperbolics: An example, *SIAM J. Numer. Anal.* **10**, 890–899.

ECKHAUS, W. (1972), Boundary layers in linear elliptic singular perturbation problems, *SIAM Rev.* **14**, 225–270.

ECKHAUS, W. (1973), *Matched Asymptotic Expansions and Singular Perturbations* (North-Holland, Amsterdam).

ECKHAUS, W. (1979), *Asymptotic Analysis of Singular Perturbations* (North-Holland, Amsterdam).

ECKHAUS, W. and E.M. DE JAEGER (1966), Asymptotic solutions of singular perturbation problems for linear differential equations of elliptic type, *Arch. Rational Mech. Anal.* **23**, 26–86.

ERIKSSON, K. (1985a), Improved accuracy by adapted mesh refinements in the finite element method, *Math. Comp.* **44**, 321–344.

ERIKSSON, K. (1985b), Higher order local rate of convergence by mesh refinement in the finite element method, *Math. Comp.* **45**, 109–142.

ERIKSSON, K. and C. JOHNSON (1988), An adaptive finite element method for linear elliptic problems, *Math. Comp.* **50**, 361–384.

ERIKSSON, K. and V. THOMÉE (1984), Galerkin methods for singular boundary value problems in one space dimension, *Math. Comp.* **42**, 345–367.

FICHERA, G. (1978), *Numerical and Quantitative Analysis*, Survey and Reference Works in Mathematics 3 (Pitman, San Francisco, CA).

FREHSE, J. and R. RANNACHER (1976), Eine L^1-Fehlerabschätzung diskreter Grundlösungen in der Methode der finiten Elemente, Tagungsband "Finite Elemente", Bonner Mathematische Schriften (Universität Bonn, Bonn).

FRENCH, D. (1987), The finite element method for a degenerate elliptic equation, *SIAM J. Numer. Anal.* **24**, 788–815.

GALLAGHER, R.H. (1975), Survey and evaluation of the finite element method in linear fracture mechanics analysis, in: *Computational Fracture Mechanics, Second National Congress on Pressure Vessels and Piping* (ASME, New York) 637–653.

GENIS, A. (1984), On finite element methods for the Euler–Poisson–Darboux equation, *SIAM J. Numer. Anal.* **21**, 1080–1106.

GRISVARD, P. (1976), Behavior of the solution of an elliptic boundary value problem in a polygonal or polyhedral domain, in: B. HUBBARD, ed., *Numerical Solution of Partial Differential Equations* **3** (Academic Press, New York) 207–274.

GRISVARD, P. (1985), *Elliptic Problems in Non-Smooth Domains* (Pitman, Boston, MA).

GRISVARD, P., W. WENDLAND and J.R. WHITEMAN, eds. (1985), *Singularities and Constructive Methods for Their Treatment*, Lecture Notes in Mathematics **1121** (Springer, Berlin).

GÜSMAN, B. (1981), Bounds of Galerkin projections on splines with highly nonuniform meshes, *SIAM J. Numer. Anal.* **18**, 1109–1119.

HALE, J., X. LIN and G. RAUGEL (1988), Upper semicontinuity of attractors in dynamical systems, *Math. Comp.* **50**, 89–123.

HAVERKAMP, R. (1984), Eine Aussage zur L_∞-Stabilität und zur genauen Konvergenzordnung der H_0^1-Projection, *Numer. Math.* **44**, 393–405.

HELFRICH, H.-P. (1974), Fehlerabschätzungen für das Galerkin Verfahren zur Lösung von Evolutionsgleichungen, *Manuscripta Math.* **13**, 219–235.

HELLEN, T.K. (1980), Numerical methods in fracture mechanics, in: G.C. CHELL, ed., *Developments in Fracture Mechanics* **1** (Applied Science Publishers, Barking, England) 145–181.

HEYWOOD, J.G. and R. RANNACHER (1988), Finite element approximation of the nonstationary Navier–Stokes problem, Part III, *SIAM J. Numer. Anal.* **25**, 489–512.

HILBERT, S. (1973), A mollifier useful for approximations in Sobolev spaces and some applications to approximating solutions of differential equations, *Math. Comp.* **27**, 81–89.

HUANG, M. and V. THOMÉE (1981), Some convergence estimates for semidiscrete type schemes for time-dependent nonselfadjoint parabolic equations, *Math. Comp.* **37**, 327–346.

HUANG, M. and V. THOMÉE (1982), On the backward Euler method for parabolic equations with rough initial data, *SIAM J. Numer. Anal.* **19**, 599–603.

HUGHES, T.J.R. and A.N. BROOKS (1979), A multidimensional upwind scheme with no crosswind diffusion, in: T.J.R. HUGHES, ed., *Finite Element Methods for Convection Dominated Flows*, AMD **34** (ASME, New York) 19–35.

HUGHES, T.J.R., L.P. FRANCA and M. MALLET (1986), A new finite element formulation for computational fluid dynamics: I. Symmetric forms of the compressible Euler and Navier–Stokes equations and the second law of thermodynamics, *Comput. Meths. Appl. Mech. Engrg.* **54**, 223–234.

HUGHES, T.J.R., M. MALLET and A. MIZUKAMI (1986), A new finite element formulation for computational fluid dynamics: II. Beyond SUPG, *Comput. Meths. Appl. Mech. Engrg.* **54**, 341–355.

HUGHES, T.J.R. and M. MALLET (1986a), A new finite element formulation for computational fluid dynamics: III. The generalized streamline operator for multidimensional advective-diffusive systems, *Comput. Meths. Appl. Mech. Engrg.* **58**, 305–328.

HUGHES, T.J.R. and M. MALLET (1986b), A new finite element formulation for computational fluid dynamics: IV. A discontinuity-capturing operator for multidimensional advective-diffusive systems, *Comput. Meths. Appl. Mech. Engrg.* **58**, 329–336.

JAMET, P. (1976), Estimations d'erreur pour des éléments finis droits presque dégénérés, *RAIRO Anal. Numér.* **10**, 43–61.

JESPERSSEN, D. (1978), Ritz–Galerkin methods for singular boundary value problems, *SIAM J. Numer. Anal.* **15**, 813–834.

JOHNSON, C., S. LARSSON, V. THOMÉE and L.B. WAHLBIN (1987), Error estimates for spatially discrete approximations of semilinear parabolic equations with nonsmooth initial data, *Math. Comp.* **49**, 331–357.

JOHNSON, C. and NÄVERT, U. (1981), An analysis of some finite element methods for advection–diffusion problems, in: O. AXELSSON, L.S. FRANK and A. VAN DER SLUIS, eds., *Analytical and Numerical Approaches to Asymptotic Problems in Analysis* (North-Holland, Amsterdam) 99–116.

JOHNSON, C., U. NÄVERT and J. PITKÄRANTA (1984), Finite element methods for linear hyperbolic problems, *Comput. Meths. Appl. Mech. Engrg.* **45**, 285–312.

JOHNSON, C. and J. PITKÄRANTA (1986), An analysis of the discontinuous Galerkin method for a scalar hyperbolic equation, *Math. Comp.* **46**, 1–26.

JOHNSON, C. and J. SARANEN (1986), Streamline diffusion methods for the incompressible Euler and Navier-Stokes equations, *Math. Comp.* **47**, 1–18.

JOHNSON, C., A.H. SCHATZ and L.B. WAHLBIN (1987), Crosswind smear and pointwise errors in streamline diffusion finite element methods, *Math. Comp.* **49**, 25–38.

JUNCOSA, M.L. and D.M. YOUNG (1957), On the Crank–Nicolson procedure for solving parabolic partial differential equations, *Proc. Cambridge Philos. Soc.* **53**, 448–461.

KELLOGG, R.B. (1971), Interpolation between subspaces of a Hilbert space, Tech. Note BN-719, University of Maryland, College Park, MD.

KELLOGG, R.B. (1972), Higher order singularities for interface problems, in: A.K. AZIZ, ed., *The Mathematical Foundations of the Finite Element Method* (Academic Press, New York) 589–602.

KONDRAT'EV, V.A. (1967), Boundary problems for elliptic equations in domains with conical or angular points, *Trans. Moscow Math. Soc.* **16**, 227–313.

KREISS, H. and J. OLIGER (1973), *Methods for the Approximate Solution of Time Dependent Problems*, GARP (Global Atmospheric Research Programme) Publication Series **10** (World Meteorological Organization, Geneva).

KREISS, H.O., V. THOMÉE, and O. WIDLUND (1970), Smoothing of initial data and rates of convergence for parabolic difference equations, *Comm. Pure Appl. Math.* **23**, 241–259.

KRIZEK, P. and P. NEITTAANMÄKI (1984), Superconvergence phenomenon in the finite element method arising from averaging gradients, *Numer. Math.* **45**, 105–116.

KRIZEK, P. and P. NEITTAANMÄKI (1987), On superconvergence techniques, *Acta Appl. Math.* **9**, 175–198.

LAASONEN, P. (1967), On the discretization error of the Dirichlet problem in a plane region with corners, *Ann. Acad. Sci. Fenn. Ser. A. I Math.* **408**, 1–16.

LARSSON, S. (1985), On reaction-diffusion equations and their approximation by finite element methods, Thesis, Chalmers University of Technology, Göteborg.

LEONARD, B.P. (1979), A survey of finite differences of opinion on numerical muddling of the incomprehensible defective confusion equation, in: T.J.R. HUGHES, ed., *Finite Element Methods for Convection Dominated Flows*, AMD **34** (ASME, New York) 1–17.

LESAINT, P. and P.A. RAVIART (1974), On the finite element method for solving the neutron transport equation, in: C. DE BOOR, ed., *Mathematical Aspects of Finite Elements in Partial Differential Equations* (Academic Press, New York) 89–123.

LESAINT, P. and M. ZLAMAL (1979), Superconvergence of the gradient of finite element solutions, *RAIRO Anal. Numér.* **13**, 139–168.

LEVINE, N. (1985), Superconvergent recovery of the gradient from piecewise linear finite-element approximations, *IMA J. Numer. Anal.* **5**, 407–427.

LIN Qun, LU Tao and SHEN Shumin (1983), Asymptotic expansion for finite element approximations, Institute of Mathematical Sciences, Chengdu Branch of Academia Sinica.

LIN Qun and XU Jinchao (1985), Linear finite elements with high accuracy, *J. Comput. Math.* **3**.

LIONS, J.L. (1973), *Perturbations Singulières dans les Problèmes aux Limites et en Contrôle Optimal*, Lecture Notes in Mathematics **323** (Springer, New York).

LÖFSTRÖM, J. (1970), Besov spaces in the theory of approximations, *Ann. Mat. Pura Appl.* (*4*) **85**, 93–184.

LORENTZ, G.G. (1966), *Approximation of Functions* (Holt, Rinehart and Winston, New York).

LOUIS, A. (1979), Acceleration of convergence for finite element solutions of the Poisson equation, *Numer. Math.* **33**, 43–53.

LUSKIN, M. and R. RANNACHER (1982a), On the smoothing property of the Galerkin method for parabolic equations, *SIAM J. Numer. Anal.* **19**, 93–113.

LUSKIN, M. and R. RANNACHER (1982b), On the smoothing property of the Crank–Nicolson scheme, *Applicable Anal.* **14**, 117–135.

MERIGOT, M. (1971), Régularité des dérivées de la solution du problème de Dirichlet dans un secteur plan, *C.R. Acad. Sci. Paris*, **273**, 356–359.

MERIGOT, M. (1972), Régularité des dérivées de la solution du problème de Dirichlet dans un secteur plan, *Matematiche* **28**, 1–36.

MERIGOT, M. (1974a), Etude du problème $\Delta u = f$ dans un polygone plane: Inéqualités à priori, *Boll. Un. Mat. Ital.* (*4*) **10**, 577–597.

MERIGOT, M. (1974b), Potentiels et régularité L_p pour les problèmes elliptiques dans un secteur plan, *C.R. Acad. Sci. Paris* **278**, 1487–1489.

NAKAO, M.T. (1986), Superconvergence of the gradient of Galerkin approximations for elliptic problems, *RAIRO Modél. Math. Anal. Numér.* **21**, 679–695.

NATTERER, F. (1975), Über die punktweise Konvergenz finiter Elemente, *Numer. Math.* **25**, 67–77.

NATTERER, F. (1977), Uniform convergence of Galerkin's method for splines on highly nonuniform meshes, *Math. Comp.* **31**, 457–468.

NÄVERT, U. (1982), A finite element method for convection-diffusion problems, Thesis, Chalmers University of Technology and the University of Göteborg, Göteborg.

NIIJIMA, K. (1980a), On a three point difference scheme for a singular perturbation problem without a first derivative term, I, *Mem. Numer. Math.* **7**.

NIIJIMA, K. (1980b), On a three point difference scheme for a singular perturbation problem without a first derivative term, II, *Mem. Numer. Math.* **7**.

NITSCHE, J. (1972a), Interior error estimates of projection methods, in: *Proceedings Conference on Differential Equations and Their Applications* (EQUADIFF 3), J.E. Purkyně University, Brno, 235–239.

NITSCHE, J.A. (1972b), On Dirichlet problems using subspaces with nearly zero boundary conditions, in: A.K. AZIZ, ed., *The Mathematical Foundations of the Finite Element Method* (Academic Press, New York) 603–627.

NITSCHE, J. (1975), L_∞ convergence for finite element approximation, in: *Proceedings Second Conference on Finite Elements*, Rennes, France.

NITSCHE, J. (1976a), Zur lokalen Konvergenz von Projectionen auf finite Elemente, in: *Approximation Theory*, Lecture Notes in Mathematics **556** (Springer, Berlin) 329–346.

NITSCHE, J. (1976b), Der Einfluss von Randsingularitäten beim Ritzschen Verfahren, *Numer. Math.* **25**, 263–278.

NITSCHE, J. (1977), L_∞ convergence of finite element approximations, in: *Mathematical Aspects of Finite Element Methods*, Lecture Notes in Mathematics **606** (Springer, New York) 261–274.

NITSCHE, J. (1979), Interior error estimates for semidiscrete Galerkin approximations for parabolic equations, in: *Proceedings Conference on Progress in the Theory and Practice of the Finite Element Method*, Göteborg.

NITSCHE, J. and A. SCHATZ (1972), On local approximation properties of L_2-projection on spline-subspaces, *Applicable Anal.* **2**, 161–168.

NITSCHE, J.A. and A.H. SCHATZ (1974), Interior estimates for Ritz–Galerkin methods, *Math. Comp.* **28**, 937–958.

NOBLE, M.E. (1954), Coefficient properties of Fourier series with a gap condition, *Math. Ann.* **128**, 55–62 (Correction, *Math. Ann.* **128**, 256).

OGANESYAN, L.A. and L.A. RUKHOVETS (1969), Study of the rate of convergence of variational difference schemes for second order elliptic equations in a two-dimensional field with a smooth boundary, *U.S.S.R. Comput. Math. and Math. Phys.* **9**, 158–183.

OLIGER, J. (1978), Methods for time dependent partial differential equations, in: *Numerical Analysis: Proceedings of Symposia in Applied Mathematics* **22** (AMS, Providence, RI) 87–108.

RAITHBY, G.D. and K.E. TORRANCE (1974), Upstream-weighted differencing schemes and their application to elliptic problems involving fluid flow, *Comput. & Fluids* **2**, 191–206.

RANNACHER, R. and R. SCOTT (1982), Some optimal error estimates for piecewise linear finite element approximations, *Math. Comp.* **38**, 437–445.

RAUGEL, G. (1978), Résolution numérique de problèmes elliptiques dans des domaines avec coins, *C. R. Acad. Sci. Paris Ser A–B* **286** (18), A791–794.

RICE, J.R. (1969), On the convergence of nonlinear spline approximation, in: I.J. SCHOENBERG, ed., *Approximations with Special Emphasis on Spline Functions* (Academic Press, New York) 349–365.

RICHTER, G.R. and R.S. FALK (1984), An analysis of a finite element method for hyperbolic equations, in: R. VICHNEVETSKY and R. STEPLEMAN, eds., *Advances in Computer Methods for Partial Differential Equations*, VI (IMACS, New Brunswick, NJ) 297–300.

RIEMANN, B. (1854), Über die Darstellbarkeit einer Funktion durch eine trigonometrische Reihe, Collected Works, 227–271 (Leipzig, 2nd ed., 1892).

RISCH, U. (1986), Ein hybrides upwind FEM Verfahren und dessen Anwendung auf swach gekoppelte elliptische Differentialgleichungssysteme mit dominanter Konvektion, Dissertation, Magdeburg.

ROSSER, J.B. (1975), Calculation of potential in a sector, MRC Rept. 1535, Madison, WI.

ROSSER, J.B. (1977), Harmonic functions on region with reentrant corners, MRC Rept. 1796, Madison, WI.

SAMMON, P.H. (1982), Convergence estimates for semidiscrete parabolic equation approximations, *SIAM. J. Numer. Anal.* **19**, 68–92.

SAMMON, P.H. (1983), Fully discrete approximation methods for parabolic problems with nonsmooth initial data, *SIAM J. Numer. Anal.* **20**, 437–470.

SCHATZ, A.H. (1980a), The finite element method on polygonal domains, in: *Seminar on Numerical Analysis and Its Applications to Continuum Physics* (Colecao ATAS, Rio de Janeiro) 57–64.

SCHATZ, A.H. (1980b), A weak discrete maximum principle and stability of the finite element method in L^∞ on plane polygonal domains, *Math. Comp.* **34**, 77–91.

SCHATZ, A.H., V. THOMÉE and L.B. WAHLBIN (1980), Maximum norm stability and error estimates in parabolic finite element equations, *Comm. Pure Appl. Math.* **33**, 265–304.

SCHATZ, A.H. and L.B. WAHLBIN (1977), Interior maximum norm estimates for finite element methods, *Math. Comp.* **31**, 414–442.

SCHATZ, A.H. and L.B. WAHLBIN (1978), Maximum norm estimates in the finite element method on plane polygonal domains. Part 1, *Math. Comp.* **32**, 73–109.

SCHATZ, A.H. and L.B. WAHLBIN (1979), Maximum norm estimates in the finite element method on plane polygonal domains, Part 2, Refinements, *Math. Comp.* **33**, 465–492.

SCHATZ, A.H. and L.B. WAHLBIN (1981), On a local asymptotic error estimate in finite elements and its use: numerical examples, in: R. VICHNEVETSKY and R.S. STEPLEMAN, eds., *Advances in Computer Methods for Partial Differential Equations*, IV (IMACS, New Brunswick, NJ) 14–17.

SCHATZ, A.H. and L.B. WAHLBIN (1982), On the quasi-optimality in L_∞ of the $\mathring{H}^1$-projection into finite element spaces, *Math. Comp.* **38**, 1–21.

SCHATZ, A.H. and L.B. WAHLBIN (1983), On the finite element for singularly perturbed reaction-diffusion problems in two and one dimensions, *Math. Comp.* **40**, 47–89.

SCHREIBER, R. (1980), Finite element methods of high order accuracy for singular two-point boundary value problems with non-smooth solutions, *SIAM J. Numer. Anal.* **17**, 547–566.

SCOTT, R. (1973), Finite element convergence for singular data, *Numer. Math.* **21**, 317–327.

SCOTT, R. (1975), Interpolated boundary conditions in the finite element method, *SIAM J. Numer. Anal.* **12**, 404–427.

SCOTT, R. (1976), Optimal L^∞ estimates for the finite element method on irregular meshes, *Math. Comp.* **30**, 681–697.

SENDOV, B. (1969), Some questions of the theory of approximations of functions and sets in the Hausdorff metric, *Russian Math. Surveys* **24**, 143–183.

SMOLLER, J. (1983), *Shock Waves and Reaction-Diffusion Equations* (Springer, New York).

STEIN, E.M. (1970), *Singular Integrals and Differentiability Properties of Functions* (Princeton University Press, Princeton, NJ).

STRANG, G. (1973), Approximation in the finite element method, *Numer. Math.* **19**, 81–98.

STRANG, G. and G.J. FIX (1973), *An Analysis of the Finite Element Method* (Prentice-Hall, Englewood Cliffs, NJ).

THATCHER, R.W. (1976), The use of infinite grid refinements at singularities in the solution of Laplace's equation, *Numer. Math.* **25**, 163–178.

THOMÉE, V. (1972), Spline approximation and difference schemes for the heat equation, in: A.K. AZIZ, ed., *The Mathematical Foundations of the Finite Element Method with Applications to Partial Differential Equations* (Academic Press, New York) 711–746.

THOMÉE, V. (1977), Higher order local approximations to derivatives in the finite element method, *Math. Comp.* **31**, 652–660.

THOMÉE, V. (1979), Some interior estimates for semidiscrete Galerkin approximations for parabolic equations, *Math. Comp.* **33**, 37–62.

THOMÉE, V. (1980), Negative norm estimates and superconvergence in Galerkin methods for parabolic problems, *Math. Comp.* **34**, 93–113.

THOMÉE, V. (1984), *Galerkin Finite Element Methods for Parabolic Problems*, Lecture Notes in Mathematics **1054** (Springer, Berlin).

THOMÉE, V. and L.B. WAHLBIN (1974), Convergence rates of parabolic difference schemes for non-smooth data, *Math. Comp.* **28**, 1–13.

THOMÉE, V. and L.B. WAHLBIN (1975), On Galerkin methods in semilinear parabolic problems, *SIAM J. Numer. Anal.* **12**, 378–389.

THOMÉE, V. and L.B. WAHLBIN (1983), Maximum norm stability and error estimates in Galerkin methods for parabolic equations in one space variable, *Numer. Math.* **41**, 345–371.

WAHLBIN, L.B. (1974a), A dissipative Galerkin method for the numerical solution of first order hyperbolic equations, in: C. DE BOOR, ed., *Mathematical Aspects of Finite Elements in Partial Differential Equations* (Academic Press, New York) 147–169.

WAHLBIN, L.B. (1974b), A dissipative Galerkin method applied to some quasilinear hyperbolic equations, *RAIRO Anal. Numér.* **8**, 109–117.

WAHLBIN, L.B. (1978), Maximum norm error estimates in the finite element method with isoparametric quadratic elements and numerical integration, *RAIRO Anal. Numér.* **12**, 173–202.

WAHLBIN, L.B. (1980), A remark on parabolic smoothing and the finite element method, *SIAM. J. Numer. Anal.* **17**, 33–38.

WAHLBIN, L.B. (1981), A brief survey of parabolic smoothing and how it affects a numerical solution: finite differences and finite elements, in: *Lectures on the Numerical Solution of Partial Differential Equations*, University of Maryland Lecture Notes **20** (University of Maryland Press, College Park, MD).

WAHLBIN, L.B. (1984), On the sharpness of certain local estimates for $\mathring{H}^1$ projections into finite element spaces: Influence of a reentrant corner, *Math. Comp.* **42**, 1–8.

WAHLBIN, L.B. (1985), A comparison of the local behavior of spline L^2 projections, Fourier series and Legendre series, in: P. GRISVARD, W. WENDLAND and J.R. WHITEMAN, eds., *Singularities and Constructive Methods for Their Treatment*, Lecture Notes in Mathematics **1121** (Springer, New York).

WENDLAND, W.L. (1979), *Elliptic Systems in the Plane* (Pitman, San Francisco, CA).

WHEELER, J.A. (1973), Simulation of heat transfer from a warm pipeline buried in permafrost, in: *Proceedings 74th National Meeting of the American Institute of Chemical Engineers.*

WHEELER, M.F. (1973), A priori L^2 error estimates for Galerkin approximations to parabolic partial differential equations, *SIAM. J. Numer. Anal.* **10**, 723–759.

ZLAMAL, M. (1977), Some superconvergence results in the finite element method, in: *Mathematical Aspects of Finite Element Methods*, Lecture Notes in Mathematics **606** (Springer, Berlin) 353–362.

ZLAMAL, M. (1978), Superconvergence and reduced integration in the finite element method, *Math. Comp.* **32**, 663–685.

Subject Index

Mixed and Hybrid Methods

J.E. Roberts

INRIA
B.P. 105
78153 Le Chesnay Cedex, France

J.-M. Thomas

Laboratoire de Mathématiques Appliquées
Université de Pau et C.N.R.S.; URA 1204
avenue de l'Université
64000 Pau, France

HANDBOOK OF NUMERICAL ANALYSIS, VOL. II
Finite Element Methods (Part 1)
Edited by P.G. Ciarlet and J.L. Lions

Contents

CHAPTER I

Introduction

1. Preliminary remarks

The terms "mixed finite element method", "hybrid finite element method", and even "mixed-hybrid finite element method" come essentially from the vocabulary of structural mechanics as do the terms "conforming finite element method" and "equilibrium finite element method"; cf. in particular the historical account of hybrid and mixed elements given by PIAN [1983]. Consider for simplicity the linear elasticity problem describing the displacement $\boldsymbol{u}$ of an elastic body Ω. A finite element method based on the variational formulation associated with the principle of minimization of the potential energy form is called a displacement method as the procedure yields an approximation of the displacement field $\boldsymbol{u}$. However, in structural mechanics one is often more interested in the stress tensor $\boldsymbol{\sigma}$ than in the displacement. Methods have been devised for computing an approximation of $\boldsymbol{\sigma}$ from the approximation of $\boldsymbol{u}$. Alternatively the stress may be characterized as the minimum of the complementary energy form on the space of symmetric tensors satisfying a relation expressing the equilibrium between the internal and boundary forces acting on Ω, and finite element methods based on the associated variational form are termed equilibrium methods. The drawback of these methods is that it is not usually easy to construct an approximation space of tensors satisfying the equilibrium relation and having the required amount of regularity, C^0 in this instance. Mixed methods and hybrid methods were devised to avoid this difficulty.

With the aid of Lagrangian multipliers the above constrained minimization problem may be turned into a saddle point problem. The Hellinger–Reissner principle characterizes the pair $(\boldsymbol{\sigma}, \boldsymbol{u})$ as the saddle point of a Lagrangian where $\boldsymbol{\sigma}$ is allowed to vary over the space of symmetric tensors subject only to regularity constraints, the equilibrium relation now being incorporated in the Lagrangian. Finite element methods based on the corresponding variational principle are called mixed methods. Mixed methods were first introduced in the literature in HERMANN [1967] for the plate problem. We shall say, more generally, that a finite element method is a mixed method if it involves the simultaneous approximation of two or more vector fields defined on the physical domain. These will in general be the principal unknown and an expression involving one or more of its derivatives.

An alternative approach is taken in the development of hybrid methods. The idea is to retain the equilibrium requirement for functions in the space in which the stress is approximated but to relax the regularity requirement. Given a finite element mesh on the domain Ω, a mesh-dependent Lagrangian is constructed for which the unique saddle point is $(\boldsymbol{\sigma}, \boldsymbol{\varphi})$ where $\boldsymbol{\sigma}$ varies over a space of tensors satisfying the equilibrium condition but defined only on the union of the interiors of the elements. The Lagrangian multiplier $\boldsymbol{\varphi}$ is a vector field defined on the union of the boundaries of the elements which turns out to be none other than the trace of the displacement $\boldsymbol{u}$. Finite element methods based on the associated variational formulation are hybrid methods. It is equally possible, starting from the original displacement formulation, to define a Lagrangian, again mesh-dependent, by introducing a multiplier λ to relax the regularity requirement on the space in which the displacement $\boldsymbol{u}$ is sought. The space of multipliers is again a space of vector fields on the union of the boundaries of the elements, and $\boldsymbol{u}$ is looked for in a space of vector fields defined on the interiors of the elements. There is a unique saddle point $(\boldsymbol{u}, \lambda)$ for this Lagrangian and, moreover, the multiplier λ is the normal constraint $\boldsymbol{\sigma}\cdot\boldsymbol{\nu}$ on the boundaries of the elements, $\boldsymbol{\nu}$ denoting a field of unit normal vectors on the union of the boundaries. Methods obtained from such a Lagrangian are also termed hybrid methods. Again we shall adopt a general terminology saying that the method is a hybrid method if it involves the simultaneous approximation of a vector field defined on the union of the elements of the discretization and another defined on the union of the boundaries of the elements. The first reference for hybrid methods is JONES [1964].

An aperçu of these methods is given in CIARLET [1978, Chapter 7].

Besides structural mechanics mixed and hybrid methods have been found useful in many fields. In fluid mechanics for the incompressible Stokes problem, for instance, the role of the displacement is played by the fluid pressure, that of the stress by the fluid velocity, that of the equilibrium condition by the incompressibility condition.

The above ideas will be developed in detail in Sections 2 and 3 in the context of a simpler model problem. First, however, we need to describe the functional framework in which we shall work.

Let Ω be a bounded, open subset of $\mathbb{R}^n$, with $n=2$ or $n=3$, having a boundary Γ, $\Gamma=\partial\Omega$, which is piecewise C^1. For each natural number m denote by $H^m(\Omega)$ the Sobolev space of order m of scalar valued functions on Ω, defined recursively by

$$H^0(\Omega)=L^2(\Omega),$$

$$H^m(\Omega)=\Big\{v\in H^{m-1}(\Omega)\colon \forall\alpha=(\alpha_1,\ldots,\alpha_n)\in\mathbb{N}^n \text{ with } \sum_{i=1}^{n}\alpha_i=m, \qquad (1.1)$$

$$\partial^\alpha v=\frac{\partial^m v}{\partial x_1^{\alpha_1}\cdots\partial x_n^{\alpha_n}}\in L^2(\Omega)\Big\} \quad \text{for all } m\geqslant 1.$$

We put

$$\|v\|_{0,\Omega}=\left(\int_\Omega v^2(x)\,\mathrm{d}x\right)^{1/2},$$

$$|v|_{m,\Omega}=\left(\sum_{|\alpha|=m}\|\partial^\alpha v\|_{0,\Omega}^2\right)^{1/2} \quad \text{for all } m\geqslant 1, \tag{1.2}$$

$$\|v\|_{m,\Omega}=(\|v\|_{m-1,\Omega}^2+|v|_{m,\Omega}^2)^{1/2} \quad \text{for all } m\geqslant 1.$$

The product space

$$(H^m(\Omega))^n=\{\boldsymbol{q}=(q_i)_{1\leqslant i\leqslant n}\colon q_i\in H^m(\Omega) \text{ for all } i=1,\ldots,n\}$$

will be equipped with the following seminorm and norm:

$$|\boldsymbol{q}|_{m,\Omega}=\left(\sum_{i=1}^n |q_i|_{m,\Omega}^2\right)^{1/2}, \qquad \|\boldsymbol{q}\|_{m,\Omega}=\left(\sum_{i=1}^n \|q_i\|_{m,\Omega}^2\right)^{1/2}. \tag{1.3}$$

We shall make use of the following result from functional analysis:

THEOREM 1.1. *The mapping $v\to v|_\Gamma$ defined a priori for functions v regular on $\bar\Omega$ can be extended to a continuous linear mapping called the trace map of $H^1(\Omega)$ into $L^2(\Gamma)$. In other words there exists a constant C, depending only on Ω, such that we have*

$$\|v\|_{0,\Gamma}\leqslant C\|v\|_{1,\Omega} \quad \textit{for all } v\in H^1(\Omega). \tag{1.4}$$

It can be shown that on the kernel $H_0^1(\Omega)$ of the trace mapping the correspondence $v\to|v|_{1,\Omega}$ defines a norm equivalent to the norm $v\to\|v\|_{1,\Omega}$; i.e. the Friedrichs–Poincaré inequality holds on $H_0^1(\Omega)$. The image of the trace mapping, denoted $H^{1/2}(\Gamma)$, is a Hilbert space with norm

$$\|\psi\|_{1/2,\Gamma}=\inf_{\{v\in H^1(\Omega):v|_\Gamma=\psi\}}\|v\|_{1,\Omega}. \tag{1.5}$$

Thus for each vector $\boldsymbol{q}=(q_i)_{1\leqslant i\leqslant n}\in(H^1(\Omega))^n$ the n traces $q_i|_\Gamma$ are defined and belong to $L^2(\Gamma)$, and in particular, the linear combination $\boldsymbol{q}\cdot\boldsymbol{\nu}=\sum_{i=1}^n q_i|_\Gamma \nu_i$, where $\boldsymbol{\nu}=(\nu_i)_{1\leqslant i\leqslant n}$ denotes the unit exterior normal vector to Ω, is in $L^2(\Gamma)$. Imposing less regularity on each component q_i, we can still define the normal trace as an element of a space of distributions on Γ, cf. Theorem 1.2 below. First, though, we need to introduce two more spaces. The dual space of $H^{1/2}(\Gamma)$ is denoted $H^{-1/2}(\Gamma)$ and is a Hilbert space with norm

$$\|\mu\|_{-1/2,\Gamma}=\sup_{\{\psi\in H^{1/2}(\Gamma):\|\psi\|_{1/2,\Gamma}=1\}}\langle\mu,\psi\rangle_\Gamma, \tag{1.6}$$

where $\langle\cdot,\cdot\rangle_\Gamma$ is the duality pairing between $H^{1/2}(\Gamma)$ and $H^{-1/2}(\Gamma)$. The space $H(\text{div};\Omega)$ is defined by

$$H(\text{div};\Omega)=\left\{\boldsymbol{q}=(q_i)_{1\leqslant i\leqslant n}\in(L^2(\Omega))^n\colon \operatorname{div}\boldsymbol{q}=\sum_{i=1}^n\frac{\partial q_i}{\partial x_i}\in L^2(\Omega)\right\} \tag{1.7}$$

and is a Hilbert space with norm

$$\|\boldsymbol{q}\|_{H(\text{div};\Omega)}=\{\|\boldsymbol{q}\|_{0,\Omega}^2+\|\operatorname{div}\boldsymbol{q}\|_{0,\Omega}^2\}^{1/2}. \tag{1.8}$$

THEOREM 1.2. *The mapping $\boldsymbol{q} \to \boldsymbol{q} \cdot \boldsymbol{v}$ defined a priori from $(H^1(\Omega))^n$ into $L^2(\Gamma)$ can be extended to a continuous linear mapping from $H(\mathrm{div};\Omega)$ onto $H^{-1/2}(\Gamma)$. Further we have the following characterization of the norm on $H^{-1/2}(\Gamma)$:*

$$\|\mu\|_{-1/2,\Gamma} = \inf_{\{\boldsymbol{q}\in H(\mathrm{div};\Omega):\, \boldsymbol{q}\cdot\boldsymbol{v}=\mu\}} \|\boldsymbol{q}\|_{H(\mathrm{div};\Omega)}. \tag{1.9}$$

A demonstration of the first part of this theorem can be found in TEMAM [1977, Theorem 1.2, p. 9]. The characterization of the norm was given by THOMAS [1977, Chapter I]. To avoid, whenever possible, working in this space $H^{-1/2}(\Gamma)$ which contains all the functions of $L^2(\Gamma)$, we define, with the aid of Theorem 1.2, the space

$$\mathscr{H}(\mathrm{div};\Omega) = \{\boldsymbol{q}\in H(\mathrm{div};\Omega):\, \boldsymbol{q}\cdot\boldsymbol{v}\in L^2(\Gamma)\} \tag{1.10}$$

which is a Hilbert space with norm

$$\|\boldsymbol{q}\|_{\mathscr{H}(\mathrm{div};\Omega)} = \{\|\boldsymbol{q}\|^2_{H(\mathrm{div};\Omega)} + \|\boldsymbol{q}\cdot\boldsymbol{v}\|^2_{0,\Gamma}\}^{1/2}. \tag{1.11}$$

We shall make use of the following version of Green's formula:

$$\int_\Omega (v\,\mathrm{div}\,\boldsymbol{q} + \mathbf{grad}\,v\cdot\boldsymbol{q})\,\mathrm{d}x = \int_\Gamma v\,\boldsymbol{q}\cdot\boldsymbol{v}\,\mathrm{d}\sigma \quad \text{for all } v\in H^1(\Omega) \text{ and } \boldsymbol{q}\in\mathscr{H}(\mathrm{div};\Omega). \tag{1.12}$$

REMARK 1.1. The reader not familiar with the notation $H^{1/2}(\Gamma)$ or $H^{-1/2}(\Gamma)$ may find it somewhat surprising. It is natural in that these spaces can also be defined as Sobolev spaces of order s, for s not necessarily an integer. The study of such spaces is not within the scope of this work but may be found in LIONS and MAGENES [1968, Chapter I] or in ADAMS [1975, Chapter 7].

Other examples of functional spaces will be used in the following, and corresponding trace theorems will be stated. However, before ending this section we would like to give necessary and sufficient conditions for a locally regular scalar function to belong to $H^1(\Omega)$ and for a locally regular vector function to belong to $H(\mathrm{div};\Omega)$. More precisely, let $\mathscr{T}_h$ be a decomposition of $\bar{\Omega}$ into compact sets T, $\bar{\Omega} = \bigcup_{T\in\mathscr{T}_h} T$, such that the interiors of the sets T of $\mathscr{T}_h$ are pairwise disjoint and such that the boundary ∂T of an element T of $\mathscr{T}_h$ is piecewise C^1. For the sake of simplicity we shall make the following abuses of notation; we shall write

$$\begin{array}{llll} H^m(T) & \text{for } H^m(\mathring{T}), & \|\cdot\|_{m,T} & \text{for } \|\cdot\|_{m,\mathring{T}}, \\ H(\mathrm{div};T) & \text{for } H(\mathrm{div};\mathring{T}), & \|\cdot\|_{H(\mathrm{div};T)} & \text{for } \|\cdot\|_{H(\mathrm{div};\mathring{T})}, \end{array}$$

etc.

THEOREM 1.3. *Let $\mathscr{T}_h$ be such a decomposition of $\bar{\Omega}$, $\bar{\Omega} = \bigcup_{T\in\mathscr{T}_h} T$. A function $v\in L^2(\Omega)$, whose restriction $v|_T$ may be identified with a function $v_T\in H^1(T)$ for each $T\in\mathscr{T}_h$, belongs to $H^1(\Omega)$ if and only if for each interface $T' = T_1\cap T_2$ with $T_1, T_2\in\mathscr{T}_h$, the* *traces of v_{T_1} and of v_{T_2} on T' coincide:*

$$v_{T_1}|_{T'} = v_{T_2}|_{T'} \quad \text{for all } T' = T_1 \cap T_2 \text{ with } T_1, T_2 \in \mathscr{T}_h. \tag{1.13}$$

Similarly a function $\boldsymbol{q} \in (L^2(\Omega))^n$, whose restriction $\boldsymbol{q}|_T$ may be identified with a function $\boldsymbol{q}_T \in \mathscr{H}(\text{div}; T)$ for each $T \in \mathscr{T}_h$, belongs to $\mathscr{H}(\text{div}; \Omega)$ if and only if for each interface $T' = T_1 \cap T_2$ with $T_1, T_2 \in \mathscr{T}_h$, the normal trace of $\boldsymbol{q}_{T_1}$ coincides with the negative of that of $\boldsymbol{q}_{T_2}$:

$$\begin{aligned}&(\boldsymbol{q}_{T_1} \cdot \boldsymbol{v}_{T_1})|_{T'} + (\boldsymbol{q}_{T_2} \cdot \boldsymbol{v}_{T_2})|_{T'} = 0\\ &\qquad \text{for all } T' = T_1 \cap T_2 \text{ with } T_1, T_2 \in \mathscr{T}_h,\end{aligned} \tag{1.14}$$

where $\boldsymbol{v}_T$ is the unit exterior normal vector of T.

Clearly we have $\boldsymbol{v}_{T_1} = -\boldsymbol{v}_{T_2}$ on $T' = T_1 \cap T_2$ so that if $\boldsymbol{v}$ is any vector orthogonal to T' we have

$$\boldsymbol{q}_{T_1} \cdot \boldsymbol{v} = \boldsymbol{q}_{T_2} \cdot \boldsymbol{v} \quad \text{on } T' = T_1 \cap T_2.$$

2. Conforming and equilibrium variational formulations for a model boundary value problem

Many physical phenomena can be modelled by a system of first-order partial differential equations on a domain $\Omega \subset \mathbb{R}^n$ of the form

$$p_i = \sum_{j=1}^{n} a_{ij} \frac{\partial u}{\partial x_j} \quad \text{in } \Omega, \quad 1 \leqslant i \leqslant n, \tag{2.1}$$

where the vector function $\boldsymbol{p} = (p_i)_{1 \leqslant i \leqslant n}$ satisfies the equilibrium relation

$$\operatorname{div} \boldsymbol{p} + f = 0 \quad \text{in } \Omega, \tag{2.2}$$

and the boundary conditions are of the form

$$(\boldsymbol{p} \cdot \boldsymbol{v} - g) + t(u - \bar{u}) = 0 \quad \text{on } \Gamma, \tag{2.3}$$

where Γ is the boundary of Ω; cf. DUVAUT and LIONS [1972, pp. 16–17] for example. For a review of such boundary conditions, cf. DAUTRAY and LIONS [1985, pp. 230–240]. The coefficients a_{ij} of the system, defined on Ω and the function t defined on Γ are known functions dependent on the physical problem. The function f defined on Ω and the functions g and $\bar{u}$ on Γ are data functions. The function t is allowed to be infinite, $t = +\infty$, on a part Γ_0 of Γ thus implying that $u = \bar{u}$ on Γ_0. The steady state heat equation with convection terms neglected is an example of such a model; u denotes the temperature, $\boldsymbol{p}$ is the heat flow, and $\boldsymbol{p} \cdot \boldsymbol{v}$, $\boldsymbol{v}$ a unit vector normal to Γ, is the flux across Γ. Linear models for elasticity problems involve an analogous system relating the displacement field $\boldsymbol{u} = (u_i)_{1 \leqslant i \leqslant n}$ and the stress field $\boldsymbol{\sigma} = (\sigma_{ij})_{1 \leqslant i,j \leqslant n}$. Hooke's law, which gives the stress field in terms of the linearized displacement field $\varepsilon_{ij}(u) = \frac{1}{2}((\partial u_i/\partial x_j) + (\partial u_j/\partial x_i))$, is the analogue of (2.1).

For these examples corresponding to fundamental problems of engineering, we see that it is at least as important to calculate an approximation of $\boldsymbol{p}$, or even the moments of $\boldsymbol{p}\cdot\boldsymbol{v}_\Gamma$ on Γ or of $\boldsymbol{p}\cdot\boldsymbol{v}$ on a surface contained in Ω with normal vector $\boldsymbol{v}$, as to calculate an approximation of u or of the trace $u|_\Gamma$ or of the trace of u on a surface lying in Ω. On the other hand, from a strictly theoretical point of view, an apparently obvious approach would seem to be to begin by simplifying the study of system (2.1), (2.2) by eliminating the unknown $\boldsymbol{p}$, thus considering u as the solution of the second order partial differential equation

$$-\sum_{i,j=1}^{n} \frac{\partial}{\partial x_i}\left(a_{ij}\frac{\partial u}{\partial x_j}\right)=f \quad \text{in } \Omega. \tag{2.4}$$

The classical methods for the numerical approximation of problem (2.1)–(2.3) are based on the following approach: one determines a function u_h satisfying (2.4) in a more or less weak sense—using for instance standard finite difference or finite element methods. Once the function u_h has been calculated, if it is necessary to obtain an approximation of $\boldsymbol{p}$, the simplest procedure is to consider the function $\boldsymbol{p}_h$ whose ith component, $1\leqslant i\leqslant n$, is given by $(p_h)_i=\sum_{j=1}^n a_{ij}(\partial u_h/\partial x_j)$. This naive treatment has the inconvenience of producing a function $\boldsymbol{p}_h$ which satisfies only in a very weak sense the equilibrium condition and for which the two normal traces $\boldsymbol{p}_h\cdot\boldsymbol{v}_{T'}$ across an interface $T'\subset\Omega$ do not agree. Experience shows that it is better, a priori, to choose a method which solves directly for the desired quantities rather than to try to obtain them in an indirect fashion as in the example above. Our objective here is to propose several such methods.

We shall consider in this section, by way of introduction, the case in which the coefficient functions a_{ij} are constant in Ω, and the matrix (a_{ij}) is symmetric and positive definite. Thus the relations (2.1) are invertible:

$$\frac{\partial u}{\partial x_i}=\sum_{j=1}^{n} A_{ij}p_j,\quad 1\leqslant i\leqslant n, \tag{2.5}$$

and the inverse matrix (A_{ij}) is also symmetric and positive-definite. For the moment we shall restrict our attention to the two following model problems:

HOMOGENEOUS DIRICHLET PROBLEM. Given f in $L^2(\Omega)$, find $\boldsymbol{p}\in H(\text{div};\Omega)$ and $u\in H_0^1(\Omega)$ satisfying (2.1) and (2.2).

HOMOGENEOUS NEUMANN PROBLEM. Given f in $L^2(\Omega)$ with $\int_\Omega f\,\mathrm{d}x=0$, find $\boldsymbol{p}\in H_0(\text{div};\Omega)$ and $u\in H^1(\Omega)/\mathbb{R}$ satisfying (2.1) and (2.2), where for the sake of simplicity we suppose that Ω is connected.

Just as $H_0^1(\Omega)$ denotes the kernel of the trace mapping of $H^1(\Omega)$ into $L^2(\Gamma)$, $H_0(\text{div};\Omega)$ denotes the kernel of the normal trace mapping of $H(\text{div};\Omega)$ into $L^2(\Gamma)$. A specific representative of the equivalence class of u in the quotient space $H^1(\Omega)/\mathbb{R}$, for u a solution of the homogeneous Neumann problem, may be chosen by requiring further that a relation such as $\int_\Omega u\,\mathrm{d}x=0$ or such as $\int_\Gamma u\,\mathrm{d}x=0$ be satisfied. (Using the standard abuse of notation we write u for an element of $H^1(\Omega)$ as well as for its equivalence class in $H^1(\Omega)/\mathbb{R}$.)

The solution $(\boldsymbol{p}, u)$ in the case of the homogeneous Dirichlet or the homogeneous Neumann problem may be characterized as the unique solution of a minimization problem. For the homogeneous Dirichlet problem put

$$W^f = \{\boldsymbol{q} \in H(\text{div}; \Omega): \text{div}\, \boldsymbol{q} + f = 0\}, \qquad V = H_0^1(\Omega), \tag{2.6D}$$

and for the homogeneous Neumann problem

$$W^f = \{\boldsymbol{q} \in H_0(\text{div}; \Omega): \text{div}\, \boldsymbol{q} + f = 0\}, \qquad V = H^1(\Omega)/\mathbb{R}. \tag{2.6N}$$

Then in either case $(\boldsymbol{p}, u)$ is the unique element of $W^f \times V$ satisfying (2.1) or equivalently (2.5). On $W^f \times V$ define the functional

$$I(\boldsymbol{q}, v) = \tfrac{1}{2} \int_\Omega \sum_{i=1}^{n} \left(\left(q_i - \sum_{j=1}^{n} a_{ij} \frac{\partial v}{\partial x_j} \right) \left(\sum_{k=1}^{n} A_{ik} q_k - \frac{\partial v}{\partial x_i} \right) \right) \mathrm{d}x. \tag{2.7}$$

For each pair $(\boldsymbol{q}, v)$ in $W^f \times V$ we have

$$I(\boldsymbol{q}, v) = \tfrac{1}{2} \int_\Omega \sum_{i,k=1}^{n} \left(\left(q_i - \sum_{j=1}^{n} a_{ij} \frac{\partial v}{\partial x_j} \right) A_{ik} \left(q_k - \sum_{l=1}^{n} a_{kl} \frac{\partial v}{\partial x_l} \right) \right) \mathrm{d}x \geqslant 0,$$

with the equality being realized, as $[A]$ is positive-definite, only in case $(\boldsymbol{q}, v)$ satisfies the relation $q_i = \sum_{j=1}^{n} a_{ij}(\partial v/\partial x_j)$ for all i, $i = 1, \ldots, n$; i.e. if $(\boldsymbol{q}, v)$ satisfies (2.1). Thus the solution $(\boldsymbol{p}, u)$ of (2.1)–(2.2) with the homogeneous Dirichlet or the homogeneous Neumann condition is the unique element of $W^f \times V$ satisfying

$$I(\boldsymbol{p}, u) = 0 \tag{2.8}$$

or in other words the unique element minimizing I on $W^f \times V$. This presentation was inspired by that in BARLOW [1986].

With this characterization of the solution $(\boldsymbol{p}, u)$ of (2.1)–(2.3), it is natural to try to construct an approximation in the following manner. Let V_h be a finite-dimensional subspace of V and W_h^f a finite-dimensional, affine subspace of $H(\text{div}; \Omega)$ contained in W^f. We consider the following problem: find $(\boldsymbol{p}, u_h)$ satisfying

$$\begin{aligned} &(\boldsymbol{p}_h, u_h) \in W_h^f \times V_h, \\ &I(\boldsymbol{p}_h, u_h) = \inf_{(\boldsymbol{q}_h, v_h) \in W_h^f \times V_h} I(\boldsymbol{q}_h, v_h). \end{aligned} \tag{2.9}$$

Problem (2.9) has a unique solution $(\boldsymbol{p}_h, u_h)$; the uniqueness is an immediate consequence of the characterizations given by (2.15) and (2.16) below. Furthermore, the calculation of the value $I(\boldsymbol{p}_h, u_h) > 0$ furnishes, a posteriori, a control of the error committed in the approximation.

Formulation (2.9) thus seems to provide a natural coupling between the approximations $\boldsymbol{p}_h$ of $\boldsymbol{p}$ and u_h of u. In fact it does not at all. To see this one has only to note that, with the choice (2.6D) as well as with the choice (2.6N), we have

$$I(\boldsymbol{q}, v) = \mathscr{I}_*(\boldsymbol{q}) + \mathscr{I}^f(v), \tag{2.10}$$

with

$$\mathcal{J}^f(v)=\tfrac{1}{2}\int_\Omega \sum_{i,j=1}^n a_{ij}\frac{\partial v}{\partial x_j}\frac{\partial v}{\partial x_i}\,dx-\int_\Omega fv\,dx, \tag{2.11}$$

$$\mathcal{J}_*(q)=\tfrac{1}{2}\int_\Omega \sum_{i,j=1}^n A_{ij}q_jq_i\,dx. \tag{2.12}$$

Hence if (p,u) is the solution of (2.8), then u is characterized as being the unique solution of the minimization problem: find u satisfying

$$u\in V,\qquad \mathcal{J}^f(u)=\inf_{v\in V}\mathcal{J}^f(v); \tag{2.13}$$

and p as being that of the minimization problem: find p satisfying

$$p\in W^f,\qquad \mathcal{J}_*(p)=\inf_{q\in W^f}\mathcal{J}_*(q). \tag{2.14}$$

Since $I(p,u)=0$, we have the relation

$$\mathcal{J}^f(u)+\mathcal{J}_*(p)=0 \tag{2.15}$$

known in mechanics as the complementary energy principle.

The solution (p_h,u_h) of (2.9) is characterized with the aid of the following two minimization problems: find u_h satisfying

$$u_h\in V_h,\qquad \mathcal{J}^f(u_h)=\inf_{v_h\in V_h}\mathcal{J}^f(v_h); \tag{2.16}$$

find p_h satisfying

$$p_h\in W_h^f,\qquad \mathcal{J}_*(p_h)=\inf_{q_h\in W_h^f}\mathcal{J}_*(q_h). \tag{2.17}$$

Since the approximate solution $u_h\in V_h$ is an element of V, we say that u_h is obtained by a conforming approximation method. Similarly, since the approximate solution $p_h\in W_h^f$ is an element of W^f we say that p_h is obtained by an equilibrium approximation method. If $\bar{\mathcal{J}}$ is the energy defined by $\bar{\mathcal{J}}=\mathcal{J}^f(u)=-\mathcal{J}_*(p)$, cf. (2.15), we remark that the conforming approximation permits us to calculate an approximation $\mathcal{J}^f(u_h)$ of $\bar{\mathcal{J}}$ which is of necessity excessive, whereas the equilibrium approximation permits us to calculate an approximation $-\mathcal{J}_*(p_h)$ of $\bar{\mathcal{J}}$ which is necessarily too small:

$$-\mathcal{J}_*(p_h)\leqslant-\mathcal{J}_*(p)=\bar{\mathcal{J}}=\mathcal{J}^f(u)\leqslant\mathcal{J}^f(u_h). \tag{2.18}$$

REMARK 2.1. The possibility to thus obtain an a posteriori bound on the energy $\bar{\mathcal{J}}$ is what motivated the development of equilibrium approximation methods, cf. FRAEIJS DE VEUBEKE [1965, 1973, 1975], FRAEIJS DE VEUBEKE and HOGGE [1972].

REMARK 2.2. Let g be a function in $L^2(\Gamma)$ satisfying $\int_\Gamma g\,ds=0$. If (p,u) is a solution of the system

$$\begin{aligned} &\boldsymbol{p}=\mathbf{grad}\, u && \text{in } \Omega, \\ &\operatorname{div} \boldsymbol{p}=0 && \text{in } \Omega, \\ &\boldsymbol{p}\cdot\boldsymbol{\nu}=g && \text{on } \Gamma, \end{aligned}$$

for Ω a connected open subset of $\mathbb{R}^n$, then as above one can verify that the function u, defined to within a constant, minimizes on $H^1(\Omega)/\mathbb{R}$ the functional

$$\mathscr{J}_g(v)=\tfrac{1}{2}\int_\Omega |\mathbf{grad}\, v|^2\,\mathrm{d}x-\int_\Gamma gv\,\mathrm{d}\sigma, \tag{2.19}$$

whereas the vector function $\boldsymbol{p}$ minimizes the functional

$$\mathscr{J}_*(\boldsymbol{q})=\tfrac{1}{2}\int_\Omega |\boldsymbol{q}|^2\,\mathrm{d}x \tag{2.20}$$

on the affine manifold of vector functions $\boldsymbol{q}$ with $\operatorname{div}\boldsymbol{q}=0$ on Ω and $\boldsymbol{q}\cdot\boldsymbol{\nu}=g$ on Γ. If u minimizes $\mathscr{J}_g(v)$ then we have $\operatorname{div}(\mathbf{grad}\, u)=0$; and if $\boldsymbol{p}$ minimizes $\mathscr{J}_*(\boldsymbol{q})$, we have $\mathbf{rot}\,\boldsymbol{p}=0$ since then $\boldsymbol{p}$ has a scalar potential. These principles are known in fluid mechanics as *Dirichlet's principle* and *Kelvin's principle*, respectively; cf. SERRIN [1959, Section 24]. In the standard terminology of fluid mechanics, the vector variables representing the flow rate are denoted $\boldsymbol{u}$ or $\boldsymbol{v}$, while the scalar variable is the scalar potential of the velocity field or the pressure and is generally denoted φ, ψ, or p.

REMARK 2.3. We shall see later on, cf. Remark 3.1, that the two minimization problems (2.13) and (2.14) (or similarly (2.19) and (2.20)) are dual problems in the sense of the mathematical theory of duality in convex analysis. Usually (2.13) is considered to be the primal problem and (2.14) its dual. In keeping with standard practice we shall say that a formulation is primal, respectively dual, when the elimination of certain variables yields the conforming formulation (2.13), respectively the equilibrium formulation (2.14).

To conclude this presentation of conforming and equilibrium approximations we give the variational formulations associated with these minimization problems:

Conforming variational formulation. Find a solution u of

$$\begin{aligned} &u\in V, \\ &\sum_{i,j=1}^n \int_\Omega a_{ij}\frac{\partial u}{\partial x_j}\frac{\partial v}{\partial x_i}\,\mathrm{d}x=\int_\Omega fv\,\mathrm{d}x \quad \text{for all } v\in V, \end{aligned} \tag{2.21}$$

which leads one to seek an approximation u_h of u satisfying

$$
\begin{gathered}
u_h \in V_h, \\
\sum_{i,j=1}^{n} \int_{\Omega} a_{ij} \frac{\partial u_h}{\partial x_j} \frac{\partial v_h}{\partial x_i}\, \mathrm{d}x = \int_{\Omega} f v_h \, \mathrm{d}x \quad \text{for all } v_h \in V_h.
\end{gathered}
\tag{2.22}
$$

Equilibrium variational formulation. Find a solution $\boldsymbol{p}$ of

$$
\begin{gathered}
\boldsymbol{p} \in W^f, \\
\sum_{i,j=1}^{n} \int_{\Omega} A_{ij} p_j q_i \, \mathrm{d}x = 0 \quad \text{for all } \boldsymbol{q} \in W^0,
\end{gathered}
\tag{2.23}
$$

which leads one to seek an approximation $\boldsymbol{p}_h$ of $\boldsymbol{p}$ satisfying

$$
\begin{gathered}
\boldsymbol{p}_h \in W_h^f, \\
\sum_{i,j=1}^{n} \int_{\Omega} A_{ij} p_{hj} q_{hi} \, \mathrm{d}x = 0 \quad \text{for all } \boldsymbol{q}_h \in W_h^0,
\end{gathered}
\tag{2.24}
$$

where W^0, respectively W_h^0, is W^f, respectively W_h^f, for $f \equiv 0$ and hence is a subspace of $H(\text{div}; \Omega)$ or $H_0(\text{div}; \Omega)$ depending on whether the problem considered is the homogeneous Dirichlet problem or the homogeneous Neumann problem.

3. First examples of mixed and hybrid formulations

The actual construction of a solution $\boldsymbol{p}_h$ of problem (2.24) (equilibrium formulation) poses several practical difficulties. If $\mathscr{T}_h$ is a triangulation of $\bar{\Omega}$, $\bar{\Omega} = \bigcup_{T \in \mathscr{T}_h} T$, and W_h^0 is a finite dimensional subspace of W^0 associated with $\mathscr{T}_h$, then for a test function $\boldsymbol{q}_h$ to belong to W_h^0 it must satisfy the following two constraints:

(i) The normal traces of $\boldsymbol{q}_h$ must be "continuous across the interface" between any two abutting finite elements T_1 and T_2 of $\mathscr{T}_h$, i.e. $\boldsymbol{q}_h \cdot \boldsymbol{\nu}_{T_1}$ and $-\boldsymbol{q}_h \cdot \boldsymbol{\nu}_{T_2}$ must coincide on $T' = T_1 \cap T_2$ (cf. the second part of Theorem 1.3).

(ii) Within each finite element T of $\mathscr{T}_h$ the divergence of $\boldsymbol{q}_h$ must vanish; i.e. $\text{div}(\boldsymbol{q}_h|_T) \equiv 0$ for all $T \in \mathscr{T}_h$; thus, in particular $\int_{\partial T} \boldsymbol{q}_h \cdot \boldsymbol{\nu}_T \, \mathrm{d}\sigma = 0$.

We do not intend to attack head on the practical problem of constructing a basis for the subspace W_h^0. To circumvent this difficulty we shall free ourselves of one of the constraints that should be satisfied by the test functions by using the technique of Lagrangian multipliers.

3.1. *The dual mixed formulation*

The affine subspace defined by (2.6D) in the case of the homogeneous Dirichlet problem and by (2.6N) in the case of the homogeneous Neumann problem may be characterized as being the set of functions $\boldsymbol{q} \in W$ satisfying the constraint div $\boldsymbol{q}+f=0$ in Ω, where W is the space defined by

$$W=H(\operatorname{div};\Omega), \tag{3.1D}$$

$$W=H_0(\operatorname{div};\Omega). \tag{3.1N}$$

Thus we are led to substitute for the minimization problem (2.14) the saddle point problem of the Lagrangian $\mathscr{L}_*(\cdot,\cdot)$ defined on $W\times L^2(\Omega)$ by

$$\mathscr{L}_*(\boldsymbol{q},v)=\mathscr{J}_*(\boldsymbol{q})+\int_\Omega (\operatorname{div}\boldsymbol{q}+f)v\,\mathrm{d}x. \tag{3.2}$$

Thus we seek $(\boldsymbol{p},u)\in W\times L^2(\Omega)$ satisfying

$$\mathscr{L}_*(\boldsymbol{p},v)\leqslant\mathscr{L}_*(\boldsymbol{p},u)\leqslant\mathscr{L}_*(\boldsymbol{q},u)\quad\text{for all } (\boldsymbol{q},v)\in W\times L^2(\Omega). \tag{3.3}$$

The solution is characterized as being the solution of the system

$$\begin{aligned}&(\boldsymbol{p},u)\in W\times L^2(\Omega),\\ &\sum_{i,j=1}^n\int_\Omega A_{ij}p_jq_i\,\mathrm{d}x+\int_\Omega u\operatorname{div}\boldsymbol{q}\,\mathrm{d}x=0\quad\text{for all } \boldsymbol{q}\in W,\\ &\int_\Omega(\operatorname{div}\boldsymbol{p})v\,\mathrm{d}x=-\int_\Omega fv\,\mathrm{d}x\quad\text{for all } v\in L^2(\Omega).\end{aligned} \tag{3.4}$$

If $(\boldsymbol{p},u)$ satisfies (3.4) it is clear that $\boldsymbol{p}$ satisfies (2.23).

Denote, for the moment, by $\hat{u}$ the solution of (2.21). Using Green's formula together with relation (2.5), we see that

$$\begin{aligned}\sum_{i,j=1}^n\int_\Omega A_{ij}p_jq_i\,\mathrm{d}x&=\int_\Omega(\mathbf{grad}\,\hat{u})\cdot\boldsymbol{q}\,\mathrm{d}x\\ &=-\int_\Omega\hat{u}\operatorname{div}\boldsymbol{q}\,\mathrm{d}x\quad\text{for all } \boldsymbol{q}\in W.\end{aligned}$$

Hence the Lagrange multiplier associated with the constraint div $\boldsymbol{q}=f$ is none other than

$$u=\hat{u}. \tag{3.5}$$

3.2. *The primal mixed formulation*

In virtue of (3.5), the solution $(\boldsymbol{p},u)$ actually belongs to $W\times V$. Using Green's formula we establish that the solution $(\boldsymbol{p}, u)$ satisfies

$$\sum_{i,j=1}^{n} \int_{\Omega} A_{ij} p_j q_i \, dx - \int_{\Omega} (\mathbf{grad}\, u) \cdot \boldsymbol{q} \, dx = 0 \quad \text{for all } \boldsymbol{q} \in W,$$

$$-\int_{\Omega} \boldsymbol{p} \cdot \mathbf{grad}\, v \, dx = -\int_{\Omega} f v \, dx \quad \text{for all } v \in V.$$

As W is dense in $(L^2(\Omega))^n$, the pair $(\boldsymbol{p}, u)$ is the solution of the system

$$\begin{aligned} &(\boldsymbol{p}, u) \in (L^2(\Omega))^n \times V, \\ &\sum_{i,j=1}^{n} \int_{\Omega} A_{ij} p_j q_i \, dx - \int_{\Omega} (\mathbf{grad}\, u) \cdot \boldsymbol{q} \, dx = 0 \quad \text{for all } \boldsymbol{q} \in (L^2(\Omega))^n, \\ &-\int_{\Omega} \boldsymbol{p} \cdot \mathbf{grad}\, v \, dx = -\int_{\Omega} f v \, dx \quad \text{for all } v \in V. \end{aligned} \tag{3.6}$$

This formulation, (3.6), can be interpreted as being a characterization of the saddle point of the functional $\mathscr{L}^f(\cdot,\cdot)$ defined on $(L^2(\Omega))^n \times V$ by

$$\mathscr{L}^f(\boldsymbol{q}, v) = \bar{\mathscr{J}}_*(\boldsymbol{q}) - \int_{\Omega} \boldsymbol{q} \cdot \mathbf{grad}\, v \, dx + \int_{\Omega} f v \, dx, \tag{3.7}$$

where $\bar{\mathscr{J}}_*$ is the continuous extension of $\mathscr{J}_*$ to the space $(L^2(\Omega))^n$. This functional can be found in the literature written in the form

$$\begin{aligned} \mathscr{L}^f(\boldsymbol{q}, v) = & -\tfrac{1}{2} \sum_{i,j=1}^{n} \int_{\Omega} A_{ij} q_j q_i \, dx \\ & + \sum_{i=1}^{n} \int_{\Omega} \left(\sum_{j=1}^{n} A_{ij} q_j - \frac{\partial v}{\partial x_i} \right) q_i \, dx + \int_{\Omega} f v \, dx. \end{aligned} \tag{3.7'}$$

The elimination of $\boldsymbol{p}$ from formulation (3.6) gives us again the conforming variational formulation (2.21) of the problem. This type of method is analyzed in BABUŠKA, ODEN and LEE [1977]. Such methods have proved useful in theoretical studies such as the development of new models for nonlinear elasticity, cf. CIARLET and DESTUYNDER [1979a, 1979b], DESTUYNDER [1986].

REMARK 3.1. We have on the one hand

$$\sup_{v \in V} \mathscr{L}^f(\boldsymbol{q}, v) = \begin{cases} \mathscr{J}_*(\boldsymbol{q}), & \text{if } \boldsymbol{q} \in W^f, \\ +\infty, & \text{otherwise,} \end{cases}$$

and on the other

$$\inf_{q\in W} \mathcal{L}^f(q,v) = -\tfrac{1}{2}\sum_{i,j=1}^{n}\int_\Omega a_{ij}\frac{\partial v}{\partial x_j}\frac{\partial v}{\partial x_i}\,\mathrm{d}x + \int_\Omega fv\,\mathrm{d}x.$$

We deduce that

$$\inf_{q\in W}\sup_{v\in V} \mathcal{L}^f(q,v) = \inf_{q\in W^f} \mathcal{J}_*(q) = \mathcal{J}_*(p)$$

and that

$$\sup_{v\in V}\inf_{q\in W} \mathcal{L}^f(q,v) = -\inf_{v\in V} \mathcal{J}^f(v) = -\mathcal{J}^f(u).$$

Then using the complementary energy principle (2.15), we obtain

$$\begin{aligned}\inf_{q\in W^f} \mathcal{J}_*(q) &= \inf_{q\in W}\sup_{v\in V} \mathcal{L}^f(q,v)\\ &= \sup_{v\in V}\inf_{q\in W} \mathcal{L}^f(q,v) = -\inf_{v\in V} \mathcal{J}^f(v).\end{aligned} \tag{3.8}$$

In this form we recognize the minimization problems (2.13) and (2.14) as dual problems in the sense of optimization theory, cf. for example CIARLET [1982, p. 222].

3.3. *The dual hybrid formulation*

To a decomposition $\mathcal{T}_h$ of $\bar{\Omega}$, $\bar{\Omega} = \bigcup_{T\in\mathcal{T}_h} T$, is associated the affine subspace Y^f of $(L^2(\Omega))^n$ defined by

$$\begin{aligned}Y^f = \{q\in(L^2(\Omega))^n\colon \forall T\in\mathcal{T}_h,\ \operatorname{div}(q|_T) + f|_T = 0,\\ \text{in the distributional sense}\}.\end{aligned} \tag{3.9}$$

In other words, Y^f is the set of vector functions $q\in(L^2(\Omega))^n$ satisfying for each $T\in\mathcal{T}_h$

$$\int_T q\cdot\mathbf{grad}\, v_T\,\mathrm{d}x = \int_T f v_T\,\mathrm{d}x \quad \text{for all } v_T\in H_0^1(T).$$

For each $T\in\mathcal{T}_h$, the restriction $q|_T$ of the function $q\in Y^f$ to T belongs to the space $H(\operatorname{div}; T)$. Further, the normal trace $q|_T\cdot\nu_T$ is defined, cf. Theorem 1.2, to be an element of $H^{-1/2}(\partial T)$, the dual space of $H^{1/2}(\partial T)$; we denote by $\langle\cdot,\cdot\rangle_{\partial T}$ this duality. We ascertain that the affine subspace W^f may be characterized as being the set of functions $q\in Y^f$ satisfying the constraint

$$\sum_{T\in\mathcal{T}_h} \langle q|_T\cdot\nu_T, v\rangle_{\partial T} = 0 \quad \text{for all } v\in V.$$

This leads us to associate to the decomposition $\mathcal{T}_h$ of $\bar{\Omega}$ the Lagrangian $\mathcal{D}_*(\cdot,\cdot)$ defined on $Y^f\times V$ by

$$\mathcal{D}_*(q,v) = \mathcal{J}_*(q) - \sum_{T\in\mathcal{T}_h} \langle q|_T\cdot\nu_T, v\rangle_{\partial T}. \tag{3.10}$$

(The choice of $\mathcal{D}_*(\boldsymbol{q}, v)$ as given in (3.10) instead of as

$$\mathcal{J}_*(\boldsymbol{q}) + \sum_{T\in\mathcal{T}_h} \langle \boldsymbol{q}|_T \cdot \boldsymbol{\nu}_T, v\rangle_{\partial T}$$

is made simply to obtain (3.13) below.) Thus we seek an element $(\boldsymbol{p}, u) \in \boldsymbol{Y}^f \times V$ satisfying

$$\mathcal{D}_*(\boldsymbol{p}, v) \leqslant \mathcal{D}_*(\boldsymbol{p}, u) \leqslant \mathcal{D}_*(\boldsymbol{q}, u) \quad \text{for all } (\boldsymbol{q}, v) \in \boldsymbol{Y}^f \times V. \tag{3.11}$$

A pair $(\boldsymbol{p}, u)$ is such an element if and only if it is a solution of the system

$$(\boldsymbol{p}, u) \in \boldsymbol{Y}^f \times V,$$
$$\sum_{i,j=1}^{n} \int_\Omega A_{ij} p_j q_i \,\mathrm{d}x + \sum_{T\in\mathcal{T}_h} \langle \boldsymbol{q}|_T \cdot \boldsymbol{\nu}_T, u\rangle_{\partial T} = 0 \quad \text{for all } \boldsymbol{q} \in \boldsymbol{Y}^0, \tag{3.12}$$
$$\sum_{T\in\mathcal{T}_h} \langle \boldsymbol{p}|_T \cdot \boldsymbol{\nu}_T, v\rangle_{\partial T} = 0, \quad \text{for all } v \in V,$$

where $\boldsymbol{Y}^0$ is $\boldsymbol{Y}^f$ for $f \equiv 0$. If $(\boldsymbol{p}, u)$ is a solution of problem (3.12), then $\boldsymbol{p}$ is the solution of (2.22). It is clear that u is not uniquely defined as an element of V; only its traces on the interfaces are determined by the first equation of (3.12). However, if $\hat{u}$ is the solution of (2.20), we can choose

$$u = \hat{u}. \tag{3.13}$$

REMARK 3.2. In fact with the dual hybrid formulation the Lagrangian multiplier should not be considered as a function in V defined on all Ω but as an element of the product space $\Pi_{T\in\mathcal{T}_h} H^{1/2}(\partial T)$ satisfying certain matching conditions at the interfaces. This will be the point of view taken for the analysis of the method given in Section 5. A variation of considerable practical interest consists of taking the multiplier to be an element of $\Pi_{T\in\mathcal{T}_h} L^2(\partial T)$ satisfying again certain matching conditions at the interfaces.

3.4. *The primal hybrid formulation*

To a decomposition $\mathcal{T}_h$ of $\bar{\Omega}$ we associate in a manner similar to the above a subspace X of $L^2(\Omega)$ defined by

$$X = \{v \in L^2(\Omega): \forall T \in \mathcal{T}_h,\ v|_T \in H^1(T)\}. \tag{3.14}$$

The space V may be characterized as being the space of functions $v \in X$ satisfying the constraint

$$\sum_{T\in\mathcal{T}_h} \langle \boldsymbol{q}\cdot\boldsymbol{\nu}|_T, v_T\rangle_{\partial T} = 0 \quad \text{for all } \boldsymbol{q} \in \boldsymbol{W}.$$

Thus we replace the minimization problem (2.13) by the saddle point problem for the Lagrangian $\mathcal{D}^f(\cdot,\cdot)$ defined on $X \times \boldsymbol{W}$ by

$$\mathcal{D}^f(v, \boldsymbol{q}) = \bar{\mathcal{J}}^f(v) - \sum_{T\in\mathcal{T}_h} \langle \boldsymbol{q}\cdot\boldsymbol{\nu}|_T, v_T\rangle_{\partial T}, \tag{3.15}$$

where $\bar{\mathscr{J}}^f$ is the extension of $\mathscr{J}^f$ to X defined as follows:

$$\bar{\mathscr{J}}^f(v)=\tfrac{1}{2}\sum_{T\in\mathscr{T}_h}\left(\int_T\sum_{i,j=1}^n a_{ij}\frac{\partial v}{\partial x_j}\frac{\partial v}{\partial x_i}\,\mathrm{d}x\right)-\int_\Omega fv\,\mathrm{d}x. \tag{3.16}$$

Thus we seek a solution $(\boldsymbol{p},u)$ of the system

$$\begin{aligned}&(\boldsymbol{p},u)\in W\times X,\\&\sum_{T\in\mathscr{T}_h}\left\{\int_T\sum_{i,j=1}^n a_{ij}\frac{\partial u}{\partial x_j}\frac{\partial v}{\partial x_i}\,\mathrm{d}x-\langle\boldsymbol{p}\cdot\boldsymbol{\nu}_T,v_T\rangle_{\partial T}\right\}=\int_\Omega fv\,\mathrm{d}x\quad\text{for all } v\in X,\\&\sum_{T\in\mathscr{T}_h}\langle\boldsymbol{q}\cdot\boldsymbol{\nu}_T,u_T\rangle_{\partial T}=0\quad\text{for all }\boldsymbol{q}\in W.\end{aligned} \tag{3.17}$$

The solution $(\boldsymbol{p},u)$ of problem (3.17) is such that u is a solution of problem (2.21) and if $\hat{\boldsymbol{p}}$ is the solution of (2.23) we may choose

$$\boldsymbol{p}=\hat{\boldsymbol{p}},$$

as indeed only the normal traces of $\boldsymbol{p}$ on the interfaces are determined by (3.17).

REMARK 3.3. The hybrid formulations have been presented here as subordinate to a decomposition $\mathscr{T}_h$ of $\bar{\Omega}$ which will, of course, coincide with the triangulation of $\bar{\Omega}$ by elements T on which will be constructed approximations of functions. One can, however, generalize this type of hybrid formulation to one subordinate to an arbitrary decomposition of $\bar{\Omega}$ into subdomains. A particular case is that for which the decomposition is the simplest possible, a decomposition into a single "subdomain". In this case the primal hybrid formulation of the homogeneous Dirichlet problem becomes: find a saddle point $(u,\lambda)\in H^1(\Omega)\times H^{-1/2}(\Gamma)$ of the Lagrangian $\mathscr{L}^f(\cdot,\cdot)$ defined by $\mathscr{L}^f(v,\mu)=\mathscr{J}^{-f}(v)-\langle\mu,v\rangle_\Gamma$ for all $(v,\mu)\in H^1(\Omega)\times H^{-1/2}(\Gamma)$. It was for this type of formulation, adapted to a nonhomogeneous Dirichlet problem, that the first mathematical analysis of a finite element method with Lagrangian multipliers was given, cf. BABUŠKA [1973].

4. General orientation

The elementary examples of mixed formulations and of hybrid formulations presented in Section 3 may all be written in the following general form: find a pair (φ,λ) satisfying

$$\begin{aligned}&\varphi\in W,\qquad \lambda\in M,\\&a(\varphi,\psi)+b(\psi,\lambda)=f(\psi)\quad\text{for all }\psi\in W,\\&b(\varphi,\mu)=g(\mu)\qquad\qquad\text{for all }\mu\in M,\end{aligned} \tag{4.1}$$

where W and M are Hilbert spaces, $a(\cdot,\cdot)$ is a symmetric bilinear form on $W\times W$, $b(\cdot,\cdot)$ is a bilinear form on $W\times M$, f is a linear form on W, and g is a linear form on

M. Then given a finite-dimensional subspace W_h of W and a finite-dimensional subspace M_h of M we consider the problem: find a pair (φ_h, λ_h) satisfying

$$\begin{aligned} &\varphi_h \in W_h, \quad \lambda_h \in M_h, \\ &a(\varphi_h, \psi_h) + b(\psi_h, \lambda_h) = f(\psi_h) \quad \text{for all } \psi_h \in W_h, \\ &b(\varphi_h, \mu_h) = g(\mu_h) \quad \text{for all } \mu_h \in M_h. \end{aligned} \tag{4.2}$$

This finite-dimensional problem may then be posed in the obvious manner as a problem of solving a linear system of N_h equations in N_h unknowns, where

$$N_h = \dim W_h + \dim M_h. \tag{4.3}$$

The matrix of this system is symmetric but cannot be positive-definite. Thus the first problem is to find sufficient conditions, bearing on the pair (W_h, M_h) of subspaces, for the matrix of the associated linear system to be invertible. In a more general way Chapter III treats the numerical analysis of this type of problem in an abstract setting. The a priori estimates of the error committed in approximating the solution (φ, λ) of (4.1) by the solution (φ_h, λ_h) of (4.2) are obtained as a function of the error of approximation of (φ, λ) in $W_h \times M_h$.

In the examples given in Section 3, the spaces W and M appearing in the mixed or hybrid formulations are either $L^2(\Omega)$, $H^1(\Omega)$, or closed subspaces of these spaces for the space of scalar functions and either $(L^2(\Omega))^n$, $H(\text{div}; \Omega)$, or closed subspaces of these spaces for the space of vector functions. Chapter II contains the major results concerning interpolation by finite elements of scalar functions in $H^1(\Omega)$ and of vector functions in $H(\text{div}; \Omega)$.

Chapters II and III are independent. Both are fundamental for the analysis of the examples developed in the subsequent chapters.

Chapters IV and V treat essentially the approximation of the model problem governed by the first-order system of partial differential equations

$$p_i = \sum_{j=1}^{n} a_{ij} \frac{\partial u}{\partial x_j}, \quad 1 \leqslant i \leqslant n, \tag{4.4}$$

$$\operatorname{div} \boldsymbol{p} + f = 0, \tag{4.5}$$

posed on a bounded, open domain $\Omega \subset \mathbb{R}^n$ with $n=2$ or $n=3$ with boundary Γ piecewise C^1. The coefficients a_{ij} are supposed to be bounded and measurable in Ω and to satisfy the uniform ellipticity hypothesis

$$\sum_{i,j=1}^{n} a_{ij}(x)\xi_i\xi_j \geqslant \alpha_p |\xi|^2 \quad \text{a.e. on } \Omega \quad \text{for all } \xi \in \mathbb{R}^n, \tag{4.6}$$

for some $\alpha_p > 0$. Unlike in Sections 2 and 3, we shall not suppose the symmetry of $[a]$, i.e. that $a_{ij} = a_{ji}$. Also we shall not maintain throughout the following the notation particular to Sections 2 and 3.

With the above hypotheses concerning the coefficients a_{ij} the matrix $[a]$ may be inverted to obtain

$$\frac{\partial u}{\partial x_i} = \sum_{j=1}^{n} A_{ij} p_j, \quad 1 \leqslant i \leqslant n; \tag{4.7}$$

the coefficients A_{ij} are then bounded and measurable in Ω and satisfy the uniform ellipticity hypothesis

$$\sum_{i,j=1}^{n} A_{ij}(x)\xi_i\xi_j \geqslant \alpha_d|\xi|^2 \quad \text{a.e. on } \Omega \quad \text{for all } \xi \in \mathbb{R}^n, \tag{4.8}$$

for some $\alpha_d > 0$.

Chapter IV is concerned with mixed methods. Primal mixed methods are treated rather rapidly, the dual mixed methods in some more detail. Several algorithms for the solution of the linear system are also given here. In Chapter V hybrid methods are studied. Primal hybrid and dual hybrid methods as well as the method of hybridization of mixed methods are included. In Chapter VI some further examples of mixed methods and of hybrid methods are mentioned, and some extensions and variants of the theory in the literature are cited.

CHAPTER II

Finite Element Type Interpolation of Scalar and Vectorial Functions

5. Lagrangian interpolation: An H^1 approximation of scalar functions

In the examples of mixed and hybrid formulations presented in Section 3, an essential role is played by two Hilbert spaces: the space $H^1(\Omega)$ of scalar functions $v \in L^2(\Omega)$ such that $\mathbf{grad}\, v \in (L^2(\Omega))^n$ and the space $H(\mathrm{div}; \Omega)$ of vector functions $\boldsymbol{q} \in (L^2(\Omega))^n$ such that $\mathrm{div}\, \boldsymbol{q} \in L^2(\Omega)$. In this chapter we shall describe several choices of "finite element type" subspaces of $H^1(\Omega)$ and of $H(\mathrm{div}; \Omega)$.

For this general presentation we shall use the most elementary setting for finite element methods (some generalizations will be presented in Section 7):

(i) Ω will denote a bounded, open, polyhedral subset of $\mathbb{R}^n$, $n=2$ or 3;

(ii) $\mathcal{T}_h$ will be a triangulation of $\bar{\Omega}$ by n-simplexes T of diameter no greater than h (T is a triangle for $n=2$, a tetrahedron for $n=3$), with

$$\bar{\Omega} = \bigcup_{T \in \mathcal{T}_h} T. \tag{5.1}$$

For the definition of a triangulation, cf. CIARLET [1978, p. 38] for example. The boundary of an n-simplex T, i.e. the union of its $n+1$ $(n-1)$-dimensional faces, will be denoted by ∂T, and in a general manner we shall use the symbol T' to represent an arbitrary $(n-1)$-dimensional face of T and we shall write $T' < T$; T' will be an edge of T when $n=2$, a triangular face when $n=3$. We shall also use the following notation:

$$\begin{aligned}
\operatorname{mes} T &= \text{the Euclidian measure of } T \text{ in } \mathbb{R}^n \\
&\qquad (\text{geometric area if } n=2, \text{ geometric volume if } n=3), \\
h_T &= \text{the diameter of } T, \text{ which here, with a triangulation by simplexes,} \\
&\qquad \text{is just the length of the longest edge,} \\
\rho_T &= \text{the radius of the circle inscribed in } T \text{ if } n=2, \text{ of the sphere in-} \\
&\qquad \text{scribed in } T \text{ if } n=3,
\end{aligned} \tag{5.2}$$

and

$$h = \max_{T \in \mathcal{T}_h} h_T. \tag{5.3}$$

Let $\mathcal{H}$ be a set of positive numbers. A family of triangulations $\{\mathcal{T}_h \colon h \in \mathcal{H}\}$ will said to be *regular* if

$$\inf_{h\in\mathscr{H}} h=0, \quad \text{and} \quad \inf_{h\in\mathscr{H}} \min_{T\in\mathscr{T}_h} \rho_T/h_T>0. \tag{5.4}$$

This definition is equivalent to that given by CIARLET [1978, p. 124]. We shall say that the family $\{\mathscr{T}_h: h\in\mathscr{H}\}$ is *uniformly regular* if

$$\inf_{h\in\mathscr{H}} h=0, \quad \text{and} \quad \inf_{h\in\mathscr{H}} \{(\min_{T\in\mathscr{T}_h} \rho_T)/(\max_{T\in\mathscr{T}_h} h_T)\}>0. \tag{5.5}$$

Thus a family of triangulations $\{\mathscr{T}_h: h\in\mathscr{H}\}$ is uniformly regular if and only if it is regular and it satisfies the *inverse hypothesis*: there exists a constant C, independent of h, such that for each $h\in\mathscr{H}$

$$\min_{T\in\mathscr{T}_h} h_T \geqslant Ch, \tag{5.6}$$

cf. CIARLET [1978, p. 140].

REMARK 5.1. The regularity hypothesis (5.4) is quite natural for the theory of interpolation by finite elements. However, in the chapters that follow we shall be rather abusive of the inverse hypothesis (5.6), using it often just to simplify the exposition of certain results.

For each natural number k, we denote by $P_k(T)$ the space of restrictions to T of polynomial functions of degree k in n variables; we have

$$\dim P_k(T)=\frac{(n+k)!}{n!k!}=\begin{cases}\frac{1}{2}(k+1)(k+2), & \text{if } n=2,\\ \frac{1}{6}(k+1)(k+2)(k+3), & \text{if } n=3.\end{cases} \tag{5.7}$$

To an n simplex T and a positive integer k is associated $\Sigma_T^{(k)}$, the principal lattice of order k of the triangle T, i.e. the set of points of T each of whose barycentric coordinates, relative to the vertices of T, is a multiple of $1/k$. We have

$$\operatorname{card} \Sigma_T^{(k)}=\frac{(n+k)!}{n!k!}. \tag{5.8}$$

The Lagrangian interpolant $\mathscr{L}_T^{(k)}v$ on T of order k, $k\geqslant 1$, of a function $v\in C^0(T)$ is by definition the unique function in $P_k(T)$ such that

$$\mathscr{L}_T^{(k)}v(a)=v(a) \quad \text{for all } a\in\Sigma_T^{(k)}. \tag{5.9}$$

Given a triangulation $\mathscr{T}_h$ of $\bar{\Omega}$ and a function $v\in C^0(\bar{\Omega})$, for a positive integer k, we shall denote by $\mathscr{L}_h^{(k)}v$ the function on $\bar{\Omega}$ which coincides with the polynomial function $\mathscr{L}_T^{(k)}(v|_T)$ on T for each simplex $T\in\mathscr{T}_h$. One says that $\mathscr{L}_h^{(k)}v$ is the Lagrangian interpolant of order k on $\mathscr{T}_h$ of the function v. It is essential to remark that $\mathscr{L}_h^{(k)}v$ is, for each positive integer k, a continuous function on $\bar{\Omega}$. Thus one can easily verify that the mapping $\mathscr{L}_h^{(k)}$ is surjective from $C^0(\bar{\Omega})$ onto the space $L_h^{(k)}$ of simplicial Lagrangian interpolants defined by

$$L_h^{(k)}=\{v\in C^0(\bar{\Omega}): \forall T\in\mathscr{T}_h,\ v|_T\in P_k(T)\}. \tag{5.10}$$

For each k, $k\geqslant 1$, the space $L_h^{(k)}$ is a finite-dimensional subspace of $H^1(\Omega)$. In dimensions $n=2$ and $n=3$ one may define $\mathscr{L}_T^{(k)}v$ for $v\in H^2(T)\subset C^0(T)$ but not for $v\in H^1(T)$.[1] With the aid of the following notation, given for each positive integer m,

$$H^m(\mathscr{T}_h)=\{v\in L^2(\Omega)\colon \forall T\in\mathscr{T}_h,\ v|_T\in H^m(T)\}, \tag{5.11}$$

one may characterize $L_h^{(k)}$ as being the fixed point space of the operator $\mathscr{L}_h^{(k)}$ on $H^1(\Omega)\cap H^2(\mathscr{T}_h)$:

$$L_h^{(k)}=\{v\in H^1(\Omega)\cap H^2(\mathscr{T}_h)\colon v=\mathscr{L}_h^{(k)}v\}. \tag{5.12}$$

Thus we have:

THEOREM 5.1. *For each positive integer k and each triangulation $\mathscr{T}_h$ of $\bar{\Omega}$, the mapping $\mathscr{L}_h^{(k)}$ defines a projection of $H^1(\Omega)\cap H^2(\mathscr{T}_h)$ onto $L_h^{(k)}$.*

We also have:

THEOREM 5.2. *Suppose we are given a regular family $\{\mathscr{T}_h\colon h\in\mathscr{H}\}$ of triangulations of $\bar{\Omega}$. For each positive integer k, there exists a constant C, independent of h, such that*

$$\begin{aligned}&|v-\mathscr{L}_h^{(k)}v|_{1,\Omega}+h^{-1}\,\|v-\mathscr{L}_h^{(k)}v\|_{0,\Omega}\leqslant Ch^m|v|_{m+1,\Omega},\\ &\quad \textit{for each } v\in H^{m+1}(\Omega),\quad 1\leqslant m\leqslant k.\end{aligned} \tag{5.13}$$

We shall not give here the demonstrations of these results fundamental for the study of the convergence of finite element methods but refer the reader to RAVIART and THOMAS [1983, pp. 79–103], for example, for the details of the proofs.

We would like now to make precise the idea of interpolation of the trace on the set $\partial\mathscr{T}_h=\bigcup_{T\in\mathscr{T}_h}\partial T$ of a scalar function v continuous on $\bar{\Omega}$. For a triangulation $\mathscr{T}_h$ of $\bar{\Omega}$ and a positive integer k, the space $L_{\partial\mathscr{T}_h}^{(k)}$ is defined by

$$L_{\partial\mathscr{T}_h}^{(k)}=\left\{\psi=(\psi_T)_{T\in\mathscr{T}_h}\in\prod_{T\in\mathscr{T}_h}L^2(\partial T)\colon \exists v\in L_h^{(k)}\text{ s.t. }\forall T\in\mathscr{T}_h,\ v|_{\partial T}=\psi_T\right\}. \tag{5.14}$$

The space $L_{\partial\mathscr{T}_h}^{(k)}$ may also be characterized as being the space to which $\psi=(\psi_T)_{T\in\mathscr{T}_h}$ belongs if and only if it satisfies the following properties:

$$\psi_T\in P_k(\partial T)\quad\text{for each } T\in\mathscr{T}_h, \tag{5.15a}$$

$$\begin{aligned}&\psi_{T_1}=\psi_{T_2}\text{ on } T'\quad\text{whenever } T' \text{ is a face common to the}\\ &\qquad\qquad\text{simplexes } T_1 \text{ and } T_2 \text{ of } \mathscr{T}_h,\end{aligned} \tag{5.15b}$$

where $P_k(\partial T)$ is defined to be the space of restrictions to the boundary ∂T of T of polynomials of degree not greater than k in n variables or equivalently the space of functions in n variables, polynomial of degree not greater than k on each face T' of T and continuous on ∂T. One may easily verify that

$$\dim P_k(\partial T)=\begin{cases}\dim P_k(T), & \text{if } k\leqslant n,\\ \dim P_k(T)-\dim P_{k-n-1}(T), & \text{if } k>n.\end{cases} \tag{5.16}$$

[1] We have adopted the notational convention $H^m(A)=H^m(\mathring{A})$, $\mathring{A}$ the interior of A.

Let $\mathscr{L}^{(k)}_{\partial\mathscr{T}_h}$ be the mapping which associates to an element $\psi=(\psi_T)_{T\in\mathscr{T}_h}\in\Pi_{T\in\mathscr{T}_h}C^0(\partial T)$ the element $(\mathscr{L}^{(k)}_{\partial\mathscr{T}}\psi_T)_{T\in\mathscr{T}_h}$, where $\mathscr{L}^{(k)}_{\partial\mathscr{T}}\psi_T$, the kth-order Lagrangian interpolant of ψ_T, is the function coinciding with $\mathscr{L}^{(k)}_{T'}(\psi_T|_{T'})$ on each face T' of T. As the elements of $L^{(k)}_h$ are of class C^0 (cf. CIARLET [1978, p. 95] for example), for each function $v\in H^1(\Omega)\cap H^2(\partial\mathscr{T}_h)$, we have

$$\mathscr{L}^{(k)}_{\partial T}(v|_{\partial T})=(\mathscr{L}^{(k)}_T(v|_T))|_{\partial T}\quad \text{for all } T\in\mathscr{T}_h,$$

or in other words, if we put

$$\psi_{(v)}=(v|_{\partial T})_{T\in\mathscr{T}_h}, \tag{5.17}$$

then we have

$$\mathscr{L}^{(k)}_{\partial\mathscr{T}_h}\psi_{(v)}=((\mathscr{L}^{(k)}_h v)|_{\partial T})_{T\in\mathscr{T}_h}. \tag{5.18}$$

THEOREM 5.3. *For each positive integer k and each triangulation $\mathscr{T}_h$ of $\bar{\Omega}$, the mapping $v\to\mathscr{L}^{(k)}_{\partial\mathscr{T}_h}\psi_{(v)}$, with $\psi_{(v)}$ given by (5.17), is a surjection of $H^1(\Omega)\cap H^2(\mathscr{T}_h)$ onto $L^{(k)}_{\partial\mathscr{T}_h}$. It is a bijection of $L^{(k)}_h$ onto $L^{(k)}_{\partial\mathscr{T}_h}$ only in case $k\leqslant n$.*

PROOF. In light of Theorem 5.1, it suffices to show that the mapping $v\to\mathscr{L}^{(k)}_{\partial\mathscr{T}_h}\psi_{(v)}$ is a surjection of $L^{(k)}_h$ onto $L^{(k)}_{\partial\mathscr{T}_h}$ and is also a bijection only if $k\leqslant n$. For $k\leqslant n$, the claim follows immediately from the facts that according to (5.16) the space $P_k(T)$ and $P_k(\partial T)$ have the same dimension and that for each ψ, $\psi=(\psi_T)_{T\in\mathscr{T}_h}$, belonging to $L^{(k)}_{\partial\mathscr{T}_h}$, there exists a unique function v in $L^{(k)}_h$ whose restriction to the boundary of T is equal to ψ_T for each T in $\mathscr{T}_h$.

For $k>n$, the dimension of $P_k(T)$ is greater than that of $P_k(\partial T)$, so a bijection from $L^{(k)}_h$ onto $L^{(k)}_{\partial\mathscr{T}_h}$ cannot exist. To show surjectivity in this case, we have only to remark that given $\psi=(\psi_T)_{T\in\mathscr{T}_h}\in L^{(k)}_{\partial\mathscr{T}_h}$, the function $v\in L^{(k)}_h$, for example, defined by its values at the lattice points $a\in\bigcup_{T\in\mathscr{T}_h}\Sigma^{(k)}_T$ as follows:

$$v(a)=\begin{cases}\psi_T(a), & \text{if } a\in\partial T \text{ for some } T\in\mathscr{T}_h,\\ 0, & \text{otherwise,}\end{cases}$$

defines an antecedent of ψ. □

Before announcing an error bound for interpolation in $L^{(k)}_{\partial\mathscr{T}_h}$, we would like to define a norm on the space $\Pi_{T\in\mathscr{T}_h}H^{1/2}(\partial T)$ which will be seen to be well-adapted to the analysis of dual hybrid methods. Thus, given a triangulation $\mathscr{T}_h$ of $\bar{\Omega}$, we put

$$\begin{aligned}|||v|||_{1,T}&=\{|v|^2_{1,T}+h^{-2}\|v\|^2_{0,T}\}^{1/2} && \text{for all } v\in H^1(T),\ T\in\mathscr{T}_h,\\ |||\psi|||_{1/2,\partial T}&=\inf_{\{v\in H^1(T):v|_{\partial T}=\psi\}}|||v|||_{1,T} && \text{for all } \psi\in H^{1/2}(\partial T),\ T\in\mathscr{T}_h,\\ |||\psi|||_{1/2,\partial\mathscr{T}_h}&=\Big\{\sum_{T\in\mathscr{T}_h}|||\psi_T|||^2_{1/2,\partial T}\Big\}^{1/2} && \text{for all } \psi=(\psi_T)_{T\in\mathscr{T}_h}\in\prod_{T\in\mathscr{T}_h}H^{1/2}(\partial T).\end{aligned} \tag{5.19}$$

We may now state:

THEOREM 5.4. *Suppose $\{\mathscr{T}_h: h\in\mathscr{H}\}$ is a regular family of triangulations of Ω. For* *each positive integer k, there exists a constant C, independent of h, such that if* $v \in H^{m+1}(\Omega)$, $1 \leqslant m \leqslant k$, *and* $\psi = \psi_{(v)}$ *is determined by* (5.17), *then*

$$|||\psi - \mathscr{L}^{(k)}_{\partial\mathscr{T}_h}\psi|||_{1/2,\partial\mathscr{T}_h} \leqslant Ch^m |v|_{m+1,\Omega}.$$

PROOF. With the norms defined in (5.19) we have

$$|||\psi - \mathscr{L}^{(k)}_{\partial\mathscr{T}_h}\psi|||^2_{1/2,\partial\mathscr{T}_h} \leqslant \|v - \mathscr{L}^{(k)}_h v\|^2_{1,\Omega} + h^{-2}\|v - \mathscr{L}^{(k)}_h v\|^2_{0,\Omega},$$

so the demonstration is completed on applying Theorem 5.2. □

6. Interpolation of the moments of a vectorial function: An H(div) approximation of vector valued functions

We retain here the assumptions made in Section 5 concerning the domain $\Omega \subset \mathbb{R}^n$, $n = 2$ or 3, and the triangulation $\mathscr{T}_h$, $\bar{\Omega} = \bigcup_{T\in\mathscr{T}_h} T$. Given a positive integer k, we shall associate to each n-simplex $T \in \mathscr{T}_h$ a subspace $D_k(T)$ of the product space $(P_k(T))^n$ and an interpolation operator $\mathscr{E}^{(k)}_T$ defined on a subspace $\mathscr{H}(\text{div}; T)$ of $H(\text{div}; T)$ and having values in the space $D_k(T)$. Denoting by $\mathscr{H}(\text{div}; \mathscr{T}_h)$ the space defined by

$$\mathscr{H}(\text{div}; \mathscr{T}_h) = \{\boldsymbol{q} \in (L^2(\Omega))^n : \forall T \in \mathscr{T}_h,\ \boldsymbol{q}|_T \in \mathscr{H}(\text{div}; T)\}, \tag{6.1}$$

we can associate to each vector function $\boldsymbol{q} \in \mathscr{H}(\text{div}; \mathscr{T}_h)$ the function $\mathscr{E}^{(k)}_h \boldsymbol{q} \in (L^2(\Omega))^n$ determined by

$$(\mathscr{E}^{(k)}_h \boldsymbol{q})|_T = \mathscr{E}^{(k)}_T(\boldsymbol{q}|_T) \quad \text{for all } T \in \mathscr{T}_h. \tag{6.2}$$

The spaces $D_k(T)$ and the operators $\mathscr{E}^{(k)}_T$ will be constructed in such a way that the following two hypotheses are satisfied

$$\begin{aligned} &\operatorname{div} \boldsymbol{q} = 0 \;\Rightarrow\; \operatorname{div}(\mathscr{E}^{(k)}_T \boldsymbol{q}) = 0 \\ &\quad \text{for all } T \in \mathscr{T}_h, \text{ and for all } \boldsymbol{q} \in \mathscr{H}(\text{div}; T), \end{aligned} \tag{6.3}$$

and

$$\mathscr{E}^{(k)}_h \boldsymbol{q} \in H(\text{div}; \Omega) \quad \text{for all } \boldsymbol{q} \in H(\text{div}; \Omega) \cap \mathscr{H}(\text{div}; \mathscr{T}_h). \tag{6.4}$$

Property (6.3) is a local property of the interpolation of vector functions in $\mathscr{H}(\text{div}; T)$. Its analogue for the approximation of scalar functions in $H^1(T)$ is that the interpolant of a function whose gradient vanishes also has a trivial gradient; i.e., the interpolant of a constant function is a constant. However we did not think it necessary to state such a natural requirement.

Property (6.4) is a continuity condition for the interpolant of a vector function having a certain amount of regularity on $\bar{\Omega}$. Here of course the scalar analogue, that the interpolant $\mathscr{L}^{(k)}_h v$ be in $H^1(\Omega)$ whenever $v \in H^1(\Omega) \cap H^2(\mathscr{T}_h)$, is satisfied as the space $L^{(k)}_h$ is a subspace of $C^0(\bar{\Omega})$.

REMARK 6.1. Almost reflexively, a mathematician's first idea is to try to reduce the study of interpolation of vector functions to that of scalar functions as given in the preceding section by choosing

$$D_k(T)=(P_k(T))^n,$$
$$\mathscr{H}(\mathrm{div},T)=(H^2(T))^n,$$
$$\mathscr{E}_T^{(k)}=(\mathscr{L}_T^{(k)})^{\otimes n}\quad \text{(kth-order Lagrangian interpolation on T carried out component by component).}$$

One may check that condition (6.4) is satisfied; however, (6.3) is not. To see this consider in two dimensions, on the reference triangle $\hat{T}$ with vertices (0, 0), (1, 0) and (0, 1), the function $\boldsymbol{q}=(q_1,q_2)$ given by

$$q_1(x_1,x_2)=\tfrac{1}{2}x_1^2-x_1x_2,$$
$$q_2(x_1,x_2)=\tfrac{1}{2}x_2^2-x_1x_2.$$

This function $\boldsymbol{q}$ is divergence-free. Yet its Lagrangian interpolant in $(P_1(T))^2$ is none other than the function $x\to(\frac{1}{2}x_1,\frac{1}{2}x_2)$ in $\hat{T}$ whose divergence is the constant function $x\to 1$.

We shall give two families of examples of $H(\mathrm{div})$ approximation; the first examples, the original examples due to THOMAS [1977, Chapter III], cf. also RAVIART and THOMAS [1977] for the two-dimensional case and NEDELEC [1980] for the three-dimensional case, we present in some detail; the second examples due to BREZZI, DOUGLAS and MARINI [1985] in the two-dimensional case, and to BREZZI, DOUGLAS, DURAN and FORTIN [1987] in the three-dimensional case will be described more briefly. Yet another family of examples of $H(\mathrm{div})$ approximation may be found in NEDELEC [1986].

For each positive integer k we put

$$D_k=(P_{k-1})^n\oplus \boldsymbol{x}P_{k-1} \tag{6.5}$$

and define $D_k(T)$ to be the space of restrictions to T of the functions of D_k. In other words, the vector function $\boldsymbol{q}=(q_1,\ldots,q_n)$ belongs to $D_k(T)$ if and only if there exists $n+1$ scalar polynomial functions $q_i^*\in P_{k-1}(T)$, $i=0,1,\ldots,n$, such that

$$q_i(x)=q_i^*(x)+x_iq_0^*(x)\quad \text{for all } x\in T,\quad i=1,\ldots,n. \tag{6.6}$$

The expression (6.6) for $\boldsymbol{q}$ may be made unique by further requiring that q_0^* be homogeneous of degree $k-1$. Thus we have

$$\dim D_k(T)=(n+1)\dim P_{k-1}(T)-\dim P_{k-2}(T), \tag{6.7}$$

which together with (5.7) gives us

$$\dim D_k(T)=(n+k)\,\frac{(n+k-2)!}{(n-1)!(k-1)!},$$

or more specifically

$$\dim D_k(T)=\begin{cases}k(k+2), & \text{if } n=2,\\ \tfrac{1}{2}k(k+1)(k+3), & \text{if } n=3.\end{cases}$$

For each nonnegative integer l we denote by $\mathscr{M}_l(T)$ the space of linear forms on $(C^0(T))^n$ defined by

$$\mathscr{M}_l(T)=\left\{\left(q\to\int_T q\cdot r\,\mathrm{d}x\right)\in\mathscr{L}((C^0(T))^n,\mathbb{R}):r\in(P_l)^n\right\},\tag{6.8}$$

i.e. an element of $\mathscr{M}_l(T)$ assigns to the vector function q on T one of its moments on T of order no greater than l. In keeping with classical notation we shall write $\mathscr{M}_l(T)=\{0\}$ for l a negative integer, in particular for $l=-1$. Similarly for each nonnegative integer l and each face T' of T we denote by $\mathscr{M}_l(T')$ the space of linear forms on $(C^0(T))^n$ defined by

$$\mathscr{M}_l(T')=\left\{\left(q\to\int_{T'} q\cdot\nu_T w\,\mathrm{d}\sigma\right)\in\mathscr{L}((C^0(T))^n,\mathbb{R}):w\in P_l\right\},\tag{6.9}$$

where ν_T is the unit exterior normal to ∂T, constant on each face T' of T. Thus an element of $\mathscr{M}_l(T')$ assigns to the vector function q on T one of the moments on T' of order no greater than l of its exterior normal trace $q\cdot\nu_T$. Now we define, for k a positive integer,

$$\mathscr{M}_T^{(k)}=\left(\bigcup_{T'<T}\mathscr{M}_{k-1}(T')\right)\cup\mathscr{M}_{k-2}(T),\tag{6.10}$$

where $T'<T$ means T' is an $(n-1)$-dimensional face of T. Without change of notation we shall consider henceforth that the forms in $\mathscr{M}_T^{(k)}$ are defined on the space $\mathscr{H}(\mathrm{div};T)$,

$$\mathscr{H}(\mathrm{div};T)=\{q\in H(\mathrm{div};T):q\cdot\nu_T\in L^2(\partial T)\}.\tag{6.11}$$

The dimension of the space $\mathscr{M}_T^{(k)}$ is given by

$$\dim\mathscr{M}_T^{(k)}=(n+1)\dim P_{k-1}(\mathbb{R}^{n-1})+n\dim P_{k-2}(T),$$

or more explicitly, by

$$\dim\mathscr{M}_T^{(k)}=(n+k)\frac{(n+k-2)!}{(n-1)!(k-1)!}.\tag{6.12}$$

Thus we see that

$$\dim D_k(T)=\dim\mathscr{M}_T^{(k)},\quad k=1,2,\ldots.\tag{6.13}$$

THEOREM 6.1. *Let T be an n simplex and k a positive integer. For each $q\in\mathscr{H}(\mathrm{div};T)$ there exists one and only one function, which we shall denote $\mathscr{E}_T^{(k)}q$, in $D_k(T)$ for which*

$$m(\mathscr{E}_T^{(k)}q)=m(q)\quad\textit{for all } m\in\mathscr{M}_T^{(k)}.\tag{6.14}$$

The mapping $\mathscr{E}_T^{(k)}:\mathscr{H}(\mathrm{div};T)\to D_k(T)$ thus defined further satisfies the commuting *diagram property*

$$\begin{array}{ccc} \mathscr{H}(\mathrm{div};T) & \xrightarrow{\mathrm{div}} & L^2(T) \\ \Big\downarrow \mathscr{E}_T^{(k)} & & \Big\downarrow \Pi_T^{(k-1)} \\ D_k(T) & \xrightarrow{\mathrm{div}} & P_{k-1}(T) \end{array}$$

i.e.

$$\operatorname{div}(\mathscr{E}_T^{(k)}\boldsymbol{q}) = \Pi_T^{(k-1)}(\operatorname{div}\boldsymbol{q}) \quad \textit{for all } \boldsymbol{q} \in \mathscr{H}(\mathrm{div};T), \tag{6.15}$$

where $\Pi_T^{(k-1)}$ denotes the orthogonal projection in $L^2(T)$ onto $P_{k-1}(T)$.

PROOF. Suppose for the moment that the operator $\mathscr{E}_T^{(k)}$ from $\mathscr{H}(\mathrm{div};T)$ to $D_k(T)$ satisfying (6.14) is well defined and let $\boldsymbol{q} \in \mathscr{H}(\mathrm{div};T)$. Then it is clear that the divergence of any element of $D_k(T)$, and in particular of $\mathscr{E}_T^{(k)}\boldsymbol{q}$, is a polynomial of degree not exceeding $k-1$. Thus using Green's formula with (6.14) we have

$$\int_T w \operatorname{div}(\mathscr{E}_T^{(k)}\boldsymbol{q})\,\mathrm{d}x = \int_T w \operatorname{div}\boldsymbol{q}\,\mathrm{d}x \quad \text{for all } w \in P_{k-1}(T)$$

or equivalently (6.15).

To show that, given $\boldsymbol{q} \in \mathscr{H}(\mathrm{div};T)$, (6.14) does indeed determine a unique element of $D_k(T)$, in view of (6.13) it suffices to show that if $\boldsymbol{p} \in D_k(T)$ and $m(\boldsymbol{p})=0$ for each $m \in \mathscr{M}_T^{(k)}$, then $\boldsymbol{p}$ must be trivial. It is easy to check that the normal trace on any face T' of T of any element of $D_k(T)$ is a polynomial of degree at most $k-1$. Thus, the normal trace of $\boldsymbol{p}$ on T' must be trivial. The divergence of $\boldsymbol{p}$, a priori a polynomial of degree at most $k-1$ on T, can then be shown to vanish by using Green's formula. It follows that $\boldsymbol{p}$ belongs to the subspace $(P_{k-1}(T))^n$ of $D_k(T)$; to see this simply apply Euler's identity for homogeneous functions: $\boldsymbol{x}\cdot\mathbf{grad}\, f(\boldsymbol{x}) = \alpha f(\boldsymbol{x})$ whenever f is homogeneous of degree α. Next we show that if $\boldsymbol{v}'$ is a vector normal to a face T' of T that $\boldsymbol{p}\cdot\boldsymbol{v}'$ vanishes on T. As $\boldsymbol{p}\cdot\boldsymbol{v}'$ vanishes on T', we have that $\boldsymbol{p}\cdot\boldsymbol{v}' = bw$, where b is the first degree polynomial whose zeroes form the $(n-1)$-dimensional hyperplane containing T', and $w \in P_{k-2}$. Now $w\boldsymbol{v}' \in (P_{k-2})^n$ and we have

$$\int_T bww\,\mathrm{d}x = \int_T \boldsymbol{p}\cdot(w\boldsymbol{v}')\,\mathrm{d}x = 0.$$

As b is either always positive or always negative on T, w and hence $\boldsymbol{p}\cdot\boldsymbol{v}'$ vanish on T. The theorem now follows on noting that for any n of the $n+1$ faces of T, the corresponding normal vectors form a basis for $\mathbb{R}^n$. □

For each $T \in \mathscr{T}_h$ we have now constructed an interpolation operator $\mathscr{E}_T^{(k)}$ from

$\mathscr{H}(\text{div}; T)$ into $D_k(T) \subset (P_k(T))^n$ satisfying (6.15) and a fortiori (6.3). To a triangulation $\mathscr{T}_h$ of $\bar{\Omega}$ we associate the space $\mathscr{H}(\text{div}; \mathscr{T}_h)$ defined by (6.1) together with (6.11):

$$\begin{aligned}&\mathscr{H}(\text{div}; \mathscr{T}_h)\\ &\quad = \{\boldsymbol{q} \in (L^2(\Omega))^n : \forall T \in \mathscr{T}_h, \boldsymbol{q}|_T \in H(\text{div}; T) \text{ and } \boldsymbol{q}|_T \cdot \boldsymbol{v}_T \in L^2(\partial T)\},\end{aligned} \tag{6.16}$$

and the equilibrium interpolation operator $\mathscr{E}_h^{(k)}$ of order k on $\mathscr{H}(\text{div}; \mathscr{T}_h)$ constructed from the operators $\mathscr{E}_T^{(k)}$ with the aid of (6.2).

To see that $\mathscr{E}_h^{(k)}$ satisfies (6.4), we note that if $\boldsymbol{q}$ is sufficiently regular in the sense that $\boldsymbol{q} \in H(\text{div}; \Omega) \cap \mathscr{H}(\text{div}; \mathscr{T}_h)$ then Theorem 1.3 implies

$$\begin{aligned}&\boldsymbol{q}|_{T_1} \cdot \boldsymbol{v}_{T_1} + \boldsymbol{q}|_{T_2} \cdot \boldsymbol{v}_{T_2} = 0 \quad \text{on } T'\\ &\quad \text{for all } T' = T_1 \cap T_2 \text{ with } T_1, T_2 \in \mathscr{T}_h.\end{aligned} \tag{6.17}$$

Thus, for each $T' = T_1 \cap T_2$ with $T_1, T_2 \in \mathscr{T}_h, \boldsymbol{r} = \mathscr{E}_h^{(k)} \boldsymbol{q}$ satisfies

$$\int_{T'} (\boldsymbol{r}|_{T_1} \cdot \boldsymbol{v}_{T_1} + \boldsymbol{r}|_{T_2} \cdot \boldsymbol{v}_{T_2}) w \, ds = 0 \quad \text{for all } w \in P_{k-1}.$$

The restrictions $\boldsymbol{r}|_T \in D_k(T)$, $T \in \mathscr{T}_h$, are polynomial of degree at most k and their normal traces on ∂T are piecewise polynomial of degree at most $k-1$; indeed, on each face T' of T, $T \in \mathscr{T}_h$, $\boldsymbol{x} \cdot \boldsymbol{v}_T$ is constant. With this observation of central importance we see that $\boldsymbol{r}$ satisfies the continuity condition (6.17) and hence $\boldsymbol{r} = \mathscr{E}_h^{(k)} \boldsymbol{q} \in H(\text{div}; \Omega)$; i.e. (6.4) holds.

We denote by $E_h^{(k)}$ the finite-dimensional space of equilibrium interpolants

$$E_h^{(k)} = \{\boldsymbol{q} \in H(\text{div}; \Omega) : \forall T \in \mathscr{T}_h, \boldsymbol{q}|_T \in \boldsymbol{D}_k(T)\}, \tag{6.18}$$

which, as we have just seen, is the space of functions $\boldsymbol{q} \in (L^2(\Omega))^n$ for which the restriction $\boldsymbol{q}|_T$ may be identified with a function in $D_k(T)$ for each $T \in \mathscr{T}_h$ and for which (6.17) holds. In analogy with (5.12), $E_h^{(k)}$ may also be characterized as a space of fixed points as follows:

$$E_h^{(k)} = \{\boldsymbol{q} \in H(\text{div}; \Omega) \cap \mathscr{H}(\text{div}; \mathscr{T}_h) : \mathscr{E}_h^{(k)}(\boldsymbol{q}) = \boldsymbol{q}\}. \tag{6.18$'$}$$

Thus we have:

THEOREM 6.2. *For each positive integer k, and each triangulation $\mathscr{T}_h$ of $\bar{\Omega}$, the mapping $\mathscr{E}_h^{(k)}$ is a projection from $H(\text{div}; \Omega) \cap \mathscr{H}(\text{div}; \mathscr{T}_h)$ onto $E_h^{(k)}$.*

The next theorem gives a bound for the error of interpolation by $\mathscr{E}_h^{(k)}$:

THEOREM 6.3. *Suppose $\{\mathscr{T}_h : h \in \mathscr{H}\}$ is a regular family of triangulations of $\bar{\Omega}$ and k is a positive integer. There exists a constant C, independent of h, such that*

$$\begin{aligned}&\|\boldsymbol{q} - \mathscr{E}_h^{(k)} \boldsymbol{q}\|_{0,\Omega} \leqslant C h^l |\boldsymbol{q}|_{l,\Omega}\\ &\quad \textit{for all } \boldsymbol{q} \in (H^l(\Omega))^n, \quad 1 \leqslant l \leqslant k,\end{aligned} \tag{6.19}$$

and

$$\|\operatorname{div}(\boldsymbol{q}-\mathscr{E}_h^{(k)}\boldsymbol{q})\|_{0,\Omega} \leqslant Ch^l|\operatorname{div}\boldsymbol{q}|_{l,\Omega}$$
$$\textit{for all } \boldsymbol{q}\in(H^1(\Omega))^n \textit{ with } \operatorname{div}\boldsymbol{q}\in H^l(\Omega),\quad 0\leqslant l\leqslant k. \tag{6.20}$$

PROOF. Estimate (6.20) is a classical corollary of property (6.15) identifying $\operatorname{div}(\mathscr{E}_T^{(k)}\boldsymbol{q})$ as the image under the orthogonal L^2 projection onto $L_h^{(k)}$ of $\operatorname{div}\boldsymbol{q}$, and we shall not give the details of its demonstration. The argument justifying (6.19) is more technical: for each $T\in\mathscr{T}_h$ we must find a bound for the quantity

$$\|\boldsymbol{q}-\mathscr{E}_T^{(k)}\boldsymbol{q}\|_{0,T}=\left[\sum_{i=1}^n \|q_i-(\mathscr{E}_T^{(k)}\boldsymbol{q})_i\|_{0,T}^2\right]^{1/2};$$

however, this cannot be done component by component as the interpolation space is not a simple tensor product of n copies of the same space of polynomials. Thus it is necessary to work directly with the subspaces of vector functions of n variables with values in $\mathbb{R}^n$. We employ the technique of reducing the question of the error bound on T to that on the reference triangle $\hat{T}$ as in CIARLET and RAVIART [1972]. For each $T\in\mathscr{T}_h$, there exists an affine transformation $F\colon \hat{T}\to T=F(\hat{T})$, bijective from $\hat{T}$ onto T. We denote by DF the linear tangent mapping and by J the Jacobian of the transformation. We have here $|J|=(\operatorname{mes} T)/(\operatorname{mes}\hat{T})$. To each scalar function $\hat{v}\in H^1(\hat{T})$, classically is associated the function $v\in H^1(T)$, defined by

$$v(x)=\hat{v}(\hat{x})\quad \text{for each } x=F(\hat{x})\in T. \tag{6.21}$$

Here, to each vector function $\hat{\boldsymbol{q}}\in H(\operatorname{div};\hat{T})$ we associate the function $\boldsymbol{q}$, that we shall show belongs to $H(\operatorname{div}; T)$, defined by

$$\boldsymbol{q}(x)=\frac{1}{|J|}\mathrm{D}F\,\hat{\boldsymbol{q}}(\hat{x})\quad \text{for each } x=F(\hat{x})\in T. \tag{6.22}$$

The transformation (6.22) was constructed in such a way that

$$\int_T \boldsymbol{q}\cdot\mathbf{grad}\, v\,\mathrm{d}x=\int_{\hat{T}}\hat{\boldsymbol{q}}\cdot\mathbf{grad}\,\hat{v}\,\mathrm{d}\hat{x}. \tag{6.23}$$

Using Green's formula we deduce, cf. THOMAS [1976, Lemma 5.1], that

$$\int_T v\operatorname{div}\boldsymbol{q}\,\mathrm{d}x=\int_{\hat{T}}\hat{v}\operatorname{div}\hat{\boldsymbol{q}}\,\mathrm{d}\hat{x}, \tag{6.24}$$

and we have, with $\hat{\boldsymbol{\nu}}$ denoting of course the exterior unit normal of $\partial\hat{T}$,

$$\int_{\partial T} v\boldsymbol{q}\cdot\boldsymbol{\nu}\,\mathrm{d}\sigma=\int_{\partial\hat{T}}\hat{v}\hat{\boldsymbol{q}}\cdot\hat{\boldsymbol{\nu}}\,\mathrm{d}\hat{\sigma}, \tag{6.25}$$

for all sufficiently regular functions $\hat{v}$ and $\hat{q}$. Thus $\hat{\boldsymbol{q}}\to\boldsymbol{q}$ is a bijective affine transformation from $H(\mathrm{div};\hat{T})$ onto $H(\mathrm{div};T)$ such that

$$(\operatorname{mes} T)\operatorname{div} \boldsymbol{q} = (\operatorname{mes} \hat{T})\operatorname{div} \hat{\boldsymbol{q}} \tag{6.26}$$

and on each face $T' = F(\hat{T}')$ of T

$$(\operatorname{mes} T')\boldsymbol{q}\cdot\boldsymbol{v} = (\operatorname{mes} \hat{T}')\hat{\boldsymbol{q}}\cdot\hat{\boldsymbol{v}}. \tag{6.27}$$

Furthermore, it is immediately verified that this transformation takes $D_k(\hat{T})$ onto $D_k(T)$:

$$\boldsymbol{q} \in D_k(T) \Leftrightarrow \hat{\boldsymbol{q}} \in D_k(\hat{T}). \tag{6.28}$$

We conclude from (6.26), (6.27) and (6.28) that the interpolant in $D_k(T)$ of a function $\boldsymbol{q}$, image of the function $\hat{\boldsymbol{q}}$ defined on $\hat{T}$, is none other than the image of the interpolant in $D_k(\hat{T})$ of $\hat{\boldsymbol{q}}$:

$$(\mathscr{E}_T^{(k)}\boldsymbol{q})\cdot F = \frac{1}{|J|}\mathrm{D}F(\mathscr{E}_{\hat{T}}^{(k)}\hat{\boldsymbol{q}}). \tag{6.29}$$

This property allows us to reduce the problem of obtaining an error bound for interpolation on T to that of obtaining an error bound for interpolation on $\hat{T}$, the reference n-simplex. As the subspace $(P_{k-1})^n$ of D_k is invariant under the interpolation operator, we know that there exists a constant $\hat{C}$ depending only on $\hat{T}$ and k such that

$$\|\hat{\boldsymbol{q}} - \mathscr{E}_{\hat{T}}^{(k)}\hat{\boldsymbol{q}}\|_{0,\hat{T}} \leqslant \hat{C}|\hat{\boldsymbol{q}}|_{l,\hat{T}} \quad \text{for all } \hat{\boldsymbol{q}} \in (H^l(\hat{T}))^n \text{ with } 1 \leqslant l \leqslant k.$$

With this bound and change of variable formulas we obtain for each T in $\mathscr{T}_h$ a local bound for $\|\boldsymbol{q} - \mathscr{E}_T^{(k)}\boldsymbol{q}\|_{0,T}$ and then the global bound (6.19). □

REMARK 6.2. The introduction of the spaces D_k is fundamental for our construction of an $H(\mathrm{div};\Omega)$ approximation. We have the strict inclusions $(P_{k-1})^n \subset D_k \subset (P_k)^n$, and we have chosen to index by k the space D_k strictly between $(P_{k-1})^n$ and $(P_k)^n$ though in the literature one also finds this same space indexed by $k-1$. Our choice was made with consideration for the error bounds. As Lagrangian interpolation, $\mathscr{L}_h^{(k)}$, by functions locally in P_k leads to error bounds in $H^1(\Omega)$ of order h^k, so the interpolation $\mathscr{E}_h^{(k)}$, with $\mathscr{E}$ as in equilibrium, by functions locally in D_k leads to an error in $H(\mathrm{div};\Omega)$ of order h^k.

REMARK 6.3. The transformation $\hat{\boldsymbol{q}} \to \boldsymbol{q}$ introduced during the demonstration of Theorem 6.3 in (6.22) is none other than the transformation known in mechanics as the Piola transformation of a vector field defined on a manifold of $\mathbb{R}^n$, cf. for example MARSDEN and HUGHES [1983, pp. 116–118].

Next we present a few results concerning the approximation of the normal traces on the set $\partial\mathscr{T}_h = \bigcup_{T\in\mathscr{T}_h} \partial T$ of sufficiently regular vector functions $\boldsymbol{q}$, for example $\boldsymbol{q} \in (H^1(\Omega))^n$. Given a triangulation $\mathscr{T}_h$ of $\bar{\Omega}$ and a positive integer k, we denote by $E_{\partial\mathscr{T}_h}^{(k)}$ the space defined by

$$E^{(k)}_{\partial\mathcal{T}_h}=\left\{\mu=(\mu_T)_{T\in\mathcal{T}_h}\in\prod_{T\in\mathcal{T}_h}L^2(\partial T)\colon \exists \boldsymbol{q}\in E^{(k)}_h \text{ s.t. } \forall T\in\mathcal{T}_h, \boldsymbol{q}|_{\partial T}\cdot\boldsymbol{\nu}_T=\mu_T\right\},\tag{6.30}$$

where $E^{(k)}_h$ is the space defined by (6.18). The space $E^{(k)}_{\partial\mathcal{T}_h}$ may also be characterized as the space of functions $\mu=(\mu_T)_{T\in\mathcal{T}_h}$ satisfying the following two properties:

$$\mu_T\in D_k(\partial T)\quad \text{for each } T\in\mathcal{T}_h,\tag{6.31a}$$

$$\mu_{T_1}+\mu_{T_2}=0\quad \text{on each } T',\ T'=T_1\cap T_2 \text{ with } T_1,T_2\in\mathcal{T}_h,\tag{6.31b}$$

where by $D_k(\partial T)$ we mean the space

$$D_k(\partial T)=\{\mu\in L^2(\partial T)\colon \exists \boldsymbol{q}\in D_k(T), \boldsymbol{q}\cdot\boldsymbol{\nu}_T=\mu\},\tag{6.32}$$

or equivalently the space of functions polynomial of degree at most $k-1$ on each face T' of T with no continuity constraint at the intersection of two faces. Thus we have

$$\dim D_k(\partial T)=(n+1)\frac{(n+k-2)!}{(n-1)!(k-1)!}.\tag{6.33}$$

Let $\mathscr{E}^{(k)}_{\partial\mathcal{T}_h}$ be the operator which associates to each element $\mu=(\mu_T)_{T\in\mathcal{T}_h}\in\Pi_{T\in\mathcal{T}_h}L^2(\partial T)$ the element $(\mathscr{E}^{(k)}_{\partial T}\mu_T)_{T\in\mathcal{T}_h}$ where $\mathscr{E}^{(k)}_{\partial T}$ is the orthogonal projection from $L^2(\partial T)$ onto $D_k(\partial T)$, $T\in\mathcal{T}_h$. Associating to each vector function $\boldsymbol{q}\in\mathscr{H}(\mathrm{div};\mathcal{T}_h)\cap H(\mathrm{div};\Omega)$ the element

$$\mu_{(\boldsymbol{q})}=(\boldsymbol{q}\cdot\boldsymbol{\nu}_T)_{T\in\mathcal{T}_h},\tag{6.34}$$

we clearly have

$$\mathscr{E}^{(k)}_{\partial\mathcal{T}_h}\mu_{(\boldsymbol{q})}=((\mathscr{E}^{(k)}_h\boldsymbol{q})\cdot\boldsymbol{\nu}_T)_{T\in\mathcal{T}_h}\in E^{(k)}_{\partial\mathcal{T}_h}.\tag{6.35}$$

Hence we deduce:

THEOREM 6.4. *For each positive integer k and each triangulation $\mathcal{T}_h$ of $\bar\Omega$, the mapping $\boldsymbol{q}\to\mathscr{E}^{(k)}_{\partial\mathcal{T}_h}\mu_{(\boldsymbol{q})}$, with $\mu_{(\boldsymbol{q})}$ given by (6.34), is a surjection of $H(\mathrm{div};\Omega)\cap\mathscr{H}(\mathrm{div};\mathcal{T}_h)$ onto $E^{(k)}_{\partial\mathcal{T}_h}$. It is a bijection of $E^{(k)}_h$ onto $E^{(k)}_{\partial\mathcal{T}_h}$ only in case $k=1$.*

Here we shall give only one result concerning error bounds for interpolation in $E^{(k)}_{\partial\mathcal{T}_h}$. However, beforehand, in a manner analogous to (5.19), we define several norms that we shall find useful. Given a triangulation $\mathcal{T}_h$ of $\bar\Omega$, we put

$$|||\boldsymbol{q}|||_{H(\mathrm{div};T)}=\{\|\boldsymbol{q}\|^2_{0,T}+h^2\|\mathrm{div}\,\boldsymbol{q}\|^2_{0,T}\}^{1/2}\quad \text{for all } \boldsymbol{q}\in H(\mathrm{div};T),\ T\in\mathcal{T}_h,$$

$$|||\mu|||_{-1/2,\partial T}=\inf_{\{\boldsymbol{q}\in H(\mathrm{div};T):\boldsymbol{q}\cdot\boldsymbol{\nu}_T=\mu\}}|||\boldsymbol{q}|||_{H(\mathrm{div};T)}\quad \text{for all } \mu\in H^{-1/2}(\partial T),\ T\in\mathcal{T}_h,$$

$$|||\mu|||_{-1/2,\partial\mathcal{T}_h}=\left\{\sum_{T\in\mathcal{T}_h}|||\mu_T|||^2_{-1/2,\partial T}\right\}^{1/2}\quad \text{for all } \mu=(\mu_T)\in\prod_{T\in\mathcal{T}_h}H^{-1/2}(\partial T).\tag{6.36}$$

The following result will be needed for our analysis of primal hybrid methods:

THEOREM 6.5. *Suppose $\{\mathcal{T}_h: h\in\mathcal{H}\}$ is a regular family of triangulations of $\bar{\Omega}$. For each positive integer k, there exists a constant C, independent of h, such that if $q\in(H^l(\Omega))^n$, $1\leqslant l\leqslant k$, and $\mu=\mu(q)$ is determined by* (6.34), *then*

$$|||q-\mathscr{E}^{(k)}_{\partial\mathcal{T}_h}\mu|||_{-1/2,\partial\mathcal{T}_h}\leqslant Ch^l|q|_{l,\Omega}. \tag{6.37}$$

To obtain this result we employ the bound (6.19) of order l and the bound (6.20) of order $(l-1)$.

The second family of examples, is defined for integers k, $k\geqslant 2$. One chooses for interpolation space on the simplex T the space

$$D^*_k(T)=(P_{k-1}(T))^n, \tag{6.38}$$

where we have kept the indexing convention of Remark 6.2 instead of that of BREZZI, DOUGLAS and MARINI [1985]. For $q\in\mathcal{H}(\mathrm{div};T)$ the interpolant $\mathscr{E}^{*(k)}_T q$ is defined to be the unique element of $D^*_k(T)$ such that

$$m(\mathscr{E}^{*(k)}_T q)=m(q)\quad\text{for all } m\in\mathcal{M}^{*(k)}_T,$$

where the set of moments $\mathcal{M}^{*(k)}_T$ used to define the interpolant is the union

$$\mathcal{M}^{*(k)}_T=\left(\bigcup_{T'<T}\mathcal{M}_{(k-1)}(T')\right)\cup\mathcal{M}^{\mathrm{grad}}_{k-2}(T)\cup\mathcal{M}^{\perp}_{k-1}(T),$$

with $M_l(T')$ defined by (6.9),

$$\mathcal{M}^{\mathrm{grad}}_l(T)=\left\{\left(q\to\int_T q\cdot\mathbf{grad}\,w\,\mathrm{d}x\right)\in\mathcal{L}((C^0(T))^n,\mathbb{R}): w\in P_l\right\}$$

and

$$\mathcal{M}^{\perp}_l(T)=\left\{\left(q\to\int_T q\cdot w\,\mathrm{d}x\right)\in\mathcal{L}((C^0(T))^n,\mathbb{R}):\right.$$

$$\left. w\in(P_l)^n,\ w\cdot\nu_T=0 \text{ on } \partial T, \text{ and } \operatorname{div} w=0\right\}.$$

Define the operator $\mathscr{E}^{*(k)}_h$ by (6.2) and the space $E^{*(k)}_h$ by analogy with (6.18) or equivalently (6.18′). It can be shown that (6.15) still holds with $\mathscr{E}^{(k)}_T$ replaced by $\mathscr{E}^{*(k)}_T$ and $\Pi^{(k-1)}_T$ replaced by $\Pi^{(k-2)}_T$, and thus so do (6.3) and (6.4) if $\mathscr{E}^{(k)}_h$ is also replaced by $\mathscr{E}^{*(k)}_h$. In fact the analogues of Theorems 6.1 and 6.2 obtained by replacing $\mathscr{E}^{(k)}_T$, $\mathscr{E}^{(k)}_h$, $D_k(T)$, and $E^{(k)}_h$ by $\mathscr{E}^{*(k)}_T$, $\mathscr{E}^{*(k)}_h$, $D^*_k(T)$ and $E^{*(k)}_h$ respectively remain valid. One can also demonstrate the following analogue of Theorem 6.3: if $\{\mathcal{T}_h: h\in\mathcal{H}\}$ is a regular family of triangulations of $\bar{\Omega}$, then

$$\|q-\mathscr{E}^{*(k)}_h q\|_{0,\Omega}\leqslant Ch^l|q|_{l,\Omega}\quad\text{for all } q\in(H^l(\Omega))^n \text{ with } 1\leqslant l\leqslant k, \tag{6.39}$$

and

$$\|\operatorname{div}(q-\mathscr{E}_h^{*(k)}q)\|_{0,\Omega}\leqslant Ch^l|\operatorname{div} q|_{l,\Omega}$$
$$\text{for all } q\in(H^1(\Omega))^n \text{ with } \operatorname{div} q\in H^l(\Omega) \text{ for } 0\leqslant l\leqslant k-1. \tag{6.40}$$

REMARK 6.4. The space $E_h^{*(k)}$,

$$E_h^{*(k)}=\{q\in H(\operatorname{div};\Omega)\cap\mathscr{H}(\operatorname{div};\mathscr{T}_h):\ q=\mathscr{E}_h^{*(k)}q\}, \tag{6.41}$$

is clearly a proper subspace of $E_h^{(k)}$. Hence fewer degrees of freedom are needed to determine the interpolants in $E_h^{*(k)}$ than in $E_h^{(k)}$. The interpolation errors in these two approximation spaces are of the same order when measured in the L^2 norm $\|\cdot\|_{0,\Omega}$; however, interpolation in $E_h^{*(k)}$ is less precise by one order than that in $E_h^{(k)}$ when measured in the $H(\operatorname{div};\Omega)$ norm unless $|\operatorname{div} q|_{k-1,\Omega}=0$, i.e. unless $\operatorname{div} q\in P_{k-2}$. Moreover one may verify that

$$D_k^*(T)=\{q\in D_k(T):\operatorname{div} q\in P_{k-2}(T)\}.$$

REMARK 6.5. In fact one of the most important distinctions between the approximations in $E_h^{(k)}$ and in $E_h^{*(k)}$ lies not in the interpolations themselves but in the fact that for the dual mixed method for the elliptic problem the $E_h^{(k)}$ interpolation for the vector variable is used in conjunction with a locally P_{k-1} interpolation for the scalar variable whereas when interpolation in $E_h^{*(k)}$ is used for the vector variable the natural domain for approximating the scalar variable is the space of locally P_{k-2} functions. Thus, for example, when a piecewise constant approximation is sufficient for the scalar variable but more precision than that provided by the $E_h^{(1)}$ interpolant is desired for the vector variable, one has the option of approximating the vector function in $E_h^{*(2)}$ while retaining the locally P_0 approximation of the scalar function instead of using the full P_1, $E_h^{(2)}$ approximation. The gain in accuracy thus obtained, however, is, a priori, only in the $(L^2(\Omega))^n$ norm.

7. Examples of finite element type approximation for nontriangular geometric forms

Using the families of examples of finite element type approximation developed in Sections 5 and 6, we can construct approximations of order k, for each positive integer k, of functions in $H^1(\Omega)$ and of functions in $H(\operatorname{div};\Omega)$ in the academic setting Ω, a polyhedral domain triangulated by n-simplexes. To simplify vocabulary we fix $n=2$. Thus for the preceding theory Ω was supposed to be an open, polygonal domain triangulated by triangles. Here we shall indicate a few generalizations related to the geometry of the triangulation: rectangular finite elements, curved finite elements such as curvilinear triangles, convex quadrilaterals....

We begin with the case of rectangular finite elements. We shall describe families of examples of approximation of functions in $H^1(\Omega)$ and of functions in $H(\operatorname{div};\Omega)$ when Ω is an open set of $\mathbb{R}^2$ "triangulated" in the form $\bar{\Omega}=\bigcup_{T\in\mathscr{T}_h}T$ by rectangles T having sides parallel to the axes. These families will be the rectangular analogues of the triangular families defined in the two preceding sections.

We first consider the approximation of H^1 functions. For any two nonnegative integers k and l, denote by $P_{k,l}$ the space of polynomials in two variables, of degree at most k in the first variable and of degree at most l in the second variable, and by $P_{k,l}(T)$ the space of restrictions to the rectangle T of the polynomials in $P_{k,l}$. To a rectangle T and a positive integer k is associated the set $\Sigma_T^{(k)}$ of points of the grid on T obtained by subdividing each edge of T into k equal parts. The Lagrangian interpolant $\mathscr{L}_T^{(k)}v$ on T of order k of a function $v \in C^0(T)$ is the unique function in $P_{k,k}(T)$ which coincides with v at each point of $\Sigma_T^{(k)}$. For $\mathscr{T}_h$ a triangulation of $\bar{\Omega}$ by rectangles, we define the interpolation operator $\mathscr{L}_h^{(k)}$ from $H^1(\Omega)\cap H^2(\mathscr{T}_h)$ onto $L_h^{(k)}$, where $L_h^{(k)}$ is defined by

$$L_h^{(k)} = \{v \in H^1(\Omega) \colon \forall T \in \mathscr{T}_h, v|_T \in P_{k,k}(T)\}, \tag{7.1}$$

by requiring that $\mathscr{L}_h^{(k)}v$ agree on T with $\mathscr{L}_T^{(k)}(v|_T)$ for each $T \in \mathscr{T}_h$. The error estimate for Lagrangian interpolation given by Theorem 5.2 remains valid when the family of triangulations is a regular family of triangulations by rectangles; i.e. if $\{\mathscr{T}_h \colon h \in \mathscr{H}\}$ is a family of triangulations $\mathscr{T}_h$ of $\bar{\Omega}$ by rectangles T of diameter $h_T \leqslant h$ which is regular in the sense that the ratio of the width to the length of each rectangle T, $T \in \mathscr{T}_h$, $h \in \mathscr{H}$, is bounded below by a positive constant independent of h, then there exists a constant C such that

$$\begin{aligned} &|v - \mathscr{L}_h^{(k)}v|_{1,\Omega} + h^{-1}\|v - \mathscr{L}_h^{(k)}v\|_{0,\Omega} \leqslant Ch^m|v|_{m+1,\Omega} \\ &\quad \text{for all } v \in H^{m+1}(\Omega) \text{ with } 1 \leqslant m \leqslant k. \end{aligned} \tag{7.2}$$

The demonstration for this case may be found in RAVIART and THOMAS [1983, pp. 79–103], for example.

For the approximation of vector functions on a rectangle T we define the space $D_k(T)$ to be the product space $P_{k,k-1}(T) \times P_{k-1,k}(T)$. The equilibrium interpolant $\mathscr{E}_T^{(k)}\boldsymbol{q}$ of order k, $k \geqslant 1$, on T of the vector function $\boldsymbol{q} \in \mathscr{H}(\text{div}; T)$ is the unique function in $D_k(T)$ such that

$$\int_{T'} (\mathscr{E}_T^{(k)}\boldsymbol{q}) \cdot \boldsymbol{\nu}_T w \, \mathrm{d}\sigma = \int_{T'} \boldsymbol{q} \cdot \boldsymbol{\nu}_T w \, \mathrm{d}\sigma \quad \text{for all } w \in P_{k-1}(T') \text{ with } T' < T, \tag{7.3}$$

and

$$\int_T (\mathscr{E}_T^{(k)}\boldsymbol{q}) \cdot \boldsymbol{r} \, \mathrm{d}x = \int_T \boldsymbol{q} \cdot \boldsymbol{r} \, \mathrm{d}x \quad \text{for all } \boldsymbol{r} \in P_{k-2,k-1}(T) \times P_{k-1,k-2}(T). \tag{7.4}$$

For $\mathscr{T}_h$ a triangulation of $\bar{\Omega}$ by rectangles we define the interpolation operator $\mathscr{E}_h^{(k)}$ from $H(\text{div}; \Omega) \cap \mathscr{H}(\text{div}; \mathscr{T}_h)$ onto $E_h^{(k)}$, where $E_h^{(k)}$ is defined by

$$E_h^{(k)} = \{\boldsymbol{q} \in H(\text{div}; \Omega) \colon \forall T \in \mathscr{T}_h, \boldsymbol{q}|_T \in D_k(T)\}, \tag{7.5}$$

by requiring that $\mathscr{E}_h^{(k)}\boldsymbol{q}$ agree on T with $\mathscr{E}_T^{(k)}(\boldsymbol{q}|_T)$ for each $T \in \mathscr{T}_h$. Using Green's formula one can show that on each rectangle T

$$\operatorname{div}(\mathscr{E}_T^{(k)} q) = \Pi_T^{(k-1)}(\operatorname{div} q) \quad \text{for all } q \in \mathscr{H}(\operatorname{div}; T), \tag{7.6}$$

where $\Pi_T^{(k-1)}$ is the orthogonal L^2 projection of $L^2(T)$ onto $P_{k-1,k-1}(T)$. The error estimate given by Theorem 6.3 remains valid in the rectangular case: if $\{\mathscr{T}_h : h \in \mathscr{H}\}$ is a regular family of triangulations of $\bar{\Omega}$ by rectangles then there is a constant C such that

$$\|q - \mathscr{E}_h^{(k)} q\|_{0,\Omega} \leqslant Ch^l |q|_{l,\Omega} \tag{7.7}$$
$$\text{for all } q \in (H^l(\Omega))^2 \text{ with } 1 \leqslant l \leqslant k,$$

and

$$\|\operatorname{div}(q - \mathscr{E}_h^{(k)} q)\|_{0,\Omega} \leqslant Ch^l |\operatorname{div} q|_{l,\Omega}$$
$$\text{for all } q \in (H^1(\Omega))^2 \text{ with } \operatorname{div} q \in H^l(\Omega), \quad 0 \leqslant l \leqslant k. \tag{7.8}$$

For the demonstration in the case $n=2$ treated here, see THOMAS [1977, Chapter III]; for the generalization to the case $n=3$, see NEDELEC [1980].

The analogue for rectangular elements of the second family described in Section 6 is obtained by choosing for $k \geqslant 2$ for interpolation space on the rectangle T the space

$$D_k^*(T) = (P_{k-1}(T))^2 + (\mathbf{rot}\, x_1^k x_2) + (\mathbf{rot}\, x_1 x_2^k) \tag{7.9}$$

and for interpolation operator $\mathscr{E}_T^{*(k)}$ on $\mathscr{H}(\operatorname{div}; T)$ the mapping associating to $q \in \mathscr{H}(\operatorname{div}; T)$ the unique function in $D_k^*(T)$ such that

$$\int_{T'} (\mathscr{E}_T^{*(k)} q) \cdot \nu_T w \, \mathrm{d}\sigma = \int_{T'} q \cdot \nu_T w \, \mathrm{d}\sigma \tag{7.10}$$
$$\text{for all } w \in P_{k-1}(T') \text{ with } T' < T,$$

and

$$\int_T (\mathscr{E}_T^{*(k)} q) \cdot r \, \mathrm{d}x = \int_T q \cdot r \, \mathrm{d}x \quad \text{for all } r \in (P_{k-2}(T))^2. \tag{7.11}$$

Then to a triangulation $\mathscr{T}_h$ of $\bar{\Omega}$ by rectangles we associate the interpolation space $E_h^{*(k)}$ and interpolation operator $\mathscr{E}_h^{*(k)}$ from $H(\operatorname{div}; \Omega) \cap \mathscr{H}(\operatorname{div}; \mathscr{T}_h)$ onto $E_h^{*(k)}$ defined by analogy with the preceding example. The analogue of (7.6) holds in this case as do the error bounds (6.39) and (6.40). For the details of the construction and the demonstration of the error bounds see BREZZI, DOUGLAS and MARINI [1985, Section 5]. The extension to the three-dimensional case is given by BREZZI, DOUGLAS, DURAN and FORTIN [1987, Section 3].

REMARK 7.1. In the rectangular case the difference in size of the spaces $D_k(T)$ and

$D_k^*(T)$ is more striking than in the triangular case. For T a rectangle,

$$\dim D_k(T) = 2k^2 + 2k,$$
$$\dim D_k^*(T) = k^2 + k + 2,$$

and indeed, if NE is the number of elements in $\mathcal{T}_h$,

$$\dim E_h^{(k)} - \dim E_h^{*(k)} = (k^2 + k - 2)\text{NE} = \begin{cases} 4\text{NE}, & \text{if } k = 2, \\ 10\text{NE}, & \text{if } k = 3. \end{cases}$$

However, at the level of construction of the spaces $E_h^{(k)}$ and $E_h^{*(k)}$, we note that the degrees of freedom which assure that the continuity condition (6.17) holds, i.e. those which assure the connections between the finite elements and thus contribute to the size of the algebraic system to be solved, are exactly the same for the two methods.

The situation is not the same in three dimensions. If $D_k(T)$ and $D_k^*(T)$ denote the local approximation spaces on a rectangular solid T as defined by NEDELEC [1980] and by BREZZI, DOUGLAS, DURAN and FORTIN [1987] respectively, we have

$$\dim D_k(T) = 3k^2(k+1),$$
$$\dim D_k^*(T) = \tfrac{1}{2}k(k^2 + 3k + 8).$$

Given a triangulation $\mathcal{T}_h$ of $\bar{\Omega}$ by rectangular solids, the degrees of freedom needed to guarantee the required continuity in the corresponding finite-dimensional subspaces of $H(\text{div}; \Omega)$, $E_h^{(k)}$ and $E_h^{*(k)}$, are the moments of the normal component with respect to

$P_{k-1,k-1}(T')$ (i.e. k^2 moments) for elements of $E_h^{(k)}$,
$P_{k-1}(T')$ (i.e. $\frac{1}{2}k(k+1)$ moments) for elements of $E_h^{*(k)}$.

Here the gain in calculation time with interpolation by $\mathscr{E}_h^{*(k)}$, the operator associated with $E_h^{*(k)}$, instead of $\mathscr{E}_h^{(k)}$, the operator associated with $E_h^{(k)}$, is clear. If NF is the number of two-dimensional faces T' of elements T of $\mathcal{T}_h$ with $T' \subset \Omega$, then the difference in calculation time can be estimated in terms of the number

$$(3k^2 - 3k)\text{NF} = \begin{cases} 6\text{NF}, & \text{if } k = 2, \\ 18\text{NF}, & \text{if } k = 3. \end{cases}$$

Nevertheless, concerning the precision of the interpolation in three dimensions as well as in two dimensions, we note as in the simplicial case (cf. Remark 6.5) that while the order of approximation is the same for $\mathscr{E}_h^{(k)}$ and $\mathscr{E}_h^{*(k)}$ in $(L^2(\Omega))^n$, the accuracy in $H(\text{div}; \Omega)$ of $\mathscr{E}_h^{*(k)}$ is of one order less than that of $\mathscr{E}_h^{(k)}$.

REMARK 7.2. Another family of examples of $H(\text{div})$ approximation on rectangles in two dimensions and in three dimensions may be found in BREZZI, DOUGLAS, FORTIN and MARINI [1987]. For rectangular elements in three dimensions as well as for prisms, other families are given in NEDELEC [1986].

We cannot develop here a theory of approximation as general as that constructed above for finite elements of such an elementary form as a triangle or a rectangle for the case of finite elements of other geometric forms. However, we shall describe an example where the triangulation $\mathcal{T}_h$ of $\bar{\Omega}$ consists of convex quadrilaterals. Thus for $T\in\mathcal{T}_h$ there exists a bijection $F\in(P_{1,1})^2$ such that $T=F(\hat{T})$, with $\hat{T}$ the reference square having vertices (0, 0), (1, 0), (1, 1) and (0, 1). For each point $\hat{x}\in\hat{T}$, the linear tangent mapping $DF(\hat{x})$ and its Jacobian $J(\hat{x})=\det(DF(\hat{x}))$ are affine functions of the coordinates $(\hat{x}_1, \hat{x}_2)$.

For a scalar function $v\in H^2(T)$ we may define a function $\mathcal{L}_T^{(1)}v$ in the space

$$\{w\in H^1(T)\colon w\circ F\in P_{1,1}(\hat{T})\} \tag{7.12}$$

by interpolating the values of v at the four vertices of the quadrilateral T.

Similarly for a vector function $\boldsymbol{q}\in\mathcal{H}(\mathrm{div}; T)$ we may interpolate, on each of the four edges, the average value of the normal trace to define a function $\mathcal{E}_T^{(1)}\boldsymbol{q}$ in the space

$$\{\boldsymbol{r}\in H(\mathrm{div}; T)\colon |J|DF^{-1}(\boldsymbol{r}\circ F)\in P_{2,1}(\hat{T})\times P_{1,2}(\hat{T})\}. \tag{7.13}$$

Even though the spaces defined in (7.12) and (7.13) are spaces of polynomial functions only if T is a parallelogram, we note that in any case the trace of $\mathcal{L}_T^{(1)}v$ on each edge T' of T is an affine function and the normal trace of $\mathcal{E}_T^{(1)}\boldsymbol{q}$ on T' is constant. Furthermore, $|J|\mathrm{div}(\mathcal{E}_T^{(1)}\boldsymbol{q})$ is constant on T and we have

$$|J|\mathrm{div}(\mathcal{E}_T^{(1)}\boldsymbol{q})=\frac{1}{\mathrm{mes}\ \hat{T}}\int_T \mathrm{div}\ \boldsymbol{q}\ \mathrm{d}x. \tag{7.14}$$

Proceeding as in each of the preceding examples, by patching together the maps $\mathcal{L}_T^{(1)}$, $T\in\mathcal{T}_h$, we define an operator $\mathcal{L}_h^{(1)}$ on $H^1(\Omega)\cap H^2(\mathcal{T}_h)$ having as image the finite-dimensional subspace $L_h^{(1)}$; and using the maps $\mathcal{E}_T^{(1)}$, $T\in\mathcal{T}_h$, we define the operator $\mathcal{E}_h^{(1)}$ on $H(\mathrm{div}; \Omega)\cap\mathcal{H}(\mathrm{div}; \mathcal{T}_h)$ having as image a finite-dimensional subspace $E_h^{(1)}$ of $H(\mathrm{div}; \Omega)$.

Error estimates for the interpolation operators $\mathcal{L}_h^{(1)}$ and $\mathcal{E}_h^{(1)}$ are given by THOMAS [1977, Theorems III.4.2 and III.4.4], cf. BERNARDI [1986, Chapter 6] and JENSEN [1979], for analogous results on curved domains.

CHAPTER III

Abstract Theory of Approximation of Solutions of Problems with Lagrangian Multipliers

8. General remarks

In this chapter we develop an abstract theory for the approximation of solutions of problems with Lagrangian multipliers. The main purpose for the construction of such a theory is to unify the analysis of the mixed and hybrid methods presented in Chapter I, and hence to provide a framework for generalizations and extensions of the analysis to that of similar problems.

Even though these methods were conceived of in the context of approximation of problems of minimization of a quadratic functional with affine constraints, the properties of constrained optimization are not at all fundamental to the analysis of the problems studied here. On the contrary, more than one familiar with conforming approximations of minimization problems seems to have been misguided by this approach. In the exposition of the abstract theory given in this chapter, we will make no restrictions which allow the class of problems studied here to be interpreted as being minimization problems on an affine manifold. The expression "with Lagrangian multiplier" will be used in the same formal sense as is the expression "variational problem".

Let us now give the general framework for this study. Let W and M be Hilbert spaces with inner products $((\cdot,\cdot))_W$ and $((\cdot,\cdot))_M$ respectively. The norms associated with these inner products are denoted $\|\cdot\|_W$ and $\|\cdot\|_M$. Further, let $a(\cdot,\cdot)$ and $b(\cdot,\cdot)$ be bilinear forms defined on $W\times W$ and on $W\times M$ respectively. Given the linear forms $f(\cdot)$ defined on W and $g(\cdot)$ defined on M, we consider the following problem: find a pair (φ,λ) satisfying

$$\begin{aligned}
&(\varphi,\lambda)\in W\times M,\\
&a(\varphi,\psi)+b(\psi,\lambda)=f(\psi) \quad \text{for all } \psi\in W,\\
&b(\varphi,\mu)=g(\mu) \quad \text{for all } \mu\in M.
\end{aligned} \tag{8.1}$$

We introduce the product space $X=W\times M$ which is a Hilbert space with norm

$$\|v\|_X=(\|\psi\|_W^2+\|\mu\|_M^2)^{1/2} \quad \text{for all } v=(\psi,\mu)\in X.$$

Clearly one may also formulate (8.1) in the following manner: find $u=(\varphi,\lambda)$ satisfying

$$u\in X,\qquad k(u,v)=l(v)\quad \text{for all } v\in X, \tag{8.2}$$

where $k(\cdot,\cdot)$ is a bilinear form on $X\times X$ given by

$$\begin{aligned} &k(u,v)=a(\varphi,\psi)+b(\psi,\lambda)+b(\varphi,\mu)\\ &\qquad \text{for all } u=(\varphi,\lambda)\in X \text{ and for all } v=(\psi,\mu)\in X, \end{aligned} \tag{8.3}$$

and where $l(\cdot)$ is a linear form defined on X by

$$l(v)=f(\psi)+g(\mu)\quad \text{for all } v=(\psi,\mu)\in X. \tag{8.4}$$

All questions of continuity aside, it is evident that the analysis of the variational problem (8.2) cannot be carried out with the aid of the Lax–Milgram theorem since the bilinear form $k(\cdot,\cdot)$ is not X-elliptic: for each $v=(0,\mu)\in X$, we have $k(v,v)=0$. One may show, nevertheless, the following result:

THEOREM 8.1. *Suppose that the bilinear form $k(\cdot,\cdot)$ is continuous on $X\times X$ and that it satisfies*

$$\inf_{\{u\in X:\|u\|_X=1\}}\ \sup_{\{v\in X:\|v\|_X=1\}} k(u,v)>0 \tag{8.5a}$$

and

$$\sup_{\{u\in X:\|u\|_X=1\}} k(u,v)>0\quad \text{for all } v\in X \text{ with } v\neq 0. \tag{8.5b}$$

Then for each continuous linear form $l(\cdot)$ on X, there exists a unique solution of problem (8.2).

This theorem is a special case of the theorem due to BABUŠKA [1971] given in the following section.

REMARK 8.1. Hypothesis (8.5b) is superfluous when the bilinear form $k(\cdot,\cdot)$ is symmetric because, in this case, (8.5a) implies (8.5b).

9. Analysis and approximation of the solution of a variational problem with Lagrangian multipliers (theory of Babuška)

We consider the two Hilbert spaces X and Y equipped with the norms $\|\cdot\|_X$ and $\|\cdot\|_Y$, and let $k(\cdot,\cdot)$ be a bilinear form on $X\times Y$. We have

THEOREM 9.1. *Suppose that $k(\cdot,\cdot)$ is a continuous, bilinear form on $X\times Y$ which satisfies*

$$\inf_{\{u\in X:\|u\|_X=1\}}\ \sup_{\{v\in Y:\|v\|_Y=1\}} k(u,v)>0 \tag{9.1a}$$

and

$$\sup_{\{u\in X:\|u\|_X=1\}} k(u,v)>0 \quad \text{for all } v\in Y \text{ with } v\neq 0. \tag{9.1b}$$

Then for each continuous linear form $l(\cdot)$ on Y, the problem: find u satisfying

$$u\in X, \qquad k(u,v)=l(v) \quad \text{for all } v\in Y, \tag{9.2}$$

has a unique solution.

PROOF. Define the continuous linear operator $\mathscr{K}$ from X to Y by

$$((\mathscr{K}u,v))_Y=k(u,v) \quad \text{for all } u\in X \text{ and for all } v\in Y.$$

Hypothesis (9.1a) implies that the operator $\mathscr{K}$ is injective and that its image is closed in Y; hypothesis (9.1b) expresses the fact that the image of $\mathscr{K}$ is dense in Y. □

REMARK 9.1. Conditions (9.1a) and (9.1b) together give a necessary and sufficient condition for problem (9.2) to have a unique solution given a continuous, linear form $l(\cdot)$. In the case $X=Y$, the ellipticity condition,

$$\inf_{\{v\in X:\|v\|_X=1\}} k(v,v)>0,$$

obviously implies conditions (9.1a) and (9.1b); thus it is a sufficient condition for problem (9.2) when $Y=X$ to have a unique solution. In this sense Theorem 9.1 generalizes the Lax–Milgram theorem.

In the next section we shall come back to and treat in detail the "particular case" where $X=W\times M$ and $k(\cdot,\cdot)$ is of the form (8.3), i.e. the case that motivates our study. First, however, we give a general result concerning the approximation of the solution of problem (9.2). Given finite-dimensional subspaces X_h and Y_h of X and Y respectively, we seek the "approximate solution" $u_h\in X_h$, an approximation of the solution u of problem (9.2), defined by

$$u_h\in X_h, \qquad k(u_h,v_h)=l(v_h) \quad \text{for all } v_h\in Y_h. \tag{9.3}$$

Applying Theorem 9.1 in the finite-dimensional case, we obtain:

THEOREM 9.2. *Suppose that $k(\cdot,\cdot)$ is a bilinear form on $X_h\times Y_h$ which satisfies*

$$\inf_{\{u_h\in X_h:\|u_h\|_X=1\}} \sup_{\{v_h\in Y_h:\|v_h\|_Y=1\}} k(u_h,v_h)>0 \tag{9.4a}$$

and

$$\sup_{\{u_h\in X_h:\|u_h\|_X=1\}} k(u_h,v_h)>0 \quad \text{for all } v_h\in Y \text{ with } v_h\neq 0. \tag{9.4b}$$

Then for each linear form $l(\cdot)$ on Y_h, problem (9.3) has a unique solution.

REMARK 9.2. The space X_h being finite-dimensional, condition (9.4a) is equivalent to

$$\sup_{\{v_h \in Y_h:\, \|v_h\|_Y = 1\}} k(u_h, v_h) > 0 \quad \text{for all } u_h \in X_h \text{ with } u_h \neq 0.$$

Furthermore, when the hypotheses of Theorem 9.2 are realized, one has necessarily that

$$\dim Y_h = \dim X_h. \tag{9.5}$$

Conversely, when we know, a priori, that (9.5) is satisfied (for example when $X_h = Y_h$), it suffices to check that either one or the other of conditions (9.4a) and (9.4b) holds.

REMARK 9.3. It is important to note that it is not sufficient that hypotheses (9.1a) and (9.1b) be satisfied in order that the corresponding conditions (9.4a) and (9.4b) be satisfied as well. Herein lies the charm of the analysis!

REMARK 9.4. Even when $X = Y$ and the bilinear form $k(\cdot, \cdot)$ is X-elliptic, it can be useful to consider examples for which (9.5) holds with $X_h \neq Y_h$. Such is, for instance, the case in a Petrov–Galerkin type approach to a diffusion-convection problem, cf. for example MIZUKAMI and HUGHES [1985].

We conclude this section with an estimation of the error committed in approximating the solution u of problem (9.2) by the solution u_h of problem (9.3):

THEOREM 9.3. *Suppose that $k(\cdot, \cdot)$ is a continuous, bilinear form on $X \times Y$ and let K be a number such that*

$$\sup_{\{v \in Y:\, \|v\|_Y = 1\}} k(u, v) \leqslant K \|u\|_X \quad \textit{for all } u \in X. \tag{9.6}$$

Suppose that the choice of subspaces X_h and Y_h is such that

$$\dim Y_h = \dim X_h \tag{9.7}$$

and that there exists a number $\kappa > 0$ such that

$$\sup_{\{v_h \in Y_h:\, \|v_h\|_Y = 1\}} k(u_h, v_h) \geqslant \kappa \|u_h\|_X \quad \textit{for all } u_h \in X_h. \tag{9.8}$$

Then for each linear form $l(\cdot)$ on Y, if u satisfies (9.2) and u_h is the solution to (9.3), we have the error bound

$$\|u - u_h\|_X \leqslant (1 + (K/\kappa)) \inf_{w_h \in X_h} \|u - w_h\|_X. \tag{9.9}$$

This result is due to BABUŠKA [1971]; (cf. also BABUŠKA and AZIZ [1972, Theorem 6.2.1, p. 186]); a simple demonstration in the case $X_h = Y_h$ can be found in RABIER and THOMAS [1985, Exercise 3.13], for example. For a given problem of the form (9.2), the constants K and κ which appear in (9.6) and (9.8), respectively, obviously depend on the choice of norm on X and on Y and on the choice of the subspaces X_h and

Y_h—nothing prevents one from making the choice of the norms as a function of the choice of the finite-dimensional subspaces. We will see some such examples later on.

For the moment we shall settle for a simple criterion for convergence when the index h, indicating the choice of finite-dimensional subspace X_h as well as a real parameter, tends toward 0. The set of parameters will be denoted $\mathscr{H}$.

THEOREM 9.4. *Suppose that the norms $\|\cdot\|_X$ and $\|\cdot\|_Y$ on the spaces X and Y, respectively, are chosen independently of the parameter $h \in \mathscr{H}$ and that condition (9.6) is satisfied. Suppose that for each $h \in \mathscr{H}$, we have been able to choose the finite-dimensional subspaces X_h and Y_h to be of the same dimension and such that condition (9.8) is realized with a constant $\kappa > 0$ independent of h. Suppose further that there exist a subspace $\mathscr{X}$ of X dense in X and a mapping r_h of $\mathscr{X}$ into X_h such that*

$$\lim_{h\to 0} \|w - r_h(w)\|_X = 0 \quad \textit{for all } w \in \mathscr{X}.$$

Then, the variational approximation method converges in the sense that we have

$$\lim_{h\to 0} \|u - u_h\|_X = 0,$$

where u is the solution of (9.2) and u_h the solution of (9.3).

The demonstration of this theorem is identical to that showing the convergence of the approximate solutions of an elliptic variational problem, cf. for example RAVIART and THOMAS [1983, Theorem 3.1–3.3].

10. Analysis and approximation of the solution of a variational problem with Lagrangian multipliers (theory of Brezzi)

Equipped with the general results of the preceding section, we proceed with the analysis of a problem of the form (8.1) which we shall call a problem with Lagrangian multipliers: find a pair (φ, λ) satisfying

$$\begin{aligned} &(\varphi, \lambda) \in W \times M, \\ &a(\varphi, \psi) + b(\psi, \lambda) = f(\psi) \quad \text{for all } \psi \in W, \\ &b(\varphi, \mu) = g(\mu) \quad \text{for all } \mu \in M. \end{aligned} \tag{10.1}$$

In particular if a pair (φ, λ) satisfies (10.1) with $g(\cdot) = 0$, then the first argument φ is a solution of the variational problem: find φ such that

$$\varphi \in V, \qquad a(\varphi, \psi) = f(\psi) \quad \text{for all } \psi \in V, \tag{10.2}$$

where V is the subspace of W defined by

$$V = \{\psi \in W: \forall \mu \in M,\ b(\psi, \mu) = 0\}. \tag{10.3}$$

The following theorem is due to BREZZI [1974a, 1974b]:

THEOREM 10.1. *Suppose that the bilinear form $a(\cdot,\cdot)$ is continuous on $W \times W$ and V-elliptic, i.e.*

$$\inf_{\{\psi \in V:\, \|\psi\|_W = 1\}} a(\psi, \psi) > 0. \tag{10.4a}$$

Suppose that the bilinear form $b(\cdot,\cdot)$ is continuous on $W \times M$ and that it satisfies

$$\inf_{\{\mu \in M:\, \|\mu\|_M = 1\}} \sup_{\{\psi \in W:\, \|\psi\|_W = 1\}} b(\psi, \mu) > 0. \tag{10.4b}$$

Then for each pair of continuous linear forms $f(\cdot)$ on W and $g(\cdot)$ on M, problem (10.1) has a unique solution.

PROOF. The original demonstration of this theorem is given in BREZZI [1974b]; a complete demonstration may also be found in GIRAULT and RAVIART [1979, Chapter I, Theorem 4.1] for example. We shall indicate here a variant of the demonstration of this fundamental result. Let $V^\perp$ denote the subspace of W orthogonal to V; denote by ψ^0 and $\psi^\perp$ the orthogonal projections of $\psi \in W$ in V and $V^\perp$, respectively. With this notation, problem (10.1) may be written: find $\varphi^\perp$, φ^0, and λ such that

$$\begin{aligned}
&\varphi^\perp \in V^\perp, \qquad \varphi^0 \in V, \qquad \lambda \in M,\\
&a(\varphi^\perp, \psi^\perp) + a(\varphi^0, \psi^\perp) + b(\psi^\perp, \lambda) = f(\psi^\perp) \quad \text{for all } \psi^\perp \in V^\perp,\\
&a(\varphi^\perp, \psi^0) + a(\varphi^0, \psi^0) = f(\psi^0) \quad \text{for all } \psi^0 \in V,\\
&b(\varphi^\perp, \mu) = g(\mu) \quad \text{for all } \mu \in M.
\end{aligned} \tag{10.5}$$

Note that we have the equality

$$\sup_{\{\psi \in V^\perp:\, \|\psi\|_W = 1\}} b(\psi, \mu) = \sup_{\{\psi \in W:\, \|\psi\|_W = 1\}} b(\psi, \mu) \quad \text{for all } \mu \in M,$$

and that we have the characterization of V as a subset of W:

$$\psi \in V \quad \text{iff} \quad \sup_{\{\mu \in M:\, \|\mu\|_M = 1\}} b(\psi, \mu) = 0.$$

Using hypothesis (10.4b) one may then establish with the aid of Theorem 9.1 that the linear operator $\mathscr{B}^*$ from M into $V^\perp$ defined by

$$((\psi, \mathscr{B}^* \mu))_W = b(\psi, \mu) \quad \text{for all } \psi \in W,\ \mu \in M,$$

is an isomorphism of M onto $V^\perp$. Thus the linear operator $\mathscr{B}$ from $V^\perp$ into M defined by

$$((\mathscr{B}\psi, \mu))_M = ((\psi, \mathscr{B}^* \mu))_W \quad \text{for all } \psi \in W,\ \mu \in M$$

is an isomorphism of $V^\perp$ onto M.

Finally, using hypothesis (10.4a) one may show with the aid of the Lax–Milgram theorem that the linear operator $\mathscr{A}$ from V into V defined by

$$((\mathscr{A}\varphi, \psi))_W = a(\varphi, \psi) \quad \text{for all } \varphi \in V,\ \psi \in W,$$

is an automorphism. Thus the linear system (10.5), to which is associated a matrix of operators of the form

$$\begin{pmatrix} \times & \times & \mathscr{B}^* \\ \times & \mathscr{A} & 0 \\ \mathscr{B} & 0 & 0 \end{pmatrix}$$

has a unique solution. □

REMARK 10.1. Condition (10.4b) is often called the *inf–sup condition*; some authors refer to it as the *Brezzi condition*, others as the *Babuška–Brezzi condition*. The current trend seems to be to use the expression *LBB condition*, where LBB stands for LADYZHENSKAYA [1949], BABUŠKA [1971], BREZZI [1974].

REMARK 10.2. A necessary and sufficient condition for problem (10.1) to have a unique solution $(\varphi, \lambda) \in W \times M$ given the linear form $f(\cdot)$ on W and $g(\cdot)$ on M is that the inf–sup condition be satisfied and that the operator $\mathscr{A} \in \mathscr{L}(V; V)$ associated with the bilinear form $a(\cdot,\cdot)$ be an isomorphism. This result, though presented differently, can be found in GIRAULT and RAVIART [1979, Chapter I, Theorem 4.1].

With the goal of approximating the solution of (10.1), we take now finite-dimensional subspaces of W and of M,

$$W_h \subset W, \qquad M_h \subset M, \tag{10.6}$$

and define the discrete problem: find (φ_h, λ_h) satisfying

$$\begin{aligned} &(\varphi_h, \lambda_h) \in W_h \times M_h, \\ &a(\varphi_h, \psi_h) + b(\psi_h, \lambda_h) = f(\psi_h) \quad \text{for all } \psi_h \in W_h, \\ &b(\varphi_h, \mu_h) = g(\mu_h) \qquad\qquad\quad \text{for all } \mu_h \in M_h. \end{aligned} \tag{10.7}$$

As in the continuous case, we introduce the subspace V_h of W_h given by

$$V_h = \{\psi_h \in W_h \colon \forall \mu_h \in M_h,\ b(\psi_h, \mu_h) = 0\}. \tag{10.8}$$

One should be careful to note that in general V_h is not a subspace of the space V defined in (10.3). Transposing to the finite-dimensional case the results of Theorem 10.1, we obtain

THEOREM 10.2. *Suppose that*

the bilinear form $a(\cdot,\cdot)$ *is* V_h*-elliptic* (10.9a)

and

the bilinear form $b(\cdot,\cdot)$ *satisfies*

$$\inf_{\{\mu_h \in M_h \colon \|\mu_h\|_M = 1\}} \ \sup_{\{\psi_h \in W_h \colon \|\psi_h\|_W = 1\}} b(\psi_h, \mu_h) > 0. \tag{10.9b}$$

Then for each pair of linear forms $f(\cdot)$ on W_h and $g(\cdot)$ on M_h, problem (10.7) has a unique solution.

REMARK 10.3. Condition (10.9b) is called the *discrete inf–sup condition* or again the *discrete LBB condition*. It says that the choice of the subspace W_h of W and that of the subspace M_h of M cannot be made independently one from the other, that there is a compatibility relation between the two subspaces. If we introduce the bases $(\psi_h^{(j)})_{j=1,\ldots,\dim W_h}$ and $(\mu_h^{(i)})_{i=1,\ldots,\dim M_h}$ of the subspaces W_h and M_h respectively, the discrete inf–sup condition may be interpreted as expressing the fact that the matrix $[B_h]$ whose i,j entry is $[B_h]_{i,j}=b(\psi_h^{(j)},\mu_h^{(i)})$ has rank equal to the dimension of M_h:

$$\mathrm{rg}[B_h]=\dim M_h.$$

In this form, the condition has been called the *rank condition* (cf. BABŬSKA, ODEN and LEE [1978]). Clearly a necessary condition for this condition to hold is that

$$\dim M_h \leqslant \dim W_h.$$

REMARK 10.4. Obviously if the bilinear form $a(\cdot,\cdot)$ is V-elliptic and $V_h\subset V$ (as will be the case for dual mixed methods), then $a(\cdot,\cdot)$ is V_h-elliptic, or better yet if $a(\cdot,\cdot)$ is W-elliptic, so one no longer needs to have $V_h\subset V$ (as will be the case for dual hybrid and primal mixed methods), then $a(\cdot,\cdot)$ is V_h-elliptic. In other cases it is necessary to verify directly the V_h-ellipticity of $a(\cdot,\cdot)$, something which can be rather technical (this will be the situation for primal hybrid methods).

To verify the discrete inf–sup condition one can use the following result (cf. FORTIN [1977]), the proof of which is immediate:

THEOREM 10.3. *Suppose there is a Hilbert space $\mathscr{W}\subset W$, an operator $\mathscr{R}_h$ from M_h into $\mathscr{W}$ satisfying*

$$\frac{b(\mathscr{R}_h\mu_h,\mu_h)}{\|\mathscr{R}_h\mu_h\|_{\mathscr{W}}}\geqslant C \sup_{\{\psi\in W:\|\psi\|_W=1\}} b(\psi,\mu_h)\quad \text{for all } \mu_h\in M_h, \tag{10.10}$$

for some $C\geqslant 0$, and a continuous linear operator π_h from $\mathscr{W}$ into W_h such that the following diagram commutes:

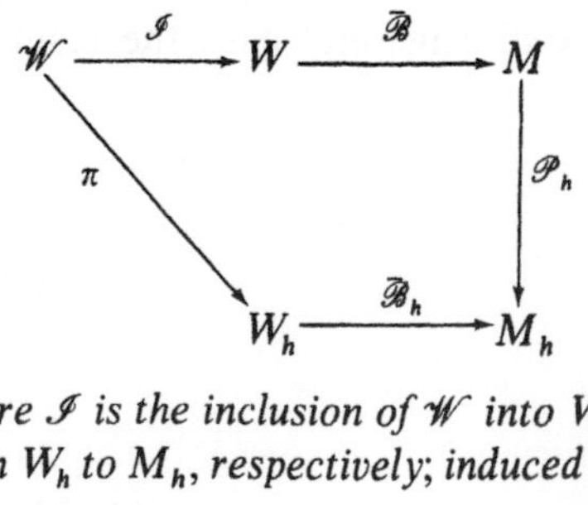

(10.11)

where $\mathscr{I}$ is the inclusion of $\mathscr{W}$ into W, $\bar{\mathscr{B}}$ and $\bar{\mathscr{B}}_h$ are the mappings from W to M and from W_h to M_h, respectively; induced by $b(\cdot,\cdot)$[1] *and $\mathscr{P}_h$ is the orthogonal projection of*

[1] The symbols $\mathscr{B}$ and $\mathscr{B}_h$ have been reserved for the restrictions of $\bar{\mathscr{B}}$ to $V^\perp$ and of $\bar{\mathscr{B}}_h$ to $V_h^\perp$, respectively.

M onto M_h. Then the inf–sup condition of Theorem 10.1 implies the discrete inf–sup condition of Theorem 10.2.

We give an estimate of the error made in approximating the solution of problem (10.1) by that of problem (10.7). This estimate was obtained by BREZZI [1974a, 1974b].

THEOREM 10.4. *Suppose that $a(\cdot,\cdot)$ and $b(\cdot,\cdot)$ are continuous bilinear forms on $W\times W$ and on $W\times M$ respectively. Let A and B be numbers such that*

$$a(\varphi,\psi)\leqslant A\|\varphi\|_W\|\psi\|_W \quad \text{for all } \varphi\in W,\ \psi\in W, \tag{10.12}$$

$$b(\psi,\mu)\leqslant B\|\psi\|_W\|\mu\|_M \quad \text{for all } \psi\in W,\ \mu\in M. \tag{10.13}$$

Suppose further that the subspaces W_h and M_h have been chosen such that there exist positive numbers α and β for which we have

$$a(v_h, v_h)\geqslant \alpha\|v_h\|_W^{1/2} \quad \text{for all } v_h\in V_h, \tag{10.14}$$

$$\inf_{\{\mu_h\in M_h:\|\mu_h\|_M=1\}}\ \sup_{\{\psi_h\in W_h:\|\psi_h\|_W=1\}} b(\psi_h,\mu_h)\geqslant\beta. \tag{10.15}$$

Then there exists a constant C, depending only on A, B, α, and β, such that for each pair of linear forms $f(\cdot)$ on W and $g(\cdot)$ on M, if (φ,λ) satisfies (10.1) and (φ_h,λ_h) satisfies (10.6), we have

$$\|\varphi-\varphi_h\|_W+\|\lambda-\lambda_h\|_M\leqslant C\left\{\inf_{\psi_h\in W_h}\|\varphi-\psi_h\|_W+\inf_{\mu_h\in M_h}\|\lambda-\mu_h\|_M\right\}. \tag{10.16}$$

PROOF. Let $V_h^\perp$ be the subspace of W_h orthogonal to V. We introduce, as in the proof of Theorem 10.1, the operator $\mathscr{B}_h\in\mathscr{L}(V_h^\perp, M)$ and its adjoint $\mathscr{B}_h^*\in\mathscr{L}(M, V_h^\perp)$. The discrete inf–sup condition (10.15) signifies that the operators $\mathscr{B}_h$ and $\mathscr{B}_h^*$ are invertible or more precisely that we have

$$\|\mathscr{B}_h^{-1}\|_{\mathscr{L}(M_h,V_h^\perp)}=\|\mathscr{B}_h^{*-1}\|_{\mathscr{L}(V_h^\perp,M_h)}\leqslant 1/\beta.$$

Similarly the condition of V_h-ellipticity (10.14) implies that the operator $\mathscr{A}_h\in\mathscr{L}(V_h, V_h)$ associated with the bilinear form $a(\cdot,\cdot)$ is invertible and that we have

$$\|\mathscr{A}_h^{-1}\|_{\mathscr{L}(V_h,V_h)}\leqslant 1/\alpha.$$

It is now easy to check that the bilinear form defined with the aid of (9.5) on $X\times X$, where $X=W\times M$ (and $Y=X$), satisfies (9.6) with a constant K depending only on A and B and for the choice $X_h=W_h\times M_h$ (and $Y_h=X_h$) satisfies (9.8) with a constant $\kappa>0$ which depends only on A, α and β. An application of Theorem 9.3 then leads us to the estimation (10.16). □

REMARK 10.5. This error estimate is optimal in the product space $X=W\times M$. This does not mean that the estimation of a single term, $\|\varphi-\varphi_h\|_W$ for example, is necessarily optimal. Say for instance that $g(\cdot)=0$ and V_h is a subspace of V. Then we obviously have

$$\|\varphi-\varphi_h\|_W \leqslant C \inf_{\psi_h \in V_h} \|\varphi-\psi_h\|_W.$$

This sort of refinement of the theory has been developed by FALK and OSBORN [1980]. We will point out such refinements in the examples treated hereafter.

REMARK 10.6. We can deduce from Theorem 10.4 a convergence theorem similar to Theorem 9.4 for a family of pairs of subspaces (W_h, M_h) indexed by $h\in\mathscr{H}$. On the other hand we point out that the approximation result of Theorem 10.4 is still of interest if the chosen norms or even the spaces W and M themselves change with the parameter h as will indeed be the case for hybrid methods. The error bound (10.16) holds with constant C independent of h as long as the constants A, B, α and β may be taken independently of h.

REMARK 10.7. Theorem 10.3 gives sufficient conditions for concluding that the inf–sup condition for $b(\cdot,\cdot)$ on $W\times M$ is inherited by $b(\cdot,\cdot)$ on $W_h\times M_h$. However, given a family of pairs of subspaces (W_h, M_h) indexed by $h\in\mathscr{H}$, to obtain convergence results (cf. Remark 10.6) we need to know that the discrete inf–sup condition holds uniformly in the sense that there exists β independent of h such that for each $h\in\mathscr{H}$

$$\inf_{\{\mu_h\in M_h:\|\mu_h\|_M=1\}} \; \sup_{\{\psi_h\in W_h:\|\psi_h\|_W=1\}} b(\psi_h,\mu_h)\geqslant\beta>0.$$

This will be the case if the norms of R_h and π_h of Theorem 10.3 are independent of h.

REMARK 10.8. With the methods used here to analyze numerically a problem with Lagrangian multipliers of the form (10.1) whose solution is approximated by the solution of a finite-dimensional problem of the form (10.7), it is easy to find sufficient conditions for analyzing a problem of the following form: find φ and λ satisfying

$$\begin{aligned}
&\varphi\in W, \qquad \lambda\in M,\\
&a(\varphi,\psi)+b(\psi,\lambda)=f(\psi) \quad \text{for all } \psi\in W,\\
&b(\varphi,\mu)+d(\lambda,\mu)=g(\mu) \quad \text{for all } \mu\in M.
\end{aligned} \tag{10.17}$$

This generalization introduces a supplementary bilinear form $d(\cdot,\cdot)$ on $M\times M$ which we suppose is continuous, and we let D be a constant such that

$$d(\lambda,\mu)\leqslant D\|\lambda\|_M\|\mu\|_M \quad \text{for all } \lambda\in M,\ \mu\in M. \tag{10.18}$$

Given finite-dimensional subspaces $W_h\subset W$ and $M_h\subset M$, the corresponding discrete problem associated with (10.17) is written as follows: find φ_h and λ_h satisfying

$$\begin{aligned}
&\varphi_h\in W_h, \qquad \lambda_h\in M_h,\\
&a(\varphi_h,\psi_h)+b(\psi_h,\lambda_h)=f(\psi_h) \quad \text{for all } \psi_h\in W_h,\\
&b(\varphi_h,\mu_h)+d(\lambda_h,\mu_h)=g(\mu_h) \quad \text{for all } \mu_h\in M_h.
\end{aligned} \tag{10.19}$$

If in addition to the hypotheses of Theorem 10.4 we suppose that

$$a(\psi_h, \psi_h) \geqslant 0 \quad \text{for all } \psi_h \in W_h,$$
$$d(\mu_h, \mu_h) \leqslant 0 \quad \text{for all } \mu_h \in M_h,$$

and that one of the two forms $a(\cdot,\cdot)$ and $d(\cdot,\cdot)$ is symmetric, then problem (10.19) has a unique solution (φ_h, λ_h). Moreover, if (φ, λ) denotes the solution of (10.17), an error bound analogous to that given by (10.16) may be obtained with a constant C depending only on A, B, D, α and β. Another generalization of Theorem 10.4 is obtained by assuming in addition to the hypotheses of Theorem 10.4 that there is a positive number δ, $0<\delta\leqslant 1$, such that

$$d(\lambda_h, \mu_h) \leqslant (1-\delta)\frac{\alpha}{\sqrt{\alpha^2+A^2}}\frac{\beta^2}{A}\|\lambda_h\|_M\|\mu_h\|_M \quad \text{for all } \lambda_h \in M_h,\ \mu_h \in M_h.$$

The conclusion is the same as in the above generalization except that here the constant C depends also on δ.

The demonstration of the first generalization relies on the fact that if H is a Hilbert space with inner product $((\cdot,\cdot))_H$ and $\mathscr{U} \in \mathscr{L}(H,H)$ is positive in the sense that $((\mathscr{U}x, x))_H \geqslant 0$, for all $x \in H$, then $\mathscr{I}+\mathscr{U}$ is invertible in $\mathscr{L}(H,H)$ and $\|(\mathscr{I}+\mathscr{U})^{-1}\|_{\mathscr{L}(H,H)} \leqslant 1$. To show that the second generalization is valid one uses that if H is a Hilbert space and $\mathscr{U} \in \mathscr{L}(H,H)$ has norm in $\mathscr{L}(H,H)$ less than $1-\theta$ for some θ in $(0,1]$, then $\mathscr{I}+\mathscr{U}$ is invertible in $\mathscr{L}(H,H)$ and $\|(\mathscr{I}+\mathscr{U}^{-1})\|_{\mathscr{L}(H,H)} \leqslant 1/\theta$.

REMARK 10.9. An extension of the above results of BREZZI [1974b] to the case of the following generalized saddle point problem:

$$(\varphi, \lambda) \in W \times M,$$
$$a(\varphi, \psi) + b_1(\psi, \lambda) = f(\psi) \quad \text{for all } \psi \in W,$$
$$b_2(\varphi, \mu) = g(\mu) \quad \text{for all } \mu \in M,$$

for which the bilinear forms $b_1(\cdot,\cdot)$ and $b_2(\cdot,\cdot)$ on $W \times M$ are distinct, can be deduced from the more extensive generalization given in NICOLAIDES [1982], see also BERNARDI, CANUTO and MADAY [1988].

11. Effects of numerical integration

Under the hypotheses of Theorem 10.2, problem (10.7), the problem with Lagrangian multiplier, is reduced, once the bases of W_h and of M_h have been chosen, to the resolution of an invertible linear system of order N, with $N = \dim W_h + \dim M_h$. Several algorithms adapted to the resolution of such linear systems will be presented in Section 17. For the moment we remark only that even the formation of the linear system can turn out to be practically impossible or simply too expensive in calculation time. Thus we are led to replace problem (10.7) by the following problem:

find $(\varphi_h^*, \lambda_h^*)$ satisfying

$$(\varphi_h^*, \lambda_h^*) \in W_h \times M_h,$$
$$a_h(\varphi_h^*, \psi_h) + b_h(\psi_h, \lambda_h^*) = f_h(\psi_h) \quad \text{for all } \psi_h \in W_h, \tag{11.1}$$
$$b_h(\varphi_h^*, \mu_h) = g_h(\mu_h) \quad \text{for all } \mu_h \in M_h;$$

where $a_h(\cdot,\cdot)$ and $b_h(\cdot,\cdot)$ are bilinear forms on $W_h \times W_h$ and $W_h \times M_h$, respectively, and $f_h(\cdot)$ and $g_h(\cdot)$ are linear forms on W_h and M_h, respectively. When $g_h(\cdot)=0$, if $(\varphi_h^*, \lambda_h^*)$ is the solution of (11.1), then the first argument φ_h^* is a solution of the variational problem: find a solution φ_h^* such that

$$\varphi_h^* \in V_h^*, \qquad a_h(\varphi_h^*, \psi_h) = f_h(\psi_h) \quad \text{for all } \psi_h \in V_h^*; \tag{11.2}$$

where the subspace V_h^* is defined by

$$V_h^* = \{\psi_h \in W_h \colon \forall \mu_h \in M_h,\ b_h(\psi_h, \mu_h) = 0\}. \tag{11.3}$$

REMARK 11.1. For all of the examples that we shall consider, the physical parameters of the modelled problem actually play a role only in the expression of the bilinear form $a(\cdot,\cdot)$. The form $b(\cdot,\cdot)$ is independent of these parameters; the same is true, a fortiori, for the subspace V defined by (10.3).

Furthermore, the spaces W_h and M_h are constructed in such a way that the exact calculation of the term $b(\psi_h, \mu_h)$ poses no practical difficulty since it always reduces to the calculation of an integral on an affine manifold of a polynomial function. In the following we shall point out (cf. Remark 11.2) where it can be of some interest to replace the bilinear form $b(\cdot,\cdot)$ by another $b_h(\cdot,\cdot)$.

Defining X_h to be the product space $W_h \times M_h$ and putting

$$\begin{aligned} &k_h(u_h, v_h) = a_h(\varphi_h, \psi_h) + b_h(\psi_h, \lambda_h) + b_h(\varphi_h, \mu_h),\\ &l_h(v_h) = f_h(\psi_h) + g_h(\mu_h), \end{aligned} \tag{11.4}$$

for $u_h = (\varphi_h, \lambda_h)$ and $v_h = (\psi_h, \mu_h)$, we can write, as in Section 9, problem (11.1) in the form: find u_h^* satisfying

$$u_h^* \in X_h, \qquad k_h(u_h^*, v_h) = l_h(v_h) \quad \text{for all } u_h \in X_h.$$

Then in a slightly more general setting, this leads us to study the approximation of the solution u of problem (9.3) by the solution u_h^* of the following problem: find u_h^* satisfying

$$u_h^* \in X_h, \qquad k_h(u_h^*, v_h) = l_h(v_h) \quad \text{for all } v_h \in Y_h, \tag{11.5}$$

where $k_h(\cdot,\cdot)$ is a bilinear form on $X_h \times Y_h$ and $l_h(\cdot)$ is a linear form on Y_h, with $X_h \subset X$ and $Y_h \subset Y$. The analysis is based on the following result:

THEOREM 11.1. *Suppose that $k(\cdot,\cdot)$ is a bilinear form, continuous on $X \times Y$ and let K be a number such that* (9.6) *is satisfied. We suppose that the choice of subspaces X_h*

and Y_h is such that

$$\dim Y_h = \dim X_h,$$

and that there exists a number $\kappa_h > 0$ such that

$$\sup_{\{v_h \in Y_h : \|v_h\|_Y = 1\}} k_h(u_h, v_h) \geqslant \kappa_h \|u_h\|_X \quad \text{for all } u_h \in X_h. \tag{11.6}$$

Then there exists a constant C, depending only on K and κ_h, such that if u is a solution of (9.3) *and u_h^* is the solution of* (11.5), *we have*

$$\|u - u_h^*\|_X \leqslant C \left\{ \inf_{w_h \in X_h} \left(\|u - w_h\|_X + \sup_{v_h \in Y_h} \frac{k(w_h, v_h) - k_h(w_h, v_h)}{\|v_h\|_Y} \right) + \sup_{v_h \in Y_h} \frac{l(v_h) - l_h(v_h)}{\|v_h\|_Y} \right\}. \tag{11.7}$$

This is, of course, in the case $k_h(\cdot,\cdot) = k(\cdot,\cdot)$ and $l_h(\cdot) = l(\cdot)$, the result (9.9) given in Theorem 9.3. It generalizes the results of STRANG [1972] developed for the elliptic variational problem. The demonstration may be obtained by simply adapting the demonstration of Strang's result given by CIARLET [1977, pp. 186–187] for example.

We return now to the discrete problem (11.1), Theorem 10.4 may be generalized to the following result:

THEOREM 11.2. *Suppose that $a(\cdot,\cdot)$ and $b(\cdot,\cdot)$ are continuous bilinear forms on $W \times W$ and on $W \times M$ respectively, and let A and B be numbers such that* (10.12) *and* (10.13) *are satisfied. Let A_h be a number such that*

$$a_h(\varphi_h, \psi_h) \leqslant A_h \|\varphi_h\|_W \|\psi_h\|_W \quad \text{for all } \varphi_h \in W_h,\ \psi_h \in W_h. \tag{11.8}$$

Suppose further that there exist numbers $\alpha_h > 0$ and $\beta_h > 0$ such that

$$a_h(v_h, v_h) \geqslant \alpha_h \|v_h\|_W^2, \quad \text{for all } v_h \in V_h^*, \tag{11.9}$$

and

$$\inf_{\{\mu_h \in M_h : \|\mu_h\|_M = 1\}} \ \sup_{\{\psi_h \in W_h : \|\psi_h\|_W = 1\}} b_h(\psi_h, \mu_h) \geqslant \beta_h. \tag{11.10}$$

Then there exists a constant C dependent only on A, B, A_h, α_h and β_h such that if (φ, λ) satisfies (10.1) and $(\varphi_h^*, \lambda_h^*)$ *is the solution of* (11.1) *then we have*

$$\begin{aligned} &\|\varphi - \varphi_h^*\|_W + \|\lambda - \lambda_h^*\|_M \\ &\quad \leqslant C \left\{ \inf_{\psi_h \in W_h} \left(\|\varphi - \psi_h\|_W + \sup_{\eta_h \in W_h} \frac{a(\psi_h, \eta_h) - a_h(\psi_h, \eta_h)}{\|\eta_h\|_W} \right) \right. \\ &\qquad + \inf_{\mu_h \in M_h} \left(\|\lambda - \mu_h\|_M + \sup_{\eta_h \in W_h} \frac{b(\eta_h, \mu_h) - b_h(\eta_h, \mu_h)}{\|\eta_h\|_W} \right) \\ &\qquad \left. + \sup_{\eta_h \in W_h} \frac{f(\eta_h) - f_h(\eta_h)}{\|\eta_h\|_W} + \sup_{v_h \in M_h} \frac{g(v_h) - g_h(v_h)}{\|v_h\|_{M_h}} \right\}. \end{aligned} \tag{11.11}$$

We may obtain this result by applying Theorem 11.1. With the given hypotheses and the definition of the bilinear form $k_h(\cdot,\cdot)$ given in (11.4), condition (9.6) is satisfied with K depending only on A and B and condition (11.6) is satisfied with $\kappa_h > 0$ depending only on A_h, α_h and β_h. A detailed demonstration of this type of result was given by THOMAS [1977, Chapter I, Section 3].

REMARK 11.2. The replacement of the bilinear form $b(\cdot,\cdot)$ by the bilinear form $b_h(\cdot,\cdot)$ entails the replacement of the subspace V_h defined by (10.8) by the subspace V_h^* defined by (11.3). The ellipticity hypothesis (11.9) can certainly be satisfied without (10.14)'s being satisfied, even in the case where $a_h(\cdot,\cdot)=a(\cdot,\cdot)$ on $W_h \times M_h$. Also, the discrete inf–sup condition (11.10) may hold without (10.15)'s holding. On the contrary, the converses are valid; (10.14) implies (11.9) and (10.15) implies (11.10). However, we stress that a careless choice of numerical integration technique can turn a well-posed discrete problem of the form (10.7) into an ill-posed discrete problem of the form (11.1). One should be aware that by replacing $b(\cdot,\cdot)$ by $b_h(\cdot,\cdot)$ one is actually changing the numerical method, and this must be done with some care.

CHAPTER IV

Mixed Finite Element Methods

12. Examples of primal mixed methods for a Dirichlet problem

A primal mixed variational formulation of the model problem, find a solution $(\boldsymbol{p}, u)$ of the system (4.4)–(4.5) with the homogeneous Dirichlet boundary condition $u=0$ on Γ, is obtained by characterizing the pair $(\boldsymbol{p}, u)$ as the solution of

$$\begin{aligned}
&(\boldsymbol{p}, u)\in(L^2(\Omega))^n \times H_0^1(\Omega),\\
&a(\boldsymbol{p}, \boldsymbol{q})+b(\boldsymbol{q}, u)=0 \quad \text{for all } \boldsymbol{q}\in(L^2(\Omega))^n,\\
&b(\boldsymbol{p}, v)=-\int_\Omega fv \,\mathrm{d}x \quad \text{for all } v\in H_0^1(\Omega),
\end{aligned} \tag{12.1}$$

where

$$a(\boldsymbol{p}, \boldsymbol{q})=\sum_{i,j=1}^{n}\int_\Omega A_{ij}p_j q_i \,\mathrm{d}x, \tag{12.2}$$

and

$$b(\boldsymbol{q}, v)=-\int_\Omega \boldsymbol{q}\cdot\mathbf{grad}\, v \,\mathrm{d}x. \tag{12.3}$$

We are here in the context of Chapter III with

$$W=(L^2(\Omega))^n, \qquad \|\boldsymbol{q}\|_W=\|\boldsymbol{q}\|_{0,\Omega}, \tag{12.4}$$

and

$$M=H_0^1(\Omega), \qquad \|v\|_M=|v|_{1,\Omega}. \tag{12.5}$$

With the given hypotheses concerning the coefficients a_{ij}, the bilinear form $a(\cdot,\cdot)$ is continuous on $W\times W$ and W-elliptic with constant of ellipticity $\alpha=\alpha_d$, cf. (4.8). The bilinear form $b(\cdot,\cdot)$ is continuous on $W\times M$ and we clearly have

$$|v|_{1,\Omega}=1 \;\Rightarrow\; \sup_{\{\boldsymbol{q}\in W:\|\boldsymbol{q}\|_{0,\Omega}=1\}} b(\boldsymbol{q}, v)=1. \tag{12.6}$$

Thus Theorem 10.1 guarantees that, for each $f\in L^2(\Omega)$, problem (12.1) has a unique solution.

In order to construct a primal mixed approximation to this solution, we introduce a finite-dimensional subspace W_h of $(L^2(\Omega))^n$ and a finite-dimensional subspace M_h of $H_0^1(\Omega)$. From Theorems 10.2 and 10.4 we deduce:

THEOREM 12.1. *Suppose that the subspaces W_h and M_h are compatible in the sense that the discrete inf–sup condition is satisfied:*

$$\beta_h = \inf_{\{v_h \in M_h : |v_h|_{1,\Omega} = 1\}} \sup_{\{q_h \in W_h : \|q_h\|_{0,\Omega} = 1\}} b(q_h, v_h) > 0. \tag{12.7}$$

Then the problem of finding a pair (p_h, u_h) satisfying

$$\begin{aligned} &(p_h, u_h) \in W_h \times M_h, \\ &a(p_h, q_h) + b(q_h, u_h) = 0 \quad \text{for all } q_h \in W_h, \\ &b(p_h, v_h) = -\int_\Omega f v_h \, \mathrm{d}x \quad \text{for all } v_h \in M_h, \end{aligned} \tag{12.8}$$

has a unique solution. Furthermore, for each family of pairs of subspaces (W_h, M_h) satisfying (12.7) *uniformly in the sense that $\beta = \inf \beta_h > 0$, there exists a constant C, independent of h, such that*

$$\|p - p_h\|_{0,\Omega} + |u - u_h|_{1,\Omega} \leq C \left\{ \inf_{q_h \in W_h} \|p - q_h\|_{0,\Omega} + \inf_{v_h \in M_h} |u - v_h|_{1,\Omega} \right\}. \tag{12.9}$$

Given a triangulation $\mathcal{T}_h$ by n-simplexes T of a domain $\bar{\Omega}$, supposed to be polyhedral, and a positive integer k, we take finite-dimensional subspaces $W_h = W_h^{(k)}$ and $M_h = M_h^{(k)}$ given as follows:

$$W_h = \{ q \in (L^2(\Omega))^n : \forall T \in \mathcal{T}_h, q|_T \in (P_{k-1}(T))^n \}, \tag{12.10}$$

and

$$M_h = \{ v \in H_0^1(\Omega) : \forall T \in \mathcal{T}_h, v|_T \in P_k(T) \}. \tag{12.11}$$

(With the notation (5.10) introduced in Chapter II we have $M_h = H_0^1(\Omega) \cap L_h^{(k)}$.) Such a choice of subspaces is proposed in REDDY and ODEN [1975]; cf. also BABUŠKA, ODEN and LEE [1977, Section 4].

THEOREM 12.2. *For each positive integer k, the subspaces W_h and M_h defined by* (12.10) *and* (12.11) *satisfy the discrete inf–sup condition* (12.7). *Moreover, if $\{\mathcal{T}_h : h \in \mathcal{H}\}$ is a regular family of triangulations, there is a constant C, independent of h, such that if the solution (p, u) of* (12.1) *belongs to $(H^k(\Omega))^n \times H^{k+1}(\Omega)$ we have the error estimate*

$$\|p - p_h\|_{0,\Omega} + |u - u_h|_{1,\Omega} \leq C h^k (|p|_{k,\Omega} + |u|_{k+1,\Omega}). \tag{12.12}$$

In virtue of Theorem 12.1, the demonstration of Theorem 12.2 consists essentially of establishing the existence of a constant $\beta > 0$ such that for each function $v_h \in M_h$, $v_h \neq 0$, there exists a function $q_h \in W_h$, $q_h \neq 0$, such that

$$b(q_h, v_h) \geq \beta \|q_h\|_{0,\Omega} |v_h|_{1,\Omega}.$$

With the choice of subspaces given in (12.10) and (12.11), we clearly have that for each $v_h \in M_h$, **grad** v_h belongs to W_h. The compatibility condition is thus satisfied with $\beta=1$. The proof is now terminated with an application of classical results giving error bounds for interpolation in $(L^2(\Omega))^n$ and in $H_0^1(\Omega)$. (As $L_h^{(k)}u$ belongs to M_h, we may use Theorem 5.2.)

REMARK 12.1. When the coefficients a_{ij} are sufficiently regular, for example when $a_{ij} \in C^k$, the regularity hypothesis $u \in H^{k+1}(\Omega)$ implies the regularity $\boldsymbol{p} \in (H^k(\Omega))^n$.

REMARK 12.2. Even with no regularity hypothesis for the solution $(\boldsymbol{p}, u)$ one can show that the primal mixed method applied in the context of Theorem 12.2 is convergent:

$$\lim_{h\to 0} \boldsymbol{p}_h = \boldsymbol{p} \quad \text{in } (L^2(\Omega))^n, \qquad \lim_{h\to 0} u_h = u \quad \text{in } H_0^1(\Omega);$$

cf. Remark 10.6.

REMARK 12.3. With $W_h = W_h^{(k)}$ as given in (12.10) and with $M_h = M_h^{(k')}$ as given in (12.11) but with k' not necessarily equal to k, the compatibility hypothesis is satisfied once $k' \leqslant k$. Error estimates optimal in $W \times M$ are obtained when $k'=k$.

REMARK 12.4. Let $\hat{T}$ be the reference n-simplex and for each $T \in \mathscr{T}_h$, let F_T denote an affine bijection of $\hat{T}$ onto T with Jacobian J_T. The spaces W_h and M_h given by (12.10) and (12.11) may also be defined as follows:

$$W_h = \{\boldsymbol{q} \in (L^2(\Omega))^n : \forall T \in \mathscr{T}_h, J_T \mathrm{D}F_T^{-1}(\boldsymbol{q}|_T \circ F_T) \in (P_{k-1}(\hat{T}))^n\}, \tag{12.10'}$$

$$M_h = \{v \in H_0^1(\Omega) : \forall T \in \mathscr{T}_h, v|_T \circ F_T \in P_k(\hat{T})\}. \tag{12.11'}$$

It is with these definitions that the theory is extended to the case of curved finite elements where T is the image of $\hat{T}$ under a bijection not required to be affine. However, we must note that for $v_h \in M_h$, **grad** v_h need no longer belong to W_h. To show the compatibility relation in this case we introduce for $v_h \in M_h$, the function $\boldsymbol{q}_h \in W_h$ defined on each $T \in \mathscr{T}_h$ by

$$(\boldsymbol{q}_h|_T) \circ F_T = (1/J_T)(\mathrm{D}F_T)(\mathrm{D}F_T)^{\mathrm{T}}(\mathbf{grad}\, v_h|_T) \circ F_T$$

and apply the relations (6.23).

These primal mixed methods may be easily adapted to the case in which T_h is a triangulation by rectangles having sides parallel to the axes for $n=2$, by rectangular solids having faces parallel to the coordinate planes for $n=3$. For a regular family of such triangulations one may show:

THEOREM 12.3. (Stated here only for the case $n=2$, T a rectangle.) *For each positive integer k, with $W_h = W_h^{(k)}$ and $M_h = M_h^{(k)}$ defined as follows:*

$$W_h = \{\boldsymbol{q} \in (L^2(\Omega))^2 : \forall T \in \mathscr{T}_h, \boldsymbol{q}|_T \in P_{k-1,k}(T) \times P_{k,k-1}(T)\} \tag{12.13}$$

and

$$M_h=\{v\in H_0^1(\Omega): \forall T\in\mathcal{T}_h,\ v|_T\in P_{k,k}(T)\}, \tag{12.14}$$

the inf–sup condition, condition (12.7), is uniformly satisfied. Moreover, if the solution (p,u) *of* (12.1) *belongs to* $(H^k(\Omega))^n\times H^{k+1}(\Omega)$ *and* (p_h,u_h) *is the solution of* (12.8) *for* W_h *and* M_h *as in* (12.13) *and* (12.14), *there exists a constant C, independent of h, such that*

$$\|p-p_h\|_{0,\Omega}+|u-u_h|_{1,\Omega}\leqslant Ch^k(|p|_{k,\Omega}+|u|_{k+1,\Omega}). \tag{12.15}$$

13. Examples of dual mixed methods for a Dirichlet problem

A dual mixed variational formulation of the model problem, find a solution (p,u) of system (4.4)–(4.5) with the homogeneous Dirichlet boundary condition $u=0$ on Γ, is given by characterizing the pair (p,u) as the solution of

$$\begin{aligned}&(p,u)\in H(\mathrm{div};\Omega)\times L^2(\Omega),\\ &a(p,q)+b(q,u)=0\quad\text{for all } q\in H(\mathrm{div};\Omega),\\ &b(p,v)=-\int_\Omega fv\,\mathrm{d}x\quad\text{for all } v\in L^2(\Omega),\end{aligned} \tag{13.1}$$

where

$$a(p,q)=\sum_{i,j=1}^{n}\int_\Omega A_{ij}p_jq_i\,\mathrm{d}x \tag{13.2}$$

and

$$b(q,v)=\int_\Omega v\,\mathrm{div}\,q\,\mathrm{d}x. \tag{13.3}$$

We are again in the context of Chapter III with now

$$W=H(\mathrm{div};\Omega),\qquad \|q\|_W=\|q\|_{H(\mathrm{div};\Omega)}, \tag{13.4}$$

and

$$M=L^2(\Omega),\qquad \|v\|_M=\|v\|_{0,\Omega}. \tag{13.5}$$

The subspace V of functions $q\in W$ for which $b(q,v)=0$ for all $v\in M$ is none other than

$$V=\{q\in H(\mathrm{div};\Omega):\ \mathrm{div}\,q=0\}. \tag{13.6}$$

With the given hypotheses concerning the coefficients a_{ij}, the bilinear form $a(\cdot,\cdot)$ is continuous on $W\times W$ and V-elliptic with constant of ellipticity $\alpha=\alpha_d$, cf. (4.8). The bilinear form $b(\cdot,\cdot)$ is continuous on $W\times M$ and the inf–sup condition is satisfied:

$$\inf_{\{v\in M:\|v\|_{0,\Omega}=1\}}\ \sup_{\{q\in W:\|q\|_{H(\mathrm{div};\Omega)}=1\}} b(q,v)>0. \tag{13.7}$$

To establish (13.7) we shall use the auxiliary problem. For each $v \in L^2(\Omega)$ denote by φ_v the unique solution in $H_0^1(\Omega)$, of $-\Delta\varphi_v = v$ on Ω. Then $\boldsymbol{q}_v = \mathbf{grad}\,\varphi_v$ belongs to $H(\mathrm{div};\Omega)$ and $b(\boldsymbol{q}_v, v) = \|v\|_{0,\Omega}^2$. Since

$$\|\boldsymbol{q}_v\|_{H(\mathrm{div};\,\Omega)} = (|\varphi_v|_{1,\Omega}^2 + \|v\|_{0,\Omega}^2)^{1/2} \leqslant C\|v\|_{0,\Omega},$$

inequality (13.7) follows. Thus we can apply Theorem 10.1 to obtain the existence and uniqueness of the solution $(\boldsymbol{p}, u)$ of (13.1).

In order to construct a dual mixed approximation of the solution $(\boldsymbol{p}, u)$ of (13.1), we introduce a finite-dimensional subspace W_h of $H(\mathrm{div};\Omega)$ and a finite-dimensional subspace M_h of $L^2(\Omega)$. From Theorems 10.2 and 10.4 we deduce:

THEOREM 13.1. *Suppose that the subspaces W_h and M_h are compatible in the sense that the discrete inf–sup condition is satisfied:*

$$\beta_h = \inf_{\{v_h \in M_h : \|v_h\|_{0,\Omega} = 1\}} \; \sup_{\{\boldsymbol{q}_h \in W_h : \|\boldsymbol{q}_h\|_{H(\mathrm{div};\Omega)} = 1\}} b(\boldsymbol{q}_h, v_h) > 0. \tag{13.8}$$

Then the problem of finding a pair $(\boldsymbol{p}_h, u_h)$ satisfying

$$\begin{aligned} &(\boldsymbol{p}_h, u_h) \in W_h \times M_h,\\ &a(\boldsymbol{p}_h, \boldsymbol{q}_h) + b(\boldsymbol{q}_h, u_h) = 0 \quad \textit{for all } \boldsymbol{q}_h \in W_h,\\ &b(\boldsymbol{p}_h, v_h) = -\int_\Omega f v_h \,\mathrm{d}x \quad \textit{for all } v_h \in M_h, \end{aligned} \tag{13.9}$$

has a unique solution. Furthermore, for each family of pairs of subspaces (W_h, M_h) such that for each $\boldsymbol{q}_h \in W_h$ we have

$$\left(\int_\Omega v_h \,\mathrm{div}\,\boldsymbol{q}_h \,\mathrm{d}x = 0 \quad \textit{for all } v_h \in M_h\right) \Rightarrow \mathrm{div}\,\boldsymbol{q}_h = 0 \tag{13.10}$$

and such that (13.8) *is satisfied uniformly in the sense that $\beta = \inf \beta_h > 0$, there exists a constant C, independent of h, such that*

$$\begin{aligned} &\|\boldsymbol{p} - \boldsymbol{p}_h\|_{H(\mathrm{div};\,\Omega)} + \|u - u_h\|_{0,\Omega}\\ &\quad \leqslant C\left\{\inf_{\boldsymbol{q}_h \in W_h} \|\boldsymbol{p} - \boldsymbol{q}_h\|_{H(\mathrm{div};\,\Omega)} + \inf_{v_h \in M_h} \|u - v_h\|_{0,\Omega}\right\}. \end{aligned} \tag{13.11}$$

Hypothesis (13.10) expresses the fact that the space V_h,

$$V_h = \{\boldsymbol{q}_h \in W_h : \forall v_h \in M_h, b(\boldsymbol{q}_h, v_h) = 0\}, \tag{13.12}$$

is a subspace of the space V defined by (13.6). This hypothesis assures the uniform ellipticity of $a(\cdot,\cdot)$ on V_h, where the norm on V_h is the $H(\mathrm{div};\Omega)$ norm.

Given a triangulation $\mathcal{T}_h$ by n-simplexes T of a domain $\bar{\Omega}$, supposed to be polyhedral, and a positive integer k, we take finite-dimensional subspaces $W_h = W_h^{(k)}$ and $M_h = M_h^{(k)}$ given as follows:

$$W_h=\{q\in H(\mathrm{div};\Omega)\colon \forall T\in\mathscr{T}_h,\ q|_T\in D_k(T)\},\tag{13.13}$$

and

$$M_h=\{v\in L^2(\Omega)\colon \forall T\in\mathscr{T}_h,\ v|_T\in P_{k-1}(T)\},\tag{13.14}$$

where $D_k(T)$ denotes the space of restrictions to T of the functions in the space D_k introduced in (6.5). (With the notation (6.17) introduced in Chapter II we have $W_h=E_h^{(k)}$.) These examples were proposed and analyzed in RAVIART and THOMAS [1977] in the two-dimensional case and in NEDELEC [1980] in the three-dimensional case. In the literature the space $W_h\times M_h$ or $W_h^{(k)}\times M_h^{(k)}$ is referred to as the Raviart–Thomas space of index k (or sometimes $k-1$; cf. Remark 6.2) or, in the three-dimensional case, as the Raviart–Thomas–Nedelec space.

THEOREM 13.2. *For each positive integer k, the subspaces W_h and M_h defined by* (13.13) *and* (13.14) *satisfy the discrete inf–sup condition* (13.8). *Moreover, if $\{\mathscr{T}_h\colon h\in\mathscr{H}\}$ is a regular family of triangulations, there is a constant C independent of h such that if the solution $(\boldsymbol{p},u)$ of* (13.1) *is such that $(\boldsymbol{p},u)\in(H^k(\Omega))^n\times H^k(\Omega)$ and* $\operatorname{div}\boldsymbol{p}\in H^k(\Omega)$ *and if $(\boldsymbol{p}_h,u_h)$ is the solution of* (13.9) *associated with* (13.13) *and* (13.14), *then we have the error estimate*

$$\|\boldsymbol{p}-\boldsymbol{p}_h\|_{H(\mathrm{div};\,\Omega)}+\|u-u_h\|_{0,\Omega}\leqslant Ch^k(|u|_{k,\Omega}+|\boldsymbol{p}|_{k,\Omega}+|\operatorname{div}\boldsymbol{p}|_{k,\Omega}).\tag{13.15}$$

PROOF. With W_h and M_h as given in (13.13) and (13.14) we clearly have that div $\boldsymbol{q}_h\in M_h$ for each $\boldsymbol{q}_h\in W_h$ and thus that $V_h\subset V$. To establish the discrete inf–sup condition we apply Theorem 10.3. Take for subspace of $W=H(\mathrm{div};\Omega)$ the space $\mathscr{W}=(H^1(\Omega))^n$ and for continuous linear operator π_h from $\mathscr{W}$ into W_h, the operator $\pi_h=\mathscr{E}_h^{(k)}$, given by (6.2) together with (6.14). Then (6.15) guarantees that (10.11) is satisfied. Let R be any continuous linear map from $L^2(\Omega)$ into $(H^1(\Omega))^n$ satisfying

$$\operatorname{div}(Rv)=v,\tag{13.16}$$

and let R_h from M_h into $\mathscr{W}=(H^1(\Omega))^n$ be defined by

$$R_hv_h=Rv_h\quad\text{for all } v_h\in M_h.$$

We note that with sufficient regularity the function $\boldsymbol{q}_v\in H(\mathrm{div};\Omega)$ used to establish the inf–sup condition (13.7) belongs to $(H^1(\Omega))^n$ and Rv could be taken to be $\boldsymbol{q}_v$. To avoid having to invoke regularity results for Dirichlet problems in polyhedral domains which would necessitate certain restrictions on the geometry of Ω, if suffices to extend the functions $v\in L^2(\Omega)$ by zero to functions $\tilde v\in L^2(\mathscr{O})$ where $\mathscr{O}$ is a ball containing Ω, to solve the Dirichlet problem $\Delta\varphi_v=\tilde v$ in $\mathscr{O}$, $\varphi_{v|\partial\mathscr{O}}=0$, and to define Rv to be the restriction to Ω of **grad** φ_v.

To verify (10.10) we note there is a constant $C_{\mathscr{O}}$ depending only on the ball $\mathscr{O}$ containing Ω, such that

$$\|R_hv_h\|_{1,\Omega}\leqslant C_{\mathscr{O}}\|v_h\|_{0,\Omega}.$$

Thus

$$\frac{b(R_h v_h, v_h)}{\|R_h v_h\|_{\mathscr{W}}} = \frac{\|v_h\|_{0,\Omega}^2}{\|R_h v_h\|_{1,\Omega}} \geqslant \frac{1}{C_{\mathscr{O}}} \|v_h\|_{0,\Omega}.$$

Since

$$\sup_{\{\psi \in W: \|\psi\|_W = 1\}} b(\psi, v_h) \leqslant \|\operatorname{div} \psi\|_{0,\Omega} \|v_h\|_{0,\Omega} \leqslant \|v_h\|_{0,\Omega},$$

we have that (10.10) holds with constant $C = 1/C_{\mathscr{O}}$, and (13.8) now follows from Theorem 10.3.

Next suppose that $\{\mathscr{T}_h: h \in \mathscr{H}\}$ is a regular family of triangulations. We have just seen that R_h may be defined with norm independent of h, and Theorem 6.3 guarantees that for the operator $\pi_h = \mathscr{E}_h^{(k)}$ there is a constant C independent of h such that

$$\|\pi_h \boldsymbol{q}\|_{H(\operatorname{div};\, \Omega)} \leqslant C\{h|\boldsymbol{q}|_{1,\Omega} + \|\operatorname{div} \boldsymbol{q}\|_{0,\Omega}\}.$$

Thus the discrete inf–sup condition is satisfied uniformly and we have the error bound (13.11). Then to estimate the error of approximation $\inf_{\boldsymbol{q}_h \in W_h} \|\boldsymbol{p} - \boldsymbol{q}_h\|_{H(\operatorname{div};\, \Omega)}$, we take $\boldsymbol{q}_h = \mathscr{E}_h^{(k)}(\boldsymbol{p})$ and use (6.19) and (6.20). □

REMARK 13.1. When the coefficients a_{ij} are sufficiently regular, for example when $a_{ij} \in C^k$, the regularity hypothesis $\boldsymbol{p} \in (H^k(\Omega))^n$ and $\operatorname{div} \boldsymbol{p} \in H^{(k)}(\Omega)$ is enough to give the error bound (13.15). If we suppose only that $\boldsymbol{p} \in (H^k(\Omega))^n$ we may easily adapt the last part of the demonstration of Theorem 13.2 to obtain

$$\|\boldsymbol{p} - \boldsymbol{p}_h\|_{H(\operatorname{div};\, \Omega)} + \|u - u_h\|_{0,\Omega} \leqslant Ch^{k-1} \|\boldsymbol{p}\|_{k,\Omega},$$

or with a more careful study we obtain

$$\|\boldsymbol{p} - \boldsymbol{p}_h\|_{0,\Omega} \leqslant Ch^k \|\boldsymbol{p}\|_{k,\Omega}. \tag{13.17}$$

If in addition we suppose the regularity of the Dirichlet problem in Ω, a hypothesis which is satisfied once the open polyhedral domain Ω is convex, we have

$$\|u - u_h\|_{0,\Omega} \leqslant Ch^k \|u\|_{k^*,\Omega} \quad \text{with } k^* = \max(k, 2). \tag{13.18}$$

These results are demonstrated in FALK and OSBORN [1980, Section 3].

REMARK 13.2. Even with no regularity hypothesis for the solution $(\boldsymbol{p}, u)$ one may show that the dual mixed method applied in the context of Theorem 13.2 is convergent:

$$\lim_{h \to 0} \boldsymbol{p}_h = \boldsymbol{p} \quad \text{in } H(\operatorname{div}; \Omega), \qquad \lim_{h \to 0} u_h = u \quad \text{in } L^2(\Omega).$$

REMARK 13.3. With $W_h = W_h^{(k)}$ as given in (13.13) and with $M_h = M_h^{(k')}$ as given in (13.4) but with k' not necessarily equal to k, the compatibility hypothesis is satisfied once $k' \leqslant k$. Error estimates, optimal in $W \times M$, are obtained when $k' = k$. If $1 \leqslant k' < k$ the subspace V_h is no longer a subspace of V; however, the method remains convergent.

REMARK 13.4. Let $\hat{T}$ be the reference n-simplex and for each $T\in\mathcal{T}_h$, let F_T denote an affine bijection of $\hat{T}$ onto T with Jacobian J_T. The spaces W_h and M_h given by (13.13) and (13.14) may also be defined as follows:

$$W_h=\{q\in H(\mathrm{div};\Omega)\colon \forall T\in\mathcal{T}_h, |J_T|DF_T^{-1}(q|_T\circ F_T)\in D_k(\hat{T})\}, \tag{13.13'}$$

$$M_h=\{v\in L^2(\Omega)\colon \forall T\in\mathcal{T}_h, v|_T\circ F_T\in P_{k-1}(\hat{T})\}. \tag{13.14'}$$

It is with these definitions that the theory is extended to the case of curved finite elements where T is the image of $\hat{T}$ under a C^1 bijection not required to be affine. To show the compatibility relation in this case, we construct for $v_h\in M_h$, the function $\boldsymbol{q}_h\in W_h$ such that

$$\mathrm{div}\,\boldsymbol{q}_h=\theta_h v_h,$$

where $\theta_h\in L^\infty(\Omega)$ is defined on each $T\in\mathcal{T}_h$ by

$$\theta_h(x)=\sup_{y\in T}\frac{|J_T(y)|}{|J_T(x)|},$$

for each $x\in T$.

REMARK 13.5. In the context of Theorem 13.2 it is obvious that for $\boldsymbol{q}_h\in W_h$ we have $\mathrm{div}\,\boldsymbol{q}_h\in M_h$. Moreover, the proof of Theorem 13.2 shows that for each $v_h\in M_h$ there exists a function $\boldsymbol{q}_h=\pi_h R_h v_h\in W_h$ such that $\mathrm{div}\,\boldsymbol{q}_h=v_h$. Thus in this case we have the characterization

$$M_h=\{v_h\in L^2(\Omega)\colon \exists \boldsymbol{q}_h\in W_h \text{ with } \mathrm{div}\,\boldsymbol{q}_h=v_h\}. \tag{13.19}$$

Such is not the case for the generalizations described in Remarks 13.3 and 13.4.

For the dual mixed method analyzed above, given a triangulation $\mathcal{T}_h$ of $\bar{\Omega}$ by n-simplexes, the vector functions in $W=H(\mathrm{div};\Omega)$ are approximated in the space W_h defined by (13.13) which is the space $E_h^{(k)}$ defined by (6.17) in Chapter II. An alternative method is obtained by instead approximating the vector functions by functions in the space $E_h^{*(k)}$ also defined in Chapter II. For each integer k, $k\geqslant 2$, take $W_h=W_h^{(k)}$ and $M_h=M_h^{(k)}$ as follows:

$$W_h=\{q\in H(\mathrm{div};\Omega)\colon \forall T\in\mathcal{T}_h, \boldsymbol{q}|_T\in D_k^*(T)\}, \tag{13.20}$$

$$M_h=\{v\in L^2(\Omega)\colon \forall T\in\mathcal{T}_h, v|_T\in P_{k-2}(T)\}, \tag{13.21}$$

where $D_k^*(T)=(P_{k-1}(T))^n$; cf. (7.1). These approximation spaces were introduced and the corresponding methods analyzed in BREZZI, DOUGLAS and MARINI [1985] in dimension $n=2$ and in BREZZI, DOUGLAS, DURAN and FORTIN [1987] in dimension $n=3$. In the literature the space $W_h\times M_h$ or $W_h^{(k)}\times M_h^{(k)}$ is referred to as the Brezzi–Douglas–Marini space of index k (or sometimes $k-1$, cf. Remark 6.2), or in the three-dimensional case the Brezzi–Douglas–Duran–Fortin space.

As for the Raviart–Thomas(–Nedelec) spaces it is clear that if $\boldsymbol{p}_h \in W_h$, then $\operatorname{div} \boldsymbol{p}_h \in M_h$, and one may show that the discrete inf–sup condition (13.8) is satisfied. In fact we again have the characterization (13.19). For a regular Dirichlet problem in Ω we have the error estimates

$$\|\boldsymbol{p}-\boldsymbol{p}_h\|_{0,\Omega} \leqslant Ch^k \|\boldsymbol{p}\|_{k,\Omega},$$

$$\|u-u_h\|_{0,\Omega} \leqslant Ch^{k-1} \|u\|_{k-1,\Omega},$$

$$\|\operatorname{div}(\boldsymbol{p}-\boldsymbol{p}_h)\|_{0,\Omega} \leqslant Ch^{k-1} \|\operatorname{div} \boldsymbol{p}\|_{k-1,\Omega}.$$

For each of the dual mixed methods presented above for the case where $\mathcal{T}_h$ is a triangulation by n-simplexes, there is a variant adapted to the case where $\mathcal{T}_h$ is a triangulation by rectangles with sides parallel to the axes when $n=2$ or by rectangular solids with sides parallel to the coordinate planes when $n=3$. The description of these methods and the corresponding analyses may be found in the references cited above for the associated method for a triangulation by simplexes. Here we shall simply state for the case $n=2$ the analogue of Theorem 13.2 when $\mathcal{T}_h$ is a triangulation of $\bar{\Omega}$ by rectangles with sides parallel to the axes.

THEOREM 13.3. *For each positive integer k, the subspaces* $W_h = W_h^{(k)}$ *and* $M_h = M_h^{(k)}$ *defined by*

$$W_h = \{\boldsymbol{q} \in H(\operatorname{div}; \Omega): \forall T \in \mathcal{T}_h, \boldsymbol{q}|_T \in D_k^{\square}(T)\}, \tag{13.22}$$

and

$$M_h = \{v \in L^2(\Omega): \forall T \in \mathcal{T}_h, v|_T \in P_{k-1,k-1}(T)\}, \tag{13.23}$$

with $D_k^{\square} = P_{k,k-1} \times P_{k-1,k}$, *satisfy the discrete inf–sup condition* (13.8). *Moreover, if* $\{\mathcal{T}_h: h \in \mathcal{H}\}$ *is a regular family of triangulations, there is a constant C, independent of h, such that if the solution* $(\boldsymbol{p}, u)$ *of* (13.1) *is such that* $(\boldsymbol{p}, u) \in (H^k(\Omega))^n \times H^k(\Omega)$ *and* $\operatorname{div} \boldsymbol{p} \in H^k(\Omega)$ *and if* $(\boldsymbol{p}_h, u_h)$ *is the solution of* (13.9) *associated with* (13.22) (13.23), *we have the error estimate*

$$\|\boldsymbol{p}-\boldsymbol{p}_h\|_{H(\operatorname{div};\,\Omega)} + \|u-u_h\|_{0,\Omega} \leqslant Ch^k(|u|_{k,\Omega} + |\boldsymbol{p}|_{k,\Omega} + |\operatorname{div} \boldsymbol{p}|_{k,\Omega}).$$

Other dual mixed methods are presented and analyzed in BREZZI, DOUGLAS, FORTIN and MARINI [1987] and in NEDELEC [1986]; cf. Remark 7.2.

14. Examples of mixed methods for a Neumann problem

We consider the model problem, given $f \in L^2(\Omega)$ with $\int_\Omega f \,\mathrm{d}x = 0$, find u (to within a constant) and $\boldsymbol{p}$ satisfying (4.4) and (4.5) with the homogeneous boundary condition $\boldsymbol{p} \cdot \boldsymbol{\nu} = 0$ on Γ.

A primal mixed variational formulation of this problem is obtained by characterizing the pair $(\boldsymbol{p}, u)$ as the solution of

$$(\boldsymbol{p}, u) \in (L^2(\Omega))^n \times H^1(\Omega)/\mathbb{R},$$

$$a(\boldsymbol{p}, \boldsymbol{q}) + b(\boldsymbol{q}, u) = 0 \quad \text{for all } \boldsymbol{q} \in (L^2(\Omega))^n, \tag{14.1}$$

$$b(\boldsymbol{p}, v) = -\int_\Omega f v \,\mathrm{d}x \quad \text{for all } v \in H^1(\Omega)/\mathbb{R},$$

where the bilinear forms $a(\cdot,\cdot)$ and $b(\cdot,\cdot)$ are given by (12.2) and (12.3). Recalling the notation for the abstract problem of Section 10, here we put

$$W = (L^2(\Omega))^n, \qquad \|\boldsymbol{q}\|_W = \|\boldsymbol{q}\|_{0,\Omega}, \tag{14.2}$$

and

$$M = H^1(\Omega)/\mathbb{R}, \qquad \|v\|_M = |v|_{1,\Omega}. \tag{14.3}$$

With W_h and M_h denoting finite-dimensional subspaces of $(L^2(\Omega))^n$ and $H^1(\Omega)/\mathbb{R}$, respectively, we obtain the analogue of Theorem 12.1, with no modification of the statement, and also the analogue of Theorem 12.2 if we specify the subspaces W_h and M_h as follows:

$$W_h = \{\boldsymbol{q} \in (L^2(\Omega))^n : \forall T \in \mathcal{T}_h, \boldsymbol{q}|_T = (P_{k-1}(T))^n\} \tag{14.4}$$

and

$$M_h = \{v \in H^1(\Omega)/\mathbb{R} : \forall T \in \mathcal{T}_h, v|_T \in P_k(T)\}, \tag{14.5}$$

for some $k \geqslant 1$. In the case of a triangulation by rectangles, with the appropriate modification of the subspaces W_h and M_h, the analogue of Theorem 12.3 is obtained.

Thus for primal mixed methods there is no difficulty in adapting the theory for the Dirichlet problem to obtain that for the Neumann problem. The construction of examples may even be easier for the Neumann problem as the functions of M_h need no longer satisfy, a priori, the zero trace condition on Γ. The situation for dual mixed methods is quite the opposite.

A dual mixed variational formulation of the homogeneous Neumann problem is obtained by characterizing the pair $(\boldsymbol{p}, u)$ as the solution of

$$(\boldsymbol{p}, u) \in H_0(\mathrm{div}; \Omega) \times L^2(\Omega)/\mathbb{R},$$

$$a(\boldsymbol{p}, \boldsymbol{q}) + b(\boldsymbol{q}, u) = 0 \quad \text{for all } \boldsymbol{q} \in H_0(\mathrm{div}; \Omega), \tag{14.6}$$

$$b(\boldsymbol{p}, v) = -\int_\Omega f v \,\mathrm{d}x \quad \text{for all } v \in L^2(\Omega)/\mathbb{R},$$

where the bilinear forms $a(\cdot,\cdot)$ and $b(\cdot,\cdot)$ are given by (13.2) and (13.3). For convenience, $L^2(\Omega)/\mathbb{R}$ will be identified with the subspace of functions $v \in L^2(\Omega)$ such that $\int_\Omega v \,\mathrm{d}x = 0$. Here the spaces W and M and their norms are defined by

$$W = H_0(\mathrm{div}; \Omega), \qquad \|\boldsymbol{q}\|_W = \|\boldsymbol{q}\|_{H(\mathrm{div};\Omega)}, \tag{14.7}$$

and

$$M=\left\{v\in L^2(\Omega): \int_\Omega v\,dx=0\right\}, \qquad \|v\|_M=\|v\|_{0,\Omega}. \tag{14.8}$$

The inf–sup condition analogous to (13.7) is also established with the aid of the dual problem. For each $v\in M$ denote by φ_v the unique element of $H^1(\Omega)/\mathbb{R}$ satisfying $\Delta\varphi_v=v$ in Ω and $\partial\varphi_v/\partial\nu=0$ on Γ. The function $\boldsymbol{q}_v=\mathbf{grad}\,\varphi_v$ belongs to W and satisfies $\operatorname{div}\boldsymbol{q}_v=v$ in Ω.

With W_h and M_h denoting finite-dimensional subspaces of $H_0(\operatorname{div};\Omega)$ and M respectively, a demonstration similar to that given for Theorem 3.1 shows that if the discrete inf–sup condition is satisfied, there exists a unique solution of the following problem: find $(\boldsymbol{p}_h,u_h)$ satisfying

$$\begin{aligned}&(\boldsymbol{p}_h,u_h)\in W_h\times M_h,\\ &a(\boldsymbol{p}_h,\boldsymbol{q}_h)+b(\boldsymbol{q}_h,u_h)=0 \quad \text{for all } \boldsymbol{q}_h\in W_h,\\ &b(\boldsymbol{p}_h,v_h)=-\int_\Omega f v_h\,dx \quad \text{for all } v_h\in M_h.\end{aligned} \tag{14.9}$$

To show that the discrete inf–sup condition holds for a pair of subspaces W_h and M_h, however, is a more delicate matter here than for the Dirichlet problem. It suffices, of course, to be able to associate to each $v_h\in M_h$ a $\boldsymbol{q}_h\in W_h$ with $\operatorname{div}\boldsymbol{q}_h=v_h$, but this is made more complicated here by the requirement that $q_h\in W_h\subset W$ have a vanishing normal trace on Γ. Nonetheless, one may show that the discrete inf–sup condition is satisfied when, for a triangulation $\mathcal{T}_h$ by n-simplexes, the subspaces W_h and M_h are defined as follows:

$$W_h=\{\boldsymbol{q}\in H_0(\operatorname{div};\Omega): \forall T\in\mathcal{T}_h, \boldsymbol{q}|_T\in D_k(T)\} \tag{14.10}$$

and

$$M_h=\left\{v\in L^2(\Omega): \int_\Omega v\,dx=0 \text{ and } \forall T\in\mathcal{T}_h, v|_T\in P_{k-1}(T)\right\}, \tag{14.11}$$

for some positive integer k. We give the demonstration for the case $k=1$:

Let $v_h\in M_h$; thus v_h is constant on each n-simplex T and has zero average over Ω. Denote by Φ_h the space of functions $\varphi_h\in H^1(\mathcal{T}_h)$ with $\int_\Omega\varphi_h\,dx=0$, with $\varphi_h|_T$, for each $T\in\mathcal{T}_h$, an affine function, and with φ_h continuous across the interior faces at the barycenters of the faces T' common to two n-simplexes of $\mathcal{T}_h$. For each $v_h\in M_h$, introduce the nonconforming solution $\varphi_h(v_h)$ of the dual Neumann problem: find φ_h satisfying

$$\begin{aligned}&\varphi_h\in\Phi_h,\\ &\sum_{T\in\mathcal{T}_h}\int_T \mathbf{grad}\,\varphi_h\cdot\mathbf{grad}\,\psi_h\,dx=\int_\Omega v_h\psi_h\,dx \quad \text{for all } \psi_h\in\Phi_h.\end{aligned} \tag{14.12}$$

Let $\boldsymbol{q}_h = \boldsymbol{q}_h(v_h)$ be the function of $(L^2(\Omega))^n$ whose restriction to T, for each $T \in \mathcal{T}_h$, coincides with the function of $D_1(T)$ having normal trace on ∂T equal to that of $\boldsymbol{\varphi}_h|_T$. One can verify that $\boldsymbol{q}_h$ belongs to W_h and that $\operatorname{div} \boldsymbol{q}_h = v_h$. One may further establish that, for a uniformly regular family of triangulations $\{\mathcal{T}_h : h \in \mathcal{H}\}$, there exists a constant C, independent of h, such that

$$\|\psi_h\|_{0,\Omega} \leqslant C \left\{ \sum_{T \in \mathcal{T}_h} |\psi_h|^2_{1,T} \right\}^{1/2}, \tag{14.13}$$

a result which amounts to a generalization of the Friedrichs–Poincaré inequalities to the nonconforming case. One thus obtains the existence of a (new) constant C, independent of h, such that for the preceding construction we have

$$\|\boldsymbol{q}_h(v_h)\|_{H(\operatorname{div};\, \Omega)} \leqslant C \|v_h\|_{0,\Omega}, \tag{14.14}$$

thereby showing that the discrete inf–sup condition is satisfied uniformly. Hence we may deduce that there is a constant C, independent of h, such that

$$\|\boldsymbol{p} - \boldsymbol{p}_h\|_{H(\operatorname{div};\, \Omega)} + \|u - u_h\|_{0,\Omega} \leqslant Ch\{|u|_{1,\Omega} + |\boldsymbol{p}|_{1,\Omega} + |\operatorname{div} \boldsymbol{p}|_{1,\Omega}\}.$$

To verify the discrete inf–sup condition we have avoided using the regularity of the Neumann problem on polyhedral domains. Thus the above argument may be extended to the case of a problem with a Dirichlet condition on part of the boundary and a Neumann condition on the remaining part. For this type of problem, an extension of the above result to the case $k > 1$ may be found in THOMAS [1977, Chapter V, Section 4].

The uniform regularity hypothesis for the family of triangulations can be removed using the regularity of the solution of the Neumann problem for the Laplace equation in a polyhedral domain; cf. THOMAS [1989].

15. Examples of mixed finite element methods for a Robin's boundary condition

In this section we discuss the adaptation of mixed formulations to the following model problem: for a given function $f \in L^2(\Omega)$ and function $g \in L^2(\Gamma)$, find a solution $(\boldsymbol{p}, u)$ of system (4.4)–(4.5) with the Robin's boundary condition

$$\boldsymbol{p} \cdot \boldsymbol{\nu} + tu = g \quad \text{on } \Gamma, \tag{15.1}$$

where t is a continuous, strictly positive function on Γ. (Thus t is bounded above and away from 0.)

A primal mixed variational formulation of this problem is obtained by characterizing the pair $(\boldsymbol{p}, u)$ as the solution of

$$\begin{aligned} &(\boldsymbol{p}, u) \in (L^2(\Omega))^n \times H^1(\Omega), \\ &a(\boldsymbol{p}, \boldsymbol{q}) + b(\boldsymbol{q}, u) = 0 \qquad \text{for all } \boldsymbol{q} \in (L^2(\Omega))^n, \\ &b(\boldsymbol{p}, v) + d(u, v) = -\int_\Omega fv \, dx - \int_\Gamma gv \, d\sigma \quad \text{for all } v \in H^1(\Omega); \end{aligned} \tag{15.2}$$

where the bilinear forms $a(\cdot,\cdot)$ and $b(\cdot,\cdot)$ are defined as for the Dirichlet and Neumann problems by (12.2) and (12.3) and where the bilinear form $d(\cdot,\cdot)$ is defined on $H^1(\Omega)\times H^1(\Omega)$ by

$$d(u,v)=-\int_\Gamma tuv\,\mathrm{d}\sigma. \tag{15.3}$$

Problem (15.2) is the first example we have encountered of the generalization (10.17) developed in the abstract setting in Remark 10.8 with here

$$W=(L^2(\Omega))^n \tag{15.4}$$

and

$$M=H^1(\Omega). \tag{15.5}$$

The numerical analysis of a primal mixed finite element method derived from formulation (15.2) is not difficult when to a triangulation $\mathscr{T}_h$ of $\bar{\Omega}$ by n-simplexes T are associated the finite-dimensional subspaces

$$W_h=\{\boldsymbol{q}\in(L^2(\Omega))^n:\ \forall T\in\mathscr{T}_h,\ \boldsymbol{q}|_T\in(P_{k-1}(T))^n\} \tag{15.6}$$

and

$$M_h=\{v\in H^1(\Omega):\ \forall T\in\mathscr{T}_h,\ v|_T\in P_k(T)\}, \tag{15.7}$$

where k is a positive integer. We remark nonetheless that in this situation the inf–sup condition can never be satisfied as the functions constant belong to M_h.

With the dual mixed variational formulation, the solution $(\boldsymbol{p},u)$ of system (4.4)–(4.5) with boundary condition (15.1) is characterized as the solution of

$$(\boldsymbol{p},u)\in\mathscr{H}(\mathrm{div};\Omega)\times L^2(\Omega),$$

$$a(\boldsymbol{p},\boldsymbol{q})+b(\boldsymbol{q},u)=\int_\Gamma (1/t)g\boldsymbol{q}\cdot\boldsymbol{\nu}\,\mathrm{d}\sigma \quad \text{for all } \boldsymbol{q}\in\mathscr{H}(\mathrm{div};\Omega), \tag{15.8}$$

$$b(\boldsymbol{p},v)=-\int_\Omega fv\,\mathrm{d}x \qquad \text{for all } v\in L^2(\Omega),$$

where the bilinear form $b(\cdot,\cdot)$ is defined as for the Dirichlet and Neumann problems by (13.3) and where the bilinear form $a(\cdot,\cdot)$ is now given by

$$a(\boldsymbol{p},\boldsymbol{q})=\sum_{i,j=1}^{n}\int_\Omega A_{ij}p_jq_i\,\mathrm{d}x+\int_\Gamma (1/t)\boldsymbol{p}\cdot\boldsymbol{\nu}\,\boldsymbol{q}\cdot\boldsymbol{\nu}\,\mathrm{d}\sigma. \tag{15.9}$$

We thus obtain a problem of form (10.1) with now

$$W=\mathscr{H}(\mathrm{div};\Omega) \tag{15.10}$$

and

$$M = L^2(\Omega). \tag{15.11}$$

There exists a continuous linear operator R from $L^2(\Omega)$ into $\mathscr{H}(\text{div};\Omega)$ satisfying $\text{div}(Rv) = v$ for each $v \in L^2(\Omega)$. (One may use, for example, the operator R from $L^2(\Omega)$ into $(H^1(\Omega))^n$ constructed in the proof of Theorem 13.2.) Hence the inf–sup condition is satisfied, the norms for the spaces W and M being given by $\|\cdot\|_W = \|\cdot\|_{\mathscr{H}(\text{div};\Omega)}$ and $\|\cdot\|_M = \|\cdot\|_{0,\Omega}$.

For a triangulation $\mathscr{T}_h$ of Ω by n-simplexes T, take for finite-dimensional subspaces W_h and M_h

$$W_h = \{\boldsymbol{q} \in \mathscr{H}(\text{div};\Omega) : \forall T \in \mathscr{T}_h, \boldsymbol{q}|_T \in D_k(T)\} \tag{15.12}$$

and

$$M_h = \{v \in L^2(\Omega) : \forall T \in \mathscr{T}_h, v|_T \in P_{k-1}(T)\}, \tag{15.13}$$

where k is a positive integer. One can show with no new difficulties beyond those already treated for the problem with Dirichlet boundary condition that the problem, find a pair $(\boldsymbol{p}_h, u_h)$ satisfying

$$(\boldsymbol{p}_h, u_h) \in W_h \times M_h,$$

$$a(\boldsymbol{p}_h, \boldsymbol{q}_h) + b(\boldsymbol{q}_h, u_h) = \int_\Gamma (1/t) g \boldsymbol{q}_h \cdot \boldsymbol{\nu} \, \mathrm{d}\sigma \quad \text{for all } \boldsymbol{q}_h \in W_h, \tag{15.14}$$

$$b(\boldsymbol{p}_h, v_h) = -\int_\Omega f v_h \, \mathrm{d}x \qquad \text{for all } v_h \in M_h,$$

admits a unique solution $(\boldsymbol{p}_h, u_h)$. If $\{\mathscr{T}_h : h \in \mathscr{H}\}$ is a regular family of triangulations, the discrete inf–sup condition is satisfied uniformly and if further the solution $(\boldsymbol{p}, u)$ is sufficiently regular, then we have

$$\|\boldsymbol{p} - \boldsymbol{p}_h\|_{\mathscr{H}(\text{div};\Omega)} + \|u - u_h\|_{0,\Omega} = \mathrm{O}(h^k). \tag{15.15}$$

16. Other examples of mixed finite element methods

All of the examples presented above treat only elliptic problems governed by second-order partial differential equations of the form

$$-\sum_{i,j=1}^{n} \frac{\partial}{\partial x_i}\left(a_{ij}\frac{\partial u}{\partial x_j}\right) = f.$$

The generalization of these methods to problems governed by an equation of the form

$$-\sum_{i,j=1}^{n} \frac{\partial}{\partial x_i}\left(a_{ij}\frac{\partial u}{\partial x_j}\right) + \sum_{j=1}^{n} b_j \frac{\partial u}{\partial x_j} + cu = f$$

poses no difficulty for the primal mixed formulation. The case is quite different for the dual mixed formulation. A mathematical analysis of this problem can be found in DOUGLAS and ROBERTS [1982, 1985]. A generalization of these methods to the quasi-linear problem where the coefficients in the governing equation depend on u is analyzed by MILNER [1985]. Several methods adapted to treat numerically the case where the convection term is large in comparison with the diffusion term have been studied by JAFFRE [1984], JAFFRE and ROBERTS [1985], JOLY [1982], and THOMAS [1987]. Another generalization of mixed methods to problems governed by an equation of the form

$$-\sum_{i,j=1}^{n} \frac{\partial}{\partial x_i}\left(a_{ij}\frac{\partial u}{\partial x_j}\right)=f$$

with coefficients a_{ij} which can degenerate has been studied by LE ROUX [1982].

Estimations of the error made by a mixed method approximation have been given in norms other than the L^2 norm and the $H(\mathrm{div})$ norm. Estimates for $u-u_h$ in the L^∞ norm have been given by SCHOLZ [1977, 1983], and then by DOUGLAS and ROBERTS [1985], GASTALDI and NOCHETTO [1987] and KWON and MILNER [1988]. Negative norm estimates in $H^{-s}(\Omega)$, for s a nonnegative integer, are given for $u-u_h$, $p-p_h$, and $\mathrm{div}(p-p_h)$ by DOUGLAS and ROBERTS [1985]. These estimates have been exploited by DOUGLAS and MILNER [1985] to obtain superconvergence results.

Mixed methods have been "generalized" to fourth-order problems such as those governed by an equation of the form $\Delta^2 u=f$, where we write "generalized" with quotation marks to indicate that these methods were introduced, for the most part, before those presented for the second-order problem and without the use of the general abstract theory for problems with Lagrangian multipliers. We cite in particular the work of BALASUNDARAM and BHATTACHARYYA [1984, 1986]; BRAMBLE and FALK [1983]; BREZZI, MARINI, QUARTERONI and RAVIART [1980]; BREZZI and RAVIART [1978]; CIARLET and RAVIART [1974]; HELLAN [1967]; HERRMANN [1967, 1983]; JOHNSON [1973]; MERCIER [1974]; MIYOSHI [1973]; MONK [1987]; QUARTERONI [1980a]; SCAPOLLA [1980]; and SCHOLZ [1978]. A precise analysis of the best known of these methods can be found in FALK and OSBORN [1980]. An excellent exposition is also given in CIARLET [1978, Chapter 7].

A generalization to problems governed by an equation of the form $(-\Delta)^m u=f$ for an integer m, $m\geqslant 2$ is proposed by BRAMBLE and FALK [1985].

17. Solution of the linear system

We conclude this chapter with some remarks concerning the solution of the algebraic system resulting from a mixed method or, in particular, from a dual mixed method. We shall restrict our attention to the case of the model problem (3.1)–(3.4) as in Sections 12–14. We assume that the continuous problem is written as a problem with Lagrangian multipliers in the form (8.1), and that finite-dimensional subspaces W_h and M_h of the spaces W and M where the solution is sought have been introduced in such a way that (10.9a) and (10.9b) of Theorem 10.2 are satisfied. We thus suppose that the finite-dimensional problem is of the form (10.7): find (ψ_h, λ_h) satisfying

$$\begin{aligned} &\varphi_h \in W_h, \qquad \lambda_h \in M_h, \\ &a(\varphi_h, \psi_h) + b(\psi_h, \lambda_h) = (f, \psi_h) \quad \text{for all } \psi_h \in W_h, \\ &b(\varphi_h, \mu_h) = (g, \mu_h) \qquad\qquad\quad \text{for all } \mu_h \in M_h; \end{aligned} \tag{17.1}$$

and we further suppose that $a(\cdot,\cdot)$ is symmetric. Once bases for W_h and M_h have been chosen, we may write the algebraic problem associated with (17.1) as

$$\begin{pmatrix} A & B^* \\ B & 0 \end{pmatrix} \begin{pmatrix} \varphi_h \\ \lambda_h \end{pmatrix} = \begin{pmatrix} f_h \\ g_h \end{pmatrix}, \tag{17.2}$$

where A is a symmetric, $\dim W_h \times \dim W_h$ matrix such that $(A\psi_h, \psi'_h) = a(\psi_h, \psi'_h)$ for all (ψ_h, ψ'_h) in $W_h \times W_h$, and B is a $\dim M_h \times \dim W_h$ matrix such that $(B\psi_h, \mu_h) = b(\psi_h, \mu_h)$ for all (ψ_h, μ_h) in $W_h \times M_h$. The functions f_h and g_h are the orthogonal projections of f onto W_h and g onto M_h, respectively.

Theorem 10.2 guarantees that the matrix L,

$$L = \begin{pmatrix} A & B^* \\ B & 0 \end{pmatrix}, \tag{17.3}$$

of order $N \times N$, $N = \dim W_h \times \dim M_h$, is invertible; however, it is not positive-definite. Hence direct solution of the system (17.2) is generally not feasible. On the other hand A is here positive-definite. So, theoretically it is always possible to eliminate φ_h from the system,

$$\varphi_h = A^{-1}(f_h - B^* \lambda_h), \tag{17.4}$$

to obtain

$$BA^{-1}B^* \lambda_h = -g_h + BA^{-1} f_h. \tag{17.5}$$

However, there is no reason, a priori, why A^{-1} should be sparse.

For the primal mixed method, the space W is $(L^2(\Omega))^n$. So, among the degrees of freedom determining an element of W_h there is none needed to enforce a continuity requirement. Thus A (and hence A^{-1}) is block diagonal, and the structure of the linear system (17.5) is the same as that associated with the conforming problem (2.21).

For the dual mixed method the situation is different. Thus while we cannot give here an exhaustive treatment of the solution of linear systems of the form (17.2), we would like to mention briefly some techniques that have been used when the system (17.2) results from a dual mixed method.

REMARK 17.1. The algebraic system associated with the primal or dual hybrid methods discussed in the following chapter can, under restrictions similar to those indicated above, take the form (17.2). However, as with the primal mixed method, the form of A is such that the linear system that must be solved is the same as that for a more standard problem, again as for a conforming problem for the dual hybrid method, as for a nonconforming problem for the primal hybrid method. Thus no section analogous to this one will be given in Chapter V.

17.1. Conjugate gradient method

Even for the dual mixed method where A^{-1} is full, the system (17.5) can be solved using conjugate gradient iteration as A^{-1} never actually has to be computed. Whenever $A^{-1}\psi$ is needed one can solve the system $A\psi'=\psi$ using, for example, Gaussian elimination as A itself is sparse. However, conjugate gradient iteration without preconditioning is known to converge slowly, and in general there seems to be no obvious preconditioner, though in BROWN [1982] and in EWING and WHEELER [1983] several preconditioners are proposed and tested for the case of a triangulation by rectangles. Moreover, conjugate gradient iteration is known to be especially sensitive to round-off error. Recently BRAMBLE and PASCIAK [1988] have proposed a new preconditioning technique.

It seems appropriate to mention that there has been much research concerning the use of the conjugate gradient method and its variants to solve matrix equations involving matrices that are not necessarily symmetric or positive-definite. Thus one may consider applying such a method to solve equation (17.2) even though L is not positive-definite. In JOLY [1984] several variants of the conjugate gradient method are described and their performance in solving equations of the form (17.2) coming from the implementation of a dual mixed method are compared. As pointed out in JOLY [1984], for such systems the role of the preconditioner is no longer simply to speed up the convergence but in fact to guarantee the convergence.

17.2. Penalty method

This method consists of replacing (17.1) by a more regular problem: find $(\varphi_{h,\varepsilon}, \lambda_{h,\varepsilon})$ satisfying

$$\begin{aligned} &\varphi_{h,\varepsilon}\in W_h, \qquad \lambda_{h,\varepsilon}\in M_h,\\ &a(\varphi_{h,\varepsilon}, \psi_h)+b(\psi_h \lambda_{h,\varepsilon})=(f, \psi_h) \quad \text{for all } \psi_h\in W_h,\\ &b(\varphi_{h,\varepsilon}, \mu_h)-\varepsilon(\lambda_{h,\varepsilon}, \mu_h)=(g, \mu_h) \quad \text{for all } \mu_h\in M_h. \end{aligned} \tag{17.6}$$

The corresponding algebraic problem then becomes

$$\begin{pmatrix} A & B^* \\ B & -\varepsilon I \end{pmatrix}\begin{pmatrix} \varphi_{h,\varepsilon} \\ \lambda_{h,\varepsilon} \end{pmatrix}=\begin{pmatrix} f_h \\ g_h \end{pmatrix}, \tag{17.7}$$

and we can eliminate $\lambda_{h,\varepsilon}$,

$$\lambda_{h,\varepsilon}=(1/\varepsilon)(B\varphi_{h,\varepsilon}-g_h), \tag{17.8}$$

to obtain

$$(A+(1/\varepsilon)B^*B)\varphi_{h,\varepsilon}=f_h+(1/\varepsilon)B^*g_h. \tag{17.9}$$

For our model problem the matrix $(A+(1/\varepsilon)B^*B)$ is symmetric, positive-definite, and sparse, hence (17.9) can be solved by known techniques. In BERCOVIER [1978, Theorem 3.1] such penalized problems are studied and it is shown that, under the more general hypotheses that (10.9a) (10.9b) of Theorem 10.2 and

$$\inf_{\{\psi_h \in W_h : \|\psi_h\|_W = 1\}} \left\{ |a(\psi_h, \psi_h)| + \sup_{\{\mu_h \in M_h : \|\mu_h\|_M = 1\}} b(\psi_h, \mu_h) \right\} > 0 \tag{17.10}$$

hold, there is a unique solution $(\varphi_{h,\varepsilon}, \lambda_{h,\varepsilon})$ of (17.6), and if (φ_h, λ_h) is the solution of (17.1) then there is a positive constant C, independent of ε, such that

$$\|\varphi_h - \varphi_{h,\varepsilon}\|_W + \|\lambda_h - \lambda_{h,\varepsilon}\|_M \leqslant C\varepsilon. \tag{17.11}$$

The choice of ε may thus be made independently of h; though, as pointed out in BERCOVIER [1978], for each h there is an optimal ε.

This method is surely the simplest to implement, but it has one crucial flaw. The equation

$$\boldsymbol{B}\varphi_h = g_h$$

is not satisfied by the solution $\varphi_{h,\varepsilon}$. Thus the method, at least for our model problem, is mostly of historical interest.

17.3. *Augmented Lagrangian method*

Under the hypothesis that $a(\cdot,\cdot)$ is symmetric and W_h-elliptic, (φ_h, λ_h) is a solution of (17.1) if and only if (φ_h, λ_h) is a saddle point of the Lagrangian Γ defined on $W_h \times M_h$ as follows:

$$\Gamma(\psi_h, \mu_h) = \tfrac{1}{2}a(\psi_h, \psi_h) + b(\psi_h, \mu_h) - f(\psi_h) - g(\mu_h). \tag{17.12}$$

The classical iterative method of Uzawa to obtain the saddle point of Γ is described by the algorithm:

Initialize: Let $\lambda_h^{(0)}$ be an arbitrarily chosen element of M_h.

Calculate φ_h: Once $\lambda_h^{(n)}$ is known, $\varphi_h^{(n)}$ is defined to be the element of W_h minimizing $\Gamma(\psi_h, \lambda_h^{(n)})$; i.e. $\varphi_h^{(n)}$ is the solution of

$$\boldsymbol{A}\varphi_h^{(n)} = f_h - \boldsymbol{B}^*\lambda_h^{(n)}. \tag{17.13}$$

Advance λ_h: Let $\lambda_h^{(n+1)}$ be defined by

$$\lambda_h^{(n+1)} = \lambda_h^{(n)} + \rho_n(\boldsymbol{B}\varphi_h^{(n)} - g_h). \tag{17.14}$$

This method in general converges quite slowly. The idea of the augmented Lagrangian method is to "penalize" the Lagrangian Γ to obtain a Lagrangian Γ_r for which Uzawa's method converges more rapidly. For $r>0$, put

$$\Gamma_r(\psi_h, \mu_h) = \Gamma(\psi_h, \mu_h) + \tfrac{1}{2}r\|\boldsymbol{B}\psi_h - g_h\|^2. \tag{17.15}$$

It is easy to see that (φ_h, λ_h) is a saddle point of Γ_r if and only if it is a saddle point of Γ. Thus this method is not really a penalty method as Γ does not have to tend toward ∞ to obtain the actual solution, and the augmented Lagrangian method does not have the defect of the earlier described penalty method. Applying Uzawa's method to Γ_r one obtains the algorithm:

Initialize: Let $\lambda_h^{(0)}$ be an arbitrarily chosen element of M_h.

Calculate φ_h: Once $\lambda_h^{(n)}$ is known, $\varphi_h^{(n)}$ is defined to be the element of W_h minimizing $\Gamma_r(\psi_h, \lambda_h^{(n)})$; i.e. $\varphi_h^{(n)}$ is the solution of

$$A_r\varphi_h^{(n)} = (A + rB^*B)\varphi_h^{(n)} = f_h - B^*\lambda_h^{(n)} + rB^*g_h. \tag{17.16}$$

Advance λ_h: Let $\lambda_h^{(n+1)}$ be defined by

$$\lambda_h^{(n+1)} = \lambda_h^{(n)} + \rho_n(B\varphi_h^{(n)} - g_h). \tag{17.17}$$

The augmented Lagrangian method originated with the work of HESTENES [1969] and POWELL [1969] and is studied extensively by FORTIN and GLOWINSKI [1983] where the following two results are shown (Theorems 2.1 and 2.2) under the hypothesis that A is symmetric and positive-definite:

(i) There is a positive constant α such that if $\alpha \leqslant \rho_n \leqslant 2r$ for each n, then the sequence $\varphi_h^{(n)}$ converges to the solution φ_h.

(ii) If $\hat{\lambda}_h$ is the unique element of Im B satisfying (17.1) with φ_h, then the sequence $\lambda_h^{(n)}$ converges to $\hat{\lambda}_h + \lambda^\circ$, where λ° is the componant of $\lambda_h^{(0)}$ in Ker B^*.

REMARK 17.2. The algorithm of Uzawa, and hence that of the augmented Lagrangian, can be interpreted in the following manner: replace problem (17.1) by the virtual time-dependent problem

$$\begin{aligned} &a(\varphi_h(t), \psi_h) + b(\psi_h, \lambda_h(t)) = (f, \psi_h) \quad \text{for all } \psi_h \in W_h, \\ &b(\varphi_h(t), \mu_h) - \left(\frac{\mathrm{d}}{\mathrm{d}t}\lambda_h(t), \mu_h\right) = (g, \mu_h) \quad \text{for all } \mu_h \in M_h, \end{aligned} \tag{17.18}$$

and look for the solution (φ_h, λ_h) as the steady state solution of (17.18). Introducing an explicit time discretization with nth time step $\rho_n = t^{n+1} - t^n$, one obtains

$$\begin{aligned} &a(\varphi_h^{(n)}, \psi_h) + b(\psi_h, \lambda_h^{(n)}) = (f, \psi_h) \quad \text{for all } \psi_h \in W_h, \\ &b(\varphi_h^{(n)}, \mu_h) - \left(\frac{\lambda_h^{(n+1)} - \lambda_h^{(n)}}{\rho_n}, \mu_h\right) = (g, \mu_h) \quad \text{for all } \mu_h \in M_h, \end{aligned}$$

which gives the algebraic system

$$\begin{aligned} &A\varphi_h^{(n)} + B^*\lambda_h^{(n)} = f_h, \\ &\rho_n B\varphi_h^{(n)} - (\lambda_h^{(n+1)} - \lambda_h^{(n)}) = \rho_n g_h \end{aligned}$$

for Uzawa's method and

$$\begin{aligned} &(A + rB^*B)\varphi_h^{(n)} + B^*\lambda_h^{(n)} = f_h + rB^*g_h, \\ &\rho_n B\varphi_h^{(n)} - (\lambda_h^{(n+1)} - \lambda_h^{(n)}) = \rho_n g_h \end{aligned}$$

for the augmented Lagrangian method.

REMARK 17.3. Uzawa's algorithm can be viewed as a gradient type algorithm for the minimization of the functional J_* on M_h:

$$\begin{aligned} J_*(\mu_h) &= -\min_{\psi_h \in W_h} \Gamma(\psi_h, \mu_h) \\ &= \tfrac{1}{2}(BA^{-1}B^*\mu_h, \mu_h) - (BA^{-1}f_h, \mu_h) + \tfrac{1}{2}(A^{-1}f_h, f_h) + (g_h, \mu_h), \end{aligned}$$

in that eliminating $\varphi_h^{(n)}$ in (17.13) and (17.14),

$$\varphi_h^{(n)} = A^{-1}(f_h - B^*\lambda_h^{(n)}),$$

we obtain the algorithm

Initialize: Let $\lambda_h^{(0)}$ be an arbitrarily chosen element of M_h.
Advance λ_h: Once $\lambda_h^{(n)}$ is known, put

$$\lambda_h^{(n+1)} = \lambda_h^{(n)} - \rho_n(BA^{-1}B^*\lambda_h^{(n)} - BA^{-1}f_h + g_h).$$

Thus the augmented Lagrangian algorithm can be viewed as a gradient type algorithm for the minimization of the dual functional J_{*r} on M_h:

$$\begin{aligned} J_{*r}(\mu_h) &= -\min_{\psi_h \in W_h} \Gamma_r(\psi_h, \mu_h) \\ &= \tfrac{1}{2}(BA_r^{-1}B^*\mu_h, \mu_h) - (BA_r^{-1}(f_h + rB^*g_h), \mu_h) \\ &\quad + \tfrac{1}{2}(A_r^{-1}(f_h + rB^*g_h), f_h + rB^*g_h) + (g_h, \mu_h). \end{aligned}$$

However, A_r^{-1}, like A^{-1}, is, a priori, full. Similarly, one may eliminate $\lambda_h^{(n)}$ to obtain an algorithm involving $\varphi_h^{(n)}$, but again the matrix A_r^{-1} appears.

Uzawa's algorithm indicates no specific method for solving the equation determining $\varphi_h^{(n)}$. As A and A_r are positive-definite and sparse, either a direct method such as Gaussian elimination or an iterative method may be used to determine $\varphi_h^{(n)}$ in (17.13) or (17.16). The Arrow–Hurwitz algorithm is an algorithm that may be interpreted as being a variant of the Uzawa algorithm in which the method for calculating $\varphi_h^{(n)}$ is specified:

Initialize: Let $\lambda_h^{(0)}$ and $\varphi_h^{(0)}$ be arbitrarily chosen elements of M_h and W_h, respectively.
Advance φ_h: Once $\lambda_h^{(n)}$ and $\varphi_h^{(n)}$ are known, $\varphi_h^{(n+1)}$ is defined by

$$\varphi_h^{(n+1)} = \varphi_h^{(n)} - \omega_n(A\varphi_h^{(n)} + B^*\lambda_h^{(n)} - f_h). \tag{17.19}$$

Advance λ_h: Let $\lambda_h^{(n+1)}$ be defined by

$$\lambda_h^{(n+1)} = \lambda_h^{(n)} + \rho_n(B\varphi_h^{(n)} - g_h). \tag{17.20}$$

Thus a variant of the augmented Lagrangian method is obtained by replacing A by A_r in (17.19).

REMARK 17.4. If ω_n is taken to be equal to ρ_n, then the algorithm of Arrow–Hurwitz is also associated with the virtual time-dependent problem

$$a(\varphi_h(t), \psi_h) + (\mathrm{d}\varphi_h(t)/\mathrm{d}t, \psi_h) + b(\psi_h, \lambda_h(t)) = (f, \psi_h) \quad \text{for all } \psi_h \in W_h,$$
$$b(\varphi_h(t), \mu_h) - (\mathrm{d}\lambda_h(t)/\mathrm{d}t, \mu_h) = (g, \mu_h) \quad \text{for all } \mu_h \in M_h,$$

of which the solution of (17.1) is the steady state solution.

17.4. Alternating-direction method

These methods can be used both in two dimensions and in three dimensions when the triangulation $\mathcal{T}_h$ of the domain Ω is a triangulation by rectangles. Here again one introduces a virtual time-dependent problem to obtain the solution of the original problem as its steady state. The idea is, heuristically speaking, to reduce the solution of the two- (or three-)dimensional problem to the solution of one-dimensional problems by advancing half (or a third of) a time step in the x_1-direction, then half (or a third of) a time step in the x_2-direction (and in the three-dimensional case, a third of a time step in the x_3-direction).

For example, in the two-dimensional case, for the Raviart–Thomas elements the subspace W_h of $H(\text{div};\Omega)$ in which the vectors are approximated is locally $P_{k+1,k} \times P_{k,k+1}$ and we suppose that the basis $\{w_j\}_{j=1}^{2m}$ has been chosen such that $W_{h,1} = \text{span}\{w_j\}_{j=1}^{m}$ is locally $P_{k+1,k} \times \{0\}$, and $W_{h,2} = \text{span}\{w_j\}_{j=m+1}^{2m}$ is locally $\{0\} \times P_{k,k+1}$. Thus the vector $w_h \in W_h$ may be written uniquely as $w_h = w_{h,1} + w_{h,2}$ where $w_{h,1} \in W_{h,1}$ and $w_{h,2} \in W_{h,2}$, and the matrices $\boldsymbol{A}$ and $\boldsymbol{B}$ are block diagonal:

$$\boldsymbol{A} = \begin{pmatrix} \boldsymbol{A}_1 & 0 \\ 0 & \boldsymbol{A}_2 \end{pmatrix}, \qquad \boldsymbol{B} = \begin{pmatrix} \boldsymbol{B}_1 & 0 \\ 0 & \boldsymbol{B}_2 \end{pmatrix}.$$

As in Remark 17.4 a fictitious time-dependent problem is introduced:

$$a(\varphi_h(t), \psi_h) + b(\psi_h, \lambda_h(t)) = (f, \psi_h) \quad \text{for all } \psi_h \in W_h,$$
$$b(\varphi_h(t), \mu_h) - (\mathrm{d}\lambda_h(t)/\mathrm{d}t, \mu_h) = (g, \mu_h) \quad \text{for all } \mu_h \in M_h.$$

The two-step iterative procedure is obtained by introducing an implicit time discretization as follows:

$$a(\varphi_{h,1}^{(n^*)} + \varphi_{h,2}^{(n)}, \psi_h) + b(\psi_h, \lambda_h^{(n+(1/2))}) = (f, \psi_h) \quad \text{for all } \psi_h \in W_{h,1},$$

$$b(\varphi_{h,1}^{(n^*)} + \varphi_{h,2}^{(n)}, \mu_h) - \left(\frac{\lambda_h^{(n+(1/2))} - \lambda_h^{(n)}}{\rho_n}, \mu_h\right) = (g, \mu_h) \quad \text{for all } \mu_h \in M_h,$$

$$a(\varphi_{h,1}^{(n^*)} + \varphi_{h,2}^{(n+1)}, \psi_h) + b(\psi_h, \lambda_h^{(n+1)}) = (f, \psi_h) \quad \text{for all } \psi_h \in W_{h,2},$$

$$b(\varphi_{h,1}^{(n^*)} + \varphi_{h,2}^{(n+1)}, \mu_h) - \left(\frac{\lambda_h^{(n+1)} - \lambda_h^{(n+(1/2))}}{\rho_n}, \mu_h\right) = (g, \mu_h) \quad \text{for all } \mu_h \in M_h,$$

which gives the algebraic systems

$$A_1\varphi_{h,1}^{(n^*)}+B_1^*\lambda_h^{(n+(1/2))}=f_{h,1}-A_2\varphi_{h,2}^{(n)},$$

$$\rho_n B_1\varphi_{h,1}^{(n^*)}-\lambda_h^{(n+(1/2))}=\rho_n(g_h-B_2\varphi_{h,2}^{(n)})-\lambda_h^{(n)},$$

and

$$A_2\varphi_{h,2}^{(n+1)}+B_2^*\lambda_h^{(n+1)}=f_{h,2}-A_1\varphi_{h,1}^{(n^*)},$$

$$\rho_n B_2\varphi_{h,2}^{(n+1)}-\lambda_h^{(n+1)}=\rho_n(g_h-B_1\varphi_{h,1}^{(n^*)})-\lambda_h^{(n+(1/2))}.$$

From the first pair of the above equations $\lambda_h^{(n+1/2)}$ can be eliminated to obtain

$$(A_1+\rho_n B_1^*B_1)\varphi_{h,1}^{(n^*)}=f_{h,1}+\rho_n B_1^*g_h-(A_2+\rho_n B_1^*B_2)\varphi_{h,2}^{(n)}-B_1^*\lambda_h^{(n)},$$

and from the second pair, eliminating $\lambda_h^{(n+1)}$, one obtains

$$(A_2+\rho_n B_2^*B_2)\varphi_{h,2}^{(n+1)}=f_{h,2}+\rho_n B_2^*g_h-(A_1+\rho_n B_2^*B_1)\varphi_{h,1}^{(n^*)}-B_2^*\lambda_h^{(n+(1/2))}.$$

Thus the algorithm may be given as follows:

Initialize: Let $\lambda_h^{(0)}$ be an arbitrarily chosen element of M_h. The initialization $\varphi_{h,2}^{(0)}$ of $\varphi_{h,2}$ is obtained as the solution of

$$A_2\varphi_{h,2}^{(0)}=f_{h,2}-B_2^*\lambda_h^{(0)}. \tag{17.21}$$

Advance a half time step in the x_1-direction: Once $\lambda_h^{(n)}$ and $\varphi_{h,2}^{(n)}$ are known,

$$(A_1+\rho_n B_1^*B_1)\varphi_{h,1}^{(n^*)}=f_{h,1}+\rho_n B_1^*g_h-(A_2+\rho_n B_1^*B_2)\varphi_{h,2}^{(n)}-B_1^*\lambda_h^{(n)}. \tag{17.22}$$

Let $\lambda_h^{(n+(1/2))}$ be defined by

$$\lambda_h^{(n+(1/2))}=\lambda_h^{(n)}+\rho_n(B_1\varphi_{h,1}^{(n^*)}+B_2\varphi_{h,2}^{(n)}-g_h). \tag{17.23}$$

Advance a half time step in the x_2-direction: Once $\lambda_h^{(n+(1/2))}$ and $\varphi_{h,1}^{(n^*)}$ are known,

$$(A_2+\rho_n B_2^*B_2)\varphi_{h,2}^{(n+1)}$$
$$=f_{h,2}+\rho_n B_2^*g_h-(A_1+\rho_n B_2^*B_1)\varphi_{h,1}^{(n^*)}-B_2^*\lambda_h^{(n+(1/2))}. \tag{17.24}$$

Let $\lambda_h^{(n+1)}$ be defined by

$$\lambda_h^{(n+1)}=\lambda_h^{(n+(1/2))}+\rho_n(B_1\varphi_{h,1}^{(n^*)}+B_2\varphi_{h,2}^{(n+1)}-g_h). \tag{17.25}$$

Correct: Finally to obtain $\varphi_{h,1}^{(n+1)}$, solve

$$A_1\varphi_{h,1}^{(n+1)}=f_{h,1}-B_1^*\lambda_h^{(n+1)}. \tag{17.26}$$

This procedure without the correction step was introduced by BROWN [1982], and as given here by DOUGLAS, DURAN and PIETRA [1986, 1987]. It is an adaptation of the original alternating-direction procedure of PEACEMAN and RACHFORD [1955]. Also in DOUGLAS, DURAN and PIETRA [1986, 1987] an alternating-direction scheme for the three-dimensional case is introduced. This procedure is based on the scheme of DOUGLAS [1962]. As with the classical alternating-direction iterative schemes, only limited convergence results have been obtained; cf. the above references.

For the Brezzi–Douglas–Marini elements there is no natural decomposition of W_h as for the Raviart–Thomas elements. Thus the above algorithm is not applicable. However, an alternating-direction procedure for these elements involving the introduction of two virtual time variables is given for two dimensions in DOUGLAS and PIETRA [1986] and for three dimensions in DOUGLAS, DURAN and PIETRA [1986, 1987].

17.5. *Mixed-hybrid method*

As pointed out earlier, the reason that the algebraic system derived from the dual mixed method poses special problems is that due to the degrees of freedom imposing the inter-element continuity requirements on the elements ψ_h of W_h, the matrix A^{-1} is likely to be full. A technique often used to overcome this problem is to replace the dual mixed formulation by a mixed-hybrid formulation, cf. FRAEIJS DE VEUBEKE [1965]. The idea is to eliminate the inter-element continuity requirements from the space W_h thereby obtaining a space $\hat{W}_h$ and to impose instead the desired continuity on the solution $\hat{\varphi}_h \in \hat{W}_h$ via Lagrangian multipliers. Problem (17.1) is thus replaced by a problem of the following form: find $\hat{\varphi}_h$, $\hat{\lambda}_h$, and η_h satisfying

$$\begin{aligned}
&\hat{\varphi}_h \in \hat{W}_h, \qquad \hat{\lambda}_h \in M_h, \qquad \eta_h \in N_h,\\
&a(\hat{\varphi}_h, \psi_h) + b(\psi_h, \hat{\lambda}_h) + c(\psi_h, \eta_h) = (f, \psi_h) \quad \text{for all } \psi_h \in \hat{W}_h,\\
&b(\hat{\varphi}_h, \mu_h) = (g, \mu_h) \quad \text{for all } \mu_h \in M_h,\\
&c(\hat{\varphi}_h, \tau_h) = 0 \quad \text{for all } \tau_h \in N_h,
\end{aligned} \tag{17.27}$$

where $N_h \subset \Pi_{T' < T \in \mathcal{T}_h} L^2(T')$ is the space of Lagrangian multipliers, and the bilinear form $c(\cdot,\cdot)$ on $\hat{W}_h \times N_h$ is such that the induced linear map C from $\hat{W}_h$ into N_h has as kernel the elements of $\hat{W}_h$ which belong in fact to W_h. The linear system thus obtained is

$$\begin{pmatrix} \hat{\boldsymbol{A}} & \hat{\boldsymbol{B}}^* & \boldsymbol{C}^* \\ \hat{\boldsymbol{B}} & 0 & 0 \\ \boldsymbol{C} & 0 & 0 \end{pmatrix} \begin{pmatrix} \hat{\varphi}_h \\ \hat{\lambda}_h \\ \eta_h \end{pmatrix} = \begin{pmatrix} f_h \\ g_h \\ 0 \end{pmatrix}, \tag{17.28}$$

where $\hat{\boldsymbol{A}}$ is now block diagonal as well as symmetric and positive-definite, and $\hat{\varphi}_h$ and $\hat{\lambda}_h$ are equal as functions to φ_h and λ_h, respectively. Eliminating $\hat{\varphi}_h$ from the system,

$$\hat{\varphi}_h = \hat{\boldsymbol{A}}^{-1}(f_h - \hat{\boldsymbol{B}}^* \hat{\lambda}_h - \boldsymbol{C}^* \eta_h), \tag{17.29}$$

we obtain

$$\begin{pmatrix} \hat{\boldsymbol{B}}\hat{\boldsymbol{A}}^{-1}\hat{\boldsymbol{B}}^* & \hat{\boldsymbol{B}}\hat{\boldsymbol{A}}^{-1}\boldsymbol{C}^* \\ \boldsymbol{C}\hat{\boldsymbol{A}}^{-1}\hat{\boldsymbol{B}}^* & \boldsymbol{C}\hat{\boldsymbol{A}}^{-1}\boldsymbol{C}^* \end{pmatrix} \begin{pmatrix} \hat{\lambda}_h \\ \eta_h \end{pmatrix} = \begin{pmatrix} \hat{\boldsymbol{B}}\hat{\boldsymbol{A}}^{-1} f_h - g_h \\ \boldsymbol{C}\hat{\boldsymbol{A}}^{-1} f_h \end{pmatrix}, \tag{17.30}$$

and the space N_h and the form $c(\cdot,\cdot)$ have been constructed in such a way that this system is symmetric and positive-definite. But now we further note that $\hat{\boldsymbol{B}}\hat{\boldsymbol{A}}^{-1}\boldsymbol{B}^*$ is also block diagonal. Thus we have

$$\hat{\lambda}_h = (\hat{B}\hat{A}^{-1}\hat{B}^*)^{-1}\{\hat{B}\hat{A}^{-1}f_h - g_h - \hat{B}\hat{A}^{-1}C^*\eta_h\} \tag{17.31}$$

and

$$\begin{aligned}\{C\hat{A}^{-1}C^* - C\hat{A}^{-1}\hat{B}^*(\hat{B}\hat{A}^{-1}\hat{B}^*)^{-1}\hat{B}\hat{A}^{-1}C^*\}\eta_h \\ = \{C\hat{A}^{-1} - C\hat{A}^{-1}\hat{B}^*(\hat{B}\hat{A}^{-1}\hat{B}^*)^{-1}\hat{B}\hat{A}^{-1}\}f_h \\ + C\hat{A}^{-1}B^*(B\hat{A}^{-1}B^*)^{-1}g_h.\end{aligned} \tag{17.32}$$

This system is symmetric, positive-definite, and sparse. Thus we can obtain η_h by solving (17.32), and then $\hat{\lambda}_h$ and $\hat{\varphi}_h$ are given by (17.31) and (17.29), respectively. It is interesting to note that η_h often has a physical significance. For our model problem η_h represents the value of φ_h on the faces T' and in fact in certain cases can be used together with $\hat{\varphi}_h$ to obtain a new approximation that converges more rapidly than $\hat{\varphi}_h$ or equivalently φ_h; cf. ARNOLD and BREZZI [1985] and BREZZI, DOUGLAS and MARINI [1985]. Mixed hybrid methods will be taken up again in Section 21.

CHAPTER V

Hybrid Finite Element Methods

18. Examples of primal hybrid finite element methods

We shall develop several examples of hybrid formulations for the model problem: find $\boldsymbol{p}$ and u satisfying the system of first-order partial differential equations (4.4)–(4.5) and the Robin's boundary condition (15.1): $\boldsymbol{p}\cdot\boldsymbol{\nu}+tu=g$ on Γ, with t a continuous, strictly positive function on Γ. We shall indicate how to adapt the method to treat a problem with Dirichlet or Neumann boundary conditions.

To a triangulation $\mathcal{T}_h$ of the domain $\bar{\Omega}$, we have associated the spaces $H^1(\mathcal{T}_h)$ and $\mathcal{H}(\mathrm{div};\mathcal{T}_h)$, cf. (5.11) and (6.16). The space $H^1(\mathcal{T}_h)$ is a subspace of $L^2(\Omega)$ isomorphic to the product space $\Pi_{T\in\mathcal{T}_h}H^1(T)$, and the space $\mathcal{H}(\mathrm{div};\mathcal{T}_h)$ is a subspace of $(L^2(\Omega))^n$ isomorphic to the product space $\Pi_{T\in\mathcal{T}_h}\mathcal{H}(\mathrm{div};T)$. The hybrid formulations, both primal and dual, will make use of a Lagrangian multiplier belonging to the space $L^2(\partial\mathcal{T}_h)$ which is defined to be the product space

$$L^2(\partial\mathcal{T}_h)=\prod_{T\in\mathcal{T}_h}L^2(\partial T). \tag{18.1}$$

For the primal hybrid formulation, the subspace of multipliers will be

$$E(\partial\mathcal{T}_h)=\{\mu=(\mu_T)_{T\in\mathcal{T}_h}\in L^2(\partial\mathcal{T}_h):\\ \exists \boldsymbol{q}\in H(\mathrm{div};\Omega)\text{ s.t. }\forall T\in\mathcal{T}_h,\ \mu_T=\boldsymbol{q}\cdot\boldsymbol{\nu}_T\text{ on }\partial T\}, \tag{18.2}$$

where for this definition $\boldsymbol{q}\cdot\boldsymbol{\nu}_T$ denotes the normal trace on ∂T of the restriction to T of the vector function $\boldsymbol{q}$. Thus $\boldsymbol{q}\cdot\boldsymbol{\nu}_T$ is a priori an element of $H^{-1/2}(\partial T)$, cf. Remark 18.1 below.

Theorems 1.2 and 1.3 allow us to characterize $E(\partial\mathcal{T}_h)$ as being the subspace

$$E(\partial\mathcal{T}_h)=\{\mu=(\mu_T)_{T\in\mathcal{T}_h}\in L^2(\partial\mathcal{T}_h):\\ \forall T'=T_1\cap T_2\text{ with }T_1,T_2\in\mathcal{T}_h,\ \mu_{T_1}+\mu_{T_1}=0\text{ on }T'\}. \tag{18.3}$$

For the dual hybrid formulation the subspace of multipliers will be

$$L(\partial\mathcal{T}_h)=\{\psi=(\psi_T)_{T\in\mathcal{T}_h}\in L^2(\partial\mathcal{T}_h):\\ \exists v\in H^1(\Omega)\text{ s.t. }\forall T\in\mathcal{T}_h,\ \psi_T=v\text{ on }\partial T\}, \tag{18.4}$$

where in the expression "$\psi_T=v$ on ∂T", v denotes the trace on ∂T of the restriction to T of the function v. This trace is an element of $H^{1/2}(\partial T)$.

With Theorems 1.1 and 1.3 we may characterize $L(\partial\mathcal{T}_h)$ as being the subspace

$$L(\partial\mathcal{T}_h)=\{\psi=(\psi_T)_{T\in\mathcal{T}_h}\in\prod_{T\in\mathcal{T}_h}H^{1/2}(\partial T):\quad \forall T'=T_1\cap T_2 \text{ with } T_1,T_2\in\mathcal{T}_h, \psi_{T_1}=\psi_{T_2} \text{ on } T'\}. \tag{18.5}$$

REMARK 18.1. In fact the natural space of Lagrangian multipliers for the primal hybrid formulation is

$$E(\partial\mathcal{T}_h)=\{\mu=(\mu_T)_{T\in\mathcal{T}_h}\in\prod_{T\in\mathcal{T}_h}H^{-1/2}(\partial T):\quad \exists \boldsymbol{q}\in H(\mathrm{div};\Omega) \text{ s.t. } \forall T\in\mathcal{T}_h, \mu_T=\boldsymbol{q}\cdot\boldsymbol{\nu}_T \text{ in } H^{-1/2}(\partial T)\}.$$

The choice (18.2) while more restrictive is less cumbersome and leads to no restriction for examples of finite element methods.

REMARK 18.2. The first examples of hybrid formulations given in Section 3 were described as having as Lagrangian multipliers functions defined on all Ω. Such a description is not in general well adapted for the numerical analysis of hybrid finite element methods. For hybrid formulations the Lagrangian multipliers are defined only on $\partial\mathcal{T}_h$. It is sometimes possible to extend the multipliers to functions defined on all Ω while taking into account complementary information when necessary. Such a procedure is a postprocessing technique.

A primal hybrid formulation of problem (4.4)–(4.5) with the Robin's boundary condition (15.1) is given as follows: find a pair (u,λ) satisfying

$$(u,\lambda)\in H^1(\mathcal{T}_h)\times E(\partial\mathcal{T}_h),$$

$$a(u,v)+b(v,\lambda)=\int_\Omega fv\,\mathrm{d}x \qquad \text{for all } v\in H^1(\mathcal{T}_h), \tag{18.6}$$

$$b(u,\mu)+d(\lambda,\mu)=-\sum_{T\in\mathcal{T}_h}\int_{\partial T\cap\Gamma}(1/t)g\mu_T\,\mathrm{d}\sigma \quad \text{for all } \mu\in E(\partial\mathcal{T}_h),$$

where here

$$a(u,v)=\sum_{T\in\mathcal{T}_h}\sum_{i,j=1}^{n}\int_T a_{ij}\frac{\partial u}{\partial x_j}\frac{\partial v}{\partial x_i}\,\mathrm{d}x, \tag{18.7}$$

$$b(v,\mu)=-\sum_{T\in\mathcal{T}_h}\int_{\partial T}v|_T\,\mu_T\,\mathrm{d}\sigma \tag{18.8}$$

(in this expression, $v|_T$ denotes the trace on ∂T of the restriction of v to T), and

$$d(\lambda,\mu)=-\sum_{T\in\mathcal{T}_h}\int_{\partial T\cap\Gamma}(1/t)\lambda_T\mu_T\,\mathrm{d}\sigma. \tag{18.9}$$

It is easy to show that problem (18.6) has at most one solution. If the solution $(\boldsymbol{p}, u)$ of (15.2), which is also the solution of (15.8) and thus belongs to $\mathscr{H}(\mathrm{div};\Omega)\times H^1(\Omega)$, satisfies the regularity hypothesis that $\boldsymbol{p}\cdot\boldsymbol{v}_T\in L^2(\partial T)$ for each $T\in\mathscr{T}_h$, then the pair $(u, (\boldsymbol{p}\cdot\boldsymbol{v}_T)_{T\in\mathscr{T}_h})$ is a solution of (18.6). Thus we have

$$\lambda_T=\boldsymbol{p}\cdot\boldsymbol{v}_T \quad \text{on } \partial T \quad \text{for each } T\in\mathscr{T}_h. \tag{18.10}$$

REMARK 18.3. The spaces $H^1(\mathscr{T}_h)$ and $L^2(\partial\mathscr{T}_h)$ are supplied with Hilbert space structures in a natural way. With $W=H^1(\mathscr{T}_h)$ and with $M=E(\partial\mathscr{T}_h)$ having the norm induced by that of $L^2(\partial\mathscr{T}_h)$, we note that the bilinear form $b(\cdot,\cdot)$ does not satisfy the inf–sup condition on $W\times M$.

Toward the end of constructing an approximation (u_h, λ_h) of the solution (u, λ) of problem (18.6) we suppose that for each $T\in\mathscr{T}_h$ we have two finite-dimensional spaces $P_T\subset H^1(T)$ and $X_T\subset L^2(\partial T)$, and we put

$$W_h=\{v\in H^1(\mathscr{T}_h): \forall T\in\mathscr{T}_h, v|_T\in P_T\} \tag{18.11}$$

and

$$M_h=\{\mu=(\mu_T)_{T\in\mathscr{T}_h}\in E(\partial\mathscr{T}_h): \forall T\in\mathscr{T}_h, \mu_T\in X_T\}. \tag{18.12}$$

We then seek (u_h, λ_h) satisfying

$$\begin{aligned}
&(u_h, \lambda_h)\in W_h\times M_h,\\
&a(u_h, v_h)+b(v_h, \lambda_h)=\int_\Omega f v_h\,\mathrm{d}x \qquad \text{for all } v_h\in W_h,\\
&b(u_h, \mu_h)+d(\lambda_h, \mu_h)=-\sum_{T\in\mathscr{T}_h}\int_{\partial T\cap\Gamma}(1/t)g\mu_{hT}\,\mathrm{d}\sigma \quad \text{for all } \mu_h\in M_h.
\end{aligned} \tag{18.13}$$

REMARK 18.4. To treat the analogous problem with a Dirichlet boundary condition $u=\bar{u}$ on Γ, where $\bar{u}$ is a given function in $H^{1/2}(\Gamma)$ we would keep the same choice of subspaces W_h and M_h and look for (u_h, λ_h) satisfying

$$\begin{aligned}
&(u_h, \lambda_h)\in W_h\times M_h,\\
&a(u_h, v_h)+b(v_h, \lambda_h)=\int_\Omega f v_h\,\mathrm{d}x \qquad \text{for all } v_h\in W_h,\\
&b(u_h, \mu_h)=-\sum_{T\in\mathscr{T}_h}\int_{\partial T\cap\Gamma}\bar{u}\mu_{hT}\,\mathrm{d}\sigma \quad \text{for all } \mu_h\in M_h.
\end{aligned}$$

By contrast, to treat a homogeneous Neumann problem such as described in (14.1) or (14.6) we would introduce the subspace M_{0h} defined by

$$M_{0h}=\{\mu_h\in M_h: \forall T\in\mathscr{T}_h, \mu_{hT}=0 \text{ on } \partial T\cap\Gamma\}$$

and look for (u_h, λ_h) satisfying

$$(u_h, \lambda_h) \in W_h \times M_{0h},$$

$$a(u_h, v_h) + b(v_h, \lambda_h) = \int_\Omega f v_h \, dx \quad \text{for all } v_h \in W_h,$$

$$b(u_h, \mu_h) = 0 \qquad\qquad \text{for all } \mu_h \in M_{0h}.$$

The bilinear form $a(\cdot,\cdot)$ is not $H^1(\mathcal{T}_h)$-elliptic and will not be W_h-elliptic. However, we shall introduce a hypothesis to make it V_h-elliptic, where

$$V_h = \{v_h \in W_h : \forall \mu_h \in M_h, b(v_h, \mu_h) = 0\}. \tag{18.14}$$

THEOREM 18.1. *We suppose that the two subspaces W_h and M_h are compatible in the sense that the discrete inf–sup condition is satisfied:*

$$\{\mu_h \in M_h : \forall v_h \in W_h, b(v_h, \mu_h) = 0\} = \{0\}. \tag{18.15}$$

We suppose further that for each interface $T' = T_1 \cap T_2$ with $T_1, T_2 \in \mathcal{T}_h$, there exists a function $\chi_{T'} \in L^2(T')$ with $\int_{T'} \chi_{T'} \, d\sigma \neq 0$ such that the function $\Lambda = (\Lambda_T)_{T \in \mathcal{T}_h}$ in $E(\partial\mathcal{T}_h)$ defined by $\Lambda_{T_1} = \chi_{T'}$ on T', $\Lambda_{T_2} = -\chi_{T'}$ on T', and $\Lambda_T = 0$ on all other faces for all $T \in \mathcal{T}_h$, belongs to the subspace M_h. Similarly we suppose that for each boundary face $T' \subset \Gamma$ of an element $T_0 \in \mathcal{T}_h$, there exists a function $\chi_{T'} \in L^2(T')$ with $\int_{T'} \chi_{T'} \, d\sigma \neq 0$ such that the function $\Lambda = (\Lambda_T)_{T \in \mathcal{T}_h}$ in $E(\partial\mathcal{T}_h)$ defined by $\Lambda_{T_0} = \chi_{T'}$ on T' and $\Lambda_T = 0$ on all other faces for all $T \in \mathcal{T}_h$, belongs to the subspace M_h. Then problem (18.13) has a unique solution.

PROOF. Each function $v_h \in V_h$ satisfying $a(v_h, v_h) = 0$ is a function whose restriction to T, for each $T \in \mathcal{T}_h$, is constant. The first part of the additional hypothesis guarantees that such a function may be identified with a function constant on each component of Ω. The second part forces such a function to vanish on $\Gamma = \partial\Omega$. We conclude that the form $a(\cdot,\cdot)$ is V_h-elliptic and, in light of Remark 10.8, that the theorem follows. □

It is important to note that the compatibility condition (18.15) that should be satisfied by the subspaces W_h and M_h given in the forms (18.11) and (18.12) is satisfied as soon as, for each $T \in \mathcal{T}_h$, the spaces P_T and X_T satisfy the local compatibility condition

$$\left\{\mu \in X_T : \forall v \in P_T, \int_{\partial T} \mu v \, d\sigma = 0\right\} = \{0\}. \tag{18.16}$$

By contrast, the additional hypotheses of Theorem 18.1 demand that the functions belonging to V_h have a minimum amount of continuity at the interfaces of the triangulation. In the example that follows this continuity hypothesis will be trivially satisfied. (We will in fact choose $\chi_{T'}$ to be a constant function.)

THEOREM 18.2. *Let Ω be an open polygonal domain in $\mathbb{R}^2$ triangulated by triangles T. Let k be a positive integer, and put*

$$k^*=\begin{cases} k, & \text{if } k \text{ is odd,}\\ k+1, & \text{if } k \text{ is even.}\end{cases}$$

For each triangle $T\in\mathcal{T}_h$, define the subspace $X_T\subset L^2(\partial T)$ by

$$X_T=D_k(\partial T). \tag{18.17}$$

(With the definition of the space $E^{(k)}_{\partial\mathcal{T}_h}$ given by (6.30), we have $M_h=E^{(k)}_{\partial\mathcal{T}_h}$.) Suppose that the subspace $P_T\subset H^1(T)$ is chosen such that the following property is satisfied:

$$\forall w\in P_{k^*}(\partial T),\ \exists v\in P_T \text{ s.t. } v=w \text{ on } \partial T. \tag{18.18}$$

Then problem (18.13) *has a unique solution.*

PROOF. The elements of the space $D_k(\partial T)$ defined in (6.32) are polynomials of degree $\leqslant k-1$ on each edge. It is clear that condition (18.16) is satisfied if each element of $P_{k+1}(\partial T)$ vanishing at the vertices of T is the trace of a function $v\in P_T$. Even though we have $\dim P_k(\partial T)=\dim D_k(\partial T)$, we know (cf. RAVIART and THOMAS [1977b, Lemma 4]) that the subspace of $D_k(\partial T)$ consisting of all those functions $\mu\in D_k(\partial T)$ satisfying $\int_{\partial T}\mu w\,\mathrm{d}\sigma=0$ for each $w\in P_k(\partial T)$ is trivial only if k is odd. It is of dimension one when k is even. □

Before giving an error estimate we need to make precise our choice of norms. For the sake of simplicity we suppose here that the triangulation is uniformly regular. To each function $v\in H^1(\mathcal{T}_h)$ we associate the function $\widehat{\delta v}\in L^2(\partial\mathcal{T}_h)$, constant on each edge $T'\in\partial\mathcal{T}_h$, defined by

$$\widehat{\delta v}=\begin{cases} \dfrac{1}{\operatorname{mes} T'}\displaystyle\int_{T'}(v|_{T_1}-v|_{T_2})\,\mathrm{d}\sigma, & \text{if } T'=T_1\cap T_2;\ T_1,T_2\in\mathcal{T}_h\\ & \text{(the choice of } T_1 \text{ being arbitrary)}\\ \dfrac{1}{\operatorname{mes} T'}\displaystyle\int_{T'}v|_T\,\mathrm{d}T, & \text{if } T'=T\cap\Gamma.\end{cases} \tag{18.19}$$

Once the hypotheses of Theorem 18.1 are satisfied, we clearly have $\widehat{\delta v}=0$ if $v\in H^1_0(\Omega)\oplus V_h$. The space $L^2(\partial\mathcal{T}_h)$ is given the norm

$$\|\mu\|_{0,\partial\mathcal{T}_h}=\Big\{\sum_{T\in\mathcal{T}_h}\|\mu_T\|^2_{0,\partial T}\Big\}^{1/2}, \tag{18.20}$$

the space $H^1(\mathcal{T}_h)$ the norm

$$[v]_{1,\mathcal{T}_h}=\Big\{\sum_{T\in\mathcal{T}_h}|v|^2_{1,T}+h^{-1}\|\widehat{\delta v}\|^2_{0,\partial\mathcal{T}_h}\Big\}^{1/2}, \tag{18.21}$$

and the space $E(\partial\mathcal{T}_h)$ the norm $|||\cdot|||_{-1/2,\partial\mathcal{T}_h}$ defined in (6.36). One can show (by adapting, for example, the demonstration of THOMAS [1977, Theorem V.4.3]) that the Friedrichs–Poincaré inequality,

$$\|v\|_{0,\Omega} \leqslant C[v]_{1,\mathcal{T}_h} \quad \text{for all } v \in H^1(\mathcal{T}_h), \tag{18.22}$$

is satisfied with a constant C independent of h. (Note that the two norms on $H^1(\mathcal{T}_h)$, $v \to [v]_{1,\mathcal{T}_h}$ and $v \to |||v|||_{1,\mathcal{T}_h} = (\sum_{T\in\mathcal{T}_h}(|v|^2_{1,T} + h^{-2}\|v\|^2_{0,T}))^{1/2}$, are equivalent uniformly in h.) One may also verify that there exist constants c and C_k independent of h such that

$$|||\mu|||_{-1/2,\partial\mathcal{T}_h} \leqslant ch^{1/2}\|\mu\|_{0,\partial\mathcal{T}_h} \quad \text{for all } \mu \in L^2(\partial\mathcal{T}_h), \tag{18.23}$$

and

$$\|\mu_h\|_{0,\partial\mathcal{T}_h} \leqslant C_k h^{-1/2} |||\mu_h|||_{-1/2,\partial\mathcal{T}_h} \quad \text{for all } \mu_h \in E^{(k)}_{\partial\mathcal{T}_h}. \tag{18.24}$$

THEOREM 18.3. *We suppose the hypotheses of Theorem 18.2 except here we fix for each* $T \in \mathcal{T}_h$

$$X_T = D_k(\partial T), \qquad P_T = P_{k*}(T). \tag{18.25}$$

Then there exists a constant C independent of h such that we have

$$[u - u_h]_{1,\mathcal{T}_h} + |||\lambda - \lambda_h|||_{-1/2,\partial\mathcal{T}_h} \leqslant Ch^k\{|u|_{k+1,\Omega} + |\boldsymbol{p}|_{k,\Omega}\} \tag{18.26}$$

once the solution $(\boldsymbol{p}, u)$ of (15.2) *and of* (15.8) *belongs to $(H^k(\Omega))^2 \times H^{k+1}(\Omega)$. ($\lambda$ is expressed as a function of $\boldsymbol{p}$ in* (18.10).)

PROOF. It is clear that the bilinear form $a(\cdot,\cdot)$ satisfies

$$a(u, v) \leqslant A[u]_{1,\mathcal{T}_h}[v]_{1,\mathcal{T}_h} \quad \text{for all } u, v \in H^1(\mathcal{T}_h),$$

$$a(v_h, v_h) \geqslant \alpha_p [v_h]^2_{1,\mathcal{T}_h}, \quad \text{for all } v_h \in V_h,$$

as $v_h \in V_h$ implies $\widehat{\delta v_h} = 0$. We can also establish the existence of a constant $\beta, \beta > 0$, such that

$$\sup_{v_h \in W_h} \frac{b(v_h, \mu_h)}{[v_h]_{1,\Omega}} \geqslant \beta |||\mu_h|||_{-1/2,\partial\mathcal{T}_h} \quad \text{for all } \mu_h \in M_h.$$

Finally we show the existence of a constant B_0 such that

$$b(v, \mu) \leqslant B_0[v]_{1,\mathcal{T}_h} |||\mu|||_{-1/2,\partial\mathcal{T}_h} \quad \text{for all } v \in H^1(\mathcal{T}_h), \mu \in E_0(\partial\mathcal{T}_h),$$

where $E_0(\partial\mathcal{T}_h)$ is the subspace of functions $\mu \in E(\partial\mathcal{T}_h)$ such that $\int_{T'} \mu \, d\sigma = 0$ on each edge $T' \subset \Gamma$. Now, for a Dirichlet boundary condition or for a Neumann boundary condition, using Theorem 10.4, we obtain the existence of a constant C, independent of h, such that

$$[u - u_h]_{1,\mathcal{T}_h} + |||\lambda - \lambda_h|||_{-1/2,\partial\mathcal{T}_h}$$
$$\leqslant C\{\inf_{v_h \in W_h} [u - v_h]_{1,\mathcal{T}_h} + \inf_{\{\mu_h \in M_h : \lambda - \mu_h \in E_0(\partial\mathcal{T}_h)\}} |||\lambda - \mu_h|||_{-1/2,\partial\mathcal{T}_h}\}, \tag{18.27}$$

and then using (6.37) deduce the inequality (18.26).

To handle the case of Robin's boundary condition we introduce the direct sum decomposition of $M=E(\partial\mathcal{T}_h)$ in the form $M=M_0\oplus\bar{M}$, where M_0 is the subspace of all $\mu\in M$ such that for each $T\in\mathcal{T}_h$ with $T\cap\Gamma\neq\emptyset$, $\mu_T=0$ on each edge T' contained in the boundary Γ. Problem (18.6), respectively (18.13), may be put in the abstract form (10.1), respectively (10.7), posed in $(H^1(\mathcal{T}_h)\times\bar{M})\times M_0$, with for bilinear forms

$$(u,\bar{\lambda}),(v,\bar{\mu}) \to a(u,v)+b(v,\bar{\lambda})-b(u,\bar{\mu})-d(\bar{\lambda},\bar{\mu})$$

and

$$(v,\bar{\mu}),\mu_0 \to b(v,\mu_0).$$

We then deduce as above

$$[u-u_h]_{1,\mathcal{T}_h}+\|\bar{\lambda}-\bar{\lambda}_h\|_{0,\partial\mathcal{T}_h}+|||\lambda_0-\lambda_{0h}|||_{-1/2,\partial\mathcal{T}_h}$$
$$\leqslant Ch^k\{|u|_{k+1,\Omega}+|p|_{k,\Omega}\}. \qquad \square$$

REMARK 18.5. If the integer k is even, then $k^*=k+1$, and the choice $P_T=P_{k+1}(T)$ is not optimal. We can obtain the same results by taking P_T to be the space generated by $P_k(T)$ together with an appropriately chosen element of $P_{k+1}(T)$. For more details see RAVIART and THOMAS [1977b].

The primal hybrid finite element method corresponding to the subspaces M_h and W_h associated with the spaces X_T and P_T of (18.25) by (18.11) and (18.12) is convergent for each positive integer k, in the following sense: we can show, in the absence of any hypothesis concerning the regularity of the solution u, that

$$\lim_{h\to 0}\|u-u_h\|_{0,\Omega}=0.$$

Approximation methods based on a principle analogous to that which was the basis for the development of primal hybrid finite element methods have been studied by BABUŠKA [1973], BRAMBLE [1981], FALK [1976], and PITKÄRANTA [1979, 1980a, 1981]. For a problem with a nonhomogeneous Dirichlet boundary condition $u=\bar{u}$ on Γ, we dualize the boundary condition using a multiplier $\lambda\in H^{-1/2}(\Gamma)$. The discrete inf–sup condition is established by means of compactness arguments; thus the results, notably existence and uniqueness, are assured only for "h sufficiently small".

19. Primal hybrid methods and nonconforming methods

We consider the model Dirichtlet problem

$$-\Delta u=f \text{ in } \Omega, \qquad u=\bar{u} \text{ on } \Gamma,$$

where f is a given function in $L^2(\Omega)$, $\bar{u}$ a given function in $H^{1/2}(\Gamma)$, and Ω a bounded, open domain $\mathbb{R}^2$ with polygonal boundary Γ. For a triangulaton $\mathcal{T}_h$ of Ω by triangles, the simplest primal hybrid finite element method for solving this problem is given as follows: find (u_h, λ_h) satisfying

$$(u_h, \lambda_h) \in W_h \times M_h,$$

$$a(u_h, v_h) + b(v_h, \lambda_h) = \int_\Omega f v_h \, \mathrm{d}x \qquad \text{for all } v_h \in W_h, \tag{19.1}$$

$$b(u_h, \mu_h) = -\sum_{T \in \mathscr{T}_h} \int_{\partial T \cap \Gamma} \bar{u} \mu_{hT} \, \mathrm{d}\sigma \quad \text{for all } \mu_h \in M_h,$$

with

$$W_h = \{v \in H^1(\mathscr{T}_h): \forall T \in \mathscr{T}_h, v|_T \in P_1(T)\},$$
$$M_h = \{\mu = (\mu_T)_{T \in \mathscr{T}_h} \in E(\partial \mathscr{T}_h): \forall T \in \mathscr{T}_h, \mu_T \in D_1(\partial T)\},$$

$$a(u, v) = \sum_{T \in \mathscr{T}_h} \int_T \mathbf{grad}\, u \cdot \mathbf{grad}\, v \, \mathrm{d}x,$$

$$b(v, \mu) = -\sum_{T \in \mathscr{T}_h} \int_{\partial T \cap \Gamma} \mu_T v|_T \, \mathrm{d}\sigma.$$

Eliminating the Lagrangian multiplier, we obtain a characterization of u_h as the solution of the variational problem

$$u_h \in V_h^{\bar{u}}, \qquad a(u_h, v_h) = \int_\Omega f v_h \, \mathrm{d}x \quad \text{for all } v_h \in V_h, \tag{19.2}$$

with

$$V_h^{\bar{u}} = \{v_h \in W_h: \forall \mu_h \in M_h, b(v_h, \mu_h) = -\sum_{T \in \mathscr{T}_h} \int_{\partial T \cap \Gamma} \bar{u} \mu_{hT} \, \mathrm{d}\sigma\},$$

and

$$V_h = \{v_h \in W_h: \forall \mu_h \in M_h, b(v_h, \mu_h) = 0\}.$$

It is a simple matter to check that $V_h^{\bar{u}}$ is the set of functions in W_h that are affine on each triangle $T \in \mathscr{T}_h$, are continuous at the midpoint of each edge T' common to two triangles T_1 and T_2 of $\mathscr{T}_h$ and have for value at the midpoint of each boundary edge $T' \subset \Gamma$ the average value of $\bar{u}$ on T'. The set of midpoints of the three sides of a triangle is P_1-unisolvent. Thus a function $v_h \in V_h$ is uniquely determined by its values at the midpoints of the edges common to two triangles of $\mathscr{T}_h$. It is not difficult to give a basis of V_h associated with this choice of degrees of freedom for a function in V_h. Thus, problem (19.2) can be solved directly. Such a procedure is said to be a *nonconforming finite element method* since V_h is not a subspace of $V = H_0^1(\Omega)$, the space of test functions for the primal variational formulation.

If th. solution u belongs to $H^2(\Omega)$ we can obtain, using the results of the previous section, the error bound

$$\left\{\sum_{T\in\mathcal{T}_h} |u-u_h|_{1,T}^2\right\}^{1/2} + \|u-u_h\|_{0,T} \leqslant Ch|u|_{2,\Omega}, \tag{19.3}$$

an estimate that is classically obtained in a direct manner, cf. STRANG [1972] or STRANG and FIX [1973, Chapter 4.2].

After having calculated the solution of (19.2), to obtain the solution of (19.1) we have only to determine $\lambda_h \in M_h$ satisfying

$$b(v_h, \lambda_h) = \int_\Omega f v_h \, dx - a(u_h, v_h) \quad \text{for each } v_h \in W_h, \tag{19.4}$$

i.e. the functions λ_{hT}, constant on each edge of T, are determined by

$$\int_{\partial T} v\lambda_{hT} \, d\sigma = \int_T fv \, dx - \int_T \mathbf{grad}\, u_h \cdot \mathbf{grad}\, v \, dx \quad \text{for each } v \in P_1(T).$$

Thus once the nonconforming approximation u_h of u has been calculated, a post-processing procedure furnishes directly an approximation of the flux of **grad** u across each edge T'. If T' is the edge common to T_1 and $T_2 \in \mathcal{T}_h$, then the values λ_{hT_1} and λ_{hT_2} obtained from (19.4) satisfy

$$\lambda_{hT_1} + \lambda_{hT_2} = 0 \quad \text{on } T' = T_1 \cap T_2.$$

The flux across T' from T_1 into T_2 is minus that from T_2 into T_1.

More elaborate examples of nonconforming finite elements derived from primal hybrid finite element methods for second-order elliptic problems on an open set Ω in $\mathbb{R}^2$ may be found in RAVIART and THOMAS [1977b]. For those examples the continuity of the functions of V_h is imposed at the Gauss points of the edges T'. The example treated above, where the functions v_h are locally affine, is readily generalized to the three-dimensional case, where Ω is an open set of $\mathcal{R}^3$. However, the situation quickly becomes more difficult for higher orders of approximation. An example with $k=2$ is given in FORTIN [1985].

In fact nonconforming finite element methods were first developed for elliptic problems more complicated than the model problem above, such as:

(1) A fourth-order elliptic problem where the space V is a subspace of $(H^2(\Omega))^2$ and is thus made up of functions of class C^1 on $\bar{\Omega}$. Examples and analyses of such methods may be found in CIARLET [1974; 1978, pp. 362–380], CIAVALDINI and NEDELEC [1974], LASCAUX and LESAINT [1975] and ZHONG-CI SHI [1984a].

(2) A Stokes problem formulated as a variational problem on a subspace of the space V of divergence-free functions in $(H^1(\Omega))^n$. Analyses of such methods are given in CROUZEIX and RAVIART [1973], FORTIN [1981], HECHT [1981] and TEMAM [1977, pp. 172–181].

(3) A problem of linear elasticity formulated as a variational problem on a subspace of $V=(H^1(\Omega))^n$. The motivation in this case is to obtain models less rigid than those provided by conforming methods, cf. in particular PIAN [1971, 1972], FRAEIJS

DE VEUBEKE [1974b]. The most popular nonconforming finite element for this type of problem is Wilson's brick (WILSON et al. [1973]). An analysis of the method may be found in LESAINT [1976], LESAINT and ZLÁMAL [1980], ZHONG-CI SHI [1984b] or as a particular case of primal hybrid methods in THOMAS [1977, Chapter VI].[1] Other interpretations using hybrid methods have been proposed in IRONS [1972], PIAN and TONG [1986].

We conclude this section by pointing out that the problem of finding necessary and sufficient conditions for convergence of a nonconforming finite element method has been the subject of many discussions between theoreticians and engineers, cf. in particular the response of IRONS and LOIKKANEN [1983] to STUMMEL [1979, 1980a, 1980b]. When a nonconforming method can be interpreted as a primal hybrid method, possibly with reduced numerical integration, the study of its convergence is based in particular on the verification of the discrete inf–sup condition.

20. Examples of dual hybrid finite element methods

Let $\mathscr{T}_h$ be a triangulation of the domain $\bar{\Omega}$. Associated to this triangulation are the spaces $\mathscr{H}(\mathrm{div};\mathscr{T}_h)$ defined by (6.16) and $L(\partial\mathscr{T}_h)$ introduced in (18.4) and (18.5). Given a function $f\in L^2(\Omega)$ denote by $Q^f(\mathscr{T}_h)$ the affine manifold defined by

$$Q^f(\mathscr{T}_h)=\{\boldsymbol{q}\in\mathscr{H}(\mathrm{div};\mathscr{T}_h)\colon \forall T\in\mathscr{T}_h,\ \mathrm{div}(\boldsymbol{q}|_T)+f|_T=0\}, \tag{20.1}$$

or more precisely defined as the space of vector functions $\boldsymbol{q}\in\mathscr{H}(\mathrm{div};\mathscr{T}_h)$ which satisfy $\int_\Omega \boldsymbol{q}\cdot\mathbf{grad}\, v\,\mathrm{d}x=\int_\Omega fv\,\mathrm{d}x$, for each function $v\in H^1(\Omega)$ whose trace on $\partial\mathscr{T}_h$ is trivial. In particular, $Q^0(\mathscr{T}_h)$ is the vector space of functions $\boldsymbol{q}\in\mathscr{H}(\mathrm{div};\mathscr{T}_h)$ for which $\boldsymbol{q}|_T$ is divergence free for each $T\in\mathscr{T}_h$.

A dual mixed-hybrid formulation of our model problem (4.4)–(4.5) with Robin's boundary condition (15.1) is given as follows: find a pair $(\boldsymbol{p},\varphi)$ satisfying

$$\begin{aligned}
&(\boldsymbol{p},\varphi)\in Q^f(\mathscr{T}_h)\times L(\partial\mathscr{T}_h),\\
&a(\boldsymbol{p},\boldsymbol{q})+b(\boldsymbol{q},\varphi)=0 \qquad \text{for all } \boldsymbol{q}\in Q^0(\mathscr{T}_h),\\
&b(\boldsymbol{p},\psi)+d(\varphi,\psi)=-\sum_{T\in\mathscr{T}_h}\int_{\partial T\cap\Gamma} g\psi_T\,\mathrm{d}\sigma \qquad \text{for all } \psi\in L(\partial\mathscr{T}_h),
\end{aligned} \tag{20.2}$$

with here

$$a(\boldsymbol{p},\boldsymbol{q})=\sum_{i,j=1}^{n}\int_\Omega A_{ij}p_jq_i\,\mathrm{d}x, \tag{20.3}$$

$$b(\boldsymbol{q},\psi)=-\sum_{T\in\mathscr{T}_h}\int_{\partial T}\psi_T\boldsymbol{q}\cdot\boldsymbol{\nu}_T\,\mathrm{d}\sigma, \tag{20.4}$$

[1] An analysis of the method using Wilson's brick is also given in Sections 32–34 of Ciarlet's article in this volume (pp. 214–228).

(in this last expression, $\boldsymbol{q}\cdot\boldsymbol{\nu}_T$ denotes the normal trace on ∂T of the restriction of $\boldsymbol{q}$ to T), and

$$d(\varphi,\psi)=-\sum_{T\in\mathscr{T}_h}\int_{\partial T\cap\Gamma} t\varphi_T\psi_T\,\mathrm{d}\sigma. \tag{20.5}$$

After noting that, for each $T\in\mathscr{T}_h$, the normal trace on the boundary of T is a surjective mapping of $\{\boldsymbol{q}\in\mathscr{H}(\mathrm{div};T)\colon \operatorname{div}\boldsymbol{q}=0\}$ onto $\{\mu\in L^2(\partial T)\colon \int_{\partial T}\mu\,\mathrm{d}\sigma=0\}$, it is easy to verify that problem (20.2) has at most one solution. One shows that if the solution $(\boldsymbol{p},u)$ of (15.2) and of (15.8) belongs to $(H(\mathrm{div};\Omega)\cap\mathscr{H}(\mathrm{div};\mathscr{T}_h))\times H^1(\Omega)$ (this is the same regularity hypothesis as that required for the existence of a solution to the primal hybrid problem (18.6)), then the pair $(\boldsymbol{p},(u|_{\partial T})_{T\in\mathscr{T}_h})$ is a solution of (20.2). One thus has

$$\varphi_T=u \quad \text{on } \partial T \quad \text{for all } T\in\mathscr{T}_h. \tag{20.6}$$

REMARK 20.1. If we equip the spaces $\mathscr{H}(\mathrm{div};\mathscr{T}_h)$ and $L(\partial\mathscr{T}_h)$ with Hilbert space structures, then $b(\cdot,\cdot)$ considered as a bilinear form on $W\times M$, with $W=Q^0(\mathscr{T}_h)$ a subspace of $\mathscr{H}(\mathrm{div};\mathscr{T}_h)$ and $M=L(\partial\mathscr{T}_h)$, does not satisfy the inf–sup condition. (This situation is analogous to the case of the primal hybrid formulation, cf. Remark 18.3.)

To construct an approximation of the solution $(\boldsymbol{p},\varphi)$ of problem (20.2), we take for each $T\in\mathscr{T}_h$ two finite-dimensional subspaces, $Q_T\subset\mathscr{H}(\mathrm{div};T)$ and $Y_T\subset H^{1/2}(\partial T)$. To a given function $f\in L^2(\Omega)$, we associate the function $f_h\in L^2(\Omega)$ whose restriction to T, for each $T\in\mathscr{T}_h$, is the orthogonal projection in $L^2(T)$ of $f|_T$ into the finite-dimensional space $\{\operatorname{div}\boldsymbol{q}\colon \boldsymbol{q}\in Q_T\}$. We then put

$$W_h^f=\{\boldsymbol{q}\in\mathscr{H}(\mathrm{div};\mathscr{T}_h)\colon \forall T\in\mathscr{T}_h,\ \boldsymbol{q}|_T\in Q_T\}\cap Q^{f_h}(\mathscr{T}_h), \tag{20.7}$$

$$W_h=W_h^0=\{\boldsymbol{q}\in\mathscr{H}(\mathrm{div};\mathscr{T}_h)\colon \forall T\in\mathscr{T}_h,\ \boldsymbol{q}|_T\in Q_T\}\cap Q^0(\mathscr{T}_h), \tag{20.8}$$

and

$$M_h=\{\psi=(\psi_T)_{T\in\mathscr{T}_h}\in L(\partial\mathscr{T}_h)\colon \forall T\in\mathscr{T}_h,\ \psi_T\in Y_T\}. \tag{20.9}$$

We seek a pair $(\boldsymbol{p}_h,\varphi_h)$ satisfying

$$\begin{aligned}
&(\boldsymbol{p}_h,\varphi_h)\in W_h^f\times M_h,\\
&a(\boldsymbol{p}_h,\boldsymbol{q}_h)+b(\boldsymbol{q}_h,\varphi_h)=0 \qquad \text{for all } \boldsymbol{q}_h\in W_h,\\
&b(\boldsymbol{p}_h,\psi_h)+d(\varphi_h,\psi_h)=-\sum_{T\in\mathscr{T}_h}\int_{\partial T\cap\Gamma} g\psi_{hT}\,\mathrm{d}\sigma \quad \text{for all } \psi_h\in M_h.
\end{aligned} \tag{20.10}$$

REMARK 20.2. The analogous problem with a Neumann boundary condition $\boldsymbol{p}\cdot\boldsymbol{\nu}=g$, where g is a given function in $L^2(\Gamma)$ satisfying $\int_\Omega f\,\mathrm{d}x+\int_\Gamma g\,\mathrm{d}\sigma=0$, is formulated as follows: find a pair $(\boldsymbol{p}_h, \varphi_h)$ satisfying

$$\begin{aligned}
&(\boldsymbol{p}_h, \varphi_h) \in \boldsymbol{W}_h^f \times M_h,\\
&a(\boldsymbol{p}_h, \boldsymbol{q}_h) + b(\boldsymbol{q}_h, \varphi_h) = 0 \qquad \text{for all } \boldsymbol{q}_h \in \boldsymbol{W}_h,\\
&b(\boldsymbol{p}_h, \psi_h) = -\sum_{T \in \mathscr{T}_h} \int_{\partial T \cap \Gamma} g\psi_{hT}\, \mathrm{d}\sigma \quad \text{for all } \psi_h \in M_h,
\end{aligned}$$

where $\boldsymbol{W}_h^f$, $\boldsymbol{W}_h$, and M_h are as defined in (20.7), (20.8), and (20.9).

To treat the homogeneous Dirichlet problem we seek a solution $(\boldsymbol{p}_h, \varphi_h)$ of the following problem:

$$\begin{aligned}
&(\boldsymbol{p}_h, \varphi_h) \in \boldsymbol{W}_h^f \times M_{0h},\\
&a(\boldsymbol{p}_h, \boldsymbol{q}_h) + b(\boldsymbol{q}_h, \varphi_h) = 0 \quad \text{for all } \boldsymbol{q}_h \in \boldsymbol{W}_h,\\
&b(\boldsymbol{p}_h, \psi_h) = 0 \qquad\qquad\qquad \text{for all } \psi_h \in M_{0h},
\end{aligned}$$

where M_{0h} is the subspace of M_h defined by

$$M_{0h} = \{\psi_h \in M_h : \forall T \in \mathscr{T}_h, \psi_{hT} = 0 \text{ on } \partial T \cap \Gamma\}. \tag{20.11}$$

By construction of the function f_h, the affine manifold $\boldsymbol{W}_h^f$ is not empty. Applying Theorem 10.2 with Remark 10.8 we obtain with no difficulty the following theorem:

THEOREM 20.1. *Problem* (20.10) *has a unique solution if and only if the subspaces* $\boldsymbol{W}_h$ *and* M_h *are compatible in the sense that the discrete inf–sup condition is satisfied*:

$$\{\psi_h \in M_{0h} : \forall \boldsymbol{q}_h \in \boldsymbol{W}_h, b(\boldsymbol{q}_h, \psi_h) = 0\} = \{0\}, \tag{20.12}$$

where M_{0h} *is the subspace of* M_h *defined in* (20.11).

The global condition (20.12) is satisfied once, for each $T \in \mathscr{T}_h$, the choice of spaces Q_T and Y_T satisfy the local compatibility condition

$$\left\{\psi \in Y_T : \forall \boldsymbol{q} \in Q_T \text{ with } \operatorname{div} \boldsymbol{q} = 0, \int_{\partial T} \psi \boldsymbol{q} \cdot \boldsymbol{\nu}_T \, \mathrm{d}\sigma = 0\right\} = P_0(\partial T). \tag{20.13}$$

Next we develop an example where the above theory is applied.

THEOREM 20.2. *Let* Ω *be an open polygonal domain in* $\mathbb{R}^2$ *triangulated by triangles* T. *Let* k *be a positive integer, and put*

$$k^* = \begin{cases} k, & \text{if } k \text{ is odd},\\ k+1, & \text{if } k \text{ is even}. \end{cases}$$

For each triangle $T \in \mathscr{T}_h$ *define the subspace* $Y_T \subset H^{1/2}(\partial T)$ *by*

$$Y_T = P_k(\partial T). \tag{20.14}$$

(With the definition of the space $L^{(k)}_{\partial\mathcal{T}_h}$ given by (5.14), we have $M_h = L^{(k)}_{\partial\mathcal{T}_h}$.) Suppose that the space $Q_T \subset \mathcal{H}(\mathrm{div}, T)$ is such that the following property is satisfied:

$$\forall \mu \in D_{k^*}(\partial T), \exists \boldsymbol{q} \in Q_T \text{ s.t. } \boldsymbol{q}\cdot\boldsymbol{\nu}_T = \mu \text{ on } \partial T. \tag{20.15}$$

Then problem (20.10) *has a unique solution.*

To establish this theorem it clearly suffices to show that the local compatibility condition (20.13) is satisfied. A demonstration can be found in THOMAS [1976, Lemma 4.1]. We point out that the subspace of all $\psi \in P_k(\partial T)$ which satisfy $\int_{\partial T} \psi\mu \, d\sigma = 0$ for each $\mu \in D_k(\partial T)$ with $\int_{\partial T} \mu \, d\sigma = 0$ reduces to $P_0(\partial T)$ only when k is odd. Otherwise, when k is even, it is a subspace of dimension 2.

We shall give error estimates using the norm $\|\cdot\|_{0,\Omega}$ on $(L_2(\Omega))^2$ for vector functions (with this method, on each $T \in \mathcal{T}_h$, $\mathrm{div}(\boldsymbol{p} - \boldsymbol{p}_h) = -(f - f_h)$ and is considered to be known or estimated a priori) and the norm $|||\cdot|||_{1/2,\partial\mathcal{T}_h}$ given in (5.19) for the traces on $\partial\mathcal{T}_h$ of functions in $H^1(\Omega)$. We suppose we have a uniformly regular family of triangulations, where the hypothesis of uniformness is made simply for ease of exposition. We note the existence of constants c and C_k independent of h such that

$$\|\psi\|_{0,\partial\mathcal{T}_h} \leqslant c h^{1/2} |||\psi|||_{1/2,\partial\mathcal{T}_h} \quad \text{for all } \psi \in H^{1/2}(\partial\mathcal{T}_h), \tag{20.16}$$

$$|||\psi_h|||_{1/2,\partial\mathcal{T}_h} \leqslant C_k h^{-1/2} \|\psi_h\|_{0,\partial\mathcal{T}_h} \quad \text{for all } \psi_h \in L^{(k)}_{\partial\mathcal{T}_h}. \tag{20.17}$$

THEOREM 20.3. *We suppose the hypotheses of Theorem* 20.2 *except here we fix for each* $T \in \mathcal{T}_h$

$$\begin{aligned} Y_T &= P_k(\partial T),\\ Q_T &= D_{k^*}(T). \end{aligned} \tag{20.18}$$

Then there exists a constant C independent of h such that we have

$$\|\boldsymbol{p} - \boldsymbol{p}_h\|_{0,\mathcal{T}_h} + |||\varphi - \varphi_h|||_{1/2,\partial\mathcal{T}_h} \leqslant C h^k \{|\boldsymbol{p}|_{k,\Omega} + |u|_{k+1,\Omega}\} \tag{20.19}$$

once the solution $(\boldsymbol{p}, u)$ *of* (15.2) *and of* (15.8) *belongs to* $(H^k(\Omega))^2 \times H^{k+1}(\Omega)$ (φ *is expressed as a function of u in* (20.6)).

PROOF. To apply the general theory developed in Chapter III we begin by making the change of variables $\boldsymbol{p}$ to $\boldsymbol{p} - \mathscr{E}^{(k)}_h \boldsymbol{p}$, and $\boldsymbol{p}_h$ to $\boldsymbol{p}_h - \mathscr{E}^{(k)}_h \boldsymbol{p}$. Thus we have

$$\boldsymbol{p} - \mathscr{E}^{(k)}_h \boldsymbol{p} \in Q^0(\mathcal{T}_h), \qquad \boldsymbol{p}_h - \mathscr{E}^{(k)}_h \boldsymbol{p} \in W_h \subset Q^0(\mathcal{T}_h),$$

where $\mathscr{E}^{(k)}_h \boldsymbol{p}$ is the equilibrium interpolant of order k described in Section 6. The continuity properties and W_h-ellipticity with constants independent of h relative to the norm $\|\cdot\|_{0,\Omega}$ are obvious. Further, we can establish the existence of constants B and $\beta_0 > 0$, independent of h, such that

$$b(\boldsymbol{q}, \psi) \leqslant B \|\boldsymbol{q}\|_{0,\Omega} |||\psi|||_{1/2,\partial\mathcal{T}_h} \qquad \text{for all } \boldsymbol{q} \in Q^0(\mathcal{T}_h),\ \psi \in L(\partial\mathcal{T}_h),$$

and

$$\sup_{q_h \in W_h} \frac{b(q_h, \psi_h)}{\|q_h\|_{0,\Omega}} \geqslant \beta_0 [\psi_h]_{1/2,\partial\mathcal{T}_h} \qquad \text{for all } \psi_h \in M_h,$$

where the seminorm $[\cdot]_{1/2,\partial\mathcal{T}_h}$ is defined by

$$[\psi]_{1/2,\partial\mathcal{T}_h} = \left\{ \sum_{T \in \mathcal{T}_h} \left\| \psi_T - \frac{1}{\operatorname{mes} \partial T} \left(\int_{\partial T} \psi_T \, d\sigma \right) \right\|_{1/2,\partial T}^2 \right\}^{1/2}.$$

Using the Friedrichs–Poincaré inequality we can show that $[\cdot]_{1/2,\partial\mathcal{T}_h}$ is a norm on M_{0h} equivalent (uniformly in h) to the norm $|||\cdot|||_{1/2,\partial\mathcal{T}_h}$. Thus there exists $\beta > 0$, independent of h, such that

$$\sup_{q_h \in W_h} \frac{b(q_h, \psi_h)}{\|q_h\|_{0,\Omega}} \geqslant \beta |||\psi_h|||_{1/2,\partial\mathcal{T}_h} \qquad \text{for all } \psi_h \in M_{0h}.$$

The theorem then follows easily. □

REMARK 20.3. When the integer k is even, we can obtain the same result while taking the space Q_T to be the space generated by $D_k(T)$ together with an appropriately chosen element of $D_{k+1}(T)$. For details and other examples cf. THOMAS [1976].

The dual hybrid finite element method corresponding to the choice of subspaces M_h and W_h associated with (20.18) by (20.8) and (20.9) is convergent for each integer $k \geqslant 1$ in the following sense (with no regularity hypothesis for the solution, one can show that):

$$\lim_{h \to 0} \|p - p_h\|_{0,\Omega} = 0.$$

The principle of the dual hybrid finite element methods given above for a second-order elliptic problem corresponds to that developed by specialists in structural mechanics and termed assumed stress hybrid finite element model. We cite first of all the work of PIAN [1964, 1971, 1972, 1983]; cf. also PIAN and TONG [1969, 1986], TONG [1983], and the recent articles of PIAN and his collaborators where can be found a treatment of nonaffine-equivalent finite elements. Other examples are given by ATLURI [1971], ATLURI, TONG and MURAKAWA [1983], PUNCH and ATLURI [1984], SPILKER and MUNIR [1980a, 1980b] and WOLF [1972a, 1972b] among many other publications on the subject. For a mathematical analysis of such methods we refer to the work of BREZZI [1975, 1977], of KIKUCHI [1973] and of QUARTERONI [1979].

REMARK 20.4. The construction of the examples given above followed this strategy: given a space M_h, find a space W_h large enough for the compatibility condition to be satisfied. Under reasonable hypotheses we obtain asymptotic orders of optimal error (relative to the given W_h). The construction of examples proposed by engineers often follows another strategy: given two spaces W_h and M_h^*, the choice of spaces being consistent with the physics of the problem, we note the existence of "mechanisms" (... the homogeneous problem does not have the trivial solution as unique solution). One then tries to eliminate these mechanisms, that is to reduce M_h^* to a space M_h small enough for the compatibility condition with W_h to be satisfied. This technique does not lead to satisfactory asymptotic orders of convergence.

21. Hybridization of equilibrium methods

The dual hybrid finite element method just constructed yields—when the compatibility hypothesis (20.12) is satisfied—an approximation $\boldsymbol{p}_h$ of $\boldsymbol{p}$ satisfying div $\boldsymbol{p}_h + f_h = 0$ in the interior of each finite element T; however, the reciprocity of the normal traces on the interfaces is not realized pointwise along the entire interface. In particular, the solution $\boldsymbol{p}_h$ is not in $H(\text{div}; \Omega)$.

A variant introduced by FRAEIJS DE VEUBEKE [1965] yields an approximated solution $\boldsymbol{p}_h \in \mathscr{H}(\text{div}; \Omega)$ satisfying div $\boldsymbol{p}_h + f_h = 0$ in Ω. The mathematical analysis can be found in THOMAS [1977, Chapter VIII], cf. also RAVIART and THOMAS [1979]. These results are completed by a postprocessing technique developed in ARNOLD and BREZZI [1985].

We thus reconsider the approximation of the solution $(\boldsymbol{p}, \varphi)$ of the model problem with the dual hybrid formulation (20.2)–(20.5). On the one hand, for each $T \in \mathscr{T}_h$, we take a finite-dimensional subspace $Q_T \subset \mathscr{H}(\text{div}; T)$ and as before, cf. (20.7), (20.8), define the spaces

$$W_h^f = \{\boldsymbol{q} \in \mathscr{H}(\text{div}; \mathscr{T}_h): \forall T \in \mathscr{T}_h, \boldsymbol{q}|_T \in Q_T\} \cap Q^{f_h}(\mathscr{T}_h) \tag{21.1}$$

and

$$W_h = W_h^0 = \{\boldsymbol{q} \in \mathscr{H}(\text{div}; \mathscr{T}_h): \forall T \in \mathscr{T}_h, \boldsymbol{q}|_T \in Q_T\} \cap Q^0(\mathscr{T}_h). \tag{21.2}$$

On the other hand, for each $T \in \mathscr{T}_h$, we put

$$Y_T = \{\psi \in L^2(\partial T): \exists \boldsymbol{q} \in Q_T \text{ s.t. } \boldsymbol{q} \cdot \boldsymbol{v}_T = \psi\} \tag{21.3}$$

and define

$$\begin{aligned} M_h = \{\psi = (\psi_T)_{T \in \mathscr{T}_h} \in L^2(\partial \mathscr{T}_h): \\ \forall T \in \mathscr{T}_h, \psi_T \in Y_T, \forall T' = T_1 \cap T_2,\ T_1, T_2 \in \mathscr{T}_h, \psi_{T_1} = \psi_{T_2} \text{ on } T'\}. \end{aligned} \tag{21.4}$$

We seek $(\boldsymbol{p}_h, \varphi_h)$ satisfying (cf. (20.10))

$$\begin{aligned} &(\boldsymbol{p}_h, \varphi_h) \in W_h^f \times M_h, \\ &a(\boldsymbol{p}_h, \boldsymbol{q}_h) + b(\boldsymbol{q}_h, \varphi_h) = 0 \qquad \text{for all } \boldsymbol{q}_h \in W_h, \\ &b(\boldsymbol{p}_h, \psi_h) + d(\varphi_h, \psi_h) = -\sum_{T \in \mathscr{T}_h} \int_{\partial T \cap \Gamma} g \psi_{hT} \, d\sigma \quad \text{for all } \psi_h \in M_h. \end{aligned} \tag{21.5}$$

REMARK 21.1. Such an approximation method is not a dual hybrid approximation in the sense of the preceding section because the spaces Y_T are not subspaces of $H^{1/2}(\partial T)$, and thus the space M_h of multipliers is not a subspace of the space $L(\partial \mathcal{T}_h)$ defined in (18.4) or (18.5). Whereas examples of finite-dimensional subspaces of $L^2(\partial T)$ and of $H^{-1/2}(\partial T)$ are identical, the examples of finite-dimensional subspaces of $L^2(\partial T)$ and of $H^{1/2}(\partial T)$ are fundamentally different: the functions in $H^{1/2}(\partial T)$ which are locally continuous are globally continuous on ∂T.

It is clear that for each choice of spaces $Q_T \subset \mathscr{H}(\mathrm{div}; T)$ with Y_T associated to Q_T by (21.3), problem (21.5) has a unique solution $(\boldsymbol{p}_h, \varphi)$.

Suppose that the coefficient $t>0$ is constant on each face T' contained in the boundary Γ of Ω (this hypothesis is made solely for simplicity of exposition). Then the elimination of the multiplier φ_h in (21.5) yields a characterization of $\boldsymbol{p}_h$ as the solution of the variational problem

$$\begin{gathered}\boldsymbol{p}_h \in V_h^f,\\ \sum_{i,j=1}^{n} \int_T A_{ij} p_{hj} q_{hi}\,\mathrm{d}x + \int_\Gamma (1/t)(\boldsymbol{p}_h \cdot \boldsymbol{\nu})(\boldsymbol{q}_h \cdot \boldsymbol{\nu})\,\mathrm{d}\sigma = \int_\Gamma (1/t) g \boldsymbol{q}_h \cdot \boldsymbol{\nu}\,\mathrm{d}\sigma \qquad (21.6)\\ \text{for all } \boldsymbol{q}_h \in V_h,\end{gathered}$$

with

$$\begin{aligned} V_h^f &= \{\boldsymbol{q} \in \mathscr{H}(\mathrm{div}; \Omega) \colon \mathrm{div}\, \boldsymbol{q} + f = 0 \text{ on } \Omega\} \\ &\quad \cap \{\boldsymbol{q} \in (L^2(\Omega))^n \colon \forall T \in \mathcal{T}_h, \boldsymbol{q}|_T \in Q_T\}, \qquad (21.7)\\ V_h = V_h^0 &= \{\boldsymbol{q} \in \mathscr{H}(\mathrm{div}; \Omega) \colon \mathrm{div}\, \boldsymbol{q} = 0 \text{ on } \Omega\} \\ &\quad \cap \{\boldsymbol{q} \in (L^2(\Omega))^n \colon \forall T \in \mathcal{T}_h, \boldsymbol{q}|_T \in Q_T\}. \qquad (21.8)\end{aligned}$$

The approximation $\boldsymbol{p}_h$ of $\boldsymbol{p}$ thus obtained is an equilibrium approximation (cf. Section 2); i.e. $\boldsymbol{p}_h$ is a vector function in $\mathscr{H}(\mathrm{div}; \Omega)$ that satisfies the equilibrium equation

$$\mathrm{div}\, \boldsymbol{p}_h + f_h = 0 \quad \text{in } \Omega. \qquad (21.9)$$

Conversely, to an equilibrium formulation (21.6)–(21.8) can be associated a hybrid method defined by (21.1)–(21.5). Contrary to what one might suspect at first glance, it is simpler to solve numerically (21.5) than (21.6). The numerical solution of (21.5) can be carried out by first eliminating the principal variable $\boldsymbol{p}_h$. (This elimination is performed at the level of the n-simplex T.) Then to calculate φ_h, one must solve a linear system of the same structure as the one which results from a nonconforming finite element method, cf., in particular, FREIJS DE VEUBEKE [1974a].

REMRK 21.2. When the given function f is trivial, or sufficiently simple that one can numerically reduce the problem to the case $f=0$, to solve (21.6) (equilibrium formulation) it suffices to know how to construct a basis for V_h. In dimension $n=2$, using stream functions we can treat directly problem (21.6), cf. HASLINGER and HLAVACEK [1975, 1976a, 1976b], HLAVACEK [1980], and HLAVACEK and KRIZEK [1984]. We have not chosen this presentation because it is not easily generalized in dimension $n=3$ to the case where f is arbitrary and the boundary condition is a Robin's condition.

When $\mathcal{T}_h$ is a triangulation of $\bar{\Omega}$ by n-simplexes T, if we take for spaces Q_T the spaces $D_k(T)$, where k is a given integer, $k \geqslant 1$, the described method is to take for spaces Y_T the spaces $D_k(\partial T)$. Thus here we have

$$W_h^f = \{\boldsymbol{q} \in \mathscr{H}(\mathrm{div}; \mathcal{T}_h): \forall T \in \mathcal{T}_h, \boldsymbol{q}|_T \in D_k(T)\} \cap Q^{f_h}(\mathcal{T}_h), \tag{21.10}$$

and

$$W_h = W_h^0 = \{\boldsymbol{q} \in \mathscr{H}(\mathrm{div}; \mathcal{T}_h): \forall T \in \mathcal{T}_h, \boldsymbol{q}|_T \in (P_{k-1}(T))^n\} \cap Q^0(\mathcal{T}_h), \tag{21.11}$$

where for the characterization of W_h we have used the fact that each vector function in D_k that is divergence-free belongs in fact to $(P_{k-1}(T))^n$. The associated space V_h is then none other than

$$V_h = \{\boldsymbol{q} \in \mathscr{H}(\mathrm{div}; \Omega): \mathrm{div}\, \boldsymbol{q} = 0 \text{ on } \Omega \text{ and } \\ \text{s.t. } \forall T \in \mathcal{T}_h, \boldsymbol{q}|_T \in (P_{k-1})^n\}, \tag{21.12}$$

i.e. with the notation of (6.17)

$$V_h = E_h^{(k)} \cap \{\boldsymbol{q} \in (L^2(\Omega))^n: \mathrm{div}\, \boldsymbol{q} = 0 \text{ on } \Omega\}.$$

For simplicity of exposition, to give error estimates we suppose that the family of triangulations $\mathcal{T}_h$ is uniformly regular. With $Q_T = D_k(T)$ and $Y_T = D_k(\partial T)$ for each $T \in \mathcal{T}_h$, there is no difficulty in showing that there is a constant $\beta_0 > 0$, independent of h, such that

$$\sup_{\boldsymbol{q}_h \in W_h} \frac{b(\boldsymbol{q}_h, \psi_h)}{\|\boldsymbol{q}_h\|_{0,\Omega}} \geqslant \beta_0 h^{-1/2} [\psi_h]_{0,\partial\mathcal{T}_h} \quad \text{for all } \psi_h \in M_h,$$

where the seminorm $[\cdot]_{0,\partial\mathcal{T}_h}$ is defined by

$$[\psi]_{0,\partial\mathcal{T}_h} = \left\{ \sum_{T \in \mathcal{T}_h} \left\| \psi_T - \frac{1}{\mathrm{mes}\, \partial T} \left(\int_{\partial T} \psi_T \, \mathrm{d}\sigma \right) \right\|_{0,\partial T}^2 \right\}^{1/2}.$$

Thus we have:

THEOREM 21.1. *Suppose that, for each* $T \in \mathcal{T}_h$,

$$Q_T = D_k(T), \qquad Y_T = D_k(\partial T). \tag{21.13}$$

Then there exists a constant C, *independent of* h, *such that if* $(\boldsymbol{p}_h, \varphi_h)$ *is the solution of* (21.5), *then*

$$\|\boldsymbol{p} - \boldsymbol{p}_h\|_{0,\mathcal{T}_h} + h^{-1/2} \|\varphi - \varphi_h\|_{0,\partial\mathcal{T}_h} \leqslant C h^k \{|\boldsymbol{p}|_{k,\Omega} + |u|_{k+1,\Omega}\}, \tag{21.14}$$

provided, of course, that the solution $(\boldsymbol{p}, u)$ *of* (15.2) *and of* (15.8) *belongs to* $(H^k(\Omega))^n \times H^{k-1}(\Omega)$. ($\varphi$ *is expressed as a function of u in* (20.6).)

Once the solution $(\boldsymbol{p}_h, \varphi_h)$ of problem (21.5), with the spaces Q_T and Y_T as given in (21.13), has been obtained, we can easily calculate the function $u_h \in L^2(\Omega)$ whose restriction to each $T \in \mathcal{T}_h$ is the polynomial of degree at most $k-1$ defined by

$$\int_T u_h \operatorname{div} \boldsymbol{q} \,\mathrm{d}x = -\sum_{i,j=1}^{n} \int_T A_{ij} p_{hj} q_i \,\mathrm{d}x + \int_{\partial T} \varphi_{hT} \boldsymbol{q} \cdot \boldsymbol{\nu}_T \,\mathrm{d}\sigma$$

for each $\boldsymbol{q} \in D_k(T)$.

Using that $(\boldsymbol{p}_h, \varphi_h)$ satisfies (21.5) we remark that this function u_h satisfies

$$\int_\Omega u_h \operatorname{div} \boldsymbol{q}_h \,\mathrm{d}x = -\sum_{i,j=1}^{n} \int_\Omega {}_{ij} p_{hj} q_{hi} \,\mathrm{d}x + \int_\Gamma (1/t)(g - \boldsymbol{p}_h \cdot \boldsymbol{\nu}) \boldsymbol{q}_h \cdot \boldsymbol{\nu} \,\mathrm{d}\sigma$$

for each $\boldsymbol{q}_h \in \mathscr{H}(\operatorname{div}; \Omega)$ such that $\boldsymbol{q}_h|_T \in D_k(T)$ for each $T \in \mathcal{T}_h$.

The pair $(\boldsymbol{p}_h, u_h)$ is the unique solution of problem (15.14) (dual mixed formulation) obtained with (15.12) and (15.13). Thus we know u_h is an approximation of u of order k relative to the norm $\|\cdot\|_{0,\Omega}$.

Now for each $T \in \mathcal{T}_h$, we have both the approximation $\varphi_{hT} \in D_k(\partial T)$ of $\varphi_T = u|_{\partial T}$ and the approximation $u_h|_T \in P_{k-1}(T)$ of $u|_T$. These two approximations provide complementary information about $u|_T$ and it is reasonable to try to use all of this information to obtain a better approximation of $u|_T$. We take an example in dimension $n=2$ when k is odd. The relations

$$\int_{\partial T} (u_h^* - \varphi_{hT}) \psi \,\mathrm{d}\sigma = 0 \quad \text{for all } \psi \in D_k(\partial T),$$

and

$$\int_T (u_h^* - u_h) v \,\mathrm{d}x = 0 \qquad \text{for all } v \in P_{k-3}(T) \text{ (if } k \geqslant 3)$$

define a unique function $u_h^* \in L^2(\Omega)$ whose restriction to each triangle $T \in \mathcal{T}_h$ is a polynomial of degree at most k. Under the standard regularity hypotheses, one can show (ARNOLD and BREZZI [1985, Theorem 2.2]) the order of approximation is thus improved:

$$\|u - u_h^*\|_{0,\Omega} = \mathrm{O}(h^{k+1}).$$

For similar results see also STENBERG [1988c].

CHAPTER VI

Extensions and Variations

22. Other examples of mixed and hybrid methods

The mixed and hybrid finite element methods given in the preceding sections were developed almost exclusively for second-order elliptic problems. Formulations of such methods for fourth-order elliptic problems were noted. The mathematical framework constructed for the numerical analysis of these methods for approximating the solutions of partial differential equations apparently ignores the physical origin of the model. We have not even sought to exploit the supplementary properties that might result from the fact that in most cases we are concerned with the solutions of constrained minimization problems. This does not however imply that the elaboration of such approximation techniques can be successfully carried out without taking into account the physical properties of the underlying problem. There is no lack of examples in computational mechanics in the recent scientific literature. We give here only a brief overview.

Stokes problem is a fundamental problem of fluid mechanics. The primary unknown u is the velocity vector for the fluid displacement. It can be characterized as being the solution of a minimization problem in a subspace of the space

$$V=\{v=(v_i)\colon v_i\in H^1(\Omega),\ 1\leqslant i\leqslant n,\ \text{s.t. div } v=0\}$$

The Lagrangian multiplier associated with the constraint that the velocity field be divergence-free is identified with a pressure. The approximation of the solution of this problem has been and will continue to be the object of many publications. The state of the art in this domain can be found in the book of GIRAULT and RAVIART [1986], and it does not seem opportune to duplicate here the bibliography of this work.

The system of linear elasticity constitutes another privileged domain for the application of the theory of mixed and hybrid methods. The variational formulation of the particle displacement being taken as primal formulation, the dual formulation called the equilibrium formulation, is a variational formulation of the stress equation. In the case of no volume forces this equilibrium formulation is a minimization principle in a subspace of the space

$$S=\{\boldsymbol{\sigma}=(\sigma_{ij})\colon \sigma_{ij}\in L^2(\Omega),\ \sigma_{ij}=\sigma_{ji},\ 1\leqslant i,\ j\leqslant n,\ \mathbf{div}\,\boldsymbol{\sigma}=\mathbf{0}\},$$

where $\mathbf{div}\,\boldsymbol{\sigma}=\mathbf{0}$ means $\Sigma_{j=1}^{3}\partial\sigma_{ij}/\partial x_j=0$ for $i=1,2,3$. With some simplifying restrictions concerning the data and geometry of the problem under consideration, it is possible to use Airy functions to reduce the study of the equilibrium formulation to that of the two-dimensional biharmonic problem. The degree of sophistication necessary for constructing equilibrium finite element models is more easily understood with this analogy. Examples can be found in JOHNSON and MERCIER [1978, 1979], KRIZEK [1982, 1983], and KRIZEK and NEITTAAMÄKI [1986]. Analyses of mixed and hybrid formulations for the system of elasticity are proposed by AMARA and THOMAS [1978, 1979], ARNOLD, BREZZI and DOUGLAS [1985], ARNOLD, DOUGLAS and GUPTA [1984], PITKÄRANTA and STENBERG [1983], and STENBERG [1986, 1988a, 1988b]. The principal source of difficulty in the approximation of the stress formulation comes from the necessity of respecting, rigorously or not, the symmetry relations that should be satisfied by the stresses. The article of ARNOLD and FALK [1988] proposes a new formulation with which one obtains distinct approximations of σ_{ij} and σ_{ji}.

Other recent articles concerning the numerical treatment of problems of solid mechanics with mixed and hybrid methods are referenced in the bibliography. While some of these articles may appear to lack a certain degree of mathematical rigor, experience has shown that the examples described therein cannot, nevertheless, be neglected. A survey of such examples in fluid and solid mechanics can be found in the book of HUGHES [1987, Chapters 4 and 5].

Another domain where quite a few models using mixed or hybrid finite elements have been constructed is the subdomain of fluid mechanics dealing with fluid flow in porous media and in particular petroleum reservoir engineering. These reservoir models are based on nonlinear equations too complex to be described here in a few lines. We refer the interested reader to the book of CHAVENT and JAFFRE [1986], where in Chapter V some models directly inspired by the mixed methods developed above may be found. Other models are described and analyzed in CHAVENT, COHEN, DUPUY, JAFFRE and RIBERA [1984], DARLOW, EWING and WHEELER [1984], DOUGLAS [1984], DOUGLAS, EWING and WHEELER [1983], DOUGLAS and ROBERTS [1983], EYMARD, GALLOUET and JOLY [1988], JAFFRE and ROBERTS [1985], and WHEELER and GONZALES [1984].

We conclude this section by mentioning that mixed finite elements have also been used for the equations of electromagnetism: BENDALI [1984a, 1984b], BOSSAVIT [1988], NEDELEC [1980].

23. Extensions and variations of the theory

The methods of Lagrangian multipliers that have been presented here concern the approximation of elliptic problems. These methods have been extended to the study of eigenvalue problems; cf. CANUTO [1978], KIKUCHI [1987], and MERCIER, OSBORN, RAPPAZ and RAVIART [1981]. They have also been extended to the study of problems of evolution: parabolic problems as in JOHNSON and THOMÉE [1981] and in THOMÉE [1984, Chapter 13], see also SQUEFF [1987]; hyberbolic problems as in CANUTO [1981], in DOUGLAS and GUPTA [1986], in GEVECI [1988], in QUARTERONI [1980b] and more recently in DUPONT, GLOWINSKI, KINTON and WHEELER [1989].

Certain mixed methods have been successful only because of the use of reduced numerical integration. Some examples can be found in JOHNSON and PITKÄRANTA [1982], MALKUS and HUGHES [1978], NOOR and PETERS [1983], and SHIMODAIRA [1985]. Other examples of reduced integration are proposed in RAVIART and THOMAS [1977b] to obtain the interpretation of certain primal hybrid methods as nonconforming methods.

In the preceding studies the approximation of elliptic problems of constrained minimization has been approached exclusively with the technique of dualization of the constraints using Lagrangian multipliers. The technique of penalization of the constraints constitutes an alternative way of attacking these minimization problems. These techniques can fortunately be combined. The resulting technique is called the augmented Lagrangian technique for which we refer the reader to the book of FORTIN and GLOWINSKI [1983]. Even though the penalization technique was not initially retained in the mixed and hybrid formulations of the problem, it can nonetheless be useful for the resolution of the linear system resulting from the mixed hybrid formulation; cf. BERCOVIER [1978], FELIPPA [1986]; cf. also Section 17.

Other finite element methods such as nodal methods can be interpreted as being mixed or hybrid methods; cf. HENNART [1985, 1986a, 1986b], and HENNART, JAFFRE and ROBERTS [1988]. The so-called Trefftz methods can also be understood as being variants of the dual mixed hybrid formulations; cf. JIROUSEK and LAN GUEZ [1986] and ZIELINSKI and ZIENKIEWICZ [1985].

As an extension of the classical abstract theory of Ritz–Galerkin, with the development of the theory of internal approximation of variational problems which has been seen to be fundamental for the analysis of "standard" finite element methods, the abstract theory of Babuşka–Brezzi for the approximation of problems with Lagrangian multipliers has become popular because of its applications to the analysis of mixed and hybrid finite element methods. Other applications of the Babuška–Brezzi theory have been proposed more recently; cf. for example the analysis of spectral methods proposed in the articles of BERNARDI, CANUTO and MADAY [1988], and of BERNARDI, MADAY and METIVET [1987] or the analysis of modal decomposition methods in the articles of STOLARSKI and BELYTSCHKO [1986a, 1986b].

Domain decomposition methods for the solution of partial differential equations are currently enjoying a resurgence in popularity for their use in constructing algorithms well adapted to use on multiprocessing machines. The use of mixed and hybrid formulations seem adequate for this type of method; cf. GLOWINSKI and WHEELER [1988] for a first example in this direction.

References

ADAMS, R.A. (1975), *Sobolev Spaces* (Academic Press, London).

AMARA, M. and J.-M. THOMAS (1978), Approximation par éléments finis équilibres du système de l'élasticité linéaire, *C.R. Acad. Sci. Paris Sér. A* **286**, 1147–1150.

AMARA, M. and J.-M. THOMAS (1979), Equilibrium finite elements for the linear elastic problem, *Numer. Math.* **33**, 367–383.

ARNOLD, D.N. (1981), Discretization by finite elements of a model parameter dependent problem, *Numer. Math.* **37**, 405–421

ARNOLD, D.N. and F. BREZZI (1985). Mixed and nonconforming finite element methods: Implementation, postprocessing and error estimates, *RAIRO Modél. Math. Anal. Numér.* **19**, 7–32.

ARNOLD, D.N., F. BREZZI and J. DOUGLAS Jr (1985), Peers: A new mixed finite element for plane elasticity, *Japan J. Appl. Math.* **1**, 347–367.

ARNOLD, D.N., J. DOUGLAS Jr and C.P. GUPTA (1984), A family of higher order mixed finite element methods for plane elasticity, *Numer. Math.* **45**, 1–22.

ARNOLD, D.N. and R. FALK (1988), A new mixed formulation for elasticity, *Numer. Math.* **53**, 13–30.

ATLURI, S.N. (1971), A new assumed stress hybrid finite element model for solid continua, *AIAA J.* **9**, 1647–1649.

ATLURI, S.N., P. TONG and H. MURAKAWA (1983), Recent studies in hybrid and mixed finite element methods in mechanics, in: S.N. ATLURI, R.H. GALLAGHER and O.C. ZIENKIEWICZ, eds., *Hybrid and Mixed Finite Element Methods* (Wiley, Chichester) 51–71.

BABUŠKA, I. (1971), Error bounds for finite element method, *Numer. Math.* **16**, 322–333.

BABUŠKA, I. (1973), The finite element method with Lagrange multipliers, *Numer. Math.* **20**, 179–192.

BABUŠKA, I. and A.K. AZIZ (1972), Survey lectures on the mathematical foundations of the finite element method, in: A.K. AZIZ, ed., *The Mathematical Foundations of the Finite Element Method with Applications to Partial Differential Equations* (Academic Press, New York) 3–359.

BABUŠKA, I. and A. MILLER (1984), The postprocessing approach in the finite element method, Part 1: Calculation of displacements, stresses and other higher derivatives of the displacements, *Internat. J. Numer. Methods Engrg.* **20**, 1085–1109.

BABUŠKA, I., J.T. ODEN and J.K. LEE (1977), Mixed-hybrid finite element approximations of second-order elliptic boundary-value problem, Part 1, *Comput. Methods Appl. Mech. Engrg.* **11**, 175–206.

BABUŠKA, I., J.T. ODEN and J.K. LEE (1978), Mixed-hybrid finite element approximations of second-order elliptic boundary-value problem, Part 2: Weak-hybrid methods, *Comput. Methods Appl. Mech. Engrg.* **14**, 1–22.

BABUŠKA, I. and J.E. OSBORN (1983), Generalized finite element methods: Their performance and their relation to mixed methods, *SIAM J. Numer. Anal.* **20**, 510–536.

BABUŠKA, I., J.E. OSBORN, J. PITKÄRANTA (1980), Analysis of mixed methods using mesh dependent norms, *Math. Comp.* **35**, 1039–1062.

BALASUNDARAM, S. and P.K. BHATTACHARYYA (1984), A mixed finite element method for fourth order elliptic equations with variable coefficients, *Comput. Math. Appl.* **10**, 245–256.

BALASUNDARAM, S. and P.K. BHATTACHARYYA (1986), A mixed finite element method for fourth order partial differential equations, *Z. Angew. Math. Mech.* **66**, 489–499.

BARLOW, J. (1986), A different view of the assumed stress hybrid method, *Internat. J. Numer. Methods Engrg.* **22**, 11–16.

BENDALI, A. (1984a), Numerical analysis of the exterior boundary value problem for the time-harmonic

Maxwell equations by a boundary finite element method, Part 1: The continuous problem, *Math. Comp.* **43**, 29–46.

BENDALI, A. (1984b), Numerical analysis of the exterior boundary value problem for the time-harmonic Maxwell equations by a boundary finite element method, Part 2: The discrete problem, *Math. Comp.* **43**, 47–68.

BERCOVIER, M. (1978), Pertubation of mixed variational problems. Application to mixed finite element methods, *RAIRO Anal. Numér.* **12**, 211–236.

BERNARDI, C. (1986), Contributions à l'Analyse Numérique de Problèmes Non Linéaires, Thèse d'Etat, Université Pierre et Marie Curie, Paris.

BERNARDI, C., C. CANUTO and Y. MADAY (1988), Generalized inf–sup condition for Chebyshev approximation of the Stokes problem, *SIAM J. Numer. Anal.* **25**, 1237–1271.

BERNARDI, C., Y. MADAY and B. METIVET (1987), Calcul de la pression dans la résolution spectrale du problème de Stokes, *Rech. Aérospat.* **1**, 1–21

BOSSAVIT, A. (1988), Mixed finite elements and the complex of Whitney forms, in: J.R. WHITEMAN, ed., *MAFELAP* 87, *The Mathematics of Finite Elements and Applications* (Academic Press, New York) 137–144.

BRAMBLE, J.H. (1981), The Lagrange multiplier method for Dirichlet's problem, *Math. Comp.* **37**, 1–11.

BRAMBLE, J.H. and R.S. FALK (1983), Two mixed finite element methods for the simply supported plate Problem, *RAIRO Anal. Numér.* **17**, 337–384.

BRAMBLE, J.H. and R.S. FALK (1985), A mixed-Lagrange multiplier finite element method for the polyharmonic equation, *RAIRO Modél. Math. Anal. Numér.* **19**, 519–557.

BRAMBLE, J.H. and E. PASCIAK (1988), A preconditioning technique for indefinite systems resulting from mixed approximations of elliptic problems, *Math. Comp.* **50**, 1–17.

BREZZI, F. (1974a), Théorèmes d'existence, d'unicité et d'approximation numérique pour des problèmes de point-selle, *C.R. Acad. Sci. Paris Sér. A* **278**, 839–842.

BREZZI, F. (1974b), On the existence, uniqueness and approximation of saddle-point problems arising from Lagrangian multipliers, *RAIRO Anal. Numér.* **8**, 129–151.

BREZZI, F. (1974c), Sur une méthode hybride pour l'approximation du problème de la torsion d'une barre élastique, *Istit. Lombardo (Rend. Sci.) A* **108**, 274–300.

BREZZI, F. (1975), Sur la méthode des éléments finis hybrides pour le problème biharmonique, *Numer. Math.* **24**, 103–131.

BREZZI, F. (1977), Hybrid method for fourth order elliptic equations, in: I. GALLIGANI and E. MAGENES, eds., *Mathematical Aspects of Finite Element Methods*, Lecture Notes in Mathematics **606** (Springer, Berlin) 35–46.

BREZZI, F. (1979), Non-standard finite elements for fourth order elliptic problems, in: R. GLOWINSKI, E.Y. RODIN and O.C. ZIENKIEWICZ, eds., *Energy Methods in Finite Element Analysis* (Wiley, Chichester) 193–211.

BREZZI, F., J. DOUGLAS Jr, R. DURAN and M. FORTIN (1987), Mixed finite elements for second order elliptic problems in three variables, *Numer. Math.* **51**, 237–250.

BREZZI, F., J. DOUGLAS Jr, M. FORTIN and L.D. MARINI (1987), Efficient rectangular mixed finite elements in two and three variables, *RAIRO Modél. Math. Anal. Numér.* **21**, 581–604.

BREZZI, F., J. DOUGLAS Jr, and L.D. MARINI (1985), Two families of mixed finite elements for second order elliptic problems, *Numer. Math.* **47**, 217–235.

BREZZI, F. and M. FORTIN (1986), Numerical approximation of Mindlin Reissner plates, *Math. Comp.* **47**, 151–158.

BREZZI, F. and L.D. MARINI (1975), On the numerical solution of plate bending problems by hybrid methods, *RAIRO Anal. Numér.* **9**, 5–50.

BREZZI, F., L.D. MARINI, A. QUARTERONI and P.A. RAVIART (1980), On an equilibrium finite element method for plate bending problems, *Calcolo* **17**, 271–291.

BREZZI, F. and P.-A. RAVIART (1978), Mixed finite element methods for 4th order elliptic equations, in: J. MILLER, ed., *Topics in Numerical Analysis* **III** (Academic Press, New York) 35–56.

BROWN, D.C. (1982), Alternating-direction iterative schemes for mixed finite element methods for second order elliptic problems, Thesis, University of Chicago, Chicago, IL.

CANUTO, C. (1978), Eigenvalue approximations by mixed methods, *RAIRO Anal. Numér.* **12**, 27–50.

CANUTO, C. (1981), A hybrid finite element method to compute the free vibration frequencies of a clamped plate, *RAIRO Anal. Numér*, **15**, 101–118.

CAREY, G.F., S.S. CHOW and M.K. SEAGER (1985), Approximate boundary–flux calculations, *Comput. Methods Appl. Mech. Engrg.* **50**, 107–120.

CHAVENT, G., G. COHEN, M. DUPUY, J. JAFFRE and I. RIBERA (1984), Simulation of two dimensional waterflooding using mixed finite elements, *Soc. Petroleum Engrs. J.* **24**, 382–390.

CHAVENT, G. and J. JAFFRE (1986), *Mathematical Models and Finite Elements for Reservoir Simulation*, Studies in Mathematics and Its Applications **17** (North-Holland, Amsterdam).

CHEUNG, Y.K. and CHEN WAJNI (1988), Isoparametric hybrid hexahedral elements for threedimensional stress analysis, *Internat. J. Numer. Methods Engrg.* **26**, 677–693.

CIARLET, P.G. (1974), Conforming and nonconforming finite element methods for solving the plate problem, in: G.A. WATSON, ed., *Numerical Solution of Differential Equations*, Lecture Notes in Mathematics **363** (Springer, Berlin) 156–176.

CIARLET, P.G. (1978), *The Finite Element Method for Elliptic Problems*, Studies in Mathematics and Its Applications **4** (North-Holland, Amsterdam).

CIARLET, P.G. (1982), *Introduction à l'Analyse Numérique Matricielle et à l'Optimisation* (Masson, Paris).

CIARLET, P.G. and P. DESTUYNDER (1979a), A justification of the two-dimensional linear plate model, *J. Mécanique* **18**, 315–344.

CIARLET, P.G. and P. DESTUYNDER (1979b), A justification of a nonlinear model in plate theory, *Comput. Methods Appl. Mech. Engrg.* **17**, 227–258.

CIARLET, P.G. and P.-A. RAVIART (1972), General Lagrange and Hermite interpolation in $\mathbb{R}^n$ with applications to finite element methods, *Comput. Methods Appl. Mech. Engrg.* **1**, 217–249.

CIARLET, P.G. and P.-A. RAVIART (1974), A mixed finite element method for the biharmonic equation, in C. DE BOOR, ed., *Mathematical Aspects of Finite Elements in Partial Differential Equations* (Academic Press, New York) 125–145.

CIAVALDINI, J.F. and J.-C. NEDELEC (1974), Sur l'élément de Fraeijs de Veubeke et Sander, *RAIRO Anal. Numér.* **8**, 29–45.

DARLOW, B., R.E. EWING and M.F. WHEELER (1984), Mixed finite element methods for miscible displacement problems in porous media, *Soc. Petroleum Engrs. J.* **24**, 391–398.

DAUTRAY, R. and J.-L. LIONS (1984), *Analyse Mathématique et Calcul Numérique pour les Sciences et les Techniques* **1**, Collection Commissariat à l'Energie Atomique (Masson, Paris).

DAY, M.L. and T.Y. YANG (1982), A mixed variational principle for finite element analysis, *Internat. J. Numer. Methods Engrg.* **18**, 1213–1230.

DESTUYNDER, P. (1986), *Une Théorie Asymptotique des Plaques Minces en Elasticité Linéaire*, Recherches en Mathématiques Appliquées **2** (Masson, Paris).

DESTUYNDER, P. and J.-C. NEDELEC (1986), Approximation numérique du cisaillement transverse dans les plaques minces en flexion, *Numer. Math.* **48**, 281–302.

DOUGLAS Jr, J. (1962), Alternating-direction methods for three space variables, *Numer. Math.* **4**, 41–63.

DOUGLAS Jr, J. (1984), Numerical methods for the flow of miscible fluids in porous media, in: R.W. LEWIS, P. BETTESS and E. HINTON, eds., *Numerical Methods in Coupled Systems* (Wiley, Chichester).

DOUGLAS Jr, J., R. DURAN and P. PIETRA (1986), Alternating-direction iteration for mixed finite element methods, in: R. GLOWINSKI and J.L. LIONS, eds., *Computing Methods in Applied Sciences and Engineering*, **VII** (North-Holland, Amsterdam).

DOUGLAS Jr, J., R. DURAN and P. PIETRA (1987), Formulation of alternating-direction iterative methods for mixed methods in three space, in: E.L. ORTIZ, ed., *Numerical Approximation of Partial Differential Equations*, Mathematics Studies **133** (North-Holland, Amsterdam) 21–30.

DOUGLAS Jr, J., R.E. EWING and M.F. WHEELER (1983), The approximation of the pressure by a mixed method in the simulation of miscible displacement, *RAIRO Anal. Numér.* **17**, 17–33.

DOUGLAS Jr, J. and C.P. GUPTA (1986), Superconvergence for a mixed finite element method for elastic wave propagation in a plane domain, *Numer. Math.* **49**, 189–202.

DOUGLAS Jr, J. and A. MILNER (1985), Interior and super convergence estimates for mixed methods for second order elliptic problems, *RAIRO Modél. Math. Anal. Numér.* **19**, 397–428.

DOUGLAS Jr, J. and P. PIETRA (1986), A description of some alternating-direction iterative techniques for mixed finite element methods, in: W.E. FITZGIBBON, ed., *Mathematical and Computational Methods in*

Seismic Exploration and Reservoir Modeling (SIAM, Philadelphia, PA) 37–53.

DOUGLAS Jr, J. and J.E. ROBERTS (1982), Mixed finite element methods for second order elliptic problems, *Math. Appl. Comput.* **1**, 91–103.

DOUGLAS Jr, J. and J.E. ROBERTS (1983), Numerical methods for a model for compressible miscible displacement in porous media, *Math. Comp.* **41**, 441–459.

DOUGLAS Jr, J. and J.E. ROBERTS (1985), Global estimates for mixed methods for second order elliptic problems, *Math. Comp.* **44**, 39–52.

DUPONT, T., R. GLOWINSKI, W. KINTON and M.F. WHEELER (1989), Mixed finite element methods for time dependent problems: Application to control, Research Rept. 54, Department of Mathematics, University of Houston, Houston, TX.

DUVAUT, G. and J.-L. LIONS (1972), *Les Inéquations en Mécanique et en Physique* (Dunod, Paris).

EWING, R.E. and M.F. WHEELER (1983), Computational aspects of mixed finite element methods, in: R.S. STEPLEMAN, ed., *Numerical Methods for Scientific Computing* (North-Holland, Amsterdam) 163–172.

EYMARD, R., T. GALLOUET and P. JOLY (1988), Hybrid finite element technics for oil recovery simulation, Publications du Laboratoire d'Analyse Numérique, Université Pierre et Marie Curie, Paris.

FALK, R.S. (1976), A Ritz method based on complementary variational principle, *RAIRO Anal. Numér.* **10**, 39–48.

FALK, R.S. (1978), Approximation of the biharmonic equation by a mixed finite element method, *SIAM J. Numer. Anal.* **15**, 556–567.

FALK, R.S. and J.E OSBORN (1980), Error estimates for mixed methods, *RAIRO Anal. Numér.* **14**, 249–277.

FELIPPA, C.A. (1986), Penalty-function iterative procedures for mixed finite element formulations, *Internat. J. Numer. Methods Engrg.* **22**, 267–279.

FIX, G.J. (1976), Hybrid finite element methods, *SIAM Rev.* **18**, 460–484.

FIX, G.J., M.D. GUNZBURGER and R.A. NICOLAIDES (1979), On finite element methods of least squares type, *Comput. Methods Appl. Mech. Engrg.* **10**, 175–198.

FIX, G.J., M.D. GUNZBURGER and R.A. NICOLAIDES (1981), On mixed finite element methods for first order elliptic systems, *Numer. Math.* **37**, 29–48.

FORTIN, M. (1977), Analysis of the convergence of mixed finite element methods, *RAIRO Anal. Numér.* **11**, 341–354.

FORTIN, M. (1985), A three-dimensional quadratic nonconforming element, *Numer. Math.* **46**, 269–279.

FORTIN, M. and R. GLOWINSKI (1983), *Résolution de Problèmes aux Limites par des Méthodes de Lagrangien Augmenté* (Dunod, Paris).

FORTIN, M. and R. GLOWINSKI (1983), *Augmented Lagrangian Methods: Application to the Numerical Solution of Boundary-Value Problems*, Studies in Mathematics and Its Applications **15** (North-Holland, Amsterdam).

FRAEIJS DE VEUBEKE, B. (1965), Displacement and equilibrium models in the finite element method, in: O.C. ZIENKIEWICZ and G.S. HOLISTER, eds., *Stress Analysis* (Wiley, New York) 145–197 (Chapter 9).

FRAEIJS DE VEUBEKE, B. (1973), Diffusive equilibrium models, University of Calgary Lecture Notes, International Research Seminar on the Theory and Applications of the Finite Element Methods, Calgary, Alta.

FRAEIJS DE VEUBEKE, B. (1974a), Finite element method in aerospace engineering problems in: R. GLOWINSKI and J.-L. LIONS, eds., *Computing Methods in Applied Sciences and Engineering*, Lecture Notes on Computer Science **16** (Springer, Berlin) 224–258.

FRAEIJS DE VEUBEKE, B. (1974b), Variational principles and the patch test, *Internat. J. Numer. Methods Engrg.* **8**, 783–801.

FRAEIJS DE VEUBEKE, B. (1975), Stress function approach, in: *World Congress on Finite Element Methods in Structural Mechanics* **1**, Bournemouth, England, J.1–J.51.

FRAEIJS DE VEUBEKE, B. and M.A. HOGGE (1972), Dual analysis for heat conduction problems by finite elements, *Internat. J. Numer. Methods Engrg.* **5**, 65–82.

FRAEIJS DE VEUBEKE, B. and G. SANDER (1968), An equilibrium model for plate bending, *Internat. J. Solids Structures* **4**, 447–468.

GASTALDI, L. and R. NOCHETTO (1987), Optimal L^∞-estimates for nonconforming and mixed finite element methods of lowest order, *Numer. Math.* **50**, 587–611.

GEVECI, T. (1988), On the application of mixed finite element methods to the wave equations, *RAIRO Modél. Math. Anal. Numér.* **22**, 243–250.

GIRAULT, V. and P.-A. RAVIART (1979), *Finite Element Approximation of the Navier–Stokes Equations*, Lecture Notes in Mathematics **749** (Springer, Berlin).

GIRAULT, V. and P.-A. RAVIART (1986), *Finite Element Methods for Navier–Stokes Equations: Theory and Algorithms*, Lecture Notes in Computational Mathematics **5** (Springer, Berlin).

GLOWINSKI, R. and M.F. WHEELER (1988), Domain decomposition and mixed finite element methods for elliptic problems, in: R. GLOWINSKI, G. GOLUB, G. MEURANT and J. PÉRIAUX, eds., *First International Symposium on Domain Decomposition Methods for Partial Differential Equations* (SIAM, Philadelphia, PA) 144–172.

HASLINGER, J. and I. HLAVACEK (1975), Curved element in a mixed finite element method, *Apl. Mat.* **20**, 233–252.

HASLINGER, J. and I. HLAVACEK (1976a), Convergence of an equilibrium finite element method based on the dual variational formulation, *Apl. Mat.* **21**, 43–65.

HASLINGER, J. and I. HLAVACEK (1976b), A mixed finite element method close to the equilibrium model, *Numer. Math.* **26**, 85–97.

HASLINGER, J. and P. NEITTAANMÄKI (1984), On different finite element methods for approximating the gradient of the solution to the Helmholtz equation, *Comput. Methods Appl. Mech. Engrg.* **42**, 131–148.

HECHT, F. (1981), Construction d'une base d'un élément fini P_1 non conforme à divergence nulle dans $\mathbb{R}^3$. *RAIRO Anal. Numér.* **15**, 119–150.

HELLAN K. (1967), Analysis of elastic plates in flexure by a simplified finite element method, *Acta Polytech. Scand. Civil Engrg. Ser.* **46**.

HENNART, J.P. (1985), Nodal schemes, mixed-hybrid finite elements and block-centered finite differences, Rapport de Recherche 386, INRIA Rocquencourt.

HENNART, J.P. (1986a), A general family of nodal schemes, *SIAM J. Sci. Statist. Comput.* **3**, 264–287.

HENNART, J.P. (1986b), A general finite element framework for nodal methods, in: J.R. WHITEMAN, ed., *MAFELAP 1984, The Mathematics of Finite Elements and Applications* (Academic Press, London) 309–316.

HENNART, J.P., J. JAFFRE and J.E. ROBERTS (1988), A constructive method for deriving finite elements of nodal type, *Numer. Math.* **53**, 701–738.

HERRMANN, L.R. (1967), Finite element bending analysis for plates, *J. Engrg. Mech. Div. ASCE* **93** (5), 13–26.

HERRMANN, L.R. (1983), Mixed Finite elements for couple-stress analysis, in: S.N. ATLURI, R.H. GALLAGHER and O.C. ZIENKIEWICZ, eds., *Hybrid and Mixed Finite Element Methods* (Wiley, Chichester) 1–17.

HLAVACEK, I. (1979), Convergence of an equilibrium finite element model for plane elastostatics, *Apl. Mat.* **24**, 427–457.

HLAVACEK, I. (1980), The density of solenoidal functions and the convergence of a dual finite element method, *Apl. Mat.* **25**, 39–55.

HLAVACEK, I. and M. KRIZEK (1984), Internal finite element approximations in the dual variational methods for second order elliptic problems with curved boundaries, *Apl. Mat.* **29**, 52–69.

HUGHES, T.J.R. (1987), *The Finite Element Method, Linear Static and Dynamic Finite Element Analysis* (Prentice-Hall, Englewood Cliffs, NJ).

IRONS, B.M. (1972), An assumed-stress version of the Wilson 8-nodes isoparametric brick, Computer Rept. CNME/56/, University of Wales, Swansea.

IRONS, B.M. (1984), Alternative ways for the formulation of hybrid stress elements, *Internat. J. Numer. Methods Engrg.* **20**, 780–782.

IRONS, B.M. and M. LOIKKANEN (1983), An engineers' defence of the patch test, *Internat. J. Numer. Methods Engrg.* **19**, 1391–1401.

IRONS, B.M. and A. RAZZAQUE (1972), Experience with the patch test for convergence of finite elements, in: A.K. AZIZ, ed., *The Mathematical Foundations of the Finite Element Method with Applications to Partial Differential Equations* (Academic Press, New York) 557–587.

JAFFRE, J. (1984), Elements finis mixtes et décentrage pour les équations de diffusions-convection, *Calcolo* **23**, 171–197.

JAFFRE, J. and J.E. ROBERTS (1985), Upstream weighting and mixed finite elements in the simulation of miscible displacements, *RAIRO Modél. Math. Anal. Numér.* **19**, 443–460.

JENSEN, C. (1979), A mixed finite element method with curved elements, Research Rept. 79.07 R, Department of Computer Sciences, Chalmers University of Technology and University of Göteborg, Sweden.

JIROUSEK, J. and LAN GUEX (1986), The hybrid-Trefftz finite element model and its application to plate bending, *Internat. J. Numer. Methods Engrg.* **23**, 651–693.

JOHNSON, C. (1973), On the convergence of the mixed finite element method for plate bending problems, *Numer. Math.* **21**, 43–62.

JOHNSON, C. and B. MERCIER (1978), Some equilibrium finite element methods for two-dimensional elasticity problems, *Numer. Math.* **30**, 103–116.

JOHNSON, C. and B. MERCIER (1979), Some equilibrium finite element methods for two-dimensional problems in continuum mechanics, in: R. GLOWINSKI, E.Y. RODIN and O.C. ZIENKIEWICZ, eds., *Energy Methods in Finite Element Analysis* (Wiley, Chichester) 213–224.

JOHNSON, C. and J. PITKÄRANTA (1982), Analysis of some finite element methods related to reduced integration, *Math. Comp.* **38**, 375–400.

JOHNSON, C. and V. THOMÉE (1981), Error estimates for some mixed finite element methods for parabolic type problems, *RAIRO Anal. Numér.* **15**, 41–78.

JOLY, P. (1982), La méthode des éléments finis mixtes appliquée au problème de diffusion-convection, Thèse de 3ème cycle, Université Pierre et Marie Curie, Paris.

JOLY, P. (1984), Résolution de systèmes linéaires non symétriques par des méthodes de gradient conjugué, Publications du Laboratoire d'Analyse Numérique, Université Pierre et Marie Curie, Paris.

JOLY, P. (1986), Présentation de synthèse des méthodes de gradient conjugué, *RAIRO Modél. Math. Anal. Numér.* **20**, 639–665.

KIKUCHI, F. (1973), Some considerations of the convergence of hybrid stress method, in: Y. YAMADA and R.H. GALLAGHER, eds., *Theory and Practice in Finite Element Structural Analysis* (University of Tokyo Press, Tokyo) 25–42.

KIKUCHI, F. (1983), On a mixed method related to the discrete Kirchhoff assumption, in: S.N. ATLURI, R.H. GALLAGHER and O.C. ZIENKIEWICZ, eds., *Hybrid and Mixed Finite Element Methods* (Wiley, Chichester) 137–154.

KIKUCHI, F. (1987), Mixed and penalty formulations for finite element analysis of an eigenvalue problem in electromagnetism, *Comput. Methods Appl. Mech. Engrg.* **64**, 509–521.

KRIZEK, M. (1982), An equilibrium finite element method in three-dimensional elasticity, *Apl. Mat.* **27**, 46–75.

KRIZEK, M. (1983), Conforming equilibrium finite element methods for some elliptic plane problems, *RAIRO Anal. Numér.* **17**, 35–65.

KRIZEK, M. and P. NEITTAANMÄKI (1986), Internal FE approximation of spaces of divergence-free functions in three-dimensional domains, *Internat. J. Numer. Methods Fluids* **6**, 811–817.

KWON, Y. and F.A. MILNER (1988), L^∞-error estimates for semilinear second-order elliptic equations, *SIAM J. Numer. Anal.* **25**, 46–53.

LADYZHENSKAYA, O.-A. (1969) *The Mathematical Theory of Viscous Flow* (Gordon and Breach, New York, 2nd ed.).

LASCAUX, P. and P. LESAINT (1975), Some nonconforming finite elements for the plate bending problem, *RAIRO Anal. Numér.* **9**, 9–53.

LEE, S.W. and J.J. RHIU (1986), A new approach to the formulation of mixed finite element models for structural analysis, *Internat. J. Numer. Methods Engrg.* **21**, 1629–1641.

LEE, S.W., S.C. WONG and J.J. RHIU (1985), Study of a nine node mixed formulation finite element for thin plates and shells, *Comput. Structures* **21**, 1325–1334.

LE ROUX, M.-N. (1982), A mixed finite element method for a weighted elliptic problem, *RAIRO Anal. Numér.* **16**, 243–273.

LESAINT, P. (1976), On the convergence of Wilson's nonconforming element for solving the elastic problem, *Comput. Methods Appl. Mech. Engrg.* **7**, 1–6.

LESAINT, P. and M. ZLÁMAL (1980), Convergence of the nonconforming Wilson element for arbitrary quadrilateral meshes, *Numer. Math.* **36**, 33–52.

LI, Z.C. and G.P. LIANG (1981), On the simplified hybrid-combined method, *Math. Comp.* **41**, 13–25.

LIONS, J.-L. and E. MAGENES (1968), *Problèmes aux Limites non Homogènes et Applications* **1** (Dunod, Paris).

MALKUS, D.S. and T.J.R. HUGHES (1978), Mixed finite element methods, reduced and selective integration techniques: A unification of concepts, *Comput. Methods Appl. Mech. Engrg.* **15**, 63–81.

MARINI, L.D. (1985), An inexpensive method for the evaluation of the solution of the lowest order Raviart–Thomas mixed method, *SIAM J. Numer. Anal.* **22**, 493–496.

MARSDEN, J.E. and T.J.R. HUGHES (1983), *Mathematical Foundations of Elasticity* (Prentice-Hall, Englewood Cliffs, NJ).

MERCIER, B. (1974), Numerical solution of the biharmonic problem by mixed finite elements of class $\mathscr{C}^0$, *Boll. Un. Mat. Ital.* **10**, 133–149.

MERCIER, B., J. OSBORN, J. RAPPAZ and P.-A. RAVIART (1981), Eigenvalue approximation by mixed and hybrid methods, *Math. Comp.* **36**, 427–453.

MIGNOT, A.L. and C. SURRY (1981), A mixed finite element family in plane elasticity, *Appl. Math. Modelling* **5**, 259–262.

MILNER, F. (1985), Mixed finite element methods for quasilinear second order elliptic problems, *Math. Comp.* **44**, 303–320.

MIRZA, F.A. and M.D. OLSON (1980), The mixed finite element method in plane elasticity, *Internat. J. Numer. Methods Engrg.* **15**, 273–289.

MIYOSHI, T. (1973), A finite element method for the solution of fourth order partial differential equations, *Kumamoto J. Sci. (Math.)* **9**, 87–116.

MIZUKAMI, A. (1986), A mixed finite element method for boundary flux computation, *Comput. Methods Appl. Mech. Engrg.* **57**, 239–243.

MIZUKAMI, A. and T.J.R. HUGHES (1985), A Petrov–Galerkin finite element method for convection-dominated flows: An accurate upwinding technique for satisfying the maximum principle, *Comput. Methods Appl. Mech. Engrg.* **50**, 181–193.

MONK, P. (1987), A mixed finite element method for the biharmonic equation, *SIAM J. Numer. Anal.* **24**, 737–749.

MORLEY, L.S.D. (1968), The triangular equilibrium element in the solution of the plate bending problems, *Aero. Quart.* **19**, 255–269.

NEDELEC, J.-C. (1980), Mixed finite elements in $\mathbb{R}^3$, *Numer. Math.* **35**, 315–341.

NEDELEC, J.-C. (1986), A new family of mixed finite element in $\mathbb{R}^3$, *Numer. Math.* **50**, 57–81.

NEITTAANMÄKI, P. and J. SARANEN (1981), On finite element approximation of the gradient for solution of Poisson equation, *Numer. Math.* **37**, 333–337.

NICOLAIDES, R.A. (1982), Existence, uniqueness and approximation for generalized saddle point problems, *SIAM J. Numer. Anal.* **19**, 349–357.

NOOR, A.K. and J.M. PETERS (1983), Mixed models and reduced selective integration displacement models for vibration analysis of shells, in: S.N. ATLURI, R.H. GALLAGHER and O.C. ZIENKIEWICZ, eds., *Hybrid and Mixed Finite Element Methods* (Wiley, Chichester) 537–564.

ODEN, J.T. (1973), Some contributions to the mathematical theory of mixed finite element approximations, in: Y. YAMADA and R.H. GALLAGHER, eds., *Theory and Practice in Finite Element Structural Analysis* (University of Tokyo Press, Tokyo) 3–23.

ODEN, J.T. and J.K. LEE (1975), Theory of mixed and hybrid finite element approximations in linear elasticity, in: *Proceedings IUTAM, IUM Symposium on Applications of Methods of Functional Analysis to Problems of Mechanics*, Marseille.

ODEN, J.T. and J.N. REDDY (1975), Some observations on properties of certain mixed finite element approximations, *Internat. J. Numer. Methods Engrg.* **9**, 933–949.

ODEN, J.T. and J.N. REDDY (1976a), On mixed finite element approximations, *SIAM J. Numer. Anal.* **13**, 392–404.

ODEN, J.T. and J.N. REDDY (1976b), *An Introduction to the Mathematical Theory of Finite Elements* (Wiley, New York).

PEACEMAN, D.W. and H.H. RACHFORD Jr (1955), The numerical solution of parabolic and elliptic differential equations, *J. SIAM* **3**, 28–41.

PIAN, T.H.H. (1964), Derivation of element stiffness matrices by assumed stress distributions, *AIAA J.* **2**, 1333–1336.

PIAN, T.H.H. (1971), Formulations of finite element methods for solid continua, in: R.H. GALLAGHER, Y. YAMADA and J.T. ODEN, eds., *Recent Advances in Matrix Methods of Structural Analysis and Design* (University of Alabama Press, Tuscaloosa, AL), 49–83.

PIAN, T.H.H. (1972), Finite element formulation by variational principles with relaxed continuity requirements, in: A.K. AZIZ, ed., *The Mathematical Foundations of the Finite Element Method with Applications to Partial Differential Equations* (Academic Press, New York) 671–687.

PIAN, T.H.H. (1982), On the equivalence between incompatible displacement element and hybrid stress element, *Appl. Math. Mech.* (*English Ed.*) **3** (6), 773–776.

PIAN, T.H.H. (1983), Reflections and remarks on hybrid and mixed finite element methods, in: S.N. ATLURI, R.H. GALLAGHER and O.C. ZIENKIEWICZ, eds., *Hybrid and Mixed Finite Element Methods* (Wiley, Chichester) 565–570.

PIAN, T.H.H. (1985), Finite element based on consistently assumed stresses and displacements, *Finite Elements Anal. Design* **1**, 131–140.

PIAN, T.H.H. and D.P. CHEN (1982), Alternative ways for formulation of hybrid stress elements, *Internat. J. Numer. Methods Engrg.* **18**, 1676–1684.

PIAN, T.H.H., D.P. CHEN and D. KANG (1983), A new formulation of hybrid-mixed finite elements, *Comput. Structures* **16**, 81–87.

PIAN, T.H.H., D. KANG and C. WANG (1987), Hybrid plate elements based on balanced stresses and displacements, in: T.J.R. HUGHES and E. HINTON, eds., *State-of-the-Art Texts on Finite Element Methods in Plate and Shell Structural Analysis* (to appear).

PIAN, T.H.H., M.-S. LI and D. KANG (1986), Hybrid stress elements based on natural isoparametric coordinates, in: *Proceedings Invitational China–American Workshop on Finite Element Methods*, Chengde, People's Republic of China.

PIAN, T.H.H. and K. SUMIHARA (1984), Rational approach for assumed stress finite elements, *Internat. J. Numer. Methods Engrg.* **22**, 173–181.

PIAN, T.H.H. and P. TONG (1969), Basis of finite element methods for solid continua, *Internat. J. Numer. Methods Engrg.* **1**, 3–28.

PIAN, T.H.H. and P. TONG (1986), Relations between incompatible displacement model and hybrid stress model, *Internat. J. Numer. Methods Engrg.* **22**, 173–181.

PINSKY, P.M. and R.V. JASTI (1989), A mixed finite element formulation for Reissner–Mindlin plates based on the use of bubble functions, *Internat. J. Numer. Methods Engrg.* **28**, 1677–1702.

PITKÄRANTA, J. (1979), Boundary subspaces for the finite element method with Lagrange multipliers, *Numer. Math.* **33**, 273–289.

PITKÄRANTA, J. (1980a), Local stability conditions for the Babuška method of Lagrange multipliers, *Math. Comp.* **35**, 1113–1129.

PITKÄRANTA, J. (1980b), A conforming finite element method with Lagrange multipliers for the biharmonic problem, *RAIRO Anal. Numér.* **14**, 309–324.

PITKÄRANTA, J. (1981), The finite element method with Lagrange multipliers for domains with corners, *Math. Comp.* **37**, 13–30.

PITKÄRANTA, J. and R. STENBERG (1983), Analysis of some mixed finite element methods for plane elasticity equations, *Math. Comp.* **41**, 399–423.

POWELL, M.J.D. (1969), A method for nonlinear constraints in minimization problems, in: R. FLETCHER, ed., *Optimization* (Academic Press, London) 283–298.

PUNCH, E.F. and S.N. ATLURI (1984), Development and testing of stable, invariant, isoparametric curvilinear 2- and 3-D hybrid-stress elements, *Comput. Methods Appl. Mech. Engrg.* **47**, 331–356.

QUARTERONI, A. (1979), Error estimates for the assumed stresses hybrid methods in the approximation of fourth order elliptic equations, *RAIRO Anal. Numér.* **13**, 355–367.

QUARTERONI, A. (1980a), On mixed methods for fourth order problems, *Comput. Methods Appl. Mech. Engrg.* **24**, 13–24.

QUARTERONI, A. (1980b), Mixed approximations of evolution problems, *Comput. Methods Appl. Mech. Engrg.* **24**, 137–163.

RABIER, P. and J.-M. THOMAS (1985), *Introduction à l'Analyse Numérique des Equations aux Dérivées Partielles, Exercices* (Masson, Paris).

RANNACHER, R. (1979), On nonconforming and mixed finite element methods for plate bending problems: The linear case, *RAIRO Anal. Numér.* **13**, 369–387.

RAVIART, P.-A. (1984), Mixed finite element methods, in: D.F. GRIFFITHS, ed., *The Mathematical Basis of Finite Element Methods, with Applications to Partial Differential Equations* (Clarendon Press, Oxford), 123–156.

RAVIART, P.-A. and J.-M. THOMAS (1977a), A mixed finite element method for second order elliptic problems, in: I. GALLIGANI and E. MAGENES, eds., *Mathematical Aspects of Finite Element Methods* Lecture Notes in Mathematics **606** (Springer, Berlin) 292–315.

RAVIART, P.-A. and J.-M. THOMAS (1977b), Primal hybrid finite element methods for second order elliptic equations, *Math. Comp.* **31**, 391–413.

RAVIART, P.-A. and J.-M. THOMAS (1979), Dual finite element models for second order elliptic problems, in: R. GLOWINSKI, E.Y. RODIN and O.C. ZIENKIEWICZ, eds., *Energy Methods in Finite Element Analysis* (Wiley, Chichester) 175–191.

RAVIART, P.-A. and J.-M. THOMAS (1983), *Introduction à l'Analyse Numérique des Equations aux Dérivées Partielles* (Masson, Paris).

REDDY, J.N. and J.T. ODEN (1973), Convergence of mixed finite-element approximations of a class of linear boundary-value problems, *J. Structural Mech.* **2**, 83–108.

REDDY, J.N. and J.T. ODEN (1975), Mixed finite element approximations of linear boundary-value problems, *Quart. Appl. Math.* **33**, 255–280.

REISSNER, E. (1984), On a certain mixed variational theorem and a proposed application, *Internat. J. Numer. Methods Engrg.* **20**, 1366–1368.

RUBINSTEIN, R., E. PUNCH and S.N. ATLURI (1983), An analysis of and remedies for, kinematic modes in hybrid stress finite elements: selection of stable, invariant stress fields, *Comput. Methods Appl. Mech. Engrg.* **38**, 63–92.

SAMUELSSON, A. (1979), The global constant strain condition and the patch test, in: R. GLOWINSKI, E.Y. RODIN and O.C. ZIENKIEWICZ, eds., *Energy Methods in Finite Element Analysis* (Wiley, Chichester) 47–58.

SANDER, G. and P. BECKERS (1977), The influence of the choice of connectors in finite element method, *Internat. J. Numer. Methods Engrg.* **11**, 1491–1505.

SARIGUL, N. and R.H. GALLAGHER (1989), Assumed stress function finite element method: two-dimensional elasticity, *Internat. J. Numer. Methods Engrg.* **28**, 1577–1598.

SCAPOLLA, T. (1980), A mixed finite element method for the biharmonic problem, *RAIRO Anal. Numér.* **14**, 55–79.

SCHOLZ, R. (1976), Approximation von Sattelpunkten mit finiten Elementen, *Tagungsband, Math. Schriften* **89**, 53–66.

SCHOLZ, R. (1977), L^∞-convergence of saddle-point approximations for second order problems, *RAIRO Anal. Numér.* **11**, 209–216.

SCHOLZ, R. (1978), A mixed method for fourth order problems using linear finite elements, *RAIRO Anal. Numér.* **12**, 85–90.

SCHOLZ, R. (1983), Optimal L_∞-estimates for a mixed finite element method for second order elliptic and parabolic problems, *Calcolo* **20**, 355–377.

SERRIN, J. (1959), Mathematical principles of classical fluid mechanics, in: S. FLÜGGE, ed., Handbuch der Physik **VIII/1**: *Strömungs Mechanik* **I** (C. TRUESDELL, coed.) (Springer, Berlin) 125–260.

SHIMODAIRA, H. (1985), Equivalence between mixed models and displacement models using reduced integration, *Internat. J. Numer. Methods Engrg.* **21**, 89–104.

SPILKER, R.L. (1982), Invariant 8-node hybrid-stress elements for thin and moderately thick plates, *Internat. J. Numer. Methods Engrg.* **18**, 1153–1178.

SPILKER, R. L. (1983), Hybrid-stress reduced Mindlin isoparametric elements for analysis of thin plates, *J. Structural Mech.* **11**, 49–66.

SPILKER, R. L. and T. BELYTSCHKO, (1983), Bilinear Mindlin plate elements in: S.N. ATLURI, R.H.

GALLAGHER and O.C. ZIENKIEWICZ, eds., *Hybrid and Mixed Finite Element Methods* (Wiley, Chichester) 117–136.

SPILKER, R. L., S.M. MASKERI and E. KANIA (1981), Plane isoparametric hybrid-stress elements, *Internat. J. Numer. Methods Engrg.* **17**, 1469–1496.

SPILKER, R. L. and N.I. MUNIR (1980a), The hybrid-stress model for thin plates, *Internat, J. Numer. Methods Engrg.* **15**, 1239–1260.

SPILKER, R. L. and N.I. MUNIR (1980b), A hybrid-stress quadratic serendipity displacement Mindlin plate bending element, *Comput. Structures* **12**, 11–21.

SPILKER, R. L. and N.I. MUNIR (1980c), A serendipity cubic-displacement hybrid-stress element for thin and moderately thick plates, *Internat. J. Numer. Methods Engrg.* **15**, 1261–1278.

SQUEFF, M.C. (1987), Superconvergence of mixed finite element methods for parabolic equations, *RAIRO Modél. Math. Anal. Numér.* **21**, 327–352.

STANLEY, G.M., K.C. PARK and T.J.R. HUGHES (1986), Treatment of large rotations of thin shells using mixed finite element methods, in: *First World Congress on Computational Mechanics*, Austin, TX.

STEIN, E. and R. AHMAD (1977), An equilibrium method for stress calculation using finite element displacement models, *Comput. Methods Appl. Mech. Engrg.* **10**, 175–198.

STENBERG, R. (1986), On the construction of optimal mixed finite element methods for the linear elasticity problem, *Numer. Math.* **48**, 447–462.

STENBERG, R. (1988a), A family of mixed finite elements for the elasticity problem, *Numer. Math.* **53**, 513–538.

STENBERG, R. (1988b), Two low-order mixed methods for the elasticity problem, in: J.R. WHITEMAN, ed., *MAFELAP 1987, The Mathematics of Finite Elements and Applications* **VI** (Academic Press, London) 271–280.

STENBERG, R. (1988c), Postprocessing schemes for some mixed finite elements, Rapport de Recherche 800 INRIA, Rocquencourt.

STOLARSKI, H. and T. BELYTSCHKO (1986a), On the equivalence of mode decomposition and mixed finite elements based on the Hellinger–Reissner principle, Part I: Theory, *Comput. Methods Appl. Mech. Engrg.* **58** 249–263.

STOLARSKI, H. and T. BELYTSCHKO (1986b), On the equivalence of mode decomposition and mixed finite elements based on the Hellinger–Reissner principle, Part II: Applications, *Comput. Methods Appl. Mech. Engrg.* **58**, 265–284.

STRANG, G. (1972), Variational crimes in the finite element method, in: A.K. AZIZ, ed., *The Mathematical Foundations of the Finite Element Method with Applications to Partial Differential Equations* (Academic Press, New York) 689–710.

STRANG, G. and G.J. FIX (1973), *An Analysis of the Finite Element Method* (Prentice-Hall, Englewood Cliffs, NJ).

STUMMEL, F. (1979), The generalized patch test, *SIAM J. Numer. Anal.* **16**, 449–471.

STUMMEL, F. (1980a), The limitations of the patch test, *Internat. J. Numer. Methods Engrg.* **15**, 177–188.

STUMMEL, F. (1980b), Basic compactness properties of nonconforming and hybrid finite element spaces, *RAIRO Anal. Numér.* **14**, 81–115.

TAYLOR, R.L., P.J. BERESFORD and E.L. WILSON (1976), A nonconforming element for stress analysis, *Internat. J. Numer. Methods Engrg.* **10**, 1211–1219.

TEMAM, R. (1979), *Navier–Stokes Equations*, Studies in Mathematics and Its Applications **2** (North-Holland, Amsterdam).

THOMAS, J.-M. (1976) Méthode des éléments finis hybrides duaux pour les problèmes du second ordre, *RAIRO Anal. Numér.* **10**, 51–79.

THOMAS, J.-M. (1977), Sur l'analyse numérique des méthodes d'eléments finis hybrides et mixtes, Thèse d'Etat, Université Pierre et Marie Curie, Paris.

THOMAS, J.-M. (1987), Mixed finite elements methods for convection-diffusion problems, in: E.L. ORTIZ, ed., *Numerical Approximation of Partial Differential Equations*, Mathematics Studies **133** (1989), (North- Holland, Amsterdam) 241–250.

THOMAS, J.-M. (1989), De la convergence des méthodes d'éléments finis mixtes, *Les Annales de l'Enit* **3**, 5–18.

THOMÉE, V. (1984), *Galerkin Finite Element Methods for Parabolic Problems*, Lecture Notes in Mathematics **1054** (Springer, Berlin).

TIAN, Z.S. and T.H.H. PIAN (1985), Axisymmetric solid elements by a rational hybrid stress method, *Comput. Structures* **20**, 141–149.

TONG, P. (1969), An assumed stress hybrid finite element method for incompressible materials, *Internat. J. Solids Structures* **5**, 455–461.

TONG, P. (1982), A family of hybrid plate elements, *Internat. J Numer. Methods Engrg.* **15**, 1771–1812.

TONG, P. and T.H.H. PIAN, (1969), A variational principle and the convergence on a finite element method based on assumed stress distribution, *Internat. J. Solids Structures* **5**, 463–472.

WHEELER, M.F and R. GONZALES (1984), Mixed finite element methods for petroleum reservoir engineering problems, in: R. GLOWINSKI and J.-L. LIONS, eds., *Computing Methods in Applied Sciences and Engineering* **VI** (North-Holland, Amsterdam) 639–657.

WILSON, E.L., R.L. TAYLOR, W.P. DOHERTY and J. GHABOUSSI (1973), Incompatible displacement models, in: S.J. FENVES et al., eds., *Numerical and Computer Methods in Structural Mechanics* (Academic Press, New York) 43–57.

WOLF, J.P. (1972a), Structural averaging of stresses in the hybrid stress model, *AIAA J.* **10**, 843–845.

WOLF, J.P. (1972b), Generalized hybrid stress finite element models, *AIAA J.* **11**, 386–388.

ZHONG-CI SHI (1984a), On the convergence properties of the quadrilateral elements of Sander and Beckers, *Math. Comp.* **42**, 493–504.

ZHONG-CI SHI (1984b), A convergence condition for the quadrilateral Wilson element, *Numer. Math.* **44**, 349–361.

ZHONG-CI SHI (1984c), An explicit analysis of the Stummel's patch test examples, *Internat. J. Numer. Methods Engrg.* **20**, 1233–1246.

ZIELINSKI, A.P. and O.C. ZIENKIEWICZ (1985), Generalized finite element analysis with T-complete boundary solutions functions, *Internat. J. Numer. Methods Engrg.* **21**, 509–528.

ZIENKIEWICZ, O.C., S. QU, R.L. TAYLOR and S. NAKAZAWA, (1986), The patch test for mixed formulations, *Internat. J. Numer. Methods Engrg.* **23**, 1873–1883.

ZIENKIEWICZ, O.C., LI XI-KUI and S. NAKASAWA (1984), Iterative solution of mixed problems and the stress recovery procedures, Research Rept. **476**, Institute for Numerical Methods in Engineering, University College of Swansea, Wales.

List of Some Special Symbols

Spaces

Norms and seminorms

Subject Index

Eigenvalue Problems

I. Babuška*

Institute for Physical Science and Technology
and Department of Mathematics
University of Maryland
College Park, MD 20742, USA

J. Osborn**

Department of Mathematics
University of Maryland
College Park, MD 20742, USA

*Partially supported by the Office of Naval Research under contract N00014-85-K-0169 and by the National Science Foundation under grant DMS-85-16191.

**Partially supported by the National Science Foundation under grant DMS-84-10324.

HANDBOOK OF NUMERICAL ANALYSIS, VOL. II
Finite Element Methods (Part 1)
Edited by P.G. Ciarlet and J.L. Lions

Contents

CHAPTER I

Introduction and Preliminaries

1. Examples of eigenvalue problems

In this section we present several model eigenvalue problems arising in physics and engineering. Specifically, we will discuss briefly some important physical interpretations of eigenvalues and eigenfunctions. Some of the model problems we discuss here will serve as illustrative examples in connection with the approximation methods considered in Chapter III. We will attempt to provide a clear understanding of the fundamental ideas, but will not present a detailed treatment. For a more complete discussion of the material in this section we refer to COURANT and HILBERT [1953].

1.1. One-dimensional problems

1.1.1. The longitudinal vibration of an elastic bar

We are interested in studying the small, longitudinal vibrations of a longitudinally loaded, elastically supported, elastic bar with masses attached to its ends. The bar is shown in Fig. 1.1.

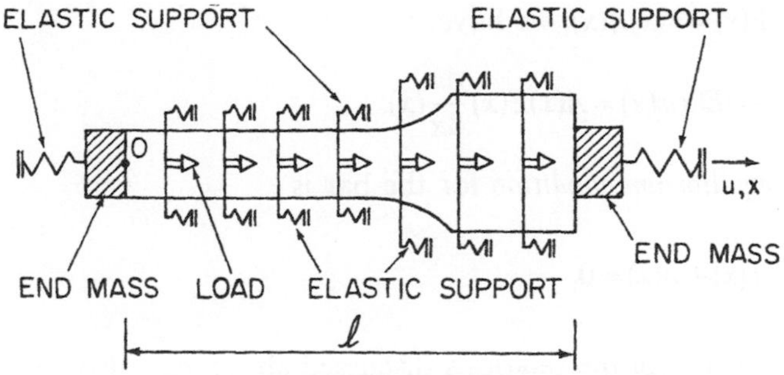

FIG. 1.1. Elastic bar.

We now derive the governing differential equation and boundary conditions for the problem. First we consider the *static problem*. Suppose

$f(x)$, $0<x<l$, represents the *external longitudinal load*, with positive $f(x)$ denoting a force directed to the right,

$u(x)$, $0<x<l$, denotes the *displacement* of the cross-section of the bar originally at x, with positive $u(x)$ denoting the displacement to the right, so that the position of a point originally at x is $x+u(x)$,

$\varepsilon(x)$, $0<x<l$, denotes the *strain in the x-direction*, i.e., the relative change in the length of the fibers in the bar ($\varepsilon(x)$ will be positive if it describes extension),

$\sigma(x)$, $0<x<l$, denotes the *normal stress* in the cross-section at x, i.e., the force per unit area exerted by the portion of the bar to the right of x on the portion to the left of x ($\sigma(x)$ will be positive if it describes tension),

$A(x)$, $0<x<l$, denotes the *area* of the cross-section at x,

$E(x)$, $0<x<l$, denotes the *modulus of elasticity* of the bar at x,

$F(x)$, $0<x<l$, denotes the *internal force* acting on the cross-section at x, i.e., the force exerted by the portion of the bar to the right of x on the portion to the left, with positive $F(x)$ denoting a force directed to the right,

$\rho(x)$, $0<x<l$, denotes the *load due to the (continuous) elastic support*, which is assumed to be of the form

$$\rho(x)=-c(x)u(x),$$

where $c(x)>0$ is the spring constant of the support (the negative sign indicates that the force is directed opposite to the displacement), and

$m(x)$, $0<x<l$, denotes the *specific mass* at x, i.e., the mass per unit volume at x.

The strain $\varepsilon(x)$ and the displacement $u(x)$ are related by

$$\varepsilon(x)=\frac{\mathrm{d}u}{\mathrm{d}x}(x).$$

This relation is valid for small displacements, i.e., when $|\varepsilon(x)|\ll 1$. The relation between stress and strain is described by the *constitutive law* of the material. We are assuming the linear relation given by *Hooke's Law*:

$$\sigma(x)=E(x)\varepsilon(x).$$

Thus, since $F(x)=\sigma(x)A(x)$, we have

$$F(x)=A(x)E(x)\varepsilon(x)=A(x)E(x)\frac{\mathrm{d}u}{\mathrm{d}x}(x).$$

Now the *equilibrium condition* for the bar is

$$\frac{\mathrm{d}F}{\mathrm{d}x}(x)+f(x)+\rho(x)=0,$$

which, with the use of the relations discussed above, can also be written as

$$\frac{\mathrm{d}}{\mathrm{d}x}\left(A(x)E(x)\frac{\mathrm{d}u}{\mathrm{d}x}(x)\right)+c(x)u(x)=f(x),\quad 0<x<l. \tag{1.1}$$

This is the governing differential equation.

We consider the three most important types of boundary conditions:

Dirichlet type:

$$u(0)=a_1, \qquad u(l)=a_2. \tag{1.2a}$$

Here the displacements of the end points of the bar are given.

Neumann type:

$$-F(0)=-\left(AE\frac{du}{dx}\right)(0)=b_1, \qquad F(l)=\left(AE\frac{du}{dx}\right)(l)=b_2. \tag{1.2b}$$

Here the forces at the ends of the bar are given. The different signs at 0 and l are used to express the outer normal derivative at the ends of the bar.

Newton type:

$$-\left(AE\frac{du}{dx}\right)(0)+\gamma_1 u(0)=c_1, \qquad \left(AE\frac{du}{dx}\right)(l)+\gamma_2 u(l)=c_2 \tag{1.2c}$$

where $\gamma_1, \gamma_2 > 0$.

Here γ_2 is the spring constant of a spring attached to the bar at $x=l$ and $-\gamma_2 u(l)$ is the force exerted on the right end of the bar by the spring. We are thus specifying the sum of the internal force and the spring force on the right end of the bar. The condition at $x=0$ has a similar interpretation.

Equation (1.1) together with one of (1.2a)–(c) determine the displacement $u(x)$ in the static case. We now turn to the *dynamic case*.

We assume the external load depends on the time t and is represented by $f(x,t)$ and suppose a_i, b_i, c_i in the boundary conditions depend on t:

$$a_i = a_i(t), \quad b_i = b_i(t), \quad c_i = c_i(t), \qquad i=1,2.$$

We further suppose the bar is subject to a damping force represented by R. If $u=u(x,t)$ is the displacement at time t, then from Newton's second law we have

$$\begin{aligned}&\frac{\partial}{\partial x}\left(A(x)E(x)\frac{\partial u}{\partial x}(x,t)\right)+c(x)u(x,t)\\ &\quad =f(x,t)-m(x)A(x)\frac{\partial^2 u}{\partial t^2}(x,t)-R, \qquad 0<x<l, \quad t>0.\end{aligned} \tag{1.3}$$

We next give the boundary conditions in the dynamic case. The Dirichlet conditions are nearly the same as in the static case, while the Neumann and Newton conditions require modification because of the forces exerted on the ends of the bar by the attached masses.

Dirichlet type:

$$u(0,t)=a_1(t), \quad u(l,t)=a_2(t), \quad t\geqslant 0. \tag{1.4a}$$

Neumann type:

$$\left(-AE\frac{\partial u}{\partial x}\right)(0,t)=-m_1\frac{\partial^2 u}{\partial t^2}(0,t)+b_1(t),$$
$$t\geqslant 0, \tag{1.4b}$$
$$\left(AE\frac{\partial u}{\partial x}\right)(l,t)=-m_2\frac{\partial^2 u}{\partial t^2}(l,t)+b_2(t),$$

where m_1 and m_2 are the masses attached to the left and right ends of the bar, respectively.

Newton type:

$$\left(-AE\frac{\partial u}{\partial x}\right)(0,t)+\gamma_1 u(0,t)=-m_1\frac{\partial^2 u}{\partial t^2}(0,t)+c_1(t),$$
$$t\geqslant 0. \tag{1.4c}$$
$$\left(AE\frac{\partial u}{\partial x}\right)(l,t)+\gamma_2 u(l,t)=-m_2\frac{\partial^2 u}{\partial t^2}(l,t)+c_2(t),$$

We remark that we can impose boundary conditions of different types at the two ends. For example, we could impose a Newton type condition at 0 and a Dirichlet type at l.

Finally in this (dynamic) case we need to impose initial conditions. We specify the initial position and velocity:

$$u(x,0)=\chi_1(x),\quad \frac{\partial u}{\partial t}(x,0)=\chi_2(x),\qquad 0\leqslant x\leqslant l. \tag{1.5}$$

Consider now equations (1.3), with $f=R=0$, and one of the conditions (1.4a)–(c), with $a_1=a_2=b_1=b_2=c_1=c_2=0$. If we seek *separated solutions* of the form

$$u(x,t)=v(x)w(t),$$

in which the spatial variable x and the temporal variable t are separated, from (1.3) we find that

$$\left[-\frac{d}{dx}\left(A(x)m(x)\frac{dv}{dx}(x)\right)+c(x)v(x)\right]w(t)=-m(x)A(x)v(x)\frac{d^2 w}{dt^2}(t)$$

or

$$\frac{-\dfrac{d}{dx}\left(A(x)E(x)\dfrac{dv}{dx}(x)\right)+c(x)v(x)}{m(x)A(x)v(x)}=\frac{-\dfrac{d^2 w}{dt^2}(t)}{w(t)}, \tag{1.6}$$

$$0<x<l,\quad t\geqslant 0.$$

Imposing the boundary conditions (1.4a)–(c) on $u=vw$ we find

$$v(0)w(t)=0, \quad v(l)w(t)=0, \qquad t\geqslant 0; \tag{1.7a}$$

$$\frac{-\left(AE\dfrac{\mathrm{d}v}{\mathrm{d}x}\right)(0)}{m_1 v(0)}=\frac{-\dfrac{\mathrm{d}^2 w}{\mathrm{d}t^2}(t)}{w(t)}, \quad \frac{\left(AE\dfrac{\mathrm{d}v}{\mathrm{d}x}\right)(l)}{m_2 v(l)}=\frac{-\dfrac{\mathrm{d}^2 w}{\mathrm{d}t^2}(t)}{w(t)}, \qquad t\geqslant 0; \tag{1.7b}$$

$$\left.\begin{aligned} &\frac{-\left(AE\dfrac{\mathrm{d}v}{\mathrm{d}x}\right)(0)+\gamma_1 v(0)}{m_1 v(0)}=\frac{-\dfrac{\mathrm{d}^2 w}{\mathrm{d}t^2}(t)}{w(t)},\\ &\frac{\left(AE\dfrac{\mathrm{d}v}{\mathrm{d}x}\right)(l)+\gamma_2 v(l)}{m_2 v(l)}=\frac{-\dfrac{\mathrm{d}^2 w}{\mathrm{d}t^2}(t)}{w(t)}, \end{aligned}\right\} \qquad t\geqslant 0. \tag{1.7c}$$

It is immediate that both sides of equation (1.6) equal a constant, which we denote by λ. We are thus led to seek a number λ and a function $v(x)\neq 0$ so that

$$-\frac{\mathrm{d}}{\mathrm{d}x}\left(A(x)E(x)\frac{\mathrm{d}v}{\mathrm{d}x}(x)\right)+c(x)v(x)=\lambda m(x)A(x)v(x), \quad 0<x<l. \tag{1.8}$$

From (1.7a)–(c) we get boundary conditions for v:

Dirichlet type:

$$v(0)=v(l)=0. \tag{1.9a}$$

Neumann type:

$$-\left(AE\frac{\mathrm{d}v}{\mathrm{d}x}\right)(0)=\lambda m_1 v(0), \qquad \left(AE\frac{\mathrm{d}v}{\mathrm{d}x}\right)(l)=\lambda m_2 v(l). \tag{1.9b}$$

Newton type:

$$-\left(AE\frac{\mathrm{d}v}{\mathrm{d}x}\right)(0)+\gamma_1 v(0)=\lambda m_1 v(0), \qquad \left(AE\frac{\mathrm{d}v}{\mathrm{d}x}\right)(l)+\gamma_2 v(l)=\lambda m_2 v(l). \tag{1.9c}$$

The problem of finding λ and $v(x)\neq 0$ satisfying (1.8) and a boundary condition (1.9) of Dirichlet, Neumann, or Newton type is called an *eigenvalue problem*. λ is called an *eigenvalue* and $v(x)$ a corresponding *eigenfunction*, or *eigenvector*, of the problem, and (λ, v) is often called an *eigenpair*. If λ is present in one or both of the boundary conditions, the problem is referred to as a *Steklov-type* eigenvalue problem. Since in the general theory of eigenvalue problems it is necessary to consider complex eigenfunctions and eigenvalues, we will use complex notation here. We note, however, that for the specific class of problems treated in this section, this generality is not necessary.

For the sake of definiteness, let us suppose we have a Newton type boundary condition at 0 and a Dirichlet type at l, and further assume that $m_1 = 0$. Thus we are considering the initial boundary value problem

$$-\frac{\partial}{\partial x}\left(AE\frac{\partial u}{\partial x}\right)+cu=-mA\frac{\partial^2 u}{\partial t^2}, \qquad 0<x<l, \quad t>0; \tag{1.3'}$$

$$-\left(AE\frac{\partial u}{\partial x}\right)(0,t)+\gamma_1 u(0,t)=0; \tag{1.4c'}$$

$$u(l,t)=0, \quad t\geqslant 0; \tag{1.4a'}$$

$$u(x,0)=\chi_1(x), \quad \frac{\partial u}{\partial t}(x,0)=\chi_2(x), \qquad 0<x<l. \tag{1.5'}$$

The corresponding eigenvalue problem is

$$\frac{\mathrm{d}}{\mathrm{d}x}\left(AE\frac{\mathrm{d}u}{\mathrm{d}x}\right)+cv=\lambda mAv, \quad 0<x<l,$$
$$-\left(AE\frac{\mathrm{d}u}{\mathrm{d}x}\right)(0)+\gamma_1 v(0)=0, \qquad v(l)=0. \tag{1.10}$$

It is known that problems of this type have a sequence of eigenvalues

$$0<\lambda_1\leqslant\lambda_2\leqslant\cdots\uparrow+\infty \tag{1.11}$$

and corresponding eigenfunctions

$$v_1(x),\ v_2(x),\ldots. \tag{1.12}$$

The eigenfunctions satisfy

$$\int_0^l m(x)A(x)v_i(x)\bar{v}_j(x)\,\mathrm{d}x=\delta_{ij}, \tag{1.13}$$

where δ_{ij} is the Kronecker symbol, i.e., they are orthonormal; in addition they are complete in L_2, i.e., any function $h(x)\in L_2$ can be written as

$$h(x)=\sum_{j=1}^{\infty} c_j v_j(x), \tag{1.14}$$

where

$$c_j=\int_0^l mAh\bar{v}_j\,\mathrm{d}x \tag{1.15}$$

and the convergence is in the L_2 norm. Regarding (1.11)–(1.15), see (4.10)–(4.14).

Corresponding to each λ_j we solve

$$\frac{d^2w}{dt^2}(t)+\lambda_j w(t)=0, \quad t>0 \tag{1.16}$$

(cf. (1.6)), obtaining

$$w(t)=w_j(t)=a_j \sin\sqrt{\lambda_j}(t+\theta_j),$$

where a_j and θ_j are arbitrary. Thus the separated solutions are given by

$$a_j v_j(x) \sin\sqrt{\lambda_j}(t+\theta_j), \quad j=1,2,\ldots. \tag{1.17}$$

It is immediate that

$$u(x,t)=\sum_{j=1}^{\infty} a_j v_j(x) \sin\sqrt{\lambda_j}(t+\theta_j) \tag{1.18}$$

is a solution of (1.3′), (1.4c′), (1.4a′) for arbitrary a_j and θ_j, provided the series converges appropriately. It remains to satisfy the initial conditions (1.5′). For this, a_j and θ_j must satisfy

$$u(x,0)=\sum_j a_j \sin\sqrt{\lambda_j}\theta_j v_j(x)=\chi_1(x),$$

$$\frac{\partial u}{\partial t}(x,0)=\sum_j a_j\sqrt{\lambda_j} \cos\sqrt{\lambda_j}\theta_j v_j(x)=\chi_2(x).$$

From the complete orthonormality of the $v_j(x)$ we see that these two equations uniquely determine a_j and θ_j. Thus (1.18), with this choice for a_j and θ_j, is the unique solution of (1.3′), (1.4c′), (1.4a′), (1.5′).

The simple motions given in (1.17) are called the eigenvibrations of (1.3′), (1.4c′), (1.4a′). All the points x of the jth eigenvibrations vibrate with the same (circular) frequency (defined to be the number of vibrations per 2π seconds) and phase displacement $\sqrt{\lambda_j}\theta_j$ and the point x vibrates with amplitude proportional to $v_j(x)$. Thus $\sqrt{\lambda_j}$ is the frequency with which the jth eigenvibration vibrates and $v_j(x)$ gives the basic shape of the eigenvibration. The amplitude factor a_j and θ_j are determined by the initial position and velocity of the eigenvibration, whereas λ_j and $v_j(x)$ are determined by the physical process itself, as represented by (1.3′), (1.4c′), and (1.4a′). We have seen that any motion of (1.3′), (1.4c′), (1.4a′) can be written as a sum or superposition of eigenvibrations.

So far we have been dealing with free vibrations, i.e., we have assumed $f(x,t)$ and R in (1.3) are zero. Now we briefly consider the case when $f\neq 0$ and $R=0$, i.e., the case of forced vibrations. If we write

$$f(x,t)=\sum_{j=1}^{\infty} f_j(t)v_j(x)m(x)A(x),$$

then we easily see that $u(x,t)=\sum_{j=1}^{\infty} a_j(t)v_j(x)$ is a solution if

$$a_j''(t)+\lambda_j a_j(t)=f_j(t).$$

If, now, $f_j(t)=\sin\sqrt{\lambda_j}(t+\theta_j)$, then we see that $a_j(t)$, and hence $u(x,t)$, will be unbounded as $t\to\infty$. This phenomenon is called *resonance* and f is called a *resonant load*; the *resonant frequencies* are $\sqrt{\lambda_j}$, $j=1,2,\ldots$.

The damping term R could be defined in various ways. For example, we could take R to be $\mu(\partial u/\partial t)$, for a constant μ, which would lead to a term of the form $\mu(\partial u/\partial t)$ in equation (1.3).

Eigenvalue problems similar to (1.8) and (1.9) or (1.10) arise in a number of other situations. We now briefly mention some of them.

1.1.2. *The transverse vibration of a string*

We are interested here in the small, transverse vibration of a homogeneous string that is stretched between two points a distance l apart. Gravity is assumed to be negligible and the particles of the string are assumed to move in a plane. We denote the density of the string by r and the tension by p. We restrict our attention to the case of free vibrations.

If the particles of the string are identified with the numbers $0\leqslant x\leqslant l$ and if $u(x,t)$ denotes the vertical displacement of the particle x at time t, then u satisfies

$$\begin{aligned}&-p\frac{\partial^2 u(x,t)}{\partial x^2}=-r\frac{\partial^2 u(x,t)}{\partial t^2}, \qquad 0<x<l, \quad t>0,\\ &u(0,t)=u(l,t)=0, \quad t\geqslant 0.\end{aligned} \tag{1.19}$$

We see that (1.19) is a very special case of (1.3) and (1.4a). The associated eigenvalue problem is

$$\begin{aligned}&-C^2v''(x)=\lambda v(x), \quad 0<x<l,\\ &v(0)=v(l)=0,\end{aligned} \tag{1.20}$$

where $C^2=p/r$. It is easily seen that the eigenvalues and eigenfunctions of (1.20) can be given explicitly; they are

$$\lambda_k=(k^2C^2\pi^2)/l^2 \tag{1.21}$$

and

$$v_k(x)=\sqrt{2/l}\,\sin(k\pi x/l), \quad k=1,2,\ldots . \tag{1.22}$$

The entire discussion of the elastic bar—i.e., the discussion of separation of variables, of eigenvalues and eigenfunctions, and of eigenvibrations—applies to this problem. We note that it is possible to find the eigenvalues and eigenfunctions explicitly only in very special situations, roughly, just in the case of eigenvalue problems for differential equations with constant coefficients in one dimension. In general, one must resort to approximation methods. The discussion of such methods is the main topic of this article.

1.1.3. Characterization of the optimal constant in the Poincaré inequality

The Poincaré inequality states that there is a constant C such that

$$\int_0^l [u(x)]^2\,\mathrm{d}x \leqslant C \int_0^l [u'(x)]^2\,\mathrm{d}x \tag{1.23}$$

for all functions $u(x)$ having a square integrable first derivative and vanishing at 0 and l. Let us consider the problem of finding the minimal constant C. We are thus interested in

$$C = \sup_{\{u:u(0)=u(l)=0\}} \frac{\int_0^l u^2\,\mathrm{d}x}{\int_0^l (u')^2\,\mathrm{d}x}. \tag{1.24}$$

Using the elementary methods of the calculus of variations we find that the function u achieving the supremum in (1.24) satisfies

$$C\int_0^l u'v'\,\mathrm{d}x = \int_0^l uv\,\mathrm{d}x$$

for all v having square integrable first derivatives and vanishing at 0 and l. By integration by parts we then find

$$-u'' = (1/C)u, \quad 0<x<l, \qquad u(0)=u(l)=0. \tag{1.25}$$

Thus $1/C$ is the lowest eigenvalue of the eigenvalue problem (1.25), and the optimal u in (1.24) (which achieves equality in (1.21)) is an associated eigenfunction.

1.2. Higher-dimensional problems

1.2.1. The vibrating membrane

Consider the small, transverse vibration of a thin membrane stretched over a bounded region Ω in the plane and fixed along its edges $\Gamma = \partial\Omega$. The vertical displacement $u(x, y, t)$ of the point (x, y) in Ω at time t satisfies

$$\begin{aligned} -\Delta u = -\frac{\partial^2 u}{\partial x^2} - \frac{\partial^2 u}{\partial y^2} &= -\frac{\partial^2 u}{\partial t^2}, \quad (x,y)\in\Omega, \quad t>0, \\ u(x,y,t) &= 0, \quad (x,y)\in\partial\Omega, \qquad t\geqslant 0. \end{aligned} \tag{1.26}$$

As with the vibrating elastic bar or the vibrating string, if we seek separated solutions of the form $u(x, y, t) = v(x, y)w(t)$, we are led to the eigenvalue problem of finding λ and

$v(x, y) \neq 0$ satisfying

$$\begin{aligned} -\Delta v &= \lambda v, \quad (x, y) \in \Omega \\ v(x, y) &= 0, \quad (x, y) \in \partial\Omega, \end{aligned} \tag{1.27}$$

and for each eigenpair (λ, v) of (1.27), to the differential equation

$$\frac{d^2 w}{dt^2}(t) + \lambda w(t) = 0, \quad t \geqslant 0, \tag{1.28}$$

for $w(t)$ (cf. (1.16)).

It is known that (1.27) has an infinite sequence of eigenvalues

$$0 < \lambda_1 \leqslant \lambda_2 \leqslant \cdots \uparrow +\infty$$

and corresponding eigenfunctions

$$v_1(x, y), v_2(x, y), \ldots .$$

The eigenfunctions are complete and orthonormal in $L^2(\Omega)$. $a_j v_j(x, y) \sin \sqrt{\lambda_j}(t + \theta_j)$, $j = 1, 2, \ldots$, are called eigenvibrations. $\sqrt{\lambda_j}$ is the frequency and $v_j(x, y)$ is the shape of the jth eigenvibration. All solutions of (1.26) can be obtained as a superposition of eigenvibrations (cf. (1.18)). We note that if, instead of fixing the membrane on Γ, we allowed it to move freely in the vertical direction, then we should have the Neumann boundary condition $\partial u/\partial n = 0$, where $\partial/\partial n$ denotes the outer normal derivative, instead of the Dirichlet condition $u = 0$. The approximation of the eigenpairs of a membrane is discussed in Sections 10.2, 11.2 and 12.

1.2.2. The problem of heat conduction

Consider the problem of heat conduction in a body occupying a region Ω in three-dimensional space. We suppose the temperature distribution throughout Ω is known at time zero, the temperature is held at zero on $\partial\Omega$ for all time, and that we want to determine the temperature $u(x, y, z, t)$ at the point $(x, y, z) \in \Omega$ at time $t > 0$. From the fundamental law of heat conduction we know that

$$\begin{aligned} &-\frac{\partial}{\partial x}\left(p(x, y, z)\frac{\partial u}{\partial x}\right) - \frac{\partial}{\partial y}\left(p(x, y, z)\frac{\partial u}{\partial y}\right) - \frac{\partial}{\partial z}\left(p(x, y, z)\frac{\partial u}{\partial z}\right) \\ &\quad = -r(x, y, z)\frac{\partial u}{\partial t}, \qquad (x, y, z) \in \Omega, \quad t > 0, \end{aligned} \tag{1.29}$$

$$\begin{aligned} &u(x, y, z, t) = 0 \qquad u(x, y, z) \in \partial\Omega, \quad t \geqslant 0, \\ &u(x, y, z, 0) = f(x, y, z), \quad (x, y, z) \in \Omega, \end{aligned}$$

where

$f(x, y, z)=$ the temperature distribution at $t=0$,

$p(x, y, z)=$ the thermal conductivity of the material at (x, y, z),

$r(x, y, z)=$ density of the material times the specific heat of the material.

If we seek separated solutions

$$u(x, y, z, t)=v(x, y, z)w(t)$$

of the differential equation and the boundary conditions in (1.29) we are led to the eigenvalue problem

$$\begin{aligned} &-\frac{\partial}{\partial x}\left(p\frac{\partial v}{\partial x}\right)-\frac{\partial}{\partial y}\left(p\frac{\partial v}{\partial y}\right)-\frac{\partial}{\partial z}\left(p\frac{\partial v}{\partial z}\right)=\lambda rv, \quad (x, y, z)\in\Omega, \\ &v(x, y, z)=0, \quad (x, y, z)\in\partial\Omega, \end{aligned} \tag{1.30}$$

and for each eigenpair (λ, v) of (1.30) we are led to the equation

$$w'+\lambda w=0, \quad t>0, \tag{1.31}$$

for $w(t)$ (cf. (1.16) and (1.28)). Problem (1.30) has eigenvalues

$$0<\lambda_1\leqslant\lambda_2\leqslant\cdots\uparrow\infty$$

and eigenfunctions

$$v_1, v_2, \ldots$$

satisfying

$$\int_\Omega v_i\bar{v}_j r\,\mathrm{d}x\,\mathrm{d}y\,\mathrm{d}z=\delta_{ij}.$$

Corresponding to each λ_j, from (1.31) we find $w(t)=w_j(t)=a_j\mathrm{e}^{-\lambda_j t}$. Thus the separated solutions are given by

$$a_j v_j(x, y, z)\mathrm{e}^{-\lambda_j t}, \quad j=1, 2, \ldots,$$

and the solution of (1.29) is

$$u(x, y, z, t)=\sum_{j=1}^{\infty}\left(\int_\Omega f\bar{v}_j r\,\mathrm{d}x\,\mathrm{d}y\,\mathrm{d}z\right)v_j(x, y, z)\mathrm{e}^{-\lambda_j t} \tag{1.32}$$

(cf. (1.18)). We note that from (1.32) and the positivity of the eigenvalues, one can show that $\lim_{t\to\infty} u(x, y, z, t)=0$ and that the rate at which the temperature u decays to zero is largely determined by λ_1.

1.2.3. The vibration of an elastic solid

The vibration of an elastic solid Ω, the three-dimensional generalization of the elastic bar, is governed by the Navier–Lamé equations

$$(\lambda+\mu)\frac{\partial\theta}{\partial x}+\mu\Delta u=-X+\rho\frac{\partial^2 u}{\partial t^2},$$

$$(\lambda+\mu)\frac{\partial\theta}{\partial y}+\mu\Delta v=-Y+\rho\frac{\partial^2 v}{\partial t^2},\qquad (x,y,z)\in\Omega,\quad t>0, \tag{1.33}$$

$$(\lambda+\mu)\frac{\partial\theta}{\partial z}+\mu\Delta w=-Z+\rho\frac{\partial^2 w}{\partial t^2},$$

where $u(x,y,z,t)$, $v(x,y,z,t)$, and $w(x,y,z,t)$ are the x-, y- and z-components of the displacement of the point $(x,y,z)\in\Omega$ at time t, $\theta=\partial u/\partial x+\partial v/\partial y+\partial w/\partial z$, X, Y, and Z are the components of the external force per unit volume acting at (x,y,z), $\lambda>0$, and $\mu>0$ are the Lamé elastic constants, and ρ is the density of the material.

As in the case of the bar, boundary conditions of various types may be prescribed. For example, the Dirichlet boundary conditions prescribe the values of u, v and w on $\Gamma=\partial\Omega$. Neumann conditions are more complicated. Let n be the unit outer normal to Γ, let n_x, n_y and n_z be the x-, y- and z-components of n, and let

$$\frac{\partial}{\partial n}=n_x\frac{\partial}{\partial x}+n_y\frac{\partial}{\partial y}+n_z\frac{\partial}{\partial z}$$

be the outer normal derivative. Then define

$$X_n=\lambda\theta n_x+\mu\frac{\partial u}{\partial n}+\mu\left[\frac{\partial u}{\partial x}n_x+\frac{\partial v}{\partial x}n_y+\frac{\partial w}{\partial x}n_z\right], \tag{1.34a}$$

$$Y_n=\lambda\theta n_y+\mu\frac{\partial v}{\partial n}+\mu\left[\frac{\partial u}{\partial y}n_x+\frac{\partial v}{\partial y}n_y+\frac{\partial w}{\partial y}n_z\right], \tag{1.34b}$$

$$Z_n=\lambda\theta n_z+\mu\frac{\partial w}{\partial n}+\mu\left[\frac{\partial u}{\partial z}n_x+\frac{\partial v}{\partial z}n_y+\frac{\partial w}{\partial z}n_z\right]. \tag{1.34c}$$

The Neumann conditions then consist in prescribing X_n, Y_n and Z_n on the boundary. One can also mix the boundary conditions in various ways, e.g., impose Dirichlet conditions on one part of the boundary and Neumann conditions on the remainder of the boundary or prescribe X_n, Y_n and w on Γ.

The eigenvalue problem associated with (1.33) is given by

$$-(\lambda+\mu)\frac{\partial\theta}{\partial x}-\mu\Delta u=\omega\rho u,$$

$$-(\lambda+\mu)\frac{\partial\theta}{\partial y}-\mu\Delta v=\omega\rho v,\qquad (x,y,z)\in\Omega, \tag{1.35}$$

$$-(\lambda+\mu)\frac{\partial\theta}{\partial z}-\mu\Delta w=\omega\rho w,$$

where we have denoted the eigenvalue parameter by ω (to avoid confusion with the

Lamé constants μ and λ), and where u, v, w and θ denote functions of x, y and z only, i.e., the separation of variables has been written as $u(x, y, z, t)=u(x, y, z)T(t)$, etc. For boundary conditions we can consider any of those mentioned above. If we consider Dirichlet conditions ($u=v=w=0$ on Γ) we refer to the clamped solid and if we consider Neumann conditions ($X_n=Y_n=Z_n=0$ on Γ) we refer to the free solid.

The approximation of the eigenvalues of the free L-shaped panel (a two-dimensional analogue of the elastic solid) is treated in detail in Section 10.1.

1.2.4. The Steklov eigenvalue problem

The Steklov eigenvalues of the differential operator $-\Delta+I$ are those numbers λ such that for some nonzero u,

$$-\Delta u+u=0 \quad \text{in } \Omega,$$
$$\partial u/\partial n=\lambda u \quad \text{on } \Gamma=\partial\Omega.$$

Problems of this type, in which the eigenvalue parameter appears in the boundary condition, arise in a number of applications (cf. (1.9b) and (1.9c)).

1.2.5. The problem of stability of a nonlinear problem

Consider the quasilinear parabolic problem

$$\frac{\partial u}{\partial t}-\Delta u+u\frac{\partial u}{\partial x}=0, \quad (x, y)\in\Omega, \quad t>0,$$

$$u(x, y, t)=\varphi(x, y), \quad (x, y)\in\partial\Omega, \quad t>0.$$

Suppose $\tilde{u}(x, y)$ is a stationary solution, i.e., suppose

$$-\Delta\tilde{u}+\tilde{u}\frac{\partial\tilde{u}}{\partial x}=0, \quad (x, y)\in\Omega,$$

$$\tilde{u}(x, y)=\varphi(x, y), \quad (x, y)\in\partial\Omega.$$

Then we consider a nearby time-dependent solution

$$u(x, y, t)=\tilde{u}(x, y)+w(x, y, t)$$

and ask whether $\tilde{u}$ is a *stable stationary solution*, i.e., whether

$$\lim_{t\to\infty} u(x, y, t)=\tilde{u}(x, y)$$

or, equivalently,

$$\lim_{t\to\infty} w(x, y, t)=0.$$

We easily see that w satisfies

$$\begin{aligned}&\frac{\partial w}{\partial t}+Lw+Nw=0, \quad (x, y)\in\Omega, \quad t>0\\&w=0, \quad (x, y)\in\partial\Omega,\end{aligned} \tag{1.36}$$

where

$$Lw = -\Delta w + \tilde{u}\frac{\partial w}{\partial x} + \frac{\partial \tilde{u}}{\partial x} w$$

and

$$Nw = w\frac{\partial w}{\partial x}.$$

Conditions ensuring $w \to 0$ as $t \to \infty$ can be given in terms of the eigenvalues of

$$Lw = \lambda w \quad \text{in } \Omega,$$
$$w = 0 \quad \text{on } \partial\Omega.$$

In fact, if all the eigenvalues of this problem have positive real parts, then $\tilde{u}$ is asymptotically stable in the L_2 norm, i.e., there is a constant $\delta > 0$ such that if

$$\| w(\cdot, \cdot, 0) \|_{L_2(\Omega)} < \delta,$$

then

$$\| w(\cdot, \cdot, t) \|_{L_2(\Omega)} \to 0 \quad \text{as } t \to \infty.$$

If the term N in (1.36) is neglected, then this result is similar to that mentioned at the end of the discussion of heat conduction. Note that L is a nonselfadjoint operator and its eigenvalues will, in general, be complex (cf. Section 3). For further details on this type of stability results see PRODI [1962].

2. Sobolev spaces

The natural setting for a discussion of eigenvalue problems and their approximation is the theory of linear operators on a Hilbert space. In this section we will sketch the definitions and basic properties of the function spaces we will make use of. These are mainly the Sobolev and Besov spaces.

Let Ω be a bounded open subset of $\mathbb{R}^n$ and denote by $x = (x_1, \ldots, x_n)$ a point in $\mathbb{R}^n$. For each integer $m \geqslant 0$, the real (complex) Sobolev space $H^m(\Omega)$ is defined by

$$H^m = H^m(\Omega) = \{u: \partial^\alpha u \in L_2(\Omega) \forall |\alpha| \leqslant m\}, \tag{2.1}$$

where $L_2(\Omega)$ denotes the usual space of real- (complex-) valued square integrable functions on Ω equipped with the inner product

$$(u, v) = (u, v)_{L_2(\Omega)} = \int_\Omega u\bar{v}\, dx \tag{2.2}$$

and norm

$$\|u\| = \|u\|_{L_2(\Omega)} = \left(\int_\Omega |u|^2\, dx\right)^{1/2} \tag{2.3}$$

On $H^m(\Omega)$ we have the inner product

$$((u, v))_m = ((u, v))_{m,\Omega} = \sum_{|\alpha| \leqslant m} \int_\Omega \partial^\alpha u \overline{\partial^\alpha v}\, dx \tag{2.4}$$

and norm

$$\|u\|_{H^m(\Omega)} = \|u\|_m = \|u\|_{m,\Omega} = \Big(\sum_{|\alpha| \leqslant m} \int_\Omega |\partial^\alpha u|^2 \, dx \Big)^{1/2}. \tag{2.5}$$

With this inner product, $H^m(\Omega)$ is a Hilbert space. Here $\alpha = (\alpha_1, \ldots, \alpha_n)$, with α_i a nonnegative integer, $|\alpha| = \Sigma_i \alpha_i$, and $\partial^\alpha u = \partial^{|\alpha|} u / \partial x_1^{\alpha_1} \cdots \partial x_n^{\alpha_n}$. We also have the semi-inner product

$$(u, v)_{H^m(\Omega)} = (u, v)_m = (u, v)_{m,\Omega} = \sum_{|\alpha| = m} \int_\Omega \partial^\alpha u \overline{\partial^\alpha v}\, dx \tag{2.6}$$

and seminorm

$$|u|_{H^m(\Omega)} = |u|_m = |u|_{m,\Omega} = \Big(\sum_{|\alpha| = m} \int_\Omega |\partial^\alpha u|^2 \, dx \Big)^{1/2} \tag{2.7}$$

It is immediate that $H^0(\Omega) = L_2(\Omega)$ and $\|u\|_{0,\Omega} = |u|_{0,\Omega} = \|u\|_{L_2(\Omega)}$. If $\Gamma = \partial\Omega$ is Lipschitz continuous, then $C^m(\bar{\Omega})$ is dense in $H^m(\Omega)$. (Γ is called Lipschitz continuous if it can be locally represented by a Lipschitz continuous function; see NEČAS [1967] for further details.)

$H_0^1(\Omega)$ is defined as the closure in $H^1(\Omega)$ of $C_0^\infty(\Omega)$, the space of infinitely differentiable functions on Ω which vanish near Γ. The Poincaré inequality, which states that

$$|u|_{0,\Omega} \leqslant C |u|_{1,\Omega} \quad \forall u \in H_0^1(\Omega), \tag{2.8}$$

shows that $|\cdot|_{1,\Omega}$ is a norm on $H_0^1(\Omega)$. $H_0^m(\Omega)$ is the closure in $H^m(\Omega)$ of $C_0^\infty(\Omega)$.

If Γ is Lipschitz continuous, then we can define the space $L_2(\Gamma)$, which consists of functions u defined on Γ for which $\|u\|_{L_2(\Gamma)} = (\int_\Gamma |u|^2 \, ds)^{1/2} < \infty$, where ds denotes the surface area. $L_2(\Gamma)$ is a Hilbert space with inner product $(u, v)_{L_2(\Gamma)} = \int_\Gamma u\bar{v}\, ds$. It is also known that a function $u \in H^1(\Omega)$ has a well-defined restriction to Γ, denoted by $\operatorname{tr} u$, in the sense of trace; $u = \operatorname{tr} u$ satisfies

$$\|u\|_{L_2(\Gamma)} \leqslant C \|u\|_{1,\Omega} \quad \forall u \in H^1(\Omega) \tag{2.9}$$

and

$$H_0^1(\Omega) = \{u \in H^1(\Omega) \colon u = 0 \text{ on } \Gamma \text{ in the sense of trace}\}.$$

Furthermore, a function $u \in C^1(\bar{\Omega})$ is in $H_0^1(\Omega)$ if and only if $u = 0$ for all $x \in \Gamma$. We note that if Γ is Lipschitz continuous, then the normal vector n is defined almost everywhere on Γ. The outer normal derivative $\partial u/\partial n$ is defined for $u \in H^2(\Omega)$ and

$$H_0^2(\Omega)=\left\{u \in H^2(\Omega): u=\frac{\partial u}{\partial n}=0 \text{ on } \Gamma\right\}.$$

We shall occasionally make use of the vector-valued Sobolev spaces $\mathbb{H}^m(\Omega)$ which are defined by

$$\mathbb{H}^m(\Omega)=\{(u_1(x),\ldots,u_k(x)): u_j(x) \in H^m(\Omega), j=1,\ldots,k\} \tag{2.10}$$

and

$$\|u\|^2_{\mathbb{H}^m(\Omega)}=\|u_1\|^2_{m,\Omega}+\cdots+\|u_k\|^2_{m,\Omega}. \tag{2.11}$$

In the study of eigenvalue problems, central use will be made of Rellich's theorem (cf. AGMON [1965]), which states that every bounded sequence in $H^m(\Omega)$ has a subsequence which converges in $H^j(\Omega)$ if $j<m$, provided Ω is a bounded open set in $\mathbb{R}^n$ with a Lipschitz continuous boundary.

So far we have defined the Sobolev space $H^m(\Omega)$ only for m an integer. We will sometimes use $H^m(\Omega)$, for m fractional, and also the Besov spaces, so we now turn to their definition, using the K-method.

For $u \in H^m(\Omega)$ and $0<t<\infty$ set

$$K(u,t)=\inf_{\substack{v\in H^m, w\in H^{m+1}\\ v+w=u}}\{\|v\|_{m,\Omega}+t\|w\|_{m+1,\Omega}\}. \tag{2.12}$$

Then for $m<k<m+1$ define

$$\|u\|_{H^k(\Omega)}=\|u\|_k=\|u\|_{k,\Omega}=\left(\int_0^\infty [t^{-\theta}K(u,t)]^2\frac{dt}{t}\right)^{1/2} \tag{2.13}$$

and

$$\|u\|_{\hat{H}^k(\Omega)}=\sup_{0<t<\infty}\{t^{-\theta}K(u,t)\}, \tag{2.14}$$

where $\theta=k-m$. The space

$$H^k(\Omega)\equiv\{u \in H^m(\Omega): \|u\|_{H^k(\Omega)}<\infty\} \tag{2.15}$$

is the Sobolev space with fractional order k and

$$\hat{H}^k(\Omega)\equiv\{u \in H^m(\Omega): \|u\|_{\hat{H}^k(\Omega)}<\infty\} \tag{2.16}$$

is a Besov space, the one often denoted by $B^k_{2,\infty}$.

In order to fix these ideas and to obtain a fact we will use in the sequel (cf. Sections 10.1 and 10.2), we now consider the function

$$u=r^\alpha \quad \text{for } (r,\theta)\in S=\{(r,\theta): 0<r<1, 0\leqslant\theta\leqslant\theta_0\leqslant 2\pi\},$$

where $-1<\alpha<0$, (r,θ) being polar coordinates, and prove that $u \in \hat{H}^{\alpha+1}(S)$.

THEOREM 2.1. *For* $-1<\alpha<0$, *we have*

$$u=r^{\alpha}\in \hat{H}^{1+\alpha}(S).$$

PROOF. Let $\varphi(x)$, $0<x<\infty$, be a function having derivatives of all orders and satisfying

$$\varphi(x)=\begin{cases}0 & \text{for } 0<x<\frac{1}{2},\\ 1 & \text{for } 1<x<\infty.\end{cases}$$

For $0<\delta\leqslant 1$, define

$$v=[1-\varphi(r/\delta)]u, \qquad w=\varphi(r/\delta)u.$$

Then we obviously have $u=v+w$. Now

$$\|v\|^2_{\hat{H}^0(S)}\leqslant \theta_0\int_0^{\delta} r^{2\alpha+1}\,\mathrm{d}r=\frac{\delta^{2\alpha+2}\theta_0}{2\alpha+2}\leqslant C\delta^{2\alpha+2}$$

and

$$\begin{aligned}\|w\|^2_{\hat{H}^1(S)}&=\int_S\left[|w|^2+\left|\frac{\partial w}{\partial x_1}\right|^2+\left|\frac{\partial w}{\partial x_2}\right|^2\right]\mathrm{d}x_1\,\mathrm{d}x_2\\ &=\int_0^{\theta_0}\int_0^1\left[|w|^2+\left|\frac{\partial w}{\partial r}\right|^2+r^{-2}\left|\frac{\partial w}{\partial \theta}\right|^2\right]r\,\mathrm{d}r\,\mathrm{d}\theta\\ &=\theta_0\int_0^1\left[|w|^2+\left|\frac{\partial w}{\partial r}\right|^2\right]r\,\mathrm{d}r\\ &\leqslant C\left[\int_{\delta/2}^1 r^{2\alpha+1}\,\mathrm{d}x+\alpha^2\int_{\delta/2}^1 r^{2\alpha-1}\,\mathrm{d}r+\delta^{-2}\int_{\delta/2}^{\delta} r^{2\alpha+1}\,\mathrm{d}r\right]\\ &\leqslant C\delta^{2\alpha},\end{aligned}$$

with C independent of δ. Hence

$$K(u,t)\leqslant C[\delta^{\alpha+1}+t\delta^{\alpha}]$$

and thus

$$t^{-(\alpha+1)}K(u,t)\leqslant C[\delta^{\alpha+1}t^{-(\alpha+1)}+t^{1-(\alpha+1)}\delta^{\alpha}].$$

If $0<t<1$, let $\delta=t$ to get

$$t^{-(\alpha+1)}K(u,t)\leqslant 2C$$

and hence

$$\sup_{0<t<1} \{t^{-(\alpha+1)}K(u, t)\} \leqslant 2C.$$

If $t \geqslant 1$, we obviously have

$$K(u, t) = \|u\|_{H^0(\Omega)} \leqslant C'$$

and hence

$$\sup_{1\leqslant t<\infty} \{t^{-(\alpha+1)}K(u, t)\} \leqslant C'.$$

Therefore

$$\|u\|_{\hat{H}^{(1+\alpha)}(S)} = \sup_{0<t<\infty} \{t^{-(1+\alpha)}K(u, t)\} \leqslant C'' < \infty$$

and hence $u \in \hat{H}^{1+\alpha}(S)$, as was to be proved.

In a similar way, one can also prove that $r^\alpha \in \hat{H}^{1+\alpha}(S)$ for $\alpha > 0$, not an integer. Finally we note that $r^\alpha \notin H^{1+\alpha}(S)$, but $r^\alpha \in H^{1+\alpha-\varepsilon}(S)$ for any $\varepsilon > 0$. □

For a complete discussion of the Sobolev and Besov spaces we refer to ADAMS [1975], NEČAS [1967], and BUTZER and BERENS [1967].

REMARK 2.1. The definition of the Sobolev spaces with fractional order m has a very simple interpretation. For u to be in $\hat{H}^{1+\alpha}(S)$ means that for any $0<t<\infty$, u can be split into the sum of a smooth function and a nonsmooth function in a natural way. We have employed this natural splitting in the proof of Theorem 2.1.

So far we have considered only one special family of Sobolev spaces or Sobolev-type spaces. Several other families are important in various situations. For example, if $\Omega \subset \mathbb{R}^2$ with $0 \in \partial\Omega$, and if $0<\beta<1$ and $m \geqslant l \geqslant 1$, we can define

$$H_\beta^{m,l}(\Omega) = \{u \in H^{l-1}(\Omega)\colon (\partial^\alpha u) r^{\beta+|\alpha|-l} \in L_2(\Omega) \text{ for } l \leqslant |\alpha| \leqslant m\} \tag{2.17}$$

and

$$\|u\|^2_{H_\beta^{m,l}(\Omega)} = \|u\|^2_{H^{l-1}(\Omega)} + \sum_{|\alpha|=l}^{m} \|(\partial^\alpha u) r^{\beta+|\alpha|-l}\|^2_{L_2(\Omega)}, \tag{2.18}$$

where $r = (x_1^2 + x_2^2)^{1/2}$. Spaces of this kind are called weighted Sobolev spaces. For more details we refer to KUFNER [1985]. Consider the function $u = r^\gamma$, with $0<\gamma<1$. One can show that $u \in H_\beta^{m,l}(\Omega)$, where $\Omega = \{(r, \theta)\colon 0<r<1\}$, for $\beta > 1-\gamma$, $m \geqslant 2$, and $l=2$. In fact, since $|\partial^\alpha u| \leqslant C(\alpha) r^{\gamma-|\alpha|}$, we have $|\partial^\alpha u| r^{\beta+|\alpha|-2} \leqslant C r^{\gamma+\beta-2}$, and we see that $u \in H_\beta^{m,l}(\Omega)$ for m, l and β as given.

We will also have occasion to use countably normed spaces constructed from Sobolev spaces. For example, consider the space

$$\mathscr{B}_\beta^2(\Omega) = \{u \in H_\beta^{2,2}(\Omega)\colon \|(\partial^\alpha u) r^{|\alpha|-2+\beta}\|_{L_2(\Omega)} \leqslant C d^{|\alpha|} \alpha! \text{ for } |\alpha| > 2, \text{ with } C \text{ and } d \text{ independent of } \alpha\}. \tag{2.19}$$

It is easy to see that all functions $u \in \mathscr{B}^2_\beta(\Omega)$ are analytic in $\bar{\Omega}\backslash\{0\}$. The function r^γ considered above belongs to $\mathscr{B}^2_\beta(\Omega)$ for $\beta > 1-\gamma$. We have here only considered weights with respect to the origin. More generally, one can consider weights with respect to the vertices of domains with piecewise smooth boundaries. An important reason for introducing these spaces is to characterize the solution (eigenfunctions) of a problem as precisely as possible by embedding it (them) in as small a space as possible. There are other classes of function spaces that are important in various contexts, but we will not go further in this direction.

REMARK 2.2. We have followed the usual custom of using the same notation for real and complex function spaces. It will be clear from the context which version we are using. See Remark 4.1.

3. Variational formulation of eigenvalue problems

In Section 1 the eigenvalue problems were stated in classical form, i.e., we were seeking an eigenvalue λ and a corresponding nonzero eigenfunction $u(x)$ such that the eigenvalue equation and boundary condition were satisfied in the classical pointwise sense. These problems can alternately be given as variational formulations. Since finite element approximation methods are most naturally defined in terms of variational formulations we now briefly indicate how eigenvalue problems can be cast in variational form. We will do this by discussing second-order elliptic eigenvalue problems in two dimensions in some detail. We begin by describing this type of problem.

Consider the problem: Seek a real or complex number λ and a nonzero real- or complex-valued function $u(x)$ satisfying

$$\begin{aligned} (Lu)(x) &= \lambda(Mu)(x), \quad x \in \Omega, \\ (Bu)(x) &= 0, \quad x \in \Gamma = \partial\Omega, \end{aligned} \tag{3.1}$$

where Ω is a bounded, open, connected set in $\mathbb{R}^2$ with Lipschitz continuous boundary Γ, and

$$Lu(x) = -\sum_{i,j=1}^{2} \partial_j(a_{ij}(x)\partial_i u) + \sum_{i=1}^{2} b_i(x)\partial_i u + c(x)u, \quad \partial_i = \partial/\partial x_i, \tag{3.2}$$

where $a_{ij}(x) = a_{ji}(x)$, $b_i(x)$, and $c(x)$ are given real or complex functions on Ω,

$$Mu(x) = d(x)u(x), \tag{3.3}$$

where $d(x)$ is a given real function which is bounded below by a positive constant on Ω, and

$$(Bu)(x) = u(x) \quad \text{or} \quad (Bu)(x) = -\sum_{i,j=1}^{2} a_{ij} n_j \partial_i u, \tag{3.4}$$

where $n(x) = (n_1, n_2)$ is the exterior unit normal to $\Gamma = \partial\Omega$ at x. L is assumed to be uniformly strongly elliptic in Ω, i.e., there is a positive constant a_0 such that

$$\operatorname{Re} \sum_{i,j=1}^{2} a_{ij}(x)\xi_i\xi_j \geqslant a_0 \sum_{i=1}^{2} \xi_i^2 \qquad \forall x \in \Omega \quad \text{and} \quad \forall(\xi_1,\xi_2)\in\mathbb{R}^2. \tag{3.5}$$

In addition, a_{ij}, b_i, c and d are assumed to be bounded and measurable. (A portion of the theory of eigenvalue problems can be developed under the more general hypothesis that $d(x)$ is merely assumed to be a bounded, measurable, complex function, but we will not pursue this direction.)

(λ, u) is called an eigenpair of the second-order differential operator L (relative to the zeroth-order differential operator M). If $Bu=u$, the boundary condition $Bu=0$ is the Dirichlet condition, and if

$$Bu = -\sum_{i,j=1}^{2} a_{ij}n_j\partial_i u = \frac{\partial u}{\partial v}$$

$$= \text{the conormal derivative of } u,$$

then $Bu=0$ yields the Neumann condition.

It is immediate that all of the examples discussed in Section 1—except the Steklov-type eigenvalue problems and the problem of the vibration of an elastic solid—are of the form (3.1) or its one- or higher-dimensional analogues. In any case, our discussion of approximation methods will be in terms of an abstract framework that will cover all the examples.

Let

$$L^*v(x) = -\sum_{i,j=1}^{2} \partial_i(\overline{a_{ij}}\partial_j v) - \sum_{i=1}^{2} \partial_i(\bar{b}_i v) + \bar{c}v \tag{3.6}$$

and

$$\frac{\partial v}{\partial v^*} = -\sum_{i,j=1}^{2} \overline{a_{ij}}n_i\partial_j v - \sum_{i=1}^{2} \bar{b}_i n_i v. \tag{3.7}$$

L^* is called the formal adjoint of L. It is an immediate consequence of the divergence theorem that

$$\begin{aligned}\int_\Omega Lu\bar{v}\,\mathrm{d}x &= \int_\Omega \left(\sum_{i,j=1}^{2} a_{ij}\partial_i u\overline{\partial_j v} + \sum_{i=1}^{2} b_i\partial_i u\bar{v} + cu\bar{v}\right)\mathrm{d}x + \int_\Gamma \frac{\partial u}{\partial v}\bar{v}\,\mathrm{d}s \\ &= \int_\Omega u\overline{L^*v}\,\mathrm{d}x + \int_\Gamma \frac{\partial u}{\partial v}\bar{v}\,\mathrm{d}x - \int_\Gamma u\overline{\frac{\partial v}{\partial v^*}}\,\mathrm{d}s\end{aligned} \tag{3.8}$$

for all smooth functions u and v. Hence we have

$$\int_\Omega Lu\bar{v}\,\mathrm{d}x = \int_\Omega u\overline{L^*v}\,\mathrm{d}x \tag{3.9}$$

if either $u=v=0$ on Γ or $\partial u/\partial\nu=\partial v/\partial\nu^*=0$ on Γ.

If a_{ij} and c are real and $b_i=0$, then $L^*=L$ and $\partial/\partial\nu=\partial/\partial\nu^*$. In this case we say L, M, B or, more briefly, L is *formally selfadjoint*, and we have

$$\int_\Omega d\left(\frac{1}{d}L\right)u\bar{v}\,\mathrm{d}x=\int_\Omega du\overline{\left(\frac{1}{d}L\right)v}\,\mathrm{d}x \tag{3.10}$$

if either $u=v=0$ on Γ or $\partial u/\partial\nu=\partial v/\partial\nu=0$ on Γ. All of the examples treated in Section 1 are formally selfadjoint except the operator arising in the stability analysis of the nonlinear initial boundary value problem.

Now we turn to the derivation of a variational formulation for (3.1). Suppose $(\lambda, u(x))$ satisfies (3.1) in the classical sense, i.e., the differential equation and the boundary condition hold pointwise, and consider first the case of the Dirichlet boundary condition: $u(x)=0$ for $x\in\Gamma$. Multiplying (3.1) by v, integrating over Ω, and using (3.3) and (3.8) we find that

$$\begin{aligned}\lambda b(u,v)&\equiv\lambda\int_\Omega du\bar{v}\,\mathrm{d}x=\int_\Omega Lu\bar{v}\,\mathrm{d}x\\&=\int_\Omega\left(\sum_{i,j=1}^{2}a_{ij}\partial_i u\overline{\partial_j v}+\sum_{i=1}^{2}b_i\partial_i u\bar{v}+cu\bar{v}\right)\mathrm{d}x+\int_\Gamma\frac{\partial u}{\partial\nu}\bar{v}\,\mathrm{d}s\\&=\int_\Omega\left(\sum_{i,j=1}^{2}a_{ij}\partial_i u\overline{\partial_j v}+\sum_{i=1}^{2}b_i\partial_i u\bar{v}+cu\bar{v}\right)\mathrm{d}x\\&\equiv a(u,v)\quad\text{for all } v\in C^1(\bar{\Omega})\text{ that vanish on }\Gamma.\end{aligned} \tag{3.11}$$

$a(u, v)$ and $b(u, v)$, as defined in (3.11), are bilinear forms (sometimes referred to as sesquilinear forms in the complex case) in u and v. They are clearly defined for $u,v\in C^1(\bar{\Omega})$ and, in fact, $a(u, v)$ is defined for $u, v\in H^1(\Omega)$ and $b(u, v)$ for $u,v\in H^0(\Omega)=L_2(\Omega)$. Furthermore, using the fact that a_{ij}, b_i, c and d are bounded, it follows from Schwarz's inequality that a is bounded on $H^1(\Omega)$ and b is bounded on $H^0(\Omega)$, i.e., that

$$|a(u,v)|\leqslant C_1\|u\|_{1,\Omega}\|v\|_{1,\Omega}\quad\forall u,v\in H^1(\Omega), \tag{3.12}$$

$$|b(u,v)|\leqslant C_2\|u\|_{0,\Omega}\|v\|_{0,\Omega}\quad\forall u,v\in H^0(\Omega). \tag{3.13}$$

We note one further property of the form $a(u, v)$:

$$\operatorname{Re} a(u,u)\geqslant\begin{cases}\tfrac{1}{2}a_0\|u\|_{1,\Omega}^2\quad\forall u\in H^1(\Omega),\\ \qquad\text{if }\operatorname{Re} c(x)\geqslant\tfrac{1}{2}a_0+\tfrac{1}{2}b^2/a_0\text{ for all }x\in\Omega,\\ \qquad\text{where } b=\max\limits_{\substack{x\in\Omega\\ i=1,2}}|b_i(x)|,\\ a_0|u|_{1,\Omega}^2\geqslant C\|u\|_{1,\Omega}^2\quad\forall u\in H_0^1(\Omega),\\ \qquad\text{if } b_i(x)=0,\ i=1,2,\ \operatorname{Re} c(x)\geqslant 0.\end{cases} \tag{3.14}$$

a_0 here is the ellipticity constant in (3.5); the result follows directly from (3.5).

Since the eigenfunction u vanishes on Γ, $u \in H_0^1(\Omega)$. Thus, using (3.12), (3.13), and the fact that $\{v \in C^1(\bar{\Omega}): v=0 \text{ on } \Gamma\}$ is dense in $H_0^1(\Omega)$, it follows from (3.11) that the eigenpair (λ, u) satisfies

$$\begin{aligned} &u \in H_0^1(\Omega), \qquad u \neq 0,\\ &a(u, v) = \lambda b(u, v) \quad \forall v \in H_0^1(\Omega). \end{aligned} \tag{3.15}$$

(3.15) is called a variational formulation of (3.1). We have shown that if (λ, u) is an eigenpair in the classical sense then it is an eigenpair in the variational sense. We now show that the converse is true, provided Γ, a_{ij}, b_i, c and d are sufficiently smooth.

Suppose (λ, u) satisfies (3.15) and suppose in addition Ω is a bounded open set with Lipschitz continuous boundary Γ and that $u \in C^2(\bar{\Omega})$. Then from the equation in (3.15) and from (3.8) we have

$$\begin{aligned} \int_\Omega Lu\bar{v}\, \mathrm{d}x &= a(u, v) + \int_\Omega \frac{\partial u}{\partial \nu}\bar{v}\, \mathrm{d}s\\ &= a(u, v) = \lambda b(u, v)\\ &= \lambda \int_\Omega du\bar{v}\, \mathrm{d}x \quad \forall v \in C^1(\bar{\Omega}) \text{ that vanish on } \Gamma. \end{aligned} \tag{3.16}$$

Since $\{v \in C^1(\bar{\Omega}): v=0 \text{ on } \Gamma\}$ is dense in $L_2(\Omega)$ we see from (3.16) that

$$Lu(x) = \lambda Mu(x), \quad x \in \Omega.$$

Also, since Γ is Lipschitz continuous and $u \in C^2(\bar{\Omega}) \cap H_0^1(\Omega)$ we know that $u(x)=0$ for all $x \in \Gamma$. Thus we see that (λ, u) is an eigenpair in the classical sense.

We next present conditions that guarantee that $u \in C^2(\bar{\Omega})$. From (3.15) we see that u is a weak solution of the boundary value (source) problem

$$Lu = f \ \text{ in } \Omega, \qquad u = 0 \ \text{ on } \Gamma,$$

where $f = \lambda\, du$. Using standard regularity results for elliptic equations we find that $u \in C^2(\bar{\Omega})$ provided Γ, a_{ij}, b_i, c and d are sufficiently smooth. In the two-dimensional case we are discussing it is sufficient to assume

(i) Γ is of class C^4,
(ii) $a_{ij}, b_i \in C^3(\bar{\Omega})$,
(iii) $c, d \in C^2(\bar{\Omega})$.

In the general n-dimensional case it is sufficient to assume

(i) Γ is of class C^k,
(ii) $a_{ij}, b_i \in C^{k-1}(\bar{\Omega})$,
(iii) $c, d \in C^{k-2}(\bar{\Omega})$, where $k = [\frac{1}{2}n] + 3$.

For these results we refer to AGMON [1965, Theorems 3.9 and 9.8].

Eigenvalue problems on domains with corners arise in many applications but are not covered by the above results because of the requirement that Γ be smooth.

Nevertheless, when Ω has corners, analogous results in a generalized form involving weighted Sobolev spaces can be proved for problems with smooth coefficients (see GRISVARD [1985] and BABUŠKA and GUO [1987]). Furthermore these results provide information on the behavior of u near the corners that is important in assessing the accuracy of eigenvalue approximations. This matter will be taken up in Section 10. We now briefly outline the extent to which the eigenpair (λ, u) of (3.15) satisfies (3.1) in the classical sense in the case in which Ω is a polygon and $L=-\Delta$ and $d(x)=1$. From regularity results for elliptic equations we conclude that $u\in C^{\infty}(\bar{\Omega}\backslash\{\text{vertices of } \Omega\})$. Thus we see that $Lu(x)=\lambda Mu(x)$ for all $x\in\Omega$ and $u(x)=0$ for $x\in\Gamma\backslash\{\text{vertices of } \Omega\}$. u fails, however, to be an eigenfunction in the classical sense in that $u\notin C^2$ at any vertex of Ω with interior angle larger than π.

Under the hypothesis sketched above, the classical and variational formulations of (3.1) are equivalent. For the remainder of this article, we will take the point of view that our eigenvalue problems are given in variational form. Thus we will consider problems of the form (3.15), or problems that are generalizations of the form (3.15); see Section 8.

Consider now the case of the Neumann boundary condition: $\partial u(x)/\partial\nu=0$ for $x\in\Gamma$. Suppose (λ, u) satisfies (3.1) in the classical sense. Then, using (3.8) we find

$$\begin{aligned}\lambda b(u, v) &= a(u, v)+\int_{\Gamma}\frac{\partial u}{\partial\nu}\bar{v}\,\mathrm{d}s\\ &= a(u, v)\quad\text{for all } v\in C^1(\bar{\Omega}),\end{aligned}$$

and thus, using the fact that $C^1(\bar{\Omega})$ is dense in $H^1(\Omega)$, we see that (λ, u) satisfies

$$\begin{aligned}&u\in H^1(\Omega),\qquad u\neq 0,\\ &a(u, v)=\lambda b(u, v)\quad\forall v\in H^1(\Omega).\end{aligned}\tag{3.17}$$

(3.17) is a variational form for (3.1) with the Neumann condition. Now suppose (λ, u) satisfies (3.17) and assume $u\in C^2(\bar{\Omega})$. From (3.17) and (3.8) we obtain

$$\begin{aligned}\int_{\Omega}Lu\bar{v}\,\mathrm{d}x &= a(u, v)+\int_{\Gamma}\frac{\partial u}{\partial\nu}\bar{v}\,\mathrm{d}x=\lambda b(u, v)+\int_{\Gamma}\frac{\partial u}{\partial\nu}\bar{v}\,\mathrm{d}s\\ &=\lambda\int_{\Omega}du\bar{v}\,\mathrm{d}x+\int_{\Gamma}\frac{\partial u}{\partial\nu}\bar{v}\,\mathrm{d}s\quad\forall v\in C^1(\bar{\Omega}).\end{aligned}\tag{3.18}$$

Taking $v\in C^1(\bar{\Omega})$ which vanish on Γ we find that

$$Lu(x)=\lambda d(x)u(x)\quad\forall x\in\Omega.$$

Thus (3.18) reduces to

$$\int_{\Gamma}\frac{\partial u}{\partial\nu}\bar{v}\,\mathrm{d}s=0\quad\forall v\in C^1(\bar{\Omega}),$$

which implies that $\partial u/\partial \nu = 0$ on Γ. Thus we have shown that (λ, u) satisfies (3.1) in the classical sense. As with the Dirichlet condition, the analysis is valid under appropriate smoothness assumptions on Γ, a_{ij}, b_i, c and d. We will not state these in detail.

Note that the Neumann boundary condition is not explicitly stated in (3.17). It is, however, implicitly contained in (3.17). We refer to the Neumann condition as a *natural boundary condition*, in contrast to the Dirichlet condition which is referred to as an *essential boundary condition*, and which is explicitly contained in the variational formulation (3.15). The fact that the Neumann condition is natural has important implications for the approximation of eigenvalues; see Remark 10.4.

In summary, for (3.1) we get one of the following forms:

PROBLEM 3.1. *Dirichlet boundary condition.*

Seek $\lambda, u \neq 0$ satisfying

$$u \in H_0^1(\Omega),$$
$$a(u, v) = \lambda b(u, v) \quad \forall v \in H_0^1(\Omega).$$

PROBLEM 3.2. *Neumann boundary condition.*

Seek $\lambda, u \neq 0$ satisfying

$$u \in H^1(\Omega),$$
$$a(u, v) = \lambda b(u, v) \quad \forall v \in H^1(\Omega).$$

We will sometimes refer to (λ, u) as an eigenpair of the form a relative to the form b. Regarding the forms a and b we assume (3.12)–(3.14) hold.

In a similar way, many other problems—including all of the examples discussed in Section 1—can be given variational formulations. This is done for a number of problems in Chapter III. We mention in particular the eigenvalue problems corresponding to the vibration of a free L-shaped panel (a two-dimensional analogue of the elastic solid).

Finally we wish to make one further point regarding variational formulations of eigenvalue problems, namely, that a given eigenvalue problem can often be given a variety of different variational formulations and that some of these may lead to more effective finite element methods than others. We illustrate the possibility of various variational formulations by considering the simple model problem

$$\begin{aligned} &-(a(x)u')' + cu = \lambda u, \quad 0 < x < 1, \\ &u(0) = u(1) = 0. \end{aligned} \tag{3.19}$$

This has already been cast into the variational form:

$$\begin{aligned} &\text{Seek } \lambda, u \neq 0 \text{ satisfying} \\ &u \in H_0^1(0, 1), \\ &a(u, v) = \lambda b(u, v) \quad \forall v \in H_0^1(0, 1), \end{aligned} \tag{3.20}$$

where

$$a(u, v)=\int_0^1 (au'\,\bar{v}'+cu\bar{v})\,\mathrm{d}x$$

and

$$b(u, v)=\int_0^1 u\bar{v}\,\mathrm{d}x$$

are bounded bilinear forms in $H_0^1 \times H_0^1$. An alternate formulation is

$$\begin{aligned}&\text{Seek } \lambda,\ 0\neq u\in L_2(0, 1) \text{ satisfying}\\ &a_1(u, v)=\lambda b_1(u, v)\quad \forall v\in H^2(0, 1)\cap H_0^1(0, 1),\end{aligned}\tag{3.21}$$

where

$$a_1(u, v)=\int_0^1 u[-(a\bar{v}')'+c\bar{v}]\,\mathrm{d}x$$

and

$$b_1(u, v)=\int_0^1 u\bar{v}\,\mathrm{d}x$$

are bounded bilinear forms on $L_2\times[H^2(0, 1)\cap H_0^1(0, 1)]$. Formulations (3.20) and (3.21) are equivalent in the sense that (λ, u) is an eigenpair of one if and only if it is an eigenpair of the other.

Another formulation is obtained as follows. If we let $\sigma=au'$, then (3.19) can be written as a first-order system of equations,

$$\begin{aligned}&-\sigma'+cu=\lambda u,\\ &u'-\sigma/a=0,\\ &u(0)=u(1)=0.\end{aligned}\tag{3.22}$$

System (3.22) can then be given the variational formulation,

$$\begin{aligned}&\text{Seek } \lambda,\ (\sigma, u)\in L_2(0, 1)\times H_0^1(0, 1) \text{ satisfying}\\ &a_2(\sigma, u, \psi, v)=\lambda b_2(\sigma, u, \psi, v)\quad \forall(\psi, v)\in L_2(0, 1)\times H_0^1(0, 1),\end{aligned}\tag{3.23}$$

where

$$a_2(\sigma, u, \psi, v)=\int_0^1 \left(\sigma\bar{v}'+cu\bar{v}+u'\bar{\psi}-\frac{\sigma\bar{\psi}}{a}\right)\mathrm{d}x$$

and

$$b_2(\sigma, u, \psi, v) = \int_0^1 u\bar{v}\,dx.$$

a_2 and b_2 are bounded bilinear forms on $L_2 \times H_0^1$. Formulations (3.19) and (3.22) are equivalent in the sense that if (λ, u) is an eigenpair of (3.19) and $\sigma = au'$, then $(\lambda, (u, \sigma))$ is an eigenpair of (3.22), and if $(\lambda, (\sigma, u))$ is an eigenpair of (3.22), then (λ, u) is one of (3.19) and $\sigma = au'$. (3.22) and (3.23) are called *mixed formulations* of the eigenvalue problem (3.19); see Section 11. We can also write (3.19) in the form

$$\begin{aligned} &(\sigma, u) \in L_2(0, 1) \times H_0^1(0, 1), \qquad (\sigma, u) \neq (0, 0), \\ &A(\sigma, \psi) + \overline{B(\psi, u)} = 0 \quad \forall \psi \in L_2(0, 1), \\ &B(\sigma, v) - \int_0^1 cu\bar{v}\,dx = \int_0^1 -\lambda\, u\bar{v}\,dx \quad \forall v \in H_0^1, \end{aligned} \tag{3.24}$$

where

$$A(\sigma, \psi) = \int_0^1 \frac{\sigma\bar{\psi}}{a}\,dx$$

and

$$B(\sigma, v) = -\int_0^1 \sigma\bar{v}'\,dx.$$

In Chapter III we will consider further examples of variational formulations and show how they can be used to define a variety of finite element methods.

4. Properties of eigenvalue problems

In this section we discuss the basic properties of eigenvalue problems. As in Section 3 this discussion will be in terms of second-order elliptic eigenvalue problems.

We thus consider the problem (3.1) in variational form:

$$\begin{aligned} &\text{Seek } \lambda,\ 0 \neq u \in H \text{ satisfying} \\ &a(u, v) = \lambda b(u, v) \quad \forall v \in H, \end{aligned} \tag{4.1}$$

where $H = H_0^1(\Omega)$ for Dirichlet boundary conditions and $H = H^1(\Omega)$ for Neumann conditions. The forms $a(\cdot,\cdot)$ and $b(\cdot,\cdot)$ are assumed to satisfy

$$|a(u, v)| \leqslant C_1 \|u\|_{1,\Omega} \|v\|_{1,\Omega}, \quad \forall u, v \in H, \tag{4.2}$$

$$|b(u, v)| \leqslant C_2 \|u\|_{0,\Omega} \|v\|_{0,\Omega}, \quad \forall u, v \in H, \tag{4.3}$$

and

$$\operatorname{Re} a(u, u) \geq \alpha \|u\|_{1,\Omega}^2 \quad \forall u \in H, \tag{4.4}$$

where $\alpha > 0$. Sufficient conditions for (4.2)–(4.4) to hold were given in Section 3; cf. (3.12)–(3.14).

For the study of (4.1) it is useful to introduce the operator $T: H^0(\Omega) \to H$ defined by

$$Tf \in H, \qquad a(Tf, v) = b(f, v) \quad \forall v \in H. \tag{4.5}$$

T is the solution operator for the boundary value (source) problem

$$Lu = df \ \text{ in } \Omega, \qquad Bu = 0 \ \text{ on } \Gamma, \tag{4.6}$$

i.e., $u = Tf$ solves (4.6). Thus T is the inverse of the differential operator L, considered on functions that satisfy the boundary conditions. It follows immediately from (4.2)–(4.4) and the Riesz representation theorem in the special case in which $a(\cdot, \cdot)$ is an inner product on H or the Lax–Milgram theorem (LAX and MILGRAM [1954]) in the general case, that (4.5) has a unique solution Tf for each $f \in H^0(\Omega)$ and that

$$\|Tf\|_{1,\Omega} \leq (C_2/\alpha) \|f\|_{0,\Omega} \quad \forall f \in H^0(\Omega), \tag{4.7}$$

i.e., $T: H^0(\Omega) \to H$ is bounded. In Section 2 we noted that H is compactly embedded in $H^0(\Omega)$, provided Γ is Lipschitz continuous (Rellich's theorem). From this fact and (4.7) we see that $T: H^0(\Omega) \to H^0(\Omega)$ is a compact operator. We can also view T as an operator on H; we will, in fact, mainly consider T on H. Another application of Rellich's theorem shows that $T: H \to H$ is compact.

It follows immediately from (4.1) and (4.5) that (λ, u) is an eigenpair of (4.1) (or of L) if and only if

$$Tu = (1/\lambda)u, \quad u \neq 0,$$

i.e., if and only if $(\mu = \lambda^{-1}, u)$ is an eigenpair of T. Through this correspondence, properties of the eigenvalue problem (4.1) can be derived from the spectral theory for compact operators. A complete development of this theory can be found in DUNFORD and SCHWARTZ [1958, 1963]. We now give a brief sketch of it; a slightly more complete treatment is given in Section 6. We present this theory under the assumption that the space H is complex. This leads to the simplest general statement of the theory. In the special case in which T is selfadjoint, H can be taken to be real or complex. We will specialize to the selfadjoint case later.

Denote by $\rho(T)$ the resolvent set of T, i.e., the set

$$\begin{aligned}\rho(T) = \{z\colon & z \in \mathbb{C} \equiv \text{the complex numbers}, \\ & (z - T)^{-1} \text{ exists as a bounded operator on } H\},\end{aligned}$$

and by $\sigma(T)$ the spectrum of T, i.e., the set $\sigma(T) = \mathbb{C} \setminus \rho(T)$. $\sigma(T)$ is countable with no nonzero limit points; nonzero numbers in $\sigma(T)$ are eigenvalues; and if zero is in $\sigma(T)$, it may or may not be an eigenvalue. Let $0 \neq \mu \in \sigma(T)$. The space $N(\mu - T)$ of eigenvectors corresponding to μ is finite-dimensional; its dimension is called the (geometric) multiplicity of μ.

Now suppose L is formally selfadjoint. Then it follows immediately from their definitions that $a(u, v)$ and $b(u, v)$ satisfy

$$a(u, v) = a(v, u) \quad \forall u, v \in H, \tag{4.8a}$$

$$b(u, v) = b(v, u) \quad \forall u, v \in H^0(\Omega), \tag{4.8b}$$

i.e., a and b are symmetric (or Hermitian) forms. Thus from (4.2)–(4.4) we see that $a(u, v)$ is an inner product on H that is equivalent to $((u, v))_{1,\Omega}$. In a similar way we see that $b(u, v)$ is an inner product on $H^0(\Omega)$ that is equivalent to $(u, v)_{0,\Omega}$ (recall that $d(x)$ is bounded above and is bounded below by a positive constant). It follows from (4.8) that

$$a(Tu, v) = a(u, Tv) \quad \forall u, v \in H, \tag{4.9a}$$

$$b(Tu, v) = b(u, Tv) \quad \forall u, v \in H^0(\Omega), \tag{4.9b}$$

i.e., T, considered as an operator on H, is selfadjoint with respect to $a(u, v)$, and, considered as an operator on $H^0(\Omega)$, is selfadjoint with respect to $b(u, v)$. (We have previously noted in (3.10) that $b((1/d)Lu, v) = b(u, (1/d)Lv)$ if $u = v = 0$ on Γ or if $\partial u/\partial \nu = \partial v/\partial \nu = 0$ on Γ, provided L is formally selfadjoint.)

From the fact that T is selfadjoint on H it follows that the eigenvalues of T are real and the eigenfunctions corresponding to distinct eigenvalues are orthogonal with respect to $a(u, v)$. We noted above that T is compact on H and it follows from (4.5) that T is positive definite. Thus T will have a countably infinite sequence of eigenvalues

$$0 \downarrow \cdots \leqslant \mu_2 \leqslant \mu_1$$

and associated eigenfunctions

$$u_1, u_2, \ldots,$$

which satisfy

$$a(u_i, u_j) = \lambda_i b(u_i, u_j) = \delta_{ij}.$$

It is further known that the eigenfunctions are complete in $L_2(\Omega)$, i.e., that

$$u = \sum_{j=1}^{\infty} c_j u_j \quad \forall u \in L_2(\Omega), \tag{4.10}$$

where

$$c_j = \lambda_j b(u, u_j) = \lambda_j \int_\Omega d u \bar{u}_j \, \mathrm{d}x, \tag{4.11}$$

and convergence is in either the L_2 norm or the norm $\|u\|_b = \sqrt{b(u, u)}$. Equation (4.10) converges in the H norm for $u \in H$.

Now the spectral properties of (4.1) (or of L) can be inferred from these facts by recalling that the eigenvalues of (4.1) (or L) are the reciprocals of those of T and that (4.1) and T have the same eigenfunctions. Thus, if L is formally selfadjoint, then (4.1) (or L) has eigenvalues

$$0<\lambda_1\leqslant\lambda_2\leqslant\cdots\uparrow+\infty \tag{4.12}$$

and corresponding eigenfunctions

$$u_1, u_2, \ldots \tag{4.13}$$

satisfying

$$a(u_i, u_j)=\lambda_i b(u_i, u_j)=\delta_{ij}. \tag{4.14}$$

In the sequence $\lambda_1, \lambda_2, \ldots$, eigenvalues are repeated according to their (geometric) multiplicity. The properties of eigenvalues and eigenfunctions used in Section 1 in the discussion of separation of variables (cf. (1.11)–(1.15)) all follow from the properties we have sketched here.

Although our discussion has been in terms of second-order elliptic problems, it is immediate that the results hold for any eigenvalue problem in variational form provided the bilinear forms are symmetric and satisfy (4.2)–(4.4). We will refer to this as the selfadjoint, positive-definite case. In Section 8 this, as well as a more general, class of variationally formulated eigenvalue problems is discussed.

REMARK 4.1. The eigenvalues of selfadjoint eigenvalue problems are real and the eigenfunctions may be taken to be real. Thus these problems may be formulated in terms of real function spaces. Nonselfadjoint eigenvalue problems, on the other hand, may have complex eigenvalues and complex eigenfunctions, and are formulated in terms of complex spaces.

We end this section with a discussion of the regularity of the eigenfunctions of the second-order elliptic operator L. L is not assumed to be formally selfadjoint here.

THEOREM 4.1. *Suppose for* $k\geqslant 2$,

(i) $\Gamma=\partial\Omega$ *is of class* C^k,

(ii) $a_{ij},b_i\in C^{k-1}(\bar{\Omega})$, *and*

(iii) $c,d\in C^{k-2}(\bar{\Omega})$.

Then all eigenfunctions of L *(see* (3.2)) *lie in* $H^k(\Omega)$ *and*

$$\|u_j\|_{k,\Omega}\leqslant C\lambda_j^{k/2}\|u_j\|_{0,\Omega}, \quad j=1,2,\ldots.$$

PROOF. This result is a direct consequence of standard results on the regularity of solutions of elliptic boundary value problems. In particular, we refer to AGMON [1965, Theorem 9.8]. □

THEOREM 4.2. *Suppose*

(i) Γ *is of class* C^∞, *and*

(ii) $a_{ij},b_i,c,d\in C^\infty(\bar{\Omega})$.

Then $u_j\in C^\infty(\bar{\Omega})$ *for* $j=1,2,\ldots$.

PROOF. This result follows directly from Theorem 4.1. □

THEOREM 4.3. *Suppose*
(i) $\Gamma = \partial\Omega$ *is analytic, and*
(ii) a_{ij}, b_i, c, d *are analytic on* $\bar{\Omega}$.
Then u_j *is analytic on* $\bar{\Omega}$ *for each* j.

PROOF. For a proof of this result see MORREY [1966, Section 5.7]. □

In practice most of the domains of interest have piecewise analytic boundaries. Let us mention a result for such domains.

THEOREM 4.4. *Suppose*
(i) $\Omega \subset \mathbb{R}^2$,
(ii) Γ *is piecewise analytic, and*
(iii) a_{ij}, b_i, c, d *are analytic on* $\bar{\Omega}$.
Then every eigenfunction of L *is analytic in* $\bar{\Omega}\setminus\bigcup$*(vertices), and moreover, belongs to the space* $\mathscr{B}^2_\beta(\Omega)$, *for properly chosen* β.

PROOF. This theorem follows from the results in BABUŠKA and GUO [1988a]. □

REMARK 4.2. Assume that $Lu = -\Delta u$, Ω is a polygon, and the boundary conditions are of Dirichlet type. If Ω is a convex polygon, then the eigenfunctions $u \in H^2(\Omega)$, and if Ω is a nonconvex polygon, then $u \in \hat{H}^k(\Omega) \cap H^1_0(\Omega)$, where $k = (\pi/\alpha) + 1$, with α the maximal interior angle of the vertices of Ω.

For a comprehensive treatment of regularity results for problems on domains with corners, we refer to GRISVARD [1985].

5. A brief overview of the finite element method for eigenvalue approximation

In this section we give a brief overview of the use of finite element methods for approximating eigenvalues and eigenfunctions of differential operators. We will restrict the discussion to a simple model problem in one dimension and its approximation by the simplest type of finite element method.

Consider the selfadjoint eigenvalue problem

$$\begin{aligned} &(Lu)(x) = -(a(x)u')' + c(x)u = \lambda d(x)u, \quad 0 < x < l \\ &u(0) = u(1) = 0, \end{aligned} \tag{5.1}$$

where $a \in C^1[0, l]$, $c, d \in C^0[0, l]$, and

$$0 < a_0 \leqslant a(x), \quad 0 \leqslant c(x), \quad 0 < d_0 \leqslant d(x) \qquad \text{for } 0 \leqslant x \leqslant l$$

(cf. (3.1)–(3.4)). As indicated in Section 3, this problem has the variational characterization

$$u \in H_0^1(0, l), \qquad a(u, v) = \lambda b(u, v) \quad \forall v \in H_0^1(0, l), \tag{5.2}$$

where

$$a(u, v) = \int_0^l a(x)u'v' \, dx$$

and

$$b(u, v) = \int_0^l duv \, dx.$$

(5.1) (or (5.2)) has a sequence of eigenvalues

$$0 < \lambda_1 \leqslant \lambda_2 \leqslant \cdots \uparrow + \infty$$

and corresponding eigenfunctions

$$u_1, u_2, \ldots$$

satisfying

$$\lambda_i \int_0^l d(x) u_i u_j \, dx = \delta_{ij}.$$

On $[0, l]$ consider an arbitrary mesh

$$\Delta = \{0 = x_0 < x_1 < \cdots < x_n = l\},$$

where $n = n(\Delta)$ is a positive integer, and let

$$S_h = \{u : u \in C[0, l],\ u(0) = u(l) = 0,$$
$$u \text{ is linear on } I_j,\ j = 1, \ldots, n\},$$

where $h_j = x_j - x_{j-1}$ and $I_j = (x_{j-1}, x_j)$ for $j = 1, \ldots, n$ and $h = h(\Delta) = \max_j h_j$. S_h is an $(n-1)$-dimensional subspace of $H_0^1(0, l)$. The pairs (λ, u) have been characterized in (5.2) as eigenpairs of the bilinear form $a(u, v)$ relative to the form $b(u, v)$ over the space $H_0^1(0, l) \times H_0^1(0, l)$. We now consider eigenpairs of $a(u, v)$ relative to $b(u, v)$ over the space $S_h \times S_h$, i.e., we consider the eigenvalue problem,

$$\begin{aligned} &\text{Seek } \lambda_h,\ 0 \neq u_h \in S_h \text{ satisfying} \\ &a(u_h, v) = \lambda_h b(u_h, v) \quad \forall u \in S_h, \end{aligned} \tag{5.3}$$

and then view the eigenpairs of (5.3) as approximations to those of (5.2). (λ_h, u_h) is called a finite element (Galerkin) approximation to (λ, u). A wide variety of finite element methods for eigenvalue problems will be introduced and analyzed in

Chapter III. Here we will outline the general features of these methods by examining the method (5.3) as it applies to (5.1).

Since S_h is finite-dimensional, (5.3) is equivalent to a generalized matrix eigenvalue problem. In fact, if $\phi_1,\ldots,\phi_{n-1}$ is a basis for S_h, then it is easily seen that $(\lambda_h, u_h=\Sigma_{j=1}^{n-1} z_j\phi_j)$ is an eigenpair of (5.3) if and only if

$$Az=\lambda_h Bz, \quad z\neq 0, \tag{5.4}$$

where $z=(z_1,\ldots, z_{n-1})^{\mathrm{T}}$ and

$$A=(A_{ij}) \quad \text{with } A_{ij}=a(\phi_j, \phi_i),$$
$$B=(B_{ij}) \quad \text{with } B_{ij}=b(\phi_j, \phi_i).$$

Problem (5.3) (respectively, (5.4)) has eigenvalues

$$0<\lambda_{1,h}\leqslant\lambda_{2,h}\leqslant\cdots\leqslant\lambda_{n-1,h}$$

and corresponding eigenfunctions $u_{1,h},\ldots, u_{n-1,h}$ (respectively, $z_{j,h}=(z_{j,1,h},\ldots, z_{j,n-1,h})^{\mathrm{T}}$, $j=1,\ldots, n-1$), satisfying

$$\lambda_{i,h}\int_0^l du_{i,h}u_{j,h}\,\mathrm{d}x=\delta_{ij}$$

(respectively, $\lambda_{i,h}z_{i,h}^{\mathrm{T}}Bz_{j,h}=\delta_{ij}$). We further note that if we choose as basis functions the usual hat functions determined by

$$\phi_i(x_j)=\delta_{ij},$$

then A and B are sparse; in fact, they are tridiagonal. We easily see that the three nonzero diagonals are given by

$$A_{i,i+1}=-h_i^{-1}h_{i+1}^{-1}\int_{x_i}^{x_{i+1}} a(x)\,\mathrm{d}x+h_i^{-1}h_{i+1}^{-1}\int_{x_i}^{x_{i+1}}(x_{i+1}-x)(x-x_i)c(x)\,\mathrm{d}x, \tag{5.5a}$$

$$A_{ii}=h_i^{-2}\int_{x_{i-1}}^{x_i} a(x)\,\mathrm{d}x+h_{i+1}^{-2}\int_{x_i}^{x_{i+1}} a(x)\,\mathrm{d}x$$

$$+h_i^{-2}\int_{x_{i-1}}^{x_i}(x-x_{i-1})^2c(x)\,\mathrm{d}x+h_{i+1}^{-2}\int_{x_{i-1}}^{x_{i+1}}(x_{i+1}-x)^2c(x)\,\mathrm{d}x, \tag{5.5b}$$

$$A_{i-1,i}=-h_{i-1}^{-1}h_i^{-1}\int_{x_{i-1}}^{x_i} a(x)\,\mathrm{d}x+h_{i-1}^{-1}h_i^{-1}\int_{x_{i-1}}^{x_i}(x-x_i)(x-x_{i-1})c(x)\,\mathrm{d}x; \tag{5.5c}$$

$$B_{i,i+1}=h_i^{-1}h_{i+1}^{-1}\int\limits_{x_i}^{x_{i+1}}(x_{i+1}-x)(x-x_1)d(x)\,\mathrm{d}x, \tag{5.6a}$$

$$B_{i,i}=h_i^{-2}\int\limits_{x_{i-1}}^{x_i}(x-x_{i-1})^2 d(x)\,\mathrm{d}x+h_{i+1}^{-2}\int\limits_{x_i}^{x_{i+1}}(x_{i+1}-x)^2 d(x)\,\mathrm{d}x, \tag{5.6b}$$

$$B_{i-1,i}=h_{i-1}^{-1}h_i^{-1}\int\limits_{x_{i-1}}^{x_i}(x_i-x)(x-x_{i-1})d(x)\,\mathrm{d}x. \tag{5.6c}$$

Now we specialize (5.1) to the vibrating string problem discussed in Section 1, i.e., we let $a(x)=p=$ the tension of the string, $c(x)=0$, and $d(x)=r=$ the density of the string. We also suppose the mesh is uniform, i.e., we let $x_i=iln^{-1}$; we then have $h=h_i=ln^{-1}$. It is easily seen from (5.5)–(5.6) that

$$A=ph^{-1}\begin{pmatrix} 2 & -1 & & & \\ -1 & 2 & -1 & & \\ & -1 & 2 & -1 & \\ & & & \ddots & \\ & & & \cdot & -1 \\ & & & -1 & 2 \end{pmatrix} \tag{5.7}$$

and

$$B=\tfrac{1}{6}rh\begin{pmatrix} 4 & 1 & & & \\ 1 & 4 & 1 & & \\ & 1 & 4 & 1 & \\ & & & \ddots & \\ & & & \cdot & 1 \\ & & & 1 & 4 \end{pmatrix} \tag{5.8}$$

If the integrals defining the B_{ij} are approximated by the trapezoid quadrature rule, then instead of the matrix B we would obtain the matrix

$$\tilde{B}=rhI \tag{5.9}$$

and instead of (5.4) we would have

$$Az=\tilde{\lambda}\tilde{B}z. \tag{5.10}$$

We finally note that the eigenvalues and eigenvectors of (5.4) and (5.10) can, in this special case, be explicitly found. The eigenvalues of (5.4) are given by

$$\begin{aligned}\lambda_{j,h} &= 6h^{-2}(1-\cos(j\pi h/l))(2+\cos(j\pi h/l))^{-1}pr^{-1},\\ j &= 1,2,\ldots,n-1,\end{aligned} \tag{5.11}$$

and those of (5.10) by

$$\tilde{\lambda}_{j,h} = 2h^{-2}(1-\cos(j\pi h/l))pr^{-1}, \quad j=1,2,\ldots,n-1. \tag{5.12}$$

The unnormalized eigenvectors of both problems are given by

$$z_{j,h} = (z_{j,1,h},\ldots,z_{j,n-1,h})^{\mathrm{T}}, \tag{5.13}$$

where

$$z_{j,k,h} = \sin(j\pi kh/l), \quad j,k=1,2,\ldots,n-1. \tag{5.14}$$

The eigenvalues and eigenfunctions of (5.1), in this case, are given by

$$\lambda_j = j^2\pi^2 p/l^2 r, \quad j=1,2,3,\ldots \tag{5.15}$$

and

$$u_j(x) = \sqrt{2/lr}\,\sin(j\pi x/l), \quad j=1,2,\ldots\,. \tag{5.16}$$

From (5.11) and (5.15) we see that

$$\lambda_{j,h}-\lambda_j = \frac{j^4\pi^4 p}{12rl^4}h^2 + \frac{j^6\pi^6 p}{360rl^6}h^4 + \cdots = \mathrm{O}(h^2) \tag{5.17}$$

and from (5.12) and (5.15) we see that

$$\lambda_j-\tilde{\lambda}_{j,h} = \frac{j^4\pi^4 p}{12rl^4}h^2 - \frac{j^6\pi^6 p}{360rl^6}h^4 + \cdots = \mathrm{O}(h^2). \tag{5.18}$$

From (5.13), (5.14), and (5.16) we see that, neglecting the normalizing factors, the eigenvector $z_{j,h}$ consists of the values of $u_j(x)$ at $x=x_1, x_2,\ldots, x_{n-1}$.

Equation (5.17) shows that the eigenvalue error $\lambda_{j,h}-\lambda_j$ is $\mathrm{O}(h^2)$. Thus the small eigenvalues of (5.3) (or of (5.4)) approximate the eigenvalues of (5.2), but the larger ones do not since $\lambda_{j,h}-\lambda_j$ is small only if j^2h is small. If, for example, $j\simeq n^{1/2}$, then j^2h is of order one and we would not expect $\lambda_{j,h}-\lambda_j$ to be small. Thus only a small percentage of the eigenvalues of (5.4) are of interest. This observation influences the selection of numerical methods for the extraction of the eigenvalues of (5.4). We also note that (5.17) and (5.18) show that $\tilde{\lambda}_{j,h}\leqslant\lambda_j\leqslant\lambda_{j,h}$ for h small. It is known that $\lambda_j\leqslant\lambda_{j,h}$ for all h; cf. (8.42).

5.1. *A physical interpretation of the finite element eigenvalue problem* (5.10)

We consider here the vibration of a weightless elastic string loaded with several point masses. Suppose we have a weightless elastic string of length l loaded with $n-1$ particles of mass m at distances $ln^{-1}, 2ln^{-1},\ldots,(n-1)ln^{-1}$ from one end and fixed at both ends. Gravity is assumed to be negligible and the particles are assumed to move in a plane. We shall study the small free vibrations of this system of $n-1$ degrees of freedom.

Let p denote the tension in the string and let $h=ln^{-1}$. If $q_i(t)$ denotes the vertical displacement of the ith particle, the particles being numbered from the left (see Fig. 5.1), then the equation of motion for the ith particle is easily seen to be

$$-mq_i''(t)=-p\frac{q_{i-1}-2q_i+q_{i+1}}{h}, \quad i=1, 2, \ldots, n-1, \tag{5.19}$$

where we assume $q_0=q_n=0$.

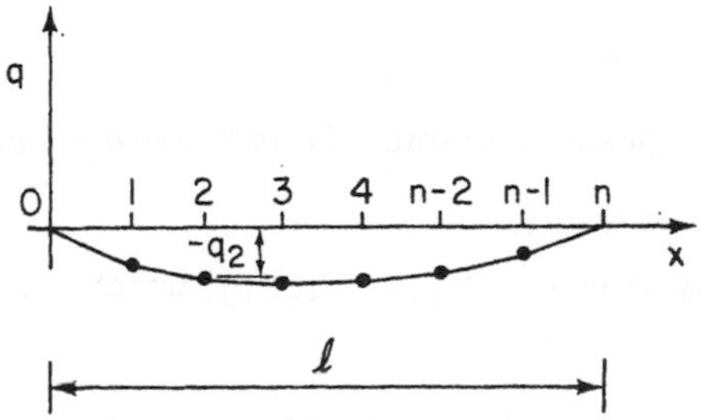

FIG. 5.1. Elastic string with point masses.

If we seek separated solutions of the form

$$q_1(t)=z_1q(t),$$
$$\vdots$$
$$q_{n-1}(t)=z_{n-1}q(t),$$

or, in vector form,

$$\boldsymbol{q}(t)=\boldsymbol{z}q(t),$$

in which the (discrete) spatial variable j and the temporal variable t are separated, we find that

$$-mz_iq''(t)=-p\frac{z_{i-1}-2z_i+z_{i+1}}{h}q(t)$$

or

$$-\frac{p(z_{i-1}-2z_i+z_{i+1})/h}{mz_i}=-\frac{q''(t)}{q(t)} \quad \text{for all } i \text{ and } t.$$

Both members of this equation must equal a constant, which we denote by λ. We are thus led to seek (λ, $z\neq 0$) such that

$$p(-z_{i-1}+2z_i-z_{i+1})/h=\lambda mz_i, \quad i=1,\ldots, n-1,$$

i.e., to seek eigenpairs (λ, z) of the matrix

$$ph^{-1}\begin{pmatrix} 2 & -1 & & & \\ -1 & 2 & -1 & & \\ & -1 & 2 & -1 & \\ & & & \ddots & -1 \\ & & & -1 & 2 \end{pmatrix} \tag{5.20}$$

relative to the matrix mI, and, for each eigenvalue λ, solutions to the differential equation

$$q''(t)+\lambda q(t)=0, \quad t>0. \tag{5.21}$$

The matrix (5.20) is positive-definite. Thus it has $n-1$ eigenvalues

$$0<\lambda_{1,h}\leqslant\lambda_{2,h}\leqslant\cdots\leqslant\lambda_{n-1,h}$$

and corresponding eigenvectors $z_{1,h},\ldots,z_{n-1,h}$, which satisfy

$$\lambda_{i,h}mz_{i,h}^{\mathrm{T}}z_{j,h}=\delta_{ij}.$$

$z_{1,h},\ldots,z_{n-1,h}$ thus form an orthonormal basis (i.e., are complete) in $(n-1)$-dimensional space. Corresponding to $\lambda_{j,h}$, the solutions of (5.21) are given by

$$q(t)=q_j(t)=a_j\sin\sqrt{\lambda_{j,h}}(t+\theta_j),$$

where a_j and θ_j are arbitrary. Thus the separated solutions are given by

$$z_{j,h}a_j\sin\sqrt{\lambda_{j,h}}(t+\theta_j), \quad j=1,\ldots,n-1. \tag{5.22}$$

As with the vibrating string, it is easily seen that all solutions of (5.19) can be written as the superposition of the separated solutions (5.22). These simple motions are called the eigenvibrations. The components of the jth eigenvibration all vibrate with some circular frequency $\sqrt{\lambda_{j,h}}$ and phase displacement $\sqrt{\lambda_{j,h}}\theta_j$, and the components are proportional to the components of $z_{j,h}$. Thus $\sqrt{\lambda_{j,h}}$ is the frequency and $z_{j,h}$ the shape of the jth eigenvibration.

A complete discussion of the vibration of a weightless elastic string loaded with several point masses can be found in COURANT and HILBERT [1953] and SYNGE and GRIFFITH [1959].

We now draw a parallel with the finite element problem (5.10). It follows immediately from (5.7), (5.9), and (5.20) that the eigenvalue problem that we obtained, i.e., the problem of finding the eigenpairs of the matrix in (5.20) relative to mI, is identical to the eigenvalue problem (5.10) provided $m=rh=rln^{-1}$. We have thus arrived at the following physical interpretation of (5.10): Consider the problem of a vibrating string with density r and tension p. Divide the total mass rl of the string into $n-1$ particles of mass $m=rln^{-1}$, which are placed at the points $x_1,\ldots,x_{n-1}$, and two particles of mass $rl(2n)^{-1}$, which are placed at x_0 and x_n. Then the eigenvalue problem corresponding to this system is identical to the problem (5.10) arrived at by approximating (5.2) by the finite element method (5.3), and then approximating the matrix B by $\tilde{B}$ via the trapezoid rule. Thus the finite element eigenvalue problem (5.10) is the same as the eigenvalue problem that arises when the mass of the string is "lumped" as indicated above.

The matrix A in (5.7) is called the *stiffness matrix* and B in (5.8) is called the *mass matrix*. Because of the physical analogy we have noted, $\tilde{B}$ is called the *lumped mass matrix* and, in contrast, B is sometimes referred to as the *consistent mass matrix*.

CHAPTER II

Abstract Spectral Approximation Results

In this chapter we present the abstract spectral approximation results we will use in the sequel.

6. Survey of spectral theory for compact operators

Since the differential operators we consider have compact inverses, our approximation results will be developed for the class of compact operators. We turn now to a survey of the spectral theory for compact operators. A complete development of this theory can be found in DUNFORD and SCHWARTZ [1963, Section XI.9].

Let $A: X \to X$ be a compact operator on a complex Banach space X with norm $\|\cdot\|_X = \|\cdot\|$. We denote by $\rho(A)$ the *resolvent set of* A, i.e., the set

$$\rho(A) = \{z\colon z \in \mathbb{C} \equiv \text{the complex numbers}, \\ (z-A)^{-1} \text{ exists as a bounded operator on } X\},$$

and by $\sigma(A)$ the *spectrum of* A, i.e., the set $\sigma(A) = \mathbb{C} \backslash \rho(A)$. For any $z \in \rho(A)$, $R_z(A) = (z-A)^{-1}$ is the *resolvent operator*. $\sigma(A)$ is countable with no nonzero limit points; nonzero numbers in $\sigma(A)$ are eigenvalues; and if zero is in $\sigma(A)$, it may or may not be an eigenvalue.

Let $\mu \in \sigma(A)$ be nonzero. There is a smallest integer α, called the *ascent of* $\mu - A$, such that $N((\mu-A)^\alpha) = N((\mu-A)^{\alpha+1})$, where N denotes the null space. $N((\mu-A)^\alpha)$ is finite-dimensional and $m = \dim N((\mu-A)^\alpha)$ is called the *algebraic multiplicity of* μ. The vectors in $N((\mu-A)^\alpha)$ are called *generalized eigenvectors of* A *corresponding to* μ. The *order of a generalized eigenvector* u is the smallest integer j such that $u \in N((\mu-A)^j)$. The generalized eigenvectors of order 1, i.e., the vectors in $N(\mu-A)$, are, of course, the *eigenvectors of* A *corresponding to* μ. The *geometric multiplicity of* μ is equal to $\dim N(\mu-A)$, and is less than or equal to the algebraic multiplicity. The ascent of $\mu-A$ is one and the two multiplicities are equal if X is a Hilbert space and A is selfadjoint; in this case the eigenvalues are real. If μ is an eigenvalue of A and f is a corresponding eigenvector, we will often refer to (μ, f) as an *eigenpair of* A.

Throughout this section we will consider a compact operator $T: X \to X$ and a family of compact operators $T_h: X \to X$, $0 < h \leqslant 1$, such that $T_h \to T$ in norm as $h \downarrow 0$.

Let μ be a nonzero eigenvalue of T with algebraic multiplicities m. Let Γ be a circle in the complex plane centered at μ which lies in $\rho(T)$ and which encloses no other points of $\sigma(T)$. The *spectral projection associated with T and* μ is defined by

$$E = E(\mu) = \frac{1}{2\pi i} \int_\Gamma R_z(T)\, dz.$$

E is a projection onto the space of generalized eigenvectors associated with μ and T, i.e., $R(E) = N((\mu - T)^\alpha)$, where R denotes the range. For h sufficiently small, $\Gamma \subset \rho(T_h)$ and the spectral projection

$$E_h = E_h(\mu) = \frac{1}{2\pi i} \int_\Gamma R_z(T_h)\, dz$$

exists, E_h converges to E in norm, and $\dim R(E_h(\mu)) = \dim R(E(\mu)) = m$. E_h is the *spectral projection associated with* T_h *and the eigenvalues of* T_h *which lie in* Γ and is a projection onto the direct sum of the spaces of generalized eigenvectors corresponding to these eigenvalues, i.e.,

$$R(E_h) = \sum_{\substack{\mu(h) \in \sigma(T_h) \\ \mu(h) \text{ inside } \Gamma}} N((\mu(h) - T_h)^{\alpha_{\mu(h)}}),$$

where $\alpha_{\mu(h)}$ is the ascent of $\mu(h) - T_h$. Thus, counting according to algebraic multiplicities, there are m eigenvalues of T_h in Γ; we denote these by $\mu_1(h), \ldots, \mu_m(h)$. Furthermore, if Γ' is another circle centered at μ with an arbitrarily small radius, we see that $\mu_1(h), \ldots, \mu_m(h)$ are all inside of Γ' for h sufficiently small, i.e., $\lim_{h \to 0} \mu_j(h) = \mu$ for $j = 1, \ldots, m$.

$R(E)$ and $R(E_h)$ are invariant subspaces for T and T_h, respectively, and $TE = ET$ and $T_h E_h = E_h T_h$. $\{R_z\{T_h\}: z \in \Gamma, h \text{ small}\}$ is bounded.

If μ is an eigenvalue of T with algebraic multiplicity m, then μ is an eigenvalue with algebraic multiplicity m of the adjoint operator T' on the dual space X'. The ascent of $\mu - T'$ will be α. E' will be the projection operator associated with T' and μ; likewise E_h' will be the projection operator associated with T_h' and $\mu_1(h), \ldots, \mu_m(h)$. If $f \in X$ and $f' \in X'$, we denote the value of the linear functional f' at f by $\langle f, f' \rangle$.

T' here is the Banach adjoint. If $X = H$ is a Hilbert space, we would naturally work with the Hilbert adjoint T^*, which acts on H. Then μ would be an eigenvalue of T if and only if $\bar{\mu}$ is an eigenvalue of T^*.

Given two closed subspaces M and N of X, we define

$$\delta(M,N) = \sup_{\substack{x \in M \\ \|x\| = 1}} \operatorname{dist}(x, N) \quad \text{and} \quad \hat{\delta}(M,N) = \max(\delta(M,N), \delta(N,M)).$$

$\hat{\delta}(M,N)$ is called the *gap between M and N*. The gap provides a natural way in which to formulate results on the approximation of generalized eigenvectors. We will need the following:

THEOREM 6.1. *If* $\dim M = \dim N < \infty$, *then* $\delta(N,M) \leqslant \delta(M,N)[1 - \delta(M,N)]^{-1}$.

For a discussion of this result and the result that $\delta(N,M)=\delta(M,N)$ if $X=H$ is a Hilbert space and $\hat{\delta}(M, N)<1$, we refer to KATO [1958].

7. Fundamental results on spectral approximation

In this section we present estimates which show how the eigenvalues and generalized eigenvectors of T are approximated by those of T_h. Estimates for this type of approximation were obtained by VAINIKKO [1964, 1967, 1970], BRAMBLE and OSBORN [1973], and OSBORN [1975]; our presentation follows OSBORN [1975]. We refer also to CHATELIN [1973, 1975, 1981], GRIGORIEFF [1975], CHATELIN and LEMORDANT [1978], STUMMEL [1977], and to the excellent and comprehensive monograph of CHATELIN [1983]. Let μ be a nonzero eigenvalue of T with algebraic multiplicity m and assume the ascent of $\mu-T$ is α. Let $\mu_1(h),\ldots, \mu_m(h)$ be the eigenvalues of T_h that converge to μ.

THEOREM 7.1. *There is a constant C independent of h, such that*

$$\hat{\delta}(R(E), R(E_h))\leqslant C\|(T-T_h)|_{R(E)}\| \tag{7.1}$$

for small h, where $(T-T_h)|_{R(E)}$ denotes the restriction of $T-T_h$ to $R(E)$.

PROOF. For $f\in R(E)$ with $\|f\|=1$ we have $\|f-E_hf\|=\|(E-E_h)f\|\leqslant\|E-E_h\|$. Thus, since E_h converges to E in norm, $\lim_{h\to 0}\delta(R(E), R(E_h))=0$. Using Theorem 6.1, with $M=R(E)$ and $N=R(E_h)$, we thus have

$$\begin{aligned}\delta(R(E_h), R(E))&\leqslant\delta(R(E), R(E_h))[1-\delta(R(E), R(E_h))]^{-1}\\&\leqslant 2\delta(R(E), R(E_h))\end{aligned}$$

and hence

$$\hat{\delta}(R(E), R(E_h))\leqslant 2\delta(R(E), R(E_h)) \tag{7.2}$$

for small h.

Now for $f\in R(E)$ we have

$$\begin{aligned}\|f-E_hf\|=\|Ef-E_hf\|&=\left\|\frac{1}{2\pi i}\int_\Gamma[R_z(T)-R_z(T_h)]f\,dz\right\|\\&=\left\|\frac{1}{2\pi i}\int_\Gamma R_z(T_h)(T-T_h)R_z(T)f\,dz\right\|\end{aligned}$$

and hence, recalling that $R(E)$ is invariant for T and thus for $R_z(T)$,

$$\|f-E_hf\|\leqslant\frac{1}{2\pi}\text{length}(\Gamma)\sup_{z\in\Gamma}\|R_z(T_h)\|\,\|(T-T_h)|_{R(E)}\|\sup_{z\in\Gamma}\|R_z(T)\|\,\|f\|. \tag{7.3}$$

As noted above, $\sup_{z\in\Gamma}\|R_z(T_h)\|$ is bounded in h. Thus from (7.2) and (7.3) we have

$$\hat{\delta}(R(E), R(E_h)) \leqslant C\|(T-T_h)|_{R(E)}\|,$$

where

$$C=\frac{1}{\pi}\text{length}(\Gamma)\sup_{\substack{z\in\Gamma\\0<h}}\|R_z(T_h)\|\sup_{z\in\Gamma}\|R_z(T)\|. \qquad \square$$

REMARK 7.1. The proof of Theorem 7.1 also shows that

$$\|(E-E_h)|_{R(E)}\| \leqslant C\|(T-T_h)|_{R(E)}\|.$$

Although each of the eigenvalues $\mu_1(h), \ldots, \mu_m(h)$ is close to μ for small h, their arithmetic mean is generally a closer approximation to μ (cf. BRAMBLE and OSBORN [1973]). Thus we define

$$\hat{\mu}(h)=\frac{1}{m}\sum_{j=1}^{m}\mu_j(h).$$

Our next theorem gives an estimate for $\mu-\hat{\mu}(h)$.

THEOREM 7.2. *Let $\phi_1, \ldots, \phi_m$ be any basis for $R(E)$ and let $\phi'_1, \ldots, \phi'_m$ be the dual basis in $R(E')$, as defined in the proof to follow. Then there is a constant C, independent of h, such that*

$$|\mu-\hat{\mu}(h)| \leqslant \frac{1}{m}\sum_{j=1}^{m}|\langle(T-T_h)\phi_j, \phi'_j\rangle| + C\|(T-T_h)|_{R(E)}\| \, \|(T'-T'_h)|_{R(E')}\|. \qquad (7.4)$$

PROOF. For small h, the operator $E_h|_{R(E)}: R(E)\to R(E_h)$ is one-to-one since $\|E-E_h\|\to 0$ and $E_h f=0$, $f\in R(E)$ implies

$$\|f\|=\|Ef-E_hf\|\leqslant\|E-E_h\| \, \|f\|,$$

and $E_h|_{R(E)}$ is onto since

$$\dim R(E_h)=\dim R(E)=m.$$

Thus $(E_h|_{R(E)})^{-1}: R(E_h)\to R(E)$ is defined; we write E_h^{-1} for $(E_h|_{R(E)})^{-1}$. For h sufficiently small and $f\in R(E)$ with $\|f\|=1$ we have

$$1-\|E_hf\|=\|Ef\|-\|E_hf\|\leqslant\|E-E_h\|\leqslant\tfrac{1}{2}$$

and hence $\|E_hf\|\geqslant\frac{1}{2}\|f\|$. This implies $\|E_h^{-1}\|$ is bounded in h. We note that $E_hE_h^{-1}$ is the identity on $R(E_h)$ and $E_h^{-1}E_h$ is the identity on $R(E)$. Now we define

$$\hat{T}_h=E_h^{-1}T_hE_h|_{R(E)}: R(E)\to R(E).$$

Using the fact that $R(E_h)$ is invariant for T_h we see that $\sigma(\hat{T}_h)=\{\mu_1(h), \ldots, \mu_m(h)\}$ and that the algebraic (geometric, respectively) multiplicity of any $\mu_j(h)$ as an eigenvalue of $\hat{T}_h$ is equal to its algebraic (geometric, respectively) multiplicity as an eigenvalue of T_h. Letting $\hat{T}=T|_{R(E)}$, we likewise see that $\sigma(\hat{T})=\{\mu\}$. Thus trace $\hat{T}=m\mu$ and trace $\hat{T}_h=m\hat{\mu}(h)$ and, since $\hat{T}$ and $\hat{T}_h$ act on the same space, we can write

$$\mu-\hat{\mu}(h)=\frac{1}{m}\operatorname{trace}(\hat{T}-\hat{T}_h). \tag{7.5}$$

Let $\phi_1,\ldots,\phi_m$ be a basis for $R(E)$ and let $\phi'_1,\ldots,\phi'_m$ be the dual basis to $\phi_1,\ldots,\phi_m$. Then from (7.5) we get

$$\mu-\hat{\mu}(h)=\frac{1}{m}\operatorname{trace}(\hat{T}-\hat{T}_h)=\frac{1}{m}\sum_{j=1}^{m}\langle(\hat{T}-\hat{T}_h)\phi_j,\phi'_j\rangle. \tag{7.6}$$

Here each ϕ'_j is an element of $R(E)'$, the dual space of $R(E)$, but we can extend each ϕ'_j to all of X as follows. Since $X=R(E)\oplus N(E)$, any $f\in X$ can be written as $f=g+h$ with $g\in R(E)$ and $h\in N(E)$. Define $\langle f,\phi'_j\rangle=\langle g,\phi'_j\rangle$. Clearly ϕ'_j, so extended, is bounded, i.e., $\phi'_j\in X'$. Now $\langle f,(\mu-T')^\alpha\phi'_j\rangle=\langle(\mu-T)^\alpha f,\phi'_j\rangle$ vanishes for all f. This follows from the observation that it obviously vanishes for $f\in R(E)=N((\mu-T)^\alpha)$ and it vanishes for $f\in N(E)$ since $N(E)$ is invariant for $\mu-T$. Thus we have shown that $\phi'_1,\ldots,\phi'_m\in R(E')$.

Using the facts that $T_hE_h=E_hT_h$ and $E_h^{-1}E_h$ is the identity on $R(E)$, we have

$$\begin{aligned}\langle(\hat{T}-\hat{T}_h)\phi_j,\phi'_j\rangle&=\langle T\phi_j-E_h^{-1}T_hE_h\phi_j,\phi'_j\rangle\\&=\langle E_h^{-1}E_h(T-T_h)\phi_j,\phi'_j\rangle\\&=\langle(T-T_h)\phi_j,\phi'_j\rangle+\langle(E_h^{-1}E_h-I)(T-T_h)\phi_j,\phi'_j\rangle.\end{aligned} \tag{7.7}$$

Let $L_h=E_h^{-1}E_h$. L_h is the projection on $R(E)$ along $N(E_h)$. Hence L'_h is the projection on $N(E_h)^\perp=R(E'_h)$ along $R(E)^\perp=N(E')$. Thus

$$\langle(E_h^{-1}E_h-I)(T-T_h)\phi_j,\phi'_j\rangle=\langle(L_h-I)(T-T_h)\phi_j,(E'-E'_h)\phi'_j\rangle. \tag{7.8}$$

From (7.8), the boundedness of L_h, and Remark 7.1 (applied to T' and $\{T'_h\}$) we have

$$\begin{aligned}&|\langle(E_h^{-1}E_h-I)(T-T_h)\phi_j,\phi'_j\rangle|\\&\quad\leqslant\Big(\sup_h\|L_h-I\|\Big)\|(T-T_h)|_{R(E)}\|\;\|(E'-E'_h)|_{R(E)}\|\;\|\phi_j\|\;\|\phi'_j\|\\&\quad\leqslant C\|(T-T_h)|_{R(E)}\|\,\|(T'-T'_h)|_{R(E')}\|.\end{aligned} \tag{7.9}$$

Finally, (7.6), (7.7), and (7.9) yield the desired result. □

REMARK 7.2. Our treatment of the term $\langle(E_h^{-1}E_h-I)(T-T_h)\phi_j,\phi'_j\rangle$, which differs from that in OSBORN [1975], was suggested by DESCLOUX, NASSIF and RAPPAZ [1978b].

REMARK 7.3. A slight modification of the proof of Theorem 7.2 shows that for any $1\leqslant\bar{i},\bar{j}\leqslant m$, $|\langle(T-T_h)\phi_{\bar{j}},\phi'_{\bar{i}}\rangle|$ is bounded by $C\delta_h$, where

$$\delta_h=\sum_{i,j=1}^{m}|\langle(T-T_h)\phi_j,\phi'_i\rangle|+\|(T-T_h)|_{R(E)}\|\;\|(T'-T'_h)|_{R(E')}\|.$$

Noting that $\langle(\hat{T}-\hat{T}_h)\phi_j, \phi_i'\rangle$ is a matrix representation of $\hat{T}-\hat{T}_h$, we see that

$$\|\hat{T}-\hat{T}_h\| \leq C\delta_h. \tag{7.10}$$

Since it is immediate that

$$|\mu-\hat{\mu}(h)| = \frac{1}{m}|\text{trace}(\hat{T}-\hat{T}_h)| \leq \|\hat{T}-\hat{T}_h\|,$$

from (7.10) we get

$$|\mu-\hat{\mu}(h)| \leq C\delta_h, \tag{7.11}$$

an estimate that is similar to, and of equal use in applications as, (7.4).

We also have

$$\begin{aligned}\left|\mu^{-1}-\left(\sum_j \mu_j(h)^{-1}\right)\Big/ m\right| &= \frac{1}{m}|\text{trace}(\hat{T}^{-1}-\hat{T}_h^{-1})| \\ &\leq \|\hat{T}^{-1}-\hat{T}_h^{-1}\| = \|\hat{T}^{-1}(\hat{T}-\hat{T}_h)\hat{T}_h^{-1}\| \\ &\leq \|\hat{T}^{-1}\|\,\|\hat{T}_h-\hat{T}\|\,\|\hat{T}_h^{-1}\| \leq C\|\hat{T}_h-\hat{T}\|.\end{aligned}$$

Hence we see that

$$\left|\mu^{-1}-\left(\sum_j \mu_j(h)^{-1}\right)\Big/ m\right| \leq C\delta_h. \tag{7.12}$$

It is also know that

$$|\mu-\mu_j(h)|^\alpha \leq C\|\hat{T}-\hat{T}_h\| \tag{7.13}$$

for any $1 \leq j \leq m$. Hence

$$|\mu-\mu_j(h)|^\alpha \leq C\delta_h. \tag{7.14}$$

Equation (7.14) is established directly in Theorem 7.3. We note, however, that the proof of Theorem 7.3 is closely related to one of the ways of proving (7.13).

REMARK 7.4. It follows immediately from (7.4) that

$$|\mu-\hat{\mu}(h)| \leq C\|(T-T_h)|_{R(E)}\|.$$

However, the second term on the right-hand side of (7.4) is of higher order than $\|(T-T_h)|_{R(E)}\|$, namely of order

$$\|(T-T_h)|_{R(E)}\|\,\|(T'-T_h')|_{R(E)}\|.$$

We will also see that in a large variety of applications, $\Sigma_{j=1}^m\langle(T-T_h)\phi_j, \phi_j'\rangle$ is of higher order than $\|(T-T_h)|_{R(E)}\|$.

In addition to estimating $\mu-\hat{\mu}(h)$ we may estimate $\mu-\mu_j(h)$ for each j.

THEOREM 7.3. *Let α be the ascent of $\mu-T$. Let $\phi_1,\ldots,\phi_m$ be any basis for $R(E)$ and let $\phi'_1,\ldots,\phi'_m$ be the dual basis. Then there is a constant C such that*

$$|\mu-\mu_j(h)|\leqslant C\left\{\sum_{i,k=1}^{m}|\langle(T-T_h)\phi_i,\phi'_k\rangle|+\|(T-T_h)|_{R(E)}\|\,\|(T'-T'_h)|_{R(E')}\|\right\}^{1/\alpha},\tag{7.15}$$

$$j=1,\ldots,m.$$

PROOF. For each h, $\mu_j(h)$ is one of the eigenvalues of $\hat{T}_h$. Let $\hat{T}_h w_h=\mu_j(h)w_h$, $\|w_h\|=1$. We can choose $w'_h\in N((\mu-T')^\alpha)$ in such a way that $\langle w_h,w'_h\rangle=1$ and the norms $\|w'_h\|$ are bounded in h. First, using the Hahn–Banach theorem, choose $w'_h\in R(E)'$ such that $\langle w_h,w'_h\rangle=1$ and $\|w'_h\|=1$; then extend w'_h to all of X as in the proof of Theorem 7.2. w'_h, so extended will be in $R(E')$ and satisfy $\|w'_h\|\leqslant\|E\|$. Now, noting that $(T'-\mu)^\alpha w'_h=0$, we have

$$\begin{aligned}
&|\mu-\mu_j(h)|^\alpha\\
&\quad=|\langle(\mu-\mu_j(h))^\alpha w_h,w'_h\rangle|\\
&\quad=|\langle((\mu-\mu_j(h))^\alpha-(\mu-T)^\alpha)w_h,w'_h\rangle|\\
&\quad=\left|\left\langle\sum_{k=0}^{\alpha-1}(\mu-\mu_j(h))^k(\mu-T)^{\alpha-1-k}(\mu_j(h)-T)w_h,w'_h\right\rangle\right|\\
&\quad\leqslant\sum_{k=0}^{\alpha-1}|\mu-\mu_j(h)|^k|\langle(\mu_j(h)-T)w_h,(\mu-T')^{\alpha-1-k}w'_h\rangle|\\
&\quad\leqslant\sum_{k=0}^{\alpha-1}|\mu-\mu_j(h)|^k\max_{\substack{\phi'\in R(E')\\ \|\phi'\|=1}}|\langle(\mu_j(h)-T)w_h,\phi'\rangle|\|\mu-T'\|^{\alpha-1-k}\|w'_h\|.
\end{aligned}\tag{7.16}$$

For any $\phi'\in R(E')$ with $\|\phi'\|=1$,

$$\begin{aligned}
&|\langle(\mu_j(h)-T)w_h,\phi'\rangle|\\
&\quad=|\langle(\hat{T}_h-T)w_h,\phi'\rangle|\\
&\quad=|\langle E_h^{-1}E_h(T_h-T)w_h,\phi'\rangle|\\
&\quad=|\langle(T-T_h)w_h,\phi'\rangle+\langle(E_h^{-1}E_h-I)(T-T_h)w_h,\phi'\rangle|\\
&\quad\leqslant|\langle(T_h-T)w_h,\phi'\rangle|+C\|(T-T_h)|_{R(E)}\|\,\|(T'-T'_h)|_{R(E')}\|.
\end{aligned}\tag{7.17}$$

There is obviously a constant C' such that

$$|\langle(T_h-T)w_h,\phi'\rangle|\leqslant C'\sum_{i,k=1}^{m}|\langle(T_h-T)\phi_i,\phi'_k\rangle|\tag{7.18}$$

for all $w_h\in R(E)$ and $\phi'\in R(E')$ with $\|w_h\|=\|\phi'\|=1$. From (7.16)–(7.18) we get the desired result. □

Theorem 7.1 shows how the generalized eigenvectors of T are approximated by those of T_h. Our next result concerns the proximity of certain elements of $R(E_h)$ to certain elements of $R(E)$. It shows, for example, that eigenvectors of T_h are close to eigenvectors of T.

THEOREM 7.4. *Let $\mu(h)$ be an eigenvalue of T_h such that $\lim_{h\to 0}\mu(h)=\mu$. Suppose for each h that w_h is a unit vector satisfying $(\mu(h)-T_h)^k w_h=0$ for some positive integer $k\leqslant\alpha$. Then, for any integer l with $k\leqslant l\leqslant\alpha$, there is a vector u_h such that $(\mu-T)^l u_h=0$ and*

$$\|u_h-w_h\|\leqslant C\|(T-T_h)|_{R(E)}\|^{(l-k+1)/\alpha}. \tag{7.19}$$

PROOF. Since $N((\mu-T)^l)$ is finite-dimensional, there is a closed subspace M of X such that $X=N((\mu-T)^l)\oplus M$. For $y\in R((\mu-T)^l)$, the equation $(\mu-T)^l x=y$ is uniquely solvable in M. Thus $(\mu-T)^l|_M\colon M\to R((\mu-T)^l)$ is one-to-one and onto. Hence $(\mu-T)^l|_M^{-1}\colon R((\mu-T)^l)\to M$ exists and, by the closed graph theorem, is bounded. Thus there is a constant C such that

$$\|f\|\leqslant C\|(\mu-T)^l f\|\quad\text{for all } f\in M.$$

Set $u_h=Pw_h$, where P is the projection on $N((\mu-T)^l)$ along M. Then $(\mu-T)^l u_h=0$ and $w_h-u_h\in M$, and hence

$$\|w_h-u_h\|\leqslant C\|(\mu-T)^l(w_h-u_h)\|. \tag{7.20}$$

By Theorem 7.1 there are vectors $\tilde{u}_h\in R(E)$ such that

$$\|w_h-\tilde{u}_h\|\leqslant C'\|(T-T_h)|_{R(E)}\|.$$

Hence there is a constant C'' such that

$$\begin{aligned}
&\|[(\mu-T)^l-(\mu-T_h)^l]w_h\|\\
&\quad=\left\|\sum_{j=0}^{l-1}(\mu-T_h)^j(T-T_h)(\mu-T)^{l-j-1}[(w_h-\tilde{u}_h)+\tilde{u}_h]\right\|\\
&\quad\leqslant C''\|(T-T_h)|_{R(E)}\|.
\end{aligned}\tag{7.21}$$

Since $k\leqslant l$,

$$\begin{aligned}
\|(\mu-T_h)^l w_h\|&=\left\|\sum_{j=0}^{l-1}\binom{l}{j}(\mu-\mu(h))^j(\mu(h)-T_h)^{l-j}w_h\right\|\\
&=\left\|\sum_{j=l-k+1}^{l}\binom{l}{j}(\mu-\mu(h))^j(\mu(h)-T_h)^{l-j}w_h\right\|\\
&\leqslant C'''|\mu-\mu(h)|^{l-k+1}.
\end{aligned}\tag{7.22}$$

Combining (7.20)–(7.22) we get

$$\begin{aligned}\|w_h-u_h\| &\leqslant C\|(\mu-T)^l(w_h-u_h)\|\\ &\leqslant C\|(\mu-T)^l w_h\|\\ &= C\|[(\mu-T)^l-(\mu-T_h)^l]w_h+(\mu-T_h)^l w_h\|\\ &\leqslant C[C''\|(T-T_h)|_{R(E)}\|+C'''|\mu-\mu(h)|^{l-k+1}].\end{aligned}$$

The result now follows immediately from Theorem 7.3. □

REMARK 7.5. If $X=H$ is a Hilbert space, we let T^* and T_h^* denote the Hilbert adjoints of T and T_h, respectively. In Theorems 7.2 and 7.3 we would let $\phi_1,\ldots,\phi_m$ be an orthonormal basis for $R(E)$ and let $\phi_j^*=E^*\phi_j$. Then $\phi_1^*,\ldots,\phi_m^*\in N((\bar{\mu}-T^*)^\alpha)$ and trace

$$(\hat{T}-\hat{T}_h)=\sum_{j=1}^{m}((\hat{T}-\hat{T}_h)\phi_j,\phi_j^*),$$

where $(\cdot,\cdot)=(\cdot,\cdot)_H$ denotes the inner product on H, and with only minor modifications all the results of this section remain valid.

We end this section by specializing the results in Theorems 7.1–7.4 to the case where $X=H$ is a Hilbert space and T and T_h are selfadjoint. If μ is a nonzero eigenvalue of T, then, as noted above, the ascent α of $\mu-T$ is one and the algebraic and geometric multiplicities of μ are equal. Likewise the eigenvalues $\mu_j(h)$ of T_h which converge to μ have equal algebraic and geometric multiplicities. μ and $\mu_j(h)$ are, of course, real.

Thus, under the present hypotheses, Theorems 7.2 and 7.3 give the estimate

$$|\mu-\mu_j(h)|\leqslant C\left\{\sum_{i,j=1}^{m}|((T-T_h)\phi_i,\phi_j^*)|+\|(T-T_h)|_{R(E)}\|^2\right\},$$

$$j=1,\ldots,m.$$

Now consider Theorem 7.4 in the selfadjoint case. Suppose $\mu(h)$ is an eigenvalue of T_h converging to μ. If w_h is a unit eigenvector of T_h corresponding to $\mu(h)$, then it follows immediately from Theorem 7.1 and the definition of $\hat{\delta}(R(E),R(E_h))$ that there is an eigenvector u_h of T corresponding to μ such that

$$\|u_h-w_h\|\leqslant C\|(T-T_h)|_{R(E)}\|.$$

This is Theorem 7.4 in the case $\alpha=1$. We further note that one may assume $\|u_h\|=1$. From Theorem 7.1 we can also conclude that if u is a unit eigenvector of T corresponding to μ then there is a unit eigenvector $w_h\in R(E_h)$ of T_h such that

$$\|u-w_h\|\leqslant C\|(T-T_h)|_{R(E)}\|.$$

Compare the discussion of the Ritz method near the end of Section 8.

REMARK 7.6. In the selfadjoint case one may assume the Hilbert space H is real (cf. Remark 4.1). Starting with a real space H we can in the usual way obtain a complex space by complexifying. Then the contour integrals

$$\frac{1}{2\pi i}\int_\Gamma R_z(T)\,dz \quad\text{and}\quad \frac{1}{2\pi i}\int_\Gamma R_z(T_h)\,dz,$$

which are the fundamental tools in the analysis, can be introduced and the results derived. The results will be in the complex context but can immediately be translated to the real context.

REMARK 7.7. Results for noncompact operators T which parallel those in this section were proved by DESCLOUX, NASSIF, and RAPPAZ [1978]. See Remark 13.4.

8. Spectral approximation of variationally formulated eigenvalue problems

As explained in Section 3, eigenvalue problems can be given variational formulations. For the most part, we will consider eigenvalue problems formulated in this manner. In this section we will first sketch the functional analysis framework for variationally formulated eigenvalue problems and then discuss their approximation. Results of the type presented in this section, specifically Theorems 8.1 and 8.3, were proved by BABUŠKA and AZIZ [1973, Chapter 12] and FIX [1973] for the case of an eigenvalue with multiplicity one; in the general case they were proved by KOLATA [1978]. Our treatment is similar to Kolata's.

Let H_1 and H_2 be complex Hilbert spaces with inner products and norms $(\cdot,\cdot)_1$ and $\|\cdot\|_1$ and $(\cdot,\cdot)_2$ and $\|\cdot\|_2$, respectively. Let $a(\cdot,\cdot)$ be a bilinear (or sesquilinear) form on $H_1 \times H_2$ satisfying

$$|a(u,v)| \leqslant C_1\|u\|_1\|v\|_2 \qquad \forall u\in H_1,\quad \forall v\in H_2, \tag{8.1}$$

$$\inf_{\substack{u\in H_1\\ \|u\|_1=1}} \sup_{\substack{v\in H_2\\ \|v\|_2=1}} |a(u,v)| = \alpha > 0, \tag{8.2}$$

and

$$\sup_{v\in H_1} |a(u,v)| > 0 \qquad \forall v\in H_2 \quad \text{with } v\neq 0. \tag{8.3}$$

The Riesz representation theorem and (8.1) imply that there is a bounded linear map A from H_1 to H_2 such that $a(u,v)=(Au,v)_2$ for all $u\in H_1$, $v\in H_2$. The adjoint A' is a bounded map from H_2 to H_1 satisfying $a(u,v)=(u,A'v)_1$ for all $u\in H_1$, $v\in H_2$. Equations (8.1), (8.2), and (8.3) imply that A is an isomorphism of H_1 onto H_2. In fact, in the presence of (8.1), equations (8.2) and (8.3) hold if and only if A is an isomorphism, cf. BABUŠKA [1971] and BABUŠKA and AZIZ [1973, Chapter 5]. Using the fact that A is an isomorphism if and only if A' is an isomorphism we see that in the presence of (8.1), (8.2) and (8.3) hold if and only if

$$\inf_{\substack{v\in H_2\\ \|v\|_2=1}} \sup_{\substack{u\in H_1\\ \|u\|_1=1}} |a(u,v)| = \alpha > 0 \tag{8.4}$$

and

$$\sup_{v\in H_2} |a(u, v)| > 0 \qquad \forall u \in H_2 \quad \text{with } u \neq 0. \tag{8.5}$$

(8.2) and (8.3) (or (8.4) and (8.5)) are called the inf–sup conditions.

Suppose $\|\cdot\|'_1$ is a second norm on H_1 which is compact with respect to $\|\cdot\|_1$, i.e., every sequence in H_1 which is bounded in $\|\cdot\|_1$ has a subsequence with is Cauchy in $\|\cdot\|'_1$. Let $b(u, v)$ be a bilinear form on $H_1 \times H_2$ satisfying

$$|b(u, v)| \leqslant C_2 \|u\|'_1 \|v\|_2 \qquad \forall u \in H_1, \quad v \in H_2. \tag{8.6}$$

We remark that in many applications the form $b(u, v)$ is defined on $W_1 \times W_2$, where

$H_1 \subset W_1$ with a compact imbedding,

$H_2 \subset W_2$ with a bounded imbedding,

and satisfies

$$|b(u, v)| \leqslant C_2 \|u\|_{W_1} \|v\|_{W_2} \qquad \forall u \in W_1, \quad u \in W_2. \tag{8.7}$$

If $\|\cdot\|'_1 = \|\cdot\|_{W_1}$, then it is immediate that $\|\cdot\|'_1$ is compact with respect to $\|\cdot\|_1$ and that (8.6) holds.

It is shown in BABUŠKA [1971] and BABUŠKA and AZIZ [1973, Chapter 5] that (8.1)–(8.3) imply there are unique bounded operators $T: H_1 \to H_1$ and $T_*: H_2 \to H_2$ satisfying

$$\begin{aligned} a(Tu, v) &= b(u, v) \qquad && \forall u \in H_1, \quad \forall v \in H_2, \\ a(u, T_* v) &= b(u, v) && \forall u \in H_1, \quad \forall v \in H_2. \end{aligned} \tag{8.8}$$

Furthermore

$$\|Tu\|_1 \leqslant (C_2/\alpha)\|u\|'_1 \quad \forall u \in H_1. \tag{8.9}$$

If u_j is a bounded sequence in H_1, then, since $\|\cdot\|'_1$ is compact with respect to $\|\cdot\|_1$, we know there is a subsequence u_{j_l} that is Cauchy in $\|\cdot\|'_1$. It then follows immediately from (8.9), applied to $u_{j_l} - u_{j_k}$, that Tu_{j_l} is Cauchy, and hence convergent, in H_1. Thus $T: H_1 \to H_1$ is compact. It is immediate that $a(Tu, v) = a(u, T_* v)$. The operator T_* is related to T^*, the usual adjoint of T on H_1, by the transformation $T^* = A' T_* A'^{-1}$. T^* and T_* are compact.

A complex number λ is called an *eigenvalue of the form a relative to the form b* if there is a nonzero vector $u \in H_1$, called an associated *eigenvector*, satisfying

$$a(u, v) = \lambda b(u, v) \quad \forall v \in H_2. \tag{8.10}$$

It is easily seen from (8.8) that (λ, u) satisfies (8.10) if and only if $\lambda Tu = u$, i.e., if and only if (λ^{-1}, u) is an eigenpair of the compact operator T. (8.10) is referred to as a *variationally posed eigenvalue problem* (cf. (3.15)). The notions of ascent, generalized eigenvector, and algebraic and geometric multiplicities are defined in terms of T. The generalized eigenvectors of T corresponding to λ can, however, be characterized in terms of the forms $a(\cdot,\cdot)$ and $b(\cdot,\cdot)$. u^j is a generalized eigenvector of order $j > 1$ if and only if $a(u^j, v) = \lambda b(u^j, v) + \lambda a(u^{j-1}, v)$ for all $v \in H_2$, where u^{j-1} is a generalized eigenvector of order $j-1$. Since $T_* = A'^{-1} T^* A'$, it is immediate that $\sigma(T_*) = \sigma(T^*)$ and that $N((\lambda^{-1} - T_*)^j) = A'^{-1}\{N((\overline{\lambda^{-1}} - T^*)^j)\}$. From this we see that the generalized eigenvectors of T_* have a similar characterization to those of T, namely, $a(u, v^j) = \lambda b(u, v^j) + \lambda a(u, v^{j-1})$ for all $u \in H_1$. In particular, $(\overline{\lambda^{-1}}, v)$ is an eigenpair of T_* if and only if $a(u, v) = \lambda b(u, v)$ for all $u \in H_1$, i.e., (λ, v) is an *adjoint eigenpair* of (8.10).

In order to construct approximations to the eigenvalues and eigenvectors of (8.10) we select finite dimensional subspaces $S_{1,h} \subset H_1$ and $S_{2,h} \subset H_2$, indexed by a parameter h, that satisfy

$$\inf_{\substack{u \in S_{1,h} \\ \|u\|_1 = 1}} \sup_{\substack{v \in S_{2,h} \\ \|v\|_2 = 1}} |a(u, v)| \geq \beta = \beta(h) > 0 \tag{8.11}$$

and

$$\sup_{u \in S_{1,h}} |a(u, v)| > 0 \qquad \text{for each } v \in S_{2,h} \quad \text{with } v \neq 0. \tag{8.12}$$

We also assume

$$\forall u \in H_1, \quad \lim_{h \to 0} \beta(h)^{-1} \inf_{\chi \in S_{1,h}} \|u - \chi\|_1 = 0. \tag{8.13}$$

We note that if $\dim S_{1,h} = \dim S_{2,h}$, then (8.12) follows from (8.11). We assume $\dim S_{1,h} = \dim S_{2,h}$ for the remainder of this article. $S_{1,h}$ and $S_{2,h}$ are referred to as test and trial spaces, respectively, and, if they consist of piecewise polynomial functions they are called finite element (approximation) spaces. Equation (8.11) is referred to as the discrete inf–sup condition.

We then consider eigenpairs of the form a relative to the form b, but now restricted to $S_{1,h} \times S_{2,h}$, i.e., pairs (λ_h, u_h), where λ_h is a number and $0 \neq u_h \in S_{1,h}$, satisfying

$$a(u_h, v) = \lambda_h b(u_h, v) \quad \forall v \in S_{2,h}, \tag{8.14}$$

and use λ_h and u_h as approximations to λ and u, respectively. Equation (8.14) is called a *variational approximation method* or *Galerkin method* for (8.10) in general and, if $S_{1,h}$ and $S_{2,h}$ consist of piecewise polynomial functions, it is called a *finite element method*. Since $N = \dim S_{1,h} = \dim S_{2,h} < \infty$, (8.14) is equivalent to a matrix eigenvalue problem. In fact, if $\phi_1, \ldots, \phi_N$ and $\psi_1, \ldots, \psi_N$ are bases for $S_{1,h}$ and $S_{2,h}$, respectively, then $(\lambda_h, u_h = \Sigma_{j=1}^N z_j \phi_j)$ is an eigenpair of (8.14) if and only if

$$Az = \lambda_h Bz, \tag{8.15}$$

where $z = (z_1, \ldots, z_N)^{\mathrm{T}}$,

$$A = (A_{ij}), \quad A_{ij} = a(\phi_j, \psi_i),$$

$$B = (B_{ij}), \quad B_{ij} = b(\phi_j, \psi_i).$$

(λ_h, u_h) is an eigenpair of (8.14) if and only if (λ_h^{-1}, u_h) is an eigenpair of the compact operator $T_h : H_1 \to S_{1,h}$ defined by

$$a(T_h u, v) = b(u, v) \qquad \forall u \in H_1, \quad v \in S_{2,h}. \tag{8.16}$$

The operator T_h can be written as $P_h T$, where P_h is the projection of H_1 onto $S_{1,h}$ defined by

$$a(P_h u, v) = a(u, v) \qquad \forall u \in H_1, \quad v \in S_{2,h}. \tag{8.17}$$

Using the central result in BABUŠKA [1971] and BABUŠKA and AZIZ [1973, Chapter 6], it follows from (8.1)–(8.3), (8.11), and (8.17) that

$$\|u - P_h u\|_1 \leqslant \left(1 + \frac{C_1}{\beta(h)}\right) \inf_{\chi \in S_{1,h}} \|u - \chi\|_1.$$

Thus from (8.13) we see that $P_h \to I$ pointwise. Since T is compact, $T_h = P_h T \to T$ in norm on H_1.

Let λ be an eigenvalue of (8.10) with algebraic multiplicity m, by which we mean that λ^{-1} is an eigenvalue of T with algebraic multiplicity m. Let α = ascent of $\lambda^{-1} - T$. Since $T_h \to T$ in norm, m eigenvalues $\lambda_1(h), \ldots, \lambda_m(h)$ of (8.14) will converge to λ. The $\lambda_j(h)$ are counted according to the algebraic multiplicities of the $\mu_j(h) = \lambda_j(h)^{-1}$ as eigenvalues of T_h. Let

$$M = M(\lambda) = \{u\colon u \text{ a generalized eigenvector of (8.10) corresponding to } \lambda,\ \|u\|_1 = 1\}, \tag{8.18}$$

$$M^* = M^*(\lambda) = \{v\colon v \text{ a generalized adjoint eigenvector of (8.10) corresponding to } \lambda,\ \|v\|_2 = 1\}, \tag{8.19}$$

$$M_h = M_h(\lambda) = \{u\colon u \text{ in the direct sum of the generalized eigenspace of (8.14) corresponding to the eigenvalues } \lambda_j(h) \text{ that converge to } \lambda,\ \|u\|_1 = 1\}, \tag{8.20}$$

and define

$$\varepsilon_h = \varepsilon_h(\lambda) = \sup_{u \in M(\lambda)} \inf_{\chi \in S_{1,h}} \|u - \chi\|_1, \tag{8.21}$$

$$\varepsilon_h^* = \varepsilon_h^*(\lambda) = \sup_{v \in M^*(\lambda)} \inf_{\eta \in S_{2,h}} \|v - \eta\|_2. \tag{8.22}$$

Let $\bar{M}(\lambda) = R(E)$ and $\bar{M}_h(\lambda) = R(E_h)$.

We now state and prove four results which correspond to Theorems 7.1–7.4. Let α denote the ascent of $\lambda^{-1} - T$.

THEOREM 8.1. *There is a constant C such that*

$$\delta(\bar{M}(\lambda), \bar{M}_h(\lambda)) \leqslant C\beta(h)^{-1} \varepsilon_h. \tag{8.23}$$

THEOREM 8.2. *There is a constant C such that*

$$\left|\lambda - \left(\frac{1}{m} \sum_{j=1}^{m} \lambda_j(h)^{-1}\right)^{-1}\right| \leqslant C\beta(h)^{-1} \varepsilon_h \varepsilon_h^*. \tag{8.24}$$

THEOREM 8.3. *There is a constant C such that*

$$|\lambda - \lambda_j(h)| \leqslant C[\beta(h)^{-1} \varepsilon_h \varepsilon_h^*]^{1/\alpha}. \tag{8.25}$$

THEOREM 8.4. *Let $\lambda(h)$ be an eigenvalue of* (8.14) *such that* $\lim_{h\to 0}\lambda(h)=\lambda$. *Suppose for each h that w_h is a unit vector satisfying $(\lambda(h)^{-1}-T)^k w_h=0$ for some positive integer $k\leqslant\alpha$. Then, for any integer l with $k\leqslant l\leqslant\alpha$, there is a vector u_h such that $(\lambda^{-1}-T)^l u_h=0$ and*

$$\|u_h-w_h\|_1\leqslant C(\beta(h)^{-1}\varepsilon_h)^{(l-k+1)/\alpha} \tag{8.26}$$

PROOFS. The eigenvalues and generalized eigenvectors of (8.10) and (8.14) have been characterized in terms of the compact operators T and T_h and we know that $T_h\to T$ in norm. Thus we can apply the results in Section 7, with $X=H_1$ and T and T_h as defined in (8.8) and (8.16), to estimate the eigenvalue and eigenvector errors. Note that

$$M=R(E)\cap(\text{unit sphere in } H_1),$$

where E is the spectral projection associated with T and λ^{-1}, and

$$M_h(\lambda)=R(E_h)\cap(\text{unit sphere in } H_1)$$

where E_h is the spectral projection associated with T_h and $\lambda_j^{-1}(h)$, $j=1,\ldots,m$. Consider first the proofs of Theorems 8.1 and 8.4. These results will follow immediately from Theorems 8.1 and 8.4, respectively, if we show that

$$\|(T-T_h)|_{R(E)}\|\leqslant C\beta(h)^{-1}\varepsilon_h. \tag{8.27}$$

From BABUŠKA [1971] and BABUŠKA and AZIZ [1973, Chapter 6] and (8.1)–(8.3), (8.6), (8.8), (8.11) and (8.16) we have

$$\|(T-T_h)u\|_1\leqslant\left(1+\frac{C_2}{\beta(h)}\right)\inf_{\chi\in S_{1,h}}\|Tu-\chi\|_1.$$

Since $M=R(E)$ is invariant for T, for $u\in R(E)$ we obtain

$$\inf_{\chi\in S_{1,h}}\|Tu-\chi\|_1\leqslant\varepsilon_h\|T\|\,\|u\|_1.$$

(8.27) follows from these two estimates.

Now consider the proofs of Theorems 8.2 and 8.3. The left-hand side of (8.24) is bounded by

$$C\left\{\sum_{i,j=1}^{m}|((T-T_h)\phi_j,\phi_j^*)_1|+\|(T-T_h)|_{R(E)}\|\,\|(T^*-T_h^*)|_{R(E^*)}\|\right\}.$$

We now show that this quantity can be bounded by $C\beta(h)^{-1}\varepsilon_h\varepsilon_h^*$. For $u\in H_1$ with $\|u\|_1=1$ and for $v^*\in R(E^*)$ with $\|v^*\|_1=1$ we have

$$\begin{aligned}((T-T_h)u,v^*)_1&=a((T-T_h)u,A'^{-1}v^*)\\&=a((T-T_h)u,A'^{-1}v^*-\eta)\\&\leqslant C_1\|(T-T_h)u\|_1\,\|A'^{-1}v^*-\eta\|_2,\quad\forall\eta\in S_{2,h}.\end{aligned}$$

We have here used the definition of the operator A, (8.8), and (8.16). Recalling the

A'^{-1} maps $R(E^*)=N((\overline{\lambda^{-1}}-T^*)^\alpha)$ onto $N((\overline{\lambda^{-1}}-T_*)^\alpha)=M^*(\lambda)$, we get

$$((T-T_h)u, v^*)_1 \leqslant C_1 \|(T-T_h)u\|_1 \|A'^{-1}\| \varepsilon_h^*.$$

From this it is immediate that

$$\|(T^*-T_h^*)v^*\|_1 = \sup_{\substack{u\in H_1\\ \|u\|_1=1}} |(u,(T^*-T_h^*)v^*)_1| \leqslant C_1 \|T-T_h\| \, \|A'^{-1}\| \varepsilon_h^* \tag{8.28}$$

$$\forall v^* \in R(E^*) \quad \text{with } \|v^*\|_1 = 1,$$

$$\begin{aligned}((T-T_h)\phi_i, \phi_j^*)_1 &= ((T-T_h)\phi_i, E^*\phi_j)_1 \\ &\leqslant C_1 \|E^*\| \, \|(T-T_h)|_{R(E)}\| \, \|A'^{-1}\| \varepsilon_h^* \leqslant C\beta(h)^{-1}\varepsilon_h \varepsilon_h^*. \end{aligned} \tag{8.29}$$

Now, using (8.27)–(8.29) we get

$$\sum_{i,j=1}^{m} |((T-T_h)\phi_j, \phi_j^*)_1| + \|(T-T_h)|_{R(E)}\| \, \|(T^*-T_h^*)|_{R(E^*)}\| \leqslant C\beta(h)^{-1}\varepsilon_h\varepsilon_h^*.$$

Thus Theorem 8.2 follows from Theorem 7.2 and Theorem 8.3 from Theorem 7.3. □

REMARK 8.1. The proof we have given for (8.24), together with (7.12), shows that

$$|\lambda - \hat{\lambda}(h)| \leqslant C\beta(h)\varepsilon_h\varepsilon_h^*, \tag{8.30}$$

where

$$\hat{\lambda}(h) = \frac{1}{m}\sum_{j=1}^{m} \lambda_j(h). \tag{8.31}$$

We end this section by specializing the results to the Ritz method for selfadjoint, positive-definite problems and then presenting a lower bound for the eigenvalue error. Suppose $H_1 = H_2 = H$, $\|\cdot\|_H = \|\cdot\|$ is a real Hilbert space. Let $a(\cdot,\cdot)$ be a symmetric bilinear form on H satisfying (8.1) and

$$a(u,u) \geqslant \alpha\|u\|^2 \quad \forall u \in H, \tag{8.32}$$

with α a positive constant. Note that (8.1) and (8.32) imply that $a(u,u)^{1/2}$ and $\|u\|$ are equivalent norms; $a(u,u)^{1/2}$ is often called the energy norm of u. Let b be a symmetric bilinear form on W satisfying (8.7), with $W_1 = W_2 = W \supset H$ compactly, and satisfying

$$b(u,u) > 0 \quad \forall u \in H, \quad u \neq 0. \tag{8.33}$$

Equation (8.32) implies (8.2) and (8.3) are satisfied. We note that (8.1) and (8.32) show that $a(\cdot,\cdot)$ is equivalent to the given inner product $(\cdot,\cdot)=(\cdot,\cdot)_H$ on H. We will now take $a(\cdot,\cdot)$ to be the inner product on H and take $\|\cdot\|_a = \sqrt{a(\cdot,\cdot)}$ to be the norm. We see that $T^* = T_* = T$. Thus T is selfadjoint; it is also positive-definite. It is, of course, compact. Let $S_{1,h} = S_{2,h} = S_h \subset H$ be a family of finite-dimensional spaces satisfying (8.13). In this case the variational approximation method (8.14) is called the Ritz method. Inequalities (8.11), with $\beta(h) = \alpha$, and (8.12) follow from (8.32).

Under these hypotheses, the problem (8.10) has a countable sequence of eigenvalues

$$0<\lambda_1\leqslant\lambda_2\leqslant\lambda_3\leqslant\cdots\uparrow+\infty$$

and corresponding eigenvectors

$$u_1, u_2, u_3, \ldots,$$

which can be assumed to satisfy

$$a(u_i, u_j)=\lambda_j b(u_i, u_j)=\delta_{ij} \tag{8.34}$$

(cf. Section 4). In the sequence $\{\lambda_j\}$, the λ_j are repeated according to geometric multiplicity. Furthermore, the λ_j can be characterized as various extrema of the *Rayleigh quotient*

$$R(u)=a(u, u)/b(u, u).$$

We state these characterizations now.

Minimum principle:

$$\begin{aligned} \lambda_1 &=\min_{u\in H} R(u)=R(u_1),\\ \lambda_k &= \min_{\substack{u\in H\\ a(u,u_i)=0\\ i=1,\ldots,k-1}} R(u)=R(u_k),\quad k=2,3,\ldots. \end{aligned} \tag{8.35}$$

Minimum–maximum principle:

$$\lambda_k=\min_{\substack{V_k\subset H\\ \dim V_k=k}}\ \max_{u\in V_k} R(u)=\max_{u\in U_k\equiv \mathrm{sp}(u_1,\ldots,u_k)} R(u),\quad k=1,2,\ldots. \tag{8.36}$$

Maximum–minimum principle:

$$\begin{aligned} \lambda_k &= \max_{z_1,\ldots,z_{k-1}\in H}\ \min_{\substack{u\in H\\ a(u,z_i)=0\\ i=1,\ldots,k-1}} R(u)\\ &= \min_{\substack{u\in H\\ a(u,u_i)=0\\ i=1,\ldots,k-1}} R(u),\quad k=1,2,\ldots \end{aligned} \tag{8.37}$$

Likewise (8.14) (with $S_{1,h}=S_{2,h}=S_h$) has a finite sequence of eigenvalues

$$0<\lambda_{1,h}\leqslant\lambda_{2,h}\leqslant\cdots\leqslant\lambda_{N,h},\quad N=\dim S_h,$$

and corresponding eigenvectors

$$u_{1,h},\ldots, u_{N,h},$$

which can be taken to satisfy

$$a(u_{i,h}, u_{j,h})=\lambda_{j,h} b(u_{i,h}, u_{j,h})=\delta_{ij}. \tag{8.38}$$

For the $\lambda_{k,h}$ we also have extremal characterizations.

Minimum principle:

$$\begin{aligned}\lambda_{1,h}&=\min_{u\in S_h} R(u)=R(u_{1,h}),\\ \lambda_{k,h}&=\min_{\substack{u\in S_h\\ a(u,u_{i,h})=0\\ i=1,\dots,k-1}} R(u)=R(u_{k,h}),\quad k=2,\dots,N.\end{aligned}\tag{8.39}$$

Minimum–maximum principle:

$$\lambda_{k,h}=\min_{\substack{V_{k,h}\in S_h\\ \dim V_{k,h}=k}}\ \max_{u\in V_{k,h}} R(u)=\max_{u\in U_{k,h}\equiv \mathrm{sp}(u_{1,h},\dots,u_{k,h})} R(u),\quad k=1,2,\dots,N.\tag{8.40}$$

Maximum–minimum principle:

$$\begin{aligned}\lambda_{k,h}&=\max_{z_{1,h},\dots,z_{k-1,h}\in H}\ \min_{\substack{u\in S_h\\ a(u,z_{i,h})=0\\ i=1,\dots,k-1}} R(u)\\ &=\min_{\substack{u\in S_h\\ a(u,u_{i,h})=0\\ i=1,\dots,k-1}} R(u),\quad k=1,\dots,N.\end{aligned}\tag{8.41}$$

It follows directly from the minimum and the minimum–maximum principles that

$$\lambda_k\leqslant\lambda_{k,h},\quad k=1,2,\dots,N=\dim S_h.\tag{8.42}$$

For a comprehensive treatment of such extremal characterizations of eigenvalues and their applications we refer to COURANT and HILBERT [1953], WEINSTEIN and STENGER [1972], and WEINBERGER [1974].

If λ_k has geometric multiplicity q, i.e., if $\lambda_k=\lambda_{k+1}=\cdots=\lambda_{k+q-1}$, then $\lambda_{k,h},\dots,\lambda_{k+q-1,h}\downarrow\lambda_k$, and combining (8.40) with Theorems 8.2 and 8.3 we see that

$$\lambda_k\leqslant\lambda_{j,h}\leqslant\lambda_k+C\varepsilon_h^2(\lambda_k),\quad j=k,\dots,k+q-1.\tag{8.43}$$

(Recall that the ascent α of $\lambda_k^{-1}-T$ is one.) Regarding the approximation of eigenvectors, from Theorems 8.1 and 8.4 we see that if $w_h=u_{j,h}$, $j=k,\dots,k+q-1$, then there is a unit eigenvector $u=u_h$ of (8.10) corresponding to λ_k such that

$$\|u-w_h\|_1\leqslant C\varepsilon_h,\tag{8.44}$$

and if u is a unit eigenvector of (8.10) corresponding to λ_k, then there is a unit vector w_h in $\mathrm{sp}(u_{k,h},\dots,u_{k+q-1,h})$ such that

$$\|u-w_h\|_1\leqslant C\varepsilon_h.\tag{8.45}$$

If λ_k is simple, i.e., its geometric multiplicity is one, we have

$$\|u_k-u_{k,h}\|_1\leqslant C\varepsilon_h.\tag{8.46}$$

To be more precise, if $u_1,u_2,\dots$ satisfy (8.34), then $u_{1,h},u_{2,h},\dots,u_{N,h}$ can be chosen so that (8.38) and (8.46) hold. Regarding these applications of Theorems 8.1–8.4, see the discussion of the selfadjoint case at the end of Section 7.

We now state some refinements of (8.43)–(8.46) due to Chatelin. Consider λ_k, which we suppose has multiplicity of q. Let $E=E(\lambda_k^{-1})$ be the orthogonal projection of H onto $\operatorname{span}\{u_k, \ldots, u_{k+q-1}\}$ and let $E_h = E_h(\lambda_k^{-1})$ be the orthogonal projection of H onto $\operatorname{span}\{u_{k,h}, \ldots, u_{k+q-1,h}\}$. (These are the spectral projections introduced in Section 6.) CHATELIN [1975, 1983] showed that

$$\|u - E_h u\|_1 = r_h^{(a)} \inf_{\chi \in S_h} \|u - \chi\|_a, \quad \forall u \in \operatorname{span}\{u_k, \ldots, u_{k+q-1}\}, \tag{8.47a}$$

$$\|u_{j,h} - Eu_{j,h}\|_1 = r_h^{(b)} \inf_{\chi \in S_h} \|Eu_{j,h} - \chi\|_a, \quad j = k, \ldots, k+q-1, \tag{8.47b}$$

$$(\lambda_{j,h} - \lambda_k)/\lambda_k = r_h^{(c)} \inf_{\chi \in S_h} \|Eu_{j,h} - \chi\|_a^2, \quad j = k, \ldots, k+q-1, \tag{8.47c}$$

where $r_h^{(l)} \to 1$ as $h \to 0$, for $l = \text{a, b, c}$. Regarding $r_h^{(l)}$, BABUŠKA and OSBORN [1988] have shown that

$$|r_h^{(l)} - 1| \leqslant d\eta^2(h), \quad l = \text{a, b, c},$$

where

$$\eta(h) = \sup_{b(u,u)=1} \inf_{\chi \in S_h} \|Tu - \chi\|_a.$$

Note that (8.47c) provides a lower, as well as an upper, bound for the eigenvalue error. For example, if λ_k is simple, we have

$$C_1 \varepsilon_h^2 \leqslant \lambda_{k,h} - \lambda_k \leqslant C_2 \varepsilon_h^2,$$

where C_1 and C_2 are positive constants, showing that the eigenvalue error is of the same order as ε_h^2. If λ_k is a multiple eigenvalue and $\lambda(h)$ is any approximate eigenvalue with $\lambda(h) \downarrow \lambda$, then one has

$$\lambda(h) - \lambda \geqslant C \inf_{u \in M(\lambda_k)} \inf_{\chi \in S_h} \|u - \chi\|_1^2. \tag{8.48}$$

KOLATA [1978] also proved this type of lower bound for the eigenvalue error.

PIERCE and VARGA [1972] established improved estimates for the eigenvector error in the norm $\|\cdot\|_b = \sqrt{b(\cdot,\cdot)}$. BABUŠKA and OSBORN [1989] have also proved such results, as well as refinements for the case of multiple eigenvalues.

9. An additional result for multiple eigenvalues

In the previous section we presented error estimates for variationally formulated problems and at the end of the section we specialized these results to the Ritz method for selfadjoint, positive-definite problems. Then we stated some refinements of these results due to Chatelin. In this section we present a result of BABUŠKA and OSBORN [1987] for multiple eigenvalues.

As at the end of Section 8, we assume that $a(\cdot,\cdot)$ is a symmetric bilinear form on H satisfying (8.1) and (8.32), that $b(\cdot,\cdot)$ is a symmetric bilinear form on W satisfying

(8.7), with $W_1 = W_2 = \dot{W} \supset H$ compactly, and (8.33). We take $a(\cdot,\cdot)$ and $\|\cdot\|_a = \sqrt{a(\cdot,\cdot)}$ to be the inner product and norm on H and set $\|\cdot\|_b = \sqrt{b(\cdot,\cdot)}$. Then, as stated in Section 8, the eigenvalue problem (8.10), i.e., the problem

$$\begin{aligned} &u \in H, \qquad u \neq 0 \\ &a(u, v) = \lambda b(u, v) \quad \forall v \in H \end{aligned} \tag{9.1}$$

has a countable sequence of eigenvalues

$$0 < \lambda_1 \leqslant \lambda_2 \leqslant \cdots \uparrow +\infty$$

and corresponding eigenvectors

$$u_1, u_2, \ldots,$$

which can be chosen to satisfy

$$a(u_i, u_j) = \lambda_j b(u_i, u_j) = \delta_{ij}. \tag{9.2}$$

Furthermore, any $u \in H$ can be written as

$$u = \sum_{j=1}^{\infty} a_j u_j \quad \text{with } a_j = a(u, u_j), \tag{9.3}$$

with convergence in $\|\cdot\|_a$ (cf. (4.10) and (4.11)). We assume $S_h \subset H$ is a family of finite-dimensional spaces satisfying (8.13) with $\beta(h) = \alpha$. The eigenvalue problem (8.14) with $S_{1,h} = S_{2,h} = S_h$, i.e., the problem

$$\begin{aligned} &u_h \in S_h, \qquad u_h \neq 0 \\ &a(u_h, v) = \lambda_h b(u_h, v) \quad \forall v \in S_h \end{aligned} \tag{9.4}$$

has a finite sequence of eigenvalues

$$0 < \lambda_{1,h} \leqslant \lambda_{2,h} \leqslant \cdots \leqslant \lambda_{N,h}, \quad N = \dim S_h,$$

and corresponding eigenvectors

$$u_{1,h}, \ldots, u_{N,h},$$

which can be chosen to satisfy

$$a(u_{i,h}, u_{j,h}) = \lambda_{j,h} b(u_{i,h}, u_{j,h}) = \delta_{ij}. \tag{9.5}$$

The λ_k and $\lambda_{k,h}$ satisfy the extremal principles stated in Section 8.

Our analysis makes use of the following lemma that expresses a fundamental property of eigenvalue and eigenvector approximation.

LEMMA 9.1. *Suppose (λ, u) is an eigenpair of (9.1) with $\|u\|_b = 1$, suppose w is any vector in H with $\|w\|_b = 1$, and let $\tilde{\lambda} = a(w, w)$. Then*

$$\tilde{\lambda} - \lambda = \|w - u\|_a^2 - \lambda \|w - u\|_b^2. \tag{9.6}$$

(*Note that we have assumed u and w are normalized with respect to $\|\cdot\|_b$ here, whereas in (9.2) and (9.5) we assumed u_i and $u_{i,h}$ are normalized with respect to $\|\cdot\|_a$.*)

PROOF. By an easy calculation,

$$\|w-u\|_a^2-\lambda\|w-u\|_b^2$$
$$=\|w\|_a^2-2a(w,u)+\|u\|_a^2-\lambda\|w\|_b^2+2\lambda b(w,u)-\lambda\|u\|_b^2.$$

Then, since

$$\|w\|_b=\|u\|_b=1,$$
$$\|w\|_a^2=\bar{\lambda}, \qquad \|u\|_a^2=\lambda,$$

and

$$a(w,u)=\lambda b(w,u),$$

we get the desired result. □

For $i=1,2,\ldots$ suppose λ_{k_i} is an eigenvalue of (9.1) with multiplicity q_i, i.e., suppose

$$\lambda_{k_i-1}<\lambda_{k_i}=\lambda_{k_i+1}=\cdots=\lambda_{k_i+q_i-1}<\lambda_{k_i+q_i}=\lambda_{k_{i+1}}.$$

Here $k_1=1$, k_2 is the lowest index of the second distinct eigenvalue, k_3 is the lowest index of the third distinct eigenvalue, etc. Let

$$\begin{aligned}&\varepsilon_h(i,j)=\inf_{u\in M(\lambda_{k_i})}\inf_{\chi\in S_h}\|u-\chi\|_a, \quad j=1,\ldots,q_i,\\ &a(u,u_{k_i,h})=\cdots=a(u,u_{k_i+j-2,h})=0,\end{aligned} \tag{9.7}$$

where $M(\lambda_{k_i})$ is defined in (8.18). The restrictions $a(u,u_{k_i,h})=\cdots=a(u,u_{k_i+j-2,h})=0$ are considered vacuous if $j=1$.

Note that $\varepsilon_h(i,j)\leqslant\varepsilon_h(\lambda_i)$, where $\varepsilon_h(\lambda_i)$ is defined in (8.21). We now estimate the eigenvalue and eigenvector errors for the Galerkin (Ritz) method (9.4) in terms of the approximability quantities $\varepsilon_h(i,j)$.

THEOREM 9.1. *There are constants C and h_0 such that*

$$\begin{aligned}&\lambda_{k_i+j-1,h}-\lambda_{k_i+j-1}\leqslant C\varepsilon_h^2(i,j)\\ &\quad\forall 0<h\leqslant h_0, \quad j=1,\ldots,q_i, \quad i=1,2,\ldots,\end{aligned} \tag{9.8}$$

and such that the eigenvectors $u_1,u_2,\ldots$ of (9.1) *can be chosen so that*

$$\begin{aligned}&\|u_{k_i+j-1,h}-u_{k_i+j-1}\|_a\leqslant C\varepsilon_h(i,j),\\ &\quad\forall 0<h\leqslant h_0, \quad j=1,\ldots,q_i, \quad i=1,2,\ldots,\end{aligned} \tag{9.9}$$

and so that (9.2) *holds.*

REMARK 9.1. Equations (9.8) and (9.9) should be compared and contrasted with estimates (8.47a)–(c) of CHATELIN [1975, 1983] and (8.48) of KOLATA [1978].

PROOF OF THEOREM 9.1. *Overview of the proof.* The complete details of the proof, which proceeds by induction, are given below. Here we provide an overview. In

Step A we give the proof for $i=1$. The proof is very simple in this case and rests entirely on the minimum principle (8.39) and Lemma 9.1.

The central part of the proof is given in Step B. There we prove the theorem for $i=2$, proving first the eigenvalue estimate (9.8) and then the eigenvector estimate (9.9). In particular, in Steps B.1 and B.2, estimates (9.8) and (9.9), respectively, are proved for $j=1$. We further note that the argument used in Step B proves the main inductive step in our proof, yielding the result for $i=\underline{i}+1$ on the assumption that it is true for $i\leqslant\underline{i}$. To be somewhat more specific, the argument in Step B.1 proves (9.8) directly for any $i\geqslant 2$ (and $j=1$) and that in Step B.2 proves (9.9) for $i=\underline{i}+1$ (and $j=1$) under the assumption that $\|u_{l,h}-u_l\|_a\to 0$ as $h\to 0$ for $l\leqslant k_{\underline{i}+1}-1$ (cf. (9.30)).

Details of the proof. Throughout the proof we use the fact that $\varepsilon_h(i,j)$ can also be expressed as

$$\varepsilon_h(i,j)=\inf_{\substack{u\in M(\lambda_{k_i})\\ a(u,u_{k_i,h})=\cdots=a(u,u_{k_i+j-2,h})=0}}\ \inf_{\substack{\chi\in S_h\\ a(\chi,u_{k_i,h})=\cdots=a(\chi,u_{k_i+j-2,h})=0}}\|u-\chi\|_a \tag{9.7'}$$

Step A. Here we prove the theorem for $i=1$.

Step A.1. Suppose λ_{k_1} ($k_1=1$) is an eigenvalue of (9.1) with multiplicity q_1, i.e., suppose $\lambda_1=\lambda_2=\cdots=\lambda_{q_1}<\lambda_{q_1+1}$. In this step we estimate $\lambda_{1,h}-\lambda_1$, the error between λ_1 and the approximate eigenvalue among $\lambda_{1,h},\ldots,\lambda_{q_1,h}$ that is closest to λ_1, i.e., we prove (9.8) for $i=j=1$. Note that

$$\varepsilon_h(1,1)=\inf_{u\in M(\lambda_1)}\ \inf_{\chi\in S_h}\|u-\chi\|_a$$

is the error in the approximation by elements of S_h of the most easily approximated eigenvector associated with λ_1.

From the definition of $\varepsilon_h(1,1)$ we see that there is a $\bar{u}_h\in M(\lambda_1)$ and an $s_h\in S_h$ such that

$$\|\bar{u}_h-s_h\|_a=\varepsilon_h(1,1). \tag{9.10}$$

Let

$$\bar{\bar{u}}=\frac{\bar{u}_h}{\sqrt{b(\bar{u}_h,\bar{u}_h)}},\qquad \tilde{s}_h=\frac{s_h}{\sqrt{b(s_h,s_h)}}.$$

By the minimum principle (8.39) we have

$$\lambda_{1,h}-\lambda_1\leqslant a(\tilde{s}_h,\tilde{s}_h)-\lambda_1. \tag{9.11}$$

Now apply Lemma 9.1 with $(\lambda,u)=(\lambda_1,\bar{\bar{u}}_h)$, $w=\tilde{s}_h$, and $\bar{\lambda}=a(\tilde{s}_h,\tilde{s}_h)$. We obtain

$$\begin{aligned} a(\tilde{s}_h,\tilde{s}_h)-\lambda_1&=\|\tilde{s}_h-\bar{\bar{u}}_h\|_a^2-\lambda_1\|\tilde{s}_h-\bar{\bar{u}}_h\|_b^2\\ &\leqslant\|\tilde{s}_h-\bar{\bar{u}}_h\|_a^2\leqslant C\|s_h-\bar{u}_h\|_a^2. \end{aligned} \tag{9.12}$$

Equations (9.10)–(9.12) yield the desired result.

Step A.2. In this step we prove (9.9) for $i=j=1$. Let $u_1,u_2,\ldots$ be eigenvectors of

(9.1) satisfying (9.2). Write

$$u_{1,h}=\sum_{j=1}^{\infty} a_j^{(1)} u_j. \tag{9.13}$$

(cf. (9.3)). From (9.5) and (9.11)–(9.13) we have

$$\begin{aligned}\left(1-\frac{\lambda_1}{\lambda_{q_1+1}}\right)\sum_{j=q_1+1}^{\infty}(a_j^{(1)})^2 &\leqslant \left|\sum_{j=q_1+1}^{\infty}(a_j^{(1)})^2\left(1-\frac{\lambda_1}{\lambda_j}\right)\right| \\ &= \left|\sum_{j=1}^{\infty}(a_j^{(1)})^2\left(1-\frac{\lambda_1}{\lambda_j}\right)\right| \\ &= |a(u_{1,h},u_{1,h})-\lambda_1 b(u_{1,h},u_{1,h})| \\ &= (\lambda_{1,h}-\lambda_1)\lambda_{1,h}^{-1} \\ &\leqslant C\varepsilon_h^2(1,1).\end{aligned}$$

Hence

$$\begin{aligned}\left\|u_{1,h}-\sum_{j=1}^{q_1} a_j^{(1)} u_j\right\|_a &= \left[\sum_{j=q_1+1}^{\infty}(a_j^{(1)})^2\right]^{1/2} \\ &\leqslant C\left(1-\frac{\lambda_1}{\lambda_{q_1+1}}\right)^{-1/2}\varepsilon_h(1,1).\end{aligned} \tag{9.14}$$

Redefining u_1 to be

$$\left(\sum_{j=1}^{q_1} a_j^{(1)} u_j\right)\Bigg/\left\|\sum_{j=1}^{q_1} a_j^{(1)} u_j\right\|_a,$$

we easily see that $\|u_1\|_a=1$, so that (9.2) still holds, and from (9.14) we obtain

$$\|u_{1,h}-u_1\|_a \leqslant C\varepsilon_h(1,1), \tag{9.15}$$

as desired. Note that u_1 may depend on h.

Step A.3. Suppose $q_1 \geqslant 2$. From (9.7′) we see that

$$\varepsilon_h(1,2)=\inf_{\substack{u\in M(\lambda_1)\\ a(u,u_{1,h})=0}}\ \inf_{\substack{\chi\in S_h\\ a(\chi,u_{1,h})=0}} \|u-\chi\|_a. \tag{9.16}$$

Choose $\bar{u}_h \in M(\lambda_1)$ with $a(\bar{u}_h, u_{1,h})=0$ and $s_h \in S_h$ with $a(s_h, u_{1,h})=0$ so that

$$\|\bar{u}_h - s_h\|_a = \varepsilon_h(1,2), \tag{9.17}$$

and let

$$\tilde{\bar{u}}_h=\frac{\bar{u}_h}{\sqrt{b(\bar{u}_h,\bar{u}_h)}}, \qquad \tilde{s}_h=\frac{s_h}{\sqrt{b(s_h,s_h)}}.$$

Since $a(s_h, u_{1,h})=0$, from the minimum principle (8.39), Lemma 9.1, and (9.17), we have

$$\lambda_{2,h}-\lambda_2 \leqslant \|\tilde{s}_h-\tilde{\tilde{u}}_h\|_a^2 \leqslant C\varepsilon_h^2(1,2). \tag{9.18}$$

This is (9.8) for $i=1$ and $j=2$.

Step A.4. In Step A.2 we redefined u_1. Now redefine $u_2,\ldots,u_{q_1}$ so that $u_1,\ldots,u_{q_1}$ are a-orthogonal. Write

$$u_{h,2}=\sum_{j=1}^{\infty} a_j^{(2)}u_j.$$

Now, proceeding as in Step A.2 and using (9.18), we have

$$\begin{aligned}\left(1-\frac{\lambda_2}{\lambda_{q_1+1}}\right)\sum_{j=q_1+1}^{\infty}(a_j^{(2)})^2 &\leqslant \left|\sum_{j=1}^{\infty}(a_j^{(2)})^2\left(1-\frac{\lambda_2}{\lambda_j}\right)\right| \\ &=|a(u_{2,h},u_{2,h})-\lambda_2 b(u_{2,h},u_{2,h})| \\ &=(\lambda_{2,h}-\lambda_2)\lambda_{2,h}^{-1} \\ &\leqslant C\varepsilon_h^2(1,2).\end{aligned}$$

Thus

$$\left\|u_{2,h}-\sum_{j=1}^{q_1} a_j^{(2)}u_j\right\|_a \leqslant C\varepsilon_h(1,2). \tag{9.19}$$

But by (9.15),

$$\begin{aligned}a_1^{(2)}&=a(u_{2,h},u_1)\\ &=a(u_{2,h},u_1-u_{1,h})\\ &\leqslant \|u_{2,h}\|_a\|u_1-u_{1,h}\|_a\\ &\leqslant C\varepsilon_h(1,1)\\ &\leqslant C\varepsilon_h(1,2).\end{aligned} \tag{9.20}$$

Combining (9.19) and (9.20) we get

$$\begin{aligned}\left\|u_{2,h}-\sum_{j=2}^{q_1} a_j^{(2)}u_j\right\|_a &\leqslant \left\|u_{2,h}-\sum_{j=1}^{q_1} a_j^{(2)}u_j\right\|_a + \|a_1^{(2)}u_1\|_a \\ &\leqslant C\varepsilon_h(1,2).\end{aligned}$$

Redefining u_2 to be

$$\left(\sum_{j=2}^{q_1} a_j^{(2)}u_j\right)\Bigg/\left\|\sum_{j=2}^{q_1} a_j^{(2)}u_j\right\|_a,$$

we see that $\|u_2\|_a=1$ and $a(u_1,u_2)=0$, so that (9.2) holds and

$$\|u_{2,h}-u_2\|_a\leqslant C\varepsilon_h(1,2), \tag{9.21}$$

which is (9.9) for $i=1, j=2$.

Step A.5. Continuing in the above manner we obtain the proof of (9.8) and (9.9) for $i=1$ and $j=1,\ldots,q_1$.

Step B. Here we prove Theorem 9.1 for $i=2$.

Step B.1. Suppose λ_{k_2} $(k_2=q_1+1)$ is an eigenvalue of (9.1) of multiplicity q_2. In this step we estimate $\lambda_{k_2,h}-\lambda_{k_2}$, the error between λ_{k_2} and the approximate eigenvalue among $\lambda_{k_2,h},\ldots,\lambda_{k_2+q_2-1,h}$ that is closest to λ_{k_2}. Note that

$$\varepsilon_h(2,1)=\inf_{u\in M(\lambda_{k_2})}\ \inf_{\chi\in S_h}\|u-\chi\|_a. \tag{9.22}$$

Introduce next the operators $T,T_h: H\to H$ defined in (8.8) and (8.16), respectively, i.e., the operators defined by

$$Tf\in H,\qquad a(Tf,v)=b(f,v)\quad \forall v\in H$$

and

$$T_hf\in S_h,\qquad a(T_hf,v)=b(f,v)\quad \forall v\in S_h.$$

It follows from (8.1), (8.7), (8.32), and (8.33) that T and T_h are defined and compact on H. Furthermore

$$\|(T-T_h)f\|_a\leqslant C\inf_{\chi\in S_h}\|Tf-\chi\|_a. \tag{9.23}$$

We now suppose the space H and the bilinear forms a and b have been complexified in the usual manner. Let Γ be a circle in the complex plane centered at $\mu_{k_2}=\lambda_{k_2}^{-1}$, with sufficiently small radius. Then for h sufficiently small, $\Gamma\subset\rho(T_h)$ and $\operatorname{Int}(\Gamma)\cap\sigma(T_h)=\{\mu_{k_2,h},\ldots,\mu_{k_2+q_2-1,h}\}$, where $\mu_{k_2+i}=\lambda_{k_2+i}^{-1}$. Also, as we have seen in Section 6, $E(\mu_{k_2})$, the spectral projections associated with T and μ_{k_2}, and $E_h(\mu_{k_2})$, the spectral projection associated with T_h and $\mu_{k_2+i,h}$, $i=0,\ldots,q_2-1$, respectively, can be written as

$$E(\mu_{k_2})=\frac{1}{2\pi \mathrm{i}}\int_\Gamma (z-T)^{-1}\,\mathrm{d}z \tag{9.24}$$

and

$$E_h(\mu_{k_2})=\frac{1}{2\pi \mathrm{i}}\int_\Gamma (z-T_h)^{-1}\,\mathrm{d}z. \tag{9.25}$$

Let $u\in R(E(\mu_{k_2}))$. Then $v_h=E_h(\mu_{k_2})u\in R(E_h(\mu_{k_2}))$, and from the formulas (9.24) and (9.25) we obtain

$$\begin{aligned}
\|u-v_h\|_a &= \|(E(\mu_{k_2})-E_h(\mu_{k_2}))u\|_a \\
&= \left\|\frac{1}{2\pi i}\int_\Gamma (z-T_h)^{-1}(T-T_h)(z-T)^{-1}u\,dz\right\| \\
&= \left\|\frac{1}{2\pi i}\int_\Gamma (z-T_h)^{-1}(T-T_h)\,\frac{u}{z-\mu_{k_2}}\,dz\right\| \\
&\leqslant \frac{1}{2\pi}\,[2\pi\,\mathrm{rad}(\Gamma)]\,\sup_{z\in\Gamma}\|(z-T_h)^{-1}\|\frac{1}{\mathrm{rad}(\Gamma)}\|(T-T_h)u\|_a \\
&= (\mu_{k_2+q_2-1,h}-\mu_{k_2}+\mathrm{rad}(\Gamma))^{-1}\|(T-T_h)u\|_a \\
&\leqslant C\|(T-T_h)u\|_a.
\end{aligned} \tag{9.26}$$

Equations (9.23) and (9.26) yield

$$\begin{aligned}
\|u-v_h\|_a &\leqslant C\inf_{\chi\in S_h}\|Tu-\chi\|_a \\
&= C\inf_{\chi\in S_h}\|\mu_{k_2}u-\chi\|_a \leqslant C\inf_{\chi\in S_h}\|u-\chi\|_a.
\end{aligned} \tag{9.27}$$

This is an eigenvector estimate; it shows that starting from any $u\in R(E(\mu_{k_2}))$ we can construct a $v_h=v_h(u)\in R(E_h(\mu_{k_2}))$ that is close to u. We now use (9.27) to prove the desired eigenvalue estimate.

By the minimum principle (8.39) we have

$$\lambda_{k_2,h}-\lambda_{k_2}=\inf_{\substack{v\in S_h\\ \|v\|_b=1\\ a(v,u_{i,h})=0,\\ i=1,\dots,k_2-1}} a(v,v)-\lambda_{k_2}. \tag{9.28}$$

Since $v_h\in R(E_h(\mu_{k_2}))$, we know that $a(v_h,u_{i,h})=0, i=1,\dots,k_2-1$. Thus, from (9.28) we find

$$\lambda_{k_2,h}-\lambda_{k_2}\leqslant a\left(\frac{v_h}{\|v_h\|_b},\frac{v_h}{\|v_h\|_b}\right)-\lambda_{k_2}.$$

Combining this with Lemma 9.1 and (9.27) we obtain

$$\begin{aligned}
\lambda_{k_2,h}-\lambda_{k_2} &\leqslant \left\|\frac{v_h}{\|v_h\|_b}-\frac{u}{\|u\|_b}\right\|_a^2-\lambda_{k_2}\left\|\frac{v_h}{\|v_h\|_b}-\frac{u}{\|u\|_b}\right\|_b^2 \\
&\leqslant C\|v_h-u\|_a^2\leqslant C\inf_{\chi\in S_h}\|u-\chi\|_a^2
\end{aligned}$$

for $u\in R(E(\mu_{k_2}))$ with $\|u\|_a=1$. Hence, using (9.22),

$$\lambda_{k_2,h}-\lambda_{k_2}\leqslant C\inf_{u\in M(\lambda_{k_2})}\inf_{\chi\in S_h}\|u-\chi\|_a^2=C\varepsilon_h^2(2,1). \tag{9.29}$$

which is (9.8) for $i=2, j=1$.

Comment on inequality (9.29). A careful examination of the proof of (9.29) shows that C depends only on $\mu_{k_1}-\mu_{k_2}$ and $\mu_{k_2}-\mu_{k_3}$, but is independent of h, and that (9.29) is valid for $h\leqslant h_0$, where h_0 is such that $h\leqslant h_0$ implies $\Gamma\subset\rho(T_h)$,

$$\mathrm{Int}(\Gamma)\cap\sigma(T_h)=\{\mu_{k_2,h},\ldots,\mu_{k_2+q_2-1,h}\},$$

and $\mu_{k_2}-\mu_{k_2+q_2-1,h}$ is small, say $\mu_{k_2}-\mu_{k_2+q_2-1,h}\leqslant\frac{1}{2}\,\mathrm{rad}(\Gamma)$. Note that if we were considering a family of problems depending on a parameter τ, we could bound $C=C(\tau)$ above, independent of τ, provided $\mu_{k_1}(\tau)-\mu_{k_2}(\tau)$ and $\mu_{k_2}(\tau)-\mu_{k_3}(\tau)$ were bounded away from 0, and we could bound $h_0(\tau)$ away from 0 if $\Gamma(\tau)\subset\rho(T_h(\tau))$,

$$\mathrm{Int}(\Gamma(\tau))\cap\sigma(T_h(\tau))=\{\mu_{k_2,h}(\tau),\ldots,\mu_{k_2+q_2-1,h}(\tau)\},$$

and $\mu_{k_2}(\tau)-\mu_{k_2+q_2-1,h}(\tau)\leqslant\frac{1}{2}\,\mathrm{rad}(\Gamma(\tau))$, uniformly in τ.

Step B.2. Suppose, as in Step B.1, that λ_{k_2} has multiplicity q_2. We have shown in Step A.5 that we can choose the eigenvectors $u_1,u_2,\ldots$ of (9.1) so that (9.2) holds and so that

$$\|u_{j,h}-u_j\|_a\leqslant C\varepsilon_h(1,j),\quad j=1,\ldots,q_1=k_2-1. \tag{9.30}$$

Write

$$u_{k_2,h}=\sum_{j=1}^{\infty}a_j^{(k_2)}u_j. \tag{9.31}$$

From (9.31) we have

$$\begin{aligned}\left|\sum_{j=1}^{\infty}[a_j^{(k_2)}]^2\left(1-\frac{\lambda_{k_2}}{\lambda_j}\right)\right|&=|a(u_{k_2,h},u_{k_2,h})-\lambda_{k_2}b(u_{k_2,h},u_{k_2,h})|\\&=(\lambda_{k_2,h}-\lambda_{k_2})\lambda_{k_2,h}^{-1},\end{aligned}$$

which, together with (9.29), yields

$$\left|\sum_{j=1}^{k_2-1}[a_j^{(k_2)}]^2\left(1-\frac{\lambda_{k_2}}{\lambda_j}\right)+\sum_{j=k_2+q_2}^{\infty}[a_j^{(k_2)}]^2\left(1-\frac{\lambda_{k_2}}{\lambda_j}\right)\right|\leqslant C\varepsilon_h^2(2,1). \tag{9.32}$$

Note that the first term inside the absolute value is negative and the second is positive. In addition

$$C_1\leqslant|1-\lambda_{k_2}/\lambda_j|\leqslant C_2\quad\forall j\neq k_2,k_2+1,\ldots,k_2+q_2-1,$$

with C_1,C_2 positive numbers. Hence from (9.32) we obtain

$$\sum_{j=1}^{k_2-1}[a_j^{(k_2)}]^2\leqslant D_1\varepsilon_h^2(2,1)+D_2\sum_{j=k_2+q_2}^{\infty}[a_j^{(k_2)}]^2 \tag{9.33}$$

and

$$\sum_{j=k_2+q_2}^{\infty}[a_j^{(k_2)}]^2\leqslant D_3\varepsilon_h^2(2,1)+D_4\sum_{j=1}^{k_2-1}[a_j^{(k_2)}]^2. \tag{9.34}$$

Write

$$u_{i,h}-u_i=\sum_{j=1}^{\infty} b_{i,j}u_j, \quad i=1,\ldots,k_2-1=q_1. \tag{9.35}$$

Then, by (9.30),

$$\sum_{j=1}^{\infty} b_{i,j}^2=\|u_{i,h}-u_i\|_a^2 \leqslant C\varepsilon_h^2(1,i), \quad i=1,\ldots,k_2-1. \tag{9.36}$$

Next we wish to find constants $\alpha_1,\ldots,\alpha_{k_2-1}$ so that

$$a\left(u_i, \sum_{j=1}^{k_2-1}\alpha_j u_{j,h}\right)=a_i^{(k_2)}, \quad i=1,\ldots,k_2-1. \tag{9.37}$$

Using (9.35), these equations can be written as

$$a\left(u_i, \sum_{j=1}^{k_2-1}\left(\alpha_j u_j+\alpha_j\sum_{l=1}^{\infty} b_{j,l}, u_l\right)\right)=\alpha_i+\sum_{j=1}^{k_2-1} b_{j,i}\alpha_j=a_i^{(k_2)},$$

$$i=1,\ldots,k_2-1. \tag{9.38}$$

Since (8.13) implies $\varepsilon_h^2(2,1)\to 0$ as $h\to 0$, from (9.36) we see that the $b_{j,i}$ are small for $h\leqslant h_0$, with h_0 sufficiently small, and hence the system (9.38) is uniquely solvable, and, moreover, there is a constant L, depending only on k_2, such that

$$\left(\sum_{j=1}^{k_2-1}\alpha_i^2\right)^{1/2} \leqslant L\left[\sum_{j=1}^{k_2-1}(a_i^{(k_2)})^2\right]^{1/2} \tag{9.39}$$

Now, from (9.30) we obtain

$$\begin{aligned}
|a_j^{(k_2)}| &= |a(u_{k_2,h}, u_j)| \\
&= |a(u_{k_2,h}, u_j-u_{j,h})| \\
&\leqslant \|u_{k_2,h}\|_a\|u_j-u_{j,h}\|_a \\
&= \|u_j-u_{j,h}\|_a \\
&\leqslant C\varepsilon_h(i,j), \quad j=1,\ldots,k_2-1.
\end{aligned}$$

Letting

$$\rho_{k_2}^2(h)=\sum_{j=1}^{k_2-1}\varepsilon_h^2(1,j), \tag{9.40}$$

we see that

$$\left[\sum_{j=1}^{k_2-1}(a_j^{(k_2)})^2\right]^{1/2} \leqslant C\rho_{k_2}(h), \tag{9.41}$$

and thus, from (9.39)

$$\left(\sum_{j=1}^{k_2-1}\alpha_i^2\right)^{1/2} \leqslant LC\rho_{k_2}(h)\leqslant C\rho_{k_2}(h). \tag{9.42}$$

Now let

$$\psi = u_{k_2,h} - \sum_{j=1}^{k_2-1} \alpha_j u_{j,h}. \tag{9.43}$$

Then $\psi \in S_h$. Furthermore, from (9.35) and (9.37) we get

$$a(u_i, \psi) = \begin{cases} 0, & i \leqslant k_2 - 1 \\ a_i^{(k_2)} - \sum_{j=1}^{k_2-1} \alpha_j b_{j,i}, & i \geqslant k_2. \end{cases} \tag{9.44}$$

From (9.42) and (9.43),

$$\begin{aligned} |\|\psi\|_a - 1| &= |\|\psi\|_a - \|u_{k_2,h}\|_a| \\ &\leqslant \|\psi - u_{k_2,h}\|_a \\ &\leqslant \left(\sum_{j=1}^{k_2-1} \alpha_j^2\right)^{1/2} \\ &\leqslant C\rho_{k_2}(h). \end{aligned} \tag{9.45}$$

Using (9.29), (9.44), and (9.45), and the fact that $\rho_{k_2}(h) \to 0$ as $h \to 0$, we get

$$\begin{aligned} C\varepsilon_h^2(2, 1) &\geqslant \frac{\lambda_{k_2,h} - \lambda_{k_2}}{\lambda_{k_2,h}} \\ &\geqslant a\left(u_{k_2,h}, \frac{\psi}{\|\psi\|_a}\right) - \lambda_{k_2} b\left(u_{k_2,h}, \frac{\psi}{\|\psi\|_a}\right) \\ &= C'\left[\sum_{l=k_2+q_2}^{\infty} a_l^{(k_2)}\left(a_l^{(k_2)} - \sum_{i=1}^{k_2-1} \alpha_i b_{i,l}\right)\left(1 - \frac{\lambda_{k_2}}{\lambda_l}\right)\right], \end{aligned} \tag{9.46}$$

where $C' > 0$ and is independent of h. Combining (9.36), (9.39), (9.40), and (9.46) we obtain

$$\begin{aligned} &\sum_{l=k_2+q_2}^{\infty} (a_l^{(k_2)})^2 \\ &\leqslant C\left[\varepsilon_h^2(2, 1) + \sum_{l=k_2+q_2}^{\infty} |a_l^{(k_2)}| \sum_{i=1}^{k_2-1} |\alpha_i| \, |b_{i,l}|\right] \\ &\leqslant C\left[\varepsilon_h^2(2, 1) + \sum_{i=1}^{k_2-1} |\alpha_i| \sum_{l=k_2+q_2}^{\infty} |a_l^{(k_2)}| \, |b_{i,l}|\right] \\ &\leqslant C\left[\varepsilon_h^2(2, 1) + \sum_{i=1}^{k_2-1} |\alpha_i| \left(\sum_{l=k_2+q_2}^{\infty} |a_l^{(k_2)}|^2\right)^{1/2} \left(\sum_{l=k_2+q_2}^{\infty} |b_{i,l}|^2\right)^{1/2}\right]. \end{aligned}$$

$$\leqslant C\left[\varepsilon_h^2(2,1)+\sum_{i=1}^{k_2-1}|\alpha_i|\left(\sum_{l=k_2+q_2}^{\infty}|a_l^{(k_2)}|^2\right)^{1/2}\max_{i=1,\dots,k_2-1}\varepsilon_h(1,i)\right]$$

$$\leqslant C\left[\varepsilon_h^2(2,1)+\varepsilon_h(1,k_2-1)\sqrt{k_2-1}\left(\sum_{i=1}^{k_2-1}|\alpha_i|^2\right)^{1/2}\left(\sum_{l=k_2+q_2}^{\infty}|a_l^{(k_2)}|^2\right)^{1/2}\right]$$

$$\leqslant C\left[\varepsilon_h^2(2,1)+\varepsilon_h(1,k_2-1)\sqrt{k_2-1}\,L\left(\sum_{i=1}^{k_2-1}[a_i^{(k_2)}]^2\right)^{1/2}\left(\sum_{l=k_2+q_2}^{\infty}|a_l^{(k_2)}|^2\right)^{1/2}\right]$$

$$\leqslant C\left[\varepsilon_h^2(2,1)+\varepsilon_h(1,k_2-1)\left(\sum_{i=1}^{k_2-1}[a_i^{(k_2)}]^2\right)^{1/2}\left(\sum_{l=k_2+q_2}^{\infty}|a_l^{(k_2)}|^2\right)^{1/2}\right]. \tag{9.47}$$

(9.47) is a quadratic inequality in

$$\left(\sum_{l=k_2+q_2}^{\infty}[a_l^{(k_2)}]^2\right)^{1/2},$$

whose solution yields

$$\sum_{l=k_2+q_2}^{\infty}[a_l^{(k_2)}]^2\leqslant C\varepsilon_h^2(1,k_2-1)\sum_{i=1}^{k_2-1}[a_i^{(k_2)}]^2+C\varepsilon_h^2(2,1). \tag{9.48}$$

Combining (9.33) and (9.48) we get

$$\sum_{i=1}^{k_2-1}(a_i^{(k_2)})^2\leqslant D_1\varepsilon_h^2(2,1)+D_2C\varepsilon_h(1,k_2-1)\sum_{i=1}^{k_2-1}(a_i^{(k_2)})^2+D_2C\varepsilon_h^2(2,1),$$

and thus, since $\varepsilon_h(1,k_2-1)$ is small for h small,

$$\sum_{i=1}^{k_2-1}(a_i^{(k_2)})^2\leqslant D_5\varepsilon_h^2(2,1). \tag{9.49}$$

Next, combining (9.34) and (9.49), we get

$$\sum_{l=k_2+q_2}^{\infty}(a_l^{(k_2)})^2\leqslant D_6\varepsilon_h^2(2,1). \tag{9.50}$$

Finally, from (9.31), (9.49), and (9.50) we have

$$\left\|u_{k_2,h}-\sum_{j=k_2}^{k_2+q_2-1}a_j^{(k_2)}u_j\right\|_a=\left[\sum_{j=1}^{k_2-1}(a_j^{(k_2)})^2+\sum_{j=k_2+q_2}^{\infty}(a_j^{(k_2)})^2\right]^{1/2}$$

$$\leqslant C\varepsilon_h(2,1).$$

Redefining u_{k_2} to be

$$\left(\sum_{j=k_2}^{k_2+q_2-1}a_j^{(k_2)}u_j\right)\bigg/\left\|\sum_{j=k_2}^{k_2+q_2-1}a_j^{(k_2)}u_j\right\|_a$$

we see that $\|u_{k_2}\|_a=1$, so that (9.2) holds, and

$$\|u_{k_2,h}-u_{k_2}\|_a\leqslant C\varepsilon_h(2,1). \tag{9.51}$$

This is (9.9) for $i=2$, $j=1$.

Comment on estimate (9.51). In the proof of (9.51) we used (9.30), which was proved in Step A. A careful examination of the proof of (9.51) shows that we did not use the full strength of (9.30), but only the weaker fact that $\|u_{j,h}-u_j\|_a\to 0$ as $h\to 0$ for $j\leqslant k_2-1$. (Cf. "Overview of the proof".)

Step B.3. Suppose $q_2\geqslant 2$. In Step B.1 we estimated $\lambda_{k_2,h}-\lambda_{k_2}$. In this step we estimate $\lambda_{k_2+1,h}-\lambda_{k_2+1}$.

We proceed by modifying problems (9.1) and (9.4) by restricting them to the spaces

$$H^{k_2,h}=\{u\in H: a(u,u_{k_2,h})=0\},$$
$$S_h^{k_2,h}=\{u\in S_h: a(u,u_{k_2,h})=0\},$$

respectively, i.e., we consider the problems $(9.1)^{k_2,h}$ and $(9.4)^{k_2,h}$ obtained by replacing H and S_h by $H^{k_2,h}$ and $S_h^{k_2,h}$ in (9.1) and (9.4), respectively. $(9.4)^{k_2,h}$ has the same eigenpairs $(\lambda_{j,h},u_{j,h})$ as does (9.4), except that the pair $(\lambda_{k_2,h},u_{k_2,h})$ has been eliminated. $(9.1)^{k_2,h}$ has eigenpairs $(\lambda_j^{k_2,h},u_j^{k_2,h})$, which in general depend on h. Nevertheless,

$$\lambda_{k_2+l}^{k_2,h}=\lambda_{k_2+1},\quad l=0,\ldots,q_2-2, \tag{9.52}$$

i.e., λ_{k_2}, the eigenvalue under consideration, is an eigenvalue of multiplicity q_2-1 for problem $(9.1)^{k_2,h}$. Its eigenspace is

$$\bar{M}^{k_2,h}(\lambda_{k_2})=\{u\in M(\lambda_{k_2}): a(u,u_{h,k_2})=0\}.$$

We can now apply the argument used in Step B.1 to problems $(9.1)^{k_2,h}$ and $(9.4)^{k_2,h}$ and, using (9.7′), we obtain (cf. (9.29))

$$\lambda_{k_2+1,h}-\lambda_{k_2+1}\leqslant C\varepsilon_{2,2}^2(h)\quad\text{for } h<h_0. \tag{9.53}$$

Since $u_{k_2,h}$ depends on h, the problems $(9.1)^{k_2,h}$ and $(9.4)^{k_2,h}$ depend on h. It follows from the "Comment on inequality (9.29)" with $\tau=h$ that we can apply the argument in Step B.1 obtaining C and h_0 that are independent of h. To see this, note that $\mu_{k_2}(h)=\mu_{k_2}$, by (9.52), $\mu_{k_3}(h)\leqslant\mu_{k_3}$, by the minimum principle, and $\mu_{k_1}(h)\to\mu_{k_1}$, since $\mu_{k_1}-\mu_{k_1}(h)\leqslant\mu_{k_1}-\mu_{k_1,h}$, by the minimum–maximum principle, and $\mu_{k_1,h}\to\mu_{k_1}$ (cf. (9.51)), and hence that $\mu_{k_1}(h)-\mu_{k_2}(h)$ and $\mu_{k_2}(h)-\mu_{k_3}(h)$ are bounded away from 0. Then note that

$$\Gamma(h)=\Gamma\subset\rho(T_h(h))=\rho(T_h)\cup\{\mu_{k_2,h}\},$$
$$\operatorname{Int}(\Gamma(h))\cap\sigma(T_h(h))=\operatorname{Int}(\Gamma)\cap(\sigma(T_h)-\{\mu_{k_2,h}\})=\{\mu_{k_2+1,h},\ldots,\mu_{k_2+q_2-1,h}\},$$

and

$$\mu_{k_2}(h)-\mu_{k_2+q_2-1,h}(h)=\mu_{k_2}-\mu_{k_2+q_2-1,h}\leqslant\tfrac{1}{2}\operatorname{rad}(\Gamma(h))=\tfrac{1}{2}\operatorname{rad}(\Gamma)$$

for all h.

Step B.4. Suppose $q_2\geqslant 2$ as in Step B.3. Here we show that u_{k_2+1} can be chosen so that $\|u_{k_2+1,h}-u_{k_2+1}\|_a \leqslant C\varepsilon_h(2,2)$. We know that

$$\|u_{j,h}-u_j\|_a \leqslant \begin{cases} C\varepsilon_h(1,j), & j=1,\dots,q_1, \\ C\varepsilon_h(2,1), & j=q_1+1=k_2 \end{cases} \tag{9.54}$$

(cf. (9.15), (9.21), and (9.51)). Assume that $u_{k_2+1},\dots,u_{k_2+q_2-1}$ have been redefined so that (9.2) holds. Write

$$u_{k_2+1,h}=\sum_{j=1}^{\infty} a_j^{(k_2+1)} u_j.$$

If we apply the argument used in Step B.2 to $u_{k_2+1,h}$, i.e., if we let k_2 be replaced by k_2+1 and use (9.53) instead of (9.29), we obtain

$$\left\| u_{k_2+1,h} - \sum_{j=k_2}^{k_2+q_2-1} a_j^{(k_2)} u_j \right\|_a \leqslant C\varepsilon_h(2,2).$$

But, by (9.54),

$$\begin{aligned} |a_{k_2}^{(k_2+1)}| &= |a(u_{k_2+1,h}, u_{k_2})| = |a(u_{k_2+1,h}, u_{k_2}-u_{k_2,h})| \\ &\leqslant \|u_{k_2}-u_{k_2,h}\|_a \leqslant C\varepsilon_h(2,1) \leqslant C\varepsilon_h(2,2) \end{aligned}$$

and hence

$$\left\| u_{k_2+1,h} - \sum_{j=k_2+1}^{k_2+q_2-1} a_j^{(k_2)} u_j \right\|_a \leqslant C\varepsilon_{2,2}(h).$$

Redefining u_{k_2+1} to be

$$\left(\sum_{j=k_2+1}^{k_2+q_2-1} a_j^{(k_2)} u_j\right) \Bigg/ \left\| \sum_{j=k_2+1}^{k_2+q_2-1} a_j^{(k_2)} u_j \right\|_a ,$$

we see that $\|u_{k_2+1}\|_a=1$, $a(u_{k_2+1},u_j)=0$, $j=1,\dots,k_2$, so that (9.2) holds, and

$$\|u_{k_2+1,h}-u_{k_2+1}\|_a \leqslant C\varepsilon_{2,2}(h),$$

which is (9.9) for $i=j=2$.

Step B.5. Continuing in this manner we prove (9.8) and (9.9) for $i=2$ and $j=1,\dots,q_2$.

Step C. Repeating the argument of Step B we get (9.8) and (9.9) for $i=3,4,\dots$. This completes the proof. □

REMARK 9.2. BABUŠKA and AZIZ [1973], FIX [1973], CHATELIN [1975], and KOLATA [1978] proved the estimate

$$\lambda_{k_i+j-1,h}-\lambda_{k_i+j-1} \leqslant C\varepsilon_h^2(\lambda_i), \quad j=1,\dots,q_i, \tag{9.55}$$

where $\varepsilon_h(\lambda_i)$ is defined in (8.21). Estimate (9.55) is weaker than (9.8). For $j=1,\dots,q_i-1$, (9.8) shows higher rates of convergence for certain problems; see the discussion of multiple eigenvalues in Section 10.2. BIRKHOFF, DE BOOR, SWARTZ, and WENDROFF [1966] estimated $\lambda_{k_i+j-1,h}-\lambda_{k_i+j-1}$ in terms of the sum of the squares of the a-norm distances between S_h and the unit eigenvectors associated with all the eigenvalues λ_l not exceeding λ_{k_i}.

REMARK 9.3. In BABUŠKA and OSBORN [1989], simplified proofs of (9.8) and (9.9), as well as refinements of these estimates, are given. These proofs are based, in part, on the estimate (8.47a) of Chatelin. BABUŠKA and OSBORN [1989] also contains improved eigenvector estimates in the norm $\|\cdot\|_b$. Cf. the comments at the end of Section 8.

CHAPTER III

Applications

In this chapter we apply the abstract results developed in Chapter II to several representative problems.

10. The Ritz method for second-order problems

10.1. Vibrations of a free L-shaped panel

We consider the problems of the plane strain vibration of an *L-shaped panel* Ω with free boundary. The specific shape of the panel is shown in Fig. 10.1.

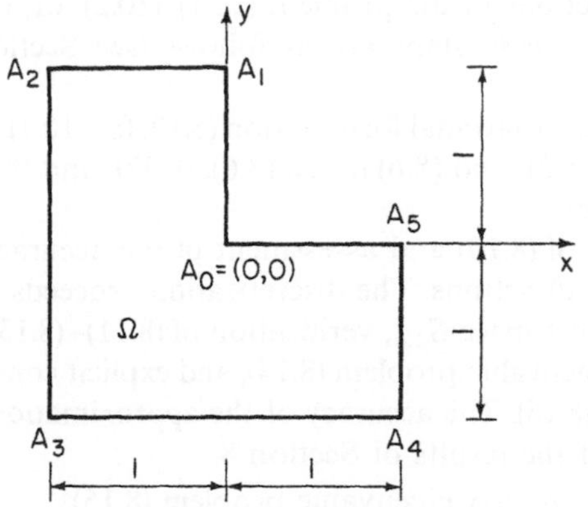

Fig. 10.1. The L-shaped panel Ω.

The equations governing the vibration of an elastic solid were discussed in Section 1 (see (1.33)–(1.35)). Corresponding to the L-shaped panel we have the eigenvalue problem

$$\begin{aligned} &-(\lambda+\mu)\frac{\partial\theta}{\partial x}-\mu\Delta u=\omega\rho u, \\ & \qquad\qquad\qquad\qquad\qquad\qquad (x, y)\in\Omega, \\ &-(\lambda+\mu)\frac{\partial\theta}{\partial y}-\mu\Delta v=\omega\rho v, \end{aligned} \tag{10.1}$$

where $\theta=(\partial u/\partial x)+(\partial v/\partial y)$. We obtain (10.1) from (1.35) by assuming that $u(x, y, z)$ and $v(x, y, z)$ are independent of z and that $w(x, y, z)=0$. The boundary conditions describing the traction-free boundary are

$$X_n = Y_n = 0, \quad (x, y)\in \Gamma = \partial\Omega, \tag{10.2}$$

where

$$\begin{aligned} X_n &= \lambda\theta n_x + \mu\frac{\partial u}{\partial n} + \mu\left(\frac{\partial u}{\partial x}n_x + \frac{\partial v}{\partial x}n_y\right), \\ Y_n &= \lambda\theta n_y + \mu\frac{\partial v}{\partial n} + \mu\left(\frac{\partial u}{\partial y}n_x + \frac{\partial v}{\partial y}n_y\right). \end{aligned} \tag{10.3}$$

Equations (10.2), with X_n and Y_n given in (10.3) are the Neumann conditions discussed in connection with the elastic solid specialized to the L-shaped panel.

We now consider the specific case in which

$$\nu = \frac{\lambda}{2(\lambda+\mu)} = 0.3, \qquad E = \frac{\mu(3\lambda+2\mu)}{\lambda+\mu} = 1$$

(i.e., in which $\lambda=\frac{15}{26}$ and $\mu=\frac{5}{13}$). ν is called *Poisson's ratio* and E is called *Young's modulus of elasticity*. $G=\mu$ is called the *modulus of rigidity*. Note that $0\leqslant\nu<\frac{1}{2}$ for any material.

We now discuss the basic steps in the finite element approximation of the eigenvalues and eigenfunctions of the problem (10.1), (10.2), or, more generally, of any eigenvalue problem. These steps are as follows (see Sections 10.1.1–10.1.3 and Section 10.1.1′):

(1) Derivation of a variational formulation (8.10) for (10.1), (10.2) and verification of conditions (8.1)–(8.3), and (8.6) ((8.1), (8.6), (8.32), and (8.33) in the selfadjoint, positive-definite case).

(2) Discretization of (8.10) and assessment of the accuracy of the approximate eigenvalues and eigenfunctions. The discretization proceeds by the selection of the trial space $S_{1,h}$ and test space $S_{2,h}$, verification of (8.11)–(8.13), consideration of the finite-dimensional eigenvalue problem (8.14), and explicit construction of the matrix eigenvalue problem (8.15). The accuracy of the approximation is assessed by means of the application of the results of Section 8.

(3) Solution of the matrix eigenvalue problem (8.15).

The accuracy of the approximation method (8.14) depends in a crucial way on the trial and test spaces $S_{1,h}$ and $S_{2,h}$, and their rational selection is strongly influenced by the available information on the eigenfunctions, typically information regarding their regularity. Thus, also of importance is:

(1′) Analysis of the regularity of the eigenfunctions.

REMARK 10.1. The approximation methods we will discuss in this section are referred to as Ritz methods, by which we mean that the eigenvalue problems under consideration are selfadjoint and positive-definite and that the test and trial space are equal ($S_{1,h}=S_{2,h}=S_h$); see the discussion in Section 8.

10.1.1. Variational formulation

We begin by casting our problem in the variational form

$$u\in H, \qquad a(u,v)=\omega b(u,v) \quad \forall v\in H, \tag{10.4}$$

where H is an appropriately chosen Hilbert space and a and b are appropriately selected bilinear forms. This process was explained in Section 3. We typically proceed as follows. Multiplying the first equation in (10.1) by ϕ, the second by ψ, adding the resulting equations together, and integrating over Ω, we obtain

$$\begin{aligned}&\int_\Omega \left\{\left[-(\lambda+\mu)\frac{\partial\theta}{\partial x}-\mu\Delta u\right]\phi+\left[-(\lambda+\mu)\frac{\partial\theta}{\partial y}-\mu\Delta v\right]\psi\right\}\,\mathrm{d}x\,\mathrm{d}y\\ &\quad=\omega\int_\Omega \rho(u\phi+v\psi)\,\mathrm{d}x\,\mathrm{d}y.\end{aligned} \tag{10.5}$$

Now, integration by parts shows that

$$\begin{aligned}&\int_\Omega \left\{\left[-(\lambda+\mu)\frac{\partial\theta}{\partial x}-\mu\Delta u\right]\phi+\left[-(\lambda+\mu)\frac{\partial\theta}{\partial y}-\mu\Delta v\right]\psi\right\}\,\mathrm{d}x\,\mathrm{d}y\\ &\quad=\int_\Omega (\lambda+\mu)\theta\frac{\partial\phi}{\partial x}\,\mathrm{d}x\,\mathrm{d}y-\int_\Gamma (\lambda+\mu)\theta\phi n_x\,\mathrm{d}s+\int_\Omega \mu\nabla u\cdot\nabla\phi\,\mathrm{d}x\,\mathrm{d}y\\ &\qquad-\int_\Gamma \mu\frac{\partial u}{\partial n}\phi\,\mathrm{d}s+\int_\Omega (\lambda+\mu)\theta\frac{\partial\psi}{\partial y}\,\mathrm{d}x\,\mathrm{d}y-\int_\Gamma (\lambda+\mu)\theta\psi n_y\,\mathrm{d}s\\ &\qquad+\int_\Omega \mu\nabla v\cdot\nabla\psi\,\mathrm{d}x\,\mathrm{d}y-\int_\Gamma \mu\frac{\partial v}{\partial n}\psi\,\mathrm{d}s\\ &\quad=\int_\Omega \left\{(\lambda+2\mu)\left(\frac{\partial u}{\partial x}+\frac{\partial v}{\partial y}\right)\left(\frac{\partial\phi}{\partial x}+\frac{\partial\psi}{\partial y}\right)\right.\\ &\qquad\left.+\mu\left[\left(\frac{\partial u}{\partial y}+\frac{\partial v}{\partial x}\right)\left(\frac{\partial\phi}{\partial y}+\frac{\partial\psi}{\partial x}\right)-2\frac{\partial u}{\partial x}\frac{\partial\psi}{\partial y}-2\frac{\partial v}{\partial y}\frac{\partial\phi}{\partial x}\right]\right\}\,\mathrm{d}x\,\mathrm{d}y\\ &\qquad-\int_\Gamma (X_n\phi+Y_n\psi)\,\mathrm{d}s.\end{aligned} \tag{10.6}$$

Combining (10.5) and (10.6) we see that if $(\omega, (u, v))$ satisfies (10.1) and (10.2), then

$$\int_\Omega \left\{(\lambda+2\mu)\left(\frac{\partial u}{\partial x}+\frac{\partial v}{\partial y}\right)\left(\frac{\partial \phi}{\partial x}+\frac{\partial \psi}{\partial y}\right)\right.$$

$$\left.+\mu\left[\left(\frac{\partial u}{\partial y}+\frac{\partial v}{\partial x}\right)\left(\frac{\partial \phi}{\partial y}+\frac{\partial \psi}{\partial x}\right)-2\frac{\partial u}{\partial x}\frac{\partial \psi}{\partial y}-2\frac{\partial v}{\partial y}\frac{\partial \phi}{\partial x}\right]\right\}\,\mathrm{d}x\,\mathrm{d}y$$

$$=\omega\int_\Omega \rho(u\phi+v\psi)\,\mathrm{d}x\,\mathrm{d}y \tag{10.7}$$

for all smooth (ϕ, ψ), and, conversely, if (10.7) holds for all smooth (ϕ, ψ), then (10.1) and (10.2) hold, provided u and v are smooth $(u,v \in H^2(\Omega))$.

From (10.7) we see how to choose H, a, and b in (10.4). Let

$$H=H^1(\Omega)\times H^1(\Omega), \qquad \|(u, v)\|_H^2=\|u\|_{1,\Omega}^2+\|v\|_{1,\Omega}^2 \tag{10.8}$$

and on H define the bilinear form

$$a(u, v; \phi, \psi)$$

$$=\int_\Omega \left\{(\lambda+2\mu)\left(\frac{\partial u}{\partial x}+\frac{\partial v}{\partial y}\right)\left(\frac{\partial \phi}{\partial x}+\frac{\partial \psi}{\partial y}\right)\right.$$

$$\left.+\mu\left[\left(\frac{\partial u}{\partial y}+\frac{\partial v}{\partial x}\right)\left(\frac{\partial \phi}{\partial y}+\frac{\partial \psi}{\partial x}\right)-2\frac{\partial u}{\partial x}\frac{\partial \psi}{\partial y}-2\frac{\partial v}{\partial y}\frac{\partial \phi}{\partial x}\right]\right\}\,\mathrm{d}x\,\mathrm{d}y. \tag{10.9}$$

It is immediate that (8.1) is satisfied and that a is symmetric. Let us remark that $a(u, v; u, v)$ has the physical meaning of the (double) strain energy and that $\sqrt{a(u, v; u, v)}$ is referred to as the energy norm of (u, v). Recall from Section 8 that b is to be defined on a space $W \supset H$. Let

$$W=L_2(\Omega)\times L_2(\Omega), \qquad \|(u, v)\|_W^2=\|u\|_{0,\Omega}^2+\|v\|_{0,\Omega}^2 \tag{10.10}$$

and define

$$b(u, v; \phi, \psi)=\int_\Omega \rho(u\phi+v\psi)\,\mathrm{d}x\,\mathrm{d}y. \tag{10.11}$$

It is immediate that b is symmetric and satisfies (8.7) and (8.33) and that $H \subset W$, compactly. It remains to consider (8.32). Note that since a and b are symmetric, $H^1(\Omega)$ and $L_2(\Omega)$ may be taken to be real.

We begin by expressing $a(u, v; \phi, \psi)$ in terms of the Poisson ratio ν and the modulus of rigidity G:

$$a(u, v; \phi, \psi)$$

$$=\frac{2G}{1-2\nu}\int_\Omega \left\{(1-\nu)\left(\frac{\partial u}{\partial x}\frac{\partial \phi}{\partial x}+\frac{\partial v}{\partial y}\frac{\partial \psi}{\partial y}\right)\right.$$

$$+\nu\left(\frac{\partial u}{\partial x}\frac{\partial \psi}{\partial y}+\frac{\partial v}{\partial y}\frac{\partial \phi}{\partial x}\right)+\tfrac{1}{2}(1-2\nu)\left(\frac{\partial u}{\partial y}+\frac{\partial v}{\partial x}\right)\left(\frac{\partial \phi}{\partial y}+\frac{\partial \psi}{\partial x}\right)\Bigg\}\,\mathrm{d}x\,\mathrm{d}y. \quad (10.12)$$

From (10.12) we have

$$\begin{aligned} &a(u, v; u, v)\\ &=\frac{2G}{1-2\nu}\int_\Omega\Bigg\{(1-\nu)\left[\left(\frac{\partial u}{\partial x}\right)^2+\left(\frac{\partial v}{\partial y}\right)^2\right]+2\nu\frac{\partial u}{\partial x}\frac{\partial v}{\partial y}\\ &\quad+\tfrac{1}{2}(1-2\nu)\left(\frac{\partial u}{\partial x}+\frac{\partial v}{\partial x}\right)^2\Bigg\}\,\mathrm{d}x\,\mathrm{d}y\\ &\geqslant\frac{2G}{1-2\nu}\int_\Omega\Bigg\{(1-2\nu)\left[\left(\frac{\partial u}{\partial x}\right)^2+\left(\frac{\partial v}{\partial y}\right)^2\right]+\tfrac{1}{2}(1-2\nu)\left(\frac{\partial u}{\partial y}+\frac{\partial v}{\partial x}\right)^2\Bigg\}\,\mathrm{d}x\,\mathrm{d}y. \quad (10.13)\end{aligned}$$

Recalling that $0\leqslant\nu<\frac{1}{2}$, we see from (10.13) that

$$a(u, v; u, v)\geqslant 0 \quad \forall u,v$$

(as was to be expected from the physical interpretation), and that $a(u, v; u, v)=0$ if and only if

$$\begin{aligned} u&=\bar{u}_{c_1,c_2,c_3}=c_1+c_2y,\\ v&=\bar{v}_{c_1,c_2,c_3}=c_3-c_2x \end{aligned} \quad (10.14)$$

for some c_1, c_2, c_3. These displacements, which are characterized as having no strain energy, are the "rigid body motions," i.e., translations and rotations. Thus (8.32) does not hold with H defined by (10.8), but the above considerations suggest that it might hold if H is replaced by a smaller space that did not include the rigid body motions. In fact, if we define

$$\tilde{H}=\left\{(u, v)\in H:\ \int_\Omega \rho(u\bar{u}_{c_1,c_2,c_3}+v\bar{v}_{c_1,c_2,c_3})\,\mathrm{d}x\,\mathrm{d}y=0\ \forall c_1,c_2,c_3\right\}, \quad (10.15)$$

then it can be shown (see NEČAS and HLAVÁČEK [1970] and KNOPS and PAYNE [1971]) that

$$\begin{aligned} &a(u, v; u, v)\geqslant\alpha\|(u, v)\|_H^2\\ &\quad=\alpha(\|u\|_{1,\Omega}^2+\|v\|_{1,\Omega}^2)\quad \forall (u, v)\in\tilde{H}, \end{aligned} \quad (10.16)$$

where α is a positive constant. This is (8.32).

We thus restrict $a(u, v; \phi, \psi)$ to $\tilde{H}$ and $b(u, v; \phi, \psi)$ to

$$\tilde{W}=\left\{(u, v)\in W:\ \int_\Omega \rho(u\bar{u}_{c_1,c_2,c_3}+v\bar{v}_{c_1,c_2,c_3})\,\mathrm{d}x\,\mathrm{d}y=0\ \forall c_1,c_2,c_3\right\}. \quad (10.17)$$

For the eigenvalue problem (10.1), (10.2) we therefore have the variational formulation

$$\begin{aligned}&0\neq(u,v)\in\tilde{H},\\&a(u,v;\ \phi,\psi)=\tilde{\omega}b(u,v;\ \phi,\psi)\quad\forall(\phi,\psi)\in\tilde{H}.\end{aligned}\tag{10.18}$$

Thus, with a, b, H, $\tilde{H}$, W, and $\tilde{W}$ chosen as in (10.8)–(10.11), (10.15)–(10.17), we see that a and b are symmetric and conditions (8.1), (8.7), (8.32), and (8.33) are satisfied. Equation (10.18) is a selfadjoint, positive-definite problem of the type studied at the end of Section 8.

As stated in Section 8, (10.18) has a countable sequence of eigenvalues

$$0<\omega_1\leqslant\omega_2\leqslant\cdots\uparrow+\infty$$

and corresponding eigenfunctions

$$(u_1,\ v_1),\ (u_2,\ v_2),\ldots,$$

which can be chosen to satisfy

$$a(u_i,\ v_i;\ u_j,\ v_j)=\omega_j b(u_i,\ v_i;\ u_j,\ v_j)=\omega_j\delta_{ij}.$$

When implementing our approximation method it is simpler to consider the eigenvalue problem on the space H instead of on $\tilde{H}$, i.e., to consider (10.18) with $\tilde{H}$ replaced by H. Then $\omega_0=0$ will be a triple eigenvalue with eigenfunctions (u, v) given by $(1, 0)$, $(0, 1)$, and $(-y, x)$. These eigenpairs and their approximations are then ignored. If the rigid body motions are included in the space $S_{1,h}$ and $S_{2,h}$, then $\omega_0=0$ is also a triple approximate eigenvalue with the rigid body motions again the corresponding approximate eigenfunctions. If this is not the case, then dealing with $\tilde{H}$ and H does not lead to the same approximate eigenvalues and eigenfunctions. It is easy to analyze the case in which the rigid body motions are not in $S_{1,h}$ and $S_{2,h}$, but we will not do so. Alternately, the validity of (10.16) or (8.32) can be ensured by considering

$$\tilde{a}(u,\ v;\ \phi,\ \psi)=a(u,\ v;\ \phi,\ v)+b(u,\ v;\ \phi,\ \psi)$$

instead of $a(u,\ v;\ \phi,\ \psi)$. Then the triple eigenvalue 1 would be the lowest eigenvalue. Usually the first alternative is used.

10.1.1′. Regularity of the eigenfunctions

We have seen in Section 8 that the accuracy of the approximate eigenvalues and eigenfunctions depends on the degree to which the exact eigenfunctions and adjoint eigenfunctions can be approximated by elements in the spaces $S_{1,h}$ and $S_{2,h}$, respectively (see (8.23)–(8.26)). In the selfadjoint, positive-definite case this reduces to the degree to which the eigenfunctions can be approximated by S_h (see (8.44)–(8.46)). Since the approximability of the eigenfunctions depends on their regularity, it is essential to determine the basic regularity properties of the eigenfunctions.

The eigenfunctions $(u_i,\ v_i)$ of (10.18) have the following properties:

– The functions u_i and v_i are analytic in $\bar{\Omega}\setminus\bigcup A_j$, where A_i are the vertices of Ω. This follows from the general theory of elliptic equations (see MORREY [1966, Section 6.6]).

– The functions u_i and v_i are singular at the vertices of A_j, and the character of the singularity is known. The strength of the singularity at A_j depends on the interior angle at A_j. For the domain we are considering, the strongest singularity is at the vertex A_0. The leading terms of u_i and v_i at a vertex have the form

$$u^* = C_1 r^\sigma F_1(\theta), \qquad v^* = C_2 r^\sigma F_2(\theta), \tag{10.19}$$

where (r, θ) are the polar coordinates with origin at the vertex, σ depends on the interior angle and on λ and μ, and $F_1(\theta)$ and $F_2(\theta)$ are analytic functions of θ. The value of σ is characterized as the root of a nonlinear equation and, in general, can be real or complex. For our example of the L-shaped domain, $\sigma = \sigma_0 = 0.544481\ldots$ for the vertex A_0. For a more complete discussion of the singularities of solutions of elliptic equations in polygonal domains we refer to KONDRATÉV [1968], MERIGOT [1974], and GRISVARD [1985]. Using their results, any eigenfunction can be written as $(u, v) = (u^1, v^1) + (u^2, v^2)$, where $u^2, v^2 \in H^k(\Omega)$, where k is an integer which is greater than or equal to 3, and (u^1, v^1) is a linear combination of functions of the type on the right-hand side of (10.19) with $\sigma \geqslant \sigma_0$ and with coordinates centered in the vertices of Ω. Application of the method used in the proof of Theorem 2.1 shows that, for our domain, u^* and v^* and thus u_i and v_i are contained in $\hat{H}^{\sigma_0+1}(\Omega)$, with $\sigma_0 = 0.544481\ldots$. This statement of the regularity or smoothness of the eigenfunctions is the strongest that can be made in terms of Sobolev spaces (without weights).

– It is also possible to show that $u, v \in \mathscr{B}^2_\beta(\Omega)$, for any $\beta > \sigma_0 (\beta = \sigma_0 + \varepsilon)$; see Theorem 4.4. Of course, the space $\mathscr{B}^2_\beta(\Omega)$ is much smaller than $\hat{H}^{\sigma_0+1}(\Omega)$ and hence we can make a more effective choice for S_h if we use $\mathscr{B}^2_\beta(\Omega)$ instead of $\hat{H}^{\sigma_0+1}(\Omega)$.

10.1.2. Discretization of (8.10) *and assessment of the accuracy of the approximate eigenvalues and eigenfunctions*

The discretization of (8.10) is accomplished by selecting the trial and test spaces $S_{1,h}$ and $S_{2,h}$ satisfying (8.11)–(8.13), considering the finite-dimensional eigenvalue problem (8.14), and deriving the matrix eigenvalue problem (8.15) from which the approximate eigenvalues are obtained. The selection of $S_{1,h}$ and $S_{2,h}$ is the most important part of this process. It is influenced by three considerations.

(a) The spaces $S_{1,h}$ and $S_{2,h}$ have to satisfy (8.11) and (8.12). Note, however, that if the problem under consideration is selfadjoint and positive-definite, from (8.32) we see that (8.11) and (8.12) hold for $S_{1,h} = S_{2,h} = S_h$, for any S_h. Our problem (10.18) is selfadjoint and positive-definite and we will take $S_{1,h} = S_{2,h} = S_h$.

(b) $S_{1,h}$ should accurately approximate the eigenfunctions of (10.18) and $S_{2,h}$ should accurately approximate the adjoint eigenfunctions. Usually we also require that the rigid body motion functions are included in $S_{1,h}$ and $S_{2,h}$. If this is not the case, then we have to assume that the rigid body motion functions are very well approximated. If they are not well approximated, although there will be no change in the asymptotic rate of convergence, the accuracy will deteriorate, especially with long domains (such as long beams), for which the rigid body motions for some parts of the domain could be relatively large.

(c) The matrices A and B in (8.15) should be reasonably sparse, since sparsity is strongly related to computational complexity. Sparsity is achieved by choosing finite element spaces for $S_{1,h}$ and $S_{2,h}$. These spaces then have bases consisting of functions with local supports, and, as a consequence, A and B will be sparse. We note that the sparseness of A and B is not required for the validity of the results of Section 8 and, in fact, in certain applications one does use non finite element type trial and test spaces, spaces consisting of global polynomials or trigonometric polynomials, e.g.

We now describe some typical choices for S_h for the L-shaped panel.

10.1.2.1. The h-version on a uniform mesh. Let Ω be covered by a mesh of uniform squares I_{ij} of size h as shown in Fig. 10.2. Then for $p=1,2,\ldots$ let

$$\tilde{S}_h^p=\{u: u\in H^1(\Omega),\ u|_{I_{ij}}=\sum_{(m,n)\in\mathscr{D}(p)} A_{m,n}x^m y^n\ \forall I_{ij}\}, \tag{10.20}$$

where

$$\mathscr{D}(p)=\{(m,n): 0\leqslant m,n, \text{ and } m+n\leqslant p \text{ or } (m,n)=(1,p) \text{ or } (p,1)\}. \tag{10.21}$$

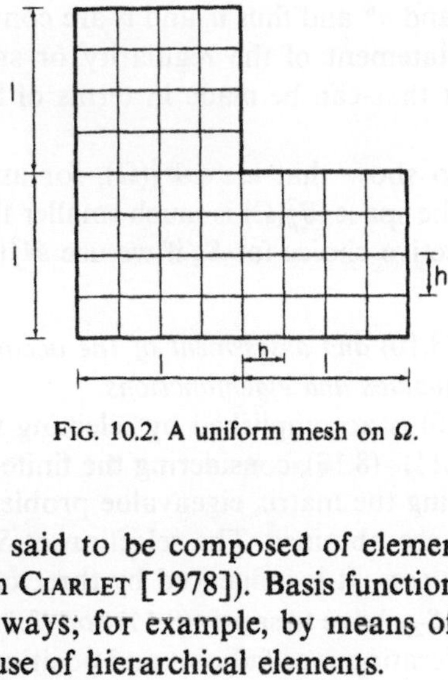

FIG. 10.2. A uniform mesh on Ω.

Spaces of this type are said to be composed of elements of type Q'_p (the cases $p=1,2,3$ are discussed in CIARLET [1978]). Basis functions for these spaces can be constructed in various ways; for example, by means of Lagrange elements (see CIARLET [1978]) or by use of hierarchical elements.

Regarding the approximation properties of the family $\{\tilde{S}_h^p\}_{0<h}$, it can be shown (see CIARLET [1978]) that if $k\geqslant 1$ is an integer, then

$$\inf_{\chi\in S_h^p}\|u-\chi\|_{1,\Omega}\leqslant C(p)h^{\mu-1}\|u\|_{k,\Omega} \quad \text{for any } u\in H^k(\Omega) \text{ and any } h>0, \tag{10.22}$$

where

$$\mu=\min(k,\ p+1) \tag{10.23}$$

and $C(p)$ depends on p, k and Ω, but is independent of u and h. Equation (10.22) is optimal in the sense that h^μ on the right-hand side cannot be replaced by a higher power of h when the mesh is uniform. If k is not an integer, then we have

$$\inf_{\chi \in S_h^p} \|u-\chi\|_{1,\Omega} \leqslant C(p) h^{\mu-1} \|u\|_{\tilde{H}^k(\Omega)}, \tag{10.24}$$

with μ given by (10.23). Note that we have not said anything about the dependence of $C(p)$ on p. The proof in CIARLET [1978] suggests that $C(p)$ grows with p, and thus could lead to the conclusion that it is improper to use $p > k-1$. However, this conclusion is not justified because, in fact, $C(p) \leqslant Cp^{-(k-1)}$ (see BABUŠKA and SURI [1987b]).

We will now derive (10.24)–(10.23) from (10.22)–(10.23) using the method outlined in Section 2 (cf. (2.12)–(2.16) and Theorem 2.1).

Suppose $m < k < m+1$. Since

$$\|u\|_{\tilde{H}^k(\Omega)} = \sup_{0<t<\infty} \{t^{-\theta} K(u,t)\},$$

where $\theta = k-m$, we see that

$$K(u,t) \leqslant t^\theta \|u\|_{\tilde{H}^k(\Omega)}.$$

Let $\varepsilon > 0$. Then for any $t > 0$ there exist $v_t \in H^m(\Omega)$ and $w_t \in H^{m+1}(\Omega)$ such that $u = v_t + w_t$ and

$$\|v_t\|_{m,\Omega} + t\|w_t\|_{m+1,\Omega} \leqslant K(u,t) + t^\theta \varepsilon \leqslant t^\theta (\|u\|_{\tilde{H}^k(\Omega)} + \varepsilon).$$

Therefore, using (10.22)–(10.23) we can choose $\chi_{1,t}, \chi_{2,t} \in \tilde{S}_h^p$ so that

$$\begin{aligned}\|v_t - \chi_{1,t}\|_{1,\Omega} &\leqslant C(p) h^{\mu_1 - 1} \|v_t\|_{m,\Omega} \\ &\leqslant C(p) h^{\mu_1 - 1} t^\theta (\|u\|_{\tilde{H}^k(\Omega)} + \varepsilon),\end{aligned}$$

where $\mu_1 = \min(m, p+1)$, and

$$\begin{aligned}\|w_t - \chi_{2,t}\|_{1,\Omega} &\leqslant C(p) h^{\mu_2 - 1} \|w_t\|_{m+1,\Omega} \\ &\leqslant C(p) h^{\mu_2 - 1} t^{\theta - 1} (\|u\|_{\tilde{H}^k(\Omega)} + \varepsilon),\end{aligned}$$

where $\mu_2 = \min(m+1, p+1)$. Letting $\chi_t = \chi_{1,t} + \chi_{2,t}$, we thus have

$$\|u - \chi_t\|_{1,\Omega} \leqslant C(p)(h^{\mu_1 - 1} t^\theta + h^{\mu_2 - 1} t^{\theta - 1})(\|u\|_{\tilde{H}^k(\Omega)} + \varepsilon) \quad \text{for any } t > 0. \tag{10.25}$$

If $p \geqslant m$, select $t = h$ in (10.25) to obtain

$$\begin{aligned}\inf_{\chi \in \tilde{S}_h^p} \|u-\chi\|_{1,\Omega} &\leqslant C(p) h^{\theta + \mu_1 - 1} (\|u\|_{\tilde{H}^k(\Omega)} + \varepsilon) \\ &= C(p) h^{k-1} (\|u\|_{\tilde{H}^k(\Omega)} + \varepsilon) = C(p) h^{\mu - 1} (\|u\|_{\tilde{H}^k(\Omega)} + \varepsilon),\end{aligned}$$

where $\mu = \min(k, p+1)$. If $p < m$, let $t = 1$ in (10.25) to get

$$\begin{aligned}\inf_{\chi \in \tilde{S}_h^p} \|u-\chi\|_{1,\Omega} &\leqslant C(p) h^{\mu_1 - 1} (\|u\|_{\tilde{H}^k(\Omega)} + \varepsilon) \\ &= C(p) h^p (\|u\|_{\tilde{H}^k(\Omega)} + \varepsilon) = C(p) h^{\mu - 1} (\|u\|_{\tilde{H}^k(\Omega)} + \varepsilon),\end{aligned}$$

with $\mu=\min(k, p+1)$. Letting $\varepsilon=\|u\|_{\hat{H}^k(\Omega)}$ in these estimates yields (10.24)–(10.23).

Now define

$$S_{1,h}=S_{2,h}=S_h=\tilde{S}_h^p\times\tilde{S}_h^p. \tag{10.26}$$

We remark that the rigid body motions belong to S_h (cf. (10.14)). Since (10.18) is selfadjoint and positive-definite, it satisfies (8.32), and hence (8.11), with $\beta(h)=\alpha$, and (8.12) hold. (10.22) and (10.24) show that S_h accurately approximates the exact eigenfunctions. Thus (10.22) and (10.24), together with a density argument, show that (8.13) is satisfied. If an appropriate basis is chosen for S_h, the matrices A and B in (9.15) can be calculated and they will be sparse. Thus the issues raised above in (a), (b) and (c) have been addressed.

Now consider the problem (8.14) with this choice for $S_{1,h}$ and $S_{2,h}$ and denote its eigenvalues and eigenfunctions by

$$0<\omega_{1,h}\leqslant\cdots\leqslant\omega_{N,h}$$

and

$$(u_{1,h}, v_{1,h}),\ldots,(u_{N,h}, v_{N,h}),$$

where $N=\dim S_h$. To assess the accuracy of these approximate eigenpairs, the results of Section 8 will be applied. All of the hypotheses for these results have now been shown to be satisfied for our problem and approximation procedure.

THEOREM 10.1. *Let $S_{1,h}$ and $S_{2,h}$ be selected as in* (10.26). *Suppose ω_k is an eigenvalue of* (10.18) *with multiplicity q, i.e., suppose $\omega_{k-1}<\omega_k=\omega_{k+1}=\cdots=\omega_{k+q-1}<\omega_{k+q}$. Then*

$$|\omega_{j,h}-\omega_k|\leqslant C(p)h^{1.088962\ldots},\quad j=k,\ldots,k+q-1. \tag{10.27}$$

If $(w_h, z_h)=(u_{j,h}, v_{j,h}), j=k,\ldots,k+q-1$, then there is a unit eigenfunction $(u, v)=(u_h, v_h)$ of (10.18) *such that*

$$\|u-w_h\|_{1,\Omega}+\|v-z_h\|_{1,\Omega}\leqslant C(p)h^{0.544481\ldots}, \tag{10.28a}$$

and if (u, v) is a unit eigenfunction of (10.18) *corresponding to ω_k, then there is a unit vector*

$$(w_h, z_h)\in\operatorname{sp}\{(u_{k,h}, v_{k,h}),\ldots,(u_{k+q-1,h}, v_{k+q-1,h})\}$$

such that

$$\|u-w_h\|_{1,\Omega}+\|v-z_h\|_{1,\Omega}\leqslant C(p)h^{0.544481\ldots}. \tag{10.28b}$$

If ω_k is simple, the eigenfunction estimates reduce to

$$\|u_{k,h}-u_k\|_{1,\Omega}+\|v_{k,h}-v_k\|_{1,\Omega}\leqslant C(p)h^{0.544481\ldots}. \tag{10.28c}$$

PROOF. We saw in Section 10.1.1′ that u_j and v_j are in $\hat{H}^{\sigma_0+1}$, with $\sigma_0=0.544481\ldots$.

Thus from (10.22)–(10.24) we have

$$\varepsilon_h = \sup_{j=k,\ldots,k+q-1} \inf_{\chi=(\chi_1,\chi_2)\in S_h} \|(u_j, v_j)-(\chi_1,\chi_2)\|_{H^1(\Omega)\times H^1(\Omega)}$$

$$\leqslant C(p)h^{\sigma_0}$$

Equations (10.27) and (10.28) follow from this estimate and (8.44)–(8.46). □

To show the effectiveness of estimates (10.27)–(10.28) we would have to know the exact eigenfunctions and eigenvalues. Because these are not available we consider instead the quantity

$$Q(p,h) = \inf_{\chi=(\chi_1,\chi_2)\in S_h} a(u^*-\chi_1, v^*-\chi_2; u^*-\chi_1, v^*-\chi_2), \tag{10.29}$$

where u^* and v^* are given in (10.19). $Q(p,h)$ can be computed numerically. Figure 10.3 shows the graph of

$$\|e\|_{\mathrm{E,R}} = [Q(p,h)/a(u^*, v^*; u^*, v^*)]^{1/2}$$

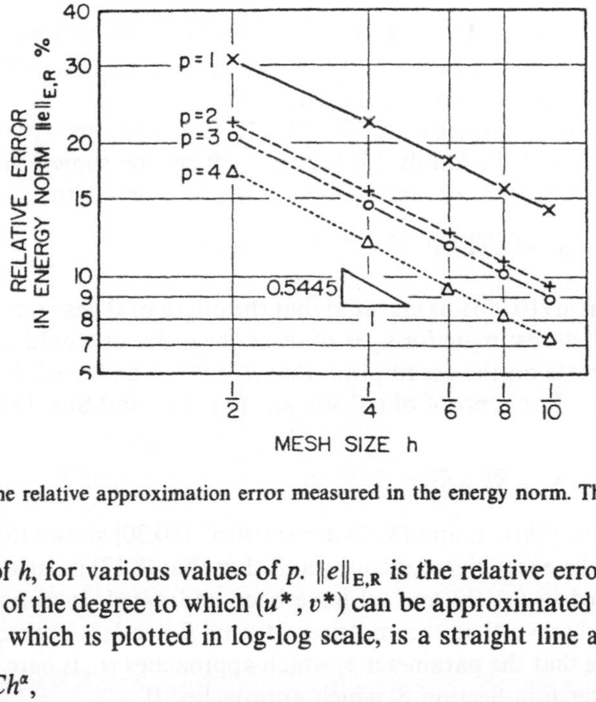

FIG. 10.3. The relative approximation error measured in the energy norm. The *h*-version.

as a function of h, for various values of p. $\|e\|_{\mathrm{E,R}}$ is the relative error in the energy norm measure of the degree to which (u^*, v^*) can be approximated by functions in S_h. The graph, which is plotted in log-log scale, is a straight line and thus

$$\|E\|_{\mathrm{E,R}} = Ch^{\alpha},$$

where α is the slope of the line. We see that the slope is very close to the theoretically predicted $\alpha = 0.544481\ldots$. Increasing p decreases the constant C but does not affect the slope α.

From an analysis of Fig. 10.3 we can draw several conclusions:

– To achieve an accuracy of 5% (respectively, 3%) with elements of degree $p=1$

we would require N to be about 25,000 (respectively, N to be about 170,000) and with elements of degree 2 we would require N to be about 19,000 (respectively, N to be about 124,000). This shows that a uniform or quasiuniform mesh is completely unacceptable for our problem.

– Because the rate of convergence for eigenvalues is twice that for eigenfunctions, we see that the eigenvalues are much cheaper to compute than the eigenfunctions. Roughly speaking, we see that for eigenvalue calculations the required number of unknowns would be about $N=160$ (respectively, about $N=400$) for $p=1$ and about $N=140$ (respectively, $N=350$) for $p=2$.

– While (10.22) qualitatively characterizes the error behavior, it does not give all the desired quantitative information because C and $\|u\|_{k,\Omega}$ are not known. More precise quantitative information can be gained only by a posteriori analysis. We will not, however, be able to pursue this direction. For a survey of results on a posteriori assessment of the quality of finite element computations, we refer to NOOR and BABUŠKA [1987]. A posteriori error analysis is used also in connection with adaptive approaches, in which the goal is to let the computer construct the mesh required to achieve the desired accuracy.

10.1.2.2. The p-version. In the h-version of the finite element method accuracy is achieved by letting $h\downarrow 0$, while p is held fixed. In the p-version of the finite element method, one, in contrast, fixes h and lets $p\uparrow\infty$.

Let $\tilde{S}_h^p$ again be defined by (10.20)–(10.21). Regarding the (p-version) approximation properties of the family $\{\tilde{S}_h^p\}_{p=1,2,\ldots}$, it can be shown that if $u=u_1+u_2$, where $u_1\in H^k(\Omega)$, with $k\geqslant 2$, and $u_2=Kr^\sigma F(\theta)$, with $\sigma>0$, then

$$\inf_{\chi\in\tilde{S}_h^p}\|u-\chi\|_{1,\Omega}\leqslant C(h)[Kp^{-2\sigma}+p^{-(k-1)}\|u_1\|_{k,\Omega}]. \tag{10.30}$$

We remark that in (10.30) it is essential that the origin of Ω lies on an element vertex; for in this case, the estimate for u_2 is of twice the order as would be obtained if we based our estimate on the assumption that $u_2\in\hat{H}^{\sigma+1}$ and used the h-version with a uniform mesh. For a proof of (10.30), see BABUŠKA and SURI [1987a].

Define

$$S_{1,p}=S_{2,p}=S_p=\tilde{S}_h^p\times\tilde{S}_h^p. \tag{10.31}$$

Then (8.11), with $\beta(h)=\alpha$, and (8.12) are satisfied. (10.30) shows that S_p accurately approximates the exact eigenfunctions and thus that (8.13) is satisfied. We see that the issues raised in (a), (b) and (c) have been addressed. In connection with (c), however, we observe that the matrices A and B are less sparse than with the h-version. Note that the parameter p, which approaches ∞, is here playing the role of the parameter h in Section 8, which approached 0.

Now consider the problem (8.14) with this choice for $S_{1,p}$ and $S_{2,p}$ and denote the eigenvalues and eigenfunctions by

$$0<\omega_{1,p}\leqslant\cdots\leqslant\omega_{N,p}$$

and

$$(u_{1,p}, v_{1,p}), \ldots, (u_{N,p}, v_{N,p}),$$

where $N = \dim S_p$. As with the h-version, the accuracy of the approximate eigenpairs may be assessed with the results of Section 8.

THEOREM 10.2. *Let $S_{1,p}$ and $S_{2,p}$ be chosen as in (10.31). Suppose ω_k is an eigenvalue of (10.18) with multiplicity q. Then*

$$|\omega_{j,p} - \omega_k| \leqslant C(h)p^{-2.177924\ldots}, \quad j = k, \ldots, k+q-1, \tag{10.32}$$

and

$$\|u_{k,p} - u_k\|_{1,\Omega} + \|v_{k,p} - v_k\|_{1,\Omega} \leqslant C(h)p^{-1.088962\ldots}. \tag{10.33}$$

Note that we have given the eigenfunction estimate the simplified form it has when ω_k is simple; it would have to be modified if ω_k has multiplicity greater than 1. See the statement of Theorem 10.1.

PROOF. Suppose ω_k has multiplicity q and let w be either component of one of the eigenfunctions corresponding to ω_k. We have seen that w can be written in the form $w = w^1 + w^2$, where $w^2 \in H^k(\Omega)$, with $k \geqslant 3$, and w^2 is a sum of terms of the type (10.19) with $\sigma \geqslant \sigma_0$ and with coordinate centers at the vertices of Ω. Because $\sigma_0 = 0.544481\ldots$ in (10.19), from (10.30) we have

$$\varepsilon_p \leqslant C(h)p^{-1.088962\ldots}.$$

(10.32) and (10.33) follow from this estimate and (8.44)–(8.46). □

To illustrate the performance of the p-version we consider, as with the h-version, the relative error in the energy norm measure of the degree to which (u^*, v^*) can be approximated by S_p (cf. (10.29)). Figure 10.4 presents the graph of $\|e\|_{E,R}$ as a function of p, for various values of h. Again the log-log scale is used. We see that

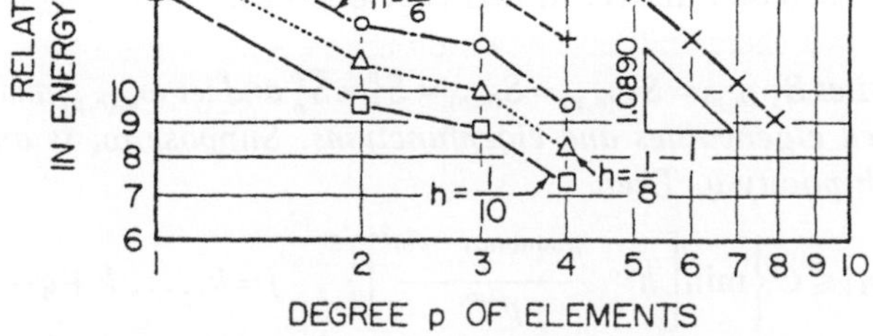

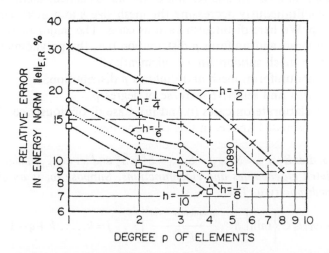

FIG. 10.4. The relative approximation error measured in the energy norm. The p-version.

the slope is close to the theoretically predicted $-1.088962\ldots$. This is valid only for $p \geqslant 3$, but recall that all our results are of an asymptotic nature.

To assess the relative effectiveness of the h- and p-versions, to understand, in particular, their dependence on the choice of S_h and S_p is not easy. Here we content ourselves with a brief assessment in terms of the number of degrees of freedom: $N = 2 \dim \tilde{S}_h^p = \dim S_h = \dim S_p$. In Fig. 10.5, the relative error in the energy norm measure of the accuracy is plotted as a function of N. Since $N \simeq h^{-2}$ and $N \simeq p^2$, the rates of convergence shown in Fig. 10.5 are half those shown in Figs. 10.3 and 10.4. We see that with respect to degrees of freedom, the p-version with $h = \frac{1}{2}$ performs better than the h-version with $p = 1$, 2, 3, or 4.

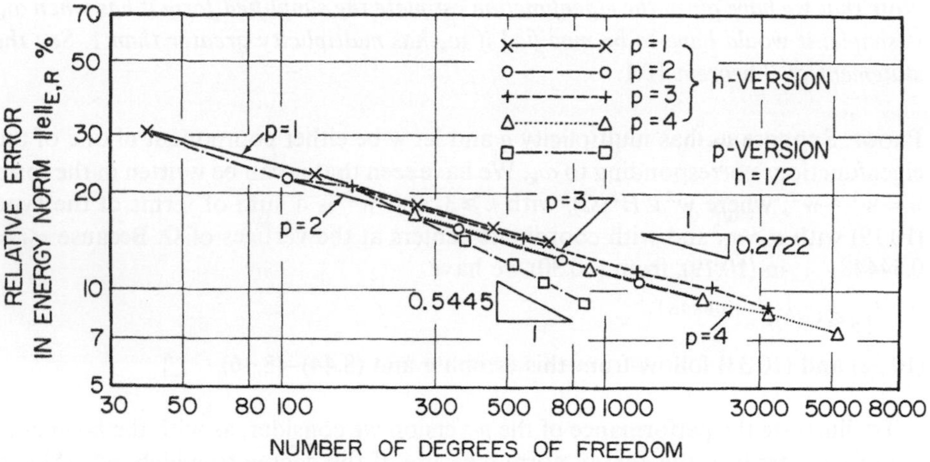

FIG. 10.5. The relative error in the energy norm in dependence on N.

10.1.2.3. The h-p version. In this version of the finite element method accuracy is achieved by simultaneously decreasing the mesh size h and increasing the polynomial degree p. We here distinguish various cases. The major ones are:

(i) uniform mesh and uniform p distribution (i.e., the polynomial degree p is the same on each mesh subdomain, i.e., element);

(ii) refined (nonuniform) mesh and uniform p distribution;

(iii) refined mesh and selective increase of degree p.

We will now elaborate on cases (i) and (ii). Case (i) obviously combines the h- and p-versions discussed above. In this case one has:

THEOREM 10.3. *Let $S_{1,(h,p)} = S_{2,(h,p)} = S_{(h,p)} = \tilde{S}_h^p \times \tilde{S}_h^p$ and let $\omega_{k,(h,p)}$ and $(u_{k,(h,p)}, v_{k,(h,p)})$ be the associated eigenvalues and eigenfunctions. Suppose ω_k is an eigenvalue of* (10.18) *with multiplicity q. Then*

$$|\omega_{j,(h,p)} - \omega_k| \leqslant C\left\{\min\left[h^{\sigma_0}, \frac{h^{\min(\sigma_0, p-\sigma_0)}}{p^{2\sigma_0}}\right]\right\}^2, \quad j = k, \ldots, k+q-1, \tag{10.34}$$

and

$$\|u_{k,(h,p)}-u_k\|_{1,\Omega}+\|v_{k,(h,p)}-v_k\|_{1,\Omega}\leqslant C\left\{\min\left[h^{\sigma_0},\frac{h^{\min(\sigma_0,p-\sigma_0)}}{p^{2\sigma_0}}\right]\right\}, \tag{10.35}$$

where $\sigma_0=0.544481\ldots$ *and* C *is independent of* h *and* p.

PROOF. The basic approximation results for this type of approximation were proved by BABUŠKA and SURI [1987b]. (10.34) and (10.35) follow directly from these results and (8.44)–(8.46). □

In case (ii) we will consider only geometric meshes with ratio factor 0.15; see Fig. 10.6. This ratio is close to optimal. The space $S_{(h,p)}=\tilde{S}_h^p\times\tilde{S}_h^p$ is now more

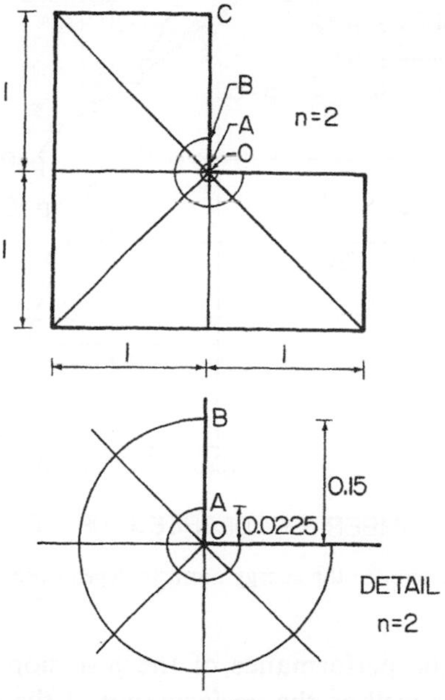

FIG. 10.6. The strongly refined mesh with $n=2$ layers.

complicated. $\tilde{S}_h^p$ is defined by

$$\begin{aligned}\tilde{S}_h^p=\{u\in H^1(\Omega)\colon u|_{I_{ij}} &\text{ is the image of a polynomial}\\ &\text{in a square } S=\{(\xi,\eta)\colon |\xi|,|\eta|\leqslant 1\}\\ &\text{or a triangle } T=\{(\xi,\eta)\colon 0\leqslant\eta\leqslant\xi, 0\leqslant\xi\leqslant 1\},\\ &\text{for all subdomains } I_{ij} \text{ in the mesh}\}.\end{aligned}$$

For a more detailed description of $\tilde{S}_h^p$ see BABUŠKA and GUO [1988b, 1989] and SZABO [1986]. For a thorough discussion of the h-p version in the one-dimensional setting and the optimality of the meshes and degree distributions. We refer to GUI and BABUŠKA [1986]. For the two-dimensional setting see GUI [1988].

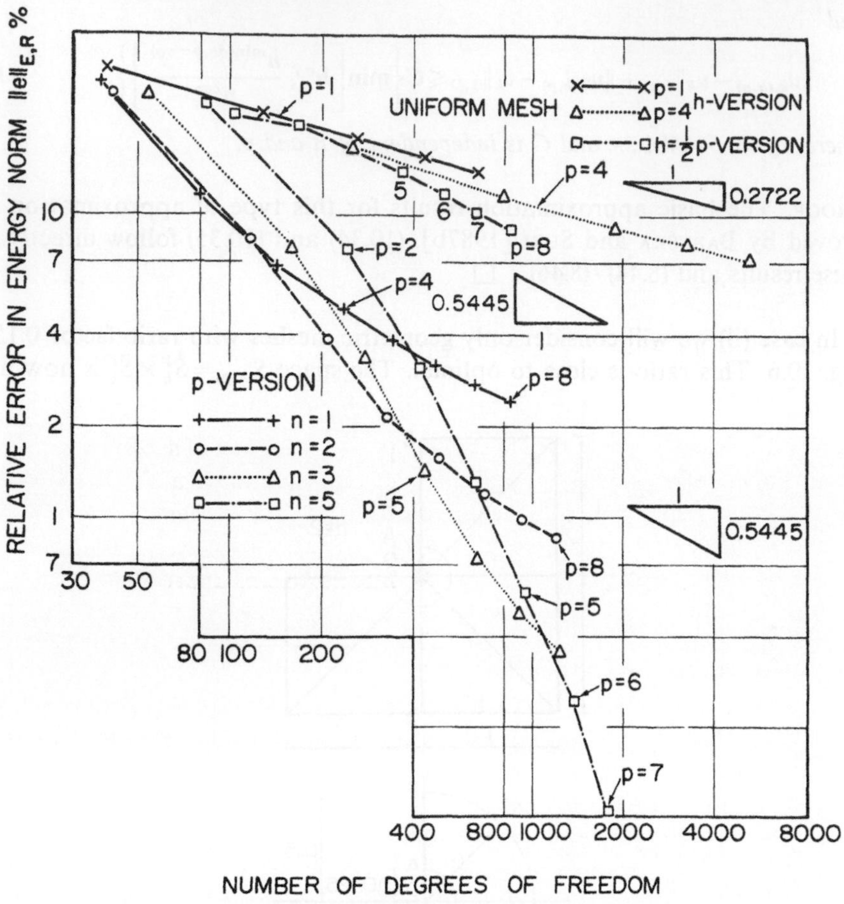

FIG. 10.7. The relative error in the energy norm in dependence on N for various meshes.

Figure 10.7 shows the performance of the p-version on meshes with various numbers of layers n, as well as the performance of the p- and the h-versions for uniform meshes. We typically see a reverse S curve for the accuracy of the p-version on a geometric mesh. The first part of the curve is convex and then it is concave, the slope approaching $N^{-\sigma_0}$. The h-p version appears as the envelope of the p-version on geometric meshes with various numbers of layers. This envelope shows the optimal relation between the number of layers and the polynomial degree. In BABUŠKA and GUO [1988b], it is shown that if $u \in \mathscr{B}^2_\beta(\Omega)$, with $0<\beta<1$, then a geometric mesh and a proper selection of the degree p leads to the exponential rate

$$\|e\|_{E,R} \leqslant Ce^{-\alpha\sqrt[3]{N}}.$$

Using this approximation result and the results of Section 8, we obtain:

THEOREM 10.4. *Suppose the components u_i and v_i of the eigenfunctions belong to* $\mathscr{B}^2_\beta$

(in our case $\beta = 0.544481\ldots + \varepsilon$). Then with a proper choice of geometric mesh and the degree p we have

$$|\omega_{k,p} - \omega_k| \leqslant Ce^{-2\alpha\sqrt[3]{N}} \tag{10.36}$$

and

$$\|u_{k,p} - u_k\|_{1,\Omega} + \|v_{k,p} - v_k\|_{1,\Omega} \leqslant Ce^{-\alpha\sqrt[3]{N}}, \tag{10.37}$$

where α depends on the ratio of the mesh, the relation of p and the number of layers, and the domain, but is independent of N.

PROOF. (10.36) and (10.37) follow directly from the results of Section 8 and the above estimate for $\|e\|_{E,R}$. □

Figure 10.7 clearly shows the effect of the proper selection of meshes and element degrees on the effectiveness of the finite element method. It also shows that the optimal choice depends on the required accuracy. The design of the mesh and selection of the degree of the elements is a delicate task. Various approaches to deal with this problem are in the research phase. One promising approach is to apply the principles of artificial intelligence (expert systems). For further details we refer to BABUŠKA and RANK [1987]. Figure 10.7 shows only the dependence of the accuracy on the number of degrees of freedom N. It is also essential to judge the complexity of the method with respect to such factors as number of operations, computer architecture, user's interaction, etc. For a detailed study of computer time, accuracy, and performance for various numbers of degrees of freedom, we refer to BABUŠKA and SCAPOLLA [1987]. We can see directly from Fig. 10.7 that the proper mesh design leading to an accuracy of 5% has two layers (the ratio of the sizes of elements is of order 50) and $p=3$. For an accuracy of 3%, the optimal p is 3 or 4 and the number of layers is 2 or 3 (which leads to size ratios from 50 to 300). The number of degrees of freedom is 200–300 (compared with 25,000–170,000 for a uniform mesh and $p=1$).

10.1.3. Solution of the matrix eigenvalue problem

We have seen that the approximation procedure developed in Section 8 leads from the eigenvalue problem (8.10) or (10.18) to the generalized matrix eigenvalue problem (8.15), and that the matrices A and B in (8.15) are sparse if the bases for the trial and test spaces are properly chosen. We have proved convergence for each fixed eigenvalue, but convergence does not occur for a fixed percentage of the available eigenvalues. Nevertheless, from the error estimates in Section 8 we know that the low eigenvalues of (8.15) give reasonable approximations to the corresponding exact eigenvalues. The size of the matrix problem will thus be much larger than the number of eigenvalues we are attempting to calculate. The matrix eigenvalue solver, a crucial component of the complete computational procedure, should therefore be designed to effectively find the low eigenvalues of large, sparse, generalized matrix problems. An appropriate version of the Lanczos algorithm is suitable for this class of matrix problems and is often used in practice. We refer to the monographs by PARLETT [1980] and CULLUM and WILLOUGHBY [1985]. Because the extraction of the eigenvalues is very expensive, various "tricks" are used in engineering practice to reduce the sizes of the matrices under consideration. We will not go further in this direction.

REMARK 10.2. It should be emphasized that, generally, the goal of the computation is to find, in addition to the eigenpairs, certain functionals of the eigenfunctions (u, v)—for example, the stress intensity factors, which are combinations of the derivatives of u and v. We will not pursue this direction since it lies beyond the scope of this article. We refer, e.g., to BABUŠKA and MILLER [1984] and SZABO and BABUŠKA [1986].

REMARK 10.3. The complete computational resolution of an eigenvalue problem is influenced by a wide range of factors. Some of the most important of these—the smoothness of the eigenfunctions and the approximation properties of the trial and test space, for example—have been discussed in detail. Others—the accuracy of the matrix eigenvalue solver and the relation between the accuracy of the matrix solver and the error $\omega_{k,h}-\omega_k$, for example—have not been mentioned or have only been mentioned briefly. While these latter factors are important, we will not be able to pursue them. We also note that the important function of a posteriori analysis of computed data has not been discussed. Likewise we have not discussed any adaptive approaches. For some ideas on the assessment of the quality of the finite element computations we refer to the survey paper of NOOR and BABUŠKA [1987].

10.2. *Vibration of a membrane*

We consider here the eigenvalue problem associated with the small, transverse vibration of a membrane stretched over a bounded region Ω in the plane and fixed along the edge $\Gamma=\partial\Omega$, i.e., the eigenvalue problem

$$-\Delta u=\lambda u \quad \text{in } \Omega, \qquad u=0 \quad \text{on } \Gamma \tag{10.38}$$

(cf. Section 1.2, in particular, (1.27)). We turn now to a discussion of the basic steps (1), (1′), (2) and (3) (cf. Section 10.1 above) in the finite element approximation of the eigenpairs of (10.38). The discussion can be brief since these steps are similar for the two problems (10.1)–(10.2) and (10.38), in fact for any eigenvalue problem.

10.2.1. *Variational formulation*

Problem (10.38) is a special case of problem (3.1) and the variational formulation (3.15) of (3.1) was derived in Section 3. Thus we see that the variational formulation of (10.38) is given by

$$\begin{gathered} u\in H_0^1(\Omega), \qquad u\neq 0 \\ \int_\Omega \left(\frac{\partial u}{\partial x}\frac{\partial v}{\partial x}+\frac{\partial u}{\partial y}\frac{\partial v}{\partial y}\right)\mathrm{d}x\,\mathrm{d}y=\lambda\int_\Omega uv\,\mathrm{d}x \quad \forall v\in H_0^1(\Omega). \end{gathered} \tag{10.39}$$

Let

$$a(u, v) = \int_\Omega \left(\frac{\partial u}{\partial x}\frac{\partial v}{\partial x} + \frac{\partial u}{\partial y}\frac{\partial v}{\partial y} \right) dx\, dy = \int_\Omega \nabla u \cdot \nabla v \, dx\, dy$$

be defined for $u, v \in H = H_0^1(\Omega)$, and let

$$b(u, v) = \int_\Omega uv \, dx$$

be defined for $u, v \in W = L_2(\Omega)$. Then (10.39) has the form of (8.10), and a and b are symmetric forms, (8.1), (8.7), (8.32), and (8.33) are satisfied, and $H \subset W$, compactly. All of this can be easily seen for the concrete problem we are considering; it also follows from the more general discussion in Section 3. Problem (10.39) is a selfadjoint, positive-definite problem. It has eigenvalues

$$0 < \lambda_1 \leqslant \lambda_2 \leqslant \cdots \uparrow +\infty$$

and corresponding eigenfunctions

$$u_1, u_2, \ldots,$$

which can be chosen to satisfy

$$\int_\Omega \nabla u_i \cdot \nabla v_j \, dx\, dy = \lambda_j \int_\Omega u_i u_j \, dx\, dy = \delta_{ij}.$$

10.2.1′. Regularity of the eigenfunctions

From Theorems 4.1–4.4 we obtain the following regularity results for the eigenfunctions u_i of (10.39) (or (10.38)).

- For $k \geqslant 2$, if $\Gamma = \partial\Omega$ is of class C^k, then $u_i \in H^k(\Omega)$.
- If Γ is of class C^∞, then $u_i \in C^\infty(\bar{\Omega})$.
- If Γ is analytic, then u_i is analytic in $\bar{\Omega}$.
- If Ω is a curved polygon with analytic sides and with vertices $A_0, A_1, \ldots$, then u_i is analytic in $\bar{\Omega} \setminus \bigcup A_j$. u_i is singular at the vertices; the strengths of the singularities depend on the interior vertex angles. Moreover, $u \in \mathscr{B}_\beta^2(\Omega)$ for properly chosen β.

10.2.2. Discretization of (8.10) *and assessment of the accuracy of the approximate eigenpairs*

Suppose Ω is a polygon. By a *triangulation* or *mesh* on $\bar{\Omega}$ we will mean a finite family $\tau = \{T_i\}_{i=1}^{M(\tau)}$ satisfying:

- each T_i is a closed triangle,
- $\bar{\Omega} = \bigcup_{i=1}^{M(\tau)} T_i$,
- for any T_i and $T_j \in \tau$, $T_i \cap T_j = \emptyset$ or a common vertex or a common side.

For $0 < \alpha < \pi$, a triangulation τ is said to be *α-regular* if the minimal angle of every triangle $T \in \tau$ is greater than or equal to α. For any τ, let

$$h = h(\tau) = \max_{i=1,\ldots,M(\tau)} \operatorname{diam} T_i$$

and

$$\underline{h}(\tau) = \min_{i=1,\ldots,M(\tau)} \operatorname{diam} T_i.$$

τ is said to be q-quasiuniform if

$$h(\tau)/\underline{h}(\tau) \leqslant q.$$

We will often view triangulations $\tau = \tau_h$ as parameterized by $h = h(\tau)$ and consider families $\gamma = \{\tau\} = \{\tau_h\}$ of triangulations that are α-regular. An example of a $\frac{1}{4}\pi$-regular, 1-quasiuniform triangulation of the domain $\Omega = \{(x, y){:}\ -1 < x < 1, -1 < y < 1\}$ is shown in Fig. 10.8. It is called a uniform triangulation.

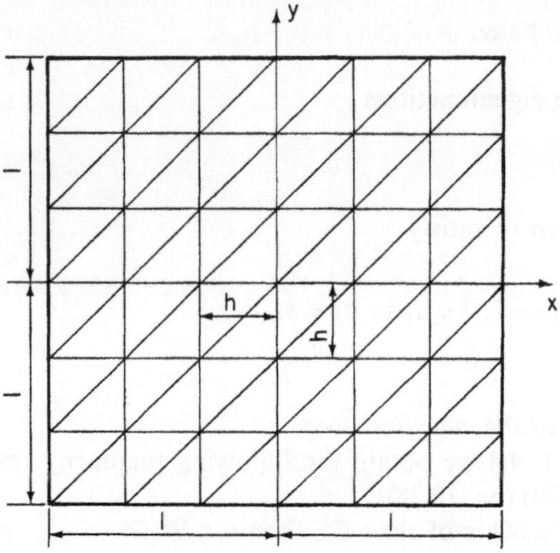

FIG. 10.8. A uniform triangulation.

For τ a triangulation of Ω and $p = 1, 2, \ldots$ let

$$S^p(\tau) = \{u{:}\ u \in H^1(\Omega),\ u|_T = \text{a polynomial of degree } p \text{ for each } T \in \tau\}$$

and let

$$S_0^p(\tau) = S^p(\tau) \cap H_0^1(\Omega).$$

Regarding the approximation properties of $S^p(\tau)$ and $S_0^p(\tau)$, if $k \geqslant 1$ (integer or non-integer) and $p \geqslant 1$ and if $\gamma = \{\tau\}$ is a family of α-regular triangulations of Ω, then

$$\inf_{\chi \in S^p(\tau)} \|u - \chi\|_{1,\Omega} \leqslant C(h(\tau)^{\mu-1}/p^{k-1})\|u\|_{k,\Omega} \tag{10.40a}$$

for any $u \in H^k(\Omega)$ and for any $\tau \in \gamma$,

and

$$\inf_{\chi \in S_0^p(\tau)} \|u-\chi\|_{1,\Omega} \leqslant C(h(\tau)^{\mu-1}/p^{k-1})\|u\|_{k,\Omega}$$
$$\text{for any } u \in H^k(\Omega) \cap H_0^1(\Omega) \text{ and for any } \tau \in \gamma, \tag{10.40b}$$

where

$$\mu = \min(k, p+1). \tag{10.41}$$

The constant C in (10.40) is independent of p, τ and u, but depends on Ω, k and α. For a complete proof of these estimates we refer to BABUŠKA and SURI [1987b].

Now define

$$S_{1,(h,p)} = S_{2,(h,p)} = S_{(h,p)} = S_0^p(\tau). \tag{10.42}$$

Since (10.39) is selfadjoint and positive-definite and satisfies (8.32), we see that (8.11) and (8.12) are satisfied. Equation (10.40b) shows that $S_{(h,p)}$ accurately approximates the exact eigenfunctions and thus that (8.13) is satisfied. If a suitable basis is chosen for $S_{(h,p)}$, then the matrices A and B in (8.15) will be sparse. The issues raised in Section 10.1.2(a)–(c) above have now been addressed for this choice for $S_{(h,p)}$. Note that in using the notation $S_{(h,p)}$ we are identifying $h = h(\tau)$ with τ. An alternate, and more explicit, notation would be $S_{(\tau,p)}$.

Now consider problem (8.14) with $S_{(h,p)}$ defined as in (10.42) and denote the approximate eigenvalues and eigenfunctions by

$$\lambda_{1,(h,p)} \leqslant \cdots \leqslant \lambda_{N,(h,p)}$$

and

$$u_{1,(h,p)}, \ldots, u_{N,(h,p)},$$

where $N = \dim S_{(h,p)}$. To assess the accuracy of these approximate eigenpairs, we apply the results of Section 8, obtaining:

THEOREM 10.5. *Let $S_{(h,p)}$ be selected as in* (10.42) *and suppose λ_j is an eigenvalue of* (10.39) *with multiplicity q. Then*

$$|\lambda_{l,(h,p)} - \lambda_j| \leqslant C(h^{2\mu-2}/p^{2k-2}), \quad l = j, \ldots, j+q-1, \tag{10.43}$$

and

$$\|u_{j,(h,p)} - u_j\|_{1,\Omega} \leqslant C(h^{\mu-1}/p^{k-1}), \tag{10.44}$$

where $k \geqslant 1$ is such that the eigenfunctions corresponding to λ_j are in $H^k(\Omega)$ and $\mu = \min(k, p+1)$. Note that we have given the eigenfunction estimate the simple form it has when λ_j is simple; it would have to be modified in the general case. See the statement of Theorem 10.1.

PROOF. Suppose λ_j has multiplicity q. Then (10.39) has eigenfunctions $u_j, \ldots, u_{j+q-1}$ associated with λ_j; by assumption, these eigenfunctions are in $H^k(\Omega)$. Thus, by (10.40)–(10.41), we have

$$\begin{aligned}\varepsilon_h &= \max_{l=j,\dots,j+q-1} \inf_{\chi\in S_{(h,p)}} \|u_l-\chi\|_{1,\Omega}\\ &\leqslant C(h^{\mu-1}/p^{k-1}) \max_{l=j,\dots,j+q-1} \|u_l\|_{k,\Omega}\\ &= C(h^{\mu-1}/p^{k-1}).\end{aligned} \tag{10.45}$$

(10.43) and (10.44) follow directly from this estimate and (8.44)–(8.46). □

REMARK 10.4. If our membrane is free instead of fixed along its edge, then we would have considered the Neumann boundary condition $\partial u/\partial n=0$. In this situation the eigenvalue problem would have the variational foundation

$$\begin{aligned}&u\in H, \qquad u\neq 0,\\ &a(u,v)=\lambda b(u,v),\end{aligned} \tag{10.46}$$

where a and b are as above, but

$$H=\left\{u\colon u\in H^1(\Omega), \int_\Omega u\,\mathrm{d}x\,\mathrm{d}y=0\right\},$$

$$\|u\|_H=\|u\|_{1,\Omega}.$$

We would choose

$$S_{(h,p)}=\left\{u\colon u\in S^p(\tau), \int_\Omega u\,\mathrm{d}x\,\mathrm{d}y=0\right\}$$

for the trial and test space. Then all of the hypotheses in Section 8 are satisfied, approximation results similar to (10.40) can be proved, and for the approximate eigenpairs, the error estimates (10.43) and (10.44) follow. We note in particular that the Neumann boundary condition is only implicitly stated in (10.46), i.e., is natural, and thus that the boundary condition need not be imposed on the trial and test functions. This fact makes implementation easier, especially for domains with curved boundaries. See the discussion of natural and essential boundary conditions in Section 3.

10.2.3. Solution of the matrix eigenvalue problem

The comments made in Section 10.1.3 apply here as well.

EXAMPLE 10.1. Consider the membrane eigenvalue problem (10.38) on the domain Ω shown in Fig. 10.1. The eigenfunctions will be analytic on $\bar{\Omega}$, except possibly at A_0. Because of the various symmetries in Ω, the leading term in the singularities in the eigenfunctions at A_0 are as follows:

(a) for the first (simple) eigenfunction,

$$u=Cr^{2/3}\sin\tfrac{2}{3}\theta+\cdots,$$

(b) for the second (simple) eigenfunction,

$$u = Cr^{4/3}\sin\tfrac{4}{3}\theta + \cdots,$$

(c) for the third (simple) eigenfunction, u is analytic on $\bar{\Omega}$, in fact,

$$u = (\tfrac{2}{3})^{1/2}\pi^{-1}\sin \pi x \sin \pi y.$$

We see that the higher eigenfunctions can be smoother than the lower ones. The regularity of the eigenfunctions for problems with piecewise analytic data has been discussed in BABUŠKA, GUO and OSBORN [1989].

BABUŠKA, GUO and OSBORN [1989] also contains the results of detailed numerical computations for finite element approximation of the eigenvalue problem mentioned above; we will now report some of these numerical results. Strongly refined meshes similar to those introduced in Section 10.1 (see Fig. 10.6) were used. The sequence of meshes with n layers is shown in Fig. 10.9; the associated finite element spaces are denoted by $S_{(n,p)}$.

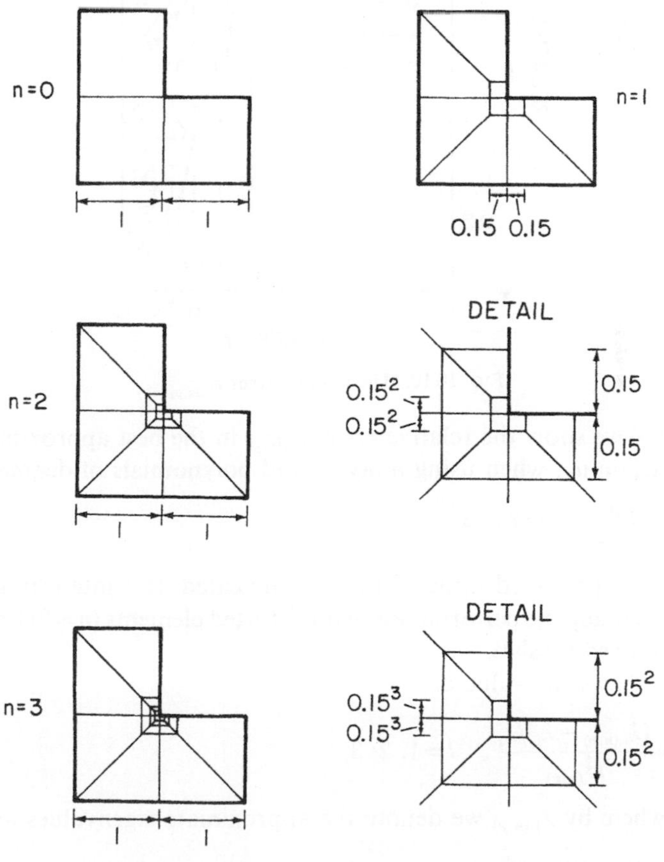

FIG. 10.9. Meshes for the finite element spaces.

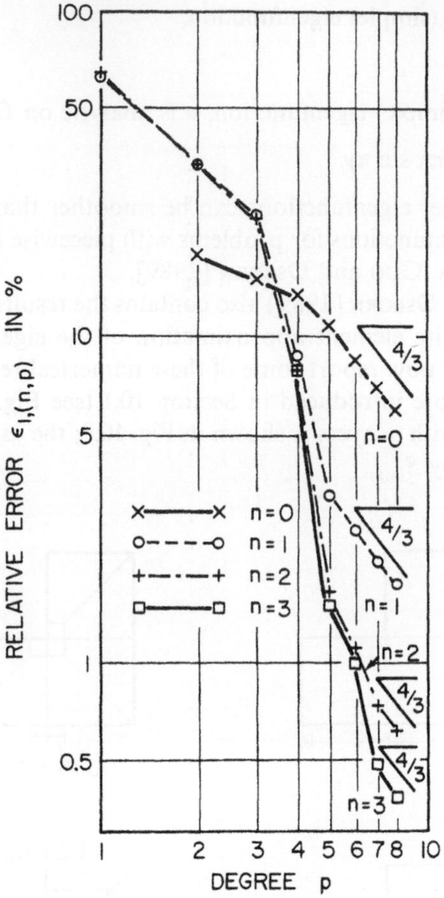

FIG. 10.10. The relative error $\varepsilon_{1,(n,p)}$.

In Fig. 10.10 we show the relative error $\varepsilon_{1,(n,p)}$ in the best approximation of the first eigenfunction u_1, when using n layers and polynomials of degree p, i.e.,

$$\varepsilon_{1,(n,p)} = \inf_{\chi \in S(n,p)} \|u_1 - \chi\|_{1,\Omega}.$$

The theoretically predicted slope of $\frac{4}{3}$ is also indicated. It is interesting to observe that for low accuracy the performance of undistorted elements ($n=0$) leads to better results than refined meshes.

Table 10.1 shows the value of

$$C_{i,(n,p)} = \frac{(\lambda_{i,(n,p)} - \lambda_i)/\lambda_i}{\varepsilon_{i,(n,p)}^2}, \quad i=1, 2, 3,$$

(cf. (8.47c)), where by $\lambda_{i,(n,p)}$ we denote the approximate eigenvalues and

$$\varepsilon_{i,(n,p)} = \inf_{\chi \in S(n,p)} \|u_i - \chi\|_{1,\Omega}.$$

TABLE 10.1. The values of $C_{i,(n,p)}$.

	p	k = 0	1	2	3
	1	—	1.322	1.349	2.346
	2	0.912	0.988	0.978	0.978
	3	0.967	0.947	0.932	0.932
	4	0.962	0.945	0.929	0.927
λ_1	5	0.977	0.991	0.965	0.954
	6	0.984	0.995	0.973	0.962
	7	0.992	0.998	0.997	0.992
	8	0.995	0.999	0.998	0.996
	1	—	1.795	1.777	1.777
	2	0.867	0.919	0.921	0.978
	3	0.787	0.892	0.893	0.932
	4	0.780	0.886	0.886	0.927
λ_2	5	0.871	0.932	0.931	0.934
	6	0.948	0.932	0.929	0.962
	7	0.966	0.963	0.962	0.991
	8	0.975	0.974	0.975	0.995
	1	—	2.692	2.740	2.740
	2	—	1.126	1.128	1.128
	3	1.150	1.121	1.124	1.124
	4	0.831	0.902	0.902	0.901
λ_3	5	0.831	0.899	0.899	0.898
	6	0.905	0.904	0.905	0.904
	7	0.905	0.928	0.928	0.928
	8	0.942	0.946	0.946	0.946

Figure 10.11 shows the relative error in $\lambda_{1,(n,p)}$ in dependence on the number of degrees of freedom N. We see that to get relatively high accuracy for the eigenvalues is not difficult. In our case, with $N=2$ we already have an accuracy of 5%. Note that an unrefined mesh ($N=0$) is more effective than a refined mesh ($N>0$) for low accuracies (see Figs. 10.10 and 10.11). Thus, while it is worthwhile, in general, to refine the mesh at appropriate places, in some cases requiring only low accuracy, there is no advantage in refinement.

Multiple eigenvalues. The results proved in this subsection and in Section 10.1 cover the case of multiple eigenvalues. Recall that the estimates for $|\lambda_{j,(h,p)}-\lambda_j|$ and $\|u_{j,(h,p)}-u_j\|_{1,\Omega}$ are in terms of

$$\varepsilon_h = \max_{l=1,\ldots,q} \inf_{\chi\in S_{1,(h,p)}} \|u_{i_l}-\chi\|_{1,\Omega},$$

where q is the multiplicity of λ_j and $u_{i_1},\ldots,u_{i_q}$ are the corresponding eigenfunctions. We now make some comments on multiple eigenvalues and then make an application of the refined error estimates for multiple eigenvalues proved in Section 9.

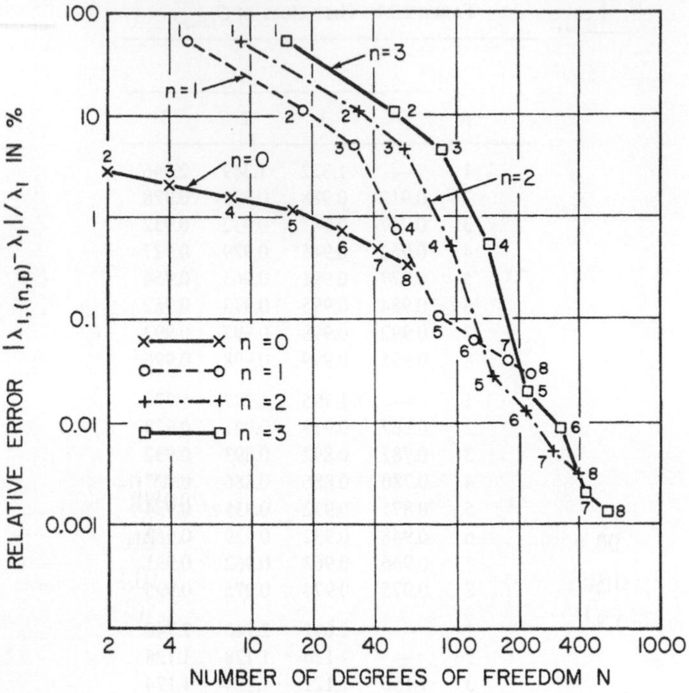

FIG. 10.11. Relative error in $\lambda_{1,(n,p)}$.

The eigenvalues and eigenfunctions of the membrane problem on a square, i.e., the problem

$$-\Delta u = \lambda u \quad \text{in } \Omega, \qquad u = 0 \quad \text{on } \Gamma, \tag{10.47}$$

where

$$\Omega = \{(x, y) : |x|, |y| < \pi\},$$

are easily seen to be given by

$$\lambda_{k,l} = k^2 + l^2$$

and

$$u_{k,l} = \sin kx \sin ly, \quad k, l = 1, 2, \ldots .$$

Hence we see that there are multiple eigenvalues. Problem (10.47) is typical of problems with symmetries ((10.47) is symmetric with respect to x and y), and we thus see that multiple eigenvalues are common in applications.

For $i = 1, 2, \ldots$, let λ_{k_i} be an eigenvalue of (10.47) of multiplicity q_i, i.e., suppose

$$\lambda_{k_i - 1} < \lambda_{k_i} = \lambda_{k_i + 1} = \cdots = \lambda_{k_i + q_i - 1} < \lambda_{k_i + q_i} = \lambda_{k_i + 1}.$$

Note that we are here using the notation introduced in Section 10, whereby

$k_1=1, k_2$ is the lowest index of the second distinct eigenvalue, etc. Suppose now that $q_i>1$, i.e., that λ_{k_i} is multiple. Let $\{S_h\}$ be any family of finite-dimensional subspaces of $H_0^1(\Omega)$ satisfying (9.14). Recall from Section 7 that the q_i approximate eigenvalues

$$\lambda_{k_i,h},\ldots,\lambda_{k_i+q_i-1,h}$$

converge to λ_{k_i}. While these approximate eigenvalues may be equal, i.e., we may have one distinct eigenvalue with multiplicity q_i, consideration of the situation in which we choose S_h to be $S^1(\tau)$, where τ is the triangulation shown in Fig. 10.8, shows that they may not be equal, since some of the symmetries present in (10.47) are not present in the discrete problem. Nevertheless, Theorem 10.5 provides estimates for each of the errors $|\lambda_{k_i+j-1,h},\ldots,\lambda_{k_i+j-1}|, j=1,\ldots,q_i$. As we have seen the estimates are

$$|\lambda_{k_i+j-1,h}-\lambda_{k_i+j-1}|$$
$$\leqslant C\varepsilon_h^2=C\Big[\sup_{u\in M(\lambda_{k_i})}\ \inf_{\chi\in S^1(\tau)}\|u-\chi\|_{1,\Omega}\Big]^2,\quad j=1,\ldots,q_i, \tag{10.48}$$

which suggest that the error in λ_{h,k_i+j-1} depends on the degree to which $S^1(\tau)$ can approximate all of the eigenfunctions corresponding to λ_{k_i}.

Recall that in Section 9 (Theorem 9.1) we proved refined estimates, namely,

$$|\lambda_{k_i+j-1,h}-\lambda_{k_i+j-1}|\leqslant C\varepsilon_{i,j}^2(h),\quad j=1,\ldots,q_i, \tag{10.49}$$

where

$$\varepsilon_{i,j}(h)=\inf_{u\in M(\lambda_{k_i})}\ \inf_{\chi\in S^1(\tau)}\|u-\chi\|_{1,\Omega},$$

$$a(u,u_{k_1,h})=\cdots=a(u,u_{k_1+j-2,h})=0.$$

Now for the specific problem (10.47), all eigenfunctions have the same smoothness properties and $S^1(\tau)$ with τ given in Fig 10.8, will approximate them all with the same asymptotic accuracy and (10.48) and (10.49) would each lead to the same estimate in terms of h. The multiplicative constants in the estimates could, however, be different. We further note that there are eigenvalue problems for which the different eigenfunctions corresponding to a multiple eigenvalue have strikingly different approximability properties. For such problems (10.49) would provide a striking improvement over (10.48).

As an example of such a problem, consider

$$-\left[\frac{1}{\varphi'(x)}u'(x)\right]'=\lambda\varphi'(x)u,\quad x\in I=(-\pi,\pi),$$

$$u(-\pi)=u(\pi), \tag{10.50}$$

$$\left[\frac{1}{\varphi'}u'\right](-\pi)=\left[\frac{1}{\varphi'}u'\right](\pi),$$

where

$$\varphi(x)=\pi^{-\alpha}|x|^{1+\alpha}\operatorname{sgn} x, \quad 0<\alpha<1.$$

It is easy to check that the eigenvalues and eigenfunctions are as shown in Table 10.2. We see that $\lambda_1=\lambda_2$, $\lambda_3=\lambda_4$, etc.

TABLE 10.2. Eigenvalues and eigenfunctions of the eigenvalue problem (10.50).

i	λ_i	u_i
0	0.0	1
1	1.0	$\cos\varphi(x)$
2	1.0	$\sin\varphi(x)$
3	4.0	$\cos 2\varphi(x)$
4	4.0	$\sin 2\varphi(x)$
⋮	⋮	⋮

We cast this problem into the variational form (10.1) by choosing

$$H=\left\{u(x):\ \|u\|=\left(\int_{-\pi}^{\pi}\frac{(u')^2}{\varphi'}\,dx\right)^{1/2}<\infty,\ u(-\pi)=u(\pi),\ \int_{-\pi}^{\pi}\varphi' u\,dx=0\right\},$$

$$a(u,v)=\int_{-\pi}^{\pi}u'v'\frac{1}{\varphi'}\,dx \quad\text{and}\quad b(u,v)=\int_{-\pi}^{\pi}uv\varphi'\,dx.$$

With these choices (8.10) is equivalent to (10.50), with the understanding that the eigenpair $(\lambda_0,u_0)=(0,1)$ of (10.50) is not present in (8.10). Note that $\|u\|=\|u\|_a$. Let $\|u\|_b=(b(u,u))^{1/2}$. The assumptions made in Section 9 are clearly satisfied. Our approximation is defined by (9.4) with

$$\begin{aligned}S_{1,h}&=S_{2,h}=S_h\\&=\{u\in H: u \text{ linear on } (-\pi+jh,\ -\pi+(j+1)h),\ j=0,1,\ldots,n-1\},\end{aligned}$$

where $h=2\pi/n$ and n is an even integer.

Now with this choice for $\{S_h\}$ it is easily seen that

$$\inf_{\chi\in S_h}\|\cos\varphi(x)-\chi\|_a^2\simeq Ch^2 \tag{10.51}$$

and

$$\inf_{\chi\in S_h}\|\sin\varphi(x)-\chi\|_a^2\simeq Ch^{1+\alpha} \tag{10.52}$$

Hence from Theorem 10.1 we would expect $\lambda_{1,h}$ and $\lambda_{2,h}$, the two approximate eigenvalues that converge to the double eigenvalue $\lambda_1=\lambda_2$, to have different convergence rates.

From Table 10.3 we can find the errors in $\lambda_{i,h}$, $i=1,2,3,4$, for $\alpha=0.4$. These errors

TABLE 10.3. Numerical solution of the eigenvalue problem (10.50) for $\alpha=0.4$.

n	i	$\lambda_{i,h}$	$K(i)$		$C_1^{(i)}$		$C_2^{(i)}$		$\dfrac{\lambda_{i+1,h}-\lambda_{i+1}}{\lambda_{i,h}-\lambda_i}$
8	1	1.0716754	0.2704	0	0.5637791	0	−0.1124891	−16	1.5562955
	2	1.1115481	0.3423	0	−0.4151973	−13	−0.5636998	0	
	3	5.0394692	0.1075	+1	0.5558919	0	0.1317809	−12	1.1943249
	4	5.2414639	0.1191	+1	0.5022638	−13	0.5516234	0	
16	1	1.0175850	0.1329	0	0.5641633	0	0.1596754	−12	2.0041570
	2	1.0352431	0.1881	0	−0.8916589	−12	0.5641519	0	
	3	4.2691915	0.5259	0	0.5636643	0	0.1124328	−13	1.2575063
	4	4.3385100	0.5869	0	−0.2689727	−12	0.5637697	0	
32	1	1.0043740	0.6618	−1	0.5641879	0	0.6411454	−11	2.6003887
	2	1.0113741	0.1067	0	0.1323421	−10	0.5641830	− 0	
	3	4.0666055	0.2589	0	0.5641561	0	0.1970954	−10	1.4067517
	4	4.0936974	0.3067	0	−0.7375504	−10	0.5641613	0	
64	1	1.0010921	0.3305	−1	0.5641895	0	0.7729760	− 9	3.5190001
	2	1.0038431	0.6202	−1	0.8670648	− 9	0.5641883	0	
	3	4.0166006	0.1289	0	0.5641875	0	0.3641341	−10	1.6437659
	4	4.0272875	0.1653	0	0.1415775	− 8	0.5641858	0	
128	1	1.0002729	0.1651	−1	0.5641895	0	0.4535626	− 7	4.9215830
	2	1.0013431	0.3665	−1	0.3251219	− 7	0.5641893	0	
	3	4.0041468	0.6440	−1	0.5641895	0	0.4409247	− 7	2.0.107071
	4	4.0083380	0.9135	−1	−0.9705611	− 8	0.5641890	0	
256	1	1.0000682	0.8255	−2	0.5641896	0	0.8070959	− 5	7.0542522
	2	1.0004811	0.2193	−1	0.7269570	− 6	0.5641895	0	
	3	4.0010365	0.3217	−1	0.5641896	0	0.6435344	− 6	2.5706705
	4	4.0026645	0.5162	−1	−0.2601000	− 6	0.5641895	0	

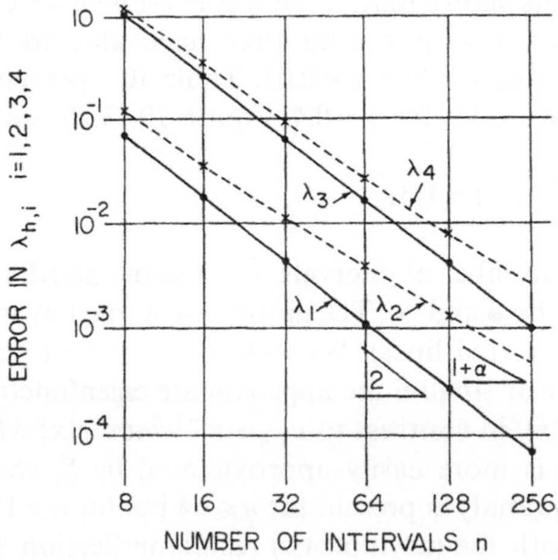

FIG. 10.12. The error in the eigenvalues $\lambda_{1,h}, \lambda_{2,h}$ and $\lambda_{3,h}, \lambda_{4,h}$ for $\alpha=0.4$ in dependence on the number of intervals n.

are plotted in Fig. 10.12 in log-log scale. We clearly see the different rates of convergence, specifically seeing the rates h^2 and $h^{1+\alpha}=h^{1.4}$ for the errors in $\lambda_{i,h}$, for $i=1$, 3 and $i=2$, 4, respectively, as suggested by (10.51) and (10.52). It should be noted that the estimates presented in Theorem 10.1 are of an asymptotic nature in that they provide information only for small h (or large n), i.e., for h (or n) in the asymptotic range. From Fig. 10.12 we see that for $\alpha=0.4$ we are in the asymptotic range quite quickly, say for $n\geqslant 16$.

Consider $u_{1,h}$ and $u_{2,h}$, the approximate eigenfunctions corresponding to $\lambda_{1,h}$ and $\lambda_{2,h}$, respectively, normalized by $\|\cdot\|_b=1$. The results of Section 9 suggest that $u_{1,h}$ should be close to $C\cos\varphi(x)$ and $u_{2,h}$ close to $C\sin\varphi(x)$ (cf. (10.51) and (10.52)), where C is such that $C\sin\varphi(x)$ and $C\cos\varphi(x)$ are normalized by $\|\cdot\|_b=1$, i.e., $C=\pi^{-1/2}$. To illustrate this point we have computed $C_1^{(i)}$ and $C_2^{(i)}$, $i=1,2,3,4$, so that

$$K(i)=\begin{cases}\|u_{i,h}-C_1^{(i)}\cos\varphi(x)-C_2^{(i)}\sin\varphi(x)\|_a, & i=1,2,\\ \|u_{i,h}-C_1^{(i)}\cos 2\varphi(x)-C_2^{(i)}\sin 2\varphi(x)\|_a, & i=3,4,\end{cases}$$

is minimal. We would expect that

$$C_1^{(2)}, C_1^{(4)}, C_2^{(1)}, C_2^{(3)}\simeq 0 \tag{10.53}$$

and

$$C_1^{(1)}=C_2^{(2)}=C_1^{(3)}=C_2^{(4)}\simeq C=0.564189583\ldots. \tag{10.54}$$

Table 10.3 shows some of the results for $\alpha=0.4$. We see clearly the results predicted in (10.53) and (10.54). Table 10.3 also shows that $K(1)<K(2)$ and $K(3)<K(4)$, as we would expect.

The last column in Table 10.3 and Figure 10.8 show that the ratios

$$\frac{\lambda_{i+1,h}-\lambda_{i+1}}{\lambda_{i,h}-\lambda_i},\quad i=1,3,$$

increase as $h\to 0$. This shows that in the whole h-range we considered, the approximate eigenvalues converging to a multiple eigenvalue are well separated.

Consider next the case when $\alpha=0.01$. Table 10.4 presents the same results for $\alpha=0.01$ as Table 10.3 does for $\alpha=0.4$. Figure 10.13 shows the graph of

$$\log\frac{\lambda_{i+1,h}-\lambda_{i+1}}{\lambda_{i,h}-\lambda_i},\quad i=1,3,$$

as a function of the number of intervals n in a semilogarithmic scale. The computed values are indicated by • and ×. The graphs are formed by interpolation (solid lines) and extrapolation (dotted lines). We note three related phenomena that did not occur with $\alpha=0.4$. For small n the approximate eigenfunction associated with $\lambda_{1,h}$ is $u_{1,h}\simeq\pi^{-1/2}\sin\varphi(x)$, in contrast to $u_{1,h}\simeq\pi^{-1/2}\cos\varphi(x)$ when $\alpha=0.4$. We remark that $\pi^{-1/2}\cos\varphi(x)$ is more easily approximated by S_h than is $\pi^{-1/2}\sin\varphi(x)$ for all $0<\alpha<1$. This anomaly is present for $n\leqslant 64$ but for $n\geqslant 128$ we get results which are in agreement with the (asymptotic) results in Section 9. For $\lambda_{3,h}$ and $\lambda_{4,h}$ we have to take $n\geqslant 256$ to get results which agree with the asymptotic theory.

For $\alpha=0.01$ we see that $K(2)<K(1)$ for small n ($n\leqslant 64$) and $K(2)>K(1)$ for large

TABLE 10.4. Numerical solution of the eigenvalue problem (4.1) for $\alpha=0.01$.

n	i	$\lambda_{i,h}$	$K(i)$		$C_1^{(i)}$		$C_2^{(i)}$		$\dfrac{\lambda_{i+1,h}-\lambda_{i+1}}{\lambda_{i,h}-\lambda_i}$
8	1	1.0520268	0.2338	0	0.8181940	−11	0.5634386	0	1.0171143
	2	1.0529172	0.2268	0	0.5645965	0	−0.2916448	−11	
	3	4.8576239	0.9593	0	−0.9346720	−13	0.5597529	0	1.0164293
	4	4.8717141	0.9615	0	0.5604533	0	0.1167277	−11	
16	1	1.0128661	0.1223	0	0.8717399	−10	0.5635957	0	1.0111689
	2	1.0130098	0.1052	0	0.5647369	0	−0.8480131	−9	
	3	4.2088367	0.4650	0	0.2507177	−10	0.5636658	0	1.0087030
	4	4.2106542	0.4577	0	0.5642694	0	−0.3101833	−10	
32	1	1.0032139	0.7274	−1	−0.9345818	−9	0.5636031	0	1.0068764
	2	1.0032360	0.3568	−1	0.5647430	0	0.1273043	−7	
	3	4.0515675	0.2384	0	0.3745461	−9	0.5638178	0	1.0057284
	4	4.0518629	0.2205	0	0.5644172	0	−0.4115544	−9	
64	1	1.0008063	0.5369	−1	−0.1311961	−5	0.5636032	0	1.0017363
	2	1.0008077	0.3398	−1	0.5647430	0	0.2462939	−7	
	3	4.0128623	0.1343	0	0.2743681	−7	0.5638240	0	1.0035997
	4	4.0129086	0.9792	−1	0.5644235	0	0.3196172	−8	
128	1	1.0002018	0.4196	−1	0.5647430	0	0.3356056	−5	1.0064420
	2	1.0002031	0.4775	−1	0.7414162	−6	0.5636032	0	
	3	4.0032196	0.9166	−1	0.2379072	−6	0.5638239	0	1.0010560
	4	4.0032230	0.9745	−2	0.5644235	0	0.1197135	−5	
256	1	1.0000504	0.4372	−1	0.5647429	0	0.1061527	−4	1.0218254
	2	1.0000515	0.4614	−1	−0.1553659	−4	0.5636031	0	
	3	4.0008054	0.5011	−1	0.5644234	0	−0.2123278	−4	1.0031040
	4	4.0008079	0.7741	−1	0.1165012	−5	0.5638238	0	

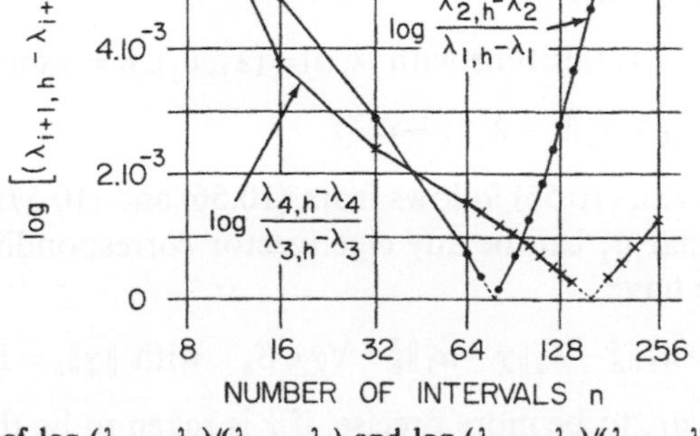

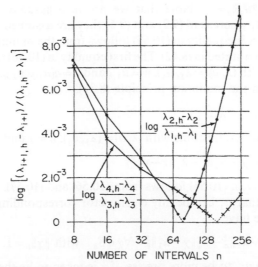

FIG. 10.13. The graphs of log $(\lambda_{2,h}-\lambda_2)/(\lambda_{1,h}-\lambda_1)$ and log $(\lambda_{4,h}-\lambda_4)/(\lambda_{3,h}-\lambda_3)$ for $\alpha=0.01$ in dependence on the number of intervals n.

n and $K(4)<K(3)$ for small n ($n\leqslant 128$) and $K(4)>K(3)$ for large n. Recall that $K(2)>K(1)$ and $K(4)>K(3)$ for all n when $\alpha=0.4$.

Finally we note that when $\alpha=0.01$ the ratio

$$\frac{\lambda_{i+1,h}-\lambda_{i+1}}{\lambda_{i,h}-\lambda_i}, \quad i=1,3,$$

first decreases as n increases, then for some n the two eigenvalue errors become equal, and then the ratio increases again. This is in contrast to the case for $\alpha=0.4$, in which the ratio increased over the whole range of n values. We further note that the value $\bar{n}$ for which the eigenvalue errors are equal—$\bar{n}\simeq 70$ for $i=1$ and $\bar{n}\simeq 160$ for $i=2$ (see Fig. 10.9)—marks a transition in each of these situations from $u_{1,h}\simeq \pi^{-1/2}\sin\varphi(x)$ to $u_{1,h}\simeq\pi^{-1/2}\cos\varphi(x)$ and $u_{3,h}\simeq\pi^{-1/2}\sin 2\varphi(x)$ to $u_{3,h}\simeq\pi^{-1/2}\cos 2\varphi(x)$, from $K(2)<K(1)$ and $K(4)<K(3)$ to $K(2)>K(1)$ and $K(4)>K(3)$, and from $(\lambda_{i+1,h}-\lambda_{i+1})/(\lambda_{i,h}-\lambda_i)$, $i=1,3$, decreasing to increasing.

We have thus seen that for $\alpha=0.4$ the numerical results are in concert with the (asymptotic) results in Section 9 for the whole range of n considered, while for $\alpha=0.01$ they are in disagreement for small n, but are in agreement for large n. We now make an observation that further illuminates these two phases of error behavior—the pre-asymptotic and the asymptotic. Toward this end we note that if (λ_1, u_1), with $\|u_1\|_b=1$, and $(\lambda_{1,h}, u_{1,h})$, with $\|u_{1,h}\|_b=1$, are first eigenpairs of (9.1) and (9.4), respectively, then

$$\begin{aligned} 0\leqslant\lambda_{1,h}-\lambda &= \|u_{1,h}-u\|_a^2-\lambda_1\|u_{1,h}-u_1\|_b^2 \\ &= \inf_{\substack{\chi\in S_h\\ \|\chi\|_b=1}} [\|\chi-u_1\|_a^2-\lambda_1\|\chi-u_1\|_b^2]. \end{aligned} \tag{10.55}$$

If λ_1 is a multiple eigenvalue, then the u_1 in (10.56) can be any corresponding eigenvector with $\|u_1\|_b=1$. (Note that we are here assuming u_1 and $u_{1,h}$ have $\|\cdot\|_b$-length equal 1, whereas in (9.2) and (9.5) they are assumed to have $\|\cdot\|_a$-length equal to 1.) The first inequality in (10.55) follows from the minimum principle (8.35) and has already been stated in (8.42). The first equality in (10.56) follows immediately from Lemma 9.1 with $(\lambda, u)=(\lambda_1, u_1)$, $w=u_{1,h}$ and $\tilde{\lambda}=a(u_{1,h}, u_{1,h})=\lambda_{1,h}$. If $\chi\in S_h$ with $\|\chi\|_b=1$, then from the minimum principle (8.35),

$$\lambda_{1,h}-\lambda_1\leqslant a(\chi,\chi)-\lambda_1. \tag{10.56}$$

Again from Lemma 9.1, this time with $(\lambda, u)=(\lambda_1, u_1)$, $w=\chi$ and $\tilde{\lambda}=a(\chi,\chi)$, we have

$$a_0(\chi,\chi)-\lambda_1=\|\chi-u_1\|_a^2-\lambda_1\|\chi-u_1\|_b^2. \tag{10.57}$$

The second equality in (10.55) follows from (10.56) and (10.57). It is clear from the above discussion that u_1 can be any eigenvector corresponding to λ_1.

From (10.55) we have

$$\lambda_{1,h}-\lambda_1\leqslant\|\chi-u_1\|_a^2-\lambda_1\|\chi-u_1\|_b^2 \quad \forall\chi\in S_h \quad \text{with } \|\chi\|_b=1. \tag{10.58}$$

If χ is $\|\cdot\|_a$-close to u_1, to be more precise, if χ is taken to be the a-projection of u_1 onto S_h (cf. (8.17)), then the second term at the right-hand side of (10.58) is negligible with respect to the first term. This follows from the compactness assumption made in Section 9. On the other hand, if $\|u_1-\chi\|_a$ is not small, $\lambda_{1,h}-\lambda_1$ may still be small because of cancellation between the two terms on the right-hand side of (10.58). Regarding the case $\alpha=0.01$, this explains why for h large (the pre-asymptotic phase), we can have $u_{1,h}\simeq\pi^{-1/2}\sin\varphi(x)$ and $K(1)>K(2)$, and yet have $\lambda_{h,1}$, the approximate eigenvalue associated with $u_{1,h}$, closer to λ_1 than is $\lambda_{2,h}$, the approximate eigenvalue associated with $u_{2,h}\simeq\pi^{-1/2}\cos\varphi(x)$, while for h small (the asymptotic phase), we have $u_{1,h}\simeq\pi^{-1/2}\cos\varphi(x)$, $K(1)<K(2)$ and $\lambda_{1,h}$ closer to λ_1 than is $\lambda_{2,h}$, showing that the eigenvalue error, $\lambda_{1,h}-\lambda_i$, is governed by $\inf_{\chi\in S_h}\|\chi-u_1\|_a^2$.

The analysis of example (10.50) we have presented is taken from BABUŠKA and OSBORN [1987].

10.3. *Eigenvalue problems for general second-order elliptic operators*

We consider here the approximation of the eigenpairs of general second-order elliptic operators. This problem is, in large part, similar to those discussed in Sections 10.1 and 10.2; we will thus be brief, discussing in detail only those issues that have a treatment in this case that differs from that for the L-shaped panel or the membrane, or those issues that did not arise with those problems.

Consider the eigenvalue problem

$$\begin{aligned}&(Lu)(x)=\lambda(Mu)(x), \quad x\in\Omega,\\ &(Bu)(x)=0, \quad x\in\Gamma=\partial\Omega,\end{aligned} \tag{10.59}$$

where Ω is a polygonal domain in $\mathbb{R}^2$, L is given in (3.2), M in (3.3), and B in (3.4), L is assumed to be uniformly strongly elliptic (cf. (3.5)), a_{ij} b_i, c and d to be bounded and measurable, and d to be bounded below by a positive constant (cf. Section 2).

In Section 2 we saw that (10.59) has the variational form (8.10) (cf. (3.15) and (3.17)), with $H_1=H_2=H_0^1(\Omega)$ in the case of Dirichlet boundary conditions and $H_1=H_2=H^1(\Omega)$ in the case of Neumann conditions, and a and b given in (3.11). Equations (8.1)–(8.3) hold, (8.2) and (8.3) being a consequence of (3.5), provided

$$\operatorname{Re} c(x)\geqslant\tfrac{1}{2}a_0+\tfrac{1}{2}b^2/a_0, \tag{10.60}$$

where

$$b=\max_{\substack{x\in\Omega\\ i=1,2}}|b_i(x)|$$

(cf. 3.14). We remark that (10.60) can be easily achieved. It does not hold for the given operator L, L can be modified, by adding an appropriate multiple of $d(x)$ to $c(x)$, so that it does hold. This change shifts the eigenvalues and leaves the eigenfunctions unchanged. We also see that (8.7) is satisfied with $W_1=W_2=L^2(\Omega)$. Thus (10.59) has the form of the problem analyzed in Section 8.

We remark that in this subsection, since we are not imposing any selfadjointness assumptions, the spaces H and $S_{1,(h,p)}=S_{2,(h,p)}$ must be taken to be complex and the eigenvalue parameter λ must be considered complex.

As we have seen, the selection of the trial and test spaces $S_{1,(h,p)}$ and $S_{2,(h,p)}$ is guided by the regularity properties of the exact eigenfunctions and adjoint eigenfunctions. In general, determining this regularity and then using it to choose effective trial and test spaces is a delicate task. The regularity can depend on the coefficients in the differential equation, e.g., on where they have discontinuities and where they are smooth, and on the domain, as we have seen with the L-shaped panel. We will not go further in this direction, but will instead assume the eigenfunctions belong to $H^{k_1}(\Omega)$ and the adjoint eigenfunctions to $H^{k_2}(\Omega)$, and select trial and test spaces so as to reflect this assumption.

REMARK 10.5. For eigenvalue problems with rough coefficients, which arise in the analysis of vibrations in structures with rapidly changing material properties (such as composite materials) it is known that the eigenfunctions do not lie in any high-order Sobolev space. Nevertheless, for one-dimensional problems, their regularity can be understood and, based on this understanding, one can select trial and test spaces that lead to very accurate and robust approximations. These trial and test spaces are not of the usual polynomial type, but instead closely reflect the coefficients. For details see BABUŠKA and OSBORN [1983, 1985] and BABUŠKA, CALOZ and OSBORN [1990]. Cf. also Section 11.3.

REMARK 10.6. The mathematical study of the use of regularity information for the optimal selection of trial and test functions belongs to the area of complexity and information-based approaches. See, e.g., WOZNIAKOWSKI [1985].

Based on the information that the eigenfunctions lie in $H^{k_1}(\Omega)$ and the adjoint eigenfunctions in $H^{k_2}(\Omega)$, with $k_1, k_2 \geqslant 1$, it is appropriate to discretize (8.10) by choosing

$$S_{(h,p)} = S_{1,(h,p)} = S_{2,(h,p)} = \begin{cases} S_0^p(\tau) & \text{for Dirichlet conditions,} \\ S^p(\tau) & \text{for Neumann conditions,} \end{cases}$$

as in Section 10.2, where $\tau \in \gamma$ and $\gamma = \{\tau\} = \{\tau_h\}$ is a family of α-regular triangulations of Ω. Equation (8.11), with $\beta(h) = \frac{1}{2}a_0$, and (8.12) follow from (3.5). Equation (8.13) follows from (10.40).

Equation (8.14) (or (8.15)) can now be considered and from it we get eigenpairs $(\lambda_{(h,p)}, u_{(h,p)})$ which serve as approximations to the eigenpairs (λ, u) of (10.59) (or (8.10)). The errors in the approximate eigenpairs can be estimated with the results of Section 8.

Let λ be an eigenvalue of (10.59) (or (8.10)) with algebraic multiplicity m (by which we mean that λ^{-1} is an eigenvalue with algebraic multiplicity m of the compact operator T introduced in (8.8)). Recall that

$M(\lambda)$ = the unit ball (with respect to $H^1(\Omega)$) in the space of generalized eigenfunctions associated with λ,

and

$M^*(\lambda)$ = the unit ball in the space of generalized adjoint eigenvectors associated with λ.

From (10.40b) in the case of the Dirichlet problem and (10.40a) in the case of the Neumann problem we have

$$\varepsilon_{(h,p)}(\lambda)=\sup_{u\in M(\lambda)}\inf_{\chi\in S_{h,p}}\|u-\chi\|_{1,\Omega}$$
$$\leqslant C(h^{\mu_1-1}/p^{k_1-1})\sup_{u\in M(\lambda)}\|u\|_{k_1,\Omega},$$

where $\mu_1=\min(p+1,k_1)$, and

$$\varepsilon^*_{(h,p)}(\lambda)=\sup_{v\in M^*(\lambda)}\inf_{\eta\in S_{h,p}}\|v-\eta\|_{1,\Omega}$$
$$\leqslant C(h^{\mu_2-1}/p^{k_2-1})\sup_{v\in M^*(\lambda)}\|u\|_{k_2,\Omega},$$

where $\mu_2=\min(p+1,\ k_2)$.

Let $\lambda_{1,(h,p)},\ldots,\lambda_{m,(h,p)}$ be the eigenvalues of (8.14) that converges to λ, let

$M_{h,p}(\lambda)=\{u$: u in the direct sum of the generalized eigenspaces of (8.14) corresponding to the eigenvalues $\lambda_{1,(h,p)},\ldots,\lambda_{m,(h,p)}$, $\|u\|_{1,\Omega}=1\}$,

and let $\alpha=$ ascent of $(\lambda^{-1}-T)$.

Applying Theorem 8.2 we have

$$\left|\lambda-\left(\frac{1}{m}\sum_{j=1}^{m}\lambda^{-1}_{j,(h,p)}\right)\right|^{-1}$$
$$\leqslant C\varepsilon_{(h,p)}(\lambda)\varepsilon^*_{(h,p)}(\lambda)$$
$$\leqslant C(h^{\mu_1+\mu_2-2}/p^{k_1+k_2-2})\sup_{u\in M(\lambda)}\|u\|_{k_1,M}\sup_{v\in M^*(\lambda)}\|v\|_{k_2,\Omega}. \tag{10.61}$$

In light of Remark 8.1 we also have

$$\left|\lambda-\frac{1}{m}\sum_{j=1}^{m}\lambda_{j,(h,p)}\right|$$
$$\leqslant C(h^{\mu_1+\mu_2-2}/p^{k_1+k_2-2})\sup_{u\in M(\lambda)}\|u\|_{k_1,m}\sup_{v\in M^*(\lambda)}\|v\|_{k_2,\Omega}. \tag{10.62}$$

From Theorem 8.3 we obtain

$$|\lambda-\lambda_{j,(h,p)}|$$
$$\leqslant C((h^{\mu_1+\mu_2-2}/p^{k_1+k_2-2})\sup_{u\in M(\lambda)}\|u\|_{k_1,M}\sup_{v\in M^*(\lambda)}\|v\|_{k_2,\Omega})^{1/\alpha}. \tag{10.63}$$

Regarding eigenfunction estimates, we apply Theorems 8.1 and 8.2. From Theorem 8.1 we have

$$\delta(M(\lambda),M_{(h,p)}(\lambda))\leqslant C(h^{\mu_1-1}/p^{k_1-1})\sup_{u\in M(\lambda)}\|u\|_{k_1,\Omega}. \tag{10.64}$$

Let $\lambda_{(h,p)}$ be an eigenvalue of (10.1) (or 8.10)) such that $\lim_{h\to 0,p\to\infty}\lambda(h,p)=\lambda$ and let $w_{(h,p)}$ be a unit vector satisfying

$$(\lambda(h,p)^{-1}-T)^{l_1}w_{(h,p)}=0 \quad \text{for some positive integer } l_1 \leqslant \alpha.$$

Then, from Theorem 8.4, for any integer l_2 with $l_1 \leqslant l_2 \leqslant \alpha$, there is a vector $u_{(h,p)}$ such that $(\lambda^{-1}-T)^{l_2}u_{(h,p)}=0$ and

$$\|u_{(h,p)}-w_{(h,p)}\|_{1,\Omega} \leqslant C((h^{\mu_1-1}/p^{k_1-1}) \sup_{u\in M(\lambda)} \|u\|_{k_1,\Omega})^{(l_2-l_1+1)/\alpha}. \tag{10.65}$$

REMARK 10.7. In this section we have considered triangular meshes. One could also consider quadrilateral meshes, which are a generalization of the type of mesh employed in Section 10.1, or curvilinear meshes. Since these generalizations properly belong to approximation theory we will not pursue them. We refer the reader to CIARLET [1978], BABUŠKA and GUO [1988b], and SZABO [1986].

REMARK 10.8. We have mentioned here only estimates based on the information that $u\in H^k(\Omega)$. If we know, e.g., that $u\in \mathscr{B}^2_\beta(\Omega)$, then we can say more, provided a proper mesh is selected.

REMARK 10.9. The approximate eigenvalues $\lambda_{j,(h,p)}$ here, as in any finite element method, are defined by the eigenvalue problem (8.14), which involves integrals over the domain Ω. In practice these integrals often must be evaluated (approximated) by quadrature formulas. For estimates of the eigenvalue error due to this quadrature error we refer to FIX [1972] and BANERJEE and OSBORN [1990]. In BANERJEE and OSBORN [1990] it is shown that to preserve optimal order for eigenvalue error more quadrature points than for source problems are required. We note that the use of a finite element method in conjunction with a quadrature method often leads to a finite difference method for eigenvalue approximation. For example, if we approximate the eigenvalues of

$$-\Delta u=\lambda u \quad \text{in } \Omega, \qquad u=0 \quad \text{on } \Gamma$$

with the finite element method corresponding to $p=1$ and a uniform triangulation (cf. Fig. 10.8) and evaluate the resulting integrals with an appropriate quadrature formula, we obtain the standard five-point difference eigenvalue approximation for the Laplacian (cf. Section 5). This observation is due to COURANT [1927, 1943]. For further results on finite difference methods we refer to POLYA [1952], HERSCH [1955, 1963], WEINBERGER [1956, 1958, 1974], HUBBARD [1961, 1962], KUTTLER [1970], and KREISS [1972].

REMARK 10.10. Since the eigenvalue $\lambda_{j,(h,p)}$ are defined by a Ritz method, they are upper bounds for the exact eigenvalues λ_j:

$$\lambda_j \leqslant \lambda_{j,(h,p)}$$

(cf. (8.42)). If we could derive a lower bound $\tilde{\lambda}_{j,(h,p)}$, then one would have bracketed λ_j. Much attention has been directed to the derivation of lower bounds. WEINSTEIN [1935, 1937, 1953, 1963] developed the method of intermediate problems. Many authors have contributed to the development of this and other related variational methods. We mention D.H. WEINSTEIN [1934], ARONSZAJN and WEINSTEIN [1942], ARONSZAJN [1948, 1949–50], WEINBERGER [1952, 1956, 1959, 1960], BAZLEY [1959], BAZLEY and FOX [1961, 1963]. In addition we mention the monographs by COLLATZ [1948], WEINSTEIN and STENGER [1972], and WEINBERGER [1974].

REMARK 10.11. Most books and monographs that treat finite element methods contain a section or chapter on eigenvalue problems. For a survey of books and monographs on finite element methods we refer to NOOR [1985]. Of the more mathematically oriented of these, we mention STRANG and FIX [1973], ODEN and REDDY [1976], and ODEN and CAREY [1982].

11. Approximation by mixed methods

In Section 3 we saw, in terms of an example, how eigenvalue problems can be given mixed formulations. Mixed formulations can be discretized and thereby lead to approximation methods referred to as mixed finite element methods. In this section we discuss three such methods. We begin by presenting an abstract result designed for the analysis of mixed methods.

REMARK 11.1. Mixed methods for source problems have received considerable attention. We mention HERMANN [1967], GLOWINSKI [1973], JOHNSON [1973], ODEN [1973], BREZZI [1974], CIARLET and RAVIART [1974], MERCIER [1974], SCHOLZ [1976], RAVIART and THOMAS [1977], BREZZI and RAVIART [1978], FALK [1978], BABUŠKA, OSBORN and PITKÄRANTA [1980], and FALK and OSBORN [1980].

11.1. An abstract result

Let V, W, H and G be four real Hilbert spaces with inner products and norms $(\cdot,\cdot)_V$, $\|\cdot\|_V$, $(\cdot,\cdot)_W$, $\|\cdot\|_W$, $(\cdot,\cdot)_H$, $\|\cdot\|_H$ and $(\cdot,\cdot)_G$, $\|\cdot\|_G$, respectively. We assume $V\subset H$ and $W\subset G$. Let $A(\sigma,\psi)$ and $B(\psi,u)$ be bilinear forms on $H\times H$ and $V\times W$, respectively, that satisfy

$$|A(\sigma,\psi)|\leqslant C_1\|\sigma\|_H\|\psi\|_H \quad \forall\sigma,\psi\in H \tag{11.1a}$$

and

$$|B(\psi,u)|\leqslant C_2\|\psi\|_V\|u\|_W \qquad \forall\psi\in V,\quad u\in W. \tag{11.1b}$$

We assume $A(\sigma,\psi)$ is symmetric and satisfies

$$A(\sigma,\sigma)>0 \quad \forall 0\neq\sigma\in H, \tag{11.2a}$$

and assume

$$\sup_{\psi\in V}|B(\psi,u)|>0 \quad \forall 0\neq u\in W. \tag{11.2b}$$

We then consider the following eigenvalue problem:

$$\begin{aligned}&(\sigma, u)\in V\times W, \qquad (\sigma, u)\neq(0,0),\\&A(\sigma,\psi)+B(\psi,u)=0 \quad \forall\psi\in V,\\&B(\sigma,v)=-\lambda(u,v)_G \qquad \forall v\in W.\end{aligned} \tag{11.3}$$

A discretization of (11.3) is obtained by selecting finite-dimensional spaces $V_h \subset V$ and $W_h \subset W$ and considering the approximate eigenvalue problem

$$\begin{aligned}&(\sigma_h, u_h)\in V_h\times W_h, \qquad (\sigma_h, u_h)\neq(0,0),\\&A(\sigma_h,\psi)+B(\psi,u_h)=0 \quad \forall\psi\in V_h,\\&B(\sigma_h,v)=-\lambda_h(u_h,v)_G \qquad \forall v\in W_h.\end{aligned} \tag{11.4}$$

We then view $(\lambda_h, (\sigma_h, u_h))$ as an approximation to $(\lambda, (\sigma, u))$. Given bases for V_h and W_h, (11.4) becomes a matrix eigenvalue problem.

REMARK 11.2. If we let

$$a((\sigma,u),(\psi,v))=A(\sigma,\psi)+B(\psi,u)+B(\sigma,v),$$

$$b((\sigma,u)),(\psi,v))=-(u,v)_G,$$

and

$$H=V\times W,$$

then (11.3) can be written as

$$\begin{aligned}&(\sigma,u)\in H, \qquad (\sigma,u)\neq(0,0)\\&a((\sigma,u),(\psi,v))=\lambda b((\sigma,u),(\psi,v)) \quad \forall(\psi,v)\in H,\end{aligned} \tag{11.5}$$

which has the form of (8.10). Also (11.4) has the form of (8.14) with $S_{1,h}=S_{2,h}=V_h\times W_h$. Problems (11.3) and (11.5) do not, however, satisfy all of the hypotheses of the results in Section 8. We thus need an alternative analysis. This will be provided by Theorem 11.1, which is based on the results of Section 7. Note that even though the methods considered in this and the next section are not covered by the results of Section 8, it is still useful to discuss them, to the extent possible, in terms of the basic steps (1), (1′), (2′), and (3) introduced in Section 10.

In order to estimate the error in the approximate eigenpairs $(\lambda_h, (\sigma_h, u_h))$ we consider the associated source and approximate source problems:

Given $g\in G$, find $(\sigma,u)\in V\times W$ satisfying

$$\begin{aligned}&A(\sigma,\psi)+B(\psi,u)=0 \quad \forall\psi\in V,\\&B(\sigma,v)=-(g,v)_G \qquad \forall v\in W.\end{aligned} \tag{11.6}$$

Given $g\in G$, find $(\sigma_h,u_h)\in V_h\times W_h$ satisfying

$$\begin{aligned}&A(\sigma_h,\psi)+B(\psi,u_h)=0 \quad \forall\psi\in V_h,\\&B(\sigma_h,v)=-(g,v)_G \qquad \forall v\in W_h.\end{aligned} \tag{11.7}$$

We assume (11.6) and (11.7) are uniquely solvable for each $g \in G$. We then introduce the corresponding component solution operators:

$$S: G \to V, \qquad Sg = \sigma, \tag{11.8a}$$

$$S_h: G \to V, \qquad S_h g = \sigma_h, \tag{11.8b}$$

$$T: G \to G, \qquad Tg = u, \tag{11.8c}$$

$$T_h: G \to G, \qquad T_h g = u_h, \tag{11.8d}$$

where (σ, u) and (σ_h, u_h) are defined by (11.6) and (11.7), respectively. (Note that the T introduced here is different from that introduced in (8.8).)

The eigenpairs $(\lambda, (\sigma, u))$ of (11.3) can be characterized in terms of the operator T. Before establishing this we note that $\lambda = A(\sigma, \sigma)/(u, u)_G$, which shows that $\lambda > 0$. This follows from (11.3) and the observation that both components u and σ of an eigenvector are nonzero. Now, if $(\lambda, (\sigma, u))$ is an eigenpair of (11.3), then $\lambda Tu = u, u \neq 0$, and if $\lambda Tu = u, u \neq 0$, then there is a $\sigma \in V (\sigma = S(\lambda u))$ such that $(\lambda, (\sigma, u))$ is an eigenpair of (11.3). Thus λ is an eigenvalue of (11.3) if and only if λ^{-1} is an eigenvalue of T. The correspondence between eigenvectors is given by $u \leftrightarrow (\sigma, u)$. In a similar way the approximate eigenvalues defined by (11.4) can be characterized in terms of the eigenvalues of T_h. λ_h is an eigenvalue of (11.4) if and only if λ_h^{-1} is an eigenvalue of T_h; the correspondence between the eigenpairs is given by $u_h \leftrightarrow (\sigma_h, u_h)$.

We assume

$$\|T - T_h\|_{GG} \to 0 \quad \text{as } h \to 0, \tag{11.9}$$

where, for an operator $A: D(A) (\subset X) \to Y$, we let

$$\|A\|_{XY} = \sup_{w \in D(A)} \|Aw\|_Y / \|w\|_X.$$

(In particular, we assume T is a bounded operator on G.) Since $\dim R(T_h) < \infty$ for each h, the T_h are compact and (11.9) thus implies T is compact. We also note that T is selfadjoint on G. This is seen as follows. Let $v = Tf$ in the second equation in (11.6) to obtain

$$B(Sg, Tf) = -(g, Tf)_G.$$

Again consider (11.6), but with g replaced by f, and let $\psi = Sg$ in the first equation to get

$$A(Sf, Sg) + B(Sg, Tf) = 0.$$

From these two equations we have

$$(g, Tf)_G = A(Sf, Sg) \quad \forall f, g \in G. \tag{11.10}$$

Using (11.10) and the symmetry of A we get

$$(Tg, f)_G = (f, Tg)_G = A(Sg, Sf) = A(Sf, Sg) = (g, Tf)_G,$$

showing T is selfadjoint. In a similar way we see T_h is selfadjoint.

We now apply Theorems 7.3 and 7.4 to the operator T and family of operators $\{T_h\}$ on the space G. By virtue of the correspondence between the eigenpairs of T and T_h and those of (11.3) and (11.4) we will thereby obtain estimates for the errors in $(\lambda_h, (\sigma_h, u_h))$. The hypotheses have all been shown to be satisfied; cf. Remarks 7.5 and 7.6. Let λ^{-1} be an eigenvalue of multiplicity m. Since $\|T-T_h\|_{GG}\to 0$ we know that m eigenvalues $\lambda_{1,h}^{-1},\ldots,\lambda_{m,h}^{-1}$ of T_h converge to λ^{-1}. Since T and T_h are self-adjoint the relevant ascents are one and all eigenvalues have equal geometric and algebraic multiplicities. Let $\bar{M}(\lambda^{-1})$ be the eigenspace of T corresponding to λ^{-1}. Recall that $\bar{M}=\bar{M}(\lambda^{-1})=R(E)$, the range of the spectral projection E associated with T and λ^{-1}. We have denoted this space by $\bar{M}$ to distinguish it from the set M of normalized eigenvectors introduced in Section 8.

THEOREM 11.1. *Under the assumptions made above, there is a constant C such that*

$$|\lambda-\lambda_{l,h}|\leqslant C\{\|(S-S_h)|_{\bar{M}}\|_{GH}^2 + \|(S-S_h)|_{\bar{M}}\|_{GV}\,\|(T-T_h)|_{\bar{M}}\|_{GW}+\|(T-T_h)|_{\bar{M}}\|_{GG}^2\},\quad l=1,\ldots,m. \tag{11.11}$$

PROOF. Let $u_1,\ldots,u_m$ be an orthonormal basis for $\bar{M}(\lambda^{-1})$. From Theorem 7.3 with $\alpha=1$ we have

$$|\lambda^{-1}-\lambda_{l,h}^{-1}|\leqslant C\left\{\sum_{i,j=1}^{m}|((T-T_h)u_i,u_j)_G|+\|(T-T_h)|_{\bar{M}}\|_{GG}^2\right\},\quad l=1,\ldots,m. \tag{11.12}$$

For $g,f\in G$ we estimate $((T-T_h)g,f)_G$. Adding the two equations in (11.6) and recalling the definitions of Tg and Sg in (11.8) we find

$$(g,v)_G=-A(Sg,\psi)-B(\psi,Tg)-B(Sg,v)\quad \forall(\psi,v)\in V\times W.$$

Setting $v=(T-T_h)f$ and $\psi=(S-S_h)f$ yields

$$(g,(T-T_h)f)_G = -A(Sg,(S-S_h)f)-B((S-S_h)f,Tg)-B(Sg,(T-T_h)f). \tag{11.13}$$

Next note that substraction of the equations (11.7) from (11.6) (with g replaced by f) gives

$$A((S-S_h)f,\psi)+B(\psi,(T-T_h)f)+B((S-S_h)f,v)=0\quad \forall(\psi,v)\in V_h\times W_h. \tag{11.14}$$

Now, combining (11.13) and (11.14) and using (11.1) we have

$$\begin{aligned}
&|(g,(T-T_h)f)_G|\\
&\quad=|A((S-S_h)f,Sg-\psi)+B((S-S_h)f,Tg-v)+B(Sg-\psi,(T-T_h)f)|\\
&\quad\leqslant C_1\|(S-S_h)f\|_H\|Sg-\psi\|_H\\
&\qquad+C_2\|(S-S_h)f\|_V\|Tg-v\|_W+C_2\|Sg-\psi\|_V\|(T-T_h)f\|_W.
\end{aligned}$$

Setting $\psi = S_h g$ and $v = T_h g$ gives

$$\begin{aligned}
&|((T\,T_h)g,f)_G| \\
&\quad \leqslant C_1 \|(S-S_h)f\|_H \|(S-S_h)g\|_H \\
&\quad + C_2 \|(S-S_h)f\|_V \|(T-T_h)g\|_W + C_2 \|(S-S_h)g\|_V \|(T-T_h)f\|_W. \qquad (11.15)
\end{aligned}$$

Letting $g = u_i$ and $f = u_j$ in (11.15) yields

$$\begin{aligned}
&|((T-T_h)u_i, u_j)_G| \\
&\quad \leqslant C_1 \|(S-S_h)|_M\|^2_{GH} + 2C_2 \|(S-S_h)_{\overline{M}}\|_{GV} \|(T-T_h)|_{\overline{M}}\|_{GW}. \qquad (11.16)
\end{aligned}$$

(11.11) follows immediately from (11.12) and (11.16). □

THEOREM 11.2. *Under the assumptions made above, there is a constant C such that*

$$\|u-u_h\|_G \leqslant C\|(T-T_h)|_M\|_{GG}. \qquad (11.17)$$

PROOF. This result is an immediate consequence of Theorems 7.1 and 7.4. Note that we have given this estimate the simplified form it has when λ is simple, and it would have to be modified in the general case. Cf. the statement of Theorem 10.1 and (8.44)–(8.46). □

Theorems 11.1 and 11.2 were proved by OSBORN [1979] and by MERCIER, OSBORN, RAPPAZ and RAVIART [1981].

11.2. A mixed method for the vibrating membrane

We consider, as in Section 10.2, the vibrating membrane problem

$$-\Delta u = \lambda u \quad \text{in } \Omega,\ u = 0 \quad \text{on } \Gamma = \partial\Omega, \qquad (11.18)$$

where Ω is a convex polygon in $\mathbb{R}^2$, but we will here give it a mixed variational formulation. Otherwise we will proceed in a parallel way, discussing in turn the steps (1), (1′), (2) and (3) introduced in Section 10.1. We will clearly see how the variational formulation influences the entire approximation method.

Before proceeding with the variational formulation, we introduce an additional function space. Let

$$\begin{aligned}
H(\operatorname{div}, \Omega) = \Big\{ \sigma = (\sigma_1, \sigma_2) \colon & \sigma_1, \sigma_2 \in H^0(\Omega) \text{ and there exists} \\
& z = \operatorname{div} \sigma \in H^0(\Omega) \text{ such that} \\
& \int_\Omega \sigma \cdot \nabla\phi \,\mathrm{d}x\,\mathrm{d}y = -\int_\Omega z\phi \,\mathrm{d}x\,\mathrm{d}y, \forall \phi \in C_0^\infty(\Omega) \Big\},
\end{aligned}$$

$$\|\sigma\|^2_{H(\operatorname{div}, \Omega)} = \int_\Omega [\sigma_1^2 + \sigma_2^2 + (\operatorname{div} \sigma)^2]\,\mathrm{d}x\,\mathrm{d}y.$$

11.2.1. Variational formulation

Suppose (λ, u) is an eigenpair of (11.18), by which we will mean

$$0 \neq u \in H_0^1(\Omega),$$
$$\int_\Omega \nabla u \cdot \nabla u \,\mathrm{d}x\,\mathrm{d}y = \lambda \int_\Omega uv \,\mathrm{d}x\,\mathrm{d}y \quad \forall v \in H_0^1(\Omega), \tag{11.19}$$

i.e., we will assume (11.18) to have the variational formulation considered in Section 10.2. We now derive a mixed variational formulation for (11.18). Introduce the auxiliary variable

$$\sigma = \nabla u. \tag{11.20}$$

From (11.19) we see that $\sigma \in H(\operatorname{div}, \Omega)$ and

$$\operatorname{div} \sigma = -\lambda u. \tag{11.21}$$

From (11.21) we get

$$\int_\Omega v \operatorname{div} \sigma \,\mathrm{d}x\,\mathrm{d}y = -\lambda \int_\Omega uv \,\mathrm{d}x\,\mathrm{d}y \quad \forall v \in H^0(\Omega) \tag{11.22}$$

and from (11.20) and the definition of $H(\operatorname{div}, \Omega)$ we have

$$\int_\Omega \sigma \cdot \psi \,\mathrm{d}x\,\mathrm{d}y = \int_\Omega \nabla u \cdot \psi \,\mathrm{d}x\,\mathrm{d}y = -\int_\Omega u \operatorname{div} \psi \,\mathrm{d}x\,\mathrm{d}y$$
$$\forall \psi \in (\operatorname{div}, \Omega). \tag{11.23}$$

Combining (11.22) and (11.23) we obtain

$$(\sigma, u) \in H(\operatorname{div}, \Omega) \times H^0(\Omega), \qquad (\sigma, u) \neq (0, 0),$$
$$\int_\Omega \sigma \cdot \psi \,\mathrm{d}x\,\mathrm{d}y + \int_\Omega u \operatorname{div} \psi \,\mathrm{d}x\,\mathrm{d}y = 0 \quad \forall \psi \in H(\operatorname{div}, \Omega), \tag{11.24}$$
$$\int_\Omega v \operatorname{div} \sigma \,\mathrm{d}x\,\mathrm{d}y = -\lambda \int_\Omega uv \,\mathrm{d}x\,\mathrm{d}y \quad \forall v \in H^0(\Omega).$$

Now suppose $(\lambda, (\sigma, u))$ satisfies (11.24). Let $\bar{u}$ be the solution to

$$\Delta \bar{u} = \lambda u \quad \text{in } \Omega, \qquad \bar{u} = 0 \quad \text{on } \Gamma, \tag{11.25}$$

and let

$$\bar{\sigma} = \nabla \bar{u}.$$

Then, by the argument used above,

$$(\bar{\sigma}, \bar{u}) \in H(\mathrm{div}, \Omega) \times H^0(\Omega),$$

$$\int_\Omega \bar{\sigma} \cdot \psi \,\mathrm{d}x\,\mathrm{d}y + \int_\Omega \bar{u} \,\mathrm{div}\, \psi \,\mathrm{d}x\,\mathrm{d}y = 0 \quad \forall \psi \in H(\mathrm{div}, \Omega), \tag{11.26}$$

$$\int_\Omega v \,\mathrm{div}\, \sigma \,\mathrm{d}x\,\mathrm{d}y = -\lambda \int_\Omega uv \,\mathrm{d}x\,\mathrm{d}y \quad \forall v \in H^0(\Omega).$$

Subtraction of the equations in (11.26) from those in (11.24) yields

$$(\sigma - \bar{\sigma}, u - \bar{u}) \in H(\mathrm{div}, \Omega) \times H^0(\Omega),$$

$$\int_\Omega (\sigma - \bar{\sigma}) \cdot \psi \,\mathrm{d}x\,\mathrm{d}y + \int_\Omega (u - \bar{u}) \mathrm{div}\, \psi \,\mathrm{d}x\,\mathrm{d}y = 0 \quad \forall \psi \in H(\mathrm{div}, \Omega), \tag{11.27}$$

$$\int_\Omega v \,\mathrm{div}(\sigma - \bar{\sigma}) \,\mathrm{d}x\,\mathrm{d}y = 0 \quad \forall v \in H^0(\Omega).$$

In (11.27), if in the second equation we take v arbitrary in $H^0(\Omega)$ we get $\mathrm{div}(\sigma - \bar{\sigma}) = 0$, and if we take $\psi = \sigma - \bar{\sigma}$ in the first equation we obtain $0 = \int_\Omega (\sigma - \bar{\sigma}) \cdot (\sigma - \bar{\sigma}) \,\mathrm{d}x\,\mathrm{d}y$, which implies

$$\sigma = \bar{\sigma}. \tag{11.28}$$

Then the first equation in (11.27) implies

$$\int_\Omega (u - \bar{u}) \mathrm{div}\, \psi \,\mathrm{d}x\,\mathrm{d}y = 0 \quad \forall \psi \in H(\mathrm{div}, \Omega). \tag{11.29}$$

Let w satisfy $\Delta w = u - \bar{u}$ and let $\psi = \nabla w$ in (11.29). Since $\mathrm{div}\ \psi = u - \bar{u}$, this choice leads to

$$u = \bar{u}. \tag{11.30}$$

Equations (11.25), (11.28) and (11.30) show that (λ, u) is an eigenpair of (11.18) (or (11.19)), and that $\sigma = \nabla u$.

In summary, if (λ, u) is an eigenpair of (11.18) and $\sigma = \nabla u$, then $(\lambda, (u, \sigma))$ satisfies (11.24), and if $(\lambda, (\sigma, u))$ satisfies (11.24), then (λ, u) is an eigenpair of (11.18) and $\sigma = \nabla u$. (11.24) is the desired mixed formulation.

It is immediate that (11.24) has the form of (11.3) with

$$V = H(\mathrm{div}, \Omega), \qquad W = G = H^0(\Omega), \qquad H = \mathbb{H}^0(\Omega),$$

$$A(\sigma, \psi) = \int_\Omega \sigma \cdot \psi \,\mathrm{d}x\,\mathrm{d}y, \qquad B(\psi, u) = \int_\Omega u \,\mathrm{div}\, \psi \,\mathrm{d}x\,\mathrm{d}y.$$

Furthermore, A is symmetric and (11.1) and (11.2) hold. The symmetry of A and (11.1) and (11.2a) are trivial. To prove (11.2b), let w solve $\Delta w = u$ and set $\bar{\psi} = \nabla w$. Then $\operatorname{div} \bar{\psi} = u$ and we have

$$\sup_{\psi \in H(\operatorname{div},\Omega)} \left| \int_\Omega u \operatorname{div} \psi \, dx \, dy \right|$$

$$\geqslant \left| \int_\Omega u \operatorname{div} \bar{\psi} \, dx \, dy \right| = \int_\Omega u^2 \, dx \, dy > 0 \quad \text{for } 0 \neq u \in H^0(\Omega),$$

which proves (11.2b).

From the fact that (11.18) has a sequence of positive eigenvalues and from the correspondence between the eigenpairs of (11.18) and (11.24) we see that (11.24) has a sequence of eigenvalues

$$0 < \lambda_1 \leqslant \lambda_2 \leqslant \cdots \uparrow \infty$$

and corresponding eigenfunctions

$$(\sigma_1, u_1), (\sigma_2, u_2), \ldots,$$

with $\sigma_j = \nabla u_j$ and with the (λ_j, u_j) being the eigenpairs of (11.18).

11.2.1'. Regularity of the eigenfunctions

If (σ, u) is an eigenfunction of (11.24), then u is an eigenfunction of (11.18) and $\sigma = \nabla u$. Thus the regularity of (σ, u) can be inferred from the regularity of the eigenfunction of (11.18), which was discussed in Section 11.2.

11.2.2. Discretization of (11.24) *and assessment of the accuracy of the approximate eigenvalues and eigenfunctions*

We will use a discretization of the general form of (11.4). It thus remains to select the subspaces $V_h \subset H(\operatorname{div}, \Omega)$ and $W_h \subset H^0(\Omega)$. This will be done with an eye toward ensuring (11.9) holds and the terms on the right-hand side of (11.11) in Theorem 11.1 are small. A mixed method approximation of the associated source problem (cf. (11.6) and (11.7)) has been proposed and analyzed in RAVIART and THOMAS [1977]. We will take their choice of trial and test functions. The source problem has also been analyzed by FALK and OSBORN [1980].

Let $\hat{T}$ be the unit triangle in the (ξ, η)-plane whose vertices are $\hat{a}_1 = (1, 0)$, $\hat{a}_2 = (0, 1)$ and $\hat{a}_3 = (0, 0)$. Then with $p \geqslant 0$ an even integer and $\hat{T}$ associate the space $\hat{Q}^{(p+1)}$ of all functions $\hat{\psi} = (\hat{\psi}_1, \hat{\psi}_2)$ of the form

$$\hat{\psi}_1 = \operatorname{pol}_p(\xi, \eta) + \alpha_0 \xi^{p+1} + \alpha_1 \xi^p \eta + \cdots + \alpha_{p/2} \xi^{p/2+1} \eta^{p/2},$$
$$\hat{\psi}_2 = \operatorname{pol}_p(\xi, \eta) + \beta_0 \eta^{p+1} + \beta_1 \xi \eta^p + \cdots + \beta_{p/2} \xi^{p/2} \eta^{p/2+1},$$

where $\operatorname{pol}_p(\xi, \eta)$ denotes an arbitrary polynomial of degree p and where

$$\sum_{i=0}^{p/2} (-1)^i (\alpha_i - \beta_i) = 0,$$

and with $p \geqslant 1$ an odd integer and $\hat{T}$ associate the space $\hat{Q}^{(p+1)}$ of all $\hat{\psi}$ of the form

$$\hat{\psi}_1 = \mathrm{pol}_p(\xi, \eta) + \alpha_0 \xi^{p+1} + \alpha_1 \xi^p \eta + \cdots + \alpha_{(p+1)/2} \xi^{(p+1)/2} \eta^{(p+1)/2},$$
$$\hat{\psi}_2 = \mathrm{pol}_p(\xi, \eta) + \beta_0 \eta^{p+1} + \beta_1 \xi \eta^p + \cdots + \beta_{(p+1)/2} \xi^{(p+1)/2} \eta^{(p+1)/2},$$

where

$$\sum_{i=0}^{(p+1)/2} (-1)^i \alpha_i = \sum_{i=0}^{(p+1)/2} (-1)^i \beta_i = 0.$$

We remark that for $\hat{\psi} \in \hat{Q}^{(p+1)}$, $\hat{\psi}_1$ and $\hat{\psi}_2$ are polynomials of degree $p+1$. With a general triangle T in the (x, y)-plane, we associate the space $Q_T^{(p+1)}$ defined by

$$Q_T^{(p+1)} = \{\psi : \psi(x, y) = (1/J_T) B_T \hat{\psi}(F_T^{-1}(x, y)),\ \hat{\psi} \in Q^{(p+1)}\},$$

where $F_T(\xi, \eta) = B_T(\xi, \eta) + b_T$ is the linear transformation mapping $\hat{T}$ onto T and $J_T = \det(B_T)$.

Let $\gamma = \{\tau\} = \{\tau_h\}$ be a family of α-regular triangularizations of $\bar{\Omega}$. Then for $p \geqslant 0$ an integer let

$$V_h = \{\psi \in H(\mathrm{div}, \Omega) : \psi|_T \in Q_T^{(p+1)}\ \forall T \in \tau_h\} \tag{11.31a}$$

and

$$W_h = \{u \in H^0(\Omega) : u|_T \text{ polynomial of degree } p\ \forall T \in \tau_h\}. \tag{11.31b}$$

Now we consider (11.4) with this choice for V_h and W_h. (11.4) will have eigenvalues

$$\lambda_{1,h} \leqslant \cdots \leqslant \lambda_{N,h}$$

and corresponding eigenfunctions

$$(\sigma_{1,h}, u_{1,h}), \ldots, (\sigma_{N,h}, u_{N,h}),$$

where $N = \dim(V_h \times W_h)$. It remains to derive error estimates by applying Theorems 11.1 and 11.2.

THEOREM 11.3. *Let V_h and W_h be selected as in* (11.31). *Suppose the eigenfunctions of* (11.18) *belong to $H^{p+2}(\Omega)$. Then*

$$|\lambda_{k,h} - \lambda_k| \leqslant C(p) h^{2p+2} \tag{11.32}$$

and

$$\|u_{k,h} - u_k\|_{0,\Omega} \leqslant C(p) h^{p+1} \tag{11.33}$$

PROOF. We begin by showing that all of the hypotheses of Theorems 11.1 and 11.2 are satisfied. We have already noted that A is symmetric and that (11.1) and (11.2) are satisfied for the problem (11.24).

The source problem (11.6) is uniquely solvable for each $g \in G = H^0(\Omega)$. In fact the unique solution is (σ, u), where

$$-\Delta u = g, \quad u \in H_0^1(\Omega)$$

and

$$\sigma = \nabla u$$

(cf. the discussion in Section 11.2.1). To see that (11.7) is uniquely solvable for each $g \in G$ it is sufficient to show that $g=0$ implies σ_h and u_h are zero. Now $g=0$ implies, using the second equation in (11.7), that $B(\sigma_h, v)=0$, $\forall v \in W_h$. Setting $\psi=\sigma_h$ in the first equation and using this fact shows that $A(\sigma_h, \sigma_h)=0$ which, together with (11.2a), shows that $\sigma_h=0$. Then, using the first equation in (11.6) again we get $B(\psi, u_h)=0$, $\forall \psi \in V_h$. For our specific problem this is $\int_\Omega u_h \operatorname{div} \psi \,dx\,dy=0$, $\forall \psi \in V_h$. It is shown in RAVIART and THOMAS [1977, Theorem 4] that corresponding to any $u_h \in W_h$ there is a $\psi \in V_h$ such that $\operatorname{div} \psi_h = u_h$. Using this ψ we thus have $\int_\Omega |u_h|^2 \,dx\,dy=0$ which implies $u_h=0$.

It remains to check (11.9). FALK and OSBORN [1980, Section 3(d)] have shown that

$$\|Tg - T_h g\|_{0,\Omega} \leqslant \begin{cases} Ch^2 \|Tg\|_{2,\Omega} & \text{for } p \geqslant 1 \\ Ch\|Tg\|_{2,\Omega} & \text{for } p=0 \end{cases}$$

$$\leqslant Ch\|g\|_{0,\Omega} \qquad \text{for } p \geqslant 0, \tag{11.34}$$

which proves (11.9).

We now apply Theorems 11.1 and 11.2. From RAVIART and THOMAS [1977, Theorem 5] we have

$$\|(S-S_h)g\|_{H^0(\Omega)} = \|(S-S_h)g\|_{H(\operatorname{div},\Omega)} \leqslant Ch^{p+1}(\|Tg\|_{p+2,\Omega} + \|g\|_{p+1,\Omega})$$

and

$$\|(T-T_h)g\|_{0,\Omega} \leqslant Ch^{p+1}(\|Tg\|_{p+2,\Omega} + \|g\|_{p+1,\Omega}).$$

If $g \in \bar{M}(\lambda_k^{-1})$, then $Tg = \lambda_k^{-1} g$ and g is an eigenfunction of (11.18) corresponding to λ_k and by our hypotheses, $\|g\|_{p+2,\Omega} < \infty$. Thus

$$\|(S-S_h)|_{\bar{M}}\|_{H^0(\Omega),H^0(\Omega)} \leqslant Ch^{p+1}, \tag{11.35a}$$

$$\|(S-S_h)_{\bar{M}}\|_{H^0(\Omega),H(\operatorname{div},\Omega)} \leqslant Ch^{p+1}, \tag{11.35b}$$

$$\|(T-T_h)|_{\bar{M}}\|_{H^0(\Omega),H^0(\Omega)} \leqslant Ch^{p+1}. \tag{11.35c}$$

(11.32) follows immediately from Theorem 11.1 and estimates (11.35). (11.33) follows immediately from Theorem 11.2 and (11.35c). □

REMARK 11.3. Theorems 11.1 and 11.2 estimate the errors in mixed method approximation of eigenpairs in terms of error estimates for the corresponding source problems. For our problem, these were mainly provided by the results of RAVIART and THOMAS [1977]. Note, however, the estimate (11.34)—the estimate that ensures the approximate eigenvalues converge—is not proved in RAVIART and THOMAS [1977].

11.2.3. Solution of matrix eigenvalue problem
The matrix problem (11.4) with V_h and W_h given in (11.31) is large and sparse, but is not positive-definite.

11.3. A mixed method for the vibrating plate

The eigenvalue problem

$$\Delta^2 u = \lambda u \quad \text{in } \Omega, \qquad u = \partial u/\partial n = 0 \quad \text{in } \Omega \tag{11.36}$$

arises in connection with the small, transverse vibration of a clamped plate. A commonly used variational formulation of (11.36) is

$$\begin{gathered} u \in H_0^2(\Omega), \qquad u \neq 0, \\ \int_\Omega \Delta u \Delta v \,\mathrm{d}x\,\mathrm{d}y = \lambda \int_\Omega uv \,\mathrm{d}x\,\mathrm{d}y \quad \forall v \in H_0^2(\Omega). \end{gathered} \tag{11.37}$$

A finite element method based on (11.37) would require trial and test space that were subspaces of $H_0^2(\Omega)$, and this would require C^1-elements, i.e., piecewise polynomials that are C^1 across interelement boundaries. In order to avoid this requirement we will use a different variational formulation for (11.37), one that permits the use of C^0-elements. We do, however, use (11.37) to show that (11.36) has a sequence of eigenvalues

$$0 < \lambda_1 \leqslant \lambda_2 \leqslant \cdots \uparrow \infty$$

and corresponding eigenfunctions

$$u_1, u_2, \ldots,$$

which can be chosen so that

$$\int_\Omega \Delta u_i \Delta u_j \,\mathrm{d}x\,\mathrm{d}y = \lambda_j \int_\Omega u_i u_j \,\mathrm{d}x\,\mathrm{d}y = \delta_{ij}.$$

11.3.1. Variational formulation
Introduce the auxiliary variable $\sigma = -\Delta u$. Then (11.36) can be written as a second-order system:

$$\begin{aligned} \sigma + \Delta u &= 0 \quad \text{in } \Omega, \\ -\Delta\sigma &= \lambda u \quad \text{in } \Omega, \\ u = \partial u/\partial n &= 0 \quad \text{on } \Gamma. \end{aligned}$$

Multiplying the first equation by ψ, the second by v, integrating over Ω, and integrating by parts leads to

$$0=\int_\Omega \sigma\psi\,\mathrm{d}x\,\mathrm{d}y+\int_\Omega \Delta u\psi\,\mathrm{d}x\,\mathrm{d}y$$

$$=\int_\Omega \sigma\psi\,\mathrm{d}x\,\mathrm{d}y-\int_\Omega \nabla u\cdot\nabla\psi+\int_\Gamma (\partial u/\partial n)\psi\,\mathrm{d}s$$

$$=\int_\Omega \sigma\psi\,\mathrm{d}x\,\mathrm{d}y-\int_\Omega \nabla u\cdot\nabla\psi\,\mathrm{d}x\,\mathrm{d}y\quad \forall\psi\in H^1(\Omega)$$

and

$$\lambda\int_\Omega uv\,\mathrm{d}x\,\mathrm{d}y=-\int_\Omega \Delta\sigma v\,\mathrm{d}x\,\mathrm{d}y$$

$$=\int_\Omega \nabla\sigma\cdot\nabla v\,\mathrm{d}x\,\mathrm{d}y-\int_\Gamma (\partial\sigma/\partial n)v\,\mathrm{d}s$$

$$=\int_\Omega \nabla\sigma\cdot\nabla v\,\mathrm{d}x\,\mathrm{d}y\quad \forall v\in H_0^1(\Omega).$$

Thus we arrive at the variational formulation

$$\begin{aligned}&(\sigma,u)\in H^1(\Omega)\times H_0^1(\Omega),\qquad (\sigma,u)\neq(0,0)\\&\int_\Omega \sigma\psi\,\mathrm{d}x\,\mathrm{d}y-\int_\Omega \nabla u\cdot\nabla\psi\,\mathrm{d}x\,\mathrm{d}y=0\quad \forall\psi\in H^1(\Omega),\\&-\int \nabla\sigma\cdot\nabla v\,\mathrm{d}x\,\mathrm{d}y=-\lambda\int_\Omega uv\,\mathrm{d}x\,\mathrm{d}y\quad \forall v\in H_0^1(\Omega).\end{aligned}\tag{11.38}$$

We derived (11.38) formally from (11.36). One can, however, easily make the argument rigorous with the aid of a well-known regularity result: If w is the solution to

$$\Delta^2 w=f\quad\text{in }\Omega,\qquad w=\partial w/\partial n=0\quad\text{on }\Gamma,$$

where Ω is a convex polygon and $f\in H^0(\Omega)$, then $w\in H^3(\Omega)$ and $\|w\|_{3,\Omega}\leqslant C\|f\|_{0,\Omega}$, cf. GRISVARD [1985] and KELLOGG and OSBORN [1975]. We assume Ω is a convex polygon in the remainder of this subsection. Using this result we can show that if (λ, u) is an eigenpair of (11.36) and $\sigma=-\Delta u$, then $(\lambda,(\sigma,u))$ is an eigenpair of (11.38), and if $(\lambda,(\sigma,u))$ is an eigenpair of (11.38), then (λ,u) is an eigenpair of (11.36) and

$\sigma=-\Delta u$. Problem (11.38) has the form of (11.3) with

$$V=H^1(\Omega), \qquad W=H_0^1(\Omega), \qquad H=G=H^0(\Omega),$$

$$A(\sigma,\psi)=\int_\Omega \sigma\psi \,\mathrm{d}x\,\mathrm{d}y, \qquad B(\psi,u)=\int_\Omega \nabla\psi\cdot\nabla u\,\mathrm{d}x\,\mathrm{d}y.$$

It is easily seen that A is symmetric and that (11.1) and (11.2) are satisfied.

(11.38) has eigenvalues

$$0<\lambda_1\leqslant\lambda_2\leqslant\cdots\uparrow\infty$$

and corresponding eigenfunctions

$$(\sigma_1,u_1),(\sigma_2,u_2),\ldots,$$

with $\sigma_j=-\Delta u_j$.

11.3.1′. Regularity of the eigenfunctions

If (σ,u) is an eigenfunction of (11.38) then, as we have seen above, u is an eigenfunction of (11.36) and $\sigma=-\Delta u$, and hence the regularity of (σ,u) can be inferred from the regularity properties of (11.36). For results on this later regularity question we refer to GRISVARD [1985] and KELLOGG and OSBORN [1975].

11.3.2. Discretization of (11.38) *and assessment of the accuracy of the approximate eigenpairs*

As in Section 11.2, our discretization will be via (11.4). For our specific problem, a mixed method for the associated source problem has been studied by GLOWINSKI [1973], CIARLET and RAVIART [1974], MERCIER [1974], and FALK and OSBORN [1980]. We will use the same trial and test spaces employed in those papers.

Let $\gamma=\{\tau\}=\{\tau_h\}$ be a family of α-regular, q-quasiuniform triangulations of $\bar{\Omega}$. Then for $p=2,3,\ldots$, let

$$V_h=S^p(\tau_h) \tag{11.39a}$$

and

$$W_h=S_0^p(\tau_h)\cap H_0^1(\Omega). \tag{11.39b}$$

We then consider (11.4) with these choices. We will have approximate eigenvalues and eigenfunctions

$$\lambda_{1,h}\leqslant\cdots\leqslant\lambda_{N,h}$$

and

$$(\sigma_{1,h},u_{1,h}),\ldots,(\sigma_{N,h},u_{N,h}),$$

where $N=\dim(V_h\times W_h)$.

THEOREM 11.4. *Let V_h and W_h be as in* (11.39) *with $p\geqslant 2$ and suppose the eigenfunctions* *of* (11.36) *belong to* $H^{p+1}(\Omega)$. *Then*

$$|\lambda_{k,h}-\lambda_k|\leqslant C(p)h^{2p-2} \tag{11.40}$$

and

$$\|u_{k,h}-u_k\|_{0,\Omega}\leqslant C(p)h^p. \tag{11.41}$$

PROOF. The symmetry of A and the validity of (11.1) and (11.2) for problem (11.38) have already been noted.

The source problem (11.6) is uniquely solvable for each $g\in G=H^0(\Omega)$. The unique solution is (σ, u), where

$$\Delta^2 u=g,\quad u\in H_0^2(\Omega)$$

and

$$\sigma=-\Delta u$$

(cf. the derivation of (11.38)). The unique solvability of (11.7) is easily checked.

FALK and OSBORN [1980, Section 3(a)] have shown that

$$\|Tg-T_hg\|_{0,\Omega}\leqslant Ch^2\|Tg\|_{3,\Omega}.$$

This, together with the regularity result mentioned above, gives

$$\|(T-T_h)g\|_{0,\Omega}\leqslant Ch^2\|g\|_{0,\Omega},$$

which proves (11.9).

Thus, all of the hypotheses for Theorems 11.1 and 11.2 have been verified for the problem under consideration. Using the results in FALK and OSBORN [1980, Section 3(a)], we have

$$\|(S-S_h)g\|_{0,\Omega}\leqslant Ch^{p-1}\|Tg\|_{p+1,\Omega},$$

$$\|(S-S_h)g\|_{1,\Omega}\leqslant Ch^{p-2}\|Tg\|_{p+1,\Omega},$$

$$\|(T-T_h)g\|_{0,\Omega}\leqslant Ch^p\|Tg\|_{p+1,\Omega},$$

$$\|(T-T_h)g\|_{1,\Omega}\leqslant Ch^p\|Tg\|_{p+1,\Omega},$$

from which we obtain

$$\|(S-S_h)|_{\overline{M}}\|_{H^0(\Omega),H^0(\Omega)}\leqslant Ch^{p-1}, \tag{11.42a}$$

$$\|(S-S_h)|_{\overline{M}}\|_{H^0(\Omega),H^1(\Omega)}\leqslant Ch^{p-2}, \tag{11.42b}$$

$$\|(T-T_h)|_{\overline{M}}\|_{H^0(\Omega),H^0(\Omega)}\leqslant Ch^p, \tag{11.42c}$$

$$\|(T-T_h)|_{\overline{M}}\|_{H^0(\Omega),H_0^1(\Omega)}\leqslant Ch^p. \tag{11.42d}$$

(11.40) follows immediately from Theorem 11.1 and (11.42), and (11.41) follows from Theorem 11.2 and (11.42c). □

REMARK 11.4. The estimates obtained in this subsection were first obtained by

CANUTO [1978]. We note, however, that the estimation techniques used here will yield an improvement over the estimates of Canuto in the case when the eigenfunctions have low regularity. Our method of proof does not yield any estimates for $p=1$. For this case, see ISHIHARA [1978].

11.3.3. Solution of matrix eigenvalue problem
See Section 11.2.3.

For further results in eigenvalue approximation by mixed methods, and also by hybrid methods, we refer to MERCIER, OSBORN, RAPPAZ, and RAVIART [1981], MERCIER and RAPPAZ [1978], and ISHIHARA [1977].

REMARK 11.5. We have seen in this section and in Section 10 that there are various methods available for the approximate calculation of the eigenvalues of a specific problem. For example, we have analyzed two methods for the membrane problem. Furthermore, this discussion, together with that in Section 3, shows that there are many more possibilities. Clearly the rational choice of a method for any particular concrete problem is important. The effective choice of a method is complex, depending on many aspects of the underlying problem.

11.4. A mixed method for a problem in one dimension with rough coefficients

Consider the problem

$$\begin{aligned} &-(a(x)u')' = \lambda b(x)u, \quad 0<x<1, \\ &u(0)=u(1)=0. \end{aligned} \tag{11.43}$$

This is a special case of the problem (1.8), (1.9a) discussed in Section 1. We will be especially interested here in the case in which the coefficients $a(x)$ and $b(x)$ are rough functions. Such problems arise in the analysis of the vibrations of structures with rapidly varying material properties, of composite materials, for example. In Section 3, we gave (11.43) the mixed formulation (3.26) (or (3.27)). In this subsection we analyze a mixed method based on (3.27).

Hence we consider the problem

$$\begin{aligned} &(\sigma, u) \in L_2(0,1) \times H_0^1(0,1), \qquad (\sigma, u) \neq (0,0), \\ &\int_0^1 (\sigma\psi/a)\,\mathrm{d}x - \int_0^1 u'\psi\,\mathrm{d}x = 0 \quad \forall \psi \in L_2(0,1), \\ &-\int_0^1 \sigma v'\,\mathrm{d}x = -\lambda \int_0^1 buv\,\mathrm{d}x \quad \forall v \in H_0^1(0,1), \end{aligned} \tag{11.44}$$

where $a(x)$ is of bounded variation and $b(x)$ is measurable and

$$0<a_0 \leqslant a(x) \leqslant a_1, \qquad 0 \leqslant b_0 \leqslant b(x) \leqslant b_1.$$

(11.44) is of the form (11.3) with

$$V=H=L_2(0,1), \qquad W=H_0^1(0,1),$$

$$G=L_2(0,1) \quad \text{with } (u,v)_G=\int_0^1 buv\,\mathrm{d}x,$$

$$A(\sigma,\psi)=\int_0^1 \frac{\sigma\psi}{a}\,\mathrm{d}x, \qquad B(\sigma,v)=-\int_0^1 \sigma v'\,\mathrm{d}x.$$

$(\lambda,(\sigma,u))$ is an eigenpair of (11.44) if and only if (λ,u) is an eigenpair of (11.43) and $\sigma=au'$. We discretize (11.44) by letting $\tau=\{T_i\}_{i=1}^{M(\tau)}$ be a mesh on [0, 1], defining

$$V_h=\{\sigma:\sigma|_{T_i}=\text{ constant}, i=1,\ldots,M(\tau)\} \tag{11.45a}$$

and

$$W_h=\{v: v\in H_0^1(0,1), v|_{T_i}=\text{linear polynomial}, i=1,\ldots,M(\tau)\}, \tag{11.45b}$$

with

$$h=h(\tau)=\max_{i=1,\ldots,M(\tau)} \operatorname{diam} T_i,$$

and considering (11.4). The eigenpairs $(\lambda_h,(\sigma_h,u_h))$ of (11.4) are then considered as approximations to the eigenpairs $(\lambda,(\sigma,u))$ of (11.44). Although this approximation method satisfies the hypotheses of Theorem 11.1, a direct application of that result does not yield the best possible estimate. We will employ an analysis that is parallel to, but different than, that used in the proof of Theorem 11.1.

The analysis begins by introducing the operators $T,T_h:G\to G$ and $S,S_h:G\to V$ that are defined in (11.8). λ is an eigenvalue of (11.44) if and only if λ^{-1} is an eigenvalue of T; the correspondence between the eigenfunctions is given by $(\sigma,u)\leftrightarrow u$. Likewise λ_h is an eigenvalue of (11.4) if and only if λ_h^{-1} is an eigenvalue of T_h, with the correspondence between eigenfunctions given by $(\sigma_h,u_h)\leftrightarrow u_h$. $\|T-T_h\|_{GG}\to 0$, as will be shown later, so we may apply Theorem 7.2 to T and T_h on the space G. Let λ^{-1} be an eigenvalue of T. The eigenvalues of a problem of the type (11.43) are simple and hence λ^{-1} is a simple eigenvalue of T. Thus one eigenvalue λ_h of (11.4) converges to λ. By Theorem 7.2 we have

$$|\lambda-\lambda_h|\leqslant C\{|((T-T_h)u,u)_G|+\|(T-T_h)u\|_G^2\}, \tag{11.46}$$

where u is any eigenfunction of T corresponding to λ^{-1} with $\|u\|_G=1$. We now proceed to analyze $((T-T_h)u,u)_G$.

From (11.6) we have

$$((T-T_h)u,u)_G=\int_0^1 bu(T-T_h)u\,\mathrm{d}x$$

$$= -B(Su, (T-T_h)u)$$

$$= A(Su, (S-S_h)u) + B((S-S_h)u, Tu) - B(Su, (T-T_h)u), \tag{11.47}$$

and from (11.6) and (11.7) we get

$$0 = A((S-S_h)u, \xi) + B((S-S_h)u, \eta) + B(\xi, (T-T_h)u), \quad \forall \eta \in W_h, \quad \xi \in V_h. \tag{11.48}$$

Combining (11.47) and (11.48) we get

$$((T-T_h)u, u)_G = A((S-S_h)u, Su+\xi) + B((S-S_h)u, Tu+\eta) + B(\xi - Su, (T-T_h)u),$$

$$\forall \eta \in W_h, \quad \xi \in V_h,$$

which, letting $\eta = -T_h u$ and $\xi = S_h u$, yields

$$((T-T_h)u, u)_G = A((S-S_h)u, (S+S_h)u)$$
$$= 2A((S-S_h)u, Su) - A((S-S_h)u, (S-S_h)u). \tag{11.49}$$

Now, again using (11.6) and (11.7) we get

$$a((S-S_h)u, Su) = -B((S-S_h)u, Tu)$$
$$= -B((S-S_h)u, Tu - \Sigma_h Tu),$$

where $\Sigma_h Tu$ is the W_h-interpolant of Tu, and hence, using

$$\int_0^1 S_h u\,[Tu - \Sigma_h Tu]'\,\mathrm{d}x = 0$$

and (11.6) we have

$$A((S-S_h)u, Su) = -B(Su, Tu - \Sigma_h Tu) = \int_0^1 bu[Tu - \Sigma_n Tu]\,\mathrm{d}x. \tag{11.50}$$

Finally, combining (11.49) and (11.50) we get

$$((T-T_h)u, u)_G = 2\int_0^1 bu[Tu - \Sigma_h Tu]\,\mathrm{d}x - \int_0^1 \frac{|(S-S_h)u|^2}{a}\,\mathrm{d}x$$

$$= 2\lambda^{-1}\int_0^1 bu(u - \Sigma_h u)\,\mathrm{d}x - \int_0^1 \frac{|(S-S_h)u|^2}{a}\,\mathrm{d}x. \tag{11.51}$$

Now, using (11.50) and (11.46) we get

$$|\lambda-\lambda_h|\leqslant C\left\{\left|\int_0^1 bu\,(u-\Sigma_h u)\,\mathrm{d}x\right|+a_0^{-1}\,\|(S-S_h)u\|_{L_2}+\|(T-T_h)u\|_{L_2}^2\right\}. \tag{11.52}$$

It remains to estimate the three terms on the right-hand side of (11.52).

Recall that $\Sigma_h u$ is the W_h-interpolant of u. By a result of PROSDORF and SCHMIDT [1981] we know that

$$\|u-\Sigma_h u\|_{L^1}\leqslant Ch^2\,V_0^1(u'), \tag{11.53}$$

where $V_0^1(u')$ denotes the variation of u'. Recall that u is an eigenfunction of (11.43) with $\|u\|_G=1$. Since $a(x)$ is of bounded variation, u' will be of bounded variation; in fact

$$V_0^1(u')\leqslant C, \tag{11.54}$$

where $C=C(a_0,a_1,b_0,b_1,V_0^1(a),\lambda)$ depends on $a_0,a_1,b_0,b_1,V_0^1(a)$ and λ. Also

$$\|u\|_{L^\infty}\leqslant C. \tag{11.55}$$

Using Hölders inequality, together with (11.53)–(11.55), we get

$$\left|\int_0^1 bu(u-\Sigma_h u)\,\mathrm{d}x\right|\leqslant\|bu\|_{L^\infty}\|u-\Sigma_h u\|_{L^1}\leqslant Ch^2V(a), \tag{11.56}$$

where $C=C(a_0,a_1,b_0,b_1,V_0^1(a),\lambda)$.

Next we consider $\|(S-S_h)u\|_{L_2}$ and $\|(T-T_h)u\|_{L_2}$. It is easily seen that the results in FALK and OSBORN [1980] imply

$$\|(S-S_h)u\|_{L_2},\ \|(T-T_h)u\|_{L_2}\leqslant C(a_0,a_1,b_0,b_1,\lambda)h. \tag{11.57}$$

Note that (11.57) shows that $\|T-T_h\|_{GG}\to 0$.

Finally, combining (11.52), (11.56), and (11.57) we have:

THEOREM 11.5. *Suppose λ is an eigenvalue of* (11.43) (*or of* (11.44)) *and let λ_h be the approximate eigenvalue defined by* (11.4) *with V_h and W_h defined by* (11.45). *Then*

$$|\lambda-\lambda_h|\leqslant C(a_0,a_1,b_0,b_1,V_0^1(a),\lambda)h^2. \tag{11.58}$$

The striking feature of estimate (11.58) is that the constant C depends on the bounds a_0,a_1,b_0 and b_1 and on $V_0^1(a)$, but is otherwise independent of $a(x)$ and $b(x)$. This shows that the approximation method is effective for problems with rough coefficients (cf. discussion of alternate variational formulations at the end of Section 3). In fact, the rate of convergence indicated by (11.58) is the same as that for the usual Ritz method for problems with smooth coefficients. Estimate (11.58) was proved by BANERJEE [1988]. The use of mixed methods for eigenvalue approximation for problems with rough coefficients was first suggested by NEMAT-NASSER [1972, 1974]. Rate of convergence estimates for several such mixed methods were derived by BABUŠKA and OSBORN [1978].

REMARK 11.6. It is of interest to note that the variable σ_h can be eliminated from (11.4) in the present context (i.e., with the choices for V, W, H, G, A, B, V_h and W_h we have made in this subsection) leading to the problem

$$u_h \in W_h, \tag{11.59}$$

$$\sum_{i=1}^{M} \int_{T_i} a_\tau u_h' \bar{v}' \,\mathrm{d}x = \lambda_h \int_0^1 b u_h \bar{v} \,\mathrm{d}x \quad \forall v \in W_h,$$

where a^τ is a step function with

$$a_\tau|_{T_i} = \left(\frac{\int_{T_i} \mathrm{d}x/a}{\operatorname{diam} T_i}\right)^{-1}, \quad i=1,\ldots, M(\tau).$$

Thus (11.59) differs from the usual Ritz method only in that the coefficient enters the calculation through its harmonic averages over the subintervals of the mesh instead of through its averages.

12. Methods based on one-parameter families of variational formulations

In our treatment of the membrane problem in Section 10.2, the trial and test functions satisfied the essential boundary condition $u=0$ (cf. (10.42)). In fact, if one bases the approximation method on the usual variational formulation (10.39), one must impose the boundary condition on the trial and test functions. To avoid this, methods have been developed that use test and trial functions that are not required to satisfy essential boundary conditions. (See the discussion of essential and natural boundary conditions in Section 3.) In this section we discuss two such methods. They are both based on a one-parameter family of variational formulations. We will be rather brief and will not explicitly discuss each of the steps (1), (1′), (2), and (3) of finite element approximation outlined in Section 10.

12.1. The least squares method

Consider, as in Sections 10.2 and 11.2, the membrane problem

$$-\Delta u=\lambda u \quad \text{in } \Omega, \qquad u=0 \quad \text{on } \Gamma=\partial\Omega, \tag{12.1}$$

where Ω is a bounded, open set with boundary Γ, which, for the sake of simplicity, we assume to be of class C^∞. Note that we are not assuming Ω to be a polygon. Problem (12.1) has eigenvalues

$$0<\lambda_1 \leqslant \lambda_2 \leqslant \cdots \uparrow \infty$$

and eigenfunctions

$$u_1, u_2, \ldots .$$

We begin by introducing the least squares method for the corresponding source problem,

$$-\Delta w=f \quad \text{in } \Omega, \qquad w=0 \quad \text{on } \Gamma, \tag{12.2}$$

which is usually given the variational formulation (see Remark 12.1 for the reason for using complex functions here),

$$\begin{gathered} w \in H_0^1(\Omega), \\ \int_\Omega \nabla w \cdot \overline{\nabla v}\, \mathrm{d}x\, \mathrm{d}y = \int_\Omega f \bar{v}\, \mathrm{d}x\, \mathrm{d}y \quad \forall v \in H_0^1(\Omega). \end{gathered} \tag{12.3}$$

We now give (12.2) a different variational formulation. w solves (12.2) if and only if

$$\begin{gathered} w \in H^2(\Omega), \\ \int \Delta w \overline{\Delta v}\, \mathrm{d}x\, \mathrm{d}y + \rho \int_\Gamma w \bar{v}\, \mathrm{d}x\, \mathrm{d}y = -\int_\Omega f\, \overline{\Delta v}\, \mathrm{d}x\, \mathrm{d}y, \\ \forall v \in H^2(\Omega), \quad \forall 0<h\leqslant 1, \end{gathered} \tag{12.4}$$

where $\rho=\rho_h \geqslant 1$ is a parameter that approaches ∞ as $h \to 0$. To pass from (12.2) to (12.4) is immediate. To go from (12.4) to (12.2) we proceed as follows. First take $v \in H^2(\Omega)$ to satisfy

$$\Delta v = \Delta w + f \quad \text{in } \Omega, \qquad v=0 \quad \text{on } \Gamma.$$

This choice for v in (12.4) yields $-\Delta w=f$ in Ω. The equation in (12.4) then becomes

$$\rho \int_\Omega w \bar{v}\, \mathrm{d}x = 0 \quad \forall v \in H^2(\Omega),$$

which implies $w=0$ on Γ. In (12.4) the boundary conditions $w=0$ is not explicitly imposed. This is the major advantage of the formulation (12.4) over (12.3) for our purposes. We note that w can also be characterized by an extremal property: the solution w of (12.4) is the unique minimizer of the functional

$$\int_\Omega |-\Delta v - f|^2\, \mathrm{d}x\, \mathrm{d}y + \rho \int_\Gamma |v|^2\, \mathrm{d}s$$

over $v \in H^2(\Omega)$.

In order to discretize (12.4) we suppose we have a family $\gamma=\{\tau\}=\{\tau_h\}$ of triangulations of $\overline{\Omega'}$, where Ω' is some fixed rectangle containing $\bar{\Omega}$. Then let

$$\begin{aligned} S_h &= S^{p,2}(\tau_h) \\ &= \{u \in H^2(\Omega') : u|_T = \text{polynomial of degree } p\ \forall T \in \tau_h\} \end{aligned}$$

and let S_h consist of the restrictions of functions in $S^{p,2}(\tau_h)$ to Ω. The family S_h satisfies the following approximation result: If $p \geqslant 5$, then

$$\inf_{\chi \in S_h} \sum_{j=0}^{2} h^j \|v-\chi\|_{j,\Omega} \leqslant Ch^t \|v\|_{t,\Omega} \quad \text{for } 2 \leqslant t \leqslant p+1. \tag{12.5}$$

See CIARLET [1978] for a proof of (12.5). Then we define an approximate solution w_h to w by letting w_h be the unique solution to

$$\begin{aligned} &w_h \in S_h, \\ &\int_\Omega \Delta w_h \overline{\Delta v}\, dx\, dy + \rho \int_\Gamma w_h \bar{v}\, ds = -\int_\Omega f \overline{\Delta v}\, dx\, dy \quad \forall v \in S_h. \end{aligned} \tag{12.6}$$

w_h is called the least squares approximation to w since it can be alternately characterized as the unique minimizer of

$$\int_\Omega |-\Delta v - f|^2\, dx\, dy + \rho \int |v|^2\, ds$$

over $v \in S_h$. BRAMBLE and SCHATZ [1970] proposed and analyzed this method for $\rho = \rho_h = h^{-3}$. They also showed $\rho = h^{-3}$ to be the optimal choice for ρ.

Now we return to the eigenvalue problem (12.1). Proceeding in a similar way we see that (12.1) has the variational formulation

$$\begin{aligned} &u \in H^2(\Omega), \\ &\int_\Omega \Delta u \overline{\Delta v}\, dx\, dy + \rho \int_\Gamma u \bar{v}\, ds = -\lambda \int_\Omega u \overline{\Delta v}\, dx\, dy \quad \forall v \in H^2(\Omega). \end{aligned} \tag{12.7}$$

Problem (12.7) is then discretized by

$$\begin{aligned} &\lambda_h \text{ complex}, \quad 0 \neq u_h \in S_h \\ &\int_\Omega \Delta u_h \overline{\Delta v}\, dx\, dy + \rho \int_\Gamma u \bar{v}\, ds = -\lambda_h \int_\Omega u_h \overline{\Delta v}\, dx\, dy \quad \forall v \in S_h. \end{aligned} \tag{12.8}$$

Problem (12.8) has eigenpairs $(\lambda_{j,h}, u_{j,h})$, $j=1,\dots,N$, where $N = \dim S_h$.

If for $f \in H^0(\Omega)$ we define $Tf = w$ and $T_h f = w_h$, where w and w_h are defined by (12.2) (or (12.4)) and (12.6), respectively, then we easily see that (λ, u) is an eigenpair of (12.1) if and only if $(\mu = \lambda^{-1}, u)$ is an eigenpair of T and (λ_h, u_h) is an eigenpair of (12.8) if and only if $(\mu_h = \lambda_h^{-1}, u_h)$ is an eigenpair of T_h. We will estimate the error in (μ_h, u_h), and thus in (λ_h, u_h), by applying the results in Section 7. T and T_h are clearly compact on $H^0(\Omega)$. We will show $\|T - T_h\| \to 0$ in the next paragraph.

In order to apply Theorem 7.3 on $H^0(\Omega)$ we need estimates for $((T-T_h)u, u)$, $\|(T-T_h)u\|_{0,\Omega}$ and $\|(T-T_h^*)u\|_{0,\Omega}$, where u is an eigenfunction of (12.1) corresponding to the eigenvalue λ (or μ) we are approximating. These estimates are all contained in BRAMBLE and SCHATZ [1970] (and also in BAKER [1973]) for the choice $\rho=h^{-3}$. In their Corollary 4.1 take $\gamma=\frac{3}{2}$, $\lambda=t-2$, $g=0$, $l=-s$ and $r=p+1$ to get

$$|((T-T_h)\phi,\psi)_{0,\Omega}|\leqslant Ch^{s+t}\|\phi\|_{t-2,\Omega}\|\psi\|_{s,\Omega} \quad \text{for } 0\leqslant s\leqslant p-3,\quad 2\leqslant t\leqslant p+1. \tag{12.9}$$

Taking $s=0$ and $t=2$ in (12.9) shows that $\|T-T_h\|\to 0$. Now take $s=p-3$ and $t=p+1$ to obtain

$$|((T-T_h)\phi,\psi)_{0,\Omega}|\leqslant Ch^{2p-2}\|\phi\|_{p-1,\Omega}\|\phi\|_{p-3,\Omega}, \tag{12.10}$$

take $s=0$ and $t=p+1$ to obtain

$$|((T-T_h)\phi,\psi)_{0,\Omega}|\leqslant Ch^{p+1}\|\phi\|_{p-1,\Omega}\|\psi\|_{0,\Omega},$$

and hence

$$\|((T-T_h)\phi)\|_{0,\Omega}\leqslant Ch^{p+1}\|\phi\|_{p-1,\Omega}, \tag{12.11}$$

and take $s=p-3$ and $t=2$ to obtain

$$|((T-T_h)\phi,\psi)_{0,\Omega}|=|(\phi,(T-T_h^*)\psi)_{0,\Omega}|\leqslant Ch^{p-1}\|\phi\|_{0,\Omega}\|\psi\|_{p-3,\Omega},$$

and hence

$$\|((T-T_h^*)\psi)\|_{0,\Omega}\leqslant Ch^{p-1}\|\psi\|_{p-3,\Omega}. \tag{12.12}$$

THEOREM 12.1. *Suppose the approximate eigenpairs* $(\lambda_{j,h},u_{j,h})$ *are defined by* (12.8) *with* $\rho=h^{-3}$ *and suppose the eigenfunctions of* (12.1) *belong to* $H^{p-1}(\Omega)$. *Then*

$$|\lambda_{k,h}-\lambda_k|\leqslant Ch^{2p-2} \tag{12.13}$$

and

$$\|u_{k,h}-u_k\|_{0,\Omega}\leqslant Ch^{p+1}. \tag{12.14}$$

PROOF. Let λ_k be any eigenvalue of (12.1) and suppose its geometric multiplicity is q, i.e., the geometric multiplicity of $\mu_k=\lambda_k^{-1}$ is q. Since T is selfadjoint, the ascent is one and the algebraic multiplicity of μ_k is also q. q of the $\lambda_{j,h}$ will converge to λ_k. Let $\lambda_{k,h}$ be one of them. Theorem 7.3 can now be applied and (12.13) follows directly from (7.15) and (12.10)–(12.12) since all of the eigenfunctions of (12.1) corresponding to λ_k belong to $H^{p-1}(\Omega)$. Estimate (12.14) follows from Theorems 7.1 and 7.4 and (12.11). □

REMARK 12.1. Even though (12.1) is selfadjoint, (12.8) is a nonselfadjoint (finite-dimensional) problem. Thus one needs the general (not necessarily selfadjoint) theory in Section 7 to analyze the least squares method. The nonselfadjointness of (12.8) is the reason we have used complex function spaces in this analysis.

12.2. The penalty method

We will once more consider the membrane eigenvalue problem (10.38) and assume the boundary Γ of Ω is of class C^∞ (cf. also (12.1)). In Section 10 we gave this problem the variational formulation

$$u \in H_0^1(\Omega), \qquad a(u, v) = \lambda b(u, v) \quad \forall v \in H_0^1(\Omega), \tag{12.15}$$

where

$$a(u, v) = \int_\Omega \nabla u \cdot \nabla v \, \mathrm{d}x \, \mathrm{d}y \tag{12.16a}$$

and

$$b(u, v) = \int_\Omega uv \, \mathrm{d}x \, \mathrm{d}y. \tag{12.16b}$$

Let us replace the boundary condition $u=0$ on Γ in (10.38) by $u+\psi^{-1}(\partial u/\partial n)=0$, i.e., let us consider the problem

$$\begin{aligned} -\Delta u &= \lambda u \quad \text{in } \Omega, \\ u+\psi^{-1}\frac{\partial u}{\partial n} &= 0 \quad \text{on } \Gamma, \end{aligned} \tag{12.17}$$

where $\psi = \psi_h \geqslant 1$ is a parameter that approaches $+\infty$ as $h \to 0$. It is easily seen that (12.17) has the variational form

$$u \in H^1(\Omega), \qquad a_\psi(u, v) = \lambda b(u, v) \quad \forall v \in H^1(\Omega), \tag{12.18}$$

where

$$a_\psi(u, v) = \int_\Omega \nabla u \cdot \nabla u \, \mathrm{d}x \, \mathrm{d}y + \psi \int_\Omega uv \, \mathrm{d}s. \tag{12.19}$$

Note that in (12.18), in contrast to (12.15), we have not imposed any constraint on either u or v. This is the case since $u+\psi^{-1}(\partial u/\partial n)=0$ is a natural boundary condition (cf. Section 3).

We now estimate the error between the eigenvalues and eigenvectors of (12.15) and (12.18). Toward this end consider the corresponding source problems:

$$-\Delta u_0 = f \quad \text{in } \Omega, \qquad u=0 \quad \text{on } \Gamma \tag{12.20}$$

and

$$-\Delta u = f \quad \text{in } \Omega, \qquad u+\psi^{-1}\frac{\partial u}{\partial n} = 0 \quad \text{on } \Gamma. \tag{12.21}$$

We view (12.21) as an approximation to (12.20). Denoting by u_0 (respectively u_ψ) the solution of (12.20) (respectively (12.21)), we are interested in estimating $u_\psi - u_0$. It is shown in BABUŠKA and AZIZ [1973, Section 7.2] that

$$u_\psi = u_0 - \psi^{-1}\xi + \zeta, \tag{12.22}$$

where ξ is the solution of the problem

$$-\Delta\xi+\xi=0 \quad \text{in } \Omega, \qquad \xi=\partial u_0/\partial n \quad \text{in } \Omega \tag{12.23}$$

and ζ is the solution to

$$\zeta\in H^1(\Omega), \qquad a_\psi(\zeta,v)=\psi^{-1}a(\xi,v) \quad \forall v\in H^1(\Omega). \tag{12.24}$$

From (12.16), (12.19) and (12.24) we have

$$\|\zeta\|^2_{H^1(\Omega)}\leqslant a_\psi(\zeta,\zeta)=\psi^{-1}a(\xi,\zeta)=\psi^{-1}\|\xi\|_{H^1(\Omega)}\|\zeta\|_{H^1(\Omega)}$$

and hence

$$\|\zeta\|_{H^1(\Omega)}\leqslant\psi^{-1}\|\xi\|_{H^1(\Omega)}. \tag{12.25}$$

From (12.23) and (12.25) we obtain

$$\|u_\psi-u_0\|_{H^1(\Omega)}\leqslant 2\psi^{-1}\|\xi\|_{H^1(\Omega)}. \tag{12.26}$$

From (12.23) and regularity results for elliptic boundary value problems we get

$$\|\xi\|_{H^1(\Omega)}\leqslant C\|f\|_{H^0(\Omega)}. \tag{12.27}$$

Combining (12.26) and (12.27) yields

$$\|u_\psi-u_0\|_{H^1(\Omega)}\leqslant C\psi^{-1}\|f\|_{H^0(\Omega)}. \tag{12.28}$$

If we now introduce the operators T and T_ψ on $H^0(\Omega)$ by

$$Tf=u_0 \quad \text{and} \quad T_\psi f=u_\psi,$$

then (12.28) implies that

$$\|(T-T_\psi)f\|_{H^0(\Omega)}\leqslant C\psi^{-1}\|f\|_{0,\Omega}. \tag{12.29}$$

It is immediate that (λ,u) is an eigenpair of (12.15) if and only if $(\mu=\lambda^{-1},u)$ is an eigenpair of T; likewise (λ_ψ,u_ψ) is an eigenpair of (12.18) if and only if $\mu_\psi=(\lambda_\psi^{-1},u_\psi)$ is an eigenpair of T_ψ. It thus follows immediately from Theorems 7.1–7.4 and (12.29) that

$$|\lambda_j-\lambda_{\psi,j}|\leqslant C\psi^{-1} \tag{12.30}$$

and

$$\|u_j-u_{\psi,j}\|_{H^1(\Omega)}\leqslant C\psi^{-1}, \tag{12.31}$$

where (λ_j,u_j) and $(\lambda_{\psi,j},u_{\psi,j})$ denote the eigenpairs of (12.15) and (12.18), respectively. Note that (12.30) and (12.31) are estimates of the same order for both the eigenvalue and eigenvector errors. This is in contrast to approximations we have analyzed previously in this article. An analysis of a one-dimensional model problem shows that, for the type of approximations we are considering, the eigenvalue and eigenvector error is, indeed, of the same order.

Next we consider the problem (12.18) and approximate it by a finite element method, letting the resulting eigenpairs be $(\lambda_{\psi,j,h},u_{\psi,j,h})$. Since u and v in (12.18) are taken in $H^1(\Omega)$, we need not impose any boundary condition on the trial and test space S_h. If one now analyzes the error in the finite element approximation of (12.18), selects ψ so that the error in passing from (12.15) to (12.18) is of the same magnitude of that incurred in the finite element approximation of (12.18), and then combines the error estimates (12.30) and (12.31) with those for the finite element approximations of (12.18), one obtains estimates for the difference between (λ_j, u_j) and $(\lambda_{\psi,j,h}, u_{\psi,j,h})$. We stress that the $(\lambda_{\psi,j,h}, u_{\psi,j,h})$ are calculated from a matrix eigenvalue problem corresponding to trial and test spaces that are not required to satisfy the essential boundary condition for the membrane problem (12.1). The approximation method we have outlined is referred to as the penalty method.

We refer the reader to BABUŠKA and AZIZ [1973, Section 7.2] for a detailed analysis of the penalty method for the source problem. Estimates for the errors in eigenvalue approximation can be easily derived from the corresponding source problem estimates by means of Theorems 7.1–7.4. Because this application of these error estimates to the eigenvalue problem is similar to those discussed above and raises no new issues, we will not give a formal statement of the results.

REMARK 12.2. If Ω is a polygon, then the choice $\psi = \infty$ corresponds to satisfying the boundary condition on $\partial\Omega$, i.e., constraining $S^p(\tau)$ to be $\mathring{S}^p(\tau)$, and the resulting method is identical with that discussed in Section 10.2. If Γ is not polygonal, then $\psi = \infty$ will lead to the constraint $S^p(\tau) = \tilde{\mathring{S}}^p(\tau)$, where $\tilde{\mathring{S}}^p(\tau)$ consists of those $u \in S^p(\tau)$ which are zero on every triangle which intersects Γ. The finite element solution then solves the problem on $\tilde{\Omega}$ instead of Ω, where $\tilde{\Omega}$ consists of the union of all triangles which do not intersect Γ. Sometimes the mesh is constructed so that $\Omega\backslash\tilde{\Omega}$ is as small as possible by interpolating Γ by straight lines.

REMARK 12.3. In practical computation (codes) the penalty method (or some equivalent method) is often also used when Ω is a polygon by taking ψ to be very large (say $\psi = 10^8$). This is just a way of imposing the essential boundary conditions in the code.

REMARK 12.4. The least squares method and penalty method are seldom used as a way to treat essential boundary conditions on a curved boundary because of the difficulty in the computation of $a_\psi(u, v)$, which requires area integrations over triangles which intersect the boundary. The usual approach is to use curvilinear elements, which allow exact satisfaction of the boundary condition in a similar way as when the domain is polygonal (cf. Remark 12.3).

Let us end this section by noting some similarities and differences in the least squares and penalty methods.

– Both methods circumvent essential boundary conditions by reformulating the original problem in terms of a one-parameter family of variational formulations. In both methods, the optimal value of the parameter depends on the mesh, i.e., on h.

– With the least squares method, the optimal value of the parameter $(p = h^{-3})$ is independent of the solution. This is related to the fact that the alternate variational formulation characterizes the solution exactly for any value of the parameter. In the case of the penalty method, the optimal value of the parameter depends on the mesh and the smoothness of the solution or the eigenfunction. This is related to the fact that the exact solution does not exactly satisfy the one-parameter family of formulations for any value of the parameter $\psi \neq +\infty$.

– The least squares method employs C^1-elements (i.e., subspaces of $H^2(\Omega)$), whereas the penalty method employs C^0-elements (i.e., subspaces of $H^1(\Omega)$). As we have previously noted, C^0-elements are easier to construct than C^1-elements.

13. Concluding remarks

REMARK 13.1. We have illustrated the application of the general theory that was presented in Chapter II by considering several important model problems. It should be clear from the analysis of these model problems how to treat a wide variety of problems. We have seen, however, that the application of the general theory to a concrete problem may require subtle analysis.

REMARK 13.2. In Sections 10, 11 and 12 we have illustrated the main approach to finite element approximation of eigenvalue problems. We have seen that there are many available methods and that their basic theoretical properties can be established as an application of the results in Chapter II. Nevertheless, the implementation of these methods raises many other important questions; although we cannot address these questions in detail, we now mention some of them.

(1) Which method is most effective for a specific problem? What is the goal of the computation? We remark that sometimes high accuracy is achieved for eigenvalue approximation, but that only low accuracy is obtained for the approximation of other important quantities such as the stresses, moments, or shear forces.

(2) What types of meshes or adaptive mesh procedures are desirable? How should the quality of the computed results be assessed a posteriori? For a survey of results in this direction, see NOOR and BABUŠKA [1987].

(3) Which matrix eigenvalue solvers should be used? What is the influence of the computer architecture?

These questions are, of course, not restricted to eigenvalue computation. They also arise with finite element computation of source problems.

REMARK 13.3. The Ritz method, which was discussed in Section 10, is most easily analyzed with the results of Section 8, specifically with (8.44)–(8.46). Note that because of (8.32), (8.11) is satisfied with $\beta = \alpha$ and thus the major requirement on S_h is that it have good approximation properties.

REMARK 13.4. We have seen in Remark 11.2 that mixed methods for eigenvalue approximation have the form of (8.10) and (8.14). Thus, if a method satisfies the hypotheses of Section 8, specifically (8.1), (8.2), (8.3), (8.6), (8.11) and (8.13), then the method can also be analyzed with the results of Section 8. Most mixed methods, however, fail to satisfy at least one of these hypotheses, and we thus cannot rely on the results of Section 8. We now comment on two of the examples discussed in Section 11 in regard to which results in Chapter II their analysis is based on.

(1) Consider first the mixed method discussed in Section 11.2 for the membrane problem. It is easily seen that the variational formulation (11.24) satisfies (8.1), (8.2), and (8.3), but that it does not satisfy (8.6). In Section 8, assumption (8.6) is used to show that the operator T defined by (8.8) is compact. For our example, for $(f, g) \in H(\text{div}, \Omega) \times H^0(\Omega)$,

$$T(f, g) = (\sigma, u),$$

where u solves

$$u \in H_0^1(\Omega), \qquad -\Delta u = g \quad \text{in } \Omega$$

and $\sigma = \nabla u$, and, by noting in particular the dependence of σ on g, we see that $T: H(\text{div}, \Omega) \times H^0(\Omega) \to H(\text{div}, \Omega) \times H^0(\Omega)$ is not compact. Since T is not compact, T_h, as defined by (8.16), cannot converge to T in norm. Because of these facts, the results of Section 7 do not apply (to this T). The analysis that we used for this problem (cf. Theorem 11.1) is based on Section 7 and circumvents this difficulty by using a different operator, namely $T: H^0(\Omega) \to H^0(\Omega)$ defined by $Tg = u$ (cf. 11.8(c)).

As mentioned in Remark 7.7, results for noncompact operators which parallel those in Section 7 have been proved by DESCLOUX, NASSIF and RAPPAZ [1978a, 1978b], and one can, if fact, use them to derive the estimates we obtained in Section 11.2, specifically (11.32) and (11.33). We will not present the details of this analysis but will comment briefly on the applicability of the results of DESCLOUX, NASSIF, and RAPPAZ [1978] to our problem.

For their results, T is not required to be compact and T_h is assumed to converge to T in the sense that

$$\begin{aligned} &\inf_{(\chi, \eta) \in S_h = V_h \times W_h} \|(\sigma, u) - (\chi, \eta)\|_H \to 0 \\ &\qquad \text{for each } (\sigma, u) \in H(\text{div}, \Omega) \times H^0(\Omega) \end{aligned} \tag{13.1}$$

and

$$\|T_h - T\|_h = \sup_{\substack{(f, g) \in V_h \times W_h \\ \|(f, g)\|_{H(\text{div}, \Omega) \times H^0(\Omega)}}} \|(T_h - T)(f, g)\|_{H(\text{div}, \Omega) \times H^0(\Omega)} \to 0 \quad \text{as } h \to 0. \tag{13.2}$$

With V_h and W_h defined as in (11.31), (13.1) follows from the approximation result in RAVIART and THOMAS [1977].

We now verify (13.2), which is central point in any application of the results of DESCLOUX, NASSIF and RAPPAZ [1978]. For $(f, g) \in V_h \times W_h$, let $(\sigma, u) = T(f, g)$ and $(\sigma_h, u_h) = T_h(f, g)$, where T and T_h are defined by (8.8) and (8.16), respectively, for the problem discussed in Section 11.2. We know that $u \in H^1(\Omega)$, $-\Delta u = g$ and

$\sigma=\nabla u$, and hence $\operatorname{div}\sigma=g$. Also, if $g\in W_h$ it is easily seen that $\operatorname{div}\sigma_h=-g$. Thus

$$\begin{aligned}
&\|(T_h-T)(f,g)\|_{H(\operatorname{div},\Omega)\times H^0(\Omega)}\\
&\quad=\|(\sigma_h,u_h)-(\sigma,u)\|_{H(\operatorname{div},\Omega)\times H^0(\Omega)}\\
&\quad=(\|\operatorname{div}\sigma_h-\operatorname{div}\sigma\|_{0,\Omega}^2+\|\sigma_h-\sigma\|_{H^0(\Omega)}^2+\|u_h-u\|_{0,\Omega}^2)^{1/2}\\
&\quad=(\|\sigma_h-\sigma\|_{H^0(\Omega)}^2+\|u_h-u\|_{0,\Omega}^2)^{1/2}\quad\text{for } (f,g)\in V_h\times W_h.
\end{aligned}\tag{13.3}$$

From the results in FALK and OSBORN [1980] we have

$$\|\sigma_h-\sigma\|_{H^0(\Omega)}\leqslant Ch\|u\|_{2,\Omega}\leqslant Ch\|g\|_{0,\Omega}\tag{13.4a}$$

and

$$\|u_h-u\|_{0,\Omega}\leqslant Ch^2\|u\|_{2,\Omega}\leqslant Ch^2\|g\|_{0,\Omega}.\tag{13.4b}$$

Combining (13.3) and (13.4) we get

$$\begin{aligned}
&\|(T_h-T)(f,g)\|_{H(\operatorname{div},\Omega)\times H^0(\Omega)}\\
&\quad\leqslant Ch\|g\|_{0,\Omega}\\
&\quad\leqslant Ch\|(f,g)\|_{H(\operatorname{div},\Omega)\times H^0(\Omega)}\quad\text{for } (f,g)\in V_h\times W_h.
\end{aligned}\tag{13.5}$$

(13.2) follows directly from (13.5).

(2) Consider next the method discussed in Section 11.3 for the vibrating plate problem. The variational formulation (11.38) for the problem does not satisfy (8.2) and (8.11). Note that the method was analyzed by means of Theorem 11.1 which is based on Theorem 7.3.

REMARK 13.5. The fact that many mixed approximation methods fail to satisfy the usual hypotheses (cf. BABUŠKA [1971], BABUŠKA and AZIZ [1979] and BREZZI [1974]) for variational approximation methods is an issue for the approximation of source problems as well as eigenvalue problems. The abstract results in FALK and OSBORN [1980] have as their main application the analysis of mixed methods which fail to satisfy the usual hypotheses for variational approximation methods. In this connection see also BABUŠKA, OSBORN and PITKÄRANTA [1980], where problem (11.38) is reformulated in terms of alternate spaces with alternate (mesh-dependent) norms so as to satisfy the usual hypotheses.

Acknowledgement

The authors would like to thank Professor Christopher Beattiz for his careful reading of the manuscript.

References

ADAMS, R.A. (1975), *Sobolev Spaces* (Academic Press, New York).

AGMON, S. (1965), *Lectures on Elliptic Boundary Value Problems*, Van Nostrand Mathematical Studies **2** (Van Nostrand, Princeton, NJ).

ARONSZAJN, N. (1948), The Rayleigh–Ritz and A. Weinstein methods for approximation of eigenvalues, I: Operators in a Hilbert space; II: Differential operators, *Proc. Nat. Acad. Sci. U.S.A.* **34**, 474–480, 594–601.

ARONSZAJN, N. (1949–50), The Rayleigh–Ritz and the Weinstein methods for approximation of eigenvalues, Tech. Reps. 1–4, Oklahoma A&M College, Stillwater, OK.

ARONSZAJN, N. and A. WEINSTEIN (1942), On the unified theory of eigenvalues of plates and membranes, *Amer. J. Math.* **64**, 623–645.

BABUŠKA, I. (1971), Error bounds for finite element method, *Numer. Math.* **16**, 322–333.

BABUŠKA, I. and A. AZIZ (1973), Survey lectures on the mathematical foundations of the finite element method, in: A.K. AZIZ, ed., *The Mathematical Foundations of the Finite Element Method with Application to Partial Differential Equations* (Academic Press, New York) 5–359.

BABUŠKA, I., G. LALOZ and J. OSBORN (1990), Finite element method for second order partial differential equations with rough coefficients.

BABUŠKA, I. and B.Q. GUO (1988a), Regularity of the solution of elliptic problems with piecewise analytic data, Part 1: Boundary value problems for linear elliptic equations of the second order, *SIAM J. Math. Anal.* **19**, 172–203.

BABUŠKA, I. and B.Q. GUO (1988b), The *h-p* version of finite element method with curved boundary, *SIAM J. Math. Anal.* **25**, 837–861.

BABUŠKA, I. and B.Q. GUO (1989), The *h-p* version of finite element method for problems with non-homogeneous essential boundary condition, *Comput. Methods Appl. Math. Engrg.* **74**, 1–28.

BABUŠKA, I., B.Q. GUO and J. OSBORN (1989), Regularity and numerical solution of eigenvalue problems with piecewise analytic data.

BABUŠKA, I. and A. MILLER (1984), The post-processing approach in the finite element method, Part 2: The calculation of stress intensity factors, *Internat. J. Numer. Math. Engrg.* **20**, 1111–1129.

BABUŠKA, I. and J. OSBORN (1978), Numerical treatment of eigenvalue problems for differential equations with discontinuous coefficients, *Math. Comp.* **32**, 991–1023.

BABUŠKA, I. and J. OSBORN (1983), Generalized finite element methods and their relation to mixed methods, *SIAM J. Numer. Anal.* **20**, 510–536.

BABUŠKA, I. and J. OSBORN (1985), Finite element methods for the solution of problems with rough data, in: P. GRISVARD, W. WENDLAND and J.R. WHITMAN, eds., *Singularities and Constructive Methods for Their Treatment*, Lecture Notes in Mathematics **1121** (Springer, Berlin) 1–18.

BABUŠKA, I. and J. OSBORN (1987), Estimates for the errors in eigenvalue and eigenvector approximation by Galerkin methods, with particular attention to the case of multiple eigenvalues, *SIAM J. Numer. Anal.* **24**, 1249–1276.

BABUŠKA, I. and J. OSBORN (1989), Finite element–Galerkin approximation of the eigenvalues and eigenvectors of selfadjoint problems, *Math. Comp.* **52**, 275–297.

BABUŠKA, I., J. OSBORN and J. PITKÄRANTA (1980), Analysis of mixed methods using mesh dependent norms, *Math. Comp.* **35**, 1039–1062.

BABUŠKA, I. and E. RANK (1987), An expert-system-like feedback approach in the *h-p* version of finite element method, *Finite Elements Anal. Design* **3**, 127–147.

BABUŠKA, I. and T. SCAPOLLA (1987), The computational aspects of the *h*, *p* and *h-p* versions of finite element method, in: R. VICHNEVETSKY and R. STEPLEMAN, eds., *Advances in Computer Methods for Partial Differential Equations* **VI** (IMACS, New Brunswick, NJ), 233–240.

BABUŠKA, I. and M. SURI (1987a), The optimal convergence rate of the *p*-version of finite element method, *SIAM J. Numer. Anal.* **24**, 750–776.

BABUŠKA, I. and M. SURI (1987b), The *h-p* version of the finite element method with quasiuniform meshes, *RAIRO Modél. Math. Anal. Numér.* **21**, 119–238.

BAKER, G. (1973), Simplified proofs of error estimates for the least squares method for Dirichlet's problem, *Math. Comp.* **27**, 229–235.

BANERJEE, U. (1987), Approximation of eigenvalues of differential equations with non-smooth coefficients, *RAIRO Modél. Math. Anal. Numér.* **22**, 29–51.

BANERJEE, U. and J. OSBORN (1990), Estimates of the effect of numerical integration in finite element eigenvalue approximation.

BAZLEY, N.W. (1959), Lower bounds for eigenvalues with applications to the helium atom, *Proc. Nat. Acad. Sci. U.S.A.* **45**, 850–853.

BAZLEY, N.W. (1961), Lower bounds for eigenvalues, *J. Math. Mech.* **10**, 289–308.

BAZLEY, N.W. and D.W. FOX (1961a), Truncation in the method of intermediate problems for lower bounds to eigenvalues, *J. Res. Nat. Bur. Standards* **65B**, 105–111.

BAZLEY, N.W. and D.W. FOX (1961b), Lower bounds for eigenvalues of Schrödinger's equation, *Phys. Rev.* **124**, 483–492.

BAZLEY, N.W. and D.W. FOX (1963), Lower bounds for energy levels of molecular systems, *J. Math. Phys.* **4**, 1147–1153.

BIRKHOFF, G., C. DE BOOR, B.J. SWARTZ and B. WENDROFF (1966), Rayleigh–Ritz approximation by piecewise cubic polynomials, *SIAM J. Numer. Anal.* **3**, 188–203.

BRAMBLE, J.H. and J.E. OSBORN (1973), Rate of convergence estimates for nonselfadjoint eigenvalue approximations, *Math. Comp.* **27**, 525–549.

BRAMBLE, J.H. and A.H. SCHATZ (1970), Rayleigh–Ritz–Galerkin methods for Dirichlet's problem using subspaces without boundary conditions, *Comm. Pure Appl. Math.* **23**, 653–675.

BREZZI, F. (1974), On the existence uniqueness and approximation of saddle-point problems arising from Lagrangian multipliers, *RAIRO R2* **8**, 129–151.

BREZZI, F. and P. RAVIART (1977), Mixed finite element methods for 4th order elliptic equations, in: J. MILLER, ed., *Topics in Numerical Analysis* **III** (Academic Press, New York) 33–56.

BUTZER, P.L. and H. BERENS (1967), *Semi-Groups of Operators and Approximation*, Lecture Notes in Mathematics **145** (Springer, Berlin).

CANUTO, C. (1978), Eigenvalue approximations by mixed methods, *RAIRO Anal. Numér.* **12**, 27–50.

CHATELIN, F. (1973), Convergence of approximate methods to compute eigenelements of linear operators, *SIAM J. Numer. Anal.* **10**, 939–948.

CHATELIN, F. (1975), La méthode de Galerkin, Ordre de convergence des éléments propres, *C.R. Hebd. Séances Acad. Ser. A* **278**, 1213–1215.

CHATELIN, F. (1981), The spectral approximation of linear operators with applications to the computation of eigenelements of differential and integral operators, *SIAM Rev.* **23**, 495–522.

CHATELIN, F. (1983), *Spectral Approximations of Linear Operators* (Academic Press, New York).

CHATELIN, F. and J. LEMORDANT (1978), Error bounds in the approximation of eigenvalues of differential and integral operators, *J. Math. Anal. Appl.* **62**, 257–271.

CIARLET, P. (1978), *The Finite Element Method for Elliptic Problems* (North-Holland, Amsterdam).

CIARLET, P. and P. RAVIART (1974), A mixed finite element method for the biharmonic equation, in: C. DE BOOR, ed., *Symposium on Mathematical Aspects of Finite Elements in Partial Differential Equations* (Academic Press, New York) 125–143.

COLLATZ, L. (1948), *Eigenwertprobleme und Ihre Numerische Behandlung* (Chelsea, New York).

COURANT, R. (1927), Über direkte Methoden in der Variationsrechnung und über verwandte Fragen, *Math. Ann.* **97**, 711–736.

COURANT, R. (1943), Variational methods for the solution of problems of equilibrium and vibrations, *Bull. Amer. Math. Soc.* **49**, 1–23.

COURANT, R. and D. HILBERT (1953), *Methods of Mathematical Physics* **1** (Wiley-Interscience, New York).

CULLUM, J. and R.A. WILLOUGHBY (1985), *Lanczos Algorithms for Large Symmetric Eigenvalue Computations*, **I**: *Theory*; **II**: *Programs* (Birkhäuser, Boston, MA).

DESCLOUX, J., N. NASSIF and J. RAPPAZ (1978a), On spectral approximation, Part 1: The problem of convergence, *RAIRO Anal. Numér.* **12**, 97–112.

DESCLOUX, J., N. NASSIF and J. RAPPAZ (1978b), On spectral approximation, Part 2: Error estimates for the Galerkin method, *RAIRO Anal. Numér.* **12**, 113–119.

DUNFORD, N. and J.T. SCHWARTZ (1958), *Linear Operators*, **I**: *General Theory* (Wiley-Interscience, New York).

DUNFORD, N. and J.T. SCHWARTZ (1963), *Linear Operators*, **II**: *Spectral Theory, Selfadjoint Operators in Hilbert Spaces* (Wiley-Interscience, New York).

FALK, R. (1978), Approximation of the biharmonic equation by a mixed finite element method, *SIAM J. Numer. Anal.* **15**, 556–567.

FALK, R. and J. OSBORN (1980), Error estimates for mixed methods, *RAIRO Anal. Numer.* **14**, 249–277.

FIX, G.J. (1972), Effects of quadrature errors in finite element approximation of steady state, eigenvalue, and parabolic problems, in: A.K. AZIZ, ed., *The Mathematical Foundations of the Finite Element Method with Applications to Partial Differential Equations* (Academic Press, New York) 525–556.

FIX, G.J. (1973), Eigenvalue approximation by the finite element method, *Adv. in Math.* **10**, 300–316.

GLOWINSKI, R. (1973), Approximations externes par éléments finis de Lagrange d'ordre un et deux, du problème de Dirichlet pour l'opérateur biharmonique, Méthodes iteratives de résolutions des problèmes approchés, in: J.J.H. MILLER, ed., *Topics in Numerical Analysis* (Academic Press, New York) 123–171.

GRIGORIEFF, R.D. (1975a), Diskrete Approximation von Eigenwertproblemen, I: Qualitative Konvergenz, *Numer. Math.* **24**, 355–374.

GRIGORIEFF, R.D. (1975b), Diskrete Approximation von Eigenwertproblemen, II: Konvergenzordnung, *Numer. Math.* **24**, 415–433.

GRIGORIEFF, R.D. (1975c), Diskrete Approximation von Eigenwertproblemen, III: Asymptotische Entwicklung, *Numer. Math.* **25**, 79–97.

GRISVARD, P. (1985), *Elliptic Problems in Nonsmooth Domain* (Pitman, Boston, MA).

GUI, W. (1988), Hierarchical elements, local mappings and the *h-p* version of the finite element method, Part I, Part II, *J. Comp. Math.* **6**, 54–68, 142–156.

GUI, W. and I. BABUŠKA (1986), The *h, p, h-p* versions of finite element methods in one dimension, *Numer. Math.* **49**, 577–683.

GUO, B.Q. and I. BABUŠKA (1986), The *h-p* version of finite element method, Part 1: The basic approximation results, *Comp. Mech.* **1**, 21–41; Part 2: General results and application, *Comp. Mech.* **1**, 203–220.

HERRMANN, L. (1967), Finite element bending analysis for plates, *J. Engrg. Mech. Div. ASCE* **93**, 49–83.

HERSCH, J. (1955), Equations différentielles et fonctions des cellules, *C.R. Acad. Sci. Paris* **240**, 1602–1605.

HERSCH, J. (1963), Lower bounds for all eigenvalues by cell functions: A refined form of H.F. Weinberger's method, *Arch. Rational Mech. Anal.* **12**, 361–366.

HUBBARD, B.E. (1961), Bounds for eigenvalues of the free and fixed membrane by finite difference methods, *Pacific J. Math.* **11**, 559–590.

HUBBARD, B.E. (1962), Eigenvalues of the non-homogeneous rectangular membrane by finite difference methods, *Arch. Rational Mech. Anal.* **9**, 121–133.

ISHIHARA, K. (1977), Convergence of the finite element method applied to the eigenvalue problem $\Delta u+\lambda u=0$, *Publ. Inst. Math. Sci. Kyoto Univ.* **13**, 48–60.

ISHIHARA, K. (1978a), The buckling of plates by the mixed finite element method, *Mem. Numer. Math.* **5**, 73–82.

ISHIHARA, K. (1978b), A mixed finite element method for the biharmonic eigenvalue problem of plate bending, *Publ. Res. Inst. Math. Sci. Kyoto Univ.* **14**, 399–414.

JOHNSON, C. (1973), On the convergence of a mixed finite element method for plate bending problems, *Numer. Math.* **21**, 43–62.

KATO, T. (1958), Perturbation theory for nullity, deficiency and other quantities of linear operators, *J. Anal. Math.* **6**, 261–322.

KATO, T. (1966), *Perturbation Theory for Linear Operators*, Lecture Notes in Mathematics **132** (Springer, Berlin).

KELLOGG, R.B. and J.E. OSBORN (1975), A regularity result for the Stokes problem in a convex polygon, *J. Funct. Anal.* **21**, 397–431.

KNOPS, R.J. and L.E. PAYNE (1971), *Uniqueness Theorems in Linear Elasticity* (Springer, Berlin).

KOLATA, W. (1978), Approximation of variationally posed eigenvalue problems, *Numer. Math.* **29**, 159–171.

KONDRATÉV, V.A. (1968), Boundary problems for elliptic equations with conical or angular points, *Trans. Moscow Math. Soc.* **16** (1967) (translated by American Mathematical Society, Providence, RI).

KREISS, H.O. (1972), Difference approximation for boundary and eigenvalue problems for ordinary differential equations, *Math. Comp.* **26**, 605–624.

KUFNER, A. (1985), *Weighted Sobolev Spaces* (Wiley, New York).

KUTTLER, J.R. (1970a), Finite difference approximations for eigenvalues of uniformly elliptic operators, *SIAM J. Numer. Anal.* **7**, 206–232.

KUTTLER, J.R. (1970b), Upper and lower bounds for eigenvalues by finite differences, *Pacific J. Math.* **35**, 429–440.

LAX, P.D. and A.N. MILGRAM (1954), Parabolic equations, in: Annals of Mathematics Studies **33** (Princeton University Press, Princeton, NJ) 167–190.

MERCIER, B. (1974), Numerical solution of the biharmonic problems by mixed finite elements of class C^0, *Boll. Un. Mat. Ital.* **10**, 133–149.

MERCIER, B., J. OSBORN, J. RAPPAZ and D. RAVIART (1981), Eigenvalue approximation by mixed and hybrid methods, *Math. Comp.* **36**, 427–453.

MERCIER, B. and J. RAPPAZ (1978), Eigenvalue approximation via non-conforming and hybrid finite element methods, Rapport du Centre de Mathématiques Appliquées, École Polytechnique, Palaiseau, France.

MERIGOT, M. (1974), Regularité des functions propres du Laplacien dans un cone, *C.R. Acad. Sci. Paris Ser. A* **279**, 503–505.

MORREY, C.B. (1966), *Multiple Integrals in Calculus of Variations* (Springer, Berlin).

NEČAS, J. (1967), *Les Méthodes Directes en Théorie des Équations Elliptiques* (Masson, Paris).

NEČAS, J. and I. HLAVÁČEK (1970), On inequalities of Korn's type, II: Applications to linear elasticity, *Arch. Rational Mech. Anal.* **36**, 312–334.

NEMAT-NASSER, S. (1972a), General variational methods for elastic waves in composites, *J. Elasticity* **2**, 73–90.

NEMAT-NASSER, S. (1972b), Harmonic waves in layered composites, *J. Appl. Mech.* **39**, 850–852.

NEMAT-NASSER, S. (1974), *General Variational Principles in Non-linear and Linear Elasticity with Applications*, Mechanics Today **1** (Pergamon Press, New York) 214–261.

NOOR, A.K. (1985), Books and monographs on finite element technology, *Finite Elements Anal. Design* **1**, 101–111.

NOOR, A.K. and I. BABUŠKA (1987), Quality assessment and control of finite element solutions, *Finite Elements Anal. Design* **3**, 1–26.

ODEN, J. (1973), Some contributions to the mathematical theory of mixed finite element approximations, in: *Theory and Practice in Finite Element Structural Analysis* (University of Tokyo Press, Tokyo) 3–23.

ODEN, J.T. and G.T. CAREY (1982), *Finite Elements Mathematical Aspects* **IV** (Prentice-Hall, Englewood Cliffs, NJ).

ODEN, J.T. and J.N. REDDY (1976), *An Introduction to the Mathematical Theory of Finite Elements* (Wiley, New York).

OSBORN, J.E. (1975), Spectral approximation for compact operators, *Math. Comp.* **26**, 712–725.

OSBORN, J.E. (1979), Eigenvalue approximations by mixed methods, in: R. VICHNEVETSKY and R. STEPLEMAN, eds., *Advances in Computer Methods for Partial Differential Equations* **III** (IMACS, New Brunswick, NJ) 158–161.

OSBORN, J.E. (1981), The numerical solution of differential equations with rough coefficients, in: R. VICHNEVETSKY and R. STEPLEMAN, eds., *Advances in Computer Methods for Partial Differential Equations* **IV** (IMACS, New Brunswick, NJ) 9–13.

PARLETT, B.N. (1980), *The Symmetric Eigenvalue Problem* (Prentice-Hall, Englewood Cliffs, NJ).

PIERCE, J.G. and R.S. VARGA (1972), Higher order convergence results for the Rayleigh–Ritz method applied to eigenvalue problems, 2: Improved bounds for eigenfunctions, *Numer. Math.* **19**, 155–169.

POLYA, G. (1952), Sur une interprétation de la méthode des différences finies qui peut fournir des bornes supérieures ou inférieures, *C.R. Acad. Sci. Paris* **235**, 995–997.

PRODI, G. (1962), Theoremi di tipo locale per il sistema de Navier-Stokes e stabilitá della soluzioni stationarie, *Rend. Sem. Mat. Univ. Padova* **32**, 374–397.

PROSDORF, S. and G. SCHMIDT (1981), A finite element collocation method for singular integral equations, *Math. Nachr.* **100**, 33–60.

RAVIART, P.A. and J.M. THOMAS (1977), *A Mixed Finite Element Method for Second Order Elliptic Problems*, Lecture Notes in Mathematics **606** (Springer, Berlin) 292–315.

SCHOLZ, R. (1976), Approximation von Sattelpunkten mit finiten Elementen, Tagungsband, *Bonn. Math. Schr.* **89**, 53–66.

STRANG, G. and G.J. FIX (1973), *An Analysis of the Finite Element Method* (Prentice-Hall, Englewoo Cliffs, NJ).

STUMMEL, F. (1977), Approximation methods for eigenvalue problems in elliptic differential equations, in: E. BOHL, L. COLLATZ and K.P. HADELER, eds. *Numerik und Anwendungen von Eigenwertaufgaben und Verzweigungsproblemen* (Birkhäuser, Basel) 133–165.

SYNGE, J.L. and B.A. GRIFFITH (1959), *Principles of Mechanics* (McGraw-Hill, New York, 34th ed.).

SZABO, B.A. (1985), PROBE: Theoretical Manual, NOETIC Tech., St. Louis, MO.

SZABO, B.A. (1986), Mesh design for the p-version of the finite element method, *Comput. Methods Appl. Math. Engrg.* **55**, 181–197.

SZABO, B.A. and I. BABUŠKA (1986), Computation of the amplitude of stress, singular terms for cracks and reentrant corners, in: T.A. CRUSE, ed., *Fracture Mechanics, 19th Symposium ASTM*, STP 1969 (American Society Test. and Mat., Philadelphia, PA) 101–126.

VAĬNIKKO, G.M. (1964), Asymptotic error bounds for projection methods in the eigenvalue problem, *U.S.S.R. Comput. Math. and Math. Phys.* **4**, 9–36.

VAĬNIKKO, G.M. (1967), Rapidity of convergence of approximation methods in the eigenvalue problem, *U.S.S.R. Comput. Math. and Math. Phys.* **7**, 18–32.

VAĬNIKKO, G.M. (1970), On the rate of convergence of certain approximation methods of Galerkin type in an eigenvalue problem, *Amer. Math. Soc. Transl.* **36**, 249–259.

WEINBERGER, H.F. (1952a), Error estimation in the Weinstein method for eigenvalues, *Proc. Amer. Math. Soc.* **3**, 643–646.

WEINBERGER, H.F. (1952b), An optimum problem in the Weinstein method for eigenvalues, *Pacific J. Math.* **2**, 413–418.

WEINBERGER, H.F. (1956), Upper and lower bounds for eigenvalues by finite difference methods, *Comm. Pure Appl. Math.* **9**, 613–623; also in: *Proceedings Conference on Partial Differential Equations*, Berkeley, CA (1955).

WEINBERGER, H.F. (1958), Lower bounds for higher eigenvalues by finite difference methods, *Pacific J. Math.* **8**, 339–368.

WEINBERGER, H.F. (1959), A theory of lower bounds for eigenvalues, Tech Note BN-183, Institute for Fluid Dynamics and Applied Mathematics, University of Maryland, College Park, MD.

WEINBERGER, H.F. (1960), Error bounds in the Rayleigh Ritz approximation of eigenvectors, *J. Res. Nat. Bur. Standards* **64B**, 217–225.

WEINBERGER, H.F. (1974), *Variational Methods for Eigenvalue Approximation*, Regional Conference Series in Applied Mathematics **15** (SIAM, Philadelphia, PA).

WEINSTEIN, A. (1935), Sur la stabilité des plaques encastrées, *C.R. Acad. Sci. Paris* **200**, 107–109.

WEINSTEIN, A. (1937), Étude des spectres des équations aux dérivées partielles de la théorie des plaques élastiques, *Mém. Sci. Math.* **88**.

WEINSTEIN, A. (1953), Variational methods for the approximation and exact computation of eigenvalues, *Nat. Bur. Standards Appl. Math. Ser.* **29**, 83–89.

WEINSTEIN, A. (1963), On the Sturm–Liouville theory and the eigenvalues of intermediate problems, *Numer. Math.* **5**, 238–245.

WEINSTEIN, A. and W. STENGER (1972), *Methods for Intermediate Problems for Eigenvalues* (Academic Press, New York).
WEINSTEIN, D.H. (1934), Modified Ritz method, *Proc. Nat. Acad. Sci. U.S.A.* **20**, 529–532.
WOZNIAKOWSKI, H.A. (1985), A survey of information-based complexity, *J. Complexity* **1**, 11–44.

Subject Index

Evolution Problems

Hiroshi Fujita

Department of Mathematics,
School of Science and Technology,
Meiji University,
Higashimata, Tamaku,
Kawasaki-shi 214, Japan

Takashi Suzuki

Department of Mathematics,
Faculty of Science,
Tokyo Metropolitan University,
Fukakusa, Setagayaku,
Tokyo 158, Japan

HANDBOOK OF NUMERICAL ANALYSIS, VOL. II
Finite Element Methods (Part 1)
Edited by P.G. Ciarlet and J.L. Lions

Contents

CHAPTER I

Elliptic Boundary Value Problems and Finite Element Methods: A Review and Remarks

The purpose of this article is to present a mathematical study of the finite element method applied to evolution problems such as parabolic or hyperbolic initial value problems. Our analysis is basically operator-theoretical, namely, from the viewpoint of evolution equations in Banach spaces. Therefore, our arguments on approximation, as well as those on the original problem in the continuous version, heavily depend on the study of elliptic boundary value problems and their finite element approximations. Those elliptic problems have been described in detail in the first article of this volume. However it seems to be convenient for later quotation to review here some fundamental facts together with some advanced results concerning them which are directly connected with our study of evolution problems.

1. Review of elliptic boundary value problems

For the sake of simplicity, we suppose that $\Omega \subset \mathbb{R}^2$ is a polygon or a bounded domain with smooth boundary $\partial\Omega$, and consider the Poisson equation

$$-\Delta u = f \quad \text{in } \Omega \tag{1.1}$$

with the Dirichlet boundary condition

$$u = 0 \quad \text{on } \partial\Omega. \tag{1.2}$$

By a well-known argument, this boundary value problem is reduced to a weak form, that is, the variational problem in $V = H_0^1(\Omega)$. Find $u \in V$ such that

$$a(u, v) = (f, v), \quad v \in V. \tag{1.3}$$

Here, $(,)$ and $a(,)$ denote the L^2 inner product and the Dirichlet form respectively. Namely,

$$(f, g) = \int_\Omega f(x)\overline{g(x)}\,\mathrm{d}x, \quad f, g \in L^2(\Omega),$$

$$a(u, v) = \int_\Omega \nabla u(x) \cdot \nabla v(x)\, dx, \quad u,v \in H^1(\Omega).$$

Actually in this article, unless otherwise stated we assume that functions are complex-valued and that Hilbert spaces are complex spaces.

We note that $u \in H_0^1(\Omega)$ implies $u|_{\partial\Omega} \equiv$ trace of u to $\partial\Omega = 0$. Then, supposing that u is smooth, we have by Green's formula that

$$a(u, v) = \int_\Omega \nabla u \cdot \overline{\nabla v}\, dx = -\int_\Omega \Delta u\, \bar{v}\, dx = (f, v),$$

since $v \in H_0^1(\Omega)$. Hence the identity (1.1) follows formally from the arbitrariness of v.

Henceforth, various generic constants are denoted indifferently by C. If it depends on some parameters, say $\alpha, \beta, \ldots$, we shall denote it by $C_{\alpha,\beta}, \ldots$. Then by means of Poincaré's inequality

$$\|v\|_{L^2} \leqslant C_\Omega \|\nabla v\|_{L^2}, \quad v \in V, \tag{1.4}$$

where $a(\,,\,)$ may be regarded as an inner product in $V = H_0^1(\Omega)$. Let V^* be the adjoint space of V. Then, regarding $f \in L^2(\Omega)$ as an element in V^* through ${}_{V^*}\langle f, v\rangle_V = (f, v)$ for $v \in V$, we can verify the unique solvability of (1.3) from Riesz's representation theorem. In the case that Ω has a smooth boundary or is a convex polygon, the solution $u \in V$ belongs to $H^2(\Omega)$ when $f \in X = L^2(\Omega)$ and becomes a strong solution of (1.1) with (1.2). (See, AGMON [1965], LIONS and MAGENES [1968], or GRISVARD [1985], for instance.)

We proceed to the case of $V = H^1(\Omega)$ in (1.1). Supposing that u is smooth, we obtain by Green's formula that

$$a(u, v) = \int_\Omega \nabla u \cdot \overline{\nabla v}\, dx = \int_{\partial\Omega} (\partial u/\partial n) \cdot \bar{v}\, dS - \int_\Omega \Delta u \cdot \bar{v}\, dx = (f, v).$$

Here $\partial/\partial n$ is the differentiation along the outer unit normal vector $n = (n_1, n_2)$ on $\partial\Omega$ and dS denotes the surface element of $\partial\Omega$. Taking $v \in H_0^1(\Omega)$, we have $\int_{\partial\Omega} (\partial u/\partial n) \cdot \bar{v}\, dS = 0$ so that the identity of (1.3) follows. Therefore,

$$\int_{\partial\Omega} (\partial u/\partial n) \cdot \bar{v}\, dS = 0$$

holds for every $v \in H^1(\Omega)$, hence $\partial u/\partial n = 0$ on $\partial\Omega$. In other words, the Neumann problem which is composed of

$$-\Delta u = f \quad \text{in } \Omega, \tag{1.1$'$}$$

$$\partial u/\partial n = 0 \quad \text{on } \partial\Omega \tag{1.2$'$}$$

is reduced to the variational problem (1.3) for $V = H'(\Omega)$.

Unfortunately, since Poincaré's inequality (1.4) does not hold for $V=H^1(\Omega)$, $a(\,,)$ is not an inner product in $H^1(\Omega)$. Actually, the Neumann problem (1.1′) with (1.2′) is not uniquely solvable. For instance, any constant function $u\equiv c\in\mathbb{R}$ satisfies (1.1′) with (1.2′) for $f=0$.

To generalize the consideration above, let $a_{ij}(x)=a_{ji}(x)$, $b_j(x)$ and $c(x)$ be real smooth functions on $\bar{\Omega}$ and suppose that the uniform ellipticity

$$\sum_{i,j=1}^{2} a_{ij}(x)\xi_i\xi_j \geqslant \delta_1|\xi|^2, \qquad \xi=(\xi_1,\xi_2)\in\mathbb{R}^2, \quad x\in\bar{\Omega}, \tag{1.5}$$

holds with a constant $\delta_1>0$. We put

$$\mathscr{A}=\mathscr{A}(x,D)=-\sum_{i,j=1}^{2}\frac{\partial}{\partial x_i}a_{ij}(x)\frac{\partial}{\partial x_j}+\sum_{j=1}^{2}b_j(x)\frac{\partial}{\partial x_j}+c(x).$$

Then, the boundary value problem for

$$\mathscr{A}u=f\in X \quad \text{in } \Omega \tag{1.6}$$

with

$$u=0 \quad \text{on } \partial\Omega \tag{1.7}$$

is reduced to the variational problem (1.3) if we take $V=H_0^1(\Omega)$ and

$$a(u,v)=\int_{\Omega}\left\{\sum_{i,j=1}^{2}a_{ij}(x)\frac{\partial}{\partial x_j}u\,\overline{\frac{\partial}{\partial x_i}v}+\sum_{j=1}^{2}b_j(x)\frac{\partial}{\partial x_j}u\cdot\bar{v}+c(x)u\bar{v}\right\}\mathrm{d}x.$$

Similarly, if $\sigma(\xi)$ is a smooth function on $\partial\Omega$ and if we write

$$\frac{\partial}{\partial\nu_{\mathscr{A}}}=\sum_{i,j=1}^{2}n_i(x)a_{ij}(x)\frac{\partial}{\partial x_j}$$

to denote the outer conormal differentiation associated with $\mathscr{A}$, then the boundary value problem (1.6) with

$$\partial u/\partial\nu_{\mathscr{A}}+\sigma u=0 \quad \text{on } \partial\Omega \tag{1.7′}$$

is reduced to the variational problem (1.3) with $V=H^1(\Omega)$ and

$$a(u,v)=\int_{\Omega}\left\{\sum_{i,j=1}^{2}a_{ij}(x)\frac{\partial}{\partial x_j}u\,\overline{\frac{\partial}{\partial x_i}v}+\sum_{j=1}^{2}b_j(x)\frac{\partial}{\partial x_j}u\cdot\bar{v}+c(x)u\bar{v}\right\}\mathrm{d}x+\int_{\partial\Omega}\sigma u\bar{v}\,\mathrm{d}S,$$

$$u,v\in V.$$

The sesquilinear form $a(\,,)$ which appears above is bounded in the sense that

$$|a(u,v)|\leqslant C\|u\|_V\cdot\|v\|_V, \quad u,v\in V. \tag{1.8}$$

Here we note that from the trace theorem (e.g. LIONS and MAGENES [1968], MIZOHATA [1973]) the following inequality follows

$$\|v|_{\partial\Omega}\|_{L^2(\partial\Omega)} \leqslant \varepsilon\|v\|_{H^1(\Omega)} + C_\varepsilon\|v\|_{L^2(\Omega)}, \quad v \in H^1(\Omega)$$

for each $\varepsilon > 0$ with a constant $C_\varepsilon > 0$. Therefore, by the uniform ellipticity (1.5), it holds furthermore that for each δ, $0 < \delta < \delta_1$, there exists a $\lambda \in \mathbb{R}$ such that

$$\operatorname{Re} a(u, u) \geqslant \delta\|u\|_V^2 - \lambda\|u\|_X^2, \quad u \in V. \tag{1.9}$$

If $b_j \equiv 0$, $c \geqslant 0$ and $V = H_0^1(\Omega)$, or if $b_j \equiv 0$, $c > 0$, $\sigma \geqslant 0$ and $V = H^1(\Omega)$, then we may take $\lambda = 0$ in (1.9) In this case the variational problem (1.3) is uniquely solvable by Lax–Milgram's theorem (see YOSIDA [1964]).

Recall that V^* denotes the adjoint space of V. Then, the boundedness (1.8) of $a(\,,)$ implies the well-definedness of a bounded operator $A: V \to V^*$ through the identity

$$a(u, v) = {}_{V^*}\langle Au, v\rangle_V, \quad u, v \in V.$$

On the other hand, identifying X^* with X through Riesz's representation theorem, we get a triple of Hilbert spaces $V \subset X \subset V^*$ with continuous and dense inclusions. Setting $D(A) = \{u \in V: Au \in X\}$, we write the restriction of $A: V \to V^*$ to $D(A)$ again by A, which is regarded as an operator in X. Then, (1.3) is simply expressed as an abstract equation in X, that is

$$Au = f. \tag{1.10}$$

Again, when Ω has a smooth boundary or is a convex polygon, it is known that

$$D(A) = H^2(\Omega) \cap H_0^1(\Omega)$$

for the boundary condition (1.7) and

$$D(A) = H^2(\Omega) \cap \{\partial v/\partial v_{\mathscr{A}} + \sigma v = 0 \text{ on } \partial\Omega\}$$

for the boundary condition (1.7′) (see AGMON [1965], LIONS and MAGENES [1968], GRISVARD [1985]). Therefore, in those cases we obtain the strong solution $u \in H^2(\Omega)$ of (1.6) with (1.7) (or (1.7′)), provided that $\lambda = 0$ in (1.9).

The operator A in X which arises as above from the sesquilinear form $a(\,,)$ on V is called an m-sectorial operator in view of the following property (see KATO [1966]). For simplicity, let us suppose $\lambda = 0$ in (1.9) and specify the constant C in (1.8) as C_1. Taking θ in $0 < \theta < \frac{1}{2}\pi$ so that $\cos\theta = \delta/C_1$, we set:

$$\Sigma_\theta = \{z \in \mathbb{C}: 0 \leqslant |\arg z| \leqslant \theta\}.$$

Then, we have $\Sigma_\theta \supset \sigma(A) =$ "the spectrum of A", namely,

$$\mathbb{C}\backslash\Sigma_\theta \subset \rho(A) \equiv \text{the resolvent set of } A.$$

Furthermore, for each θ_1, $\theta < \theta_1 < \frac{1}{2}\pi$, the estimate

$$\|(z - A)^{-1}\| \leqslant \frac{1}{\sin(\theta_1 - \theta)} \cdot \frac{1}{|z|}, \quad z \in G_{\theta_1} \equiv \{z \in \mathbb{C}: \theta_1 \leqslant |\arg z| \leqslant \pi\}$$

holds, which implies that $-A$ generates a holomorphic semigroup $\{e^{-tA}\}_{t\geqslant 0}$ in X as is discussed in detail in the next chapter.

Here, we recall the following notation, commonly used in the semigroup theory: A densely defined closed operator A in a Banach space X is said to be of type (θ, M), $0<\theta<\pi$ and $M\geqslant 1$, if

$$\mathbb{C}\backslash\Sigma_\theta \subset \rho(A), \qquad \|\lambda(\lambda-A)^{-1}\|\leqslant M, \quad \lambda<0,$$
$$\|(\lambda-A)^{-1}\|\leqslant M_\varepsilon/|\lambda|, \quad \pi\geqslant|\arg\lambda|\geqslant\theta+\varepsilon>\theta,$$

with a constant $M_\varepsilon>0$. In this terminology, an m-sectorial operator A is of type (θ, M) for some θ, $0<\theta<\frac{1}{2}\pi$.

2. Review of finite element methods for elliptic problems

As in the preceding section, let $V\subset X\subset V^*$ be a triple of Hilbert spaces and let $a(\,,)$ be a sesquilinear form on V. Assume that $a(\,,)$ is bounded:

$$|a(u, v)|\leqslant C\|u\|_V\|v\|_V, \quad u,v\in V, \tag{2.1}$$

and is "strongly coercive", that is, (1.9) holds for $\lambda = 0$:

$$\operatorname{Re} a(u, v)\geqslant\delta\|u\|_V^2, \quad u\in V, \tag{2.2}$$

where $\delta>0$ is constant. Then for any given $f\in X$ the variational problem,

Find $u\in V$ such that

$$a(u, v)=(f, v), \quad v\in V, \tag{2.3}$$

is uniquely solvable.

The Ritz–Galerkin method is the most classical approximating method for (2.3) and consists of the following procedure. Namely, we prepare a family of finite-dimensional subspaces $\{V_h\}_{h>0}$ of V with the property that $V_h\to V$ as $h\downarrow 0$ in some sense. Then, the variational problem,

Find the $u_h\in V_h$ such that

$$a(u_h, \chi)=(f, \chi), \quad \chi\in V_h, \tag{2.4}$$

is regarded as an approximation of (2.3). The unique solvability of (2.4) follows from the same reasoning as for (2.3). Let X_h be the space V_h with the topology induced by the inner product in X and let P_h: $X\to X_h$ be the orthogonal projection. Then, the variational problem (2.4) is equivalent to the equation

$$A_h u_h = P_h f \tag{2.5}$$

in X_h, which is of finite dimension.

Let us refer to the stability of the approximate solution u_h and to the estimate of the error $e_h = u_h - u$. To this end, we introduce the so-called Ritz operator $R_h: V \to V_h$ through the relation

$$a(R_h u, \chi) = a(u, \chi), \qquad u \in V, \quad \chi \in V_h. \tag{2.6}$$

The well-definedness of R_h again follows from the Lax–Milgram theorem in the same way as for the unique solvability of (2.4). If $u \in V$ is the solution of (1.2), then the approximate solution u_h of (2.4) is nothing but $R_h u$. Therefore, we have $R_h A^{-1} = A_h^{-1} P_h$. Furthermore, in the case of symmetric $a(\,,)$, namely, if $a^* = a$, where

$$a^*(u, v) = \overline{a(v, u)}, \quad u, v \in V,$$

the operator $R_h: V \to V_h$ is the orthogonal projection with respect to the inner product $a(\,,)$. Thus, in this case, we have

$$a(R_h u, R_h u) \leqslant a(u, u)$$

and

$$a(R_h u - u, R_h u - u) \leqslant a(\chi - u, \chi - u), \quad \chi \in V_h.$$

Again by the boundedness (2.1) and the strong coerciveness (2.2) of $a(\,,)$ we get the stability of approximate solutions

$$\|R_h u\|_V \leqslant C\|u\|_V \tag{2.7}$$

and the error estimate

$$\|R_h v - u\|_V \leqslant C \inf_{\chi \in V_h} \|\chi - u\|_V. \tag{2.8}$$

Actually, in the general case that $a^* \neq a$, these relations also hold true. In fact, from (2.2) and (2.3) we obtain

$$\begin{aligned} \delta\|R_h u\|_V^2 &\leqslant \operatorname{Re} a(R_h u, R_h u) = \operatorname{Re} a(u, R_h u) \\ &\leqslant C\|R_h u\|_V \cdot \|u\|_V \end{aligned}$$

and

$$\begin{aligned} \delta\|R_h u - u\|_V^2 &\leqslant \operatorname{Re} a(R_h u - u, R_h u - u) = \operatorname{Re} a(R_h u - u, \chi - u) \\ &\leqslant C\|R_h u - u\|_V \cdot \|\chi - u\|_V, \quad \chi \in V_h. \end{aligned}$$

Hence (2.7) and (2.8) follow.

Now, we proceed to the finite element approximation, and set $X = L^2(\Omega)$, $V = H_0^1(\Omega)$ or $H^1(\Omega)$. Let $f \in X$ and $a(\,,)$ be the sesquilinear form associated with an elliptic boundary value problem in the preceding section. The finite element method is a kind of Ritz–Galerkin approximation for the variational problem (2.3). In the present article, we mainly exemplify the method through the simplest finite elements.

Namely, in the case that Ω is a polygon, we divide it into small triangles with size parameter $h > 0$. Let τ_h be the triangulation. Then each $T \in \tau_h$ is called an element. $\rho(T)$ and $\sigma(T)$ mean the radii of the inscribed and the outscribed circles of T, respectively. The family of triangulations $\{\tau_h\}_{h>0}$ is said to be regular if there exists a constant $\nu>0$ such that

$$\rho(T)/\sigma(T)\geqslant \nu, \qquad T\in\tau_h, \quad h>0.$$

V_h denotes the totality of continuous piecewise functions in V which are linear on each $T\in\tau_h$. Obviously,

$$V_h\subset V \tag{2.9}$$

and

$$V_h \text{ is of finite dimension.} \tag{2.10}$$

For each continuous function $u\in V$, there exists a unique element in V_h whose values at the vertices of $T\in\tau_h$ are equal to those of u. That function is called the interpolant of u and is denoted by $\pi_h u$. The following inequality is due to BRAMBLE and HILBERT [1970]:

$$\|\pi_h u-u\|_{L^2(T)}+h\|\nabla(\pi_h u-u)\|_{L^2(T)}\leqslant Ch^2\|u\|_{H^2(T)} \tag{2.11}$$
$$u\in V\cap H^2(\Omega), \quad T\in\tau_h.$$

When Ω has a curved boundary, we must modify $\pi_h u$ near the boundary. Such a modification has been given by ZLÁMAL [1973], which assures us of the existence of V_h and π_h subject to (2.9)–(2.11).

From the general theory developed above the stability (2.7) of the approximate solution $u_h=R_h u$ and the estimate (2.8) of the error $e_h=u_h-u$ follows. In particular, we have

$$\|R_h u-u\|_{H^1}\leqslant Ch\|u\|_{H^2}, \quad u\in H^2(\Omega)\cap V. \tag{2.12}$$

When Ω has a smooth boundary or is a convex polygon, the elliptic estimate

$$\|A^{-1}f\|_{H^2}\leqslant C\|f\|_{L^2}, \quad f\in X, \tag{2.13}$$

holds true. In this case, the following L^2 error estimate can be obtained:

$$\|R_h u-u\|_{L^2}\leqslant Ch^2\|u\|_{H^2} \quad u\in H^2(\Omega)\cap V. \tag{2.14}$$

In fact, putting $e_h=R_h u-u$, we have

$$\begin{aligned}\|e_h\|_{L^2}^2&=a(e_h, A^{*-1}e_h)=a(R_h u-u, A^{*-1}e_h)\\ &=a(R_h u-u, A^{*-1}e_h-\pi_h A^{*-1}e_h)\leqslant C\|R_h u-u\|_V\cdot\|(\pi_h-1)A^{*-1}e_h\|_V\\ &\leqslant Ch^2\|u\|_{H^2}\|A^{*-1}e_h\|_{H^2}\leqslant Ch^2\|u\|_{H^2}\|e_h\|_{L^2}.\end{aligned}$$

The argument here is sometimes referred to as Nitsche's trick (CIARLET [1978]). Similarly we can show that

$$\|R_h u-u\|_{L^2}\leqslant Ch\|u\|_{H^1}, \quad u\in V. \tag{2.15}$$

As we have seen in the preceding section, the relation $R_h A^{-1}=A_h^{-1}P_h$ holds so that (2.14) implies

$$\|A_h^{-1}P_h-A^{-1}\|\leqslant Ch^2. \tag{2.16}$$

We should note that this inequality is weaker than (2.14) in the case of $V=H^1(\Omega)$ where $D(A)$ involves the boundary condition: $D(A)=H^2(\Omega)\cap\{\partial u/\partial\nu_{\mathscr{A}}+\sigma u=0$ on $\partial\Omega\}$.

Henceforth, the family of triangulations $\{\tau_h\}_{h>0}$ is always supposed to be regular.

3. Inverse assumptions and their consequences

We say that the family $\{\tau_h\}_{h>0}$ of triangulations satisfies the inverse assumption when there exists a constant $\nu>0$ such that

$$\rho(T)\geqslant \nu h \qquad T\in\tau_h, \quad h>0$$

(see CIARLET [1978]). Under this assumption, the following inequality holds for $1\leqslant p\leqslant\infty$:

$$\|\chi\|_{L^\infty(T)}\leqslant Ch^{-2/p}\|\chi\|_{L^p(T)}, \qquad T\in\tau_h, \quad \chi\in V_h, \quad h>0. \tag{3.1}$$

We note that its converse always holds true. Furthermore, we have

$$\|\nabla\chi\|_{L^2(T)}\leqslant Ch^{-1}\|\chi\|_{L^2(T)}, \qquad T\in\tau_h, \quad \chi\in V_h, \quad h>0, \tag{3.2}$$

and, hence $\|\nabla\chi\|_{L^2}\leqslant Ch^{-1}\|\chi\|_{L^2}$.

From these inequalities we can deduce the following propositions:

PROPOSITION 3.1. *There exists a constant $\beta>0$ such that*

$$\|A_h\|\leqslant\beta h^{-2}. \tag{3.3}$$

PROOF. In the case that $a(\,,)$ is strongly coercive and symmetric, we have for $\chi\in V_h$ that

$$\|A_h^{1/2}\chi\|_X^2=(A_h\chi,\chi)=a(\chi,\chi),$$

so that

$$C^{-1}\|\chi\|_V\leqslant\|A_h^{1/2}\chi\|_X\leqslant C\|\chi\|_V, \qquad \chi\in V_h, \quad h>0,$$

where $C>0$ is independent of $h>0$. Therefore for $u,v\in V_h$ we get from (3.2) that

$$|(A_hu,v)|=|(A_h^{1/2}u, A_h^{1/2}v)|$$
$$\leqslant C\|u\|_V\cdot\|v\|\leqslant Ch^{-2}\|u\|_X\|v\|_X,$$

hence (3.3) follows.

In the general case, we break $a(\,,)$ as

$$a(u,v)=a^0(u,v)+(Bu,v), \quad u,v\in V,$$

where $a^0(\,,)$ is strongly coercive and symmetric and $B: V\to X$ is bounded. Then, we can write

$$A_h=A_h^0+B_h,$$

where A_h^0 is associated with $a^0|_{V_h\times V_h}$ and $B_h=P_hB|_{V_h}$. The inequality (3.3) follows from

$$\|A_h^0\|\leqslant Ch^{-2},$$
$$\|B_h\chi\|_X\leqslant C\|\chi\|_V\leqslant Ch^{-1}\|\chi\|_X,\quad \chi\in V_h. \qquad \square$$

PROPOSITION 3.2. *When Ω has a smooth boundary or is a convex polygon, the L^2 orthogonal projection $P_h: X\to X_h$ satisfies the estimate*

$$\|P_hv\|_V\leqslant C\|v\|_V,\quad v\in V. \tag{3.4}$$

PROOF. Since $\|P_h\|_{X\to X}\leqslant 1$, we have only to show that

$$\|\nabla(P_h-1)v\|_{L^2}\leqslant C\|v\|_{H^1},\quad v\in V.$$

To this end let $R_h: V\to V_h$ be the Ritz operator associated with

$$a(u,v)=\int_\Omega(\nabla u\cdot\overline{\nabla v}+u\bar{v})\,\mathrm{d}x,\quad u,v\in V.$$

Since $P_h|_{V_h}=$identity, we have for $\chi=R_hv\in V_h$ that

$$\begin{aligned}\|\nabla(P_h-1)v\|_{L^2}&=\|\nabla(P_h-1)(v-\chi)\|_{L^2}\leqslant\|\nabla P_h(v-\chi)\|_{L^2}+\|\nabla(v-\chi)\|_{L^2}\\&\leqslant Ch^{-1}\|P_h(v-\chi)\|_{L^2}+\|v\|_{H^1}+\|\chi\|_{H^1}\\&\leqslant Ch^{-1}\|v-\chi\|_{L^2}+\|v\|_{H^1}+\|\chi\|_{H^1}.\end{aligned}$$

As we have seen in the preceding section, the inequalities

$$\|\chi\|_{H^1}\leqslant\|v\|_{H^1},\qquad \|\chi-v\|_{L^2}\leqslant Ch\|v\|_{H^1}$$

hold. Hence (3.4) has been established. $\square$

The following lemma is due to Descloux. For the proof, see THOMÉE [1984, p. 64], for instance.

LEMMA 3.1. *When $\{\tau_h\}_{h>0}$ satisfies the inverse assumption, there exists a constant $C>0$ such that for each $T_0\in\tau_h$ and a subdomain $\Omega_0\subset\Omega$ with $\bar{T}_0\cap\bar{\Omega}_0=\emptyset$, the function $v\in L^2(\Omega)$ with $\operatorname{supp} v\subset T_0$ satisfies the inequality*

$$\|P_hv\|_{L^2(\Omega_0)}\leqslant\exp(-C\operatorname{dist}(\Omega_0,T_0)h^{-1})\|v\|_{L^2}. \tag{3.5}$$

Lemma 3.1 implies the L^q boundedness of the L^2 orthogonal projection $P_h: X\to X_h$:

PROPOSITION 3.3. *For $1\leqslant q\leqslant\infty$, the estimate*

$$\|P_h\|_{L^q\to L^q}\leqslant C \tag{3.6}$$

holds true.

PROOF. First, we note the identity

$$_{L^1}\langle P_h u, v\rangle_{L^\infty} = {}_{L^1}\langle u, P_h v\rangle_{L^\infty}$$

for $u \in L^2(\Omega)$ and $v \in L^\infty(\Omega)$. If (3.6) is shown for $q=\infty$, the equality above implies that (3.6) holds also for $q=1$, and hence for $1 \leqslant q \leqslant \infty$ by Riesz–Thorin's interpolation theorem.

Suppose that v is real-valued and let the maximum of $P_h v$ be attained on $\bar{T}_0$, $T_0 \in \tau_h$. For each $T \in \tau_h$, we set

$$v_T = \begin{cases} v & \text{on } T, \\ 0 & \text{otherwise.} \end{cases}$$

Then, v is decomposed as $v = \sum_{T \in \tau_h} v_T$, so that

$$\|P_h v\|_{L^\infty} = \|P_h v\|_{L^\infty(T_0)} \leqslant \sum_T \|P_h v_T\|_{L^\infty(T_0)}.$$

By virtue of Descloux's lemma, we have

$$\begin{aligned} \|P_h v_T\|_{L^\infty(T_0)} &\leqslant Ch^{-1}\|P_h v_T\|_{L^2(T_0)} \\ &\leqslant Ch^{-1}\exp(-C\,\mathrm{dist}(T_0, T)h^{-1})\|v_T\|_{L^2(T)} \\ &\leqslant C\exp(-C\,\mathrm{dist}(T_0, T)h^{-1})\|v_T\|_{L^\infty}. \end{aligned}$$

Now, starting from $R_0 = T_0$ we define subsets $R_j \subset \Omega$, $j = 0, 1, 2, \ldots$, recursively as follows. Namely, R_k is the union of (closed) triangles in τ_h which are not in but adjacent to $\bigcup_{l<k} R_l$. Then, the points of R_k have a distance to T_0 which is of order kh. Furthermore, the number of triangles of R_k is bounded by the number of triangles of $\bigcup_{l \leqslant k} R_l$, which is bounded by Ck^2. In this way we have

$$\begin{aligned} \|P_h v\|_{L^\infty} &\leqslant C\left(\sum_k \sum_{T \ni R_k} \mathrm{e}^{-Ck}\right)\|v\|_{L^\infty} \\ &\leqslant C\left(\sum_k k^2 \mathrm{e}^{-Ck}\right)\|v\|_{L^\infty} = C\|v\|_{L^\infty}. \end{aligned}$$

For other inequalities related to the inverse assumption we refer to CIARLET [1978]. □

4. Some L^∞ estimates

We have shown that the Ritz operator $R_h : V \to V_h$ associated with the finite element method described in Section 2 satisfies

$$\|R_h u\|_{H^1} \leqslant C\|u\|_{H^1}, \quad u \in V, \tag{4.1}$$

and, under the elliptic estimate (2.13), also

$$\|R_h u - u\|_{H^j} \leqslant Ch^{k-j}\|u\|_{H^k}, \quad v \in H^k(\Omega) \cap V$$

for $j=0$, 1 and $k=1,2$. Unfortunately, the L^2 stability

$$\|R_h u\|_{L^2} \leqslant C\|u\|_{L^2}$$

cannot be expected. In fact, if this were true, we should have from $R_h A^{-1} = A_h^{-1} P_h$ that $\|\overline{A_h^{-1} P_h A}\|_{X \to X} \leqslant C$ and hence the following type of estimate by taking the adjoint operator:

$$\|AA_h^{-1} P_h\|_{X \to X} \leqslant C,$$

which is impossible because of $V_h \not\subset D(A)$.

The following inequality has been obtained by NITSCHE [1975] and SCOTT [1976] under the inverse assumption for the case $V = H_0^1(\Omega)$ and $a(u,v) = \int_\Omega \nabla u \cdot \overline{\nabla v}\, \mathrm{d}x$.

$$\|R_h u\|_{L^\infty} \leqslant C|\log h| \|u\|_{L^\infty}, \quad u \in V. \tag{4.2}$$

Since $R_h|_{V_h} =$identity, (4.2) implies the error estimate

$$\begin{aligned}\|R_h u - u\|_{L^\infty} &= \|(R_h - 1)(u - \chi)\|_{L^\infty} \\ &\leqslant C\{|\log h| + 1\}\|u - \chi\|_{L^\infty}, \quad \chi \in V_h.\end{aligned}$$

The inequality

$$\|\pi_h u - u\|_{L^\infty} \leqslant Ch^2 \|u\|_{W^{2,\infty}}, \quad u \in V \cap W^{2,\infty}(\Omega),$$

is known to hold so that we have

$$\|R_h u - u\|_{L^\infty} \leqslant Ch^2 |\log h| \|u\|_{W^{2,\infty}}. \tag{4.3}$$

In obtaining this result NITSCHE [1975] adopts the weighted norms while SCOTT [1976] makes use of the discrete Green function. We shall refer to these methods later in Section 16.

Concerning the optimality of (4.3), we note the following. First, in the case that $\Omega \subset \mathbb{R}^2$ is a ball: $\Omega = \{x\colon |x| < R\}$, and f is radial, the Dirichlet problem (1.1) with (1.2) is reduced to a two-point boundary value problem on $[0, R]$ with singular coefficients. Then, it is known that $h^2|\log h|$ is the optimal rate of convergence in L^∞ norm for the finite element approximation by piecewise linear trial functions for this reduced one-dimensional problem (JESPERSON [1978], FRIED [1980]). Next, due to HAVERKAMP [1984], the rate $h^2|\log h|$ is optimal generally in the two-dimensional problem. Namely, there exist a domain $\Omega \subset \mathbb{R}^2$, a triangulation $\{\tau_h\}$, and a constant $c > 0$ such that for each h with $h \downarrow 0$ there exists a $u \in C_0^2(\bar{\Omega})$ such that

$$\|R_h u - u\|_{L^\infty} \geqslant ch^2 |\log h| \|u\|_{W^{2,\infty}}. \tag{4.4}$$

We should note that the boundedness of the Ritz operator implies its convergence estimate, but the converse does not necessarily hold as is seen at the beginning of this section with respect to the L^2 norm. Thus, the boundedness is a problem on its own.

A related topic is the discrete maximum principle due to CIARLET and RAVIART [1973]. See also CIARLET [1978] for relevant works. The discrete maximum principle will be also dealt with in Chapter V of the present article.

Recently SUZUKI and FUJITA [1986] have obtained by means of Stampacchia's method the inequality

$$\|R_h u\|_{L^\infty} \leqslant C_p \|u\|_{W^{1,p}}, \quad u \in V \cap W^{1,p}(\Omega) \tag{4.5}$$

for $p>2$, which may be regarded as a discrete Sobolev inequality. Actually, under the inverse assumption RANNACHER and SCOTT [1982] have shown the estimate

$$\|R_h u\|_{W^{1,p}} \leqslant C \|u\|_{W^{1,p}}, \quad u \in V \cap W^{1,p}(\Omega) \tag{4.6}$$

for $2 \leqslant p \leqslant \infty$ in the case that

$$a(u, v) = \int_\Omega \nabla u \cdot \overline{\nabla v}\, dx \quad \text{with } V = H_0^1(\Omega)$$

and Ω has a smooth boundary or is a convex polygon. In that case, (4.5) is implied by (4.6), because of Sobolev's inequality

$$\|v\|_{L^\infty} \leqslant C_p \|v\|_{W^{1,p}}, \quad p>2.$$

Incidentally, (4.5) holds for general $a(,)$ without any elliptic estimates or inverse assumptions.

5. Topics on spectra and fractional powers of approximate operators

Let A be the elliptic operator considered in Section 1 and A_h be its finite element approximation described in Section 2. We shall show some properties of the spectra and the fractional powers of A_h which are uniform in h.

5.1. Spectra of approximate operators

The following theorem can be proven by the method of AGMON [1965]:

THEOREM 5.1. *The spectrum $\sigma(A_h)$ of A_h is contained in a parabolic region in the complex plane independent of $h>0$. Further, for each θ, $0<\theta<\frac{1}{2}\pi$, there exist constants $C>0$ and $M \geqslant 1$ independent of h such that $A_h - C$ is of type (θ, M) in X_h, where X_h is the set V_h endowed with the L^2 inner product.*

PROOF. As in Section 3, we split $a(,)$ as

$$a(u, v) = a^0(u, v) + (Bu, v), \quad u, v \in V,$$

where a^0 is strongly coercive and symmetric and $B: V \to X$ is bounded. Let A_h^0 be the self-adjoint operator in X_h associated with $a^0|_{V_h \times V_h}$. Then

$$\|(\lambda - A_h^0)^{-1}\|_{X_h \to X_h} \leqslant \frac{1}{|\operatorname{Im} \lambda|}. \tag{5.1}$$

Therefore,

$$\|A_h^0(\lambda-A_h^0)^{-1}v\|_X \leqslant \left(1+\frac{|\lambda|}{|\text{Im }\lambda|}\right)\|v\|_X \leqslant 2\frac{|\lambda|}{|\text{Im }\lambda|}\|v\|_X, \quad v\in X_h. \tag{5.2}$$

By means of Heinz's inequality, we have

$$\|A_{0h}^{1/2}(\lambda-A_{0h})^{-1}\|_{X_h\to X_h} \leqslant \sqrt{2}\frac{|\lambda|^{1/2}}{|\text{Im }\lambda|}. \tag{5.3}$$

On the other hand, the relation

$$A_h = A_h^0 + B_h$$

holds with $B_h = P_h B|_{V_h}$ so that $B_h\colon V_h \to X_h$ is uniformly bounded.

Let us introduce positive constants C_1 and C_2 such that

$$\|B_h\|_{V_h\to X_h} \leqslant C_1, \quad \|A_h^{0-1/2}\|_{V_h\to X_h} \leqslant C_2.$$

As we have seen in the proof of Proposition 3.1, these constants can be taken uniformly in h. Then, we have for $v\in V_h$ that

$$\begin{aligned}\|(\lambda-A_h^0)^{-1}B_h v\|_V &\leqslant C_2\|A_h^{0\,1/2}(\lambda-A_h^0)^{-1}B_h v\|_X\\ &\leqslant \sqrt{2}C_1C_2\frac{|\lambda|^{1/2}}{|\text{Im }\lambda|}\|v\|_V = M\frac{|\lambda|^{1/2}}{|\text{Im }\lambda|}\|v\|_V.\end{aligned}$$

From the identity

$$(\lambda-A_h^0)^{-1}(\lambda-A_h) = 1-(\lambda-A_{0h})^{-1}B_h,$$

the inequality $M\cdot|\lambda|^{1/2}/|\text{Im }\lambda| \leqslant \frac{1}{2}$ implies $\lambda\in\rho(A_h)$ and

$$\|(\lambda-A_h)^{-1}\chi\|_V \leqslant 2\|(\lambda-A_h^0)^{-1}\chi\|_V, \quad \chi\in V_h. \tag{5.4}$$

We also have

$$(\lambda-A_h^0)^{-1} = (\lambda-A_h)^{-1}-(\lambda-A_h^0)^{-1}B_h(\lambda-A_h)^{-1}.$$

Therefore,

$$\begin{aligned}\|(\lambda-A_h)^{-1}\chi\|_X &\leqslant \|(\lambda-A_h^0)^{-1}\|_{X_h\to X_h}\{\|\chi\|_X+\|B_h(\lambda-A_h)^{-1}\chi\|_X\}\\ &\leqslant \frac{1}{|\text{Im }\lambda|}\{\|\chi\|_X + C_1\|(\lambda-A_h)^{-1}\chi\|_V\}\\ &\leqslant \frac{1}{|\text{Im }\lambda|}\{\|\chi\|_X + 2C_1\|(\lambda-A_h^0)^{-1}\chi\|_V\}\\ &\leqslant \frac{1}{|\text{Im }\lambda|}\{\|\chi\|_X + 2C_1C_2\|(A_h^0)^{1/2}(\lambda-A_h^0)^{-1}\chi\|_V\}\\ &\leqslant \frac{1}{|\text{Im }\lambda|}\left\{\|\chi\|_X + 2C_1C_2\sqrt{2}\frac{|\lambda|^{1/2}}{|\text{Im }\lambda|}\|\chi\|_V\right\}\\ &\leqslant \frac{2}{|\text{Im }\lambda|}\|\chi\|_X,\end{aligned}$$

when

$$\frac{|\lambda|^{1/2}}{|\mathrm{Im}\,\lambda|}\leqslant\frac{1}{2M},$$

from which the assertion follows. □

5.2. Negative powers of approximate operators

We suppose that the sesquilinear form $a(,)$ is strongly coercive. Then, the approximate operator A_h in X_h defined in Section 2 is uniformly of type (θ, M) for $0<\theta<\frac{1}{2}\pi$. Consequently, the negative powers $A_h^{-\alpha}, \alpha\geqslant 0$, are defined through Dunford integrals:

$$A_h^{-\alpha}=\frac{1}{2\pi\sqrt{-1}}\int_\Gamma \lambda^{-\alpha}(\lambda-A)^{-1}\,\mathrm{d}\lambda,$$

Γ being the positively oriented boundary of $G_{\theta_1}=\{z\in\mathbb{C}: \theta_1\leqslant|\arg z|\leqslant\pi\}$ with $\theta<\theta_1<\frac{1}{2}\pi$ (for instance, see TANABE [1979]). The following theorem is due to USHIJIMA [1979].

THEOREM 5.2. *For each* α, $0<\alpha<1$, *the inequality*

$$\|A_h^{-\alpha}P_h-A^{-\alpha}\|\leqslant C_\alpha h^{2\alpha} \tag{5.5}$$

holds true.

PROOF. The estimate

$$\|(\lambda-A_h)^{-1}P_h-(\lambda-A)^{-1}\|\leqslant Ch^2,\quad \lambda\in G_{\theta_1},$$

will be shown in Section 7. On the other hand, we have

$$\|(\lambda-A_h)^{-1}P_h-(\lambda-A)^{-1}\|$$
$$\leqslant\|(\lambda-A_h)^{-1}P_h\|+\|(\lambda-A)^{-1}\|\leqslant C\frac{1}{|\lambda|},\quad \lambda\in G_{\theta_1},$$

hence

$$\|(\lambda-A_h)^{-1}P_h-(\lambda-A)^{-1}\|\leqslant Ch^{2\beta}\frac{1}{|\lambda|^{1-\beta}}.$$

for $0\leqslant\beta\leqslant 1$. We divide Γ into $\Gamma_1=\{\lambda\in\Gamma: |\lambda|\leqslant h^{-2}\}$ and $\Gamma_2=\{\lambda\in\Gamma: |\lambda|\geqslant h^{-2}\}$. Then in dealing with

$$A_h^{-\alpha}P_h-A^{-\alpha}=\frac{1}{2\pi\sqrt{-1}}\left(\int_{\Gamma_1}+\int_{\Gamma_2}\right)\lambda^{-\alpha}\{(\lambda-A_h)^{-1}P_h-(\lambda-A)^{-1}\}\,\mathrm{d}\lambda$$
$$=\mathrm{I}+\mathrm{II},$$

we can estimate

$$\|\mathrm{I}\| \leqslant C \int_0^{h^{-2}} \rho^{-\alpha-1+\beta}\, \mathrm{d}\rho\, h^{2\beta} = C_\alpha h^{-2(-\alpha+\beta)} h^{2\beta} = C_\alpha h^{2\alpha}$$

with $\beta > \alpha$, and

$$\|\mathrm{II}\| \leqslant C \int_{h^{-2}}^{\infty} \rho^{-\alpha-1+\beta}\, \mathrm{d}\rho\, h^{2\beta} = C_\alpha h^{-2(-\alpha+\beta)} h^{2\beta} = C_\alpha h^{2\alpha}$$

with $\beta < \alpha$. Hence (5.5) follows. □

REMARK 5.1. The estimate

$$\|(P_h - 1)A^{-\alpha}\| \leqslant Ch^{2\alpha}, \quad 0 \leqslant \alpha \leqslant 1,$$

holds by the interpolation theory so that

$$\|A_h^{-\alpha} P_h - P_h A^{-\alpha}\| \leqslant C_\alpha h^{2\alpha}, \quad 0 < \alpha < 1,$$

holds also true.

5.3. *A discrete elliptic Sobolev inequality*

Let us suppose that the elliptic estimate

$$\|A^{-1} f\|_{H^2} \leqslant C \|f\|_{L^2}, \quad f \in X,$$

holds true. Then, interpolation theory yields

$$\|A^{-\alpha} f\|_{H^\alpha} \leqslant C \|f\|_{L^2}, \quad f \in X,$$

for $0 \leqslant \alpha \leqslant 1$. By virtue of Sobolev's inequalities

$$\|f\|_{L^\infty} \leqslant C_p \|f\|_{W^{1,p}}, \quad f \in W^{1,p}(\Omega) \cap V$$

with $p > 2$ and

$$\|f\|_{W^{1,p}} \leqslant C(\alpha, p, \Omega)\, \|f\|_{H^{2\alpha}}, \quad f \in V \cap H^{2\alpha}(\Omega),$$

with $\alpha > \frac{1}{2}$ and p close to 2, we arrive at

$$\|f\|_{L^\infty} \leqslant C_\varepsilon \|A^{1/2+\varepsilon} f\|_{L^2}, \quad f \in D(A^{1/2+\varepsilon})$$

where $\varepsilon > 0$. SUZUKI and FUJITA [1986] have shown that a similar inequality holds for A_h uniformly in h. Namely:

THEOREM 5.3. *Under the inverse assumption, for each $\varepsilon > 0$, there exists a constant $C_\varepsilon > 0$ such that*

$$\|\chi\|_{L^\infty} \leqslant C_\varepsilon \|A_h^{1/2+\varepsilon} \chi\|_{L^2}, \quad \chi \in V_h, \tag{5.6}$$

holds true.

PROOF. We recall the relation

$$R_h A^{-1} = A_h^{-1} P_h$$

in Section 2 and the estimate

$$\|R_h u\|_{L^\infty} \leqslant C_p \|u\|_{W^{1,p}}, \quad u \in V \cap W^{1,p}(\Omega),$$

for $p > 2$ of Section 4. Then, we have

$$\|A_h^{-1} P_h f\|_{L^\infty} \leqslant C_p \|A^{-1} f\|_{W^{1,p}}, \quad f \in X,$$

or

$$\|\chi\|_{L^\infty} \leqslant C_p \|A^{-1} A_h \chi\|_{W^{1,p}}, \quad \chi \in V_h.$$

Taking $\alpha > \frac{1}{2}$, we can bound the right-hand side from above by

$$C_\alpha \|A^{\alpha-1} A_h \chi\|_{L^2}.$$

Therefore, the desired estimate (5.5) is reduced to

$$\|A^{-\beta} A_h^\beta P_h\|_{X \to X} \leqslant C_\beta$$

for $0 \leqslant \beta < \frac{1}{2}$. In the adjoint form, the inequality is given as

$$\|A_h^\beta P_h A^{-\beta}\|_{X \to X} \leqslant C_\beta. \tag{5.7}$$

In order to prove (5.7), let us suppose first that the sesquilinear form $a(,)$ is symmetric: $a^* = a$. Then, the relations $D(A^{1/2}) = V$ and

$$C^{-1} \|f\|_V \leqslant \|A^{1/2} f\|_X \leqslant C \|f\|_V, \quad f \in V,$$

hold. Similarly, the relation

$$C^{-1} \|\chi\|_V \leqslant \|A_h^{1/2} \chi\|_X \leqslant C \|\chi\|_V, \quad \chi \in V_h,$$

holds uniformly in h as we have seen in Section 3. Therefore, (5.7) holds for $\beta = \frac{1}{2}$ by virtue of Proposition 3.2. Inequality (5.7) is obvious for $\beta = 0$. Therefore, it holds for the full range $0 \leqslant \beta \leqslant \frac{1}{2}$ by the interpolation theory.

As for the general case, we set $a^0 = \frac{1}{2}(a + a^*)$. Then the m-sectorial operator A_h^0 in X_h associated with $a^0|_{V_h \times V_h}$ is self-adjoint so that

$$\|(A_h^0)^\beta P_h (A_h^0)^{-\beta}\| \leqslant C$$

holds for $0 \leqslant \beta \leqslant \frac{1}{2}$. On the other hand, the inequalities

$$\|A_h^\beta (A_h^0)^{-\beta}\| \leqslant C_\beta, \quad \|(A_h^0)^\beta A_h^{-\beta}\| \leqslant C_\beta$$

are obtained for $0 \leqslant \beta < \frac{1}{2}$ according to KATO [1961a]. Here, $C_\beta > 0$ is independent of $h > 0$ and the proof has been completed. □

CHAPTER II

The Semigroup Theory and the Finite Element Method

6. The evolutionary problem and the semigroup of operators

The purpose of the present chapter is to study the finite element method for initial value problems along the line of the semigroup theory, namely, the exponential functions of operators. Here in this section we describe the way in which the semigroup is applied as an abstract method in dealing with initial value problems for parabolic and hyperbolic equations.

As in the preceding chapter, let us suppose again that Ω is a bounded domain in $\mathbb{R}^2$ with boundary $\partial\Omega$ smooth or a convex polygon. The most typical initial value problem for a parabolic equation is composed of the heat equation

$$\partial u/\partial t = \Delta u, \qquad t>0, \quad x\in\Omega, \tag{6.1}$$

the initial condition

$$u|_{t=0} = u_0(x), \quad x\in\Omega, \tag{6.2}$$

and the boundary condition. To fix the idea, let the boundary condition be the homogeneous Dirichlet condition

$$u=0, \qquad t>0, \quad x\in\partial\Omega. \tag{6.3}$$

It is well known that the initial value problem (to be exact, the initial boundary value problem) is uniquely solvable if the initial function u_0 is not very singular. Furthermore, even if u_0 is not continuous, for instance, if $u_0\in L^2(\Omega)$, then the solution $u(t, x)$ is quite smooth for $t>0$. This fact is sometimes referred to as the smoothing property of the heat equation, and is reflected by the semigroup of operators arising from the initial value problems for parabolic equations that we now start to describe.

We regard $X=L^2(\Omega)$ as a (complex) Hilbert space and denote $-\Delta$ associated with the boundary condition (6.3) by A, as in the preceding chapter. Then,

$$D(A)=H^2(\Omega)\cap H_0^1(\Omega) \tag{6.4}$$

and the operator $A: D(A)\to X$ is self-adjoint in X. By means of A, the initial value problem (6.1)–(6.3) is formulated as an abstract Cauchy problem for the unknown function $u\colon [0,\infty)\to X$. Namely,

$$du/dt + Au = 0, \quad t>0, \tag{6.5}$$

with

$$u(0) = u_0. \tag{6.6}$$

If we introduce e^{-tA}, the semigroup (of operators) generated by $-A$, then the solution of (6.5) and (6.6) is simply given by

$$u(t) = e^{-tA}u_0. \tag{6.7}$$

A general theory concerning generation of e^{-tA} for given A has been established by Hille and Yosida (see, for instance, YOSIDA [1964, Chapter 9] and KATO [1966, Chapter 9]). However, since the present A admits of a complete orthonormal system of eigenfunctions, we can define e^{-tA} through the eigenfunction expansion. Namely, let $\{\varphi_n\}_{n=0}^{\infty}$ be the complete orthonormal system of eigenfunctions of A associated with the eigenvalues $\{\lambda_n\}_{n=0}^{\infty}$;

$$A\varphi_n = \lambda_n\varphi_n, \qquad (\varphi_n, \varphi_m) = (\varphi_n, \varphi_m)_{L^2(\Omega)} = \delta_{nm}.$$

Then any function v in X can be expanded as

$$v = \sum_{n=0}^{\infty} (v, \varphi_n)\varphi_n = \sum_{n=0}^{\infty} \beta_n\varphi_n, \tag{6.8}$$

where $\beta_n = (v, \varphi_n)$, $n = 0, 1, 2, \ldots$.

Now, the bounded operator e^{-tA} in X may be defined by

$$e^{-tA}v = \sum_{n=0}^{\infty} \beta_n e^{-t\lambda_n}\varphi_n, \quad t\geqslant 0 \tag{6.9}$$

for v represented as in (6.8). Actually, from (6.9) we have

$$\|e^{-tA}v\|^2 = \sum_{n=0}^{\infty} |\beta_n e^{-t\lambda_n}|^2 = \sum_{n=0}^{\infty} e^{-2t\lambda_n}|\beta_n|^2$$

$$\leqslant \sum_{n=0}^{\infty} |\beta_n|^2 = \|v\|^2.$$

This implies that e^{-tA} is a contraction;

$$\|e^{-tA}\| \leqslant 1. \tag{6.10}$$

Since $e^{-tA}\varphi_n = e^{-t\lambda_n}\varphi_n$ from (6.9), it is easy to see that $e^{-tA}\cdot e^{-sA}v = e^{-(t+s)A}v$ for any $v\in X$. On the other hand, $e^{-tA}|_{t=0} = I$ (the identity) is obvious. Thus the family of operators $\{e^{-tA}\}_{t\geqslant 0}$ has the following semigroup property:

$$U(t)\cdot U(s) = U(t+s), \qquad t\geqslant 0, \quad s\geqslant 0,$$

$$U(0) = I.$$

In view of (6.10), we can say that $\{e^{-tA}\}_{t\geqslant 0}$ is a contraction semigroup. Moreover, it is easy to see that $e^{-tA}v$ is continuous in t, particularly, $e^{-tA}v \to v$ as $t \to 0+$. In other words, $\{e^{-tA}\}_{t\geqslant 0}$ forms a strongly continuous semigroup, which is referred to as a (C_0) semigroup. The condition for v in (6.8) to belong to $D(A)$ is

$$\sum_{n=0}^{\infty} |\beta_n \lambda_n|^2 < +\infty, \tag{6.11}$$

which is equivalent to the convergence of

$$\sum_{n=0}^{\infty} \beta_n \lambda_n \varphi_n$$

in X. By virtue of

$$0 < \lambda e^{-t\lambda} = (1/t)(t\lambda e^{-t\lambda}) \leqslant 1/t, \tag{6.12}$$

it is easy to see from (6.11) that $e^{-tA}v$ belongs to $D(A)$ for any $v \in X$ and $t>0$. Actually we have

$$\begin{aligned} \|Ae^{-tA}v\|^2 &= \|\sum_{n=0}^{\infty} \beta_n \lambda_n e^{-t\lambda_n}\varphi_n\|^2 \\ &= \sum_{n=0}^{\infty} |\beta_n|^2 |\lambda_n e^{-t\lambda_n}|^2 \\ &\leqslant \frac{1}{t^2} \sum_{n=0}^{\infty} |\beta_n|^2 = \frac{\|v\|^2}{t^2}, \end{aligned}$$

hence it follows that

$$\|Ae^{-tA}\| \leqslant \frac{1}{t}, \quad t>0, \tag{6.13}$$

which implies a certain smoothness in x of $e^{-tA}v$. Finally, for $t>0$ we have

$$\begin{aligned} \frac{d}{dt} e^{-tA}v &= \frac{d}{dt} \sum_{n=0}^{\infty} \beta_n e^{-t\lambda_n}\varphi_n = \sum_{n=0}^{\infty} \beta_n(-\lambda_n)e^{-t\lambda_n}\varphi_n \\ &= -\sum_{n=0}^{\infty} \beta_n \lambda_n e^{-t\lambda_n}\varphi_n = -Ae^{-tA}v. \end{aligned}$$

Namely, if we set $u(t) = e^{-tA}v$, $v \in X$, then we have $dv/dt = -Av$. In view of this we see that (6.7) holds true, i.e., $e^{-tA}u_0$ solves the abstract Cauchy problem (6.5) and (6.6). From (6.13) it follows also that

$$\|A^2 e^{-tA}\| = \|Ae^{-tA/2} \cdot Ae^{-tA/2}\| \leqslant \|Ae^{-tA/2}\|^2 \leqslant 4/t^2.$$

Thus e^{-tA} carries any $v \in X$ into $D(A^2)$. Similarly, we can show that for any $n \in \mathbb{N} = \{1, 2, 3, \ldots\}$ and $t>0$, $A^n e^{-tA}$ is bounded and is subject to

$$\|A^n e^{-tA}\| \leqslant M_n/t^n, \quad t>0, \tag{6.14}$$

for some constant M_n. Consequently, for any $v \in X$, $u(t) = e^{-tA}v$ is differentiable any times and we have

$$(\mathrm{d}/\mathrm{d}t)^n \mathrm{e}^{-tA} v = (-A)^n \mathrm{e}^{-tA} v.$$

From what was stated above, we notice the smoothing property of e^{-tA} for $t>0$. Also we note that the argument above can be applied with the same result to the case where the boundary condition (6.3) is replaced by the Neumann condition

$$\partial u/\partial n = 0. \tag{6.15}$$

We proceed to the general case where A is the L^2 realization of

$$\begin{aligned}\mathscr{A} &= \mathscr{A}(x, D)\\ &= -\sum_{i,j=1}^{2} \frac{\partial}{\partial x_j} a_{ij}(x) \frac{\partial}{\partial x_j} + \sum_{j=1}^{2} b_j(x) \frac{\partial}{\partial x_j} + c(x)\end{aligned} \tag{6.16}$$

introduced in Section 1. As for the boundary condition, we adopt either the Dirichlet condition (6.3) or the Robin condition

$$\frac{\partial}{\partial \nu_{\mathscr{A}}} u + \sigma u = 0 \tag{6.17}$$

in (1.7). Then the operator $A: D(A) \to X$ is defined through the sesquilinear form $a(\,,)$. By the argument in Section 1, A is an m-sectorial operator:

$$\begin{aligned} a(u, v) = &\int_{\Omega} \left\{ \sum_{i,j=1}^{2} a_{ij}(x) \frac{\partial}{\partial x_j} u \overline{\frac{\partial}{\partial x_i} v} + \sum_{j=1}^{2} b_j(x) \frac{\partial}{\partial x_j} u\bar{v} + c(x) u\bar{v} \right\} \mathrm{d}x \\ &+ \int_{\partial\Omega} \sigma u \bar{v} \, \mathrm{d}s, \end{aligned} \tag{6.18}$$

where $u, v \in V = H^1(\Omega)$. Actually,

$$D(A) = \{u \in H^2(\Omega): \partial u/\partial \nu_{\mathscr{A}} + \sigma u = 0 \text{ on } \partial\Omega\} \tag{6.19}$$

as indicated in Section 1. When we consider the case of the Dirichlet boundary condition $u=0$ on $\partial\Omega$, the last term in (6.18) disappears, for we take $V = H_0^1(\Omega)$. Then

$$\begin{aligned} D(A) &= H^2(\Omega) \cap H_0^1(\Omega) \\ &= \{u \in H^2(\Omega): u = 0 \text{ on } \partial\Omega\}. \end{aligned} \tag{6.20}$$

By considering $u(t) = \mathrm{e}^{-t\lambda} u(t)$ instead of u, we replace A by $A + \lambda I$ for some suitable $\lambda \in \mathbb{R}$. Therefore, we may assume that

$$\operatorname{Re} a(u, u) \geqslant \delta \|u\|_V^2, \quad u \in V, \tag{6.21}$$

holds true. Hence follows that $\|\mathrm{e}^{-tA}\| \leqslant 1$, namely, e^{-tA} is a contraction semigroup.

Thus, for any $u_0 \in X = L^2(\Omega)$, $u(t) = \mathrm{e}^{-tA} u_0$, $t \geqslant 0$, solves

$$\mathrm{d}u/\mathrm{d}t + Au = 0, \quad t > 0, \qquad u(0) = u_0. \tag{6.22}$$

In other words, $u=e^{-tA}u_0$ gives the unique solution of the initial value problem for the equation

$$\partial u/\partial t+\mathscr{A}(x,D)u=0 \tag{6.23}$$

with the corresponding boundary condition and the initial condition $u(0,x)=u_0(x)$. Furthermore, e^{-tA} is a holomorphic semigroup in the sense that

$$\|A^{\alpha}e^{-tA}\|\leqslant M_{\alpha}/t^{\alpha},\quad t>0, \tag{6.24}$$

holds for each $\alpha>0$ and with $M_{\alpha}>0$ depending on α. As to the details of this fact, we refer to YOSIDA [1964, Chapter 9] or KATO [1966, Chapter 9]. However, in view of an analogous argument in dealing with the finite element approximation in the Section 7, we here indicate an outline of the proof. First of all, with a constant $c=c_0>0$, $a(\,,)$ satisfies (1.8). Namely,

$$|a(u,v)|\leqslant c_0\|u\|_V\cdot\|v\|_V. \tag{6.25}$$

Then, defining a sector $\Sigma_0\subset\mathbb{C}$ by

$$\Sigma_0=\{z\in\mathbb{C}: 0\leqslant|\arg z|\leqslant\theta_0\}$$

for θ_0 with $0<\theta_0<\frac{1}{2}\pi$ and $\cos\theta_0=\delta/c_0$, we see that the numerical range $v(A)$ of A is included in Σ_0. That is,

$$v(A)\equiv\{(Au,u)=a(u,u): u\in D(A),\|u\|=1\}\subset\Sigma_0. \tag{6.26}$$

From (6.26), it follows that $\mathbb{C}\backslash\Sigma_0$ is included in $\rho(A)$. Furthermore, if we fix θ_1 with $0<\theta_0<\theta_1<\frac{1}{2}\pi$ and assume that

$$\theta_1\leqslant|\arg z|\leqslant\pi,$$

then we have

$$\|(z-A)^{-1}\|\leqslant\frac{1}{\sin(\theta_1-\theta_0)}\frac{1}{|z|} \tag{6.27}$$

by the following consideration. In fact, if we put

$$d(z)=\operatorname{dist}(z,\partial\Sigma_0)=\inf\{|z-\zeta|:\zeta\in\Sigma_0\},$$

then

$$d(z)\geqslant|z|\sin(\theta_1-\theta_0).$$

On the other hand,

$$|z-(Au,u)/\|u\|^2|\geqslant\operatorname{dist}(z,v(A))\geqslant\operatorname{dist}(z,\Sigma_0)=d(z)$$

yields

$$|((z-A)u,u)|\geqslant d(z)\|u\|^2,\quad u\in D(A), \tag{6.28}$$

hence $\|(z-A)u\|\geqslant d(z)\|u\|$ and (6.27) follow. A useful consequence of (6.27) is the following integral representation (Dunford integral) of e^{-tA}:

$$e^{-tA}=\frac{1}{2\pi i}\int_{\Gamma}(z-A)^{-1}e^{-tz}\,dz, \tag{6.29}$$

where Γ is the boundary of

$$\Sigma_1=\{z: 0\leqslant|\arg z|\leqslant\theta_1\},$$

with the sense that it starts from $+\infty e^{i\theta_1}$, passes through the origin and goes to $+\infty e^{-i\theta_1}$. Obviously, we may deform Γ to some extent in $\mathbb{C}\backslash\Sigma_0$ without affecting the integral representation (6.26). Note that (6.29) is a kind of Cauchy integral representation. In fact, it follows that

$$A^{\alpha}e^{-tA}=\frac{1}{2\pi i}\int_{\Gamma}z^{\alpha}(z-A)^{-1}e^{-tz}\,dz. \tag{6.30}$$

By virtue of (6.27), we can deduce (6.24) as

$$\|A^{\alpha}e^{-tA}\|\leqslant\frac{1}{2\pi}\cdot 2\int_0^{\infty}r^{\alpha}\frac{1}{\sin(\theta_1-\theta_0)}\frac{1}{r}e^{-tr\cos\theta_1}\,dr$$

$$=\frac{1}{\pi}\frac{1}{t^{\alpha}}\int_0^{\infty}\lambda^{\alpha-1}e^{-\lambda\cos\theta_1}\,d\lambda.$$

Before concluding this section, we note that, in terms of e^{-tA}, the solution of the inhomogeneous equation

$$du/dt+Au=f(t),\quad 0\leqslant t\leqslant T, \tag{6.31}$$

with $u(0)=u_0$ is formally given by

$$u(t)=e^{-tA}u_0+\int_0^t e^{-(t-s)A}f(s)\,ds, \tag{6.32}$$

where f: $[0, T]\to X$ is a given continuous function. For instance, $u=u(t)$ in (6.32) is shown to be the strong solution of (6.31) in the case that f is Hölder continuous.

REMARK 6.1. The hyperbolic equation can be dealt with by means of a semigroup, which is strongly continuous but not holomorphic. This will be described in Section 15 when we discuss approximation for hyperbolic problems.

7. Semidiscrete approximation for parabolic equations

In this section we consider the finite element approximation applied to the initial value problem for the equation

$$\partial u/\partial t = \mathscr{A}(x, D)u, \qquad t>0, \quad x\in\Omega, \tag{7.1}$$

where $\mathscr{A}$ is the elliptic differential operator given by (6.16). All assumptions made in Section 6 concerning Ω are kept here. The boundary condition is either the Robin condition (6.17) or the Dirichlet condition

$$u=0, \qquad t>0, \quad x\in\Omega. \tag{7.2}$$

To fix the idea, we describe our results for the case of (7.2) for the time being. The initial value $u_0 = u_0(x)$ is taken from $L^2(\Omega)$. Thus in terms of the operator A introduced before, the problem to be approximated is written as (6.22) which we reproduce here:

$$\mathrm{d}u/\mathrm{d}t + Au = 0, \quad t>0, \qquad u(0)=u_0. \tag{7.3}$$

We denote this initial value problem in $X = L^2(\Omega)$ by (IVP). Since the solution u of (IVP) is given by $u = \mathrm{e}^{-tA}u_0$, we may say that our aim is to approximate e^{-tA}.

The approximation which is discussed here is the so-called semidiscrete finite element approximation. Namely, we discretize the space variable $x\in\Omega$ according to the procedure of the finite element method, while the time variable is dealt with as a continuous variable. Actually, we triangulate the domain Ω, which is now assumed to be a convex polygon for the sake of simplicity, and introduce the finite-dimensional space V_h in accordance with Chapter I. All notations made in Chapter I remain unchanged. Thus the family of triangulations $\{\tau_h\}_{h>0}$, h being the largest diameter of triangles in τ_h, is assumed to be regular. We introduce the inverse assumption when necessary. Unless otherwise stated, V_h is composed of elementwise linear functions vanishing on the boundary, for we are considering the case of the Dirichlet boundary condition.

The semidiscrete approximate solution u_h is the function $u_h : [0, \infty) \to V_h$ determined by the following conditions:

$$\frac{\mathrm{d}}{\mathrm{d}t}(u_h(t), v_h) + a_h(u_h(t), v_h) = 0, \quad v_h \in V_h, \tag{7.4}$$

for $t>0$ and

$$(u_h(0), v_h) = (u_0, v_h) \quad v_h \in V_h. \tag{7.5}$$

Here $(\,,)$ is the L^2 inner product and $a_h(\,,)$ is the restriction of $a(\,,)$ onto $V_h \times V_h$. Recalling that the operator $A_h : V_h \to V_h$ is defined by

$$a_h(v_h, w_h) = (A_h v_h, w_h)_{L^2(\Omega)},$$

we can rewrite (7.4) as

$$\mathrm{d}u_h/\mathrm{d}t + A_h u_h = 0, \quad t>0. \tag{7.6}$$

On the other hand, by means of the L^2-orthogonal projection $P_h: X \to V_h$, the condition (7.5) is equivalent to

$$u_h(0) = P_h u_0. \tag{7.7}$$

We denote the initial value problem (7.6) and (7.7) for u_h by $(\mathrm{IVP})_h$. The solution of $(\mathrm{IVP})_h$ is given by

$$u_h = \mathrm{e}^{-tA_h} P_h u_0, \quad t \geqslant 0, \tag{7.8}$$

although for practical computation one sets

$$u_h(t) = \sum_{j=1}^{N} \xi_j^h(t) \varphi_h^{(j)}, \quad t \geqslant 0, \tag{7.9}$$

with the standard basis function $\varphi_h^{(j)}, j = 1, 2, \ldots, N$, and determine $\xi_j^h(t)$ through the ordinary differential equation

$$M_h (\mathrm{d}/\mathrm{d}t) \vec{\xi} + K_h \vec{\xi} = 0, \tag{7.10}$$

where $\vec{\xi} = (\xi_1^h, \xi_2^h, \ldots, \xi_N^h)^{\mathrm{T}}$ is an N-vector-valued unknown function, and where M_h and K_h are $N \times N$ matrices defined, respectively, by

$$M_h = (M_{ij}(h)), \quad M_{ij}(h) = (\varphi_h^{(i)}, \varphi_h^{(j)})_{L^2}, \tag{7.11}$$

and

$$K_h = (K_{ij}(h)), \quad K_{ij}(h) = a_h(\varphi_h^{(i)}, \varphi_h^{(j)}). \tag{7.12}$$

Since our analysis here is an abstract one, we mostly use the operator-theoretical representation of (7.8) for the semidiscrete approximate solution u_h. Thus we have to study operator-theoretical properties of A_h in the subspace V_h of $X = L^2(\Omega)$, which are uniform in h. In this direction, we have already shown Theorem 5.1 in the preceding chapter. However, we note that the crucial fact can be easily seen from a consideration of the numerical range as in the case of operator A. Actually, we firstly note that e^{-tA_h} is a contraction semigroup and that the numerical range

$$\nu(A_h) = \{(A_h v_h, v_h) : v_h \in V_h, \ \|v_h\|_{L^2} = 1\}$$

of A_h is included in the sector Σ_0 of the preceding section, for $(A_h v_h, v_h) = a_h(v_h, v_h) = a(u, u)$. Therefore, by the same argument which has led to (6.27), we have

$$\|(z - A_h)^{-1}\| \leqslant \frac{1}{\sin(\theta_1 - \theta_0)} \frac{1}{|z|} \tag{7.13}$$

for $z \in \mathbb{C}$ subject to $\theta_0 < \theta_1 \leqslant |\arg z| \leqslant \pi$.

Moreover, we have

$$\mathrm{e}^{-tA_h} = \frac{1}{2\pi\mathrm{i}} \int_{\Gamma} (z - A_h)^{-1} \mathrm{e}^{-tz} \, \mathrm{d}z, \tag{7.14}$$

where the path Γ is the same as in (6.29). Consequently, we can write

$$\mathrm{e}^{-tA_h}P_h=\frac{1}{2\pi\mathrm{i}}\int_\Gamma (z-A_h)^{-1}P_h\mathrm{e}^{-tz}\,\mathrm{d}z.$$

A corollary of (7.14) is the estimate

$$\|A_h^\alpha \mathrm{e}^{-tA_h}\|\leqslant M_\alpha/t^\alpha,\qquad t>0,\quad \alpha>0, \tag{7.15}$$

with M_α independent of h.

The main result of this section is the following theorem which asserts that the error $\|u_h(t)-u(t)\|_{L^2}$ in the semidiscrete approximation is of the optimal order $\mathrm{O}(h^2)$ for $t>0$ even if the initial function u_0 is not smooth, which may be regarded as a consequence of the smoothing property of parabolic equations.

THEOREM 7.1. *Let u and u_h be the solutions of* (IVP) *and* $(\text{IVP})_h$, *respectively. Then for any* $u_0\in L^2(\Omega)$,

$$\|u_h(t)-u(t)\|_{H^1}\leqslant C(h/t)\|u_0\|_{L^2},\quad t>0, \tag{7.16}$$

and

$$\|u_h(t)-u(t)\|_{L^2}\leqslant C(h^2/t)\|u_0\|_{L^2},\quad t>0. \tag{7.17}$$

REMARK 7.1. The proof below is adapted from FUJITA and MIZUTANI [1976], while these estimates have been obtained for the case of self-adjoint A by HELFRICH [1974] without resort to "complex analysis". See Section 12 concerning Helfrich's method.

Before going into the proof of the theorem, we have to make a little more refined consideration of the numerical range of the operators A and A_h, or more exactly, the numerical range of the sesquilinear forms $a(\,,)$ and $a_h(\,,)$. We recall that the acute angle θ_0 was defined by $\cos\theta_0=\delta/c_0$ with the constants δ and c_0 in (6.21) and (6.25), respectively and that θ_1 is another acute angle with $\theta_0<\theta_1<\frac{1}{2}\pi$. Now we choose a positive number δ' such that

$$\cos\theta_1<\delta'/c_0<\delta/c_0=\cos\theta_0$$

and define the angle θ' by

$$\theta_0<\theta'<\theta_1,\qquad \cos\theta'=\delta'/c_0.$$

Also we put $\gamma_0=\delta-\delta'$. Then from (6.21) and (6.25) we have

$$\begin{aligned}&\operatorname{Re}a(v,v)-\gamma_0\|v\|_{H^1}^2\\&\quad\geqslant\delta'\|v\|_{H^1}^2\geqslant\frac{\delta'}{c_0}\cdot c_0\|v\|_{H^1}^2\geqslant\cos\theta'|a(v,v)|,\quad v\in V.\end{aligned} \tag{7.18}$$

Therefore, if ζ is a point in the numerical range of $a(\,,)$, namely, if $\zeta=a(v,v)/\|v\|^2$ for some $v\in V$, then we have

$$\operatorname{Re}\zeta-\gamma_0\mu(v)\geqslant\cos\theta'|\zeta|, \tag{7.19}$$

where $\mu(v)=\|v\|_{H^1}^2/\|v\|^2$. From this, we see that $\zeta=\zeta(v)$ is contained in the shifted sector $\Sigma'=\{z: |\arg(z-\gamma_0\mu(v))|\leqslant\theta'\}$. Now we have:

LEMMA 7.1. *There exists a positive constant δ_1 such that for any $v\in V$ and for any z in $\Sigma_1=\{z: \theta_1\leqslant|\arg z|\leqslant\pi\}$, the inequality*

$$|z|\,\|v\|_{L^2}^2+\|v\|_{H^1}^2\leqslant\delta_1|z\,\|v\|_{L^2}^2-a(v,v)| \tag{7.20}$$

holds true.

PROOF. We note that the distance $\operatorname{dist}(z,\Sigma')$ between z and the shifted sector above satisfies

$$\operatorname{dist}(z,\Sigma')\geqslant|z|\sin(\theta_1-\theta')+\mu(v)\sin\theta'.$$

Consequently, we have

$$\begin{aligned}
&|z\|v\|_{L^2}^2-a(v,v)|\\
&\quad=\|v\|_{L^2}^2|z-\zeta(v)|\geqslant\|v\|_{L^2}^2\operatorname{dist}(z,\Sigma')\\
&\quad\geqslant\|v\|_{L^2}^2\{|z|\sin(\theta_1-\theta')+(2\|v\|_{H^1}^2/\|v\|_{L^2}^2)\sin\theta'\}\\
&\quad\geqslant(1/\delta_1)\{|z|\,\|v\|_{L^2}^2+\|v\|_{H^1}^2\}
\end{aligned}$$

with $\delta_1=(\min\{\sin(\theta_1-\theta'),\sin\theta'\})^{-1}$. This proves the lemma. □

PROOF OF THEOREM 7.1. Writing $f=u_0$ for the time being, we put $w=(z-A)^{-1}f$ and $w_h=(z-A_h)^{-1}P_hf$ for $z\in\Gamma$, respectively. Then the error $e_h(t)=u_h(t)-u(t)$ is represented by

$$e_h(t)=\frac{1}{2\pi\mathrm{i}}\int_\Gamma \mathrm{e}^{-tz}\{w_h-w\}\,\mathrm{d}z. \tag{7.21}$$

On the other hand, w and w_h satisfy

$$(zw-Aw,\varphi)_{L^2}=(f,\varphi)_{L^2},\quad \varphi\in V,$$

and

$$(zw_h-A_hw_h,\varphi_h)_{L^2}=(f,\varphi_h)_{L^2},\quad \varphi_h\in V_h,$$

respectively. Namely,

$$z(w,\varphi)_{L^2}-a(w,\varphi)_{L^2}=(f,\varphi)_{L^2}, \tag{7.22}$$

$$z(w_h,\varphi_h)_{L^2}-a(w_h,\varphi_h)_{L^2}=(f,\varphi_h)_{L^2}. \tag{7.23}$$

Since $V_h\subset V$, we have from (7.22) and (7.23)

$$z(w-w_h,\varphi_h)_{L^2}+a(w-w_h,\varphi_h)_{L^2}=0,\quad \varphi_h\in V_h. \tag{7.24}$$

Henceforth, we write $r_h = w_h - w$. Then by virtue of (7.20) and (7.24), we have

$$|z|\,\|r_h\|_{L^2}^2 + \|r_h\|_{H^1}^2 \leqslant \delta_1 |z\|r_h\|_{L^2}^2 - a(r_h, r_h)|$$
$$= \delta_1 |z(r_h, w-\chi_h)_{L^2} - a(r_h, w-\chi_h)|$$

for any $\chi_h \in V_h$. Furthermore, by (6.25) we have

$$|z|\,\|r_h\|_{L^2}^2 + \|r_h\|_{H^1}^2$$
$$\leqslant \delta_1 \{|z|^{1/2}\|r_h\|_{L^2}\cdot|z|^{1/2}\|w-\chi_h\|_{L^2} + c_0\|r_h\|_{H^1}\cdot\|w-\chi_h\varphi_h\|_{H^1}\}. \tag{7.25}$$

On the other hand, we recall

$$\inf_{\chi_h \in V_h} \|w-\chi_h\|_{L^2} \leqslant Ch\|w\|_{H^1},$$
$$\inf_{\chi_h \in V_h} \|w-\chi_h\|_{H^1} \leqslant Ch\|w\|_{H^2}.$$

In order to estimate $\|w\|_{H^1}$, we use (7.20) again, obtaining

$$|z|\,\|w\|_{L^2}^2 + \|w\|_{H^1}^2 \leqslant \delta_1 |((z-A)w, w)_{L^2}| \leqslant \delta_1 \|f\|_{L^2}\cdot\|w\|_{L^2}, \tag{7.26}$$

hence it follows that

$$|z|\,\|w\|_{L^2}^2 \leqslant \delta_1 \|f\|_{L^2}\|w\|_{L^2},$$
$$\|w\|_{L^2} \leqslant C\|f\|_{L^2}/|z|. \tag{7.27}$$

Substituting (7.27) into (7.26), we obtain $\|w\|_{H^1}^2 \leqslant C\|f\|_{L^2}^2/|z|$ and end up with

$$\|w\|_{H^1} \leqslant C\|f\|_{L^2}/|z|^{1/2}. \tag{7.28}$$

Estimation of $\|w\|_{H^2}$ is made as follows:

$$\begin{aligned}\|w\|_{H^2} &\leqslant C\|Aw\|_{L^2}\\ &= C\|A(z-A)^{-1}f\|_{L^2}\\ &= C\|((A-z)+z)(z-A)^{-1}f\|_{L^2}\\ &\leqslant C\|f\|_{L^2} + |z|\,\|(z-A)^{-1}f\|_{L^2}\\ &\leqslant C\|f\|_{L^2}.\end{aligned}$$

Here we have used (6.27). From what we have obtained above, we get

$$\inf_{\chi_h \in V_h} \|w-\chi_h\|_{L^2} \leqslant Ch\|f\|_{L^2}/|z|^{1/2},$$
$$\inf_{\chi_h \in V_h} \|w-\chi_h\|_{H^1} \leqslant Ch\|f\|_{L^2}.$$

Substituting these into (7.25), we have

$$|z|\,\|r_h\|_{L^2}^2 + \|r_h\|_{H^1}^2 \leqslant \{|z|^{1/2}\|r_h\|_{L^2} + \|r_h\|_{H^1}\}\cdot h\|f\|_{L^2}.$$

This yields

$$|z|\,\|r_h\|_{L^2}^2 + \|r_h\|_{H^1}^2 \leqslant Ch^2\|f\|_{L^2}^2. \tag{7.29}$$

Hence we have $\|r_h\|_{H^1} \leqslant Ch\|f\|_{L^2}$, and in view of (7.21), we have

$$\|e_h(t)\|_{H^1} \leqslant \frac{1}{2\pi}\int_{\Gamma} |e^{-tz}|\, \|w_h - w\|_{H^1} |dz|$$

$$= \frac{1}{2\pi}\cdot 2\int_0^\infty e^{-tr\cos\theta_1}\cdot Ch\|f\|_{L^2}\, dr$$

$$= Ch\|f\|_{L^2}/t = Ch\|u_0\|_{L^2}/t.$$

This proves (7.16). To obtain (7.17) we need

$$\|w - w_h\|_{L^2} = \|r_h\|_{L^2} \leqslant Ch^2\|f\|_{L^2} \tag{7.30}$$

which is deduced from (7.24) by Nitsche's trick as follows; namely, for $g \in L^2(\Omega)$ we put

$$v = (z - A^*)^{-1} g, \qquad v_h = (z - A_h^*)^{-1} P_h g,$$

and note

$$\|r_h\|_{L^2} \leqslant \sup_{g \in L^2} \frac{|(r_h, g)|_{L^2}}{\|g\|_{L^2}} = \sup_{g \in L^2} \frac{|z(r_h, v)_{L^2} - a(r_h, v)|}{\|g\|_{L^2}}$$

and, moreover,

$$\begin{aligned}
&|z(r_h, v)_{L^2} - a(r_h, v)| \\
&\quad = |z(r_h, v - v_h)_{L^2} - a(r_h, v - v_h)| \\
&\quad \leqslant |z|\, \|r_h\|_{L^2}\|v - v_h\|_{L^2} + c_0\|r_h\|_{H^1}\cdot\|v - v_h\|_{H^1} \\
&\quad \leqslant C(h|z|^{1/2}\|f\|_{L^2}\cdot h\|v\|_{H^1} + h\|f\|_{L^2}\cdot h\|v\|_{H^2}) \\
&\quad \leqslant Ch^2(|z|^{1/2}\|f\|_{L^2}\cdot |z|^{-1/2}\|g\|_{L^2} + \|f\|_{L^2}\|g\|_{L^2}),
\end{aligned}$$

because the estimates obtained above for A and A_h hold also true for their adjoints. Thus (7.30) has been verified. Substituting it into

$$\|e_h\|_{L^2} \leqslant \frac{1}{2\pi}\int_{\Gamma} |e^{-tz}|\, \|w_h - w\|_{L^2} |dz|,$$

we immediately obtain (7.17), which establishes the theorem. □

REMARK 7.2. Similar results hold true for the case of the Robin boundary condition, particularly, for the Neumann boundary condition.

REMARK 7.3. If we are concerned only with the convergence, we can relax the assumption on Ω considerably.

REMARK 7.4. The inequality (7.30) is represented as

$$\|(\lambda-A_h)^{-1}P_h-(\lambda-A)^{-1}\|\leqslant Ch^2,\quad \lambda\in G_{\theta_1}. \tag{7.31}$$

8. Fully discrete approximation for parabolic equations

In this section we consider approximation methods for (IVP) (7.3), where the space variable $x\in\Omega$ is discretized just as in the preceding section and the time variable t is discretized by a uniform mesh, namely, as $t=n\tau$, $\tau>0$, $n=0,1,2,\ldots$. For the sake of simplicity, we approximate du/dt in (7.3) by the simplest backward or forward difference quotient. Thus the approximation which we are going to discuss might be called the difference finite element approximation.

8.1. Backward difference approximation

Firstly, we deal with the following approximation with backward difference for the time variable:

$$\frac{u_h^\tau(t+\tau)-u_h^\tau(t)}{\tau}+A_hu_h^\tau(t)=0,\qquad t=n\tau,\quad n=0,1,2,\ldots, \tag{8.1}$$

$$u_h^\tau(0)=P_hu_0,$$

where A_h and P_h mean the same as before. Thus the approximate solution u_h^τ is a function from the discrete time $t=n\tau$ into V_h. Obviously, it is expressed as

$$u_h^\tau(t)=(I+\tau A_h)^{-n}P_hu_0,\quad t=n\tau, \tag{8.2}$$

although for actual computation one discretizes the time variable in (7.9) and (7.10). The approximate problem (8.1) is denoted by $(\mathrm{IVP})^{\mathrm{B}}_{\tau,h}$.

Our main concern here is to study the rate of convergence of u_h^τ to u. However, before doing so, we mention the stability of $(\mathrm{IVP})^{\mathrm{B}}_{\tau,h}$.

THEOREM 8.1. *The fully discrete approximation* $(\mathrm{IVP})^{\mathrm{B}}_{\tau,h}$ *with backward difference in t is absolutely stable. In fact*

$$\|u_h^\tau(t)\|_{L^2}\leqslant\|u_0\|_{L^2}\quad t=n\tau. \tag{8.3}$$

This theorem is obvious since $(I+\tau A_h)^{-1}$ is a contraction because of

$$\mathrm{Re}(A_hv_h,v_h)_{L^2}=\mathrm{Re}\ a(v_h,v_h)\geqslant 0.$$

As to the rate of convergence, which reflects the smoothing property of parabolic equations, we claim:

THEOREM 8.2. *Let* $e_h^\tau(t)=u_h^\tau(t)-u(t)$ *be the error for* $(\mathrm{IVP})^{\mathrm{B}}_{\tau,h}$. *Then we have*

$$\|e_h^\tau(t)\|_{L^2}\leqslant C(h^2+\tau)\|u_0\|_{L^2}/t,\qquad t=n\tau,\quad n=1,2,\ldots \tag{8.4}$$

PROOF. Putting

$$\varepsilon^{(1)} = u_h(t) - u(t) = \mathrm{e}^{-tA_h} P_h u_0 - \mathrm{e}^{-tA} u_0,$$

$$\varepsilon^{(2)} = (I + \tau A_h)^{-n} P_h u_0 - \mathrm{e}^{-tA_h} P_h u_0 \equiv K_h^\tau P_h u_0.$$

We notice $e_h^\tau = \varepsilon^{(1)} + \varepsilon^{(2)}$. According to Theorem 7.1,

$$\|\varepsilon^{(1)}\| \leqslant Ch^2 \|u_0\|/t, \quad t > 0, \tag{8.5}$$

while we can estimate $\|K_h^\tau\| = \|(I + \tau A_h)^{-n} - \mathrm{e}^{-tA_h}\|$, as follows, to obtain

$$\|K_h^\tau\| \leqslant C\tau/t, \quad t = n\tau > 0. \tag{8.6}$$

In fact, we can write

$$K_h^\tau = \int_0^\tau \frac{\mathrm{d}}{\mathrm{d}s}((I + sA_h)^{-n} \mathrm{e}^{-n(\tau - s)A_h})\,\mathrm{d}s$$

$$= n \int_0^\tau sA_h^2 (I + sA_h)^{-n-1} \mathrm{e}^{-n(\tau - s)A_h}\,\mathrm{d}s$$

$$= n \int_0^\tau sA_h^{3/2} (I + sA_h)^{-(n+1)} \cdot A_h^{1/2} \mathrm{e}^{-n(\tau - s)A_h}\,\mathrm{d}s$$

and use the inequalities

$$\|A_h^\alpha (I + sA_h)^{-k}\| \leqslant C_\alpha (ks)^{-\alpha}, \qquad k > \alpha > 0, \quad s > 0, \tag{8.7}$$

which is an analogue of (7.15) and can be proved by means of the Dunford integral (for details, see FUJITA and MIZUTANI [1976]). Namely, we substitute

$$\|sA_h^{3/2}(I + sA_h)^{-(n+1)}\| \leqslant C((n+1)s)^{-3/2}$$
$$\leqslant C(ns)^{-3/2},$$

$$\|A_h^{1/2}\, \mathrm{e}^{-n(\tau - s)A_h}\| \leqslant C(n(\tau - s))^{-1/2}$$

into the integral representation of K_h^τ above and obtain

$$\|K_h^\tau\| \leqslant n \int_0^\tau s \cdot C(ns)^{-3/2} \cdot 1 \cdot C(n(\tau - s))^{-1/2}\,\mathrm{d}s$$

$$= \frac{C}{n} \int_0^\tau s^{-1/2} (\tau - s)^{-1/2}\,\mathrm{d}s = C\frac{\tau}{n\tau}$$

$$= C\tau/t,$$

which proves (8.6). Hence, in view of $\|P_h\|=1$, we get $\|\varepsilon^{(2)}\|\leqslant C\tau\|u_0\|/t$, which yields (8.4) when combined with (8.5). □

8.2. Forward difference approximation

We proceed to the forward difference approximation and define the approximate solution $u_h^\tau: t=n\tau \mapsto V_h$ by

$$\frac{u_h^\tau(t+\tau)-u_h^\tau(t)}{\tau}+A_h u_h^\tau(t)=0, \qquad t=n\tau, \quad n=0,1,2,\ldots, \tag{8.8}$$

$$u_h^\tau(0)=P_h u_0. \tag{8.9}$$

The discrete initial value problem (8.9)–(8.10) is denoted by $(\mathrm{IVP})_{\tau,h}^{\mathrm{F}}$. Formally, the solution of $(\mathrm{IVP})_{\tau,h}^{\mathrm{F}}$ is given by

$$u_h^\tau(t)=(I+\tau A_h)^n P_h u_0, \qquad t=n\tau, \quad n=0,1,2,\ldots. \tag{8.10}$$

Like in the usual difference approximation, the stability consideration is more crucial with the forward difference scheme. In this connection we claim:

THEOREM 8.3. *Suppose that the condition*

$$\tau\|A_h\|\leqslant 2\cos\theta_0 \tag{8.11}$$

is satisfied as $\tau,h\to 0$, *where* θ_0 *is the acute angle defined by* $\cos\theta_0=\delta/c_0$. *Then* $(\mathrm{IVP})_{\tau,h}^{\mathrm{F}}$ *is stable. Precisely we have*

$$\|u_h^\tau(t)\|_{L^2}\leqslant 2\|u_0\|_{L^2}, \qquad t=n\tau, \quad n=0,1,2,\ldots. \tag{8.12}$$

Under the inverse assertion, in particular, $(\mathrm{IVP})_{\tau,h}^{\mathrm{F}}$ *is stable if* τ/h^2 *is sufficiently small.*

PROOF. Putting $S_h^\tau=I-\tau A_h$, we consider the numerical range of S_h^τ. Let $\varphi_h\in V_h$ with $\|\varphi_h\|_{L^2}=1$. Then

$$(S_h^\tau\varphi_h,\varphi_h)_{L^2}=1-\tau\zeta \qquad \text{with } \zeta=(A_h\varphi_h,\varphi_h)_{L^2}=a(\varphi_h,\varphi_h)$$

is seen to satisfy $|(S_h^\tau\varphi_h,\varphi_h)|\leqslant 1$ in view of

$$|\tau\zeta|=|\tau(A_h\varphi_h,\varphi_h)_{L^2}|\leqslant\tau\|A_h\|\leqslant 2\cos\theta_0$$

and $|\arg(\tau\zeta)|=|\arg\zeta|\leqslant\theta_0$. Thus according to a theorem on the numerical range of iterated operators (see, for instance KATO [1966]), the numerical range of $(S_h^\tau)^n$ stays in the unit disk, which implies

$$\|(S_h^\tau)^n\|\leqslant 2, \quad n=0,1,2,\ldots.$$

This gives (8.12), since $u_h^\tau(t)=(S_h^\tau)^n P_h u_0$.

Finally, (8.11) is obvious under the inverse assumption for small τ/h^2, since $\|A_h\|\leqslant C/h^2$ holds true then. □

As to the rate of convergence of $(\mathrm{IVP})_{\tau,h}^{\mathrm{F}}$, we have:

THEOREM 8.4. *Assume in* $(\mathrm{IVP})^{\mathrm{F}}_{\tau,h}$ *that the condition*

$$\sup_{\tau,h\to 0} \tau\|A_h\| < 2\cos\theta_0 \tag{8.13}$$

is satisfied (which is a little more stringent than (8.11)). *Then for the error*

$$e(t) = u_h^\tau(t) - u(t) = (I - \tau A_h)^n P_h u_0 - \mathrm{e}^{-tA} u_0, \qquad t = n\tau, \quad n = 1, 2, \ldots$$

we have

$$\|e(t)\|_{L^2} \leqslant C(h^2 + \tau)\|u_0\|_{L^2}/t. \tag{8.14}$$

PROOF. We again split the error as $e(t) = \varepsilon^{(1)} + \varepsilon^{(2)}$ with

$$\varepsilon^{(1)} = \mathrm{e}^{-tA_h} P_h u_0 - \mathrm{e}^{-tA} u_0$$

and

$$\varepsilon^{(2)} = (I - \tau A_h)^n P_h u_0 - \mathrm{e}^{-tA_h} P_h u_0 \equiv K_h^\tau P_h u_0. \tag{8.15}$$

Since $\|\varepsilon^{(1)}\|_{L^2} \leqslant Ch^2 \|u_0\|_{L^2}/t$ according to Theorem 7.1, it suffices to show

$$\|K_h^\tau\| \leqslant C\tau/t, \qquad t = n\tau, \quad n = 1, 2, \ldots. \tag{8.16}$$

We can derive the integral representation of K_h^τ as

$$\begin{aligned} -K_h^\tau &= -\int_0^\tau \frac{\mathrm{d}}{\mathrm{d}s}((I - sA_h)^n \mathrm{e}^{-n(\tau - s)A_h})\,\mathrm{d}s \\ &= n\int_0^\tau sA_h^2 (I - sA_h)^{n-1} \mathrm{e}^{-n(\tau - s)A_h}\,\mathrm{d}s \\ &= n\int_0^\tau sA_h^{3/2} (I - sA_h)^{n-1} \cdot A_h^{1/2} \mathrm{e}^{-n(\tau - s)A_h}\,\mathrm{d}s. \end{aligned}$$

In order to estimate the integral above, we make use of (7.15) and its analogue

$$\|A_h^\alpha (I - sA_h)^n\| \leqslant C\alpha (ns)^{-\alpha}, \qquad \alpha > 0, \quad 0 < s \leqslant \tau, \tag{8.17}$$

to be proved below. We then have

$$\begin{aligned} \|K_h^\tau\| &\leqslant n\int_0^\tau s \cdot C \cdot ((n-1)s)^{-3/2} \cdot C(n(\tau - s))^{-1/2}\,\mathrm{d}s \\ &= Cn^{1/2}(n-1)^{-3/2} \int_0^\tau s^{-1/2} (\tau - s)^{-1/2}\,\mathrm{d}s \\ &\leqslant Cn^{-1} = Ct/\tau, \end{aligned}$$

which yields the required estimate of $\|\varepsilon^{(2)}\|_{L^2}$.

Finally we give the proof of (8.17). We choose positive constants κ and μ such that

$$\tau\|A_h\|(1+\kappa)\leqslant\mu<2\cos\theta_1, \tag{8.18}$$

which is possible because of (8.13). Then we introduce a positively oriented contour $\tilde{\Gamma}$ which is composed of the following two portions (as sets):

$$\Gamma^{(1)}=\{re^{\pm i\theta_1}: 0\leqslant r\leqslant R\}, \qquad \Gamma^{(2)}=\{Re^{i\theta}: -\theta_1\leqslant\theta\leqslant\theta_1\},$$

where $R=\mu/s$. We put $F_n(\lambda)=(n\lambda)^\alpha(1-\lambda)^n$ and represent $F_n(sA_h)$ by the following Dunford integral:

$$F_n(sA_h)=\frac{1}{2\pi i}\int_\Gamma F_n(sz)(z-A_h)^{-1}\,dz=\frac{1}{2\pi i}(I^{(1)}+I^{(2)}),$$

where $I^{(j)}$ means the contribution to the integral form $\Gamma^{(j)}$, $j=1,2$. For $z=re^{\pm i\theta_1}\in\Gamma^{(1)}$ we have

$$|1-sz|^2=1+s^2r^2-2sr\cos\theta_1, \quad 0\leqslant sr\leqslant\mu.$$

Since $\mu<2\cos\theta_1$, there exists a positive constant γ which depends only on θ_1 and μ such that

$$|1-sz|\leqslant 1-\gamma sr, \quad 0\leqslant sr\leqslant\mu.$$

Thus we have

$$\|I^{(1)}\|\leqslant C\int_0^R (nsr)^\alpha(1-\gamma sr)^n\,\frac{dr}{r}$$

$$\leqslant C\int_0^\infty (ns\gamma)^\alpha e^{-\gamma ns\gamma}\,\frac{dr}{r}=C\int_0^\infty \xi^{\alpha-1}e^{-\gamma\xi}\,d\xi=C_\alpha^{(1)}. \tag{8.19}$$

On the other hand, for $z=Re^{i\theta}\in\Gamma^{(2)}$, we have

$$\|(z-A_h)^{-1}\|\leqslant\frac{1}{|z|}\,\frac{1}{1-\|A_h\|/|z|}$$

$$=\frac{1}{R}\,\frac{1}{1-\|A_h\|/R}=\frac{1}{R}\,\frac{1}{1-1/(1+\kappa)}$$

$$=\frac{1}{R}\,\frac{1+\kappa}{\kappa},$$

$$|1-sz|\leqslant|1-\mu e^{i\theta_1}|=(1+\mu^2-2\mu\cos\theta_1)^{1/2}\equiv\delta_1<1.$$

By virtue of these estimates, we have

$$\|I^{(2)}\| \leqslant C \int_{\theta_1}^{\theta_2} (nsR)^{\alpha}\delta_1^n \frac{(1+\kappa)}{R\kappa} R\,\mathrm{d}\theta$$
$$\leqslant C(n\mu)^{\alpha}\delta_1^n \leqslant C_{\alpha}^{(2)},$$

since $n^{\alpha}\delta_1^n \to 0$ as $n\to\infty$ by $0<\delta_1<1$. Combining this result with (8.19), we get to $\|F_n(sA_h)\| \leqslant C_{\alpha}$ which implies (8.17). □

9. Approximation for inhomogeneous equations

Let us consider the semidiscrete approximation applied to the inhomogeneous equation

$$\mathrm{d}u/\mathrm{d}t + Au = f(t), \quad 0\leqslant t\leqslant T, \tag{9.1}$$

with the initial condition $u(0)=u_0 \in L^2(\Omega)=X$. A standard smoothness assumption for f: $[0, T]\to X$ is the Hölder continuity. Namely, we assume that

$$[f]_{\theta} = \sup_{\substack{t,s\in[0,T]\\ t\neq s}} \frac{\|f(t)-f(s)\|_{L^2}}{|t-s|^{\theta}} < +\infty \tag{9.2}$$

for some θ, $0<\theta<1$. Then a unique solution $u\in C^{1+\theta}([0, T]\to X)$ exists and is expressed by

$$u(t) = \mathrm{e}^{-tA}u_0 + \int_0^t \mathrm{e}^{-(t-s)A} f(s)\,\mathrm{d}s \tag{9.3}$$

as mentioned in Section 6.

Now we define our approximate solution u_h: $[0, T]\to V_h$ by

$$\begin{aligned} &\mathrm{d}u_h/\mathrm{d}t + A_h u_h = P_h f(t), \quad 0\leqslant t\leqslant T,\\ &u_h(0) = P_h u_0. \end{aligned} \tag{9.4}$$

Then u_h can be expressed by

$$u_h(t) = \mathrm{e}^{-tA_h}u_0 + \int_0^t \mathrm{e}^{-(t-s)A_h} P_h f(s)\,\mathrm{d}s. \tag{9.5}$$

Our aim is to estimate $\|u_h(t)-u(t)\|$. To this end, let us introduce the error operator

$$E_h(t) = \mathrm{e}^{-tA_h}P_h - \mathrm{e}^{-tA} \tag{9.6}$$

acting on X. First of all, we note

$$\|E_h(t)\| \leqslant 2, \tag{9.7}$$

since e^{-tA} and e^{-tA_h} are both contractions under our assumptions. Furthermore, according to Theorem 7.1, the estimate

$$\|E_h(t)\| \leqslant Ch^2/t, \quad t>0, \tag{9.8}$$

holds true. We split the error $e_h(t)=u_h(t)-u(t)$ as follows:

$$\begin{aligned} e_h(t) &= E_h(t)u_0 + \int_0^t E_h(t-s)f(s)\,\mathrm{d}s \\ &= \varepsilon^{(1)}(t)+\varepsilon^{(2)}(t)+\varepsilon^{(3)}(t), \end{aligned}$$

where

$$\varepsilon^{(1)}(t) = E_h(t)u_0,$$

$$\varepsilon^{(2)}(t) = \int_0^t E_h(t-s)(f(s)-f(t))\,\mathrm{d}s,$$

$$\varepsilon^{(3)}(t) = \int_0^t E_h(t-s)f(t)\,\mathrm{d}t.$$

Then obviously we have

$$\|\varepsilon^{(1)}(t)\| \leqslant Ch^2\|u_0\|/t,$$

$$\begin{aligned} \|\varepsilon^{(2)}(t)\| &\leqslant \int_0^t Ch^2(t-s)^{-1}\|f(s)-f(t)\|\,\mathrm{d}s \\ &\leqslant Ch^2[f]_\theta \int_0^t (t-s)^{-1}(t-s)^\theta\,\mathrm{d}s \\ &\leqslant C_\theta h^2[f]_\theta t^\theta. \end{aligned}$$

Incidentally, here and hereafter we may write simply $\|\cdot\|$ for $\|\cdot\|_{L^2}$ when no confusion arises. In order to estimate $\|\varepsilon^{(3)}(t)\|$, we compute

$$\begin{aligned} \varepsilon^{(3)}(t) &= \int_0^t e^{-(t-s)A}f(t)\,\mathrm{d}s - \int_0^t e^{-(t-s)A_h}P_hf(t)\,\mathrm{d}s \\ &= [A^{-1}e^{-(t-s)A}f(t)]_{s=0}^{s=t} - [A_h^{-1}e^{-(t-s)A_h}P_hf(t)]_{s=0}^{s=t} \\ &= (A_h^{-1}e^{-tA_h}P_h - A^{-1}e^{-tA})f(t) - (A_h^{-1}P_h - A^{-1})f(t) \end{aligned}$$

and note that

$$\|A_h^{-1}P_h - A^{-1}\| \leqslant Ch^2 \tag{9.9}$$

according to Section 2. Thus it is only necessary to estimate the operator norm of

$$A_h^{-1}e^{-tA_h}P_h - A^{-1}e^{-tA} = \frac{1}{2\pi i}\int_\Gamma z^{-1}e^{-tz}[(z-A_h)^{-1}P_h - (z-A)^{-1}]\,d\lambda. \tag{9.10}$$

Here Γ is a positively oriented contour which is obtained by deforming the original one in the following manner: $\Gamma = \Gamma^{(1)} \cup \Gamma^{(2)}$ with

$$\Gamma^{(1)} = \left\{\frac{1}{t}e^{i\theta}: \theta_1 \leqslant |\theta| \leqslant \pi\right\}, \qquad \Gamma^{(2)} = \left\{re^{\pm i\theta_1}: \frac{1}{t} \leqslant r < +\infty\right\}.$$

Let $I^{(j)}$, $j=1,2$, stand for the contributions to the integral in (9.10) from $\Gamma^{(j)}$, $j=1,2$, respectively. Then we have

$$\|I^{(1)}\| \leqslant \frac{1}{2\pi}\cdot 2\int_{\theta_1}^{\pi} te^{-\cos\varphi}\cdot Ch^2 \cdot \frac{1}{t}\,d\varphi = Ch^2$$

by virtue of (7.25). Furthermore, we have

$$\|I^{(2)}\| \leqslant \frac{1}{2\pi}\cdot 2\int_{1/t}^{\infty} \frac{1}{r}e^{-tr\cos\theta_1}\cdot Ch^2\,dr$$

$$= Ch^2\int_1^{\infty}\frac{1}{\rho}e^{-\rho\cos\theta_1}\,d\rho = Ch^2.$$

Thus we have

$$\|\varepsilon^{(3)}(t)\| \leqslant Ch^2\|f(t)\|.$$

Summing the estimates of $\|\varepsilon^{(j)}(t)\|$, $j=1,2,3$, we have:

THEOREM 9.1. *As to the error $e_h(t) = u_h(t) - u(t)$ committed by the semidiscrete finite element approximation for the inhomogeneous initial value problem* (9.4), *we have*

$$\|e_h(t)\| \leqslant Ch^2\left(\frac{\|u_0\|}{t} + [f]_\theta t^\theta + \|f(t)\|\right). \tag{9.11}$$

10. Remarks on approximation with higher accuracy

As in the case of elliptic boundary value problems (e.g., CIARLET [1978]), one may use a more sophisticated finite element space V_h for V and try to obtain approximate solutions with higher accuracy. In order to have a look into this direction, we again consider the same (IVP)

$$du/dt + Au = 0, \quad t > 0, \qquad u(0) = u_0 \tag{10.1}$$

as above. We study its semidiscrete approximation $(\mathrm{IVP})_h$

$$du_h/dt + A_h u_h = 0, \qquad u_h(0) = P_h u_0 \tag{10.2}$$

of the same form as in Section 7, but assume that the finite-dimensional subspace V_h has the following properties: there exists an integer $s \geqslant 2$ such that if $v \in H^s(\Omega) \cap V$, then for some v_h in V_h we have

$$\|v - v_h\|_{L^2} \leqslant Ch^{s-1} \|v\|_{s-1}, \tag{10.3}$$

$$\|v - v_h\|_{H^1} \leqslant Ch^{s-1} \|v\|_s, \tag{10.4}$$

where C is independent of v and h. If $s > 2$, we suppose that the conditions (10.3) and (10.4) with s replaced by 2 hold true as well. Let us confirm that all assumptions on A and $a(\,,)$ remain unchanged. Then we have:

THEOREM 10.1. *As to the error*

$$e_h(t) = u_h(t) - u(t) = e^{-tA_h} P_h u_0 - e^{-tA} u_0,$$

we have

$$\|u_h(t) - u(t)\|_{L^2} \leqslant C\left(\frac{2}{\sqrt{t}}\right)^s \|u_0\|_{L^2}, \quad t > 0, \tag{10.5}$$

for any $u_0 \in L^2(\Omega)$.

Concerning the details of the proof we refer to FUJITA and MIZUTANI [1978]. Here we just mention the outline. The first step is to show, for $w = (z - A)^{-1} f$ and z with $\theta_1 \leqslant |\arg z| \leqslant \pi$, the inequality

$$\begin{aligned} &\|w\|_s + |z|^{1/2} \|w\|_{s-1} + \cdots + |z|^{s/2} \|w\|_0 \\ &\quad \leqslant C(\|f\|_{s-2} + |z|^{1/2} \|f\|_{s-3} + \cdots + |z|^{(s-2)/2} \|f\|_0), \end{aligned} \tag{10.6}$$

where $\|\cdot\|_j$ stands for $\|\cdot\|_{H^j}$, $j = 0, 1, \ldots, s$. As the second step, we recall that (7.20) still holds true, i.e.,

$$|z| \|v\|_0^2 + \|v\|_1^2 \leqslant \delta_1 |z\|v\|_0^2 - a(v, v)| \tag{10.7}$$

for any $v \in V$. Making use of (10.3), (10.4), (10.6) and (10.7), we can show that if $f \in H^{s-2}(\Omega)$, then we have for $r_h = (z - A)^{-1} f - (z - A_h)^{-1} P_h f$ that

$$\|r_h\|_0 \leqslant Ch^s(\|f\|_{s-2} + |z|^{1/2} \|f\|_{s-3} + \cdots + |z|^{(s-2)/2} \|f\|_0). \tag{10.8}$$

By means of (10.8) we can estimate the integral below which expresses the error

$e_h(t)=u_h(t)-u(t)$:

$$e_h(t)=\frac{1}{2\pi i}\int_\Gamma e^{-tz}\{(z-A)^{-1}-(z-A_h)^{-1}P_h\}u_0\,dz.$$

Thus we obtain

$$\|u_h(t)-u(t)\|\leqslant Ch^s t^{-1}(\|u_0\|_{s-2}+t^{-1/2}\|u_0\|_{s-3}+\cdots+t^{-(s-2)/2}\|u_0\|_0)$$

for $u_0\in H^{s-2}(\Omega)$. In order to relax the assumption $u_0\in H^{s-2}(\Omega)$ and obtain (10.5), we follow the idea due to HELFRICH (see Section 12 and FUJITA and MIZUTANI [1978]).

As an example of the fully discrete approximation with higher accuracy, we consider the Crank–Nicolson scheme which defines the approximate solution $u_h^\tau(t)$, $t=n\tau$, $n=0,1,2,\ldots$, by

$$\frac{u_h^\tau(t+\tau)-u_h^\tau(t)}{\tau}+\tfrac{1}{2}A_h u_h^\tau(t+\tau)+\tfrac{1}{2}A_h u_h^\tau(t)=0 \tag{10.9}$$

with $u_h^\tau(0)=P_h u_0$. Consequently, u_h^τ is expressed as

$$u_h^\tau(t)=(I-\tfrac{1}{2}\tau A_h)^n(I+\tfrac{1}{2}\tau A_h)^{-n}P_h u_0.$$

A condition for the stability is

$$\sup_{\tau,h\to 0}\tau\|A_h\|<+\infty,$$

which is always met if the inverse assumption is satisfied and τ/h^2 stays bounded. Moreover, let us assume that V_h is subject to (10.3) and (10.4) with $s=4$. Then we have:

THEOREM 10.2. *Under the assumptions stated above, the error* $e_h^\tau=u_h^\tau(t)-u(t)$ *committed by the fully discrete Crank–Nicolson approximation is established as*

$$\|e_h^\tau(t)\|_{L^2}\leqslant C\frac{h^4+\tau^2}{t^2}\|u_0\|_{L^2},\quad t=n\tau>0. \tag{10.10}$$

As for the details of the proof, which is similar to that of Theorem 8.4 and is reduced to estimation of the corresponding Dunford integral, we refer to FUJITA and MIZUTANI [1978].

REMARK 10.1. Various works have been done to analyze approximations with higher accuracy. In this connection see RAVIART and THOMAS [1983] and its bibliography. See also Section 17.

CHAPTER III

Evolution Equations and Error Analysis by Real Methods

In the case of temporally inhomogeneous parabolic equations, the complex method developed in the preceding chapter does not work so well by itself. Here, we adopt first the method of Helfrich and then that of energy, in order to extend the error estimates obtained in the preceding chapter to the temporally inhomogeneous case. Also, we mention the discretization of hyperbolic equations.

11. Generation theory on evolution equations

In the present chapter, we study temporally inhomogeneous parabolic equations; that is, $\Omega \subset \mathbb{R}^2$ is as before, and $\mathscr{A} = \mathscr{A}(t, x, D)$ denotes a second-order elliptic differential operator with time-dependent real smooth coefficients:

$$\mathscr{A} = \mathscr{A}(t, x, D) = -\sum_{i,j=1}^{2} \frac{\partial}{\partial x_i} a_{ij}(t, x) \frac{\partial}{\partial x_j} + \sum_{j=1}^{2} b_j(t, x) \frac{\partial}{\partial x_j} + c(t, x). \tag{11.1}$$

Uniform ellipticity

$$\sum_{i,j=1}^{2} a_{ij}(t, x)\xi_i \xi_j \geqslant \delta_1 |\xi|^2, \quad \xi = (\xi_1, \xi_2) \in \mathbb{R}^2, \tag{11.2}$$

is assumed, $\delta_1 > 0$ being a constant. We consider the parabolic equation

$$\partial u/\partial t + \mathscr{A}(t, x, D)u = 0, \qquad 0 < t \leqslant T, \quad x \in \Omega, \tag{11.3}$$

with the boundary condition either

$$u|_{\partial\Omega} = 0, \quad 0 < t \leqslant T, \tag{11.4}$$

or

$$\partial u/\partial \nu_{\mathscr{A}} + \sigma u|_{\partial\Omega} = 0, \quad 0 < t \leqslant T, \tag{11.4'}$$

and with the initial condition

$$u|_{t=0} = u_0(x), \quad x \in \Omega. \tag{11.5}$$

In (11.4′), $\sigma=\sigma(t,\xi)$ is a smooth function on $[0,T]\times\partial\Omega$, and $\partial/\partial\nu_{\mathscr{A}}$ denotes the differentiation along the outer conormal vector $\nu_{\mathscr{A}}$:

$$\partial/\partial\nu_{\mathscr{A}}=\sum_{i,j=1}^{2} n_i a_{ij}(t,x)(\partial/\partial x_j), \tag{11.6}$$

where $n=(n_1,n_2)$ is the outer unit normal on $\partial\Omega$.

Assuming $u_0\in X=L^2(\Omega)$, we can reduce the equation (11.3) with (11.4) (or (11.4′)) and (11.5) to the evolution equation

$$\mathrm{d}u/\mathrm{d}t+A(t)u=0,\quad 0<t\leqslant T, \tag{11.7}$$

with

$$u(0)=u_0 \tag{11.8}$$

in $X=L^2(\Omega)$. Namely, let $V=H_0^1(\Omega)$ or $H^1(\Omega)$ according to the boundary condition (11.4) or (11.4′) and put

$$\begin{aligned} a_t(u,v)=&\sum_{i,j=1}^{2}\int_\Omega a_{ij}(t,x)\frac{\partial}{\partial x_j}u\cdot\overline{\frac{\partial}{\partial x_i}v}\,\mathrm{d}x+\sum_{j=1}^{2}\int_\Omega b_j(t,x)\frac{\partial}{\partial x_j}u\cdot\bar v\,\mathrm{d}x\\ &+\int_\Omega c(t,x)u\cdot\bar v\,\mathrm{d}x+\int_{\partial\Omega}\sigma(t,x)u\cdot\bar v\,\mathrm{d}S \end{aligned} \tag{11.9}$$

for $u,v\in V$, where $\mathrm{d}S$ denotes the arc element of $\partial\Omega$. An m-sectorial operator $A(t)$ in X can be defined through the relation

$$a_t(u,v)=(A(t)u,v),\qquad u\in D(A(t))\subset V,\quad v\in V \tag{11.10}$$

as before. As is described in Chapter I, the relation

$$D(A(t))=H_0^1(\Omega)\cap H^2(\Omega) \tag{11.11}$$

holds for the case $V=H_0^1(\Omega)$ and

$$D(A(t))=\{v_0\in H^2(\Omega)\colon \partial v/\partial\nu_{\mathscr{A}}+\sigma v|_{\partial\Omega}=0\} \tag{11.11′}$$

for the case $V=H^1(\Omega)$.

The generation theory of evolution operators $\{U(t,s)\}_{T>t>s>0}$ in X due to SOBOLEVSKII, KATO, TANABE and others assures us of the unique solvability of (11.7). Namely, a unique solution

$$u=u(t)\in C^1((s,T]\to X)\cap C^0((s,T]\to D(A))\cap C^0([s,T]\to X)$$

of

$$\mathrm{d}u/\mathrm{d}t+A(t)u=0,\quad s<t\leqslant T, \tag{11.7′}$$

with

$$u(s)=u_0\in X \tag{11.8′}$$

exists and is given by

$$u(t)=U(t, s)u_0. \tag{11.12}$$

It is worthwhile to give a short summary of these theories here. In fact, our error analysis on the finite element approximation of (11.7) is based on a certain stability of approximate solutions as well as on a certain smoothness of original ones, both of which can be established by re-examining these theories in our contexts.

First we note that the crucial assumptions of the generation theory are that

(1) each $-A(t)$ generates a holomorphic semigroup with certain estimates uniform in $t\in[0, T]$,

(2) $A(t)$ is smooth in $t\in[0, T]$ in some sense or other.

Then, the evolution operator $\{U(t, s)\}$ is constructed by Levi's method. Actually, condition (1) enables us to give a first approximation of $\{U(t, s)\}$, while condition (2) makes an iteration scheme converge.

As for condition (1), we can verify it in the same way as in the preceding chapter. In fact, we have the boundedness of the sesquilinear form $a_t(\,,\,)$:

$$|a_t(u, v)|\leqslant C\|u\|\,\|v\|, \quad u,v\in V, \tag{11.13}$$

as well as its coerciveness

$$\operatorname{Re}\, a_t(u, v)\geqslant \delta\|u\|_V^2-\lambda\|u\|_X^2, \quad u\in V, \tag{11.14}$$

with constants δ and $\lambda\in\mathbb{R}$ by (11.2). As in the preceding chapter, we may suppose $\lambda=0$ in (11.14′), just by taking $v=e^{-\lambda t}u(t)$ instead of $u=u(t)$, namely

$$\operatorname{Re}\, a_t(u, u)\geqslant \delta\|u\|_V^2, \quad u\in V. \tag{11.14′}$$

Then, the relation

$$G_\theta\equiv\{z\in\mathbb{C}: \pi\geqslant|\arg z|>\theta\}\subset\rho(A(t)) \tag{11.15}$$

holds for some θ, $0<\theta<\frac{1}{2}\pi$, as well as the estimate

$$\|(\lambda-A(t))^{-1}\|\leqslant C_\varepsilon/|\lambda|, \quad \lambda\in G_{\theta+\varepsilon}, \tag{11.16}$$

for each $\varepsilon>0$. These relations are uniform in $t\in[0, T]$. Therefore, each $-A(t)$ generates a holomorphic semigroup $\{e^{-sA(t)}\}_{s\geqslant 0}$ uniformly bounded in $t\in[0, T]$. Actually, we have

$$\|e^{-sA(t)}\|\leqslant 1, \qquad 0\leqslant s<\infty, \quad 0\leqslant t\leqslant T. \tag{11.17}$$

The requirement of condition (2), on the other hand, depends on the generation theory.

11.1. *Generation theory of Tanabe–Sobolevskii* (TANABE [1960], SOBOLEVSKII [1961a])

In the case of $V=H_0^1(\Omega)$, the relation (11.11) holds. Hence $D(A(t))$ is independent of t. Furthermore, from an integration by parts and the elliptic estimate by AGMON, DOUGLIS and NIRENBERG [1959], the inequality

$$\|A(t)A(s)^{-1}-1\|\leqslant C|t-s|^\alpha, \quad t, s\in[0, T], \tag{11.18}$$

follows with an α in $0<\alpha\leqslant 1$. In this case, the evolution operators $\{U(t,s)\}_{T\geqslant t\geqslant s\geqslant 0}$ are constructed by

$$U(t,s)=\mathrm{e}^{-(t-s)A(s)}+\int_s^t \mathrm{e}^{-(t-r)A(r)}R(r,s)\,\mathrm{d}r, \tag{11.19}$$

where $R=R(t,s)$ is the unique solution of the integral equation

$$R(t,s)-\int_s^t R_1(t,r)R(r,s)\,\mathrm{d}r=R_1(t,s) \tag{11.20}$$

of Volterra type with

$$R_1(t,s)=-(A(t)-A(s))\mathrm{e}^{-(t-s)A(s)}. \tag{11.21}$$

Furthermore, for the evolution operators constructed in this way, the estimates

$$\|A(t)U(t,s)\|,\|\overline{U(t,s)A(s)}\|\leqslant C(t-s)^{-1}, \qquad 0\leqslant s<t\leqslant T, \tag{11.22}$$

$$\|U(t,s)\|,\|A(t)U(t,s)A(s)^{-1}\|\leqslant C, \qquad 0\leqslant s\leqslant t\leqslant T, \tag{11.23}$$

$$\|A(t)[U(t,s)-U(r,s)]A(s)^{-1}\|\leqslant C_\theta(t-r)^\theta(r-s)^{-\theta}, \quad 0\leqslant s<r<t\leqslant T, \tag{11.24}$$

can be derived, where $0\leqslant\theta<\alpha$.

11.2. Generation theory of Fujie–Tanabe (FUJIE and TANABE [1973])

The m-sectorial operator $A(t)$ in X defined through the equality (11.10) may be regarded as that in V^*, which is denoted by $\hat{A}(t)$. Then, the domain of $\hat{A}(t)$ is independent of t, that is, $D(\hat{A}(t))=V$. The relation (11.15) and the estimate (11.16) hold for $\hat{A}(t)$ in $\hat{X}=V^*$. The coefficients a_{ij}, b_j, c and σ are smooth so that the inequality

$$|a_t(u,v)-a_s(u,v)|\leqslant C|t-s|^\alpha\|u\|_V\|v\|_V, \quad u,v\in V, \tag{11.25}$$

holds with $0<\alpha\leqslant 1$, which implies (11.18) for $\hat{A}(t)$ in $\hat{X}=V^*$. Therefore, by the preceding theory, $\hat{A}(t)$ generates an evolution operator $\{\hat{U}(t,s)\}_{T\geqslant t\geqslant s\geqslant 0}$ in $\hat{X}=V^*$. It is shown that in the case of $\alpha>\frac{1}{2}$,

$$U(t,s)=\hat{U}(t,s)|_X \tag{11.26}$$

is a bounded operator in X and becomes the desired evolution operator generated by $-A(t)$. Furthermore, this $\{U(t,s)\}$ satisfies (11.22) and the first inequality of (11.23) in X.

11.3. Generation theory of Kato–Sobolevskii (KATO [1961b], SOBOLEVSKII [1961b])

From the theory of fractional powers of m-sectorial operators due to KATO [1961a], the domain of $A(t)^\rho$ is independent of t for $0\leqslant\rho<\frac{1}{2}$. Furthermore, it is shown that

(11.25) implies

$$\|A(t)^\rho A(s)^{-\rho}-1\|\leqslant C|t-s|^\alpha, \quad t,s\in[0,T]. \tag{11.27}$$

From these facts, the evolution operators $\{U(t,s)\}_{T\geqslant t\geqslant s\geqslant 0}$ can be constructed in another way, provided that $\alpha+\rho>1$, namely, under the assumption $\alpha>\frac{1}{2}$ again.

For simplicity, we consider the case $\rho=1/m$ according to KATO [1961b], where m is an integer. By taking an appropriate approximation of $A(t)$, say, the modified Yosida approximation: $A_\lambda(t)=A(t)(1+\lambda A(t)^\rho)^{-m}$, where $\lambda\downarrow 0$, we can reduce the theory to the case where each $A(t)$ is bounded in X. In fact, we can show that (11.27) implies

$$\|A_\lambda(t)^\rho A_\lambda(s)^{-\rho}-1\|\leqslant C|t-s|^\alpha, \quad t,s\in[0,T], \tag{11.27$'$}$$

$C>0$ being independent of $\lambda>0$. The existence of the evolution operator $\{U_\lambda(t,s)\}_{T\geqslant t\geqslant s\geqslant 0}$ for the approximate operator $A_\lambda(t)$ is obvious from the boundedness of $A_\lambda(t)$. All we have to do is to derive some a priori estimates on A_λ, and make them to pass to the limit:

$$U(t,s)=s-\lim_{\lambda\downarrow 0} U_\lambda(t,s).$$

Then, $\{U(t,s)\}$ become the desired evolution operators.

For instance, the estimate

$$\|A(t)^\beta U(t,s)\|\leqslant C_\beta(t-s)^{-\beta}, \quad 0\leqslant s<t\leqslant T, \tag{11.28}$$

follows from

$$\|A_\lambda(t)^\beta U_\lambda(t,s)\|\leqslant C_\beta(t-s)^{-\beta}, \tag{11.28$'$}$$

where $0\leqslant\beta<\alpha+\rho$. We shall give an outline of the proof of (11.28′). Henceforth, we drop the suffix λ for simplicity of writing. First, setting

$$D(t,s)=A(t)^\rho A(s)^{-\rho}-1, \tag{11.29}$$

we have

$$\|D(t,s)\|\leqslant C|t-s|^\alpha, \tag{11.30}$$

which is nothing but (11.27′). Furthermore, Sobolevskii's identity

$$A(t)-A(s)=\sum_{p=1}^{m} A(t)^{1-p\rho}D(t,s)A(s)^{p\rho} \tag{11.31}$$

holds by $\rho=1/m$. Therefore, we have

$$\begin{aligned} U(t,s)-e^{-(t-s)A(t)} &= \int_s^t \frac{\partial}{\partial r}[e^{-(t-r)A(t)}U(r,s)]\,dr \\ &= \sum_{p=1}^{m}\int_s^t e^{-(t-r)A(t)}A(t)^{1-p\rho}D(t,r)A(r)^{p\rho}U(r,s)\,dr. \end{aligned} \tag{11.32}$$

We introduce a few notations. For operator-valued functions $K_l = K_l(t, s)$, $l = 1,2$, on $D = \{(t, s): T \geqslant t \geqslant s \geqslant 0\}$, we define another $K = K_1 * K_2$ by

$$(K_1 * K_2)(t, s) = \int_s^t K_1(t, r) K_2(r, s)\, dr. \tag{11.33}$$

Furthermore, for $a > 0$ and $M > 0$ we say that $K \in Q(a, M)$ if the inequality

$$\|K(t, s)\| \leqslant M(t-s)^{a-1} \tag{11.34}$$

holds. Then, $K_l \in Q(a_l, M_l)$, $l = 1, 2$, implies that $K_1 * K_2 \in Q(a_1 + a_2, B(a_1, a_2) M_1 M_2)$, where $B(a, b)$ denotes the beta function:

$$B(a, b) = \int_0^1 (1-x)^{a-1} x^{b-1}\, dx. \tag{11.35}$$

Now, we put

$$W(t, s) = U(t, s) - e^{-(t-s)A(t)}, \tag{11.36}$$

$$Y_q(t, s) = A(t)^{q\rho} W(t, s). \tag{11.37}$$

Then, the equality (11.32) reads

$$Y_q = \sum_{p=1}^{m} H_{q,p} * Y_p + Y_{q,0}, \tag{11.38}$$

where

$$H_{q,p}(t, s) = A(t)^{1 - p\rho + q\rho} e^{-(t-s)A(t)} D(t, s), \tag{11.39}$$

$$Y_{q,0} = \sum_{p=1}^{m} H_{q,p} * Y_{p,-1} \tag{11.40}$$

with

$$Y_{p,-1}(t, s) = A(t)^{p\rho} e^{-(t-s)A(t)}. \tag{11.41}$$

For a technical reason, we take

$$Z_q = Y_q - Y_{q,0} \tag{11.42}$$

and transform (11.38) into a system of integral equations for Z_q, $q = 1, \ldots, m$:

$$Z_q = \sum_{p=1}^{m} H_{q,p} * Z_p + Z_{q,0}, \tag{11.43}$$

where

$$Z_{q,0} = \sum_{p=1}^{m} H_{q,p} * Y_{p,0}. \tag{11.44}$$

Consequently, $Z_q, q=1,\ldots, m$, can be generated by an iteration scheme as

$$Z_q=\sum_{i=0}^{\infty} Z_{q,i} \tag{11.45}$$

with

$$Z_{q,i+1}=\sum_{p=1}^{m} H_{q,p}*Z_{p,i}, \quad i=0,1,\ldots. \tag{11.46}$$

From the definition (11.39), we have

$$H_{q,p}\in Q(\alpha-q\rho+p\rho, M_1) \tag{11.47}$$

with a constant $M_1>0$, because

$$\|A(r)^\kappa e^{-sA(r)}\|\leqslant c_\kappa s^{-\kappa}, \quad 0<s<\infty \tag{11.48}$$

holds for $\kappa\geqslant 0$ by (11.16). Furthermore, we can show that

$$Z_{q,0}\in Q(1+\alpha-q\rho, M_0) \tag{11.49}$$

for some $M_0>0$. Then, we get

$$Z_{q,i}\in Q(1+(i+1)\alpha-q\rho, M_i) \tag{11.50}$$

with $M_{i+1}/M_i=mM_0M_1B(\alpha+\rho-1,(i+1)\alpha)$ by induction. Thus

$$Z_q\in Q(1+\alpha-q\rho, C) \tag{11.51}$$

follows from (11.45) with a constant $C>0$. We can deduce from (11.51) an estimate on Y_q with not necessarily integral q, provided that $\alpha-q\rho+\rho>0$. In fact, (11.42) makes sense for nonintegral $q\geqslant 0$, while then $Z_{q,0}$ is again given by (11.44) and Z_q is to be defined by (11.43) (in which p takes integral values as before). Since Z_q in (11.43) has been estimated by (11.51), (11.42) gives an estimate of Y_q, because

$$Y_{q,0}\in Q(1-q\rho, C) \tag{11.52}$$

can be shown. In this way, we obtain an estimate $Y_q\in Q(1-q\rho, C)$, provided that $\alpha-q\rho+\rho>0$. Writing $q\rho=\beta$, we thus arrive at the estimate (11.28′) in view of (11.48).

11.4. *Generation theory of Kato–Tanabe* (KATO and TANABE [1962])

Let us define another sesquilinear form $\dot{a}_t(\,,)$ on V by

$$\dot{a}_t(u,v)=\sum_{i,j=1}^{2}\int_\Omega \frac{\partial}{\partial t}a_{ij}(t,x)\frac{\partial}{\partial x_j}u\cdot\overline{\frac{\partial}{\partial x_i}v}+\sum_{j=1}^{2}\int_\Omega \frac{\partial}{\partial t}b_j(t,x)\frac{\partial}{\partial x_j}u\cdot\bar{v}\,dx$$

$$+\int_\Omega \frac{\partial}{\partial t}c(t,x)u\cdot\bar{v}\,dx+\int_{\partial\Omega}\frac{\partial}{\partial t}\sigma(t,\xi)u\cdot\bar{v}\,dS. \tag{11.53}$$

Then, we have

$$|\dot{a}_t(u, v)| \leqslant C\|u\|_V\,\|v\|_V, \quad u,v \in V. \tag{11.54}$$

$$|\dot{a}_t(u, v)-\dot{a}_s(u, v)| \leqslant C|t-s|^\alpha\,\|u\|_V\,\|v\|_V, \quad u,v \in V, \tag{11.55}$$

$$\lim_{t\to s}\ \sup_{\substack{u,v\in V\\ \|u\|_V,\|v\|_V\leqslant 1}} \left|\frac{1}{t-s}(a_t-a_s)(u, v)-\dot{a}_s(u, v)\right| = 0 \tag{11.56}$$

for some α in $0<\alpha\leqslant 1$. From these relations, we can show that $A(t)^{-1}$ is strongly C^1 in t and that the inequalities

$$\left\|\frac{\mathrm{d}}{\mathrm{d}t}A(t)^{-1}-\frac{\mathrm{d}}{\mathrm{d}s}A(s)^{-1}\right\| \leqslant C|t-s|^\alpha, \quad t,s\in[0,T], \tag{11.57}$$

$$\left\|\frac{\partial}{\partial t}(\lambda-A(t))^{-1}\right\| \leqslant C_\varepsilon/|\lambda|, \quad \lambda\in G_{\theta+\varepsilon}, \tag{11.58}$$

hold, where $\varepsilon>0$. (See KATO and TANABE [1962] or SUZUKI [1982].) In this case, the evolution operators $\{U(t,s)\}_{T\geqslant t\geqslant s\geqslant 0}$ may be given by means of

$$U(t,s)=\mathrm{e}^{-(t-s)A(t)}+\int_s^t \mathrm{e}^{-(t-r)A(t)}R(r,s)\,\mathrm{d}r, \tag{11.59}$$

where $R=R(t,s)$ is the unique solution of the integral equation

$$R(t,s)-\int_s^t R_1(t,r)R(r,s)\,\mathrm{d}r = R_1(t,s), \tag{11.60}$$

where

$$R_1(t,s)=\frac{1}{2\pi\sqrt{-1}}\int_\Gamma \mathrm{e}^{-(t-s)\lambda}\frac{\partial}{\partial t}(\lambda-A(t))^{-1}\,\mathrm{d}\lambda, \tag{11.61}$$

Γ being the positively oriented boundary of $\Sigma_{\theta+\varepsilon}$ for some $\varepsilon>0$. Furthermore, the estimates (11.22) and the first inequality of (11.23) are also derived by this scheme.

This theory of generation is particularly remarkable in the sense that any assumptions on the domains of $A(t)$ are not made. Consequently, it is no wonder that a little stronger assumption on the smoothness in t of $A(t)$ is imposed.

In each generation theory stated so far, the construction yields estimates on the smoothness of evolution operators at the same time. For further details on these theories, see also TANABE [1979].

12. Semidiscrete approximation of temporally inhomogeneous parabolic equations

As we have seen in the preceding section, the parabolic equation (11.3) with (11.4) (or (11.4′)) and (11.5) is reduced to the evolution equation (11.7) with (11.8). In the same way as in the preceding chapter, this equation is discretized with respect to the space variables $x=(x_1, x_2)$. Namely, let us triangulate Ω into small elements with the size parameter $h>0$ and let $V_h \subset V$ be the space of piecewise linear trial functions. X_h means V_h equipped with the L^2 topology. The m-sectorial operator in X_h associated with $a_t|_{V_h \times V_h}$ is denoted by $A_h(t)$. Finally, $P_h : X \to X_h$ is the orthogonal projection. Then, the semidiscrete finite element approximation of (11.7) with (11.8) is given by

$$\mathrm{d}u_h/\mathrm{d}t + A_h(t)u_h = 0, \quad 0 \leqslant t \leqslant T, \tag{12.1}$$

with

$$u_h(0) = P_h u_0 \tag{12.2}$$

in X_h. According to the generation theory described in the preceding section, $-A_h(t)$ generates the evolution operator $\{U_h(t,s)\}_{T \geqslant t \geqslant s \geqslant 0}$. Furthermore, the inequalities

$$\|A_h(t)U_h(t,s)\|, \|U_h(t,s)A_h(s)\| \leqslant C(t-s)^{-1}, \quad 0 \leqslant s < t \leqslant T, \tag{12.3}$$

$$\|U_h(t,s)\| \leqslant C, \quad 0 \leqslant s \leqslant t \leqslant T, \tag{12.4}$$

hold uniformly in h. In the present section, putting

$$e_h(t) = u_h(t) - u(t), \tag{12.5}$$

we shall show the estimate

$$\|e_h(t)\|_X \leqslant C(h^2/t)\|u_0\|_X, \quad 0 < t \leqslant T, \tag{12.6}$$

and extend the similar result in Section 7. To this end, we employ the method of HELFRICH [1974, 1975].

First, we introduced the error operator $E_h \equiv E_h(t,s)$ by

$$E_h(t,s) = U_h(t,s)P_h - U(t,s). \tag{12.7}$$

Obviously,

$$e_h(t) = E_h(t,0)u_0. \tag{12.8}$$

Thus we have only to derive

$$\|E_h(t,s)\| \leqslant Ch^2/(t-s), \quad 0 \leqslant s < t \leqslant T. \tag{12.9}$$

From the equality

$$-\frac{\partial}{\partial r}[U_h(t,r)P_h E_h(r,s)] = U_h(t,r)[A_h(r)P_h - P_h A(r)]U(r,s)$$

follows the identity

$$P_h E_h(t,s) = \int_s^t U_h(t,r)[A_h(r)P_h - P_h A(r)]U(r,s)\,\mathrm{d}r. \tag{12.10}$$

Now, we introduce the Ritz operator $R_h(t) : V \to V_h$ through the relation

$$a_t(R_h(t)v, \chi) = a_t(v, \chi), \qquad v \in V, \quad \chi \in V_h. \tag{12.11}$$

Then,

$$A_h(t)R_h(t)v = P_h A(t)v,$$

as is shown in Section 2. Therefore, the equality

$$E_h(t,s) = (1-P_h)E_h(t,s) + P_h E_h(t,s) = E_h^1(t,s) + E_h^2(t,s) + E_h^3(t,s) \tag{12.12}$$

follows with

$$E_h^1(t,s) = (1-U_h(t,s)P_h)(R_h(t)-1)U(t,s), \tag{12.13}$$

$$E_h^2(t,s) = \int_s^t U_h(t,r)A_h(r)[R_h(r)-R_h(t)]U(t,s)\,\mathrm{d}r, \tag{12.14}$$

$$E_h^3(t,s) = \int_s^t U_h(t,r)A_h(r)P_h(R_h(r)-1)[U(r,s)-U(t,s)]\,\mathrm{d}r. \tag{12.15}$$

It suffices to show $\|E_h^l(t,s)\| \leqslant Ch^2/(t-s)$ for $l=1,2$ and 3.

(1) *Estimation of* $E_h^1(t,s)$. In Section 2, we have shown the estimates

$$\|(R_h(t)-1)v\|_V \leqslant Ch\|v\|_{H^2}, \qquad v \in V \cap H^2(\Omega), \tag{12.16}$$

$$\|(R_h(t)-1)v\|_X \leqslant Ch^2\|v\|_{H^2}, \quad v \in V \cap H^2(\Omega). \tag{12.17}$$

Therefore, from the elliptic estimate we obtain

$$\begin{aligned}\|E_h^1(t,s)\| &\leqslant (1+\|U_h(t,s)\|\cdot\|P_h\|)\cdot\|(R_h(t)-1)A(t)^{-1}\|\cdot\|A(t)U(t,s)\| \\ &\leqslant Ch^2/(t-s).\end{aligned} \tag{12.18}$$

(2) *Estimation of* $E_h^2(t,s)$. We shall show the inequality

$$\|(R_h(t)-R_h(s))v\|_X \leqslant Ch^2|t-s|^\alpha\|v\|_{H^2}, \quad v \in V \cap H^2(\Omega), \tag{12.19}$$

according to SUZUKI [1979b, 1982], from which we get

$$\begin{aligned}&\|E_h^2(t,s)\| \\ &\leqslant \int_s^t \|U_h(t,r)A_h(r)\|\cdot\|(R_h(r)-R_h(t))A(t)^{-1}\|\cdot\|A(t)U(t,s)\|\,\mathrm{d}r \\ &\leqslant C\int_s^t (t-s)^{-1+\theta}h^2\,\mathrm{d}r(t-s)^{-1} = Ch^2(t-s)^{-1+\theta} \leqslant Ch^2/(t-s).\end{aligned} \tag{12.20}$$

To this end, we introduce the adjoint sesquilinear form $a_t^*(\,,)$ by $a_t^*(u,v) = \overline{a_t(v,u)}$, $u,v \in V$, and denote by $\hat{R}_h(t)$ the Ritz operator associated with $a_t^*(\,,)$:

$$a_t^*(\hat{R}_h(t)v,\chi) = a_t^*(v,\chi), \qquad v \in V, \quad \chi \in V_h. \tag{12.21}$$

The inequalities

$$\|(\hat{R}_h(t)-1)v\|_V \leqslant Ch\|v\|_{H^2} \quad v\in V\cap H^2(\Omega), \tag{12.22}$$

$$\|(\hat{R}_h(t)-1)v\|_X \leqslant Ch^2\|v\|_{H^2} \quad v\in V\cap H^2(\Omega) \tag{12.23}$$

hold as in (12.16) and (12.17), respectively. Now, setting

$$z=(R_h(t)-R_h(s))v\in V_h, \tag{12.24}$$

we obtain

$$\begin{aligned}
\|z\|_X^2 &= a_t(z, A(t)^{*-1}z)=a_t(z, \hat{R}_h(t)A(t)^{*-1}z)\\
&= a_t((1-R_h(s))v, \hat{R}_h(t)A(t)^{*-1}z)\\
&= (a_t-a_s)((1-R_h(s))v, \hat{R}_h(t)A(t)^{*-1}z)\\
&= (a_t-a_s)((1-R_h(s))v, (\hat{R}_h(t)-1)A(t)^{*-1}z)\\
&\quad +(a_t-a_s)((1-R_h(s))v, A(t)^{*-1}z)\\
&= (a_t-a_s)((1-R_h(s))v, (\hat{R}_h(s)-1)A(t)^{*-1}z)\\
&\quad +a_s((1-R_h(s))v, (A(s)^{*-1}-A(t)^{*-1})z)\\
&= (a_t-a_s)((1-R_h(s))v, (\hat{R}_h(s)-1)A(t)^{*-1}z)\\
&\quad +a_s((1-R_h(s))v, (1-R_h(s))(A(s)^{*-1}-A(t)^{*-1})z)\\
&\leqslant C|t-s|^\alpha\|(1-R_h(s)v\|_V\cdot\|(\hat{R}_h(s)-1)A(t)^{*-1}z\|_V\\
&\quad +C\|(1-R_h(s)v\|_V\cdot\|(1-R_h(s))(A(s)^{*-1}-A(t)^{*-1})z\|_V\\
&\leqslant C|t-s|^\alpha h^2\|v\|_{H^2}\|A(t)^{*-1}z\|_{H^2}\\
&\quad +Ch^2\|v\|_{H^2}\cdot\|(A(s)^{*-1}-A(t)^{*-1})z\|_{H^2}.
\end{aligned}$$

Now, from the elliptic estimate follow

$$\|A(t)^{*-1}z\|_{H^2}\leqslant C\|z\|_X, \qquad \|(A(s)^{*-1}-A(t)^{*-1})z\|_{H^2}\leqslant C|t-s|^\alpha\|z\|_X,$$

and the desired inequality (12.19) has been established.

(3) *Estimation of* $E_h^3(t,s)$, *Case* 1. In the case of $V=H_0^1(\Omega)$, a duality argument due to Helfrich is effective (FUJITA and SUZUKI [1979]). Namely, in this case the estimate (11.24) holds so that for $u_0\in D(A(s))\equiv D$ we have

$$\begin{aligned}
\|E_h^3(t,s)u_0\| &\leqslant \int_s^t \|U_h(t,r)A_h(r)\|\cdot\|(R_h(r)-1)A(t)^{-1}\|\\
&\qquad \cdot\|A(t)[U(t,s)-U(r,s)]A(s)^{-1}\|\cdot\|A(s)u_0\|\,\mathrm{d}s\\
&\leqslant C_\theta\int_0^t (t-r)^{-1}h^2(t-r)^\theta(r-s)^{-\theta}\,\mathrm{d}r\,\|A(s)u_0\|\\
&= Ch^2\|A(s)u_0\|.
\end{aligned}$$

Similarly, from the second estimate of (11.23) we can show that

$$\|E_h^l(t,s)u_0\| \leqslant Ch^2\|A(s)u_0\|$$

for $l=1,2$. Hence

$$\|E_h(t,s)A(s)^{-1}\| \leqslant Ch^2. \tag{12.25}$$

Now, by virtue of the semigroup property of evolution operators, which means

$$U_h(t,r)U_h(r,s)=U_h(t,s), \qquad U(t,r)U(r,s)=U(t,s),$$

where $0\leqslant s\leqslant r\leqslant t\leqslant T$, the identity

$$E_h(t,s)=U_h(t,s_0)P_hE_h(s_0,s)+E_h(t,s_0)U(s_0,s) \tag{12.26}$$

holds with $s_0=\frac{1}{2}(t+s)$. The second term of the right-hand side of (12.26) is estimated as

$$\begin{aligned}&\|E_h(t,s_0)U(s_0,s)\|\\&\quad\leqslant \|E_h(t,s_0)A(s_0)^{-1}\|\cdot\|A(s_0)U(s_0,s)\|\\&\quad\leqslant Ch^2/(t-s)\end{aligned} \tag{12.27}$$

by (12.25). On the other hand, by (12.11) we have

$$\begin{aligned}&\|U_h(t,s)P_hE_h(s_0,s)\|\\&\quad=\|U_h(t,s_0)A_h(s_0)R_h(s_0)A(s_0)^{-1}E_h(s_0,s)\|\\&\quad\leqslant\|U_h(t,s_0)A_h(s_0)P_h(R_h(s_0)-1)A(s_0)^{-1}E_h(s_0,s)\|\\&\qquad+\|U_h(t,s_0)A_h(s_0)P_hA(s_0)^{-1}E_h(s_0,s)\|\\&\quad\leqslant C(t-s)^{-1}\{\|(R_h(s_0)-1)A(s_0)^{-1}\|\cdot\|E_h(s_0,s)\|\\&\qquad+\|A(s_0)^{-1}E_h(s_0,s)\|\}.\end{aligned} \tag{12.28}$$

Since $\|E_h(s_0,s)\|\leqslant C$ by (12.4), the desired estimate (12.9) is reduced to

$$\|A(t)^{-1}E_h(t,s)\|\leqslant Ch^2. \tag{12.29}$$

We set

$$\hat{U}(t,s)=U(T-s,T-t)^*, \qquad \hat{U}_h(t,s)=U_h(T-s,T-t)^*.$$

Then, $\{\hat{U}(t,s)\}_{T\geqslant t\geqslant s\geqslant 0}$ and $\{\hat{U}_h(t,s)\}_{T\geqslant t\geqslant s\geqslant 0}$ are nothing but the evolution operators generated by $\hat{A}(t)\equiv A(T-t)^*$ and $\hat{A}_h(t)=A_h(T-t)^*$ in X and X_h respectively. Furthermore, the relation

$$\hat{E}_h(t,s)\equiv\hat{U}_h(t,s)P_h-\hat{U}(t,s)=E_h(T-s,T-t)^*$$

holds. Therefore, in the same way as in (12.25), the inequality

$$\|\hat{E}_h(t,s)\hat{A}(s)^{-1}\|\leqslant Ch^2 \tag{12.30}$$

is obtained so that

$$\|A(t)^{-1}E_h(t,s)\| = \|E_h(t,s)^* A(s)^{*-1}\| \leqslant Ch^2.$$

(4) *Estimation of* $E_h^3(t,s)$, *Case* 2. In the case of $V=H^1(\Omega)$, $D(A(t))$ varies as t changes, and we cannot adopt the estimate

$$\|A(t)U(t,s)A(s)^{-1}\| \leqslant C$$

or

$$\|A(t)[U(t,s)-U(r,s)]A(s)^{-1}\| \leqslant C_\theta (t-r)^\theta (r-s)^{-\theta}, \quad 0\leqslant s<r<t<T.$$

The following argument is due to SUZUKI [1979b, 1982]. Namely, by a telescoping, we have

$$(t-s)E_h^3(t,s) = \sum_{l=1}^{5} F_h^l(t,s) \tag{12.31}$$

with

$$F_h^1(t,s) = \int_s^t (r-s)U_h(t,r)A_h(r)\cdot P_h(R_h(r)-1)[U(r,s)-U(t,s)]\,\mathrm{d}r, \tag{12.32}$$

$$F_h^2(t,s) = \int_s^t (t-r)U_h(t,r)A_h(r)\cdot (R_h(r)-R_h(s))[U(r,s)-U(t,s)]\,\mathrm{d}r, \tag{12.33}$$

$$F_h^3(t,s) = \int_s^t (t-r)[U_h(t,r)A_h(r)-U_h(t,s)A_h(s)]\cdot P_h(R_h(s)-1)[U(r,s)-U(t,s)]\,\mathrm{d}r, \tag{12.34}$$

$$F_h^4(t,s) = -U_h(t,s)A_h(s)\cdot P_h(R_h(s)-1)\int_s^t (r-s)[U(r,s)-U(t,s)]\,\mathrm{d}r \tag{12.35}$$

$$F_h^5(t,s) = U_h(t,s)A_h(s)\cdot P_h(R_h(s)-1)(t-s)\int_s^t [U(r,s)-U(t,s)]\,\mathrm{d}r. \tag{12.36}$$

We have only to derive $\|F_h^l(t,s)\| \leqslant Ch^2$ for $l=1,2,\ldots,5$.

By virtue of the construction of the evolution operators $\{U_h(t,s)\}$ and $\{U(t,s)\}$ of KATO and TANABE [1962] described in the preceding section and by the elliptic estimate (AGMON, DOUGLIS and NIRENBERG [1959]), we can show that

$$\begin{aligned}&\|U(t,s)-U(r,s)\|_{L^2\to H^2}\\ &\quad \leqslant C_\kappa\{(t-r)^\beta (r-s)^{-\beta-1} + (t-r)^\kappa (r-s)^{-1}\}, \quad 0\leqslant s<r<t\leqslant T,\end{aligned} \tag{12.37}$$

$$\| U_h(t,r)A_h(r) - U_h(t,s)A_h(s) \|$$
$$\leqslant C_\kappa\{(t-r)^{-1-\beta}(r-s)^\beta + (t-r)^{-1}(r-s)^\kappa\}, \quad 0 \leqslant s < r < t \leqslant T, \tag{12.38}$$

$$\left\| \int_s^t [U(t,s) - U(r,s)]\,\mathrm{d}r \right\|_{L^2 \to H^2} \leqslant C \tag{12.39}$$

for $0 \leqslant \beta \leqslant 1$ and $0 \leqslant \kappa < \alpha$ (see SUZUKI [1982]). From these inequalities, we can derive the following estimates:

$$\begin{aligned}\| F_h^1(t,s) \| &\leqslant \int_s^t (r-s)\| U_h(t,r)A_h(r) \| \\ &\qquad \cdot \|(R_h(r)-1)\|_{H^2 \to L^2} \cdot \| U(r,s) - U(t,s) \|_{L^2 \to H^2}\,\mathrm{d}r \\ &\leqslant C_\kappa h^2 \int_s^t (r-s)(t-r)^{-1} \\ &\qquad \cdot \{(t-r)^\beta (r-s)^{-\beta-1} + (t-r)^\kappa (r-s)^{-1}\}\,\mathrm{d}r \\ &\leqslant Ch^2, \quad 0 < \beta < 1. \end{aligned} \tag{12.40}$$

$$\begin{aligned}\| F_h^2(t,s) \| &\leqslant \int_s^t (t-r)\| U_h(t,r)A_h(r) \| \\ &\qquad \cdot \|(R_h(r)-R_h(s)\|_{H^2 \to L^2} \cdot \| U(r,s) - U(t,s) \|_{L^2 \to H^2}\,\mathrm{d}r \\ &\leqslant C_\kappa h^2 \int_s^t (r-s)\{(t-r)^\beta (r-s)^{-\beta-1} + (t-r)^\kappa (r-s)^{-1}\}\,\mathrm{d}r \\ &\leqslant Ch^2, \quad 0 < \beta < 1. \end{aligned} \tag{12.41}$$

$$\begin{aligned}\| F_h^3(t,s) \| &\leqslant \int_s^t (t-r)\| U_h(t,r)A_h(r) - U_h(t,s)A_h(s) \| \\ &\qquad \cdot \| R_h(s) - 1 \|_{H^2 \to L^2} \cdot \| U(t,s) - U(r,s) \|_{L^2 \to H^2}\,\mathrm{d}r \\ &\leqslant C_\kappa h^2 \int_s^t (t-r)\{(t-r)^{-1-\beta}(r-s)^\beta + (t-r)^{-1}(r-s)^\kappa\} \\ &\qquad \cdot \{(t-r)^\gamma (r-s)^{-1-\gamma} + (t-r)^\kappa (r-s)^{-1}\}\,\mathrm{d}r \\ &\leqslant Ch^2, \quad 1 \geqslant \beta \geqslant \kappa > \gamma > 0. \end{aligned} \tag{12.42}$$

$$\|F_h^4(t,s)\| \leqslant \|U_h(t,s)A_h(s)\| \cdot \|R_h(s)-1\|_{H^2\to L^2}$$

$$\cdot \int_s^t \|U(t,s)-U(r,s)\|_{L^2\to H^2} \cdot (r-s)\,\mathrm{d}r$$

$$\leqslant C_\kappa (t-s)^{-1} h^2$$

$$\cdot \int_s^t \{(t-r)^\beta (r-s)^{-\beta-1} + (t-r)^\kappa (r-s)^{-1}\}(r-s)\,\mathrm{d}r$$

$$\leqslant Ch^2, \quad 0<\beta<1. \tag{12.43}$$

$$\|F_h^5(t,s)\| \leqslant \|U_h(t,s)A_h(s)\| \cdot \|R_h(s)-1\|_{H^2\to L^2}$$

$$\cdot \left\| \int_s^t [U(r,s)-U(t,s)]\,\mathrm{d}r \right\|_{L^2\to H^2}$$

$$\leqslant Ch^2. \tag{12.44}$$

Summing up these estimates, we obtain $\|E_h^3(t,s)\| \leqslant Ch^2/(t-s)$.

13. Fully discrete approximation of temporally inhomogeneous parabolic equations

We obtain fully discrete approximations by discretizing the semidiscrete equation

$$\mathrm{d}u_h/\mathrm{d}t + A_h(t)u_h = 0, \quad 0 \leqslant t \leqslant T \tag{13.1}$$

with

$$u_h(0) = P_h u_0, \tag{13.2}$$

furthermore with respect to the time variable t. In the present section, we adopt the backward difference method with the mesh length $\tau>0$, $T=N\tau$, that is,

$$(u_h^\tau(t+\tau) - u_h^\tau(t))/\tau + A_h(t+\tau)u_h^\tau(r+\tau) = 0$$
$$t = n\tau, \quad n = 0, 1, \ldots, N \tag{13.3}$$

in X_h with

$$u_h^\tau(0) = P_h u_0. \tag{13.4}$$

Under the assumption (11.14′), the scheme is uniquely solvable and

$$e_h^\tau(t) = u_h^\tau(t) - u_h(t), \quad t = n\tau, \tag{13.5}$$

denotes the error. According to SUZUKI [1979b, 1982], we shall derive

$$\|e_h^\tau(t)\|_X \leqslant C(\tau/t)\|u_0\|_X, \tag{13.6}$$

and extend a similar estimate in Section 8. Combining (13.6) with (12.6), we obtain

$$\|u_h^\tau(t)-u(t)\|_X \leqslant C((h^2+\tau)/t)\|u_0\|_X, \quad t=n\tau. \tag{13.7}$$

Henceforth, we set $t_n = n\tau$ and

$$U_h^\tau(t_n, t_j) = \begin{cases} (1+\tau A_h(t_n))^{-1}(1+\tau A_h(t_{n-1}))^{-1}\cdots(1+\tau A_h(t_{j+1}))^{-1}, \\ \qquad n>j, \\ 1, \quad n=j. \end{cases} \tag{13.8}$$

Then we have

$$u_h^\tau(t) = U_h^\tau(t,0)P_h u_0, \quad t=t_n. \tag{13.9}$$

On the other hand, we have

$$u_h(t) = U_h(t,0)P_h u_0. \tag{13.10}$$

Henceforth, we drop the suffix h for simplicity of writing. Then, by virtue of (13.1) and (13.3), we get for $t=t_n$ that

$$\begin{aligned} &e^\tau(t+\tau)-e^\tau(t) \\ &\quad = \int_t^{t+\tau} [A(r)u(r)-A(t+\tau)u^\tau(t+\tau)]\,\mathrm{d}r \\ &\quad = \int_t^{t+\tau} [A(r)u(r)-A(t+\tau)u(t+\tau)]\,\mathrm{d}r - \tau A(t+\tau)e^\tau(t+\tau). \end{aligned}$$

Hence

$$\begin{aligned} e^\tau(t+\tau) &= (1+\tau A(t+\tau))^{-1}e^\tau(t) \\ &\qquad + (1+\tau A(t+\tau))^{-1}\int_t^{t+\tau}[A(r)u(r)-A(t+\tau)u(t+\tau)]\,\mathrm{d}r. \end{aligned}$$

On the other hand, we have $e^\tau(0)=0$ so that

$$\begin{aligned} e^\tau(t_n) &\equiv E^\tau(t_n)Pu_0 \\ &= -\sum_{k=1}^{n}\int_{t_{k-1}}^{t_k} (1+\tau A(t_n))^{-1}(1+\tau A(t_{n-1}))^{-1}\cdots(1+\tau A(t_k))^{-1} \\ &\qquad \cdot [A(t_k)U(t_k,0)-A(r)U(r,0)]\,\mathrm{d}r\, Pu_0, \end{aligned} \tag{13.11}$$

$E^\tau(t)$ being the error operator, $E^\tau(t) = U^\tau(t,0)-U(t,0)P$. From the identity (13.11), we see that the error estimate (13.6) can be reduced to the inequalities on the stability of the approximated solution and on the smoothness of the original one, that is, those on $U^{\tau}(t_n, t_{j+1})$ and on $A(t)U(t,s)-A(r)U(r,s)$, respectively. Actually, we can show the following:

LEMMA 13.1. *For each* $0 \leqslant \beta < \frac{4}{3}$, *the inequality*

$$\|U^{\tau}(t_n, t_j)A(t_{j+1})^{\beta}\| \leqslant C_{\beta}(t_n-t_j)^{-\beta}, \quad n-j > \beta, \tag{13.12}$$

holds true.

LEMMA 13.2. *For each* β, $0<\beta<\frac{1}{2}$, *the equality*

$$\begin{aligned} &A(t)U(t,s)-A(r)U(r,s) \\ &\quad = A(t)[\mathrm{e}^{-(t-s)A(t)} - \mathrm{e}^{-(r-s)A(t)}] + A(t)^{\beta}Z_{\beta}(t,r,s) \end{aligned} \tag{13.13}$$

holds for $0 \leqslant s < r < t \leqslant T$ *with* $Z_{\beta}(t,r,s)$ *subject to the estimate*

$$\|Z_{\beta}(t,r,s)\| \leqslant C_{\beta}(t-r)(r-s)^{\beta-1}. \tag{13.14}$$

For the moment, let us admit these lemmas. Then, the operator E^{τ} in the right-hand side of (13.11) splits as

$$\begin{aligned} -E^{\tau}(t_n) = {} & \sum_{k=1}^{n} \int_{t_{k-1}}^{t_k} (1+\tau A(t_n))^{-1} \cdots (1+\tau A(t_k))^{-1} A(t_k) \\ & \qquad \cdot [\mathrm{e}^{-t_k A(t_k)} - \mathrm{e}^{-rA(t_k)}]\, \mathrm{d}r \\ & + \sum_{k=1}^{n} \int_{t_{k-1}}^{t_k} (1+\tau A(t_n))^{-1} \cdots (1+\tau A(t_k))^{-1} A(t_k)^{\beta} Z_{\beta}(t_k, r, 0)\, \mathrm{d}r \\ = {} & (1+\tau A(t_n))^{-1} A(t_n) \int_{t_{n-1}}^{t_n} [\mathrm{e}^{-t_n A(t_n)} - \mathrm{e}^{-rA(t_n)}]\, \mathrm{d}r \\ & + \sum_{k=1}^{n-1} \int_{t_{k-1}}^{t_k} U^{\tau}(t_n, t_{k-1}) A(t_k) [\mathrm{e}^{-t_k A(t_k)} - \mathrm{e}^{-rA(t_k)}]\, \mathrm{d}r \\ & + \sum_{k=1}^{n} \int_{t_{k-1}}^{t_k} U^{\tau}(t_n, t_{k-1}) A(t_k)^{\beta} Z_{\beta}(t_k, r, 0)\, \mathrm{d}r. \end{aligned} \tag{13.15}$$

Let Γ be the positively oriented boundary of Σ_{θ_1} with $\theta_1 > \theta$. Noting that relation (11.15) and estimate (11.16) hold for $A(t) = A_h(t)$ uniformly in h, we obtain for $\kappa > -1$

that

$$\|A(r)^{\kappa}[\mathrm{e}^{-tA(r)}-\mathrm{e}^{-sA(r)}]\|$$

$$=\left\|\frac{1}{2\pi \mathrm{i}}\int_{\Gamma}\lambda^{\kappa}\mathrm{e}^{-s\lambda}[\mathrm{e}^{-(t-s)\lambda}-1](\lambda-A(r))^{-1}\,\mathrm{d}r\right\|$$

$$\leqslant C\int_0^{\infty}\mu^{\kappa}\mathrm{e}^{-s\mu\cos\theta_1}(t-s)\mu\,\frac{\mathrm{d}\mu}{\mu}$$

$$=C_{\kappa}(t-s)s^{-\kappa-1},\quad 0<s\leqslant t<\infty. \tag{13.16}$$

Then, supposing $n\geqslant 2$ we can estimate the first term of the right-hand side of (13.15) as

$$\|(1+\tau A(t_n))^{-1}A(t_n)\|\cdot\int_{t_{n-1}}^{t_n}\|\mathrm{e}^{-t_nA(t_n)}-\mathrm{e}^{-rA(t_n)}\|\,\mathrm{d}r$$

$$\leqslant C\tau^{-1}\int_{t_{n-1}}^{t_n}(t_n-r)r^{-1}\,\mathrm{d}r\leqslant C\tau^{-1}\cdot\tau^2(t_{n-1})^{-1}\leqslant Cn^{-1}.$$

On the other hand, by Lemma 13.2, the third term of the right-hand side of (13.15) is estimated as

$$\sum_{k=1}^{n}\int_{t_{k-1}}^{t_k}\|U^{\tau}(t_n,t_{k-1})A(t_k)^{\beta}\|\cdot\|Z_{\beta}(t_k,r,0)\|\,\mathrm{d}r$$

$$\leqslant C_{\beta}\sum_{k=1}^{n}\int_{t_{k-1}}^{t_k}(t_n-t_{k-1})^{-\beta}(t_k-r)r^{-1+\beta}\,\mathrm{d}r$$

$$\leqslant C_{\beta}\sum_{k=1}^{n}(n-k+1)^{-\beta}\tau^{-\beta}\tau^2(k\tau)^{\beta-1}$$

$$=C_{\beta}\tau\sum_{k=1}^{n}(n-k+1)^{-\beta}k^{\beta-1}.$$

We here recall the elementary inequality

$$B^N(a,b)\equiv\frac{1}{N}\sum_{k=1}^{N-1}\left(1-\frac{k}{N}\right)^{a-1}\left(\frac{k}{N}\right)^{b-1}$$

$$\leqslant B(a,b)\equiv\int_0^1(1-x)^{a-1}x^{b-1}\,\mathrm{d}x \tag{13.17}$$

for $0<a\leqslant b$ and $a\leqslant 1$. In fact, $f(x)=(1-x)^{a-1}x^{b-1}$ is monotonically increasing in $[0,1)$ when $b\geqslant 1\geqslant a$, while $f(x)$ is convex in $(0,1)$ when $a,b\leqslant 1$. Then, we have

$$\sum_{k=1}^{n}\int_{t_{k-1}}^{t_k}\|U^\tau(t_n,t_{k-1})A(t_k)^\beta\|\cdot\|Z_\beta(t_k,r,0)\|\,dr\leqslant C\tau.$$

In this way, the desired estimate $\|E^\tau(t,0)\|\leqslant C/n$, $t=t_n$, is reduced to an inequality concerning the second term of the right-hand side of (13.15), that is,

$$n\left\|\sum_{k=1}^{n-1}\int_{t_{k-1}}^{t_k}U^\tau(t_n,t_{k-1})A(t_k)[e^{-t_kA(t_k)}-e^{-rA(t_k)}]\,dr\right\|\leqslant C. \tag{13.18}$$

In fact, taking β in $0<\beta<\frac{1}{3}$, we have

$$\begin{aligned}
&n\sum_{k=1}^{n-1}\int_{t_{k-1}}^{t_k}U^\tau(t_n,t_{k-1})A(t_k)[e^{-t_kA(t_k)}-e^{-rA(t_k)}]\,dr\\
&=\sum_{k=1}^{n-1}(n-k+1)U^\tau(t_n,t_{k-1})A(t_k)^{1+\beta}\\
&\qquad\cdot\int_{t_{k-1}}^{t_k}A(t_k)^{-\beta}[e^{-t_kA(t_k)}-e^{-rA(t_k)}]\,dr\\
&\quad+\sum_{k=2}^{n-1}(k-1)U^\tau(t_n,t_{k-1})A(t_k)^{1-\beta}\\
&\qquad\cdot\int_{t_{k-1}}^{t_k}A(t_k)^{\beta}[e^{-t_kA(t_k)}-e^{-rA(r_k)}]\,dr.
\end{aligned}\tag{13.19}$$

The first term of the right-hand side of (13.19) is estimated by Lemma 13.1 and (13.16) as

$$\begin{aligned}
&\sum_{k=1}^{n-1}(n-k+1)\|U^\tau(t_n,t_{k-1})A(t_k)^{1+\beta}\|\cdot\int_{t_{k-1}}^{t_k}\|A(t_k)^{-\beta}[e^{-t_kA(t_k)}-e^{-rA(t_k)}]\|\,dr\\
&\leqslant C_\beta\sum_{k=1}^{n-1}(n-k+1)^{-\beta}\tau^{-1-\beta}\int_{t_{k-1}}^{t_k}(t_k-r)r^{\beta-1}\,dr\\
&\leqslant C_\beta\sum_{k=1}^{n-1}(n-k+1)^{-\beta}\tau^{-1-\beta}\cdot\tau^2(k\tau)^{\beta-1}
\end{aligned}$$

$$=C_\beta \sum_{k=1}^{n} (n-k+1)^{-\beta} k^{\beta-1} \leqslant C.$$

Similarly, the second term of the right-hand side of (13.19) is estimated as

$$\sum_{k=2}^{n-1} (k-1)\|U^\tau(t_n, t_{k-1})A(t_k)^{1-\beta}\| \cdot \int_{t_{k-1}}^{t_k} \|A(t_k)^\beta[\mathrm{e}^{-t_k A(t_k)} - \mathrm{e}^{-rA(t_k)}]\|\, \mathrm{d}r$$

$$\leqslant C_\beta \sum_{k=2}^{n-1} (k-1)(t_n - t_{k-1})^{-1+\beta} \cdot \int_{t_{k-1}}^{t_k} (t_k - r) r^{-\beta-1}\, \mathrm{d}r$$

$$\leqslant C_\beta \sum_{k=2}^{n-1} (k-1)(n-k+1)^{-1+\beta} \cdot \tau^{-1+\beta} \cdot \tau^2 \cdot ((k-1)\tau)^{-\beta-1}$$

$$\leqslant C_\beta \sum_{k=2}^{n-1} (k-1)^{-\beta}(n-k+1)^{\beta-1} \leqslant C.$$

In this way, (13.18) has been established.

In the remainder of the present section, we shall give outlines of the proof of Lemmas 13.1 and 13.2.

Stability of the approximated solution (Proof of Lemma 13.1)

First, we note that the inequality (13.12) is reduced to

$$\|A(t_n)^\beta U^\tau(t_n, t_j)\| \leqslant C_\beta (t_n - t_j)^{-\beta}, \quad 0 \leqslant \beta < \tfrac{4}{3}, \tag{13.20}$$

by considering its adjoint form. Now, we recall that the generation theorem of KATO [1961b] described in Section 11 has yielded the continuous version of (13.20), that is,

$$\|A(t)^\beta U(t,s)\| \leqslant C_\beta (t-s)^{-\beta}. \tag{13.20'}$$

All we have to do is to trace his computations in the contexts of the discrete version.

In fact, we note that the inequality (11.27) holds for $A(t) = A_h(t)$ uniformly in h by the theory of KATO [1961b]. Sobolevskii's identity (11.31) holds for $A(t) = A_h(t)$ so that we obtain

$$U^\tau(t_n, t_j) - (1+\tau A(t_j))^{-(n-j)}$$

$$= \sum_{k=j}^{n-1} [(1+\tau A(t_n))^{-(n-k-1)} U^\tau(t_{k+1}, t_j) - (1+\tau A(t_n))^{-(n-k)} U^\tau(t_k, t_j)]$$

$$= \sum_{k=j}^{n-1} (1+\tau A(t_n))^{-(n-k)} [(1+\tau A(t_n)) - (1+\tau A(t_{k+1}))] U^\tau(t_{k+1}, t_j)$$

$$= \tau \sum_{k=j}^{n-1} (1+\tau A(t_n))^{-(n-k)} [A(t_n) - A(t_{k+1})] U^\tau(t_{k+1}, t_j)$$

$$= \sum_{p=1}^{n} \tau \sum_{k=j}^{n-2} (1+\tau A(t_n))^{-(n-k)} A(t_n)^{1-p\rho} \cdot D(t_n, t_{k+1}) A(t_{k+1})^{p\rho} U^\tau(t_{k+1}, t_j). \tag{13.21}$$

Let us introduce the following notations: For operator-valued functions $K_l = K_l(t_n, t_j)$, $l=1,2$, on $D^\tau = \{(t_n, t_j): N \geqslant n \geqslant j \geqslant 0\}$, another function $K = K_1 *^\tau K_2$ is defined by

$$(K_1 *^\tau K_2)(t_n, t_j) = \tau \sum_{k=j}^{n-2} K_1(t_n, t_{k+1}) K_2(t_{k+1}, t_j).$$

Furthermore, we set

$$W^\tau(t_n, t_j) = U^\tau(t_n, t_j) - (1+\tau A(t_n))^{-(n-j)}, \tag{13.22}$$

$$Y_q^\tau(r_n, t_j) = A(t_n)^{q\rho} W^\tau(t_n, t_j). \tag{13.23}$$

Then, the equality (13.21) reads:

$$Y_q^\tau = \sum_{p=1}^{m} H_{q,p}^\tau *^\tau Y_p^\tau + Y_{q,0}^\tau, \tag{13.24}$$

where

$$H_{q,p}^\tau(t_n, t_j) = A(t_n)^{1-p\rho+q\rho}(1+\tau A(t_n))^{-(n-j+1)} D(t_n, t_j), \tag{13.25}$$

$$Y_{q,0}^\tau = \sum_{p=1}^{m} H_{q,p}^\tau *^\tau Y_{p,-1}^\tau \tag{13.26}$$

with

$$Y_{p,-1}^\tau(t_n, t_j) = A(t_n)^{p\rho}(1+\tau A(t_n))^{-(n-j)}. \tag{13.27}$$

Note the elementary inequality (13.17). Then, the desired inequality (13.20) can be derived in a similar way to that of KATO [1961b] (see SUZUKI [1982], for details).

Smoothness of the original solution (*Proof of Lemma 13.2*)

We restate the construction of the evolution operator $\{U(t,s)\}_{T \geqslant t \geqslant s \geqslant 0}$ by KATO and TANABE [1962] described in Section 11, that is,

$$U(t,s) = \mathrm{e}^{-(t-s)A(t)} + W(t,s) \tag{13.28}$$

with

$$W(t,s) = \int_s^t \mathrm{e}^{-(t-r)A(t)} R(r,s)\,\mathrm{d}r, \tag{13.29}$$

where $R = R(t,s)$ is the unique solution of

$$R(t,s)=R_1(t,s)+\int_s^t R_1(t,r)R(r,s)\,\mathrm{d}s \tag{13.30}$$

for

$$R_1(t,s)=\frac{1}{2\pi \mathrm{i}}\int_\Gamma \mathrm{e}^{-(t-s)\lambda}\,\frac{\partial}{\partial t}(\lambda-A(t))^{-1}\,\mathrm{d}\lambda. \tag{13.31}$$

From these relations, we have

$$\begin{aligned}
&A(t)U(t,s)-A(r)U(r,s)\\
&\quad=A(t)[\mathrm{e}^{-(t-s)A(t)}-\mathrm{e}^{-(r-s)A(t)}]\\
&\qquad+A(t)^\beta[A(t)^{1-\beta}\mathrm{e}^{-(r-s)A(t)}-A(r)^{1-\beta}\mathrm{e}^{-(r-s)A(r)}]\\
&\qquad+A(t)^\beta(1-A(t)^{-\beta}A(r)^\beta)A(r)^{1-\beta}\mathrm{e}^{-(r-s)A(r)}\\
&\qquad+A(t)^\beta[A(t)^{1-\beta}W(t,s)-A(r)^{1-\beta}W(r,s)]\\
&\qquad+A(t)^\beta(1-A(t)^{-\beta}A(r)^\beta)A(r)^{1-\beta}W(r,s),
\end{aligned}$$

so that

$$Z_\beta(t,r,s)=\sum_{l=1}^{4}Z_\beta^l(t,r,s), \tag{13.32}$$

where

$$Z_\beta^1(t,r,s)=A(t)^{1-\beta}\mathrm{e}^{-(r-s)A(t)}-A(r)^{1-\beta}\mathrm{e}^{-(r-s)A(r)}, \tag{13.33}$$

$$Z_\beta^2(t,r,s)=(1-A(t)^{-\beta}A(r)^\beta)A(r)^{1-\beta}\mathrm{e}^{-(r-s)A(r)}, \tag{13.34}$$

$$Z_\beta^3(t,r,s)=A(t)^{1-\beta}W(t,s)-A(r)^{1-\beta}W(r,s), \tag{13.35}$$

$$Z_\beta^4(t,r,s)=(1-A(t)^{-\beta}A(r)^\beta)A(r)^{1-\beta}W(r,s). \tag{13.36}$$

By the Dunford integral, we have

$$Z_\beta^1(t,r,s)=\frac{1}{2\pi\mathrm{i}}\int_\Gamma \lambda^{1-\beta}\mathrm{e}^{-(r-s)\lambda}[(\lambda-A(t))^{-1}-(\lambda-A(s))^{-1}]\,\mathrm{d}\lambda, \tag{13.37}$$

while

$$\|(\lambda-A(t))^{-1}-(\lambda-A(s))^{-1}\|\leqslant C_\varepsilon|t-s|/|\lambda|,\quad \lambda\in\Sigma_{\theta+\varepsilon} \tag{13.38}$$

follows from (11.58). Therefore, we obtain

$$\begin{aligned}
\|Z_\beta^1(t,r,s)\|&\leqslant C\int_0^\infty \mu^{1-\beta}\mathrm{e}^{-(r-s)\mu\cos\theta_1}\,\frac{\mathrm{d}\mu}{\mu}\,|t-r|\\
&=C_\beta(t-r)(r-s)^{\beta-1},\quad \theta_1>\theta.
\end{aligned} \tag{13.39}$$

Next,

$$\|1-A(t)^{-\beta}A(s)^{\beta}\| \leqslant C_{\beta}|t-s|, \quad 0\leqslant\beta<\tfrac{1}{2} \tag{13.40}$$

follows from (11.27), and consequently

$$\begin{aligned}\|Z_{\beta}^{2}(t,r,s)\| &\leqslant \|1-A(t)^{-\beta}A(r)^{\beta}\|\cdot\|A(r)^{1-\beta}e^{-(r-s)A(r)}\|\\ &\leqslant C_{\beta}(t-r)(r-s)^{\beta-1}.\end{aligned} \tag{13.41}$$

The estimate

$$\|R(t,s)\| \leqslant C \tag{13.42}$$

has been shown in KATO and TANABE [1962] so that

$$\begin{aligned}\|A(t)^{1-\beta}W(t,s)\| &= \left\|\int_{s}^{t} A(t)^{1-\beta}e^{-(t-r)A(t)}R(r,s)\,dr\right\|\\ &= C_{\beta}\int_{s}^{t}(t-r)^{\beta-1}\,dr\\ &= C_{\beta}(t-s)^{\beta}.\end{aligned} \tag{13.43}$$

Therefore,

$$\|Z_{\beta}^{4}(t,r,s)\| \leqslant C_{\beta}(t-r)(r-s)^{\beta}. \tag{13.44}$$

Thus, the proof of Lemma 13.2 has been reduced to

$$\begin{aligned}\|Z_{\beta}^{3}(t,r,s)\| &= \|A(t)^{1-\beta}W(t,s)-A(r)^{1-\beta}W(r,s)\|\\ &\leqslant C_{\beta}(t-r)(r-s)^{\beta-1}, \quad 0<\beta<\tfrac{1}{2}.\end{aligned} \tag{13.45}$$

We now recall (13.29) and obtain

$$A(t)^{1-\beta}W(t,s)-A(r)^{1-\beta}W(r,s)=\sum_{l=5}^{7}Z_{\beta}^{l}(t,r,s), \tag{13.46}$$

where

$$Z_{\beta}^{5}(t,r,s)=\int_{r}^{t}A(t)^{1-\beta}e^{-(t-z)A(t)}[R(z,s)-R(t,s)]\,dz, \tag{13.47}$$

$$\begin{aligned}Z_{\beta}^{6}(t,r,s)=\int_{s}^{r}&[A(t)^{1-\beta}e^{-(t-z)A(t)}-A(r)^{1-\beta}e^{-(r-z)A(r)}]\\ &\cdot[R(z,s)-R(r,s)]\,dz,\end{aligned} \tag{13.48}$$

$$Z_\beta^7(t,r,s)=\int_r^t A(t)^{1-\beta}e^{-(t-z)A(t)}\,dz\cdot R(t,s)$$

$$+\int_s^r [A(t)^{1-\beta}e^{-(t-z)A(t)}-A(r)^{1-\beta}e^{-(r-z)A(r)}]\,dz\cdot R(r,s). \tag{13.49}$$

The estimate

$$\|R(t,s)-R(r,s)\|\leqslant C_\gamma(t-r)^\gamma(r-s)^{-\gamma},\quad 0\leqslant s<r<t\leqslant T, \tag{13.50}$$

is shown for $0<\gamma<1$ by KATO and TANABE [1962]. Therefore, we get

$$\|Z_\beta^5(t,r,s)\|\leqslant \int_r^t \|A(t)^{1-\beta}e^{-(t-z)A(t)}\|\cdot\|R(z,s)-R(t,s)\|\,dz$$

$$\leqslant C_{\beta,\gamma}\int_r^t (t-z)^{\beta-1}(t-z)^\gamma(z-s)^{-\gamma}\,dz$$

$$\leqslant C_{\beta,\gamma}\int_r^t (t-z)^{\beta+\gamma+1}(z-r)^{-\beta+1+\gamma}\,dz$$

$$=C_\beta(t-r)(r-s)^{\beta-1} \tag{13.51}$$

by taking γ in $1-\beta<\gamma<1$. Next, from (13.16) and (13.39) we obtain

$$\|A(t)^{1-\beta}e^{-(t-z)A(t)}-A(r)^{1-\beta}e^{-(r-z)A(r)}\|$$
$$\leqslant \|A(t)^{1-\beta}e^{1(t-z)A(t)}-A(r)^{1-\beta}e^{-(r-z)A(r)}\|$$
$$+\|A(r)^{1-\beta}[e^{-(t-z)A(r)}-e^{-(r-z)A(r)}]\|$$
$$\leqslant C_\beta(t-r)(r-z)^{\beta-2},$$

so that

$$\|Z_\beta^6(t,r,s)\|\leqslant \int_s^r \|A(t)^{1-\beta}e^{(t-z)A(t)}-A(r)^{1-\beta}e^{-(r-z)A(r)}\|$$
$$\cdot\|R(r,s)-R(z,s)\|\,dz$$

$$\leqslant C_{\beta,\gamma}\int_s^r (t-r)(r-z)^{\beta-2}(r-z)^\gamma(z-s)^{-\gamma}\,dz$$

$$=C_\beta(t-r)(r-s)^{\beta-1} \tag{13.52}$$

holds when γ is taken in $1-\beta<\gamma<1$.

Finally, we have

$$\begin{aligned} Z_\beta^7(t,r,s) &= [A(t)^{-\beta}e^{-(t-z)A(t)}]_{z=r}^{z=t}\cdot R(t,s) \\ &\quad + [A(t)^{-\beta}e^{-(t-z)A(t)} - A(r)^{-\beta}e^{-(r-z)A(r)}]_{z=s}^{z=r}\cdot R(t,s) \\ &= \sum_{l=8}^{11} Z_\beta^l(t,r,s), \end{aligned} \tag{13.53}$$

where

$$Z_\beta^8(t,r,s) = (1-e^{-(t-r)A(t)})[A(t)^{-\beta}R(t,s) - A(r)^{-\beta}R(r,s)], \tag{13.54}$$

$$Z_\beta^9(t,r,s) = -e^{-(t-r)A(t)}[A(r)^{-\beta} - A(t)^{-\beta}]R(r,s), \tag{13.55}$$

$$Z_\beta^{10}(t,r,s) = -(A(t)^{-\beta}e^{-(t-s)A(t)} - A(t)^{-\beta}e^{-(r-s)A(t)})R(r,s), \tag{13.56}$$

$$Z_\beta^{11}(t,r,s) = -(A(t)^{-\beta}e^{-(r-s)A(t)} - A(r)^{-\beta}e^{-(r-s)A(r)})R(r,s). \tag{13.57}$$

The inequality

$$\|A(t)^{-\beta} - A(r)^{-\beta}\| \leqslant C_\beta|t-s|, \quad 0<\beta<\tfrac{1}{2} \tag{13.58}$$

follows from (11.27), and we get

$$\|Z_\beta^9(t,r,s)\| \leqslant C_\beta(t-r)(r-s)^{\beta-1}. \tag{13.59}$$

Furthermore, the inequalities

$$\|Z_\beta^{10}(t,r,s)\| \leqslant C_\beta(t-r)(r-s)^{\beta-1}, \tag{13.60}$$

$$\|Z_\beta^{11}(t,r,s)\| \leqslant C_\beta(t-r)(r-s)^{\beta-1} \tag{13.61}$$

follow from (13.16) and (13.39), respectively. Therefore, the proof of Lemma 13.2 is now reduced to

$$\begin{aligned} &\|A(t)^{-\beta}R(t,s) - A(r)^{-\beta}R(r,s)\| \leqslant C_\beta(t-r)(r-s)^{\beta-1} \\ &0\leqslant s<r<t\leqslant T, \end{aligned} \tag{13.62}$$

for $0<\beta<\frac{1}{2}$.

By virtue of the integral equation (13.30), this inequality is furthermore reduced to

$$\begin{aligned} &\|A(t)^{-\beta}R_1(t,s) - A(r)^{-\beta}R_1(r,s)\| \leqslant C_\beta(t-r)(r-s)^{\beta-1} \\ &0\leqslant s<r<t\leqslant T, \end{aligned} \tag{13.63}$$

$R_1 = R_1(t,s)$ being the right-hand side of (13.31). The estimate (13.63) is established by means of the spectral decomposition in the case that the $A(t)$ are self-adjoint. In the general case, we take the "real part" of $A(t)$ and compare it with $A(t)$ as in Section 5. Then, (13.63) follows (see SUZUKI [1982], for details).

14. Error analysis by energy methods

In 1979, the second author obtained the estimate

$$\|u_h^\tau(t)-u(t)\|_X \leqslant C((h^2+\tau)/t)\|u_0\|_X \tag{14.1}$$

by the methods described above in Sections 12 and 13 (SUZUKI [1979a]). This method does not employ the Hilbert space structure of the problem and does work in the framework of Banach spaces as well. On the other hand, error analysis on the fully discrete approximation of temporally inhomogeneous parabolic equations has been done independently by Sammon, Luskin, Rannacher, Thomée, Huang, Biocchi, Brezzi and others (BAIOCCHI and BREZZI [1983], HUANG and THOMÉE [1981, 1982], LUSKIN and RANNACHER [1982], SAMMON [1982, 1983], e.g.). The methods of these authors are characterized by systematic use of the Hilbert space structure of the problem. In fact, by energy methods, LUSKIN and RANNACHER [1982a] and HUANG and THOMÉE [1982] succeeded in deriving the optimal rate of convergence, that is (14.1), for the backward difference finite element approximation. In the present section, we introduce the methods of LUSKIN and RANNACHER [1982a] and HUANG and THOMÉE [1982] for the semidiscrete approximation and the fully discrete one, respectively.

14.1. *A priori estimates*

First, we shall show several a priori estimates on the evolution equation

$$\mathrm{d}u/\mathrm{d}t+A(t)u=f(t), \quad 0<t\leqslant T, \tag{14.2}$$

in a Hilbert space X with the initial condition

$$u(0)=u_0. \tag{14.3}$$

Here $A(t)$ is an m-sectorial operator associated with a sesquilinear form $a_t(,)$ on V, where $V\subset X\subset V^*$ denotes the triple of Hilbert spaces described in Section 1. As before, we suppose that $a_t(,)$ satisfies

$$|a_t(u,v)|\leqslant C\|u\|_V\cdot\|v\|_V, \quad u,v\in V, \tag{14.4}$$

and

$$\operatorname{Re} a_t(u,u)\geqslant \delta\|u\|_V^2, \quad u\in V, \tag{14.5}$$

without loss of generality, where C and δ are positive constants. Furthermore, we assume that $a_t(u,v)$, $u,v\in V$, is sufficiently smooth and put

$$\dot a_t(u,v)=(\partial/\partial t)a_t(u,v),$$
$$\ddot a_t(u,v)=(\partial^2/\partial t^2)a_t(u,v), \quad \text{etc.}$$

Then, (14.2) is equivalent to its weak form, namely,

$$(u_t,v)+a_t(u,v)=(f(t),v), \quad v\in V, \tag{14.6}$$

$(\,,\,)$ being the inner product in X. Therefore, we also have

$$(u_{tt}, v)+a_t(u_t, v)= -\dot{a}_t(u, v)+(f_t(t), v), \quad v\in V. \tag{14.7}$$

PROPOSITION 14.1. *The solution* $u=u(t)$ *of* (14.1) *with* (14.2) *satisfies the inequality*

$$\|u(t)\|_X^2+\int_0^t \|u(s)\|_V^2\,\mathrm{d}s\leqslant C\|u_0\|_X^2+C\int_0^t \|f(s)\|_{V^*}^2\,\mathrm{d}s. \tag{14.8}$$

PROOF. Putting $v=u(t)$ in (14.6), we have

$$\tfrac{1}{2}(\mathrm{d}/\mathrm{d}t)\|u(t)\|_X^2+\delta\|u(t)\|_V^2\leqslant\|f(t)\|_{V^*}\cdot\|u(t)\|_V$$

by (14.5). Therefore, from Schwarz's inequality,

$$(\mathrm{d}/\mathrm{d}t)\|u(t)\|_X^2+\|u(t)\|_V^2\leqslant C\|f(t)\|_{V^*}^2$$

follows. Hence (14.8) holds. □

We recall the adjoint a_t^* and the real part a_t^0 of a_t, that is,

$$a_t^*(u,v)=\overline{a_t(v,u)}, \quad u,v\in V, \qquad a_t^0=\tfrac{1}{2}(a_t+a_t^*).$$

Henceforth we suppose that the identity

$$a_t(u,v)=a_t^0(u,v)+(Bu,v), \quad u,v\in V, \tag{14.9}$$

holds with a bounded operator $B=B(t): V\to X$. This assumption is satisfied when $a_t(\,,\,)$ is associated with a second-order elliptic operator with real coefficients as we have seen in Chapter I.

PROPOSITION 14.2. *The solution* $u=u(t)$ *of* (14.2) *with* (14.3) *satisfies the inequality*

$$\|u(t)\|_V^2+\int_0^t \|u_t(s)\|_X^2\,\mathrm{d}s\leqslant C\|u_0\|_V^2+C\int_0^t \|f(s)\|_X^2\,\mathrm{d}s. \tag{14.10}$$

PROOF. Putting $v=u_t$ in (14.6), we have

$$\|u_t\|_X^2+a_t(u,u_t)=(f,u_t)\leqslant\tfrac{1}{2}\|f\|_X^2+\tfrac{1}{2}\|u_t\|_X^2.$$

On the other hand, by means of (14.9) we obtain

$$a_t(u,u_t)=\tfrac{1}{2}(\mathrm{d}/\mathrm{d}t)a_t(u,u)-\tfrac{1}{2}\dot{a}_t(u,v)+\tfrac{1}{2}(Bu,u_t),$$

so that

$$\|u_t\|_X^2+(\mathrm{d}/\mathrm{d}t)a_t(u,u)\leqslant C\|f\|_X^2+C\|u\|_V^2. \tag{14.11}$$

The inequality (14.10) follows from (14.5), (14.7) and (14.11). □

PROPOSITION 14.3. *The solution $u=u(t)$ of* (14.2) *with* (14.3) *satisfies the inequality*

$$\|u_t(t)\|_X^2+\int_0^t \|u_t(s)\|_V^2 \leqslant C\|u_t(0)\|_X^2+C\|u_0\|_X^2+C\int_0^t \{\|f(s)\|_{V^*}^2+\|f_t(s)\|_{V^*}^2\}\,\mathrm{d}s. \tag{14.12}$$

PROOF. Putting $v=u_t$ in (14.7), we have

$$\begin{aligned}\tfrac{1}{2}(\mathrm{d}/\mathrm{d}t)\|u_t\|_X^2+\delta\|u_t\|_V^2 &\leqslant C\|u\|_V\cdot\|u_t\|_V+\|f_t\|_{V^*}\cdot\|u_t\|_V\\ &\leqslant \tfrac{1}{2}\delta\|u_t\|_V^2+C\|u\|_V^2+C\|f_t\|_{V^*}^2,\end{aligned}$$

so that

$$\|u_t(t)\|_X^2+\int_0^t \|u_t(s)\|_V^2\,\mathrm{d}s \leqslant \|u_t(0)\|_X^2+C\int_0^t \|u(s)\|_V^2\,\mathrm{d}s+C\int_0^t \|f_t(s)\|_{V^*}^2\,\mathrm{d}s. \tag{14.13}$$

The inequality (14.12) follows from (14.8) and (14.13). □

PROPOSITION 14.4. *The solution $u=u(t)$ of* (14.2) *with* (14.3) *satisfies the inequality*

$$\int_0^t s^2\|u_t(s)\|_V^2\,\mathrm{d}s \leqslant C\|u_0\|_X^2+C\int_0^t \{\|f(s)\|_{V^*}^2+\|f_t(s)\|_{V^*}^2\}\,\mathrm{d}s. \tag{14.14}$$

PROOF. Putting $v=t^2u_t(t)$ in (14.7), we have

$$\begin{aligned}&\tfrac{1}{2}(\mathrm{d}/\mathrm{d}t)(t^2\|u_t\|_X^2)+\delta t^2\|u_t\|_V^2\\ &\quad\leqslant -t^2\dot{a}_t(u,u_t)+t\|u_t\|_X^2+t^2(f_t,u_t)\\ &\quad\leqslant \tfrac{1}{4}\delta t^2\|u_t\|_V^2+C\|u\|_V^2+t\|u_t\|_X^2+\tfrac{1}{4}\delta t^2\|u_t\|_V^2+C\|f_t\|_{V^*}^2\end{aligned}$$

so that

$$(\mathrm{d}/\mathrm{d}t)(t^2\|u_t\|_X^2)+t^2\|u_t\|_V^2 \leqslant C\|u\|_V^2+t\|u_t\|_X^2+C\|f_t\|_{V^*}^2.$$

Therefore, we get

$$\int_0^t s^2\|u_t(s)\|_V^2\,\mathrm{d}s \leqslant C\int_0^t \{\|u(s)\|_V^2+s\|u_t(s)\|_X^2+\|f_t(s)\|_{V^*}^2\}\,\mathrm{d}s. \tag{14.15}$$

Next, putting $v=tu_t(t)$ in (14.6), we have

$$\begin{aligned}t\|u_t\|_X^2+(\mathrm{d}/\mathrm{d}t)(ta_t(u,u)) &= a_t(u,u)+\dot{a}_t(u,u)+ta_t(u_t,u)+t(f,u_t)\\ &\leqslant \varepsilon t^2\|u_t\|_V^2+C_\varepsilon\|u\|_V^2+C_\varepsilon\|f\|_{V^*}^2\end{aligned}$$

for each $\varepsilon>0$. Therefore,

$$\int_0^t s\|u_t(s)\|_X^2\,ds \leqslant \varepsilon\int_0^t s^2\|u_t(s)\|_V^2\,ds + C_\varepsilon\int_0^t \{\|u(s)\|_V^2+\|f(s)\|_{V^*}^2\}\,ds \tag{14.16}$$

holds. Taking $\varepsilon>0$ small enough and combining (14.15) with (14.16), we have

$$\int_0^t s^2\|u_t(s)\|_V^2\,ds \leqslant C\int_0^t \{\|u(s)\|_V^2+\|f(s)\|_{V^*}^2+\|f_t(s)\|_{V^*}^2\}\,ds. \tag{14.17}$$

The inequalities (14.8) and (14.17) imply (14.14). □

PROPOSITION 14.5. *The solution $u=u(t)$ of* (14.2) *and* (14.3) *satisfies the inequality*

$$t\|u_t(t)\|_V^2+\int_0^t \|u_{tt}(s)\|_X^2\,ds$$
$$\leqslant C\|u_t(0)\|_X^2+C\|u_0\|_X^2+C\int_0^t \{\|f_t(s)\|_X^2+\|f(s)\|_{V^*}^2\}\,ds. \tag{14.18}$$

PROOF. Putting $v=u_{tt}(t)$ in (14.7), we have

$$\|u_{tt}\|_X^2+a_t(u_t,u_{tt})=-\dot a_t(u,u_{tt})+(f_t,u_{tt}).$$

Here, we note

$$\tfrac12(\mathrm d/\mathrm dt)a_t(u_t,u_t)=a_t(u_t,u_{tt})+\tfrac12\dot a_t(u_t,u_t)+\tfrac12(Bu_t,u_{tt})$$
$$\leqslant a_t(u_t,u_{tt})+C\|u_t\|_V^2+\tfrac14\|u_{tt}\|_X^2$$

as well as

$$-\dot a_t(u,u_{tt})=-(\mathrm d/\mathrm dt)\dot a_t(u,u_t)+\ddot a_t(u,t_t)+a_t(u_t,u_t)$$
$$\leqslant -(\mathrm d/\mathrm dt)\dot a_t(u,u_t)+C\|u_t\|_V^2+C\|u\|_V^2$$

and

$$(f_t,u_{tt})\leqslant 4\|f_t\|_X^2+\tfrac14\|u_{tt}\|_X^2.$$

Summing these relations, we obtain

$$\|u_{tt}\|_X^2+(\mathrm d/\mathrm dt)a_t(u_t,u_t)+(\mathrm d/\mathrm dt)\dot a_t(u,u_t)$$
$$\leqslant C\|u_t\|_V^2+C\|u\|_V^2+C\|f_t\|_X^2. \tag{14.19}$$

Multiplying with t on both sides, we get

$$t\|u_{tt}\|_X^2+(\mathrm d/\mathrm dt)t\{a_t(u_t,u_t)+\dot a_t(u,u_t)\}$$
$$\leqslant C(t+1)\{\|u_t\|_V^2+\|u\|_V^2+\|f_t\|_X^2\},$$

which implies

$$ta_t(u_t, u_t) + \int_0^t s\|u_{tt}(s)\|_X^2 \leqslant C \int_0^t \{\|u_t(s)\|_V^2 + \|u(s)\|_V^2 + \|f_t(s)\|_X^2\}\, ds. \tag{14.20}$$

The desired inequality (14.18) follows from (14.8), (14.12) and (14.20). □

14.2. Error analysis of semidiscrete approximations

Let

$$du/dt + A(t)u = 0, \quad 0 < t \leqslant T, \tag{14.21}$$

in $X = L^2(\Omega)$ be the evolution equation described in Section 11 with the initial condition

$$u(0) = u_0. \tag{14.22}$$

We want to make an error analysis by the energy method concerning the semidiscrete finite element approximation

$$du_h/dt + A_h(t)u_h = 0, \quad 0 \leqslant t \leqslant T, \tag{14.23}$$

in X_h with

$$u_h(0) = P_h u_0, \tag{14.24}$$

described in Section 12. We recall that $R_h = R_h(t)\colon V \to V_h$ denotes the Ritz operator associated with $a_t(\,,)$:

$$a_t(R_h(t)v, \chi) = a_t(v, \chi), \qquad v \in V, \quad \chi \in V_h. \tag{14.25}$$

In the present subsection, we only consider the case $V = H_0^1(\Omega)$, for simplicity.
First, we note the following:

LEMMA 14.1. *Any V-valued C^1 function $v = v(t)$ satisfies*

$$\|\partial_t(v - R_h v)\|_{H^j} \leqslant Ch^{k-j}\{\|v\|_{H^k} + \|v_t\|_{H^{k-j}}\} \tag{14.26}$$

for $k = 1, 2$ and $j = 0, 1$.

PROOF. From relation (14.25), we have

$$a_t(\partial_t(v - R_h v), \chi) = -\dot{a}_t(v - R_h v, \chi), \qquad \chi \in V_h,$$

so that

$$\begin{aligned}
&\|\partial_t(v - R_h v)\|_V^2 \\
&\quad \leqslant a_t(\partial_t(v - R_h v), \partial_t(v - R_h v)) \\
&\quad = a_t(\partial_t(v - R_h v), \partial_t v - \chi) - \dot{a}_t(v - R_h v, \chi) \\
&\quad \leqslant C\|\partial_t(v - R_h v)\|_V \cdot \|\partial_t v - \chi\|_V + \|v - R_h v\|_V \cdot \|\chi\|_V, \quad \chi \in V_h.
\end{aligned}$$

We recall

$$\|(R_h(t)-1)v\|_{H^j} \leqslant Ch^{k-j}\|v\|_{H^k}, \qquad k=1,2, \quad j=0,1, \tag{14.27}$$

and take $\chi = R_h(t)\partial_t v(t)$. Then, we get (14.26) for $j=1$ and $k=1,2$. The remaining case $j=0$ is obtained by Nitsche's trick and the elliptic estimate of $A(t)$. □

For $u_0 \in V$, $\bar{u}_h = \bar{u}_h(t) \in X_h$ denotes the solution of

$$\mathrm{d}\bar{u}_h/\mathrm{d}t + A_h(t)\bar{u}_h = 0, \quad 0 \leqslant t \leqslant T, \tag{14.28}$$

with

$$\bar{u}_h(0) = R_h(0)u_0. \tag{14.29}$$

The following lemma holds for the error $\bar{e}_h = \bar{u}_h - u$:

LEMMA 14.2. *We have*

$$\int_0^t \|\bar{e}_h(s)\|_X^2 \,\mathrm{d}s \leqslant Ch^2 \|u_0\|_V^2. \tag{14.30}$$

PROOF. From (14.21) and (14.28), we get

$$(\partial_t \bar{e}_h, \chi) + a_t(\bar{e}_h, \chi) = 0, \quad \chi \in V_h. \tag{14.31}$$

Making use of (14.31) with $\chi = R_h(t)\bar{e}_h(t)$, we have

$$\begin{aligned}
&\tfrac{1}{2}(\mathrm{d}/\mathrm{d}t)\|\bar{e}_h\|_X^2 + a_t(\bar{e}_h, \bar{e}_h) \\
&\quad = (\partial_t \bar{e}_h, (1-R_h)\bar{e}_h) + a_t(\bar{e}_h, (1-R_h)\bar{e}_h) \\
&\quad = -(\partial_t \bar{e}_h, (1-R_h)u) - a_t(\bar{e}_h, (1-R_h)u) \\
&\quad \leqslant C_\varepsilon h^2\{\|u\|_{H^2}^2 + \|\partial_t \bar{e}_h\|_X^2\} + \varepsilon\|\bar{e}_h\|_V^2
\end{aligned}$$

for each $\varepsilon > 0$ by (14.27). From the coerciveness (14.5) of a_t, we have

$$(\mathrm{d}/\mathrm{d}t)\|\bar{e}_h\|_X^2 + \|\bar{e}_h\|_V^2 \leqslant Ch^2\{\|u\|_{H^2}^2 + \|\partial_t \bar{e}_h\|_X^2\},$$

taking $\varepsilon > 0$ small enough. Hence

$$\begin{aligned}
&\int_0^t \|\bar{e}_h(s)\|_V^2 \,\mathrm{d}x \\
&\quad \leqslant Ch^2 \int_0^t \{\|u(s)\|_{H^2}^2 + \|\partial_t \bar{e}_h(s)\|_X^2\}\,\mathrm{d}s + \|\bar{e}_h(0)\|_X^2 \\
&\quad \leqslant Ch^2 \int_0^t \{\|u(s)\|_{H^2}^2 + \|u_t(s)\|_X^2 + \|\partial_t \bar{u}_h(s)\|^2\}\,\mathrm{d}s + Ch^2\|u_0\|_V^2.
\end{aligned} \tag{14.32}$$

By the elliptic estimate, we have

$$\|u(s)\|_{H^2} \leqslant C\|A(s)u(s)\|_X = C\|u_t(s)\|_X,$$

while the estimate in Proposition 14.2 is valid for $u = u_h$ uniformly in h. Therefore, the right-hand side of (14.32) is further estimated from above by

$$Ch^2\{\|u_0\|_V^2 + \|R_h(0)u_0\|_V^2\} \leqslant Ch^2\|u_0\|_V^2. \qquad \square$$

REMARK 14.1. Under the inverse assumption, the L^2 orthogonal projection $P_h: X \to X_h$ satisfies $\|P_h\|_{V\to V} \leqslant C$ (Proposition 3.2). Then, the estimate

$$\int_0^t \|e_h(s)\|_V^2 \, ds \leqslant Ch^2\|u_0\|_V^2 \tag{14.30'}$$

follows in a similar manner to the above calculation, where $e_h = u_h - u$.

However, such an estimate can be verified without assuming the inverse assumption. That is,

LEMMA 14.3. *In the case of* $u_0 \in V \cap H^{k-j}(\Omega)$, $k=1,2$, $j=0,1$, *the inequality*

$$\int_0^t \|e_h(s)\|_X^2 \, dx \leqslant Ct^{1-j}h^{2k}\|u_0\|_{H^{k-j}}^2 \tag{14.33}$$

holds true.

PROOF. We fix $t>0$ and consider the dual problems

$$(v, \partial_s w) - a_s(v, w) = (v, e_h(s)), \quad v \in V, \tag{14.34}$$

with

$$w(t) = 0 \tag{14.35}$$

in X and

$$(\chi, \partial_s w_h) - a_s(\chi, w_h) = (\chi, e_h(s)), \quad \chi \in V_h, \tag{14.36}$$

with

$$w_h(t) = 0 \tag{14.37}$$

in X_h, respectively. These problems have unique solutions

$$w = w(t) \in C^0([0,t) \to D(A(t)^*)) \cap C^0([0,t] \to X) \cap C^1([0,t) \to X),$$

$$w_h = w_h(t) \in C^1([0,t] \to X_h),$$

respectively. We note the appearance of inhomogeneous terms in the present case, while we repeat the argument in the proof of Lemma 14.2 with reversing the time variable t. Then, we obtain

$$\int_0^t \|w_h(s)-w(s)\|_V^2\,ds \leqslant Ch^2 \int_0^t \|e_h(s)\|_X^2\,ds.$$

Combining this inequality with the a priori estimate in Proposition 14.2, we get

$$\int_0^t \{\|\partial_t w_h(s)-\partial_t w(s)\|_X^2 + h^{-2}\|w_h(s)-w(s)\|_V^2\}\,ds$$

$$\leqslant \int_0^t \{2(\|\partial_t w_h(s)\|_X^2 + \|\partial_t w(s)\|_X^2) + h^{-2}\|w_h(s)-w(s)\|_V^2\}\,ds$$

$$\leqslant C\int_0^t \|e_h(s)\|_X^2\,ds. \tag{14.38}$$

Now, we note that

$$(\partial_t e_h, \chi) + a_t(e_h, \chi) = 0, \quad \chi \in V_h, \tag{14.39}$$

$$(\chi, \partial_s(w_h - w)) - a_s(\chi, w_h - w) = 0, \quad \chi \in V_h. \tag{14.40}$$

Taking $v = e_h(s)$ in (14.34), we get

$$\|e_h(s)\|_X^2 = (e_h(s), \partial_s w(s)) - a_s(e_h(s), w(s))$$
$$= \{(e_h, \partial_s(w-w_h)) - a_s(e_h, w-w_h)\} + (e_h, \partial_s w_h) - a_s(e_h, w_h).$$

Since $e_h = (u_h - R_h u) + (R_h u - u)$, we furthermore have

$$\|e_h(s)\|_X^2 = \{(R_h u - u, \partial_s(w-w_h)) - a_s(R_h u - u, w - w_h)\}$$
$$+ (d/ds)(e_h, w_h) - (\partial_s e_h, w_h) - a_s(e_h, w_h)$$

by (14.40). Finally, from (14.39)

$$\|e_h(s)\|_X^2 = (R_h u - u, \partial_s(w-w_h)) - a_s(R_h u - u, w-w_h) + (d/ds)(e_h, w_h).$$

Therefore, by $e_h(0) = w_h(t) = 0$

$$\int_0^t \|e_h(s)\|_X^2\,ds = \int_0^t \{(R_h u - u, \partial_s(w-w_h)) - a_s(R_h u - u, w-w_h)\}\,ds$$

$$\leqslant \varepsilon \int_0^t \{\|\partial_s w - \partial_s w_h\|_X^2 + h^{-2}\|w-w_h\|_V^2\}\,ds$$

$$+C_\varepsilon \int_0^t \{\|R_h u-u\|_X^2+h^2\|R_h u-u\|_V^2\}\,ds, \tag{14.41}$$

$\varepsilon>0$ being a constant. Taking $\varepsilon>0$ small enough, we arrive at

$$\int_0^t \|e_h(s)\|_X^2\,ds \leqslant C\int_0^t \{\|R_h u-u\|_X^2+h^2\|R_h u-u\|_V^2\}\,ds \tag{14.42}$$

by (14.38) and (14.41). Now, the desired estimate (14.33) follows with the aid of (14.27). □

As is described in Section 12, the following theorem implies

$$\|e_h(t)\|_X \leqslant C(h^2/t)\|u_0\|_X \tag{14.43}$$

by Helfrich's duality argument. Recall that $V=H_0^1(\Omega)$:

THEOREM 14.1. *If* $u_0\in V\cap H^2(\Omega)$, *the estimate*

$$\|e_h(t)\|_X \leqslant Ch^2\|u_0\|_{H^2} \tag{14.44}$$

holds true.

PROOF. Putting

$$\xi=\xi(t)=R_h(t)e_h(t)=R_h(t)u(t)-u_h(t)\in V_h,$$

we have

$$\begin{aligned}(\xi_t,\xi)+a_t(\xi,\xi)&=(\partial_t(R_h u),\xi)-(\partial_t u_h,\xi)+a_t(R_h u,\xi)-a_t(u_h,\xi)\\ &=(\partial_t(R_h u),\xi)-(\partial_t u,\xi)\\ &=-(\partial_t(u-R_h u),\xi).\end{aligned}$$

Multiplying both sides by t, we get

$$\tfrac12(d/dt)(t\|\xi\|_X^2)+ta_t(\xi,\xi)=\tfrac12\|\xi\|_X^2-t(\partial_t(u-R_h u),\xi)$$

so that

$$t\|\xi(t)\|_X^2 \leqslant C\int_0^t \|\xi(s)\|_X^2\,ds+C\int_0^t s^2\|\partial_s(u-R_h u)(s)\|_X^2\,ds.$$

Hence

$$\begin{aligned}t\|e_h(t)\|_X^2 &\leqslant t\|(u-R_h u)(t)\|_X^2\\ &\quad+C\int_0^t \{\|(u-R_h u)(s)\|_X^2+s^2\|\partial_s(u-R_h u)\|_X^2\}\,ds+C\int_0^t \|e_h(s)\|_X^2\,ds.\end{aligned}$$

By virtue of Lemma 14.1, we have

$$t\|e_h(t)\|_X^2 \leqslant Cth^4 \left\{ \max_{0\leqslant s\leqslant t} \|u(s)\|_{H^2}^2 + \int_0^t s\{\|u(s)\|_{H^2}^2 + \|u_t(s)\|_{H^2}^2\}\,\mathrm{d}s + C\int_0^t \|e_h(s)\|_X^2\,\mathrm{d}s \right\}. \tag{14.45}$$

On the other hand,

$$\int_0^t \|e_h(s)\|_X^2\,\mathrm{d}s \leqslant Cth^4\|u_0\|_{H^2}^2 \tag{14.46}$$

holds by (14.33). Furthermore, the inequalities

$$\|u(s)\|_{H^2} \leqslant C\|u_0\|_{H^2}, \quad 0\leqslant s\leqslant T, \tag{14.47}$$

$$\int_0^t s\{\|u(s)\|_{H^2}^2 + \|u_t(s)\|_{H^2}^2\}\,\mathrm{d}s \leqslant C\|u_0\|_{H^2}^2 \tag{14.48}$$

follow from Propositions 14.3 and 14.5, respectively. Summing up these inequalities (14.45)–(14.48), we obtain (14.44). □

14.3. Error analysis of fully discrete approximations

We adopt the backward difference finite element method and obtain the fully discrete approximation

$$(u_h^\tau(t+\tau)-u_h^\tau(t))/\tau + A_h(t+\tau)u_h^\tau(t+\tau)=0$$
$$t=t_n, \quad 0\leqslant n\leqslant N, \tag{14.49}$$

in X_h with

$$u_h^\tau(0)=P_h u_0 \tag{14.50}$$

as in the preceding section. We want to estimate the error $e_h^\tau(t)=u_h^\tau(t)-u_h(t)$ by the energy method.

Henceforth, we drop the suffix h. Furthermore, we put

$$\Delta_t u^\tau(t)=(u^\tau(t_j)-u^\tau(t_{j-1}))/\tau, \qquad t=t_j, \quad j\geqslant 1, \tag{14.51}$$

for simplicity of writing. First, we show the following:

LEMMA 14.4. *The inequality*

$$t^2\|e^\tau(t)\|_X^2 \leqslant C\tau^2\|u_0\|_X^2 + C\tau\sum_{j=1}^{n-1}\|e^\tau(t_j)\|_{V^*}^2, \quad t=t_n, \tag{14.52}$$

holds true.

PROOF. From the relations

$$(u_t, v)+a_t(u, v)=0, \quad v\in V, \tag{14.53}$$

$$(\Delta_t u^\tau, v)+a_t(u^\tau, v)=0, \qquad v\in V, \quad t=t_j, \tag{14.54}$$

we have

$$(\Delta_t e^\tau(t_j), v)+a_{t_j}(e^\tau(t_j), v)=(\gamma_j, v), \quad v\in V, \tag{14.55}$$

where

$$\gamma_j=\Delta_t u(t_j)-u_t(t_j)=\frac{1}{\tau}\int_{t_{j-1}}^{t_j}(s-t_{j-1})u_{tt}(s)\,\mathrm{d}s. \tag{14.56}$$

Therefore, putting

$$\tilde{e}(t_j)=t_j e^\tau(t_j), \qquad \tilde{\gamma}_j=t_j\gamma_j,$$

we get

$$(\Delta_t\tilde{e}(t_j), v)+a_{t_j}(\tilde{e}(t_j), v)=(\tilde{\gamma}_j+e(t_{j-1}), v), \quad v\in V.$$

Now, we take $v=\tilde{e}(t_j)$ to obtain

$$\frac{1}{\tau}(\tilde{e}(t_j)-\tilde{e}(t_{j-1}), \tilde{e}(t_j))+a_{t_j}(\tilde{e}(t_j), \tilde{e}(t_j))=(\tilde{\gamma}_j+e(t_{j-1}), \tilde{e}(t_j)).$$

In other words,

$$\tfrac{1}{2}\{\|\tilde{e}(t_j)\|_X^2-\|\tilde{e}(t_{j-1})\|_X^2+\tau^2\|\Delta_t\tilde{e}(t_j)\|_X^2\}+\tau a_{t_j}(\tilde{e}(t_j), \tilde{e}(t_j))$$
$$=\tau(\tilde{\gamma}_j+e(t_{j-1}), \tilde{e}(t_j))$$

so that

$$\|\tilde{e}(t_j)\|_X^2-\|\tilde{e}(t_{j-1})\|_X^2+2\tau\delta\|\tilde{e}(t_j)\|_V^2$$
$$\leqslant 2\tau(\tilde{\gamma}_j+e(t_{j-1}), \tilde{e}(t_j))$$
$$\leqslant 2\tau\{C(\|\tilde{\gamma}_j\|_{V^*}^2+\|e(t_{j-1})\|_{V^*}^2)+\delta\|\tilde{e}(t_j)\|_V^2\}.$$

Namely,

$$\|\tilde{e}(t_j)\|_X^2-\|\tilde{e}(t_{j-1})\|_X^2\leqslant 2C\tau(\|\tilde{\gamma}_j\|_{V^*}^2+\|e(t_{j-1})\|_{V^*}^2),$$

which implies

$$t^2\|e^\tau(t_n)\|_X^2=\|\tilde{e}(t_n)\|_X^2$$
$$\leqslant 2C\tau\left(\sum_{j=1}^{n}t_j^2\|\gamma_j\|_{V^*}^2+\sum_{j=1}^{n-1}\|e(t_j)\|_{V^*}^2\right). \tag{14.57}$$

Here, by definition (14.56), we get

$$\|\gamma_j\|_{V^*}^2 \leqslant \frac{1}{\tau} \int_{t_{j-1}}^{t_j} (s-t_{j-1})^2 \|u_{tt}(s)\|_{V^*}^2 \, ds.$$

Therefore,

$$t_j^2 \|\gamma_j\|_{V^*}^2 \leqslant \tau \int_{t_{j-1}}^{t_j} s\|u_{tt}(s)\|_{V^*}^2 \, ds$$

follows, because of

$$t_j(s-t_{j-1}) \leqslant \tau s, \quad t_{j-1} \leqslant s \leqslant t_j.$$

We now recall (14.7) and get

$$\|u_{tt}\|_{V^*} \leqslant C\|u_t\|_V + C\|u\|_V. \tag{14.58}$$

Therefore, the inequality

$$\begin{aligned}\tau \sum_{j=1}^{n} t_j^2 \|\gamma_j\|_{V^*}^2 &\leqslant \tau^2 \int_0^{t_n} s^2 \|u_{tt}(s)\|_{V^*}^2 \, ds \\ &\leqslant C\tau^2 \int_0^{t} \{\|u(s)\|_V^2 + s^2 \|u_t(s)\|_V^2\} \, ds \\ &\leqslant C\tau^2 \|u_0\|_X^2 \end{aligned} \tag{14.59}$$

hold by Propositions 14.1 and 14.4. The desired inequality (14.52) follows from (14.57) and (14.59). □

For the operator $T(t) \equiv A(t)^{-1}$, the relation

$$\|f\|_{V^*} \sim \|T(t)f\|_V, \quad f \in V_h, \tag{14.60}$$

holds uniformly in h. Furthermore, from $T'(t) = -A(t)^{-1} A'(t) A(t)^{-1}$, the inequality

$$\|T'(t)f\|_V \leqslant C\|f\|_{V^*} \tag{14.61}$$

follows. In terms of $T(t)$, (14.55) can be written as

$$T(t_j)\Delta_t e^\tau(t_j) + e^\tau(t_j) = T(t_j)\gamma_j.$$

Setting $\Gamma_j = T(t_j)\gamma_j$, we get

$$\Delta_t(T_j)e^\tau(t_j) + e^\tau(t_j) = \Gamma_j + (\Delta_t T(t_j))e^\tau(t_{j-1}).$$

Therefore, by taking the inner product with $T(t_j)e^\tau(t_j)$, we have

$$\tfrac{1}{2}\{(1/\tau)(\|T(t_j)e^\tau(t_j)\|_X^2-\|T(t_{j-1})e^\tau(t_{j-1})\|_X^2)+\tau\|\Delta_t(T(t_j)e^\tau(t_j))\|^2\}$$
$$+\mu\|e^\tau(t_j)\|_{V^*}^2$$
$$\leqslant\|\Gamma_j\|_{V^*}\|T(t_j)e^\tau(t_j)\|_{V^*}\|(\Delta_t T(t_j))e^\tau(t_{j-1})\|_X\|T(t_j)e^\tau(t_j)\|_X \tag{14.62}$$

as in the proof of Lemma 14.4, where $\mu>0$ is a constant related to the equivalence (14.60). By the mean value theorem, the right-hand side is furthermore estimated from above for $t_*\in(t_{j-1},t_j)$ as

$$\|\Gamma_j\|_{V^*}\cdot\|T(t_j)e^\tau(t_j)\|_V+\|T'(t_*)e(t_{j-1})\|_V\cdot\|T(t_j)e^\tau(t_j)\|_X$$
$$\leqslant C\{\|\Gamma_j\|_{V^*}\cdot\|e^\tau(t_j)\|_{V^*}+\|e(t_{j-1})\|_{V^*}\cdot\|T(t_j)e^\tau(t_j)\|_X\}$$
$$\leqslant\tfrac{1}{4}\mu\{\|e^\tau(t_j)\|_{V^*}^2+\|e(t_{j-1})\|_{V^*}^2\}+C\{\|T(t_j)e^\tau(t_j)\|_X^2+\|\Gamma_j\|_{V^*}^2\}$$

so that

$$\tfrac{1}{2}\{\|T(t_j)e^\tau(t_j)\|_X^2-\|T(t_{j-1})e^\tau(t_{j-1})\|_X^2\}$$
$$+\tfrac{1}{4}\tau\mu\{\|e^\tau(t_j)\|_{V^*}^2-\|e(t_{j-1})\|_{V^*}^2\}+\tfrac{3}{4}\tau\mu\|e^\tau(t_j)\|_{V^*}^2$$
$$\leqslant C\tau\{\|T(t_j)e^\tau(t_j)\|_X^2+\|\Gamma_j\|_{V^*}^2\}. \tag{14.63}$$

From this, we obtain

$$\|T(t_n)e^\tau(t_n)\|_X^2+\tau\sum_{j=1}^n\|e^\tau(t_j)\|_{V^*}^2\leqslant C\tau\sum_{j=1}^n\|\Gamma_j\|_{V^*}^2+C\tau\sum_{j=1}^n\|T(t_j)e^\tau(t_j)\|_X^2.$$

By virtue of a discrete version of Gronwall's inequality, we furthermore get

$$\|T(t_n)e^\tau(t_n)\|_X^2+\tau\sum_{j=1}^n\|e^\tau(t_j)\|_{V^*}^2\leqslant C\tau\sum_{j=1}^n\|\Gamma_j\|_{V^*}^2. \tag{14.64}$$

Now, in the same way as in the proof of Lemma 14.4, Schwarz's inequality gives

$$\|\Gamma_j\|_{V^*}^2=\|T(t_j)\gamma_j\|_{V^*}^2\leqslant\frac{1}{\tau}\int_{t_{j-1}}^{t_j}(s-t_{j-1})^2\|T(t_j)u_{tt}(s)\|_{V^*}^2\,ds.$$

On the other hand, $T(t)u_t+u=0$ implies

$$T(t)u_{tt}=-u_t-T'(t)u_t,$$

so that for $s\in[t_{j-1},t_j]$, $s_*\in(t_{j-1},t_j)$:

$$\|T(t_j)u_{tt}(s)\|_{V^*}\leqslant\|T(s)u_{tt}(s)\|_{V^*}+\tau\|T'(s_*)u_{tt}(s)\|_{V^*}$$
$$\leqslant\|u_t(s)\|_{V^*}+\|T'(s)u_t(s)\|_{V^*}+\|T'(s_*)u_{tt}(s)\|_{V^*}$$
$$\leqslant C(\|u_t(s)\|_{V^*}+\|u_t(s)\|_V+\tau\|u_{tt}(s)\|_{V^*}).$$

However,

$$\|u_t\|_{V^*}\leqslant\|A(t)u\|_{V^*}\leqslant C\|v\|_V$$

is obvious. Therefore, noting (14.58), we obtain

$$\|T(t_j)u_{tt}(s)\|_{V^*} \leqslant C(\|u(s)\|_V + \tau\|u_t(s)\|_V). \tag{14.65}$$

Consequently, the inequality

$$\begin{aligned}\tau \sum_{j=1}^{n} \|\Gamma_j\|_{V^*}^2 &\leqslant C \sum_{j=1}^{n} \int_{t_{j-1}}^{t_j} (s-t_{j-1})^2 \{\|u(s)\|_V^2 + \tau^2 \|u_t(s)\|_V^2\}\, \mathrm{d}s \\ &\leqslant C\tau^2 \int_0^{t_n} (\|u(s)\|_V^2 + s^2 \|u_t(s)\|_V^2)\, \mathrm{d}s \\ &\leqslant C\tau^2 \|Pu_0\|_X^2\end{aligned} \tag{14.66}$$

has been established by Propositions 14.1 and 14.4.

In this way, combining (14.52) with (14.64) and (14.66), we have proven the following:

THEOREM 14.2. *The estimate*

$$\|e^\tau(t)\|_X \leqslant C(\tau/t)\|u_0\|_X, \quad t=t_n, \tag{14.67}$$

holds true.

REMARK 14.2. As for error analysis on the schemes of higher accuracy, we refer to BAIOCCHI and BREZZI [1983], LUSKIN and RANNACHER [1982b], and SAMMON [1982, 1983]. See also Section 17.

15. Discretization of hyperbolic equations

Let us consider the hyperbolic equation

$$\partial^2 u/\partial t^2 + \mathscr{A}(x, D)u = 0, \quad 0 \leqslant t \leqslant T, \quad x \in \Omega, \tag{15.1}$$

with

$$u|_{\partial\Omega} = 0, \quad 0 \leqslant t \leqslant T, \tag{15.2}$$

or

$$\partial u/\partial \nu_{\mathscr{A}} + \sigma u|_{\partial\Omega} = 0, \quad 0 \leqslant t \leqslant T, \tag{15.2'}$$

with

$$u|_{t=0} = u_1, \qquad (\partial u/\partial t)|_{t=0} = u_0, \quad x \in \Omega. \tag{15.3}$$

Then, it is reduced to the evolution equation of second order in $X = L^2(\Omega)$,

$$\mathrm{d}^2 u/\mathrm{d}t^2 + Au = 0, \quad 0 \leqslant t \leqslant T, \tag{15.4}$$

with

$$u(0)=u_1, \qquad u'(0)=u_0. \tag{15.5}$$

Therefore, its finite element approximation can be constructed in a natural manner. In this section, we develop error analysis on the semidiscrete approximation in an abstract framework, for the case that A is self-adjoint.

To this end we recall the triple of Hilbert spaces $V \subset X \subset V^*$ given in Section 1 and suppose that $a=a(\,,\,)$ is a sesquilinear form on V, satisfying $a^*=a$,

$$|a(u,v)| \leqslant C\|u\|_V \cdot \|v\|_V, \tag{15.6}$$

$$a(u,u) \geqslant \delta \|u\|_V^2, \tag{15.7}$$

with positive constants C and δ. Let A be the self-adjoint operator in X associated with $a(\,,\,)$. We consider the evolution equation (15.4) with (15.5) of hyperbolic type, where $u_1 \in V$ and $u_0 \in X$.

We can transform the equation to a system of first order. Namely, let

$$\mathscr{X}=\begin{pmatrix} V \\ X \end{pmatrix}, \qquad \mathscr{A}=\begin{pmatrix} 0 & -1 \\ A & 0 \end{pmatrix}.$$

Then, equation (15.4) with (15.5) is reduced to

$$\mathrm{d}U/\mathrm{d}t+\mathscr{A}U=0, \quad 0 \leqslant t \leqslant T, \tag{15.8}$$

in $\mathscr{X}$ with

$$U(0)=U_0, \qquad U=\begin{pmatrix} u \\ u' \end{pmatrix}, \qquad U_0=\begin{pmatrix} u_1 \\ u_0 \end{pmatrix}. \tag{15.9}$$

The relation $D(A^{1/2})=V$ and the equivalence

$$\|f\|_V \sim \|A^{1/2}f\|_X, \quad f \in X, \tag{15.10}$$

follow from $a^*=a$, and we replace the norm $\|\cdot\|_{\mathscr{X}}$ in $\mathscr{X}$, where

$$\|U\|_{\mathscr{X}}=\|u_0\|_X+\|u_1\|_V \quad \text{for } U=\begin{pmatrix} u_1 \\ u_0 \end{pmatrix}$$

by $|||\cdot|||_0$, where

$$|||U|||_0=\|u_0\|_X+\|A^{1/2}u_1\|_X. \tag{15.11}$$

Then, $\mathscr{A}$ becomes skew-adjoint and hence generates a (C_0) semigroup $\{T(t)\}$ in $\mathscr{X}$ (YOSIDA [1964, Chapter 9], HILLE and PHILLIPS [1957], e.g.). In fact, by an easy calculation we can verify that $\mathrm{i}\mathscr{A}$ is associated with a symmetric sesquilinear form

$$\mathscr{B}=\mathscr{B}(\,,\,) \quad \text{on } \mathscr{V}=\begin{pmatrix} V \\ V \end{pmatrix}$$

such that

$$\mathscr{B}(U, W) = \mathrm{i}\{-a(u_0, w_1) + a(u_1, w_0)\}$$

$$\text{for } U = \begin{pmatrix} u_1 \\ u_0 \end{pmatrix}, \quad W = \begin{pmatrix} w_1 \\ w_0 \end{pmatrix} \in \mathscr{V}. \tag{15.12}$$

Therefore, in the case of

$$U_0 \in D(\mathscr{A}) = \begin{pmatrix} D(A) \\ V \end{pmatrix},$$

the function $U(t) = T(t)U_0$ is C^1 in $\mathscr{X}$ on $[0, T]$, contained in $D(A)$, and actually satisfies (15.8) with (15.9). Furthermore we have

$$\|T(t)\| \leqslant C, \quad 0 \leqslant t \leqslant T. \tag{15.13}$$

Now, let $V_h \subset V$, $h > 0$, be a family of finite-dimensional subspaces having the property

$$\inf_{\chi \in V_h} \|\chi - v\|_V \leqslant Ch\|Av\|, \quad v \in D(A). \tag{15.14}$$

X_h denotes the space V_h equipped with the topology induced from X. Furthermore, $P_h: X \to X_h$ is the orthogonal projection, and $R_h: V \to V_h$ is the Ritz projection associated with $a(\,,\,)$:

$$a(R_h v, \chi) = a(v, \chi), \qquad v \in V, \quad \chi \in V_h.$$

It is shown in Section 2 that the estimate

$$\|A^{j/2}(R_h v - v)\|_X \leqslant Ch^{k-j}\|A^{k/2}v\|_X \tag{15.15}$$

holds for $k = 1, 2$ and $j = 0, 1$.

Let A_h be the m-sectorial operator in X_h associated with $a|_{V_h \times V_h}$. The semidiscrete approximation of (15.4) with (15.5) is now given as

$$\mathrm{d}^2 u_h/\mathrm{d}t^2 + A_h u_h = 0, \quad 0 \leqslant t \leqslant T, \tag{15.16}$$

in X_h with

$$u_h(0) = R_h u_1, \qquad u_h'(0) = P_h u_0. \tag{15.17}$$

We can derive certain estimates on the error $e_h(t) = u_h(t) - u(t)$ by the real methods developped above. Here, we note that the equivalence

$$\|\chi\|_V \sim \|A_h^{1/2}\chi\|_X, \quad \chi \in V_h, \tag{15.10'}$$

holds uniformly in h.

15.1. Error estimate by Helfrich's method

In the same way as in (15.4) and (15.5), equation (15.16) with (15.17) is reduced to

$$dU_h/dt + \mathscr{A}_h U_h = 0, \quad 0 \leqslant t \leqslant T, \quad \text{in } \mathscr{X}_h = \begin{pmatrix} V_h \\ X_h \end{pmatrix} \tag{15.18}$$

with

$$U_h(0) = U_{0h}, \tag{15.19}$$

where

$$\mathscr{A}_h = \begin{pmatrix} 0 & -1 \\ A_h & 0 \end{pmatrix}, \qquad U_h = \begin{pmatrix} u_h \\ u_h' \end{pmatrix}, \qquad U_{0h} = \begin{pmatrix} R_h u_h \\ P_h u_0 \end{pmatrix}.$$

The operator $-\mathscr{A}_h$ also generates a (C_0) semigroup $\{T_h(t)\}$ in $\mathscr{X}_h$, and the solution $U_h = U_h(t)$ of (15.18) with (15.19) is given by

$$U_h(t) = T_h(t) U_{0h}. \tag{15.20}$$

The inequality

$$\| T_h(t) \| \leqslant C \tag{15.21}$$

holds.

Henceforth, we put

$$\mathscr{X}_l = \begin{pmatrix} D(A^{l+1/2}) \\ D(A^l) \end{pmatrix},$$

$$|||U_0|||_l = \|A^l u_0\|_X + \|A^{l+1/2} u_1\|_X \quad \text{for } U_0 = \begin{pmatrix} u_1 \\ u_0 \end{pmatrix} \in \mathscr{X}_l, \tag{15.22}$$

where $l = 0, \frac{1}{2}, 1$. This notation is compatible to $|||\cdot|||_0$ for $\mathscr{X} = \mathscr{X}_0$ introduced in (15.11). Then the inequalities

$$|||T(t)U_0|||_l \leqslant C |||U_0|||_l \tag{15.23}$$

hold true.

We can trace the computations in Section 12 to obtain certain error estimates, just noting that (i) $T(t)$ and $T_h(t)$ are merely (C_0) semigroups and (ii) $\mathscr{A}$ and $\mathscr{A}_h$ are independent of t. In fact, note that

$$\mathscr{P}_h = \begin{pmatrix} R_h & 0 \\ 0 & P_h \end{pmatrix} : \mathscr{X} \to \mathscr{X}_h$$

is the orthogonal projection and let

$$\varepsilon_h(t) = T_h(t) P_h - T(t)$$

be the error operator. Then, first we have

$$|||(1 - \mathscr{P}_h)\varepsilon_h(t) U_0|||_0$$
$$\leqslant \inf_{\chi \in \mathscr{X}_h} |||\varepsilon_h(t) U_0 - \chi|||_0$$

$$\leqslant \inf_{\chi\in\mathscr{X}_h} |||T(t)U_0-\chi|||_0 \leqslant Ch\|A^{1/2}u'(t)\|_X+\|Au(t)\|_X\}$$

$$\leqslant Ch|||U_0|||_{1/2}. \tag{15.24}$$

Next, the identity

$$-(\partial/\partial r)[T_h(t-r)\mathscr{P}_h\varepsilon_h(r)]=T_h(t-r)[\mathscr{A}_h\mathscr{P}_h-\mathscr{P}_h\mathscr{A}]T(r)$$

implies

$$\begin{aligned}\mathscr{P}_h\varepsilon_h(t)&=\int_0^t T_h(t-r)[\mathscr{A}_h\mathscr{P}_h-\mathscr{P}_h\mathscr{A}]T(r)\,\mathrm{d}r\\&=\int_0^t\left(-\frac{\partial}{\partial r}T_h(t-r)\right)\mathscr{P}_h[\mathscr{A}^{-1}-\mathscr{A}_h^{-1}\mathscr{P}_h]\mathscr{A}T(r)\,\mathrm{d}r\\&=-[T_h(t-r)\mathscr{P}_h[\mathscr{A}^{-1}-\mathscr{A}_h^{-1}\mathscr{P}_h]\mathscr{A}T(r)]_{r=0}^{r=t}\\&\quad+\int_0^t T_h(t-r)\mathscr{P}_h[\mathscr{A}^{-1}-\mathscr{A}_h^{-1}\mathscr{P}_h]\mathscr{A}^2T(r)\,\mathrm{d}r\\&=\sum_{l=1}^3\varepsilon_h^l(t),\end{aligned} \tag{15.25}$$

where

$$\varepsilon_h^1(t)=-\mathscr{P}_h[\mathscr{A}^{-1}-\mathscr{A}_h^{-1}\mathscr{P}_h]\mathscr{A}T(t),$$

$$\varepsilon_h^2(t)=T_h(t)\mathscr{P}_h[\mathscr{A}^{-1}-\mathscr{A}_h^{-1}\mathscr{P}_h]\mathscr{A},$$

$$\varepsilon_h^3(t)=\int_0^t T_h(t-r)\mathscr{P}_h[\mathscr{A}^{-1}-\mathscr{A}_h^{-1}\mathscr{P}_h]\mathscr{A}^2T(r)\,\mathrm{d}r.$$

Here, we have

$$\begin{aligned}\mathscr{A}^{-1}-\mathscr{A}^{-1}\mathscr{P}_h&=\begin{pmatrix}0&A^{-1}\\1&0\end{pmatrix}-\begin{pmatrix}0&-A_h^{-1}\\1&0\end{pmatrix}\begin{pmatrix}R_h&0\\0&P_h\end{pmatrix}\\&=\begin{pmatrix}0&-A^{-1}+A_h^{-1}P_h\\1-R_h&0\end{pmatrix},\end{aligned}$$

$$\|A_h^{-1}P_h-A^{-1}\|_{X\to V}\leqslant Ch,$$

$$\|1-R_h\|_{V\to X}\leqslant Ch,$$

so that

$$\|\mathscr{A}^{-1}-\mathscr{A}_h^{-1}\mathscr{P}_h\|_{\mathscr{X}\to\mathscr{X}}\leqslant Ch$$

holds. Consequently, we obtain

$$|||\varepsilon_h^1(t)U_0|||_0 \leqslant Ch|||\mathscr{A}U_0|||_0 = Ch|||U_0|||_{1/2},$$

$$|||\varepsilon_h^2(t)U_0|||_0 \leqslant Ch|||\mathscr{A}U_0|||_0 = Ch|||U_0|||_{1/2},$$

$$|||\varepsilon_h^3(t)U_0|||_0 \leqslant Cth|||\mathscr{A}^2U_0|||_0 = Cth|||U_0|||_1,$$

which implies the following:

THEOREM 15.1. *The "energy estimate"*

$$\|u_h(t)-u(t)\|_V + \|u_h'(t)-u'(t)\|_X \leqslant Ch\{t|||U_0|||_1 + |||U_0|||_{1/2}\} \tag{15.26}$$

holds true.

The "L^2 estimate" is obtained in the following way. We set

$$\mathscr{X}_{-1} = \begin{pmatrix} X \\ V^* \end{pmatrix}, \qquad \mathscr{X}_{-1h} = \begin{pmatrix} X_h \\ V_h^* \end{pmatrix}.$$

The (C_0) semigroups $\{T(t)\}$ and $\{T_h(t)\}$ extend to those on $\mathscr{X}_{-1}$ and $\mathscr{X}_{-1h}$, respectively, satisfying

$$\|T(t)\|_{\mathscr{X}_{-1}\to\mathscr{X}_{-1}},\ \|T_h(t)\|_{\mathscr{X}_{-1h}\to\mathscr{X}_{-1h}} \leqslant C. \tag{15.27}$$

Under the inverse assumption, (15.10) implies the equivalence

$$\|\chi\|_{V_h^*} \sim \|A^{-1/2}\chi\|_X \sim \|A_h^{-1/2}\chi\|_X, \quad \chi \in V_h.$$

In fact, we have

$$\|v\|_{V^*} \sim \|A^{-1}v\|_V, \qquad v \in V,$$

$$\|\chi\|_{V_h^*} \sim \|A_h^{-1}\chi\|_{V_h}, \quad \chi \in V_h,$$

while $\|\chi\|_{V^*} \sim \|\chi\|_{V_h^*}$ for $\chi \in V_h$, follows from $\|P_h\|_{V\to V} \leqslant C$:

$$\begin{aligned}
\|\chi\|_{V_h^*} &= \operatorname{supp}\{|(\chi, v)|: v \in V_h,\ \|v\|_V \leqslant 1\} \\
&\leqslant \|\chi\|_{V^*} = \sup\{|(\chi, v)|: v \in V,\ \|v\|_V \leqslant 1\} \\
&= \sup\{|(\chi, P_h v)|: v \in V,\ \|v\|_V \leqslant 1\} \\
&\leqslant \|\chi\|_{V_h^*} \sup\{\|P_h v\|_V: v \in V,\ \|v\|_V \leqslant 1\} \\
&\leqslant \|\chi\|_{V_h^*} \|P_h\|_{V\to V}.
\end{aligned}$$

Furthermore,

$$\|\overline{A_h^{-1/2}P_hA^{1/2}}\|_{X\to X} = \|A^{1/2}A_h^{-1/2}P_h\|_{X\to X} \leqslant C$$

holds, and hence

$$\|P_h\|_{V^*\to V^*} \leqslant C. \tag{15.28}$$

On the other hand, from a theorem by Ushijima described in Section 5, we have for

$v \in X$ and $\chi \in V_h$ that

$$\begin{aligned}\|\chi - v\|_{V^*} &\leqslant C\|A^{-1/2}(\chi - v)\|_X \\ &\leqslant C\{\|(A^{-1/2} - A_h^{-1/2}P_h)\chi\|_X + \|A_h^{-1/2}\chi - A^{-1/2}v\|_X\} \\ &\leqslant Ch\|\chi\|_X + C\|A_h^{-1/2}\chi - A^{1/2}v\|_X.\end{aligned}$$

Taking $\chi = A_h^{1/2}R_hA^{-1/2}v$, we obtain

$$\inf_{\chi \in V_h} \|x - v\|_{V^*} \leqslant Ch\|v\|_X. \tag{15.29}$$

Since $P_h|_{V_h} = \text{identity}$, the inequalities (15.28) and (15.29) yield

$$\|(P_h - 1)v\|_{V^*} \leqslant Ch\|v\|_X. \tag{15.30}$$

Therefore, first

$$\mathscr{P}_{0h} = \begin{pmatrix} P_h & 0 \\ 0 & P_h \end{pmatrix}$$

satisfies

$$\|\mathscr{P}_{0h}\|_{\mathscr{X}_{-1} \to \mathscr{X}_{-1}} \leqslant C. \tag{15.31}$$

Furthermore, from

$$\mathscr{A}^{-1} - \mathscr{A}_h^{-1}\mathscr{P}_{0h} = \begin{pmatrix} 0 & A_h^{-1}P_h - A^{-1} \\ 1 - P_h & 0 \end{pmatrix},$$

it follows that

$$\|\mathscr{A}^{-1} - \mathscr{A}_h^{-1}\mathscr{P}_{0h}\|_{\mathscr{X}_{-1} \to \mathscr{X}_{-1}} \leqslant Ch, \tag{15.32}$$

because

$$\begin{aligned}\|(A_h^{-1}P_h - A^{-1})u_0\|_X &= \|(R_h - 1)A^{-1}u_0\|_X \\ &\leqslant Ch\|A^{-1}u_0\|_V \leqslant Ch\|u_0\|_{V^*}.\end{aligned}$$

Now, for

$$\varepsilon_{0h}(t) = T_h(t)\mathscr{P}_{0h} - T(t)$$

in $\mathscr{X}_{-1}$, we repeat the computation given above for $\varepsilon_h(t)$ in $\mathscr{X}_0$. By virtue of (15.27), (15.31) and (15.32), we shall get

$$|||\varepsilon_{0h}(t)U_0|||_{-1} \leqslant Ch\{t|||U_0|||_{1/2} + |||U_0|||_0\}. \tag{15.33}$$

However, the identity

$$\varepsilon_h(t) = \varepsilon_{0h}(t) + T_h(t)(\mathscr{P}_h - \mathscr{P}_{0h}) \tag{15.34}$$

holds with

$$\mathscr{P}_h - \mathscr{P}_{0h} = \begin{pmatrix} R_h - P_h & 0 \\ 0 & 0 \end{pmatrix}.$$

Since

$$\|(R_h-P_h)v\|_X=\|P_h(R_h-1)v\|_X\leqslant\|(R_h-1)v\|_X\leqslant Ch\|v\|_V,$$

the estimate

$$\|\mathscr{P}_h-\mathscr{P}_{0h}\|_{\mathscr{X}_{-1}\to\mathscr{X}_0}\leqslant Ch \tag{15.35}$$

holds. By means of $\|T_h(t)\|_{\mathscr{X}_0\to\mathscr{X}_0}\leqslant C$, we have thus proven the following:

THEOREM 15.2. *The "L^2 estimate"*

$$\|u_h(t)-u(t)\|_X+\|A^{-1/2}(u_h'(t)-u'(t))\|_X\leqslant Ch\{t|||U_0|||_{1/2}+|||U_0|||_0\} \tag{15.36}$$

holds true.

15.2. Error estimate by the energy method

We recall the weak forms (15.4) with (15.5) and (15.16) with (15.17):

$$(u'',v)+a(u,v)=0,\quad v\in V,$$

$$u(0)=u_1,\qquad u'(0)=u_0, \tag{15.37}$$

$$(u_h'',\chi)+a(u_h,\chi)=0,\quad \chi\in V_h,$$

$$u_h(0)=R_hu_1,\qquad u_h'(0)=P_hu_0. \tag{15.38}$$

The energy equalities

$$\|u'(t)\|_X^2+\|A^{1/2}u(t)\|_X^2=\|u_0\|_X^2+\|A^{1/2}u_1\|_X^2, \tag{15.39}$$

$$\|u_h'(t)\|_X^2+\|A_h^{1/2}u_h(t)\|_X^2=\|P_hu_0\|_X^2+\|A^{1/2}R_hu_1\|_X^2 \tag{15.40}$$

follow at once from these equations.

First, we shall derive the energy estimate of the error. To this end, we put

$$f_h(t)^2=\|u_h'(t)-P_hu'(t)\|_X^2+a(u_h(t)-R_hu(t),u_h(t)-R_hu(t)).$$

Then, we have $f_h(0)=0$ and

$$\begin{aligned}f_h\cdot f_h'&=(u_h''-P_hu'',u_h'-P_hu')+a(u_h-R_hu,u_h'-R_hu')\\&=\{(u_h'',u_h'-P_hu')+a(u_h,u_h'-R_hu')\}\\&\quad-\{(u'',u_h'-P_hu')+a(u,u_h'-R_hu')\}\\&=a(u_h,(P_h-R_h)u')-a(u,(P_h-R_h)u')\\&=a(u_h-R_hu,P_h(1-R_h)u')\\&\leqslant f_h\|P_h(1-R_h)u'\|_V.\end{aligned}$$

Supposing the inverse assumption

$$\|\chi\|_V\leqslant Ch^{-1}\|\chi\|_X,\quad \chi\in V_h, \tag{15.41}$$

we can estimate the right-hand side from above furthermore by $Ch^{-1}f_h\|(1-R_h)u'\|_X$.

Hence

$$f'_h \leqslant Ch^{-1}\|(1-R_h)u'\|_X \leqslant Ch\|Au'\|_X \leqslant Ch|||U_0|||_1,$$

so that

$$f_h(t) \leqslant Cht|||U_0|||_1.$$

Therefore, we obtain

$$\begin{aligned}
&\|u_h(t)-u(t)\|_V + \|u'_h(t)-u(t)\|_X \\
&\quad\leqslant f_h(t) + \|(R_h-1)u(t)\|_V + \|u'_h(t)-P_hu'(t)\|_X \\
&\quad\leqslant f_h(t) + Ch\{\|Au(t)\|_X + \|u'(t)\|_V\} \\
&\quad\leqslant Ch\{t|||U_0|||_1 + |||U_0|||_{1/2}\},
\end{aligned} \tag{15.42}$$

which is the energy estimate given in Theorem 15.1.

The L^2 estimate is shown in the following way: Setting

$$g_h(t)^2 = \|A_h^{1/2}(u'_h - P_hu')\|_X^2 + \|u_h - R_hu\|_X^2, \tag{15.43}$$

we have $g_h(0)=0$ and

$$\begin{aligned}
g_hg'_h &= (A_h^{-1/2}(u''_h - P_hu''), A_h^{-1/2}(u'_h - P_hu')) + (u_h - R_hu, u'_h - R_hu') \\
&= \{u''_h, A_h^{-1}(u'_h - P_hu')) + a(u_h, A_h^{-1}(u'_h - R_hu'))\} \\
&\quad - \{(u'', A_h^{-1}(u'_h - P_hu')) + a(u, A_h^{-1}(u'_h - R_hu'))\} \\
&= a(u_h, A_h^{-1}(P_h - R_h)u') - a(u, A_h^{-1}(P_h - R_h)u') \\
&= a(u_h - R_hu, A_h^{-1}(P_h - R_h)u') \\
&= (u_h - R_hu, P_h(1-R_h)u') \\
&\leqslant g_h\|P_h(1-R_h)u'\| \leqslant g_h\|(1-R_h)u'\|_X.
\end{aligned} \tag{15.44}$$

Therefore,

$$g'_h \leqslant Ch\|u'\|_V \leqslant Ch|||U_0|||_{1/2},$$

so that

$$g_h(t) \leqslant Cth|||U_0|||_{1/2}. \tag{15.45}$$

We note

$$u_h - u = (u_h - R_hu) + (1-R_h)u \tag{15.46}$$

with

$$\|(1-R_h)u\|_X \leqslant Ch\|u\|_V \leqslant Ch|||U_0|||_0.$$

Furthermore,

$$A^{-1/2}(u'_h - u') = (A_h^{-1/2}P_h - A^{-1/2})(u' - u'_h) + A_h^{-1/2}(u'_h - P_hu') \tag{15.47}$$

holds with

$$\begin{aligned}\|(A_h^{-1/2}P_h - A^{-1/2})(u'-u_h')\|_X &\leqslant Ch\|u'-u_h'\|_X \\ &\leqslant Ch\{\|u'\|_X + \|u_h'\|_X\} \leqslant Ch\{|||U_0|||_0 + |||U_{0h}|||_0\} \\ &\leqslant Ch|||U_0|||_0 \end{aligned} \tag{15.48}$$

so that the L^2 estimate

$$\|u_h(t)-u(t)\|_X + \|A^{-1/2}(u_h'(t)-u'(t))\|_X \leqslant Ch\{t|||U_0|||_{1/2} + |||U_0|||_0\} \tag{15.49}$$

follows.

Furthermore, we see that the argument given above implies:

THEOREM 15.3. *The estimate*

$$\|u_h(t)-u(t)\|_X + \|A^{-1/2}(u_h'(t)-u'(t))\|_X \leqslant Ch^2\{t|||U_0|||_1 + |||U_0|||_{1/2}\} \tag{15.50}$$

holds true.

PROOF. In fact, we have

$$g_h' \leqslant \|(1-R_h)u'\|_X \leqslant Ch^2\|Au'\| \leqslant Ch^2|||U_0|||_1$$

by (15.44) and

$$\|(1-R_h)u_X \leqslant Ch^2\|Au\|_X \leqslant Ch^2|||U_0|||_{1/2}$$

in (15.46). Furthermore, from Theorem 15.1, we get

$$\begin{aligned}\|(A_h^{-1/2}P_h - A^{-1/2})(u'-u_h')\|_X &\leqslant Ch\|u'-u_h'\|_X \\ &\leqslant Ch^2\{t|||U_0|||_1 + |||U_0|||_{1/2}\}\end{aligned}$$

in (15.47). These inequalities give (15.50). □

15.3. Comments

As to the discretization for the time variable t of the semidiscrete approximation (15.16), there is a scheme called "Newmark's β's", that is,

$$D_{\tau\bar\tau}u_h^\tau(t) + A_h u_h^\tau(t) + \beta\tau^2 A_h D_{\tau\bar\tau}u_h^\tau(t) = 0,$$
$$t = t_n = n\tau, \quad n = 0, 1, \ldots, N \tag{15.51}$$

with

$$u_h^\tau(0) = R_h u_1, \qquad u_h^\tau(-1) = R_h u_1 - \tau P_h u_0. \tag{15.52}$$

Here, $\tau > 0$ is the size parameter, $\beta \geqslant 0$ is a fixed constant, and

$$D_{\tau\bar\tau}u(t) = \tau^{-2}\{u(t+\tau) - 2u(t) + u(t-\tau)\}, \quad t = t_n.$$

Taking $0<\gamma<\sqrt{4/(1-4\beta)}$ in case $0\leqslant\beta\leqslant\frac{1}{4}$, FUJII [1974] proved the following:

THEOREM 15.4. *Suppose the inverse assumption* (15.41) *and the inequality*

$$\tau/h \leqslant \kappa\gamma \tag{15.53}$$

in case $0 \leqslant \beta \leqslant \frac{1}{4}$, *where* $\kappa > 0$ *is an absolute constant. Then, the "energy estimate"*

$$\|u_h^\tau(t) - u(t)\|_V + \|D_\tau u_h^\tau(t) - u'(t)\|_X \leqslant C(h+\tau)|||U_0|||_1, \quad t \equiv t_n \leqslant T, \tag{15.54}$$

holds.

See also USHIJIMA [1975, 1979], for the proof. Theorems 15.1 and 15.2 are also shown there under the inverse assumption (15.41), by a systematic and sophisticated use of Helfrich's method.

For related works, we refer to BAKER and BRAMBLE [1979] and GEVECI [1984].

CHAPTER IV

Refinements and Generalizations

16. L^∞ estimates

Motivated by the L^∞ error analysis on the finite element approximation of elliptic equations by Nitsche and Scott, a similar study for parabolic equations has been done by Nitsche himself, Thomée and others (NITSCHE and WHEELER [1981–82], SCHATZ, THOMÉE and WAHLBIN [1980], THOMÉE and WAHLBIN [1983]). In the present section, we shall present the method of THOMÉE [1984] on semidiscrete approximations.

To fix the idea, we consider again the parabolic equation

$$\partial u/\partial t = \Delta u, \qquad 0<t<\infty, \quad x\in\Omega, \tag{16.1}$$

with the boundary condition

$$u|_{\partial\Omega}=0, \quad 0<t<\infty, \tag{16.2}$$

and with the initial condition

$$u|_{t=0}=u_0(x), \quad x\in\Omega, \tag{16.3}$$

where the bounded domain $\Omega\subset\mathbb{R}^2$ has a smooth boundary $\partial\Omega$ or is a convex polygon. Taking $X=L^2(\Omega)$, $V=H_0^1(\Omega)$ and $a(u,v)=\int_\Omega \nabla u\cdot\nabla v\,\mathrm{d}x$, $u,v\in V$, we can reduce equation (16.1) with (16.2) and (16.3) to the evolution equation

$$\mathrm{d}u/\mathrm{d}t + Au=0, \quad 0<t<\infty, \tag{16.4}$$

in X with

$$u(0)=u_0, \tag{16.5}$$

where A is the self-adjoint operator associated with $a(\,,)$. Let $V_h\subset V$ be the finite-dimensional space introduced in Section 1 and let X_h be the space V_h with the inner product induced from X. Then, as we have seen in Section 7, the semidiscrete approximation of (16.4) with (16.5) is given by the evolution equation

$$\mathrm{d}u_h/\mathrm{d}t + A_h u_h=0, \quad 0<t<\infty, \tag{16.6}$$

in X_h with

$$u_h(0)=P_h u_0, \tag{16.7}$$

where A_h is the self-adjoint operator in X_h associated with $a|_{V_h \times V_h}$ and $P_h: X \to X_h$ is the orthogonal projection. Let $\{e^{-tA}\}_{t \geq 0}$ and $\{e^{-tA_h}\}_{t \geq 0}$ be the semigroups generated by A and A_h in X and X_h, respectively. Then,

$$E_h(t) = e^{-tA_h} P_h - e^{-tA} \tag{16.8}$$

denotes the error operator. In this section, we want to give an upper bound of $\| E_h(t) u_0 \|_{L^\infty}$. To this end, we must mention the L^∞ stability of approximate solutions. Namely,

THEOREM 16.1. *The estimate*

$$\| e^{-tA_h} \chi \|_{L^\infty} \leq C |\log h| \, \| \chi \|_{L^\infty}, \quad \chi \in V_h, \tag{16.9}$$

holds true.

In order to prove Theorem 16.1, we introduce the discrete delta function $\delta_h^x \in V_h$ and the discrete fundamental solution $\Gamma_h^x(t) \in V_h$ for each $x \in \Omega$ through

$$(\delta_h^x, \chi) = \chi(x), \quad \chi \in V_h, \qquad \Gamma_h^x(t) = e^{-tA_h} \delta_h^x,$$

respectively. For $\chi \in V_h$, we have

$$(\Gamma_h^x(t), \chi) = (e^{-tA_h} \delta_h^x, \chi) = (\delta_h^x, e^{-tA_h} \chi) = (e^{-tA_h} \chi)(x),$$

so that

$$\| e^{-tA_h} \chi \|_{L^\infty} \leq \sup_{x \in \Omega} \| \Gamma_h^x(t) \|_{L^1} \cdot \| \chi \|_{L^\infty}.$$

Therefore, (16.9) is reduced to

$$\| \Gamma_h^x(t) \|_{L^1} \leq C |\log h|, \qquad t \geq 0, \quad x \in \Omega. \tag{16.10}$$

According to Nitsche, we introduce the modified distance function

$$\omega(y) = \omega_h^x(y) = (|y - x|^2 + h^2)^{1/2}.$$

Then, we have

$$\| \omega^{-1} \|_{L^2}^2 = \int_\Omega \omega^{-2} \, dx \leq C \int_0^C \frac{r \, dr}{r^2 + h^2} = C |\log h|,$$

and hence

$$\| \Gamma_h^x(t) \|_{L^1} \leq \| \omega^{-1} \|_{L^2} \cdot \| \omega \Gamma_h^x(t) \|_{L^2} \leq C |\log h|^{1/2} \| \omega \Gamma_h^x(t) \|_{L^2}.$$

Consequently, we have only to show that

$$\| \omega \Gamma_h^x(t) \|_{L^2} \leq C |\log h|^{1/2}. \tag{16.11}$$

Henceforth, we suppose the inverse assumption. Then, Descloux's lemma holds with $\| P_h \|_{L^\infty \to L^\infty} \leq C$, as is described in Section 3. Furthermore, we simply write $\| \cdot \|$

instead of $\|\cdot\|_{L^2}$. The following lemma is related to the superapproximation property of NITSCHE and SCHATZ [1974]:

LEMMA 16.1. *The estimate*

$$\|\nabla(\omega^2\chi - P_h(\omega^2\chi))\| \leqslant Ch\{\|\chi\| + \|\omega\nabla\chi\|\}, \quad \chi \in V_h, \tag{16.12}$$

holds uniformly in $x \in \Omega$.

PROOF. Let $\phi = \pi_h(\omega^2\chi)$ be the interpolant of $\omega^2\chi$. Then, for each $T \in \tau_h$ we have

$$\|\omega^2\chi - \phi\|_{L^2(T)} + h\|\nabla(\omega^2\chi - \phi)\|_{L^2(T)} \leqslant Ch^2 \sum_{|\alpha| \leqslant 2} \|D^\alpha(\omega^2\chi)\|_{L^2(T)}.$$

By virtue of Leibniz's rule and $D^\alpha\chi = 0$, $|\alpha| = 2$, we find

$$\sum_{|\alpha| \leqslant 2} \|D^\alpha(\omega^2\chi)\|_{L^2(T)} \leqslant C\{\|\chi\|_{L^2(T)} + \|\omega\nabla\chi\|_{L^2(T)}\},$$

so that

$$\|\omega^2\chi - \phi\| + h\|\nabla(\omega^2\chi - \phi)\| \leqslant Ch^2\{\|\chi\| + \|\omega\nabla\chi\|\}.$$

On the other hand, for $\psi = P_h(\omega^2\chi) \in V_h$ the inequality

$$\|\nabla(\phi - \psi)\| \leqslant Ch^{-1}\|\phi - \psi\| = Ch^{-1}\|P_h(\phi - \omega^2\chi)\| \leqslant Ch^{-1}\|\phi - \omega^2\chi\|$$

holds. The desired estimate (16.12) is an immediate consequence of these two inequalities. □

The following lemma shows that the modified distance function ω compensates the singularity of the discrete delta function δ_h^x in a certain sense.

LEMMA 16.2. *The inequality*

$$\|\omega_h^x\delta_h^x\| \leqslant C \tag{16.13}$$

holds true.

PROOF. We fix $x \in \Omega$ and set $\omega = \omega_h^x$ and $\delta = \delta_h^x$. Let

$$\Omega_j = \{y \in \Omega : 2^{j-1}h < |y - x| < 2^j h\}, \quad j \geqslant 1,$$

$$\Omega_0 = \{y \in \Omega : |y - x| < h\}.$$

Obviously, $\omega(y) \leqslant Ch2^j$, $y \in \Omega_j$, so that

$$\|\omega\delta\| \leqslant C \sum_{j \geqslant 0} h2^j \|\delta\|_{L^2(\Omega_j)}.$$

To estimate $\|\delta\|_{L^2(\Omega_j)}$, we take $\varphi \in C_0^\infty(\Omega_j)$. Supposing $x \in T \in \tau_h$, we have

$$|(\delta, \varphi)| = |(\delta, P_h\varphi)| = |(P_h\varphi)(x)| \leqslant Ch^{-1}\|P_h\varphi\|_{L^2(T)}$$

by the inverse assumption. Furthermore, from Descloux's lemma the right-hand side is furthermore estimated above by

$$Ch^{-1}\exp(-C2^j)\|\varphi\|_{L^2(\Omega_j)}.$$

Therefore,

$$\|\delta\|_{L^2(\Omega_j)} \leqslant Ch^{-1}\exp(-C2^j)$$

holds and (16.13) has been established. □

The following lemma is a kind of discrete Sobolev imbedding:

LEMMA 16.3. *The inequality*

$$\|\chi\|_{L^\infty} \leqslant C|\log h|^{1/2}\|\nabla\chi\|, \quad \chi\in V_h, \tag{16.14}$$

holds true.

PROOF. We combine Sobolev's inequality

$$\|\chi\|_{L^p} \leqslant Cp^{1/2}\|\nabla\chi\|, \quad \chi\in V_h,$$

for $p>2$ and the inequality

$$\|\chi\|_{L^\infty} \leqslant Ch^{-2/p}\|\chi\|_{L^p}, \quad \chi\in V_h,$$

which is derived from the inverse assumption. Putting $p=|\log h|$, we obtain (16.14). □

Now, we can proceed to prove Theorem 16.1.

PROOF OF THEOREM 16.1. We have to show (16.11). In fact, we have

$$\tfrac{1}{2}(\mathrm{d}/\mathrm{d}t)\|\omega\Gamma\|^2 + \|\omega\nabla\Gamma\|^2 = (\partial_t\Gamma, \omega^2\Gamma) + (\nabla\Gamma, \nabla(\omega^2\Gamma)) - 2(\nabla\Gamma, \omega\Gamma\nabla\omega)$$

while

$$(\partial_t\Gamma, \psi) + (\nabla\Gamma, \nabla\psi) = 0$$

for $\psi\in V_h$. Therefore,

$$\tfrac{1}{2}(\mathrm{d}/\mathrm{d}t)\|\omega\Gamma\|^2 + \|\omega\nabla\Gamma\|^2 = \mathrm{I} + \mathrm{II} + \mathrm{III},$$

where

$$\mathrm{I} = (\partial_t\Gamma, \omega^2\Gamma - \psi), \qquad \mathrm{II} = (\nabla\Gamma, \nabla(\omega^2\Gamma - \psi)), \qquad \mathrm{III} = -2(\nabla\Gamma, \omega\Gamma\nabla\omega).$$

Taking $\psi = P_h(\omega^2\Gamma)\in V_h$, we first have $\mathrm{I}=0$. By means of the inverse assumption and Lemma 16.1, we next get

$$\begin{aligned}\|\mathrm{II}\| &\leqslant \|\nabla\Gamma\|\cdot\|\nabla(\omega^2\Gamma - \psi)\| \\ &\leqslant Ch^{-1}\|\Gamma\|h\{\|\Gamma\| + \|\omega\nabla\Gamma\|\} \\ &= C\{\|\Gamma\|^2 + \|\Gamma\|\,\|\omega\nabla\Gamma\|\}.\end{aligned}$$

Finally,

$$\|\mathrm{III}\| \leqslant C\|\Gamma\| \cdot \|\omega\nabla\Gamma\|$$

because of $\|\nabla\omega\|_{L^\infty} \leqslant C$. Summing these inequalities, we obtain

$$(\mathrm{d}/\mathrm{d}t)\|\omega\Gamma\|^2 + \|\omega\nabla\Gamma\|^2 \leqslant C\|\Gamma\|^2,$$

so that

$$\|\omega\Gamma(t)\|^2 + \int_0^t \|\omega\nabla\Gamma\|^2 \,\mathrm{d}S \leqslant \|\omega\delta_h^x\|^2 + C\int_0^t \|\Gamma\|^2 \,\mathrm{d}S.$$

Consequently, (16.11) has been furthermore reduced to

$$\int_0^t \|\Gamma\|^2 \,\mathrm{d}S \leqslant C|\log h| \tag{16.15}$$

by virtue of Lemma 16.2.

As in Section 14, we put $T_h = A_h^{-1}$. Then Γ satisfies

$$T_h \partial_t \Gamma + \Gamma = 0, \quad t > 0, \qquad \Gamma(0) = \delta_h^x,$$

and

$$\tfrac{1}{2}(\mathrm{d}/\mathrm{d}t)(T_h\Gamma, \Gamma) + \|\Gamma\|^2 = 0$$

holds. Therefore, we have

$$\tfrac{1}{2}(T_h\Gamma, \Gamma) + \int_0^t \|\Gamma\|^2 \,\mathrm{d}s = \tfrac{1}{2}(T_h\delta_h^x)(x).$$

Now, we set

$$G_h^x = T_h\delta_h^x.$$

T_h is positive-definite and the inequality (16.15) has been reduced to

$$G_h^x(x) \leqslant C|\log h|. \tag{16.16}$$

The function $G_h^x \in V_h$ is nothing but the discrete Green function:

$$(\nabla G_h^x, \nabla\chi) = (\nabla T_h\delta_h^x, \nabla\chi) = (\delta_h^x, \chi) = \chi(x), \quad \chi \in V_h.$$

In particular, $G_h^x(x) = \|\nabla G_h^x\|^2$. Therefore, from Lemma 16.3 we obtain

$$\begin{aligned} G_h^x(x) &\leqslant C|\log h|^{1/2}\|\nabla G_h^x\| \\ &= C(|\log h| \cdot G_h^x(x))^{1/2}, \end{aligned}$$

which implies (16.16). Thus, the proof has been completed. □

By means of the real methods described in Chapter III, we can show, for instance, the following error estimates from Theorem 16.1 and Nitsche–Scott's inequality

$$\|R_h v - v\|_{L^\infty} \leqslant Ch^2|\log h|\,\|v\|_{W^{2,\infty}},$$

which is described in Section 4. We shall make use of Helfrich's method to prove the estimate presented in the following:

THEOREM 16.2. *The inequality*

$$\|u_h(t)-u(t)\|_{L^\infty} \leqslant Ch^2|\log h|^2 \left\{ \sup_{0\leqslant t\leqslant T} \|u(t)\|_{W^{2,\infty}} + \int_0^t \|u_t\|_{W^{2,\infty}}\,\mathrm{d}s \right\} \tag{16.17}$$

holds true for the solutions $u=u(t)$ *and* $u_h=u_h(t)$ *of* (16.4) *with* (16.5) *and of* (16.6) *with* (16.7), *respectively.*

PROOF. As we have shown in Sections 12 and 15, the error operator $E_h(t)$ is expressed as

$$E_h(t) = (1-\mathrm{e}^{-tA_h}P_h)(R_h-1)\mathrm{e}^{-tA}$$
$$+ \int_0^t \mathrm{e}^{-(t-r)A_h} A_h P_h (R_h-1)(\mathrm{e}^{-rA}-\mathrm{e}^{-tA})\,\mathrm{d}r.$$

Integrating by parts, we have

$$-\int_0^t \mathrm{e}^{-(t-r)A_h} A_h P_h (R_h-1)(\mathrm{e}^{-rA}-\mathrm{e}^{-tA})\,\mathrm{d}r$$
$$= [\mathrm{e}^{-(t-r)A_h} P_h (R_h-1)(\mathrm{e}^{-rA}-\mathrm{e}^{tA})]_{r=0}^{r=t} + \int_0^t \mathrm{e}^{-(t-r)A_h} P_h (R_h-1) A \mathrm{e}^{-rA}\,\mathrm{d}r$$

and write

$$E_h(t) = \sum_{l=1}^{3} E_h^l(t)$$

with

$$E_h^1(t) = (R_h-1)\mathrm{e}^{-tA},$$
$$E_h^2(t) = -\mathrm{e}^{-tA_h} P_h (R_h-1),$$
$$E_h^3(t) = \int_0^t \mathrm{e}^{-(t-r)A_h} P_h (R_h-1) A \mathrm{e}^{-rA}\,\mathrm{d}r.$$

From Nitsche–Scott's inequality it follows that

$$\|E_h^1(t)u_0\|_{L^\infty} \leqslant Ch^2|\log h|\,\|u(t)\|_{W^{2,\infty}}.$$

On the other hand, Theorem 16.1 yields

$$\|E_h^2(t)u_0\|_{L^\infty} \leqslant Ch^2|\log h|^2\,\|u_0(t)\|_{W^{2,\infty}},$$

$$\|E_h^3(t)u_0\|_{L^\infty} \leqslant Ch^2|\log h|^2 \int_0^t \|u_t(t)\|_{W^{2,\infty}}\,\mathrm{d}r.$$

Thus, (16.17) has been established. □

We can deduce other error estimates even for nonsmooth initial data u_0, taking account of the smoothing properties of $\{e^{-tA}\}$ and $\{e^{-tA_h}\}$. One simple way to establish these inequalities is to use the discrete elliptic Sobolev inequality given in Section 5.

However, we do not go into details here. Another way based on the energy method can be seen in THOMÉE [1984].

17. Rational approximation of semigroups

In Chapters II and III, we studied the fully discrete finite element approximation for parabolic equations. There, we considered the forward, backward and Crank–Nicolson schemes. However, many other methods exist as for the time discretization. In the present section, we present a general way to deal with them.

To begin with, let $\{e^{-tA}\}_{t\geqslant 0}$ be a uniformly bounded (C_0) semigroup in a Banach space X and put $r(z)=(1+z)^{-1}$, $z\in\mathbb{C}$. Then, r is a rational function with the properties

$$r(z)=e^{-z}+O(|z|^{p+1}), \quad |z|\downarrow 0, \tag{17.1}$$

for $p=1$ and

$$|r(z)|\leqslant 1 \quad \text{for } \operatorname{Re} z\geqslant 0. \tag{17.2}$$

These properties (17.1) and (17.2) are sometimes referred to as being of order p and (Dahlquist's) A-acceptability, respectively. The operator $r^n(\tau A)$ is regarded as an approximation of e^{-tA} for $t=n\tau$, $\tau>0$ being the time mesh parameter. Actually, from the identity

$$\begin{aligned}
&(\mathrm{d}/\mathrm{d}s)(r^n(sA)e^{-n(\tau-s)A}) \\
&\quad = nr^{n-1}(sA)\{r'(sA)+r(sA)\}Ae^{-n(\tau-s)A} \\
&\quad = nr^{n+1}(sA)sA^2e^{-n(\tau-s)A}
\end{aligned}$$

it follows that

$$[r^n(\tau A)-\mathrm{e}^{-n\tau A}]A^{-2}=n\int_0^{\tau} r^{n+1}(sA)s\mathrm{e}^{-n(\tau-s)A}\,\mathrm{d}s.$$

On the other hand, the uniform boundedness of $\{\mathrm{e}^{-tA}\}_{t\geq 0}$,

$$\|\mathrm{e}^{-tA}\|\leqslant C,\quad 0\leqslant t<\infty,$$

implies the stability

$$\|r^n(\tau A)\|\leqslant C$$

by a standard argument in the semigroup theory (YOSIDA [1964, Chapter 9], KATO [1966, Chapter 9], e.g.). Therefore, we obtain

$$\|[r^n(\tau A)-\mathrm{e}^{-n\tau A}]A^{-p-1}\|\leqslant Ct\tau^p,\quad t=n\tau, \tag{17.3}$$

for $p=1$. HERSH and KATO [1979] and BRENNER and THOMÉE [1979] have shown the error estimate (17.3) for a general rational function $r=r(z)$ with the properties of being of order p and A-acceptability. In the present section, we shall give an error estimate when $\{\mathrm{e}^{-tA}\}_{t\geq 0}$ is holomorphic, taking account of its smoothing effects and following the method of BAKER, BRAMBLE and THOMÉE [1977].

Before doing that, we generalize the notion of A-acceptability and introduce the following:

DEFINITION 17.1. For $0\leqslant\theta<\pi$, let

$$\Sigma_\theta=\{z\in\mathbb{C}: |\arg z|\leqslant\theta\}.$$

Then, a *rational function* $r=r(z)$ is said to have the properties

(i)$_\theta$ if $|r(z)|\leqslant 1$, $z\in\Sigma_\theta$, and $|r(\infty)|<1$,

(ii)$_\theta$ if $|r(z)|<1$, $z\in\Sigma_\theta\backslash\{0\}$,

(iii)$_\theta$ if $|r(z)|<1$, $0<|z|<\delta$, $z\in\Sigma$, for some δ.

The property $|r(z)|\leqslant 1$, $z\in\Sigma_\theta$ is called *A_θ-acceptability*. Then, A-acceptability means $A_{\pi/2}$-acceptability in this terminology.

REMARK 17.1. Each rational function of order p ($\geqslant 1$) has property (iii)$_\theta$ for $0<\theta<\frac{1}{2}\pi$. In fact, for $\varphi\in[-\theta,\theta]$ we have

$$(\partial/\partial\rho)|r(\rho\mathrm{e}^{\mathrm{i}\varphi})|^2=2\operatorname{Re} r(\rho\mathrm{e}^{\mathrm{i}\varphi})r'(\rho\mathrm{e}^{\mathrm{i}\varphi})\mathrm{e}^{\mathrm{i}\varphi}.$$

Since

$$r(0)=\mathrm{e}^{-z}|_{z=0}=1,\qquad r'(0)=(\mathrm{d}/\mathrm{d}z)\mathrm{e}^{-z}|_{z=0}=-1,$$

we get

$$\begin{aligned}(\partial/\partial\rho)|r(\rho\mathrm{e}^{\mathrm{i}\varphi})|^2|_{\rho=0}&=-2\operatorname{Re}\mathrm{e}^{\mathrm{i}\varphi}\\&=-2\cos\varphi\\&\leqslant-2\cos\theta<0.\end{aligned}$$

REMARK 17.2. Let $r=r(z)$ be the Padé approximation of e^{-z} with degrees n and m of the numerator and the dominator, respectively. Then r is of order $p=n+m$. Furthermore, it has property $(\mathrm{i})_\theta$, $(\mathrm{ii})_\theta$ or $(\mathrm{iii})_\theta$ for some θ, $0<\theta<\frac{1}{2}\pi$ for $n<m$, $n=m$ or $n>m$, respectively.

In fact, in this case we have $r(z)=R_{nm}(z)=P_{nm}(z)/Q_{nm}(z)$ with

$$P_{nm}(z)=\sum_{j=0}^{n}\frac{(n+m-j)!\,n!}{(n+m)!\,j!\,(n-j)!}(-z)^j,$$

$$Q_{nm}(z)=\sum_{j=0}^{m}\frac{(n+m-j)!\,m!}{(n+m)!\,j!\,(m-j)!}z^j.$$

The relation

$$|R_{nm}(z)-e^{-z}|\leqslant C|z|^{n+m+1},\quad |z|\downarrow 0,$$

is well-known (HITOTUMATU [1963], e.g.). Put

$$a_{nm}(j)=\frac{(n+m-j)!\,n!}{(n+m)!\,j!\,(n-j)!},\quad 0\leqslant j\leqslant n.$$

Then, in the case of $n\leqslant m$ we have

$$a_{nm}(j)\geqslant 0,\quad a_{nm}(j)\leqslant a_{mn}(j),\qquad 0\leqslant j\leqslant n,$$

so that, when $m\varphi\leqslant\frac{1}{2}\pi$,

$$\begin{aligned}|\mathrm{Re}\,P_{nm}(\rho e^{\pm i\varphi})|&=\left|\sum_{j=0}^{n}a_{nm}(j)(-\rho)^j\cos j\varphi\right|\\&\leqslant\sum_{j=0}^{n}a_{nm}(j)\rho^j\cos j\varphi\\&\leqslant\sum_{j=0}^{m}a_{mn}(j)\rho^j\cos j\varphi\\&\leqslant\mathrm{Re}\,Q_{nm}(\rho e^{\pm i\varphi}).\end{aligned}$$

Similarly, under the assumption $m\varphi\leqslant\frac{1}{2}\pi$ we have

$$\begin{aligned}|\mathrm{Im}\,P_{nm}(\rho e^{\pm i\varphi})|&=\left|\sum_{j=0}^{n}a_{nm}(j)(-\rho)^j\sin(\pm j\varphi)\right|\\&\leqslant\sum_{j=0}^{n}a_{nm}(j)\rho^j\sin j\varphi\\&\leqslant\sum_{j=0}^{m}a_{mn}(j)\rho^j\sin j\varphi\\&=\left|\sum_{j=0}^{m}a_{nm}(j)\rho^j\sin(\pm j\varphi)\right|\\&=|\mathrm{Im}\,Q_{nm}(\rho e^{\pm i\varphi})|.\end{aligned}$$

Furthermore,

$$\operatorname{Re} Q_{nm}(\rho e^{\pm i\varphi}) = \sum_{j=0}^{m} a_{nm}(j)\rho^j \cos j\varphi \neq 0,$$

if $m\varphi \leqslant \frac{1}{2}\pi$. Therefore, $|r(\rho e^{\pm i\varphi})| \leqslant 1, 0 \leqslant \varphi \leqslant \frac{1}{2}\pi/m$.

Now in the case of $n<m$, we have $r(\infty)=0$. Hence $r=R_{nm}$ satisfies $(\mathrm{i})_\theta$ for $\theta = \frac{1}{2}\pi/m$. If $n=m$, then $|r(\infty)| = 1 < +\infty$ so that $|r(z)| \leqslant 1$, $z \in \Sigma_{\pi/2m}$, by Phragmén–Lindelof's theorem. Therefore, $|r(z)|<1$, $z\in\Sigma_\theta\backslash\{0\}$, for $\theta<\frac{1}{2}\pi/m$ by the maximum principle, which means that $r=R_{nm}$ satisfies $(\mathrm{ii})_\theta$. Finally, by virtue of Remark 17.1, $r=R_{nm}$ always satisfies $(\mathrm{iii})_\theta$ for some θ in $0<\theta<\frac{1}{2}\pi$.

REMARK 17.3. Incidentally, the Padé approximation $r=R_{nm}$ of e^{-z} is $A_{\pi/2}$-acceptable if and only if $m-2\leqslant n\leqslant m$. This was conjectured by Ehle and has been proven by WANNER, HAIRER and NORSETT [1978]. In the case of $n=m$ and $n=m-1$, R_{nm} obeys a recursive formula based on the continued fraction expansion (MORI [1974]). Namely, from the expansion

$$e^z = \cfrac{1}{1-\cfrac{z}{1+\cfrac{z}{2-\cfrac{z}{3+z\cdots}}}}$$

which is expressed by

$$e^z = \frac{1}{1}-\frac{z}{1}+\frac{z}{2}-\frac{z}{3}+\frac{z}{2}-\frac{z}{5}+\cdots+\frac{z}{2}-\frac{z}{2j-1}+\cdots,$$

we get

$$e^z = 1+\frac{2z}{2-z}+\frac{z^2}{6}+\frac{z^2}{10}+\cdots+\frac{z^2}{2(2j-1)}+\cdots.$$

Accordingly, a rational approximation $H_{2k+1}(z) = G_{2k+1}(z)/F_{2k+1}(z)$ of e^z is introduced inductively with

$$F_1=1, \qquad F_3=2-z, \qquad G_1=1, \qquad G_3=2+z,$$

$$F_{2j+1}=2(2j-1)F_{2j-1}+z^2F_{2j-3}, \qquad G_{2j+1}=2(2j-1)G_{2j-1}+z^2G_{2j-3}.$$

Similarly, $H_{2k}(z)=G_{2k}(z)/F_{2k}(z)$ is defined by

$$F_0=1, \qquad F_2=1-z, \qquad G_0=0, \qquad G_2=1,$$

$$F_{2j}=\left\{2(2j-1)+\frac{2}{2j-3}z\right\}F_{2j-2}+\frac{2j-1}{2j-3}z^2F_{2j-4},$$

$$G_{2j}=\left\{2(2j-1)+\frac{2}{2j-3}z\right\}G_{2j-2}+\frac{2j-1}{2j-3}G_{2j-4}.$$

Then, the relations $H_{2k+1}(-z)=R_{kk}(z)$ and $H_{2k}(-z)=R_{k-1\,k}(z)$ can be verified.

Now, let X be a Banach space and $-A$ be the generator of a holomorphic semigroup $\{e^{-tA}\}$ of type (θ_0, M_0), $0<\theta_0<\frac{1}{2}\pi$, $M_0\geqslant 1$, that is,

$$\sigma(A)\subset\Sigma_{\theta_0}\setminus\{0\},$$

$$\|\lambda(\lambda-A)^{-1}\|\leqslant M_0, \quad \lambda>0,$$

$$\|(\lambda-A)^{-1}\|\leqslant M_\varepsilon/|\lambda|, \qquad \varepsilon>0, \quad \lambda\in\mathbb{C}\setminus\Sigma_{\theta_0+\varepsilon}.$$

Furthermore, suppose that A is bounded for simplicity. Then,

THEOREM 17.1. *The error estimate*

$$\|r^n(\tau A)-e^{-n\tau A}\|\leqslant C(\tau/t)^p, \quad 0<t=n\tau<T, \tag{17.4}$$

holds under the assumption that for some $\theta>\theta_0$ *one of the following holds:*

(a) $r(z)$ *has property* $(\mathrm{i})_\theta$,

(b) $r(z)$ *has property* $(\mathrm{ii})_\theta$ *and* $\tau\|A\|\leqslant\hat{M}_1<\infty$,

(c) $r(z)$ *has the property* $(\mathrm{iii})_\theta$ *and* $\tau\|A\|\leqslant\hat{M}_1<\delta$.

Here, the constant $C>0$ *in* (17.4) *depends only on* θ_0, θ *and* M_{θ_1}, $\theta_0<\theta_1<\theta$, *for the case* (a) *and on* $\theta_0, \theta, M_{\theta_1}$, $\theta_0<\theta_1<\theta$, *and* $\hat{M}_1$ *for the cases* (b) *and* (c).

REMARK 17.4. BAKER, BRAMBLE and THOMÉE [1977] have shown the theorem for the case that A is self-adjoint under slightly weaker assumptions on $r(z)$. Theorem 17.1(a) has been obtained by LE ROUX [1979]. The other cases can also be derived by her method.

We have to prepare some propositions for the proof of Theorem 17.1. Henceforth $\beta>0$ and $C>0$ stand for small and large absolute constants, respectively.

PROPOSITION 17.1. *If a rational function* $r=r(z)$ *is of order* p $(\geqslant 1)$ *and* A_θ*-acceptable,* $0<\theta<\frac{1}{2}\pi$, *then there exist constants* $\sigma>0$ *and* $\beta>0$ *such that*

$$|r^n(z)-e^{-nz}|\leqslant Cn|z|^{p+1}e^{-\beta n\operatorname{Re} z}, \qquad z\in\Sigma_\theta, \quad |z|\leqslant\sigma.$$

PROOF. First, we note that

$$r^n(z)-e^{-nz}=\sum_{j=1}^{n} r^{j-1}(z)(e^{-z}-r(z))e^{-(n-j+1)z}.$$

On the other hand, the inequality

$$|r(z)-e^{-z}|\leqslant C|z|^{p+1}, \quad |z|\leqslant\sigma_0,$$

holds for some $\sigma_0>0$ so that

$$|r(z)|\leqslant e^{-\operatorname{Re} z}+C(\operatorname{Re} z)^{p+1}, \qquad z\in\Sigma_\theta, \quad |z|\leqslant\sigma_0.$$

Set $f(t)=e^{-t}+Ct^{p+1}$. Then, $f=f(t)$ is real-valued, $f(0)=1$ and $f'(0)<0$. Therefore, $f(t)\leqslant e^{-\beta t}, 0\leqslant t\leqslant\sigma_1$, holds for some $\beta>0$ and $\sigma_1>0$. Namely,

$$|r(z)|\leqslant e^{-\beta \operatorname{Re} z},\qquad |z|\leqslant\min(\sigma_1,\sigma_0),\quad z\in\Sigma_\theta.$$

Therefore, if $|z|\leqslant\sigma\equiv\min(\sigma_1,\sigma_0)$ and $z\in\Sigma_\theta$, we have

$$\begin{aligned}|r^n(z)-e^{-nz}| &\leqslant \sum_{j=1}^{n} e^{-(j-1)\beta \operatorname{Re} z} C|z|^{p+1} e^{-(n-j+1)\cos\theta \operatorname{Re} z}\\ &= Cn|z|^{p+1}e^{-\beta n \operatorname{Re} z}. \qquad \square\end{aligned}$$

LEMMA 17.1. *Under the assumption of Theorem* 17.1, *the estimate*

$$\|[r^n(\tau A)-e^{-tA}]A^{-p}\|\leqslant C\tau^p,\quad t=n\tau, \tag{17.5}$$

holds, where the constant $C>0$ *depends on* θ_0,θ and $M_{\theta_1}, \theta_0<\theta_1<\theta$, *for the cases* (a) *and* (b), *and on* $\theta_0,\theta,M_{\theta_1},\theta_0<\theta_1<\theta$, *and* $\hat{M}_1$ *for the case* (c).

PROOF. We fix a constant $M>0$ such that $\tau\|A\|<M$, $\hat{M}_1<M$ and $\hat{M}_1<M<\delta$ for the cases (a), (b) and (c), respectively. We may suppose that $\sigma<M$ in Proposition 17.1. We take a path of integration Γ_0 which is divided into three parts: $\Gamma_0=\Gamma_1\cup\Gamma_2\cup\Gamma_3$, where the sets Γ_1,Γ_2 and Γ_3 are given by

$$\begin{aligned}&\Gamma_1=\{\lambda=\rho e^{\pm i\theta_1}: 0\leqslant\rho\leqslant\sigma\tau^{-1}\},\\ &\Gamma_2=\{\lambda=\rho e^{\pm i\theta_1}: \sigma\tau^{-1}\leqslant\rho\leqslant M\tau^{-1}\},\\ &\Gamma_3=\{M\tau^{-1}e^{i\phi}: |\phi|\leqslant\theta_1\}\end{aligned}$$

with $\theta_0<\theta_1<\theta$. Then, we have

$$\begin{aligned}[r^n(\tau A)-e^{-tA}]A^{-p}&=\frac{1}{2\pi i}\left(\int_{\Gamma_1}+\int_{\Gamma_2}+\int_{\Gamma_3}\right)(r^n(\tau\lambda)-e^{-t\lambda})\lambda^{-p}(\lambda-A)^{-1}\,d\lambda\\ &=\mathrm{I}+\mathrm{II}+\mathrm{III}.\end{aligned}$$

By virtue of Proposition 17.1, we have

$$\begin{aligned}\|\mathrm{I}\|&\leqslant C\int_0^\infty n(\tau\rho)^{p+1}e^{-\beta n\tau\rho\cos\theta_1}\rho^{-p}\frac{d\rho}{\rho}\\ &=C\int_0^\infty (n\tau\rho)^{p+1}e^{-\beta(n\tau\rho)\cos\theta_1}(n\tau\rho)^{-p}\frac{d\rho}{\rho}\tau^p\\ &=C\tau^p.\end{aligned}$$

From the assumption it follows that

$$\|\mathrm{II}\| \leqslant C \int_{\Gamma_2} \{|r^n(\tau\lambda)| + |e^{-t\tau}|\} |\lambda|^{-p} \|(\lambda - A)^{-1}\| \, |d\lambda|$$

$$\leqslant C \int_{\sigma\tau^{-1}}^{\infty} \rho^{-p} \frac{d\rho}{\rho} = C\tau^p.$$

Finally,

$$\|(\lambda - A)^{-1}\| \leqslant \frac{1}{|\lambda| - \|A\|} = \frac{\tau}{M - \tau\|A\|}$$

holds on Γ_3 so that

$$\|\mathrm{III}\| \leqslant C \int_{\Gamma_3} \{|r^n(\tau\lambda)| + |e^{-t\lambda}|\} |\lambda|^{-p} \|(\lambda - A)^{-1}\| \, |d\lambda|$$

$$\leqslant C \int_{-\theta_1}^{\theta_1} (M\tau^{-1})^{-p} \frac{M \, d\phi}{M - \tau\|A\|} = C_M \tau^p.$$

In the cases (a) and (b), the constants C in the estimates of $\|\mathrm{I}\|$ and $\|\mathrm{II}\|$ are independent of M, while C_M goes to zero as $M \to \infty$. Thus, the proof has been completed. □

PROPOSITION 17.2. *If $r = r(z)$ satisfies* (iii)$_\theta$, *then for each δ', $0 < \delta' < \delta$, there exists $\beta > 0$ such that*

$$|r(z)| \leqslant e^{-\beta|z|}, \qquad |z| \leqslant \delta', \quad z \in \Sigma_\theta.$$

When $r = r(z)$ satisfies (ii)$_\theta$, *δ' can be arbitrarily large.*

PROOF. As is shown in Remark 17.1, the inequality

$$(\partial/\partial\rho)|r(\rho e^{i\phi})|^2|_{\rho=0} \leqslant -2\cos\theta, \quad |\phi| \leqslant \theta, \tag{17.6}$$

holds so that for each $\beta_0, 0 < \beta_0 < \cos\theta$, there exists a $\sigma_0 > 0$ such that

$$|r(\rho e^{i\phi})| \leqslant \exp(-\beta_0 \rho), \qquad 0 \leqslant \rho \leqslant \sigma_0, \quad |\phi| \leqslant \theta.$$

Next, we set

$$\max\{|r(\rho e^{i\phi})| : \rho \in [\sigma_0, \delta'], \phi \in [-\theta, \theta]\} = 1 - \varepsilon < 1,$$

and take $\beta_1 > 0$ so that $\exp(-\beta_1 \delta') = 1 - \varepsilon$. Then, we have

$$|r(\rho e^{i\phi})| \leqslant \exp(-\beta_1 \rho), \qquad \sigma_0 \leqslant \rho \leqslant \delta', \quad |\phi| \leqslant \theta.$$

If we set $\beta = \min(\beta_0, \beta_1)$, the desired inequality holds. □

LEMMA 17.2. *In the cases* (b) *and* (c) *of Theorem* 17.1, *the estimate*

$$\| A^{\alpha} r^{n}(\tau A) \| \leqslant C_{\alpha}(n\tau)^{-\alpha} \tag{17.7}$$

holds for $\alpha > 0$. *Here the constant* $C_{\alpha} > 0$ *depends on* $\hat{M}_1$.

PROOF. We take a path of integration as in the proof of Lemma 17.1, and obtain

$$(n\tau A)^{\alpha} r^{n}(\tau A) = \frac{1}{2\pi \mathrm{i}} \Bigg(\int\limits_{\Gamma_1 + \Gamma_2} + \int\limits_{\Gamma_3} \Bigg) (n\tau\lambda)^{\alpha} r^{n}(\tau\lambda)(\lambda - A)^{-1} \, \mathrm{d}\lambda$$

$$= \mathrm{I} + \mathrm{II}.$$

From the assumption, we can take δ' of Proposition 17.2 in $M < \delta'$. Then, we have

$$\|\mathrm{I}\| \leqslant C \int\limits_{0}^{\infty} (n\tau\lambda)^{\alpha} \mathrm{e}^{-\beta n\tau\rho \cos\theta_1} \frac{\mathrm{d}\rho}{\rho} = C_{\alpha},$$

$$\|\mathrm{II}\| \leqslant C \int\limits_{-\theta_1}^{\theta_1} (n\tau M)^{\alpha} \mathrm{e}^{-\beta M n} \frac{M}{M - \hat{M}_1} \, \mathrm{d}\phi \leqslant C_{\alpha}. \qquad \square$$

By means of Helfrich's duality argument, we can now give the

PROOF OF THEOREM 17.1(b)–(c). Dividing $N \in \mathbb{N}$ into $n = l + m$ with $l, m \in \mathbb{N}$ and $0 \leqslant l - m \leqslant 1$, we have

$$r^{n}(\tau A) - \mathrm{e}^{-tA} = (r^{l}(\tau A) - \mathrm{e}^{-l\tau A}) A^{-p} r^{m}(\tau A) + \mathrm{e}^{-l\tau A} A^{-p} (r^{m}(\tau A) - \mathrm{e}^{-m\tau A}).$$

The first term of the right-hand side is estimated as

$$\| [r^{l}(\tau A) - \mathrm{e}^{-l\tau A}] A^{-p} \| \cdot \| A^{p} r^{m}(\tau A) \| \leqslant C\tau^{p} (m\tau)^{-p} \leqslant C/n^{p} = C(\tau/t)^{p}$$

by Lemmas 17.1 and 17.2. The second term is estimated similarly by the adjoint form of these lemmas. $\square$

Unfortunately, $r(\infty) = 0$ does not hold in general even in the case $(\mathrm{i})_{\theta}$. Therefore, we cannot take $\delta' = \infty$ in Proposition 17.2. Consequently, we have to give other considerations to prove Theorem 17.1(a).

LEMMA 17.3. *Let* ϕ *be a meromorphic function. Suppose that there exist functions* $f_1, f_2 : \mathbb{R}_+ \to \mathbb{R}_+$ *and a constant* $R > 0$ *such that for* θ_1, $\theta_0 < \theta_1$,

$$\int\limits_{0}^{R} f_1(r) \frac{\mathrm{d}r}{r} < \infty, \qquad \int\limits_{0}^{R} f_2(r) \frac{\mathrm{d}r}{r} < \infty,$$

$$|\phi(z)| \leqslant f_1(|z|), \quad |\phi(z)-\phi(\infty)| \leqslant f_2(|z|), \quad |z| \leqslant R, \quad |\arg z| = \theta_1,$$

Then, the inequality

$$\|\varphi(A)\| \leqslant C_R \left\{ \int_0^R f_1(r) \frac{dr}{r} + \int_0^R f_2(r) \frac{dr}{r} + |\varphi(\infty)| \right\} \tag{17.8}$$

holds true.

PROOF. Putting

$$h(z) = \varphi(z) - \frac{z}{1+z} \varphi(\infty),$$

we have

$$\varphi(A) = h(A) + \varphi(\infty) A(1+A)^{-1}.$$

Here

$$\|A(1+A)^{-1}\| \leqslant C$$

holds. Therefore, we have

$$\|\varphi(A)\| \leqslant \|h(A)\| + C|\varphi(\infty)|.$$

To estimate $\|h(A)\|$, let Γ be the path of integration given in Chapter II, that is, the positively oriented boundary of $\Sigma_{\theta_1}, \theta_0 < \theta_1 < \theta$. Then, the identity

$$h(A) = \frac{1}{2\pi i} \int_\Gamma h(z)(z-A)^{-1} \, dz$$

holds. We have

$$|h(z)| = \left| \varphi(z) - \frac{z}{1+z} \varphi(\infty) \right| \leqslant f_1(|z|) + |z| \, |\varphi(\infty)|$$

if $|z| \leqslant R$ and $|\arg z| = \theta_1$, while

$$|h(z)| = \left| \varphi(z) - \varphi(\infty) + \frac{1}{1+z} \varphi(\infty) \right| \leqslant f_2(|z|) + \frac{1}{|z|} |\varphi(\infty)|$$

holds for $|z| \geqslant R$ and $|\arg z| = \theta_1$. Therefore, we get

$$\|h(A)\| \leqslant C \left(\int_0^R \{f_1(r) + r|\varphi(\infty)|\} \frac{dr}{r} + \int_R^\infty \left\{ f_2(r) + \frac{1}{r} |\varphi(r)| \right\} \frac{dr}{r} \right)$$

$$= C \left\{ \int_0^R f_1(r) \frac{dr}{r} + \int_0^R f_2(r) \frac{dr}{r} + \left(R + \frac{1}{R} \right) |\varphi(\infty)| \right\}.$$

Hence the desired estimate (17.8) has been established. □

Now, we can give the following proof.

PROOF OF THEOREM 17.1(a): Take θ_1 in $\theta_0<\theta_1<\theta$ and $\sigma>0$ as in Proposition 17.1. Then, the function $\phi(z)=e^{-nz}-r^n(z)$ satisfies, for some $\beta>0$,

$$|\phi(z)|\leqslant Cn|z|^{p+1}e^{-\beta n|z|\cos\theta_1}, \qquad z\in\Sigma_{\theta_1},\quad |z|\leqslant\sigma. \tag{17.9}$$

On the other hand, making $\beta>0$ smaller if necessary, we may suppose that

$$\sup\{|r(z)|: z\in\Sigma_{\theta_1}, |z|\geqslant\sigma\}=e^{-\beta}<1$$

from the assumption. Furthermore,

$$|r(z)-r(\infty)|\leqslant C/|z|,\quad |z|\geqslant\sigma,$$

holds because $r=r(z)$ is rational so that

$$\begin{aligned}|r^n(z)-r^n(\infty)|&=|r(z)-r(\infty)|\left|\sum_{j=0}^{n-1} r^j(z)r^{n-1-j}(\infty)\right|\\ &\leqslant Ce^{-n\beta}/|z|,\qquad z\in\Sigma_{\theta_1},\quad |z|\geqslant\sigma.\end{aligned}$$

Therefore,

$$|\phi(z)-\phi(\infty)|\leqslant e^{-n|z|\cos\theta_1}+Ce^{-n\beta}/|z|, \qquad z\in\Sigma_{\theta_1},\quad |z|\geqslant\sigma. \tag{17.10}$$

Finally,

$$|\phi(\infty)|\leqslant e^{-n\beta}. \tag{17.11}$$

By virtue of these three inequalities (17.9)–(17.11), we obtain

$$\|\phi(\tau A)\|=\|r^n(\tau A)-e^{-n\tau A}\|\leqslant C/n^p$$

from Lemma 17.3. □

The above abstract theorem can be applied to the semidiscrete finite element approximation of parabolic equations, that is,

$$du_h/dt+A_h u_h=0,\quad 0\leqslant t\leqslant T, \tag{17.12}$$

in X_h with

$$u_h(0)=P_h u_0, \tag{17.13}$$

A_h, X_h and P_h being as before. We recall that the spectrum of A_h lies in a parabolic region in the complex plane which is uniform in h. Furthermore, for each θ_0 in $0<\theta_0<\frac{1}{2}\pi$, there exist constants $M_0>0$ and $\lambda>0$ independent of h such that $-A_h+\lambda$ is of type (θ_0, M_0). Therefore, taking $v_h=e^{-\lambda t}u_h$ instead of u_h, we can make the exponent $\theta>0$ as small as we like. On the other hand, under the inverse assumption holds the inequality

$$\|A_h\|\leqslant\beta h^{-2}$$

with an absolute constant $\beta>0$. Therefore, we have obtained:

THEOREM 17.2. *Let $r=r(z)$ be a rational function of order p ($\geqslant 1$) such that for some $\theta>0$ one of the following is satisfied*
(a) $(\mathrm{i})_\theta$,
(b) $(\mathrm{ii})_\theta$ *and* $\tau/h^2 \leqslant M_1 < \infty$,
(c) $(\mathrm{iii})_\theta$ *and* $\tau/h^2 \leqslant M_1 < \beta h$.
Then, the estimate

$$\| r^n(\tau A_h) - \mathrm{e}^{-tA_h} \| \leqslant C(\tau/t)^p, \quad t=n\tau$$

holds with a constant $C>0$ independent of h.

We can also apply Theorem 17.1 for the semidiscrete finite element approximation with higher accuracy. In that case, we will obtain a more natural result from the viewpoint of the correspondence of the rate of convergence with respect to the time discretization and the space discretization. The theorems on the backward, forward and Crank–Nicolson schemes in Sections 8 and 10 may be obtained as special cases of that general result.

18. Multi-step methods

The time discretization considered in the preceding section determines, at the nth time mesh, $u_n=u(t_n)$ from the $(n-1)$th value $u_{n-1}=u(t_{n-1})$, starting from the given initial data $u_0=u(t_0)$, where $t_n=n\tau$. In this sense, it may be called the single-step method. On the contrary, a multi-step method of order q ($\geqslant 2$) is the way to determine u_n from the values at the preceding q-steps: $u_{n-1}, u_{n-2}, \ldots, u_{n-q}$ after determining $u_1, \ldots, u_{q-1}$ from u_0 by other suitable methods.

In the present section, we adopt this kind of schemes to discretize in time the evolution equation

$$\mathrm{d}u/\mathrm{d}t + Au = 0, \quad 0 \leqslant t \leqslant T, \tag{18.1}$$

in a Banach space X with

$$u(0)=u_0. \tag{18.2}$$

Here $-A$ is the generator of a holomorphic semigroup $\{\mathrm{e}^{-tA}\}$ of type (θ, M) with $0<\theta<\frac{1}{2}\pi$ and $M\geqslant 1$. For simplicity, A is supposed to be bounded. In the scheme which we consider here, one firstly determines $u_1, \ldots, u_{q-1}$ from u_0 by a single-step approximation in use of a rational function, and then computes $u_n, n\geqslant q$ through the relation

$$\sum_{i=0}^{q} (a_i + \tau b_i A) u_{n+i} = 0, \quad n=0,1,\ldots, \tag{18.3}$$

where $a_i, b_i \in \mathbb{R}$.

Without loss of generality, we may take $a_q = 1$ in (18.3). Setting $P(\zeta)=\Sigma_{i=0}^{q} a_i\zeta^i$ and $S(\zeta)=\Sigma_{i=0}^{q} b_i\zeta^i$, we call the scheme (18.3) the multi-step method (P, S). Then:

DEFINITION 18.1. A multi-step method (P, S) is said to be of order p ($\geqslant 1$) if

$$\sum_{i=0}^{q} a_i = 0, \qquad \sum_{i=0}^{q} ia_i = \sum_{i=0}^{q} b_i,$$

$$\sum_{i=1}^{q} i^j a_i = j \sum_{i=1}^{q} i^{j-1} b_i, \quad 2 \leqslant j \leqslant p.$$

Understanding $0^0 = 1$ and $0 \cdot 0^{-1} = 0$, we can write these relations simply as

$$\sum_{i=0}^{q} i^j a_i = j \sum_{i=0}^{q} i^{j-1} b_i, \quad 0 \leqslant j \leqslant p.$$

The meaning of the above equalities is seen as follows. Regard A as the differential operator d/dx, and make the Taylor expansion around $\tau = 0$ of the finite difference operation

$$L_\tau[y] = \sum_{i=0}^{q} [a_i y(x+i\tau) + \tau b_i y'(x+i\tau)].$$

Then, the relations are obtained by making the coefficients up to pth powers of τ to be zero. The scheme described by (18.3) implies the backward, forward, Crank–Nicolson and modified Crank–Nicolson (which is of order 1) in the case of $q=1$.

Putting $w(\zeta, z) = P(\zeta) + zS(\zeta)$, we have the following definition:

DEFINITION 18.2. For $0 < \theta < \frac{1}{2}\pi$, a multi-step method (P, S) is said to have the property:

$(\mathrm{III})_\theta$ if each of the roots $\{\zeta_j : 1 \leqslant j \leqslant q\}$ of $P(\zeta) = 0$ is simple and lies in the closed unit disk $|\zeta| \leqslant 1$, and moreover, if for any ζ_j with $|\zeta_j| = 1$ the inequality $\operatorname{Re} \lambda_j / |\lambda_j| > \sin\theta$ holds for $\lambda_j = \zeta_j S(\zeta_j)/P'(\zeta_j)$,

$(\mathrm{II})_\theta$ if it has the property $(\mathrm{III})_\theta$ and each of the roots $\{\zeta_j(z) : 1 \leqslant j \leqslant q\}$ of $w(\zeta, z) = 0$ is simple and lies in the open unit disk $|\zeta| < 1$ for $z \in \Sigma_\theta \backslash \{0\}$,

(I) if it has the property $(\mathrm{II})_\theta$, each root of $S(\zeta) = 0$ is simple and lies in the open unit disk $|\zeta| < 1$, and $b_q > 0$.

We note the following:

REMARK 18.1. If (P, S) has the property $(\mathrm{III})_\theta$, then there exists a constant $\kappa > 0$ such that the condition on $\{\zeta_j(z) : 1 \leqslant j \leqslant q\}$ stated in $(\mathrm{II})_\theta$ holds for $0 < |z| < \kappa$ and $z \in \Sigma_\theta$.

In fact, $\zeta_j(z)$ is continuous in z and we may suppose that $\zeta_j(0) = \zeta_j$ by reordering the numbers if necessary. We have only to show that $|\zeta_j(z)| < 1, 0 < |z| < \kappa, z \in \Sigma_\theta$, holds for some $\kappa > 0$, assuming $|\zeta_j| = 1$. Actually, ζ_j is simple and $g(\rho) = |\zeta_j(\rho e^{i\phi})|^2, |\phi| \leqslant \theta$, is differentiable in ρ at $\rho = 0$. Therefore, we have

$$g'(0) = 2\operatorname{Re}((\zeta_j(0)\zeta_j'(0)e^{i\phi}).$$

On the other hand, we find

$$\zeta_j'(0) = -S(\zeta_j)/P'(\zeta_j)$$

from $P(\zeta_j(z)) + zS(\zeta_j(z)) = 0$ so that $g'(0) = -2\,\mathrm{Re}\,\lambda_j e^{i\phi}$. Therefore we have

$$\sup\{g'(0): |\phi| \leqslant \theta\} < 0,$$

and the assertion follows.

Let the rational function $r = r(z)$ which is used to construct the approximate solution of (18.1) with (18.2) for the values of $u_i = u(t_i)$, $1 < i \leqslant q-1$, be of order $p-1$, and let the multi-step method (P, S) for the values $u_n = u(t_n)$, $n \geqslant q$, be of order p. In this case, we have:

THEOREM 18.1. *The error estimate*

$$\| u_n - e^{-tA} u_0 \| \leqslant C(\tau/t)^p \| u_0 \|, \qquad 0 < t = n\tau \leqslant T, \quad n \geqslant q, \tag{18.4}$$

holds, if for some $\theta > \theta_0$ one of the following is satisfied:

(a) *(P, S) and r have the properties* $(\mathrm{I})_\theta$ *and* $(\mathrm{ii})_\theta$, *respectively*

(b) *(P, S) and r have the properties* $(\mathrm{III})_\theta$ *and* $(\mathrm{iii})_\theta$, *respectively, together with the condition*

$$\tau \| A \| \leqslant M_1 < \min(\tilde{\delta}, \kappa, 1/|b_q|).$$

Here the constant $\tilde{\delta} > 0$ is introduced so that r has no poles in the part $|z| < \tilde{\delta}$ of $z \in \Sigma_\theta$. Further, κ is the positive constant specified in Remark 18.1.

In the case (b), *the constant $C > 0$ in* (18.4) *depends on M_1. Furthermore,* $\min(\tilde{\delta}, \kappa, 1/|b_q|)$ *can be replaced by* $\min(\tilde{\delta}, 1/|b_q|)$ *and* $\min(\tilde{\delta}, \kappa)$, *provided that (P, S) has the property* $(\mathrm{II})_\theta$ *and that $b_q > 0$, respectively.*

LE ROUX [1979] has treated the case (a). The other case can be also proved by her method.

Before proceeding to the proof of Theorem 18.1, we introduce a few notations. We set

$$\delta_i(z) = \frac{a_i + b_i z}{a_q + b_q z}$$

for $1 \leqslant i \leqslant q$. Then, the relation (18.3) reads:

$$\sum_{i=0}^{q} \delta_i(\tau A) u_{n+i} = 0, \quad n = 0, 1, 2, \ldots .$$

Therefore, let us consider first the functions $u_n = u_n(z)$, $z \in \mathbb{C}$, satisfying

$$\sum_{i=0}^{q} \delta_i(z) u_{n+i}(z) = 0, \quad n = 0, 1, 2, \ldots . \tag{18.5}$$

Then, for these functions we have:

LEMMA 18.1. *If the* $\{\zeta_j(z): 1 \leqslant j \leqslant q\}$ *are distinct, then* $u_n = u_n(z)$ $(n \geqslant q)$ *satisfies*

$$|u_n(z)| \leqslant C \sum_{j=1}^{q} |\zeta_j(z)|^n$$

where $C>0$ *depends only on*

$$u_0, \ldots, u_{q-1}, \qquad \sup\{|\zeta_j(z)|: 1 \leqslant j \leqslant q\} \quad \textit{and} \quad \inf\{|\zeta_j(z) - \zeta_i(z)|: i \neq j\}.$$

PROOF. From (18.5) follows

$$u_{n+q}(z) = -\sum_{i=0}^{q-1} \delta_i(z) u_{n+i}(z), \tag{18.6}$$

while the relation

$$P(\zeta) + zS(\zeta) = (a_q + b_q z) \sum_{j=1}^{q} (\zeta - \zeta_j(z))$$

implies

$$\zeta^q = -\sum_{i=0}^{q-1} \sigma_i(z) \zeta^i + \prod_{j=1}^{q} (\zeta - \zeta_j(z)), \tag{18.7}$$

where $\sigma_i = \sigma_i(z)$, $i = 0, \ldots, q-1$ is determined by

$$\frac{\{P(\zeta) + zS(\zeta)\}}{(a_q + b_q z)} = \zeta^q + \sum_{i=0}^{q-1} \sigma_i(z) \zeta^i.$$

For the moment, we write $\delta_i = \delta_i(z)$, $\zeta_j = \zeta_j(z)$ and $u_n = u_n(z)$, for simplicity.

If $u_i = \zeta_j^i$ holds for $0 \leqslant i \leqslant q-1$ with some j, then we have $u_n = \zeta_j^n$, $n = q, q+1, \ldots$. In fact, we can show inductively that

$$u_{n+q} = -\sum_{i=0}^{q-1} \delta_i u_{n+i} = -\left(\sum_{i=0}^{q-1} \delta_i \zeta_j^i\right) \zeta_j^n = \zeta_j^{q+n}.$$

Therefore, in the case where

$$u_i = \sum_{j=1}^{q} \alpha_j \zeta_j^i, \quad 0 \leqslant i \leqslant q-1, \tag{18.8}$$

for some $\alpha_j \in \mathbb{C}$, $1 \leqslant j \leqslant q$, we obtain

$$u_n = \sum_{j=1}^{q} \alpha_j \zeta_j^n, \quad n = q, q+1, \ldots.$$

In particular, this yields

$$|u_n| \leqslant \max\{|\alpha_j|: 1 \leqslant j \leqslant q\} \sum_{j=1}^{q} |\zeta_j|^n.$$

The linear transformation

$$S: (\alpha_1, \ldots, \alpha_q)^{\mathrm{T}} \mapsto (u_0, \ldots, u_{q-1})^{\mathrm{T}}$$

is expressed by the matrix

$$\begin{pmatrix} 1 & \cdots & 1 \\ \zeta_1 & \cdots & \zeta_q \\ \vdots & & \vdots \\ \zeta_1^{q-1} & \cdots & \zeta_q^{q-1} \end{pmatrix},$$

whose determinant is that of Vandermonde. Since the $\{\zeta_j\colon 1\leqslant j\leqslant q\}$ are distinct, such a $\{u_0,\ldots,u_{q-1}\}$ as in (18.8) can represent an arbitrary element in $\mathbb{C}^q$. Thus the proof has been completed. □

REMARK 18.2. In the case that $\{\zeta_j(z)\colon 1\leqslant j\leqslant q\}$ has the maximum of the multiplicities $m=m(z)\geqslant 2$, then the inequality

$$|u_n(z)|\leqslant C(1+n)^{m(z)-1}\sum_{j=1}^{q}|\zeta_j(z)|^n$$

can be proved.

Let us turn to the scheme under consideration. The approximation operator is denoted by $T_n^\tau(A)$:

$$u_n=T_n^\tau(A)u_0.$$

Defining a rational function $s_n=s_n(z)$, $n=0,1,2\ldots$, inductively as

$$\begin{aligned}&s_n(z)=r^n(z),\quad 0\leqslant n\leqslant q-1,\\ &\sum_{j=0}^{q}\delta_i(z)s_{n+i}(z)=0,\quad n=0,1,2,\ldots,\end{aligned}\tag{18.9}$$

we obtain

$$T_n^\tau(A)=s_n(\tau A).$$

Therefore,

$$\sum_{i=0}^{q}\delta_i(z)(\mathrm{e}^{-(n+i)z}-s_{n+i}(z))=F_n(z)\tag{18.10}$$

holds, where

$$F_j(z)=\sum_{i=0}^{q}\delta_i(z)\mathrm{e}^{-(j+i)z},\ j\geqslant 0.\tag{18.11}$$

The following lemma represents the error operator $T_n^\tau(A)-\mathrm{e}^{-tA}$ through the operator $F_j(\tau A)$ just introduced. Let us define the rational functions $\gamma_j(z), j\in\mathbb{Z}$, inductively as

$$\gamma_j(z)=\begin{cases}0, & j<0,\\ 1, & j=0,\end{cases}$$

$$\sum_{k=0}^{q}\gamma_{j-k}(z)\delta_{q-k}(z)=0,\quad j>0. \tag{18.12}$$

LEMMA 18.2. *The identity*

$$e^{-t_{n+q}A}-T^{\tau}_{n+q}(A)$$

$$=\sum_{j=0}^{n}\gamma_{n-j}(\tau A)F_j(\tau A)-\sum_{j=1}^{q-1}\sum_{k=0}^{j}\gamma_{n-k}(\tau A)\delta_{j-k}(\tau)(e^{-t_jA}-r^j\tau A))$$

holds for $n=0,1,\ldots$.

PROOF. From the definition (18.11) of F_j, we have

$$\sum_{j=0}^{n}\gamma_{n-j}(z)F_j(z)=\sum_{j=0}^{n}\sum_{i=0}^{q}\delta_i(z)\gamma_{n-j}(z)e^{-(j+i)z}$$

$$=\sum_{j=0}^{n}\sum_{i=0}^{q}\delta_i(z)\gamma_{n-j}(z)\{e^{-(j+i)z}-s_{j+i}(z)\}$$

$$=\sum_{j=0}^{n+q}B_j(z)(e^{-jz}-s_j(z)),$$

where

$$B_j(z)=\sum_{k=0}^{j}\gamma_{n-k}(z)\delta_{j-k}(z),\quad 0\leqslant j\leqslant q,$$

$$B_{q+j}(z)=\sum_{k=0}^{q}\gamma_{n-j-k}(z)\delta_{q-k}(z).$$

Note that $\gamma_j=0$, $j<0$, by the definition.

Here, $B_j(z)=0$ for $q\leqslant j\leqslant n+q-1$ from the definition (18.12) of $\gamma_l(z)$. On the other hand, we have

$$B_{n+q}(z)=\sum_{k=0}^{q}\gamma_{-k}(z)\delta_{q-k}(z)=\delta_q(z)=1,$$

so that

$$e^{-(n+q)z}-s_{n+q}(z)$$

$$=\sum_{j=0}^{n}\gamma_{n-j}(z)P_j(z)-\sum_{j=0}^{q-1}\sum_{k=0}^{j}\gamma_{n-k}(z)\delta_{j-k}(z)(e^{-jz}-s_j(z)).\qquad\square$$

By virtue of the identity in Lemma 18.2, the error estimate

$$\|\exp(-t_{n+g}A) - T^{\tau}_{n+q}(A)\|$$

is reduced to those on $\gamma_j(z)$ and $F_j(z)$. First, we have for $\gamma_j = \gamma_j(z)$ that:

PROPOSITION 18.1. *Let (P, S) have the property $(\mathrm{III})_\theta$ and $\kappa > 0$ be the constant given in Remark 18.1. Then, for each $\kappa' < \min(\kappa, 1/|b_q|)$, there exist constants $C > 0$ and $\beta > 0$ such that*

$$|\gamma_n(z)| \leqslant \mathrm{e}^{-\beta n|z|}, \qquad |z| \leqslant \kappa', \quad z \in \Sigma_\theta.$$

PROOF. By the argument in Remark 18.1, the roots $\{\zeta_j(z): 1 \leqslant j \leqslant q\}$ of $w(\zeta, z) = P(\zeta) + zS(\zeta) = 0$ satisfy the inequalities

$$|\zeta_j(z)| \leqslant \mathrm{e}^{-\beta|z|}, \qquad |z| \leqslant \kappa', \quad z \in \Sigma_\theta.$$

for some $\beta > 0$. Therefore, we have

$$|\gamma_n(z)| \leqslant C\mathrm{e}^{-\beta n|z|}, \qquad |z| \leqslant \kappa', \quad z \in \Sigma_\theta,$$

by Lemma 18.1. Here, the constant $C > 0$ depends on $\gamma_0, \ldots, \gamma_{q-1}$, which, however, are polynomials of $\delta_0, \ldots, \delta_{q-1}$. These $\delta_0, \ldots, \delta_{q-1}$ are bounded on $|z| \leqslant \kappa' < 1/|b_q|$, hence $C > 0$ also is bounded there. Recall that $a_q = 1$. □

As for $F_j = F_j(z)$ we have:

PROPOSITION 18.2. *If (P, S) is of order $p\,(\geqslant 1)$, then for $\kappa' < 1/|b_q|$ the inequality*

$$|F_j(z)| \leqslant C|z|^{p+1}\mathrm{e}^{-j\,\mathrm{Re}\,z}, \qquad \mathrm{Re}\,z \geqslant 0, \quad |z| \leqslant \kappa',$$

holds true.

PROOF. Putting $v(t) = \mathrm{e}^{-tz}$, we have

$$F_j(z) = \sum_{i=0}^{q} \delta_i(z)\mathrm{e}^{-(j+i)z} = (a_q + b_q z)^{-1}\left\{\sum_{i=0}^{q} a_i v(j+i) - \sum_{i=0}^{q} b_i v'(j+i)\right\}.$$

Since (P, S) is of order p, the right-hand side is equal to

$$(a_q + b_q z)^{-1}\left\{\sum_{i=1}^{q} a_i \int_j^{j+1} \frac{(j+i-t)^p}{p!} v^{(p+1)}(t)\,\mathrm{d}t\right.$$

$$\left. - \sum_{i=1}^{q} b_i \int_j^{j+i} \frac{(j+i-t)^{p-1}}{(p-1)!} v^{(p+1)}\,\mathrm{d}t\right\}.$$

In fact, we have

$$A_j^i = \int_j^{j+i} \frac{(j+i-t)^p}{p!} v^{(p+1)}(t)\,dt$$

$$= \left[\frac{(j+i-t)^p}{p!} v^{(p)}(t) + \frac{(j+i-t)^{p-1}}{(p-1)!} v^{(p-1)}(t) + \cdots + (j+i-t)v'(t) + v(t) \right]_{t=j}^{t=j+i}$$

$$= -\sum_{l=0}^{p} \frac{i^l}{l!} v^{(l)}(j) + v(j+i),$$

$$B_j^i = \int_j^{j+i} \frac{(j+i-t)^{p-1}}{(p-1)!} v^{(p+1)}(t)\,dt$$

$$= \left[\frac{(j+i-t)^{p-1}}{(p-1)!} v^{(p)}(t) + \frac{(j+i-t)^{p-2}}{(p-2)!} v^{(p-2)}(t) + \cdots + (j+i-t)v''(t) + v'(t) \right]_{t=j}^{t=j+i}$$

$$= -\sum_{l=1}^{p} \frac{i^l}{(l+1)!} v^{(l)}(j) + v'(j+i),$$

which yield

$$\sum_{i=1}^{q} \{a_i A_j^i - b_i B_j^i\}$$

$$= \sum_{i=0}^{q} \{a_i A_j^i - b_i B_j^i\}$$

$$= \sum_{i=0}^{q} \{a_i v(j+i) - b_i v'(j+i)\} - \left(\sum_{i=0}^{q} a_i \right) v(j)$$

$$- \sum_{l=1}^{p} \frac{v^{(l)}(j)}{(l-1)!} \left\{ \sum_{i=0}^{q} i^l a_i - l \sum_{i=0}^{q} i^{l-1} b_i \right\}$$

$$= \sum_{i=0}^{q} \{a_i v(j+i) - bv'(j+i)\}$$

in view of Definition 18.1.

Now, into the identity

$$F_j(z)=(a_q+b_qz)^{-1}\left\{\sum_{i=1}^{q} a_i \int_j^{j+i} \frac{(j+i-t)^p}{p!} v^{(p+1)}(t)\,dt\right.$$

$$\left.-\sum_{i=1}^{q} b_i \int_j^{j+i} \frac{(j+i-t)^{p-1}}{(p-1)!} v^{(p+1)}(t)\,dt\right\}$$

we substitute $v^{(p+1)}(t)=(-z)^{p+1}e^{-tz}$, which gives the desired inequality. □

REMARK 18.3. In the case of $b_q>0$, we can take $\kappa'<\kappa$ and $\kappa'=\infty$ in Propositions 18.1 and 18.2, respectively.

Now, we can give:

PROOF OF THEOREM 18.1(b). We fix κ' in $M_1<\kappa'<\min(\bar{\delta},\kappa,1/|b_q|)$. The rational function $r=r(z)$ satisfies

$$|r^j(z)-e^{-jz}|\leqslant C|z|^p, \qquad 1\leqslant j\leqslant q-1, \quad |z|\leqslant\sigma, \quad z\in\Sigma_\theta,$$

for some $\sigma>0$. Then, in the second term of the right-hand side of the equality in Lemma 18.2 for $0\leqslant k\leqslant j\leqslant q-1$ and $j\geqslant k$, we have

$$|\gamma_{n-k}(z)\delta_{j-k}(z)(e^{-jz}-r^j(z))|\leqslant C|z|^p e^{-\beta(n-k)|z|}$$

for $|z|\leqslant\sigma_0=\min(\sigma,\kappa')$ and $z\in\Sigma_\theta$. On the other hand, for a fixed M in $M_1<M<\kappa'$, we have

$$|\gamma_{n-k}(z)(\delta_{j-k}(z)(e^{-jz}-r^j(z))|\leqslant Ce^{-\beta(n-k)|z|}$$

for $\sigma_0\leqslant|z|\leqslant M$ and $z\in\Sigma_\theta$.

We take a path of integration Γ_0 which consists of three parts (as sets): $\Gamma_0=\Gamma_1\cup\Gamma_2\cup\Gamma_3$, where

$$\Gamma_1=\{\lambda=\rho e^{\pm i\theta_1}: 0\leqslant\rho\leqslant\sigma_0\tau^{-1}\},$$

$$\Gamma_2=\{\lambda=\rho e^{\pm i\theta_1}: \sigma_0\tau^{-1}\leqslant\rho\leqslant M\tau^{-1}\},$$

$$\Gamma_3=\{M\tau^{-1}e^{i\varphi}: |\varphi|\leqslant\theta_1\}$$

with $\theta_0<\theta_1<\theta$. Then, we have

$$\gamma_{n-k}(\tau A)\delta_{j-k}(\tau A)(e^{-t_jA}-r^j(\tau A))$$

$$=\frac{1}{2\pi i}\left(\int_{\Gamma_1}+\int_{\Gamma_2}+\int_{\Gamma_3}\right)\gamma_{n-k}(\tau\lambda)\delta_{j-k}(\tau\lambda)(e^{-j\tau\lambda}-r^j(\tau\lambda))(\lambda-A)^{-1}\,d\lambda$$

$$=\mathrm{I}+\mathrm{II}+\mathrm{III},$$

of which each term admits of the following estimates:

$$\|\mathrm{I}\| \leqslant C \int_0^\infty e^{-\beta(n-k)\tau\rho}(\tau\rho)^p \frac{d\rho}{\rho}$$

$$= C(n-k)^{-p} \leqslant Cn^{-p}, \quad n = q, q+1, \ldots,$$

$$\|\mathrm{II}\| \leqslant C \int_{\sigma_0\tau^{-1}}^\infty e^{-\beta(n-k)\tau\rho} \frac{d\rho}{\rho} = C \int_{\sigma_0\beta n}^\infty e^{-\rho} \frac{d\rho}{\rho} \leqslant Cn^{-p},$$

$$\|\mathrm{III}\| \leqslant C \int_{-\theta_1}^{\theta_1} e^{-\beta(n-k)M} \, d\varphi \leqslant Cn^{-p}.$$

Thus the second term in Lemma 18.2 is estimated as

$$\left\| \sum_{j=1}^{q-1} \sum_{k=0}^{j} \gamma_{n-k}(\tau A)\delta_{j-k}(\tau A)(e^{-t_jA} - r^j(\tau A)) \right\| \leqslant Cn^{-p}.$$

Next, to estimate the first term of the right-hand side of the identity in Lemma 18.2, we note that the inequality

$$\left| \sum_{j=0}^{n} \gamma_{n-j}(z)F_j(z) \right| \leqslant Cn \, e^{-\delta n|z|} |z|^{p+1}$$

follows from Propositions 18.1 and 18.2 if $|z| \leqslant M$ and $z \in \Sigma_\theta$. As to the representation

$$\sum_{j=0}^{n} \gamma_{n-j}(\tau A)F_j(\tau A) = \frac{1}{2\pi i}\left(\int_{\Gamma_1+\Gamma_2} + \int_{\Gamma_3} \right) \sum_{j=0}^{n} \gamma_{n-j}(\tau\lambda)F_j(\tau\lambda)(\lambda - A)^{-1} \, d\lambda$$

$$= \mathrm{IV} + \mathrm{V},$$

we obtain

$$\|\mathrm{IV}\| \leqslant C \int_0^\infty n e^{-\beta n\tau\rho}(\tau\rho)^{p+1} \frac{d\rho}{\rho} = Cn^{-p},$$

$$\|\mathrm{V}\| \leqslant C \int_{-\theta_1}^{\theta_1} n e^{-\beta nM} \frac{\tau}{M - M_1} \, d\varphi \leqslant Cn^{-p}.$$

Therefore, the first term in Lemma 18.2 is also estimated as

$$\left\| \sum_{j=0}^{n} \gamma_{n-j}(\tau A)F_j(\tau A) \right\| \leqslant Cn^{-p}.$$

Thus the inequality (18.4) has been established. □

When (P, S) has the property $(\mathrm{II})_\theta$, we can replace $\min(\kappa, 1/|b_q|)$ by $1/|b_q|$ in Proposition 18.1, and accordingly, $\min(\bar{\delta}, 1/|b_q|)$ by $\min(\bar{\delta}, 1/|b_q|)$ in our theorem. Similarly, $\min(\bar{\delta}, \kappa, 1/|b_q|)$ is replaced by $\min(\bar{\delta}, \kappa)$ in the case of $b_q > 0$ by Remark 18.3.

PROOF OF THEOREM 18.1(a). If (P, S) has the property $(\mathrm{I})_\theta$, then the inequality

$$|\zeta_j(z)| \leqslant \mathrm{e}^{-\beta|z|}, \qquad |z| \leqslant \kappa', \quad z \in \Sigma_\theta,$$

holds for arbitrarily large $\kappa' > 0$ with some $\beta > 0$. Therefore, the conclusion of Proposition 18.1 holds for arbitrarily large κ'. On the other hand, by making $\beta > 0$ smaller if necessary, we have

$$|\zeta_j(z)| \leqslant \mathrm{e}^{-\beta} < 1, \qquad |z| \geqslant \kappa', \quad z \in \Sigma_\theta, \tag{18.13}$$

so that

$$|\gamma_n(z)| \leqslant C\,\mathrm{e}^{-n\beta}, \qquad |z| \geqslant \kappa', \quad z \in \Sigma_\theta, \tag{18.14}$$

is obtained by the argument in the proof of Proposition 18.1.

The first term of the right-hand side of the identity in Lemma 18.2 is estimated as follows. We fix $M > \tau\|A\|$ and represent

$$\sum_{j=0}^{n} \gamma_{n-j}(\tau A) F_j(\tau A) = \frac{1}{2\pi \mathrm{i}} \left(\int_{\Gamma_1 + \Gamma_2} + \int_{\Gamma_3} \right) \sum_{j=0}^{n} \gamma_{n-j}(\tau\lambda) F_j(\tau\lambda)(\lambda - A)^{-1}\, \mathrm{d}\lambda$$
$$= \mathrm{I} + \mathrm{II}.$$

As in the proof for the case (b), we have

$$\|\mathrm{I}\| \leqslant C n^{-p},$$

where the constant $C > 0$ is independent of M. On the other hand, in the case of $j \geqslant 1$, we get

$$\left\| \frac{1}{2\pi \mathrm{i}} \int_{\Gamma_3} \gamma_{n-j}(\tau\lambda) F_j(\tau\lambda)(\lambda - A)^{-1}\, \mathrm{d}\lambda \right\| \leqslant C \int_{-\theta_1}^{\theta_1} \mathrm{e}^{-(n-j)\beta} M^{p+1} \mathrm{e}^{-jM\cos\theta_1}\, \mathrm{d}\varphi,$$

of which the right-hand side goes to zero as $M \to \infty$.

To estimate the term for $j = 0$, we have to make the same consideration as in the preceding section. Namely, first we note

$$\lim_{|z| \to \infty, z \in \Sigma_\theta} F_0(z) = b_0/b_q$$

and obtain

$$|\gamma_n(z) F_0(z) - \gamma_n(\infty)(b_0/b_q)|$$
$$\leqslant |\gamma_n(z) - \gamma_n(\infty)|\,|F_n(z)| + |F_0(z) - F_0(\infty)|\,|\gamma_n(\infty)|.$$

Here, we can show the inequality

$$|\gamma_n(z)-\gamma_n(\infty)|\leqslant C\,e^{-\beta n}/|z|, \quad |z|\geqslant R, \quad z\in\Sigma_\theta, \tag{18.15}$$

for $R>0$, $\beta>0$ being taken smaller if necessary. To this end, we have only to derive

$$|\zeta_j(z)^n-\zeta_j(\infty)^n|\leqslant C\,e^{-\beta n}/|z|, \qquad |z|\geqslant R, \quad z\in\Sigma_\theta,$$

for the roots $\{\zeta_j(z)\colon 1\leqslant j\leqslant q\}$ of $w(\zeta,z)=P(\zeta)+zS(\zeta)=0$, by means of the proof of Lemma 18.1. In fact, (18.13) implies

$$\begin{aligned}|\zeta_j(z)^n-\zeta_j(\infty)^n| &= |\zeta_j(z)-\zeta_j(\infty)|\cdot\left|\sum_{k=0}^{n-1}\zeta_j^{n-1-k}(z)\zeta_j^k(\infty)\right|\\ &\leqslant n\,e^{-n\beta}|\zeta_j(z)-\zeta_j(\infty)|,\end{aligned}$$

where $\zeta_j(z)-\zeta_j(\infty)=O(1/|z|)$, $|z|\to\infty$, $z\in\Sigma_\theta$, because (P,S) has the property $(\mathrm{I})_\theta$. Thus (18.15) has been proven. On the other hand, $|F_0(z)|\leqslant C$ is obvious and also

$$\begin{aligned}|F_0(z)-F_0(\infty)| &= \left|\frac{a_0-b_0/b_q}{1+b_qz}+\sum_{i=1}^{q}\delta_i(z)\,e^{-iz}\right|\\ &\leqslant C/|z|+C\,e^{-|z|\cos\theta_1}.\end{aligned}$$

Therefore, we obtain

$$|\gamma_n(z)F_0(z)-\gamma_n(\infty)(b_0/b_q)|\leqslant C\,e^{-\beta n}\left\{\frac{1}{|z|}+e^{-|z|\cos\theta_1}\right\}.$$

Thus, by virtue of Lemma 17.3 the first term of the right-hand side of the identity in Lemma 18.2 is estimated as

$$\begin{aligned}\left\|\sum_{j=0}^{n}\gamma_{n-j}(\tau A)F_j(\tau A)\right\| &\leqslant C\{n^{-p}+e^{-\beta n}+|\gamma_n(\infty)F_0(\infty)|\}\\ &\leqslant Cn^{-p}\end{aligned}$$

Let us proceed to the second term in Lemma 18.2. As is shown in the proof for the case (b), the inequality

$$|\gamma_{n-k}(z)\delta_{j-k}(z)(e^{-jz}-r^j(z))|\leqslant C|z|^p\,e^{-\beta(n-k)|z|} \tag{18.16}$$

holds for $|z|\leqslant\sigma_0$ and $z\in\Sigma_\theta$ with

$$\int_0^\infty e^{-\delta n\rho}\rho^p\frac{d\rho}{\rho}=Cn^{-p}$$

On the other hand, we have

$$\begin{aligned}&|\gamma_{n-k}(z)\delta_{j-k}(z)(e^{-jz}-r^j(z))-\gamma_{n-k}(\infty)\delta_{j-k}(\infty)(-r^j(\infty))|\\ &\quad\leqslant|\gamma_{n-k}(z)-\gamma_{n-k}(\infty)||\delta_{j-k}(z)r^j(z)|+|\gamma_{n-k}(\infty)||\delta_{j-k}(z)-\delta_{j-k}(z)||r^j(z)|\\ &\qquad+|\gamma_{n-k}(\infty)\delta_{j-k}(\infty)||r^j(z)-r^j(\infty)|.\end{aligned} \tag{18.17}$$

Here,

$$|\gamma_{n-k}(z)-\gamma_{n-k}(\infty)|\leqslant C\,\mathrm{e}^{-\beta n}/|z|$$

for $|z|\geqslant\sigma_0$ and $z\in\Sigma_\theta$ as we have seen before. Furthermore, γ and δ_{j-k} satisfy

$$|\delta_{j-k}(z)-\delta_{j-k}(\infty)|\leqslant C/|z|,\quad |r(z)-r(\infty)|\leqslant C/|z|,\qquad |z|\geqslant\sigma_0,$$

respectively, because they are rational functions. Also, we recall that

$$|\gamma_n(\infty)|\leqslant C\,\mathrm{e}^{-\beta n},\qquad |\delta_{n-k}(z)|\leqslant C.$$

Summing these relations, we can give a bound from above of the right-hand side of (18.17), namely, $C\,\mathrm{e}^{-\beta n}/|z|$. Therefore, the second term in Lemma 18.2 is estimated as

$$\left\|\sum_{j=1}^{q-1}\sum_{k=0}^{j}\gamma_{n-k}(\tau A)\delta_{j-k}(\tau A)(\mathrm{e}^{-t_jA}-r^j(\tau A))\right\|\leqslant C/n^p$$

by Lemma 17.3. Thus, the proof has been completed. □

CHAPTER V

Commentary

Concluding our article, we give some commentary. First we describe some variants of the finite element method, and then show applications to nonlinear problems.

19. Variations of the finite element method

19.1. Lumping of mass

In Chapter II, we constructed a semidiscrete finite element approximation of the heat equation

$$\partial u/\partial t - \Delta u = f \quad \text{in } (0, T) \times \Omega \tag{19.1}$$

with

$$u|_{\partial\Omega} = 0, \qquad u|_{t=0} = u_0(x), \tag{19.2}$$

where $\Omega \subset \mathbb{R}^2$ is a convex polygon. Namely, taking account of its weak form

$$(\mathrm{d}/\mathrm{d}t)(u, v) + a(u, v) = (f, v), \quad v \in V = H_0^1(\Omega) \tag{19.3}$$

with

$$u|_{t=0} = u_0 \tag{19.4}$$

in $X = L^2(\Omega)$, we discretize it as

$$(\mathrm{d}/\mathrm{d}t)(u_h, \chi) + a(u_h, \chi) = (f, \chi), \quad \chi \in V_h, \tag*{(19.3)$_h$}$$

with

$$(u_h|_{t=0}, \chi) = (u_0, \chi), \quad \chi \in V_h, \tag*{(19.4)$_h$}$$

in V_h, the approximate space V constituted by piecewise linear trial functions, where

$$a(u, v) = \int_\Omega \nabla u \overline{\nabla v}\, \mathrm{d}x. \tag{19.5}$$

As a variant, there is a method for semidiscretization called lumping. Although there are several methods of lumping, we state here a procedure based on the barycenter. Let the barycenter of the triangular element $T\in\tau_h$ whose nodes are P_1, P_2 and P_3 be G_T, and let the midpoints of P_1P_2, P_2P_3 and P_3P_1 be P'_3, P'_1, and P'_2, respectively. We divide T into three subdomains by three segments $G_TP'_3$, $G_TP'_1$, and $G_TP'_2$ and assign each subdomain to the corresponding node. For example, the quadrilateral $P_1P'_3G_TP'_2$ is assigned to the node P_1.

After making this assignment for all the nodes, we combine the subdomains assigned to a particular node into a region. This region is called the barycentric region corresponding to the node. Then, for each inner node $j\in I$ in the triangulation τ_h we define a piecewise constant function $\bar{\varphi}_j$ whose value is 1 in the barycentric region corresponding to the jth inner node and is 0 otherwise. From these $\bar{\varphi}_j$, a finite-dimensional space $\bar{V}_h$ is generated. There is a natural isomorphism

$$V_h \underset{k_h}{\overset{L_h}{\rightleftarrows}} \bar{V}_a.$$

The lumped mass finite element approximation for (19.3) is described as

$$(\mathrm{d}/\mathrm{d}t)(L_hu_h, L_h\chi)+a(u_h,\chi)=(L_hf, L_h\chi), \quad \chi\in V_h, \tag{19.5$_h$}$$

in V_h with

$$(u_h|_{t=0},\chi)=(u_0,\chi), \quad \chi\in V_h. \tag{19.6$_h$}$$

Writing $u_h(t)=\Sigma_{j\in I}V_j(t)\bar{\varphi}_j$, we can reduce $(19.5)_h$ to

$$(\mathrm{d}/\mathrm{d}t)\overline{M}_hU_h+A_hU_h=F_h$$

with suitable matrices $\overline{M}_h$ and A_h and a vector F_h. Here, $\overline{M}_h$ is diagonal, which is useful in real computations. Error analysis for $(19.5)_h$ has been done by USHIJIMA [1979].

19.2. *Upwind finite element method*

It is known that for the diffusion equation

$$(\partial u/\partial t)-d\Delta u+(b\cdot\nabla)u=0 \quad \text{in } (0,T)\times\Omega \tag{19.7}$$

with

$$u|_{\partial\Omega}=0, \qquad u|_{t=0}=u_0(x) \tag{19.8}$$

involving the drift terms $b\cdot\nabla$, we obtain a fine approximate solution by combining finite element with "upwind difference", especially in the case that $\|b\|_{L^\infty}$ is large relatively to $d>0$. Among several manners of upwind difference, we describe a method employing the lumping for the elliptic equation

$$-d\Delta u+(b\cdot\nabla)u=f \quad \text{in } \Omega \tag{19.9}$$

with

$$u|_{\partial\Omega}=0 \tag{19.10}$$

according to TABATA [1979].

For simplicity, let $\Omega\subset\mathbb{R}^2$ be a convex polygon. Subject to its triangulation τ_h, we take the spaces of piecewise linear and piecewise constant trial functions V_h and $\bar{V}_h$, respectively as above. Then, the upwind finite element approximation of (19.9) is given as

$$da(u_h, \chi)+(B_h u_h, L_h\chi)=(L_h f, L_h\chi), \quad \chi\in V_h, \tag{19.9}_h$$

in V_h, where $L_h\colon V_h\to\bar{V}_h$ is the isomorphic mapping described before. Here, the upwind approximate operator $B_h\colon V_h\to\bar{V}_h$ is defined as follows: First, we say that an element $T\in\tau_h$ is upwind to its vertex $j\in T$ if the vector $b(j)$ starting from j does not cross $T\backslash\{j\}$. (Here, $b=b(x)$ is the coefficient in (19.9).) For each inner node $j\in I$ we select an element $T=T_j\in\tau_h$ which contains j as a vertex and is upwind to j. Then, for $v_h\in V_h$ we set

$$B_h v_h=\sum_{j\in I}\{(b(j)\cdot\nabla)v_h\}|_{T_j}\bar{\varphi}_j, \tag{19.11}$$

where $\{\bar{\varphi}_j\colon j\in I\}$ denotes the basis of $\bar{V}_h$ given before.

These methods are also applicable for parabolic equations and their error analysis has been made. See TABATA [1979] and IKEDA [1983].

20. Application to nonlinear problems

20.1. Nonstationary Navier–Stokes equation

The nonstationary Navier–Stokes equation

$$\partial u/\partial t-\Delta u+(u\cdot\nabla)u+\nabla p=f \quad \text{in } (0, T)\times\Omega \tag{20.1}$$

with

$$\nabla\cdot u=0, \tag{20.2}$$

$$u|_{\partial\Omega}=0, \quad u|_{t=0}=u_0(x) \tag{20.3}$$

can be treated from the semigroup-theoretical viewpoint of FUJITA and KATO [1964].

Namely, let $\Omega\in\mathbb{R}^2$ be a convex polygon and set

$$X=\{v\in L^2(\Omega)^2\colon \operatorname{div} v=0\}, \qquad v\cdot n=0 \quad \text{on } \partial\Omega,$$

$$V=\{v\in H_0^1(\Omega)^2\colon \operatorname{div} v=0\}.$$

A sesquilinear form $a(\,,)$ on $V\times V$ is defined as

$$a(w, v)=(\nabla w, \nabla v), \quad w, v\in V,$$

where $(\,,)$ denotes the natural inner product in $L^2(\Omega)^2$. A self-adjoint operator A in X is associated with $a=a(\,,)$ and the relation $D(A)=H^2(\Omega)^2 V$ holds. For each $\gamma>\frac{1}{2}$, the nonlinear operator $F\colon D(A^\gamma)\to X$ is defined through

$$(Fw, v)=((w\cdot\nabla)w, v), \quad v\in X,$$

by virtue of Sobolev's imbedding. Explicitly, $Fw = P(w\cdot\nabla)w$, where $P: H \equiv L^2(\Omega)^2 \to X$ is the orthogonal projection.

Then the system (20.1)–(20.3) is reduced to the evolution equation

$$du/dt + Au + Fu = Pf \tag{20.4}$$

with

$$u(0) = u_0 \tag{20.5}$$

in X. The self-adjoint operator $-A$, called the Stokes operator, generates a semigroup $\{e^{-tA}\}$. Hence (20.4) is furthermore reduced to the integral equation

$$u(t) = e^{-tA}u_0 - \int_0^t e^{-(t-s)A}F(u(s))\,ds + \int_0^t e^{-(t-s)A}Pf(s)\,ds. \tag{20.6}$$

Iteration works for this Volterra equation because the estimates

$$\|F(w)\|_X \leq C\|A^\gamma w\|_X \|A^{1/2}w\|_X, \quad w \in D(A^\gamma),$$

$$\|F(w)-F(v)\|_X \leq C\{\|A^\gamma w\|_X \|A^{1/2}(w-v)\|_X + \|A^{1/2}v\|_X \|A^\gamma(w-v)\|_X\},$$

$$w, v \in D(A^\gamma)$$

follow from Sobolev's imbedding. Thus, for instance if $f \equiv 0$ and $u_0 \in X$, there exists a unique solution $u = u(t) \in C([0,\infty) \to X) \subset C^1((0,\infty) \to X) \cap C^0((0,\infty) \to V)$ of (20.6). Consequently, a semidiscrete approximation of (20.1) will be obtained by discretizing the stationary Navier–Stokes equation

$$-\Delta u + (u\cdot\nabla)u + \nabla p = f \quad \text{in } \Omega \tag{20.7}$$

with

$$\nabla\cdot u = 0, \tag{20.8}$$

$$u|_{\partial\Omega} = 0, \tag{20.9}$$

or simply stationary Stokes equation

$$-\Delta u + \nabla p = f \quad \text{in } \Omega \tag{20.10}$$

with

$$\nabla\cdot u = 0, \tag{20.11}$$

$$u|_{\partial\Omega} = 0. \tag{20.12}$$

Put $H = L^2(\Omega)^2$ and $M = \{q \in L^2(\Omega): \int_\Omega q(x)\,dx = 0\}$. Then the relation

$$V = \{v \in H: (v, \nabla q) = 0 \text{ for any } q \in M\}$$

holds. Noticing this, BERCOVIER and PIRONNEAU [1979] introduced the following way of approximation for the stationary Stokes problem. Namely, let τ_h, $h > 0$, be the triangulation described in Section 19.2. We break each element $T = \triangle P_1P_2P_3 \in \tau_h$ into two four parts by taking middle points P'_1, P'_2 and P'_3 of P_2P_3, P_3P_1 and P_1P_2, respectively, and construct $\triangle P'_1P'_2P_3$, $\triangle P'_2P'_3P_1$, $\triangle P'_3P'_1P_2$ and $\triangle P'_1P'_2P'_3$. Thus we obtain a new triangulation $\tilde{\tau}_h$. The approximate space H_h of H consists of the functions that are continuous, linear on each $T\in\tilde{\tau}_h$, and 0 at the boundary $\partial\Omega$. Moreover, M_h denotes the set of functions that are continuous, linear on each $T\in\tau_h$, and 0 integral over Ω. Then we put

$$V_h=\{v_h\in H_h:(v_h,\nabla q_h)=0 \text{ for any } q_h\in M_h\}.$$

We note that $V_h\not\subset V$. A self-adjoint operator A_h in V_h is associated with the sesquilinear form $a|_{V_h\times V_h}$. It is taken as an approximation of the Stokes operator A. Such a way of approximation is sometimes referred to as the mixed finite element method.

If the condition

$$\text{each } T\in\tau_h \text{ has at least one vertex in } \Omega \tag{20.13}$$

is satisfied, then what is called Brezzi's condition (BREZZI [1974], KIKUCHI [1973]),

there exists a $\beta>0$ such that

$$\sup_{\chi\in V_h}\frac{|(\chi,\nabla q_h)|}{\|\chi\|_{V_h}}\geq\beta\|\nabla q_h\|_X,\quad q_h\in M_h \text{ as } h\downarrow 0 \tag{20.14}$$

holds. See BERCOVIER and PIRONNEAU [1979] or GLOWINSKI and PIRONNEAU [1979]. From this follows the unique existence, stability, and error estimate about the approximate solution $u_h\in V_h$ of

$$A_hu_h-Q_hf \tag*{(20.10)$_h$}$$

for (20.10), where $Q_h:H\to V_h$ denotes the orthogonal projection.

On the other hand, approximation $F_h:V_h\to V_h$ of the nonlinear term F in (20.4) is done in the following way. Namely, noting

$$(Fw,v)=((w\cdot\nabla)w,v)=b(w,w,v),\qquad w\in D(A^\gamma),\quad v\in V$$

for

$$b(u;w,v)=\tfrac{1}{2}\{((u\cdot\nabla)w,v)-((u\cdot\nabla)v,w)\},\qquad u\in D(A^\gamma),\quad v,w\in V,$$

we impose

$$(F_hw,v_h)=b(w;w,v_h)\quad\text{for any } v_h\in V_h.$$

Under these situations, a semidiscrete approximate equation

$$\mathrm{d}u_h/\mathrm{d}t+A_hu_h+F_hu_h=Q_hPf \tag*{(20.4)$_h$}$$

with

$$u_h(0)=Q_hu_0 \tag*{(20.5)$_h$}$$

in V_h for (20.4) arises, which is reduced to

$$u_h(t)=\mathrm{e}^{-tA_h}Q_hu_0-\int_0^t\mathrm{e}^{-(t-s)A_h}F_h(u_h(s))\,\mathrm{d}s+\int_0^t\mathrm{e}^{-(t-s)A_h}Q_hPf(s)\,\mathrm{d}s. \tag*{(20.6)$_h$}$$

Its error $e_h(t)\equiv u_h(t)-u(t)$ has been analyzed by HEYWOOD and RANNACHER [1982] and OKAMOTO [1982b]. For instance, the following estimates are derived by OKAMOTO [1982b], where δ is a positive constant.

THEOREM 20.1. *If* $u_0\in V$, *then*

$$\|u_h(t)-u(t)\|_X\leqslant C(\|u_0\|_V)\mathrm{e}^{-\delta t}(h^2/t^{1/2}),\quad 0<t<\infty.$$

holds.

THEOREM 20.2. *If* $u_0\in D(A)$, *then*

$$\|u_h(t)-u(t)\|_X+h\|u_h(t)-u(t)\|_V\leqslant C(\|Au_0\|_X)\mathrm{e}^{-\delta t}h^2,\quad 0<t<\infty,$$

holds.

Other mixed finite element methods for Stokes or Navier–Stokes equations are described by GIRAULT and RAVIART [1986].

20.2. Parabolic equations of blow-up type

Asymptotic behavior of the semilinear parabolic equation

$$\partial u/\partial t-\Delta u=f(u)\quad \text{in } (0,T)\times\Omega \tag{20.15}$$

with

$$u|_{\partial\Omega}=0,\qquad u|_{t=0}=u_0(x) \tag{20.16}$$

has been studied theoretically. Some equations allow only local solutions in time. Such a phenomenon is called blow-up of the solution. See, for instance, FUJITA [1966, 1969]. Using the finite difference method, NAKAGAWA [1976] gave numerical computations and their numerical analysis for such equations. Later NAKAGAWA and USHIJIMA [1977] studied similar problems for the finite element method.

Recently, WEISSLER [1984] has discovered what is called one-point blow-up for this type of equations. Motivated by this, CHEN [1986] reexamined the finite difference method for such equations and studied blow-up of approximate solutions.

20.3. Stefan problem

The Stefan problem invol·es the phenomenon of phase transition in heat equations. The most simple case is that of one space dimension and one phase. Then, the problem is to find the temperature $u=u(t,x)$ and the free boundary $s=s(t)$ through

$$\partial u/\partial t=\sigma(\partial^2u/\partial x^2),\qquad 0<x<s(t),\quad 0<t<T, \tag{20.17}$$

with

$$u|_{x=0}=g(t),\quad u|_{x=s(t)}=0,\quad u|_{t=0}=f(x),\qquad 0<x<b=s(0), \tag{20.18}$$

$$\mathrm{d}s/\mathrm{d}t=-u_x(t,s(t)),\quad 0<t<T. \tag{20.19}$$

Here, $0<x<s(t)$ represents the region of water. Ice is supposed to be located in $s(t)<x<\infty$.

For this problem, MORI [1976] proposed a scheme employing finite difference and finite element for the time and space variables respectively, and showed stability and convergence to the genuine solution of its approximate solutions. MORI [1978] studied similar problems for the two-phase case.

References

AGMON, S. (1965), *Lectures on Elliptic Boundary Value Problems* (Van Nostrand, Princeton, NJ).

AGMON, S., A. DOUGLIS and L. NIRENBERG (1959), Estimates near the boundary for solutions of elliptic partial differential equations satisfying general boundary conditions, I, *Comm. Pure Appl. Math.* **12**, 623–727.

BABA, K. and M. TABATA (1981), On a conservative upwind finite element scheme for convective diffusion equations, *RAIRO Numer. Anal.* **15**, 3–26.

BAIOCCHI, C. and F. BREZZI (1983), Optimal error estimates for linear parabolic problems under minimal regularity assumptions, Preprint.

BAKER, G.A. and J.H. BRAMBLE (1979), Semidiscrete and single step fully discrete approximations for second order hyperbolic equations, *RAIRO Numer. Anal.* **13**, 75–100.

BAKER, G.A., J.H. BRAMBLE and V. THOMÉE (1977), Single step Galerkin approximations for parabolic problems, *Math. Comp.* **31**, 818–847.

BAKER, G.A. and V.A. DOUGALIS (1980), On the L^∞-convergence of Galerkin approximations for second-order hyperbolic equations, *Math. Comp.* **34**, 401–424.

BALES, L.A. (1984), Semidiscrete and single step fully discrete approximations for second order hyperbolic equations with time-dependent coefficients, *Math. Comp.* **43**, 383–414.

BERCOVIER, H. and O. PIRONNEAU (1979), Error estimate for finite element method solution of the Stokes problem in the primitive variables, *Math. Comp.* **34**, 401–424.

BRAMBLE, J.H. and S.R. HILBERT (1970), Estimation of linear functionals on Sobolev spaces with applications to Fourier transforms and spline interpolation, *SIAM J. Numer. Anal.* **7**, 112–124.

BRAMBLE, J.H. and P.H. SAMMON (1980), Efficient higher order single step methods for parabolic problems, Part I, *Math. Comp.* **35**, 655–677.

BRAMBLE, J.H., A.H. SCHATZ, V. THOMÉE and L.B. WAHLBIN (1977), Some convergence estimates for semidiscrete Galerkin type approximations for parabolic equations, *SIAM J. Numer. Anal.* **14**, 218–241.

BRAMBLE, J.H. and V. THOMÉE (1972), Semi-discrete least square methods for a parabolic boundary value problem, *Math. Comp.* **26**, 633–648.

BRAMBLE, J.H. and V. THOMÉE (1974), Discrete time Galerkin methods for a parabolic boundary value problem, *Ann. Math. Pure Appl.* **101**, 115–152.

BRENNER, P. and V. THOMÉE (1979), On rational approximations of semigroups, *SIAM J. Numer. Anal.* **16**, 683–694.

BRENNER, P. and V. THOMÉE (1980), On rational approximations of groups of operators, *SIAM J. Numer. Anal.* **17**, 119–125.

BREZZI, F. (1974), On the existence, uniqueness and approximation of saddle point problems arising from Lagrangean multipliers, *RAIRO Numer. Anal.* **8**, 129–151.

CHEN, Y.G. (1986), Asymptotic behaviours of blowing-up solutions for finite difference analogue for $u_t = u_{xx} + u^{1+\alpha}$, *J. Fac. Sci. Univ. Tokyo, Sec. IA* **33**, 541–574.

CIARLET, P.G. (1978), *The Finite Element Method for Elliptic Problems* (North-Holland, Amsterdam).

CIARLET, P.G. and P.A. RAVIART (1973), Maximal principles and uniform convergence for the finite element method, *Comput. Methods Appl. Mech. Engrg.* **2**, 17–31.

CROUZEIX, M. (1980), Une méthode multipas implicite-explicite pour l'approximation des équations d'évolution parabolique, *Numer. Math.* **35**, 257–276.

DOUGALIS, V.A. (1979), Multistep-Galerkin methods for hyperbolic equations, *Math. Comp.* **33**, 563–584.

DOUGLAS Jr, J. and T. DUPONT (1970), Galerkin methods for parabolic equations, *SIAM J. Numer. Anal.* **7**, 575–626.

DOUGLAS Jr, J., T. DUPONT and L. WAHLBIN (1975), The stability in L^q of the L^2-projection into finite element function spaces, *Numer. Math.* **23**, 193–197.

DUPONT, T. (1973), L^2-estimates for Galerkin methods for second order hyperbolic equations, *SIAM J. Numer. Anal.* **10**, 880–897.

FIX, G. and N. NASSIF (1972), On finite element approximation in time dependent problems, *Numer. Math.* **19**, 127–135.

FRIED, I. (1980), On the optimality of the pointwise accuracy of the finite element solution, *Internat. J. Numer. Methods Engrg.* **15**, 451–456.

FUJIE, Y. and H. TANABE (1973), On some parabolic equations in Hilbert space, *Osaka J. Math.* **19**, 127–135.

FUJII, H. (1974), A note on finite element approximation for evolution equations, *Kôkyûroku RIMS Kyoto Univ.* **202**, 96–117.

FUJITA, H. (1966), On the blowing up of solutions to the Cauchy problem for $u_t = \Delta u + u^{1+\alpha}$, *J. Fac. Sci. Univ. Tokyo Sec. 1A* **13**, 109–124.

FUJITA, H. (1969), On the nonlinear equations $\Delta u + e^u = 0$ and $\partial v/\partial t = \Delta v + e^v$, *Bull. Amer. Math. Soc.* **75**, 132–135.

FUJITA, H. and T. KATO (1964), On the Navier–Stokes initial value problem, I, *Arch. Rational Mech. Anal.* **16**, 269–315.

FUJITA, H. and A. MIZUTANI (1976), On the finite element method for parabolic equations, I: Approximation of holomorphic semi-groups, *J. Math. Soc. Japan* **28**, 749–771.

FUJITA, H. and A. MIZUTANI (1978), Remarks on the finite element methods for parabolic equations with higher accuracy, in: *Functional Analysis and Numerical Analysis* (Japan–France Seminar in Tokyo and Kyoto 1976) (Japan Society for the Promotion of Science, Tokyo).

FUJITA, H. and T. SUZUKI (1979), On the finite element approximation for evolution equations of parabolic type, in: *Computing Methods in Applied Sciences and Engineering 1977*, Lecture Notes in Mathematics **704** (Springer, Berlin).

GEKELER, E. (1978), A priori error estimates of Galerkin backward differentiation methods in time-inhomogeneous parabolic problems, *Numer. Math.* **30**, 369–383.

GEVECI, T. (1984), On the convergence of Galerkin approximation schemes for second-order hyperbolic equations in energy and negative norm, *Math. Comp.* **42**, 393–416.

GIRAULT, V. and P.A. RAVIART (1986), *Finite Element Methods for Navier–Stokes Equations* (Springer, Heidelberg).

GLOWINSKI, R. and O. PIRONNEAU (1979), On a mixed finite element approximation of the Stokes problem, I: Convergence of the approximate solutions, *Numer. Math.* **33**, 397–424.

GRISVARD, P. (1985), *Elliptic Problems in Non-smooth Domains* (Pitman, London).

HAVERKAMP, R. (1984), Eine Aussage zur L_∞-Stabilität und zur genauen Konvergenzordnung der H_0^1-Projektion, *Numer. Math.* **44**, 393–405.

HELFRICH, H.P. (1974), Fehlerabschätzungen für das Galerkinverfahren zur Lösung von Evolutionsgleichungen, *Manuscripta Math.* **13**, 219–235.

HELFRICH, H.P. (1975), Lokale Konvergenz des Galerkinverfahrens bei Gleichungen vom parabolischen Typ in Hilberträumen, Thesis.

HERSH, R. and T. KATO (1979), High-accuracy stable difference schemes for well-posed initial-value problems, *SIAM J. Numer. Anal.* **16**, 670–682.

HEYWOOD, J.G. and R. RANNACHER (1982), Finite element approximation of the nonstationary Navier–Stokes problem, I: Regularity of solutions and second-order error estimates for spatial discretizations, *SIAM J. Numer. Anal.* **19**, 275–311.

HILLE, E. and R.S. PHILLIPS (1957), *Functional Analysis and Semi-groups* (American Mathematical Society, Providence, RI).

HITOTUMATU, S. (1963), *Approximation Formulae* (Takeuchi Shoten, Tokyo) (in Japanese).

HUANG, M.Y. and V. THOMÉE (1981), Some convergence estimates for semidiscrete type schemes for time dependent nonselfadjoint parabolic equations, *Math. Comp.* **37**, 327–346.

HUANG, M.Y. and V. THOMÉE (1982), On the backward Euler method for parabolic equations with rough initial data, *SIAM J. Numer. Anal.* **19**, 599–603.

IKEDA, T. (1983), *Maximum Principle in Finite Element Models for Convection-Diffusion Phenomena* (Kinokuniya/North-Holland, Tokyo/Amsterdam).

JESPERSON, D. (1978), Ritz–Galerkin methods for singular boundary value problems, *SIAM J. Numer. Anal.* **15**, 813–834.

KATO, T. (1961a), Fractional powers of dissipative operators, *J. Math. Soc. Japan* **13**, 246–274.

KATO, T. (1961b), Abstract evolution equations of parabolic type in Banach and Hilbert spaces, *Nagoya Math. J.* **5**, 93–125.

KATO, T. (1966), *Perturbation Theory for Linear Operators* (Springer, Berlin).

KATO, T. and H. TANABE (1962), On the abstract evolution equation, *Osaka Math. J.* **14**, 107–133.

KIKUCHI, F. (1973), Some considerations of convergence of hybrid stress method, in: Y. YAMADA and R.H. GALLAGHER, eds., *Theory and Practice in Finite Element Structural Analysis* (University of Tokyo Press, Tokyo).

LADYZHENSKAYA, O.A. (1963), *The Mathematical Theory of Viscous Incompressible Flow* (Gordon and Breach, New York).

LE ROUX, M.N. (1979), Semidiscretization in time for parabolic problems, *Math. Comp.* **33**, 919–931.

LIONS, J.L. and E. MAGENES (1968), *Problèmes aux Limites Non Homogènes et Applications* **1** (Gauthier-Villars, Paris).

LUSKIN, M. and R. RANNACHER (1982a), On the smoothing property of Galerkin method for parabolic equations, *SIAM J. Numer. Anal.* **19**, 93–113.

LUSKIN, M. and R. RANNACHER (1982b), On the smoothing property of the Crank–Nicolson scheme, *Applicable Anal.* **14**, 117–135.

MIZOHATA, S. (1973), *The Theory of Partial Differential Equations* (Cambridge University Press, Cambridge).

MORI, M. (1974), Approximation of exponential function of a matrix by continued fraction expansion, *Publ. RIMS Kyoto Univ.* **10**, 257–269.

MORI, M. (1976), Stability and convergence of a finite element method for solving the Stefan problem, *Publ. RIMS Kyoto Univ.* **12**, 539–563.

MORI, M. (1978), Stability of a finite element method for solving the Stefan problem in one space dimension, in: H. FUJITA, ed., *Functional Analysis and Numerical Analysis* (Japan Society for the Promotion of Science, Tokyo).

MORI, M. (1986), *The Finite Element Method and Its Applications* (Macmillan, New York).

NAKAGAWA, T. (1976), Blowing up of a finite difference solution to $u_t = u_{xx} + u^2$, *Appl. Math. Optimization* **2**, 337–350.

NAKAGAWA, T. and T. USHIJIMA (1977), Finite element analysis of the semi-linear heat equation of blow-up type, in: J.J.H. MILLER, ed., *Topics in Numerical Analysis* **III** (Academic Press, London).

NITSCHE, J.A. (1975), L^∞-convergence of finite element approximation, in: *Proceedings Second Conference on Finite Elements*, Rennes.

NITSCHE, J.A. (1976), Über L_∞-Abschätzungen von Projektionen auf Finite Elemente, *Bonn. Math. Schr.* **89**, 13–30.

NITSCHE, J.A. (1979), L_∞-convergence of finite element Galerkin approximations for parabolic problems, *RAIRO Numer. Anal.* **13**, 31–54.

NITSCHE, J.A. and A.H. SCHATZ (1974), Interior estimates for Ritz–Galerkin methods, *Math. Comp.* **28**, 937–958.

NITSCHE, J.A. and M.F. WHEELER (1981–82), L_∞-boundedness of the finite element Galerkin operation for parabolic problems, *Numer. Funct. Anal. Optim.* **4**, 325–353.

OKAMOTO, H. (1982a), On the semi-discrete finite element approximation for the nonstationary Stokes equation, *J. Fac. Sci. Univ. Tokyo Sec. IA* **29**, 241–260.

OKAMOTO, H. (1982b), On the semi-discrete finite element approximation for the nonstationary Navier–Stokes equation, *J. Fac. Sci. Univ. Tokyo Sec. IA* **29**, 613–652.

RANNACHER, R. (1984), Finite element solution of diffusion problems with irregular data, *Numer. Math.* **43**, 309–327.

RANNACHER, R. and R. SCOTT (1982), Some optimal error estimates for piecewise linear finite element approximation, *Math. Comp.* **38**, 437–445.

RAVIART, P.A. (1978), Multistep methods and parabolic equations, in: *Functional Analysis and Numerical Analysis* (Japan–France Seminar in Tokyo and Kyoto 1976) (Japan Society for the Promotion of Science, Tokyo).

RAVIART, P.A. and J.M. THOMAS (1983), *Introduction à l'Analyse Numérique des Équations aux Dérivées Partielles* (Masson, Paris).

SAMMON, P.H. (1982), Convergence estimates for semidiscrete parabolic equation approximations, *SIAM J. Numer. Anal.* **19**, 68–92.

SAMMON, P.H. (1983), Fully discrete approximation methods for parabolic problems with non-smooth initial data, *SIAM J. Numer. Anal.* **20**, 437–470.

SCHATZ, A.H. (1980), A weak maximum principle and stability of the finite element method in L_∞ on plane polygonal domains, I, *Math. Comp.* **30**, 681–697.

SCHATZ, A.H., V. THOMÉE and L.B. WAHLBIN (1980), Maximal norm stability and error estimates in parabolic finite element equations, *Comm. Pure Appl. Math.* **33**, 265–304.

SCHATZ, A.H. and L.B. WAHLBIN (1979), Maximum norm estimates in the finite element method on plane polygonal domains, Part 2: Refinement, *Math. Comp.* **33**, 465–492.

SCHATZ, A.H. and L.B. WAHLBIN (1982), On the quasi-optimality in L_∞ of the $\mathring{H}^1$-projection into finite element spaces, *Math. Comp.* **38**, 1–22.

SCOTT, R. (1976), Optimal L^∞ estimates for the finite element method on irregular meshes, *Math. Comp.* **30**, 681–697.

SOBOLEVSKII, P.E. (1961a), Parabolic type equations in Banach spaces, *Trudy Moscov. Math.* **10**, 297–350 (in Russian).

SOBOLEVSKII, P.E. (1961b), On equations of parabolic type in Banach spaces with unbounded time-dependent generators whose fractional powers are of constant domains, *Dokl. Acad. Nauk SSSR* **13**, 59–62 (in Russian).

SUZUKI, T. (1979a), On the rate of convergence of the difference finite element approximation of parabolic equations, *Proc. Japan Acad. Ser. A* **54**, 326–331.

SUZUKI, T. (1979b), On some approximation theorems for evolution equations of parabolic type: An operator theoretical approach to the finite element method, *Kôkyûroku RIMS Kyoto Univ.* **357**, 165–188 (in Japanese).

SUZUKI, T. (1982), Full-discrete finite element approximation of evolution equation $u_t + A(t)u = 0$ of parabolic type, *J. Fac. Sci. Univ. Tokyo, Sec. IA* **29**, 195–240.

SUZUKI, T. and H. FUJITA (1986), A remark on the L^∞ bounds of the Ritz operator associated with a finite element approximation, *Numer. Math.* **49**, 529–544.

TABATA, M. (1979), L^∞-analysis of the finite element method, in: H. FUJII, F. KIKUCHI, T. NAKAGAWA and T. USHIJIMA, eds., *Numerical Analysis of Evolution Equations* (Kinokuniya, Tokyo).

TANABE, H. (1960), On the equations of evolution in a Banach space, *Osaka Math. J.* **12**, 363–376.

TANABE, H. (1979), *Equations of Evolution* (Pitman, London).

TEMAM, R. (1979), *Navier–Stokes Equation* (North-Holland, Amsterdam).

THOMÉE, V. (1984), *Galerkin Finite Element Methods for Parabolic Problems* (Springer, Berlin).

THOMÉE, V. and L.B. WAHLBIN (1983), Maximum-norm stability and error estimates in Galerkin methods for parabolic equations in one space variable, *Numer. Math.* **41**, 345–371.

USHIJIMA, T. (1975), Approximation theory for semi-groups of linear operators and its application to approximation of wave equations, *Japan J. Math.* **1**, 185–224.

USHIJIMA, T. (1979a), Error estimates for the lumped mass approximation of the heat equation, *Mem. Numer. Math.* **6**, 65–82.

USHIJIMA, T. (1979b), On the finite element approximation of semi-groups of linear operators, in: H. FUJII, F. KIKUCHI, T. NAKAGAWA and T. USHIJMA, eds., *Numerical Analysis of Evolution Equations* (Kinokuniya, Tokyo).

WANNER, G., E. HAIRER and S.P. NORSETT (1978), Order stars and stability theorems, *BIT* **18**, 475–489.

WEISSLER, F.B. (1984), Single point blow up for a semilinear initial value problem, *J. Differential Equations* **55**, 204–224.

WHEELER, M.F. (1973), A priori L_2 estimates for Galerkin approximations to parabolic partial differential equations, *SIAM J. Numer. Anal.* **10**, 723–759.

YOSIDA, K. (1964), *Functional Analysis* (Springer, Berlin).

ZLÁMAL, M. (1973), Curved elements in the finite element method, I, *SIAM J. Numer. Anal.* **10**, 229–240.

ZLÁMAL, M. (1974), Finite element methods for parabolic equations, *Math. Comp.* **28**, 393–404.

ZLÁMAL, M. (1975), Finite element multistep discretizations of parabolic boundary value problems, *Math. Comp.* **29**, 350–359.

WHEELER, M.F. (1973), A priori L_2 estimates for Galerkin approximations to parabolic partial differential equations, *SIAM J. Numer. Anal.* **10**, 723–759.
YOSIDA, K. (1966), *Functional Analysis* (Springer, Berlin).
ZLÁMAL, M. (1973), Curved elements in the finite element method I, *SIAM J. Numer. Anal.* **10**, 229–240.
ZLÁMAL, M. (1974), Finite element methods for parabolic equations, *Math. Comp.* **28**, 393–404.
ZLÁMAL, M. (1975), Finite element multistep discretizations of parabolic boundary value problems, *Math. Comp.* **29**, 350–359.

Subject Index

Printed and bound by CPI Group (UK) Ltd, Croydon, CR0 4YY
03/10/2024
01040428-0017